H U M A N   A N A T O M Y

e th
d by
belo
ised
we

late

plicatio
g the ba

LEEDS BECKETT UNIVERSITY
LIBRARY
DISCARDED

LEEDS BECKETT UNIVERSITY
17 0112066 5

FOURTH EDITION

HUMAN ANATOMY

KENT M. VAN DE GRAAFF

BRIGHAM YOUNG UNIVERSITY

WCB **Wm. C. Brown Publishers**
Dubuque, Iowa•Melbourne, Australia•Oxford, England

Book Team

Editor *Colin H. Wheatley*
Developmental Editor *Kristine Noel*
Production Editor *Cathy Ford Smith*
Designers *K. Wayne Harms/Lu Ann Schrandt*
Art Editor *Mary E. Powers*
Photo Editor *John Leland*
Permissions Coordinator *Karen L. Storlie*

LEEDS METROPOLITAN
UNIVERSITY LIBRARY
1701120665
B41MV
247295 3.7.95
15.9.95
1011 VAN

Wm. C. Brown Publishers
A Division of Wm. C. Brown Communications, Inc.

Vice President and General Manager *Beverly Kolz*
Vice President, Publisher *Kevin Kane*
Vice President, Director of Sales and Marketing *Virginia S. Moffat*
Vice President, Director of Production *Colleen A. Yonda*
National Sales Manager *Douglas J. DiNardo*
Marketing Manager *Craig S. Marty*
Advertising Manager *Janelle Keeffer*
Production Editorial Manager *Renée Menne*
Publishing Services Manager *Karen J. Slaght*
Permissions/Records Manager *Connie Allendorf*

Wm. C. Brown Communications, Inc.

President and Chief Executive Officer *G. Franklin Lewis*
Corporate Senior Vice President, President of WCB Manufacturing *Roger Meyer*
Corporate Senior Vice President and Chief Financial Officer *Robert Chesterman*

Cover photos: © John Campos

Copyedited by Ann Mirels

The credits section for this book begins on page 777 and
is considered an extension of the copyright page.

Copyright © 1995 by Wm. C. Brown Communications, Inc. All
rights reserved

A Times Mirror Company

Library of Congress Catalog Card Number: 93–73841

ISBN 0–697–16065–3

No part of this publication may be reproduced, stored in a retrieval
system, or transmitted, in any form or by any means, electronic,
mechanical, photocopying, recording, or otherwise, without the
prior written permission of the publisher.

Printed in the United States of America by Wm. C. Brown Communications, Inc.,
2460 Kerper Boulevard, Dubuque, IA 52001

10 9 8 7 6 5 4 3 2

*To the memory of my
parents, John and Nell*

Brief Contents

Preface xv
Connecting the Concepts xx

Contents

Preface xv
Connecting the Concepts xx

9 Muscular System 223

10 Surface and Regional Anatomy 283

UNIT V

Integration and Coordination

11 Nervous Tissue and the Central Nervous System 329

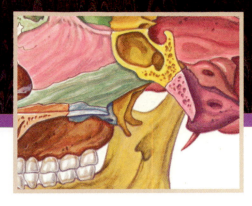

12 Peripheral Nervous System 383

13 Autonomic Nervous System 413

14 Endocrine System 431

15 Sensory Organs 461

Maintenance of the Body

16 Circulatory System 507

21 Female Reproductive System 689

22 Developmental Anatomy, Postnatal Growth, Inheritance 717

Human Anatomy is an essential foundation subject for many careers, including nursing, medicine, paramedicine, dentistry, medical technology, physical therapy, and athletic training. Since all of these careers are application oriented, the focus of this text is on presenting practical and relevant information. Numerous examples throughout the text reinforce the principle that learning anatomy helps students to become better acquainted with themselves. The fourth edition of *Human Anatomy* is intended to serve as a basic introduction to human anatomy for students majoring in biological, medical, allied-health, and physical education programs. In addition, this text can be of immense value to any student wanting to know more about the structure, function, and possible dysfunction of the human body.

Objectives

In preparing a text and its ancillaries (study guide, laboratory manual, study cards, instructors manual, and so forth), it is essential to consider both the needs of the student and the needs of the instructor. A well-written and inviting text is at the heart of an effective educational package. With this in mind, the following objectives were formulated for the Fourth edition of *Human Anatomy:*

- To provide an accurate and up-to-date text that is visually appealing, interesting, applicable, and readable—a text that will entice you to study the material and thereby enhance your appreciation of life through a better understanding of the structure, function, and magnificence of your own body.

- To identify and clearly state the basic concepts and accompanying learning objectives of the discipline of human anatomy.

- To express the beauty of the body through spectacular art. Anatomy is a visual science where exactness is essential. Over 75 new pieces of high-quality art were prepared for this edition to augment the acclaimed art program of the previous edition.

- To use a series of icons to clearly indicate the topics of the commentaries. These chapter asides point up the importance of particular technical information.

- To provide a correlated and balanced presentation of anatomy at the developmental, cellular, histological, clinical, and gross levels.

- To develop in you a sufficient anatomical vocabulary so that you are conversant with medical terminology.

- To encourage you to take proper care of your body so that you may enjoy a more healthy and productive life.

- To acquaint you with the history of the science of anatomy, from its primitive beginnings to recent advances.

Text Organization

The 22 chapters in this text are grouped into seven units that are identified by *colored tabs* on the outside page margins.

UNIT I In this unit, the stage is set for studying human anatomy by providing a historical perspective on how this science has developed over the centuries. Anatomy is an exciting and dynamic science that continues to progress through the contributions of many individuals. It is hoped that this unit will make you feel a part of the heritage of anatomy.

UNIT II In this unit, the human is described as a vertebrate animal. The various levels of body organization are also described, and the basic terminology necessary for understanding the structure and functioning of the body is defined.

UNIT III Body organization is discussed at the cellular and histological levels in this unit. Cellular chemistry is also introduced and emphasized as an integral aspect of understanding body functions.

UNIT IV Support, protection, and movement of the human body are the topics of discussion in this unit. The integumentary system provides the body with external support and protection and the skeletal system provides internal support and protection for certain organs of the body. Movement is possible at the joints of the skeleton as the associated skeletal muscles are contracted. The topics of surface anatomy and regional anatomy are discussed in chapter 10 of this unit. Photographs of excellent dissections of human cadavers are also included in this chapter.

UNIT V This unit includes the nervous system, endocrine system, and sensory organs. The concepts identified and discussed in these chapters are concerned with the integration and coordination of body functions and the perception of environmental stimuli.

UNIT VI In this unit, the structure and function of the circulatory, respiratory, digestive, and urinary systems are discussed as they regulate and maintain various body processes. The concepts and learning objectives presented in this unit emphasize those structures that directly relate to maintaining normal body functions and health.

UNIT VII The male and female reproductive systems are described in this unit and the continuance of the human species through sexual reproduction is discussed. Unit 7 provides you with an overview of the entire sequence of human life, including prenatal development and postnatal growth, development and aging.

Student Aids

You will find that each of the 22 chapters of this text includes a number of pedagogical devices that organize the material, stimulate interest in it, and facilitate the study of human anatomy. If you use these aids, your study will be more effective and enjoyable.

Chapter Introductions

The chapter introductions contain an overview of the *contents of each chapter* in outline form. Page numbers are indicated to guide you to sections dealing with particular topics. A *clinical investigation* in the form of a *case study in anatomy* also appears at the beginning of most of the chapters. These hypothetical situations are indicative of the type of application information that will be presented in the chapter. The solution to the case study is presented at the end of the chapter, following the chapter summary.

Understanding Terminology

Where each technical term first appears and is defined in the narrative, it is set off in *boldface* or *italic type,* and is often followed by a bracketed *phonetic pronunciation*. Many of the terms are further elucidated in footnotes explaining their *derivations*. In addition, the roots of each term can be identified by referring to the *glossary of prefixes and suffixes* on the inside of the front and back covers. Knowing the pronunciation, derivation, and roots of a scientific term will help you retain its meaning. The anatomical terminology used in this edition of *Human Anatomy* is current as presented in *Nomina Anatomica (NA),* sixth edition, Churchill Livingstone, Inc., 1989.

Chapter Sections

Each chapter is divided into several principal sections, and each section is prefaced by a *concept statement* and a list of *learning objectives*. A concept statement is a succinct expression of the main idea, or central theme, of

the information presented in a chapter section. The learning objectives indicate the level of competency you should attain in order to understand the concept thoroughly and be able to apply it in practical situations. The narrative that follows discusses the concept in detail and addresses the objectives. *Review questions* at the end of each chapter section test your understanding of the concept and mastery of the learning objectives.

Commentaries and Clinical Information

Short paragraphs highlighted by accompanying *topic icons* appear throughout the text. The topics covered are of practical value and reinforce the importance of learning the preceding facts. The five icons used are as follows:

 Clinical information is represented by a stethoscope.

 Information relevant to *normal aging* is represented by an hourglass.

 Developmental information of practical importance is represented by a human embryo.

 Information relevant to the body processes that maintain *homeostasis* (a state of dynamic equilibrium) is represented by a gear mechanism.

 Topics relevant to human anatomy that are quite simply of *academic interest* are represented by a mortarboard.

Selected developmental disorders, clinical procedures, and diseases or dysfunctions of specific organ systems are described in the *clinical considerations sections* at the end of each chapter. Pathological changes are illustrated in many of the discussions.

Developmental Expositions

In each body system chapter, the discussion of prenatal development is presented at the end of the chapter, just before the chapter summary. Each of these discussions includes *exhibits* and *explanations* of the morphogenic events involved in the development of a body system. Placement at the end of the chapter ensures that the anatomical terminology needed to understand the embryonic structures has been introduced.

Internal Affairs

The page of information in each body system chapter entitled "Internal Affairs" is new to this edition of the text. This information ties the functional aspects of one body system to each of the others. The page layout makes it easier to visualize how body systems make up the organism. The use of the gear mechanism icon for homeostasis indicates the interaction of body systems necessary for normal body function.

7. An updated set of *70 color slides* that depict important pathological conditions lends support to information given in the text and is free to adopters of this textbook. A narrative describing each slide is provided in the Instructor's Manual.

8. WCB Microtest is a computerized testing service for generating examinations and quizzes. It is available free to instructors who adopt this text.

9. *Study Cards for Anatomy and Physiology* by Kent M. Van De Graaff, R. Ward Rhees, and Christopher H. Creek is a boxed set of more than 300 3-by-5-inch cards. It serves as a well-organized, illustrated synopsis of the structure and function of the human body. Clinical information is presented as it applies to specific body organs and systems or physiological processes. The Study Cards offer a quick and effective way for students to review human anatomy and physiology.

10. A *Computer Review of Human Anatomy and Physiology* by S. Scott Zimmerman, Thomas V. Davis, and Kent M. Van De Graaff is contained in a 13-disk set designed for Apple® II- or IBM®-compatible computers. This computer disk set provides a graphic and innovative means for learning and reviewing human anatomy and physiology.

11. *Knowledge Map of Human Anatomy Systems* by Craig Gundy is a computer tutorial developed for the Macintosh. The computer disk set provides an anatomical review of all the body systems.

12. The WCB Anatomy & Physiology Video Series consists of
    —*Introduction to the Human Cadaver and Dissection* (30 min.; color)
    —*Introduction to Cat Dissection: Musculature* (55 min.; color)
    —*Dry Lab of Blood Drawing/Typing Projects* (30 min.; color)
    —*Internal Organs and the Circulatory System* (58 min.; color)
    —*Human Skeletal System* (51 min.; color)
    These exceptional videotapes are available free to adopters.
    —*Human Skeletal Musculature System* (90 min.; color)
    Free to qualified adopters.

13. *Anatomy and Physiology of the Heart Videodisc and "Slice of Life, vol. V," Anatomy and Physiology Videodisc.*

14. *The Coloring Review Guide to Human Anatomy* by Hogin McMurtrie and James Krall Rikel is a spiral-bound book emphasizing the connections between facts and structures through specific student activities—most notably learning through the process of color association. The Coloring Guide provides a thorough review of anatomical concepts.

15. *Atlas of the Skeletal Muscles* by Robert J. Stone and Judith A. Stone is a guide to the anatomy and actions of human skeletal muscles. The illustrations aid in visual orientation and understanding of the action of the muscles.

Acknowledgments

The fourth edition of *Human Anatomy* could not have been prepared without the support and encouragement of colleagues and students. My sincere appreciation is extended to professors Michael J. Shively, Mark Nielsen, Stuart I. Fox, Ferron L. Andersen, Richard A. Heckmann, Kerry Openshaw, and Wilford M. Hess for their professional counsel and example. Kyle M. Van De Graaff offered valuable suggestions on how to improve this edition of the text. Linda K. Anderson, Connie J. Erdmann, Jeffrey Prince, Karianne N. Prince, and Jason Kerr provided valuable input from a student perspective.

I wish to acknowledge friends who have assisted in specific ways: Dr. W. Geoff Williams accomplished the formidable task of preparing the Clinical Case Studies and their solutions that appear in each of the body systems chapters of the text. Drs. Eric J. Van De Graaff, J. Philip Freestone, Douglas W. Hacking, and Charles H. Stewart provided professional advice and reviewed the clinical portions of the text. John Van De Graaff, D.D.S., reviewed the sections pertinent to the oral cavity and clinical considerations of teeth. Drs. Gary M. Watts and Brent C. Chandler and the Department of Radiology at Utah Valley Regional Medical Center provided many of the radiographs used in the text. The photographs of surface anatomy were taken by Glenn L. Anderson, John L. Crawley, and Sheril D. Burton, Ph.D.

Quality illustrations for the fourth edition of this text were a top priority. The talented artists that contributed to this project were Christopher H. Creek and Felecia Paras. It was a pleasure working with these professionals and I appreciated their contribution.

Ann Mirels copy edited the manuscript. She was especially helpful in polishing this edition. I am grateful for her friendship, talent, and dedicated efforts. The editorial staff production people at Wm. C. Brown Publishers have been superb to work with. Acquisitions Editor Colin H. Wheatley and Developmental Editor Kris Noel worked diligently throughout the rewriting and production of this text, providing continual guidance and insight. K. Wayne Harms designed the text and the production editor, Cathy Smith, orchestrated its final production by coordinating the efforts of many.

Wm. C. Brown Publishers assembled a review panel of competent anatomists to review the manuscript and illustrations. These individuals aided my work immeasurably, and I am especially grateful for their frank criticisms, comments, and encouragements:

Donald M. Allen
University of Texas of the Permian Basin

D. Andy Anderson
Utah State University

Raymond Bernor
Howard University

Steve Cole
Iowa Central Community College

Marianne Crocker
Ozark Technical Community College

Mary Jo Fourier
Johnson County Community College

Carl D. Frailey
Johnson County Community College

Larry Ganion
Ball State University

Rebecca German
University of Cincinnati

Kathleen D. Gulino
Penn State University-Fayette Campus

Chapter Summaries

A *summary,* in outline form, at the end of each chapter reinforces your learning experience. Reading these comprehensive summaries is an excellent way for you to review the chapters and prepare for examinations.

Review Activities

At the end of each chapter the review activities consisting of *objective and essay questions,* provide you with feedback as to the depth of your understanding and learning. The correct answers are provided in Appendix A (page 758). Each correct answer is also explained, so you can effectively use these review activities to better understand the subject matter.

Illustrations and Tables

Because anatomy is a descriptive science, great care has been taken to provide you with an outstanding illustration program to maximize your learning. In addition to being aesthetically pleasing, each illustration has been checked and rechecked to ensure the conceptual clarity, precision, and accuracy of the linework, labels, and caption. All of the figures in the fourth edition are integrated with the text narrative, and most are original *full-color art.* Chapters 6, 7, and 8 on the skeletal system and articulations have been extensively illustrated with new color art. Several new color plates have also been incorporated into chapters 17, 18, and 19 on the respiratory, digestive, and urinary systems, respectively. *Color graphics* are used throughout the text to depict functional processes where complex ideas need to be clarified. Microscopic anatomy is illustrated by photomicrographs at the light microscope level and by electron micrographs from scanning and transmission electron microscopy. Carefully selected *photographs* throughout the text provide a balanced view of the gross anatomy. *Color-coding* is used in certain places within the text as a technique to aid learning. The bones of the skull, in chapter 6 for example, are color-coded with new illustrations so that each bone can be readily identified in the many renderings that are included. A set of *anatomical reference plates* follows chapter 3, where it will be of most value for review purposes. Photographs of *dissected human cadavers* are presented in chapter 10 to illustrate the complexity and structural relationships of the body that can be fully appreciated only when seen in a human specimen. Elsewhere in the text, photographs of specific organs from cadavers are used to augment the illustrations.

Numerous *tables* throughout the text summarize information and clarify complex data. These tables are placed near related textual material or at the end of chapter sections. Selected *illustrated tables* are incorporated for the first time in this edition.

Appendixes, Glossary, and Index

The appendixes, following chapter 22, contain valuable reference information. *Appendix A* contains the answers to the objective questions that appear at the end of each chapter, along with *Explanations* (see page 758). *Appendix B* provides suggestions for the preparation of laboratory demonstrations in anatomy (see page 762). *Appendix C* lists the medical and pharmacological abbreviations commonly used in clinical practice (see page 762). *Appendix D* lists the various units of measurements and their equivalents and describes how to convert unit into another (see page 763).

Provided in the *glossary* are pronunciations and definitions of more important scientific and clinical terms used in the te Synonyms and eponyms are indicated, as appropriate. The comp hensive *index* features numerous cross-references.

Human Anatomy, Fourth Edition as a Reference Book

Not only is *Human Anatomy,* fourth edition, an effective text for course of study, it is also designed to be a valuable reference book. Th text can serve as an important resource in any health-oriented cours In addition, it can also be used as a family reference book regardin the structure, function, development, senescence (aging), and possi ble dysfunctions of the human body.

Supplementary Materials

The supplementary materials that accompany the fourth edition of *Human Anatomy* are designed to assist you in your learning activities and to help anatomy instructors plan course work and presentations.

1. Since the laboratory is an integral aspect of instruction in human anatomy, two laboratory manuals are available that address the range of a laboratory's diverse needs. The third edition of *Human Anatomy Laboratory Textbook* by Kent M. Van De Graaff accommodates the use of human cadavers, as well as selected cat dissections. Clinical application is emphasized. The fourth edition of *Human Anatomy Laboratory Manual* by Kent M. Van De Graaff is specifically designed to accompany the fourth edition of the Human Anatomy textbook. This laboratory manual contains cat dissections and selected organ dissections. It emphasizes learning anatomical structures through visual observation, palpation, and knowledge of the functional relationship of one body system to another.

2. The *Instructor's Manual for the Laboratory Manual* by Jeffrey Prince and Karianne N. Prince provides the answers to the questions that appear in the laboratory reports in the Laboratory Manual.

3. A *Student Study Guide to Accompany Human Anatomy* by Kent M. Van De Graaff contains chapter concepts, focus questions, mastery quizzes, study activities and answer keys with explanations that correspond to the chapters of the text.

4. An *Instructor's Manual and Test Item File* by Jeffrey Prince and Karianne N. Prince provides instructional support in the use of the textbook. It also contains a test item file with 70 items for each chapter to help instructors make up examinations.

5. A set of *120 acetate transparencies* is available to instructors who adopt this text. The transparencies are made from selected illustrations in the text and have been chosen for their value in reinforcing lecture presentations.

6. *Student Study Art Notebook* contains full-color illustrations that match the acetate transparencies. Students use the notebook to make notes as they look at the overhead transparencies.

Kevin P. Haley
College of St. Mary

Mary Etta Hight
Marshall University

Bert H. Jacobson
Oklahoma State University

Anne Johnson
Southern Illinois University

James L. Junker
Campbell University

Ronald L. Koller
Los Angeles Pierce College

Robert T. Lynn
Professor Emeritus, Southwestern Oklahoma State University

Sherwin Mizell
Indiana University at Bloomington

Tim R. Mullican
Dakota Wesleyan University

Stephen P. Nawrocki
University of Indianapolis

Mark T. Nielsen
University of Utah

Nicholas J. Pantazis
University of Iowa

Peter K. Pedersen
Cuesta College

J. Robert Rhodes
Private Practice, Plastic Surgery

Robert R. Roales
Indiana University Kokomo

Gloria Jeanne Rogillio
Northeast Louisiana University

M. J. Shively
Utah Valley State College

Tracy L. Soltesz
Marshall University

Charles Sontag
University of Wisconsin Center-Manitowoc

Kerstin Stoedefalke
Colby-Sawyer College

Eric A. Wise
Santa Barbara City College

Finally, but most importantly, I am continually appreciative of my family for their sacrifice, encouragement, and love.

Connecting the Concepts

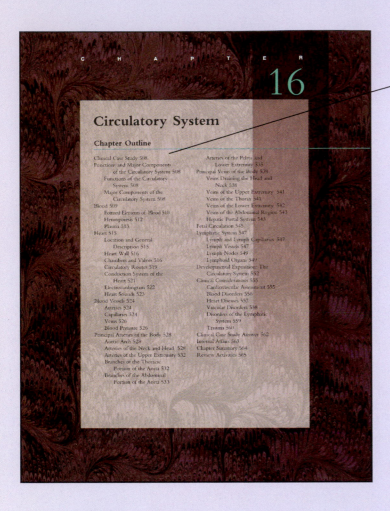

Chapter Outline

Each chapter has an outline that permits you to see at a glance how the chapter is organized and what major topics are included. The outlines include the first- and second-level heads for the chapter and page-number references for the topics discussed.

Color Coding

The outside margins of each page are color-coded to assist students in quickly locating a particular section of the book. For easy reference, these color bars correspond to those shown in the Table of Contents.

Concept Statement

A concept is a carefully worded sentence that expresses the central theme of the information presented in a chapter section; it is the unifying thought pertinent to the detail that follows.

Learning Objectives

The learning objectives for a chapter section indicate the level of competency you should attain in order to thoroughly understand the concept and apply it in practical situations.

---

432 Integration and Coordination

CLINICAL CASE STUDY

A 38-year-old female visits her family doctor complaining of chronic fatigue and weakness, especially in her legs. Upon greeting the patient, the doctor notes that although she is mildly obese, there is an unusually round contour to her face. During questioning, he learns that at her recent 20-year high school reunion, nobody recognized her because her face looked so different. He also learns that she has a chronic unrelenting headache. Physical examination yields, in addition to the facial findings, an unusual fat distribution consisting of a hump on the upper back and marked truncal obesity. Additional findings include hypertension and blindness in the lateral visual fields of both eyes (bitemporal hemianopsia). Laboratory findings are remarkable for elevated blood glucose and other evidence that glucocorticoid levels are high.

Can a correlation be drawn between the patient's visual findings and the rest of her clinical picture? Explain. In which endocrine organ does the primary disease process reside? Does that same organ produce the hormones that directly cause the patient's signs and symptoms? Explain. Do you suppose that similar findings could occur due to a tumor elsewhere in the endocrine system? *Hints:* Study figure 14.8 and note the position of the endocrine glands with respect to the optic nerve and optic chiasma. Also, study table 14.4 to review the actions of the hormones released by the endocrine glands within the brain.

UNIT V

Introduction to the Endocrine System

Hormones are regulatory chemicals secreted by the endocrine glands into the blood, which transports them to their target tissues. Feedback mechanisms in the target cells control the secretion (production) of the hormones.

Objective 1. Distinguish between endocrine and exocrine glands.

Objective 2. Compare and contrast the nervous system and the endocrine system in regulating body functions and maintaining homeostasis.

Objective 3. Define *mixed gland* and identify the endocrine glands that are mixed.

Objective 4. Describe the action of a hormone on its target cell and explain how negative feedback regulates hormonal secretion.

Objective 5. Differentiate between the three principal kinds of hormones: steroids, proteins, and amines.

Figure 14.1 Comparison of (a) an endocrine gland (e.g., the thyroid) with (b) an exocrine gland (e.g., a sweat gland in the skin). An endocrine gland is a ductless gland that releases hormones into the blood. By contrast, exocrine glands secrete their products into ducts that lead to body surfaces.

The numerous glands of the body can be classified into two types based on structure and function: *exocrine* and *endocrine*. Exocrine glands, such as sweat, salivary, or mucous glands, produce secretions that are transported through ducts to their respective destinations. Each of the exocrine glands functions within a particular system of the body. The endocrine glands constitute their own system, the endocrine system, and have several distinctions. Endocrine glands are ductless glands that secrete specific chemicals called *hormones* directly into the blood (fig. 14.1). The blood then transports these hormones to specific sites called *target cells*, where they perform precise functions.

The endocrine system functions closely with the nervous system in regulating and integrating body processes and maintaining homeostasis. The nervous system regulates body activities through the action of electrochemical impulses that are transmitted by means of neurons, resulting in rapid, but usually brief, responses. By contrast, the glands of the endocrine system secrete chemical regulators that travel through the bloodstream to their intended sites where they have a slow but relatively long-lasting effect. Neurological responses are measured in milliseconds, but hormonal action requires seconds or days to elicit a response. Some hormones may have an effect that lasts for minutes; for others, the effect may last for weeks or months.

endocrine: Gk. *endon*, within; *krinein*, to separate
exocrine: Gk. *exo*, outside; *krinein*, to separate
hormone: Gk. *hormon*, to set in motion

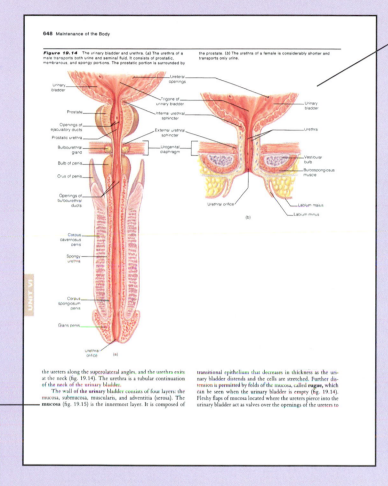

Figure 19.14 The urinary bladder and urethra. (a) The urethra of a male transports both urine and seminal fluid. It consists of prostatic, membranous, and spongy portions. The prostatic portion is surrounded by the prostate. (b) The urethra of a female is considerably shorter and transports only urine.

the ureters along the superolateral angles, and the urethra exits at the neck (fig. 19.14). The urethra is a tubular continuation of the neck of the urinary bladder.

The wall of the urinary bladder consists of four layers: the mucosa, submucosa, muscularis, and adventitia (serosa). The **mucosa** (fig. 19.15) is the innermost layer. It is composed of transitional epithelium that decreases in thickness as the urinary bladder distends and the cells are stretched. Further distension is permitted by folds of the mucosa, called **rugae**, which can be seen when the urinary bladder is empty (fig. 19.14). Fleshy flaps of mucosa located where the ureters pierce into the urinary bladder act as valves over the openings of the ureters to

Beautifully Rendered Full-Color Art

Carefully prepared, accurate illustrations are a hallmark of this text. Human anatomy is a visual science and realistic art is essential. The layout of the text accommodates placement of the art on the same page or very near the structures being described.

Boldfaced Words

New terms appear in boldface print as they are introduced and immediately defined in context. Boldfaced terms are listed in the glossary, where a phonetic pronunciation is given.

Color-coded Illustrations

Color-coding is effectively used to promote learning. For example, the bones of the skull are color-coded so that each bone can be readily identified throughout the chapter.

Figure 6.19 The floor of the cranial cavity.

Figure 6.20 A posterior view of a frontal (coronal) section of the skull.

Illustrated Tables

Selected illustrated tables clearly depict the structure and function of body organs.

Clinical Case Study

A hypothetical medical situation sets the stage for the type of application information that will be presented in the chapter. The solution to the case study is presented at the end of the chapter, following the chapter summary.

Topic Icons

Topic icons highlight information of practical application and special interest. These commentaries reinforce the importance of learning the preceding facts. The five icon images and the topics they represent are clinical information (stethoscope), aging (hourglass), developmental information (embryo), homeostasis (gear mechanism), and academic information (mortarboard).

Derivations

Many of the terms are elucidated in footnotes that indicate their Greek or Latin derivations.

Table 16.3   Layers of the Heart Wall

Layer	Characteristics	Function
Epicardium (visceral pericardium)	Serous membrane including blood capillaries, lymph capillaries, and nerve fibers	Serves as lubricative outer covering
Myocardium	Cardiac muscle tissue separated by connective tissues and including blood capillaries, lymph capillaries, and nerve fibers	Provides muscular contractions that eject blood from the heart chambers
Endocardium	Endothelial tissue and a thick subendothelial layer of elastic and collagenous fibers	Serves as protective inner lining of the chambers and valves

tricles are separated from each other by the thick, muscular **interventricular septum**. Left and right *atrioventricular valves* are located between the atria and ventricles and *semilunar valves* are located at the bases of the two large vessels leaving the heart. These valves maintain the flow of blood in one direction.

Grooved depressions on the surface of the heart indicate the partitions between the chambers and also contain **cardiac vessels** that supply blood to the muscular wall of the heart. The most prominent groove is the *coronary sulcus* that encircles the heart and marks the division between the atria and ventricles. The partition between right and left ventricles is denoted by two (anterior and posterior) *interventricular sulci*.

The following discussion follows the sequence in which blood flows through the structures of the atria, ventricles, and valves.

Right Atrium   The right atrium (*a'tre-um*) receives systemic venous blood from the **superior vena cava**, which drains the upper portion of the body, and from the **inferior vena cava**, which drains the lower portion (fig. 16.9). There is an addi-

tional venous opening into the right atrium from the *coronary sinus*, which returns venous blood from the myocardium of the heart itself.

Right Ventricle   Blood from the right atrium passes through the **right atrioventricular valve** (also called the *tricuspid valve*) to fill the right ventricle. The right atrioventricular (AV) valve is characterized by three valve leaflets, or *cusps*. Each cusp is held in position by strong tendinous cords called **chordae tendineae** that are secured to the ventricular wall by cone-shaped **papillary muscles**. These structures prevent the valves from everting, like an umbrella in a strong wind, when the ventricles contract.

Ventricular contraction causes the right atrioventricular valve to close and the blood to leave the right ventricle through the **pulmonary trunk** and to enter the capillaries of the lungs via the right and left **pulmonary arteries**. The **pulmonary valve** (also called the *pulmonary semilunar valve*) lies at the base of the pulmonary trunk, where it prevents the backflow of ejected blood into the right ventricle.

vena cava: L. *vena*, vein; *cava*, empty

chordae tendineae: L. *chorda*, string; *tendere*, to stretch

---

CLINICAL CASE STUDY

A 27-year-old woman gives birth to twin boys, followed by an apparently single placenta. After examining the two infants, the pediatrician informs the mother that one of her babies has a cleft palate but that the other is normal. She then adds that in some instances cleft palate is a genetic disorder and asks if any family members have had the problem. The mother says that she knows of none. Further examination of the placenta reveals two amnions and only one chorion.

Does the presence of two amnions and one chorion indicate monozygotic or dizygotic twins? How would you account for the fact that one baby has a cleft palate, while the other does not?

Hints: Carefully read the section on multiple pregnancy at the end of the chapter and study the examples presented in figure 22.46. Can any conclusions be drawn regarding genetic similarities between the two infants? Note that identical twins are always the same gender.

Fertilization

Upon fertilization of a secondary oocyte by a spermatozoon in the uterine tube, meiotic development is completed and a diploid zygote is formed.

Objective 1.   Define *morphogenesis, capacitation,* and *fertilization.*

Objective 2.   Describe the changes that occur in the spermatozoon and ovum prior to, during, and immediately following fertilization.

The structure of the human body forms before birth, or *prenatally*, through a process called **morphogenesis** (*mor''fo-jen'e-sis*). Through morphogenic events, the organs and systems of the body are established in a functional relationship. A comprehensive view of the structure and function of each body system requires an understanding of morphogenesis. Associated with each organ and system are sensitive stages of morphogenesis during which genetic or environmental factors may affect the normal development of the baby. Many clinical problems are congenital in nature and originate during morphogenic development.

The prenatal development of a human is a fascinating and awesome event. It begins with a single fertilized egg and culminates some 38 weeks later with a complex organization composed of billions of cells. Prenatal development can be divided into a *pre-embryonic period*, initiated by the fertilization of a secondary oocyte; an *embryonic period*, during which the body's organ systems are formed; and a *fetal period*, culminating in parturition, or the birth of the baby.

During coitus (sexual intercourse), a male ejaculates between 100 million and 500 million sperm into the female's

morphogenesis: Gk. *morphe*, form; *genesis*, beginning

vagina. This tremendous number is needed because of the high rate of sperm fatality—only about 100 survive to contact the secondary oocyte in the uterine tube. During passage through the acidic female tract, sperm gain the ability to fertilize a secondary oocyte through a process called **capacitation.** The changes occurring in sperm capacitation are not fully understood. Evidence indicates, however, that when placed in the acidic environment of the female reproductive tract, there is an enzymatic reaction resulting in a molecular alteration in the cell membrane of the sperm.

Experiments confirm that freshly ejaculated sperm are infertile and must be in the female reproductive tract for at least 7 hours before they can fertilize a secondary oocyte. During in vitro fertilization, sperm capacitation is induced artificially by treating the ejaculate with a solution of gamma globulin, free serum, follicular fluid, dextran, serum dialysate, and adrenal gland extract.

A woman usually ovulates one secondary oocyte a month, totaling approximately 400 during her reproductive years. Each ovulation releases a secondary oocyte arrested at metaphase of the second meiotic division. As the secondary oocyte enters the uterine tube, it is surrounded by a thin transparent layer of protein and polysaccharides, called the *zona pellucida*, and a layer of granulosa cells, called the *corona radiata* (fig. 22.1).

The head of each spermatozoon is capped by an organelle called an **acrosome** (figs. 22.1 and 22.2). The acrosome contains a trypsinlike protein-digesting enzyme and *hyaluronidase* (*hi''ă-loo-ron'i-dās*), which digests hyaluronic acid, an important constituent of connective tissue. When a spermatozoon meets a secondary oocyte in the uterine tube, an *acrosomal reaction* occurs that exposes the acrosome's digestive enzymes and allows the spermatozoon to penetrate through the corona radiata and the zona pellucida.

As a spermatozoon penetrates the zona pellucida, a chemical change in the zona prevents other sperm from entering. Only one spermatozoon, therefore, is allowed to fertilize a secondary oocyte. When the cell membranes of the spermatozoon and the secondary oocyte merge, the secondary oocyte becomes an **ovum.** As fertilization occurs, the ovum is stimulated to complete its second meiotic division (fig. 22.3). Like the first meiotic division, the second produces one cell that contains all of the cytoplasm and one *polar body*. The healthy cell is the mature ovum, and the second polar body, like the first, ultimately fragments and disintegrates.

At fertilization, the entire spermatozoon enters the cytoplasm of the much larger ovum (fig. 22.4). Within 12 hours, the nuclear membrane in the ovum disappears, and the *haploid number* of chromosomes (23) in the ovum is joined by the haploid number of chromosomes from the spermatozoon. A fertilized egg, or zygote (*zi'gōt*), containing the *diploid number* of chromosomes (46) is thus formed.

capacitation: L. *capacitas*, capable of
zona pellucida: L. *zone*, a girdle; *pellis*, skin
corona radiata: Gk. *korone*, crown; *radiata*, radiate
acrosome: Gk. *akron*, extremity; *soma*, body
haploid: Gk. *haploos*, single; L. *ploideus*, multiple in form
diploid: Gk. *diploos*, double; L. *ploideus*, multiple in form

Developmental Expositions

Discussions of prenatal development are presented at the ends of appropriate chapters, just before the chapter summary. These illustrated discussions explain the morphogenic events involved in the development of a body system.

Internal Affairs

The information presented in the Internal Affairs features ties the functional aspects of one body system to each of the others.

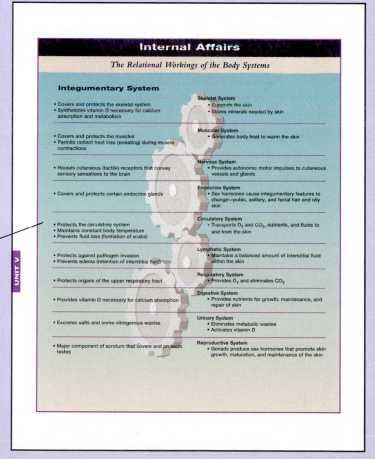

Chapter Summary

A summary, in outline form, at the end of each chapter reinforces your mastery of the chapter content.

Review Activities

At the end of each chapter, the objective and essay questions allow you to test the depth of your understanding and learning.

Answers to Objective Questions

Appendix A contains both the answers and the explanations for all of the end-of-chapter objective questions.

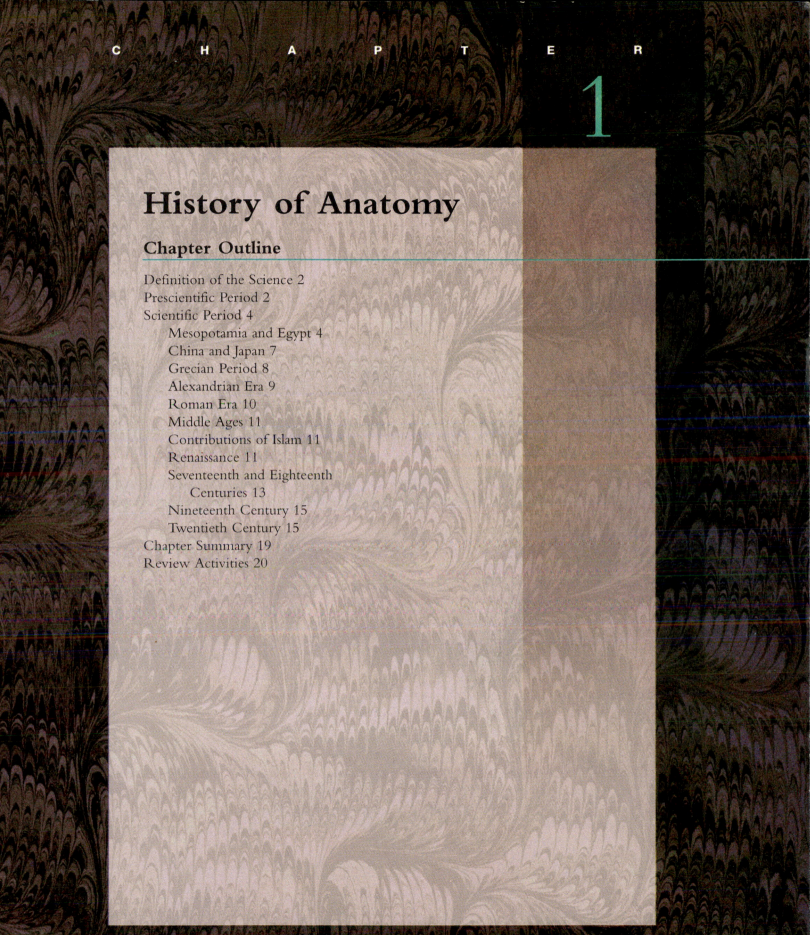

History of Anatomy

Chapter Outline

Definition of the Science

The science of human anatomy is concerned with the structural organization of the human body. The descriptive anatomical terminology is principally of Greek and Latin derivation.

Objective 1.   Define *anatomy*.

Objective 2.   Distinguish between anatomy, physiology, and biology.

Objective 3.   Explain why Greek and Latin have been used to describe structures of the human body.

Human anatomy is the science concerned with the structure of the human body. The term **anatomy** *(ah-nat'ŏ-me)* is derived from a Greek word meaning "to cut up," and in the past, the word *anatomize* was more commonly used than the word *dissect.* **Physiology** *(fiz''e-ol'ŏ-je)* is an extension beyond anatomy in that, as a science, it is concerned with the function of the body. The term *physiology* is derived from another Greek word meaning "the study of nature"—the "nature" of an organism is its function. Anatomy and physiology are both subdivisions of the science of **biology,** the study of living organisms. The anatomy of every structure of the body is adapted for performing a function, or perhaps several functions. Natural selection eliminates organisms with inappropriate structures and functions and determines which favorable structures will be passed from one generation to the next.

The dissection of human **cadavers** *(kah-dav'erz)* has served as the basis for understanding the structure and function of the human body for a long time. Each beginning student can discover and learn firsthand as the structures of the body are systematically studied. The anatomical terms that a student learns while becoming acquainted with a structure represent the work of hundreds of dedicated past anatomists who have dissected, diagrammed, described, and named the multitude of body parts.

Most of the terms that form the language of anatomy are of Greek or Latin derivation. Latin was the language of the Roman Empire, during which time an interest in scientific description was cultivated. With the decline of the Roman Empire, Latin became a "dead language," but it retained its value in nomenclature because it remained unchanged throughout history. As a consequence, anatomy is a descriptive science in which many of the terms can be understood through a knowledge of the basic prefixes and suffixes (see inside front and back covers). Although the Greeks and Romans made significant contributions to anatomical terminology, over the centuries many individuals from other cultures have also contributed to the science of human anatomy.

As a scientific field of inquiry, human anatomy has had a rich, long, and frequently troubled heritage. The history of human anatomy parallels that of medicine. In fact, interest in

anatomy: Gk. *ana,* up; *tome,* a cutting
physiology: Gk. *physis,* nature; *logos,* study
biology: Gk. *bios,* life; *logos,* study
cadaver: L. *cadere,* to fall

the structure of the body often results from the desire of the medical profession to explain a body dysfunction. Various religions, on the other hand, have at one time or another stifled the study of human anatomy through their restrictions on human dissections and their emphasis on nonscientific explanations of diseases and debilitations.

People have always had an innate interest in their own structure and physical capabilities. The Greeks esteemed physical athletic competition and expressed the beauty of the body in their sculptures. Many of the great masters of the Renaissance portrayed human figures in their art. Indeed, many of these artists were excellent anatomists because their attention to detail demanded it. Such an artistic genius was Michelangelo, who was preoccupied with the detail and beauty of the human form, capturing it in sculpture with the *David* (fig. 1.1) and in paintings like those in the Sistine Chapel.

Shakespeare's reverence for the structure of the human body found expression in his writings: "What a piece of work is a man! How noble in reason! how infinite in faculty! In form and moving, how express and admirable! In action how like an angel! In apprehension how like a god! The beauty of the world! The paragon of animals!" (*Hamlet* 2.2.315–319).

In the past, human anatomy was an academic, descriptive science primarily concerned with identifying and naming body structures. Although dissection and description form the basis of anatomy, the importance of human anatomy today is in its functional approach and clinical applications. Human anatomy is a practical, applied science that provides the foundation for understanding physical performance and body health. Studying the history of anatomy helps us to appreciate the relevant science that it is today.

1. What is the derivation and meaning of *anatomy?*
2. Explain the statement that anatomy is a science of observation and physiology is a science of experimentation and observation.
3. Why is knowing the anatomy and physiology of an organism essential to understanding its biology?
4. Discuss the value of using established Greek or Latin prefixes and suffixes to name newly described structures.

Prescientific Period

Evidence indicates that prehistoric anatomy was of survival value and provided the foundation for medicine.

Objective 4.   Discuss why understanding human anatomy is essential to the science of medicine.

Objective 5.   Define *trepanation* and *paleopathology.*

It is likely that a type of practical comparative anatomy is the oldest science. Certainly, humans have always been aware of some of their anatomical structures and how they function. Our prehistoric ancestors undoubtedly knew their own functional abilities and limitations as compared to those of other animals.

Figure 1.1 Michelangelo completed the seventeen-foot-tall *David* in 1504, sculptured from a single block of white, unflawed Carrara marble. This masterpiece captures the physical beauty of the human body in an expression of art.

Figure 1.2 Contemporary redrawings of large game mammals that were depicted on caves occupied by early *Homo sapiens* in western Europe. Presumably the location of the heart is drawn on the mammoth, and vulnerable anatomical sites are shown on the two bison. Prehistoric people needed a practical knowledge of anatomy simply for survival.

Through the trial and error of hunting, they discovered the "vital organs" of an animal, which, if penetrated with an object, would cause death (fig. 1.2). Likewise, they knew the vulnerable areas of their own bodies.

Prehistoric people knew which parts of an animal's body were useful as food, clothing, or implements. Undoubtedly, they knew that the muscles functioned in locomotion and also provided a major source of food. The skin from mammals with its associated fur served as a protective covering for the sparsely haired and vulnerable human. These people knew that the skeletal system formed a durable framework within their bodies and those of other vertebrates. They used the bones from the animals on which they fed to make a variety of tools and weapons. They knew that their own bones could be broken through accidents, and that improper healing would result in permanent disability. They knew that if an animal was wounded, it would bleed, and that excessive loss of this vital fluid would cause death. Perhaps they also realized that a severe blow on the head could cause deep sleep and debilitate an animal without killing it. Sexual differences were anatomically apparent, even though basic reproductive functions could not have been understood. The knowledge these people had was the basic, practical type, necessary for survival.

Certain surgical skills are also ancient. **Trepanation,** the drilling of a hole in the skull, or removal of a portion of a cranial bone, seems to have been practiced by several groups of prehistoric people. Trepanation was probably used as a ritualistic procedure to release evil spirits, or on some patients, perhaps, to relieve cranial pressure resulting from a head wound. Trepanated skulls have repeatedly been found in archaeological sites (fig. 1.3). Judging from the partial reossification in some of these skulls, apparently the patients occasionally survived.

What is known about prehistoric humans is conjectured through information derived from the cave drawings and

trepanation: Gk. *trypanon,* a borer

UNIT I

Figure 1.3 The surgical art of trepanation was practiced by several prehistoric cultures. Ossification around the bony edges of the wound indicates recovery from the operation.

fossils that contain paleopathological information. **Paleopathology** is the science concerned with studying diseases and causes of death in prehistoric humans. A person's age can be determined from skeletal remains, as well as certain injuries and diseases, including nutritional deficiencies. Diets and dental conditions are indicated by fossil teeth. What cannot be determined, however, is the amount of anatomical information and knowledge that was transmitted orally until humans invented symbols to record their thoughts, experiences, and history.

1. Using the example of trepanation, discuss the importance of knowing the anatomy of the skull and brain in order for the procedure to be successful.
2. List the types of data that a paleopathologist would be interested in obtaining from an Egyptian mummy.

Scientific Period

Human anatomy is a dynamic science that has had a long, exciting heritage and is currently providing the foundation for medical, biochemical, developmental, cytogenetic, and biomechanical research.

Objective 6. Discuss key historical events in the science of human anatomy.

Objective 7. List the periods of time when cadavers were used in studying human anatomy.

Objective 8. Explain why an understanding of human anatomy is important and personally applicable.

Objective 9. Discuss how a person can keep abreast of developments in anatomical research and note why this is important.

The scientific period begins with recorded anatomical observations made in early Mesopotamia on clay tablets in *cuneiform script* over 3,000 years ago and continues to the present day. Obviously, all of the past contributions to the science of anatomy cannot be mentioned; however, certain individuals and cultures had a tremendous impact and will be briefly commented upon in this section.

The history of anatomy has an interesting parallel with the history of the dissection of human cadavers, as is depicted in figure 1.4. The anatomists who made significant contributions are presented in table 1.1. Some of these contributions were in the form of books (table 1.2) that describe and illustrate the structure of the body and in some cases explain various body functions.

Mesopotamia and Egypt

Mesopotamia was the name given to the long, narrow wedge of land between the Tigris and Euphrates rivers, which is now a large part of present-day Iraq. Archaeological excavations and ancient records show that this area was settled prior to 4000 B.C. Because of the recorded information and the culture of the people, Mesopotamia is frequently called the cradle of civilization.

Many early investigations of the body were an attempt to describe basic life forces. For example, people wondered what specific organ constituted the soul, or the governing structure of the body. Some cuneiform writings from this era depict and describe body organs that were thought to serve this function. The liver, which was extensively studied in sacrificial animals (fig. 1.5), was thought to be the "guardianship of the soul and of the sentiments that make us men." This was a logical assumption because of the size of the liver and its close association with blood, which was observed to be vital for life. Even today, several European cultures attribute the liver with various emotions.

The warm blood and arrangement of blood vessels are obviously a governing system within the body, and this influenced the search for the soul. When excessive blood is lost, the body dies. Therefore, some concluded, blood must contain a vital, life-giving force. The scholars of Mesopotamia were influenced by this idea, as was Aristotle, the Greek scientist who lived centuries later. Aristotle believed that the seat of the soul was the heart and that the brain functioned in cooling the blood that flowed from the heart. The association of the heart, in song and poetry, with the emotions of love and caring has its basis in Aristotelian thought.

paleopathology: Gk. *palaios*, ancient; *pathos*, suffering disease; *logos*, to study

cuneiform: L. *cuneus*, wedge; *forma*, shape

Figure 1.4 An anatomical timeline depicting the story of cadaver dissections.

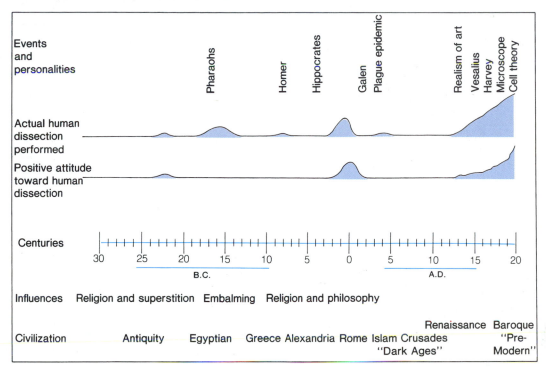

UNIT I

Table 1.1	Historical Perspective of Contributors to the Science of Human Anatomy		
Person	**Civilization**	**Lifetime or Date of Contribution**	**Contribution**
Menes	Egyptian	About 3400 B.C.	Wrote first anatomy manual
Homer	Ancient Greece	About 800 B.C.	Described the anatomy of wounds in the *Iliad*
Hippocrates	Ancient Greece	About 460–377 B.C	Father of medicine; Hippocratic oath
Aristotle	Ancient Greece	384–322 B.C.	Comparative anatomist; first recorded anatomical illustrations
Herophilus	Alexandria	About 325 B.C.	Remarkable work on aspects of the nervous system
Erasistratus	Alexandria	About 300 B.C.	Sometimes called father of physiology; concerned with body dysfunction and physiological changes
Celsus	Roman	30 B.C.–A.D. 30	Compiled information from Alexandrian school
Galen	Roman	130–201	Influential anatomical and medical writer
de' Luzzi	Renaissance	1487	Prepared dissection guide
Leonardo da Vinci	Renaissance	1452–1519	Outstanding anatomical sketches and innovative anatomical preparations
Vesalius	Renaissance	1514–64	Refuted past misconceptions by direct observation and experiment; often called father of anatomy
Harvey	Premodern (European)	1578–1657	Demonstrated function of circulatory system; applied experimental method to anatomy
Leeuwenhoek	Premodern (European)	1632–1723	Refined microscope; described various cells and tissues
Malpighi	Premodern (European)	1628–94	Regarded as father of histology
Sugita	Premodern (Japanese)	1774	Compiled a five-volume book of anatomy
Schleiden and Schwann	Modern (European)	1838–39	Formulated cell theory
Roentgen	Modern (European)	1895	Discovered X rays
Crick and Watson	Modern (England and United States)	1953	Determined the structure of DNA

Table 1.2	**Influential Anatomical Books and Publications**	

Aristotle. 384–322 B.C. *Historia animalium (History of animals), De partibus animalium (On the parts of animals),* and *De generatione animalium (On the generation of animals).* These classic works of the great Greek philosopher exerted a profound influence on biological thinking for centuries.

Celsus, Cornelius. 30 B.C.–A.D. 30. *De re medicina.* This eight-volume work was primarily a compilation of the medical data that was available from the Alexandrian school.

Galen, Claudius. 130–201. Nearly 500 medical papers on descriptive anatomy. Although Galen's writings contained numerous errors, his authoritative explanations of the structure and function of the body remained unchanged and influenced anatomists and physicians for nearly 1,500 years.

de' Luzzi, Mondino. 1487. *Anathomia.* This book was used as a dissection guide for over 225 years, during which time it underwent 40 revisions.

Vesalius, Andreas. 1543. *De humani corporis fabrica (On the structure of the human body).* The beautifully illustrated *Fabrica* boldly challenged many of the errors that had been perpetuated by Galen. In spite of the controversies this book provoked, it was well accepted and established a new standard of excellence in anatomy texts.

Fabricius ab Aquapendente, Hieronymus. 1600–1621. *De formato foetu (On the formation of the fetus)* and *De formatione ovi et pulli (On the formation of eggs and birds).* These books marked the beginning of embryological study.

Harvey, William. 1628. *Exercitatio de motu cordis et sanguinis in animalibus (On the motion of the heart and blood in animals).* Harvey advanced knowledge of circulation, and his experimental methods are still regarded as classic examples of method in science.

Descartes, René. 1637. *Discourse on method.* This philosophic thesis stimulated a mechanistic interpretation of biological data.

Linnaeus, Carolus. 1758. *Systema naturae.* The basis for the classification of living organisms was presented in this monumental work. Its anatomical value is in comparative anatomy, where the anatomy of different species is compared.

Haller, Albrecht von. 1760. *Elementa physiologiae (Physiological elements).* Some basic concepts of physiology are presented in this book, including a summary of what was then known of the functioning of the nervous system.

Sugita, Genpaku. 1774. *Kaitai shinsho (A new book of anatomy).* This book adopted the European concepts of the body and ushered in a new era of anatomy for the Japanese people.

Cuvier, Georges. 1817. *Le règne animal (The animal kingdom).* This comprehensive comparative vertebrate anatomy book had enormous influence on contemporary zoological thought.

Baer, Karl Ernst von. 1828–37. *Über entwicklungsgeschichté der thiere (On the development of animals).* This book provided the foundation for modern embryology by discussing germ layer formation.

Beaumont, William. 1833. *Experiments and observations on the gastric juice and the physiology of digestion.* Basic digestive functions are accurately described in this classic work.

Müller, Johannes. 1834–40. *Handbuch der physiologie des menschen für vorlesungen (Elements of physiology).* This book established physiology as a science concerned with the functioning of the body.

Schwann, Theodor. 1839. *Mikroskopische untersuchungen über die übereinstimmung in der struktur und dem wachstum der thiere und pflanzen (Microscopic researches into accordance in the structure and growth of animals and plants).* The concept of cellular organization of the body is presented in this classic study.

Kölliker, Albrecht von. 1852. *Mikroskopische anatomia (Microscopic anatomy).* This was the first textbook in histology. It became the foundation for the new science of histology.

Gray, Henry. 1858. *Anatomy, descriptive and surgical.* This masterpiece, better known as *Gray's anatomy,* is still in print and still contains over 200 of the original illustrations. Thousands of physicians have used this book to learn gross human anatomy.

Virchow, Rudolf. 1858. *Die cellularpathologie (Cell pathology).* Descriptions of normal and diseased tissues are presented in this book.

Darwin, Charles. 1859. *On the origin of species.* This was one of the most influential books ever published in biology. Its importance to anatomy is that it provided an explanation for the anatomical variation evident among different species.

Mendel, Gregor. 1866. *Versuche über pflanzenhybriden (Experiments with plant hybrids).* In this work, Mendel demonstrated the basic principles of heredity.

Owen, Richard. 1866. *Anatomy and physiology of the vertebrates.* Some basic concepts of structure and function, such as homologue and analogue, are presented in this book.

Balfour, Francis M. 1880. *Comparative embryology.* This book is considered the source of information for the science of modern experimental embryology.

Weismann, August. 1892. *Das keimplasma (The germplasm).* Weismann postulated the theory of meiosis, which states that a reduction in the chromosome number is necessary in the gametes of both the male and female for fertilization to occur.

Hertwig, Oskar. 1893. *Zelle and gewebe (Cell and tissue).* Important distinctions between the sciences of cytology and histology are made in this book.

Wilson, Edmund B. 1896. *The cell in development and heredity.* This book had a profound influence on the development of cytogenetics.

Pavlov, Ivan. 1897. *Le travail des glandes digestives (The work of the digestive glands).* The physiological functioning of the digestive system is described in this classic experimental work.

Sherrington, Charles. 1906. *The integrative action of the nervous system.* The basic concepts of neurophysiology were first established in this book.

Garrod, Archibald. 1909. *Inborn errors of metabolism.* Genetic defects are discussed in this pioneer book and are shown to be caused by defective genes.

Bayliss, William M. 1915. *Principles of general physiology.* This book provided a much needed synthesis of a newly emerging science.

Spemann, Hans. 1938. *Embryonic development and induction.* This masterful book provided the foundation for the science of experimental embryology.

Crick, Francis H. C., and James D. Watson. 1953. *Genetic implications of the structure of deoxyribonucleic acid.* This remarkable work provided explanations of genetic replication and control of cellular functions.

The ancient Egyptian culture neighbored Mesopotamia to the west. Here, the sophisticated science of embalming the dead in the form of mummies was perfected (fig. 1.6). No attempts were made to perform anatomical or pathological studies on the corpses, however, because embalming was strictly a religious ritual, reserved for royalty and the wealthy to prepare them for a life after death.

Several written works concerning anatomy have been discovered from ancient Egypt, but none of these influenced succeeding cultures. Menes, a king-physician during the first

---

embalm: L. *in,* in; *balsamum,* balsam

physician: Gk. *physikos,* natural

Figure 1.5 An inscribed clay model of a sheep's liver from the eighteenth or nineteenth century B.C. The people of ancient Mesopotamia regarded the liver as the source of human emotions.

Figure 1.6 Perhaps the greatest contribution of the ancient Egyptian era to anatomy and medicine is the information obtained from the mummies. Certain diseases, injuries, deformities, and occasionally cause of death can be determined from paleopathogenic examination of the mummified specimens. Shown on the right is a congenital clubfoot from a mummy of a person who lived during the Nineteenth Dynasty (about 1300 B.C.).

Egyptian dynasty of about 3400 B.C. (even before the pyramids were built), wrote what is thought to be the first manual on anatomy. Later writings (2300–1250 B.C.) attempted a systematization of the body, beginning with the head and progressing downward.

The techniques of embalming could have been a major contribution to the science of anatomy had these procedures been recorded and shared with future cultures. Apparently, however, embalmings were not well accepted by the people in ancient Egypt. In fact, embalmers were looked down upon and were frequently persecuted and even stoned. Embalming was a mystic art related more to religion than to science, and since its practice required a certain amount of mutilation of the dead body, it was regarded as demonism and antireligious. Consequently, embalming techniques that would provide embalmed cadavers as dissection specimens had to wait until centuries later to be rediscovered.

Like the people of Mesopotamia and Greece, the ancient Egyptians were concerned with a controlling spirit of the body. In fact, they even had a name for this life force—the Ba spirit—and they believed that it was associated with the bowels and the heart. Food was placed in the tomb of a mummy to feed the Ba spirit during the journey to Osiris, Egyptian god of the underworld.

China and Japan

China   The Chinese have always had an interest in the structure and function of the human body, but their studies were rooted in philosophy rather than investigation. The Chinese revered the body and abhorred its mutilation. Their only knowledge of the internal organs was from wounds and injuries. Only in recent times have the dissections of cadavers been permitted.

The ancient Chinese had an abiding belief that everything within the universe depended on the balance of the two opposing principles of *yin* and *yang*. They believed that yin and yang in the body functioned as the circulatory system. The blood was the conveyor of the yang, and the heart and vessels represented the yin. Other structures of the body were composed of lesser forces termed $z\bar{o}$ and $\int \bar{u}$.

The Chinese were great herbalists. Writings over 5,000 years old describe various herbal concoctions and potions to alleviate numerous ailments, such as diarrhea and constipation. Opium was described as an excellent painkiller.

Until recently, the Chinese have been possessive of their beliefs, and, consequently, Western cultures were not influenced by Chinese thoughts or writings. Perhaps the best known but least understood of the Chinese contributions to human anatomy and medicine is acupuncture.

Acupuncture is an ancient practice that was established to maintain a balance between the yin and the yang. Three hundred sixty-five precise meridian sites, or vital points, that correspond to the number of days in a year, were identified on the body (fig. 1.7). Needles inserted into the various sites were believed to release bad secretions and rid the tissues of obstructions. Acupuncture is still practiced in China and has gained acceptance with some medical specialists in the United States as a

Figure 1.7   An acupuncture chart from the Ming dynasty of ancient China.

In 1774, a five-volume book called *Kaitai Shinsho (A New Book of Anatomy)*, published by a Japanese physician, Genpaku Sugita, totally adopted the Dutch concepts of the body. This book marked the beginning of a modern era in anatomy and medicine for the Japanese people.

For several hundred years, Western nations were welcome in Japan. In 1603, however, the Japanese government banned all contact with the Western world because they feared the influences of Christianity on their society. Although this ban was strictly enforced and Japan became isolated, Japanese scholars continued to circulate Western anatomical and medical books. These books eventually prompted Japanese physicians to reassess what they had been taught concerning the structure of the body.

Grecian Period

Anatomy was first widely accepted as a science in ancient Greece. The writings of several Greek philosophers had a tremendous impact on future scientific thinking. During this period, the Greeks were obsessed with the physical beauty of the human body, which they expressed through exquisite sculptures.

The youth of Greece were urged to be athletic and develop their physical abilities, but at about age 18 they were directed to intellectual pursuits of science, literature, and philosophy. With this strong cultural emphasis, it was only natural that great strides were made in the sciences.

Perhaps the first written reference to the anatomy of wounds sustained in battle is contained in the *Iliad,* written by Homer in about 800 B.C. Homer's detailed descriptions of the anatomy of wounds were exceedingly accurate. However, he described clean wounds—not the type of traumatic wound that would likely be suffered on a battlefield. This has led to speculation that human dissections were conducted during this period and that anatomical structure was well understood. Victims of human sacrifice may have served as subjects for anatomical study and demonstration.

Hippocrates   Hippocrates (460–377 B.C.) was a famous Greek physician who is regarded as the father of medicine because of the sound principles of medical practice that he established (fig. 1.8). His name is memorialized in the Hippocratic oath, which many graduating medical students repeat as a promise of professional stewardship and duty to humankind (table 1.3).

Hippocrates probably had limited exposure to human dissections but was well disciplined in the popular *humoral theory* of body organization. Four body humors were recognized, and each was associated with a particular body organ: blood with the liver; choler, or yellow bile, with the gallbladder; phlegm with the lungs; and melancholy, or black bile, with the spleen. A healthy person was thought to have a balance of the four humors. The concept of humors has long since been discarded, but it dominated medical thought for over 2,000 years.

Perhaps the greatest contribution of Hippocrates was that he attributed diseases to natural causes rather than to the displeasure of the gods. His application of logic and reason to medicine was the beginning of observational medicine.

technique of anesthesia and as a cure for certain ailments. The effectiveness of acupuncture has been documented and is more than psychological. Acupuncture sites have been identified on domestic animals and have been used to a limited degree in veterinary medicine. Why acupuncture is effective remains a mystery, although it has recently been correlated with endorphin production within the brain (see chapter 11).

Japan   The advancement of anatomy in Japan was strongly influenced by the Chinese and Dutch. The earliest records of anatomical interest in Japan date back to the sixth century. Buddhist monks from Japan were trained in China where they were exposed to Chinese philosophy, and so the Chinese beliefs concerning the body were prevalent in Japan as well. By the eighteenth century, influences from the Western nations, especially the Dutch, were such that the Japanese sought to determine for themselves which version of anatomy was correct.

Figure 1.8 A fourteenth-century painting of the famous Greek physician Hippocrates. Hippocrates is referred to as the father of medicine; his creed is immortalized as the Hippocratic oath.

UNIT I

Table 1.3	**The Hippocratic Oath**

I swear by Apollo Physician and Aesculapius and Hygeia and Panacea and all the gods and goddesses, making them my witnesses, that I will fulfill according to my ability and judgment this oath and this covenant:

To hold him who has taught me this art as equal to my parents and to live my life in partnership with him, and if he is in need of money to give him a share of mine, and to regard his offspring as equal to my brothers in male lineage and to teach them this art—if they desire to learn it—without fee and covenant; to give a share of precepts and oral instruction and all the other learning to my sons and to the sons of him who has instructed me and to pupils who have signed the covenant and have taken an oath according to the medical law, but to no one else.

I will apply dietetic measures for the benefit of the sick according to my ability and judgment; I will keep them from harm and injustice.

I will neither give a deadly drug to anybody if asked for it, nor will I make a suggestion to this effect. Similarly I will not give to a woman an abortive remedy. In purity and holiness I will guard my life and my art.

I will not use the knife, not even on sufferers from stone, but will withdraw in favor of such men as are engaged in this work.

Whatever houses I may visit, I will come for the benefit of the sick, remaining free of all intentional injustice, of all mischief, and in particular of sexual relations with both female and male persons, be they free or slaves.

What I may see or hear in the course of the treatment or even outside of the treatment in regard to the life of men, which on no account one must spread abroad, I will keep to myself, holding such things shameful to be spoken about.

If I fulfill this oath and do not violate it, may it be granted to me to enjoy life and art, being honored with fame among all men for all time to come; if I transgress it and swear falsely, may the opposite of all this be my lot.

The four humors are a part of our language and medical practice even today. *Melancholy* is a term used to describe a depressed or despondent temperament in a person, while *melanous* refers to a black or sallow complexion. The prefix *melano-* means black. *Cholera* is an infectious intestinal disease that causes diarrhea and vomiting. *Phlegm* (pronounced *flem*) within the upper respiratory system is symptomatic of several respiratory disorders. *Sanguine*, which originally referred to blood, is used to describe a passionate temperament.

Aristotle   Aristotle (384–322 B.C.), a pupil of Plato, was an accomplished writer, philosopher, and zoologist (fig. 1.9). He was also a renowned teacher and was hired by King Philip of Macedonia to tutor his son, Alexander, who later became known as Alexander the Great.

Aristotle made careful investigations of all kinds of animals, including references to humans, and established a type of scientific method in obtaining data. He wrote the first known account of embryology, in which he described the development of the heart in a chick embryo. He named the aorta and contrasted the arteries and veins. Aristotle's best known zoological works are *History of Animals, On the Parts of Animals,* and *On the Generation of Animals.*

In spite of his tremendous accomplishments, Aristotle perpetuated some erroneous views of anatomy. For example, the doctrine of the humors formed the boundaries of his thought. Plato had described the brain as the seat of feeling and thought, but Aristotle disagreed. He stated that the heart was the seat of intelligence and that the function of the brain, which was bathed in fluid, was to cool the blood that was pumped from the heart.

Alexandrian Era

Alexander the Great founded Alexandria in 332 B.C. and established it as the capital of Egypt and a center of learning. In addition to a great library, there was also a school of medicine in Alexandria. The study of anatomy flourished because of the acceptance of dissections of human cadavers and human **vivisections** (dissections of living things). This seemingly barbaric practice was commonly performed on condemned criminals. People reasoned that to best understand the functions of the body, it should be studied while the subject was alive, and that a condemned man could best repay society through the use of his body for a scientific vivisection.

humor: L. *humor*, fluid
melancholy: Gk. *melan*, black; *chole*, bile
choler: Gk. *chole*, bile
phlegm: Gk. *phlegm*, inflammation
sanguine: L. *sanguis*, bloody

Aesculapius: Gk. (mythology) son of Apollo and god of medicine
Hygeia: Gk. (mythology) daughter of Aesculapius; personification of health; *hygies*, healthy
Panacea: Gk. (mythology) also a daughter of Aesculapius; assisted in temple rites and tended sacred serpents; *pan*, all, every; *akos*, remedy
vivisection: L. *vivus*, alive; *sectio*, a cutting

UNIT I

Figure 1.9   This Roman copy of a Greek sculpture is believed to be of Aristotle, the famous Greek philosopher.

Unfortunately, the scholarly contributions and scientific momentum of Alexandria did not endure. Most of the written works were destroyed when the great library was burned by the Romans as they conquered the city in 30 B.C. What is known about Alexandria was obtained from the writings of later scientists, philosophers, and historians, including Pliny, Celsus, Galen, and Tertullian. Two men of Alexandria, Herophilus and Erasistratus, made lasting contributions to the study of anatomy.

Herophilus   Herophilus (about 325 B.C.) was trained in the Hippocratic school but became a great teacher of anatomy in Alexandria. Through vivisections and dissections of human cadavers, Herophilus provided excellent descriptions of the skull, eye, various visceral organs and organ relationships, and the functional relationship of the spinal cord to the brain. Two monumental works of Herophilus are entitled *On Anatomy* and *Of the Eyes*. Herophilus regarded the brain as the seat of intelligence and described many of its structures, such as the meninges, cerebrum, cerebellum, and fourth ventricle. He was the first to identify nerves as either sensory or motor.

Erasistratus   Erasistratus (about 300 B.C.) was more interested in body functions than structure and is therefore frequently referred to as the father of physiology. Erasistratus authored a book on the causes of diseases, in which he included observations on the heart, vessels, brain, and cranial nerves. He noted the toxic effects of snake venom on various visceral organs and described changes in the liver resulting from various diseases. Although some of the writings of Erasistratus were very scientific, he also had notions that were primitive and mystical. He thought that the cranial nerves carried animal spirits and that muscles contracted because of distention by spirits. He believed that the left ventricle of the heart was filled with a vital air spirit (pneuma) that came in from the lungs and that the vessels between the lungs and heart contained pneuma rather than blood. Erasistratus thought that veins contained blood and that arteries transported pneuma but filled with blood that drained from veins when an artery was cut.

> Both Herophilus and Erasistratus were greatly criticized later in history for the vivisections they performed. Celsus (about 30 B.C.) and Tertullian (about A.D. 200) were particularly critical of the practice of vivisection. Herophilus was described as a butcher of men who had dissected as many as 600 living human beings, sometimes in public demonstrations.

Roman Era

In many respects, the Roman Empire stifled scientific advancements and set the stage for the Dark Ages. The interest and emphasis of science shifted from theoretical to practical during this time. Few dissections of cadavers were performed other than in attempts to determine the cause of death in criminal cases. Medicine was not preventive but was almost solely limited to the treatment of soldiers injured in battle. Later in Roman history, laws were established that attested to the influence of the Church on medical practice. According to Roman law, for example, no deceased pregnant woman could be buried without prior removal of the fetus from the womb so that it could be baptized.

The scientific documents that have been preserved from the Roman Empire are mostly compilations of information obtained from the Greek and Egyptian scholars. New anatomical information was scant and for the most part consisted of speculation or was derived from dissections of animals. Two important anatomists from the Roman era were Celsus and Galen.

Celsus   The major contribution of Cornelius Celsus (30 B.C.–A.D. 30) was the preservation of a portion of the medical information from the Alexandrian school. He compiled this information into an eight-volume work called *De re medicina*. Some pathological information is included in this record.

Galen   Claudius Galen (A.D. 130–201) was perhaps the best physician since Hippocrates. He was certainly the most influential writer of all times on medical subjects. For nearly 1,500

visceral: L. *viscus*, internal organ of torso

pneuma: Gk. *pneuma*, air

years, the writings of Galen served as the unquestionable authority on anatomy and medical treatment. Galen probably dissected no more than two or three human cadavers during his career, limiting his anatomical descriptions to animal dissections. He compiled nearly 500 medical papers (of which 83 have been preserved) from earlier works of others, as well as from his personal studies. He perpetuated the concept of the humors of the body and gave authoritative explanations of nearly all body functions.

Galen's works contain many errors, primarily because he wanted to present definitive answers and because his conclusions regarding human body functions were largely based on data from nonhuman animals. He did, however, provide some astute and accurate anatomical details that are still regarded as classics. He proved to be an experimentalist, demonstrating that the heart of a pig would continue to beat when spinal nerves were transected so that nerve impulses could not reach the heart. He showed that the squealing of a pig stopped when the particular nerve that innervated the vocal cords was cut. He also proved that arteries contained blood rather than pneuma.

Galen compiled a list of many medicinal plants and used medications to treat many illnesses. Although he frequently used bloodletting to balance the four humors, he cautioned against removing too much blood. He accumulated a wide variety of medical instruments and suggested their use as forceps, retractors, scissors, and splints (fig. 1.10). He was also a strong advocate of helping nature heal through hygiene, diet, rest, and exercise.

Middle Ages

The Middle Ages, frequently referred to as the Dark Ages, began with the fall of Rome to the Goths in A.D. 476 and lasted nearly 1,000 years, until Constantinople was conquered by the Turks in 1453. The totalitarian Christian Church suppressed science, and Europe sank into stagnation. Human existence continued to be miserably precarious, and people no longer felt capable of learning from personal observation but rather accepted life "on faith."

Dissections of cadavers were totally prohibited, and molesting a corpse was a criminal act frequently punishable by burning at the stake. If mysterious deaths occurred, examinations by inspection and palpation were acceptable. During the plague epidemic in the sixth century, however, a few necropsies and dissections were performed in hopes of determining the cause of this dreaded disease (see fig. 1.4).

During the Crusades soldiers cooked the bones of their dead comrades so that they could be returned home for proper burial. Even this act, however, was eventually considered sacrilegious and was strongly condemned by the Church. One of the ironies of the Middle Ages was that the peasants received far less respect and had fewer rights when they were alive than when they were dead.

epidemic: Gk. *epi*, upon; *demos*, people
necropsy: Gk. *nekros*, corpse; *opsy*, view

Figure 1.10 These surgical and gynecological instruments were found in the House of the Surgeon (about A.D. 62–79) in Pompeii and are representative of the medical equipment used throughout the Roman Empire during this time.

Contributions of Islam

The Arabic-speaking people made a profound contribution to the history of anatomy in a most unusual way. On several occasions during the Middle Ages, Arab armies conquered sections of the Middle East. As part of the spoils of their conquest, most of the anatomical writings were seized from the libraries of Alexandria. These materials were then carried back to the Arab countries and translated from Greek to Arabic.

As the Dark Ages stagnated Europe, the Christian Church attempted to destroy any scholarship of worldly knowledge that was not acceptable within Christian dogma. The study of the human body was considered heretical, and the Church banned all written anatomical material. Had it not been for the Arab invasions, the early writings of anatomy from Western civilization would never have survived. It wasn't until the thirteenth century that the Arabic translations were returned to Europe and, in turn, translated to Latin. During the translation process, any Arabic terminology that had been introduced was systematically removed so that today we find few anatomical terms of Arabic origin.

Renaissance

The period of time known as the Renaissance was characterized by a rebirth of science. It lasted from the fourteenth through the sixteenth century and was a transitional period from the Middle Ages to the modern age of science.

The Renaissance was ushered in by the great European universities established in Bologna, Salerno, Padua, Montpellier, and Paris. The first recorded human dissections at these newly established universities were the work of the surgeon William of Saliceto (1215–80), from the University of Bologna. The study of anatomy quickly spread to other universities, and by the year 1300 human dissections had become an integral part of the medical curriculum. However, the Galenic dogma that normal human anatomy was sufficiently understood persisted, so interest at this time centered on methods and techniques of dissection rather than on the study of the human body.

The development of movable type in about 1450 revolutionized the production of printed books. One of the first anatomy books to be printed in this manner was that of Jacopo Berengario of Carpi, who was a professor of surgery at Bologna. He described many anatomical structures, including the appendix, thymus, and larynx.

The most influential text of this period was written by Mondino de' Luzzi, also of the University of Bologna. This book, first published in 1487, was more of a dissection guide than a study of gross anatomy, and in spite of its numerous Galenic errors, 40 editions were published, until the time of Vesalius.

> Because of the rapid putrefaction of an unembalmed corpse, the anatomy textbooks of the early Renaissance were organized so that the more perishable portions of the body were dissected first. Dissection began with the abdominal cavity, then the chest, followed by the head, and finally the appendages. Dissection was a marathon event, frequently continuing for perhaps 4 days.

With the increased interest in anatomy during the Renaissance, obtaining cadavers for dissection became a serious problem. Medical students regularly practiced grave robbing until finally an official decree was issued that permitted the bodies of executed criminals to be used as specimens.

> Corpses were embalmed to prevent deterioration, but it was not especially effective and the stench from cadavers was apparently a continual problem. Anatomy professors lectured from a thronelike chair at some distance from the immediate area (fig. 1.11). The phrase "I wouldn't touch that with a 10-foot pole" probably originated during this time in reference to the smell of a decomposed cadaver.

The major advancements in anatomy during the Renaissance came from the artistic and scientific abilities of Leonardo da Vinci and Andreas Vesalius. Working in the fifteenth and sixteenth centuries, each produced monumental studies of the human form.

Leonardo The great Renaissance Italian Leonardo da Vinci (1452–1519) is best known for his artistic works (e.g., *Mona Lisa*) and his scientific contributions. He displayed genius as a painter, sculptor, architect, musician, and anatomist—although his contributions to anatomy were lost until the twentieth century. As a young man, Leonardo regularly observed cadaver dissections and intended to publish a textbook on anatomy with the Pavian professor Marcantonio della Torre. The untimely death of della Torre at the age of 31 halted their plans. When Leonardo died, his anatomical sketches were lost and were not discovered for over four centuries. The advancement of anatomy

Figure 1.11 The scene of a cadaver dissection, 1500, from *Fasciculus Medicinae* by Johannes de Ketham. The anatomy professor removed himself from the immediate area to a thronelike chair overlooking the proceedings. The dissections were performed by hired assistants. One of them, the ostensor, pointed to the internal structures with a wand as the professor lectured.

would have been accelerated by many years if Leonardo's notebooks had been available to the world at the time of his death.

Leonardo's illustrations incorporated a totally new perspective that showed the structural relationships of the various body organs. He was intent on accuracy, and his sketches are incredibly detailed (fig. 1.12). He experimentally determined the structure of complex body organs such as the brain and the heart. He made wax casts of the ventricles of the brain to study its structure. He constructed models of the heart valves to demonstrate their action.

Vesalius The contribution of Andreas Vesalius (1514–64) to the science of human anatomy and to modern medicine is immeasurable. Vesalius was born in Brussels to a family of physicians. He trained in Paris prior to accepting a position at the University of Padua in Italy. It was at Padua that he earned recognition as an anatomist and an artist. Vesalius participated in human dissections and initiated the use of live models to determine surface landmarks for internal structures (fig. 1.13).

Vesalius apparently had enormous energy and ambitions. He completed the masterpiece of his life, *De Humani Corporis Fabrica,* by the time he was 28 years old. The various body systems and individual organs were beautifully illustrated and described in the *Fabrica* (fig. 1.14). Vesalius was devoutly religious and considered the human body to be God's most beautiful creation. His book was especially important in that it boldly chal-

Figure 1.12 Only a master artist could achieve the detail and accuracy of the anatomical sketches of Leonardo da Vinci. "The painter who has acquired a knowledge of the nature of the sinews, muscles and tendons," Leonardo wrote, "will know exactly in the movement of any limb how many and which of the sinews are the cause of it, and which muscle by its swelling is the cause of the sinew's contracting."

Figure 1.13 A painting of the great anatomist Andreas Vesalius, as he dissects a cadaver. From his masterpiece *De Humani Corporis Fabrica*.

Figure 1.14 A plate from *De Humani Corporis Fabrica*, which Vesalius completed at the age of twenty-eight. This book, published in 1543, revolutionized the science of anatomy.

lenged hundreds of Galen's erroneous teachings. Vesalius wrote of his surprise at finding numerous anatomical errors that were taught as fact, and he refused to accept Galen's explanations on faith. Bitter controversies ensued between Vesalius and the traditional Galenic anatomists, including Vesalius's former teacher Sylvius. Vesalius became so incensed by the relentless attacks on him that he destroyed much of his unpublished work and ceased his dissections. Another point to his credit is that unlike Sylvius, Fallopius, Eustachius, and other anatomists of his time, Vesalius chose not to have his name attached to parts of the body that he described.

Although Vesalius was the greatest anatomist of his epoch, others made significant contributions and to an extent paved the way for Vesalius. Michelangelo pursued anatomy in 1495, being supplied with corpses by the friar of a local monastery. Mondino de' Luzzi and the surgeon Jacopo Berengario of Carpi also corrected many errors of Galen. Fallopius (1523–62) and Eustachius (1524–74) completed detailed dissections of specific body regions.

Seventeenth and Eighteenth Centuries

During the seventeenth and eighteenth centuries, the science of anatomy attained an unparalleled acceptance and theatrical-like status. Elaborate amphitheaters were established in various parts of Europe for public demonstrations of human dissections

Figure 1.15 *The Anatomy Lesson of Dr. Tulp,* a famous Rembrandt painting completed in 1632, depicts one of the public anatomies that were popular during this period.

(fig. 1.15). Exorbitantly priced tickets for the public anatomies were sold to the wealthy, who witnessed the dissection of a cadaver by elegantly robed anatomists who were splendid orators. The subject was chosen from condemned criminals, and the performance was scheduled during cold weather because of the perishable nature of the body.

Fortunately there were also serious, scientific-minded anatomists during this period who made several significant contributions. Two of the most important contributions were the explanation of blood flow and the development and use of the microscope.

Harvey   In 1628, the English anatomist William Harvey (1578–1657) published his outstanding work *On the Motion of the Heart and Blood in Animals.* This important research established brilliant proof of the continuous circulation of blood within contained vessels, and the technique of investigation used by Harvey is still regarded as a classic example of the scientific method for conducting research (fig. 1.16). Like Vesalius, Harvey was severely criticized for his departure from Galenic philosophy. The controversy over circulation of the blood raged for 20 years, until other anatomists finally repeated Harvey's experiments and added information.

Leeuwenhoek   Antoni van Leeuwenhoek (1632–1723) was a Dutch lens grinder who improved the microscope to where he achieved a magnification of 270 times. His many contributions included developing techniques for examining tissues and describing blood cells, spermatozoa, and skeletal muscle. Leeuwenhoek did not understand the role of sperm in fertilization, but thought that a human sperm contained a miniature human being called a *homunculus.*

The development of the microscope added an entirely new dimension to anatomy and eventually led to explanations of basic body functions. In addition, the improved microscope was invaluable for understanding the etiologies of many diseases and discovering cures for many of them. Although Leeuwenhoek improved the microscope, credit for its invention is usually given to the Dutch maker of spectacles, Zacharius Janssen. The first scientific investigation using a microscope was done by Francisco Stelluti in 1625 on the structure of a bee.

Malpighi and Others   Marcello Malpighi (1628–94), an Italian anatomist, has been referred to as the father of histology. He

homunculus: L. diminutive form of *homo,* man

Figure 1.16 In the early seventeenth century, the English physician William Harvey demonstrated that blood circulates and does not flow back and forth through the same vessels.

The term *cell* was coined in 1665 by an English physician, Robert Hooke, as he examined the structure of cork in an attempt to explain its buoyancy. What Hooke actually observed were the rigid walls that surrounded the empty cavities of dead cells. The significance of cellular structure did not become apparent until approximately 150 years after Hooke's work.

With improved microscopes, finer details were observed. In 1809, a French zoologist, Jean Lamarck, observed the jellylike substance within a living cell and speculated that this material was more important than the outside structure of a cell. Fifteen years later, René H. Dutrochet described the differences between plant and animal cells.

Two German scientists, Schleiden and Schwann, are credited with the biological principle referred to as the *cell theory.* In 1838 the botanist Matthias Schleiden suggested that each plant cell leads a double life. He believed that in some respects the cell was an independent organism, but at the same time it cooperated with the other cells that made up the whole plant. A year later, in his work with animal cells, Theodor Schwann concluded that all organisms were composed of cells that were essentially alike. This was followed 19 years later by a further biological principle that seemed to complete the explanation of cells. In 1858 the German pathologist Rudolf Virchow wrote a book entitled *Cell Pathology* in which he proposed that cells come only from other cells. The mechanism of cellular replication, however, was not understood for several more decades.

Johannes Müller (1801–58) is noted for applying the sciences of physics, chemistry, and psychology to the study of the human body. With this increased dimensional breadth to the subject, anatomy became a comparative science.

Twentieth Century

The contributions to the science of anatomy during the twentieth century have not been as astounding as they were when little was known about the structure of the body. The study of anatomy within the twentieth century became specialized, and research became more detailed and complex.

One innovation that gained momentum early in the twentieth century was the simplification and standardization of nomenclature. Because of the proliferation of scientific literature toward the end of the nineteenth century, over 30,000 terms for structures in the human body were reported, many of which were synonymous and referred to similar structures. In 1895, in an attempt to clarify the confusion, the German Anatomical Society met in Basel, Switzerland, and an official list of approximately 500 terms was approved, which was called the *Basle Nomina Anatomica (BNA)*. The terms on this list were universally accepted for use in teaching and publications.

Other conferences on nomenclature have been held throughout the century, with increased interest. The meetings have become known as the *International Congress of Anatomists.* During the meetings of the Seventh International Congress held in New York City in 1960, a resolution was passed to eliminate all proper names ("tombstone names") from anatomical terminology and use descriptive names in their place. Structures like Stensen's duct and Wharton's duct, for example, are now correctly referred to respectively as the parotid duct and

discovered the capillary blood vessels that Harvey had postulated, described the alveoli of lungs, and discussed the histological structure of the spleen and kidneys.

Many other anatomists made contributions during this 200-year period. In 1672, the Dutch anatomist de Graaf described the ovaries of the female reproductive system, and in 1775 Spallanzani showed that both the ovum and the sperm were necessary for conception. Francis Glisson (1597–1677) described the liver and associated the gallbladder, the stomach, and the intestines. Thomas Wharton (1614–73) and Niels Stensen (1638–86) separately contributed to knowledge of the salivary glands and lymph nodes within the neck and facial regions. In 1664, Thomas Willis published a summary of what was known about the nervous system.

A number of anatomical structures throughout the body are named in honor of these early anatomists. Thus we have *Graafian follicles, Stensen's* and *Wharton's ducts, Fallopian tubes, Bartholin's glands,* the *circle of Willis,* and many others. These terms are more meaningful when something is known of their origin.

Nineteenth Century

The major contribution in the nineteenth century was the formulation of the cell theory and the implications it had for a clearer understanding of the structure and functioning of the body. Once the microscope was invented, it was merely a matter of time before cells were discovered and described.

Figure 1.17 Different techniques for viewing microscopic anatomy have greatly enhanced our understanding of the structure and function of the human body. (*a*) The appearance of hair under a simple magnifying glass, (*b*) an observation of a stained section of a hair through a light microscope, and (*c*) a hair emerging from skin as viewed through an electron microscope (340× at 35-mm size).

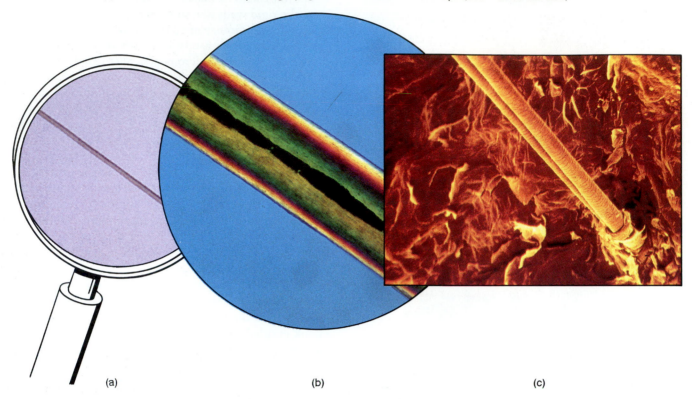

(a)   (b)   (c)

submandibular duct. Because of popular usage, however, it will be extremely difficult to eliminate all the proper names from anatomical terminology. But at least there is a trend toward descriptive simplification.

In this text, the preferred descriptive names are used in both the text narrative and the accompanying illustrations. Where the term first appears in the narrative, however, the preferred name is followed by the traditional name honoring an individual enclosed in parentheses. It will appear like this: **uterine** (fallopian) **tube.** The terminology used in this text is in accordance to the official anatomical nomenclature presented in the reference publication, *Nomina Anatomica,* Sixth Edition.

In response to the increased technology and depth of understanding in the twentieth century, new disciplines and specialties have appeared in the science of human anatomy in an attempt to categorize and use the new knowledge. The techniques of such cognate disciplines as chemistry, physics, electronics, mathematics, and computer science have been incorporated into research efforts.

There are several well-established divisions of human anatomy. The oldest, of course, is **gross anatomy,** which is the study of the structures of a cadaver that can be observed with the unaided eye. Stringent courses in gross anatomy in professional schools provide the foundation for the student's entire medical or paramedical training. Gross anatomy also forms the basis for the other specialties within anatomy. **Surface anatomy** (see chapter 10) deals with surface features of the body that can be observed or palpated (felt firmly).

Microscopic Anatomy Structures smaller than 0.1 mm (100 μm) can be seen only with the aid of a microscope. The sciences of **cytology** (study of cells), or **cellular biology,** and **histology** (study of tissues) are specialties of anatomy that have provided additional understanding of the structure and function of the human body. One can observe greater detail with the *electron microscope* than with the *light microscope* (fig. 1.17). New techniques in staining and histochemistry have aided study using the electron microscope—sometimes referred to as *ultrastructure,* or *fine structure.*

Radiological Anatomy Radiographic anatomy, or *radiology,* provides a way of observing structures within the living body. Radiology is based on the principle that substances of different densities absorb different amounts of X rays, resulting in a differential exposure on film. Radiopaque substances like barium can be ingested (swallowed) or injected into the body to produce even greater contrasts (fig. 1.18). **Angiography** involves making a radiograph after injecting a dye into the bloodstream. In *angiocardiography,* the heart and its associated vessels are x-rayed. *Cineradiography* permits the study of certain body systems through the use of motion picture radiographs. Traditional radiographs have had limitations as diagnostic tools for understanding human anatomy because of the vertical two-dimensional plane that is photographed. Since radiographs compress the body image with an overlap of organs and tissues, diagnosis is often difficult.

Figure 1.18 The versatility of radiography makes this technique one of the most important tools in diagnostic medicine and provides a unique way of observing specific anatomical structures within the body. (*a*) A radiograph of a healing fracture, (*b*) a radiograph of gallstones within a gallbladder, and (*c*) a radiograph of a stomach.

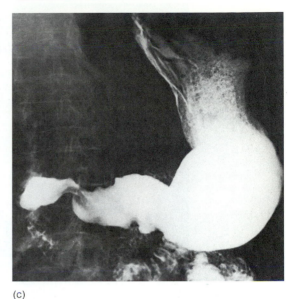

(a)   (b)   (c)

X rays were discovered in 1895 by Wilhelm Konrad Roentgen. The radiograph image that is produced on film is frequently referred to as a roentgenograph. The recent development of the computer tomographic technique has been hailed as the greatest advancement in diagnostic medicine since the discovery of X rays themselves.

The **computerized axial tomography** technique (CT, or CAT, scan) has greatly enhanced the versatility of X rays, using a computer to display a cross-sectional image similar to that which could only be obtained in an actual section through the body (fig. 1.19a).

Another technique of radiographic anatomy is the **dynamic spacial reconstructor** (DSR) scan (fig. 1.19b). The DSR functions as an electronic knife that pictorially slices an organ, such as the heart, to provide three-dimensional images. The DSR can be used to observe movements of organs, detect defects, assess the extent of a disease like cancer, or determine the extent of trauma to tissues after a stroke or a heart attack.

Magnetic resonance imaging (MRI), also called **nuclear magnetic resonance** (NMR), provides a new technique for diagnosing diseases and following the response of a disease to chemical treatment (fig. 1.19c). An MRI image is created rapidly as individual molecules of cells are focused upon simultaneously to determine their response to magnetism. MRI has the advantage of being noninvasive—that is, no chemicals are introduced into the body.

A **positron emission tomography** (PET) scan is a radiological technique used to observe the metabolic activity in organs (fig. 1.19d) after the injection of a radioactive substance, such as treated glucose, into the bloodstream.

Human anatomy will always be a relevant science. It is important to a personal understanding of body functioning as well as to the medical profession, which is concerned with the body's dysfunctions. Human anatomy is no longer confined to the isolated observation and description of structures, but has expanded to include the complexities of how the body functions. The science of anatomy is dynamic and has remained vital because anatomists have dared to explore the intricate territory of the human body.

Autopsies are an important aspect of human anatomy and medicine. An autopsy is a thorough postmortem examination of all of the organs and tissues of a body. Autopsies were routinely performed in the early part of the twentieth century but their numbers have declined significantly in the last three decades. Currently, only 15% of corpses are autopsied in the United States, which is down from over 50% 30 years ago. Autopsies are of value because (1) they often reveal the cause of death, which may confirm or disconfirm preliminary death statements; (2) they frequently uncover diseases or structural defects that remained undetected in life; (3) they check the effectiveness of a particular drug therapy for a patient or the success of a particular surgery; and (4) they are important in the training of medical students.

An interesting piece of information regarding the value of an autopsy in confirming the cause of death was revealed in a study of 2,557 autopsies conducted over a 30-year period to determine the accuracy of physicians' diagnoses of deaths (see "Autopsy," S. A. Geller, *Scientific American*, March 1983). In this study, the causes of death had been improperly or inaccurately recorded in 42% of

UNIT I

Figure 1.19 Different techniques in radiological anatomy provide unique views of human anatomy. (*a*) A CT scan through the head, (*b*) a DSR scan through the thorax (chest), (*c*) an MRI image through the head, and (*d*) a PET scan through the head.

(a)

(b)

(c)

(d)

the cases. This means that because of a decline in the number of autopsies, there may be over 1 million death certificates filed in the United States each year that are in error. This has important implications for criminal justice and medical genetics, as well as for the insurance industry.

An objective of this text is to enable students to become educated and conversant in anatomy. An excellent way to keep up with anatomy during and after completing the formal course is to subscribe to and read scientific magazines like *Science, Scientific American, Discover,* and *Science Digest.* These publications and others include many articles on recent developments in science, written so as to be easily understood by the informed layperson. Many specialties of anatomy are experiencing a rapid proliferation of exciting knowledge. It is important to become and stay informed if you are to be an educated contributor to society.

1. Discuss the impact that each of the following has had on the science of human anatomy: the humoral theory of body organization, vivisections, the Middle Ages, human dissections, movable type, embalming cadavers, the invention of the microscope, and the development of x-ray techniques.
2. Give some examples of how culture and religion influenced, positively or negatively, the science of anatomy.
3. Explain how a knowledge of anatomy can be of practical value to a person.
4. How can a student stay current in the science of human anatomy? Why is it important to do so?

CHAPTER SUMMARY

Definition of the Science (p. 2)

1. Human anatomy is the science concerned with the structure of the human body.
2. The terms of anatomy are descriptive and are generally of Greek or Latin derivation.
3. The history of human anatomy parallels that of medicine and has also been greatly influenced by various religions.

Prescientific Period (pp. 2–4)

1. Prehistoric interest in anatomy was undoubtedly limited to practical information necessary for survival.
2. Trepanation was a surgical technique that was practiced by several cultures.
3. Paleopathology is the science concerned with diseases of prehistoric people.

Scientific Period (pp. 4–19)

1. A few anatomical descriptions were inscribed in clay tablets in cuneiform writing by people who lived in Mesopotamia in about 4000 B.C.
2. Egyptians of about 3400 B.C. developed an embalming technique. It was not recorded, however, and therefore was not of value in furthering the study of anatomy.
3. The belief in a balance between yin and yang was a compelling influence in Chinese philosophy and provided the rationale for the practice of acupuncture.
4. The advancement of anatomy in Japan was largely due to the influence of the Chinese and Dutch.

5. Anatomy first found wide acceptance as a science in ancient Greece.
   (a) Hippocrates is regarded as the father of medicine because of the sound principles of medical practice he established.
   (b) The Greek philosophy of body humors dominated medical thought for over 2,000 years.
   (c) Aristotle established a type of scientific method for obtaining data, and his writings contain some basic anatomy.
6. Alexandria was a center of scientific learning from 300 to 30 B.C.
   (a) Human dissections and vivisections were performed in Alexandria.
   (b) Erasistratus is referred to as the father of physiology because of his interpretations of various body functions.
7. Theoretical data was de-emphasized during the Roman era.
   (a) Celsus's eight-volume work was a compilation of a portion of the medical data from the Alexandrian school.
   (b) Galen was an influential medical writer who established some sound principles and at the same time introduced serious errors into the literature that went unchallenged for centuries.
   (c) Science was suppressed for nearly 1,000 years during the Middle Ages, and dissections of human cadavers were prohibited.
   (d) Anatomical writings were taken from Alexandria by Arab armies and thus saved from destruction during the Dark Ages in Europe.

8. Great European universities were established during the Renaissance.
   (a) Andreas Vesalius and Leonardo da Vinci were renowned Renaissance men who produced monumental studies of the human form.
   (b) *De Humani Corporis Fabrica,* written by Vesalius, had a tremendous impact on the advancement of human anatomy. Vesalius is regarded as the father of human anatomy.
9. The two major contributions during the seventeenth and eighteenth centuries were an explanation of blood flow and the development and utilization of the microscope.
   (a) William Harvey, in 1628, correctly described the circulation of blood.
   (b) Shortly after the microscope had been perfected by Antoni van Leeuwenhoek, many investigators added new discoveries to the rapidly changing specialty of microscopic anatomy.
10. The cell theory was formulated during the nineteenth century by Matthias Schleiden and Theodor Schwann, and cellular biology became established as a science separate from anatomy.
11. A trend toward simplification and standardization of anatomical nomenclature began in the twentieth century, and many specialties within anatomy developed, including cytology, histology, electron microscopy, and radiology.

REVIEW ACTIVITIES

Objective Questions

1. *Anatomy* is derived from a Greek word meaning
   (a) to cut up.
   (b) to analyze.
   (c) functioning part.
   (d) to observe death.
2. The most important contribution of William Harvey was his research on
   (a) the continuous circulation of blood.
   (b) the microscopic structure of spermatozoa.
   (c) the detailed structure of the kidney.
   (d) the striped appearance of skeletal muscle.
3. Which of the following men would be most likely to disagree with the concept of body humors?
   (a) Galen   (c) Vesalius
   (b) Hippocrates (d) Aristotle
4. Anatomy was first widely accepted as a science in ancient
   (a) Rome.   (c) China.
   (b) Egypt.   (d) Greece.
5. The establishment of sound principles of medical practice earned this man the title of father of medicine.
   (a) Hippocrates (c) Erasistratus
   (b) Aristotle   (d) Galen
6. Which of the four body humors was believed by Hippocrates to be associated with the lungs?
   (a) black bile (c) phlegm
   (b) yellow bile (d) blood

7. The anatomical masterpiece *De Humani Corporis Fabrica* was written and illustrated by
   (a) Leonardo.
   (b) Harvey.
   (c) Vesalius.
   (d) Leeuwenhoek.
8. What event of about 1450 helped usher in the Renaissance?
   (a) the development of the microscope
   (b) an acceptance of the scientific method
   (c) the development of the cell theory
   (d) the development of movable type
9. The body organ thought by Aristotle to be the seat of intelligence was
   (a) the liver.   (c) the brain.
   (b) the heart.   (d) the intestine.
10. X rays were discovered during the late nineteenth century by
   (a) Roentgen.   (c) Schleiden.
   (b) Hooke.   (d) Müller.

Essay Questions

1. Define the terms *anatomize, trepanation, paleopathology, vivisection,* and *cadaver.*
2. Why were the techniques of embalming a corpse, which were perfected in ancient Egypt, not shared with other cultures or recorded for future generations?
3. What is acupuncture? What are some of its uses today?
4. Why is Latin an ideal language for the derivation of anatomical terms? What is the current trend in anatomical nomenclature regarding the use of proper names in referring to anatomical structures?

5. Why do you suppose the Hippocratic oath has survived for over 2,000 years as a creed for medical practice? What aspects of the oath are difficult to conform to in today's society?
6. What is meant by the humoral theory of body organization? Which great anatomists were influenced by this theory? When did the humoral theory cease to be an influence on anatomical investigation and interpretation?
7. Discuss the impact that Galen had on the advancement of anatomy and medicine. What ideological circumstances permitted the philosophies of Galen to survive for such a long period?
8. Briefly discuss the establishment of anatomy as a science during the Renaissance.
9. Who invented the microscope? What part did it play in the advancement of anatomy? What specialties of anatomical study have arisen since the introduction of the microscope?
10. Discuss the impact of the work of Andreas Vesalius on the science of anatomy.
11. Give some examples of how culture and religion influenced the science of anatomy.
12. List some techniques currently used to study anatomy and identify the specialties within which these techniques are used.

2

Body Organization and Anatomical Terminology

Chapter Outline

UNIT II

CLINICAL CASE STUDY

A 20-year-old female has been hit by a car while crossing a street. Upon arrival at the scene, paramedics find the patient to be a bit dazed but reasonably lucid, complaining of pain in her abdomen and the left side of her chest. Otherwise, her vital signs are within normal limits. Initial evaluation in the emergency room reveals a very tender abdomen and left chest. The chest radiograph demonstrates a collapsed left lung due to air in the pleural space (pneumothorax). The emergency room physician places a drainage tube into the left chest (into the pleural space) in order to treat the pneumothorax. Attention is then turned to the abdomen. Due to the finding of tenderness, a peritoneal lavage is performed. This maneuver involves penetration of the abdominal wall and insertion of a tube into the peritoneal cavity. Clear fluid is then instilled to the abdomen and siphoned out again. Fluid used in this procedure is called lavage fluid. A return of lavage fluid containing blood, fecal matter, or bile indicates injury to an abdominal organ that requires surgery. The return of lavage fluid from our patient is clear. However, the nurse states that lavage fluid is draining out of the chest tube.

Could this phenomenon be explained based on normal anatomy? Explain your answer using knowledge of the organization of, and relationships between, the various body cavities. Explain what could account for this finding in our patient. Does the absence of bile, blood, etc., in the peritoneal lavage fluid guarantee that no organ has been ruptured. If not, explain why in terms of the relationship of the various organs to the membranes within the abdomen.

Classification and Characteristics of Humans

Humans are biological organisms belonging to the phylum Chordata within the kingdom Animalia and to the family Hominidae within the class Mammalia and the order Primates.

Objective 1. List the taxonomic classification of humans.

Objective 2. List the characteristics that identify humans as chordates and as mammals.

Objective 3. Describe the characteristics that humans have but other primates do not.

The human organism, or **Homo sapiens,** as we have named ourselves, is unique in many ways. Our scientific name translates from Latin to mean man the intelligent, and indeed this is our most distinguishing feature. Through our intelligence we have built civilizations, conquered many diseases, and established cultures. We have devised a means of communicating through written symbols. We can record our own history, as well as that of other animals, and speculate on our future. We are social organisms with well-established patterns of behavioral interactions. Most of us have become so intellectually specialized that we are not self-sufficient. We need one another as much as we need the recorded knowledge of the past.

As we seek knowledge, we are challenged to learn more about ourselves. It is intriguing to study our structure and function and to realize our close relationship to other living organisms. Often, it is sobering to realize our biological imperfections and limitations.

We share many characteristics with all living animals. As human organisms, we breathe, eat and digest food, excrete bodily wastes, locomote, and reproduce our own kind. We are subject to disease, injury, pain, aging, mutations, and death. Being composed of organic materials, we will decompose after death as microorganisms consume our flesh as food. The processes by which our bodies produce, store, and utilize energy are similar to those found in all living organisms. The genetic code that regulates our development is found throughout nature. The fundamental patterns of development of many nonhuman animals also characterize the formation of the human embryo.

In the classification, or taxonomic, scheme established by biologists to organize the structural and evolutionary relationships of living organisms, each category of classification is referred to as a *taxon.* The highest taxon is the kingdom and the most specific taxon is the species. Humans are species belonging to the **animal kingdom.** *Phylogeny (fi-loj'eny)* is the origin and evolutionary development of animal species.

Phylum Chordata

Human beings belong to the phylum Chordata (*fi'lum kor-dah'tah*) along with fishes, amphibians, reptiles, birds, and other mammals. All chordates have three structures in common: a **notochord** (*no'to-kord), a **dorsal hollow nerve cord,** and **pharyngeal** (*fah-rin'je-al*) **pouches** (fig. 2.1). These chordate characteristics are well expressed during the embryonic stage of development and, to a certain extent, are present in an adult. The notochord is a flexible rod of tissue that extends the length of the back of an embryo. A portion of the notochord persists in the adult as the **nucleus pulposus,** located within each intervertebral disc (fig. 2.2). The dorsal hollow nerve cord is positioned above the notochord and develops into the **brain** and **spinal cord,** which are highly functional as the central nervous system in the adult. Pharyngeal pouches form gill openings in fishes and some amphibians. In other chordates, such as humans, embryonic pharyngeal pouches develop, but only one of the pouches persists, becoming the **auditory** (eustachian) (*yoo-sta'ke-an*) **canal,** a connection between the middle-ear cavity and **pharynx** (*far'ingks*) (throat area).

---

taxon: Gk. *taxis,* order
phylogeny: L. *phylum,* tribe; Gk. *logos,* study

Figure 2.1 A schematic diagram of the front part of the embryo of a chordate.

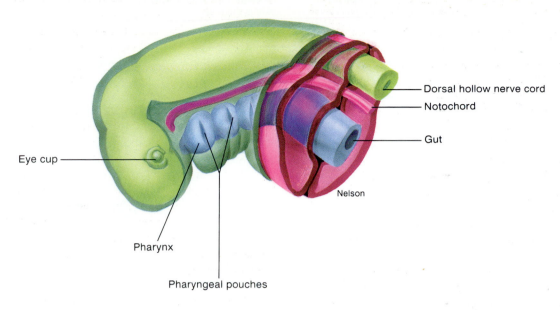

Dorsal hollow nerve cord

Notochord

Gut

Eye cup

Nelson

Pharynx

Pharyngeal pouches

UNIT II

Figure 2.2 A lateral view of the vertebral column showing the intervertebral discs and, to the right, the superior view of an intervertebral disc showing the nucleus pulposus.

Intervertebral discs

Nucleus pulposus

Intervertebral disc
(anterior view)

Creek

Vertebral
column

The function of the nucleus pulposus and the intervertebral discs is to allow flexibility between vertebrae for movement of the entire spinal column while preventing compression. Spinal nerves exit between vertebrae, and the discs maintain the spacing to avoid nerve damage. A "slipped disc," resulting from straining the back, is a misnomer. What actually occurs is a herniation, or rupture, because of a weakened wall of the nucleus pulposus, causing severe pain as a nerve is compressed.

Class Mammalia

Mammals are chordate animals with hair and mammary glands. Hair is a thermoregulatory protective covering for most mammals, and mammary glands serve for suckling the young (fig. 2.3). Other characteristics of mammals include three auditory ossicles (bones), heterodont dentition (differently shaped

heterodont: Gk. *heteros*, other; *odontos*, tooth

UNIT II

Figure 2.3 The mammary ridge and accessory nipples. (*a*) Mammary glands are positioned along a mammary ridge. (*b*) In humans, additional nipples (polythelia) occasionally develop elsewhere along the mammary ridge.

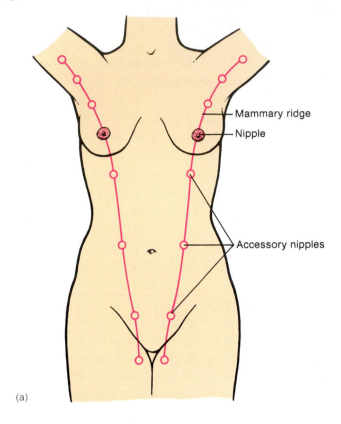

Mammary ridge

Nipple

Accessory nipples

(a)

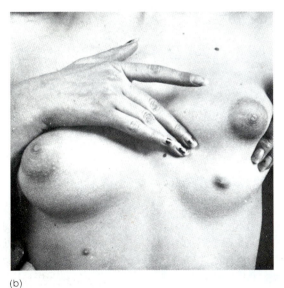

(b)

Figure 2.4 Mammals have several distinguishing characteristics, some of which are listed in the photo with their approximate location within the body.

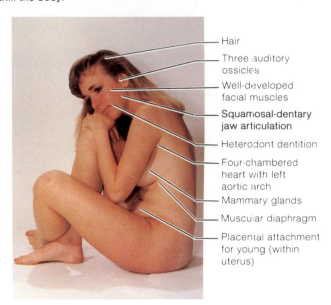

Hair

Three auditory ossicles

Well-developed facial muscles

Squamosal-dentary jaw articulation

Heterodont dentition

Four-chambered heart with left aortic arch

Mammary glands

Muscular diaphragm

Placental attachment for young (within uterus)

teeth), squamosal-dentary jaw articulation (a joint between the lower jaw and skull), an attached placenta *(plă-sen'tă)*, well-developed facial muscles, a muscular diaphragm, and a four-chambered heart with a left aortic arch (fig. 2.4).

Order Primates

There are several subdivisions of closely related groupings of mammals, called orders. Humans, along with lemurs, monkeys, and great apes, belong to the order called Primates. Members of this order have prehensile hands (fig. 2.5), digits modified for grasping, and relatively large, well-developed brains (fig. 2.6).

Family Hominidae

Humans are the sole living members of the family **Hominidae.** *Homo sapiens* is included within this family, to which all the varieties or ethnic groups of humans belong (fig. 2.7). Each "racial group" has distinguishing features that have been established in isolated populations over thousands of years. Our classification pedigree is presented in table 2.1.

With greater ease of travel and communication, many traditional cultural barriers no longer inhibit interracial marriage. This may lead to a mixing of the "gene pool" until distinct racial groups are less evident. Perhaps multiple ethnic ties in everyone's pedigree would help to reduce cultural hostility and strife.

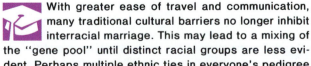

placenta: L. *placenta*, flat cake
Primates: L. *primas*, first
prehensile: L. *prehensus*, to grasp

Figure 2.5   A prehensile grip is a characteristic of primates.

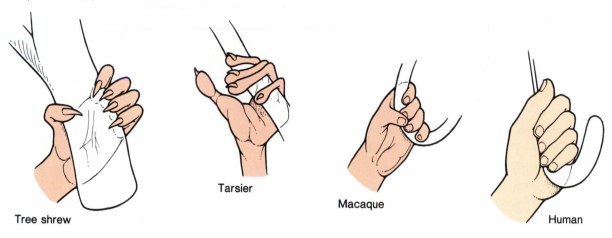

Tree shrew

Tarsier

Macaque

Human

Figure 2.6   The relative size and development of the brain of various vertebrates. The cerebrum in each representative is shaded. (Note that only mammals have a convoluted cerebrum.)

Shark

Frog

Alligator

Chicken

Cat

Chimpanzee

Human

Figure 2.7 The principal races of humans: (*a*) Mongoloid (Thailand), (*b*) Caucasoid (U.S.), (*c*) Negroid (Africa), (*d*) people of Indian subcontinent (Nepal), (*e*) Capoid (Kalahari bushman), and (*f*) Australoid (Ngatatjara man, West Australia).

(a)

(b)

(c)

(d)

(e)

(f)

Table 2.1	Classification Scheme of Human Beings	
Taxon	Designated Grouping	Characteristics
Kingdom	Animalia	Eucaryotic cells that lack walls, plastids, and photosynthetic pigments
Phylum	Chordata	Dorsal hollow nerve cord; notochord; pharyngeal pouches
Subphylum	Vertebrata	Vertebral column
Class	Mammalia	Mammary glands; hair
Order	Primates	Well-developed brain; prehensile hands
Family	Hominidae	Large cerebrum; bipedal locomotion
Genus	*Homo*	Flattened face; prominent chin and nose with inferiorly positioned nostrils
Species	*sapiens*	Largest cerebrum

Characteristics of Humans

As human beings, we have certain anatomical characteristics that are so specialized that they are diagnostic in separating us from other animals and even from other closely related mammals. We also have characteristics that are equally well developed in other animals, but when these function with the human brain, they provide remarkable capabilities. Our anatomical characteristics include the following:

1. **A large, well-developed brain.** The average adult human brain weighs between 1,350 and 1,400 g (3 lbs). This gives humans a large brain-to-body-weight ratio. But more important is the development of portions of the brain. Certain extremely specialized regions and structures within the brain account for emotion, thought, reasoning, memory, and even precise, coordinated movement.

2. **Bipedal locomotion.** Because humans stand and walk on two appendages, our style of locomotion is said to be *bipedal.* Upright posture imposes certain other diagnostic structural features such as the sigmoid (S-shaped) curvature of the spine, the anatomy of the hip and thighs, and arched feet. Some of these features may cause clinical problems in older individuals.

3. **An opposable thumb.** The human thumb joint is structurally adapted for tremendous versatility in grasping objects. Most primates have opposable thumbs.

4. **Well-developed vocal structures.** Humans, like no other animals, have developed articulated speech. The anatomical structure of the vocal organs and the well-developed brain have made this possible.

5. **Stereoscopic vision.** Although this characteristic is well developed in several other animals, it is also keen in

humans. Our eyes are directed forward so that when we focus on an object, we view it from two angles. Stereoscopic vision gives us depth perception, or a three-dimensional image.

We also differ from other animals in the number and arrangement of vertebrae (vertebral formula), the kinds and number of teeth (tooth formula), the degree of development of facial muscles, and the structural design of various body organs.

The characteristics just described account for the success of humans in acquiring culture. As bipedal animals, we have our hands free to grasp and manipulate objects with our opposable thumb. We can assimilate experiences into our highly developed brain, apply this knowledge at a later time, and even share this learning through vocal or written communication.

1. What is a chordate? Why are humans considered members of the phylum Chordata?
2. Why are humans designated as mammals and primates? What characteristics distinguish humans from other primates?
3. Which of the characteristics of humans are adaptive for social organization?

Body Organization

Structural and functional levels of organization characterize the human body, and each of its parts contributes to the total organism.

Objective 4. Identify the structures of a cell, tissue, organ, and system and explain the relationships among these structures as they constitute an organism.

Objective 5. Explain the general function of each system.

Cellular Level

The **cell** is the basic structural and functional component of life. Humans are multicellular organisms composed of between 60 and 100 trillion cells. It is at the cellular level that such vital functions of life as metabolism, growth, irritability and adaptability, repair, and reproduction are carried out.

Cells are composed of minute particles called **atoms,** which are bound together to form larger particles called **molecules** (fig. 2.8). Certain molecules are arranged into small functional structures called **organelles** *(or''gă-nelz).* Each organelle carries out a specific function within the cell. The nucleus, mitochondria, and endoplasmic reticulum are organelles. The structure of cells and the function of the organelles will be examined in more detail in chapter 3.

---

bipedal: L. *bi,* two; *pedis,* foot
sigmoid: Gk. *sigma,* shaped like the letter *s*
stereoscopic: Gk. *stereos,* solid; *skopein,* to view

---

cell: L. *cella,* small room

Figure 2.8 The levels of structural organization and complexity within the human body.

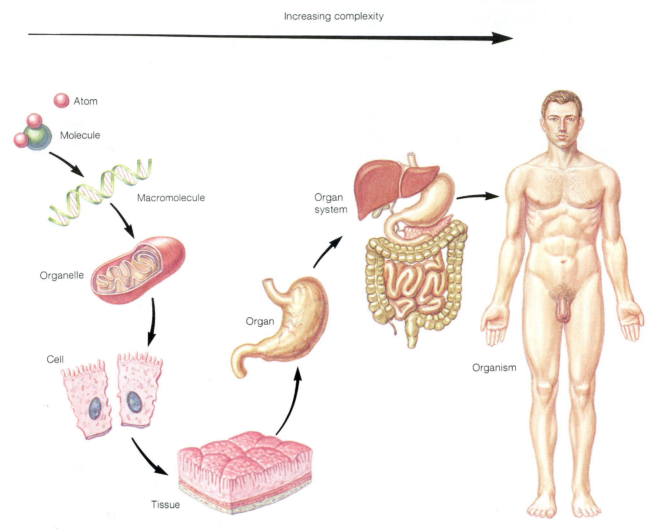

Increasing complexity

Atom

Molecule

Macromolecule

Organelle

Cell

Tissue

Organ

Organ system

Organism

UNIT II

The human body contains many distinct kinds of cells, each specialized to perform specific functions. Examples of specialized cells are bone cells, muscle cells, fat cells, blood cells, and nerve cells. Each of these cell types has a unique structure directly related to its function.

Tissue Level

Tissues are layers or aggregations of similar cells that perform specific functions (fig. 2.8). An example of a tissue is the muscle within the heart, which functions to contract and pump the blood through the body. The outer layer of skin is a tissue because it is composed of similar cells that are bound together and function to protect and contain body contents.

Organ Level

An **organ** is an aggregate of two or more tissues, integrated to perform a particular function. Organs occur throughout the body and vary greatly in size and function. Examples of organs

tissue: Fr. *tissu,* woven; from L. *texo,* to weave
organ: Gk. *organon,* instrument

are the heart, spleen, pancreas, ovary, skin, and even any bone within the body. Each organ usually has one or more primary tissues and several secondary tissues. In the stomach, for example, the inside epithelial lining is considered the primary tissue because the basic functions of secretion and absorption occur within this layer. Secondary tissues of the stomach are the supporting connective tissue and vascular, nervous, and muscle tissues.

System Level

The **systems** of the body constitute the next level of structural organization. A body system consists of various organs that have similar or related functions. Examples of systems are the circulatory system, nervous system, digestive system, and endocrine system. Certain organs may serve several systems. For example, the pancreas functions with both the endocrine and digestive systems and the pharynx serves both the respiratory and digestive systems. All the systems of the body are interrelated and function together, constituting the total organism.

system: Gk. *systema,* being together

Figure 2.9 Developmental changes from childhood to adulthood. Controlled by hormones, puberty is the period of body growth when sexual characteristics are expressed and the sexual organs become functional.

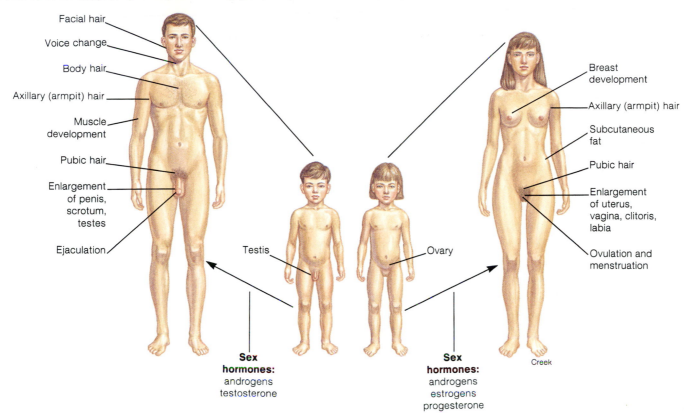

Facial hair

Voice change

Body hair

Axillary (armpit) hair

Muscle development

Pubic hair

Enlargement of penis, scrotum, testes

Ejaculation

Testis

Ovary

Sex hormones: androgens testosterone

Sex hormones: androgens estrogens progesterone

Breast development

Axillary (armpit) hair

Subcutaneous fat

Pubic hair

Enlargement of uterus, vagina, clitoris, labia

Ovulation and menstruation

Creek

UNIT II

Growth is a normal process by which an organism increases in size as a result of the accretion of cells and tissues similar to those already present within organs. Growth is an integral part of development that continues until adulthood. Normal growth depends not only on proper nutrition but on the concerted effect of several hormones (chemicals produced by endocrine glands), including insulin, growth hormone, and (during adolescence) the sex hormones. It is through the growth process that each of the body systems eventually matures (fig. 2.9). *Puberty* is the developmental transition during the growth process when sexual features become expressed in several of the body systems and the reproductive organs become functional.

In a *systematic (systemic) approach* to studying anatomy, the functional relationships of various organs within a system are emphasized. For example, the functional role of the digestive system can be better understood if all of the organs within that system are studied together. Another approach to anatomy, the *regional approach,* has merit in professional schools because the structural relationships of portions of several systems are observed simultaneously. Dissections of cadavers are usually conducted on a regional basis. Trauma or injury usually affects a region of the body, whereas a disease that affects a region may also involve an entire system.

This text uses a systematic approach to anatomy. In the chapters that follow, you will become acquainted, system by system, with the functional anatomy of the entire body. An overview of the structure and function of each of the body systems is presented in figure 2.10.

1. Construct a diagram to illustrate the levels of structural organization that characterize the body. Which of these levels are microscopic?
2. Why is the skin considered an organ?
3. Which body systems control the functioning of the others? Which are supportive of the organism? Which serve a transportive role?

Anatomical Terminology

In order to understand the science of anatomy, students must master its descriptive terminology.

Objective 6. Explain how anatomical terms are derived.

Objective 7. Describe what is meant by prefixes and suffixes.

Anatomy is a descriptive science. Analyzing anatomical terminology can be a rewarding experience in itself, as one learns something of the character of antiquity. Not only is an understanding of the roots of words of academic interest, but familiarity with technical terms reinforces the learning process. The majority of anatomical terms are derived from Greek or Latin, while some of the more recent terms are of German and French derivation. Unfortunately, some anatomical terms have been coined in honor of various anatomists or physicians. Such terms

Figure 2.10   Various systems of the human body.

Integumentary system
Function: external support
and protection of body

Skeletal system
Function: internal support and
flexible framework for body
movement; production of
blood cells

Muscular system
Function: body movement;
production of body heat

Lymphatic system
Function: body immunity;
absorption of fats; drainage
of tissue fluid

Urinary system
Function: filtration of blood;
maintenance of volume and
chemical composition
of the blood; removal of
metabolic wastes from
body

Endocrine system
Function: secretion of
hormones for
chemical regulation

Figure 2.10   continued

Nervous system
Function: control and
regulation of all
other systems of
the body

Respiratory system
Function: gaseous exchange
between external environment
and blood

Circulatory system
Function: transport of life-
sustaining materials to body
cells; removal of metabolic
wastes from cells

Digestive system
Function: breakdown and
absorption of food materials

Male reproductive system
Function: production of male
sex cells (sperm); transfer of
sperm to reproductive system
of female

Female reproductive system
Function: production of
female sex cells (ova);
receptacle of sperm from
male; site for fertilization
of ovum, implantation, and
development of embryo
and fetus; delivery of fetus

UNIT II

Table 2.2		Some Examples of Singular and Plural Word Endings
Singular Ending	**Plural Ending**	**Examples**
-a	-ae	Axilla, axillae
-ax	-aces	Thorax, thoraces
-en	-ina	Lumen, lumina
-ex	-ices	Cortex, cortices
-is	-es	Diagnosis, diagnoses
-is	-ides	Epididymis, epididymides
-ix	-ices	Appendix, appendices
-ma	-mata	Carcinoma, carcinomata
-on	-a	Mitochondrion, mitochondria
-um	-a	Cilium, cilia
-us	-i	Tarsus, tarsi
-us	-ora	Corpus, corpora
-us	-era	Viscus, viscera
-x	-ges	Pharynx, pharynges
-y	-ies	Ovary, ovaries

Source: Dr. Kenneth S. Saladin, Georgia College.

have no descriptive basis, cannot be associated with anything, and must simply be memorized.

Many Greek and Latin terms were coined more than 2,000 years ago. It is interesting to decipher the meanings of these terms and gain a glimpse into our medical heritage. Many terms referred to common plants or animals. Thus, the term *vermis* means worm, *cochlea (kok'le-ă)* means snail shell, *cancer* refers to crab, *uvula* means little grape, and even *muscle* comes from the Latin *musculus,* which means mouse. Other terms suggest the warlike environment of the Greek and Latin era. *Thyroid,* for example, means shield, *xiphos (zi'fos)* means sword, and *thorax,* breastplate. *Sella* means saddle and *stapes (sta'pēz),* the stirrup. Various tools or instruments were referred to in early anatomy. The malleus and anvil resemble miniatures of a blacksmith's implements, and tympanum refers to a drum.

You will encounter many new terms throughout your study of anatomy. You can learn these terms more easily if you know the meaning of their prefixes and suffixes. Use the glossary of prefixes and suffixes (on the inside of the front and back covers) as an aid in learning new terms. Pronouncing these terms as you learn them will also aid your retention. A guide to the relationship between the singular and plural forms of words is presented in table 2.2.

Learning the material presented in the remainder of this chapter establishes a basic foundation for anatomy, as well as for all medical and paramedical fields. Anatomy is a very precise science because of its universally accepted reference language for describing body parts and locations.

1. Explain the statement, "Anatomy is a descriptive science."
2. Refer to the glossary of prefixes and suffixes listed on the inside of the front and back covers to decipher the terms *blastocoel, hypodermic, dermatitis,* and *orchiectomy.*

Figure 2.11   Planes of reference through the body.

Planes of Reference and Descriptive Terminology

All of the descriptive planes of reference and terms of direction used in anatomy are standardized because of their reference to the body in anatomical position.

Objective 8.   Identify the planes of reference used to locate structures within the body.

Objective 9.   Describe the anatomical position.

Objective 10.   Define and be able to properly use the descriptive and directional terms that are used with reference to the body.

Planes of Reference

In order to visualize and study the structural arrangements of various organs, the body may be sectioned and diagrammed according to planes of reference. Three fundamental planes, **midsagittal** *(mid-saj'i-tal),* **coronal,** and **transverse,** are frequently used to depict structural arrangement (figs. 2.11 and 2.12).

coronal: L. *corona,* crown

Figure 2.12 The human brain sectioned along (a) a transverse plane, (b) a coronal plane, and (c) a midsagittal plane.

Cross section of brain

Frontal section of brain

Para

Midsagittal section of brain

(a)

(b)

(c)

UNIT II

As applied to the body, a *midsagittal plane* passes lengthwise through the midplane, dividing it into right and left halves. *Sagittal planes* also extend vertically and divide the body into unequal right and left portions. *Coronal,* or *frontal, planes* also pass lengthwise and divide the body into front and back portions. *Transverse planes,* also called *horizontal,* or *cross-sectional, planes,* divide the body into superior (upper) and inferior (lower) portions.

The value of the computerized tomographic X-ray (CT) scan is that it displays an image along a transverse plane similar to that which could otherwise be obtained only in an actual section through the body. Prior to the development of this X-ray technique, conventional radiographs (X-ray images) were on a vertical plane, and dimensions of body irregularities were difficult, if not impossible, to ascertain (see fig. 1.18).

Descriptive Terminology

Anatomical Position   All terms of direction that describe the relationship of one body part to another are made in reference to the **anatomical position.** In the anatomical position, the body is erect, the feet are parallel to one another and flat on the floor, the eyes are directed forward, the arms are at the sides of the body with the palms of the hands turned forward and the fingers are pointed straight down (fig. 2.13).

Directional Terms   Directional terms are used to locate the position of structures, surfaces, and regions of the body. These terms are always relative to the specimen positioned in the anatomical position. A summary of directional terms is presented in table 2.3.

Clinical Procedures   Certain clinical procedures are important in determining anatomical structure and function in a living individual. The more common of these are as follows:

1. **Observation.** Visual inspection for any revealing clinical symptoms, such as abnormal skin color, swelling, or rashes. Other observations may include needle marks on the skin, irregular breathing rates, or abnormal behavior.
2. **Palpation.** Applying the fingers with firm pressure to the surface of the body to detect surface landmarks, lumps, tender spots, or pulsations.
3. **Percussion.** Tapping sharply on various locations on the thorax or abdomen to detect resonating vibrations in determining fluid concentrations and organ densities.
4. **Auscultation.** Listening to the sounds that various organs make as they perform their functions (breathing sounds, heartbeats, digestive sounds, etc.).
5. **Reflex-response tests.** Reflex responses indicate the condition of parts of the nervous system and some associated organs. One test of a reflex mechanism involves tapping a predetermined tendon with a reflex hammer and observing the response.

---

1. Explain why a transverse plane through an organ is more important in studying specific systems than a transverse plane through the body.
2. What do we mean when we say that directional terms are relative and must be used in reference to a body structure or a body in anatomical position?
3. Write a list of statements, similar to the examples in table 2.3, that correctly express the directional terms used to describe the relative positions of various body structures.

Body Regions

The human body is divided into regions and specific local areas that can be identified on the surface. Each region contains internal organs, the location of which are anatomically and clinically important.

Figure 2.13   In the anatomical position, the body is erect, the feet parallel, the eyes directed forward, the arms to the sides with the palms directed forward, and the fingers pointed straight down.

Objective 11.   List the regions of the body and the principal local areas that make up each region.

Objective 12.   Explain why it is important to be able to describe the body areas and regions where each major internal organ is located.

The human body is divided into several regions that can be identified on the surface of the body. Learning the terminology used in reference to these regions now will make it easier to learn the names of underlying structures later. The major body regions are the **head, neck, trunk, upper extremity,** and **lower extremity** (fig. 2.14). The trunk is frequently divided into the thorax, or the pectoral region, and the abdomen.

Head

The **head** is divided into a **facial region,** which includes the eyes, nose, and mouth, and a **cranium** *(kra'ne-um),* or **cranial region,** which covers and supports the brain. Further regionalization of the head is important for identifying specific glands, muscles, nerves, and vessels located under the skin. The iden-

Table 2.3	Directional Terms for the Human Body	
Term	**Definition**	**Example**
Superior (cranial, cephalic)	Toward the head; toward the top	The thorax is superior to the abdomen.
Inferior (caudal)	Away from the head; toward the bottom	The legs are inferior to the trunk.
Anterior (ventral)	Toward the front	The navel is on the anterior side of the body.
Posterior (dorsal)	Toward the back	The kidneys are posterior to the intestine.
Medial	Toward the midline of the body	The heart is medial to the lungs.
Lateral	Away from the midline of the body	The ears are on the lateral sides of the head.
Internal (deep)	Away from the surface of the body	The brain is internal to the cranium.
External (superficial)	Toward the surface of the body	The skin is external to the muscles.
Proximal	Toward the trunk of the body	The knee is proximal to the foot.
Distal	Away from the trunk of the body	The hand is distal to the elbow.
Visceral	Related to internal organs	The lungs are covered by a thin membrane called the visceral pleura.
Parietal	Related to the body walls	The parietal pleura is the inside lining of the thoracic cavity.

tifying names for detailed surface regions are based on associated organs—such as the orbital (eye), nasal (nose), oral (mouth), mental (chin), and auricular (ear) regions.

Neck

The **neck,** referred to as the **cervix** *(ser'viks),* or **cervical region,** supports the head and permits it to move. As with the head, detailed subdivisions of the neck can be identified. Additional information concerning the neck region can be found in chapter 10.

Thorax

The **thorax,** or **thoracic** *(thŏ-ras'ik)* **region,** is commonly referred to as the chest. The **mammary region** of the thorax surrounds the nipple and in sexually mature females is enlarged as the breast. Between the mammary regions is the **sternal region.** The armpit is called the **axillary fossa,** or simply

thorax: L. *thorax,* chest
mammary: L. *mamma,* breast
axillary: L. *axilla,* armpit

Figure 2.14 Body regions. (*a*) Anterior view and (*b*) posterior view.

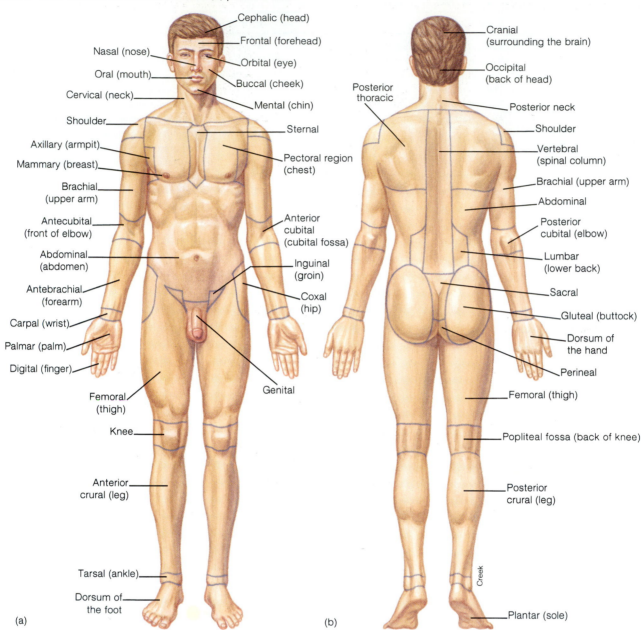

Cephalic (head)
Frontal (forehead)
Nasal (nose)
Orbital (eye)
Oral (mouth)
Buccal (cheek)
Cervical (neck)
Mental (chin)
Shoulder
Sternal
Axillary (armpit)
Mammary (breast)
Pectoral region (chest)
Brachial (upper arm)
Antecubital (front of elbow)
Anterior cubital (cubital fossa)
Abdominal (abdomen)
Inguinal (groin)
Antebrachial (forearm)
Coxal (hip)
Carpal (wrist)
Palmar (palm)
Digital (finger)
Genital
Femoral (thigh)
Knee
Anterior crural (leg)
Tarsal (ankle)
Dorsum of the foot
(a)

Cranial (surrounding the brain)
Occipital (back of head)
Posterior thoracic
Posterior neck
Shoulder
Vertebral (spinal column)
Brachial (upper arm)
Abdominal
Posterior cubital (elbow)
Lumbar (lower back)
Sacral
Gluteal (buttock)
Dorsum of the hand
Perineal
Femoral (thigh)
Popliteal fossa (back of knee)
Posterior crural (leg)
Creek
Plantar (sole)
(b)

axilla, and the surrounding area the **axillary region.** The **vertebral region,** following the vertebral column, extends the length of the back.

> The heart and lungs are contained within the thoracic cavity. Easily identified surface landmarks greatly facilitate a physical examination of them. A physician must know, for example, where the valves of the heart can best be detected and where to listen for respiratory sounds. The axilla becomes important when examining for infected lymph nodes. When fitting a patient for crutches, a physician will instruct the patient to avoid supporting the weight of the body on the axillary region because of the possibility of damaging the underlying nerves and vessels.

Abdomen

The **abdomen** is located below the thorax. The **navel,** or **umbilicus,** is an obvious landmark on the front and center of the abdomen. The abdomen has been divided into nine regions in order to describe the location of internal organs. In figure 2.15 the subdivisions of the abdomen are diagrammed, and the internal organs located within these regions are identified in table 2.4. Subdividing the abdomen into four quadrants (fig. 2.16) is a common clinical practice that enables various pains and conditions to be more easily located.

The **pelvic region** forms the lower portion of the abdomen. Within the pelvic region is the **pubic area,** which is covered with pubic hair in sexually mature individuals. The **perineum** (*per"i-ne'um*) (fig. 2.17) is the region containing

Figure 2.15 The abdomen is subdivided into nine regions. The vertical planes are positioned just medial to the nipples, the upper horizontal plane is positioned at the level of the rib cage, and the lower horizontal plane is even with the upper border of the hipbones.

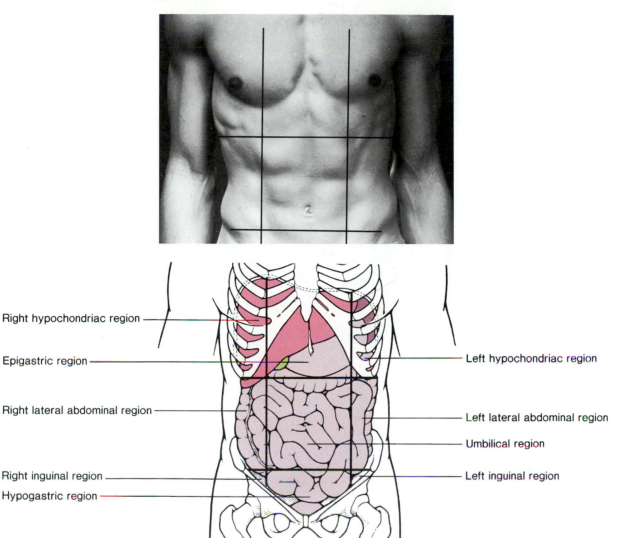

Right hypochondriac region

Epigastric region

Right lateral abdominal region

Right inguinal region

Hypogastric region

Left hypochondriac region

Left lateral abdominal region

Umbilical region

Left inguinal region

Table 2.4	Regions of the Abdomen and Pelvis	
Region	**Location**	**Internal Organs**
Right hypochondriac	Right, upper one-third of abdomen	Gallbladder; portions of liver and right kidney
Epigastric	Upper, median abdomen	Portions of liver, stomach, pancreas, and duodenum
Left hypochondriac	Left, upper one-third of abdomen	Spleen; splenic flexure of colon; portions of left kidney and small intestine
Right lateral	Right, lateral one-third of abdomen	Cecum; ascending colon; hepatic flexure; portions of right kidney and small intestine
Umbilical	Center of abdomen	Jejunum; ileum; portions of duodenum, colon, kidneys, and major abdominal vessels
Left lateral	Left, lateral one-third of abdomen	Descending colon; portions of left kidney and small intestine
Right inguinal	Right, lower one-third of abdomen	Appendix; portions of cecum and small intestine
Pubic (hypogastric)	Lower, center one-third of abdomen	Urinary bladder; portions of small intestine and sigmoid colon
Left inguinal	Left, lower one-third of abdomen	Portions of small intestine and descending and sigmoid colon

Figure 2.16   A clinical subdivision of the abdomen into four quadrants.

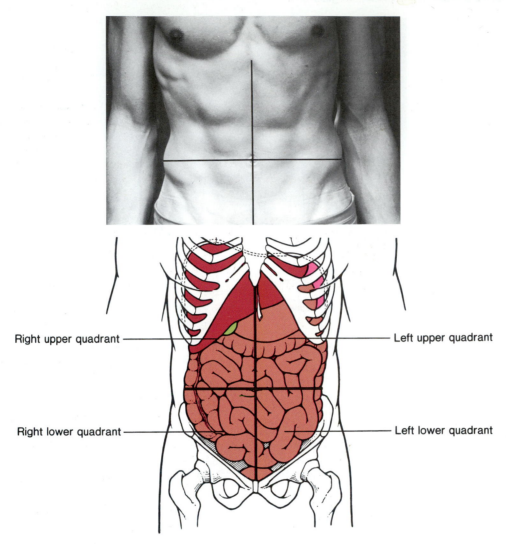

Right upper quadrant — — — — Left upper quadrant

Right lower quadrant — — — — Left lower quadrant

the external sex organs and the anal opening. The center of the back side of the abdomen, commonly called the small of the back, is the **lumbar region.** The **sacral region** is located further down at the point where the vertebral column terminates. The large hip muscles form the **buttock,** or **gluteal region.** This region is a common injection site for hypodermic needles.

Upper Extremity

The **upper extremity** is anatomically divided into the **shoulder, brachium** *(bra′ke-um)* (upper arm), **antebrachium** (forearm), and **manus** (hand) (see fig. 2.14). The shoulder is the region between the pectoral girdle and the brachium that contains the shoulder joint. The shoulder is also referred to as the **omos,** or **deltoid region.** Between the upper arm and forearm is a flexible joint called the **elbow.** The surface area of

the elbow is known as the **cubital region.** The front surface of the elbow is also known as the **cubital fossa,** an important site for intravenous injections or the withdrawal of blood. The wrist is the flexible junction between the forearm and the hand. The front of the hand is referred to as the **palm,** or **palmar region,** and the back of the hand is called the **dorsum of the hand.** Each hand has five **digits,** or fingers.

Lower Extremity

The **lower extremity** consists of the **thigh, knee, leg,** and **foot.** The thigh is commonly called the **upper leg,** or **femoral region.** The knee joint has two surfaces: the front surface is the **patellar region,** or kneecap; the back of the knee is called the **popliteal** *(pop″lĭ-te′al)* **fossa.** The leg has anterior and pos-

cubital: L. *cubitis,* elbow
popliteal: L. *poples,* ham (hamstring muscles) of the knee

Figure 2.17 A superficial view of (a) the male perineum and (b) the female perineum. The perineum consists of a urogenital triangle and an anal triangle.

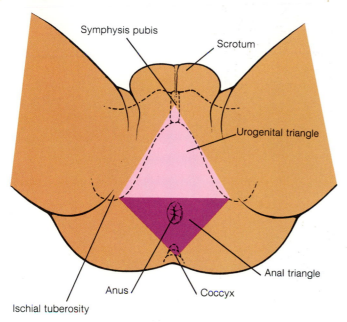

Symphysis pubis
Scrotum
Urogenital triangle
Anal triangle
Anus
Coccyx
Ischial tuberosity
(a)

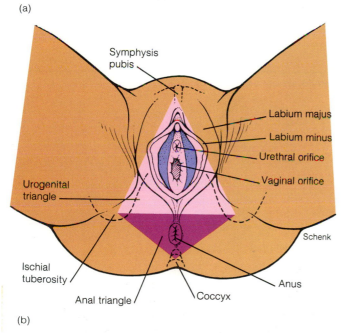

Symphysis pubis
Labium majus
Labium minus
Urethral orifice
Vaginal orifice
Urogenital triangle
Schenk
Ischial tuberosity
Anal triangle
Coccyx
Anus
(b)

1. Using yourself as a model, identify the various body regions depicted in figure 2.14. Which of these regions have surface landmarks that help distinguish their boundaries?
2. In which region of the body are intravenous injections given?
3. Distinguish between the pelvic, pubic, and perineal regions.
4. Identify the joint between the following regions: the brachium and antebrachium, the pectoral girdle and brachium, the leg and foot, the antebrachium and hand, and the thigh and leg.
5. Explain how a knowledge of the body regions has practical medical application.

Body Cavities and Membranes

For functional and protective purposes, the viscera are compartmental and supported in specific body cavities by connective and epithelial membranes.

Objective 13. Identify the various body cavities and the organs found in each.

Objective 14. Discuss the types and functions of the various body membranes.

Body Cavities

Body cavities are confined spaces within the body. They contain organs that are protected, compartmentalized, and supported by associated membranes. There are two principal body cavities: the **posterior body cavity** and the larger **anterior body cavity.** The posterior body cavity contains the brain and the spinal cord.

During development, the ventral cavity forms from a cavity called the **coelom** *(se'lom)*. The coelom is a body cavity within the trunk and is lined with a membrane that secretes a lubricating fluid. As development progresses, the coelom is partitioned by the muscular **diaphragm** into an upper **thoracic cavity,** or chest cavity, and a lower **abdominopelvic cavity** (figs. 2.18 and 2.19). Organs within the coelom are collectively called **viscera,** or **visceral** *(vis'er-al)* **organs** (fig. 2.20). Within the thoracic cavity are two **pleural** *(ploor'al)* **cavities** for the right and left lungs and a **pericardial** *(per''ĭ-kar'de-al)* **cavity** containing the heart. The area between the two lungs is known as the **mediastinum** *(me''de-ă-sti'num)*.

The abdominopelvic cavity consists of an upper **abdominal cavity** and a lower **pelvic cavity.** The abdominal cavity contains the stomach, small intestine, large intestine, liver, gallbladder, pancreas, spleen, and kidneys. The pelvic cavity is occupied by the terminal portion of the large intestine, the urinary bladder, and certain reproductive organs (uterus, uterine tubes, and ovaries in the female; seminal vesicles and prostate in the male).

terior **crural regions** (see fig. 2.14). The *shin* is a prominent bony ridge extending longitudinally along the anterior crural region, and the *calf* is the thickened muscular mass of the posterior crural region. The **ankle** is the junction between the leg and the foot. The **heel** is the back of the foot, and the **sole** of the foot is referred to as the **plantar surface.** The **dorsum of the foot** is the top surface. Each foot has five **digits,** or toes.

coelom: Gk. *koiloma,* a cavity

Figure 2.18   A midsagittal (median) section showing the body cavities.

Cranial cavity

Posterior cavity

Vertebral cavity

Thoracic cavity

Diaphragm

Abdominal cavity

Abdominopelvic cavity

Anterior cavity

Pelvic cavity

Paras

Figure 2.19   An anterior view of body cavities.

Thoracic cavity

Mediastinum

Pleural cavity

Pericardial cavity

Abdominal cavity

Abdominopelvic cavity

Pelvic cavity

Paras

UNIT II

Figure 2.20   Visceral organs of the abdominal cavity and supporting serous membranes.

Lesser omentum

Pancreas

Duodenum

Mesentery

Small intestine

Visceral peritoneum

Rectum

Diaphragm

Liver

Stomach

Large intestine

Parietal peritoneum

Greater omentum

Peritoneal cavity

Urinary bladder

Paras

Body cavities serve to confine organs and systems that have related functions. The major portion of the nervous system occupies the dorsal cavity; the principal organs of the respiratory and circulatory systems are in the thoracic cavity; the primary organs of digestion are in the abdominal cavity; and the reproductive organs are in the pelvic cavity. Not only do these cavities house and support various body organs, they also effectively compartmentalize them so that infections and diseases cannot spread from one compartment to another. For example, pleurisy of one lung membrane does not usually spread to the other, and an injury to the thoracic cavity will usually result in the collapse of only one lung rather than both.

In addition to the large ventral and dorsal cavities, there are several smaller cavities within the head. The **oral,** or **buccal,** *(buk'al)* **cavity** functions primarily with digestion and secondarily with respiration. It contains the teeth and tongue. The **nasal cavity,** which is part of the respiratory system, has two chambers created by a nasal septum. There are two **orbits,** each of which houses an eyeball and its associated muscles, vessels, and nerves. Likewise, there are two **middle-ear cavities** that contain auditory ossicles (ear bones). The location of the cavities within the head is shown in figure 2.21.

buccal: L. *bucca,* cheek or mouth
orbital: L. *orbis,* circle

Figure 2.21   Cavities within the head.

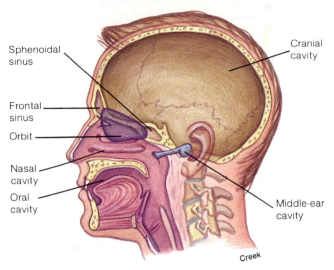

Sphenoidal sinus

Frontal sinus

Orbit

Nasal cavity

Oral cavity

Cranial cavity

Middle-ear cavity

Creek

Body Membranes

Body membranes are composed of thin layers of connective and epithelial tissue that serve to cover, separate, and support visceral organs and line body cavities. There are two basic types of body membranes: **mucous** *(myoo'kus)* **membranes** and **serous** *(se'rus)* **membranes.**

Mucous membranes secrete a thick, viscid substance called *mucus.* Generally, mucus lubricates or protects the associated organs where it is secreted. Mucous membranes line various cavities and tubes that enter or exit from the body, such as the oral and nasal cavities and the tubes of the respiratory, reproductive, urinary, and digestive systems.

Serous membranes line the thoracic and abdominopelvic cavities and cover visceral organs, secreting a watery lubricant called *serous fluid.* **Pleurae** are serous membranes associated with the lungs. Each pleura (pleura of right lung and pleura of left lung) has two parts. The **visceral pleura** adheres to the outer surface of the lung, while the **parietal** *(pă-ri'ĕ-tal)* **pleura** lines the thoracic walls and the thoracic surface of the diaphragm. The moistened space between the two pleurae is known as the **pleural cavity** (see fig. 17.20).

Pericardial membranes are the serous membranes covering the heart. A thin **visceral pericardium** covers the surface of the heart, and a thicker **parietal pericardium** is the durable covering that surrounds the heart. The space between these two membranes is called the **pericardial cavity.**

Serous membranes of the abdominal cavity are called **peritoneal** *(per''ĭ-tŏ-ne'al)* **membranes.** The **parietal peritoneum** lines the abdominal wall, and the **visceral peritoneum** covers the visceral organs. The **peritoneal cavity** is the potential space within the abdominopelvic cavity between the parietal and visceral peritoneal membranes. The *lesser omentum* and the *greater omentum* are folds of the peritoneum that extend from the stomach to store fat, cushion, and protect visceral organs of

peritoneum: Gk. *peritonaion,* stretched over

the abdominal cavity. Certain organs, such as the kidneys, adrenal glands, and a portion of the pancreas, which are within the abdominal cavity, are positioned behind the parietal peritoneum, and are therefore said to be *retroperitoneal*. **Mesenteries** (*mes'en-ter"ēz*) are double folds of peritoneum that connect the parietal to the visceral peritoneum (see figs. 2.20 and 18.3).

1. Describe the divisions and boundaries of the ventral body cavity and list the major organs contained within each division.
2. Distinguish between mucous and serous membranes and list the specific serous membranes of the thoracic and abdominopelvic cavities.
3. Explain the importance of separate and distinct body cavities.

LEEDS METROPOLITAN UNIVERSITY LIBRARY

UNIT II

CLINICAL CASE STUDY ANSWER

In normal anatomy, the thoracic cavity is effectively partitioned from the abdominal cavity by the diaphragm, peritoneum, and pleura. The phenomenon of peritoneal lavage fluid draining out of a tube properly placed in the chest can only be explained by the presence of a defect in the diaphragm. The defect in our patient is likely a traumatic rupture or laceration of the diaphragm caused by a blow to the abdomen. The blow produces sudden, upward pressure against the diaphragm, causing its rupture. The absence of bile, blood, etc., in peritoneal lavage fluid does not guarantee the absence of trauma to organs like the duodenum and pancreas. These organs do not exist within the peritoneal cavity but rather are retroperitoneal, or fully posterior to the peritoneal membrane. This keeps hemorrhage or the leaking of enzymes from these two organs out of the peritoneal space and away from lavage fluid. Other signs must therefore be relied upon, and a high index of suspicion maintained, so as not to overlook injury to these organs. Evidence of injury to any of the intra-abdominal organs or the presence of diaphragmatic rupture are indications for emergency laparotomy for repair of the structures involved.

CHAPTER SUMMARY

Classification and Characteristics of Humans (pp. 22–27)

1. Our scientific name, *Homo sapiens,* means "man the intelligent," and our intelligence is our most distinguishing feature.
2. Humans belong to the phylum Chordata because of the presence of a notochord, a dorsal hollow nerve cord, and pharyngeal pouches during the embryonic stage of human development.
3. Humans are mammals, and as such have mammalian characteristics. These include hair, mammary glands, three auditory ossicles, heterodontia, a placenta, a muscular diaphragm, and a four-chambered heart with a left aortic arch.
4. Humans are also classified within the order Primates. Primates have prehensile hands, digits modified for grasping, and well-developed brains.
5. Humans are the sole members of the family Hominidae.
6. Some of the characteristics of humans are a large, well-developed brain; bipedal locomotion; an opposable thumb; well-developed vocal structures; and stereoscopic vision.

Body Organization (pp. 27–29)

1. Cells are the fundamental structural and functional components of life.
2. Tissues are aggregations of similar cells that perform specific functions.
3. An organ is a structure consisting of two or more tissues that performs a specific function.
4. A body system is composed of a group of organs that function together.

Anatomical Terminology (pp. 29–32)

1. The majority of anatomical terms are descriptive and are of Greek or Latin derivation.
2. Learning the basic prefixes and suffixes greatly facilitates the understanding and retention of the terminology.
3. Anatomy is a foundation science for all of the medical and paramedical fields.

Planes of Reference and Descriptive Terminology (pp. 32–34)

1. The body or organs of the body may be sectioned according to planes of reference. These include a midsagittal plane that runs vertically through a structure, dividing it into right and left halves; a sagittal plane that runs vertically through a structure, dividing it into right and left portions; a coronal (frontal) plane that runs vertically through a structure, dividing it into anterior (front) and posterior (back) portions; and a transverse (cross-sectional) plane that runs horizontally through a structure, dividing it into upper and lower portions.
2. In the anatomical position, the subject stands erect and faces forward with arms at the sides and palms turned forward.
3. Directional terms are used to describe the location of one body part with respect to another part in anatomical position.
4. Clinical procedures include observation (visual inspection), palpation (feeling with firm pressure), percussion (detecting resonating vibrations), oscultation (listening to organ sounds), and reflex-response testing (determining involuntary movements).

Body Regions (pp. 34–39)

1. The head is divided into a facial region, which includes the eyes, nose, and mouth, and a cranial region, which covers and supports the brain.
2. The neck is called the cervical region and functions to support the head and permit movement.
3. The front of the thorax is subdivided into two mammary regions and one sternal region.
4. On either side of the thorax is an axillary fossa and a lateral pectoral region.
5. The abdomen may be divided into nine anatomical regions or four quadrants.
6. Regional names are given to the upper extremity and include the shoulder, brachium, antebrachium, and manus.
7. Regional names for the lower extremity include the thigh, leg, and foot.

Body Cavities and Membranes (pp. 39–42)

1. The posterior cavity, composed of the cranial and spinal cavities, encloses and protects the brain and spinal cord—the central nervous system.
2. The anterior cavity, composed of the thoracic and abdominopelvic cavities, contains the visceral organs.
3. Other body cavities include the oral, nasal, and middle-ear cavities.
4. The body has two principal types of membranes: mucous membranes, which secrete protective mucus, and serous membranes, which line the ventral cavities and cover visceral organs. Serous membranes secrete a lubricating serous fluid.
5. Serous membranes may be categorized as pleural membranes (associated with the lungs); pericardial membranes (associated with the heart); or peritoneal membranes (associated with the abdominal viscera).

REVIEW ACTIVITIES

Objective Questions

1. Which of the following is *not* a principal chordate characteristic?
   (a) a dorsal hollow nerve cord
   (b) a distinct head, thorax, and abdomen
   (c) a notochord
   (d) pharyngeal pouches
2. Prehensile hands, digits modified for grasping, and large, well-developed brains are structural characteristics of the grouping of animals referred to as
   (a) primates.   (c) mammals.
   (b) humans.   (d) chordates.

3. Layers or aggregations of similar cells that perform specific functions are called
   (a) organelles.   (c) organs.
   (b) tissues.   (d) glands.
4. Filtration and maintenance of the volume and the chemical composition of the blood are functions of
   (a) the urinary system.
   (b) the lymphatic system.
   (c) the circulatory system.
   (d) the endocrine system.
5. The cubital fossa is located in
   (a) the thorax.
   (b) the upper extremity.
   (c) the abdomen.
   (d) the lower extremity.
6. Being composed of more than one tissue type, the skin is considered
   (a) a composite tissue.   (c) an organ.
   (b) a system.   (d) an organism.
7. Which of the following is *not* a fundamental plane?
   (a) coronal   (c) vertical
   (b) transverse   (d) midsagittal
8. The external genitalia (reproductive organs) are located in
   (a) the popliteal fossa.
   (b) the perineum.
   (c) the hypogastric region.
   (d) the epigastric region.
9. The region of the thoracic cavity between the two pleural cavities is called
   (a) the midventral space.
   (b) the mediastinum.
   (c) the ventral cavity.
   (d) the median cavity.
10. The abdominal region superior to the umbilical region that contains most of the stomach is
    (a) the hypochondriac region.
    (b) the epigastric region.
    (c) the diaphragmatic region.
    (d) the inguinal region.
11. Regarding serous membranes, which of the following word pairs is *incorrect?*
    (a) visceral pleura/lung
    (b) parietal peritoneum/body wall
    (c) mesentery/heart
    (d) parietal pleura/body wall
    (e) visceral peritoneum/intestines
12. The sectional plane that divides the body into anterior and posterior portions is
    (a) sagittal.
    (b) transverse.
    (c) coronal.
    (d) cross-sectional.

13. In the anatomical position,
    (a) the arms are extended away from the body.
    (b) the palms of the hands face posteriorly.
    (c) the body is erect and the palms face anteriorly.
    (d) the body is in a fetal position.
14. Listening to sounds that functioning visceral organs make is
    (a) percussion.   (c) audiotation.
    (b) palpation.   (d) auscultation.

Essay Questions

1. Discuss the characteristics an animal must possess to be classified as a chordate; a mammal; a human.
2. Describe the relationship of the notochord to the vertebral column, the pharyngeal pouches to the ear, and the dorsal hollow nerve cord to the central nervous system.
3. Which grouping of animals are we most closely related to? What anatomical characteristics do we have in common with them?
4. Identify the four levels of complexity that constitute the human body.
5. Outline the systems of the body and identify the major organs composing each.
6. What is meant by "anatomical position"? Why is the anatomical position important in studying anatomy?
7. Define the terms *palpation, percussion,* and *auscultation.*
8. Which major region of the body contains each of the following structures or minor regions: (a) scapular region; (b) brachium; (c) popliteal fossa; (d) lumbar region; (e) cubital fossa; (f) hypochondriac region; (g) perineum; (h) axillary fossa.
9. Diagram the thoracic cavity showing the relative position of the pericardial cavity, pleural cavities, and mediastinum.
10. What is a serous membrane? Explain how the names of the serous membranes differ in accordance with each body cavity.

Gundy/Weber Software

The tutorial software accompanying Chapter 2, Body Organization and Anatomical Terminology, is Volume 1, Introduction, Tissues, Integumentary System.

UNIT II

3

Cellular Anatomy

Chapter Outline

Introduction to Cellular Anatomy

The cell is the fundamental structural and functional unit of the body. Although diverse, cells have basic structural similarities and each metabolizes in order to stay alive.

Objective 1.   Define the terms *cell*, *metabolism*, and *cytology*.

Objective 2.   Using examples, explain how cells differ from one another and how the structure of a cell determines its function.

Cells as Functional Units

Human anatomy is concerned with the structure of the human body and the relationship of its parts. The body is a masterpiece of organization. The **cell** is the basis of this organization and, as such, it is called the functional unit. As discussed in chapter 2, cellular organization forms tissues, whose organization in turn forms the organs, which in turn form systems. In order for the organs and systems to function properly, cells must function properly. Cellular function is referred to as **metabolism.** In order for cells to remain alive and metabolize, certain requirements must be met. Each cell must have access to nutrients and oxygen and be able to eliminate wastes. In addition, a constant, protective environment must be maintained. All of these requirements are achieved through organization.

Cells were first observed more than 300 years ago by the English scientist Robert Hooke. He observed the cell walls and the boxlike cavities in slices of cork and leaves and called them "little boxes or cells." Improvements in the microscope enabled further investigations and the formulation of the **cell theory,** which states that all living organisms are composed of one or more cells. The cell theory was unified in 1838 and 1839 by two German biologists, Matthias Schleiden and Theodor Schwann. Their work laid the foundation for a new science called *cytology,* which is concerned with the structure and function of cells.

A knowledge of the cellular level of organization is important for understanding the basic body processes of cellular respiration, protein synthesis, mitosis, and meiosis. An understanding of cellular structure gives meaning to the concept of tissue, organ, and system levels of functional body organization. Furthermore, many body dysfunctions and diseases originate in the cells. Although cellular structure and function have been investigated for many years, we still have much to learn about cells. The *etiologies,* or causes, of a number of complex diseases are as yet unknown. Scientists are seeking why and how the body ages. The answers will come only through a better understanding of cellular structure and function.

metabolism: Gk. *metabole*, change
cytology: L. *cella*, small room; Gk. *logos*, study of
etiology: L. *aitia*, cause; Gk. *logos*, study of

Advancements in microscopic technology have had a tremendous impact on the science of cytology. In a new process called *microtomography*, the capabilities of electron microscopy are combined with those of CT scanning to produce high-magnification, three-dimensional, microtomographic images of living cells. With this technology, living cells can be observed as they move, grow, and divide. The clinical applications are immense, as scientists can observe living, diseased (including cancerous) cells and their response to various drug treatments.

Cellular Diversity

It is amazing that from a single cell, the fertilized egg, hundreds of kinds of cells arise, composing the estimated 60 trillion to 100 trillion cells that make up an adult human. Cells vary greatly in size and shape, and their basic structure is often indicative of their function. The smallest cells are visible only through a high-powered microscope. Even the largest, an egg cell (ovum), is barely visible to the unaided eye. The sizes of cells are measured in micrometers (μm)—one micrometer equals 1/1000th of a millimeter. Using this basis of comparison, an ovum is about 140 μm in diameter and a red blood cell is about 7.5 μm in diameter. The most common type of white blood cell varies in size from 10–12 μm in diameter. Although still microscopic, some cells can be extremely long. A nerve cell (neuron), for example, may extend the entire length of an appendage and be about a meter in length.

Although a typical diagram of a cell depicts it as round or cube-shaped, the shapes of cells are actually highly variable. They can be flattened, oval, elongate, stellate, columnar, and so on (fig. 3.1). The shape of a cell is frequently an indication of its function. An oval red blood cell is adapted to transport oxygen and carbon dioxide. Thin, flattened cells may be bound together to form selectively permeable membranes. An irregularly shaped cell, such as a neuron, has a tremendous ratio of surface area to volume, which is ideal for receiving and transmitting stimuli.

The surfaces of some cells are smooth, so that substances pass over them easily. Other cells have marked depressions or grooves to facilitate absorption. Some support structures like cilia, flagella, microvilli, or gelatinous coats, which assist movement or provide adhesion. Regardless of the sizes or shapes of cells, they all have structural modifications that serve functional purposes.

1. Why is the cell considered the basic structural and functional unit of the body? List some aspects of cells that would be of concern to a cytologist.
2. What conditions are necessary for metabolism to occur?
3. Why is structural organization essential to a multicellular organism?
4. Construct a table to compare the structure and function of several kinds of cells.

Figure 3.1 Examples of the various shapes of cells within the body. (The cells are not drawn to scale.) (*a*) A nerve cell showing a cell body surrounded by numerous dendritic extensions and a portion of the axon extending below, (*b*) an epithelial cell from the lining of the lumen (hollow tube) of a blood vessel, (*c*) a smooth muscle cell from the intestinal wall, (*d*) a skeletal muscle cell, (*e*) a white blood cell, (*f*) a red blood cell, (*g*) a bone cell, and (*h*) a human sperm cell.

Cellular Chemistry

All tissues and organs are composed of cellular structures of basically the same chemicals. The most important inorganic substances in the body include water, acids, bases, and salts. The most important organic substances in the body include proteins, carbohydrates, and lipids.

Objective 3. List the common chemical elements found within cells.

Objective 4. Differentiate between inorganic and organic compounds and give examples of each.

Objective 5. Explain the importance of water in maintaining body homeostasis.

Objective 6. Differentiate between proteins, carbohydrates, and lipids.

To understand cellular structure and function, one must have a knowledge of basic cellular and general body chemistry. All of the processes that occur in the body comply with principles of chemistry. Furthermore, many of the dysfunctions of the body have a chemical basis.

Elements and Compounds

An **element** is a basic chemical substance. Four elements compose over 95% of the body. These elements and their percentages of body weight are oxygen (O) 65%, carbon (C) 18%, hydrogen (H) 10%, and nitrogen (N) 3%. Additional common elements within the body include calcium (Ca), potassium (K), sodium (Na), phosphorus (P), magnesium (Mg), and sulfur (S).

A few elements exist separately in the body, but most are linked with others to form compounds. A **compound** is a substance composed of two or more elements joined by chemical bonds. *Organic compounds* contain carbon and are formed by living organisms. *Inorganic compounds* generally lack carbon and are not formed by living organisms. Water and electrolytes (acids, bases, and salts) are examples of inorganic compounds. Proteins, carbohydrates, and lipids are organic compounds. The percentages of organic and inorganic compounds within adult males and females are compared in table 3.1.

UNIT III

Table 3.1	Compounds Within Adult Males and Females (Expressed as Percentage of Body Weight)	
Substance	Male	Female
Water	62	59
Protein	18	15
Lipid	14	20
Carbohydrates	1	1
Other (electrolytes, nucleic acids)	5	5

Table 3.2	Kinds of Electrolytes	
	Characteristic	Examples
Acid	Ionizes to release hydrogen ions (H^+)	Carbonic acid, hydrochloric acid, acetic acid, phosphoric acid
Base	Ionizes to release hydroxyl ions (OH^-) that combine with hydrogen ions	Sodium hydroxide, potassium hydroxide, magnesium hydroxide, aluminum hydroxide
Salt	Substance formed by the reaction between an acid and a base	Sodium chloride, aluminum chloride, magnesium sulfate

Water

Water is the most abundant compound within cells and in the extracellular environment. Water generally occurs within the body as a homogeneous mixture of two or more compounds called a *solution*. In this condition, the water is the *solvent,* or the liquid portion of the solution, and the *solutes* are dissolved substances suspended within the solution. Water is an excellent solvent through which many solutes can be transported through the cell membrane of a cell or from one part of the cell to another. Water is also important in maintaining a constant cellular temperature and body temperature because it absorbs and releases heat slowly. Evaporative cooling through the skin also involves water. Another function of water is as a *reactant* in the breakdown *(hydrolysis)* of food material in digestion.

> Dehydration is a condition in which fluid loss exceeds fluid intake with a resultant decrease in the volume of cellular and intracellular fluids. Rapid dehydration through vomiting, diarrhea, or excessive sweating can lead to serious medical problems by impairing cellular function. Infants are especially vulnerable because their fluid volume is so small. They can die from dehydration resulting from diarrhea within a matter of hours.

Electrolytes

Electrolytes are inorganic compounds that, when dissolved in water, break down into ions. An electrolyte is classified according to the ions it yields when it is dissolved in water. The three classes of electrolytes are *acids, bases,* and *salts,* all of which are important for normal cellular function. The functions of ions include the control of water movement through cells; the maintenance of normal acid-base *(pH)* balance; the conduction of electrical currents; the development of certain kinds of tissues, such as bones and teeth; and the normal functioning of blood, muscles, and nerves.

The three kinds of electrolytes are summarized in table 3.2.

Proteins

Proteins are nitrogen-containing organic compounds composed of amino acid molecules. An *amino acid* is an organic compound that contains an amino group ($-NH_2$) and a carboxyl group ($-COOH$). Proteins are the most abundant of the organic compounds and may exist by themselves or may be *conjugated* (joined) with other compounds; for example, with nucleic acids (RNA or DNA) to form nucleoproteins, with carbohydrates to form glycoproteins, or with lipids to form lipoproteins.

Proteins form the framework of the cell. The cell (plasma) membrane is composed primarily of protein and lipid molecules. Cellular growth, repair, and division depend on the availability of proteins. *Enzymes* are proteins or conjugated proteins that serve to catalyze chemical reactions within cells. Many *hormones* belong to a specialized group of messenger and regulator proteins produced by endocrine glands. Proteins, under certain conditions, may even be metabolized to supply cellular energy.

Carbohydrates

Carbohydrates are organic compounds that contain carbon, hydrogen, and oxygen, with a 2:1 ratio of hydrogen to oxygen. Carbohydrates include the starches and sugars and are the most abundant source of cellular energy. Excessive carbohydrate intake is converted to *glycogen* (animal starch) or to *fat* for storage in adipose tissue.

> If a person is deprived of food, the body uses the glycogen and fat reserves first and then metabolizes the protein within the cells. The gradual destruction of the cellular protein accounts for the lethargy, extreme emaciation, and ultimate death of starvation victims.

---

solution: L. *solvere,* loosen or dissolve
hydrolysis: Gk. *hydor,* water; *lysis,* dissolution
electrolyte: L. *electrum,* amber; Gk. *lysis,* a loosening
acid: L. *acidus,* sour

---

protein: Gk. *proteios,* of the first quality
hormone: Gk. *hormon,* setting in motion

Table 3.3	Location and Function of the Chemical Substances of Cells	
Substance	**Location in Cell**	**Function**
Water	Throughout	Dissolves, suspends, and ionizes materials; helps to regulate temperature
Electrolytes	Throughout	Establish osmotic gradients, pH, and membrane potentials
Proteins	Membranes, cytoskeleton, ribosomes, enzymes	Give structure, strength, contractility; catalyze, buffer
Lipids	Membranes, Golgi apparatus, inclusions	Provide reserve energy source; shape, protect, insulate
Carbohydrates	Inclusions	Preferred fuel for metabolic activity
Nucleic acids		
DNA	Nucleus, in chromosomes and genes	Controls cell activity
RNA	Nucleolus, cytoplasm	Transmits genetic information; transports amino acids
Trace materials		
Vitamins	Cytoplasm, nucleus	Work with enzymes in metabolism
Hormones	Cytoplasm, nucleus	Work with enzymes; activate or deactivate genes
Metals	Cytoplasm, nucleus	Synthesize insulin and other substances; help in maturation of blood cells

Lipids

Lipids are a third group of important organic compounds found in cells. Lipids include both fats and fat-related substances, such as phospholipids and cholesterol. Fats are important in building cell parts and supplying metabolic energy. Fats also serve to protect and insulate various parts of the body. Phospholipids and protein molecules make up the cell membrane and play an important role in regulating which substances enter or leave a cell.

Lipids, like carbohydrates, are composed of carbon, hydrogen, and oxygen atoms. Lipids, however, contain a smaller proportion of oxygen than do carbohydrates.

Table 3.3 summarizes the locations and functions of inorganic and organic substances within cells.

1. List the four most abundant elements in the body and state their relative percentages.
2. What is a compound? What two kinds of compounds exist in the body and what distinguishes each?
3. List the functions of water as related to cells and define *solvent* and *solute*.
4. Explain how electrolytes are important in maintaining homeostasis within a cell.
5. Describe a protein and discuss the functions of proteins within a cell. Explain how proteins differ from carbohydrates and lipids.

lipid: Gk. *lipos*, fat

Cellular Structure

The cell membrane separates the interior of the cell from the extracellular environment. Substances that enter or leave the cell are moved across the cell membrane. Most of the metabolic activities occur within the cytoplasmic organelles. The nucleus functions in protein synthesis and cell reproduction.

Objective 7. Describe the components of a cell.

Objective 8. Describe the composition and structure of the cell membrane and relate the structure to the functions of the membrane.

Objective 9. Distinguish between passive and active transport and describe the different ways in which each is accomplished.

Objective 10. Describe the structure and function of the endoplasmic reticulum, ribosomes, Golgi apparatus, lysosomes, and mitochondria.

Objective 11. Describe the structure and function of the nucleus.

As the basic functional unit of the body, each cell is a highly organized molecular factory. As previously discussed, cells come in a great variety of shapes and sizes. This variation, which is also apparent in the subcellular structures within different cells, reflects the variation of the functions of different cells in the body. All cells, however, share certain characteristics; for example, a cell membrane and most of the other structures listed in table 3.4. Thus, although no single cell can be considered "typical," the general structure of cells can be indicated by a single illustration (fig. 3.2).

UNIT III

Table 3.4	Structure and Function of Cellular Components	
Component	**Structure**	**Function**
Cell (plasma) membrane	Membrane composed of phospholipid and protein molecules	Gives form to cell and controls passage of materials in and out of cell
Cytoplasm	Fluid, jellylike substance between the cell membrane and the nucleus in which organelles are suspended	Serves as matrix substance in which chemical reactions occur
Endoplasmic reticulum	System of interconnected membrane-forming canals and tubules	Provides supporting framework within cytoplasm; transports materials and provides attachment for ribosomes
Ribosomes	Granular particles composed of protein and RNA	Synthesize proteins
Golgi apparatus	Cluster of flattened, membranous sacs	Synthesizes carbohydrates and packages molecules for secretion; secretes lipids and glycoproteins
Mitochondria	Double-walled membranous sacs with folded inner partitions	Release energy from food molecules and transform energy into usable ATP
Lysosomes	Single-walled membranous sacs	Digest foreign molecules and worn and damaged cells
Peroxisomes	Spherical membranous vesicles	Contain certain enzymes; form hydrogen peroxide
Centrosome	Nonmembranous mass of two rodlike centrioles	Helps to organize spindle fibers and distribute chromosomes during mitosis
Vacuoles	Membranous sacs	Store and excrete various substances within the cytoplasm
Fibrils and microtubules	Thin, hollow tubes	Support cytoplasm and transport materials within the cytoplasm
Cilia and flagella	Minute cytoplasmic extensions from cell	Move particles along surface of cell or move cell
Nuclear membrane	Membrane surrounding nucleus, composed of protein and lipid molecules	Supports nucleus and controls passage of materials between nucleus and cytoplasm
Nucleolus	Dense, nonmembranous mass composed of protein and RNA molecules	Forms ribosomes
Chromatin	Fibrous strands composed of protein and DNA molecules	Controls cellular activity for carrying on life processes

UNIT III

For descriptive purposes, a cell can be divided into three principal parts:

1. **Cell (plasma) membrane.** The outer, selectively permeable cell membrane gives form to the cell and separates the cell's internal structures from the extracellular environment.
2. **Cytoplasm** and **organelles** (or''gă-nelz). The cytoplasm is the content of a cell between the nucleus and the cell membrane. Organelles are the minute membrane-bound structures within the cytoplasm of a cell that are concerned with specific functions.
3. **Nucleus.** The nucleus is the large, generally spheroid body within a cell that contains the DNA, or genetic material, of a cell.

plasma: Gk. *plasma*, to form or mold
nucleus: L. *nucleus*, kernel or nut

Cell Membrane

The extremely thin **cell (plasma) membrane** is composed of phospholipid and protein molecules. Its thickness ranges from 65 to 100 angstroms (Å); that is, it is less than a millionth of an inch thick. The structure of the cell membrane is not fully understood, but most cytologists believe that it is composed primarily of phospholipid molecules with larger globular proteins scattered within the lipid layer (fig. 3.3). Such a structure suggests that the membrane is not static but has considerable movement. Minute openings, or pores, ranging between 7 and 10 Å in diameter extend through the membrane.

The two most important functions of the cell membrane are to enclose the components of the cell and to regulate the passage of substances into and out of the cell. A highly selective exchange of substances occurs across the membrane boundary, involving several types of passive and active processes. The kinds of movement across a cell membrane are summarized in table 3.5 and illustrated in figure 3.4.

Figure 3.2 A generalized cell and the principal organelles.

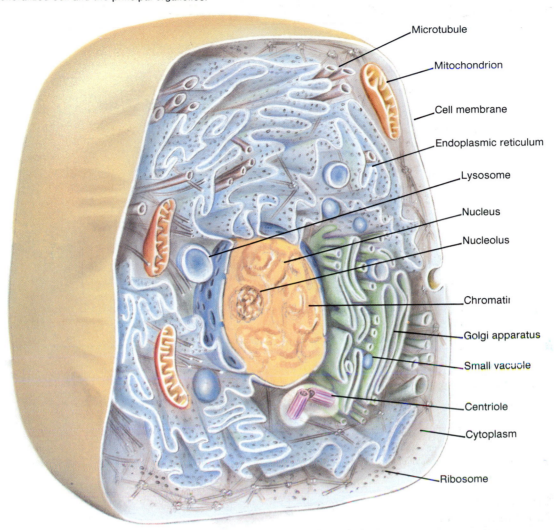

Microtubule

Mitochondrion

Cell membrane

Endoplasmic reticulum

Lysosome

Nucleus

Nucleolus

Chromatin

Golgi apparatus

Small vacuole

Centriole

Cytoplasm

Ribosome

The permeability of the cell membrane depends on the following factors:

1. **Structure of the cell membrane.** Although cell membranes on all cells are composed of phospholipids, there is evidence that their thickness and structural arrangement—both of which could affect permeability—vary considerably.
2. **Size of the molecules.** Macromolecules, such as certain proteins, are not allowed into the cell. Water and amino acids are small molecules and can readily pass through the cell membrane.
3. **Ionic charge.** The protein portion of the cell membrane may have an ionic charge. Ions with an opposite charge are attracted to and readily pass through the membrane, whereas those with a similar charge are repelled.
4. **Lipid solubility.** Substances that are easily dissolved in lipids pass into the cell with no problem, since a portion of the cell membrane is composed of lipid material.
5. **Presence of carrier molecules.** Specialized carrier molecules positioned within the cell membrane are capable of attracting and transporting substances across the membrane, regardless of size, ionic charge, or lipid solubility.

Figure 3.3 The cell membrane consists of a double layer of phospholipids. Proteins may completely or partially span the membrane. Carbohydrates are attached to the outer surface.

Extracellular side

Fibrous protein

Carbohydrate

Glycolipid

Double layer of phospholipid molecules

Cholesterol

Globular protein

Intracellular side

Table 3.5	Kinds of Movement Through Cell Membranes		
Processes	**Characteristics**	**Energy Source**	**Example**
Diffusion	Passive movement of molecules from regions of higher concentration toward regions of lower concentration	Molecular motion	Respiratory gases are exchanged in lungs
Facilitated diffusion	Use of carrier substances to speed process	Carrier energy and molecular motion	Glucose enters cell
Osmosis	Passive movement of water molecules through semipermeable membrane due to a concentration difference for water	Molecular motion	Water moves through cell wall to maintain constant turgidity of cell
Filtration	Molecules are forced by hydrostatic pressure from regions of higher pressures to regions of lower pressures	Blood pressure	Wastes are removed within kidneys
Active transport	Molecules or ions are transported through cell membrane by other molecules	Cellular energy (ATP)	Glucose and amino acids move through membranes
Pinocytosis	Membrane engulfs minute droplets of fluid from surroundings	Cellular energy	Membrane forms vacuoles containing solute and solvent
Phagocytosis	Membrane engulfs solid particles from surroundings	Cellular energy	White blood cell membrane engulfs bacterial cell

UNIT III

Figure 3.4 Examples of various kinds of movements through membranes. (*a*) Sugar molecules diffuse from compartment A to compartment B until equilibrium is achieved in 3. (*b*) Osmosis occurs as a selectively permeable membrane allows only water to diffuse through the membrane between compartments A and B, causing the level of the liquid to rise in A. (*c*) Filtration occurs as small molecules are forced through a membrane by blood pressure, leaving the larger molecules behind.

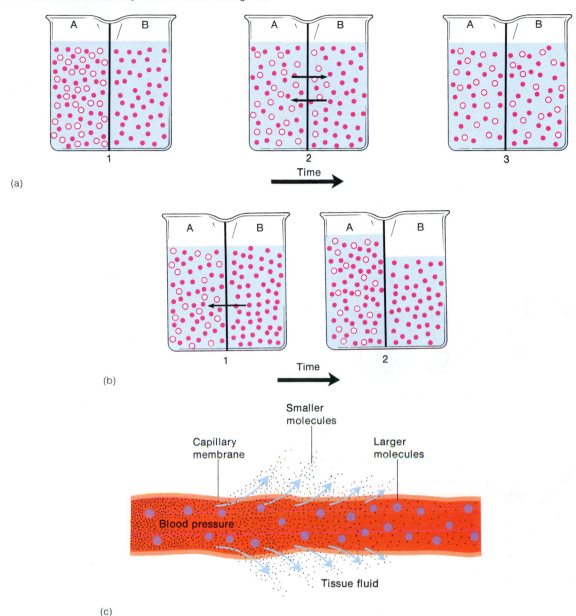

Cell membranes of certain cells are highly specialized to facilitate specific functions (fig. 3.5). The columnar cells lining the lumen (hollow portion) of the intestinal tract, for example, have numerous fine projections, or **microvilli,** that aid in the absorptive process of digestion. A single columnar cell may have as many as 3,000 microvilli on the exposed portion of the cell membrane, and a square millimeter of surface area may contain over 200 million microvilli.

Cells composing certain sensory organs have specialized cell membranes. The photoreceptors, or light-responding rods and cones of the eye, have double-layered, disc-shaped membranes called **sacs.** These structures contain pigments associated with vision. Within the spiral organ (organ of Corti) in the inner ear are **hair cells.** These tactile (touch) receptors are stimulated through mechanical vibration. Hair cells are so named because of the fine, cilialike processes that extend from the cell membrane.

Cytoplasm and Organelles

Cytoplasm refers to the material within the cell membrane but outside of the nucleus. The material within the nucleus is frequently called the **nucleoplasm. Protoplasm** refers to the cytoplasm and nucleoplasm collectively.

When observed through an electron microscope (fig. 3.6), distinct cellular components called **organelles** can be seen in the highly structured cytoplasm. The matrix of the cytoplasm

---

microvilli: Gk. *mikros,* small; *villus,* tuft of hair

Figure 3.5 Specialized cell membranes indicated by arrows. (a) Microvilli of the cells that line the lumen of the small intestine (640×), (b) rod sacs within the eye for photoreception (160×), and (c) hair cells within the spiral organ (organ of Corti) of the inner ear for tactile stimulation.

(a)

(b)

(c)

Figure 3.6 A transmission electron microscope like this one is used to observe and photograph organelles within the cytoplasm of a cell.

is a jellylike substance that is 80% to 90% water. The organelles and inorganic *colloid substances* (suspended particles) are dispersed throughout the cytoplasm. Colloid substances have similar ionic charges that space them uniformly.

Metabolic activity occurs within the organelles of the cytoplasm. Specific roles, such as heat production, cellular maintenance, repair, storage, and protein synthesis, are carried out within the organelles.

The structure and function of each of the major organelles are discussed in the following paragraphs, and a summary of this information is presented in table 3.4.

Endoplasmic Reticulum   The endoplasmic reticulum, or **ER,** is widely distributed throughout the cytoplasm as complex networks of interconnected membranes (fig. 3.7). Between the membranes are minute spaces, or **tubules,** that are connected at one end to the cell membranes. The tubules may also be connected to other organelles or to the nuclear envelope.

There are two types of endoplasmic reticulum. **Rough endoplasmic reticulum** is characterized by the presence of numerous small granules called **ribosomes,** which are attached to the outer surface of the membranous wall. **Smooth endoplasmic reticulum** lacks ribosomes. Endoplasmic reticulum provides a pathway for transportation of substances within the cell and a storage area for synthesized molecules. Smooth endoplasmic reticulum manufactures certain lipid molecules. The membranous wall of rough endoplasmic reticulum provides a site for protein synthesis within ribosomes.

A person who repeatedly uses certain drugs, such as alcohol or phenobarbital, develops a tolerance to them, so that greater quantities are required to achieve the effect they had originally. The cytological explanation for this is that repeated use causes the smooth endoplasmic reticulum to proliferate in an effort to detoxify these drugs and protect the cell. With increased amounts of smooth endoplasmic reticulum, cells can handle an increased volume of drugs.

Figure 3.7 (*a*) An electron micrograph of rough endoplasmic reticulum magnified about 100,000 times. (*b*) Rough endoplasmic reticulum has ribosomes attached to its surface.

(a)

Ribosomes

Tubule

Membrane

(b)

Walston

Ribosomes Ribosomes occur as free particles suspended within the cytoplasm or are attached to the membranous wall of rough endoplasmic reticulum. Ribosomes are small, granularlike organelles (fig. 3.7) composed of protein and RNA molecules. They function to synthesize protein molecules that may be used for building cell structures or for functioning as enzymes. Some of the proteins synthesized by ribosomes are secreted outside the cell to be used elsewhere in the body.

Golgi Apparatus The Golgi apparatus (Golgi complex) consists of several tiny membranous sacs, or vesicles, located near the nucleus (fig. 3.8a), that are continuous with the endoplasmic reticulum.

The Golgi apparatus is involved in the synthesis of carbohydrates and cellular secretions. As large carbohydrate molecules are synthesized, they combine with proteins to form compounds called glycoproteins. The glycoproteins accumulate in the channels of the Golgi complex. After a critical volume is reached, the vesicles break off from the complex and are carried to the cell membrane and released as a secretion (fig. 3.8b). Once the vesicle is outside the cell membrane, it ruptures to release its contents, thus completing the process known as *exocytosis*.

The Golgi apparatus is prominent in cells of certain secretory organs of the digestive system, such as the pancreas and the salivary glands. Pancreatic cells, for example, produce digestive enzymes that are packaged in the Golgi apparatus and secreted as droplets that flow into the pancreatic duct and are transported to the gastrointestinal (GI) tract.

Mitochondria Mitochondria are double-membraned, saclike organelles. The outer membrane is smooth, whereas the inner membrane is arranged in intricate folds called **cristae** (fig. 3.9). The folded cristae create a tremendous surface area for chemical reactions.

Mitochondria vary in size and shape. They can migrate through the cytoplasm and can reproduce themselves by budding or cleavage. Mitochondria are often called the powerhouses of cells because of their role in producing metabolic energy. Enzymes connected to the cristae control the chemical reactions that form ATP. Metabolically active cells, such as muscle cells, liver cells, and kidney cells, have a large number of mitochondria because they use energy at a high rate.

 The darker color of some cuts of meat (a chicken thigh, for example, as compared to a breast) is due to the larger amounts of myoglobin, a pigmented compound in muscle tissue that acts to store oxygen. Mitochondria are likewise more abundant in red meat. Both mitochondria and myoglobin are important for the high level of metabolic activity in red muscle tissue.

Because mitochondria are contained within ova (egg cells) but not within the head of a sperm, all of the mitochondria in a fertilized egg are derived from the mother. As cells divide during the developmental process, the mitochondria likewise replicate themselves; thus, all of the mitochondria in a fetus are genetically identical to those in the original ovum. This accounts for a unique form of inheritance that is passed only from mother to child. A rare cause of blindness—*Leber's hereditary optic neuropathy*—and perhaps genetically based neuromuscular disorders, are believed to be inherited in this manner.

Lysosomes Lysosomes vary in appearance from granularlike bodies to small vesicles to membranous spheres (fig. 3.10). They are scattered throughout the cytoplasm. Powerful enzymes enclosed within lysosomes are capable of breaking down protein and carbohydrate molecules. White blood cells contain large numbers of lysosomes and are said to be *phagocytic,* meaning that they will ingest, kill, and digest bacteria through the enzymatic activity of their lysosomes.

 The normal atrophy, or decrease in size, of the uterus following the birth of a baby is due to lysosomal digestive activity. Likewise, the secretions of lysosomes are responsible for the regression of the mammary tissue of the breasts after the weaning of an infant.

UNIT III

Golgi: from Camillo Golgi, Italian histologist, 1843–1926

mitochondrion: Gk. *mitos*, a thread; *chondros*, lump, grain
cristae: L. *crista*, crest
lysosome: Gk. *lysis*, a loosening; *somo*, body
phagocytic: Gk. *phagein*, to eat; *kytos*, a cell

UNIT III

Figure 3.8 (*a*) An electron micrograph of a Golgi apparatus. (Note the formation of vesicles at the ends of some of the flattened sacs.)

(*b*) An illustration of the processing of proteins by the rough endoplasmic reticulum and Golgi apparatus.

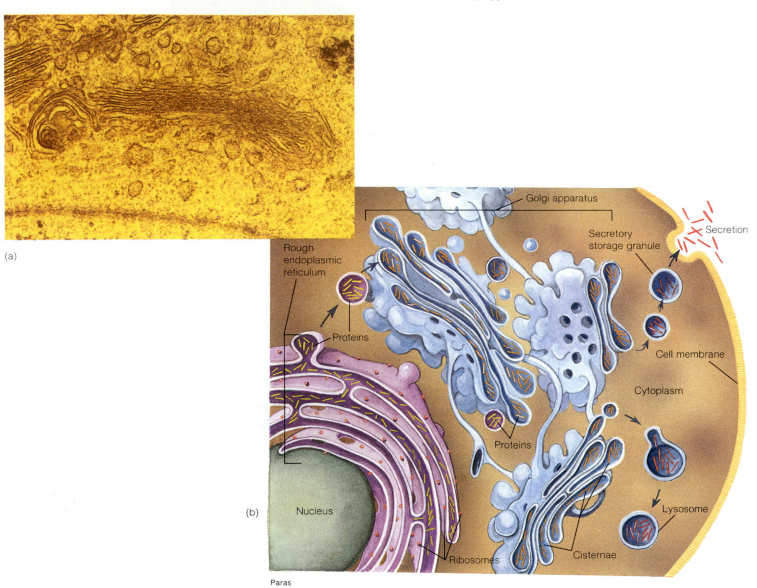

(a)

(b)

Paras

Golgi apparatus

Rough endoplasmic reticulum

Proteins

Secretory storage granule

Secretion

Cell membrane

Cytoplasm

Proteins

Nucleus

Lysosome

Ribosomes

Cisternae

Figure 3.9 (*a*) An electron micrograph of a mitochondrion magnified about 40,000 times. (*b*) A diagram showing how membranous partitions form the cristae within the organelle.

(a)

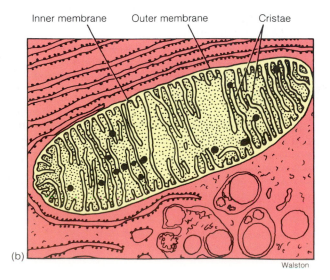

(b)

Inner membrane   Outer membrane   Cristae

Walston

Figure 3.10 (*a*) An electron micrograph of a lysosome magnified about 30,000 times. (*b*) A saclike lysosome with a diagram showing digestive phagocytic enzymes.

(a)

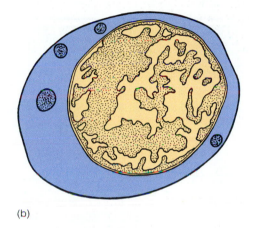

(b)

Figure 3.11 The centrioles. (*a*) A micrograph of the two centrioles in a centrosome (14,200×). (*b*) A diagram showing that the centrioles are positioned at right angles to one another.

(a)

Paras

(b)

Lysosomes also function to destroy their own worn cell parts, or the entire cell within which they reside if the cell is injured. For this reason, lysosomes are frequently called "suicide packets."

Several diseases arise from abnormalities in lysosome function. The painful inflammation of *rheumatoid arthritis*, for example, occurs when enzymes from lysosomes are released into the joint capsule and initiate digestion of the surrounding tissue.

Peroxisomes Peroxisomes are membranous sacs that resemble lysosomes in structure and, like lysosomes, they contain certain enzymes. Peroxisomes occur in most cells but are particularly abundant in the kidney and the liver. All peroxisomes contain one or more enzymes that produce hydrogen peroxide, a highly toxic substance. Hydrogen peroxide is an important compound in the white blood cells, which phagocytize diseased or worn-out cells. These organelles also contain the enzyme catalase, which breaks down excess hydrogen peroxide into water and oxygen so that there is no toxic effect on other organelles within the cytoplasm.

Centrosome and Centrioles The **centrosome** (central body) is a nonmembranous, spherical mass positioned near the nucleus. Within the centrosome is a pair of rodlike structures called **centrioles** (fig. 3.11), positioned at right angles to each other. The wall of each centriole is composed of nine evenly spaced bundles, and each bundle contains three microtubules.

Centrosomes are found only in those cells that can divide. During the mitotic (reproductive) process, the centrioles move away from one another and take positions on either side of the nucleus. They are then involved in the distribution of the chromosomes during cellular reproduction. Mature muscle and nerve cells lack centrosomes, and thus cannot divide.

Vacuoles Vacuoles are membranous sacs that usually function as storage chambers. They vary in size and are formed when a portion of the cell membrane invaginates and pinches off during a process called *vacuolation.* Vacuolation is initiated either by *phagocytosis,* in which the cell membrane engulfs solid particles, or by *pinocytosis,* in which cells take in minute droplets of liquid through the cell membrane (fig. 3.12). Vacuoles may contain liquid or solid materials that were previously outside the cell.

vacuole: L. *vacuus,* empty

Figure 3.12 Pinocytosis and phagocytosis compared. (a) During pinocytosis, the cell takes in a minute droplet of fluid from its surroundings. (b) During phagocytosis, a solid particle is engulfed and ingested through the cell membrane.

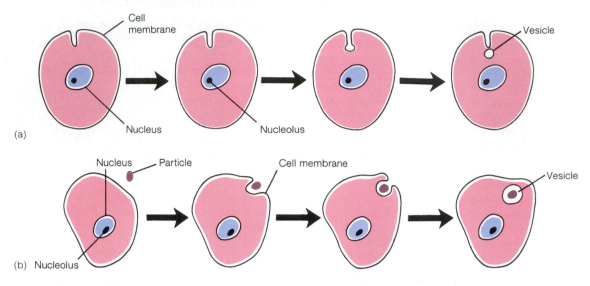

Figure 3.13 Microtubules. (a) An electron micrograph showing microtubules forming a type of cytoskeleton within a cell. (b) A diagram showing that each microtubule is composed of precisely arranged globular proteins.

Fibrils and Microtubules Both fibrils and microtubules are found throughout the cytoplasm. The fibrils are minute, rodlike structures, whereas the microtubules are fine, threadlike tubular structures of varying lengths (fig. 3.13). Both provide support to the cell by forming a type of cytoskeleton. Specialized fibrils called myofilaments are particularly abundant in muscle cells, where they aid in the contraction of these cells. Microtubules are also involved in the transportation of macromolecules throughout the cytoplasm. They are especially abundant in the cells of endocrine organs, where they aid the movement of hormones to be secreted into the blood. Microtubules in certain cells provide flexible support to cilia and flagella.

Cilia and Flagella Although cilia and flagella appear to extend from the cell membrane, they are actually cytoplasmic projections from within the cell. These projections contain cytoplasm and supportive microtubules bounded by the cell membrane (fig. 3.14). Cilia and flagella should not be confused with the previously discussed microvilli, hair cells, or stereocilia, which are specializations of cell membranes.

Cilia are numerous short projections from the exposed border of certain cells (fig. 3.15). Ciliated cells are interspersed

Figure 3.14 (*a*) Cilia and flagella are similar in the structural arrangement of their microtubules. (*b*) A spermatozoa cell has a single flagellum for propulsion. (*c*) Cilia produce a wavelike motion to move particles toward the outside of the body.

Figure 3.15 A scanning electron micrograph of ciliated cells that line the lumen of the uterine tube.

with mucus-secreting **goblet cells.** There is always a film of mucus on the free surface of ciliated cells. Ciliated cells line the lumina (hollow portions) of sections of the respiratory and reproductive systems. The function of cilia is to move the mucus and any adherent material toward the exterior of the body.

Flagella are similar to cilia in basic microtubular structure (fig. 3.14). They are somewhat longer than cilia, however, and there are only one or two flagella per cell. The only type of cell in humans that has a flagellum is a sperm cell, which uses the single structure for locomotion.

Cell Nucleus

The spherical **nucleus** is usually located near the center of the cell (fig. 3.16). It is the largest structure of the cell and contains the genetic material that determines cellular structure and controls cellular activity.

Most cells contain a single nucleus. Certain cells, however, such as skeletal muscle cells, are multinucleated. The long skeletal muscle fibers contain so much cytoplasm that several governing centers are necessary. Other cells, such as mature red blood cells, lack nuclei. These cells are limited to certain types of chemical activities and are not capable of reproduction.

The nucleus is enclosed by a double membrane called the **nuclear membrane** (fig. 3.16). The narrow space between the two walls of the nuclear membrane is called the **perinuclear**

cisterna. Minute pores located along the nuclear membrane allow substances to move between the nucleoplasm and the cytoplasm. A cell is governed by two important structures within the nucleoplasm of the nucleus:

1. **Nucleoli.** The nucleoli are small, nonmembranous, spherical bodies composed largely of protein and RNA. It is thought that they function in the production of ribosomes. As ribosomes are formed, they migrate through the nuclear membrane into the cytoplasm.
2. **Chromatin.** Chromatin is a coiled, threadlike mass. It is the genetic material of the cell and consists principally of protein and DNA molecules. When the cell begins to undergo a division, the chromatin shortens and thickens into rod-shaped chromosomes (fig. 3.17). Each chromosome has thousands of genes distributed along its threadlike structure. The genes determine the structure and function of a cell.

1. Describe the composition and specializations of the cell membrane. Discuss the importance of the selective permeability of the cell membrane.
2. Discuss the various kinds of movements across the cell membrane. Which are passive and which are active?
3. Describe the structure and function of the following cytoplasmic organelles: rough endoplasmic reticulum, a Golgi apparatus, lysosomes, and mitochondria.
4. Distinguish between the nucleus and nucleoli.
5. Distinguish between chromatin and chromosomes.

Figure 3.16 (a) The cell nucleus magnified about 20,000 times in an electron micrograph. The nucleus contains a nucleolus and masses of chromatin. (b) The double-layered nuclear envelope has pores to permit substances to pass between the nucleus and the cytoplasm.

(a)

(b)

Figure 3.17 A false-color light micrograph showing the full complement of male chromosomes arranged in numbered homologous pairs.

Cell Cycle

A cell cycle consists of growth, synthesis, and mitosis. Growth is the increase in cellular mass as the result of metabolism. Synthesis is the production of DNA and RNA to regulate cellular activity. Mitosis is the division of the nucleus and cytoplasm of a cell resulting in the formation of two daughter cells.

Objective 12.	Describe the structure of a DNA molecule.
Objective 13.	List the stages of mitosis and discuss the events of each stage.
Objective 14.	Discuss the significance of mitosis.

Cellular replication is one of the principal concepts of biology. Through the process of cellular division called **mitosis,** a multicellular organism can develop and be maintained. Mitosis enables body growth and the replacement of damaged, diseased, or worn-out cells. The process ensures that each reproduced cell has the same number and kind of chromosomes as the original parent cell.

In an average healthy adult, over 100 billion cells will die and be mitotically replaced during a 24-hour period. This represents a replacement of about 2% of the body mass each day. Some of the most mitotically active sites are the outer layer of skin, the internal lining of the digestive tract, and the liver.

Before a cell can divide, it must first duplicate its chromosomes so that the genetic traits can be passed to the succeeding generations of cells. A chromosome consists of a coiled **deoxyribonucleic acid (DNA) molecule** and is not visible until the cell begins to divide. There are 23 pairs of chromosomes in each human body (somatic) cell and approximately 20,000 genes are positioned on each chromosome.

Chromosomes are of varied lengths and shapes—some twisted, some rodlike. During mitosis, the chromosomes shorten and condense, each pair assuming a characteristic shape (fig. 3.17). On the chromosome is a **centromere** to which are attached the **spindle fibers** that direct the chromosome toward the pole during mitosis.

---

mitosis: Gk. *mitos,* thread

Structure of DNA

The DNA molecule is frequently called a *double helix* because of its resemblance to a spiral ladder (fig. 3.18). The sides of the DNA molecule are formed by alternating units of sugar, *deoxyribose,* and a phosphoric acid called the *phosphate group.* The rungs of the molecule are composed of pairs of *nitrogenous bases.* The ends of each nitrogenous base are attached to the deoxyribose-phosphate units. There are only four types of nitrogenous bases in a DNA molecule: adenine (A), thymine (T), cytosine (C), and guanine (G).

The basic structural units of the DNA molecule are called **nucleotides.** Each nucleotide consists of a molecule of deoxyribose sugar, a phosphate group, and one of the four nitrogenous bases. Thus, there is a nucleotide type for each of the four bases.

The pairing of the nitrogenous bases of the nucleotides is highly specific. The molecular configuration of each base is such that adenine always pairs with thymine and cytosine always pairs with guanine. The hydrogen bonds between these bases are relatively weak and can be easily split during cellular division (fig. 3.19). During division, the sequence of bases along the sides of the DNA molecule serves as a template that determines the sequence along each new strand.

Cell Cycle and Cell Division

A **cell cycle** is the series of changes that a cell undergoes from the time it is formed until it has completed a division and reproduced itself. **Interphase** is the first period of the cycle, from cell formation to the start of cell division (fig. 3.20). During interphase, a cell grows, carries on metabolic activities, and prepares itself for division.

Interphase is divided into **G1, S1,** and **G2 phases.** During the G1 (first growth) phase, a cell grows rapidly and is metabolically active. The length of time cells are within G1 varies considerably. It may be only hours in cells that have rapid division rates, or a matter of days or even years for other cells. At the end of G1, the centrioles replicate in preparation for their role in cell division. During the S1 (synthetic) phase, the DNA in the nucleus of a cell replicates so that the two future cells will receive identical copies of the genetic material. During the G2 (second growth) phase, the enzymes and other proteins needed for the division process are synthesized and the cell continues to grow.

The actual division of a cell is referred to as the **mitotic phase,** or simply **M phase** (fig. 3.20). The mitotic phase is further divided into mitosis and cytokinesis. **Mitosis** is the period of a cell cycle during which there is nuclear division and the duplicated chromosomes separate to form two genetically identical daughter nuclei. The process of mitosis takes place in four successive stages, each stage passing into the next without sharp structural distinctions. These stages are *prophase, metaphase, anaphase,* and *telophase* (fig. 3.21). Mitosis is followed by **cytokinesis,** during which there is division of the cytoplasm.

cytokinesis: Gk. *kytos,* a hollow; *kinesis,* movement

Figure 3.18 The double helix of a DNA molecule. Each strand of the helix contains only four kinds of organic bases (A, T, C, and G).

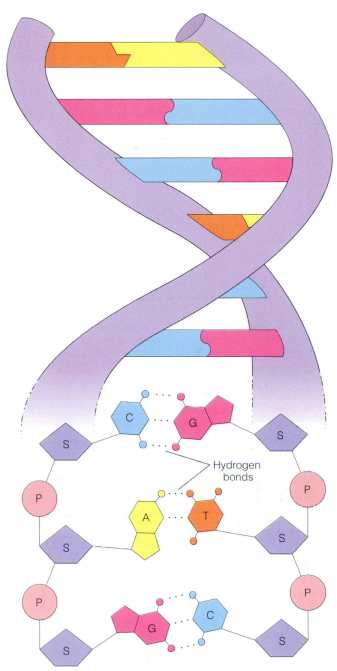

Hydrogen bonds

Figure 3.19 The replication of DNA. Each new double helix is composed of one old and one new strand. The sequence of bases of each of the new molecules is identical to that of the parent DNA because of complementary base pairing.

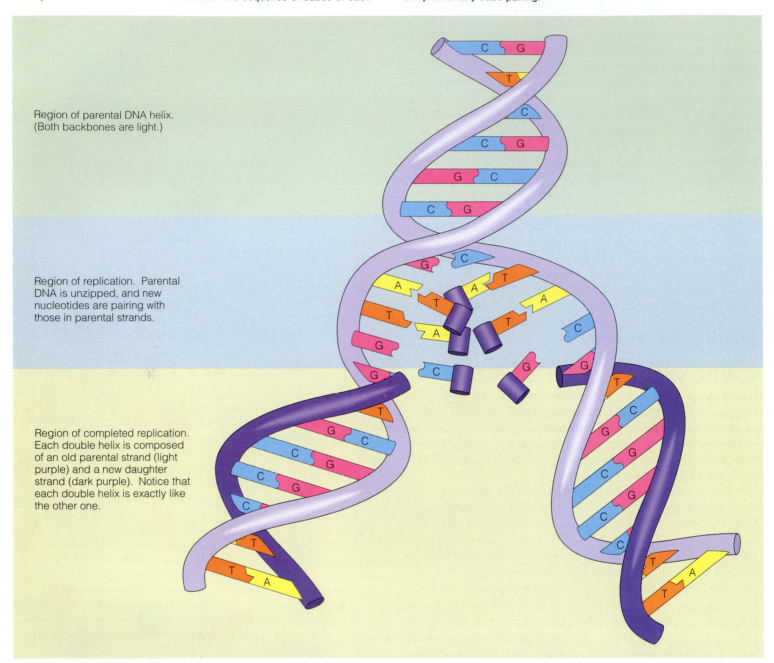

Region of parental DNA helix. (Both backbones are light.)

Region of replication. Parental DNA is unzipped, and new nucleotides are pairing with those in parental strands.

Region of completed replication. Each double helix is composed of an old parental strand (light purple) and a new daughter strand (dark purple). Notice that each double helix is exactly like the other one.

Highly specialized cells, such as muscle and nerve cells, do not replicate after a person is born. If these cells die, as the result of disease, injury, or even disuse, they are not replaced and scar tissue may form. Nerve cells are especially vulnerable to damage from alcohol, oxygen deprivation, and various drugs, which can jeopardize the health of these types of cells.

1. Explain why the DNA molecule is described as a double helix.
2. List the phases in the life cycle of a cell and describe the principal events that occur during each phase.
3. Explain why mitosis is such an important biological process.

Figure 3.20   Interphase and mitotic phase are the two principal divisions of the cell cycle. During interphase, a cell is growing, metabolizing and preparing to divide. During mitotic phase, nuclear division is followed by cytoplasmic division and the formation of two daughter cells.

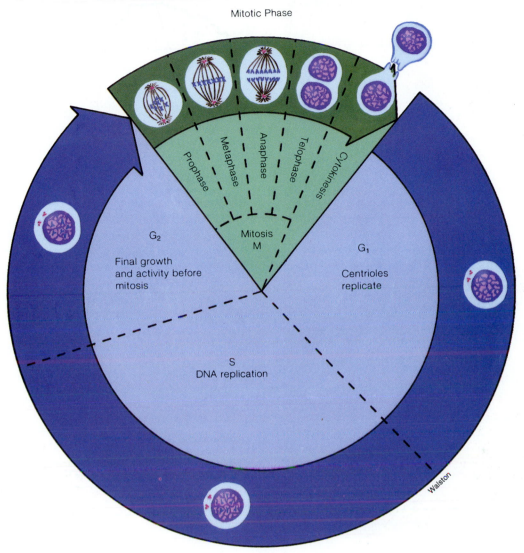

Mitotic Phase

Prophase

Metaphase

Anaphase

Telophase

Cytokinesis

Mitosis M

G₂
Final growth and activity before mitosis

G₁
Centrioles replicate

S
DNA replication

Walston

Interphase

UNIT III

Clinical Considerations

Cellular Adaptations

Apparently included within cellular specialization of structure and function is mitotic potential. Certain cells do not require further division once the organ to which they contribute becomes functional. Others, as part of their specialization, require continuous mitosis to keep an organ healthy. Thus, in the adult, it is found that some cells divide continually, some occasionally, and some not at all. For example, epidermal cells, hemopoietic cells, and cells that line the lumen of the GI tract divide continually throughout life. Cells within specialized organs, such as the liver or kidneys, divide as the need becomes apparent. Naturally occurring cellular death, disease, or trauma from surgery or injury may necessitate mitosis in these organs.

Still other cells, such as muscle or nerve cells, lose their mitotic ability as they become differentiated. Trauma to these cells frequently causes a permanent loss of function.

The factors regulating mitosis are unclear. Evidence suggests that mitotic ability is genetically controlled and, of those cells that do divide, even the number of divisions is predetermined. If this is true, it would certainly account for the aging process. Physical stress, nutrition, and hormones definitely have an effect on mitotic activity. It is thought that the reproductive activity of cells might be controlled through a feedback mechanism involving the release of a *growth-inhibiting substance*. Such a substance may slow or inhibit the cell divisions and growth of particular organs once the organ had amassed a certain number of cells or had reached a certain size.

Figure 3.21   The stages of mitosis.

(a) Interphase

The chromosomes are in an extended form and seen
   as chromatin in the electron microscope.
The nucleus is visible.

(b) Prophase

The chromosomes are seen and observed to consist
   of two chromatids joined together by a centromere.
The centrioles move apart toward opposite poles of the cell.
Spindle fibers are produced and extend from each centriole.
The nuclear membrane starts to disappear.
The nucleolus is no longer visible.

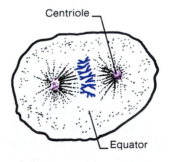

(c) Metaphase

The chromosomes are lined up at the equator of the cell.
The spindle fibers from each centriole are attached
   to the centromeres of the chromosomes.
The nuclear membrane has disappeared.

UNIT III

Figure 3.21 continued

(d) Anaphase

The centromeres split, and the sister chromatids separate as each is pulled to an opposite pole.

Nucleolus

Furrowing

Sims

(e) Telophase

The chromosomes become longer, thinner, and less distinct.
New nuclear membranes form.
The nucleolus reappears.
Cell division (cytokinesis) is nearly complete.

Except for cells located on exposed surfaces, most cells of the body are positioned in a rather homogeneous environment where continual adaptation to change is not necessary for survival. Cells do, however, have tremendous adaptability and resilience to ensure survival against conditions that might otherwise be lethal. Prolonged exposure to sunlight, for example, stimulates the synthesis of melanin and tanning of the skin. Likewise, mechanical friction to the skin increases mitotic activity and the synthesis of a fibrous protein, *keratin,* which results in the formation of a protective callus.

Cells adapt to potentially injurious stimuli by several specific mechanisms. **Hypertrophy** refers to an increase in the size of cells resulting from an increased synthesis of protein, nucleic acids, and lipids. Cellular hypertrophy can be either compensatory or hormonal. **Compensatory hypertrophy** occurs when increased metabolic demands on particular cells result in an increase in cellular mass. Examples of compensatory hypertrophy include skeletal muscle fibers enlarging as a result of ex-ercise, and cardiac (heart) muscle fibers or kidney cells enlarging because of an increased work demand. Hypertension (high blood pressure) causes cardiac cells to hypertrophy to pump blood against raised pressures. After the removal of a diseased kidney, there is a compensatory increase in the size of the cells of the remaining kidney. The remaining kidney usually increases from a normal weight of 180 gm to about 350 gm. Examples of **hormonal hypertrophy** are the increase in the size of the breasts and the increase in the size of the smooth muscles of the uterus in a pregnant woman.

Hyperplasia refers to an increase in the number of cells formed due to an increase in mitotic activity. The mechanism by which hyperplasia is induced is not well understood. The removal of a portion of the liver, for example, leads to regeneration, or hyperplasia, of the remaining liver cells to restore the loss. But the triggering mechanism for this is not known. A type of hormonally induced hyperplasia occurs in cells of the endometrium of a woman after menstruation, which restores this reproductive layer to a suitable state for implantation.

---

hypertrophy: Gk. *hyper,* over; *trophe,* nourishment

hyperplasia: Gk. *hyper,* over; *plasis,* a molding

Atrophy refers to a decrease in the size of cells and a corresponding decrease in the size of the affected organ. Atrophy can occur in the cells of any organ and may be classified as **disuse atrophy, disease atrophy,** or **aging (senile) atrophy.**

Metaplasia is a specialized cellular change in which one type of cell transforms into another. Generally, it involves the change of highly specialized cells into more generalized, protective cells. For example, excessive exposure to inhaled smoke causes the specialized ciliated columnar epithelial cells lining the bronchial airways to change into squamous epithelium, which is more resistant to injury from smoke.

Cell Trauma

As adaptable as cells are to environmental changes, many agents and factors are injurious to cells. If the trauma causes extensive cellular death, the condition may become life threatening. A person dies when a vital organ can no longer perform its necessary metabolic role in sustaining the body.

Energy deficit means that more energy is required than is available. Cells can tolerate certain mild deficits because of various reserves stored within the cytoplasm, but a severe or prolonged deficit will cause cells to die. An energy deficit occurs when the cells do not have enough glucose or oxygen to allow for glucose combustion. Examples of energy deficits are low levels of blood sugar (hypoglycemia) and the impermeability of the cell membrane to the entry of glucose (as in diabetes mellitus). Malnutrition also may result in energy deficit. Few cells can tolerate an interruption in oxygen supply. Cells of the brain and the heart have tremendous oxygen demands, and an interruption of the supply to these organs for only a matter of minutes can cause death.

Physical injury to cells, another type of cellular trauma, occurs in a variety of ways. High temperatures **(hyperthermia)** are generally less tolerable to cells than low temperatures **(hypothermia).** Respiratory rate, heart rate, and metabolism accelerate with hyperthermia. Continued hyperthermia causes protein coagulation within cells, and eventually cellular death. In frostbite, rapid or prolonged chilling causes cellular injury. In severe frostbite, ice crystals form and burst the cells.

Burns are particularly significant if they cause damage to the deeper skin layers, which interferes with the mitotic activity of cells (see chapter 5). Of immediate concern with burns, however, is the devastating effect of fluid loss and infections through traumatized cell membranes.

Accidental poisoning and suicide through drug overdose account for large numbers of deaths in the United States and elsewhere. **Drugs** and **poisons** cause cellular dysfunction by disrupting DNA replication, RNA transcription, active transport, or cell membrane activity.

Radiation is a type of cellular trauma that is cumulative in effect. When X rays are given for therapeutic purposes (radiotherapy), small doses are focused on a tumorous area over a course of many days to prevent widespread cellular injury. Some cells are more sensitive to radiation than others. Immature or mitotically active cells are highly sensitive, whereas cells that are no longer growing, such as neurons and muscle cells, are not as vulnerable to radiation injury.

Infectious agents, or **pathogens,** invade and inhabit cells, causing cellular dysfunction. Viruses and bacteria are the most common pathogens. Viruses usually destroy cells as they reproduce themselves. Bacteria, on the other hand, will frequently poison cells with their toxic metabolic wastes.

Cellular Organelles

Lysosomes are important organelles in the function of phagocytic cells. Organs like the liver, spleen, and lymph nodes have phagocytes that are immobile, or "fixed," within the organs. These are sometimes referred to as the *reticuloendothelial system.* Among other functions, the reticuloendothelial system helps to remove old red blood cells and bacteria from blood and to remove and inactivate drugs from the blood. Lysosomes also participate in *autophagy,* and by this process help to prevent the accumulation of damaged organelles and stored molecules. Genetic diseases that result in the defective production of lysosomal enzymes may produce abnormal accumulations of glycogen or lipids within particular cells. Examples of such diseases include Tay–Sachs disease and Gaucher's disease.

The **smooth endoplasmic reticulum** in liver cells and other cells contains enzymes that play a part in the inactivation of steroid hormones and many drugs. This inactivation is generally achieved by reactions that convert these compounds to more water-soluble and less active forms that can be more easily excreted by the kidneys. When people take certain drugs like alcohol and phenobarbital for a long period of time, an increasingly large dose of these compounds is required to produce a given effect. This phenomenon, called *tolerance,* is accompanied by an increase in the smooth endoplasmic reticulum, and thus an increase in the enzymes that function to inactivate these drugs.

Medical Genetics

Medical genetics is a branch of medicine concerned with diseases that have a genetic origin. Genetic factors include abnormalities in chromosome number or structure and mutant genes. Genetic diseases are a diverse group of disorders, including malformed blood cells (sickle-cell anemia), defective blood clotting (hemophilia), and mental retardation (Down syndrome).

Chromosomal abnormalities occur in approximately 0.6% of live-birth infants. The majority of them (70%) are subtle, cause no problems, and are usually undetected. Structural changes in the genetic DNA that is passed from parent to offspring by means of sex cells are called **mutations.** Mutations either occur naturally or are environmentally induced through chemicals or radiation. Natural mutations are not well understood. About 12% of all congenital malformations are caused

---

atrophy: Gk. *a*, without; *trophe*, nourishment
metaplasia: Gk. *meta*, between; *plasis*, a molding

autophagy: Gk. *autos*, self; *phagein*, to eat
mutation: L. *mutare*, to change

by mutations and probably come about through an interaction of genetic and environmental factors. Many of these problems can be predicted by knowing the genetic pedigree of prospective parents and prevented through genetic counseling. **Teratology** is the science concerned with developmental defects and the diagnosis, treatment, and prevention of malformations.

Genetic problems are occasionally caused by having too few or too many chromosomes. The absence of an entire chromosome is termed **monosomy.** Embryos with monosomy usually die. People with **Turner's syndrome** have only one X chromosome and have a better chance of survival than those who are missing one of the other chromosomes. **Trisomy,** a genetic condition in which an extra chromosome is present, occurs more frequently than monosomy. The best known among the trisomies is **Down syndrome.**

Cancer

Cancer is a complex group of diseases characterized by uncontrolled cell reproduction. The rapid proliferation of cells results in a **neoplasm,** or new cellular mass, being formed. Neoplasms, frequently called tumors, are classified as benign or malignant based on their cytological and histological features. **Benign neoplasms** usually grow slowly and are confined to a particular area. These types are usually not life threatening unless they grow to large sizes on vital organs like the brain. **Malignant neoplasms** (fig. 3.22) grow rapidly and **metastasize** (fragment and spread) easily through lymphatic or blood vessels. The original malignant neoplasm is called the *primary growth* and the new tumors, or metastatic tumors, are called *secondary growths.*

Cancer cells resemble undifferentiated or primordial cell types. Generally they do not mature before they divide and are not capable of maintaining normal cell function. Cancer causes death when a vital organ regresses because of competition from cancer cells for space and nutrients. The pain associated with cancer develops when the growing neoplasm affects sensory neurons.

The various types of cancers are classified on the basis of the tissue in which they develop. Lymphoma, for example, is a cancer of lymphoid tissue; osteogenic cancer is a type of bone cancer; myeloma is cancer of the bone marrow; and sarcoma is a general term for any cancer arising from cells of connective tissue.

The etiology (cause) of cancers is largely unknown. However, initiating factors, or **carcinogens,** such as viruses, chemicals, or irradiation may provoke cancer to develop. Cigarette smoking, for example, causes various respiratory cancers to develop. The tendency to develop other types of cancers has a genetic basis. Some researchers even think that physiological

---

teratology: Gk. *teras,* monster; *logos,* study
Turner's syndrome: from Henry H. Turner, American endocrinologist, 1892–1970
Down syndrome: from John L. H. Down, English physician, 1828–96
carcinogen: Gk, *karkinos,* cancer

Figure 3.22   A light micrograph of (*a*) cancerous tissue (see arrows) of the penis and (*b*) normal tissue of the penis (0.9✕). (Note the abnormal epithelial growth in the cancerous tissue.)

(a)

(b)

stress can promote certain types of cancerous activity. Because the causes of cancer are not well understood, emphasis is placed on early detection with prompt treatment.

Aging

Although there are obvious external indicators of aging: graying and loss of hair, wrinkling of skin, loss of teeth, and decreased muscle mass, changes due to aging within cells are not as apparent and not well understood. Certain organelles alter with age. The mitochondria, for example, may change in structure and number, and the Golgi apparatus may fragment. Also, lipid vacuoles tend to accumulate in the cytoplasm, whereas the inclusions that contain glycogen decrease.

The chromatin and chromosomes within the nucleus show changes with aging, such as clumping, shrinking, or fragmenting with repeated mitotic divisions. There is strong evidence that certain cell types have a predetermined number of mitotic divisions that are genetically controlled, thus determining the overall vitality and longevity of an organ. If this is true, identifying and genetically manipulating the "aging gene" might be possible.

Extracellular substances also change with age. Protein strands of *collagen* and *elastin* change in quality and number in aged tissues. Elastin is important in the walls of arteries, and its deterioration is thought to be associated with such vascular diseases as arteriosclerosis in aged persons.

CHAPTER SUMMARY

Introduction to Cellular Anatomy (pp. 46–47)

1. Cells are the structural and functional units of the body. Cellular function is referred to as metabolism and the study of cells is referred to as cytology.
2. The structure and function of cells depend on the specific membranes and organelles characteristic of each type of cell.

Cellular Chemistry (pp. 47–49)

1. Four elements (oxygen, carbon, hydrogen, and nitrogen) compose over 95% of the body and are linked together to form either inorganic or organic compounds.
2. Water is the most abundant inorganic compound in cells and is an excellent solvent.
   (a) Water is important in temperature control and hydrolysis.
   (b) Dehydration, a condition in which fluid loss exceeds fluid intake, may be a serious problem—especially in infants.
3. Electrolytes are inorganic compounds that form ions.
   (a) The three classes of electrolytes are acids, bases, and salts.
   (b) Electrolytes are important in maintaining pH, in conducting electrical currents, and in the development and functioning of certain tissues.
4. Proteins are organic compounds that exist by themselves or are conjugated.
   (a) Proteins provide the framework of cells and are necessary for growth, repair, and division.
   (b) Enzymes and hormones are examples of specialized proteins.
5. Carbohydrates are a group of organic compounds that includes the starches and sugars.
   (a) The composition of carbohydrates is always in a 2:1 ratio of hydrogen to oxygen.
   (b) Carbohydrates provide the most abundant source of cellular energy.

6. Lipids are organic fats and fat-related substances.
   (a) Lipids are composed of carbon, hydrogen, and oxygen.
   (b) Lipids are an important source of energy and serve to protect and insulate various parts of the body.

Cellular Structure (pp. 49–60)

1. A cell is composed of a cell membrane, cytoplasm containing organelles, and a nucleus.
2. The cell membrane, composed of phospholipid and protein molecules, encloses the contents of the cell and regulates the passage of substances.
   (a) The permeability of the cell membrane is dependent upon its structure, size of molecules, ionic charge, lipid solubility, and the presence of carrier molecules.
   (b) Cell membranes may be specialized with microvilli, sacs, hair cells, or stereocilia.
3. Cytoplasm is the material between the cell membrane and the nucleus. Nucleoplasm is the material within the nucleus. Protoplasm includes both the cytoplasm and nucleoplasm.
4. Organelles are highly structural and specialized components within the cytoplasm.
   (a) Endoplasmic reticulum provides a framework within the cytoplasm and forms a site for the attachment of ribosomes. It functions in the synthesis of lipids and proteins and in cellular transport.
   (b) Ribosomes are particles of protein and RNA that function in protein synthesis. The protein particles may be used within the cell or secreted.
   (c) The Golgi apparatus consists of membranous vesicles that synthesize glycoproteins and secrete lipids. The Golgi apparatus is prominent in secretory cells, such as those of the pancreas and salivary glands.
   (d) Mitochondria are membranous sacs that consist of an outer smooth layer and an inner folded crista. The mitochondria produce ATP and are called the "powerhouses" of a cell.
   (e) Lysosomes are spherical bodies that contain digestive enzymes. They are abundant in the phagocytic white blood cells.

   (f) Peroxisomes are membranous sacs that contain specific enzymes. They are abundant in the kidneys and liver and are important in producing hydrogen peroxide.
   (g) The centrosome is the dense area of cytoplasm near the nucleus that contains the centrioles. The paired centrioles play an important role in cell division.
   (h) Vacuoles are membranous sacs that function as storage chambers.
   (i) Fibrils and microtubules provide support in the formation of a cytoskeleton.
   (j) Cilia and flagella are projections of the cell that have the same basic structure and function in producing movement.
5. The cell nucleus is enclosed in a nuclear membrane that controls the movement of substances between the nucleoplasm and the cytoplasm.
   (a) The nucleoli are small bodies of protein and RNA within the nucleus that produce ribosomes.
   (b) Chromatin are coiled fibers of protein and DNA that form chromosomes during cell reproduction.

Cell Cycle (pp. 60–63)

1. A cell cycle consists of growth, synthesis, and mitosis.
   (a) Growth is the increase in cellular mass as the result of metabolism. Synthesis is the production of DNA and RNA to regulate cellular activity. Mitosis is the splitting of the nucleus and cytoplasm of a cell that results in two diploid cells being formed.
   (b) Mitosis permits an increase in the number of cells (body growth) and allows for the replacement of damaged, diseased, or worn-out cells.
2. A DNA molecule is in the shape of a double helix. The structural unit of the molecule is a nucleotide, which consists of a deoxyribose sugar, a phosphate, and a nitrogenous base.
3. Cell division consists of a division of the chromosomes (mitosis) and a division of the cytoplasm (cytokinesis). The stages of mitosis include prophase, metaphase, anaphase, and telophase.

UNIT III

REVIEW ACTIVITIES

Objective Questions

1. Inorganic compounds that form ions when dissociated in water are
   - (a) hydrolites.
   - (d) ionizers.
   - (b) metabolites.
   - (e) nucleic acids.
   - (c) electrolytes.
2. The four elements that compose over 95% of the body are
   - (a) oxygen, potassium, hydrogen, carbon.
   - (b) carbon, sodium, nitrogen, oxygen.
   - (c) potassium, sodium, magnesium, oxygen.
   - (d) carbon, oxygen, nitrogen, hydrogen.
   - (e) oxygen, carbon, hydrogen, sulfur.
3. Which organelle contains strong hydrolytic enzymes?
   - (a) the lysosome
   - (b) the Golgi apparatus
   - (c) the ribosome
   - (d) the vacuole
   - (e) the mitochondrion
4. Ciliated cells occur in
   - (a) the trachea.
   - (b) the ductus deferens.
   - (c) the bronchioles.
   - (d) the uterine tubes.
   - (e) all of the above.
5. Osmosis deals with the movement of
   - (a) gases.
   - (c) oxygen only.
   - (b) water only.
   - (d) both (a) and (c).
6. The phase of mitosis in which the chromosomes line up at the equator (equatorial plane) of the cell is called
   - (a) interphase.
   - (d) anaphase.
   - (b) prophase.
   - (e) telophase.
   - (c) metaphase.
7. The phase of mitosis in which the chromatids separate is called
   - (a) interphase.
   - (d) anaphase.
   - (b) prophase.
   - (e) telophase.
   - (c) metaphase.
8. The organelle that combines protein with carbohydrates and packages them within vesicles for secretion is
   - (a) the Golgi apparatus.
   - (b) the rough endoplasmic reticulum.
   - (c) the smooth endoplasmic reticulum.
   - (d) the ribosome.
9. The enlarged skeletal muscle fibers that result from an increased work demand serve to illustrate
   - (a) disuse atrophy.
   - (b) compensatory hypertrophy.
   - (c) metaplasia.
   - (d) inertia.
10. Regeneration of liver cells is an example of
    - (a) compensatory hypertrophy.
    - (b) hyperplasia.
    - (c) metaplasia.
    - (d) hypertrophy.

Essay Questions

1. Explain why understanding cellular anatomy is necessary for understanding tissue and organ functioning within the body.
2. Why is water a good fluid medium of the cell?
3. How are proteins, carbohydrates, and lipids similar? How are they different? What are enzymes and hormones?
4. Describe the cell membrane. List the kinds of movement through the cell membrane and give an example of each.
5. Describe, diagram, and list the functions of the following: (a) endoplasmic reticulum; (b) ribosome; (c) mitochondrion; (d) Golgi apparatus; (e) centrioles; (f) cilia.
6. Define *inorganic compound* and *organic compound* and give examples.
7. Define the terms *protoplasm, cytoplasm,* and *nucleoplasm.* Describe the position of the membranes associated with each of these substances.
8. Describe the structure of the nucleus and the functions of its parts.
9. What is a nucleotide? How does it relate to the overall structure of a DNA molecule?
10. Explain the relationship between DNA, chromosomes, chromatids, and genes.
11. What are the differences between mitosis and cytokinesis? Describe the major events of mitosis and discuss the significance of the mitotic process.
12. Give examples of factors that bring about cellular hypertrophy, hyperplasia, atrophy, and metaplasia.
13. Explain how cells respond to (a) energy deficit; (b) hyperthermia; (c) burns; (d) radiation; (e) pathogens.
14. Define the following genetic terms: *teratology, monosomy, trisomy,* and *mutations.*
15. In what ways do neoplasms differ from normal cells? How may a malignant neoplasm cause death?

Gundy/Weber Software

The tutorial software accompanying Chapter 3, Cellular Anatomy, is Volume 2, Viruses, Bacteria, Eukaryotic Cells.

UNIT III

4

Histology

Chapter Outline

Definition and Classification of Tissues

Histology is the specialty of anatomy that involves microscopic study of the structure of tissues. Tissues are classified into four principal types on the basis of their cellular composition and histological appearance.

Objective 1.	Define *tissue* and discuss the importance of histology.
Objective 2.	Describe the functional relationship between cells and tissues.
Objective 3.	Classify the tissues of the body into four principal types and list the distinguishing characteristics of each type.

Although cells are the structural and functional units of the body, the cells of a multicellular organism are so specialized that they do not function independently. *Tissues* are aggregations of similar cells and cell products that perform specific functions. The study of tissues is referred to as **histology** *(histol'ŏ-je)*. The various types of tissues are established during embryonic development, and as differentiation continues, organs form, each composed of a specific arrangement of tissues. Many adult organs, including the heart and muscles, contain original cells and tissues that were not replaced through mitotic activity during further growth and development. Some functional changes may occur, however, as the tissues of an organ are acted upon by hormones, or as their effectiveness diminishes with age.

Although histology is actually microscopic anatomy, it is an essential part of anatomy and physiology because it imparts an understanding of the structure and function of organs at the tissue level. Many diseases profoundly alter the tissues within an affected organ; therefore, by knowing the normal tissue structure, a physician can recognize the abnormal. In medical schools a course in histology is usually followed by a course in *pathology,* which is primarily concerned with identifying diseased tissues.

Although histologists employ many different techniques for preparing, staining, and sectioning tissues, basically only two kinds of microscopes are used to view the prepared tissues. *Light microscopy (mi-kros'kŏ-pe)* is used for the general observation of tissue structure (fig. 4.1) and *electron microscopy* permits observation of the fine details of tissue and cellular structure. Most of the histological photomicrographs in this text are at the light microscopic level to present an overview of tissue structure. A few electron micrographs from an electron microscope are used to depict the fine structural detail needed to understand a particular function.

Tissue cells are separated and bound together by a nonliving intercellular **matrix** *(ma'triks)* that the cells secrete. Matrix varies in composition from one tissue to another and may take the form of a liquid, semisolid, or solid. Blood, for

histology: Gk. *histos*, web (tissue); *logos*, study
pathology: Gk. *pathos*, suffering, disease; *logos*, study
matrix: L. *matris*, mother

Figure 4.1 The appearance of skin (*a*) magnified 25 times, as seen through a compound light microscope, and (*b*) magnified 280 times, as seen through a scanning electron microscope (SEM).

Shaft of a hair within a hair follicle

(a)

Shaft of hair emerging from the exposed surface of the skin

(b)

example, has a liquid matrix, permitting this tissue to flow through vessels. By contrast, bone cells are separated by a solid matrix, permitting this tissue to support the body.

The tissues of the body are classified into four principal kinds on the basis of structure and function: (1) *epithelial (ep''i-the'le-al) tissues* cover body and organ surfaces, line body and lumen cavities, and form glands; (2) *connective tissues* bind, support, and protect body parts; (3) *muscle tissues* contract to produce movement; and (4) *nervous tissues* initiate and transmit nerve impulses from one body part to another.

1. Define *tissue,* and explain why histology is important to the study of anatomy and physiology and medicine.
2. Cells are the functional units of the body. Explain how the matrix permits specific kinds of cells to be even more effective and functional as tissues.
3. Describe the distinguishing characteristics of the four principal kinds of tissues in the body and explain their basic functions.

Epithelial Tissue

There are two major categories of epithelia: membranous and glandular. Membranous epithelia are located throughout the body and form such structures as the outer layer of the skin, the inner lining of body cavities and lumina (hollow portions of body tubes), and the covering of visceral organs. Glandular epithelia are specialized tissues that form the secretory portion of glands.

> **Objective 4.** Compare and contrast the various types of membranous epithelia.

> **Objective 5.** Discuss the types and functions of the various body cavities.

> **Objective 6.** Define *gland* and compare and contrast the various types of glands in the body.

Characteristics of Membranous Epithelia

One side of membranous epithelia is always exposed to a body cavity, lumen, or skin surface. Some membranous epithelia are derived from ectoderm, such as the outer layer of the skin; some from mesoderm, such as the inside lining of blood vessels; and others from endoderm, such as the inside lining of the digestive tract (gastrointestinal, or GI, tract).

Membranous epithelia may be one layer or several layers thick. The upper surface of epithelia may be exposed to gases, as in the case of epithelium in the integumentary and respiratory systems; to liquids, as in the circulatory and urinary systems; or to semisolids, as in the GI tract. The deep surface of membranous epithelia is bound to underlying supportive tissue by a **basement membrane,** consisting of glycoprotein from the epithelial cells and a meshwork of collagenous and reticular fibers from the underlying connective tissue. With few exceptions, membranous epithelia are avascular (without blood vessels) and must be nourished by diffusion from underlying connective tissues. Cells forming membranous epithelia are tightly packed together, and there is little intercellular matrix between them.

Some of the functions of membranous epithelia are quite specific, but certain generalities can be made. Epithelia that cover or line surfaces provide protection from pathogens, physical injury, toxins, and desiccation. Epithelia lining the lumen of the GI tract function in absorption. The epithelium of the kidneys provides filtration, while the epithelium within the air sacs (alveoli) of the lungs allows for diffusion. Highly specialized *neuroepithelium* in the taste buds and in the nasal region serves the function of chemoreceptors.

Many membranous epithelia are exposed and, therefore, subject to trauma and destruction. For this reason, epithelial tissues have remarkable regenerative abilities. The mitotic replacement of the outer layer of skin and the lining of the GI tract, for example, is a continuous process.

Membranous epithelia are histologically classified by the number of layers of cells and the shape of the cells along the exposed surface. Epithelial tissues that are composed of a single

---

epithelium: Gk. *epi*, upon; *thelium*, to cover

Figure 4.2 (*a*) Simple squamous epithelium lines the lumina of vessels, where it permits diffusion. (*b*) A photomicrograph of this tissue and (*c*) a labeled diagram.

(a)

Paras

Red blood cells within lumen

Nucleus

Simple squamous epithelium

(b)

Red blood cells within lumen

Simple squamous epithelium (endothelium)

Nucleus

(c)

layer of cells are called *simple;* those that are layered are said to be *stratified. Squamous* cells are flattened; *cuboidal* cells are cube-shaped; and *columnar* cells are taller than they are wide.

Simple Epithelia

Simple epithelial tissues are a single layer thick and are located where diffusion, absorption, filtration, and secretion occur. The cells that constitute simple epithelia range from thin, flattened cells to tall, columnar cells, depending on function. These cells may also exhibit surface specializations like cilia and microvilli to facilitate specific surface functions.

Simple Squamous Epithelium Simple squamous (*skwa'mus*) epithelium is composed of flattened, irregularly shaped cells that are tightly bound together in a mosaiclike pattern (fig. 4.2). Each

---

squamous: L. *squamosus*, scaly

UNIT III

Developmental Exposition

The Tissues

Exhibit I The early stages of embryonic development. (*a*) Fertilization and the formation of the zygote, (*b*) the morula at about the 3rd day, (*c*) the early blastocyst at the time of implantation between the 5th and 7th day, (*d*) a blastocyst at 2 weeks, and (*e*) a blastocyst at 3 weeks showing the three primary germ layers that constitute the embryonic disc.

Explanation

Human prenatal development is initiated by the fertilization of an ovulated ovum (egg) from a female by a sperm cell from a male. The chromosomes within the nucleus of a **zygote** (*zi'gōt*) (fertilized egg) contain all the genetic information necessary for the differentiation and development of all body structures.

Within 30 hours after fertilization, the zygote undergoes a mitotic division as it moves through the uterine tube toward the uterus (see chapter 22). After several more cellular divisions, the embryonic mass consists of 16 or more cells and is called a **morula** (*mor'yoo-lah*), as shown in exhibit I. Three or 4 days after conception, the morula enters the uterine cavity where it remains

zygote: Gk. *zygotos*, yolked

morula: Gk. *morus*, mulberry

Table 4.1	Derivatives of the Germ Layers	
Ectoderm	**Mesoderm**	**Endoderm**
Epidermis of skin and epidermal derivatives: hair, nails, glands of the skin; linings of oral, nasal, anal, and vaginal cavities	Muscle: smooth, cardiac, and skeletal	Epithelium of pharynx, auditory canal, tonsils, thyroid, parathyroid, thymus, larynx, trachea, lungs, GI tract, urinary bladder and urethra, and vagina
Nervous tissue; sense organs	Connective tissue: embryonic, connective tissue proper, cartilage, bone, blood	Liver and pancreas
Lens of eye; enamel of teeth	Dermis of skin; dentin of teeth	
Pituitary gland	Epithelium of blood vessels, lymphatic vessels, body cavities, joint cavities	
Adrenal medulla	Internal reproductive organs	
	Kidneys and ureters	
	Adrenal cortex	

UNIT III

unattached for about 3 days. During this time, the center of the morula fills with fluid passing in from the uterine cavity. As the fluid-filled space develops inside the morula, two distinct groups of cells form. The single layer of cells forming the outer wall is known as the **trophoblast** and the inner aggregation of cells is known as the **embryoblast.** After further development, the trophoblast becomes a portion of the placenta and the embryoblast becomes the embryo. With the establishment of these two groups of cells, the morula becomes known as a **blastocyst** (blas'tō-sist). Implantation of the blastocyst begins between the fifth and seventh day (see chapter 22).

As the blastocyst completes implantation during the second week of development, the embryoblast undergoes marked differentiation. A slitlike space called the **amniotic** (am'ne-ot-ic) **cavity** forms within the embryoblast adjacent to the trophoblast.

The embryoblast now consists of two layers: an upper **ectoderm,** which is closer to the amniotic cavity, and a lower **endoderm,** which borders the blastocyst cavity. A short time later, a third layer called the **mesoderm** forms between the endoderm and ectoderm. These three layers constitute the **primary germ layers.**

The primary germ layers are important because all the cells and tissues of the body are derived from them (see fig. 22.10). Ectodermal cells form the nervous system; the outer layer of skin (epidermis), including hair, nails, and skin glands; and portions of the sensory organs. Mesodermal cells form the skeleton, muscles, blood, reproductive organs, dermis of the skin, and connective tissue. Endodermal cells produce the lining of the GI tract, the digestive organs, the respiratory tract and lungs, and the urinary bladder and urethra.

The derivatives of the primary germ layers are listed in table 4.1.

---

trophoblast: Gk. trophe, nourishment; blastos, germ
embryoblast: Gk. embryon, to be full, swell; blastos, germ

ectoderm: Gk. ecto, outside; derm, skin
endoderm: Gk. endo, within; derm, skin
mesoderm: Gk. meso, middle; derm, skin

Figure 4.3 (*a*) Simple cuboidal epithelium lines the lumina of ducts; for example, in the kidneys, where it permits movement of fluids and ions. (*b*) A photomicrograph of this tissue and (*c*) a labeled diagram.

(a)

(b)

(c)

Lumen

Nucleus

Cytoplasm

Cuboidal epithelial cells

Collecting tubules

Basement membrane

cell contains an oval centrally located nucleus. Simple squamous epithelium is adapted for diffusion and filtration and occurs in such places as the lining of air sacs (alveoli) within the lungs (where gaseous exchange occurs); portions of the kidney (where blood is filtered); the inside lining of the walls of blood vessels; the lining of body cavities; and the covering of the viscera. The simple squamous epithelium lining the lumina of blood and lymphatic vessels is termed **endothelium** (fig. 4.2c). That which covers visceral organs and lines body cavities is called **mesothelium.**

endothelium: Gk. *endon*, within; *thelium*, to cover
mesothelium: Gk. *meso*, middle; *thelium*, to cover

Simple Cuboidal Epithelium Simple cuboidal epithelium is composed of a single layer of tightly fitted hexagonal cells (fig. 4.3). This type of epithelium is found lining small ducts and tubules that may have excretory, secretory, or absorptive functions. It occurs on the surface of the ovaries, forms a portion of the tubules within the kidney, and lines the ducts of the salivary glands and pancreas.

Simple Columnar Epithelium Simple columnar epithelium is composed of tall, columnar cells (fig. 4.4). The height of the cells varies, depending on the site and function of the tissue. Each cell contains a single nucleus which is usually located near the basement membrane. Specialized unicellular glands called

Figure 4.4 (*a*) Simple columnar epithelium lines the lumen of the digestive tract, where it permits secretion and absorption. (*b*) A photomicrograph of this tissue and (*c*) a labeled diagram.

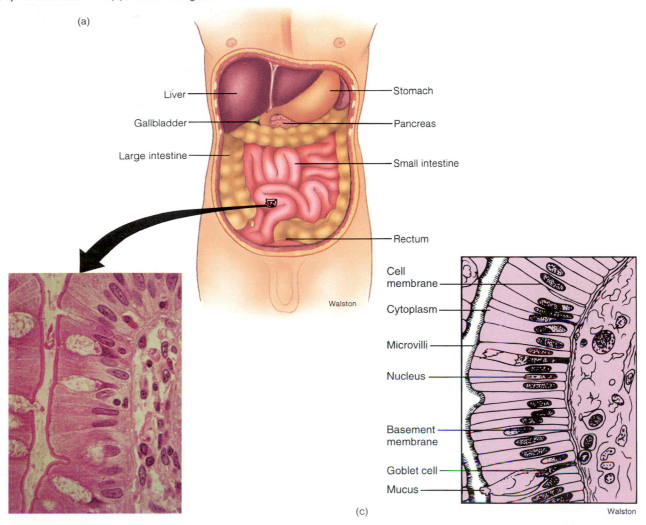

(a)

Liver

Gallbladder

Large intestine

Stomach

Pancreas

Small intestine

Rectum

Walston

Cell membrane

Cytoplasm

Microvilli

Nucleus

Basement membrane

Goblet cell

Mucus

Walston

(b)

(c)

goblet cells are dispersed throughout this tissue and secrete a lubricative and protective mucus along the free surfaces of the cells. This type of epithelium is found lining the lumen of the stomach and intestine. In the digestive system, simple columnar epithelium forms a highly absorptive surface and also secretes certain digestive substances. Within the stomach, simple columnar epithelium has a tremendous rate of mitotic activity—replacing itself every 2 or 3 days.

Simple Ciliated Columnar Epithelium Simple ciliated columnar epithelium is characterized by the presence of cilia along the free surface (fig. 4.5) whereas the simple columnar type is unciliated. Cilia produce wavelike movements that transport materials through tubes or passageways. This type of epithelium occurs in the uterine tubes of the female, where the currents generated by the cilia propel the ovum toward the uterus.

Not only do the cilia function to propel the ovum, but recent evidence indicates that sperm introduced during sexual intercourse may be moved along the return currents, or eddies, generated by ciliary movement. This greatly enhances the likelihood of fertilization.

Pseudostratified Ciliated Columnar Epithelium As the name implies, this type of epithelium appears stratified. It is actually

Figure 4.5 (a) Simple ciliated columnar epithelium lines the lumen of the uterine tube, where currents generated by the cilia propel the egg cell toward the uterus. (b) A photomicrograph of this tissue and (c) a labeled diagram.

(a)

(b)

Connective tissue

Ciliated columnar epithelium

Nucleus

Cytoplasm

Cilia

Cell membrane

Basement membrane

(c)

simple, however, since each cell is in contact with the basement membrane, although not all cells are exposed to the surface (fig. 4.6). The epithelium has a stratified appearance because the nuclei of these cells are located at different levels. Numerous goblet cells and a ciliated, exposed surface are characteristic of this epithelium. The lumina of the trachea and the bronchial tubes are lined with this tissue; hence, it is frequently called respiratory epithelium. Its function is to remove foreign dust and bacteria entrapped in mucus from the lower respiratory system.

Coughing, sneezing, or simply "clearing the throat" are protective reflex mechanisms for clearing the respiratory passages of obstruction or of inhaled particles that have been trapped in the mucus along the ciliated lining. The material that is coughed up consists of the mucus-entrapped particles.

Stratified Epithelia

Stratified epithelia are tissues consisting of two or more layers of cells. In contrast to simple epithelia, stratified epithelia are poorly suited for absorption and secretion because of their thickness. Stratified epithelia have a primarily protective function that is enhanced by a characteristic rapid mitotic activity. Stratified epithelia are classified according to the shape of the surface layer of cells, since the layer in contact with the basement membrane is cuboidal or columnar in shape.

Figure 4.6 (*a*) Pseudostratified ciliated columnar epithelium lines the lumina of most of the respiratory tract, where it traps foreign material and moves it away from the alveoli of the lungs. (*b*) A photomicrograph of this tissue and (*c*) a labeled diagram.

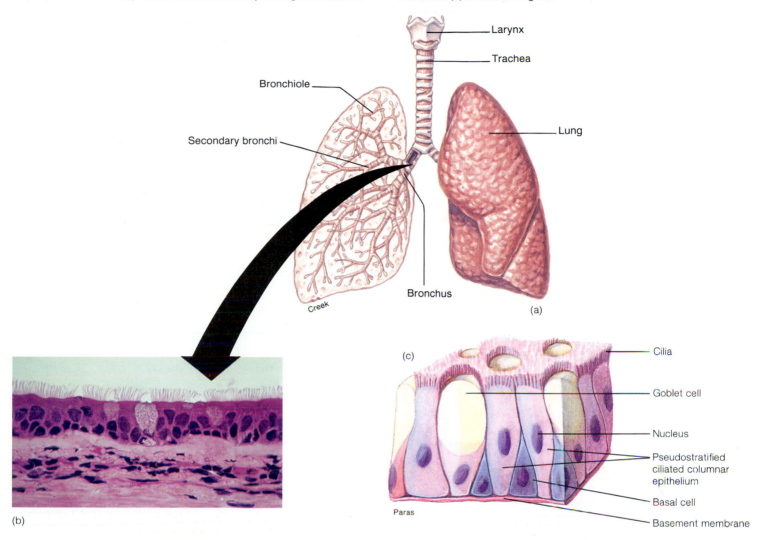

UNIT III

Stratified Squamous Epithelium   Stratified squamous epithelium is composed of a variable number of cell layers that tend to flatten near the surface (fig. 4.7). Only at the deepest layer, the **stratum basale,** does mitosis occur. The mitotic rate approximates the rate at which cells are sloughed off at the surface. As the newly produced cells grow in size, they are pushed toward the surface, where they will replace the cells that are sloughed off. Movement of the epithelial cells away from the supportive basement membrane is accompanied by progressive dehydration, flattening, and the production of the protein keratin (*ker'ă-tin*).

There are two types of stratified squamous epithelia: *keratinized* and *nonkeratinized.* Stratified squamous epithelium that is keratinized forms the outer layer, or *epidermis,* of the skin (see chapter 5). **Keratin** is a protein that strengthens the tissue. This type of epithelium is especially durable and can generally withstand physical abrasion, desiccation, and bacterial invasion. The outer layers of the stratified squamous epithelium of the skin are dead but are kept pliable by local glandular secretions.

Nonkeratinized stratified squamous epithelium lines the oral cavity and pharynx, nasal cavity, vagina, and anal canal. This type of epithelium, called *mucosa,* is well adapted to withstand moderate abrasion but not fluid loss. The cells on the free surface of this tissue remain alive and are always moistened.

Stratified squamous epithelium is the first line of defense against the entry of living organisms into the body. Stratification, along with rapid mitotic activity and keratinization within the epidermis of the skin, are important protective features. An acidic pH along the surfaces of this tissue also helps to prevent disease. The pH of the skin is between 4 and 6.8. The pH in the oral cavity ranges from 5.8 to 7.1, which tends to retard the growth of microorganisms. The pH of the anal region is about 6, and the pH along the surface of the vagina is 4 or lower.

keratin: Gk. *keras,* horn

Figure 4.7 Stratified squamous epithelium forms the outer layer of skin and the lining of body openings. In the moistened areas, such as in the vagina (*a*), it is nonkeratinized, whereas in the epidermis of the skin it is keratinized. (*b*) A photomicrograph of this tissue and (*c*) a labeled diagram.

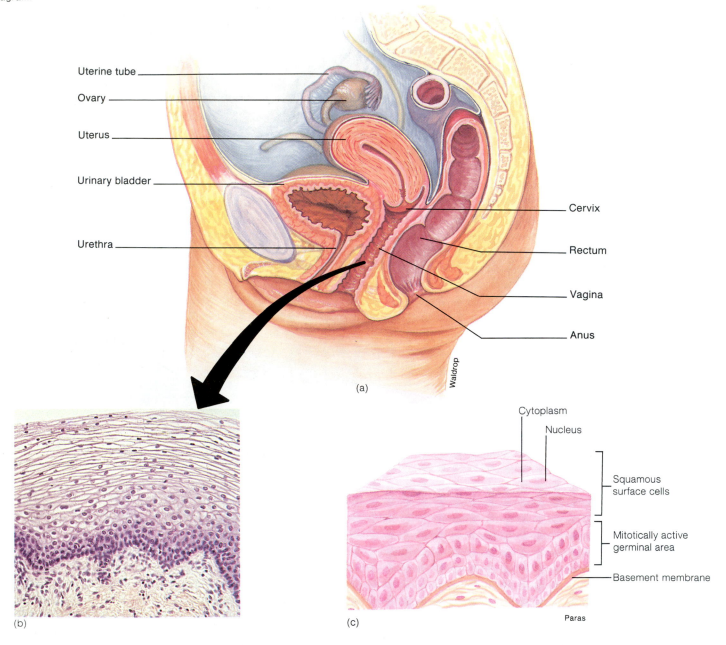

(a)

(b)

(c)

Stratified Cuboidal Epithelium   Stratified cuboidal epithelium usually consists of only two or three layers of cuboidal cells forming the lining around a lumen (fig. 4.8). This type of epithelium is confined to the linings of the larger ducts of sweat glands, salivary glands, and the pancreas. The stratification of this tissue probably provides a more robust lining than would be afforded by simple epithelium.

Transitional Epithelium   Transitional epithelium is similar to nonkeratinized stratified squamous epithelium except that the surface cells of the former are large and round rather than flat, and some may have two nuclei (fig. 4.9). Transitional epithelium is located only within the urinary system, particularly in the luminal surface of the urinary bladder and the walls of the ureters. This tissue is specialized to permit distension (stretching) of the urinary bladder.

Epithelial tissues are summarized in table 4.2.

Figure 4.8 (*a*) Stratified cuboidal epithelium lines the lumina of large ducts like the parotid duct, which drains saliva from the parotid gland. (*b*) A photomicrograph of this tissue and (*c*) a labeled diagram.

Parotid gland

Parotid duct

(a)

(b)

Lumen of parotid duct

Acinar cells

Interlobular connective tissue

(c)   Paras

UNIT III

Figure 4.9 (*a*) Transitional epithelium lines the cavity of the urinary bladder and the lumina of the ureters, where it permits distention. (*b*) A photomicrograph of this tissue and (*c*) a labeled diagram.

Ureter

Urinary bladder

Urethra

(a)

(b)

Nucleus

Cytoplasm

Basement membrane

Smooth muscle

Transitional epithelium

(c)   Paras

Table 4.2	Summary of Membranous Epithelial Tissues	
Type	**Structure and Function**	**Location**
Simple Epithelia	Single layer of cells; diffusion and filtration	Covering visceral organs; linings of lumina and body cavities
Simple squamous epithelium	Single layer of flattened, tightly bound cells; diffusion and filtration	Capillary walls; alveoli of lungs; covering visceral organs; linings of body cavities
Simple cuboidal epithelium	Single layer of cube-shaped cells; excretion, secretion, or absorption	Surface of ovaries; linings of renal tubules, salivary ducts, and pancreatic ducts
Simple columnar epithelium	Single layer of nonciliated, tall, columnar-shaped cells; protection, secretion, and absorption	Lining of most of GI tract
Simple ciliated columnar epithelium	Single layer of ciliated, columnar-shaped cells; transportive role through ciliary motion	Lining the lumen of the uterine tubes
Pseudostratified ciliated columnar epithelium	Single layer of ciliated, irregularly shaped cells, many goblet cells; protection, secretion, ciliary movement	Lining of respiratory passageways
Stratified Epithelia	Two or more layers of cells; protection, strengthening, or distension	Epidermal layer of skin; linings of body openings, ducts, and urinary bladder
Stratified squamous epithelium (keratinized)	Numerous layers containing keratin, outer layers flattened and dead; protection	Epidermis of skin
Stratified squamous epithelium (nonkeratinized)	Numerous layers lacking keratin, outer layers moistened and alive; protection and pliability	Linings of oral and nasal cavities, vagina, and anal canal
Stratified cuboidal epithelium	Usually two layers of cube-shaped cells; strengthening of luminal walls	Larger ducts of sweat glands, salivary glands, and pancreas
Transitional epithelium	Numerous layers of rounded, nonkeratinized cells; distension	Luminal walls of ureters and urinary bladder

UNIT III

Body Membranes

Body membranes are composed of thin layers of epithelial tissue and, in certain locations, epithelial tissue coupled with supporting connective tissue. Body membranes cover, separate, and support visceral organs and line body cavities. There are two basic types of body membranes: **mucous** (*myoo'kus*) **membranes** and **serous** (*se'rus*) **membranes.**

Mucous membranes secrete a thick, viscid substance called *mucus.* Generally, mucus lubricates or protects the associated organs where it is secreted. Mucous membranes line various cavities and tubes that enter or exit from the body, such as the oral and nasal cavities and the tubes of the respiratory, reproductive, urinary, and digestive systems.

Serous membranes line the thoracic and abdominopelvic cavities and cover visceral organs, secreting a watery lubricant called *serous fluid.* **Pleurae** are serous membranes associated with the lungs (see chapter 17). Each pleura (pleura of right lung and pleura of left lung) has two parts. The **visceral pleura** adheres to the outer surface of the lung, while the **parietal** (*pah-ri'ĕ-tal*) **pleura** lines the thoracic walls and the thoracic surface of the diaphragm. The moistened space between the two pleurae is known as the **pleural cavity.**

Pericardial membranes are the serous membranes of the heart (see chapter 16). A thin **visceral pericardium** covers the surface of the heart, and a thicker **parietal pericardium** is the durable covering that surrounds the heart. The space between these two membranes is called the **pericardial cavity.**

Serous membranes of the abdominal cavity are called **peritoneal** (*per''i-tŏ-ne'al*) **membranes** (see chapter 18). The **pa**-rietal peritoneum lines the abdominal wall, and the **visceral peritoneum** covers the visceral organs (fig. 4.10). The **peritoneal cavity** is the potential space within the abdominopelvic cavity between the parietal and visceral peritoneal membranes. Certain organs, such as the kidneys, adrenal glands, and a portion of the pancreas, which are within the abdominal cavity, are positioned behind the parietal peritoneum, and therefore are said to be **retroperitoneal. Mesenteries** (*mes'en-ter''ēz*) are double folds of peritoneum that connect the parietal to the visceral peritoneum.

Glandular Epithelia

Glandular epithelia are specialized tissues that have secretory functions. During the prenatal development of epithelial tissue, certain epithelial cells invade the underlying connective tissue, forming specialized secretory accumulations called **exocrine** (*ek'sŏ-krin*) **glands.** Structurally, there are two types of exocrine glands: unicellular and multicellular. *Unicellular exocrine glands* are ductless and their secretion is expelled directly to the surface of a tissue. *Multicellular exocrine glands* retain a connection to the surface in the form of a *duct* through which secretions flow. Multicellular exocrine glands within the integumentary system include sebaceous (oil) glands, sweat glands, and mammary glands. Within the digestive system, they include the salivary and pancreatic glands. Multicellular exocrine glands are classified according to the structure of the gland

---

exocrine: Gk. *exo*, outside; *krinein*, to separate

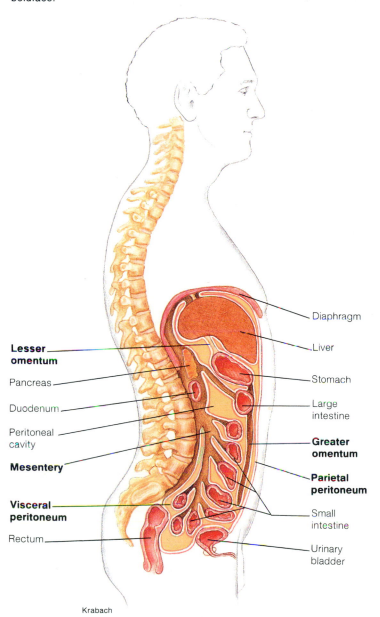

Figure 4.10 Visceral organs of the abdominal cavity and the supporting serous membranes. The serous membranes are labeled in boldface.

Diaphragm

Liver

Stomach

Large intestine

Greater omentum

Parietal peritoneum

Small intestine

Urinary bladder

Lesser omentum

Pancreas

Duodenum

Peritoneal cavity

Mesentery

Visceral peritoneum

Rectum

Krabach

Multicellular Glands   Multicellular glands, as their name implies, are composed of numerous secretory cells in addition to cells that form the walls of the ducts. Multicellular exocrine glands are divided into *simple* and *compound glands.* The ducts of the simple gland do not branch or have only a few branches, whereas those of the compound type have multiple branches (fig. 4.12). Multicellular glands are further classified according to the shape of the secretory portion. They are identified as *tubular* if the secretory portion resembles the ductule portion and as *acinar* if the secretory portion is flasklike. Multicellular glands with a secretory portion that resembles both a tube and a flask are termed *tubuloacinar.*

Multicellular exocrine glands are also classified according to the means by which they discharge the secretory product (fig. 4.13). Glands that secrete their products by exocytosis through the cell membrane of the secretory cells are called **merocrine** *(mer′ŏ-krin)* **glands.** Salivary glands, pancreatic glands, and certain sweat glands are of this type. In **apocrine** *(ap′ŏ-krin)* **glands,** the secretion accumulates on the surface of the secretory cell, and then a portion of the cell, along with the secretion, is pinched off to be discharged. Mammary glands and certain sweat glands are apocrine glands. In a **holocrine** *(hol′ŏ-krin)* **gland,** the entire secretory cell is discharged along with the secretory product. An example of a holocrine gland is a sebaceous, or oil-secreting, gland of the skin.

Glandular epithelia are summarized in table 4.3.

1. List the functions of simple squamous epithelia.
2. What are the three types of columnarlike epithelia? What do they have in common? How are they different?
3. What are the two types of stratified squamous epithelia and how do they differ?
4. Distinguish between mucous and serous membranes. List the specific serous membranes of the thoracic and abdominopelvic cavities.
5. Distinguish between unicellular and multicellular glands. Explain how multicellular glands are classified according to their mechanism of secretion.
6. In what ways are mammary glands and certain sweat glands similar?

Connective Tissue

Connective tissues are classified according to the characteristics of the matrix that binds the cells. Connective tissues provide structural and metabolic support for other tissues and organs of the body.

Objective 7.   Describe the general characteristics, locations, and functions of connective tissue.

and its means of discharging the secretory product. Exocrine glands should not be confused with endocrine glands (see chapter 14). The latter are ductless and secrete hormones into the blood.

Unicellular Glands   Unicellular glands are single-celled exocrine glands interspersed with the various columnar epithelia. A mucus-secreting goblet cell is a good example of a unicellular gland. Goblet cells are found in the epithelial linings of the respiratory, digestive, urinary, and reproductive systems, where the mucous secretion lubricates and protects the surface linings (fig. 4.11).

merocrine: Gk. *meros,* part; *krinein,* to separate
apocrine: Gk. *apo,* off; *krinein,* to separate
holocrine: Gk. *holos,* whole; *krinein,* to separate

UNIT III

Figure 4.11   A goblet cell is a unicellular gland that secretes mucus to lubricate and protect surface linings. (*a*) Goblet cells are abundant in the columnar epithelium lining the lumen of the small intestine. (*b*) A photomicrograph of a goblet cell and (*c*) a labeled diagram.

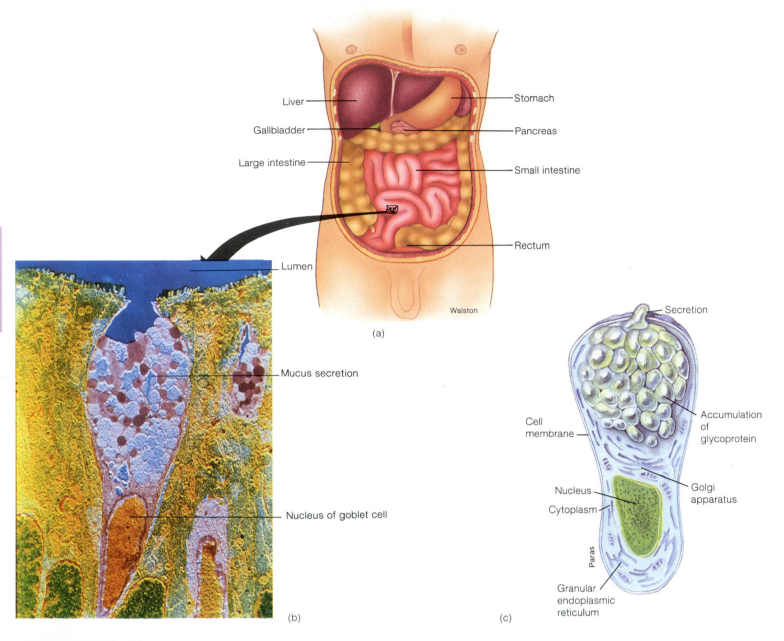

UNIT III

Objective 8.   Explain the functional relationship between embryonic and adult connective tissue.

Objective 9.   List the various ground substances, fiber types, and cells that constitute connective tissue and explain their functions.

Characteristics and Classification of Connective Tissue

Connective tissue is found throughout the body and, as the name indicates, supports or binds other tissues and provides for the metabolic needs of all body organs. Certain types of connective tissue store nutritional substances, whereas other types manufacture protective and regulatory materials.

Although connective tissue varies tremendously in structure and function, all connective tissues have similarities. With the exception of mature cartilage, connective tissues are highly vascular and well nourished. They are able to replicate and, by so doing, are responsible for the repair of body organs. Unlike epithelial tissues, which are composed of tightly fitted cells, connective tissues contain considerably more matrix (intercellular material) than cells. Connective tissues do not occur on free surfaces of body cavities or on the surface of the body, as do epithelial tissues. Furthermore, connective tissues are embryonically derived from mesoderm, whereas epithelial tissues derive from ectoderm, mesoderm, and endoderm.

Figure 4.12   Structural classification of multicellular exocrine glands. The secretory portions of the simple glands either do not branch or have only a few branches, whereas those of the compound type have multiple branches.

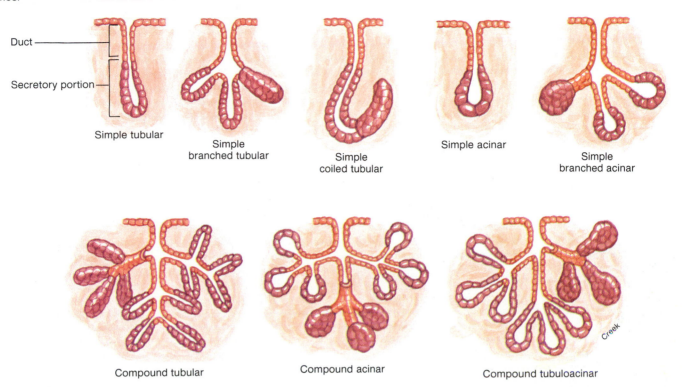

Duct

Secretory portion

Simple tubular

Simple branched tubular

Simple coiled tubular

Simple acinar

Simple branched acinar

Compound tubular

Compound acinar

Compound tubuloacinar

Figure 4.13   Secretory classification of multicellular exocrine glands: (*a*) merocrine gland, (*b*) apocrine gland, and (*c*) holocrine gland.

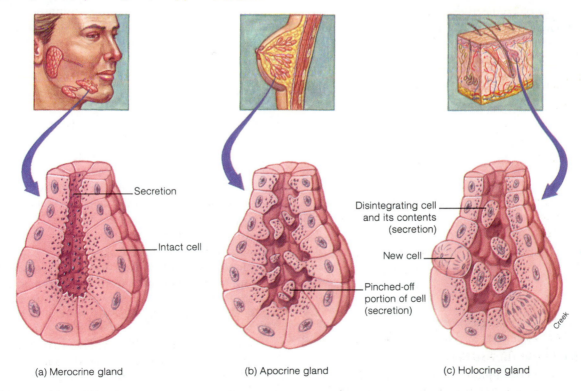

Secretion

Intact cell

Disintegrating cell and its contents (secretion)

New cell

Pinched-off portion of cell (secretion)

(a) Merocrine gland

(b) Apocrine gland

(c) Holocrine gland

UNIT III

Table 4.3	Summary of Glandular Epithelia	

Classification of Exocrine Glands by Structure

Type	Function	Example
I. Unicellular	Lubricate and protect	Goblet cells of digestive, respiratory, urinary, and reproductive systems
II. Multicellular	Protect, cool body, lubricate, aid in digestion, maintain body homeostasis	Sweat glands, digestive glands, mammary glands, sebaceous glands
A. Simple		
1. Tubular	Aid in digestion	Intestinal glands
2. Branched tubular	Protect, aid in digestion	Uterine glands, gastric glands
3. Coiled tubular	Regulate temperature	Certain sweat glands
4. Acinar	Provide additive for spermatozoa	Seminal vesicle of male reproductive system
5. Branched acinar	Condition skin	Sebaceous skin glands
B. Compound		
1. Tubular	Lubricate urethra of male, assist body digestion	Bulbourethral gland of male reproductive system, liver
2. Acinar	Provide nourishment for infant, aid in digestion	Mammary gland, salivary gland (sublingual and submandibular)
3. Tubuloacinar	Aid in digestion	Salivary gland (parotid), pancreas

Classification of Exocrine Glands by Mode of Secretion

Type	Description of Secretion	Example
Merocrine glands	Watery secretion for regulating temperature or enzymes that promote digestion	Salivary and pancreatic glands, certain sweat glands
Apocrine glands	Portion of secretory cell and secretion are discharged; provides nourishment for infant, assists in regulating temperature	Mammary glands, certain sweat glands
Holocrine glands	Entire secretory cell with enclosed secretion is discharged; conditions skin	Sebaceous glands of the skin

The classification of connective tissues is not exact, and several schemes have been devised. In general, however, they are named according to the kind and arrangement of the matrix. The following are the basic kinds of connective tissues.

- A. Embryonic connective tissue
- B. Connective tissue proper
  1. Loose (areolar)
  2. Dense regular
  3. Dense irregular
  4. Elastic
  5. Reticular
  6. Adipose
- C. Cartilage
  1. Hyaline
  2. Fibrocartilage
  3. Elastic
- D. Bone
- E. Vascular (blood) tissue

Embryonic Connective Tissue

The embryonic period of development, which lasts 6 weeks (from the beginning of the third to the end of the eighth week), is characterized by a tremendous amount of tissue differentia-tion and organ formation. At the beginning of the embryonic period, all connective tissue appears the same and is referred to as **mesenchyme** (mez'en-kīm). Mesenchyme is undifferentiated embryonic connective tissue that is derived from mesoderm. It consists of irregularly shaped cells lying in large amounts of a homogeneous, jellylike matrix (fig. 4.14). In certain periods of development, mesenchyme migrates to predisposed sites where it interacts with other tissues to form organs. Before the end of the embryonic period, once mesenchyme has migrated to the appropriate position, it differentiates, and from it all other kinds of connective tissues are formed.

Some mesenchymal-like tissue persists past the embryonic period in certain sites within the body. Good examples are the undifferentiated cells that surround blood vessels and form fibroblasts if the vessels are traumatized. Fibroblasts assist in healing wounds (see chapter 5).

Another kind of prenatal connective tissue exists only in the fetus (the fetal period is from 9 weeks to birth) and is called *mucous connective tissue,* or *Wharton's jelly.* It provides a turgid consistency to the umbilical cord.

Wharton's jelly: from Thomas Wharton, English anatomist, 1614–73

Figure 4.14 Mesenchyme is undifferentiated embryonic mesodermal connective tissue that can migrate and give rise to all other kinds of connective tissue. It is found in an early developing embryo (a). A photomicrograph of this tissue is shown in (b), and a labeled diagram in (c).

(b)

UNIT III

Connective Tissue Proper

Connective tissue proper has a loose, flexible matrix, frequently called **ground substance.** The most common cell within connective tissue proper is called a **fibroblast** (*fi'bro-blast*). Fibroblasts are large, star-shaped cells that produce collagenous (*kŏ-laj'ĕ-nus*)**, elastic, and reticular** (*rĕ-tik'yoo-lar*) fibers. **Collagenous fibers** are composed of a protein called *collagen* (*kol'ah-jen*); they are flexible, yet they have tremendous strength. **Elastic fibers** are composed of a protein called *elastin*, which gives elasticity to certain tissues. Collagenous and elastic fibers may be either sparse and irregularly arranged, as in loose connective tissue, or tightly packed, as in dense connective tissue. Tissues with loosely arranged fibers generally form packing material that cushions and protects various organs, whereas those that are tightly arranged form the binding and supportive connective tissues of the body.

Resilience in tissues that contain elastic fibers is extremely important for several physical functions of the body. Consider, for example, that elastic fibers are found in the walls of arteries and in the walls of the lower respiratory passageways. As these walls are expanded by blood moving through vessels or by inspired air, the elastic fibers must first stretch and then recoil. This maintains the pressures of the fluid or air moving through the lumina, thus ensuring adequate flow rates and rates of diffusion through capillary and lung surfaces.

Reticular fibers reinforce by forming thin, short threads that branch and join to form a delicate lattice or reticulum. Reticular fibers are common in lymphatic glands, where they form a meshlike center called the **stroma.**

Six basic types of connective tissue proper are generally recognized. These tissues are distinguished by the consistency of the ground substance and the type and arrangement of the reinforcement fibers.

Loose Connective (Areolar) Tissue   Loose connective tissue is distributed throughout the body as a binding and packing material. It binds the skin to the underlying muscles and is highly vascular, providing nutrients to the skin. Loose connective tissue surrounding muscle fibers and muscle groups is known as **fascia** (*fash'e-ă*). It also surrounds blood vessels and nerves, where it provides both protection and nourishment. Specialized cells

collagen: Gk. *kolla*, glue
elastin: Gk. *elasticus*, to drive

reticular: L. *rete*, net or netlike
stroma: Gk., *stroma*, a couch or bed
fascia: L. *fascia*, band or girdle

Figure 4.15   Loose connective tissue is packing and binding tissue
that surrounds muscles (a), nerves, and vessels and binds the skin to the
underlying muscles. (b) A photomicrograph of this tissue and (c) a labeled
diagram.

Pectoralis
major
muscle

(a)

Fascia

Paras

(b)

Elastic
fiber

Mast
cell

Fibroblast

Collagenous
fibers

(c)

called **mast cells** are dispersed throughout the loose connective tissue surrounding blood vessels. Mast cells produce *heparin (hep'ah-rin),* an anticoagulant that prevents blood from clotting within the vessels. They also produce *histamine,* which is released during inflammation and acts as a powerful vasodilator.

The cells of loose connective tissue are predominantly fibroblasts, with collagenous and elastic fibers dispersed throughout the ground substance (fig. 4.15). The irregular arrangement of this tissue provides flexibility, yet strength, in any direction. It is this tissue layer, for example, that permits the skin to move when a part of the body is rubbed.

> Much of the fluid of the body is found within loose connective tissue and is called *tissue fluid.* Sometimes excessive tissue fluid accumulates, causing a swelled condition called *edema (e-de-mă).* Edema is a symptom of a variety of dysfunctions or disease processes.

Dense Regular Connective Tissue   Dense regular connective tissue is characterized by large amounts of densely packed collagenous fibers lying parallel to the direction of force placed on this tissue during body movement. Because this tissue is silvery white in appearance, it is sometimes called dense or white fibrous connective tissue.

Dense regular connective tissue occurs where strong, flexible support is necessary (fig. 4.16). **Tendons,** which attach muscles to bones and transfer the forces of muscle contractions, and **ligaments,** which connect bone to bone across articulations, are composed of this type of tissue.

> Trauma to ligaments, tendons, and muscles are common sports-related injuries. A *strain* is an excessive stretch of the tissue composing the tendon or muscle, with no serious damage. A *sprain* is a tearing of the tissue of a ligament and may be slight, moderate,

heparin: Gk. *hepatos,* the liver

tendon: L. *tendere,* to stretch
ligament: L. *ligare,* bind

Figure 4.16 Dense regular connective tissue forms the strong and highly flexible tendons (*a*) and ligaments. (*b*) A photomicrograph of this tissue and (*c*) a labeled diagram.

Tendon of long head of biceps brachii m.

Tendon of short head of biceps brachii m.

Biceps brachii m.

(a)

Paras

(b)

Fibroblasts

Collagenous fibers

(c)

or complete. A complete tear of a major ligament is especially painful and disabling. Ligamentous tissue does not heal itself well because it has a poor blood supply. Surgical reconstruction is generally needed for the treatment of a severed ligament.

Dense Irregular Connective Tissue Dense irregular connective tissue is characterized by large amounts of densely packed collagenous fibers that are interwoven to provide tensile strength in any direction. This tissue is found in the dermis of the skin, submucosa of the GI tract, and composing the fibrous capsules of organs and joints (fig. 4.17). It also makes up the collagenous matrix of bone, called *osteoid.*

Elastic Connective Tissue Elastic connective tissue has a predominance of elastic fibers that are irregularly arranged and yellowish in color (fig. 4.18). They can be stretched to one and a half times their original lengths and will snap back to their former size. Elastic connective tissue is found in the walls of large arteries, in portions of the larynx, and in the trachea and bronchial tubes of the lungs. It is also present between the arches of vertebrae that make up the vertebral column.

Reticular Connective Tissue Reticular connective tissue is characterized by a network of reticular fibers woven through a jellylike matrix (fig. 4.19). Certain specialized cells within reticular tissue are *phagocytic (fag″o-sit′ik),* and therefore ingest foreign materials. The liver, spleen, lymph nodes, and bone marrow contain reticular connective tissue.

Adipose Connective Tissue Adipose tissue is a specialized type of loose fibrous connective tissue that contains large quantities of **adipose cells,** or **adipocytes.** Adipose cells form from mesenchyme and, for the most part, are formed prenatally and during the first year of life. Adipose cells store droplets of fat within their cytoplasm, causing them to swell and forcing their nuclei to one side (fig. 4.20).

Adipose tissue is found throughout the body but is concentrated around the kidney, in the hypodermis of the skin, on the surface of the heart, surrounding joints, and in the breasts

adipose: L. *adiposus,* fat

Figure 4.17 Dense irregular connective tissue forms joint capsules (a) that contain synovial fluid for lubricating movable joints. (b) A photomicrograph of this tissue and (c) a labeled diagram.

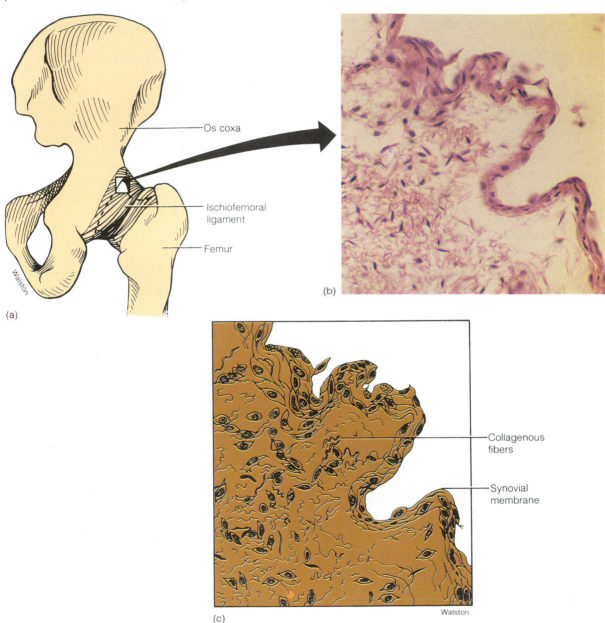

Os coxa

Ischiofemoral ligament

Femur

Walston

(a)

(b)

Collagenous fibers

Synovial membrane

Walston

(c)

of sexually mature females. Fat functions not only as a food reserve but supports and protects various organs. Fat is a good insulator against cold because it is a poor conductor of heat.

Excessive fat can be both unsightly and unhealthy, placing a strain on the heart and perhaps causing early death. For these reasons, good exercise programs and sensible diets are extremely important. Adipose tissue can also retain environmental pollutants that are ingested or absorbed through the skin. Dieting eliminates the fat stored within the tissue but not the tissue itself.

The surgical procedure of *suction lipectomy* may be used to remove small amounts of adipose tissue from certain localized body areas, such as the breasts, abdomen, buttocks, or thighs. Suction lipectomy is used for cosmetic purposes rather than as a treatment for obesity, and the risks for potentially detrimental side-effects need to be seriously considered. Potential candidates should be between 30 and 40 years old and only about 15 to 20 pounds overweight. They should also have good skin elasticity.

The characteristics, functions, and locations of connective tissue proper are summarized in table 4.4.

Figure 4.18 Elastic connective tissue permits stretching of a large artery (*a*) as blood flows through. (*b*) A photomicrograph of this tissue and (*c*) a labeled diagram.

Tunica intima (inner coat)
Endothelial cells
Elastic tissue

Paras

(a)

(b)

Fibroblast

Elastic fibers

(c)

Figure 4.19 Reticular tissue forms the stroma, or framework, of organs such as the spleen (*a*), liver, thymus, and lymph nodes. (*b*) A photomicrograph of this tissue and (*c*) a labeled diagram.

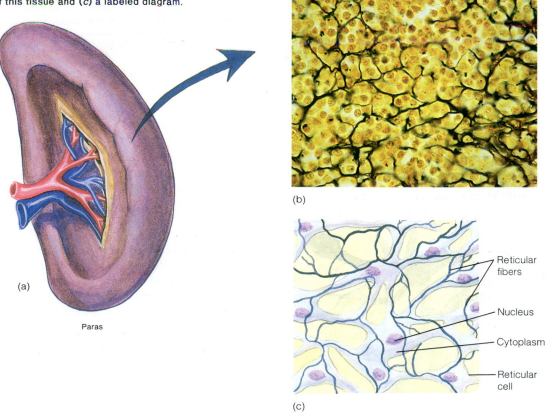

(a)

Paras

(b)

Reticular fibers

Nucleus

Cytoplasm

Reticular cell

(c)

UNIT III

Figure 4.20   Adipose tissue is abundant in the hypodermis of the skin (*a*) and around various internal organs. (*b*) A photomicrograph of this tissue and (*c*) a labeled diagram.

Table 4.4	**Summary of Connective Tissue Proper**	
Type	**Structure and Function**	**Location**
Loose connective (areolar) tissue	Predominantly fibroblast cells with lesser amounts of collagen and elastin proteins; binds organs, holds tissue fluids	Surrounding nerves and vessels, between muscles, beneath the skin
Dense regular connective tissue	Densely packed collagenous fibers lying parallel to the direction of force; provides strong, flexible support	Tendons, ligaments
Dense irregular connective tissue	Densely packed collagenous fibers arranged in a tight interwoven pattern; provides tensile strength in any direction	Dermis of skin, fibrous capsules of organs and joints
Elastic connective tissue	Predominantly irregularly arranged elastic fibers; supports, provides framework	Large arteries, lower respiratory tract, between the arches of vertebrae
Reticular connective tissue	Reticular fibers forming supportive network; stores, performs phagocytic function	Lymph nodes, liver, spleen, thymus, bone marrow
Adipose connective tissue	Adipose cells; protects, stores fat, insulates	Hypodermis of skin, surface of heart, omentum, around kidneys, back of eyeball, surrounding joints

Cartilage

Cartilage tissue consists of cartilage cells, or **chondrocytes** *(kon'dro-sītz)*, and a semisolid matrix that imparts marked elastic properties to the tissue. Cartilage is a supportive and protective connective tissue that is frequently associated with bone. It forms a precursor to one type of bone and persists at the articular surfaces on the bones of all movable joints.

The chondrocytes within cartilage may occur singly but are frequently clustered. Chondrocytes occupy cavities, called

lacunae *(lă-kyoo'ne)*, within the matrix. Most cartilage tissue is surrounded by a dense regular connective tissue called **perichondrium** *(per''ĭ-kon'dre-um)*. Cartilage at the articular surfaces of bones (articular cartilage) lacks perichondria. Because mature cartilage is avascular, it must receive nutrients through diffusion from the perichondrium and the surrounding tissue. For this reason, cartilaginous tissue has a slow rate of mitotic activity and, if damaged, heals with difficulty.

lacuna: L. *lacuna*, hole or pit

Figure 4.21 Hyaline cartilage is the most abundant cartilage within the body. It occurs in places such as the larynx (*a*), trachea, rib cage, and embryonic skeleton. (*b*) A photomicrograph of this tissue and (*c*) a labeled diagram.

(a)

Larynx

Thyroid cartilage

Cricoid cartilage

Paras

(b)

Perichondrium

Lacunae

Chondrocytes

(c)

UNIT III

There are three kinds of cartilage: hyaline *(hi'ă-līn)* cartilage, fibrocartilage, and elastic cartilage. Each is distinguished by the types, and amount, of fibers embedded within the matrix.

Hyaline Cartilage Hyaline cartilage has a homogeneous, bluish-stained matrix in which the collagenous fibers are so fine that they can be observed only with an electron microscope. When viewed through a light microscope, hyaline cartilage has a clear, glassy appearance (fig. 4.21).

Hyaline cartilage is the most abundant cartilage within the body and is commonly called *"gristle."* It covers the articular surfaces of bones, supports the tubular trachea and bronchi of the respiratory system, reinforces the nose, and forms the flexible bridge, called **costal cartilage,** between the ventral end of each of the first 10 ribs and the sternum. Most of the bones of the body form first as hyaline cartilage and later become bone in a process called *ossification.*

Fibrocartilage Fibrocartilage has its matrix reinforced with numerous collagenous fibers (fig. 4.22). It is a durable tissue adapted to withstand tension and compression. It is found at the symphysis pubis, where the two pelvic bones articulate, and between the vertebrae as intervertebral discs. It also forms the cartilaginous wedges within the knee joint, called *menisci* (see chapter 8).

By the end of the day, the intervertebral discs of the vertebral column are somewhat compacted. So a person is actually slightly shorter in the evening than in the morning following a recuperative rest. With aging comes a gradual and irreversible compression of the fibrocartilaginous discs.

Elastic Cartilage Elastic cartilage is similar to hyaline except for the presence of abundant elastic fibers, which make elastic cartilage very flexible while maintaining its strength (fig. 4.23). The numerous elastic fibers also give it a yellowish appearance. This tissue is found in the outer ear, portions of the larynx, and in the auditory canal.

The structure, function, and location of cartilage is summarized in table 4.5.

hyaline: Gk. *hyalos,* glass

Figure 4.22   Fibrocartilage is located at the symphysis pubis, within the knee joint, and between the vertebrae as the intervertebral discs (a). A photomicrograph of this tissue is shown in (b) and a labeled diagram in (c).

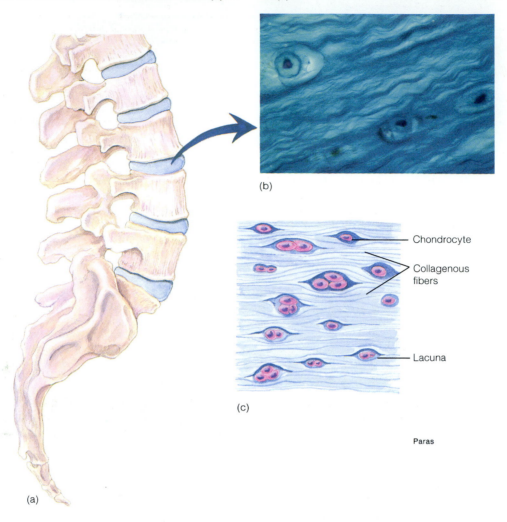

(a)

(b)

(c)

Chondrocyte

Collagenous fibers

Lacuna

Paras

Figure 4.23   Elastic cartilage gives support to the outer ear (a), auditory canal, and parts of the larynx. A photomicrograph of this tissue is shown in (b) and a labeled diagram in (c).

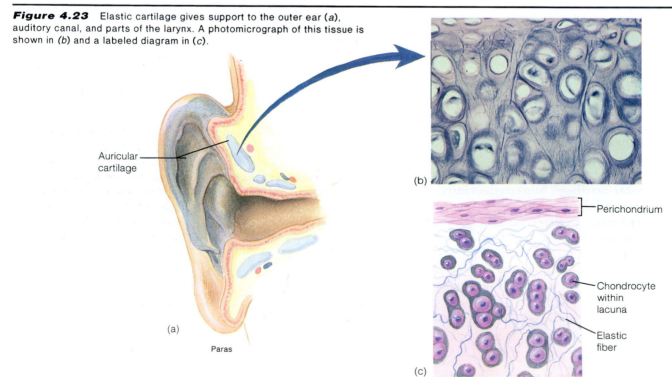

Auricular cartilage

(a)

Paras

(b)

Perichondrium

Chondrocyte within lacuna

Elastic fiber

(c)

Table 4.5	Summary of Cartilage Tissue	
Type	**Structure and Function**	**Location**
Hyaline cartilage	Homogeneous matrix with extremely fine collagenous fibers; provides flexible support, protects, is precursor to bone	Articular surfaces of bones, nose, walls of respiratory passages, fetal skeleton
Fibrocartilage	Abundant collagenous fibers within matrix; supports, withstands compression	Symphysis pubis, intervertebral discs, knee joint
Elastic cartilage	Abundant elastic fibers within matrix; supports, provides flexibility	Framework of outer ear, auditory canal, portions of larynx

Bone

Bone, or **osseous connective tissue,** is the most rigid of the connective tissues. Unlike cartilage, bone has a rich vascular supply and is the site of considerable metabolic activity. The hardness of bone is largely due to the inorganic calcium phosphate (calcium hydroxyapatite) deposited within the intercellular matrix. Numerous collagenous fibers, also embedded within the matrix, give some flexibility to bone.

When bone is placed in a weak acid, the calcium salts dissolve away and the bone becomes pliable. It retains its basic shape but can be easily bent and twisted (fig. 4.24). In calcium deficiency diseases, such as *rickets,* the osseous tissue becomes pliable and bends under the weight of the body (see fig. 5.10).

The two types of bone tissue are classified according to porosity, and most bones have both types (fig. 4.25). **Compact,** or **dense, bone tissue** is the hard outer layer, whereas **spongy,** or **cancellous, bone tissue** is the porous, highly vascular inner portion. Compact bone tissue is covered by the periosteum. It provides for attachment of muscles and protects and strengthens the bone. Spongy bone tissue makes the bone lighter and provides a space for bone marrow, where blood cells are produced.

In compact bone tissue, mature bone cells, called **osteocytes,** are arranged in concentric layers around a **central (haversian) canal,** which contains a vascular and nerve supply. Each osteocyte occupies a cavity called a **lacuna.** Radiating from each lacuna are numerous minute canals, or **canaliculi.** The canaliculi traverse the dense matrix of the bone tissue to adjacent lacunae. Nutrients diffuse through the canaliculi to reach each osteocyte. The inorganic matrix is deposited in concentric layers called **lamellae.**

Bone tissue is described in detail in chapter 6.

Vascular Connective Tissue

Blood is a highly specialized, viscous connective tissue. The cells, or **formed elements,** of vascular connective tissue are suspended in the liquid plasma matrix (fig. 4.26). Blood plays a vital role in maintaining internal body homeostasis. The three

Figure 4.24 A bone soaked in a weak acid, such as the acetic acid in vinegar, demineralizes and becomes flexible.

types of formed elements found within the blood are **erythrocytes** (red blood cells), **leukocytes** (white blood cells), and **thrombocytes** (platelets).

Formed elements and the physiology of blood are discussed in detail in chapter 16.

An injury to a portion of the body may stimulate tissue repair activity, usually involving connective tissue. A minor scrape or cut results in platelet and plasma activity of the exposed blood and the formation of a scab. The epidermis of the skin regenerates beneath the scab. A severe open wound eventuates connective tissue granulation. In this process, collagenous fibers form from surrounding fibroblasts to strengthen the traumatized area. The healed area is known as a *scar.*

haversian canal: from Clopton Havers, English anatomist, 1650–1702

erythrocyte: Gk. *erythros,* red; *kytos,* hollow (cell)
leukocyte: Gk. *leukos,* white; *kytos,* hollow (cell)
thrombocyte: Gk. *thrombos,* a clot; *kytos,* hollow (cell)

Figure 4.25 Bone (a) consists of compact and spongy bone tissues. (b) A photomicrograph of compact bone tissue and (c) a labeled diagram.

UNIT III

(a)

(b)

Lamellae

Osteocyte within a lacuna

Canaliculi

Central canal

(c)   Paras

Figure 4.26 Vascular connective tissue, or blood, is composed of a fluid matrix, called plasma, and three types of formed elements: erythrocytes (red blood cells), leukocytes (white blood cells), and thrombocytes (platelets).

Erythrocytes

Blood sample

Plasma

Cells

Leukocytes   Thrombocytes

Paras

1. List the kinds of connective tissues and describe the structure, function, and location of each.
2. Which of the previously discussed connective tissues function to protect body organs? Which type is phagocytic? Which types bind and support various structures? Which types are associated in some way with the skin?
3. What is the developmental significance of mesenchyme and how does it differ functionally from adult connective tissue?
4. Define or describe reticular fibers, fibroblasts, collagenous fibers, elastic fibers, and mast cells.

Muscle Tissue

Muscle tissues are responsible for the movement of materials through the body, the movement of one part of the body with respect to another, and for locomotion. Fibers in the three kinds of muscle tissue are adapted to contract in response to stimuli.

Objective 10. Describe the structure, location, and function of the three types of muscle tissue.

Muscle tissues are unique in possessing the property of *contractility.* The muscle cells, or *fibers,* are elongated in the direction of contraction. Movement is accomplished through the shortening of the fibers in response to a stimulus. Muscle tissue is derived from mesoderm. There are three types of muscle tissue in the body: *smooth, cardiac,* and *skeletal* (fig. 4.27).

Smooth Muscle   Smooth muscle tissue is common throughout the body, occurring in many of the systems. For example, in the wall of the GI tract it provides the motive power for mechanically digesting food and the peristaltic movements. Smooth muscle is also found in the walls of arteries, the walls of respiratory passages, and in the urinary and reproductive ducts. The contraction of smooth muscle is under autonomic (involuntary) nervous control, and is discussed in more detail in chapter 13.

Smooth muscle fibers are long, spindle-shaped cells. They contain a single nucleus and lack striations. These cells are usually grouped together in flattened sheets, forming the muscular portion of a wall around a lumen.

Cardiac Muscle   Cardiac muscle tissue makes up most of the wall of the heart. This tissue is characterized by bifurcating (branching) fibers, a centrally positioned nucleus, and transversely positioned **intercalated** *(in-ter'kă-lāt-ed)* **discs.** Intercalated discs help to hold adjacent cells together and transmit the force of contraction from cell to cell. Like skeletal muscle, cardiac muscle is striated, but unlike skeletal muscle it experiences rhythmical involuntary contractions. Cardiac muscle is further discussed in chapter 16.

Skeletal Muscle   Skeletal muscle tissue attaches to the skeleton and is responsible for voluntary body movements. Each elongated, multinucleated fiber has distinct transverse striations. Fibers of this muscle tissue are grouped into parallel fasciculi (bundles) that can be seen without a microscope in fresh muscle. Both cardiac and skeletal muscle fibers are striated and cannot replicate once tissue formation has been completed shortly after birth. Skeletal muscle tissue is further discussed in chapter 9. The three types of muscle tissue are summarized in table 4.6.

1. Identify the general characteristics of muscle tissue. What is meant by *voluntary* and *involuntary* as applied to muscle tissue?
2. Distinguish between smooth, cardiac, and skeletal muscle tissue on the bases of structure, location, and function.

Nervous Tissue

Nervous tissue is composed of neurons, which respond to stimuli and conduct impulses to and from all body organs, and neuroglia, which functionally support and physically bind neurons.

Objective 11. Describe the basic characteristics and functions of nervous tissue.

Objective 12. Distinguish between neurons and neuroglia.

Neurons   Although there are several kinds of neurons *(noor'onz)* in nervous tissue, they all have three principal components: (1) a cell body, or perikaryon; (2) dendrites; and (3) an axon (fig. 4.28). **Dendrites** function to receive stimuli and conduct impulses to the cell body. The **cell body,** or **perikaryon** *(per''ĭ-kar'e-on),* contains the nucleus and specialized organelles and microtubules. The **axon** is a cytoplasmic extension that conducts impulses away from the cell body. The term **nerve fiber** refers to any process extending from the cell body of a neuron.

Neurons derive from ectoderm and are the basic structural and functional units of the nervous system. They are specialized to respond to physical and chemical stimuli, conduct nerve impulses, and perform other functions, such as storing memory, thinking, and regulating the activity of other organs or glands. Of all the body's cells, neurons are probably the most specialized. As with muscle cells, the number of neurons is established shortly after birth; thereafter, they lack the ability to undergo mitosis, although under certain circumstances a severed portion can regenerate.

---

neuron: Gk. *neuron,* sinew or nerve
perikaryon: Gk. *peri,* around; *karyon,* nut or kernel

UNIT III

Figure 4.27 The three types of muscle tissue: (*a*) smooth muscle fibers teased apart, (*b*) cardiac muscle, and (*c*) skeletal muscle.

(a)

(b)

(c)

Neuroglia   In addition to neurons, nervous tissue is composed of neuroglia *(noo'rog le-ă)* (fig. 4.28). Neuroglial cells are about five times as abundant as neurons and have limited mitotic abilities. They do not transmit impulses but support and bind neurons together. Certain neuroglial cells are phagocytic; others assist in providing sustenance to the neurons.

Neurons and neuroglia are discussed in detail in chapter 11.

1. Compare and contrast neurons and neuroglia in terms of structure, function, and location.
2. List the structures of a neuron following the sequence of a nerve impulse passing through the cell.

neuroglia: Gk. *neuron*, nerve; *glia*, glue

Clinical Considerations

As discussed in the beginning of this chapter, the study of histology is extremely important in understanding organ and system structure and function. Histology has immense clinical importance as well. Many diseases are diagnosed through microscopic examination of tissue sections. Even in performing an autopsy, an examination of various tissues is vital in establishing the cause of death.

Several sciences are concerned with specific aspects of tissues. *Histopathology* is the study of diseased tissues. *Histochemistry* is concerned with the biochemical physiology of tissues as they function to maintain body homeostasis. *Histotechnology* studies the various ways tissues can be better stained and observed. In all of these disciplines, a thorough understanding of normal, or healthy, tissues is imperative for recognizing altered, or abnormal, tissues.

Changes in Tissue Composition

Most diseases alter tissue structure *locally,* where the disease is prevalent, although some diseases, called *general conditions,* cause changes remote from the locus of the disease. **Atrophy** (wasting of body tissue), for example, may be confined to a particular organ where the disease interferes with the metabolism of that organ; or it may involve an entire limb if nourishment or nerve impulses are decreased or prohibited. *Muscle atrophy,* for example, can be caused by a disease of the nervous system like polio, or it can be the result of a diminished blood supply to a muscle. *Senescence (sĕ-nes'ens) atrophy,* or simply *senescence,* is the natural aging of tissues and organs within the body. *Disuse*

atrophy: Gk. *a,* without; *trophe,* nourishment

Table 4.6	Summary of Muscle Tissue	
Type	**Structure and Function**	**Location**
Smooth	Elongated, spindle-shaped fiber with single nucleus; involuntary movements of internal organs	Walls of hollow internal organs
Cardiac	Branched, striated fiber with single nucleus and intercalated discs; involuntary rhythmic contraction	Heart muscle
Skeletal	Multinucleated, striated, cylindrical fiber that occurs in fasciculi; voluntary movement of skeletal parts	Associated with skeleton; spans joints of skeleton via tendons

UNIT III

Figure 4.28   Nervous tissue is found in the brain (*a*), and consists of two principal kinds of cells: (*b*) neurons and (*c*) neuroglia.

(a)

(b)

(c)

atrophy is a local atrophy that results from the inactivity of a tissue or organ. Muscular dystrophy causes a disuse atrophy that decreases muscle size and strength due to the loss of sarcoplasm within the muscle.

Necrosis *(nĕ-kro'sis)* is cellular or tissue death within the living body. It can be recognized by changes in the dead tissues. Necrosis can be caused by a number of factors, such as severe injury; physical agents (trauma, heat, radiant energy, chemical poisons); or interference with the nutrition of tissues. When histologically examined, the necrotic tissue usually appears opaque, and a whitish or yellow color is assumed. **Gangrene** is a massive necrosis of tissue accompanied by an invasion of microorganisms that live on decaying tissues.

Somatic death is the death of the body as a whole. Following somatic death, tissues undergo irreversible changes, such as **rigor mortis** (muscular rigidity), clotting of the blood, and cooling of the body. Postmortem changes occur under varying conditions at predictable rates, which is useful in estimating the approximate time of death.

Tissue Analysis

In diagnosing a disease, it is frequently important to examine tissues from a living person histologically. When this is necessary, a **biopsy** (removal of a section of living tissue) is performed. There are several techniques for biopsies. *Surgical removal* is usually done on large masses or tumors. *Curettage* involves cutting and scraping tissue, as may be done in examining for uterine cancer. In a *percutaneous needle biopsy,* a biopsy needle is inserted through a small skin incision and tissue samples are aspirated. Both normal and diseased tissues are removed for purposes of comparison.

Preparing tissues for examination involves a number of steps. **Fixation** is fundamental for all histological preparation. It is the rapid killing, hardening, and preservation of tissue to maintain its existing structure. **Embedding** the tissue in a supporting medium like paraffin wax usually follows fixation. The next step is **sectioning** the tissue into extremely thin slices, followed by **mounting** the specimen on a slide. Some tissues

---

necrosis: Gk. *nekros,* corpse
gangrene: Gk. *gangraina,* gnaw or eat

are fixed by rapid freezing and then sectioned while frozen, making embedding unnecessary. Frozen sections enable the pathologist to make a quick diagnosis during a surgical operation. These are done frequently, for example, in cases of suspected breast cancer. **Staining** is the next step. The hematoxylin and eosin (H & E) stains are routinely used on all tissue specimens. They give a differential blue and red color to the basic and acidic structures within the tissue. Other dyes may be needed to stain for specific structures.

Examination is first done with the unaided eye and then with a microscope. Practically all histological conditions can be diagnosed with low magnification ($25\times$). Higher magnification is used to clarify specific details. Further examination may be performed with an electron microscope, which makes visible the cellular structures that are the morphological bases of metabolic processes. Histological observation is the foundation of subsequent diagnosis, prognosis, treatment, and reevaluation.

Tissue Transplantation

In the last two decades, medical science has made tremendous advancements in tissue transplants. Tissue transplants are necessary for replacing nonfunctional, damaged, or lost body parts. The most successful transplant is one where tissue is taken from one place on a person's body and moved to another place, such as a skin graft from the thigh to replace burned tissue of the hand. This type of transplant is termed an **autotransplant.** **Isotransplants** are transplants between individuals who are closely related genetically. Identical twins have the best acceptance success in this type of transplant. **Homotransplants** (between individuals of the same species) and **heterotransplants** (between two different species) have low acceptance percentages because of a *tissue-rejection reaction.* When this occurs, the recipient's immune mechanisms are triggered, and the donor's tissue is identified as foreign and is destroyed. The reaction can be minimized by "matching" recipient and donor tissue. Immunosuppressive drugs also may lessen the rejection rate. These drugs act by interfering with the recipient's immune mechanisms. Unfortunately, immunosuppressive drugs may also lower the recipient's resistance to infections. New techniques involving blood transfusions from donor to recipient before a transplant are proving successful. In any event, tissue transplants are an important aspect of medical research, and significant breakthroughs are on the horizon.

CHAPTER SUMMARY

Definition and Classification of Tissues (p. 72)

1. Tissues are aggregations of similar cells that perform specific functions. The study of tissues is called histology.
2. Cells are separated and bound together by an intercellular matrix, the composition of which varies from solid to liquid.
3. The four principal types of tissues are epithelial tissue, connective tissue, muscle tissue, and nervous tissue.

Epithelial Tissue (pp. 73–83)

1. Epithelia derive from all three germ layers and may be one or several layers thick. The lower surface of membranous epithelia is supported by a basement membrane.
2. Simple epithelia vary in shape and surface characteristics. They are located where diffusion, filtration, and secretion occur.
3. Stratified epithelia consist of two or more layers of cells and are adapted for protection.
4. Transitional epithelium lines the urinary bladder and is adapted for distension.
5. The body has two principal types of membranes: mucous membranes, which secrete protective mucus, and serous membranes, which line the thoracic and abdominopelvic cavities and cover visceral organs.
6. Glandular epithelia derive from developing epithelial tissue and function as secretory exocrine glands.

Connective Tissue (pp. 83–97)

1. Connective tissues derive from mesenchymal mesoderm and, with the exception of cartilage, are highly vascular.
2. Connective tissue proper contains fibroblasts, collagenous fibers, and elastic fibers within a flexible ground substance.
3. Cartilage provides a flexible framework for many organs. It consists of a semisolid matrix of chondrocytes and various fibers.
4. Bone (osseous) tissue consists of osteocytes, collagenous fibers, and a durable matrix of mineral salts.
5. Vascular (blood) tissue consists of formed cellular elements (erythrocytes, leukocytes, and thrombocytes) suspended in a fluid plasma matrix.

Muscle Tissue (p. 97)

1. Muscle tissues (smooth, cardiac, and skeletal) are responsible for the movement of materials through the body, the movement of one part of the body with respect to another, and for locomotion.
2. Fibers in the three kinds of muscle tissues are adapted to contract in response to stimuli.

Nervous Tissue (pp. 97–99)

1. Neurons are the functional units of the nervous system. They respond to stimuli and conduct impulses to and from all body organs.
2. Neuroglia support and bind neurons. Some are phagocytic; others provide sustenance to neurons.

REVIEW ACTIVITIES

Objective Questions

1. Which of the following is *not* a principal type of body tissue?
   (a) nervous   (d) muscular
   (b) integumentary (e) epithelial
   (c) connective
2. Which of the following is a *false* statement regarding tissues?
   (a) Tissues are aggregations of similar kinds of cells that perform specific functions.
   (b) All tissues are microscopic and are studied within the science of histology.
   (c) All tissues are stationary within the body at the location of their developmental origin.
   (d) An organ within the body is composed of two or more kinds of tissues.
3. Connective tissue, muscle, and the dermis of the skin derive from embryonic
   (a) mesoderm.   (c) ectoderm.
   (b) endoderm.
4. Which statement is *false* regarding epithelia?
   (a) They are derived from mesoderm, ectoderm, and endoderm.
   (b) They are strengthened by elastic and collagenous fibers.
   (c) One side is exposed to the lumen, cavity, or external environment.
   (d) They have very few extracellular matrix-binding cells.
5. A gastric ulcer of the stomach would involve
   (a) simple cuboidal epithelium.
   (b) transitional epithelium.
   (c) simple ciliated columnar epithelium.
   (d) simple columnar epithelium.
6. Which structural and secretory designation describes mammary glands?
   (a) acinar, apocrine
   (b) tubular, holocrine
   (c) tubular, merocrine
   (d) acinar, holocrine
7. Dense regular connective tissue is in
   (a) blood vessels.   (d) the wall of
   (b) the spleen.   the uterus.
   (c) tendons.
8. The phagocytic connective tissue found in the lymph nodes, liver, spleen, and bone marrow is
   (a) reticular.   (c) mesenchyme.
   (b) loose fibrous.   (d) elastic.
9. Cartilage is slow in healing following an injury because
   (a) it is located in body areas that are under constant physical strain.
   (b) it is avascular.
   (c) its chondrocytes cannot reproduce.
   (d) it has a semisolid matrix.
10. Cardiac muscle tissue has
    (a) striations.
    (b) intercalated discs.
    (c) rhythmical involuntary contractions.
    (d) all of the above.

UNIT III

Essay Questions

1. Define *tissue.* What are the differences between cells, tissues, glands, and organs?
2. What physiological functions are epithelial tissues adapted to perform?
3. Identify the epithelial tissue
   (a) in the alveoli of the lungs;
   (b) lining the lumen of the GI tract;
   (c) in the outer layer of skin;
   (d) in the urinary bladder;
   (e) in the uterine tube; and
   (f) lining the lumina of the lower respiratory tract.
   Describe the function of the tissue in each case.

4. Why are both keratinized and nonkeratinized epithelia found within the body?
5. Describe how epithelial glands are classified according to structural complexity and secretory function.
6. Identify the connective tissue
   (a) on the surface of the heart and surrounding the kidneys;
   (b) within the lumen of the aorta;
   (c) forming the symphysis pubis;
   (d) supporting the outer ear;
   (e) forming the lymph nodes; and
   (f) forming the tendo calcaneus.
   Describe the function of the tissue in each case.

7. Compare and contrast the following: reticular fibers, collagenous fibers, elastin, fibroblasts, and mast cells.
8. What is the relationship between adipose cells and fat? Discuss the function of fat and explain the potential danger of excessive fat.
9. Discuss the mitotic abilities of each of the four principal types of tissues.
10. Define the following terms: *atrophy, necrosis, gangrene,* and *somatic death.*

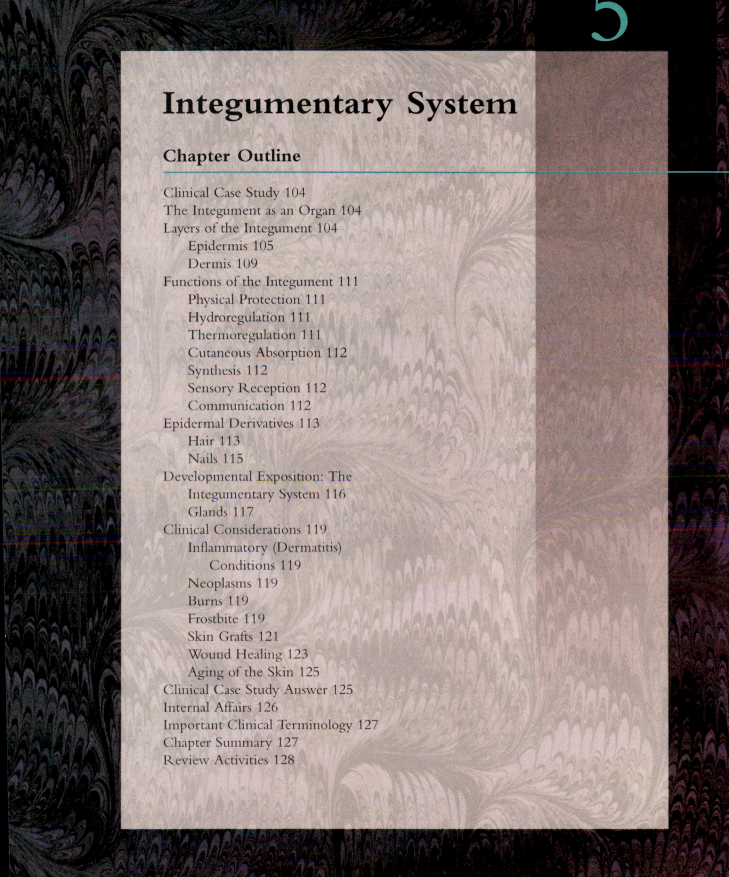

5

Integumentary System

Chapter Outline

UNIT IV

CLINICAL CASE STUDY

A 27-year-old male is involved in a gasoline explosion, sustaining burns to the face, neck, chest, and arms. Upon arrival at the emergency room, he complains of intense pain in the face and neck, which exhibit extensive blistering and erythema (redness). These findings are all curiously absent on the burned chest and arms, which have a pale, waxy appearance.

Examination reveals the skin on the patient's chest and arms to be leathery and lacking sensation. The emergency room physician comments to an observing medical student that third-degree burns are present on the skin of these regions and that excision of the burn eschar (traumatized tissue) with subsequent skin grafting will be required.

Why is the area of second-degree burn red, blistered, and painful, while the third-degree burn is pale and insensate (without sensation, including pain)? Why will the chest and arms require skin grafting while the face and neck probably will not?

Hints: Think in terms of functions of the skin and survival of the germinal cells in functioning skin. Carefully examine figures 5.1 and 5.19.

The Integument as an Organ

The integument (skin) is the largest organ of the body. Together with its epidermal modifications (hair, glands, and nails), it constitutes the integumentary system. It has adaptive modifications in certain body areas that accommodate protective or metabolic functions. The integument is a dynamic interface between the continually changing external environment and the body's internal environment and helps to maintain homeostasis.

Objective 1. Explain why the integument is considered an organ and a component of the integumentary system.

Objective 2. Describe some common clinical conditions of the integument that result from nutritional deficiencies or body dysfunctions.

We are more aware of and concerned with our integumentary system than perhaps any other system of our body. One of the first things we do in the morning is examine ourselves in a mirror and assess how to make our appearance acceptable. Periodically, we examine our skin for wrinkles and our scalp for gray hairs as signs of aging. We recognize other people to a large extent by features of the skin.

Unfortunately, the appearance of the skin frequently determines first impressions and social acceptance. For example, social rejection during teenage years, imagined or real, can be directly associated with skin problems like acne. One's image of oneself and consequent social behavior may be closely associated with physical appearance.

Even clothing styles are somewhat determined by how much skin we, or the designers, want to expose. But our skin is much more than a showpiece; it protects and regulates structures within the body.

The skin, or *integument (in-teg'yoo-ment)*, and its associated structures (hair, glands, and nails) constitute the integumentary system. Included in this system are the millions of sensory receptors and a vascular network. The skin is a dynamic interface between the body and the external environment. It protects the body from the environment even as it allows for communication with the environment.

The skin is considered an organ, since it consists of several kinds of tissues that are structurally arranged to function together. It is the largest organ of the body, covering over 7,600 sq cm (3,000 sq in.) in the average adult, and accounts for approximately 7% of a person's body weight. The skin is of variable thickness, averaging between 1.0 and 2.0 mm. It is thickest on the parts of the body exposed to wear and abrasion, such as the soles of the feet and palms of the hand, where it is about 6 mm. It is thinnest on the eyelids, external genitalia, and tympanum (eardrum), where it is approximately 0.5 mm. Even the texture varies from the rough, callous skin covering the elbows and knuckles to the soft, sensitive areas of the eyelids, nipples, and genitalia.

The general appearance and condition of the skin are clinically important because they provide clues to certain body conditions or dysfunctions. Pale skin may indicate shock, whereas red, flushed, overwarm skin may indicate fever and infection. A rash may indicate allergies or local infections. Abnormal textures of the skin may be the result of glandular or nutritional problems (table 5.1). Even chewed fingernails may be a clue to emotional problems.

1. Explain why the skin is considered an organ and why the skin plus the integumentary derivatives is considered a system.
2. Which vitamins and minerals are important for healthy skin? (See table 5.1)
3. Describe the appearance of the skin that may accompany each of the following conditions: allergy; shock; infection; dry, stiff hair; hyperpigmentation; and general dermatitis.

Layers of the Integument

The integument consists of two principal layers. The outer epidermis is stratified into four or five structural and functional layers, and the thick and deeper dermis consists of two layers. The hypodermis (subcutaneous tissue) connects the skin to underlying organs.

integument: L. *integumentum*, a covering

Table 5.1	**Conditions of the Skin and Associated Structures Indicating Nutritional Deficiencies or Body Dysfunctions**	
Condition	**Deficiency**	**Comments**
General dermatitis	Zinc	Redness and itching
Scrotal or vulval dermatitis	Riboflavin	Inflammation in genital region
Hyperpigmentation	Vitamin B$_{12}$, folic acid, or starvation	Dark pigmentation on backs of hands and feet
Dry, stiff, brittle hair	Protein, calories, and other nutrients	Usually occurs in young children or infants
Follicular hyperkeratosis	Vitamin A, unsaturated fatty acids	Rough skin due to keratotic plugs from hair follicles
Pellagrous dermatitis	Niacin and tryptophan	Lesions on areas exposed to sun
Thickened skin at pressure points	Niacin	Noted at belt area at the hips
Spoon nails	Iron	Thin nails that are concave or spoon-shaped
Dry skin	Water or thyroid hormone	Dehydration, hypothyroidism, rough skin
Oily skin (acne)		Hyperactivity of sebaceous glands

Objective 3. Describe the histological characteristics of each layer of the integument.

Objective 4. Summarize the transitional events that occur within each of the epidermal layers.

Epidermis

The **epidermis** is the superficial protective layer of the skin and is composed of stratified squamous epithelium that varies in thickness from 0.007–0.12 mm. All but the deepest layers of the epidermis are composed of dead cells. The epidermis is composed of either four or five layers, depending on its location within the body (figs. 5.1 and 5.2). The epidermis of the palms and soles has five layers because these areas are exposed to greater friction. The epidermis of all other areas of the body has only four layers. The names and characteristics of the epidermal layers are as follows.

1. **Stratum basale** (basal layer). The stratum basale is composed of a single layer of cells in contact with the dermis. Four types of cells comprise the stratum basale: *keratinocytes (ker″uh-tin′o-sīts); melanocytes (mel′ah-no-sīts); tactile cells (Merkel cells);* and, *nonpigmented granular dendrocytes (Langerhans cells).* With the exception of tactile cells, these cells are constantly dividing mitotically and moving outward to renew the epidermis. It usually takes between 6 to 8 weeks for the cells to move from the stratum basale to the surface of the skin.

   Keratinocytes are specialized cells that produce **keratin** *(ker′ă-tin),* which toughens and waterproofs the skin. As keratinocytes are pushed away from the vascular nutrient and oxygen supply of the dermis, their nuclei degenerate, their cellular content is dominated by keratin, and the process of *keratinization* is completed. By the time that keratinocytes have reached the surface of the skin, they are scalelike, dead cells filled with keratin enclosed in loose cell membranes. **Melanocytes** are specialized epithelial cells that synthesize the pigment **melanin** *(mel′ă-nin),* providing a protective barrier to the ultraviolet radiation in sunlight. **Tactile cells** are sparse as compared to keratinocytes and melanocytes. These cells aid in tactile (touch) reception. **Nonpigmented granular dendrocytes** are scattered throughout the stratum basale. They are protective *macrophagic cells* that ingest bacteria and other foreign debris.

2. **Stratum spinosum** (spiny layer). The stratum spinosum contains several stratified layers of cells. The spiny appearance of this layer is due to the changed shape of the keratinocytes. Since there is limited mitosis in the stratum spinosum, this layer and the stratum basale are collectively referred to as the **stratum germinativum** *(jer-mih″nuh-tih′vum).*

3. **Stratum granulosum** (granular layer). This layer consists of only three or four flattened rows of cells. The cells within this layer appear granular due to the process of keratinization.

---

stratum: L. *stratum*, something spread out
basale: Gk. *basis*, base
keratinocyte: Gk. *keras*, hornlike; *kytos*, cell
melanocyte: Gk. *melas*, black; *kytos*, cell

macrophagic: Gk. *makros*, large; *phagein*, to eat
spinosum: L. *spina*, thorn
germinativum: L. *germinare*, sprout or growth
granulosum: L. *granum*, grain

UNIT IV

Figure 5.1 A diagram of the skin.

Figure 5.2 A photomicrograph of the epidermis (250×).

4. **Stratum lucidum** (clear layer). The nuclei, organelles, and cell membranes are no longer visible in the cells of the stratum lucidum, and so histologically this layer appears clear. It exists only in the lips and in the thickened skin of the soles and palms.

5. **Stratum corneum** (hornlike layer). The stratum corneum *(kor′ne-um)* is composed of 25 to 30 layers of flattened, scalelike cells, which are continuously shed as flakelike residues of cells. This surface layer is cornified and is the real protective layer of the skin (fig. 5.3). *Cornification* is brought on by keratinization, and the hardening, flattening process that takes place as the cells die and are pushed to the surface.

Friction at the surface of the skin stimulates additional mitotic activity of the stratum basale, resulting in the formation of a *callus* for additional protection.

The specific characteristics of each epidermal layer are described in table 5.2.

lucidum: L. *lucidus*, light
corneum: L. *corneus*, hornlike

Tattooing colors the skin permanently because pigmented dyes are injected below the mitotic basal layer into the dermis. Because of frequent nonsterile conditions, those who administer the tattoo may introduce infections along with the dye.

Coloration of the Skin Normal skin color is caused by the expression of a combination of three pigments: *melanin, carotene,* and *hemoglobin (he''mo-glo'bin)*. **Melanin** is a brown-black pigment produced in the melanocytes of the stratum basale (fig. 5.4). All races have virtually the same number of melanocytes, but the amount of melanin produced and the degree of granular aggregation of the melanin determine whether an individual's skin color is black, brown, red, tan, or white. Melanin is a protective device that guards against the damaging effect of the ultraviolet rays in sunlight. A gradual exposure to the sunlight promotes the increased production of melanin within the melanocytes, and hence tanning of the skin. The skin of a genetically determined albino has the normal complement of melanocytes in the epidermis but lacks the enzyme *tyrosinase,* that converts the amino acid tyrosine to melanin.

Figure 5.3 A scanning electron micrograph of the surface of the skin showing the opening of a sweat gland. The fragmented appearing particles are bacteria, which are present throughout the body on the surface of the skin.

Figure 5.4 Melanocytes throughout the stratum basale (see arrow) produce melanin.

UNIT IV

Table 5.2	Layers of the Epidermis

Stratum corneum
Consists of many layers of keratinized, dead cells that are flattened and nonnucleated; cornified

Stratum lucidum
A thin, clear layer found only in the epidermis of the palms and soles

Stratum granulosum
Composed of one or more layers of granular cells that contain fibers of keratin and shriveled nuclei

Stratum spinosum
Composed of several layers of cells with centrally located, large, oval nuclei and spinelike processes; limited mitosis

Stratum basale
Consists of a single layer of cuboidal cells that undergo mitosis; contains pigment-producing melanocytes

Figure 5.5 Basic dermatoglyphic patterns in the digits: (a) arch, (b) whorl, (c) loop, and (d) combination. Individual variations are commonly understood, but it is not known why there are basic differences between men and women or why the prints of people with certain mental disorders consistently deviate from normal patterns.

(a)

(b)

(c)

(d)

Other genetic expressions of melanocytes are more common than albinism. *Freckles,* for example, are caused by aggregated patches of melanin. A lack of melanocytes in localized areas of the skin causes distinct white spots in the condition called *vitiligo (vit-ĭ-li'gō).*

Excessive exposure to sunlight can cause *basal cell carcinoma,* or *malignant melanoma.* While in sunlight, one's skin absorbs two wavelengths of ultraviolet rays known as UV-A and UV-B. The DNA within the basal skin cells may be damaged as the sun's more dangerous UV-B rays penetrate the skin. Although it was once believed that UV-A rays were harmless, findings now indicate that excessive exposure to UV-A rays may inhibit the DNA repair process that follows exposure to UV-B. Therefore, individuals who are exposed solely to UV-A rays in tanning salons are still in danger of basal cell carcinoma, since they will later be exposed to UV-B rays of sunlight when they go outdoors.

Carotene is a yellowish pigment found in the epidermal cells and fatty parts of the dermis. Carotene is abundant in the skin of people of Asian descent and, together with melanin, accounts for their yellow-tan skin.

Hemoglobin is not a pigment of the skin; rather, it is the oxygen-binding pigment found in red blood cells. Oxygenated blood flowing through the dermis gives the skin its pinkish tones.

Certain physical conditions or diseases cause symptomatic discoloration of the skin. *Cyanosis* is a bluish discoloration of the skin that appears in people with certain cardiovascular or respiratory diseases. People also become cyanotic during an interruption of breathing. In *jaundice,* the skin appears yellowish because of an excess of bile pigment in the bloodstream. Jaundice is usually symptomatic of liver dysfunction and sometimes of liver immaturity, as in a jaundiced newborn. *Erythema* is a redness of the skin generally due to vascular trauma, such as from a sunburn.

Surface Patterns The exposed surface of the skin has recognizable patterns that are either congenital or acquired. Congenital patterns called *fingerprints,* or *friction ridges,* are present on the finger and toe pads as well as on the palms and soles. The designs formed by these lines have basic similarities but are not identical in any two individuals (fig. 5.5). They are

vitiligo: L. *vitiatio,* blemish
carotene: L. *carota,* carrot (referring to orange coloration)

hemoglobin: Gk. *haima,* blood; *globus,* globe
cyanosis: Gk. *kyanosis,* dark-blue color
jaundice: L. *galbus,* yellow
erythema: Gk. *erythros,* red; *haima,* blood

UNIT IV

formed by the pull of elastic fibers within the dermis and are well established prenatally. As the name implies, friction ridges function to prevent slippage when grasping objects. Because they are precise and easy to reproduce, fingerprints are customarily used for identifying individuals.

Acquired lines include the deep *flexion creases* on the palms and the shallow *flexion lines* that can be seen on the knuckles and on the surface of other joints. Furrows on the forehead and face are acquired from continual contraction of facial muscles, such as from smiling or squinting in bright light or against the wind. Facial lines become more strongly delineated as a person ages.

The science known as *dermatoglyphics* is concerned with the classification and identification of fingerprints. Every individual's prints are unique, including those of identical twins. Fingerprints, however, are not exclusive to humans. All other primates have fingerprints, and even dogs have a characteristic "nose print" that is used for identification in the military canine corps and in certain dog kennels.

Dermis

The **dermis** is deeper and thicker than the epidermis (see fig. 5.1). Blood vessels within the dermis nourish the living portion of the epidermis, and numerous collagenous, elastic, and reticular fibers give support to the skin. The fibers within the dermis radiate in definite directions and are responsible for producing lines of tension on the surface of the skin and providing skin tone (fig. 5.6). A young person has considerably more elastic fibers in the dermis than an elderly person. The decreasing amount of elastic fiber is apparently directly associated with aging. The dermis is highly vascular and glandular and contains many nerve endings and hair follicles.

Layers of the Dermis   The dermis is composed of two layers. The upper layer, called the **stratum papillarosum** (papillary layer), is in contact with the epidermis and accounts for about one-fifth of the entire dermis. Numerous projections, called *papillae,* extend from the upper portion of the dermis into the epidermis. Papillae form the base for the friction ridges on the fingers and toes.

The deeper and thicker layer of the dermis is called the **stratum reticularosum** (reticular layer). Fibers within this layer are more dense and regularly arranged to form a tough, flexible meshwork. It is quite distensible, as is evident in pregnant women or obese individuals, but it can be stretched too far, causing "tearing" of the dermis. The repair of a strained dermal area leaves a white streak called a stretch mark, or **linea albicans.** Lineae albicantes are frequently found on the buttocks, thighs, abdomen, and breasts (fig. 5.7).

papilla: L. *papula,* swelling or pimple

Figure 5.6 Natural lines of tension within the skin. These lines are caused by the pull of elastic and collagen fibers within the dermis and are of special interest to surgeons. Incisions made parallel to the lines of tension heal more rapidly and create less scar tissue than transverse incisions.

Creek

Figure 5.7 Stretch marks (lineae albicantes) on the abdomen of a pregnant woman. Stretch marks generally fade with time but frequently leave permanent integumentary markings.

UNIT IV

UNIT IV

It is the strong, resilient reticular layer of domestic mammals that is used in making leather and suede. In the tanning process, the hide of an animal is treated with various chemicals that cause the epidermis with its hair and the papillary layer of the dermis to separate from the underlying reticular layer. The reticular layer is then softened and treated with protective chemicals to make it usable.

Innervation of the Skin   The dermis of the skin has extensive innervation. Specialized integumentary *effectors* consist of muscles or glands within the dermis that respond to motor impulses transmitted from the central nervous system to the skin by autonomic nerve fibers.

Several types of **sensory receptors** respond to various tactile (touch), pressure, temperature, tickle, or pain sensations. Some are exposed nerve endings, some form a network around hair follicles, and some extend into the papillae of the dermis. Certain areas of the body, such as the palms, soles, lips, and external genitalia, have a greater concentration of sensory receptors and are therefore more sensitive to touch. Chapter 15 contains a detailed structural and functional account of the various sensory receptors.

Vascular Supply of the Skin   Blood vessels within the dermis supply nutrients to the mitotically active stratum basale of the epidermis and to the cellular structures of the dermis, such as glands and hair follicles. Dermal blood vessels play an important role in regulating body temperature and blood pressure. Autonomic vasoconstriction or vasodilation responses can either shunt the blood away from the superficial dermal arterioles or permit it to flow freely throughout dermal vessels. Fever or shock can be detected by the color and temperature of the skin. Blushing is the result of involuntary vasodilation of dermal blood vessels.

A healthy circulating blood flow in debilitated bedridden patients is important for the prevention of bedsores, or *decubitus ulcers*. When a person lies in one position for an extended period, the dermal blood flow is restricted where the body presses against the bed. As a consequence, cells die and open wounds develop (fig. 5.8). Changing the position of the patient frequently and periodically massaging the skin to stimulate blood flow help to keep the skin healthy.

Hypodermis   The **hypodermis,** or **subcutaneous tissue,** binds the dermis to underlying organs. The hypodermis is composed primarily of loose fibrous connective tissue and adipose cells interlaced with blood vessels (see fig. 5.1). Collagenous and elastic fibers reinforce the hypodermis—particularly on the palms and soles, where the skin is firmly attached to underlying

decubitus: L. *decumbere,* lie down
ulcer: L. *ulcus,* sore
hypodermis: Gk. *hypo,* under; *derma,* skin

Figure 5.8   A decubitus ulcer on the medial surface of the ankle. The most frequent sites for decubitus ulcers are in the skin overlying a bony projection, such as the hip, ankle, heel, shoulder, or elbow.

structures. The amount of adipose in the hypodermis varies with the sex, age, region of the body, and nutritional state of the individual. The hypodermis of females is generally about 8% thicker than that of males. The hypodermis also functions to store lipids, insulate and cushion the body, and regulate temperature.

The hypodermis is of clinical importance as an injection site for subcutaneous medication. Subcutaneous injections may be administered to patients who are unconscious or uncooperative, and when oral medications are not practical. Subcutaneous medications are now available that are slow release, low dosage, and long duration. These types of medications may regulate metabolism, inhibit ovulation (birth control), or mitigate pain. Subcutaneous devices for diabetics, which have a continuous release of small dosages of insulin, are currently available.

1. List the layers of the epidermis and dermis and explain how they differ in structure and function.
2. Describe the sequence of cellular replacement within the epidermis and the processes of keratinization and cornification.
3. How do both the dermis and hypodermis function in thermoregulation?
4. What two basic types of innervation are found within the dermis?

Functions of the Integument

The integument is an extremely dynamic organ. Not only does it protect the body from pathogens and external injury, it also plays a major role in maintaining body homeostasis.

Objective 5. Discuss the role of the integument in the protection of the body from disease and external injury, the regulation of body fluids and temperature, absorption, synthesis, sensory reception, and communication.

Physical Protection

The skin is a physical barrier to most microorganisms, water, and excessive ultraviolet (UV) light. Oily secretions onto the surface of the skin form an acidic (pH 4.0–6.8) protective film that waterproofs the body and retards the growth of most pathogens. The protein keratin in the epidermis also waterproofs the skin. Cornification toughens the mostly dead cells to withstand abrasion and the penetration of microorganisms. Upon exposure to UV light, the melanocytes are stimulated to synthesize melanin, which in turn absorbs and disperses sunlight. Surface friction causes the epidermis to thicken, as a protective response, by increasing mitosis in the cells of the strata basale and spinosum, resulting in the formation of a protective *callus*.

> Regardless of skin pigmentation, everyone is susceptible to skin cancer if his or her exposure to sunlight is sufficiently intense and continuous. There are an estimated 800,000 new cases of skin cancer yearly in the United States, and approximately 9,300 of these are diagnosed as the potentially life-threatening *melanoma* (cancer of melanocytes). Melanomas (see fig. 5.18) are usually termed malignant, as they may spread rapidly. Sunscreens are advised for people who must be in direct sunlight for long periods of time.

Hydroregulation

The thickened, keratinized, and cornified epidermis of the skin is adapted for continuous exposure to the air. In addition, the outer layers are dead and scalelike, and a protein-polysaccharide basement membrane adheres the stratum basale to the dermis. Human skin is virtually waterproof, protecting the body from desiccation (dehydration) on dry land, and even from water absorption when immersed in water.

Thermoregulation

The skin plays a crucial role in the regulation of body temperature. Body heat comes from cellular metabolism, particularly in muscle cells as they maintain tone, or a degree of tension. A normal body temperature of 37° C is maintained by the antagonistic effects of sweating and shivering, which involve feedback mechanisms. Excess heat is actually lost from the body in three ways, all involving the skin: (1) through radiation from dilated blood vessels, (2) through the excretion and evaporation of perspiration, and (3) through convection and the conduction of heat directly through the skin (fig. 5.9). Sweat

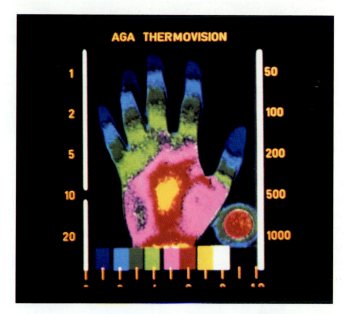

Figure 5.9 A thermogram of the hand showing differential heat radiation. Hair and body fat are good insulators. Red and yellow indicate the warmest parts of the body. Blue, green, and white indicate the coolest.

excretion increases approximately 100–150 ml/day for each 1° C elevation in body temperature. Up to 10 liters of sweat may be excreted to cool the body of a person doing hard physical work out-of-doors in the summertime.

> A serious danger of continued exposure to heat and excessive water and salt loss is *heat exhaustion*, characterized by nausea, weakness, dizziness, headache, and a decreased blood pressure. *Heat stroke* is similar to heat exhaustion, with the exception that sweating is inhibited (for reasons that are not clear) and body temperature rises. Convulsions, brain damage, and death may follow.

Excessive heat loss triggers a shivering response in muscles, which increases cellular metabolism. Not only do skeletal muscles contract, but tiny smooth muscles called **arrector pili** (ă-rek-tor pi'li), which are attached to hair follicles, are also contracted involuntarily and cause goose bumps.

> When the body's heat-producing mechanisms cannot keep pace with heat loss, *hypothermia* results. A lengthy exposure to temperatures below 20° C (68° F) and dampness may lead to this condition. This is why it is so important that a hiker, for example, dress appropriately for the weather conditions, especially on cool, rainy spring or fall days. The initial symptoms of hypothermia are numbness, paleness, delirium, and uncontrolled shivering. If the core temperature falls below 32° C (90° F), the heart loses its ability to pump blood and will go into fibrillation (erratic contractions). If the victim is not warmed, extreme drowsiness, coma, and death follow.

Figure 5.10 (*a*) A case of rickets in a child who lives in a village in Nepal, where the people reside in windowless huts. During the 5- to 6-month rainy season, the children are kept indoors. (*b*) A radiograph (X ray) of rickets in a 10-month-old child. Rickets develop from improper diets and also from lack of the UV light that is needed to synthesize vitamin D.

(a)

(b)

Cutaneous Absorption

Because of the effective protective barriers of the integument already described, cutaneous (through the skin) absorption is limited. Some gases, such as oxygen and carbon dioxide, may pass through the skin and enter the bloodstream. Small amounts of UV light, necessary for synthesis of vitamin D, are absorbed readily. Of clinical consideration is the fact that certain toxins and pesticides enter the body through cutaneous absorption.

Synthesis

The integumentary system synthesizes melanin and keratin, which remain in the skin, and vitamin D, which is used elsewhere in the body. The integumentary cells contain a compound called *dehydrocholesterol (de-hi''dro-kŏ-les'tah-rol)* from which they synthesize vitamin D in the presence of UV light. Only small amounts of UV light are necessary for vitamin D synthesis, but these amounts are very important to a growing child (fig. 5.10). Synthesized vitamin D enters the blood and helps regulate the metabolism of calcium and phosphorus, which are important in the development of strong and healthy bones. *Rickets* is a disease caused by vitamin D deficiency.

Sensory Reception

Highly specialized sensory receptors (see chapter 15) that respond to the precise stimuli of heat, cold, pressure, touch, vibration, and pain are located abundantly throughout the dermis and hypodermis of the integument. These receptors, referred to as **cutaneous receptors,** are abundant in the skin in parts of the face, the palms and fingers of the hands, the soles of the feet, and the genitalia. They are less abundant along the back and on the back of the neck and are sparse in the skin over joints, especially the elbow. Generally speaking, the thinner the skin, the greater the sensitivity.

Communication

Humans are highly social animals, and the integument plays an important role in communication. Various emotions, such as anger or embarrassment, may be reflected in changes of skin color. The contraction of specific facial muscles produces facial expressions that convey an array of emotions, including love, surprise, happiness, sadness, and despair. Secretions from certain integumentary glands frequently elicit subconscious responses from those that detect the odors.

---

1. List five modifications of the integument that are structurally or functionally protective.
2. Explain how the integument functions to regulate body fluids and temperature.
3. What substances are synthesized in the integument?

Figure 5.11   The difference between male and female in the expression of hair on the body.

Epidermal Derivatives

Hair, nails, and integumentary glands form from the epidermal layer, and are therefore of ectodermal derivation. Hair and nails are structural features of the integument and have a limited functional role. By contrast, integumentary glands are extremely important in body defense and maintenance of homeostasis.

Objective 6.   Describe the structure of hair and list the three principal types.

Objective 7.   Discuss the structure and function of nails.

Objective 8.   Compare and contrast the structure and function of the three principal kinds of integumentary glands.

Hair

Hair, or *pili,* is characteristic of all mammals, but its distribution, function, density, and texture varies across mammalian species. Humans are relatively hairless, with only the scalp, face, pubis, and axilla being densely haired. Men and women have about the same amount of hair on their bodies, but men generally appear to be more hirsute (fig. 5.11). Certain structures and regions of the body are hairless, such as the palms of the hands, soles of the feet, lips, nipples, penis, and labia minora.

Hirsutism (her'soo-tiz''em) is a condition of excessive body and facial hair, especially in women. It may be a genetic expression, as in certain ethnic groups, or occur as the result of a metabolic disorder, usually endocrine. Hirsutism occurs in some women as they experience hormonal changes during menopause. Various treatments for hirsutism include hormonal injections and electrolysis to permanently destroy selected hair follicles.

The primary function of hair is protection, even though its effectiveness is limited. Hair on the scalp and eyebrows protect against sunlight. The eyelashes and the hair in the nostrils protect against airborne particles. Hair on the scalp may also protect against mechanical injury. Some secondary functions of hair are to distinguish individuals and to serve as an ornamental sexual attractant.

Each hair consists of a diagonally positioned **shaft, root,** and **bulb** (fig. 5.12). The shaft is the visible, but dead, portion of the hair projecting above the surface of the skin. The bulb is the enlarged base of the root within the **hair follicle.** Each hair develops from stratum basale cells within the bulb of the hair, where nutrients are received from dermal blood vessels. As the cells divide, they are pushed away from the nutrient supply toward the surface, and cellular death and keratinization occur. In a healthy person, hair grows at the rate of approximately 1 millimeter every 3 days. As the hair becomes longer, however, it goes through a resting period during which it is anchored in its follicle.

The life span of a hair varies from 3 to 4 months for an eyelash to 3 to 4 years for a scalp hair. Each hair lost is replaced

pili: L. *pilus,* hair

hirsutism: L. *hirsutus,* shaggy

Figure 5.12 The structure of hair and the hair follicle. (a) A photomicrograph of the bulb and root of a hair within a hair follicle. (b) A scanning electron micrograph (280×) of a hair as it extends from a follicle. (c) A diagram of hair, a hair follicle, a sebaceous gland, and an arrector pili muscle.

(a)

Dermal papilla   Bulb   Hair follicle   Root

(b)

Hair shaft

Epidermis

Sebaceous gland

Hair follicle
External root sheath
Internal root sheath

Hair root

Hair bulb

Matrix

Dermal papilla

Arrector pili muscle

Blood vessels

(c)

Marshburn/Waldrop

by a new hair that grows from the base of the follicle and pushes the old hair out. Between 10 and 100 hairs are lost each day through replacement. Baldness results when hair is lost and not replaced. This condition may be disease-related, but it is generally inherited and most frequently occurs in males because of genetic influences combined with the action of the male sex hormone *testosterone (tes-tos′tĕ-rōn)*. No treatment is effective in reversing genetic baldness; however, flaps or plugs of skin containing healthy follicles from hairy parts of the body can be grafted to hairless regions.

Three layers can be observed in hair that is cut in cross section. An inner **medulla** *(mĕ-dul′ă)* is composed of loosely arranged cells separated by many air cells. The thick median layer called the **cortex** consists of hardened, tightly packed cells. A **cuticle** layer covers the cortex and forms the toughened outer portion of the hair. Cells of the cuticle have serrated edges that give a hair a scaly appearance when observed under a dissecting scope.

medulla: L. *medulla*, marrow
cortex: L. *cortex*, bark
cuticle: L. *cuticula*, small skin

People exposed to heavy metals, such as lead, mercury, arsenic, or cadmium, will have concentrations of these metals in their hair that are 10 times as great as those found in their blood or urine. Because of this, hair samples can be extremely important in certain diagnostic tests.

Even evidence of certain metabolic diseases or nutritional deficiencies may be detected in hair samples. For example, the hair of children with cystic fibrosis will be deficient in calcium and display excessive sodium. There is a deficiency of zinc in the hair of malnourished individuals.

Hair color is determined by the type and amount of pigment produced in the stratum basale at the base of the hair follicle. Varying amounts of melanin produce hair ranging in color from blond to brunette to black. The more abundant the melanin, the darker the hair. A pigment with an iron base (trichosiderin) produces red hair. Gray and white hair is the result of a lack of pigment production and air spaces within the layers of the shaft of the hair and generally accompanies aging. The texture of hair is determined by the cross-sectional shape; straight hair is round in cross section, wavy hair is oval, and kinky hair is flat.

Figure 5.13 A fingertip showing the associated structures of the nail. (*a*) A diagram and (*b*) a photomicrograph of the nail of a neonatal human (3.5×).

(a)

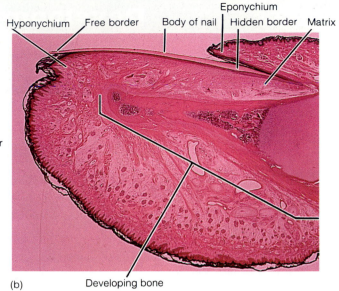

(b)

Sebaceous glands and specialized smooth muscles, called arrector pili, are attached to the hair follicle (fig. 5.12c). As mentioned previously, arrector pili muscles are involuntary, responding to thermal or psychological stimuli. When these muscles contract, the hair is pulled into a more vertical position, causing goose bumps.

Humans have three distinct kinds of hair:

1. **Lanugo.** Lanugo is a fine, silky, fetal hair that appears during the last trimester of development. Lanugo is usually not evident on the baby at birth unless the baby is born prematurely.
2. **Angora.** Angora hair grows continuously in length, as on the scalp of males and females, and on the face of males.
3. **Definitive.** Definitive hair grows to a certain length and then ceases growing. It is the most common type of hair. Eyelashes, eyebrows, pubic, and axillary hair are examples.

Anthropologists have referred to humans as the naked apes because of our relative hairlessness. The clothing that we wear over the exposed surface areas of our bodies functions to insulate and protect us, just as hair or fur does in other mammals. However, the nakedness of our skin does lead to some problems. *Skin cancer* occurs frequently in humans, particularly in regions of the skin exposed to the sun. Acne, another problem unique to humans, is partly related to the fact that hair is not present to dissipate the oily secretion from the sebaceous glands.

Nails

Nails are found on the distal dorsum of each of the fingers and toes. Both fingernails and toenails serve to protect the digits, and fingernails also aid in grasping and picking up small objects. Nails form from a hardened, transparent stratum corneum of the epidermis. The hardness of the nail is due to a dense, parallel arrangement of keratin fibrils between the cells.

Each nail consists of a **body, free border,** and **hidden border** (fig. 5.13). The platelike body of the nail rests on a **nail bed,** which is actually the stratum spinosum of the epidermis. The body and nail bed appear pinkish because of the underlying vascular tissue. The sides of the body of the nail are protected by a **nail fold,** and the furrow between the two is the **nail groove.** The free border is the distal exposed border, which is attached to the undersurface by the **hyponychium** (*hi″po-ni′ke-um*). The hidden border of the nail is attached proximally.

An **eponychium** (cuticle) covers the hidden border of the nail. The eponychium frequently splits, causing a hangnail. The growth areas of the nail are the **matrix** and the **lunula** (*loo′nyoo-lă*), the latter of which is the white half-moon-shaped area at the base of the nail. The nail grows by the transformation of the superficial cells of the matrix into nail cells. These harder, transparent cells are then pushed forward over the strata basale and spinosum of the nail bed. Fingernails grow at the rate of approximately one millimeter per week. The growth rate of toenails is somewhat slower.

The condition of a person's nails can be indicative of his or her general health and personality. Nails should appear pinkish, showing the rich vascular capillaries beneath the translucent nail. A yellowish hue may indicate certain glandular dysfunctions or nutritional deficiencies. Split nails may also be caused by nutritional deficiencies. A prominent bluish tint may indicate improper oxygenation of the blood. Spoon nails (concave body) may be the result of iron-deficiency anemia, and clubbing at the base of the nail may be caused by lung cancer. Dirty or ragged nails may indicate poor personal hygiene, and chewed nails may suggest emotional problems.

lanugo: L. *lana*, wool

hyponychium: Gk. *hypo*, under; *onyx*, nail
lunula: L. *lunula*, small moon

UNIT IV

Developmental Exposition

The Integumentary System

Exhibit I The development of the skin.

Periderm — Developing epidermis
Germinal layer —
Developing dermis
Mesenchyme —

Seven weeks

Periderm —
Hair papilla —
Melanoblast —
Developing collagenous and elastic fibers

Eleven weeks

Vernix caseosa —
Epidermis
Melanocyte —
Dermis

Waldrop

At birth

Exhibit II The development of hair and glands.

Developing epidermis — Germinal layer
Developing dermis
Hair bud — Bud of developing sweat gland

Twelve weeks

Developing sebaceous gland
Hair bulb —
Differentiated mesenchyme
Developing sweat gland

Fifteen weeks

Hair — Pores
Sebaceous gland
Arrector pili muscle
Sweat gland
Hair papilla — Blood vessels

Waldrop

At birth

Explanation

Both the ectodermal and mesodermal germ layers (see chapter 4) function in the formation of the structures of the integumentary system. The epidermis and the hair, glands, and nails of the skin develop from the ectodermal germ layer (exhibits, I, II, and III). The dermis develops from a thickened layer of undifferentiated mesoderm called *mesenchyme (mez'en-kīm'')*.

By 6 weeks the ectodermal layer has differentiated into an outer flattened *periderm* and an inner cuboidal *germinal (basal) layer* in contact with the mesenchyme. The periderm eventually sloughs off, forming the *vernix caseosa (ka''se-o'sah)*, a cheeselike protective coat that covers the skin of the fetus.

By 11 weeks, the mesenchymal cells below the germinal cells have differentiated into the distinct collagenous and elastic connective tissue fibers of the dermis. The tensile properties of these fibers cause a buckling of the epidermis and the formation of dermal papillae. During the early fetal period (about 10 weeks), specialized neural crest cells called *melanoblasts* migrate into the developing dermis and differentiate into *melanocytes*. The melanocytes soon migrate to the germinal layer of the epidermis, where they produce the pigment *melanin* which colors the epidermis.

Before hair can form, a *hair follicle* must be present. Each hair follicle begins to develop at about 12 weeks (exhibit II), as a mass of germinal cells called a *hair bud* proliferates into the underlying mesenchyme. As the hair bud becomes club-shaped,

Exhibit III The development of mammary glands. (a) Twelve weeks, (b) sixteen weeks, and (c) about twenty-eight weeks.

Developing epidermis

Developing dermis

Primary mammary bud

Twelve weeks

Mammary pit

Secondary mammary buds

Sixteen weeks

Depressed nipple

Areola

Lactiferous ducts

Lactiferous gland

Waldrop

Twenty-eight weeks

it is referred to as a *hair bulb*. The hair follicle, which physically supports and provides nourishment to the hair, is derived from specialized mesenchyme called the *hair papilla*, which is localized around the hair bulb, and from the epithelial cells of the hair bulb called the *hair matrix*. Continuous mitotic activity in the epithelial cells of the *hair bulb* results in the growth of the hair.

Sebaceous glands and sweat glands are the two principal types of integumentary glands. Both develop from the germinal layer of the epidermis (exhibit II). Sebaceous glands develop as proliferations from the sides of the developing hair follicle. Sweat glands become coiled as the secretory portion of the developing gland proliferates into the dermal mesenchyme. Mammary glands (exhibit III) are modified sweat glands that develop in the skin of the anterior thoracic region.

Glands

Although they originate in the epidermal layer, all of the glands of the skin are located in the dermis, where they receive physical support and nutritive sustenance. Glands of the skin are referred to as *exocrine*, since they either secrete or excrete substances through ducts. The glands of the skin are of three basic types: sebaceous, sudoriferous *(soo'dor-if'er-us)*, and ceruminous *(sĕ-roo'mi-nus)*.

Sebaceous   Sebaceous, or oil, glands are associated with hair follicles, since they develop from the follicular epithelium of the hair. They are simple, branched glands that are connected to hair follicles, where they secrete **sebum** onto the shaft of the hair (fig. 5.12). Sebum, which consists mainly of lipids, is dispersed along the shaft of the hair to the surface of the skin, where it lubricates and waterproofs the stratum corneum layer and also prevents the hair from becoming brittle. If the drainage pathway for sebaceous glands becomes blocked for some reason, the glands may become infected, resulting in acne. Sex hormones regulate the production and secretion of sebum, and hyperactivity of sebaceous glands can result in serious acne problems, particularly during teenage years.

Sudoriferous   Sudoriferous, or **sweat glands,** excrete perspiration, or sweat, onto the surface of the skin. Sweat glands are most numerous on the palms, soles, axillary and pubic regions, and the forehead. They are coiled and tubular shaped (fig. 5.14) and are of two types:

1. **Eccrine** *(ek'rin)* **sweat glands** are widely distributed over the body, especially on the forehead, back, palms, and soles. These glands are formed totally before birth and function in evaporative cooling (figs. 5.14 and 5.15).
2. **Apocrine** *(ap'ŏ-krin)* **sweat glands** are much larger, localized glands found in the axillary and pubic regions, where they secrete into hair follicles. Apocrine glands are not functional until puberty, and their odoriferous secretion is thought to act as a sexual attractant.

Perspiration is composed of water, salts, urea, uric acid, and traces of other elements. Perspiration is therefore valuable not only for evaporative cooling, but also for the excretion of certain wastes.

Mammary glands, found within the breasts, are specialized sudoriferous, or sweat, glands that secrete milk during lactation periods (see fig. 5.16). The breasts of the human female reach their greatest development during the childbearing years under the stimulus of pituitary and ovarian hormones.

sebum: L. *sebum*, tallow or grease
sudoriferous: L. *sudorifer*, sweat; *ferre*, to bear

UNIT IV

UNIT IV

Figure 5.14 Types of skin glands.

Figure 5.15 A photomicrograph of an eccrine sweat gland (100×). The coiled structure of the ductule portion of the gland (see arrows) accounts for its irregular appearance.

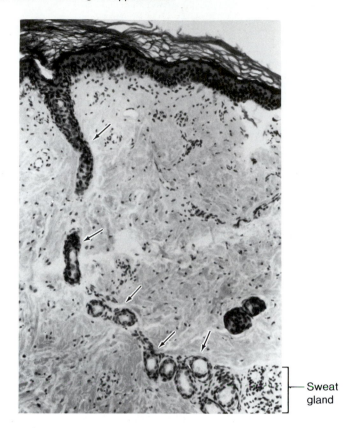

Figure 5.16 A mammary gland within the breast of a human female.

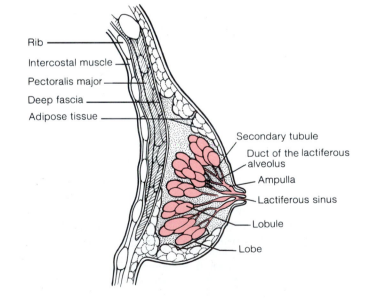

Ceruminous Glands These highly specialized glands are found only in the external auditory canal (ear canal). They secrete **cerumen** *(sĕ-roo'men),* or earwax, which is a water and insect repellent, and which also keeps the tympanum (eardrum) from drying out. Excessive amounts of cerumen may interfere with hearing.

Good routine hygiene is very important for health and social reasons. Washing away the dried residue of perspiration and sebum eliminates dirt. Excessive bathing, however, can wash off the natural sebum and dry the skin, causing it to itch or crack. The commercial lotions used for dry skin are, for the most part, refined and perfumed lanolin, which is sebum from sheep.

cerumen: L. *cera,* wax

1. Draw and label a hair. Indicate which portion is alive, and discuss what causes the cells in a hair to die. What are the three principal types of hair?
2. Describe the structure and function of nails.
3. List the types of integumentary glands and discuss the structure and function of each.
4. Are skin glands mesodermal or ectodermal in derivation? Are they epidermal or dermal in functional position?

Clinical Considerations

The skin is a buffer against the external environment and is therefore subject to a variety of disease-causing microorganisms and physical assaults. A few of the many diseases and disorders of the integumentary system are briefly discussed here.

Inflammatory (Dermatitis) Conditions

Inflammatory skin disorders are caused by immunologic hypersensitivity or infectious agents. Some people are allergic to certain foreign proteins and, because of this inherited predisposition, experience such hypersensitive reactions as asthma, hay fever, hives, drug and food allergies, and eczema. Eczema is characterized by redness, itching, and swollen vascular lesions that become dry, scaly, and crusted. **Lesions,** as applied to inflammatory conditions, are defined as more or less circumscribed pathologic changes in the tissue. Some of the more common inflammatory skin disorders and their usual sites are illustrated in figure 5.17.

There are also a number of *infectious diseases* of the skin, which is not surprising considering the highly social and communal animals that we are. Most of these diseases can now be prevented, but too frequently people fail to take appropriate precautionary measures. Infectious diseases of the skin include childhood viral infections (measles and chicken pox); bacteria, such as staphylococcus (impetigo); sexually transmitted diseases; leprosy; fungi (ringworm, athlete's foot, candida); and mites (scabies).

Neoplasms

Both benign and malignant neoplastic conditions or diseases are common in the skin. Pigmented moles (nevi), for example, are a type of benign neoplastic growth of melanocytes. Dermal cysts and benign viral infections are also common. Warts are virally caused abnormal growths of tissue that occur frequently on the hands and feet. A different type of wart, called a venereal wart, occurs in the anogenital region of sexual partners. Both types of warts are easy to treat by excision or various drugs. "Age spots" on elderly people, which appear as pigmented patches on the surface of the skin, are a benign growth of melanin pigmented basale cells. Usually no treatment is required, unless for cosmetic purposes.

---

neoplasm: Gk. *neo*, new; *plasma*, something formed
benign: L. *benignus*, good-natured
malignant: L. *malignus*, acting from malice

Skin cancer is the most common malignancy in the United States. Except for malignant melanomas (fig. 5.18), which arise from melanocytes, skin cancer is generally not life threatening. Excessive exposure to ultraviolet light from the sun is a known cause of skin cancer. The preferred treatment for this disease is complete surgical excision of the cancerous portion.

Burns

A burn is an epithelial injury caused by contact with thermal, radioactive, chemical, or electrical agents. Burns generally occur on the skin, but they can involve the linings of the respiratory and GI tracts. The extent and location of a burn is frequently less important than the degree to which it disrupts body homeostasis. Burns that have a **local effect** (local tissue destruction) are not as serious as those that have a **systemic effect.** Systemic effects directly or indirectly involve the entire body and are a threat to life. Possible systemic effects include body dehydration, shock, reduced circulation and urine production, and bacterial infections.

Burns are classified as first degree, second degree, or third degree based on their severity (fig. 5.19). In **first-degree burns,** the epidermal layers of the skin are damaged and symptoms are restricted to local effects, such as redness, pain, and edema (swelling). A shedding of the surface layers (desquamation) generally follows in a few days. A sunburn is an example. **Second-degree burns** involve both the epidermis and dermis. Blisters appear and recovery is usually complete, although slow. **Third-degree burns** destroy the entire thickness of the skin and frequently some of the underlying muscle. The skin appears waxy or charred and is insensitive to touch. As a result, ulcerating wounds develop, and the body attempts to heal itself by forming scar tissue. Skin grafts are frequently used to assist recovery.

As a way of estimating the extent of damaged skin suffered in burned patients, the *rule of nines* (fig. 5.20) is often applied. The surface area of the body is divided into regions, each of which accounts for about 9% (or a multiple of 9%) of the total skin surface. An estimation of the percentage of surface area damaged is important in treating with intravenous fluid, which replaces the fluids lost from tissue damage.

Frostbite

Frostbite is a local destruction of the skin resulting from freezing. Like burns, frostbite is classified by its degree of severity: first degree, second degree, and third degree. In **first-degree frostbite,** the skin will appear cyanotic (bluish) and swollen. Vesicle formation and hyperemia (engorgement with blood) are symptoms of **second-degree frostbite.** As the affected area is warmed, there will be further swelling, and the skin will redden and blister. In **third-degree frostbite,** there will be severe edema, some bleeding, and numbness followed by intense throbbing pain and necrosis of the affected tissue. Gangrene will follow untreated third-degree frostbite.

---

malignant: L. *malignus*, acting from malice

UNIT IV

Figure 5.17 Common inflammatory skin disorders and their usual site of occurrence.

Actinic keratosis
Spider angioma
Xanthelasma
Acne, seborrheic dermatitis, actinic keratosis
Squamous carcinoma
Impetigo
Tinea versicolor
Senile angioma
Fungal infection
Verruca vulgaris (warts)
Psoriasis
Fungal infection (between toes)
Creek

Psoriasis
Seborrheic dermatitis
Actinic keratosis
Acne
Seborrheic keratosis
Psoriasis
Actinic keratosis
Pruritus ani psoriasis
Contact dermatitis
Verruca vulgaris (warts)
Scabies
Eczema
Malignant melanoma
Verruca vulgaris (plantar warts)

Figure 5.18 Skin cancer.

Neoplasms (melanoma)

Figure 5.19 The classification of burns. (*a*) First-degree burns involve the epidermis and are characterized by redness, pain, and edema—such as with a sunburn; (*b*) second-degree burns involve the epidermis and dermis and are characterized by intense pain, redness, and blistering; and (*c*) third-degree burns destroy the entire skin and frequently expose the underlying organs. The skin is charred and numb and does not protect against fluid loss.

Loechel

Skin Grafts

If extensive areas of the stratum basale of the epidermis are destroyed in second-degree or third-degree burns or frostbite, new skin cannot grow back. In order for this type of wound to heal, a skin graft must be performed.

A **skin graft** is a segment of skin that has been excised from a *donor site* and transplanted to the *recipient site,* or *graft bed.*

An *autotransplant* is the most successful graft. This type of tissue transplant involves taking a thin sheet of healthy epidermis from a donor site of the burn or frostbite patient and moving it to the recipient site (fig. 5.21). A *heterotransplant* (between two different species) can serve as a temporary treatment to prevent infection and fluid loss.

Figure 5.20 The extent of burns, as estimated by the rule of nines. (a) Anterior and (b) posterior.

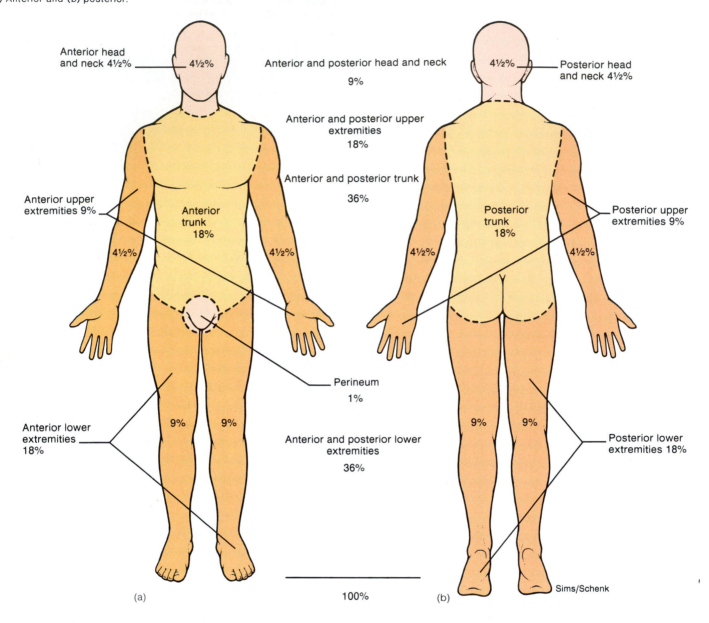

Anterior head and neck 4½%

4½%

Anterior and posterior head and neck 9%

4½%

Posterior head and neck 4½%

Anterior and posterior upper extremities 18%

Anterior and posterior trunk 36%

Anterior upper extremities 9%

Anterior trunk 18%

Posterior trunk 18%

Posterior upper extremities 9%

4½%

4½%

4½%

4½%

Perineum 1%

Anterior lower extremities 18%

9% 9%

Anterior and posterior lower extremities 36%

9% 9%

Posterior lower extremities 18%

(a)

100%

(b)

Sims/Schenk

Figure 5.21 A skin graft to the neck. (a) Traumatized skin is prepared for excision, (b) healthy skin from another body location is transplanted to graft site, and (c) one year following the successful transplant.

(a)

(b)

(c)

UNIT IV

Figure 5.22   Synthetic skin used in grafting.

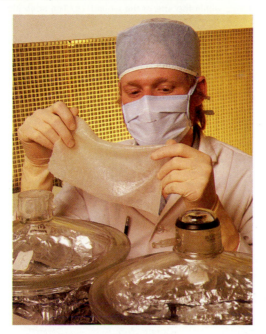

Synthetic skin fabricated from animal tissue bonded to a silicone film (fig. 5.22) may be used on a patient who is extensively burned. The completed process includes seeding the synthetic skin with basal skin cells obtained from healthy locations on the patient. This technique and treatment eliminates the problems of skin grafting, such as additional trauma, widespread scarring, and rejection, as in the case of skin obtained from a cadaver.

Wound Healing

The skin effectively protects against many abrasions, but if a wound does occur (fig. 5.23) a sequential chain of events promotes rapid healing. The process of wound healing depends on the extent and severity of the injury. Trauma to the epidermal layers stimulates an increased mitotic activity in the stratum basale, whereas injuries that extend to the dermis or subcutaneous layer elicit activity throughout the body, as well as within the wound itself. General body responses include a temporary elevation of temperature and pulse rate.

Figure 5.23   Various kinds of wounds. (*a*) Puncture, (*b*) abrasion, (*c*) laceration, and (*d*) avulsion.

(a)

(b)

(c)

(d)

Creek

UNIT IV

Figure 5.24 The process of wound healing. (*a*) A penetrating wound into the dermis ruptures blood vessels. (*b*) Blood cells, fibrinogen, and fibrin flow out of the wound. (*c*) Vessels constrict and a clot blocks the flow of blood. (*d*) A protective scab is formed from the clot, and granulation forms within the site of the wound. (*e*) The scab sloughs off as the epidermal layers are regenerated.

UNIT IV

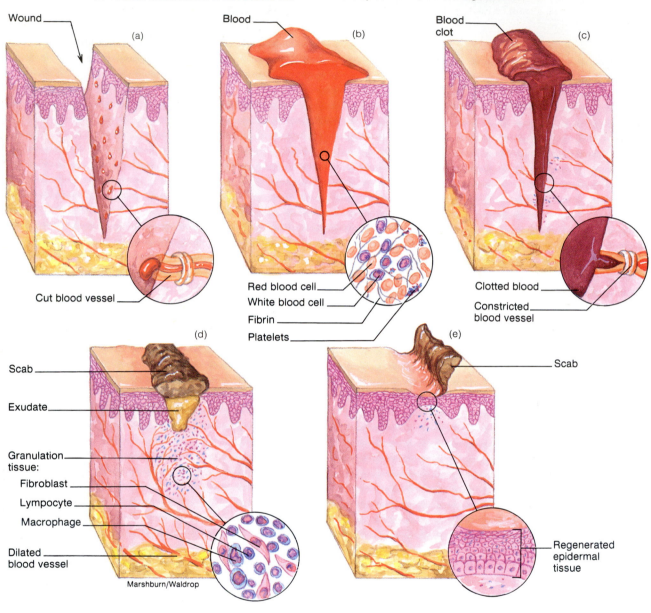

In an open wound (fig. 5.24), blood vessels are broken and bleeding occurs. Through the action of **blood platelets** (*plāt′letz*) and protein molecules, called **fibrinogen** (*fi-brin′ŏ-jen*), a clot forms and soon blocks the flow of blood. A scab forms and covers and protects the damaged area. Mechanisms are activated to destroy bacteria, dispose of dead or injured cells, and isolate the injured area. These responses are collectively referred to as *inflammation* and are characterized by redness, heat, edema, and pain. Inflammation is a response that confines the injury and promotes healing.

The next step in healing is the differentiation of binding **fibroblasts** from connective tissue at the wound margins. To-gether with new branches from surrounding blood vessels, **granulation tissue** is formed. Phagocytic cells migrate into the wound and ingest dead cells and foreign debris. Eventually the damaged area is repaired and the protective scab is sloughed off.

If the wound is severe enough, the granulation tissue may develop into **scar tissue** (fig. 5.25). Scar tissue differs from normal skin in that its collagen fibers are more dense and it has no stratified squamous or epidermal layer. Scar tissue has fewer blood vessels and may lack hair, glands, and sensory receptors. The closer the edges of a wound, the less granulation tissue develops and the less obvious a scar. This is one reason for suturing a large break in the skin.

Figure 5.25   Scars for body adornment on the face of this Buduma man from the islands of Lake Chad are created by instruments that make crescent-shaped incisions into the skin in beadlike patterns. Special ointments are applied to the cuts to retard healing and promote the forming of scars.

Figure 5.26   Aging of the skin results in a loss of elasticity and the appearance of wrinkles.

Aging of the Skin

As the skin ages, it becomes thin, dry, and inelastic. Collagenous fibers in the dermis become thicker and less elastic, and the amount of adipose tissue in the hypodermis diminishes, making it thinner. Skinfold measurements indicate that the diminution of the hypodermis begins at about the age of 45. With a loss of elasticity and a reduction in the thickness of the hypodermis, wrinkling, or the permanent infolding of the skin, becomes apparent (fig. 5.26).

During the aging of the skin, the number and activity of hair follicles, sweat glands, and sebaceous glands also diminishes. Consequently, there is a marked thinning of scalp hair and hair on the extremities, reduced sweating, and decreased sebum production. Since elderly people cannot perspire as freely, they are more likely to complain of heat and are more subject to heat exhaustion. They also become more sensitive to cold because of the loss of insulating adipose tissue and diminished circulation. A decrease in the production of sebum causes the skin to dry and crack frequently.

The integument of an elderly person is not as well protected from the sun because of thinning, and melanocytes that produce melanin gradually atrophy. The loss of melanocytes accounts for graying of the hair and pallor of the skin. After the age of 50, brown, plaquelike growths, called *seborrheic (seb''ō-re'ik) hyperkeratoses,* appear within the skin, particularly on exposed portions. Skin that has been exposed to excessive sunlight tends to develop more cutaneous carcinomas than less exposed skin.

UNIT IV

CLINICAL CASE STUDY ANSWER

The blistering and erythema characteristic of second-degree burns is a manifestation of intact and functioning blood vessels, which exist in abundance within the spared dermis. In third-degree burns, the entire dermis and its vasculature is destroyed, thus explaining the absence of these findings. Likewise, in third-degree burns, nerve endings and other nerve end organs that reside in the dermis are destroyed, resulting in a painless injury. In second-degree burns, however, significant numbers of these structures are spared and functional, thus preserving sensation, including pain. The third-degree burn areas will all require skin grafting in order to prevent infection, one of the skin's most vital functions. In second-degree burns, the spared dermis serves somewhat as a barrier to bacteria. Consequently, skin grafting is usually unnecessary, especially if sufficient numbers of skin adnexa (hair follicles, sweat glands, and so forth), which generally lie deep within the dermis, are spared. These structures serve as starting points for regeneration of surface epithelium and skin organs.

Internal Affairs

The Relational Workings of the Body Systems

Integumentary System

- Covers and protects the skeletal system
- Synthesizes vitamin D necessary for calcium absorption and metabolism

- Covers and protects the muscles
- Permits radiant and evaporative heat loss (sweating) during muscle contractions

- Houses cutaneous (tactile) receptors that convey sensory sensations to the brain

- Covers and protects certain endocrine glands

- Protects the circulatory system
- Maintains constant body temperature
- Prevents fluid loss (formation of scabs)

- Protects against pathogen invasion
- Prevents edema (retention of interstitial fluid)

- Protects organs of the upper respiratory tract

- Provides vitamin D necessary for calcium absorption

- Excretes salts and some nitrogenous wastes

- Major component of scrotum that covers and protects testes

Skeletal System
- Supports the skin
- Stores minerals needed by skin

Muscular System
- Generates body heat to warm the skin

Nervous System
- Provides autonomic motor impulses to cutaneous vessels and glands

Endocrine System
- Sex hormones cause integumentary features to change—pubic, axillary, and facial hair and oily skin

Circulatory System
- Transports O_2 and CO_2, nutrients, and fluids to and from the skin

Lymphatic System
- Maintains a balanced amount of interstitial fluid within the skin

Respiratory System
- Provides O_2 and eliminates CO_2

Digestive System
- Provides nutrients for growth, maintenance, and repair of skin

Urinary System
- Eliminates metabolic wastes
- Activates vitamin D

Reproductive System
- Gonads produce sex hormones that promote skin growth, maturation, and maintenance of the skin

Figure 5.27 The albino individual in this photograph has melanocytes within his skin, but as a result of a mutant gene he lacks the ability to synthesize melanin.

IMPORTANT CLINICAL TERMINOLOGY

acne An inflammatory condition of sebaceous glands. Acne is effected by gonadal hormones and is therefore common during puberty and adolescence. Pimples and blackheads on the face, chest, and back are expressions of this condition.

albinism A congenital, genetic deficiency of the pigment of the skin, hair, and eyes due to a metabolic block in the synthesis of melanin (fig. 5.27).

alopecia Loss of hair; baldness. Baldness is usually due to genetic factors and cannot be treated. Baldness may signify anatomical maturity.

athlete's foot (tinea pedis) A skin fungus disease of the foot.

blister A collection of fluid between the epidermis and dermis, caused by excessive friction or a burn.

boil (furuncle) A localized bacterial infection originating in a hair follicle or skin gland.

carbuncle A bacterial infection similar to a boil, except that it infects the subcutaneous tissues.

cold sore (fever blister) A lesion on the lip or oral mucous membrane, caused by type I herpes simplex virus (HSV), transmitted by oral or respiratory exposure.

comedo A plug of sebum and epithelial debris in the hair follicle and excretory duct of the sebaceous gland; also called a blackhead or whitehead.

corn A type of callus that is localized on the foot, usually over toe joints.

dandruff Common dandruff is the continual shedding of epidermal cells of the scalp. Common dandruff can be removed by normal washing and brushing of the hair. Abnormal dandruff may be caused by certain skin diseases, such as seborrhea or psoriasis.

decubitus ulcer A bedsore—an exposed ulcer caused by a continual pressure that restricts dermal blood flow to a localized portion of the skin (see fig. 5.8).

dermabrasion A procedure for removing tattoos or acne scars by high-speed sanding or scrubbing.

dermatitis An inflammation of the skin.

dermatology A specialty of medicine concerned with the study of the skin—its anatomy, physiology, histopathology, and the relationship of cutaneous lesions to systemic disease.

eczema *(eg'ze-mah)* A noncontagious inflammatory condition of the skin producing red, itching, vesicular lesions, which may be crusty or scaly.

erythema *(er''i-the'mah)* Redness of the skin caused by vasodilation from skin injury, infection, or inflammation.

furuncle A boil—a localized abscess resulting from an infected hair follicle.

gangrene Necrosis of tissue due to the obstruction of blood flow. It may be localized or extensive and may secondarily be infected with anaerobic microorganisms.

impetigo A contagious skin infection that results in lesions followed by scaly patches. It generally occurs on the face and is caused by staphylococci or streptococci.

keratosis Any abnormal growth and hardening of the stratum corneum layer of the skin.

melanoma A cancerous tumor originating from proliferating melanocytes within the epidermis of the skin.

nevus *(ne'vus)* A mole or birthmark—a congenital pigmentation of a certain area of the skin.

papilloma A benign epithelial neoplasm, such as a wart or a corn.

papule A small inflamed elevation of the skin, such as a pimple.

pruritus Itching. It may be symptomatic of systemic disorders but is generally due to dry skin.

psoriasis *(so-ri'ah-sis)* An inherited inflammatory skin disease, usually expressed as circular scaly patches of skin.

pustule A small, localized pus-filled elevation of the skin.

seborrhea *(seb''o-re'ah)* A disease characterized by an excessive activity of the sebaceous glands and accompanied by oily skin and dandruff.

urticaria (hives) A skin eruption of reddish weals, usually with extreme itching. It may be caused by an allergic reaction, stress, or contact with some external or internal precipitating factor.

wart A roughened projection of epidermal cells caused by a virus.

CHAPTER SUMMARY

The Integument as an Organ (p. 104)

1. The skin is considered an organ because it consists of several kinds of tissues that are structurally arranged to function together.
2. The appearance of the skin is clinically important because it provides clues to certain body conditions or dysfunctions.

Layers of the Integument
(pp. 104–110)

1. The stratified squamous epithelium of the epidermis is composed of five structural and functional layers: stratum basale, stratum spinosum, stratum granulosum, stratum lucidum, and stratum corneum.
   (a) Normal skin color is the result of a combination of melanin and carotene in the epidermis and hemoglobin in the blood of the dermis and hypodermis.
   (b) Fingerprints on the surface of the epidermis are individually unique; flexion creases and flexion lines are acquired.
2. The thick dermis of the skin is composed of fibrous connective tissue interlaced with elastic fibers. The two layers of the dermis are the upper papillary layer and the deeper reticular layer.
3. The hypodermis, composed of adipose and fibrous connective tissue, binds the dermis to underlying organs.

Functions of the Integument
(pp. 111–112)

1. Structural features of the skin protect the body from disease and external injury.
   (a) Keratin and an acidic oily secretion on the surface protect the skin from water and microorganisms.
   (b) Cornification of the skin protects against abrasion.
   (c) Melanin is a barrier to UV light.
2. The skin regulates body fluids and temperatures.
   (a) Fluid loss is minimal due to keratinization and cornification.
   (b) Temperature regulation is maintained by radiation, convection, and the antagonistic effects of sweating and shivering.
3. The skin permits the absorption of UV light, respiratory gases, steroids, fat-soluble vitamins, and certain toxins and pesticides.
4. The integument synthesizes melanin and keratin, which remain in the skin, and vitamin D, which is used elsewhere in the body.
5. Sensory reception in the skin is provided through cutaneous receptors throughout the dermis and hypodermis. Cutaneous receptors respond to precise sensory stimuli and are more sensitive in thin skin.
6. Certain emotions are reflected in changes in the skin.

Epidermal Derivatives
(pp. 113–119)

1. Hair is characteristic of all mammals, but its distribution, function, density, and texture varies across mammalian species.

(a) Each hair consists of a shaft, root, and bulb. The bulb is the enlarged base of the root within the hair follicle.
(b) The three layers of a hair shaft are the medulla, cortex, and cuticle.
(c) Lanugo, angora, and definitive are the three kinds of human hair.

2. Hardened, keratinized nails are found on the distal dorsum of each digit, where they protect the digits; fingernails aid in grasping and picking up small objects.
   (a) Each nail consists of a body, free border, and hidden border.
   (b) The hyponychium, eponychium, and nail fold support the nail on the nail bed.
3. Integumentary glands are exocrine, since they either secrete or excrete substances through ducts.
   (a) Sebaceous glands secrete sebum onto the shaft of the hair.
   (b) The types of sudoriferous (sweat) glands are eccrine and apocrine.
   (c) Mammary glands are specialized sudoriferous glands that secrete milk during lactation.
   (d) Ceruminous glands secrete cerumen (earwax).

REVIEW ACTIVITIES

Objective Questions

1. Hair, nails, integumentary glands, and the epidermis of the skin are derived from embryonic
   (a) ectoderm.   (c) endoderm.
   (b) mesoderm.   (d) mesenchyme.
2. Spoon-shaped nails may result when a person has a dietary deficiency of
   (a) zinc.   (c) niacin.
   (b) iron.   (d) vitamin B_{12}.
3. The epidermal layer *not* present in the thin skin of the face is the stratum
   (a) granulosum.   (c) spinosum.
   (b) lucidum.   (d) corneum.
4. Which of the following does *not* contribute to skin color?
   (a) dermal papillae
   (b) melanin
   (c) carotene
   (d) hemoglobin
5. Which of the following is *not* true of the epidermis?
   (a) It is composed of stratified squamous epithelium.
   (b) As the epidermal cells die, they undergo keratinization and cornification.
   (c) Rapid mitotic activity (cell division) within the stratum cornium accounts for the thickness of this epidermal layer.
   (d) In most areas of the body, the epidermis lacks blood vessels and nerves.

6. Integumentary glands that empty their secretions into hair follicles are
   (a) sebaceous glands.
   (b) sudoriferous glands.
   (c) eccrine glands.
   (d) ceruminous glands.
7. Fetal hair that is present during the last trimester of development is referred to as
   (a) angora.   (c) lanugo.
   (b) definitive.   (d) replacement.
8. Which of these conditions is potentially life threatening?
   (a) acne   (c) eczema
   (b) melanoma   (d) seborrhea
9. The skin of a burn victim has been severely damaged through the epidermis and into the dermis. Integumentary regeneration will be slow with some scarring, but it will be complete. Which kind of burn is this?
   (a) first degree   (c) third degree
   (b) second degree
10. The technical name for a blackhead or whitehead is
    (a) a carbuncle.   (c) a nevus.
    (b) a melanoma.   (d) a comedo.

Essay Questions

1. Discuss the development of the skin and its associated hair, glands, and nails. What role do the ectoderm and mesoderm play in integumentary development?
2. List the functions of the skin. Which of these functions occurs passively due to the structure of the skin, and which occurs dynamically due to physiological processes?
3. What are the types of tissues found in each of the three layers of skin?
4. Discuss the growth process and regeneration of the epidermis.
5. What are some physical and chemical features of the skin that make it an effective protective organ?
6. Of what practical value is it for the outer layers of the epidermis and hair to be composed of dead cells?
7. Define the following: *lines of tension, friction ridges,* and *flexion lines.* What causes each of these to develop?
8. Distinguish between a hair follicle and a hair. What other accessory structures are associated with a hair follicle and a hair?
9. Compare and contrast the structure and function of sebaceous, sudoriferous, mammary, and ceruminous glands.
10. Discuss what is meant by an inflammatory lesion. What are some frequent causes of skin lesions?

Gundy/Weber Software

The tutorial software accompanying Chapter 5, Integumentary System, is Volume 1, Introduction, Tissues, Integumentary System.

6

Skeletal System

Introduction and the Axial Skeleton

Chapter Outline

UNIT IV

CLINICAL CASE STUDY

A 68-year-old man visits his family doctor for his first physical examination in 30 years. The doctor, immediately sensing that the patient is a little disgruntled, gently tries to determine the reason. In response to the doctor's inquiry, the patient blurts out, "The nurse who measured my height is incompetent! I know for a fact I used to be six feet even when I was in the navy, and she tells me I'm 5'10''!" The doctor then performs the measurement himself, noting that although the patient's posture is excellent, he is indeed 5'10'', just as the nurse said. He explains that the spine contains some nonbony tissue, which shrivels up a bit over the years. The patient impatiently interrupts and states indignantly that he is aware of anatomic terms and principles and would like a detailed explanation. How would you explain the anatomy of the vertebral column and the changes it undergoes during the aging process?

Hints: The patient's normal posture and the fact that he has no complaints of pain indicate good health for his age. Examine figure 6.32 to understand the anatomy of the vertebral column and carefully read the accompanying discussion and the clinical considerations at the end of the chapter.

Organization of the Skeletal System

The axial and appendicular components of the skeletal system of an adult human consist of 206 individual bones arranged into a strong, flexible body framework.

Objective 1. Describe the structural organization of the skeletal system and list the bones of the axial and appendicular portions.

Osteology is the science concerned with the study of bones. Each bone is an organ that plays a part in the total functioning of the skeletal system. The skeletal system of an adult human is composed of approximately 206 bones. Actually, the number of bones differs from person to person depending on age and genetic variations. At birth, the skeleton consists of approximately 270 bones. As further bone development (ossification) occurs during infancy, the number increases. During adolescence, however, the number of bones decreases, as there is a gradual fusion of separate bones.

Some adults have extra bones within the sutures (joints) of the skull called **sutural,** or **wormian, bones.** Additional bones may develop in tendons in response to stress as the tendons repeatedly move across a joint. Bones formed this way are called **sesamoid** *(ses'ă-moid)* **bones.** Sesamoid bones, like the sutural bones, vary in number. The patellae (kneecaps) are two sesamoid bones all people have.

For the convenience of study, the skeleton is divided into the axial and appendicular portions. Anterior and posterior views of the skeleton are shown in figure 6.1. Listed in table 6.1 are the divisions of the skeleton and the number of bones in each portion.

The **axial skeleton** consists of the bones that form the axis of the body and support and protect the organs of the head, neck, and trunk:

1. **Skull.** The skull consists of two sets of bones: the cranial bones that form the cranium, or braincase, and the facial bones that support the eyes, nose, and jaws.
2. **Auditory ossicles.** Three auditory ossicles are present in the middle-ear chamber of each ear and serve to transmit sound impulses.
3. **Hyoid** *(hi'oid)* **bone.** The hyoid bone is located above the larynx and below the lower jaw. It supports the tongue and assists in swallowing.

---

sesamoid: Gk. *sesamon*, like a sesame seed
ossicle: L. *ossiculum*, little bone

Table 6.1	**Classification of the Bones of the Adult Skeleton**		
Axial Skeleton			**Appendicular Skeleton**

Axial Skeleton			Appendicular Skeleton	
Skull—22 Bones		**Auditory Ossicles**—6 Bones	**Pectoral Girdle**—4 Bones	
14 Facial Bones	*8 Cranial Bones*	malleus (2)	scapula (2)	
maxilla (2)	frontal (1)	incus (2)	clavicle (2)	
palatine (2)	parietal (2)	stapes (2)		
zygomatic (2)	occipital (1)	**Hyoid**—1 bone	**Upper Extremities**—60 Bones	
lacrimal (2)	temporal (2)	**Vertebral Column**—26	humerus (2)	carpal bones (16)
nasal (2)	sphenoid (1)	Bones	radius (2)	metacarpal bones (10)
vomer (1)	ethmoid (1)	cervical vertebra (7)	ulna (2)	phalanges (28)
inferior nasal concha		thoracic vertebra (12)	**Pelvic Girdle**—2 Bones	
(2)		lumbar vertebra (5)	os coxa (2) (each contains 3 fused bones)	
mandible (1)		sacrum (1) (5 fused bones)	**Lower Extremities**—60 Bones	
		coccyx (1) (3–5 fused bones)	femur (2)	tarsal bones (14)
		Rib Cage—25 Bones	tibia (2)	metatarsal bones (10)
		rib (24)	fibula (2)	phalanges (28)
		sternum (1)	patella (2)	

(a) an anterior view and (b) a
...on is distinguished with orange

Parietal
bone

Temporal
bone

Maxilla

Mandible

Occipital
bone

Clavicle

Scapula

Costal
cartilages

Humerus

Vertebral
column

Ulna

Os coxa

Sacrum

Coccyx

Radius

Carpal bones

Pelvic
girdle

Metacarpal
bones

Phalanges

Femur

Patella

Tibia

Fibula

Calcaneus

Tarsal bones

Metatarsal bones

Phalanges

Creek

(a)

(b)

UNIT IV

4. **Vertebral column.** The vertebral column (backbone) consists of 26 individual bones separated by cartilaginous intervertebral discs. In the pelvic region, several vertebrae are fused to form the sacrum, which provides attachment for the pelvic girdle. A few terminal vertebrae are fused to form the coccyx, the so-called tailbone.

5. **Rib cage.** The rib cage, or thoracic cage, forms the bony and cartilaginous framework of the thorax. The rib cage articulates posteriorly with the thoracic vertebrae and includes the 12 pairs of *ribs*, the flattened *sternum,* and the *costal cartilages,* which connect the ribs to the sternum on the anterior side.

The **appendicular skeleton** is composed of the bones of the upper and lower extremities and the bony girdles, which anchor the appendages to the axial skeleton:

1. **Pectoral girdle.** The paired *scapulae* and *clavicles* constitute the pectoral girdle. It is not a complete girdle, having only an anterior attachment to the axial skeleton at the sternum via the clavicles. The primary function of the pectoral girdle is to provide attachment for the muscles that move the brachium and forearm.

2. **Upper extremities.** Each upper extremity contains a proximal *humerus* within the brachium, an *ulna* and *radius* within the forearm, the *carpal bones,* the *metacarpal bones,* and the *phalanges* of the hand.

3. **Pelvic girdle.** The pelvic girdle is formed by two *ossa coxae* (hipbones), united anteriorly by the *symphysis (sim'fĭ-sis) pubis* and posteriorly by the sacrum of the vertebral column. The pelvic girdle supports the weight of the body through the vertebral column and protects the lower viscera within the pelvic cavity.

4. **Lower extremities.** Each lower extremity contains a proximal *femur* within the thigh, a *tibia* and *fibula* within the leg, the *tarsal bones,* the *metatarsal bones,* and the *phalanges* of the foot. In addition, the *patella (pah-tel'ă)* is located on the anterior surface of the knee joint between the thigh and leg.

---

1. List the bones of the body that you can palpate on yourself. Indicate which are bones of the axial skeleton and which are bones of the appendicular skeleton.
2. What are sesamoid bones and where are they found?
3. Describe the locations and functions of the pectoral and pelvic girdles.

Functions of the Skeletal System

The bones of the skeleton perform the mechanical functions of support, protection, and leverage for body movement and the metabolic functions of hemopoiesis and mineral storage.

Objective 2. Discuss the principal functions of the skeletal system and identify the body systems served by these functions.

The strength of bone comes from its inorganic components, which resist decomposition even after death. Much of what we know of prehistoric animals, including humans, has been determined from preserved skeletal remains. When we think of bone, we frequently think of a hard, dry structure. In fact, the term *skeleton* comes from a Greek word meaning "dried up." Living bone is not inert material, however; it is dynamic and adaptable in performing many body functions, including support, protection, leverage for body movement, hemopoiesis, and mineral storage.

1. **Support.** The skeleton forms a rigid framework to which the softer tissues and organs of the body are attached.

2. **Protection.** The skull and vertebral column enclose the central nervous system; the rib cage protects the heart, lungs, great vessels, liver, and spleen; and the pelvic cavity supports and protects the pelvic viscera. Even the site where red blood cells are produced is protected within the central, hollow portion of the bone.

3. **Body movement.** Bones serve as anchoring attachments for most skeletal muscles. In this capacity, the bones act as levers, with the joints functioning as pivots, when muscles contract to cause body movement.

4. **Hemopoiesis** *(he''mŏ-poi-e'sis).* The red bone marrow of an adult produces white blood cells, red blood cells, and platelets. In an infant, the spleen and liver produce red blood cells, but as the bones mature, the bone marrow assumes the performance of this formidable task. It is estimated that an average of 1 million red blood cells are produced every second by the bone marrow to replace those that are worn out and destroyed by the liver.

5. **Mineral storage.** The inorganic matrix of bone is composed primarily of the minerals calcium and phosphorus. These minerals give bone its rigidity and account for approximately two-thirds of the weight of bone. About 95% of the calcium and 90% of the phosphorus within the body are deposited in the bones and teeth. Although the concentration of these organic salts within the blood is kept within narrow limits, both of these mineral salts are essential for other body functions. Calcium is necessary for muscle contraction, blood clotting, and the movement of ions and nutrients across cell membranes. Phosphorus is required for the activities of the nucleic acids DNA and RNA, as well as for ATP utilization. If mineral salts are not present in the diet in sufficient amounts, they may be withdrawn from the bones until they are replenished through proper

nutrition. In addition to calcium and phosphorus, lesser amounts of magnesium and sodium salts are stored in bone tissue.

Vitamin D assists in the absorption of calcium and phosphorus from the small intestine into the blood. As the bones develop in a child, it is immensely important that the child's diet contains an adequate amount of these two minerals and vitamin D. If the diet is deficient in these essentials, the blood level falls below that necessary for calcification, and a condition known as *rickets* develops (see fig. 5.10). Rickets is characterized by soft bones that may result in bowlegs and malformation of the head, chest, and pelvic girdle.

In summary, the skeletal system is not an isolated body system. It is closely associated with the muscle system, since it stores the calcium needed for muscular contraction and provides an attachment for muscles as they span the movable joints. The skeletal system serves the circulatory system by producing blood cells in protected sites. Also, many of the vessels of the circulatory system are named according to the bones they parallel. The skeletal system supports and protects all of the systems of the body to varying degrees.

---

1. List the functions of the skeletal system.
2. Discuss two ways the skeletal system serves the circulatory system in the production of blood. In which two ways does it serve the muscular system?

Gross Structure of Bone

Each bone has a characteristic shape and diagnostic surface features that indicate its functional relationship to other bones, muscles, and to the body structure as a whole.

Objective 3. Classify bones according to their shapes and give an example of each type.

Objective 4. Describe the various markings on the surfaces of bones.

Objective 5. Describe the gross features of a typical long bone and list the functions of each structure.

The shape and surface features of each bone indicate its functional role in the skeleton (table 6.2). Bones that are long, for example, function as levers during body movement. Bones that support the body are massive and have large articular surfaces and processes for muscle attachment. Roughened areas on these bones may serve for the attachment of ligaments, tendons, or muscles. A flattened surface provides a placement site for a large muscle or may provide protection. Grooves around an articular end of a bone are where a tendon or nerve passes, and openings through a bone permit the passage of nerves or blood vessels.

Table 6.2	Surface Features of Bone
Structure	**Description and Example**
Articulating Surfaces	
condyle	A large, rounded, articulating knob (the occipital condyle of the occipital bone)
facet	A flattened or shallow articulating surface (the costal facet of a thoracic vertebra)
head	A prominent, rounded, articulating end of a bone (the head of the femur)
Depressions and Openings	
alveolus	A deep pit or socket (the alveoli for teeth in the maxilla)
fissure	A narrow, slitlike opening (the superior orbital fissure of the sphenoid bone)
foramen	A rounded opening through a bone (the foramen magnum of the occipital bone)
fossa	A flattened or shallow surface (the mandibular fossa of the temporal bone)
sinus	A cavity or hollow space in a bone (the frontal sinus of the frontal bone)
sulcus	A groove that accommodates a vessel, nerve, or tendon (the intertubercular sulcus of the humerus)
Nonarticulating Prominences	
crest	A narrow, ridgelike projection (the iliac crest of the os coxa)
epicondyle	A projection above a condyle (the medial epicondyle of the femur)
process	Any marked bony prominence (the mastoid process of the temporal bone)
spine	A sharp, slender process (the spine of the scapula)
trochanter	A massive process found only on the femur (the greater trochanter of the femur)
tubercle	A small rounded process (the greater tubercle of the humerus)
tuberosity	A large roughened process (the radial tuberosity of the radius)

Shapes of Bones

The bones of the skeleton are classified into four principal types on the basis of shape rather than size. The four classes are long bones, short bones, flat bones, and irregular bones (fig. 6.2).

1. **Long bones.** Long bones are longer than they are wide and function as levers. Most of the bones of the upper and lower extremities are of this type (e.g., the humerus, radius, ulna, metacarpal bones, femur, tibia, fibula, metatarsal bones, and phalanges).
2. **Short bones.** Short bones are somewhat cube-shaped and are found in confined spaces where they transfer forces (e.g., the wrist and ankle).

---

facet: Fr. *facette*, little face
trochanter: Gk. *trochanter*, runner
tuberosity: L. *tuberosus*, lump

UNIT IV

Figure 6.2 Shapes of bones. (*a*) A long bone, (*b*) a short bone, (*c*) a flat bone, and (*d*) an irregular bone.

(b)

(c)

(a)

(d)

Creek

Figure 6.3 A section through the skull showing diploe. Diploe is a layer of spongy bone sandwiched between two surface layers of compact bone. It is extremely strong yet light in weight.

Creek

Outer compact bone

Suture

Inner compact bone

Spongy bone

3. **Flat bones.** Flat bones have a broad, dense surface for muscle attachment or protection of underlying organs (e.g., the cranium, ribs, and bones of the shoulder girdle).
4. **Irregular bones.** Irregular bones have varied shapes and have many surface markings for muscle attachment or articulation (e.g., the vertebrae and certain bones of the skull).

Gross Anatomy of a Long Bone

Bone (osseous) tissue is organized as **spongy (cancellous) bone** or **compact (dense) bone.** Most bones have both types. In a flat bone of the skull, for example, the spongy bone is sandwiched between the compact bone and is called a *diploe* (*dip'lo-e*) (fig. 6.3). Because of this protective layering of bone tissue, a blow to the head may fracture the outer compact bone layer without harming the inner compact bone layer and the brain.

In a long bone from an appendage, the bone shaft, or **diaphysis** (*di-af'ĭ-sis*), consists of compact bone forming a cylinder that surrounds a central cavity called the **medullary cavity** (fig. 6.4). The medullary cavity is lined with a thin layer of connective tissue called the **endosteum** (*en-dos'te-um*) and contains **yellow bone marrow,** so named because of the large amounts of fat it contains. On each end of the diaphysis is an **epiphysis** (*ē-pif'ĭ-sis*), consisting of cancellous bone surrounded by a layer of compact bone. **Red bone marrow** is found within the porous chambers of spongy bone. In an adult, erythropoiesis, the production of red blood cells, occurs in the red bone marrow, especially that of the sternum, vertebrae, portions of the ossa coxae, and the proximal epiphyses of the femora and humeri. The red bone marrow is also responsible for the formation of white blood cells and platelets and for the phagocytosis of worn-out red blood cells. **Articular cartilage,** which is composed of thin hyaline cartilage, caps each epiphysis and facilitates joint movement. Along the diaphysis are **nutrient foramina,** small openings into the bone that allow for passage of nutrient vessels into the bone for nourishment of the living tissue.

Between the diaphysis and epiphysis is an **epiphyseal** (*ep"ĭ-fiz'e-al*) **plate** of cartilage—a region of mitotic activity that is

diploe: Gk. *diplous,* double

diaphysis: Gk. *dia,* throughout; *physis,* growth
epiphysis: Gk. *epi,* upon; *physis,* growth

Figure 6.4 A diagram of a long bone shown in longitudinal section.

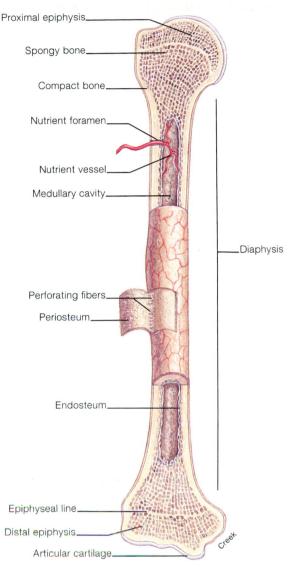

- Proximal epiphysis
- Spongy bone
- Compact bone
- Nutrient foramen
- Nutrient vessel
- Medullary cavity
- Diaphysis
- Perforating fibers
- Periosteum
- Endosteum
- Epiphyseal line
- Distal epiphysis
- Articular cartilage

Creek

responsible for *linear bone growth*. As bone growth is completed, an **epiphyseal line** replaces the plates and ossification occurs between the epiphysis and the diaphysis. A **periosteum** *(per″e-os′te-um)* of dense regular connective tissue covers the surface of the bone, except at the articulating surfaces. This highly vascular layer serves as a place for a tendon-muscle attachment and is responsible for *appositional* (width) *bone growth*. The periosteum is secured to the bone by **perforating** (Sharpey's) **fibers** (fig. 6.4) composed of bundles of collagenous fibers.

> Fracture of a long bone in a young person may be especially serious if it results in displacement of an epiphyseal plate. If such an injury is untreated, or treated improperly, longitudinal growth of the bone may be arrested or retarded, resulting in permanent shortening of the limb.

periosteum: Gk. *peri*, around; *osteon*, bone
Sharpey's fibers: from William Sharpey, Scottish physiologist and
    histologist, 1802–80

1. Using examples, discuss the function of each of the four kinds of bones as determined by shape.
2. Define each of the following surface markings on bones: *condyle, head, facet, process, crest, epicondyle, fossa, alveolus, foramen,* and *sinus.*
3. Diagram a sagittal view of a typical long bone and label the diaphysis, medullary cavity, epiphyses, articular cartilages, nutrient foramen, periosteum, and epiphyseal plates. Explain the function of each of these structures.

Bone Tissue

Bone tissue is composed of several types of bone cells embedded in a matrix of ground substance, inorganic salts, and collagenous fibers. Bone cells and ground substance give bone flexibility and strength; the inorganic salts give it hardness.

Objective 6. Identify the five types of bone cells and list the functions of each.

Objective 7. Distinguish between spongy and compact bone tissues.

Bone Cells

The bone cells of which bone tissue is composed are of five principal types. **Osteogenic** *(os″te-ŏ-jen′ik)* **cells** are found in the bone tissues in contact with the endosteum and the periosteum. These cells respond to trauma, such as a fracture, by giving rise to bone-forming cells *(osteoblasts)* and bone-destroying cells *(osteoclasts)*. **Osteoblasts** *(os′te-ŏ-blasts″)* are bone-forming cells (fig. 6.5) that synthesize and secrete unmineralized ground substance. They are abundant in areas of high metabolism within bone, such as under the periosteum and bordering the medullary cavity. **Osteocytes** *(os′te-ŏ-sītz″)* are mature bone cells (figs. 6.5 and 6.8) derived from osteoblasts that have secreted bone tissue around themselves. Osteocytes maintain healthy bone tissue by secreting enzymes and influencing bone mineral content. They also regulate the calcium release from bone tissue to blood. **Osteoclasts** *(os′te-ŏ-klasts″)* are large, multinuclear cells (fig. 6.5) that enzymatically break down bone tissue. These cells are important in bone growth, remodeling, and healing. **Bone-lining cells** are derived from osteoblasts along the surface of most bones in the adult skeleton. These cells are thought to regulate the movement of calcium and phosphate into and out of bone matrix.

Spongy and Compact Bone Tissues

Both spongy and compact bone tissues are present in most bones (fig. 6.6). **Spongy bone tissue** is deeply located compared to the compact bone tissue, and is quite porous. Minute spikes of bone tissue, called **trabeculae,** give spongy bone a latticelike appearance. Spongy bone is highly vascular and provides great strength to bone with minimal weight.

osteoblast: Gk. *osteon*, bone; *blastos*, offspring or germ
osteoclast: Gk. *osteon*, bone; *klastos*, broken

UNIT IV

UNIT IV

Figure 6.5 Types of bone cells. (*a*) Osteoblasts are important in secreting unmineralized ground substance. Osteocytes derive from osteoblasts and play a regulatory role in maintaining bone tissue.

(*b*) Osteoclasts are bone-destroying cells that help to maintain the dynamic state of bone tissue.

Figure 6.6 Compact bone tissue. (*a*) A diagram of the femur showing a cut through the compact bone into the medullary cavity. (*b*) The arrangement of the osteons within the diaphysis of the bone.

(*c*) An enlarged view of an osteon showing the osteocytes within lacunae and the concentric lamellae. (*d*) An osteocyte within a lacuna.

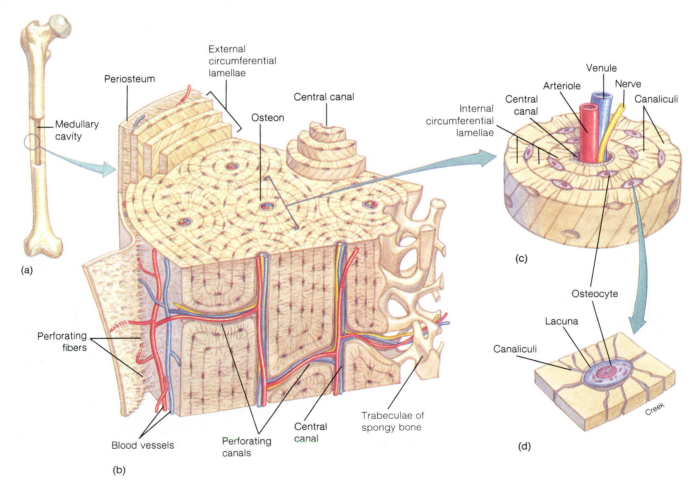

Figure 6.7 Bone tissue as seen in (a) a scanning electron micrograph and (b) a photomicrograph. The lacuna (*La*) provide spaces for the osteocytes, which are connected to one another by a canaliculus (*Ca*). Note the divisions between concentric lamellae (arrows). (*a* from: R. G. Kessel and R. H Kardon, *Tissues and Organs: A Text Atlas of Scanning Electron Microscopy,* © 1979 W. H. Freeman and Company.)

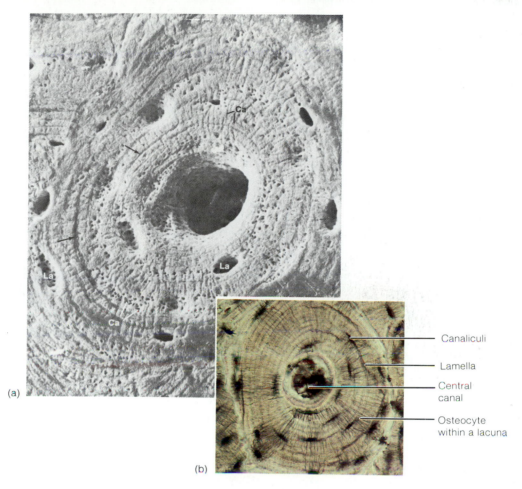

(a)

(b)

Canaliculi

Lamella

Central canal

Osteocyte within a lacuna

UNIT IV

Compact bone tissue is superficial to the spongy bone tissue and is very hard and dense. It consists of precise arrangements of microscopic cylindrical structures oriented parallel to the long axis of the bone (fig. 6.6). These columnlike structures are the **osteons** (haversian systems) of the bone tissue. The matrix of an osteon is laid down in concentric rings, called **lamellae,** around a **central** (haversian) **canal** (fig. 6.7). The central canal contains minute nutrient vessels and a nerve. Osteocytes within spaces called **lacunae** (fig. 6.8) are regularly arranged between the lamellae. The lacunae are connected by **canaliculi,** through which nutrients diffuse. Metabolic activity within bone tissue occurs at the osteon level. Between osteons there are incomplete remnants of osteons, called **interstitial systems. Perforating** (Volkmann's) **canals** penetrate compact bone, connecting osteons with blood vessels and nerves.

haversian system: from Clopton Havers, English anatomist, 1650–1702
Volkmann's canal: from Alfred Volkmann, German physiologist, 1800–1877

Figure 6.8 A photomicrograph of an osteocyte within a lacuna.

Canaliculi   Osteocyte   Lacuna

Figure 6.9 The process of endochondral ossification, beginning with (*a*) the cartilaginous model as it occurs in an embryo at 6 weeks. The bone develops (*b–e*) through intermediate stages to (*f*) adult bone.

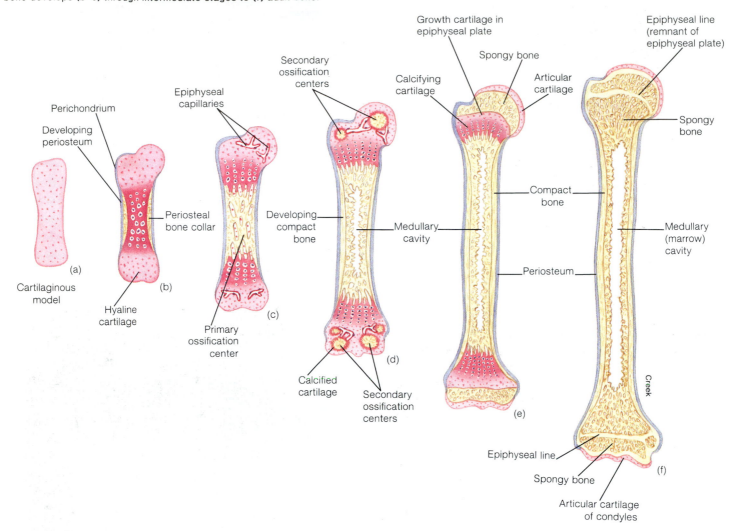

UNIT IV

1. Construct a simple table listing the location and function of each of the types of cells found within bone tissue.
2. Define an osteon and sketch the arrangement of osteons within compact bone tissue.

Bone Growth

The development of bone from embryonic to adult size depends on the orderly processes of mitotic divisions, growth, and the structural remodeling determined by genetics, hormonal secretions, and nutritional supply.

Objective 8. Describe the process of endochondral ossification as related to bone growth.

In most bone development, a cartilaginous model is gradually replaced by bone tissue during *endochondral bone formation.* As the cartilage model grows, the *chondrocytes* (cartilage cells) in the center of the shaft hypertrophy, and minerals are deposited within the matrix in a process called *calcification* (fig. 6.9). Calcification restricts the passage of nutrients to the chondrocytes, causing them to die. At the same time, some cells of the perichondrium (dense regular connective tissue surrounding cartilage) differentiate into *osteoblasts* (primordial bone cells), which secrete **osteoid,** the organic component of bone. As the perichondrium calcifies, it gives rise to a thin plate of compact bone called the **periosteal bone collar.** The periosteal bone collar is surrounded by the periosteum.

A **periosteal bud,** consisting of osteoblasts and blood vessels, invades the disintegrating center of the cartilage model from the periosteum. Once in the center, the osteoblasts secrete osteoid, thus establishing a **primary ossification center** from which ossification expands into deteriorating cartilage. This process is repeated in both the proximal and distal epiphyses, forming **secondary ossification centers** where spongy bone develops.

Once the secondary ossification centers are formed, bone tissue totally replaces cartilage tissue, except at the articular ends

Figure 6.10 A photomicrograph from an epiphyseal plate (63×).

Epiphyseal border

Reserve zone

Proliferation zone

Hypertrophic zone — Chondrocytes

Resorption zone

Ossification zone — Bone

— Red bone marrow

Diaphyseal border

Table 6.3	Average Age of Completion of Bone Ossification	
Bone	**Chronological Age of Fusion**
Scapula	18–20
Clavicle	23–31
Bones of upper extremity (brachium, forearm, hand)	17–20
Os coxa	18–23
Bones of lower extremity (thigh, leg, foot)	18–22
Vertebra	25
Sacrum	23–25
Sternum (body)	23
Sternum (manubrium, xiphoid)	30+

of the bone and at the epiphyseal plates. An epiphyseal plate contains five histological zones (fig. 6.10). The *reserve zone* borders the epiphysis and consists of small chondrocytes irregularly dispersed throughout the intercellular matrix. The chondrocytes in this zone anchor the epiphyseal plate to the bony epiphysis. The *proliferating zone* consists of larger, regularly arranged chondrocytes that are constantly dividing. The *hypertrophic zone* consists of very large chondrocytes that are arranged in columns. The linear growth of long bones is due to the cellular proliferation at the proliferating zone and the growth and maturation of these new cells within the hypertrophic zone. The *resorption zone* is the area where a change in mineral content is occurring. The *ossification zone* is a region of transformation from cartilage tissue to bone tissue. The chondrocytes within this zone die because the intercellular matrix surrounding them becomes calcified. Osteoclasts then break down the calcified matrix and the area is invaded by osteoblasts and capillaries from the bone tissue of the diaphysis. As the osteoblasts mature, osteoid is secreted and bone tissue is formed. The result of this process is a gradual increase in the length of the bone at the epiphyseal plates.

The time at which epiphyseal plates ossify varies widely from bone to bone, but generally it occurs between the ages of 18 and 20 within the long bones of most people (table 6.3). Because ossification of the epiphyseal cartilages within each bone occurs at predictable periods of development, the ages of people who are still growing can be determined by examining radiographs of their bones (fig. 6.11). Large discrepancies between bone age and chronological age may indicate a genetic or endocrine dysfunction.

Bone remodeling is a continual process throughout a person's life. The diagnostic processes on the surface of bones de-

Figure 6.11 The presence of epiphyseal plates, as seen in a radiograph of a child's hand, indicates that bones are still growing in length.

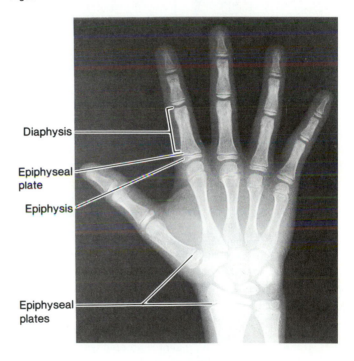

Diaphysis

Epiphyseal plate

Epiphysis

Epiphyseal plates

velop as stress is applied to the periosteum, resulting in the osteoblastic secretion of osteoid and the formation of new bone tissue. The greater trochanter of the femur, for example, develops in response to forces of stress applied to the periosteum of the bone where the tendons of muscles attach (fig. 6.12). These processes may continue to change somewhat in people who engage in rigorous physical activity, even though they have stopped growing in height.

As new bone layers are deposited on the outside surface of the bone, osteoclasts dissolve bone tissue adjacent to the medullary cavity. In this way, the size of the cavity keeps pace with the increased growth of the bone.

UNIT IV

Figure 6.12 A longitudinal section of the proximal end of a femur showing stress lines.

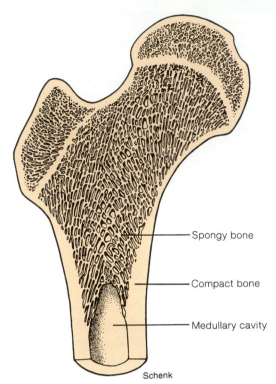

Spongy bone

Compact bone

Medullary cavity

Schenk

Bone is highly dynamic and is continually being remodeled in response to mechanical stress or even the absence of stress. The effect of the absence of stress can best be seen in the bones of bedridden or paralyzed individuals. Radiographic examination shows the loss of bone mass, or even osteoporosis. The absence of gravity that accompanies space flight may result in mineral loss from bones if an exercise regimen is not followed.

The movement of teeth in orthodontics involves bone remodeling. The teeth sockets (alveoli) are reshaped through the activity of osteoclast and osteoblast cells as stress is applied through the application of braces. The use of traction in treating certain skeletal disorders has a similar effect.

1. List the zones that are present in an epiphyseal plate and briefly describe the characteristics of each.
2. Explain the function of osteoblasts and osteoclasts in endochondral ossification and bone growth.

Skull

The human skull, consisting of 8 cranial and 14 facial bones, contains several cavities that house the brain and sensory organs. Each bone of the skull articulates with the adjacent bones and has diagnostic and functional processes, surface features, and foramina.

Objective 9. List the fontanels and discuss their functions.

Objective 10. Identify the cranial and facial bones of the skull and describe their structural characteristics.

Objective 11. Describe the location of each of the bones of the skull and identify the articulations that affix them to each other.

The skull consists of **cranial bones** and **facial bones.** The 8 bones of the cranium articulate firmly with one another to enclose and protect the brain and sensory organs. The 14 facial bones form the framework for the facial region and support the teeth. Variation in shape and density of the facial bones is a major contributor to the individuality of each human face. The facial bones, with the exception of the bone within the lower jaw, are also firmly interlocked with one another and the cranial bones.

The skull has several cavities. The **cranial cavity** is the largest, with a capacity of about 1,300–1,350 cc. The **nasal cavity** is formed by both cranial and facial bones and is partitioned into two chambers, or **nasal fossae,** by a **nasal septum** of bone and cartilage. Four sets of **paranasal sinuses** are located within the bones surrounding the nasal area and communicate via ducts into the nasal cavity. **Middle-** and **inner-ear cavities** are positioned inferior to the cranial cavity and house the organs of hearing and balance. The two **orbits** for the eyeballs are formed by facial and cranial bones. The **oral,** or **buccal, cavity** (mouth), which is only partially formed by bone, is completely within the facial region.

During fetal development and infancy, the bones of the cranium covering the brain are separated by fibrous sutures. There are also six large membranous areas of the skull that provide spaces between the developing bones (fig. 6.13). These areas are called **fontanels** (fon''tă-nelz) (soft spots) and permit the skull to undergo changes of shape, called *molding,* during parturition (childbirth). The fontanels also allow for rapid growth of the brain during infancy. Ossification of the fontanels normally has been completed by 20 to 24 months of age. A description of the six fontanels follows.

1. **Anterior (frontal) fontanel.** The anterior fontanel is diamond-shaped and is the most prominent of the six. It is located on the anteromedian portion of the skull.
2. **Posterior (occipital) fontanel.** The posterior fontanel is positioned at the back of the skull on the median line.
3. **Anterolateral fontanels.** The paired anterolateral fontanels are found on both sides of the skull, directly lateral to the anterior fontanel.
4. **Posterolateral fontanels.** The paired posterolateral fontanels are located on the posterolateral sides of the skull.

fontanel: Fr. *fontaine,* little fountain

UNIT IV

Figure 6.13   The fetal skull showing the six fontanels and the sutures. (*a*) A right lateral view and (*b*) a superior view.

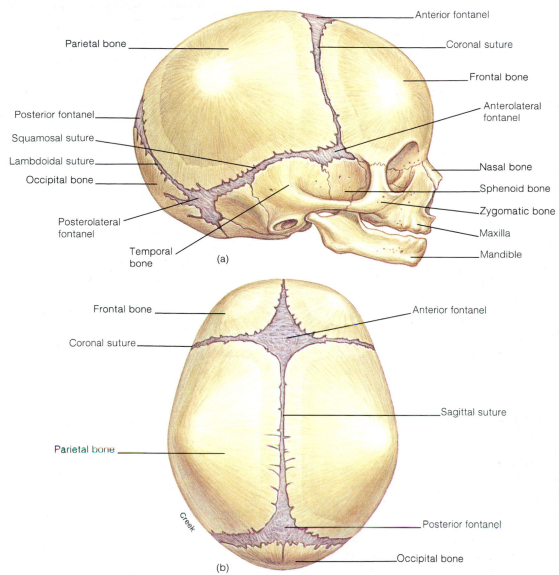

During normal parturition, the molding of the fetal skull is such that the occipital bone is usually pressed under the two parietal bones. In addition, one parietal bone overlaps the other, with the depressed one against the promontory of the mother's sacrum. In a breech (buttocks first) delivery, molding does not occur, hence, the delivery is more difficult.

A prominent **sagittal** *(saj'ĭ-tal)* **suture** extends the anteroposterior median length of the skull between the anterior and posterior fontanels. A **coronal** *(ko-rō'nal)* **suture** extends from the anterior fontanel to the anterolateral fontanel. A **lamb-doidal suture** extends from the posterior fontanel to the posterolateral fontanel. A **squamosal** *(skwah-mo'sal)* **suture** connects the posterolateral fontanel to the anterolateral fontanel.

The bones of the skull contain numerous foramina to accommodate nerves, vessels, and other structures. Table 6.4 summarizes the foramina of the skull. Figures 6.14 through 6.21 show various views of the skull. Radiographs of the skull are shown in figure 6.22.

Although the hyoid bone and the three paired auditory ossicles are not considered part of the skull, they are within the axial skeleton and will be described immediately following the discussion of the skull.

lambdoidal: Gk. *lambda*, letter λ in Greek alphabet

Table 6.4	Summary of Major Foramina of the Skull	
Foramen	**Location**	**Structures Transmitted**
Carotid canal	Petrous part of temporal bone	Internal carotid artery and sympathetic nerves
Greater palatine	Palatine bone of hard palate	Greater palatine nerve and descending palatine vessels
Hypoglossal foramen/canal	Anterolateral edge of occipital condyle	Hypoglossal nerve and branch of ascending pharyngeal artery
Incisive	Anterior region of hard palate, posterior to incisor teeth	Branches of descending palatine vessels and nasopalatine nerve
Inferior orbital fissure	Between maxilla and greater wing of sphenoid bone	Maxillary nerve of trigeminal cranial nerve, zygomatic nerve, and infraorbital vessels
Infraorbital	Inferior to orbit in maxilla	Infraorbital nerve and artery
Jugular	Between petrous portion of temporal and occipital bones, posterior to carotid canal	Internal jugular vein; vagus, glossopharyngeal, and accessory nerves
Lacerum	Between petrous portion of temporal and sphenoid bones	Branch of ascending pharyngeal artery and internal carotid artery
Lesser palatine	Posterior to greater palatine foramen in hard palate	Lesser palatine nerves
Magnum	Occipital bone	Union of medulla oblongata and spinal cord, meningeal membranes, and accessory nerves; vertebral and spinal arteries
Mandibular	Medial surface of ramus of mandible	Inferior alveolar nerve and vessels
Mental	Below second premolar on lateral side of mandible	Mental nerve and vessels
Nasolacrimal canal	Lacrimal bone	Nasolacrimal (tear) duct
Olfactory	Cribriform plate of ethmoid bone	Olfactory nerves
Optic	Back of orbit in lesser wing of sphenoid bone	Optic nerve and ophthalmic artery
Ovale	Greater wing of sphenoid bone	Mandibular nerve of trigeminal cranial nerve
Rotundum	Within body of sphenoid bone	Maxillary nerve of trigeminal cranial nerve
Spinosum	Posterior angle of sphenoid bone	Middle meningeal vessels
Stylomastoid	Between styloid and mastoid processes of temporal bone	Facial nerve and stylomastoid artery
Superior orbital fissure	Between greater and lesser wings of sphenoid bone	Four cranial nerves (oculomotor, trochlear, ophthalmic branch of trigeminal, and abducens)
Supraorbital	Supraorbital ridge of orbit	Supraorbital nerve and artery
Zygomaticofacial	Anteriolateral surface of zygomatic bone	Zygomaticofacial nerve and vessels

Figure 6.14 An anterior view of the skull.

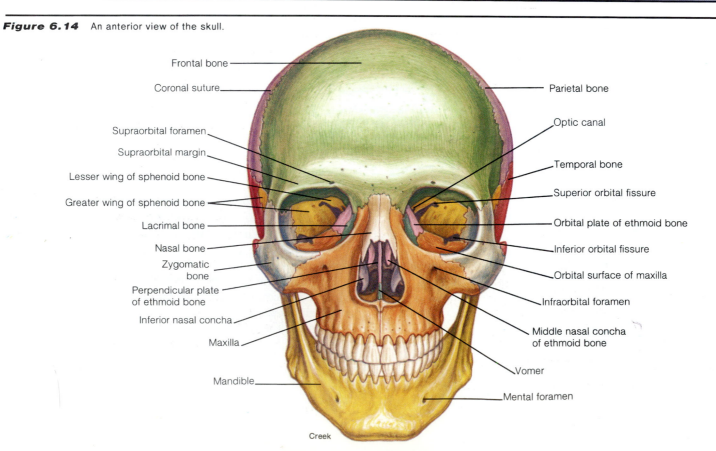

Frontal bone
Coronal suture
Parietal bone
Supraorbital foramen
Optic canal
Supraorbital margin
Lesser wing of sphenoid bone
Temporal bone
Greater wing of sphenoid bone
Superior orbital fissure
Lacrimal bone
Orbital plate of ethmoid bone
Nasal bone
Inferior orbital fissure
Zygomatic bone
Orbital surface of maxilla
Perpendicular plate of ethmoid bone
Infraorbital foramen
Inferior nasal concha
Middle nasal concha of ethmoid bone
Maxilla
Vomer
Mandible
Mental foramen

Creek

UNIT IV

Figure 6.15   A lateral view of the skull.

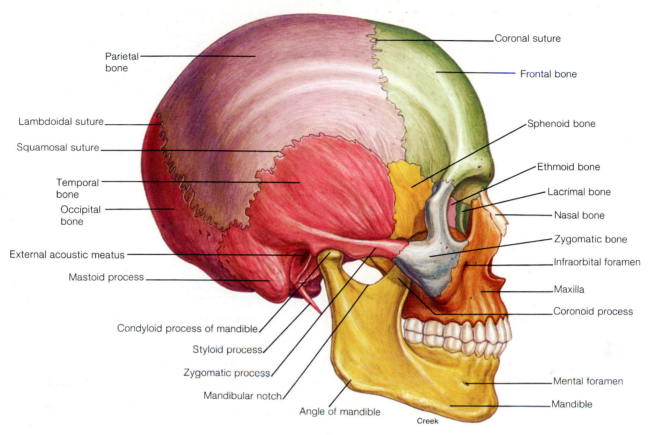

Parietal bone
Coronal suture
Frontal bone
Lambdoidal suture
Squamosal suture
Sphenoid bone
Ethmoid bone
Temporal bone
Lacrimal bone
Occipital bone
Nasal bone
External acoustic meatus
Zygomatic bone
Mastoid process
Infraorbital foramen
Maxilla
Coronoid process
Condyloid process of mandible
Styloid process
Zygomatic process
Mental foramen
Mandibular notch
Mandible
Angle of mandible
Creek

UNIT IV

Figure 6.16   An inferior view of the skull.

Incisor teeth
Premolar teeth
Canine tooth
Molar teeth
Incisive foramen
Zygomatic bone
Median palatine suture
Palatine process of maxilla
Sphenoid bone
Palatine bone
Zygomatic process
Greater palatine foramen
Vomer
Medial and lateral pterygoid processes
Mandibular fossa
Foramen ovale
External acoustic meatus
Foramen lacerum
Styloid process
Carotid canal
Mastoid process
Jugular fossa
Occipital condyle
Stylomastoid foramen
Temporal bone
Foramen magnum
Condyloid canal
Mastoid foramen
Parietal bone
Occipital bone
Superior nuchal line
External occipital protuberance
Creek

Figure 6.17 A midsagittal view of the skull.

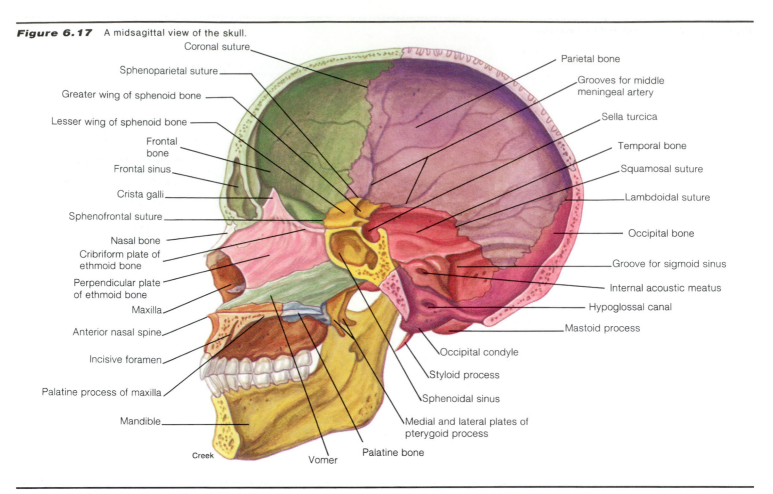

Coronal suture
Sphenoparietal suture
Greater wing of sphenoid bone
Lesser wing of sphenoid bone
Frontal bone
Frontal sinus
Crista galli
Sphenofrontal suture
Nasal bone
Cribriform plate of ethmoid bone
Perpendicular plate of ethmoid bone
Maxilla
Anterior nasal spine
Incisive foramen
Palatine process of maxilla
Mandible

Parietal bone
Grooves for middle meningeal artery
Sella turcica
Temporal bone
Squamosal suture
Lambdoidal suture
Occipital bone
Groove for sigmoid sinus
Internal acoustic meatus
Hypoglossal canal
Mastoid process
Occipital condyle
Styloid process
Sphenoidal sinus
Medial and lateral plates of pterygoid process
Palatine bone
Vomer
Creek

Figure 6.18 An inferolateral view of the skull.

Squamosal suture
Mandibular condyle
Mandibular fossa
External acoustic meatus
Mastoid process of temporal bone
Styloid process of temporal bone
Jugular foramen
Lambdoidal suture
Occipitomastoid suture
Condyloid canal
Occipital condyle
Foramen magnum

Supraorbital margin
Zygomatic arch
Coronoid process of mandible
Ramus of mandible
Mental protuberance
Angle of mandible
Digastric fossa
Mandibular foramen

UNIT IV

Figure 6.19 The floor of the cranial cavity.

Foramen cecum
Crista galli of ethmoid bone
Cribriform plate of ethmoid bone
Optic foramen
Foramen rotundum
Foramen ovale
Foramen spinosum
Temporal bone
Internal acoustic meatus
Foramen magnum
Parietal bone
Internal occipital crest

Anterior cranial fossa
Frontal bone
Sphenoid bone
Lesser wing of sphenoid bone
Greater wing of sphenoid bone
Sella turcica
Dorsum sellae
Foramen lacerum
Petrous part of temporal bone
Jugular foramen
Mastoid foramen
Posterior cranial fossa
Occipital bone

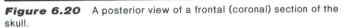

Creek

UNIT IV

Figure 6.20 A posterior view of a frontal (coronal) section of the skull.

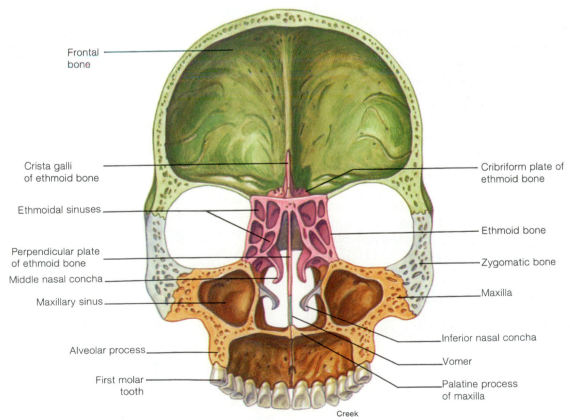

Frontal bone
Crista galli of ethmoid bone
Ethmoidal sinuses
Perpendicular plate of ethmoid bone
Middle nasal concha
Maxillary sinus
Alveolar process
First molar tooth

Cribriform plate of ethmoid bone
Ethmoid bone
Zygomatic bone
Maxilla
Inferior nasal concha
Vomer
Palatine process of maxilla

Creek

Figure 6.21   Bones of the orbit of the eye.

Frontal bone

Greater wing of sphenoid bone

Lesser wing of sphenoid bone

Greater wing of sphenoid bone

Superior orbital fissure

Orbital surface of zygomatic bone

Inferior orbital fissure

Zygomatic bone

Maxilla

Supraorbital foramen

Orbital plate of frontal bone

Anterior and posterior ethmoidal foramina

Nasal bone

Optic canal

Orbital plate of ethmoid bone

Lacrimal bone

Orbital surface of maxilla

Infraorbital foramen

Creek

Cranial Bones

The cranial bones enclose the brain and consist of one frontal, two parietals, two temporals, one occipital, one sphenoid, and one ethmoid.

Frontal Bone   The frontal bone forms the anterior roof of the cranium, the forehead, the roof of the nasal cavity, and the superior arch of the **bony orbits,** which contain the eyeballs. Table 6.5 summarizes the bones of the orbit. The frontal bone develops in two halves that grow together. Generally, they are completely fused by age 5 or 6. A complete suture sometimes persists between these two portions beyond age 6 and is referred to as a **metopic** *(mĕ-top'ik)* **suture.** The **supraorbital margin** is a prominent bony ridge over the orbit. Openings along this ridge, called *supraorbital foramina,* allow passage of small nerves and vessels.

The frontal bone contains a **frontal sinus,** which is connected to the nasal cavity (fig 6.22). This sinus, along with the other paranasal sinuses, lessens the weight of the skull and acts as a sound chamber for voice resonance.

Parietal Bone   The **coronal suture** separates the frontal bone from the parietals, and the **sagittal suture** along the superior midline separates the right and left parietal bone. The parietal bones form the upper sides and roof of the cranium (figs. 6.15

Table 6.5	Bones Forming the Orbit
Region of the Orbit	**Contributing Bones**
Roof (superior)	Frontal bone; lesser wing of sphenoid bone
Floor (inferior)	Maxilla; zygomatic bone; palatine bone
Lateral wall	Zygomatic bone
Posterior wall	Greater wing of sphenoid bone
Medial wall	Maxilla; lacrimal bone; ethmoid bone
Superior rim	Frontal bone
Lateral rim	Zygomatic bone
Medial rim	Maxilla

and 6.17). The inner concave surface of the parietal bone, as well as the inner concave surfaces of other cranial bones, is marked by shallow impressions from convolutions of the brain and vessels serving the brain.

Temporal Bone   The two temporal bones form the lower sides of the cranium (figs. 6.15, 6.16, 6.17, and 6.23). Each temporal is joined to its adjacent parietal bone by the **squamosal suture.** Structurally, each temporal has four parts.

1. **Squamous part.** The squamous part is the flattened plate of bone at the sides of the skull. Projecting forward is a **zygomatic process** that forms the posterior portion

cranium: Gk. *kranion,* skull
metopic suture: Gk. *metopon,* forehead; L. *sutura,* sew

zygomatic: Gk. *zygoma,* yolk

UNIT IV

Figure 6.22 Radiographs of the skull showing paranasal sinuses.
(*a*) An anteroposterior view and (*b*) a right lateral view.

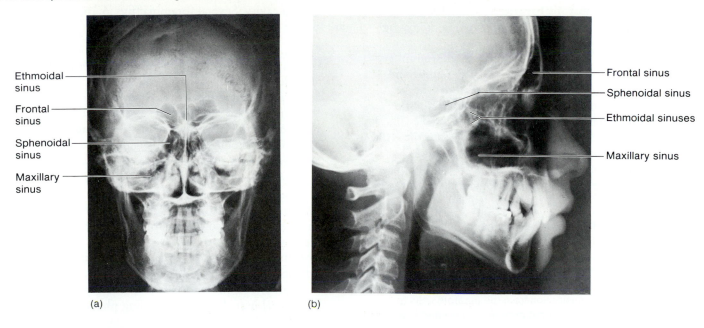

Ethmoidal sinus

Frontal sinus

Sphenoidal sinus

Maxillary sinus

Frontal sinus

Sphenoidal sinus

Ethmoidal sinuses

Maxillary sinus

(a)

(b)

Figure 6.23 The temporal bone. (*a*) A lateral view and (*b*) a medial
view.

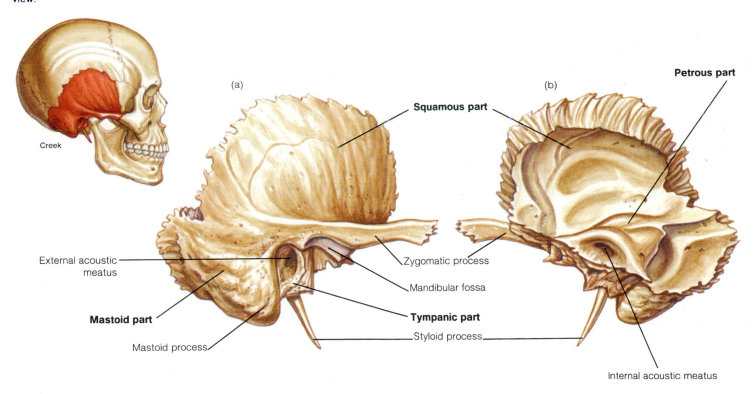

Creek

(a)

(b)

Squamous part

Petrous part

Zygomatic process

Mandibular fossa

External acoustic meatus

Mastoid part

Mastoid process

Tympanic part

Styloid process

Internal acoustic meatus

UNIT IV

of the **zygomatic arch.** On the inferior surface of the
squamous part is the **mandibular fossa,** which receives
the articular condyle of the mandible. This articulation is
referred to as the **temporomandibular** *(tem″pŏ-ro-man-
dib′yŭ-lar)* **joint.**

2. **Tympanic part.** The tympanic part of the temporal
bone contains the **external acoustic meatus,** or ear
canal, located immediately posterior to the mandibular

fossa. A thin, pointed **styloid process** (figs. 6.16, 6.17,
and 6.18) projects downward from the tympanic portion.

3. **Mastoid part.** The **mastoid process,** a rounded
projection posterior to the external acoustic meatus,

styloid: Gk. *stylos,* pillar
mastoid: Gk. *mastos,* breast

accounts for the mass of the mastoid part. The **mastoid foramen** (fig. 6.16) is directly posterior to the mastoid process. The **stylomastoid foramen,** located between the mastoid and styloid processes (fig. 6.16) is the passage for part of the facial nerve.

4. **Petrous part.** The petrous *(pet'rus)* part is viewed in figures 6.19 and 6.23. The structures of the middle and inner ear are housed in this dense, bony part of the temporal bone. The **carotid** *(kah-rot'id)* **canal** and the **jugular foramen** border on the medial side of the petrous part. The carotid canal allows blood into the brain via the internal carotid artery and the jugular foramen lets blood drain from the brain via the internal jugular vein. Three cranial nerves also pass through the jugular foramen.

> The mastoid process of the temporal bone can be easily palpated as a bony knob immediately behind the earlobe. This process contains a number of small air-filled spaces called *mastoid cells* that are clinically important because they can become infected in *mastoiditis.* A tubular communication from the mastoid cells to the middle-ear cavity may permit prolonged ear infections to spread to this region.

UNIT IV

Occipital Bone   The occipital bone forms the back and much of the base of the skull. It is fastened to the parietal bones by the **lambdoidal suture.** The **foramen magnum** is the large hole in the occipital bone through which the spinal cord attaches to the brain. On each side of the foramen magnum are the **occipital condyles** *(ok-sip'ĭ-tal kon'dīlz)* (fig. 6.16), which articulate with the atlas of the vertebral column. At the anterolateral edge of the occipital condyle is the **hypoglossal canal** (fig. 6.17), through which the hypoglossal nerve passes. A **condyloid** *(kon'dĭ-loid)* **canal** lies posterior to the occipital condyle (fig. 6.16). The **external occipital protuberance** is a prominent posterior projection on the occipital bone that can be felt as a definite bump just under the skin. The **superior nuchal** *(noo'kal)* **line** is a ridge of bone extending laterally from the occipital protuberance to the mastoid portion of the temporal bone. **Sutural bones** are small clusters of irregularly shaped bones that may be found between the joints of certain cranial bones but generally occur along the lambdoidal suture.

Sphenoid Bone   The sphenoid *(sfe'noid)* bone forms the anterior base of the cranium and can be viewed laterally and inferiorly (figs. 6.15 and 6.16). This bone resembles a butterfly with outstretched wings (fig. 6.24). It consists of a **body** with laterally projecting **greater** and **lesser wings,** which form part of the bony orbit. The body is a wedgelike central portion that contains the **sphenoidal** *(sfe-noi'dal)* **sinuses** and a prominent depression called the **sella turcica** *(sel'ă tur'si-kă)* which supports the pituitary gland. The sella turcica (meaning "Turk's saddle") is seen on the floor of the cranium (figs. 6.17 and 6.19).

A pair of **pterygoid** *(ter'ĭ-goid)* **processes (plates)** project inferiorly from the sphenoid bone to help form the lateral walls of the nasal cavity. Several foramina (figs. 6.16, 6.19, and 6.24) are located within the sphenoid bone.

1. **Optic canal.** A large opening through the lesser wing into the back of the orbit for passage of the optic nerve and the ophthalmic artery.
2. **Superior orbital fissure.** A triangular opening between the wings of the sphenoid for passage of the ophthalamic nerve off of the trigeminal cranial nerve and the oculomotor, trochlear, and the abducens cranial nerves.
3. **Foramen ovale.** An opening at the base of the pterygoid plate, through which the mandibular lateral nerve passes.
4. **Foramen spinosum.** A small opening at the posterior angle of the sphenoid bone for passage of the middle meningeal vessels.
5. **Foramen lacerum** *(las'er-um).* An opening between the sphenoid and the petrous part of the temporal bone, through which the internal carotid artery and the meningeal branch of the ascending pharyngeal artery pass.
6. **Foramen rotundum.** An opening located just posterior to the superior orbital fissure at the junction of the anterior and medial portions of the sphenoid bone. The maxillary nerve passes through this foramen.

> Located on the inferior side of the cranium, the sphenoid bone would seemingly be well protected from trauma. Actually just the opposite is true, and in fact the sphenoid is the most frequently fractured bone of the cranium. It has several broad, thin platelike extensions that are perforated and weakened with numerous foramina. A blow to almost any portion of the skull causes the buoyed, fluid-filled brain to rebound against the vulnerable sphenoid bone, often causing it to fracture.

Ethmoid Bone   The ethmoid bone is located in the anterior portion of the floor of the cranium between the orbits, where it forms the roof of the nasal cavity (figs. 6.17, 6.20, and 6.25). An inferior projection of the ethmoid bone, called the **perpendicular plate,** contributes in part to the nasal septum that separates the nasal cavity into two chambers, referred to as **nasal fossae.** A spine of the perpendicular plate, the **crista galli,** projects superiorly into the cranial cavity and serves as an attachment for the meninges covering the brain. On both lateral walls of the nasal cavity are two scroll-shaped plates of the ethmoid bone, called the **superior** and **middle nasal conchae** (fig. 6.26). At right angles to the perpendicular plate, within the floor of the cranium, is the **cribriform plate,** which has numerous perforations for the passage of olfactory nerves from the nasal cavity. The bones of the nasal cavity are summarized in table 6.6.

---

petrous: Gk. *petra,* rock
magnum: L. *magnum,* great
nuchal: Fr. *nuque,* nape of neck
sphenoid: Gk. *sphenoeides,* wedgelike

---

ethmoid: Gk. *ethmos,* sieve
crista galli: L. *crista,* crest; *galli,* cock's comb
conchae: L. *conchae,* shells
cribriform: L. *cribrum,* sieve; *forma,* like

Figure 6.24   The sphenoid bone. (*a*) A superior view and (*b*) a posterior view.

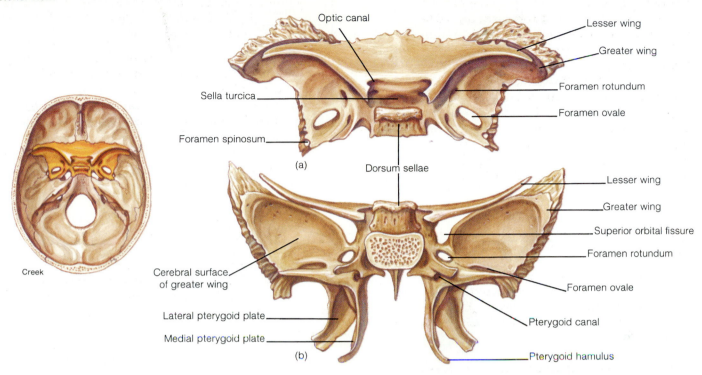

Figure 6.25   An anterior view of the ethmoid bone.

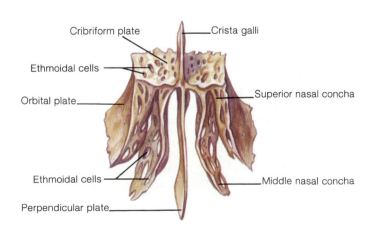

The moist, warm vascular lining within the nasal cavity is susceptible to infections, particularly if a person is not in good health. Infections of the nasal cavity can spread to several surrounding areas. The paranasal sinuses connect to the nasal cavity and are especially prone to infection. The eyes may become reddened and swollen during a nasal infection because of the connection of the nasolacrimal duct, through which tears drain from the orbit to the nasal cavity. Organisms may spread via the auditory tube from the nasopharynx to the middle ear. With prolonged nasal infections, organisms may even ascend to the meninges covering the brain, along the sheaths of the olfactory nerves, and through the cribriform plate to produce *meningitis.*

Facial Bones

The 14 bones of the skull not in contact with the brain are called **facial bones.** These bones, together with certain cranial bones (frontal and portions of the ethmoid and temporal bones), provide the basic shape of the face. Facial bones also support the teeth and provide attachments for various muscles that move the jaw and cause facial expressions. All the facial bones are paired except the vomer and mandible. The articulated facial bones can be seen in figures 6.14 through 6.21.

Maxilla   The two maxillae unite at the midline to form the upper jaw, which supports the upper teeth. **Incisors** (*in-si'zorz*), **canines** (cuspids), **premolars,** and **molars** are contained in

incisor: L. *incidere*, to cut
canine: L. *canis*, dog
molar: L. *mola*, millstone

UNIT IV

Table 6.6	Bones Forming the Nasal Cavity
Region of Nasal Cavity	**Contributing Bones**
Roof (superior)	Ethmoid bone (cribriform plate); frontal bone
Floor (inferior)	Maxilla; palatine bone
Lateral wall	Maxilla; palatine bone
Nasal septum (medial)	Ethmoid bone (perpendicular plate); vomer
Bridge	Nasal bone
Conchae	Ethmoid bone (superior and middle conchae); inferior nasal concha

sockets, or **alveoli,** within the **alveolar** (al-ve′ŏ-lar) **process** of the maxilla (fig. 6.27). The palatine process, a horizontal plate of the maxilla, forms the greater portion of the **hard palate** (pal′it), or roof of the mouth. The **incisive foramen** (fig. 6.16) is located in the anterior region of the hard palate behind the incisor teeth. An **infraorbital foramen** is located under each orbit and serves as a passageway for the infraorbital nerve and artery to the nose (figs. 6.14, 6.15, 6.21, and 6.27). A final opening within the maxilla is the **inferior orbital fissure.** It is located between the maxilla and the greater wing of the sphe-

alveolus: L. *alveus*, little cavity

Figure 6.26 The lateral wall of the nasal cavity.

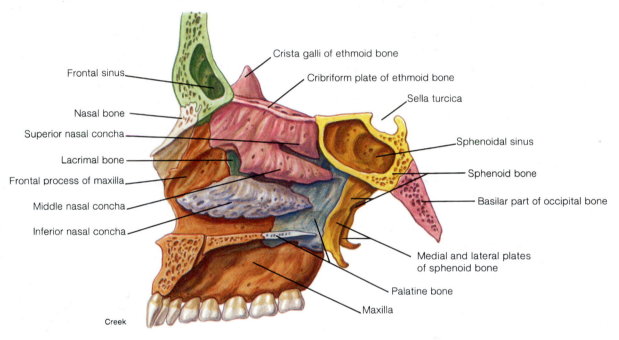

Figure 6.27 The maxilla. (*a*) A lateral view and (*b*) a medial view.

(a)   (b)

UNIT IV

noid (fig. 6.14) and is the external opening for the maxillary branch of the trigeminal nerve and infraorbital vessels. The large **maxillary sinus** located within the maxilla is one of the four paranasal sinuses (figs. 6.20 and 6.22).

If the two palatine processes fail to join during early prenatal development (about 12 weeks), a *cleft palate* results. A cleft palate may be accompanied by a *cleft lip* (harelip) lateral to the midline. These conditions can be surgically and cosmetically treated with excellent success. A more immediate problem, however, is that a newborn with a cleft palate may have a difficult time swallowing while nursing because it is unable to create the necessary suction within the oral cavity.

Palatine Bone   The L-shaped palatine bones form the posterior third of the hard palate, a portion of the orbits, and a part of the nasal cavity. The **horizontal plates** of the palatines contribute to the formation of the hard palate (fig. 6.28). On the hard palate of each palatine bone is a large **greater palatine foramen,** which permits the passage of the greater palatine nerve and descending palatine vessels (fig. 6.16). Two or more smaller **lesser palatine foramina** are positioned posterior to the greater palatine foramen. Branches of the lesser palatine nerve pass through these openings.

Zygomatic Bone   The two zygomatic bones form the cheekbones of the face. A posteriorly extending **zygomatic process** of this bone unites with that of the temporal bone to form the **zygomatic arch** (fig. 6.16). The zygomatic bone also forms the lateral margin of the orbit. A small **zygomaticofacial** (zi''gŏ-mat''ĭ-kō-fa'shal) **foramen,** located on the anterolateral surface of this bone, allows passage of the zygomatic nerves and vessels.

Lacrimal Bone   The small lacrimals (lak'rĭ-malz) are thin bones that form the anterior part of the medial wall of each orbit (fig. 6.21). Each has a **lacrimal sulcus,** a groove that helps to form the **nasolacrimal canal.** This opening permits the tears of the eye to drain into the nasal cavity.

Nasal Bone   The small, rectangular nasal bones (fig. 6.14) join in the midline to form the bridge of the nose. The nasal bones support the flexible cartilaginous plates, which are a part of the framework of the nose. Common facial injuries include fractures of the nasal bones or fragmentation of the supporting cartilages.

Inferior Nasal Concha   The two inferior nasal conchae are fragile, scroll-like bones that project horizontally and medially from the lateral walls of the nasal cavity (figs. 6.14 and 6.20). They extend into the nasal cavity just below the superior and middle nasal conchae, which are part of the ethmoid bone (fig. 6.25). The inferior nasal conchae are the largest of the three conchae, and, like the other two, are covered with mucous membranes to warm, moisten, and cleanse inhaled air.

Figure 6.28   The palatine bone. (*a*) A medial view. (*b*) The two palatine bones viewed posteriorly. The two palatine bones form the posterior portion of the hard palate.

(a)

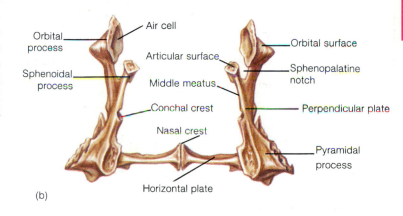

(b)

Vomer   The vomer (vo'mer) is a thin, elongated bone that forms the lower part of the nasal septum (figs. 6.16, 6.17, and 6.20). The vomer, along with the perpendicular plate of the ethmoid, supports the septal cartilage to complete the nasal septum.

Mandible   The mandible, or lower jawbone, is attached to the skull by a temporomandibular articulation, and is the only movable bone of the skull. Several muscles that close the jaw extend from the skull to the mandible (see chapter 9). The mandible of an adult supports 16 teeth within alveoli, which occlude with those of the maxilla.

vomer: L. *vomer,* plowshare
mandible: L. *mandere,* to chew

Figure 6.29 The mandible. (*a*) A lateral view and (*b*) a posterior view.

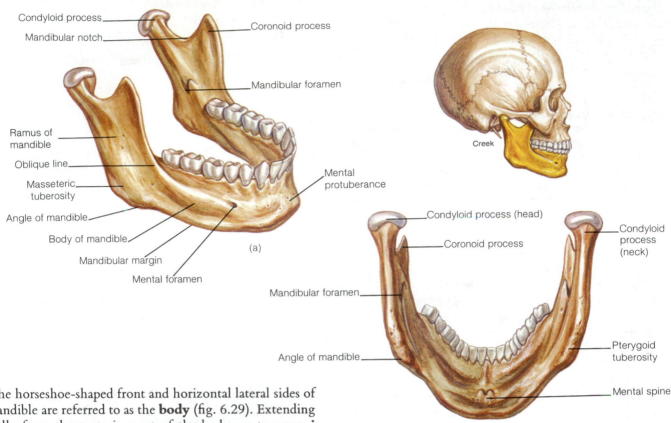

Condyloid process

Coronoid process

Mandibular notch

Mandibular foramen

Ramus of mandible

Oblique line

Masseteric tuberosity

Angle of mandible

Body of mandible

Mandibular margin

Mental foramen

Mental protuberance

(a)

Creek

Condyloid process (head)

Coronoid process

Condyloid process (neck)

Mandibular foramen

Angle of mandible

Pterygoid tuberosity

Mental spine

(b)

The horseshoe-shaped front and horizontal lateral sides of the mandible are referred to as the **body** (fig. 6.29). Extending vertically from the posterior part of the body are two **rami** *(ra'mi;* singular, *ramus)*. Each ramus has a knoblike **condyloid process,** which articulates with the mandibular fossa of the temporal bone, and a pointed **coronoid process** for the attachment of the temporalis muscle. The depressed area between these two processes is the **mandibular notch.** The angle of the mandible is where the horizontal body and vertical ramus meet at the corner of the jaw.

Two sets of foramina are found on the mandible: the **mental foramen** on the lateral side below the first molar and the **mandibular foramen** on the medial surface of the ramus. The mental nerve and vessels pass through the mental foramen and the inferior alveolar nerve and vessels are transmitted through the mandibular foramen.

Dentists use bony landmarks of the facial region to locate the nerves that traverse the foramina in order to inject anesthetics. For example, the trigeminal cranial nerve is composed of three large nerves, the lower two of which convey sensations from the teeth, gums, and jaws. The mandibular teeth can be desensitized by an injection near the mandibular foramen called a *third-division,* or *lower, nerve block.* An injection near the foramen rotundum of the skull, called a *second-division nerve block,* desensitizes all the upper teeth on one side of the maxilla.

Hyoid Bone   The hyoid bone is U-shaped and is located in the neck, just superior to the larynx (voice box). The hyoid is unique in that it does not attach directly to any other bone but is suspended from the styloid processes of the skull by the stylohyoid *(sti"lo-hi'oid)* muscles and ligaments. The hyoid has a **body,** two **lesser cornua** extending anteriorly, and two **greater cornua** (fig. 6.30), which project posteriorly to the stylohyoid ligaments. Several neck and tongue muscles attach to the hyoid bone. The hyoid may be palpated by placing a thumb and a finger on either side of the upper neck under the lateral portions of the mandible and firmly squeezing medially.

Auditory Ossicles   Three small, paired auditory ossicles, the **malleus, incus,** and **stapes,** are located within the cavities of the middle ears in the petrous part of the temporal bones (fig. 6.31). These bones transfer and amplify sound impulses through the middle ear (see p. 489).

---

ramus: L. *ramus,* branch
condyloid: L. *condylus,* knucklelike
coronoid: Gk. *korone,* like a crow's beak

---

cornu: L. *cornu,* horn
malleus: L. *malleus,* hammer
incus: L. *incus,* anvil
stapes: L. *stapes,* stirrup

UNIT IV

WANT
TO DO
WELL
ON
TESTS
AND
EXAMS?

BE
YOUR
OWN
TUTOR!

Wm. C. Brown Publishers now has developed a **TAKE-HOME TUTOR** that will help you achieve your goals in the classroom and prepare for careers in the nursing and allied health sciences. The *ANATOMY AND PHYSI-OLOGY SUPPORT PACKAGE* is a complete set of additional study aids designed to facilitate learning and achievement. The package includes:

Student Study Guide
ISBN 0-697-21100-2
For each text chapter, there is a corresponding study guide chapter—including *suggestions for review*, *objective questions*, and *essay questions*—that reviews the major chapter concepts.

Atlas of the Skeletal Muscles
ISBN 0-697-10618-7
This inexpensive atlas depicts all the human skeletal muscles in precise black-and-white drawings that show their *origin*, *insertion*, *action*, and *innervation*. Each muscle is presented on a separate page, and the spinal cord level of the nerve fibers is included in parentheses. These features aid in the visual understanding of the action of the muscles.

Study Cards for Human Anatomy and Physiology
ISBN 0-697-09731-5
This boxed set of 300 two-sided study cards provides a complete *description of terms*, *clearly labeled drawings*, *pronunciation guides*, and *clinical information* on diseases. They will accompany any anatomy and physiology textbook.

Coloring Review Guide for Anatomy and Physiology
ISBN 0-697-17109-4
This helpful manual serves as a framework for learning human anatomy and physiology. It emphasizes the *connection between facts and structures* through the specific activity of color association.

Human Skeletal Musculature Videotape
ISBN 0-697-15210-0
This videotape offers an in-depth introduction to the major groups of skeletal muscles by using a combination of *film clips*, *X rays*, *line art*, and *human cadaver photographs*. This is an excellent introduction to how the muscles provide movement to the human body.

To get your own **TAKE-HOME TUTOR**, just ask your bookstore manager. If it's unavailable through your bookstore, call toll-free 800-338-5578 (in the United States), 800-268-4178 (in Canada).

 Wm. C. Brown Publishers
A Division of Wm. C. Brown Communications, Inc.

Van De Graaff tht 16065

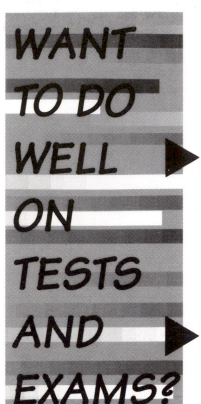

WANT TO DO WELL ON TESTS AND EXAMS?

Wm. C. Brown Publishers now has developed a **TAKE-HOME TUTOR** that will help you achieve your goals in the classroom and prepare for careers in the nursing and allied health sciences. The *ANATOMY AND PHYSI-OLOGY SUPPORT PACKAGE* is a complete set of additional study aids designed to facilitate learning and achievement. The package includes:

How to Study Science
ISBN 0-697-14474-7

This excellent workbook offers you helpful suggestions for meeting the considerable challenges of a college science course. It offers tips on *how to take notes, how to get the most out of laboratories,* as well as *how to overcome "science anxiety."* The book's unique design helps you develop critical-thinking skills, while facilitating careful note taking.

A Life Science Lexicon
ISBN 0-697-12133-X

This portable, inexpensive reference will help you quickly master the vocabulary of the life sciences. Not a dictionary, it carefully explains the *rules of word construction and derivation* and gives *complete definitions* of all important terms.

Explorations in Human Biology CD-ROM
ISBN 0-697-22959-9

This interactive software, (for use with Macintosh and IBM Windows) consists of *16 different animated modules* that cover *key topics* discussed in a human biology course.

Animated Illustrations of Physiological Processes Videotape
ISBN 0-697-21512-0

This videotape *animates 13 illustrations of complex physiological processes.* These colorful, moving visuals, make challenging concepts easier to understand.

BE YOUR OWN TUTOR!

To get your own **TAKE-HOME TUTOR**, just ask your bookstore manager. If it's unavailable through your bookstore, call toll-free 800-338-5578 (in the United States), 800-268-4178 (in Canada).

WCB **Wm. C. Brown Publishers**
A Division of Wm. C. Brown Communications, Inc.

APSPba

Figure 6.30   An anterior view of the hyoid bone.

Larynx   Hyoid

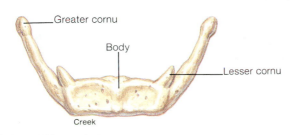

Greater cornu

Body

Lesser cornu

Creek

---

1. State which facial and cranial bones of the skull are paired and which are unpaired. Also, indicate at least two structural features for each of the bones of the skull.
2. Describe the location of each of the bones of the skull and indicate the sutures that join them.
3. What is the function of the sella turcica, foramen magnum, petrous part of the temporal bone, crista galli, and nasal conchae?
4. Which facial bones support the teeth?

Vertebral Column

The supporting vertebral column consists of vertebrae, separated by fibrocartilaginous intervertebral discs that lend flexibility and absorb the stress of movement. Vertebrae enclose and protect the spinal cord, support the skull and allow for its movement, articulate with the rib cage, and provide for the attachment of trunk muscles.

Objective 12. Identify the bones of the five regions of the vertebral column and describe the characteristic curves of each region.

Objective 13. Describe the structure of a typical vertebra.

The vertebral column is composed of 33 individual vertebrae. There are 7 **cervical,** 12 **thoracic,** 5 **lumbar,** 5 fused **sacral,**

Figure 6.31   The three auditory ossicles within cavity of the middle ear.

Creek

Superior ligament of malleus

Malleus

Incus

Tensor tympani muscle (cut)

Anterior ligament of malleus

Stapes in vestibular window

Promontory

Cochlear window

Cavity of middle ear

External acoustic meatus

Tympanic membrane

Auditory tube

UNIT IV

and 4 or 5 fused **coccygeal** *(kok-sij'e-al)* **vertebrae;** thus, the vertebral column is composed of a total of 26 movable parts. Vertebrae are separated by fibrocartilaginous intervertebral discs and are secured to each other by interlocking processes and binding ligaments. This structural arrangement provides limited movements between vertebrae but extensive movements for the entire vertebral column. Between the vertebrae are openings called **intervertebral foramina** that permit passage of spinal nerves.

Four curvatures of the vertebral column of an adult can be identified and viewed from the side (fig. 6.32). The **cervical, thoracic,** and **lumbar curves** are identified by the type of vertebrae they include. The **pelvic curve** is formed by the shape of the sacrum and coccyx *(kok'siks)*. The curves of the vertebral column play an important functional role in increasing the strength and maintaining the balance of the upper part of the body; they also make possible a bipedal stance.

The four vertebral curves are not present in a newborn. Instead, the vertebral column is somewhat anteriorly concave, and except for the cervical region, it remains this way even as an infant learns to crawl (fig. 6.33). The cervical curve begins to

UNIT IV

Figure 6.32 The vertebral column of an adult has four curves named according to the region in which they occur. The vertebrae are separated by intervertebral discs in a structural arrangement that provides flexibility.

Atlas (C1)

Axis (C2)

Cervical curve

C7

T1

Vertebrae

Thoracic curve

Intervertebral discs

T12

L1

Intervertebral foramina

Lumbar curve

L5

Sacrum

Pelvic curve

Coccyx

Creek

develop at about 3 months of age, as a baby begins holding up its head, and the curve becomes more pronounced as the baby learns to sit up. The lumbar curve develops as a child begins to walk. The thoracic and pelvic curves are called **primary curves** because they retain the anteriorly concave shape of the fetus. The cervical and lumbar curves are called **secondary curves** because they are modifications of the fetal shape that develop as adaptions to weight bearing as an infant learns to sit up and walk.

The vertebral column is commonly called the "backbone," and together with the spinal cord of the nervous system constitutes the spinal column. The vertebral column has three basic functions:

1. to support the head and upper extremities while permitting freedom of movement;
2. to provide attachment for various muscles, ribs, and visceral structures; and
3. to protect the spinal cord and permit passage of the spinal nerves.

General Structure of Vertebrae

Vertebrae show similarities in their general structure from one region to another. Figure 6.34 illustrates a typical vertebra. A vertebra is usually composed of an anterior drum-shaped **body**, adapted to withstand compression. The body is in contact with

Figure 6.33 The development of the vertebral curves. An infant is born with the two primary curves but does not develop the secondary curves until it begins sitting upright and walking. (Note the differences in the curves between the sexes.)

intervertebral discs on each end. The **vertebral arch** is affixed to the posterior surface of the body and is composed of two supporting **pedicles** *(ped'i-kulz)* and two arched **laminae.** The hollow space formed by the vertebral arch and body is the **vertebral foramen,** through which the spinal cord passes. Between the pedicles of adjacent vertebrae are the **intervertebral foramina,** through which spinal nerves emerge as they branch off from the spinal cord.

Seven processes arise from the vertebral arch: the **spinous** *(spi'nus)* **process,** two **transverse processes,** two **superior articular processes,** and two **inferior articular processes** (fig. 6.35). The first two processes serve for muscle attachment and the latter two pairs limit twisting of the vertebral column. The spinous process protrudes posteriorly and inferiorly from the vertebral arch. The transverse process extends laterally from each side of a vertebra at the point where the lamina and pedicle

---

pedicle: L. *pediculus,* small foot
lamina: L. *lamina,* thin layer

---

join. The superior articular processes of a vertebra have interlocking articulations with the inferior articular processes of the adjacent superior vertebra.

> A *laminectomy* is the surgical removal of the spinous processes and their supporting vertebral laminae in a particular region of the vertebral column. A laminectomy may be performed to relieve pressure on the spinal cord or nerve root caused by a blood clot, a tumor, or a herniated disc. It may also be performed on a cadaver to expose the spinal cord and its surrounding meninges.

Regional Characteristics of Vertebrae

Cervical Vertebrae   The seven cervical vertebrae form the flexible framework of the neck region and support the head. The bone tissue of cervical vertebrae is more dense than that found in the other vertebral regions, and, except for those in

Figure 6.34   Cervical vertebrae. (*a*) A radiograph of the cervical region, (*b*) a superior view of a typical cervical vertebra, and (*c*) the atlas and axis as they articulate.

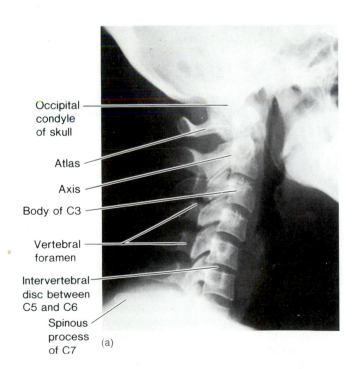

(a)

Occipital condyle of skull
Atlas
Axis
Body of C3
Vertebral foramen
Intervertebral disc between C5 and C6
Spinous process of C7

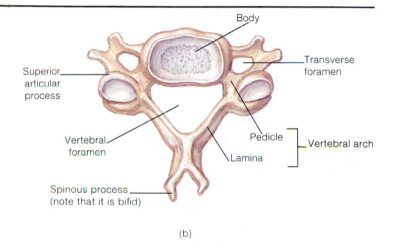

Body
Superior articular process
Transverse foramen
Vertebral foramen
Pedicle
Lamina
Vertebral arch
Spinous process (note that it is bifid)

(b)

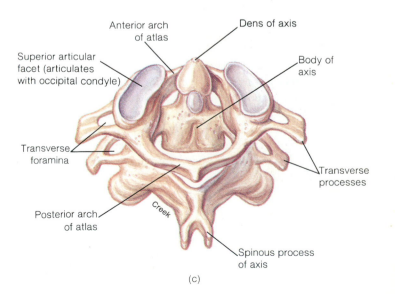

Anterior arch of atlas
Dens of axis
Superior articular facet (articulates with occipital condyle)
Body of axis
Transverse foramina
Transverse processes
Posterior arch of atlas
Spinous process of axis

(c)

Figure 6.35 Thoracic vertebrae. (*a*) Representative vertebrae in a right lateral view and (*b*) a superior view.

(a)

(b)

the coccygeal region, the cervical vertebrae are smallest. Cervical vertebrae are distinguished by the presence of a **transverse foramen** in the transverse process (fig. 6.34). The vertebral arteries pass through this opening as they transfer blood to the brain. The spinous processes of the second through the sixth cervical vertebrae are *bifid (bi'fid),* or notched, for the attachment of the strong **nuchal ligament,** that attaches to the back of the skull for added support.

The first cervical vertebra, the **atlas,** is adapted to articulate with the occipital condyles of the skull while supporting the head. The atlas has concave **superior articular surfaces** to articulate with the oval-shaped occipital condyles. This joint permits the nodding of the head in a "yes" movement. The atlas lacks a body. It has a short, rounded spinous process called the **posterior tubercle.**

The second cervical vertebra is called the **axis** and is easily identified by the presence of a peglike projection called the **dens,** or **odontoid process.** This process extends superiorly to provide a pivot for rotation with the atlas. The dens permits rotation, or the turning of the head to the side, as in a "no" movement.

"Whiplash" of the neck is a common, general term for injury to the cervical region. Muscle, bone, or ligament injury in this portion of the spinal column is relatively common in individuals involved in automobile accidents. Joint dislocation without vertebral fracture occurs commonly between the fourth and fifth or fifth and sixth cervical vertebrae, where mobility is greatest. Bilateral dislocations are particularly dangerous because of the probability of spinal cord injury. Compression fractures of the bodies of the first three cervical vertebrae are common and follow abrupt forced flexion of the neck. Fractures of this type may be extremely painful because of pinched spinal nerves.

---

atlas: from Gk. mythology, Atlas—the Titan who supported the heavens

axis: L. *axis*, axle
odontoid: Gk. *odontos,* tooth

Figure 6.36 Lumbar vertebrae. (*a*) A radiograph, (*b*) a superior view, and (*c*) a right lateral view.

(a)

(b)

(c)

UNIT IV

Thoracic Vertebrae The thoracic vertebrae serve as attachments of the ribs to form the posterior anchor of the rib cage. Thoracic vertebrae are larger than cervical vertebrae and increase in size from superior (T1) to inferior (T12). Each thoracic vertebra has a long spinous process that slopes obliquely downward and **fovea** for articulation with the ribs (fig. 6.35).

Lumbar Vertebrae The five lumbar vertebrae are easily identified by their heavy bodies and thick, blunt spinous processes (fig. 6.36) for attachment of powerful back muscles. They are the largest vertebrae of the vertebral column. Their articular processes are also distinctive in that the superior articular processes are directed medially instead of superiorly and the inferior articular processes are directed laterally instead of inferiorly.

Sacrum The wedge-shaped sacrum (fig. 6.37) consists of four or five sacral vertebrae, which become fused after age 26. The sacrum is functionally adapted to provide a strong foundation for the pelvic girdle. The sacrum has an extensive **auricular surface** on each side for the formation of a semimovable **sacroiliac** *(sak″ro-il′e-ak)* **joint** with the os coxa. A **median sacral crest** is formed along the dorsal surface by the fusion of the spinous processes. **Posterior sacral foramina** on either lateral side of the crest allow for the passage of nerves from the spinal cord. The **sacral canal** is the tubular cavity within the sacrum that is continuous with the vertebral canal. Paired **superior articular processes,** which articulate with the fifth lumbar vertebra, arise from the roughened **sacral tuberosity** along the posterior surface.

The smooth anterior surface of the sacrum forms the posterior surface of the pelvic cavity. It has four **transverse lines** denoting the fusion of the vertebral bodies. On both lateral sides of the transverse lines are the paired **pelvic foramina (ante-**

---

lumbar: L. *lumbus*, loin

sacrum: L. *sacris*, sacred

Figure 6.37 The sacrum and coccyx. (*a*) An anterior view and (*b*) a posterior view.

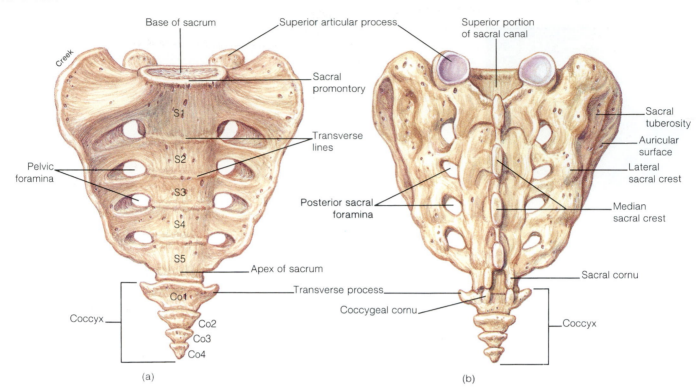

(a)

(b)

rior sacral foramina). The superior border of the anterior surface of the sacrum, called the **sacral promontory,** is an important obstetric landmark for pelvic measurements.

Coccyx   The coccyx is the so-called tailbone. It is composed of four or five fused coccygeal vertebrae, which form a triangular-shaped bone. The first vertebra of the fused coccyx has two long **coccygeal cornua,** which are attached by ligaments to the sacrum (fig. 6.37). Lateral to the cornua are the transverse processes.

The regions of the vertebral column are summarized in table 6.7.

When a person sits, the coccyx flexes anteriorly somewhat, acting as a shock absorber. An abrupt fall on the coccyx, however, may cause a painful subperiosteal bruising, fracture, or fracture-dislocation of the sacrococcygeal joint. An especially difficult childbirth can even injure the coccyx of the mother. Coccygeal trauma is painful and may require months to heal.

1. Which are the primary curves of the vertebral column and which are the secondary curves? Describe the characteristic curves of each region.
2. What is the function of the transverse foramina of the cervical vertebrae?
3. Describe the diagnostic differences between a thoracic and a lumbar vertebra. Which structures are similar and could therefore be characteristic of a typical vertebra?

coccyx: Gk. *kokkyx*, like a cuckoo's beak

Table 6.7	Regions of the Vertebral Column	
Region	**Number of Bones**	**Diagnostic Features**
Cervical	7	Transverse foramina; superior facets of atlas articulate with occipital condyle; dens of axis; spinous processes of third through fifth vertebrae are bifid
Thoracic	12	Long spinous processes that slope obliquely inferiorly; fovea for articulation with ribs
Lumbar	5	Large bodies; prominent transverse processes; short, thick spinous processes
Sacrum	4 or 5 fused vertebrae	Extensive auricular surface; median sacral crest; posterior sacral foramina; sacral promontory; sacral canal
Coccyx	4 or 5 fused vertebrae	Small, triangular; coccygeal cornua

Rib Cage

The cone-shaped and flexible rib cage consists of the thoracic vertebrae, 12 paired ribs, costal cartilages, and the sternum. It encloses and protects the thoracic viscera and is directly involved in the mechanics of breathing.

Figure 6.38   The rib cage.

Jugular notch

First rib

Costal cartilage

True ribs

Costal notches

False ribs

Floating ribs

Creek

First thoracic vertebra

Clavicular notch

Manubrium of sternum

Sternal angle

Body of sternum

Sternum

Xiphoid process

Intercostal spaces

Twelfth rib

First lumbar vertebra

| Objective 14. | Identify the parts of the rib cage, and compare and contrast the various types of ribs. |

The **sternum, ribs, costal cartilages,** and the previously described thoracic vertebrae form the **rib cage,** or **thoracic cage,** of the thorax (fig. 6.38). The rib cage is anteroposteriorly compressed and more narrow superiorly than inferiorly. It supports the pectoral girdle and upper extremities, protects and supports the thoracic and upper abdominal viscera, and plays a major role in breathing (see fig. 9.21). Certain bones of the rib cage contain active sites in the bone marrow for the production of red blood cells.

Sternum

The **sternum** (breastbone) is an elongated, flattened bony plate consisting of three separate bones: the upper **manubrium,** the central **body,** and the lower **xiphoid** *(zi'foid)* **process.** On the lateral sides of the sternum are **costal notches** where the costal

cartilages attach. A **jugular notch** is formed at the superior end of the manubrium, and a **clavicular** *(klah-vik'yoo-lar)* **notch** for articulation with the clavicle is present on both sides of the sternal notch. The manubrium articulates with the costal cartilages of the first and second ribs. The body of the sternum attaches to the costal cartilages of the second through the tenth ribs. The xiphoid process does not attach to ribs but is an attachment for abdominal muscles. The costal cartilages of the eighth, ninth, and tenth ribs fuse to form the costal margin of the rib cage. A costal angle is formed where the two costal margins come together at the xiphoid process. The **sternal angle** (angle of Louis) may be palpated as an elevation between the manubrium and body of the sternum at the level of the second rib (see fig. 6.38). The costal angle, costal margins, and sternal angle are important surface landmarks of the thorax and abdomen.

Ribs

There are twelve pairs of **ribs,** each pair being attached posteriorly to a thoracic vertebra. Anteriorly, the first seven pairs are anchored to the sternum by individual *costal cartilages* and are called **true ribs.** The remaining five pairs (ribs 8, 9, 10, 11, and 12) are termed **false ribs.** Because the last two pairs of false ribs do not attach at all to the sternum, they are referred to as **floating ribs.** The four floating ribs are embedded in the muscles of the body wall.

sternum: Gk. *sternon*, chest
manubrium: L. *manubrium*, a handle
xiphoid: Gk. *xiphos*, sword
costal: L. *costa*, rib

Developmental Exposition

The Axial Skeleton

Exhibit I Ossification centers of the skeleton of a 10-week-old fetus. (*a*) The diagram depicts endochondral ossification in red and intramembranous ossification in a stippled pattern. The cartilaginous portions of the skeleton are shown in gray. (*b*) The photograph shows the ossification centers stained with a red indicator dye.

(a)   (b)

Labels on diagram (a):
Parietal bone
Temporal bone
Occipital bone
Chondro-cranium
Vertebrae
Clavicle
Scapula
Humerus
Ribs
Ilium
Sacrum
Coccyx

Frontal bone
Zygomatic bone
Nasal bone
Maxilla
Mandible
Phalanges
Metacarpal bones
Carpal bones
Radius
Ulna
Femur
Patella
Fibula
Tibia
Phalanges
Metatarsal bones
Tarsal bones

Waldrop

Explanation

Development of Bone

Bone formation, or *ossification,* begins at about the fourth week of embryonic development, but ossification centers cannot be readily observed until about the tenth week (exhibit I). Bone tissue derives from specialized migratory cells of mesoderm (see chapter 6) known as *mesenchyme.* Some of the embryonic mesenchymal cells will transform into *chondroblasts (kon′dro-blastz)*

---

chondroblast: Gk. *chondros,* cartilage; *blastos,* offspring or germ

and develop a cartilage matrix that is later replaced by bone in a process known as **endochondral** *(en″do-kon′dral)* **ossification.** Most of the skeleton is formed in this fashion—first it goes through a hyaline cartilage stage and then it is ossified as bone.

A smaller number of mesenchymal cells develop directly into bone without first going through a cartilage stage. This type of bone formation process is referred to as **intramembranous** *(in″trah-mem′brah-nus)* **ossification.** Facial bones and certain bones of the cranium are formed this way. **Sesamoid bones** are specialized intramembranous bones that develop in tendons. The patella is an example of a sesamoid bone.

UNIT IV

Exhibit II The embryonic skull at 12 weeks is composed of bony elements from three developmental sources: the chondrocranium (colored blue-gray), the neurocranium (colored light yellow), and the viscerocranium (colored salmon).

Parietal bone

Squama of occipital bone

Occipital bone

Squama of temporal bone

Styloid cartilage

Hyoid cartilage

Thyroid cartilage

Creek

Frontal bone

Orbitosphenoid

Alisphenoid

Zygomatic bone

Nasal bone

Maxilla

Mandible

Meckel's cartilage

Cricoid cartilage

UNIT IV

Development of the Skull

The formation of the skull is a complex process that begins during the fourth week of embryonic development and continues well beyond the birth of the baby. Three factions are involved in the formation of the skull: the chondrocranium, the neurocranium, and the viscerocranium (exhibit II). The **chondrocranium** is the portion of the skull that undergoes endochondral ossification to form the bones supporting the brain. The **neurocranium** is the portion of the skull that develops through membranous ossification to form the bones covering the brain and facial region. The **viscerocranium** (splanchnocranium) is the portion that develops from the embryonic visceral arches and forms the auditory ossicles, the hyoid bone, and specific processes of the skull.

chondrocranium: Gk. *chondros*, cartilage; *kranion*, skull

viscerocranium: L. *viscera*, soft parts; Gk. *kranion*, skull

Figure 6.39 The structure of a rib.

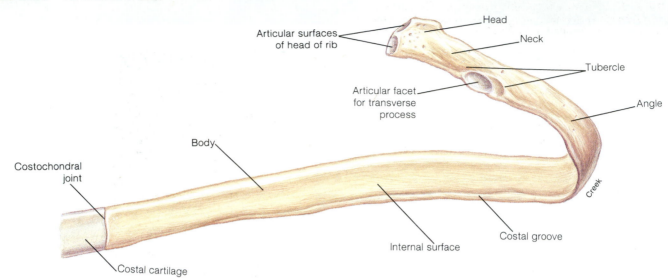

Although the structure of ribs varies, each of the first ten pairs has a **head** and a **tubercle** for articulation with a vertebra. The last two have a head but no tubercle. In addition, each of the twelve pairs has a **neck, angle,** and **body** (fig. 6.39). The head projects posteriorly and articulates with the body of a thoracic vertebra (fig. 6.40). The tubercle is a knoblike process, just lateral to the head. It articulates with the fovea on the transverse process of a thoracic vertebra. The neck is the constricted area between the head and the tubercle. The body is the main, curved part of the rib. Along the inner surface of the body is a depressed canal called the **costal groove** that protects the costal vessels and nerve. Spaces between the ribs are called **intercostal spaces** and are occupied by the intercostal muscles.

> Fractures of the ribs are relatively common injuries and most frequently occur between ribs 3 and 10. The first two pairs of ribs are protected by the clavicles, and the last two pairs move freely and will give with an impact. Little can be done to assist the healing of broken ribs other than binding them tightly to restrict movement.

1. Describe the rib cage and list its functions. What determines whether a rib is true, false, or floating?
2. Define the costal margin and the costal angle.

Clinical Considerations

Bone is a dynamic living tissue that is susceptible to hormonal or nutritional deficiency, diseases, and changes brought on by age. Since the development of bone is genetically governed, congenital conditions are possible. The hardness of bone gives it strength, yet it lacks the resiliency to avoid fracture if it undergoes excessive trauma. All of these aspects of bone provide some important and interesting clinical considerations.

Figure 6.40 Articulation of a rib with a thoracic vertebra as seen in a superior view.

Developmental Disorders

Congenital malformations account for several types of skeletal deformities. Certain bones may fail to form during osteogenesis, or they may form abnormally. **Cleft palate** and **cleft lip** are malformations of the palate and face. Cleft palates vary in severity and seem to have a mixed genetic and environmental cause. **Spina bifida** (spi'nă bif'i-dă) is a congenital defect of the vertebral column resulting from a failure of the laminae of the vertebrae to fuse, exposing the spinal cord (fig. 6.41). The lumbar area is mainly affected, and frequently only a single vertebra is involved.

Nutritional and Hormonal Disorders

Several bone disorders result from nutritional deficiencies or from excessive or deficient amounts of the hormones that regulate bone development and growth. Vitamin D has a tremen-

Figure 6.41 In spina bifida, failure of the vertebral arches to fuse permits a herniation of the meninges that cover the spinal cord through the vertebral column. This results in a condition called meningomyelocele.

dous influence on proper bone structure and function. When there is a deficiency of this vitamin, the body is unable to metabolize calcium and phosphorus. Vitamin D deficiency in children causes **rickets.** The bones of a child with rickets remain soft and are deformed from the weight of the body (see fig. 5.10).

A vitamin D deficiency in the adult causes the bones to demineralize, or to give up stored calcium and phosphorus. This condition is called **osteomalacia** (os''te-o-mă-la'shă). Osteomalacia is prevalent in women who have repeated pregnancies and poor diets and who experience relatively little exposure to sunlight.

The consequences of endocrine disorders are described in chapter 14. Since the impact of hormones on bone development is great, however, a few endocrine disorders will be briefly mentioned here. Hypersecretion of the growth hormone from the pituitary gland leads to **gigantism** in young people if it starts before ossification of their epiphyseal plates and to **acromegaly** (ak''ro-meg'ă-le) in adults. Acromegaly is characterized by hypertrophy of the bones of the face, hands, and feet. By contrast, in a child, a hyposecretion of the growth hormone can lead to **dwarfism.**

Paget's disease is a disease of disorganized metabolic processes within bone. The activity of osteoblasts and osteoclasts becomes irregular, producing areas with thickened osseous deposits and other areas where too much bone is removed. The etiology of the disease is unknown, but it is a relatively common affliction in people over the age of 50, and it occurs more frequently in males than in females.

Paget's disease: from Sir James Paget, English surgeon, 1814–99

Trauma and Injury

There are a variety of types of trauma to the skeletal system, ranging from injury to the bone itself to damage of the joints in the form of sprains or dislocations. Fractures and the healing of fractures will be discussed in chapter 7 and joint injuries will be discussed in chapter 8.

Neoplasms of Bone

Malignant bone tumors are three times more common than benign tumors. Pain is the usual symptom of either type of osseous neoplasm, although benign tumors may not have accompanying pain.

Two types of benign bone tumors are **osteoma,** which is the more frequent and which often involves the skull, and **osteoid osteoma,** which is a painful neoplasm of the long bones, usually in children.

Osteogenic sarcoma is the most virulent type of bone cancer and frequently metastasizes through the blood to the lungs. This disease usually originates in the long bones and is accompanied by aching and persistent pain.

A **bone scan** (fig. 6.42) is a diagnostic procedure frequently done on a person who has had a malignancy elsewhere in the body that may have metastasized to the bone. The patient receiving a bone scan may be injected with a radioactive substance that accumulates more rapidly in malignant tissue than normal tissue. Entire body radiographs show malignant bone areas as intensely dark dots.

Aging of the Skeletal System

Senescence affects the skeletal system by decreasing skeletal mass and density and increasing porosity and erosion (fig. 6.43). Bones become more brittle and susceptible to fracture. Articulating

UNIT IV

Figure 6.42 A bone scan of the legs of a patient suffering from arthritis in the left knee joint. In a bone scan, an arthritic joint shows up as a lighter image than a normal joint.

Figure 6.43 The geriatric skull. Note the loss of teeth and the degeneration of bone, particularly in the facial region.

UNIT IV

surfaces also deteriorate, contributing to arthritic conditions. Arthritic diseases are second to heart disease as the most common debilitation in the elderly.

Osteoporosis *(os"te-o-pŏ-ro'sis)* is the most prevalent metabolic disorder of bone in elderly people. It is characterized by marked demineralization, which weakens bone. The causes of osteoporosis include aging, prolonged inactivity, malnutrition, and an unbalanced secretion of hormones. It is most common in postmenopausal women. People with osteoporosis are prone to bone fracture, particularly at the pelvic girdle and vertebrae, as the bones become too brittle to support the body. Although there is no known cure for osteoporosis, it can be somewhat prevented in younger adults through proper diet, exercise, and good general health habits. Treatment of the disease in women through dietary calcium, exercise, and estrogens (female sex hormones) has limited positive results.

Distinct losses in height occur during middle and old age. Between the ages of 50 and 55, there is a decrease of 0.5–2 cm (0.25–0.75 in.) because of compression and shrinkage of the intervertebral discs. Elderly individuals may suffer a further major loss of height because of osteoporosis.

CLINICAL CASE STUDY ANSWER

The height (overall length) of the vertebral column is equal to the sum of the thicknesses of the vertebrae plus the sum of the thicknesses of the intervertebral discs. The body of a vertebra consists of outer compact bone and inner spongy bone. An intervertebral disc consists of a fibro-cartilage sheath called the annulus fibrosus and a mucoid center portion called the nucleus pulposus. The intervertebral discs generally change their anatomical configuration as one ages. In early adulthood, the nucleus pulposus is spongy and moist. With advanced age, however, the nucleus pulposus desiccates, resulting in a flattening of the intervertebral disc. The collective effect of this occurring in all of the discs is that the vertebral bodies lie closer together, resulting in an overall decrease in a person's height. Another cause of height loss may be the presence of undetectable compression fractures of the vertebral bodies, which are common in elderly people. This phenomenon, however, is considered pathological and not an aspect of the normal aging process. If a person has osteoporosis, there can be a marked decrease in height and perhaps more serious clinical problems as well, such as compression of spinal nerves.

Internal Affairs

The Relational Workings of the Body Systems

Skeletal System

- Provides skin with physical support

Integumentary System
- Supplies vitamin D for absorption of calcium and phosphorus

- Provides attachment sites for muscles
- Source of calcium for muscle contraction

Muscular System
- Causes bones to move at joints
- Partially responsible for bone shape and strength

- Protects central nervous system with bony encasement
- Source of calcium for neural function

Nervous System
- Sensory receptors provide sensations of body position and pain from bones and joints

- Protects endocrine glands in head and pelvis
- Source of calcium for production of certain hormones

Endocrine System
- Hormonally controls bone growth and maintenance

- Bone marrow produces blood cells
- Source of calcium for cardiac muscle contraction

Circulatory System
- Transports O_2 and CO_2, nutrients, and hormones to and from bone tissue

- Bone marrow produces and stores lymphocytes and other cells of immune system

Lymphatic System
- Maintains balanced amount of interstitial fluid within bone tissue
- Lymphocytes protect bone tissue following trauma

- Forms respiratory passageway through nasal cavity
- Protects lungs and aids in ventilation

Respiratory System
- Provides O_2 and eliminates CO_2

- Provides organs of GI tract with physical support and protection
- Stores minerals

Digestive System
- Provides nutrients for growth, maintenance, and repair of bone tissue

- Provides organs of urinary system with physical support and protection

Urinary System
- Eliminates metabolic wastes
- Activates vitamin D

- Provides organs of reproductive system with physical support and protection

Reproductive System
- Gonads produce sex hormones that promote growth and development, and maintain bone tissue

UNIT IV

UNIT IV

IMPORTANT CLINICAL TERMINOLOGY

achondroplasia (ah-kon''dro-pla'ze-ah) A genetic defect resulting in the retarded formation of cartilaginous bone during fetal development.

craniotomy Surgical cutting into the cranium to provide access to the brain.

epiphysiolysis (ep''i-fiz''e-ol'i-sis) A weakening and separation of the epiphysis from the diaphysis of a growing long bone.

laminectomy The surgical removal of the posterior arch of a vertebra, usually to repair a herniated intervertebral disc.

orthopedics A branch of medicine concerned with the diagnosis and treatment of trauma, diseases, and abnormalities involving the skeletal and muscular systems.

osteitis An inflammation of bone tissue.

osteoblastoma A benign tumor produced from bone-forming cells, most frequently in the vertebrae of young children.

osteochondritis An inflammation of bone and cartilage tissues.

osteomyelitis An inflammation of bone marrow caused by bacteria or fungi.

osteonecrosis The death of bone tissue, usually caused by obstructed arteries.

osteopathology The study of bone diseases.

osteosarcoma A malignant tumor of bone tissue.

osteotomy The cutting of a bone, usually by means of a saw or a chisel.

CHAPTER SUMMARY

Organization of the Skeletal System (pp. 130–132)

1. The axial skeleton consists of the skull, auditory ossicles, hyoid bone, vertebral column, and rib cage.
2. The appendicular skeleton consists of the bones within the pectoral girdle, upper extremities, pelvic girdle, and lower extremities.

Functions of the Skeletal System (pp. 132–133)

1. The mechanical functions of bones include the support and protection of softer body tissues and organs; in addition, certain bones function as levers during body movement.
2. The metabolic functions of bones include hemopoiesis and mineral storage.

Gross Structure of Bone (pp. 133–135)

1. Bone structure includes the shape and surface features of each bone, along with gross internal components.
2. Structurally speaking, bones may be classified as long, short, flat, or irregular.
3. The surface features of bones can be broadly classified into articulating surfaces, nonarticulating prominences, and depressions and openings.
4. A typical long bone has a diaphysis filled with marrow in the medullary cavity, epiphyses, epiphyseal plates for linear growth, and a covering of periosteum for appositional growth and the attachments of ligaments and tendons.

Bone Tissue (pp. 135–138)

1. The five types of bone cells are osteogenic cells, in contact with the endosteum and periosteum; osteoblasts (bone-forming cells); osteocytes (mature bone cells); osteoclasts (bone-destroying cells); and bone-lining cells, along the surface of most bones.
2. Compact bone consists of precise arrangements of osteons. The osteons contain osteocytes and lamellae.

Bone Growth (pp. 138–140)

1. Bone growth from embryonic to adult size is an orderly process determined by genetics, hormonal secretions, and nutritional supply.
2. Most bones develop through endochondral ossification.
3. Bone remodeling is a continual process that involves osteoclasts in bone resorption and osteoblasts in the formation of new bone tissue.

Skull (pp. 140–153)

1. The cranium encloses and protects the brain and provides for the attachment of muscles.
   - (a) Sutures are immovable joints between cranial bones.
   - (b) The eight cranial bones include the frontal, parietals, temporals, occipital, sphenoid, and ethmoid.
2. Facial bones form the basic shape of the face, support the teeth, and provide for the attachment of the facial muscles.
   - (a) The fourteen facial bones are the nasals, maxillae, zygomatics, mandible, lacrimals, palatines, inferior nasal conchae, and vomer.
   - (b) The hyoid bone is located in the neck, between the mandible and the larynx.
   - (c) The three paired auditory ossicles (malleus, incus, and stapes) are located within the middle-ear chambers of the petrous part of the temporal bones.

Vertebral Column (pp. 153–158)

1. The vertebral column consists of seven cervical, twelve thoracic, five lumbar, four or five fused sacral, and four or five fused coccygeal vertebrae.
2. Cervical vertebrae have transverse foramina; thoracic vertebrae have fovea for articulation with ribs; lumbar vertebrae have heavy bodies; sacral vertebrae are triangularly fused and articulate with the pelvic girdle; and the coccygeal vertebrae form a small triangular bone.

Rib Cage (pp. 158–162)

1. The sternum consists of a manubrium, body, and xiphoid process.
2. There are seven pairs of true ribs and five pairs of false ribs. The inferior two pairs of false ribs (pairs 11 and 12) are called floating ribs.

REVIEW ACTIVITIES

Objective Questions

1. A bone is considered to be
   - (a) a tissue.
   - (b) a cell.
   - (c) an organ.
   - (d) a system.
2. Which of the following statements is *false*?
   - (a) Bones are important in the synthesis of vitamin D.
   - (b) Bones and teeth contain about 99% of the body's calcium.
   - (c) Red bone marrow is the primary site for hemopoiesis.
   - (d) Most bones develop through endochondral ossification.
3. Match each of the following foramina to the correct bone in which it occurs.
   1. rotundum
   2. mental
   3. carotid canal
   4. olfactory
   5. magnum
   - (a) ethmoid bone
   - (b) occipital bone
   - (c) sphenoid bone
   - (d) mandible
   - (e) temporal bone

4. With respect to the hard palate, which of the following statements is *false*?
   (a) The hard palate is composed of two maxillae and two palatine bones.
   (b) The hard palate separates the oral cavity from the nasal cavity.
   (c) The mandible articulates with the posteriolateral angles of the hard palate.
   (d) The median palatine suture, incisive fossae, and greater palatine foramina are structural features of the hard palate.

5. The location of the sella turcica is immediately
   (a) superior to the sphenoidal sinus.
   (b) inferior to the frontal sinus.
   (c) medial to the petrous parts of the temporal bones.
   (d) superior to the perpendicular plate of the ethmoid bone.

6. Specialized bone cells that enzymatically reabsorb bone tissue are
   (a) osteoblasts.   (c) osteons.
   (b) osteocytes.   (d) osteoclasts.

7. The mandibular fossa is located in which structural part of the temporal bone?
   (a) the squamous part
   (b) the tympanic part
   (c) the mastoid part
   (d) the petrous part

8. The crista galli is a structural feature of which bone?
   (a) the sphenoid bone
   (b) the ethmoid bone
   (c) the palatine bone
   (d) the temporal bone

9. Transverse foramina are characteristic of
   (a) lumbar vertebrae.
   (b) sacral vertebrae.
   (c) thoracic vertebrae.
   (d) cervical vertebrae.

10. The bone disorder common in elderly people, particularly if they have prolonged inactivity, malnutrition, or an unbalanced secretion of hormones, is
    (a) osteitis.   (c) osteoporosis.
    (b) osteonecrosis. (d) osteomalacia.

Essay Questions

1. What are the functions of the skeletal system? Does each of the bones of the skeleton carry out these functions equally? Explain.
2. Explain why there are approximately 270 bones in an infant but 206 bones in a mature adult.
3. List the bones of the skull that are paired. Which are unpaired? Identify the bones of the skull that can be palpated.
4. Describe the development of the skull. What are the fontanels, where are they located, and what are their functions?
5. Which facial bones contain foramina? What structures traverse these openings?
6. Distinguish between the axial and appendicular skeletons. Describe where these two components articulate.
7. List four types of bones based on shape and give an example for each group.
8. Diagram a typical long bone. Label the epiphyses, diaphysis, epiphyseal plates, medullary cavity, nutrient foramina, periosteum, and articular cartilages.

9. List the bones that form the cranial cavity, the orbit, and the nasal cavity. Describe the location of the paranasal sinuses, the mastoidal sinus, and the inner-ear cavity.
10. Describe how bones grow in length and in circumference. How are these processes similar, and how do they differ? Explain how radiographs can be used to determine normal bone growth.
11. Explain the process of endochondral ossification of a long bone. Why is it important that a balance be maintained between osteoblast activity and osteoclast activity?
12. Describe the curvature of the vertebral column. What is meant by primary curves as compared to secondary curves?
13. List two or more characteristics by which vertebrae from each of the five regions of the vertebral column can be identified.
14. Identify the bones that form the rib cage. What functional role do the bones and the costal cartilages have in respiration?
15. Explain why a proper balance of vitamins, hormones, and minerals is essential in maintaining healthy osseous tissue. Give examples of diseases or skeletal conditions that may occur in the event of an imbalance of any of these three essential substances.

Gundy/Weber Software

The tutorial software accompanying Chapter 6, Skeletal System and the Axial Skeleton, is Volume 3, Skeletal System.

Skeletal System

Appendicular Skeleton

Chapter Outline

CLINICAL CASE STUDY

A 12-year-old boy is hit by a car while crossing a street. He is brought in stable condition to the emergency room complaining of severe pain in his right leg. Radiographs reveal a longitudinal fracture beginning on the surface of the tibial plateau, coursing 4 in. inferiorly into the anterior cortex of the tibia. The fragment of bone created by the fracture is moderately displaced. The orthopedic surgeon takes the radiographs into the waiting room and confers with the boy's parents. He explains that this particular injury is more serious in children and growing adolescents than in adults. He elaborates by stating that future growth of the bone may be jeopardized, and that surgery will be required but cannot guarantee normal growth. The parents ask, "What is it about the fracture that threatens future growth?"

How would you respond?

Hints: Review the section on bone growth in chapter 6. Carefully examine figures 6.4 and 6.9 in chapter 6 and figures 7.17 and 7.23 in this chapter.

Pectoral Girdle and Upper Extremity

The structure of the pectoral girdle and upper extremities is adaptive for freedom of movement and extensive muscle attachment.

Objective 1. Describe the bones of the pectoral girdle and the positions of articulations.

Objective 2. Identify the bones of the upper extremity and list the diagnostic features of each bone.

Pectoral Girdle

The two scapulae and two clavicles make up the pectoral (shoulder) girdle (fig. 7.1). It is not a complete girdle, having only an anterior attachment to the axial skeleton at the sternum. The primary function of the pectoral girdle is to provide attachment for the numerous muscles that move the brachium and forearm. The pectoral girdle is not weight-bearing and is therefore more delicate in structure than the pelvic girdle.

Clavicle The slender S-shaped clavicle (collarbone) binds the shoulder to the axial skeleton and positions the shoulder joint away from the trunk for freedom of movement. The articulation of the medial **sternal extremity** (fig. 7.2) of the clavicle to the manubrium is referred to as the *sternoclavicular joint.* The lateral **acromial** (ă-kro′me-al) **extremity** of the clavicle articulates with the acromion process of the scapula. This articu-

lation is referred to as the *acromioclavicular joint.* A **conoid tubercle** is present on the inferior surface of the lateral end, and a **costal tuberosity** is present on the inner surface of the medial end. Both processes serve as points of attachment for ligaments.

The long, delicate clavicle is the most commonly fractured bone in the body. Blows to the shoulder or an attempt to break a fall with an outstretched hand causes the force to be displaced to the clavicle. The most vulnerable area for a fracture of this bone is through its center, immediately proximal to the conoid tubercle. Because the clavicle is subcutaneous and not covered with muscle, a fracture can easily be palpated.

Scapula The scapula (shoulder blade) is a large, triangular flat bone positioned on the posterior aspect of the rib cage, overlying ribs 2 to 7. The **spine** of the scapula is a prominent diagonal bony ridge seen on the posterior surface (fig. 7.3). The spine increases the strength to the scapula for resistance to bending. Above the spine is the **supraspinous** (soo″prah-spi′nus) **fossa,** and below the spine is the **infraspinous fossa.** The spine broadens toward the shoulder as the **acromion** (figs. 7.3 and 7.4). The acromion serves for the attachment of several muscles, as well as for articulation with the clavicle. Inferior to the acromion is a shallow depression, the **glenoid cavity,** into which the head of the humerus fits. The **coracoid process** is a thick, upward projection that lies superior and anterior to the glenoid cavity. On the anterior surface of the scapula is a slightly concave area known as the **subscapular fossa.**

The scapula has three borders separated by three angles. The superior edge is called the **superior border.** The **medial border** is nearest to the vertebral column, positioned about 5 cm (2 in.) away. The **lateral border** is directed toward the arm. The **superior angle** is located between the superior and medial borders and the **inferior angle** is located between the medial and lateral borders. The **lateral angle** is located between the superior and medial borders. It is at the lateral angle that the scapula articulates with the head of the humerus. Along the superior border, a distinct depression called the **scapular notch** serves as a passageway for a nerve.

The anatomy of the scapula is important to know because some 15 muscles attach to its processes and fossae. Clinically, the pectoral girdle is significant because the clavicle and acromion of the scapula are frequently fractured in trying to break a fall. The acromion is palpated when locating the proper site for an intramuscular injection of the arm. This site is chosen because the musculature is quite thick and contains few nerves.

---

clavicle: L. *clavicula,* a small key
acromial: Gk. *akros,* peak; *omos,* shoulder

conoid tubercle: Gk. *konos,* cone; L. *tuberculum,* a small swelling
costal tuberosity: L. *costa,* rib; *tuberosus,* a knob
scapula: L. *scapula,* shoulder
glenoid: Gk. *glenoeides,* shallow form
coracoid process: Gk. *korakodes,* like a crow's beak

Figure 7.1 The human skeleton: (*a*) an anterior view and (*b*) a posterior view. The appendicular portion is distinguished with orange color.

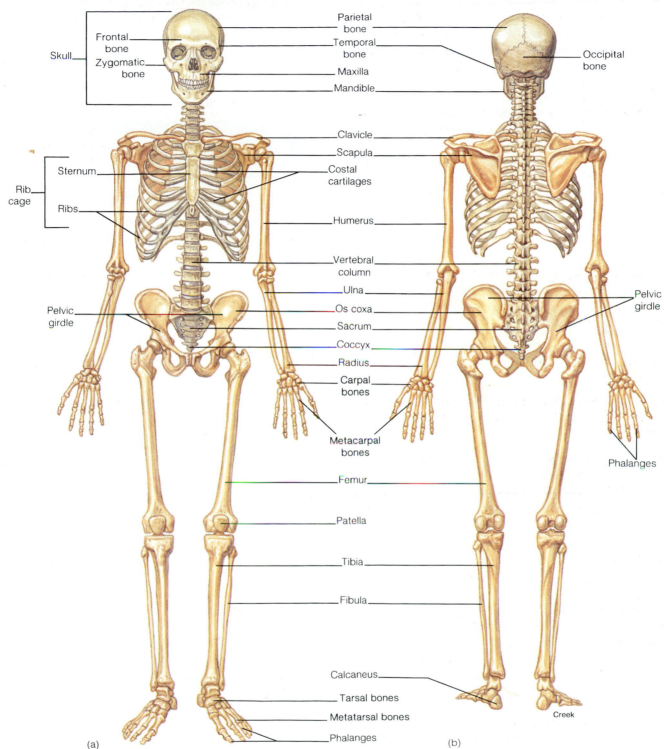

Skull
Frontal bone
Zygomatic bone
Parietal bone
Temporal bone
Maxilla
Mandible
Occipital bone
Clavicle
Scapula
Costal cartilages
Sternum
Rib cage
Ribs
Humerus
Vertebral column
Ulna
Os coxa
Sacrum
Coccyx
Radius
Carpal bones
Pelvic girdle
Pelvic girdle
Metacarpal bones
Phalanges
Femur
Patella
Tibia
Fibula
Calcaneus
Tarsal bones
Metatarsal bones
Phalanges

(a)   (b)

Creek

UNIT IV

Brachium (Upper Arm)

The brachium extends from the shoulder to the elbow and contains a single bone, the humerus.

Humerus The humerus (fig. 7.5) is the longest bone of the upper extremity. It consists of a proximal **head,** which artic-ulates with the glenoid cavity of the scapula, a **shaft** (body) and a distal end, which is modified to articulate with the two bones of the forearm. Surrounding the margin of the head is a slightly indented groove denoting the **anatomical neck.** The region where the shaft of the humerus begins to taper is referred to as the **surgical neck,** a frequent site of fractures. Lateral to the head is a large eminence, the **greater tubercle.** The **lesser tubercle** is slightly anterior to the greater and is separated from the greater by an **intertubercular (bicipital) groove,** through which the tendon from the biceps brachii muscle passes.

Along the lateral midregion of the shaft of the humerus is a roughened area, the **deltoid tuberosity,** for the attachment of the deltoid muscle.

deltoid tuberosity: Gk. *deltoeides*, shaped like the letter Δ

The distal end of the humerus has two rounded condyloid articular surfaces. The **capitulum** *(kah-pit'yoo-lum)* is the lateral rounded condyle that articulates with the radius. The

capitulum: L. *caput*, little head

Figure 7.4 This radiograph of the right shoulder shows the articulation of the scapula and humerus forming the shoulder joint.

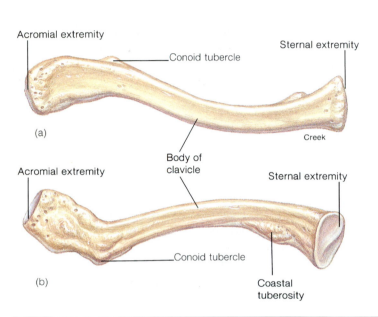

Figure 7.3 The right scapula. (*a*) An anterior view and (*b*) a posterior view.

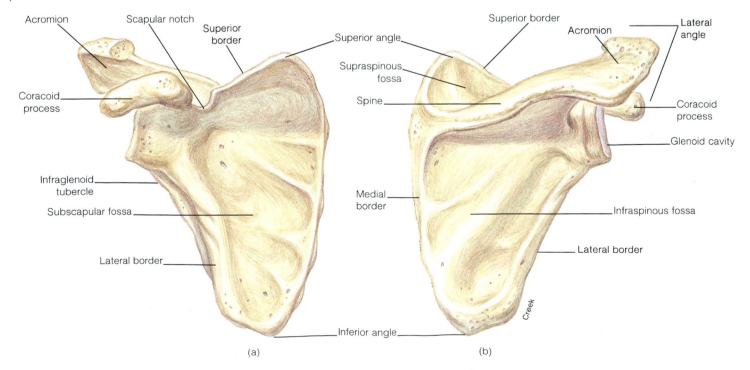

trochlea *(trok'le-ă)* of the humerus is the pulleylike medial surface that articulates with the ulna. On either side above the condyles are the **lateral** and **medial epicondyles.** The large medial epicondyle protects the ulnar nerve that passes posteriorly through the ulnar sulcus (see fig. 10.34). The **coronoid fossa** is a depression above the trochlea on the anterior surface. The **olecranon** *(o-lek'rah-non)* **fossa** is a depression on the distal posterior surface. Both fossae are adapted to receive parts of the ulna during movement of the forearm.

> The medical term for tennis elbow is *lateral epicondylitis,* which means an inflammation of the tissues surrounding the lateral epicondyle of the humerus. At least six muscles that control backward (extension) movement of the hand and fingers originate on the lateral epicondyle. Repeated strenuous contractions of these muscles, as in stroking with a tennis racket, may cause a strain on the periosteum and tendinous muscle attachments, resulting in tenderness and pain around the epicondyle. Binding usually eases the pain, but only rest can eliminate the causative factor, and recovery generally follows.

trochlea: Gk. *trochilia,* a pulley
olecranon: Gk. *olene,* ulna; *kranion,* head

Antebrachium (Forearm)

The skeletal structures of the antebrachium are the ulna on the medial side and the radius on the lateral (thumb) side (figs. 7.6 and 7.7). The ulna is more firmly connected to the humerus than the radius, and it is longer than the radius. The radius, however, contributes more significantly at the wrist joint than does the ulna.

Ulna The proximal end of the ulna articulates with the humerus and radius. A distinct depression, the **trochlear notch,** articulates with the trochlea of the humerus. The **coronoid process** forms the anterior lip of the semilunar notch, and the

Figure 7.5 The right humerus. (*a*) An anterior view and (*b*) a posterior view.

Greater tubercle
Lesser tubercle
Intertubercular groove
Head
Surgical neck
Greater tubercle
Anatomical neck
Nutrient foramen
Deltoid tuberosity
Shaft of humerus (posterior surface)
Shaft of humerus (anterior surface)
Olecranon fossa
Coronoid fossa
Lateral epicondyle
Capitulum
Medial epicondyle
Trochlea
Lateral epicondyle

(a)   (b)

UNIT IV

Figure 7.6 An anterior view of the right radius and ulna.

Figure 7.7 A posterior view of the right radius and ulna.

olecranon forms the posterior portion, or elbow. Lateral and inferior to the coronoid process is the **radial notch,** which accommodates the head of the radius.

The tapered distal end of the ulna has a knobbed part, the **head,** from which the **styloid process** projects posteromedially. The ulna articulates distally and proximally with the radius.

Radius   The radius consists of a **shaft** (body) that has a small proximal end and a large distal end. A proximal disc-shaped **head** articulates with the capitulum of the humerus and the radial notch of the ulna. The prominent **tuberosity of radius** (radial tuberosity), for attachment of the biceps brachii muscle, is on the anteromedial side of the shaft of the radius, just below the head. The distal end of the radius has a double-faceted surface for articulation with the proximal carpal bones. The distal

styloid process: Gk. *stylos*, pillar; *eidos*, resemblance

end of the radius also has the **styloid process** on the lateral tip and the **ulnar notch** on the medial side that receives the distal end of the ulna. The styloid processes on the ulna and radius provide lateral and medial stability for articulation at the wrist.

When a person falls, the natural tendency is to extend the hand to break the fall. This reflexive movement frequently results in fractured bones. Common fractures of the radius include a fracture of the head as it is driven forcefully against the capitulum, a fracture of the neck, or a fracture of the distal end *(Colles' fracture)* caused by landing on an outstretched hand.

When falling, it is less traumatic to the body to withdraw the appendages, bend the knees, and let the entire body hit the surface. Athletes learn that this is the safe way to fall.

Manus (Hand)

The hand contains 27 bones, constituting the carpus, metacarpus, and phalanges (figs. 7.8, 7.9, and 7.10).

Figure 7.8 A posterior view of the bones of the right hand. (*a*) A drawing and (*b*) a photograph.

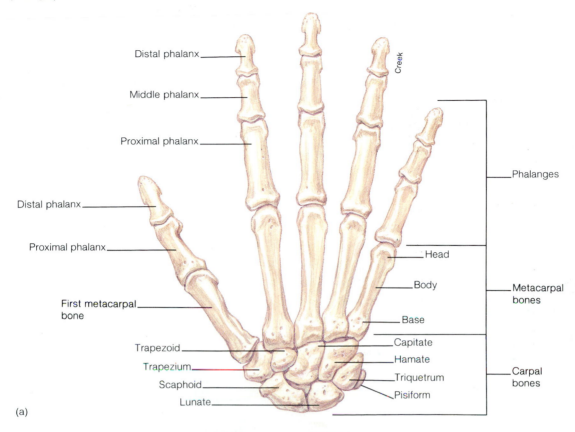

Distal phalanx

Middle phalanx

Proximal phalanx

Distal phalanx

Proximal phalanx

First metacarpal bone

Trapezoid

Trapezium

Scaphoid

Lunate

Creek

Phalanges

Head

Body

Base

Capitate

Hamate

Triquetrum

Pisiform

Metacarpal bones

Carpal bones

(a)

UNIT IV

Distal phalanx

Middle phalanx

Proximal phalanx

Hamate

Capitate

Trapezoid

Trapezium

Scaphoid

Lunate

Phalanges

Metacarpal bones

Carpal bones

Triquetrum

Figure 7.9 An anterior view of the bones of the right hand.

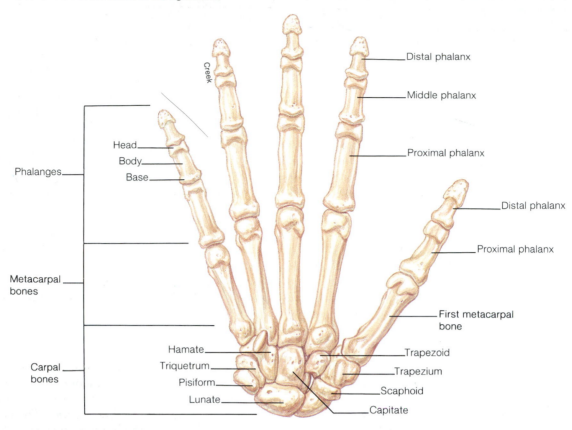

Figure 7.10 A radiograph of the right hand shown in an anteroposterior projection. (Note the presence of a sesamoid bone at the thumb joint.)

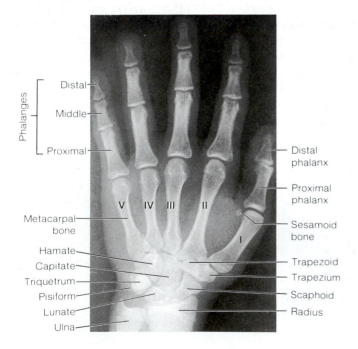

Carpus The carpus, or wrist, consists of eight carpal bones arranged in two transverse rows of four bones each. The proximal row, naming from the lateral (thumb) to medial side, consists of the **scaphoid** (navicular) **bone, lunate bone, triquetrum** (*tri-kwe′trum*) (triangular) **bone,** and **pisiform** (*pi′si-form*) **bone.** The pisiform bone forms in a tendon as a sesamoid bone. The distal row, from lateral to medial, consists of the **trapezium** (greater multangular) **bone, trapezoid** (lesser multangular) **bone, capitate bone,** and **hamate** (*ham′at*) **bone.** The scaphoid and lunate of the proximal row articulate with the distal end of the radius.

Metacarpus The metacarpus, or palm of the hand, is composed of five metacarpal bones. Each metacarpal bone consists of a proximal **base,** a **shaft,** and a distal **head** that is rounded for articulation with the base of each proximal phalanx. The heads of the metacarpal bones are distally located and form the knuckles of a clenched fist. The metacarpal bones are numbered I to V, the lateral, or thumb, side being I.

carpus: Gk. *karpos,* wrist
navicular: L. *navicula,* small ship
lunate: L. *lunare,* crescent or moon-shaped
triquetrum: L. *triquetrus,* three-cornered
pisiform: Gk. *pisos,* pea
trapezium: Gk. *trapesion,* small table
capitate: L. *capitatus,* head
hamate: L. *hamatus,* hook

Table 7.1	Bones of the Pectoral Girdle and Upper Extremities	
Name and Number	**Location**	**Diagnostic Features**
Clavicle (2)	Anterior base of neck, between sternum and scapula	S-shaped; sternal and acromial extremities
Scapula (2)	Upper back forming part of the shoulder	Triangular-shaped; spine; coracoid process and acromion
Humerus (2)	Brachium, between scapula and elbow	Longest bone of upper extremity; greater and lesser tubercles; surgical neck; deltoid tuberosity; capitulum; trochlea; coronoid and olecranon fossae
Ulna (2)	Medial side of forearm	Semilunar notch; olecranon and styloid processes
Radius (2)	Lateral side of forearm	Head; radial tuberosity; styloid process
Carpal bone (16)	Wrist	Short bones arranged in two rows of four bones each
Metacarpal bone (10)	Palm of hand	Long bones, each aligned with a digit
Phalanx (28)	Digits	Three in each digit, except two in thumb

Phalanges   The 14 phalanges are the skeletal elements of the digits. A single finger bone is called a **phalanx** *(fa'lanks)*. The phalanges of the fingers are arranged in a proximal row, a middle row, and a distal row. The thumb (pollex) has only a proximal and a distal phalanx.

A summary of the bones of the upper extremities is presented in table 7.1.

The hand is a marvel of structural organization that, despite its complexity, can withstand considerable abuse. Other than sprained fingers and dislocation, the most common bone injury is a fracture to the scaphoid bone of the wrist (about 70% of carpal fractures occur here). When immobilizing the wrist joint with a plaster cast, the wrist is positioned in the plane of relaxed function. This is the position in which the hand is about to grasp an object between the thumb and index finger.

1. Describe the structure of the pectoral girdle. Why is the pectoral girdle considered an incomplete girdle?
2. Identify the fossae and processes of the scapula.
3. Describe each of the long bones of the upper extremity.
4. Where are the styloid processes of the wrist area? What are their functions?
5. Name the bones in the proximal row of the carpus. Which of these bones articulate with the radius?

Pelvic Girdle and Lower Extremity

The structure of the pelvic girdle and lower extremities is adaptive for support and locomotion. Extensive processes and surface features on certain bones of the pelvic girdle and lower extremities accommodate massive muscles for posture and locomotion.

Objective 3.   Describe the structure of the pelvic girdle and list its functions.

Objective 4.   Describe how the male and female pelves differ structurally.

Objective 5.   Identify the bones of the lower extremity and list the diagnostic features of each bone.

Objective 6.   Describe the structural features and functions of the arches of the foot.

Pelvic Girdle

The **pelvic girdle,** or *pelvis,* is formed by two **ossa coxae** (hipbones) united anteriorly by the **symphysis pubis** (fig. 7.11). It is attached posteriorly to the sacrum of the vertebral column. The pelvic girdle and its associated ligaments support the weight of the body from the vertebral column. The pelvic girdle also supports and protects the lower viscera, including the urinary bladder, the reproductive organs, and in a pregnant woman, the developing fetus.

Clinically speaking, the basinlike pelvis is frequently divided into a **"greater,"** or **"false,"** pelvis and a **"lesser,"** or **"true,"** pelvis (see fig. 7.15). These two components are divided by the **pelvic brim,** a curved bony rim passing inferiorly from the sacral promontory to the upper margin of the symphysis pubis. The greater pelvis is the expanded portion of the pelvis, superior to the pelvic brim. The pelvic brim not only divides the two portions but surrounds the **pelvic inlet** of the lesser pelvis. The lower circumference of the lesser pelvis bounds the pelvic outlet. During parturition, a child must pass through its mother's lesser pelvis for a natural delivery. *Pelvimetry* measures the dimension of the lesser pelvis to determine whether a cesarean delivery might be necessary. Diameters may be determined by vaginal palpation or by radiographic measurements (fig. 7.12).

Each os coxa actually consists of three separate bones: the **ilium,** the **ischium** *(is'ke-um),* and the **pubis** (figs. 7.13 and 7.14). These bones are fused together in the adult. On the lateral surface of the os coxa where the three bones ossify together is a large circular depression, the **acetabulum** *(as''ĕ-tab'yŭ-lum),*

phalanx: Gk. *phalanx,* finger bone or toe bone

coxae: L. *coxae,* hips
ilium: L. *ilia,* loin
ischium: Gk. *ischion,* hip joint
pubis: L. *pubis,* genital area
acetabulum: L. *acetabulum,* vinegar cup

Figure 7.11 An anterior view of the pelvic girdle.

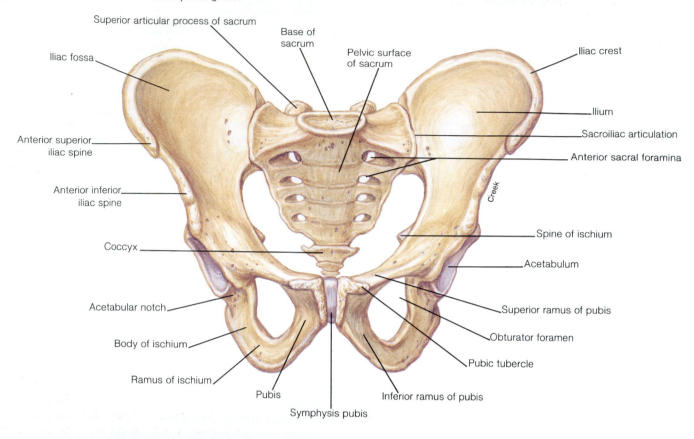

Superior articular process of sacrum

Base of sacrum

Pelvic surface of sacrum

Iliac crest

Iliac fossa

Iliac fossa

Anterior superior iliac spine

Anterior inferior iliac spine

Coccyx

Acetabular notch

Body of ischium

Ramus of ischium

Pubis

Symphysis pubis

Ilium

Sacroiliac articulation

Anterior sacral foramina

Spine of ischium

Acetabulum

Superior ramus of pubis

Obturator foramen

Pubic tubercle

Inferior ramus of pubis

Creek

UNIT IV

Figure 7.12 A radiograph of the pelvic girdle and the articulating femora.

Fifth lumbar vertebra

Ilium

Sacrum

Coccyx

Pelvic inlet

Acetabulum

Pelvic brim

Pubis

Sacral promontory

Sacroiliac joint

Anterior inferior iliac spine

Head of femur

Neck of femur

Greater trochanter of femur

Obturator foramen

Lesser trochanter

Ischium

Symphysis pubis

Figure 7.13 The lateral aspect of the right os coxa.

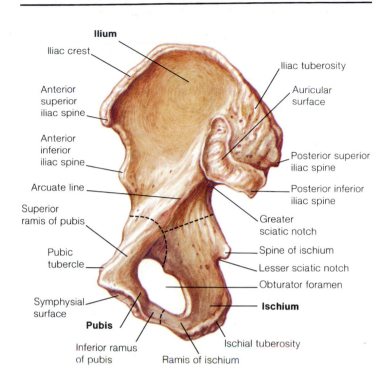

which receives the head of the femur. Although both ossa coxae are single bones in the adult, the three components are considered separately for descriptive purposes.

Ilium The ilium is the largest and uppermost of the three pelvic bones. The ilium presents a crest and four angles, or spines, which serve for muscle attachment and are important surface landmarks. The **iliac crest** forms the prominence of the hip. This crest terminates anteriorly as the **anterior superior iliac spine.** Just below this spine is the **anterior inferior iliac spine.** The posterior termination of the iliac crest is the **posterior superior iliac spine,** and just below this is the **posterior inferior iliac spine.**

Below the posterior inferior iliac spine is the **greater sciatic** *(si-at'ik)* **notch.** On the medial surface of the ilium is the roughened **auricular surface** that articulates with the sacrum. The **iliac fossa** is the smooth, concave surface on the anterior portion of the ilium. The iliacus muscle originates from this fossa. The **iliac tuberosity,** for the attachment of the sacroiliac ligament, is positioned posterior to the iliac fossa. Three roughened ridges are present on the **gluteal surface** of the posterior aspect of the ilium. These ridges serve to attach the gluteal muscles and are the **inferior, anterior,** and **posterior gluteal lines.**

Ischium The ischium is the posteroinferior component of the os coxa. This bone has several significant features. The **spine of the ischium** is the projection immediately posterior and inferior to the greater sciatic notch of the ilium and ischium. Inferior to this spine is the **lesser sciatic notch** of the ischium. The **ischial tuberosity** is the bony projection that supports the

weight of the body in the sitting position. A deep **acetabular notch** is present on the inferior portion of the acetabulum. The large **obturator foramen** is formed by the **ramus** of the ischium, together with the pubis. The obturator foramen is covered by the obturator membrane, to which several muscles attach.

Pubis The pubis is the anterior component of the os coxa. This bone consists of a **superior ramus** and an **inferior ramus** that supports the **body** of the pubis. The body contributes to the formation of the symphysis pubis—the joint between the two ossa coxae.

> The structure of the human pelvis, in its attachment to the vertebral column, permits an upright posture and locomotion on two legs (bipedal) rather than on four legs like other mammals. An upright posture may cause problems, however. The sacroiliac joint may weaken with age, causing lower back pains. The weight of the viscera may weaken the walls of the lower abdominal area and cause hernias. Some of the problems of childbirth are related to the structure of the mother's pelvis. Finally, the hip joint tends to deteriorate with age, so that many elderly people suffer from fractured hips.

Sex-Related Differences in the Pelvis Structural differences between the pelvis of an adult male and that of an adult female (fig. 7.15 and table 7.2) reflect the female's role in pregnancy and parturition.

In addition to the differences listed in table 7.2, the symphysis pubis and sacroiliac joints stretch during pregnancy and parturition.

UNIT IV

Figure 7.15 A comparison of the (a) male and (b) female pelvic girdle.

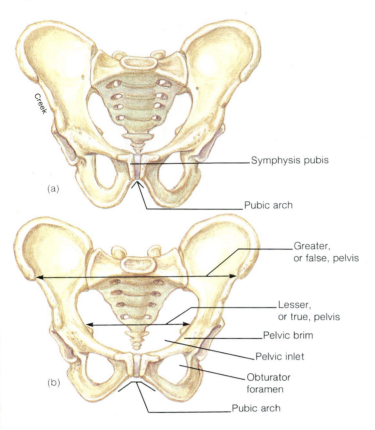

(a)

(b)

Symphysis pubis

Pubic arch

Greater, or false, pelvis

Lesser, or true, pelvis

Pelvic brim

Pelvic inlet

Obturator foramen

Pubic arch

Creek

UNIT IV

Table 7.2	Comparison of the Male and Female Pelvic Girdle	
Characteristics	Male Pelvis	Female Pelvis
General structure	More massive; prominent processes	More delicate; processes not so prominent
Pelvic inlet	Heart-shaped	Round or oval
Pelvic outlet	Narrower	Wider
Anterior superior iliac spines	Not as wide apart	Wider apart
Obturator foramen	Oval	Triangular
Acetabulum	Faces laterally	Faces more anteriorly
Symphysis pubis	Deeper, longer	Shallower, shorter
Pubic arch	Acute (less than 90°)	Obtuse (greater than 90°)

Thigh

The patella will be considered along with the femur in the discussion of the thigh, even though the femur is the only bone of the thigh.

Femur   The femur (thighbone) is the longest, heaviest, and strongest bone in the body (fig. 7.16). The proximal rounded **head** of the femur articulates with the acetabulum of the os coxa. A roughened, shallow pit, the **fovea capitis femoris,** is present in the lower center of the head of the femur. The fovea capitis femoris provides the point of attachment for the ligamentum teres, which helps to support the head of the femur against the acetabulum. The constricted region supporting the head is called the **neck** and is a common site for fractures in the elderly.

The **shaft** (body) **of the femur** has a slight medial bow so that it converges with the femur of the opposite thigh and brings the knee joints more in line with the body's plane of gravity. The degree of convergence is even greater in the female because of the wide pelvis. The shaft of the femur has several important structures for muscle attachment. On the proximolateral side of the shaft is the **greater trochanter,** and on the medial side is the **lesser trochanter.** On the anterior side, between the trochanters, is the **intertrochanteric line.** On the

posterior side, between the trochanters, is the **intertrochanteric crest.** The **linea aspera** is a vertical ridge on the posterior surface of the shaft of the femur.

The distal end of the femur is expanded for articulation with the tibia. The **medial** and **lateral condyles** are the articular processes for this joint. The depression between the condyles on the posterior aspect is called the **intercondylar fossa.** The **patellar surface** is between the condyles on the anterior side. Above the condyles on the lateral and medial sides are the **epicondyles** for ligament and tendon attachment.

Patella   The patella (kneecap) is a sesamoid bone positioned on the anterior side of the knee joint (figs. 7.17 and 7.18). It develops in response to strain in the tendon of the quadriceps femoris muscle. The patella is a triangular bone with a broad **base** and an inferiorly pointed **apex. Articular facets** on the posterior surface of this bone articulate with the medial and lateral condyles of the femur.

The functions of the patella are to protect the knee joint and to strengthen the quadriceps tendon. It also increases the leverage of the quadriceps femoris muscle as it straightens (extends) the knee joint.

The patella can be fractured by a direct blow. It usually does not fragment, however, because it is confined within the tendon. Dislocations of the patella may result from injury or may be congenital due to underdevelopment of the lateral condyle of the femur.

Leg

The tibia and fibula are the skeletal elements of the leg. The tibia is the larger and more medial of the two bones. The skeletal structure of the leg is illustrated in figure 7.18.

Tibia   The tibia (shinbone) articulates proximally with the femur at the knee joint to bear the weight of the body. On the distal end, the tibia articulates with the talus of the ankle. Two

---

femur: L. *femur*, thigh

linea aspera: L. *linea*, line; *asperare*, rough
patella: L. *patina*, small plate
tibia: L. *tibia*, shinbone, pipe, flute

Figure 7.16 The right femur. (*a*) An anterior view and (*b*) a posterior view.

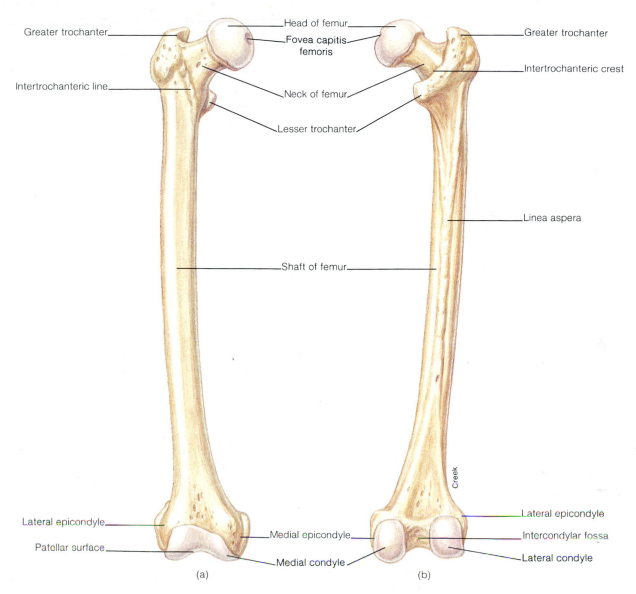

Greater trochanter

Intertrochanteric line

Head of femur
Fovea capitis femoris
Neck of femur
Lesser trochanter

Greater trochanter
Intertrochanteric crest

Linea aspera

Shaft of femur

Creek

Lateral epicondyle
Patellar surface
Medial epicondyle
Medial condyle

Lateral epicondyle
Intercondylar fossa
Lateral condyle

(a)   (b)

UNIT IV

Figure 7.17 A radiograph of the right knee region.

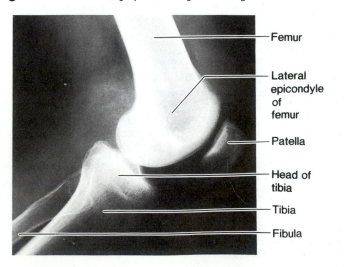

Femur

Lateral epicondyle of femur

Patella

Head of tibia

Tibia

Fibula

slightly concave surfaces on the proximal end of the tibia, the **medial** and **lateral condyles,** articulate with the condyles of the femur. Between the condyles, there is a slight upward projection called the **intercondylar eminence.** The **tibial tuberosity,** for attachment of the patellar ligament, is located on the proximoanterior part of the body of the tibia. A sharp ridge along the anterior surface of the body is called the **anterior crest.**

The **medial malleolus** (*mah-le'o-lus*) is a prominent medial knob of bone located on the distal end of the tibia. A **fibular notch,** for articulation with the fibula, is located on the distolateral end. The **lateral** and **medial epicondyles** are located above the condyles, on either side.

Fibula   The fibula is a long, narrow bone that is more important for muscle attachment than for support. The **head** of

fibula: L. *fibula,* clasp or brooch

Figure 7.18 The right tibia, fibula, and patella. (*a*) An anterior view and (*b*) a posterior view.

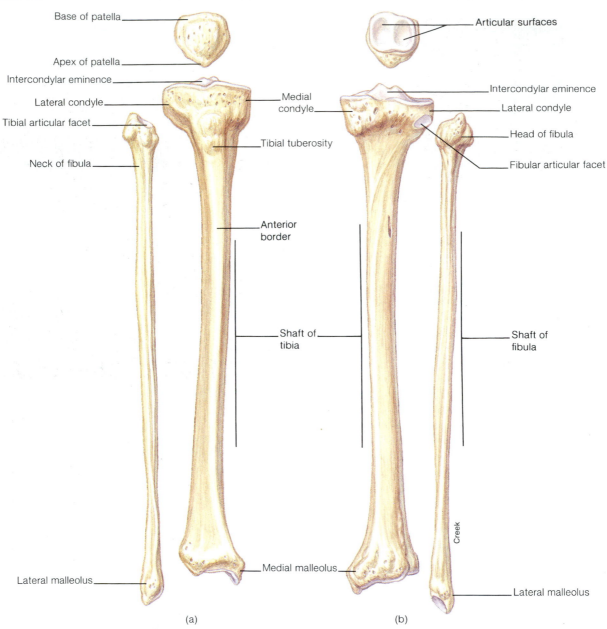

Base of patella

Apex of patella

Intercondylar eminence

Lateral condyle

Tibial articular facet

Neck of fibula

Medial condyle

Tibial tuberosity

Anterior border

Shaft of tibia

Lateral malleolus

Medial malleolus

Articular surfaces

Intercondylar eminence

Lateral condyle

Head of fibula

Fibular articular facet

Shaft of fibula

Lateral malleolus

Creek

(a)   (b)

the fibula articulates with the proximolateral end of the tibia. The distal end has a prominent knob called the **lateral malleolus.**

The lateral and medial malleoli are positioned on either side of the talus and help stabilize the ankle joint. Both processes can be seen as prominent surface features and are easily palpated. Fractures to either or both malleoli are common in skiers. These fractures, clinically referred to as *Pott's fractures,* result from a shearing force occurring at a vulnerable spot on the leg.

Pes (Foot)

The foot contains 26 bones, constituting the tarsus, metatarsus, and phalanges (figs. 7.19 and 7.20). The bones of the foot are basically similar to those of the hand. They do, however, have distinct structural differences in order to provide weight support and leverage during walking.

Tarsus   There are seven tarsal bones. The **talus** is the tarsal bone that articulates with the tibia and fibula to form the ankle joint. The **calcaneus** *(kal-ka'ne-us)* is the largest of the tarsal

malleolus: L. *malleolus,* small hammer

tarsus: G. *tarsos,* flat of the foot
talus: L. *talus,* ankle
calcaneus: L. *calcis,* heel

UNIT IV

Figure 7.19   (*a*) A superior view of the right foot. (*b*) A radiograph of the right foot. (*c*) A superior view of the bones of the right foot. (*d*) An inferior view of the bones of the right foot. (Note the presence of a sesamoid bone in (*b*) at the base of the big toe.)

Phalanges

Distal phalanx
Middle phalanx
Proximal phalanx

Metatarsal bones

I II III IV V

Medial cuneiform bone
Intermediate cuneiform bone
Lateral cuneiform bone
Navicular bone
Cuboid bone

Tarsal bones

Talus

Calcaneus

(a)

Distal phalanx
Proximal phalanx

Sesamoid bone
Metatarsal bones
Cuneiform bone
Navicular bone

Talus
Tibia
Fibula
Calcaneus

(b)

UNIT IV

Distal phalanx

Proximal phalanx

First metatarsal bone

Medial cuneiform bone

Intermediate cuneiform bone

Lateral cuneiform bone

Navicular

Talus

Distal phalanx
Middle phalanx
Proximal phalanx

Fifth metatarsal bone

Cuboid bone

Calcaneus

Tuberosity of calcaneus

Creek

(c)

Phalanges

Head

Metatarsal bones

Body

Base

Tarsal bones

(d)

bones and provides skeletal support for the heel of the foot. It has a large posterior extension, called the **tuberosity of the calcaneus,** for the attachment of the calf muscles. Anterior to the talus is the block-shaped **navicular** (*nah-vik'yŭ-lar*) **bone.** The remaining four tarsal bones form a distal series that articulate with the metatarsal bones. They are, from the medial to lateral side, the **medial, intermediate,** and **lateral cuneiform** (*kyoo-ne'i-form*) **bones** and the **cuboid bone.**

Metatarsus The skeleton of the metatarsus is composed of five metatarsal bones. The metatarsal bones are numbered I to V, with the medial, or big toe, side being I. The first metatarsal bone is larger than the others because of its weight-bearing function.

The metatarsal bones each have a **base, body,** and **head.** The proximal bases of the first, second, and third metatarsal bones articulate proximally with the cuneiform bones. The heads of the metatarsal bones articulate distally with the proximal phalanges. The proximal joints are called *tarsometatarsal joints* and the distal joints are called *metatarsophalangeal joints.* The ball of the foot is formed by the heads of the first two metatarsal bones.

Phalanges The 14 phalanges are the skeletal elements of the toes. As with the fingers of the hand, the phalanges of the toes are arranged in a proximal row, a middle row, and a distal row. The great toe (hallux) has only a proximal and a distal phalanx.

Arches of the Foot The foot has two arches that support the weight of the body and provide leverage when walking. These arches are formed by the structure and arrangement of the bones held in place by ligaments and tendons. The arches are not rigid; they yield when weight is placed on the foot and spring back as the weight is lifted (fig. 7.20).

The **longitudinal arch** is divided into medial and lateral parts. The medial part is the larger of the two. It is supported by the calcaneus bone proximally and by the heads of the first three metatarsal bones distally. The wedge, or keystone, of this part of the longitudinal arch is the talus. The shallower lateral part consists of the calcaneus, cuboid, and fourth and fifth metatarsal bones. The cuboid bone is the keystone of this part.

The **transverse arch** extends across the width of the foot and is formed by the distal part of the calcaneus, navicular, and cuboid bones and the proximal parts of all five metatarsal bones.

A weakening of the ligaments and tendons of the foot decreases the height of the longitudinal arch in a condition called *pes planus,* or flatfoot.

The bones of the lower extremities are summarized in table 7.3.

Figure 7.20 The arches of the foot. (*a*) A medial view of the right foot showing both arches and (*b*) a transverse view through the bases of the metatarsal bones showing a portion of the transverse arch.

Cuneiform bones
Cuboid bone
Talus
Calcaneus
Navicular
Transverse arch
Longitudinal arch
First metatarsal bone
Phalanges of big toe
(a)

Bases of metatarsal bones
Transverse arch
(b)

Table 7.3	Bones of the Pelvic Girdle and the Lower Extremities	
Name and Number	**Location**	**Diagnostic Features**
Os coxa (2)	Hip, part of the pelvic girdle; composed of the fused ilium, ischium, and pubis	Iliac crest, acetabulum, anterior superior iliac spine, ischial tuberosity, obturator foramen
Femur (2)	Bone of the thigh, between hip and knee	Head, fovea capitis femoris, neck, greater and lesser trochanters, lateral and medial condyles
Patella (2)	Anterior surface of knee	Triangular sesamoid bone
Tibia (2)	Medial side of leg, between knee and ankle	Medial and lateral condyles, tibial crest, medial malleolus
Fibula (2)	Lateral side of leg, between knee and ankle	Head, lateral malleolus
Tarsal bones (14)	Ankle	Large talus and calcaneus to receive weight of leg; five other wedge-shaped bones to help form arches of foot
Metatarsal bones (10)	Sole of foot	Long bones, each in line with a digit
Phalanx (28)	Digits	Three in each digit; two in big toe

Developmental Exposition

The Appendicular Skeleton

Exhibit I   The development of the appendicular skeleton. (*a*) Limb buds are apparent in an embryo by 28 days and (*b*) an ectodermal ridge is the precursor of the skeletal and muscular structures. (*c*) Mesenchymal primordial cells are present at 33 days. (*d*) Hyaline cartilaginous models of individual bones develop early in the 6th week. (*e*) Later in the 6th week, the cartilaginous skeleton of the upper extremity is well formed.

UNIT IV

Explanation

The development of the upper and lower extremities is initiated toward the end of the fourth week with the appearance of four small elevations called **limb buds** (exhibit I). The anterior pair are the arm buds, which precede the development of the posterior pair of leg buds by a few days. Each limb bud consists of a mass of undifferentiated mesoderm partially covered with a layer of ectoderm. This **apical** (*a'pĭ-kal*) **ectodermal ridge** promotes bone and muscle development.

As the limb buds elongate, migrating mesenchymal tissues differentiate into specific cartilaginous bones. Primary ossification centers soon form in each bone, and the hyaline cartilage tissue is gradually replaced by bony tissue in the process of *endochondral ossification* (see chapter 6).

Initially, the developing limbs are directed caudally, but later there is a lateral rotation in the upper extremity and a medial rotation in the lower extremity. As a result, the elbows are directed backward and the knees directed forward.

Digital rays that will form the hands and feet are apparent by the fifth week, and the individual digits separate by the end of the sixth week.

A large number of limb deformities occurred in children born between 1957 and 1962 as a result of mothers ingesting thalidomide during early pregnancy to relieve "morning sickness." It is estimated that 7,000 infants were malformed by thalidomide. The malformations ranged from *micromelia* (short limbs) to *amelia* (absence of limbs).

micromelia: Gk. *mikros*, small; *melos*, limb
amelia: Gk. *a*, without; *melos*, limb

1. Describe the structure of the pelvic girdle that reflects its weight-bearing role.
2. How can female and male pelves be distinguished? What is the clinical importance of the lesser pelvis in females?
3. Describe the structure of each of the long bones of the lower extremity and the position of each of the tarsal bones.
4. Which bones of the foot function in the formation of the arches of the foot? What are the functions of the arches?

Clinical Considerations

Developmental Disorders

Minor defects of the extremities are relatively common malformations. Extra digits, a condition called **polydactyly** (pol"e-dak'ti-le) (fig. 7.21), is the most common limb deformity. Usually an extra digit is incompletely formed and does not function. **Syndactyly** (sin-dak'ti-le), or webbed digits, is likewise a relatively common limb malformation. Polydactyly is inherited as a dominant trait, whereas syndactyly is a recessive trait.

Talipes (tal'i-pēz), or clubfoot (fig. 7.22), is a congenital malformation in which the sole of the foot is twisted medially. It is not certain if abnormal positioning or restricted movement in utero causes this condition, but both genetics and environmental conditions are involved in most cases.

Trauma and Injury

The most common type of bone injury is a **fracture.** A fracture is the cracking or breaking of a bone. Radiographs are often used to diagnose the position and extent of a fracture. Fractures may be classified in several ways, and the type and severity of the fracture varies with the age and the general health of the body. **Spontaneous,** or **pathologic, fractures,** for example, result from diseases that weaken the bones. Most fractures, however, are called **traumatic fractures** because they are caused by injuries. The following are descriptions of several kinds of traumatic fractures (fig. 7.23).

1. **Simple,** or **closed.** The fractured bone does not break through the skin.
2. **Compound,** or **open.** The fractured bone is exposed to the outside through an opening in the skin.
3. **Partial (fissured).** The bone is incompletely broken.
4. **Complete.** The fracture has separated the bone into two pieces.
5. **Capillary.** A hairlike crack occurs within the bone.
6. **Comminuted** (kom'i-noot"ed). The bone is splintered into small fragments.
7. **Spiral.** The fracture line is twisted as it is broken.

polydactyly: Gk. *polys*, many; *daktylos*, finger
syndactyly: Gk. *syn*, together; *daktylos*, finger
talipes: L. *talus*, heel; *pes*, foot

Figure 7.21 Polydactyly is the condition in which there are extra digits. It is the most common congenital deformity of the foot, although it also occurs in the hand. Syndactyly is the condition in which two or more digits are webbed together. It is a common congenital deformity of the hand, although it also occurs in the foot. Both conditions can be surgically corrected.

Figure 7.22 Talipes, or clubfoot, is a congenital malformation of a foot or both feet. The condition can be effectively treated surgically if the procedure is done at an early age.

8. **Greenstick.** In this incomplete break, one side of the bone is broken, and the other side is bowed.
9. **Impacted.** One broken end of a bone is driven into the other.
10. **Transverse.** The fracture occurs across the bone at right angles to the shaft.

Figure 7.23 Examples of types of traumatic fractures.

A *greenstick* fracture is incomplete, and the break occurs on the convex surface of the bend in the bone.

A *partial* (*fissured*) fracture involves an incomplete longitudinal break.

A *comminuted* fracture is complete and results in several bony fragments.

A *transverse* fracture is complete, and the break occurs at a right angle to the long axis of the bone.

An *oblique* fracture occurs at an angle other than a right angle to the long axis of the bone.

A *spiral* fracture is caused by twisting a bone excessively.

11. **Oblique.** The fracture occurs across the bone at an oblique angle to the long axis of the bone.
12. **Colles'.** A fracture of the distal portion of the radius.
13. **Pott's.** A fracture of either or both of the distal ends of the tibia and fibula at the level of the malleoli.
14. **Avulsion.** A portion of a bone is torn off.
15. **Depressed.** The broken portion of the bone is driven inwards, as in certain skull fractures.
16. **Displaced.** In this fracture, the bone fragments are not in anatomical alignment.
17. **Nondisplaced.** In this fracture, the bone fragments are in anatomical alignment.

When a bone fractures, medical treatment involves re-aligning the broken ends and then immobilizing them until new bone tissue is formed and the fracture is healed. The site and severity of the fracture and the age of the patient will determine the type of immobilization. The methods of immobilization include tape, splints, casts, straps, wires, and steel pins. Even with these various methods of treatment, certain fractures heal poorly. Ongoing research into these problems is producing promising results. It has been found, for example, that applying weak electrical currents to fractured bones promotes healing and reduces the time of immobilization by half.

Physicians can realign and immobilize a fracture, but the ultimate repair of the bone occurs naturally within the bone itself. Several steps occur in the repair of a fracture (fig. 7.24).

1. When a bone is fractured, the surrounding periosteum is usually torn and blood vessels in both tissues are ruptured. A blood clot called a **fracture hematoma** (*hēm-ah-to′mah*) soon forms throughout the damaged area. A disrupted blood supply to osteocytes and periosteal cells at the fracture site causes localized cellular death. This is followed by swelling and inflammation.
2. The traumatized area is "cleaned up" by the activity of phagocytic cells within the blood and osteoclasts that resorb bone fragments. As the debris is removed, fibrocartilage fills the gap within the fragmented bone, and a cartilaginous mass called a **bony callus** is formed. The bony callus becomes the precursor of bone formation in much the same way that hyaline cartilage is the precursor of developing bone.

fracture hematoma: Gk. *hema*, blood; *oma*, tumor
callus: L. *callosus*, hard

Figure 7.24 (*a–d*) Stages of the repair of a fracture. (*e*) A radiograph of a healing fracture.

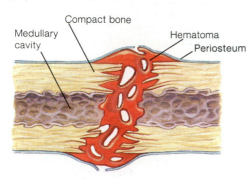

(a) Blood escapes from ruptured blood vessels and forms a hematoma.

(c) Fibrocartilage is replaced by a bony callus.

(b) Spongy bone forms in regions close to developing blood vessels; fibrocartilage forms in more distant regions.

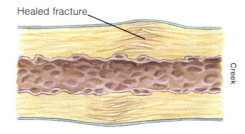

(d) Osteoclasts remove excess bony tissue, making new bone structure much like the original.

(e)

3. The remodeling of the bony callus is the final step in the healing process. The cartilaginous callus is broken down, a new vascular supply is established, and compact bone develops around the periphery of the fracture. A healed fracture line is frequently undetectable in a radiograph, except that the bone in this area is usually slightly thicker.

CLINICAL CASE STUDY ANSWER

The injury involves the cartilaginous epiphyseal growth plate, which is the site of linear growth of long bones. At cessation of growth, this plate disappears as the epiphysis and diaphysis fuse. Until this occurrence, however, disruption of the growth plate can adversely affect growth of the bone.

CHAPTER SUMMARY

Pectoral Girdle and Upper Extremity (pp. 170–177)

1. The pectoral girdle is composed of two scapulae and two clavicles. The clavicles attach the pectoral girdle to the axial skeleton at the sternum.
   (a) Diagnostic features of the clavicle include the conoid tubercle and acromial and sternal extremities.
   (b) Diagnostic features of the scapula include the spine, the acromion, and the coracoid process; the supraspinous, infraspinous, and subscapular fossae; the glenoid cavity; superior, medial, and lateral borders; and superior, inferior, and lateral angles.

2. The brachium contains the humerus, which extends from the scapula to the elbow.
   (a) Proximally, diagnostic features of the humerus include a rounded head, greater tubercle, anatomical neck, and an intertubercular groove. Distally, they include medial and lateral epicondyles, coronoid and olecranon fossae, a capitulum, and a trochlea.
   (b) The head of the humerus articulates proximally with the glenoid cavity of the scapula; distally, the trochlea and capitulum articulate with the ulna and radius, respectively.

3. The antebrachium contains the medial ulna and the lateral radius.
   (a) Proximally, diagnostic features of the ulna include the olecranon and coronoid processes and the trochlear notch. Distally, they include the styloid process and head of ulna.
   (b) Proximally, diagnostic features of the radius include the head and neck of radius and the tuberosity of radius. Distally, they include the styloid process and ulnar notch.

4. The hand contains 27 bones arranged as the carpal bones, metacarpal bones, and phalanges.

Pelvic Girdle and Lower Extremity (pp. 177–186)

1. The pelvic girdle is formed by two ossa coxae united anteriorly by the symphysis pubis.
2. The pelvis is divided into a greater pelvis, which helps to support the pelvic viscera, and a lesser pelvis, which forms the walls of the birth canal.
3. Each os coxa consists of an ilium, ischium, and pubis. Diagnostic features of the os coxa include an obturator foramen and an acetabulum, the latter of which is the socket for articulation with the head of the femur.
   (a) Diagnostic features of the ilium include an iliac crest, iliac fossa, anterior superior iliac spine, and anterior inferior iliac spine.
   (b) Diagnostic features of the ischium include the body, ramus, and ischial tuberosity.
   (c) Diagnostic features of the pubis include the ramus and pubic tubercle. The two pubic bones articulate at the symphysis pubis.
4. The thigh contains the femur, which extends from the hip to the knee where it articulates with the tibia and the patella.
   (a) Proximally, diagnostic features of the femur include the head, neck, and greater and lesser trochanters. Distally, they include the lateral and medial epicondyles, the lateral and medial condyles, and the patellar surface. The linea aspera is a roughened ridge positioned vertically along the posterior aspect of the body of the femur.
   (b) The head of the femur articulates proximally with the acetabulum of the os coxa and distally with the condyles of the tibia and the articular facets of the patella.
5. The leg contains the medial tibia and the lateral fibula.
   (a) Proximally, diagnostic features of the tibia include the lateral and medial epicondyles, intercondylar eminence, and tibial tuberosity. Distally, they include the medial malleolus. The anterior crest is a sharp ridge extending the anterior length of the tibia.
   (b) Proximally, diagnostic features of the fibula include the head. Distally, they include the lateral malleolus.
6. The foot contains 26 bones arranged as the tarsal bones, metatarsal bones, and phalanges.

REVIEW ACTIVITIES

Objective Questions

1. When in anatomical position, the subscapular fossa of the scapula faces
   (a) anteriorly.   (c) posteriorly.
   (b) medially.   (d) laterally.
2. The clavicle articulates with
   (a) the scapula and the humerus.
   (b) the humerus and the manubrium.
   (c) the manubrium and the scapula.
   (d) the manubrium, the scapula, and the humerus.
3. Which of the following bones has a conoid tubercle?
   (a) the scapula   (d) the clavicle
   (b) the humerus   (e) the ulna
   (c) the radius
4. The elbow of the ulna is formed by
   (a) the lateral epicondyle.
   (b) the olecranon.
   (c) the coronoid process.
   (d) the styloid process.
   (e) the medial epicondyle.
5. Which of the following statements concerning the carpus is *false*?
   (a) There are eight carpal bones arranged in two transverse rows of four bones each.
   (b) All of the carpal bones are considered sesamoid bones.
   (c) The navicular and the lunate bones articulate with the radius.
   (d) The trapezium, trapezoid, capitate, and hamate bones articulate with the metacarpal bones.
6. Pelvimetry is a measurement of
   (a) the os coxa.
   (b) the symphysis pubis.
   (c) the pelvic brim.
   (d) the lesser pelvis.
7. Which of the following is *not* a structural feature of the os coxa?
   (a) the obturator foramen
   (b) the acetabulum
   (c) the auricular surface
   (d) the greater sciatic notch
   (e) the linea aspera
8. A fracture across the intertrochanteric line would involve
   (a) the ilium.   (d) the fibula.
   (b) the femur.   (e) the patella.
   (c) the tibia.
9. As compared to the male pelvis, the female pelvis
   (a) is more massive.
   (b) is narrower at the pelvic outlet.
   (c) is tilted backward.
   (d) has a shallower symphysis pubis.
10. Clubfoot is a congenital malformation that is medically referred to as
    (a) talipes.   (c) pes planus.
    (b) syndactyly.   (d) polydactyly.

Essay Questions

1. Compare the pectoral and pelvic girdles in structure, articulation to the axial skeleton, and function.
2. Explain why the clavicle is more frequently fractured than the scapula.
3. List the processes of the bones of the upper and lower extremities that can be palpated. Why are these bony landmarks important to know?
4. There are basic similarities and specific differences between the bones of the hands and those of the feet. Compare and contrast these appendages, taking into account the functional role of each.
5. Define *bipedal locomotion* and discuss the adaptations of the pelvic girdle and lower extremities that allow for this type of movement.
6. What are the structural differences between male and female pelves?
7. Explain the significance of the limb buds, apical ectodermal ridges, and digital rays in limb development. When does limb development begin and when is it completed?
8. What is meant by a congenital skeletal malformation? Give two examples of such abnormalities that occur within the appendicular skeleton.
9. What are the differences between spontaneous and traumatic fractures? List some examples of traumatic fractures.
10. How does a fractured bone repair itself? Why is it important that the fracture be immobilized?

Gundy/Weber Software

The tutorial software accompanying Chapter 7, Skeletal System: The Appendicular Skeleton, is Volume 3, Skeletal System.

UNIT IV

Articulations

Chapter Outline

CLINICAL CASE STUDY

A 20-year-old male has sustained injury to his right knee during a college football game. Because of rapid swelling and intense pain, he has been taken to the emergency room at the local hospital. When the emergency room physician asks him to describe how the injury occurred, the athlete responds, "I was carrying the ball on an end run left on third down and two. As I planted my right foot to make a cut, I was hit in the knee from the side. I felt my knee give way and then I felt intense pain on the inside of my knee."

Close examination by the physician reveals marked swelling on the medial part of the knee. The doctor determines that *valgus stress* (an inward bowing stress on the knee) has caused the medial aspect of the joint to "open." What stabilizing structure is most likely injured? What cartilaginous structure is frequently injured in association with the previously mentioned structure? Is there an anatomical explanation? What other stabilizing structures within the knee are subject to athletic injury?

Hint: An impact to the one side of the knee generally results in greater trauma to the other side. Carefully read the sections in this chapter on the tibiofemoral (knee) joint and trauma to joints. In addition, examine figure 8.31 for a visual description of the knee joint.

Classification of Joints

On the basis of anatomical structure, the articulations between the bones of the skeleton are classified as fibrous joints, cartilaginous joints, or synovial joints. Fibrous joints firmly join skeletal elements with fibrous connective tissue. Cartilaginous joints firmly join skeletal elements with cartilage. Synovial joints are freely movable joints enclosed by joint capsules that contain synovial fluid.

Objective 1.   Define *arthrology* and *kinesiology.*

Objective 2.   Compare and contrast the three principal kinds of joints.

One of the functions of the skeletal system is to permit body movement. It is not the rigid bones that allow movement but the **articulations,** or **joints,** between the bones. The structure of a joint determines the range of movement it permits. Not all joints are flexible, however, and as one part of the body moves, other joints remain rigid to stabilize the body and maintain balance. The coordinated activity of all of the joints permits the sinuous, elegant movements of a gymnast or ballet dancer just as it permits all of the commonplace actions associated with walking, eating, writing, and speaking.

Arthrology is the science concerned with the study of joints. Generally speaking, an arthrologist is interested in the structure, classification, and function of joints, including any dysfunctions that may develop. **Kinesiology** *(ki-ne″se-ol′ŏ-je)* is a more practical and dynamic science concerned with the functional relationship, or biomechanics, of the skeleton, joints, muscles, and innervation as they work together to produce coordinated movement.

To study the joints, one should adopt a kinetic approach and be able to demonstrate the various movements permitted at each of the movable joints and understand the adaptive advantage of each type of movement, as well as the limitations.

The joints of the body may be classified according to structure or function. Structural classification is based on the presence or absence of a joint cavity and the kind of supportive connective tissue surrounding the joint. In classification by structure, the three types of joints are as follows:

1. **Fibrous joints.** A fibrous joint lacks a joint cavity, and fibrous connective tissue connects articulating bones.
2. **Cartilaginous joints.** A cartilaginous joint lacks a joint cavity, and cartilage binds articulating bones.
3. **Synovial joints.** A synovial joint has a joint cavity, and ligaments help to support the articulating bones.

The functional classification of joints is based on the degree of movement permitted within the joint. Using this type of classification, the three kinds of articulations are as follows:

1. **Synarthroses** *(sin′ar-thro′sēz).* Immovable joints.
2. **Amphiarthroses.** Slightly movable joints.
3. **Diarthroses** *(di″ar-thro′sēz).* Freely movable joints.

In previous editions of this text, the functional classification scheme was used in discussing the various kinds of joints of the body. In accordance with the structural classification of joints presented in the sixth edition of *Nomina Anatomica,* this chapter has been revised using a structural classification of joints.

---

1. Explain the statement that kinesiology is applied arthrology.
2. List the three types of structural joints and speculate as to which is most supportive and which is most vulnerable to trauma.

Fibrous Joints

Articulating bones in fibrous joints are tightly bound by fibrous connective tissue. These range from slightly movable joints to joints that are rigid and relatively immovable. Fibrous joints are of three types: sutures, syndesmoses, and gomphoses.

Objective 3.   Describe the structure of a suture and indicate where sutures are located.

Objective 4.   Describe the structure of a syndesmosis and indicate where syndesmoses are located.

---

arthrology: Gk. *arthron,* joint; *logos,* study

kinesiology: Gk. *kinesis,* movement; *logos,* study

Figure 8.1   A section across the skull showing a suture.

Figure 8.2   The side-to-side articulation of the ulna and radius forms a syndesmotic joint. An interosseous ligament tightly binds these bones and permits only slight movement between them.

these illustrations and make note of the bones that articulate to form the sutures identified in these figures.

A *synostosis (sin"os-to'sis)* is a sutural joint in that it is present during growth of the skull but becomes totally ossified in the adult. For example, the frontal bone forms as two separate components but the suture becomes obscured in most individuals as the skull completes its growth.

Fractures of the skull are fairly common in an adult but much less so in a child. The skull of a child is resilient to blows because of the nature of the bone and the layer of fibrous connective tissue within the sutures. The skull of an adult is much like an eggshell in its lack of resilience, and will frequently splinter on impact.

Syndesmoses

Syndesmoses *(sin"des-mo'sēz)* are fibrous joints found only in the antebrachium (forearm) and leg where adjacent bones are held together by collaginous fibers or *interosseous ligaments.* A syndesmosis is characteristic of the side-to-side joints between the tibia-fibula and the radius-ulna (fig. 8.2). Slight movement is permitted at these joints as the antebrachium or leg is rotated.

Gomphoses

Gomphoses *(gom-fo'sēz)* are fibrous joints that occur between the teeth and the supporting bones of the jaws. More specifically, a gomphosis is where the root of a tooth is attached to the periodontal ligament of the alveolus (socket) of the bone (see fig. 18.10).

Objective 5.   Describe the structure and note the locations of gomphoses. Also, discuss the importance of these joints to the profession of dentistry.

Sutures

Sutures, one type of fibrous joint, are found only between the flat bones of the skull and are characterized by a thin layer of dense regular connective tissue that binds the articulating bones (fig. 8.1). Sutures form at about 18 months of age and replace the pliable fontanels of an infant's skull (see fig. 6.13).

Sutures can be distinguished on the basis of the appearance of the articulating margin of bone. A **serrate suture** is characterized by interlocking sawlike articulations. This is the most common type of suture, an example being the sagittal suture between the two parietal bones. In a **squamous suture,** the margin of one bone overlaps that of the articulating bone. The squamous suture formed between the temporal and parietal bones is an example. In a **plane suture,** the margins of the articulating bones are fairly smooth. An example is the median palatine suture, where the paired maxillary and palatine bones articulate to form the hard palate.

The nomenclature in human anatomy is extensive and precise. There are over 30 named sutures in the skull even though just a few of the them are mentioned by name in figures 6.15, 6.16, and 6.17. Review

---

suture: L. *sutura,* sew

syndesmosis: Gk. *syndesmos,* binding together
gomphosis: Gk. *gompho,* nail or bolt

Figure 8.3 Examples of symphyses. (*a*) The symphysis pubis and (*b*) the intervertebral joints between vertebral bodies.

Fibrocartilage
(a)

Intervertebral disc

Body of vertebra

(b)

Periodontal disease occurs at gomphoses and is the inflammation and degeneration of the gum, periodontal ligaments, and alveolar bone tissue. With this condition, the teeth become loose and plaque accumulates on the roots. Periodontal disease may be caused by poor oral hygiene, compacted teeth (poor alignment), or local irritants, such as impacted food, chewing tobacco, or cigarette smoke.

1. Compare and contrast the three kinds of sutures. Give an example of each and note the location.
2. Explain how the structure of a syndesmosis permits it to move slightly.
3. Define *gomphosis* and explain why the gomphoses are of clinical importance in the field of dentistry.

Cartilaginous Joints

Cartilaginous joints allow limited movement in response to twisting, compression, or stress. The two types of cartilaginous joints are symphyses and synchondroses.

Objective 6. Describe the structure of a symphysis and indicate where symphyses occur.

Objective 7. Describe the structure of a synchondrosis and indicate where synchondroses occur.

Symphyses

The adjoining bones of a **symphysis** *(sim'fĭ-sis)* **joint** are separated by a pad of fibrocartilage. This pad cushions the joint and allows limited movement. The symphysis pubis and the intervertebral discs (fig. 8.3) are examples of symphyses. Although only limited motion is possible at each intervertebral joint, the combined movement of all the joints of the vertebral column results in extensive movement of the structure as a whole.

Synchondroses

Synchondroses *(sin''kon-dro'sēz)* are cartilaginous joints that have hyaline cartilage between the bone segments. Some of these joints are temporary, forming the epiphyseal plates (growth lines) between the diaphyses and epiphyses in the long bones of children. When growth is complete, these synchondrotic

symphysis: Gk. *symphysis*, growing together
synchondrosis: Gk. *syn*, together; *chondros*, cartilage

Figure 8.4   A radiograph of the left humerus of a 10-year-old child showing a synchondrotic joint. In a long bone, this type of joint occurs at both the proximal and distal epiphyseal plates. The mitotic activity at synchondrotic joints is responsible for bone growth in length.

Proximal epiphysis of humerus

Proximal epiphyseal plate (site of synchondrotic joint)

Diaphysis of humerus

Distal epiphyseal plate

Distal epiphysis of humerus

joints ossify. A totally ossified synchondrosis may also be referred to as a *synostosis.* Synchondroses can be clearly seen in the radiograph of a long bone of a child in figure 8.4.

A fracture of a long bone in a child may be extremely serious if it involves the mitotically active epiphyseal plate of a synchondrotic joint. If such an injury is untreated, there likely will be retarded or arrested bone growth so that the appendage will be shorter than normal.

Synchondroses that do not ossify as a person ages are those between the occipital, sphenoid, temporal, and ethmoid bones of the skull. In addition, the costochondral articulations between the ribs and the sternum are examples of synchondroses. It is interesting that elderly people often exhibit some ossification of costal cartilages of the rib cage. This may restrict movement of the rib cage and obscure an image of the lungs in a thoracic radiograph.

synostosis: Gk. *syn,* together; *osteon,* bone

1. Discuss the function of the pad of fibrocartilage in a symphysis joint and give two examples of symphyses.
2. What structural feature is characteristic of all synchondroses? Give two examples of synchondroses.

Synovial Joints

Synovial joints are freely movable joints enclosed by joint capsules that contain synovial fluid. Based on the shape of the articular surfaces and the kinds of motion they permit, synovial joints are categorized as gliding, hinge, pivot, condyloid, saddle, or ball-and-socket.

Objective 8.   Describe the structure of a synovial joint.

Objective 9.   Discuss the various kinds of synovial joints, noting where they occur and the movements they permit.

The most obvious type of articulation in the body is the freely movable synovial joint. The function of synovial joints is to provide a wide range of precise, smooth movements, at the same time maintaining stability, strength, and, in certain aspects, rigidity in the body.

Synovial joints are the most complex and varied of the three major types of joints. The range of movement of a synovial joint is limited by three factors: (1) the structure of the bones participating in the articulation (for example, the olecranon of the ulna limits the range of motion and "locks" the articulation to prevent overextension of the joint); (2) the strength and tautness of the associated ligaments, tendons, and joint capsule; and (3) the size, arrangement, and action of the muscles that span the joint. Joint motility is characterized by tremendous individual variation, most of which is related to body conditioning (fig. 8.5). "Double-jointed" is a misnomer because such a joint is not double, although it does permit extreme maneuverability.

Arthroplasty is the surgical repair or replacement of joints. Advancements in arthroplastic procedures continue as devices are sought to provide patients movement capabilities free of pain. A recent advancement in the repair of soft tissues involves the use of *artificial ligaments.* A material consisting of carbon fibers coated with a plastic called polylactic acid is sewn in and around torn ligaments and tendons. This reinforces the traumatized structures while providing a scaffolding on which the body's collagenous fibers can grow. As the healing process continues, the polylactic acid is eventually absorbed and the carbon fibers break down.

Structure of a Synovial Joint

Synovial joints are enclosed by a fibroelastic **joint capsule** that is filled with lubricating **synovial fluid,** or **synovium** (fig. 8.6). Synovial fluid is secreted by a thin **synovial membrane** that

arthroplasty: Gk. *arthron,* joint; *plasso,* to form

UNIT IV

UNIT IV

Figure 8.5 Although joint flexibility is structurally determined and limited, some individuals can achieve an extraordinary range of movement through extensive training.

Figure 8.6 A synovial joint is represented by a lateral view of the knee joint.

Femur

Bursa under lateral head of gastrocnemius muscle

Synovial membrane

Articular cartilage

Meniscus

Joint cavity filled with synovial fluid

Tibia

Suprapatellar bursa

Synovial membrane

Tendon of quadriceps femoris muscle

Patella

Subcutaneous prepatellar bursa

Infrapatellar fat pad

Subcutaneous infrapatellar bursa

Infrapatellar bursa

Patellar ligament

Creek

lines the inside of the capsule. Synovial fluid is similar to interstitial fluid (fluid surrounding cells of a tissue) and has a high concentration of hyaluronic acid—a lubricating substance. The bones that articulate in a synovial joint are capped with a smooth **articular cartilage.** The avascular articular cartilage is only about 2 mm thick and depends on the alternating compression and decompression during joint activity for the exchange of nutrients and waste products with the synovial fluid. Ligaments help to bind a synovial joint and may be located within the joint cavity or on the outside of the capsule. Tough, fibrous cartilaginous pads called **menisci**—singular, *meniscus (mě-nis′kus)*—are located within the capsule of certain synovial joints (e.g., the knee joint) and serve to cushion, as well as to guide, the articulating bones.

Many people are concerned about the cracking sounds they hear as joints move, or the popping sounds that result from "popping" or "cracking" the knuckles by forcefully pulling on the fingers. These sounds are actually quite normal. When a synovial joint is pulled upon, its volume is suddenly expanded and the pressure of the joint fluid is decreased, causing a partial vacuum within the joint. As the joint fluid is displaced within a vacuum and hits against the articular cartilage, a popping or cracking sound is heard. Similarly, displaced water in a sealed vacuum tube makes the same sound as it hits against the glass wall. Popping one's knuckles does not cause arthritis, but it can lower one's social standing.

Articulating bones of synovial joints do not come in contact with one another. Articular cartilage caps the articular surface of each bone and synovial fluid circulates through the joint during movement. Both of these joint structures serve to minimize friction and cushion the articulating bones. Should trauma or disease render either of these two joint structures nonfunctional, the two articulating bones will come in contact. Bony deposits will then form, and a type of arthritis will develop within the joint.

Located near certain synovial joints are flattened, pouch-like sacs called **bursae**—singular, *bursa (bur′sǎ)*—which are filled with synovial fluid (fig. 8.7a). These closed sacs are commonly located between muscles or within an area where a tendon passes over a bone. The functions of bursae are to cushion certain muscles and to facilitate the movement of tendons or muscles over bony or ligamentous surfaces. A **tendon sheath** (fig. 8.7b) is a modified bursa that surrounds and lubricates the tendons of certain muscles, particularly those that cross the wrist and ankle joints.

Improperly fitted shoes or inappropriate shoes can cause joint-related problems. People who wear high-heeled shoes have perpetual backaches and leg aches because their posture has to counteract the forward tilt of their bodies when standing or walking. Their knees are excessively bent and the spine is thrust forward

meniscus: Gk. *meniskos,* small moon

bursa: Gk. *byrsa,* bag or purse

Figure 8.7 Bursae and tendon sheaths are friction-reducing structures found in conjunction with synovial joints. (*a*) A bursa is a closed sac filled with synovial fluid. Bursae are commonly located between muscles or between tendons and articular capsules. (*b*) A tendon sheath encapsulates a tendon through a joint capsule. It is a closed, double-layered sac filled with synovial fluid. One layer of the sac is in contact with the tendon and the other layer is in contact with the joint capsule.

at the lumbar curvature in order to maintain balance. Tightly fitted shoes, especially ones with pointed toes, may result in the development of *hallux valgus*. This condition is a deviation of the hallux (great toe) laterally toward the other toes. Hallux valgus is generally accompanied by the formation of a *bunion* at the medial base of the proximal phalanx of the hallux. A bunion is an inflammation and an accompanying callus that develops because of pressure and rubbing of a shoe.

Kinds of Synovial Joints

Synovial joints are classified into six main types according to the shape of their articular surfaces and the kinds of motion they permit. The six types are gliding, hinge, pivot, condyloid, saddle, and ball-and-socket.

Gliding　Gliding joints allow only side-to-side and back-and-forth movements with some slight rotation. This is the

Figure 8.8 The locations of both gliding and condyloid joints. (a) Gliding joints are located between the proximal and distal carpal bones and between the individual carpal bones. (b) Condyloid joints are located between the metacarpal bones and the phalanges. (Note the diagrammatic representation of these joints showing the directions of possible movements.)

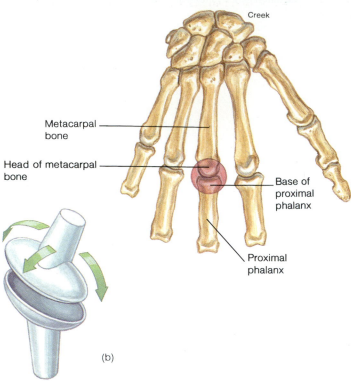

simplest type of joint movement. The articulating surfaces can be nearly flat, or one may be slightly concave and the other slightly convex (fig. 8.8). The intercarpal and intertarsal joints, the sternoclavicular joint, and the joint between the articular process of adjacent vertebrae are examples.

Hinge The structure of a hinge joint permits bending in only one plane, much like the hinge of a door. In this type of articulation, the surface of one bone is always concave and the other, convex (fig. 8.9). Hinge joints are the most common type of synovial joints and include such specific joints as the knee, the humeroulnar articulation within the elbow, and the joints between the phalanges.

Pivot The movement in a pivot joint is limited to rotation about a central axis. In this type of articulation, the articular surface on one bone is conical or rounded and fits into a depression on another bone (fig. 8.10). Examples are the proximal articulation of the radius and ulna for rotation of the forearm, as in turning a doorknob, and the articulation between the atlas and axis that enables rotational movement of the head.

Condyloid A condyloid articulation is structured so that an oval, convex articular surface of one bone fits into an elliptical, concave depression on another bone (see fig. 8.8). This permits angular movement in two directions (biaxial), as in an up-and-

Figure 8.9 A hinge joint permits only a bending movement (flexion and extension). The hinge joint of the elbow involves the distal end of the humerus articulating with the proximal end of the ulna. (Note the diagrammatic representation of this joint showing the direction of possible movement.)

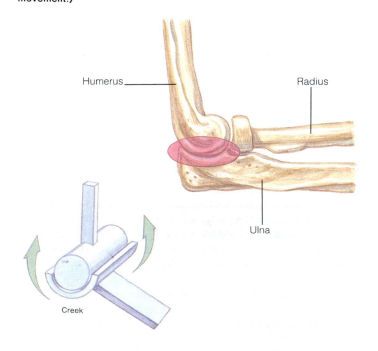

Figure 8.10 The atlas articulating with the axis forms a pivot joint that permits a rotational movement in one axis. (Note the diagrammatic representation showing the direction of possible movement.) Refer to figure 8.9 and determine which articulating bones of the elbow region form a pivot joint.

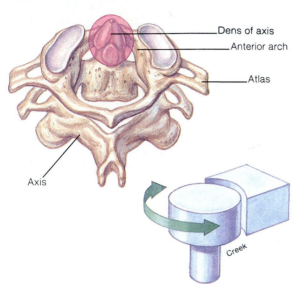

Figure 8.11 A saddle joint is formed as the trapezium articulates with the base of the first metacarpal bone. (Note the diagrammatic representation of this joint.)

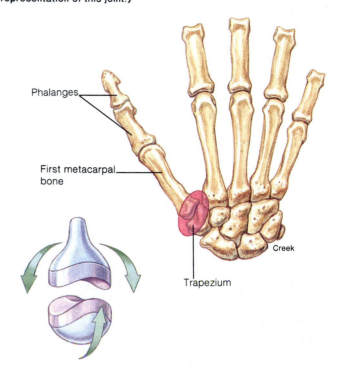

down and side-to-side motion. A condyloid joint does not permit rotational movement. The radiocarpal joint of the wrist is an example.

Saddle Each articular process of a saddle-shaped joint has a concave surface in one direction and a convex surface in another. This unique articulation is a modified condyloid joint that allows a wide range of movement. It is associated only with the thumb and is located at the articulation of the trapezium of the carpus with the first metacarpal bone (fig. 8.11).

Ball-and-Socket Ball-and-socket joints are formed by the articulation of a rounded convex surface with a cuplike cavity (fig. 8.12). This type of articulation provides the greatest range of movement of all the synovial joints. Examples are the hip and shoulder joints.

A summary of the various types of joints is presented in table 8.1.

> Synovial fluid within synovial joints lubricates the articular surfaces and provides nourishment to the articular cartilages. Trauma to the joint causes the excessive production of synovial fluid in an attempt to cushion and immobilize the joint. This leads to swelling and discomfort to the joint. The most frequent type of joint injury is a *sprain*, in which the supporting ligaments or the joint capsule are damaged to varying degrees.

1. List the structures of a synovial joint and explain the function of each.
2. What three factors limit the range of movement within synovial joints?
3. Give an example of each of the six kinds of synovial joints and describe the range of movement possible at each.

Figure 8.12 A ball-and-socket articulation illustrated by the hip joint. (Note the diagrammatic representation showing the directions of possible movement.)

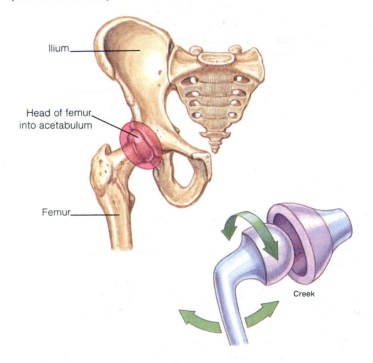

UNIT IV

Table 8.1	Types of Articulations		
Type	**Structure**	**Movements**	**Example**
Fibrous Joints	Skeletal elements joined by fibrous connective tissue		
1. Suture	Frequently serrated edges of articulating bones; separated by thin layer of fibrous tissue	None	Sutures between bones of the skull
2. Syndesmoses	Articulating bones bound by interosseous ligament	Slightly movable	Joints between tibia-fibula and radius-ulna
3. Gomphoses	Teeth bound into alveoli of bone by periodontal ligament	None	Teeth secured into alveoli (sockets)
Cartilaginous Joints	Skeletal elements joined by fibrocartilage or hyaline cartilage		
1. Symphyses	Articulating bones separated by pad of fibrocartilage	Slightly movable	Intervertebral joints; symphysis pubis and sacroiliac joint
2. Synchondroses	Mitotically active hyaline cartilage between skeletal elements	None	Epiphyseal plates within long bones
Synovial Joints	Joint capsule containing synovial membrane and synovial fluid		
1. Gliding	Flattened or slightly curved articulating surfaces	Sliding	Intercarpal and intertarsal joints
2. Hinge	Concave surface of one bone articulates with convex surface of another	Bending motion in one plane	Knee; elbow; joints of phalanges
3. Pivot	Conical surface of one bone articulates with a depression of another	Rotation about a central axis	Atlantoaxial joint; proximal radioulnar joint
4. Condyloid	Oval condyle of one bone articulates with elliptical cavity of another	Biaxial movement	Radiocarpal joint
5. Saddle	Concave and convex surface on each articulating bone	Wide range of movements	Carpometacarpal joint of thumb
6. Ball-and-socket	Rounded convex surface of one bone articulates with cuplike socket of another	Movement in all planes and rotation	Shoulder and hip joints

UNIT IV

Movements at Synovial Joints

Movements at synovial joints are produced by the contraction of skeletal muscles that span the joints and attaching to or near the bones forming the articulation. In these actions, the bones act as levers, the muscles provide the force, and the joints are the fulcra, or pivots.

Objective 10. List and discuss the various kinds of movements that are possible at synovial joints.

Objective 11. Describe the components of a lever and explain the role of synovial joints in lever systems.

Objective 12. Compare the structure of first-, second-, and third-class levers.

As previously mentioned, the range of movement at a synovial joint is determined by the structure of the individual joint and the arrangement of the associated muscle and bone. The movement at a hinge joint, for example, occurs in only one plane, whereas the structure of a ball-and-socket joint permits movement around many axes. Joint movements are broadly classified as *angular* and *circular*. Within each of these categories are specific types of movements, and certain special movements may involve several of the specific types.

Angular

Angular movements increase or decrease the joint angle produced by the articulating bones. The four types of angular movements are flexion, extension, abduction, and adduction.

Flexion   Flexion is a movement that decreases the joint angle on an anterior-posterior plane (fig. 8.13). Examples of flexion are the bending of the elbow or knee. Flexion of the elbow joint is a forward movement, whereas flexion of the knee is a backward movement. Flexion in most joints is simple to understand, such as flexion of the neck as the head is bowed or the flexing of the joints within a digit, but flexion of the ankle and shoulder joints needs further explanation. In the ankle joint, flexion occurs as the dorsum of the foot is elevated. This movement is frequently called **dorsiflexion** (fig. 8.14). Pressing the foot downward (as in rising on the toes) is **plantar flexion.** The shoulder joint is flexed when the arm is brought forward, thus decreasing the joint angle.

Extension   In extension, which is the reverse of flexion, the joint angle is increased (fig. 8.13). Extension returns a body part to the anatomical position. In an extended joint, the angle be-

flexion: L. *flectere*, to bend
extension: L. *ex*, out, away from; *tendere*, stretch

Figure 8.13 Examples of flexion and extension. Contraction of the posterior thigh muscles with the femur held rigid results in flexion at the knee joint as the leg is moved posteriorly. Contraction of the anterior hip and thigh muscles and the thigh and leg held rigid results in flexion at the hip joint as the lower extremity is moved anteriorly. Contraction of the posterior hip and thigh muscles with the thigh and leg rigid results in extension at the hip joint as the lower extremity is moved posteriorly.

Extension of the hip joint

Flexion of the hip joint

Flexion of the knee joint

Extension of the knee joint

Figure 8.14 Dorsiflexion and plantar flexion of the ankle joint.

Dorsiflexion

Plantar flexion

tween the articulating bones is 180°. The exception to this is the ankle joint, in which there is a 90° angle between the foot and the lower leg in the anatomical position. Examples of extension are the straightening of the elbow or knee joints from flexion positions. **Hyperextension** occurs when a part of the body is extended beyond the anatomical position so that the joint angle is greater than 180°. An example of hyperextension is tipping the head backward.

A common injury in runners is *patellofemoral stress syndrome,* commonly called "runner's knee." This condition is characterized by tenderness and aching pain around or under the patella. During normal knee movement, the patella glides up and down the patellar groove between the femoral condyles. In patellofemoral stress syndrome, the patella rubs laterally, causing irritation to the membranes and articular cartilage within the knee joint. Joggers frequently experience this condition from prolonged running on the slope of a road near the curb rather than on a flat surface.

Abduction Abduction is the movement of a body part away from the main axis of the body, or away from the midsagittal plane, in a lateral direction (fig. 8.15). This term usually applies to the arm or leg but can also apply to the fingers or toes, in which case the line of reference is the longitudinal axis of the limb. Examples of abduction are moving the arms sideward and away from the body or spreading the fingers apart.

Adduction Adduction, the opposite of abduction, is the movement of a body part toward the main axis of the body (fig. 8.15). In the anatomical position, the arms and legs have been adducted toward the midplane of the body.

Circular

Joints that permit **circular movement** are composed of a bone with a rounded or oval surface that articulates with a corresponding cup or depression on another bone. The two basic types of circular movements are rotation and circumduction.

Rotation Rotation is the movement of a body part around its own axis (see fig. 8.10). There is no lateral displacement during this movement. Examples are turning the head from side to side in a "no" motion and moving the forearm from a palm-up position to a palm-down position.

Supination (*soo"pi-na'shun*) is a specialized rotation of the forearm that results in the palm of the hand being turned forward (anteriorly). In the supine position, the ulna and radius of

abduction: L. *abducere*, lead away
adduction: L. *adductus*, bring to
rotation: L. *rotare*, a wheel

UNIT IV

Figure 8.15 Abduction and adduction of the lower extremity at the coxal (hip) joint in reference to the main axis of the body.

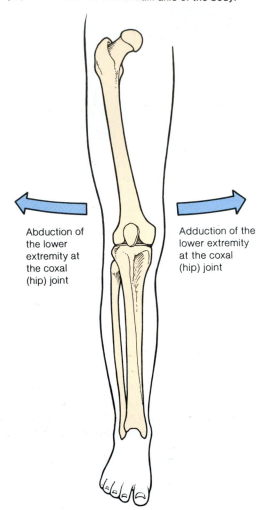

Abduction of the lower extremity at the coxal (hip) joint

Adduction of the lower extremity at the coxal (hip) joint

Figure 8.16 Supination (*a*) and pronation (*b*) of the hand. (Note the relative position of the ulna and radius in both positions.) Pronation requires medial rotation of the forearm relative to the anatomical position.

(a) Supination   (b) Pronation

Figure 8.17 Circumduction of the upper extremity at the shoulder joint.

Circumduction

the forearm are parallel and in the anatomical position. **Pronation** (*pro-na'shun*) is the opposite of supination (fig. 8.16). It is a rotational movement of the forearm that results in the palm of the hand being directed backward (posteriorly).

Circumduction Circumduction is the circular, conelike movement of a body segment. The distal extremity forms the circular movement and the proximal attachment forms the pivot (fig. 8.17). This type of motion is possible at the shoulder, wrist, trunk, hip, and ankle joints.

Special Movements

Because the terms used to describe generalized movements around axes do not apply to the structure of certain joints, other terms must be used to describe the motion of such joints.

Inversion Inversion is the movement of the sole of the foot inward or medially (fig. 8.18). The pivot axes are at the ankle and intertarsal joints.

Figure 8.18 Inversion and eversion of the foot at the intertarsal joints.

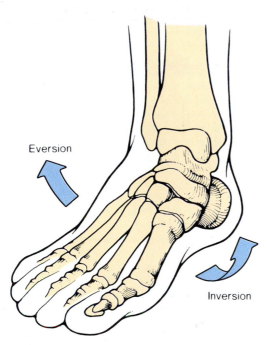

Figure 8.19 Protraction, retraction, elevation, and depression of the lower jaw.

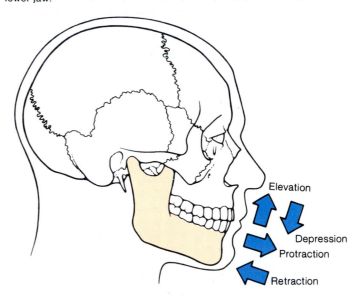

Eversion   Eversion is the opposite of inversion and is the movement of the sole of the foot outward or laterally (fig. 8.18). Both inversion and eversion are clinical terms usually used to describe developmental abnormalities.

Protraction   Protraction is the movement of part of the body forward on a plane parallel to the ground. Examples are thrusting out the lower jaw (fig. 8.19) or movement of the shoulder and upper extremity forward.

Retraction   Retraction is the pulling back of a protracted part of the body on a plane parallel to the ground (fig. 8.19). Retraction of the mandible brings the lower jaw back in alignment with the upper jaw, so that the teeth occlude.

Elevation   Elevation is a movement that results in a part of the body being lifted upward. Examples of elevation include elevating the mandible to close the mouth (fig. 8.19) or lifting the scapula to shrug the shoulder.

Depression   Depression is the opposite of elevation. Both the mandible (fig. 8.19) and shoulders are depressed when moved downward.

A visual summary of many of the movements permitted at synovial joints is presented in figures 8.20, 8.21, and 8.22.

Biomechanics of Body Movement

A lever is any rigid structure that turns about a fulcrum when force is applied. Because bones are rigid structures that can be moved at synovial joints in response to applied forces, they fit the criteria of levers. There are four basic components to a lever: (1) a rigid bar or other such structure; (2) a pivot or fulcrum; (3) an object or resistance that is moved; and (4) a force that is applied to one portion of the rigid structure.

Levers are generally associated with machines but can equally apply to other mechanical structures, such as the human body. Synovial joints are usually the fulcra (F), the muscles provide the force, or effort (E), and the bones are the rigid lever arms that move the resisting object (R).

There are three kinds of levers, determined by the arrangement of their parts (fig. 8.23).

1. In a **first-class lever,** the fulcrum is positioned between the effort and the resistance. The sequence of parts in a first-class lever is much like that of a seesaw—a sequence of resistance-pivot-effort. Scissors and hemostats are mechanical examples of first-class levers. In the body, the head at the atlantooccipital *(at-lan″to-ok-sip′i-tal)* joint is a first-class lever. The weight of the skull and facial portion of the head is the resistance, and the posterior neck muscles that contract to maintain the balance of the head on the joint are the effort.

2. In a **second-class lever,** the resistance is positioned between the fulcrum and the effort. The sequence of arrangement is pivot-resistance-effort, as in a wheelbarrow or the action of a crowbar when one end is placed under a rock and the other end lifted. Contraction of the calf muscles (E) to elevate the body (R) on the toes, with the ball of the foot acting as the fulcrum, is another example.

3. In a **third-class lever,** the effort lies between the fulcrum and the resistance. The sequence of the parts is pivot-effort-resistance, as in the action of a pair of forceps in grasping an object. The third-class lever is the most common type within the body. The flexion of the elbow

UNIT IV

Figure 8.20 A photographic summary of joint movements. (*a*) Adduction of shoulder, hip, and carpophalangeal joints; (*b*) abduction of shoulder, hip, and carpophalangeal joints; (*c*) rotation of vertebral column; (*d*) lateral flexion of vertebral column; (*e*) flexion of vertebral column; (*f*) hyperextension of vertebral column; (*g*) flexion of shoulder, hip, and knee joints of right side of body; extension of elbow and wrist joints; dorsiflexion of right ankle joint; (*h*) hyperextension of shoulder and hip joints on right side of body; plantar flexion of right ankle joint.

is an example. The effort occurs as the biceps brachii muscle is contracted to move the resistance of the forearm with the elbow joint forming the fulcrum.

Each bone-muscular interaction at a synovial joint forms some kind of lever system. The specific kind of lever is not always easy to identify. Certain joints are adapted for power at the expense of speed, whereas most are clearly adapted for speed. The specific attachment of muscles that span a joint plays an extremely important role in determining the mechanical ad-

vantage (fig. 8.24). The position of the insertion of a muscle relative to the joint is an important factor in the biomechanics of the contraction. An insertion close to the joint (fulcrum), for example, will produce a faster movement and greater range of movement than an insertion far away from the joint. An attachment far away from the joint takes advantage of the lever arm of the bone and increases power at the sacrifice of speed and range of movement.

Figure 8.21   A visual summary of some angular movements at synovial joints. (*a*) Flexion, extension, and hyperextension in the cervical region; (*b*) flexion and extension at the knee joint and dorsiflexion and plantar flexion at the ankle joint; (*c*) flexion, extension, and hyperextension at the wrist joint; (*d*) flexion, extension, and hyperextension of the hip joint and flexion and extension of the knee joint; (*e*) adduction and abduction of the arm and fingers; (*f*) posterior view of abduction and adduction of the hand at the wrist joint. (Note that the range of abduction at the wrist joint is less extensive than the range of adduction as a result of the length of the styloid process of the radius.)

(a)   (b)   (c)

(d)   (e)   (f)

UNIT IV

Figure 8.22   A visual summary of some rotational movements at synovial joints. (*a*) Rotation of the head at the cervical vertebrae—especially at the atlantoaxial joint and (*b*) rotation of the forearm (antebrachium) at the proximal radioulnar joint.

(a)

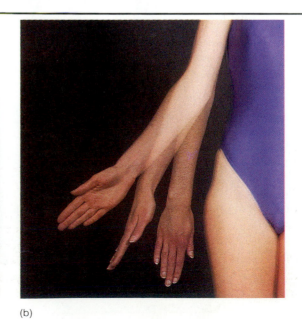

(b)

Figure 8.23 The three classes of levers. (*a*) In a first-class lever, the fulcrum (**F**) is positioned between the resistance (**R**) and the effort (**E**). (*b*) In a second-class lever, the resistance is between the fulcrum and the effort. (*c*) In a third-class lever, the effort is between the fulcrum and the resistance.

(a) (b) (c)

Figure 8.24 The position of a joint (fulcrum) relative to the length of a long bone (lever arm) and the point of attachment of a muscle (force) determines the mechanical advantage when movement occurs. (*a*) The elbow joint and extensor muscles of a human and (*b*) the elbow joint and extensor muscles of an armadillo.

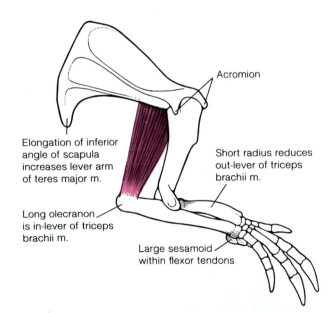

1. Describe the structure of a joint that permits rotational movement.
2. What types of joints are involved in lever systems within the body?
3. Which is the most common type of lever in the body?
4. Considering the type of synovial joint at the hip and the location of the gluteal muscles of the buttock, explain why this type of lever system is adapted for rapid, wide-ranging movements.

Specific Joints of the Body

Of the numerous joints in the body, some have special structural features that enable them to perform particular functions. Furthermore, these joints are somewhat vulnerable to trauma and are therefore clinically important.

Objective 13. Describe the structure, function, and possible clinical importance of the following joints: temporomandibular,

Figure 8.25 The temporomandibular joint. (*a*) A lateral view, (*b*) a medial view, and (*c*) a sagittal view.

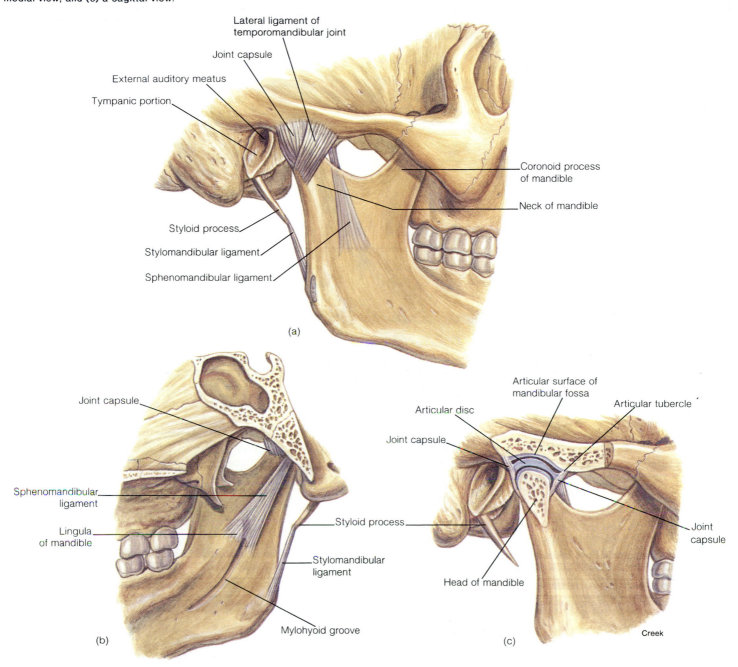

(a)

Lateral ligament of temporomandibular joint
Joint capsule
External auditory meatus
Tympanic portion
Styloid process
Stylomandibular ligament
Sphenomandibular ligament
Coronoid process of mandible
Neck of mandible

(b)

Joint capsule
Sphenomandibular ligament
Lingula of mandible
Styloid process
Stylomandibular ligament
Mylohyoid groove

(c)

Articular surface of mandibular fossa
Articular disc
Articular tubercle
Joint capsule
Joint capsule
Head of mandible

Creek

UNIT IV

sternoclavicular, humeral, elbow, meta-carpophalangeal, interphalangeal, coxal, tibiofemoral, and talocrural.

Temporomandibular Joint

The **temporomandibular joint** is the only synovial joint in the skull and represents a unique combination of a hinge joint and a gliding joint (fig. 8.25). It is formed by the mandibular condyle of the mandibular fossa and articulating with the articular tubercle of the temporal bone. An **articular disc** separates the joint cavity into superior and inferior compartments.

Three ligaments support and reinforce the temporomandibular joint. The **lateral ligament of the temporomandibular joint** is positioned on the lateral side of the joint capsule and is covered by the parotid gland. This ligament prevents the head of the mandible from being displaced posteriorly and fracturing the tympanic plate when the chin suffers a severe blow. The **stylomandibular ligament** is not directly associated with the joint but extends inferiorly and anteriorly from the styloid process to the posterior border of the ramus of the mandible. A **sphenomandibular** (*sfe″no-man-dib′yŭ-lar*) **ligament** crosses on the medial side of the joint from the spine of the sphenoid bone to the ramus of the mandible.

Figure 8.26 The sternoclavicular joint and associated ligaments. (*a*) An anterior view showing a frontal section and (*b*) a posterior view.

UNIT IV

The movements of the temporomandibular joint include depression and elevation of the mandible as a hinge joint, protraction and retraction of the mandible as a gliding joint, and lateral rotatory movements. The lateral motion is made possible by the articular disc.

The temporomandibular joint can be easily palpated by applying firm pressure to the area in front of your auricle and opening and closing your mouth. This joint is most vulnerable to dislocation when the mandible is completely depressed, as in yawning. Relocating the jaw is usually a simple task, however, and is accomplished by pressing downward on the molar teeth while pushing the mandible backward.

Temporomandibular joint (TMJ) syndrome is a recently recognized ailment that may afflict an estimated 75 million Americans. The apparent cause of TMJ syndrome is a malalignment of one or both temporomandibular joints. The symptoms of the condition vary from moderate and intermittent pain to intense and continuous pain in the head, neck, shoulders, or back. Some vertigo (disorientation of coordination) and tinnitus (ringing in ear) may be experienced.

Sternoclavicular Joint

The sternoclavicular (*ster''no-klah-vik'yŭ-lar*) joint is formed where the clavicle articulates with the manubrium of the sternum (fig. 8.26). Although a gliding joint, the sternoclavicular joint has a good range of movement because of the presence of an articular disc within the joint capsule.

Four ligaments support and give flexibility to the sternoclavicular joint. An **anterior sternoclavicular ligament** covers the anterior surface of the joint and a **posterior sternoclavicular ligament** covers the posterior surface. Both ligaments extend from the sternal end of the clavicle to the

manubrium. An **interclavicular ligament** extends between the sternal ends of both clavicles, binding them together. The **costoclavicular ligament** extends from the costal cartilage of the first rib to the costal tuberosity of the clavicle.

Of all the joints associated with the rib cage, the sternoclavicular joint is most frequently dislocated. Excessive force along the long axis of the clavicle may displace the clavicle forward and inferiorly. Injury to the costal cartilages is painful and is caused most frequently by a forceful, direct blow to the costal cartilages.

Humeral (Shoulder) Joint

The **shoulder joint** is formed by the articulation of the head of the humerus with the glenoid fossa of the scapula (fig. 8.27). It is a ball-and-socket joint and the most freely movable joint in the body. A circular band of fibrocartilage called the **glenoid labrum** passes around the rim of the shoulder joint to deepen the concavity of the glenoid cavity (fig. 8.27). The shoulder joint is protected from above by an arch formed by the acromion and coracoid process of the scapula and by the clavicle.

Although two ligaments and one retinaculum surround and support the shoulder joint, most of the stability of this joint depends on the powerful muscles and tendons that cross over it. Thus, it is an extremely mobile joint in which stability has been sacrificed for mobility. The **coracohumeral** (*kor''ah-ko-hyoo'mer-al*) **ligament** extends from the coracoid process of the scapula to the greater tubercle of the humerus. The joint capsule is reinforced with three ligamentous bands called the **glenohumeral ligaments.** The final support of the shoulder joint

labrum: L. *labrum,* lip

Figure 8.27 The humeral (shoulder) joint. (*a*) An anterior view, (*b*) a coronally sectioned anterior view, (*c*) a posterior view, and (*d*) a lateral view with the humerus removed.

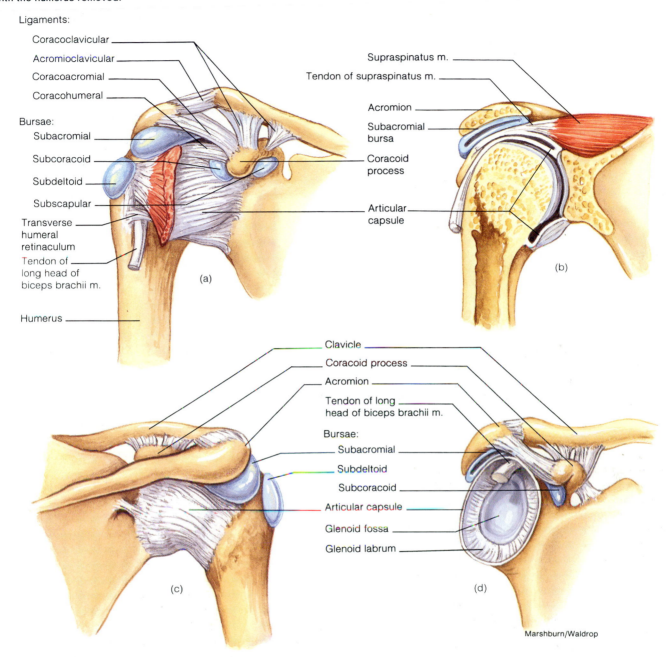

Ligaments:
Coracoclavicular
Acromioclavicular
Coracoacromial
Coracohumeral

Bursae:
Subacromial
Subcoracoid
Subdeltoid
Subscapular

Transverse humeral retinaculum

Tendon of long head of biceps brachii m.

Humerus

(a)

Supraspinatus m.
Tendon of supraspinatus m.
Acromion
Subacromial bursa
Coracoid process
Articular capsule

(b)

Clavicle
Coracoid process
Acromion
Tendon of long head of biceps brachii m.
Bursae:
Subacromial
Subdeltoid
Subcoracoid
Articular capsule
Glenoid fossa
Glenoid labrum

(c)

(d)

Marshburn/Waldrop

UNIT IV

is the **transverse humeral retinaculum,** a thin band that extends from the greater tubercle to the lesser tubercle of the humerus.

The stability of the shoulder joint is provided mainly by the tendons of the subscapularis, supraspinatus, infraspinatus, and teres minor muscles, which together form the *musculotendinous (rotator) cuff.* The cuff is fused to the underlying capsule, except in its inferior aspect. Because of the lack of inferior stability, most dislocations (subluxations) occur in this direction. The shoulder is most vulnerable to trauma when the joint is fully abducted and a sudden force from the superior direction is applied to the appendage. Degenerative changes in the musculotendinous cuff produce an inflamed, painful condition known as *pericapsulitis.*

Two major and two minor bursae are associated with the shoulder joint. The larger bursae are the **subdeltoid bursa** between the deltoid muscle and the joint capsule and the **subacromial bursa** between the acromion and joint capsule. The **subcoracoid bursa** lies between the coracoid process and the joint capsule and is frequently considered an extension of the subacromial bursa. A small **subscapular bursa** is located between the tendon of the subscapularis muscle and the joint capsule.

Figure 8.28 The right elbow region. (*a*) An anterior view, (*b*) a posterior view, (*c*) a sagittal section, (*d*) a lateral view, and (*e*) a medial view.

UNIT IV

The shoulder joint is vulnerable to dislocations from sudden jerks of the arm, especially in children before strong shoulder muscles have developed. Because of the weakness of this joint in children, parents should be careful not to force a child to follow by yanking on the arm. Dislocation of the shoulder is extremely painful and may cause permanent damage or perhaps muscle atrophy due to disuse.

Elbow Joint

There are three specific articulations in the elbow region, two of which constitute the **elbow joint** (fig. 8.28). The elbow joint is a hinge joint, formed by the trochlea of the humerus articulating with the trochlear notch of the ulna (humeroulnar joint) and the capitulum of the humerus articulating with the head of the radius (humeroradial joint). Although there are two sets

Figure 8.29 Metacarpophalangeal and interphalangeal joints. (*a*) A lateral view, (*b*) an anterior (palmar) view, and (*c*) a posterior view.

UNIT IV

of articulations at the elbow joint, there is only one joint capsule and a large **olecranon bursa** to lubricate this area. A **radial (lateral) collateral ligament** reinforces the elbow joint on the lateral side and an **ulnar (medial) collateral ligament** strengthens the medial side.

> Because so many muscles originate or insert near the elbow, it is a common site of localized tenderness, inflammation, and pain. *Tennis elbow* is a general term for musculotendinous soreness in this area. The structures most generally strained are the tendons attached to the lateral epicondyle of the humerus. The strain is caused by repeated extension of the wrist against some force, as occurs during the backhand stroke in tennis.

Metacarpophalangeal Joints and Interphalangeal Joints

The **metacarpophalangeal joints** are condyloid joints and the **interphalangeal joints** are hinge joints. These joints are formed as the heads of the metacarpal bones articulate with the proximal phalanges and as the phalanges articulate with one another (fig. 8.29). Each joint in both joint types has three ligaments. A **palmar ligament** spans each joint on the palmar, or anterior, side of the joint capsule. Each joint also has two **collateral ligaments,** one on the lateral side and one on the medial side, to further reinforce the joint capsule. There are no supporting ligaments on the posterior side.

> Athletes frequently jam a finger. It occurs when a ball forcefully strikes a distal phalanx as the fingers are extended, causing a sharp flexion at the joint between the middle and distal phalanges. No ligaments support the joint on the posterior side, but there is a tendon from the digital extensor muscles of the forearm. It is this tendon that is damaged when the finger is jammed. Treatment involves splinting the finger for a period of time. If splinting is not effective, however, surgery will be necessary to avoid a permanent crook in the finger.

Coxal (Hip) Joint

The ball-and-socket **hip joint** is formed by the head of the femur articulating with the acetabulum of the os coxa (fig. 8.30). It bears the weight of the body and is therefore much stronger and more stable than the shoulder joint. The hip joint is secured by a strong fibrous joint capsule, several ligaments, and a number of powerful muscles.

The primary ligaments of the hip joint are the anterior **iliofemoral** (*il"e-o-fem'or-al*) and **pubofemoral ligaments** and the posterior **ischiofemoral** (*is"ke-o-fem'or-al*) **ligament.** The **ligamentum capitis femoris** is located within the articular capsule and attaches the head of the femur to the acetabulum. The **transverse acetabular** (*as"ĕ-tab'yŭ-lar*) **ligament** crosses the acetabular notch and connects to the joint capsule and the ligamentum capitis femoris. The **acetabular labrum,** a fibrocartilaginous rim that rings the head of the femur as it articulates with the acetabulum, is attached to the margin of the acetabulum.

Figure 8.30 The right coxal (hip) joint. (*a*) An anterior view, (*b*) a posterior view, and (*c*) a coronal (frontal) section.

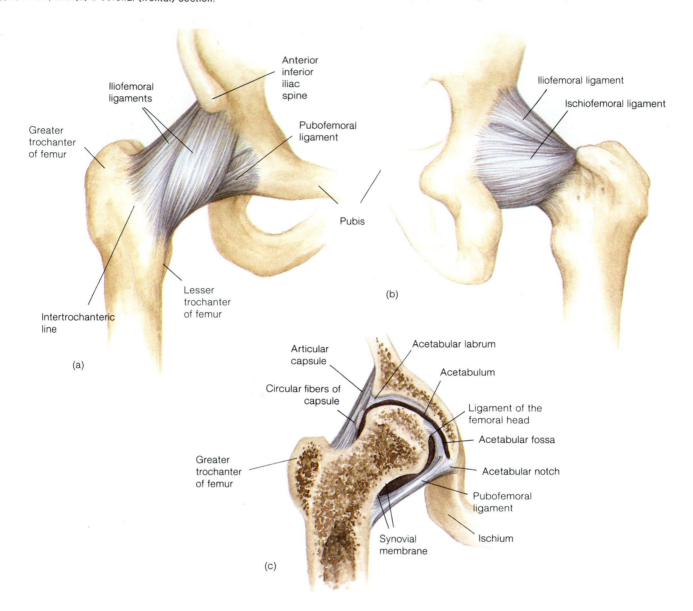

UNIT IV

Osteoarthritis is a degenerative disease of the articular cartilage of synovial joints accompanied by the formation of bony spurs in the joint cavities. Immobility results if the hip joint is severely afflicted with this disease. Fortunately, the entire hip joint can be replaced in a procedure called *hip arthroplasty*. During this surgery, the acetabulum is replaced by a low-friction polyethylene socket, which is fit into the os coxa using bone cement. The head of the femur is replaced by a stainless steel, ball-shaped prosthesis (see fig. 8.34).

Tibiofemoral (Knee) Joint

The **knee joint** is the largest, most complex, and probably the most vulnerable joint in the body. The knee joint is formed as the femur and the tibia articulate. It is a complex hinge joint that permits limited rolling and gliding movements in addition to flexion and extension. On the anterior side, the knee joint is stabilized and protected by the patella and the **patellar ligament,** forming a gliding **patellofemoral joint.**

Because of the complexity of the knee joint, only the relative positions of the ligaments, menisci, and bursae will be covered here. Although the detailed attachments and functions will not be discussed, the locations of these structures can be seen in figure 8.31.

In addition to the patella and the patellar ligament on the anterior surface, the tendinous insertion of the quadriceps femoris muscle forms two supportive bands called the **lateral** and **medial patellar retinacula** (*ret″ĭ-nak′yŭ-lă*). Four bursae are associated with the anterior aspect of the knee: the **subcutaneous prepatellar bursa,** the **suprapatellar bursa,** the **cutaneous prepatellar bursa,** and the **deep infrapatellar bursa.**

The posterior aspect of the knee is referred to as the **popliteal fossa.** The broad **oblique popliteal ligament** and the

Figure 8.31 The right tibiofemoral (knee) joint. (a) An anterior view, (b) a superficial posterior view, (c) a lateral view showing the bursae, (d) an anterior view with the knee slightly flexed and the patella removed, and (e) a deep posterior view.

(a)

Vastus lateralis muscle

Patella

Lateral patellar retinaculum

Patellar ligament

Rectus femoris muscle

Vastus medialis muscle

Medial patellar retinaculum

Bursae:

Popliteal

Semimembranosus

(b)

Popliteal ligaments:

Oblique

Arcuate

Popliteus muscle (cut)

(c)

Bursae:

Suprapatellar

Subcutaneous prepatellar

Cutaneous prepatellar

Deep infrapatellar

(d)

Femur

Patellar area on condyles

Lateral condyle of femur

Fibular collateral ligament

Lateral meniscus

Fibula

Tibia

Medial condyle of femur

Posterior cruciate ligament

Anterior cruciate ligament

Medial meniscus

Tibial collateral ligament

Transverse ligament

Patellar ligament (cut)

(e)

Marshburn/Waldrop

UNIT IV

arcuate (ar′kyoo-āt) **popliteal ligament** are superficial in position, whereas the **anterior** and **posterior cruciate** (kroo′she-āt) **ligaments** are deep within the joint. The **popliteal bursa** and the **semimembranosus bursa** are the two bursae associated with the back of the knee.

Strong **collateral ligaments** support both the medial and lateral sides of the knee joint. Two fibrocartilaginous discs called the **lateral** and **medial menisci** are located within the knee joint interposed between the distal femoral and proximal tibial condyles. The two menisci are connected by a **transverse ligament.** Several other bursae are associated with the knee joint.

cruciate: L. crucis, cross

In addition to the 4 on the anterior side and the 2 on the posterior side, there are 7 bursae on the lateral and medial sides, for a total of 13.

During normal walking, running, and supporting of the body, the knee joint functions superbly. It can tolerate even moderate stress without tissue damage. However, the knee lacks bony support to withstand sudden forceful stresses, such as commonly occur among professional athletes. Knee injuries frequently require surgery and heal with difficulty due to the avascularity of the cartilaginous tissue. Because this joint is potentially so vulnerable, it is important to realize its limitations by understanding its anatomy.

Talocrural (Ankle) Joint

There are actually two principal articulations within the **ankle joint,** both of which are hinge joints. One articulation is formed as the distal end of the tibia and its medial malleolus articulate with the talus; the other is formed as the lateral malleolus of the fibula articulates with the talus (fig. 8.32).

One joint capsule surrounds the articulations of the three bones, and four ligaments support the ankle joint on the outside of the capsule. The strong **deltoid ligament** is associated with the tibia, whereas the **lateral collateral ligaments, anterior talofibular** (ta''lo-fib'yoo-lar) **ligament, posterior talofibular ligament,** and **calcaneofibular** (kal-ka''ne-o-fib'yoo-lar) **ligament** are associated with the fibula.

The malleoli form a cap over the upper surface of the talus that prohibits side-to-side movement at the ankle joint. Unlike the condyloid joint at the wrist, the movements of the ankle are limited to flexion and extension. Dorsiflexion of the ankle is checked primarily by the tendo calcaneus, whereas plantar

Figure 8.32 The right talocrural (ankle) joint. (*a*) A lateral view, (*b*) a medial view, and (*c*) a posterior view.

UNIT IV

flexion, or ankle extension, is checked by the tension of the extensor tendons on the front of the joint and the anterior portion of the joint capsule.

Ankle sprains are a common type of locomotor injury. They vary widely in seriousness but tend to occur in certain locations. The most common cause of ankle sprain is excessive inversion of the foot, resulting in partial tearing of the anterior talofibular ligament and the calcaneofibular ligament. Less commonly, the deltoid ligament is injured by excessive eversion of the foot. Torn ligaments are extremely painful and are accompanied by immediate local swelling. Reducing the swelling and immobilizing the joint are about the only treatments for moderate sprains. Extreme sprains may require surgery and casting of the joint to facilitate healing.

A summary of the principal joints of the body and their movement is presented in table 8.2.

1. Which synovial joints have menisci?
2. What are the two types of joints found in the shoulder region? Why is the shoulder joint so vulnerable?
3. Which joints are reinforced with muscles that span the joint?
4. Describe the structure of the knee joint and indicate which structures protect and reinforce its anterior surface.

UNIT IV

Table 8.2	Principal Articulations	
Joint	**Type**	**Movement**
Most skull joints	Fibrous (suture)	Immovable
Temporomandibular	Synovial (hinge; gliding)	Elevation, depression; protraction, retraction
Atlantooccipital	Synovial (condyloid)	Flexion, extension, circumduction
Atlantoaxial	Synovial (pivot)	Rotation
Intervertebral		
bodies of vertebrae	Cartilaginous (symphysis)	Slight movement
articular processes	Synovial (gliding)	Flexion, extension, slight rotation
Sacroiliac	Synovial (gliding)	Slight gliding movement; may fuse in adults
Costovertebral	Synovial (gliding)	Slight movement during breathing
Sternocostal	Synovial (gliding)	Slight movement during breathing
Sternoclavicular	Synovial (gliding)	Slight movement when shrugging shoulders
Sternal	Cartilaginous (symphysis)	Slight movement during breathing
Acromioclavicular	Synovial (gliding)	Protraction, retraction; elevation, depression
Humeral (shoulder)	Synovial (ball-and-socket)	Flexion, extension; adduction, abduction; rotation, circumduction
Elbow	Synovial (hinge)	Flexion, extension
Proximal radioulnar	Synovial (pivot)	Rotation
Distal radioulnar (side-to-side)	Fibrous (syndesmosis)	Slight movement
Radiocarpal (wrist)	Synovial (condyloid)	Flexion, extension; adduction, abduction; circumduction
Intercarpal	Synovial (gliding)	Slight movement
Carpometacarpal		
fingers	Synovial (condyloid)	Flexion, extension; adduction, abduction
thumb	Synovial (saddle)	Flexion, extension; adduction, abduction
Metacarpophalangeal	Synovial (condyloid)	Flexion, extension; adduction, abduction
Interphalangeal	Synovial (hinge)	Flexion, extension
Symphysis pubis	Fibrous (symphysis)	Slight movement
Coxal (hip)	Synovial (ball-and-socket)	Flexion, extension; adduction, abduction; rotation; circumduction
Tibiofemoral (knee)	Synovial (hinge)	Flexion, extension; slight rotation when flexed
Proximal tibiofibular	Synovial (gliding)	Slight movement
Distal tibiofibular	Fibrous (syndesmosis)	Slight movement
Talocrural (ankle)	Synovial (hinge)	Dorsiflexion, plantar flexion; slight circumduction; inversion, eversion
Intertarsal	Synovial (gliding)	Inversion, eversion
Tarsometatarsal	Synovial (gliding)	Flexion, extension; adduction, abduction

Developmental Exposition

The Synovial Joints

Exhibit I Development of synovial joints. (*a*) At 6 weeks, different densities of mesenchyme denote where the bones and joints will form. (*b*) At 9 weeks, a basic synovial model is present. At 12 weeks, the synovial joints are formed and have either (*c*) a free joint cavity (e.g., interphalangeal joint); (*d*) a cavity containing menisci (e.g., knee joint); or (*e*) a cavity with a complete articular disc (e.g., sternoclavicular joint).

Waldrop

Explanation

The sites of developing synovial joints (freely movable joints) are discernible at 6 weeks as mesenchyme becomes concentrated in the areas where precartilage cells differentiate (exhibit I). The future joints at this stage appear as intervals of less concentrated mesenchymal cells. As cartilage cells develop within a forming bone, a thin flattened sheet of cells forms around the cartilaginous model to become the **perichondrium.** These same cells are continuous across the gap between the adjacent developing bone. Surrounding the gap, the flattened mesenchymal cells differentiate to become the **joint capsule.**

During the early part of the third month of development, the mesenchymal cells still remaining within the joint capsule begin migrating toward the epiphyses of the adjacent developing bones. The cleft eventually enlarges to become the **joint cavity.** Thin pads of hyaline cartilage develop on the surfaces of the epiphyses that contact the joint cavity. These pads become the **artic-** **ular cartilages** of the functional joint. As the joint continues to develop, a highly vascular **synovial membrane** forms on the inside of the joint capsule and begins secreting a watery *synovial fluid* into the joint cavity.

In certain developing synovial joints, the mesenchymal cells do not migrate away from the center of the joint cavity. Rather, they give rise to cartilaginous wedges called **menisci,** as in the knee joint, or to complete cartilaginous pads called **articular discs,** as in the sternoclavicular joint.

Most synovial joints have formed completely by the end of the third month. Shortly thereafter, fetal muscle contractions, known as *quickening,* cause movement at these joints. Joint movement enhances the nutrition of the articular cartilage and prevents the fusion of connective tissues within the joint.

Clinical Considerations

A synovial joint is a remarkable biologic system that acts as a self-lubricating bearing surface, able to move with almost frictionless precision under tremendous loads and impacts. Under normal circumstances and in most people, the many joints of the body perform without problems throughout life. Joints are not indestructible, however, and are subject to various forms of trauma and disease. Although not all of the diseases of joints are fully understood, medical science has made remarkable progress in the treatment of arthrological problems.

Trauma to Joints

Joints are well adapted to withstand compression and tension forces. Torsion or sudden impact to the side of a joint, however, can be devastating. These types of injuries frequently occur in athletes.

In a **strained joint,** unusual or excessive exertion stretches the tendons spanning the joint or surrounding muscles, but it causes no serious damage. Strains frequently result from not "warming up," or activating joints and muscles, prior to strenuous use. A **sprain** is a tearing of the ligaments or tendons that surround a joint. There are various grades of sprains, and the severity will determine the treatment. Severe sprains of the knee joint are frequently accompanied by damage to articular cartilages and menisci, which generally requires surgery. Sprains are usually accompanied by **synovitis** (si″nŏ-vi′tis), an inflammation of the joint capsule.

Luxation, or **joint dislocation,** is derangement of the articulating bones that compose the joint. Joint dislocation is more serious than a sprain and is usually accompanied by sprains. The shoulder and knee joints are the most vulnerable to dislocation. Self-healing of a dislocated joint, like the knee joint, may be incomplete, leaving the person with a "trick knee" that may unexpectedly give way.

Subluxation is partial dislocation of a joint. Subluxation of the hip joint is a common type of birth defect that can be treated by bracing or casting the hip joints to promote suitable bone development.

Bursitis (bur-si′tis) is an inflammation of the bursa associated with a joint. Because of its close proximity to the joint, bursitis may affect the joint capsule as well. Bursitis may be caused by excessive stress on the bursa from overexertion, or it may be a local or systemic inflammatory process. As the bursa swells, the surrounding muscles become sore and stiff. **Tendonitis** involves the tendon, in or out of a tendon sheath, and usually comes about in the same way as bursitis.

The flexible vertebral column is a marvel of mechanical engineering. Not only do the individual vertebrae articulate one with another, but together they form the portion of the axial skeleton with which the head, ribs, and pelvic girdle articulate. The vertebral column also encloses the spinal cord and provides exits for 31 pairs of spinal nerves. Considering all the articulations in the vertebral column and the physical abuse it takes, it is no wonder that back ailments are second only to headaches

as the most common physical complaint. Our way of life causes many of the problems associated with the vertebral column. Improper shoes, athletic exertion, sudden stops in vehicles, or improper lifting can all cause the back to go awry. Body weight, age, and general body condition influence a person's susceptibility to back problems.

The most common cause of back pain is *strained muscles,* generally the result of overexertion. The second most frequent back ailment is a *herniated disc.* The dislodged nucleus pulposus of a disc may push against a spinal nerve and cause excruciating pain. The third most frequent back problem is a *dislocated articular facet* between two vertebrae, caused by sudden torsion of the vertebral column. The treatment of back ailments varies from bed rest to spinal manipulation to extensive surgery.

Curvature disorders are another problem of the vertebral column. **Kyphosis** *(ki-fo′sis)* (hunchback) is an exaggeration of the thoracic curve. **Lordosis** (swayback) is an abnormal anterior convexity of the lumbar curve. **Scoliosis** (crookedness) is an abnormal lateral curvature of the vertebral column (fig. 8.33), which may be caused by one leg being longer than the other or uneven muscular development on the two sides of the vertebral column.

Diseases of Joints

Arthritis is a generalized term for over 50 different joint diseases, all of which have the symptoms of edema, inflammation, and pain. The causes of most kinds of arthritis are unknown, but certain types follow joint trauma or bacterial infections. There is evidence that some types of arthritis are the result of hormonal or metabolic disorders. The most common forms are rheumatoid *(roo′mă-toid)* arthritis, osteoarthritis, and gouty arthritis.

In **rheumatoid arthritis,** the synovial membrane thickens and becomes tender and synovial fluid accumulates. This change is usually followed by an invasion of fibrous tissue and deterioration of the articular cartilage. When the cartilage is destroyed, the exposed bone tissue is joined by the fibrous tissue and instigates ossification of the joint. It is the joint ossification that causes the crippling effect of rheumatoid arthritis. This disease affects females slightly more than males and occurs most commonly between the ages of 30 and 50.

Osteoarthritis is a degenerative joint disease that results from aging and irritation of the joints. Although osteoarthritis is far more common than rheumatoid arthritis, it is usually less damaging. Osteoarthritis is a slow, progressive disease in which the articular cartilages gradually soften and disintegrate. The affected joints seldom swell, and the synovial membrane is rarely damaged. As the articular cartilage deteriorates, ossified spurs are deposited on the exposed bone, causing pain and restricting the movement of articulating bones. Osteoarthritis most frequently affects the knee, hip, and intervertebral joints.

luxation: L. *luxus,* out of place

kyphosis: Gk. *kyphos,* hunched
lordosis: Gk. *lordos,* curving forward
scoliosis: Gk. *skoliosis,* crookedness
rheumatoid: Gk. *rheuma,* a flowing

UNIT IV

Figure 8.33 Scoliosis is a lateral curvature of the spine, usually in the thoracic region. It may be congenital, acquired, or disease related. (*a*) A posterior view of a 19-year-old woman and (*b*) a radiograph.

(a)

(b)

UNIT IV

Gouty arthritis results from a metabolic disorder in which an abnormal amount of uric acid is retained in the blood and sodium urate crystals are deposited in the joints. The salt crystals irritate the articular cartilage and synovial membrane, causing swelling, tissue deterioration, and pain. If gout is not treated, the affected joint fuses. Males have a greater incidence of gout than females, and apparently the disease is genetically determined. About 85% of gout cases affect the joints of the foot and legs. The most common joint affected is the metatarsophalangeal joint of the hallux (great toe).

Treatment of Joint Disorders

Arthroscopy (*ar-thros'kŏ-pe*) is widely used in diagnosing and, to a limited extent, treating joint disorders. Arthroscopic inspection involves making a small incision through the skin and into the joint capsule through which the tubelike arthroscopic instrument is threaded. In arthroscopy of the knee, the articular cartilage, synovial membrane, menisci, and cruciate ligaments can be observed. Samples can be extracted, and pictures taken for further evaluation.

Remarkable advancements have been made in the last 15 years in **joint prostheses** (*pros-the'sēz*) (fig. 8.34). Joint prostheses, or artificial articulations, do not take the place of normal, healthy joints, but they are a valuable option for chronically disabled arthritis patients.

CLINICAL CASE STUDY ANSWER

The mechanism of the knee injury and the location of the pain, taken together with the exam findings, indicate a complete or near-complete tear of the medial collateral ligament. Because the medial meniscus is attached to the medial collateral ligament, it is frequently torn as well in an injury of this sort. Other ligaments susceptible to athletic injury are the anterior cruciate ligament (most common) and the lateral collateral and posterior cruciate ligaments. Complete tears of these ligaments usually require surgical repair for acceptable results. Incomplete tears can often be managed nonoperatively.

gout: L. *gutta*, a drop (thought to be caused by "drops of viscous humors")
prosthesis: Gk. *pros*, in addition to; *thesis*, a setting down

Figure 8.34 Two examples of joint protheses. (*a,b*) The coxal (hip) joint and (*c,d*) the tibiofemoral (knee) joint.

(a)

(b)

(c)

(d)

Figure 8.35 Which joints of the body are being flexed as this person assumes a fetal position?

UNIT IV

IMPORTANT CLINICAL TERMINOLOGY

ankylosis Stiffening of a joint resulting in severe or complete loss of movement.

arthralgia (also *arthrodynia*) Severe pain within a joint.

arthrolith A gouty deposit in a joint.

arthrometry The measurement of the range of movement in a joint.

arthroncus Swelling of a joint due to trauma or disease.

arthropathy Any disease affecting a joint.

arthroplasty The surgical repair of a joint.

arthrosis A joint or an articulation; also, a degenerative condition of a joint.

arthrosteitis An inflammation of the bony structure of a joint.

chondritis An inflammation of the articular cartilage of a joint.

coxarthrosis A degenerative condition of the hip joint.

hemarthrosis An accumulation of blood in a joint cavity.

rheumatology The medical specialty concerned with the diagnosis and treatment of arthritis.

spondylitis An inflammation of one or several vertebrae.

synovitis The inflammation of the synovial membrane lining the inside of a joint capsule.

CHAPTER SUMMARY

Classification of Joints (p. 192)

1. Joints are formed as adjacent bones articulate. Arthrology is the science concerned with the study of joints and kinesiology is the study of movements involving certain joints.

2. Joints can be classified according to structure or function.
   (a) Structurally, joints are classified as fibrous, cartilaginous, and synovial types.
   (b) Functionally, joints are classified as synarthroses, immovable joints; amphiarthroses, slightly movable joints; and diarthroses, freely movable joints.

Fibrous Joints (pp. 192–194)

1. Articulating bones in fibrous joints are tightly bound by fibrous connective tissue. They are of three kinds: sutures, syndesmoses, and gomphoses.

2. Sutures are found only in the skull and are classified as serrate, lap, or plane.

3. Syndesmoses are found only in the antebrachium and leg where adjacent bones are held together by interosseous ligaments. Slight movement is permitted at syndesmoses.

4. Gomphoses are found only in the skull where the teeth are bound into their sockets by the periodontal ligament.

Cartilaginous Joints (pp. 194–195)

1. Cartilaginous joints have fibrocartilage or hyaline cartilage between the adjacent bones. They are of two kinds: symphyses and synchondroses.
2. The symphysis pubis and the intervertebral disc joints are examples of symphyses.
3. Synchondroses are temporary joints formed in the growth lines (epiphyseal plates) between the diaphyses and epiphyses in the long bones of children.

Synovial Joints (pp. 195–200)

1. Synovial joints are freely movable articulations between bones.
2. Each synovial joint contains a joint capsule, articular cartilages, and synovial membranes that produce lubricating synovial fluid. In addition, some synovial joints contain a meniscus or menisci to assist joint movement.
3. The range of movement of a synovial joint is determined by the structure of the articulating bones, the ligaments and tendons, and the muscles that act on the joint.

Movements at Synovial Joints (pp. 200–206)

1. Movements at synovial joints are produced by the contraction of the skeletal muscles spanning the joints and attaching to or near the bones forming the articulations. In these actions, the bones act as levers, the muscles provide the force, and the joints are the fulcra, or pivots.
2. Angular movements increase or decrease the joint angle produced by the articulating bones.
   (a) Flexion decreases the joint angle on an anterior-posterior plane; extension increases the same joint angle.
   (b) Abduction is the movement of a body part away from the main axis of the body; adduction is the movement of a body part toward the main axis of the body.
3. Circular movements can occur only in joints that are composed of a bone with a rounded surface articulating with a corresponding depression on another bone.
   (a) Rotation is the movement of a bone around its own axis.
   (b) Circumduction is a conelike movement of a body segment.

4. Special joint movements include inversion and eversion, protraction and retraction, and elevation and depression.
5. Synovial joints can be classified as first-, second-, or third-class levers.
   (a) In a first-class lever, the fulcrum is positioned between the effort and the resistance.
   (b) In a second-class lever, the resistance is positioned between the fulcrum and the effort.
   (c) In a third-class lever, the effort lies between the fulcrum and the resistance.

Specific Joints of the Body (pp. 206–215)

1. The temporomandibular joint, a combined hinge and gliding joint, is of clinical importance because of temporomandibular joint (TMJ) syndrome.
2. The humeral (shoulder) joint, a ball-and-socket joint, is vulnerable to dislocations from sudden jerks of the arm, especially in children before strong shoulder muscles have developed.
3. There are two sets of articulations at the elbow joint as the distal end of the humerus articulates with the proximal ends of the ulna and radius. Strain on the elbow joint is common during certain sports.
4. The ball-and-socket coxal (hip) joint is especially prone to osteoarthritis in elderly people.
5. The hinged tibiofemoral (knee) joint is the largest, most vulnerable joint in the body.
6. There are two hinged articulations within the talocrural (ankle) joint. Ankle sprains are common injuries of this joint.

REVIEW ACTIVITIES

Objective Questions

1. Which statement regarding joints is *false*?
   (a) Joints are the locations where two or more bones articulate.
   (b) The structural classification of joints includes fibrous, membranous, and cartilaginous types.
   (c) Arthrology is the study of joints; kinesiology is the study of the biomechanics of joint movement.
2. Synchondroses are a type of
   (a) fibrous joint.
   (b) synovial joint.
   (c) cartilaginous joint.

3. An interosseous ligament is characteristic of
   (a) a suture.
   (b) a synchondrosis.
   (c) a symphysis.
   (d) a syndesmosis.
4. Which of the following joint type–function word pairs is *incorrect*?
   (a) synchondrosis/growth at the epiphyseal plate
   (b) symphysis/movement at the intervertebral joint
   (c) suture/strength and stability in the skull
   (d) syndesmosis/movement of the jaw
5. Which of the following is a *false* statement?
   (a) Synchondroses occur only in children and young adults.
   (b) Sutures occur only in the skull.
   (c) Saddle joints occur in the thumb and in the neck, where rotational movement is possible.
   (d) Syndesmoses occur only in the antebrachium and in the leg.
6. Which of the following is *not* characteristic of all synovial joints?
   (a) articular cartilage
   (b) synovial fluid
   (c) a joint capsule
   (d) menisci
7. The atlantoaxial and the proximal radioulnar synovial joints are specifically classified as
   (a) hinge.
   (b) gliding.
   (c) pivotal.
   (d) condyloid.
8. Which of the following joints can be readily and comfortably hyperextended?
   (a) the interphalangeal joint
   (b) the coxal joint
   (c) the tibiofemoral joint
   (d) the sternocostal joint
9. Which of the following is most vulnerable to luxation?
   (a) the elbow joint
   (b) the humeral joint
   (c) the coxal joint
   (d) the tibiofemoral joint
10. A thickening and tenderness of the synovial membrane and the accumulation of synovial fluid are signs of the development of
    (a) arthroscopitis.
    (b) gouty arthritis.
    (c) osteoarthritis.
    (d) rheumatoid arthritis.

UNIT IV

UNIT IV

Essay Questions

1. What is meant by a structural classification of joints as compared to a functional classification?
2. Why is the anatomical position so important in explaining the movements that are possible at joints?
3. What are the structural components of a synovial joint that determine or limit the range of movement at that joint?
4. What are the advantages of a hinge joint over a ball-and-socket type? If ball-and-socket joints allow a greater range of movement, why are not all the synovial joints of this type?
5. What is synovial fluid? Where is it produced and what are its functions?
6. Describe a bursa and discuss its function. What is bursitis?
7. Identify four types of synovial joints found in the wrist and hand region and state the types of movement permitted by each.
8. Discuss the articulations of the pectoral and pelvic regions to the axial skeleton with regard to range of movement, ligamentous attachments, and potential clinical problems.
9. What is meant by a sprained ankle? How does a sprain differ from a strain or a luxation?
10. What occurs within the joint capsule in rheumatoid arthritis? How does rheumatoid arthritis differ from osteoarthritis?

Activity Question

Refer to figure 8.35 and identify the joints being flexed.

Gundy/Weber Software

The tutorial software accompanying Chapter 8, Articulations, is Volume 4, Articulations and the Muscular System.

9

Muscular System

Chapter Outline

UNIT IV

CLINICAL CASE STUDY

A 66-year-old man is at his doctor's office for a routine physical exam. The man's medical history reveals that he had been treated surgically for cancer of the oropharynx, 6 years earlier. The patient states that the cancer had spread to the lymph nodes in the left side of his neck. He elaborates by pointing to the involved area and states that lymph nodes, a vein, and a muscle, among other things, had been removed. On the right side, only lymph nodes had been removed, but they had not been cancerous. The patient adds, in a bewildered tone, that he can turn his head only to the left. "It seems to me Doc, that if they took the muscle out of the left side of my neck, I would be able to turn my head only to the right."

Does the patient have a valid point? If not, how would you explain the reason for his disability in terms of neck musculature?

Hints: The action of a muscle can always be explained on the basis of its points of attachment and the joint or joints it spans. Carefully examine the muscles depicted in figure 9.20 and described in table 9.5.

Organization and General Functions

The muscles of the skeletal system are adapted to contract in order to carry out the functions of motion, heat production, and posture and body support.

Objective 1. Define the term *myology* and describe the three principal functions of muscles.

Objective 2. Explain how muscles are described according to their anatomical location and cooperative function.

Myology is the study of muscles. Myology does not have much meaning, however, except as it relates to the skeletal system and joints in the performance of body movements and to the nervous system in the performance of motor control. Muscles are dynamic, and to understand them requires an applied approach.

Functions of Muscles

Muscle cells (fibers) are adapted to contract when stimulated by nerve impulses. The stimulation of several fibers is not enough to cause a noticeable effect within a muscle, but these types of isolated fiber contractions are important and occur continuously within a muscle. When many fibers of a skeletal muscle

are activated, the muscle contracts and causes body movement. Muscles perform three general functions: (1) motion, (2) heat production, and (3) posture and body support.

1. **Motion.** The most obvious function performed by the skeletal muscles is to move the body and/or its appendages, as in walking, running, writing, chewing, and swallowing. The contraction of skeletal muscle is equally important in breathing and in moving body fluids. The stimulation of isolated fibers within the muscles maintains muscle tonus and is important in the movement of venous blood and lymphatic fluid. The eyeball and even the auditory ossicles have associated skeletal muscles responsible for various movements.

   The primary impetus for blood flow is the contraction of cardiac muscles within the heart. All of the involuntary body systems (urinary, digestive, respiratory, circulatory, and so forth) contain smooth muscles for the involuntary movement of materials through the body. The act of speaking involves several different muscle groups that control air volume, tense the vocal cords, move the tongue, and purse the lips. This is a highly coordinated effort of several muscle groups controlled by several different nerve centers.

2. **Heat production.** Body temperature is remarkably consistent. Metabolism within the cells releases heat as an end product. Since muscles constitute approximately 40% of body weight and are in a continuous state of fiber activity, they are very important in the production of heat. The rate of heat production increases immensely as a person exercises strenuously.

3. **Posture and body support.** The skeletal system lends form and stability to the body, but skeletal muscles maintain posture and support around the flexible joints. Certain muscles are active postural muscles whose primary function is to work in opposition to gravity. Some postural muscles are working even when you think you are relaxed. As you are sitting, for example, the weight of your head is balanced at the atlantooccipital joint through the efforts of the muscles located at the back of the neck. If you start to get sleepy, your head will suddenly nod forward as the postural muscles relax and the weight (resistance) overcomes the effort.

Muscle tissue in the body is of three types: smooth, cardiac, and skeletal. Although these three types differ in structure and function, the following basic properties characterize all muscle tissue:

1. **Irritability.** Muscle tissue is sensitive to stimuli from nerve impulses.
2. **Contractility.** Muscle tissue responds to stimuli by contracting lengthwise, or shortening.
3. **Extensibility.** Once a stimulus has subsided and the fibers within muscle tissue are shortened and relaxed, they may be stretched back to their original length by the contraction of an opposing muscle. The fibers are then stretched in readiness for another contraction.

---

muscle: L. *mus*, mouse (from the appearance of certain muscles)
myology: Gk. *myos*, muscle; *logos*, study of

Figure 9.1 (*a*) An anterior view and (*b*) a posterior view of some of the superficial skeletal muscles.

Frontalis
Orbicularis oculi
Zygomaticus
Masseter
Orbicularis oris
Sternocleido-mastoid
Deltoid
Pectoralis major
Brachialis
Biceps brachii
Brachioradialis
Gracilis
Vastus medialis
Gastrocnemius
Soleus

Trapezius
Latissimus dorsi
Serratus anterior
External abdominal oblique
Rectus abdominis
Sartorius
Adductor longus
Vastus lateralis
Peroneus longus
Extensor digitorum longus
Tibialis anterior

Margulies/Waldrop

(a)

Brachialis
Temporalis
Occipitalis
Sternocleidomastoid
Trapezius
Deltoid
Triceps brachii
Brachio-radialis
Biceps femoris
Semitendinosus
Semimembranosus
Gastrocnemius
Tendo calcaneus

Teres major
Infraspinatus
Rhomboideus
Latissimus dorsi
External abdominal oblique
Gluteus medius
Gluteus maximus
Adductor magnus
Iliotibial tract
Gracilis
Vastus lateralis
Sartorius
Soleus
Peroneus longus

Margulies/Waldrop

(b)

UNIT IV

4. **Elasticity.** Muscle tissue has an innate tension that causes it to assume a desired shape regardless of how it might be stretched.

A histological description of each of the three muscle types was presented in chapter 4 and should be reviewed at this time. Cardiac muscle is discussed further in connection with the heart in chapter 16. Smooth muscle functions with various involuntary organs throughout the body and is discussed, when appropriate, with these organs. The remaining information presented in this chapter pertains only to skeletal muscle and the skeletal muscular system of the body.

Skeletal muscle constitutes a body system by itself and accounts for approximately 40% of body weight. Over 600 individual muscles make up the skeletal muscular system. The principal superficial muscles are shown in figure 9.1. In de-

scribing the various muscles, they are usually grouped according to anatomical location and cooperative function. The *muscles of the axial skeleton* have their attachments to the bones of the axial skeleton and include facial muscles, neck muscles, and anterior and posterior trunk muscles. The *muscles of the appendicular skeleton* include those that act on the pectoral and pelvic girdles and those that cause movement at the joints of the upper and lower extremities.

1. Discuss how the functions of muscles aid in maintaining body homeostasis.
2. What is meant by a postural muscle?
3. Distinguish between the axial and the appendicular muscles.

Structure of Skeletal Muscles

Skeletal muscle tissue and its binding connective tissue are arranged in a highly organized pattern so as to unite the forces of the contracting muscle fibers and direct them onto the structure being moved.

Objective 3. Compare and contrast the various binding connective tissues associated with skeletal muscles.

Objective 4. Explain what is meant by synergistic and antagonistic muscle groups.

Objective 5. Describe the various types of muscle fiber architecture and discuss the biomechanical advantage of each type.

Attachments

Skeletal muscles are usually long and narrow, span a joint, and are attached to a bone at either end by a tendon (fig. 9.2). As the muscle contracts, one of the bones moves relative to the other at the joint. The more fixed, or stationary, attachment is designated as the **origin** of a muscle; the movable end is its **insertion.** In muscles associated with the girdles and appendages, the origin is generally the proximal attachment and the insertion is the distal attachment. The fleshy, thickened portion of a muscle is referred to as its **belly.** The belly of a muscle is generally located on the bone proximal to the bone that is to be moved. The joint is spanned by a **tendon** from the muscle. A tendon is toughened, dense regular connective tissue that connects a muscle to the periosteum of a bone. It functions to transfer the force of contraction from the muscle across the joint and onto the bone that is to be moved. Flattened, sheetlike tendons are called **aponeuroses** (ap''ō-noo-ro'sēz) and occur where attachment is over a broad line. In certain places, especially in the wrist and ankle, the tendons are not only enclosed by protective **tendon sheaths** that lubricate the tendons with synovial fluid, but also the entire group of tendons is covered by a thin but strong band of connective tissue called a **retinaculum** (ret''ĭ-nak'yoo-lum) (see fig. 9.43).

Associated Connective Tissue

Contracting muscle fibers would not be effective if they worked as isolated units. Each fiber is bound to adjacent fibers to form bundles, and the bundles in turn are bound to other bundles. With this arrangement, the contraction in one area of a muscle works in conjunction with contracting fibers elsewhere in the muscle. The binding substances within muscles are the associated connective tissues.

Connective tissue is structurally arranged within muscle to protect, strengthen, and bind muscle fibers into bundles and bind the bundles together (fig. 9.3). The individual fibers of skeletal muscles are surrounded by a fine sheath of connective tissue

Figure 9.2 The skeletomuscular relationship. The more proximal and fixed point of muscle attachment is the origin and the distal, maneuverable point of attachment is the insertion. The contraction of muscle fibers causes one bone to move relative to another around a joint.

called **endomysium** (en''do-mis'e-um). The endomysium binds adjacent fibers and supports capillaries and nerve endings serving the muscle. Another connective tissue, **perimysium,** binds groups of fibers together into bundles called **fasciculi** (fă-sik'yŭ-li). The perimysium supports blood vessels and nerve fibers serving the various fasciculi. The entire muscle is covered by the **epimysium,** which in turn is continuous with a tendon.

Fascia (fash'e-ă) is a fibrous connective tissue of varying thickness that covers muscle and attaches to the skin. *Superficial fascia* secures the skin to the underlying muscles. The superficial fascia over the buttocks and abdominal wall is thick and laced with adipose tissue. By contrast, the superficial fascia under the skin of the dorsum of the hand and facial region is thin. *Deep fascia* is an inward extension of the superficial fascia. It lacks adipose tissue and blends with the epimysium of muscle. Deep fascia surrounds adjacent muscles, compartmentalizing and binding them into functional groups.

Muscle Groups

Just as individual muscle fibers seldom contract independently, muscles generally do not contract separately but work as functional groups. Muscles that contract together and are coordinated in accomplishing a particular movement are said to be

---

aponeurosis: Gk. *aponeurosis*, change into a tendon
retinaculum: L. *retinere*, to hold back (retain)

endomysium: Gk. *endon*, within; *myos*, muscle
perimysium: Gk. *peri*, around; *myos*, muscle
fasciculus: L. *fascis*, bundle
epimysium: Gk. *epi*, upon; *myos*, muscle
fascia: L. *fascia*, a band or girdle

Figure 9.3 The relationship between muscle tissue and connective tissue. (*a*) The fascia and tendon attach a muscle to the periosteum of a bone. (*b*) The epimysium surrounds the entire muscle, and the perimysium separates and binds the fasciculi (muscle bundles). (*c*) The endomysium surrounds and binds individual muscle fibers. (*d*) An individual muscle fiber is composed of myofibrils, which consist of myofilaments of actin and myosin.

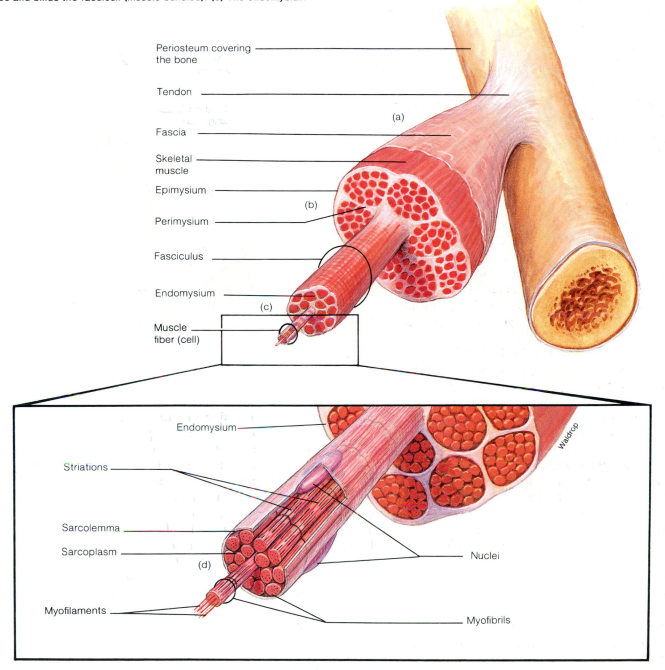

synergistic (fig. 9.4). *Antagonistic* muscles perform opposite functions and are generally located on the opposite sides of the limb. The two heads of the biceps brachii muscle together with the brachialis muscle, for example, contract to *flex* the elbow joint. The triceps brachii muscle, the antagonist to the biceps brachii and brachialis muscles, *extends* the elbow as it is contracted.

Seldom does the action of a single muscle cause a movement at a joint. Utilization of several synergistic muscles rather than one massive muscle allows for a division of labor. One muscle may be an important postural muscle, for example, whereas another may be adapted for rapid, powerful contraction. In terms of total output, the contraction of several small muscles provides more strength than the contraction of one large one.

synergistic: Gk. *synergein,* cooperate
antagonistic: Gk. *antagonistes,* struggle against

Figure 9.4 Examples of synergistic and antagonistic muscles. The two heads of the biceps brachii and the brachialis muscle are synergistic to each other, as are the three heads of the triceps brachii. The biceps brachii and the brachialis are antagonistic to the triceps brachii, and the triceps brachii is antagonistic to the biceps brachii and the brachialis muscle. When one antagonistic group contracts, the other one must relax; otherwise, movement does not occur.

Muscle Architecture

Skeletal muscles may be classified on the basis of fiber arrangement (fig. 9.5), each type having certain inherent advantages. The major types of fiber arrangements are as follows:

1. **Parallel.** Parallel muscles have relatively long excursions (contract over a great distance) and good endurance, but they are not especially strong. They are long, straplike muscles, such as the sartorius and rectus abdominis muscles.
2. **Convergent.** Convergent-fibered muscles are so named because the fibers converge at the insertion point to maximize contraction. They are fan-shaped muscles, such as the deltoid and pectoralis major muscles.
3. **Pennate.** Pennate-fibered muscles provide dexterity. They have many fibers per unit area, and hence are strong muscles. They have short excursions but generally tire quickly. The three kinds of pennate-fibered muscles are *unipennate*, *bipennate*, and *multipennate*. Examples of each of the pennate types can be found in the forearm.
4. **Sphincter.** Sphincteral muscles surround a body opening, or **orifice,** constricting it when contracted. Examples are the orbicularis oculi around the eye and orbicularis oris surrounding the mouth.

pennate: L. *pennatus*, feather
orifice: L. *orificium*, mouth; *facere*, to make

Figure 9.5 Skeletal muscle architecture.

Muscle fiber architecture can be observed on a cadaver or other dissection specimen. If you have the opportunity to learn the muscles of the body from a cadaver, observe the fiber architecture of specific muscles and try to determine the advantages afforded to each muscle by its location and action.

Blood and Nerve Supply to Skeletal Muscle

Muscle cells have a high rate of metabolic activity and therefore require extensive vascularity to receive nutrients and oxygen and to eliminate waste products. Smaller muscles generally have a single artery supplying blood and perhaps two veins returning blood (fig. 9.6). Large muscles may have several arteries and veins. The microscopic capillary exchange between arteries and veins occurs throughout the endomysium that surrounds individual fibers.

A skeletal muscle fiber cannot contract unless it is stimulated by a nerve impulse. This means that there must be extensive innervation to a muscle to ensure the connection of each muscle fiber to a nerve cell. Actually there are two nerve pathways for each muscle. A **motor (efferent) neuron** is a nerve cell that conducts nerve impulses to the muscle fiber to stimulate it to contract. A **sensory (afferent) neuron** conducts nerve impulses away from the muscle fiber to the central nervous system. Muscle fibers will atrophy if they are not periodically stimulated to contract.

The feeling of stiffness the day after strenuous exercise is due to trauma to muscle fibers and the accumulation of lactic acid within the muscle. Lactic acid is a waste product of metabolism and is removed by the venous flow from the muscle. If a person is in good physical condition, the vascularity can readily remove the metabolic wastes, and stiffness is not as likely to occur. Being in good physical condition not only improves vascularity but enlarges muscle fibers and allows them to work more efficiently over a longer duration.

Figure 9.6   The relationship of blood vessels and nerves to skeletal muscles of the axillary region. (Note the close proximity of the nerves and vessels as they pass between muscle masses.)

UNIT IV

1. Contrast the following terms: *endomysium* and *epimysium*; *fascia* and *tendon*; *aponeurosis* and *retinaculum*.
2. Discuss the biomechanical advantage of having synergistic muscles. Give some examples of synergistic muscles and state which muscles are antagonistic.
3. Which type of architecture provides dexterity and strength?

Skeletal Muscle Fibers and Mode of Contraction

Muscle fiber contraction in response to a motor impulse results from a sliding movement within the myofibrils in which the length of the sarcomeres is reduced.

Objective 6. Identify the major components of a muscle fiber and discuss the function of each part.

Objective 7. Distinguish between isotonic and isometric contractions.

Objective 8. Define *motor unit* and discuss the role of motor units in muscular contraction.

Skeletal Muscle Fibers

A skeletal muscle fiber is an elongated, cylindrical cell that may reach a length of 30 cm (12 in.) and have a diameter of 10 to 100 μm. Despite their unusual shape, muscle fibers have the same organelles that are present in other cells: mitochondria, intracellular membranes, glycogen granules, and others. Unlike most other cells in the body, however, skeletal muscle fibers are multinucleated and striated (fig. 9.7). Each fiber is surrounded by a cell membrane called a **sarcolemma,** and the cytoplasm within the cell is called **sarcoplasm.** A network of membranous channels, the **sarcoplasmic reticulum,** extends throughout the sarcoplasm (fig. 9.8). A system of **transverse tubules** (T-tubules) runs perpendicular to the sarcoplasmic reticulum and opens to the outside through the sarcolemma. Also embedded in the sarcoplasm and extending the entire length of the fiber are many parallel, threadlike structures called **myofibrils** (*mi''ō-fi'brilz*) (fig. 9.9). Each myofibril is composed of even smaller protein strands called **myofilaments.** *Thin myo-*

Figure 9.7 (*a*) A skeletal muscle fiber is composed of numerous threadlike strands of myofibrils that contain the myofilaments of actin and myosin. A skeletal muscle fiber is striated and multinucleated. (*b*) An electron micrograph of skeletal muscle fibers showing the striations and the peripheral location of the nuclei.

Sarcolemma
Sarcoplasm
Myofilaments
Myofibrils
Nucleus
Striations
(a)
Waldrop

Nuclei
Muscle fiber
(b)

filaments are about 6 nm in diameter and are composed of the protein **actin** *(ak'tin)*. *Thick myofilaments* are about 16 nm in diameter and are composed of the protein **myosin** *(mi'ŏ-sin)*.

The characteristic dark and light striations of skeletal muscle myofibrils are due to the arrangement of these myofilaments. The darker crossbands, produced by overlapping myosin myofilaments, are able to polarize visible light (anisotropic) and have been named *A bands*. The lighter crossbands, composed of thin actin myofilaments, do not polarize light (isotropic) and are called *I bands* (see figs. 9.8 and 9.9).

The I bands within a myofibril extend from the edge of one stack of thick myosin myofilaments to the edge of the next

actin: L. *actus*, motion, doing
myosin: L. *myosin*, within muscle
anisotropic: Gk. *anisos*, uneven; *tropos*, a turning
isotropic: Gk. *isos*, equal; *tropos*, a turning

stack of thick myofilaments. They are light bands because they contain only thin actin myofilaments. Each thin myofilament, however, extends partway into the A bands on each side (between the thick myofilaments of each stack). Since thick and thin myofilaments overlap at the edges of each A band, the edges are darker in appearance than the central region of the A band. These central lighter regions of the A bands are called the *H bands* (for *helle,* a German word for bright). The central H bands thus contain only myosin that is not overlapped with thin myofilaments.

At high magnification, thin dark lines can be seen in the middle of the I bands. These are labeled *Z lines* (for *Zwischenscheibe,* a German word meaning "between disc"). The arrangement of thick and thin myofilaments between a pair of Z lines forms a repeating pattern that serves as the basic subunit of skeletal muscle contraction. These subunits, from Z line to Z line,

Figure 9.8 The structural relationship of the myofibrils of a muscle fiber to the sarcolemma, transverse tubules, and sarcoplasmic reticulum. (Note the position of the mitochondria.)

Sarcolemma

Triad of the reticulum:

Terminal cisternae

Transverse tubule

Sarcoplasmic reticulum

Mitochondria

Myofibrils

A band

I band

Z line

Nucleus

Waldrop

UNIT IV

Figure 9.9 The structure of a myofibril. (*a*) Each of the many myofibrils of a skeletal muscle fiber is arranged into compartments (*b*) called sarcomeres. (*c*) The characteristic striations of a sarcomere are due to the arrangement of thin and thick myofilaments, composed of actin and myosin, respectively.

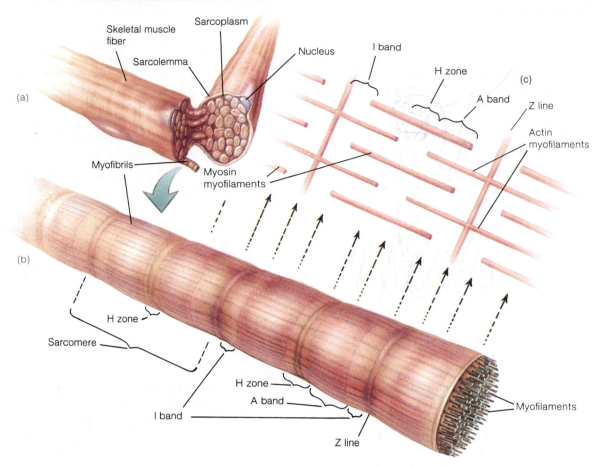

UNIT IV

are known as **sarcomeres** *(sar'kŏ-mērz)*. A longitudinal section of a myofibril thus presents a side view of successive sarcomeres (fig. 9.10a, b).

This side view is, in a sense, misleading. There are numerous sarcomeres within each myofibril that are out of the plane of the section (and out of the picture). A better appreciation of the three-dimensional structure of a myofibril can be obtained by viewing the myofibril in transverse section. In this view, it can be seen that the Z lines are actually disc shaped, and that the thin myofilaments that penetrate these Z discs surround the thick myofilaments in a hexagonal arrangement (see fig. 9.10c). If one concentrates on a single row of dark thick myofilaments in this transverse section, the alternating pattern of thick and thin myofilaments seen in longitudinal section becomes apparent.

When a muscle is stimulated to contract, it decreases in length as a result of the shortening of its individual fibers. Shortening of the muscle fibers, in turn, is produced by shortening of their myofibrils, which occurs as a result of the shortening of the distance from Z line to Z line (fig. 9.11). As the sarcomeres shorten in length, however, the A bands do *not*

shorten but instead appear closer together. The I bands—which represent the distance between A bands of successive sarcomeres—decrease in length.

Close examination reveals that the thick myosin and thin actin myofilaments remain the same length during muscle contraction. Shortening of the sarcomeres is produced not by shortening of the myofilaments, but rather by the *sliding* of thin myofilaments over thick ones. In the process of contraction, the thin myofilaments on either side of the A bands extend deeper and deeper toward the center, thereby increasing the amount of overlap with the thick myofilaments. The central H bands thus get shorter and shorter during contraction.

Isotonic and Isometric Contractions

In order for muscle fibers to shorten when they contract, they must generate a force that is greater than the opposing forces that act to prevent movement of the muscle's insertion. Flexion of the elbow joint, for example, occurs against the force of gravity and the weight of the objects being lifted. The tension

Figure 9.10 Electron micrographs of myofibrils of a muscle fiber. (*a*) At low power (1,600×), a single muscle fiber containing numerous myofibrils. (*b*) At high power (53,000×), myofibrils in longitudinal section. (Notice the sarcomeres and overlapping thick and thin myofilaments.) (*c*) The hexagonal arrangement of thick and thin filaments as seen in transverse section (arrows point to cross-bridges; SR = sarcoplasmic reticulum). (From R. G. Kessel and R. H. Kardon, *Tissues and Organs: A Text-Atlas of Scanning Electron Microscopy* © 1979 W. H. Freeman and Company.)

Kessel and Kardon

produced by the contraction of each muscle fiber separately is insufficient to overcome these opposing forces, but the combined contractions of large numbers of muscle fibers may be sufficient to overcome the opposing force and flex the elbow joint as the muscle fibers shorten in length.

The contraction that results in muscle shortening is called *isotonic contraction,* so called because the force of contraction remains relatively constant throughout the shortening process. In an *isometric contraction,* the length of the muscle remains constant because the antagonist force equals the force in the muscle being contracted. Isometric contractions occur when a person supports an object in a fixed position. An isometric contraction is converted to an isotonic contraction when increased force is generated within the muscle overcoming the resistance, resulting in muscle movement.

Neuromuscular Junction

A nerve serving a muscle is composed of both motor and sensory neurons. As the motor portion of the nerve penetrates a muscle, it splays into a number of branching neuron processes called **axons.** The terminal ends of axons contact the sarcolemma of muscle fibers by means of **motor end plates** (fig. 9.12). Each axon may split into enough branches to serve dozens of muscle fibers. The area consisting of the motor end plate and the sarcolemma of a muscle fiber is known as the **neuromuscular** (myoneural) **junction.**

isotonic: Gk. *isos,* equal; *tonos,* tension
isometric: Gk. *isos,* equal; *metron,* measure
axon: Gk. *axon,* axis

Acetylcholine (ă-sēt″l-ko′l ēn) is a neurotransmitter chemical stored in **synaptic vesicles** at the terminal ends of the axons. A nerve impulse reaching the terminal end of an axon causes the release of acetylcholine into the **neuromuscular cleft** of the neuromuscular junction. As this chemical mediator contacts the sarcolemma, it initiates physiological activity within the muscle fiber, resulting in contraction.

Motor Unit

A **motor unit** consists of a single motor neuron and the aggregation of muscle fibers innervated by the motor neuron (fig. 9.12b). When a nerve impulse travels through a motor unit, all of the fibers served by it contract simultaneously to their maximum. Most muscles have an innervation ratio of one motor neuron per 100–150 muscle fibers. Muscles capable of precise, dexterous movements, such as an eye muscle, may have an innervation ratio of 1:10. Massive muscles that are responsible for gross body movements, such as those of the thigh, may have an innervation ratio exceeding 1:500.

UNIT IV

Figure 9.11 The sliding filament model of contraction. As the myofilaments slide, the Z lines are brought closer together. The A bands remain the same length during contraction, but the I and H bands narrow progressively and may eventually become obliterated.

UNIT IV

All of the motor units controlling a particular muscle, however, are not the same size. Innervation ratios in a large thigh muscle may vary from 1:100 to 1:2,000. Neurons that innervate smaller numbers of muscle fibers have smaller cell bodies and axon diameters than neurons that have larger innervation ratios. The smaller neurons also are stimulated by lower levels of excitatory input. The small motor units, as a result, are the ones that are used most often. The larger motor units are activated only when very forceful contractions are required.

Skeletal muscles are voluntary in that they can be consciously contracted. The magnitude of the task determines the number of motor units that are activated. Performing a light task, such as lifting a book, requires few motor units, whereas lifting a table requires many. Muscles with pennate architecture have many motor units and are strong and dexterous; however, they generally fatigue more readily than muscles with fewer motor units. Being mentally "psyched up" to accomplish an athletic feat involves voluntary activation of more motor units within the muscles. Seldom does a person utilize all the motor units within a muscle, but the secretion of *epinephrine (ep''i-nef'rin)* from the adrenal gland does promote an increase in the force that can be produced when a given number of motor units are activated.

Figure 9.12 A motor end plate at the neuromuscular junction. (*a*) A neuromuscular junction is the site where the nerve fiber and muscle fiber meet. The motor end plate is the specialized portion of the sarcolemma of a muscle fiber surrounding the terminal end of the axon. (Note the slight gap between the membrane of the axon and that of the muscle fiber.) (*b*) A photomicrograph of muscle fibers and motor end plates. A motor neuron and the skeletal muscle fibers it innervates constitute a motor unit.

Motor neuron fiber
Nerve fiber branches
Muscle fiber nucleus
Motor end plate
Myofibril of muscle fiber

Mitochondria
Synaptic vesicles
Neuromuscular cleft
Folded sarcolemma
Motor end plate

Waldrop

(a)

Axon

Motor end plate

Muscle fiber

(b)

associations because they provide unfair advantages in physical competition and because they can have serious side effects, such as gonadal atrophy, induction of malignant tumors of the liver, and excessive aggressive behavior.

1. Draw three successive sarcomeres in a myofibril of a resting muscle fiber. Label the myofibril, sarcomeres, A bands, I bands, H bands, and Z lines.
2. Why do the A bands appear darker than the I bands?
3. Draw three successive sarcomeres in a myofibril of a contracted fiber. Indicate which bands get shorter during contraction and explain how this occurs.
4. Describe how the antagonistic muscles in the brachium can be exercised through both isotonic and isometric contractions.
5. Explain why motor units are considered the basic functional units of muscle contraction.

Steroids are hormones produced by the adrenal glands, testes, and ovaries. Because they are soluble in lipids, they readily pass through cell membranes and into the cytoplasm where they combine with proteins to form steroid-protein complexes that are necessary for the syntheses of specific kinds of messenger RNA molecules. Synthetic steroids were originally developed to promote weight gain in cancer and anorexic patients. It was soon determined, however, that steroids taken by bodybuilders and athletes would provide them with increased muscle mass, strength, and aggressiveness. The use of steroids is now considered illegal by most athletic

Naming of Muscles

Skeletal muscles are named on the basis of shape, location, attachment, orientation of fibers, relative position, or function.

Objective 9. Use examples to describe the various ways in which muscles are named.

One of the tasks of a myologist is to learn the names of the more than 600 skeletal muscles within the body. This task is not quite as formidable as one might imagine because most of the muscles are paired; that is, the right side is the mirror image

Figure 9.13 An anterior view of the superficial facial muscles involved in facial expression.

Galea aponeurotica

Frontalis

Orbicularis oculi — Orbital / Palpebral

Nasalis

Levator labii superioris

Zygomaticus minor

Zygomaticus major

Orbicularis oris

Risorius

Depressor anguli oris

Platysma

Depressor labii inferioris

Temporalis

Corrugator

Orbicularis oculi

Zygomaticus minor and major (cut)

Levator labii superioris (cut)

Buccinator

Masseter

Orbicularis oris

Depressor labii inferioris (cut)

Mentalis

Platysma (cut)

Sternocleidomastoid

Creek

of the left. Further simplifying the task are the descriptive names of muscles. Only the principal muscles of the body are described in the remaining pages of this chapter.

As you learn the muscles of the body, attend to the derivations of the names and be able to identify the muscles on yourself. Learn them as functional groups so that you will better understand their actions. If you can identify a muscle on yourself, you will be able to contract the muscle and describe its action. Learning anatomy in this way is easiest and most meaningful; moreover, retention is improved.

The following are some examples of how the names of muscles have been logically derived:

1. **Shape:** rhomboideus (like a rhomboid); trapezius (like a trapezoid); or denoting the number of heads of origin: triceps (three heads), biceps (two heads)
2. **Location:** pectoralis (in the chest, or pectus); intercostal (between ribs); brachii (upper arm)
3. **Attachment:** many facial muscles (zygomaticus, temporalis, nasalis); sternocleidomastoid (sternum, clavicle, and mastoid process of the skull)
4. **Size:** maximus (larger, largest); minimus (smaller, smallest); longus (long); brevis (short)
5. **Orientation of fibers:** rectus (straight); transverse (across); oblique (in an oblique direction)
6. **Relative position:** lateral, medial, internal, and external
7. **Function:** adductor, flexor, extensor, pronator, and levator (lifter)

1. Refer to chapter 2 and review the location of the following body regions: cervical, pectoral, abdominal, gluteal, perineal, brachial, antebrachial, inguinal, thigh, and popliteal.
2. Refer to chapter 8 and review the movements permitted at synovial joints.

Muscles of the Axial Skeleton

Muscles of the axial skeleton include those responsible for facial expression, mastication, eye movement, tongue movement, neck movement, and respiration and those of the abdominal wall, the pelvic outlet, and the vertebral column.

Objective 10. Locate the major muscles of the axial skeleton. Identify synergistic and antagonistic muscles and describe the action of each one.

Muscles of Facial Expression

Humans have a well-developed facial musculature (figs. 9.13 and 9.14) that provides complex facial expression as a means of social communication. Indeed, many messages can be conveyed by a person without a word spoken.

The muscles of facial expression are in a superficial position located on the scalp, face, and neck. Although highly variable

Figure 9.14 A lateral view of the superficial facial muscles involved in facial expression.

Galea aponeurotica
Periosteum
Adipose layer
Superficial and deep layers of temporal fascia
Temporalis
Occipitalis
Zygomatic arch
Masseter
Risorius
Sternocleidomastoid

Frontalis (cut)
Corrugator
Orbicularis oculi (orbital and palpebral parts)
Nasalis
Levator labii superioris
Zygomaticus minor
Zygomaticus major
Buccinator
Orbicularis oris
Depressor labii inferioris
Mentalis
Depressor anguli oris
Platysma

Creek

UNIT IV

in size and strength, these muscles all originate from the bones of the skull or flat sheetlike tendons and insert onto the hypodermis of the skin (table 9.1). They are all innervated by one of the two facial cranial nerves. The locations and points of attachments of most of the facial muscles are such that, when contracted, they cause movements around the eyes, nostrils, or mouth (fig. 9.15).

The muscles of facial expression are of clinical concern for several reasons, each of which involves the facial nerve. Located right under the skin, the many branches of the facial cranial nerve are vulnerable to trauma. Facial lacerations and fractures of the skull frequently damage branches of this nerve. The extensive pattern of motor innervation (see chapter 12) becomes apparent in stroke victims and persons suffering from *Bell's palsy*. The facial muscles on one side of the face are affected in these people and that portion of the face appears to sag.

Muscles of Mastication

The large **temporalis** and **masseter** *(mah-se'ter)* **muscles** (fig. 9.16) are powerful elevators of the mandible in synergism with the **medial pterygoid** *(ter'i-goyd)*. The primary function of the medial and lateral pterygoid muscles is to provide grinding movements of the teeth. The **lateral pterygoid** also protracts the mandible (table 9.2).

Tetanus is a bacterial disease caused by anaerobic *Clostridium tetani* being introduced into the body, usually from a puncture wound. The bacteria produce a neurotoxin that is carried to the spinal cord by sensory nerves. The motor impulses relayed back cause certain muscles to contract continuously (tetany). The muscles that move the mandible are affected first, which has given the disease the common name of *lockjaw*.

Ocular Muscles

The movements of the eyeball are controlled by six extrinsic ocular muscles (fig. 9.17). Five of these muscles arise from the margin of the optic canal at the back of the orbital cavity and insert on the outer layer (sclera) of the eyeball. Four **rectus muscles** maneuver the eyeball in the direction indicated by their names (**superior, inferior, lateral,** and **medial**), and two **oblique muscles (superior** and **inferior)** rotate the eyeball on its axis. The medial rectus on one side contracts with the medial rectus of the opposite eye when focusing on close objects. When looking to the side, the lateral rectus of one eyeball works with the medial rectus of the opposite eyeball to keep both eyes focused together. The superior oblique muscle passes through a pulleylike cartilaginous loop, the *trochlea,* before attaching to the eyeball. The ocular muscles are innervated by three cranial nerves (table 9.3).

Another muscle, the **levator palpebrae** *(le-va'tor pal'pĕ-bre)* **superioris** (fig. 9.17), is located in the ocular region but is not attached to the eyeball. It extends into the upper eyelid and raises the eyelid when contracted.

Figure 9.15 Muscles of facial expression. The actions of each of the muscles of facial expression are apparent as a person contracts them. In each of these photographs, identify the muscles that are being contracted.

(a)   (b)   (c)   (d)

(e)   (f)   (g)   (h)

UNIT IV

Table 9.1	**Muscles of Facial Expression**				
Muscle	**Derivation**	**Origin**	**Insertion**	**Action**	**Innervation**
Epicranius	*epi* = above; *crani* = skull	Galea aponeurotica and occipital bone	Skin of eyebrow and galea aponeurotica	Wrinkles forehead and moves scalp	Facial n.
Frontalis	Referring to frontal bone	Galea aponeurotica	Skin of eyebrow	Wrinkles forehead and elevates eyebrow	Facial n.
Occipitalis	Referring to occipital bone	Occipital bone and mastoid process	Galea aponeurotica	Moves scalp backward	Facial n.
Corrugator	*ruga* = a wrinkle	Fascia above eyebrow	Root of nose	Draws eyebrows toward midline	Facial n.
Orbicularis oculi	*orbis* = orbit, circular; *oculus* = eye	Bones of medial orbit	Tissue of eyelid	Closes eyes	Facial n.
Nasalis	*nasus* = nose	Maxilla and nasal cartilage	Aponeurosis of nose	Compresses nostrils	Facial n.
Orbicularis oris	*orb* = circular; *or* = mouth	Fascia surrounding lips	Mucosa of lips	Closes and purses lips	Facial n.
Levator labii superioris	*levator* = lifter; *labii* = lip; *superior* = upper	Upper maxilla and zygomatic bone	Orbicularis oris and skin above lips	Elevates upper lip	Facial n.
Zygomaticus	Pertaining to zygomatic bone	Zygomatic bone	Superior corner of orbicularis oris	Elevates corner of mouth	Facial n.
Risorius	*risor* = laughter	Fascia of cheek	Orbicularis oris at corner of mouth	Draws angle of mouth laterally	Facial n.
Depressor anguli oris	*depressor* = depress	Mandible	Inferior corner of orbicularis oris	Depresses corner of mouth	Facial n.
Depressor labii inferioris	*labii* = lip; *inferior* = lower	Mandible	Orbicularis oris and skin of lower lip	Depresses lower lip	Facial n.
Mentalis	*mentum* = chin	Mandible (chin)	Orbicularis oris	Elevates and protrudes lower lip	Facial n.
Platysma	*platy* = broad	Fascia of neck and chest	Inferior border of mandible	Depresses lower lip	Facial n.
Buccinator	*bucca* = cheek	Maxilla and mandible	Orbicularis oris	Compresses cheek	Facial n.

Figure 9.16 Muscles of mastication. (*a*) A superficial view, (*b*) a deep view, and (*c*) the deepest view, showing the pterygoid muscles. (The muscles of mastication are labeled in boldface.)

Table 9.2	Muscles of Mastication				
Muscle	**Derivation**	**Origin**	**Insertion**	**Action**	**Innervation**
Temporalis	*templus* = time, temple	Temporal fossa	Coronoid process of mandible	Elevates mandible	Trigeminal n.
Masseter	*maseter* = chew	Zygomatic arch	Lateral part of ramus of mandible	Elevates mandible	Trigeminal n.
Medial pterygoid	*medial* = toward midline; *pteryg* = winglike (referring to process of sphenoid bone)	Sphenoid bone	Medial aspect of mandible	Elevates mandible and moves mandible laterally	Trigeminal n.
Lateral pterygoid	*latus* = side	Sphenoid bone	Anterior side of mandibular condyle	Protracts mandible	Trigeminal n.

Figure 9.17 Extrinsic ocular muscles of the left eyeball. (a) An anterior view and (b) a lateral view. (The extrinsic ocular muscles are labeled in boldface.)

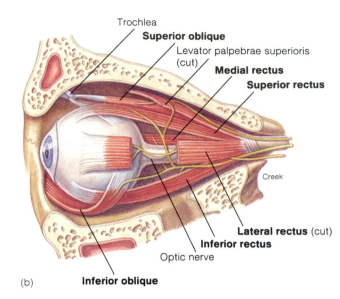

Table 9.3	Ocular Muscles	
Muscle	**Cranial Nerve Innervation**	**Movement of Eyeball**
Lateral rectus	Abducens	Lateral
Medial rectus	Oculomotor	Medial
Superior rectus	Oculomotor	Superior and medial
Inferior rectus	Oculomotor	Inferior and medial
Inferior oblique	Oculomotor	Superior and lateral
Superior oblique	Trochlear	Inferior and lateral

Muscles That Move the Tongue

The tongue is a highly specialized muscular organ that functions in speaking, manipulating food during mastication, cleansing the teeth, and swallowing. Two groups of skeletal muscles are responsible for tongue movement: intrinsic and extrinsic. The *intrinsic muscles* are confined within the tongue and are responsible for its mobility and changes of shape. *Extrinsic tongue muscles* are those that originate on structures away from the tongue and insert onto it to cause gross tongue movement (see fig. 9.18 and table 9.4). The three paired extrinsic muscles are the **genioglossus, styloglossus,** and **hyoglossus.** When the anterior portion of the genioglossus muscle is contracted, the tongue is depressed and thrust forward. If both genioglossus muscles are contracted together along their entire lengths, the superior surface of the tongue becomes transversely concave. This muscle is extremely important in infants for sucking; the tongue is positioned around the nipple with a concave groove channeled toward the pharynx.

Muscles of the Neck

Muscles of the neck either support and move the head or are attached to structures within the neck region, such as the hyoid bone and larynx. Only the more obvious neck muscles will be considered in this chapter.

Several of the muscles in this section and the sections to follow can be observed on yourself. Refer to chapter 10 to determine which muscles form important surface landmarks. These muscles are illustrated in figures 9.19 and 9.20 and are summarized in table 9.5.

Posterior Muscles   The posterior muscles include the sternocleidomastoid (originates anteriorly), trapezius, splenius capitis, semispinalis capitis, and longissimus capitis muscles.

As the name implies, the **sternocleidomastoid** *(ster"no-kli"do-mas'toid)* **muscle** originates on the sternum and clavicle and inserts on the mastoid process of the skull (fig. 9.20). When contracted on one side, it turns the head sideways in a direction opposite the side on which the muscle is located. If both sternocleidomastoid muscles are contracted, the head is pulled forward and down. The sternocleidomastoid muscle is covered by the platysma muscle (see fig. 9.14 and table 9.5).

Although a portion of the **trapezius muscle** extends over the posterior neck region, it is primarily a superficial muscle of the back and will be described later.

The **splenius capitis** is a broad muscle, positioned deep to the trapezius (fig. 9.19). It originates on the ligamentum nuchae and the spinous processes of the seventh cervical and first three thoracic vertebrae. It inserts on the back of the skull below the superior nuchal line and on the mastoid process of the temporal bone. When the splenius capitis muscle contracts on one side,

Figure 9.18 Extrinsic muscles of the tongue and deep structures of the neck. (The extrinsic muscles of the tongue are labeled in boldface.)

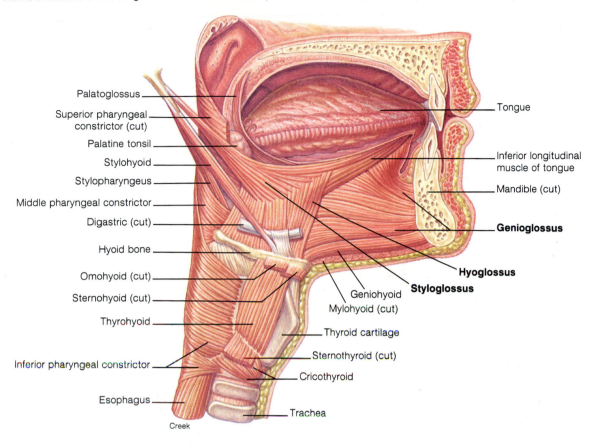

Palatoglossus
Superior pharyngeal constrictor (cut)
Palatine tonsil
Stylohyoid
Stylopharyngeus
Middle pharyngeal constrictor
Digastric (cut)
Hyoid bone
Omohyoid (cut)
Sternohyoid (cut)
Thyrohyoid
Inferior pharyngeal constrictor
Esophagus

Tongue
Inferior longitudinal muscle of tongue
Mandible (cut)
Genioglossus
Hyoglossus
Styloglossus
Geniohyoid
Mylohyoid (cut)
Thyroid cartilage
Sternothyroid (cut)
Cricothyroid
Trachea

Creek

UNIT IV

Table 9.4	Extrinsic Tongue Muscles				
Muscle	**Derivation**	**Origin**	**Insertion**	**Action**	**Innervation**
Genioglossus	*geneion* = chin; *glossus* = tongue	Mental spine of mandible	Inferior surface of tongue	Depresses and protracts tongue	Hypoglossal n.
Styloglossus	*stylo* = pertaining to styloid process of skull	Styloid process of temporal bone	Lateral side and inferior surface of tongue	Elevates and retracts tongue	Hypoglossal n.
Hyoglossus	*hyo* = pertaining to hyoid bone	Body of hyoid bone	Side of tongue	Depresses sides of tongue	Hypoglossal n.

the head rotates and extends to one side. Contracted together, these muscles extend the head at the neck. Further contraction causes hyperextension of the neck and head.

The broad, sheetlike **semispinalis capitis muscle** extends upward from the seventh cervical and first six thoracic vertebrae to insert on the occipital bone (fig. 9.19). When the two semispinalis capitis muscles contract together, they extend the head at the neck, along with the splenius capitis muscle. If one of the muscles acts alone, the head is rotated to the side.

The narrow, straplike **longissimus capitis muscle** ascends from processes of the lower four cervical and upper five thoracic vertebrae and inserts on the mastoid process of the temporal bone (fig. 9.19). This muscle extends the head at the neck, bends it to the one side, or rotates it slightly.

Suprahyoid Muscles The group of suprahyoid muscles located above the hyoid bone includes the digastric, mylohyoid, and stylohyoid muscles (fig. 9.20).

Figure 9.19 Deep muscles of the posterior neck and upper back regions.

Semispinalis capitis

Splenius capitis

Sternocleidomastoid

Levator scapulae

Splenius cervicis

Serratus posterior superior

Rhomboideus minor (cut)

Rhomboideus major (cut)

Rectus capitis posterior minor

Rectus capitis posterior major

Obliquus capitis superior

Obliquus capitis inferior

Longissimus capitis

Splenius cervicis

Levator scapulae

Scalenus medius

Scalenus posterior

Longissimus cervicis

Iliocostalis cervicis

Longissimus thoracis

Creek

UNIT IV

Table 9.5	Muscles of the Neck				
Muscle	**Derivation**	**Origin**	**Insertion**	**Action**	**Innervation**
Sternocleidomastoid	*sternum* = pertaining to sternal bone; *cleido* = clavicle; *mastoid* = mastoid process of skull	Sternum; clavicle	Mastoid process of temporal bone	Turns head to side; flexes neck	Accessory n.
Digastric	*di* = two; *gaster* = belly	Inferior border of mandible; mastoid groove	Hyoid bone	Opens mouth; elevates hyoid bone	Trigeminal n. (ant. belly); facial n. (post. belly)
Mylohyoid	*mylos* = akin to; *hyo* = hyoid bone	Inferior border of mandible	Body of hyoid bone and median raphe	Elevates hyoid bone and floor of mouth	Trigeminal n.
Stylohyoid	*stylo* = pertaining to styloid process	Styloid process of temporal bone	Body of hyoid bone	Elevates and retracts tongue	Facial n.
Hyoglossus	*hyo* = hyoid bone; *glossus* = tongue	Body of hyoid bone	Side of tongue	Depresses side of tongue	Hypoglossal n.
Sternohyoid		Manubrium	Body of hyoid bone	Depresses hyoid bone	Hypoglossal n.
Sternothyroid	*thyro* = shield shaped	Manubrium	Thyroid cartilage	Depresses thyroid cartilage	Hypoglossal n.
Thyrohyoid		Thyroid cartilage	Great cornu of hyoid bone	Depresses hyoid bone; elevates thyroid	Hypoglossal n.
Omohyoid	*omo* = shoulder	Superior border of scapula	Clavicle; body of hyoid bone	Depresses hyoid bone	Hypoglossal n.

Figure 9.20 Muscles of the anterior and lateral neck regions.

- Stylohyoid
- Posterior belly of digastric
- Sternocleidomastoid
- Semispinalis capitis
- Splenius capitis
- Common carotid artery
- Levator scapulae
- Trapezius
- Scalenus medius
- Inferior belly of omohyoid
- Brachial plexus
- Scalenus anterior
- Platysma (cut)
- Hyoglossus
- Mylohyoid
- Anterior belly of digastric
- Thyrohyoid
- Inferior constrictor
- Superior belly of omohyoid
- Sternohyoid
- Sternothyroid
- Sternocleidomastoid

Creek

UNIT IV

The **digastric muscle** is a two-bellied muscle of double origin that inserts on the hyoid bone. The anterior origin is on the mandible at the point of the chin, and the posterior origin is near the mastoid process of the temporal bone. The digastric muscle can open the mouth or elevate the hyoid.

The **mylohyoid muscle** forms the floor of the mouth. It originates on the inferior border of the mandible and inserts on the median raphe and body of the hyoid. As this muscle contracts, the floor of the mouth is elevated. It aids swallowing by forcing the food toward the back of the mouth.

The slender **stylohyoid muscle** extends from the styloid process of the skull to the hyoid bone, which it elevates as it contracts. The secondary effect of this muscle on tongue movement has already been described.

Infrahyoid Muscles Infrahyoid muscles are paired, thin, straplike muscles located below the hyoid bone. They are individually named on the basis of their origin and insertion and include the sternohyoid, sternothyroid, thyrohyoid, and omohyoid muscles (fig. 9.20).

The **sternohyoid muscle** originates on the manubrium of the sternum and inserts on the hyoid bone. It depresses the hyoid bone as it contracts.

The **sternothyroid muscle** also originates on the manubrium but inserts on the thyroid cartilage of the larynx. When this muscle contracts, the larynx is pulled downward.

The short **thyrohyoid muscle** extends from the thyroid cartilage to the hyoid bone. It elevates the larynx and lowers the hyoid bone.

The long, thin **omohyoid muscle** originates on the superior border of the scapula and inserts on the clavicle and on the hyoid bone. It acts to depress the hyoid bone.

The coordinated movements of the hyoid bone and the larynx are impressive. The hyoid bone does not articulate with any other bone, yet it has eight paired muscles attached to it. Two paired muscles involve tongue movement, one paired muscle lowers the jaw, one muscle elevates the floor of the mouth, and four paired muscles depress the hyoid or elevate the thyroid cartilage of the larynx.

Muscles of Respiration

The muscles of respiration are skeletal muscles that continually and rhythmically contract, usually involuntarily. Breathing, or *pulmonary ventilation,* is divided into two phases: *inspiration (inhalation)* and *expiration (exhalation).*

Figure 9.21 Muscles of respiration.

Muscles of Inspiration

Muscles of Expiration

Sternocleidomastoid

Scalenes

External intercostals

Internal intercostals
(interchondral part)

Diaphragm

Creek

Internal intercostals
(excluding interchondral
part)

External abdominal oblique

Internal abdominal oblique

Transversus abdominis

Rectus abdominis

During normal, relaxed inspiration, the important muscles are the **diaphragm** *(di'ă-fram)*, the **external intercostal muscles,** and the interchondral portion of the **internal intercostal muscles** (fig. 9.21). A downward contraction of the dome-shaped diaphragm causes a vertical increase in thoracic dimension. A simultaneous contraction of the external intercostals and the interchondral portion of the internal intercostals produces an increase in the lateral dimension of the thorax. In addition, the **sternocleidomastoid** and **scalenes muscles** may assist in inspiration through elevation of the first and second ribs, respectively. The intercostal muscles are innervated by the intercostal nerves, and the diaphragm receives its stimuli through the phrenic nerves.

Expiration is primarily a passive process, occurring as the muscles of inspiration are relaxed and the rib cage recoils to its original position. During forced expiration, the interosseous portion of the **internal intercostal muscles** contract, causing the rib cage to be depressed. This portion of the internal intercostal muscles lies under the external intercostals, and their fibers are directed downward and backward. The **abdominal muscles** may also contract during forced expiration, which increases pressure within the abdominal cavity and forces the diaphragm superiorly, squeezing additional air out of the lungs.

Muscles of the Abdominal Wall

The anterolateral abdominal wall is composed of four pairs of flat, sheetlike muscles: the external abdominal oblique, internal

abdominal oblique, transversus abdominis, and rectus abdominus muscles (fig. 9.22). These muscles support and protect the organs of the abdominal cavity and aid in breathing. When they contract, the pressure in the abdominal cavity increases, which can aid in defecation and in stabilizing the spine during heavy lifting.

The **external abdominal oblique muscle** is the strongest and most superficial of the three layered muscles of the lateral abdominal wall (figs. 9.22 and 9.23). Its fibers are directed inferiorly and medially. The **internal abdominal oblique muscle** lies deep to the external abdominal oblique, and its fibers are directed at right angles to the external abdominal oblique. The **transversus abdominis muscle** is the deepest of the abdominal muscles, and its fibers pass directly medially around the abdominal wall. The long, straplike **rectus abdominis muscle** is entirely enclosed in a fibrous sheath formed from the aponeuroses of the other three abdominal muscles. The *linea alba* is a band of connective tissue on the midline of the abdomen that separates the two rectus abdominis muscles. *Tendinous inscriptions* transect the rectus abdominis muscles at several points, causing the surface abdominal anatomy of a well-muscled person to appear segmented.

Refer to table 9.6 for a summary of the muscles of the abdominal wall.

Figure 9.22 Muscles of the anterolateral neck, shoulder, and trunk regions. The mammary gland is an integumentary structure positioned over the pectoralis major muscle.

Sternocleidomastoid
Splenius capitis
Levator scapulae
Trapezius
Short head of biceps brachii
Deltoid
Coracobrachialis
Long and lateral heads of triceps brachii
Subscapularis
Teres major
Latissimus dorsi
Serratus anterior
Mammary lobes
External oblique (cut)
External intercostal
Internal intercostal
External abdominal oblique (cut)
Internal abdominal oblique (cut)
Transversus abdominis
Rectus abdominis

Sternohyoid
Scalenus medius
Inferior belly of omohyoid
Pectoralis major
Mammary gland lobule
Nipple
Areola
Lactiferous sinus
Fat body
Tendinous inscription
Anterior layer of rectus sheath
Linea alba
Umbilicus

Creek

UNIT IV

Figure 9.23 Muscles of the anterior abdominal wall shown in a transverse view.

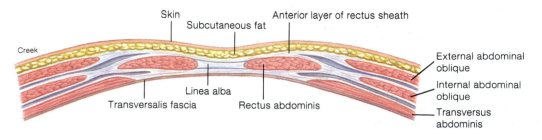

Skin
Subcutaneous fat
Anterior layer of rectus sheath
External abdominal oblique
Internal abdominal oblique
Transversus abdominis
Creek
Transversalis fascia
Linea alba
Rectus abdominis

Table 9.6	Muscles of the Abdominal Wall				
Muscle	Derivation	Origin	Insertion	Action	Innervation
External abdominal oblique	*external* = closer to surface; *oblique* = diagonal	Lower eight ribs	Iliac crest, linea alba	Compresses abdomen; lateral rotation; draws thorax downward	Intercostal, iliohypogastric, and ilioinguinal nn.
Internal abdominal oblique	*internal* = deep	Iliac crest, inguinal ligament, and lumbodorsal fascia	Linea alba, costal cartilage of last three or four ribs	Compresses abdomen; lateral rotation; draws thorax downward	Intercostal, iliohypogastric, and ilioinguinal nn.
Transversus abdominis	*transverse* = horizontal; *abdomino* = belly	Iliac crest, inguinal ligament, lumbar fascia, and costal cartilage of last six ribs	Xiphoid process, linea alba, and pubis	Compresses abdomen	Intercostal, iliohypogastric, and ilioinguinal nn.
Rectus abdominis	*rectus* = straplike; straight	Pubic crest and symphysis pubis	Costal cartilage of fifth to seventh ribs and xiphoid process of sternum	Flexes vertebral column	Intercostal nn.

Table 9.7	Muscles of the Pelvic Outlet				
Muscle	Derivation	Origin	Insertion	Action	Innervation
Levator ani	*anus* = ring	Ischial spine and pubic bone	Coccyx	Supports pelvic viscera; aids in defecation	Pudendal plexus
Coccygeus	*coccyx* = like a cuckoo's bill	Ischial spine	Sacrum and coccyx	Supports pelvic viscera; aids in defecation	Perineal branch of pudendal n.
Transversus perinei	*perineum* = pertaining to area between genitals and anus	Ischial tuberosity	Central tendon	Supports pelvic viscera	Pudendal plexus
Bulbospongiosus	*bulbus* = little bulb; *caverna* = hollow spaces	Central tendon	Males: base of penis; females: root of clitoris	Constricts urethral canal; constricts vagina	Perineal branch of pudendal n.
Ischiocavernosus	*ischio* = ischium	Ischial tuberosity	Males: pubic arch and crus of the penis Females: pubic arch and crus of the clitoris	Aids erection of penis or clitoris	Perineal branch of pudendal n.

Muscles of the Pelvic Outlet

The **pelvic outlet,** or **floor of the pelvis,** is the inferiorly positioned muscular wall that supports the pelvic viscera. It consists of the levator ani and coccygeus muscles, collectively called the *pelvic diaphragm.* The openings for the urethra and the rectum, as well as the vaginal opening in the female, pass through the pelvic diaphragm.

The **levator ani muscle** forms a thin sheet of muscle that helps to support the pelvic viscera and constrict the lower part of the rectum, pulling it forward and aiding defecation. The fan-shaped **coccygeus muscle** aids the levator ani muscle in its functions. Either or both of these muscles are occasionally stretched and torn during parturition.

The *perineal muscles* are positioned inferiorly to the pelvic diaphragm and are arranged in a superficial layer, consisting of the bulbocavernosus, ischiocavernosus, and the transversus perinei superficialis muscles, and a deep layer, consisting of the deep transversus perinei muscle and the external anal sphincter (fig. 9.24). Muscles of the deep layer, with a fascia, constitute the *urogenital diaphragm.* The **external anal sphincter muscle** of the deep layer is a funnel-shaped constrictor muscle that surrounds the anal canal. The muscles of the pelvic diaphragm and the urogenital diaphragm are similar in the male and female, but the perineal muscles of each sex markedly differ.

In males, the **bulbospongiosus muscle** of one side unites with that of the opposite side to form a muscular constriction

UNIT IV

Figure 9.24   Muscles of the pelvic outlet: (*a*) male and (*b*) female. (*c*) A superior view of the internal muscles of the female pelvic outlet.

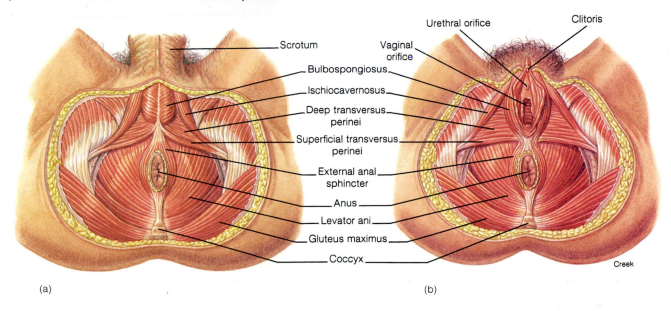

(a)

(b)

surrounding the base of the penis. When contracted, the two muscles constrict the urethral canal and assist in emptying the urethra. In females, these muscles are separated by the vaginal orifice, which they constrict as they contract. The **ischiocavernosus muscle** inserts onto the pubic arch and crus of the penis in the male and the pubic arch and crus of the clitoris of the female. This muscle aids the erection of the penis in the male and the clitoris in the female.

The muscles of the pelvic outlet are illustrated in figure 9.24 and summarized in table 9.7.

Muscles of the Vertebral Column

The muscles that move the vertebral column are strong and complex because they have to provide support and movement in resistance to the effect of gravity.

The vertebral column can be flexed, extended, abducted, adducted, and rotated. The muscle that flexes the vertebral column, the rectus abdominis, has already been described as a paired, straplike muscle of the anterior abdominal wall. The extensor muscles located on the posterior side of the vertebral column have to be stronger than the flexors because extension (such as lifting an object) is in opposition to gravity. The extensor muscles consist of a superficial group and a deep group. Only some of the muscles of the vertebral column will be described here.

The **erector spinae muscles** constitute a massive superficial muscle group that extends from the sacrum to the skull. It actually consists of three groups of muscles: **iliocostalis, longissimus,** and **spinalis** (fig. 9.25). Each of these groups, in turn, consists of overlapping slips of muscle. The iliocostalis is

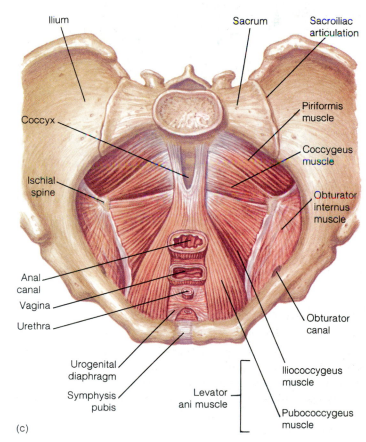

(c)

UNIT IV

Figure 9.25 Muscles of the vertebral column. The superficial neck muscles and erector spinae group of muscles are illustrated on the right and the deep neck and back muscles are illustrated on the left.

Rectus capitis posterior minor

Rectus capitis posterior major

Obliquus capitis superior

Obliquus capitis inferior

Semispinalis cervicis

Semispinalis thoracis

Levator costarum

Quadratus lumborum

Multifidus

Semispinalis capitis (medial and lateral part)

Longissimus capitis

Longissimus cervicis

First thoracic vertebra

Iliocostalis cervicis

Semispinalis cervicis

Spinalis thoracis

Longissimus thoracis

Iliocostalis thoracis

Iliocostalis lumborum

Erector spinae

Creek

UNIT IV

the most lateral group, the longissimus is intermediate in position, and the spinalis, in medial position, is positioned in contact with the spinous processes of the vertebrae.

The erector spinae muscles are frequently strained through improper lifting. A heavy object should not be lifted with the vertebral column flexed; instead, the thighs and knees should be flexed so that the pelvic and leg muscles can aid in the task.

The erector spinae muscles are also frequently strained during pregnancy. Pregnant women will try to counterbalance the effect of the protruding abdomen by hyperextending the vertebral column. This causes an exaggerated lumbar curvature, strained muscles, and a peculiar gait.

The **quadratus lumborum** is a deep muscle that originates on the iliac crest and the lower three lumbar vertebrae. It inserts on the transverse processes of the first four lumbar vertebrae and the inferior margin of the twelfth rib. When the right and left quadratus lumborum contract together, the vertebral column in the lumbar region extends. Separate contraction causes lateral flexion of the spine.

A summary of the major muscles of the back is presented in table 9.8.

Table 9.8	Muscles of the Vertebral Column				
Muscle	**Derivation**	**Origin**	**Insertion**	**Action**	**Innervation**
Quadratus lumborum	*quad* = four; *lumb* = referring to lumbar region	Iliac crest and lower three lumbar vertebrae	Twelfth rib and upper four lumbar vertebrae	Extends lumbar region; lateral flexion of vertebral column	Intercostal nerve T12 and lumbar nerves L1–L4
Erector spinae	Consists of three groups of muscles: iliocostalis, longissimus, and spinalis. The iliocostalis and longissimus are further subdivided into three groups on the basis of location along the vertebral column.				
Iliocostalis lumborum	*ilium* = flank	Crest of ilium	Lower six ribs	Extends lumbar region	Dorsal rami of lumbar nerves
Iliocostalis thoracis	*thorax* = chest	Lower six ribs	Upper six ribs	Extends thoracic region	Dorsal rami of thoracic nerves
Iliocostalis cervicis	*cervix* = neck	Angles of third to sixth rib	Transverse processes of fourth to sixth cervical vertebrae	Extends cervical region	Dorsal rami of cervical nerves
Longissimus thoracis		Transverse processes of lumbar vertebrae	Transverse processes of all the thoracic vertebrae and lower nine ribs	Extends thoracic region	Dorsal rami of spinal nerves
Longissimus cervicis		Transverse processes of upper four or five thoracic vertebrae	Transverse processes of second to sixth cervical vertebrae	Extends cervical region and lateral flexion	Dorsal rami of spinal nerves
Longissimus capitis		Transverse processes of upper five thoracic vertebrae and articular processes of lower three cervical vertebrae	Posterior margin of cranium and mastoid processes	Extends joints of cervical vertebrae	Dorsal rami of middle and lower cervical nerves
Spinalis thoracis		Spinous processes of upper lumbar and lower thoracic vertebrae	Spinous processes of upper thoracic vertebrae	Extends vertebral column	Dorsal rami of spinal nerves

UNIT IV

1. Identify the facial muscles that do the following: (a) wrinkle the forehead; (b) purse the lips; (c) protrude the lower lip; (d) produce frown lines; (e) smile; (f) frown; (g) wink; and (h) elevate the upper lip to show the teeth.
2. Describe the actions of the extrinsic muscles that move the tongue.
3. Which muscles of the neck either originate from or insert on the hyoid bone?
4. Describe the actions of the muscles of inspiration. Which muscles participate in forced expiration?
5. Which muscles of the pelvic outlet support the floor of the pelvic cavity and which are associated with the genitalia?
6. What is the extent and the compartmentalization of the erector spinae muscle?

Muscles of the Appendicular Skeleton

The muscles of the appendicular skeleton include those of the pectoral girdle, arm, forearm, wrist, hand, and fingers and those of the pelvic girdle, thigh, leg, ankle, foot, and toes.

Objective 11. Locate the major muscles of the appendicular skeleton. Identify synergistic and antagonistic muscles and describe the action of each one.

Muscles That Act on the Pectoral Girdle

The shoulder is attached to the axial skeleton only at the sternoclavicular joint; therefore, strong, straplike muscles are necessary. Furthermore, muscles that move the brachium originate on the scapula, and during brachial movement the scapula has to be held fixed. The muscles that act on the pectoral girdle originate on the axial skeleton and can be divided into anterior and posterior groups.

The anterior group of muscles that act on the pectoral girdle includes the **serratus anterior, pectoralis minor,** and **subclavius muscles** (fig. 9.26). The posterior group includes the **trapezius, levator scapulae,** and **rhomboideus muscles** (fig. 9.27). The position of these muscles is such that one muscle does not cause an action on its own, but rather several muscles contract synergistically to result in any movement of the girdle.

Treatment of advanced stages of *breast cancer* requires the surgical removal of both pectoralis major and pectoralis minor muscles in a procedure called a *radical mastectomy.* Postoperative physical therapy is primarily geared toward strengthening the synergistic muscles of this area. As the muscles that act on the brachium are learned, determine which are synergists with the pectoralis major.

The muscles that act on the pectoral girdle are summarized in table 9.9.

Figure 9.26 Muscles of the anterior trunk and shoulder regions. The superficial muscles are illustrated on the right and the deep muscles are illustrated on the left.

UNIT IV

Table 9.9	**Muscles That Act on the Pectoral Girdle**				
Muscle	**Derivation**	**Origin**	**Insertion**	**Action**	**Innervation**
Serratus anterior	*serratus* = serrated; *anterior* = front	Upper eight or nine ribs	Medial border of scapula	Pulls scapula forward and downward	Long thoracic n.
Pectoralis minor	*pectus* = chest; *minor* = lesser	Sternal ends of third, fourth, and fifth ribs	Coracoid process of scapula	Pulls scapula forward and downward	Medial pectoral n.
Subclavius	*sub* = below, under	First rib	Inferior surface of acromion of scapula	Draws clavicle downward	Spinal nerves C5, C6
Trapezius	*trapezoeides* = trapezoid shaped	Occipital bone and spines of seventh cervical and all thoracic vertebrae	Clavicle, spine, and acromion of scapula	Elevates scapula, draws head back, adducts scapula, braces shoulder	Accessory nerve and spinal nerves C3, C4
Levator scapulae	*levator* = elevator; *scapulae* = scapula	First to fourth cervical vertebrae	Superior angle of scapula	Elevates scapula	Dorsal scapular n.
Rhomboideus major	*rhomboides* = rhomboid shaped	Spines of second to fifth thoracic vertebrae	Medial border of scapula	Elevates and adducts scapula	Dorsal scapular n.
Rhomboideus minor		Seventh cervical and first thoracic vertebrae	Medial border of scapula	Elevates and adducts scapula	Dorsal scapular n.

Figure 9.27 Muscles of the posterior neck, shoulder, trunk, and gluteal regions. The superficial muscles are illustrated on the left and the deep muscles are illustrated on the right.

Splenius capitis

Sternocleidomastoid

Vertebra prominens (C7)

Trapezius

Spine of scapula

Deltoid

Infraspinatus (covered by fascia)

Teres minor (covered by fascia)

Teres major

Latissimis dorsi

External abdominal oblique

Gluteal fascia

Gluteus maximus

Semispinalis capitis

Splenius capitis

Levator scapulae

Splenius cervicis

Rhomboideus minor

Supraspinatus

Infraspinatus

Teres minor

Rhomboideus major

Teres major

Spinalis thoracis

Longissimus thoracis

Iliocostalis thoracis

Serratus posterior inferior

External abdominal oblique

Gluteus medius

Piriformis

Creek

UNIT IV

Muscles That Move the Humerus

Of the nine muscles that span the shoulder joint to insert on the humerus, only two of them, the pectoralis major and latissimus dorsi, do not originate on the scapula. These two are designated as axial muscles, whereas the remaining seven are scapular muscles. The muscles of this region are shown in figures 9.26 and 9.27 and the attachments of all the muscles that either originate or insert on the scapula are shown in figure 9.28.

In terms of their development, the pectoralis major and the latissimus dorsi muscles are not axial muscles at all. They develop in the forelimb and extend to the trunk secondarily. They are considered axial muscles only because their origins are on the axial skeleton.

Axial Muscles   The axial muscles include the pectoralis major and latissimus dorsi muscles.

UNIT IV

Figure 9.28   A posterior view of the scapula and humerus showing the areas of attachment of the associated muscles. (Points of origin are color coded red, and points of insertion are color coded blue.)

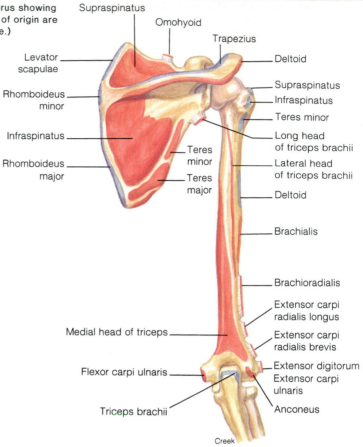

Table 9.10		Muscles That Move the Humerus			
Muscle	**Derivation**	**Origin**	**Insertion**	**Action**	**Innervation**
Pectoralis major	*pectus* = chest; *major* = greater	Clavicle, sternum, costal cartilages of second to sixth rib; rectus sheath	Crest of greater tubercle of humerus	Flexes, adducts, and rotates arm medially at shoulder joint	Medial and lateral pectoral nn.
Latissimus dorsi	*latissimus* = widest *dorsum* = back	Spines of sacral, lumbar, and lower thoracic vertebrae; iliac crest and lower four ribs	Intertubercular groove of humerus	Extends, adducts, and rotates humerus medially at shoulder joint	Thoracodorsal n.
Deltoid	*delta* = triangular	Clavicle, acromion process; spine of scapula	Deltoid tuberosity of humerus	Abducts arm and extends or flexes humerus at shoulder joint	Axillary n.
Supraspinatus	*supra* = above; *spina* = spine of scapula	Supraspinous fossa of scapula	Greater tubercle of humerus	Abducts and laterally rotates humerus at shoulder joint	Suprascapular n.
Infraspinatus	*infra* = below	Infraspinous fossa of scapula	Greater tubercle of humerus	Rotates arm laterally at shoulder joint	Suprascapular n.
Teres major		Inferior angle and lateral border of scapula	Crest of lesser tubercle of humerus	Extends humerus, or adducts and rotates arm medially at shoulder joint	Lower subscapular n.
Teres minor	*teres* = rounded; *minor* = lesser	Lateral border of scapula	Greater tubercle of humerus	Rotates arm laterally at shoulder joint	Axillary n.
Subscapularis	*sub* = below, under; *scapula* = scapula	Subscapular fossa	Lesser tubercle of humerus	Rotates arm medially at shoulder joint	Subscapular n.
Coracobrachialis	*coraco* = pertaining to coracoid process; *brachium* = arm	Coracoid process of scapula	Shaft of humerus	Flexes and adducts arm at shoulder joint	Musculocutaneous n.

Figure 9.29 Muscles of the right anterior shoulder and brachium.

Deltoid

Coracobrachialis

Pectoralis major

Humerus

Short head of biceps brachii

Long head of biceps brachii

Brachioradialis

Pectoralis minor

Subscapularis

Teres major

Latissimus dorsi

Long head of triceps brachii

Medial head of triceps brachii

Brachial artery

Median nerve

Brachialis

Bicipital aponeurosis

Creek

Coracoid process of scapula

Long head of biceps brachii

Short head of biceps brachii

Radial tuberosity

Deltoid

Coracobrachialis

Brachialis

Coronoid process of ulna

UNIT IV

The **pectoralis major muscle** is a large chest muscle (fig. 9.26) that binds the humerus to the anterior rib cage. It is the principal flexor muscle of the arm. The large, flat, triangular **latissimus dorsi muscle** covers the lower half of the thoracic region of the back (see fig. 9.27) and is the antagonist to the pectoralis major muscle. The latissimus dorsi is frequently called the "swimmer's muscle" because it powerfully extends the shoulder joint, drawing the arm downward and backward while it rotates medially. Extension of the arm is in reference to anatomical position and is therefore a backward, retracting (increasing the shoulder joint angle) movement of the arm.

A latissimus dorsi muscle, conditioned with pulsated electrical impulses, will in time come to resemble cardiac muscle tissue in that it will be nonfatigable and use oxygen at a steady rate. Following conditioning, the muscle may be used in an autotransplant to repair a surgically removed portion of a patient's diseased heart. The procedure involves detaching the latissimus dorsi muscle from its vertebral origin, leaving the blood supply and innervation intact, and slipping it into the pericardial cavity where it is wrapped around the heart like a towel. A pacemaker is required to provide the continuous, rhythmic contractions.

Scapular Muscles   The nonaxial scapular muscles include the deltoid, supraspinatus, infraspinatus, teres major, teres minor, subscapularis, and coracobrachialis muscles.

The **deltoid muscle** is a thick, powerful muscle that caps the shoulder joint (figs. 9.29 and 9.30). Although it has several functions (see table 9.10), the principal action of the deltoid muscle is abduction of the arm at the shoulder joint. Functioning together, both the pectoralis major and the latissimus dorsi muscles are antagonists to the deltoid muscle in that they cause adduction of the arm at the shoulder joint. The deltoid muscle is a common site for intramuscular injections.

The remaining six scapular muscles also help stabilize the shoulder and have specific actions at the shoulder joint (see table 9.10). The **supraspinatus muscle** laterally rotates the humerus and is synergistic with the deltoid muscle in abducting the arm at the shoulder joint. The **infraspinatus muscle** rotates the humerus laterally. The action of the **teres major muscle** is similar to that of the latissimus dorsi muscle, adducting and medially rotating the humerus. The **teres minor muscle** works with the infraspinatus muscle in laterally rotating the humerus. The **subscapularis muscle** is a strong stabilizer of the shoulder and also aids in medially rotating the

Figure 9.30 Muscles of the right posterior shoulder and brachium.

humerus. The **coracobrachialis muscle** is a synergist to the pectoralis major muscle in flexing and adducting the arm at the shoulder joint.

Four of the nine muscles that cross the shoulder joint, the supraspinatus, infraspinatus, teres minor, and subscapularis, are commonly called the *musculotendinous cuff,* or *rotator cuff*. Their distal tendons blend with and reinforce the fibrous capsule of the shoulder joint en route to their points of insertion on the humerus. This structural arrangement plays a major role in stabilizing the shoulder joint. Musculotendinous cuff injuries are common in baseball players. When throwing a baseball, there is complete abduction of the shoulder followed by a rapid and forceful rotation and flexion of the shoulder, which may strain the musculotendinous cuff.

Muscles That Act on the Forearm

The powerful muscles of the brachium are responsible for flexion and extension of the elbow joint. These muscles are the biceps brachii, brachialis, brachioradialis, and triceps brachii (figs. 9.29 and 9.30). In addition, a short triangular muscle, the anconeus, is positioned over the distal end of the triceps brachii muscle, near the elbow. A transverse section through the brachium in figure 9.31 provides a different perspective of the brachial region.

The powerful **biceps brachii muscle,** positioned on the anterior surface of the humerus, is the most familiar muscle of the arm, yet it has no attachments on the humerus. The biceps brachii muscle has a dual origin: a medial tendinous head, the **short head,** arises from the coracoid process of the scapula, and the **long head** originates on the superior tuberosity of the glenoid fossa and passes through the shoulder joint and descends in the intertubercular groove on the humerus. Both heads of the biceps brachii muscle insert on the radial tuberosity. The

Figure 9.31 The axillary region and a transverse section through the brachium.

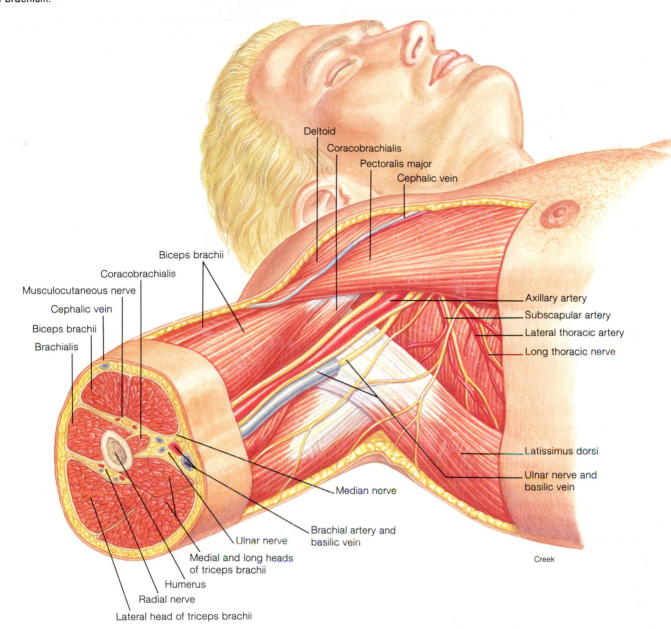

UNIT IV

Creek

brachialis muscle is located on the distal anterior half of the humerus, deep to the biceps brachii muscle. It is synergistic to the biceps brachii muscle in flexing the forearm at the elbow joint.

The **brachioradialis** is the prominent muscle positioned along the lateral (radial) surface of the forearm. It too flexes the forearm at the elbow joint.

The **triceps brachii muscle,** located on the posterior surface of the brachium, is the muscle antagonistic to the biceps

brachii muscle. It has three heads, or origins. Two of the three, the **lateral head** and **medial head,** arise from the humerus, whereas the **long head** arises from the infraglenoid tuberosity of the scapula. A common tendinous insertion attaches the triceps brachii muscle to the olecranon of the ulna. The triceps brachii muscle and the small **anconeus muscle** extend the forearm at the elbow joint.

Refer to table 9.11 for a summary of the muscles that act on the forearm.

Table 9.11	Muscles That Act on the Forearm				
Muscle	**Derivation**	**Origin**	**Insertion**	**Action**	**Innervation**
Biceps brachii	*biceps* = two heads	Coracoid process; tuberosity above glenoid fossa of scapula	Radial tuberosity	Flexes elbow joint; supinates forearm and hand at elbow joint	Musculocutaneous n.
Brachialis	*brachium* = upper arm	Anterior shaft of humerus	Coronoid process of ulna	Flexes elbow joint	Musculocutaneous, median, and radial nn.
Brachioradialis	*radialis* = pertaining to radius	Lateral supracondylar ridge of humerus	Proximal to styloid process of radius	Flexes elbow joint	Radial n.
Triceps brachii	*triceps* = three heads	Tuberosity below glenoid fossa; lateral and medial surfaces of humerus	Olecranon of ulna	Extends elbow joint	Radial n.
Anconeus	*ancon* = elbow	Lateral epicondyle of humerus	Olecranon of ulna	Extends elbow joint	Radial n.

Muscles of the Forearm That Move the Wrist, Hand, and Fingers

The muscles that cause wrist, hand, and most finger movements are positioned along the forearm (figs. 9.32 and 9.33). Several of these muscles act on two joints—the elbow and wrist. Others act on the joints of the wrist, hand, and digits. Still others produce rotational movement at the radioulnar joint. The precise actions of these muscles are complex, so only the basic movements will be described here. Most of these muscles perform four primary actions on the hand and digits: supination, pronation, flexion, and extension. Other actions of the hand include adduction and abduction.

Supination and Pronation of the Hand   The **supinator muscle** is positioned around the upper posterior portion of the radius (fig. 9.33), where it works synergistically with the biceps brachii muscle to supinate the hand. Two muscles are responsible for pronating the hand—the pronator teres and pronator quadratus. The **pronator teres muscle** is located on the upper medial side of the forearm, whereas the deep, anteriorly positioned **pronator quadratus muscle** extends between the ulna and radius on the distal fourth of the forearm. These two muscles work synergistically to rotate the palm of the hand posteriorly and position the thumb medially.

Flexion of the Wrist, Hand, and Fingers   Six of the muscles that flex the wrist, hand, and fingers will be described from lateral to medial and from superficial to deep (see figs. 9.32 and 9.33). Although four of the six arise from the medial epicondyle of the humerus (see table 9.12), their actions on the elbow joint are minimal. The brachioradialis, already described, is an obvious reference muscle for locating the muscles of the forearm that flex the joints of the hand.

The **flexor carpi radialis muscle** extends diagonally across the anterior surface of the forearm, and its distal cordlike tendon crosses the wrist under the *flexor retinaculum.*

The narrow **palmaris longus muscle** is superficial in position on the anterior surface of the forearm. It has a long, slender tendon that attaches to the *palmar aponeurosis,* where it assists in flexing the wrist.

The palmaris longus is the most variable muscle in the body. It is totally absent in approximately 8% of all people, and in 4% it is absent in one or the other forearm. Furthermore, it is absent more often in females than males, and on the left side in both sexes. Because of the superficial position of the palmaris longus muscle, its presence or absence can be readily detected on oneself by tightly clenching the fist and examining for its tendon just proximal to the wrist (see fig. 10.37).

The **flexor carpi ulnaris muscle** is positioned on the medial anterior side of the forearm, where it assists in flexing the wrist and adducting the hand.

The broad **superficial digital flexor muscle** lies directly beneath the three flexor muscles just described (see figs. 9.32 and 9.33). It has an extensive origin, involving all three of the long bones of the forearm (table 9.12). The tendon at the distal end of this muscle is united across the wrist joint but then splits to attach to the middle phalanx of digits II through V.

The **deep digital flexor muscle** lies deep to the superficial digital flexor muscle. These two muscles flex the wrist, hand, and the second, third, fourth, and fifth digits.

The **flexor pollicis longus muscle** is a deep, lateral muscle of the forearm. It flexes the joints of the thumb, assisting the grasping mechanism of the hand.

Figure 9.32 Superficial muscles of the right forearm. (*a*) An anterior view and (*b*) a posterior view.

Brachial artery

Median nerve

Lateral antebrachial cutaneous nerve

Pronator teres

Radial artery

Flexor carpi ulnaris

Brachioradialis

Flexor carpi radialis

Extensor carpi radialis longus

Palmaris longus

Extensor carpi radialis brevis

Superficial digital flexor

Radial artery

Ulnar nerve

Abductor pollicis longus

Flexor pollicis longus

Extensor pollicis brevis

Ulnar artery

Median nerve

Flexor retinaculum

Abductor digiti minimi

Abductor pollicis brevis

Flexor digiti minimi brevis

Flexor pollicis brevis

Adductor pollicis

(a)

Brachioradialis

Ulnar nerve

Extensor carpi radialis longus

Anconeus

Extensor carpi ulnaris

Flexor carpi ulnaris

Extensor carpi radialis brevis

Extensor digitorum communis

Extensor digiti minimi

Abductor pollicis longus

Extensor indicis

Extensor pollicis longus

Extensor pollicis brevis

Dorsal branch of ulnar nerve

Superficial branch of radial nerve

Extensor retinaculum

Radial artery

Dorsal interosseous

Creek

(b)

The tendons of the muscles that flex the hand can be seen on the wrist as a fist is made. These tendons are securely positioned by the flexor retinaculum (fig. 9.32a), which crosses the wrist area transversely.

Extension of the Hand The muscles that extend the hand are located on the posterior side of the forearm. Most of the primary extensor muscles of the hand can be seen superficially in figure 9.32b and will be discussed from lateral to medial.

The long, tapered **extensor carpi radialis longus muscle** is located medial to the brachioradialis muscle, where it extends the carpal joint and abducts the hand at the wrist. The **extensor carpi radialis brevis muscle** is positioned immedi-

ately medial to the extensor carpi radialis longus muscle and performs approximately the same functions. Its origin and insertion, however, are different (see table 9.12).

The bipennate-fibered **extensor digitorum communis muscle** is positioned in the center of the forearm, along the posterior surface. It originates on the lateral epicondyle of the humerus. Its tendon of insertion divides at the wrist, beneath the extensor retinaculum, into four tendons that attach to the distal tip of the medial phalanges of digits II through V.

The **extensor digiti minimi muscle** is a long, narrow muscle located on the ulnar side of the extensor digitorum communis muscle. Its tendinous insertion fuses with the tendon of the extensor digitorum communis muscle going to the fifth digit.

UNIT IV

Figure 9.33 Deep muscles of the right forearm. (*a*) Rotators, (*b*) flexors, and (*c*) extensors.

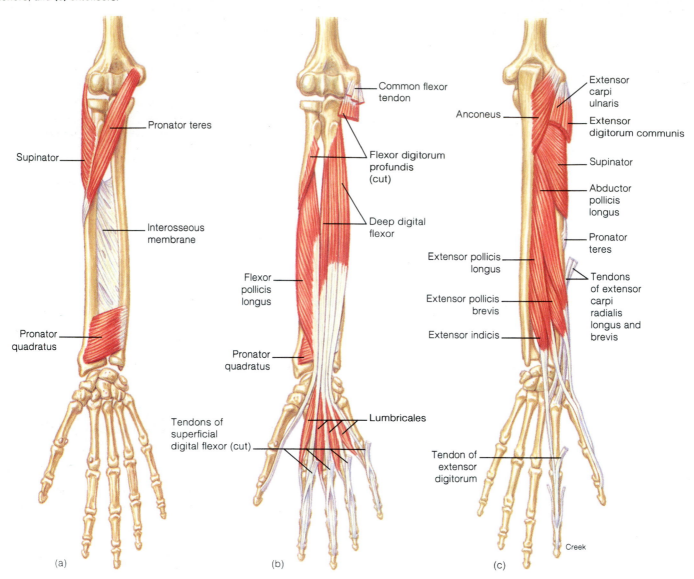

(a)

Supinator

Pronator teres

Interosseous membrane

Pronator quadratus

(b)

Common flexor tendon

Flexor digitorum profundis (cut)

Deep digital flexor

Flexor pollicis longus

Pronator quadratus

Tendons of superficial digital flexor (cut)

Lumbricales

(c)

Anconeus

Extensor carpi ulnaris

Extensor digitorum communis

Supinator

Abductor pollicis longus

Pronator teres

Tendons of extensor carpi radialis longus and brevis

Extensor pollicis longus

Extensor pollicis brevis

Extensor indicis

Tendon of extensor digitorum

Creek

The **extensor carpi ulnaris muscle** is the most medial muscle on the posterior surface of the forearm. It inserts on the base of the fifth metacarpal bone, where it functions to extend and adduct the hand.

The **extensor pollicis longus muscle** extends from the mid-ulnar region, across the lower two-thirds of the forearm, and inserts onto the base of the distal phalanx of the thumb (see fig. 9.33), where it extends the joints of the thumb and abducts the hand.

The **extensor pollicis brevis muscle** arises from the lower midportion of the radius and inserts on the base of the proximal phalanx of the thumb (fig. 9.33). The action of this muscle is similar to that of the extensor pollicis longus muscle.

As its name implies, the **abductor pollicis longus muscle** abducts the joints of the thumb and hand. It originates on the interosseous ligament, between the ulna and radius, and inserts on the base of the first metacarpal bone.

The muscles that act on the wrist, hand, and digits are summarized in table 9.12.

Notice that the joints of your hand are partially flexed even when the hand is relaxed. The muscles that extend the hand are not as strong as the muscles that flex it. This is why persons who receive strong electrical shocks through the arms will tightly flex their hands and hold on. All the muscles of the arm are stimulated to contract, but the flexors, being stronger, cause the hands to close tightly.

Muscles of the Hand

The hand is a marvelously complex structure, adapted to permit an array of intricate movements. Flexion and extension movements of the hand and phalanges are accomplished by the muscles of the forearm just described. Precise finger movements that require coordinating abduction and adduction with flexion and extension are the function of the small intrinsic muscles of the hand. These muscles and associated structures of the hand are depicted in figure 9.34. The position and actions of the muscles of the hand are listed in table 9.13.

Table 9.12	Muscles of the Forearm That Act on the Wrist, Hand, and Digits				
Muscle	**Derivation**	**Origin**	**Insertion**	**Action**	**Innervation**
Supinator	*supin* = bend back; palms up	Lateral epicondyle of humerus and crest of ulna	Lateral surface of radius	Supinates hand	Radial n.
Pronator teres	*pron* = bend forward; palms down; *teres* = long, round	Medial epicondyle of humerus	Lateral surface of radius	Pronates hand	Median n.
Pronator quadratus	*quad* = four (square)	Distal fourth of ulna	Distal fourth of radius	Pronates hand	Median n.
Flexor carpi radialis	*flexor* = decrease angle; *carpus* = wrist; *radial* = radius bone	Medial epicondyle of humerus	Base of second and third metacarpal bones	Flexes and abducts hand at wrist	Median n.
Palmaris longus	*palma* = flat of hand; *longus* = long	Medial epicondyle of humerus	Palmar aponeurosis	Flexes wrist	Median n.
Flexor carpi ulnaris		Medial epicondyle and olecranon	Carpal and metacarpal bones	Flexes and adducts wrist	Ulnar n.
Superficial digital flexor	*digitis* = finger or toe; *superficialis* = superficial	Medial epicondyle, coronoid process, and anterior border of radius	Middle phalanges of digits	Flexes wrist and digits at carpophalangeal and interphalangeal joints	Median n.
Flexor digitorum profundus	*profundus* = deep	Proximal two-thirds of ulna and interosseous membrane	Distal phalanges	Flexes wrist and digits at carpophalangeal and interphalangeal joints	Median and ulnar nn.
Flexor pollicis longus	*pollex* = thumb	Shaft of radius, interosseous membrane, and coronoid process of ulna	Distal phalanx of thumb	Flexes joints of thumb	Median n.
Extensor carpi radialis longus	*extensor* = increase joint angle	Lateral supracondylar ridge of humerus	Second metacarpal bone	Extends and abducts wrist	Radial n.
Extensor carpi radialis brevis	*brevis* = short	Lateral epicondyle of humerus	Third metacarpal bone	Extends and abducts wrist	Radial n.
Extensor digitorum		Lateral epicondyle of humerus	Posterior surfaces of digits II–V	Extends wrist and phalanges at joints of carpophalangeal and interphalangeal joints	Radial n.
Extensor digiti minimi	*minimus* = very small	Lateral epicondyle of humerus	Extensor aponeurosis of fifth digit	Extends joints of fifth digit and wrist	Radial n.
Extensor carpi ulnaris		Lateral epicondyle of humerus and olecranon	Base of fifth metacarpal bone	Extends and adducts wrist	Radial n.
Extensor pollicis longus		Middle shaft of ulna, lateral side	Base of distal phalanx of thumb	Extends joints of thumb; abducts joints of hand	Radial n.
Extensor pollicis brevis		Distal shaft of radius and interosseous membrane	Base of first phalanx of thumb	Extends joints of thumb; abducts joints of hand	Radial n.
Abductor pollicis longus		Distal radius and ulna and interosseous membrane	Base of first metacarpal bone	Abducts joints of thumb and joints of hand	Radial n.

UNIT IV

Figure 9.34 Muscles of the hand. (*a*) An anterior view and (*b*) a lateral view of the second digit (index finger).

UNIT IV

Tendons of flexor digitorum profundus

Tendons of superficial digital flexor

Palmar plates

Dorsal interosseous

Tendon of flexor pollicis longus

Tendon of deep digital flexor

Transverse head ⎤
Oblique head ⎦ Adductor pollicis

Lumbricals

Abductor digiti minimi

Opponens digiti minimi

Flexor digiti minimi brevis

Tendons of flexor digitorum profundus

Tendons of flexor digitorum superficialis

Flexor pollicis brevis

Abductor pollicis brevis

Opponens pollicis

Tendon of extensor pollicis brevis

Position of pisiform bone

Tendon of flexor carpi ulnaris

Tendon of abductor pollicis longus

Flexor retinaculum

Tendons of superficial digital flexor

Radial artery

Ulnar nerve and artery

Pronator quadratus

Tendon of palmaris longus

Median nerve

Tendon of flexor carpi radialis

Tendon of flexor pollicis longus

(a)

Extensor expansion

Collateral ligaments

Collateral ligament

Tendon of extensor digitorum

First dorsal interosseous

Vincula

First lumbrical

Fibrous digital sheath

Tendon of superficial digital flexor

Tendon of deep digital flexor

(b)

Table 9.13	Intrinsic Muscles of the Hand				
Muscle	**Derivation**	**Origin**	**Insertion**	**Action**	**Innervation**
Thenar Muscles					
Abductor pollicis brevis	*pollex* = thumb; *brevis* = short	Flexor retinaculum, scaphoid, and trapezium	Proximal phalanx of thumb	Abducts joints of thumb	Median n.
Flexor pollicis brevis		Flexor retinaculum and trapezium	Proximal phalanx of thumb	Flexes joints of thumb	Median n.
Opponens pollicis	*opponens* = against	Trapezium and flexor retinaculum	First metacarpal bone	Opposes joints of thumb	Median n.
Intermediate Muscles					
Adductor pollicis (oblique and transverse heads)		Oblique head, capitate; transverse head, second and third metacarpal bones	Proximal phalanx of thumb	Adducts joints of thumb	Ulnar n.
Lumbricales (4)	*lumbricalis* = referring to digits	Tendons of flexor digitorum profundus	Extensor expansions of digits II–V	Flexes digits at metacarpophalangeal joints; extends digits at interphalangeal joints	Median and ulnar nn.
Palmar interossei (3)	*palma* = flat of hand	Medial side of second metacarpal; lateral sides of fourth and fifth metacarpal bones	Proximal phalanges of index, ring, and little fingers and extensor digitorum communis	Adducts fingers toward middle finger at metacarpophalangeal joints	Ulnar n.
Dorsal interossei (4)		Adjacent sides of metacarpal bones	Proximal phalanges of index and middle fingers (lateral sides) plus proximal phalanges of middle and ring fingers (medial sides) and extensor digitorum communis	Abducts fingers away from middle finger at metacarpophalangeal joints	Ulnar n.
Hypothenar Muscles					
Abductor digiti minimi	*minimus* = very small	Pisiform and tendon of flexor carpi ulnaris	Proximal phalanx of digit V	Abducts joints of digit V	Ulnar n.
Flexor digiti minimi		Flexor retinaculum and hook of hamate	Proximal phalanx of digit V	Flexes joints of digit V	Ulnar n.
Opponens digiti minimi		Flexor retinaculum and hook of hamate	Fifth metacarpal bone	Opposes joints of digit V	Ulnar n.

UNIT IV

The muscles of the hand are divided into **thenar** *(the'nar),* **hypothenar** *(hi-poth'ē-nar),* and **intermediate** groups. The *thenar eminence* is the fleshy base of the thumb and is formed by three muscles: the **abductor pollicis brevis,** the **flexor pollicis brevis,** and the **opponens pollicis.** The most important of the thenar muscles is the opponens pollicis, which opposes the thumb to the palm of the hand.

The *hypothenar eminence* is the elongated, fleshy bulge at the base of the little finger. It also is formed by three muscles: the **abductor digiti minimi muscle,** the **flexor digiti minimi muscle,** and the **opponens digiti minimi muscle.**

Muscles of the intermediate group are positioned between the metacarpal bones in the region of the palm. This group includes the **adductor pollicis muscle,** the **lumbricales** *(lum'bri-kalz),* and the **palmar** and **dorsal interossei muscle.**

Muscles That Move the Thigh

The muscles that move the thigh originate from the pelvic girdle and insert on various places on the femur. These muscles stabilize a highly movable hip joint and provide support for the body during bipedal stance and locomotion. Found in this region are the most massive muscles of the body as well as some extremely small muscles. The muscles that move the thigh are divided into anterior, posterior, and medial groups.

Anterior Muscles   The anterior muscles that move the thigh are the iliacus and psoas major (figs. 9.35 and 9.36).

The triangular **iliacus muscle** arises from the iliac fossa and inserts on the lesser trochanter of the femur.

Figure 9.35 Muscles of the right anterior pelvic and thigh regions.

Psoas minor

Psoas major

Iliac crest

Iliacus

Sartorius (cut)

Rectus femoris (cut)

Tensor fasciae latae

Vastus intermedius

Vastus lateralis

Vastus medialis

Tendon of rectus femoris (cut)

Patella

Iliopsoas

Pectineus

Adductor brevis

Adductor magnus

Adductor longus (cut)

Gracilis

Femoral artery and vein

Sartorius (cut)

Creek

Figure 9.36 Anterior pelvic muscles that move the hip.

Costal part of diaphragm

Psoas major

Psoas minor

Iliacus

Iliopsoas

Central tendon of diaphragm

Twelfth rib

Quadratus lumborum

Lesser trochanter of femur

Creek

The long, thick **psoas major muscle** originates on the bodies and transverse processes of the lumbar vertebrae and inserts, along with the iliacus muscle, on the lesser trochanter (fig. 9.36). The psoas major muscle and the iliacus work synergistically in flexing and rotating the thigh and flexing the vertebral column. They are frequently referred to as a single muscle, the **iliopsoas** (il''e-ŏ-so'as) **muscle.**

Posterior and Lateral (Buttock) Muscles The posterior muscles that move the thigh include the gluteus maximus, gluteus medius, gluteus minimus, and tensor fasciae latae muscles.

The large **gluteus maximus muscle** forms much of the prominence of the buttock (figs. 9.37 and 9.40). It is a powerful extensor muscle of the thigh and is very important for bipedal stance and locomotion. The gluteus maximus muscle originates on the ilium, sacrum, coccyx, and aponeurosis of the lumbar region. It inserts on the gluteal tuberosity of the femur and the *iliotibial tract,* a thickened tendinous region of the fascia lata extending down the thigh (see fig. 9.39).

The **gluteus medius muscle** is located immediately deep to the gluteus maximus muscle (fig. 9.37). It originates on the lateral surface of the ilium and inserts on the greater trochanter of the femur. The gluteus medius muscle abducts and medially rotates the thigh. The mass of this muscle is of clinical significance as a site for intramuscular injections.

The **gluteus minimus muscle** is the smallest and deepest of the gluteal muscles (figs. 9.37 and 9.40). It also arises from the lateral surface of the ilium and inserts on the lateral surface of the greater trochanter, where it acts synergistically with the gluteus medius muscle to abduct and medially rotate the thigh.

The quadrangular **tensor fasciae latae muscle** is positioned superficially on the lateral surface of the hip (see fig. 9.39). It originates on the iliac crest and inserts on a broad lateral fascia of the thigh called the *fascia lata.* The fascia lata is continuous with the iliotibial tract. The tensor fasciae latae muscle is the principal abductor of the thigh.

A deep group of six lateral rotator muscles of the thigh is positioned directly over the posterior aspect of the hip joint. These muscles are not discussed here but are identified in figure 9.37 from superior to inferior as the **piriformis, superior gemellus, obturator internus, inferior gemellus, obturator externus,** and **quadratus femoris muscles.**

The anterior and posterior group of muscles that move the thigh at the hip joint are summarized in table 9.14.

Figure 9.37 Deep gluteal muscles.

Creek

Gluteus medius (cut)

Gluteus maximus (cut)

Gluteus minimus

Piriformis

Sciatic nerve

Posterior femoral cutaneous nerve

Obturator internus

Gracilis

Adductor magnus

Semitendinosus

Semimembranosus

Superior gemellus

Trochanteric bursa

Inferior gemellus

Obturator externus

Quadratus femoris

Gluteus maximus (cut)

Lesser trochanter

Adductor minimus (part of adductor magnus)

Adductor magnus

Iliotibial tract

Biceps femoris (long head)

Vastus lateralis

UNIT IV

Table 9.14	**Anterior and Posterior Muscles That Move the Thigh**				
Muscle	**Derivation**	**Origin**	**Insertion**	**Action**	**Innervation**
Iliacus	*iliacus* = pertaining to ilium (hip)	Iliac fossa	Lesser trochanter of femur	Flexes and rotates thigh laterally at the hip joint; flexes joints of vertebral column	Femoral n.
Psoas major	*psoa* = muscle of loin	Transverse processes of all lumbar vertebrae	Lesser trochanter with iliacus muscle	Flexes and rotates thigh laterally at the hip joint; flexes joints of vertebral column	Spinal nerves L2, L3
Gluteus maximus	*gloutos* = rump; *maximus* = greatest	Iliac crest, sacrum, coccyx, and aponeurosis of the lumbar region	Gluteal tuberosity and iliotibial tract	Extends and rotates thigh laterally at the hip joint	Inferior gluteal n.
Gluteus medius	*medius* = middle	Lateral surface of ilium	Greater trochanter	Abducts and rotates thigh medially at the hip joint	Superior gluteal n.
Gluteus minimus	*minimus* = very small	Lateral surface of lower half of ilium	Greater trochanter	Abducts and rotates thigh medially at the hip joint	Superior gluteal n.
Tensor fasciae latae	*tensor* = tightener; *fascia* = band; *latus* = wide	Anterior border of ilium and iliac crest	Iliotibial tract	Abducts thigh at the hip joint	Superior gluteal n.

Figure 9.38 Adductor muscles of the right thigh.

Anterior superior iliac spine
Sartorius (cut)
Rectus femoris (cut)
Iliopsoas (cut)
Quadratus femoris
Pectineus (cut)
Vastus intermedius
Adductor longus (cut)
Vastus lateralis (cut)
Femur
Tendon of rectus femoris
Patella
Patellar ligament
Creek

Pectineus (cut)
Adductor longus (cut)
Obturator externus
Adductor brevis (cut)
Gracilis (cut)
Adductor magnus
Adductor brevis (cut)
Vastus medialis (cut)
Adductor hiatus
Gracilis (cut)
Sartorius (cut)

Pectineus
Pubis
Adductor longus
Linea aspera on posterior surface of femur

Trochanteric fossa of femur
Rami of pubis and ischium
Obturator externus
Adductor brevis
Linea aspera on posterior surface
Gracilis
Tibia

Ischium
Adductor magnus
Medial epicondyle

Medial, or Adductor, Muscles   The medial muscles that move the thigh include the gracilis *(gras'il-is)*, pectineus, adductor longus, adductor brevis, and adductor magnus muscles (figs. 9.38, 9.39, 9.40, and 9.41).

The long, thin **gracilis muscle** is the most superficial of the medial thigh muscles. It is a two-joint muscle and can adduct the thigh or flex the leg.

The **pectineus muscle** is the uppermost of the medial muscles that move the thigh. It is a flat, quadrangular muscle that flexes and adducts the thigh.

The **adductor longus muscle** is immediately lateral to the gracilis on the upper third of the thigh, and is the most anterior of the adductor muscles. The **adductor brevis muscle** is a triangular muscle located behind and largely concealed by the adductor longus and pectineus muscles. The **adductor magnus muscle** is a large, thick muscle, somewhat triangular in shape. It is located deep to the other two adductor muscles. The adductor longus, adductor brevis, and the adductor magnus muscle are synergistic in adducting, flexing, and laterally rotating the thigh.

The muscles that adduct the thigh at the hip joint are summarized in table 9.15.

Muscles of the Thigh That Move the Leg

The muscles that move the leg originate on the pelvic girdle or thigh and are surrounded and compartmentalized by tough fascial sheets that are a continuation of the fascia lata and iliotibial tract. They are divided according to function and position into two groups: anterior extensors and posterior flexors.

Anterior, or Extensor, Muscles   The anterior muscles that move the leg are the sartorius and quadriceps femoris muscles (fig. 9.38).

The long, straplike **sartorius muscle** obliquely crosses the anterior aspect of the thigh. It can act on both the hip and knee joints to flex and rotate the thigh laterally, and also to assist in flexing the leg and rotating it medially. The sartorius muscle is the longest muscle of the body. It is frequently called the tailor's muscle because it enables one to assume a cross-legged sitting position.

The **quadriceps femoris muscle** is actually a composite of four distinct muscles that have separate origins but a common

Figure 9.39   Muscles of the right anterior thigh.

Iliacus

Gluteus medius

Inguinal ligament

Iliopsoas

Tensor fasciae latae

Vastus lateralis

Rectus femoris

Iliotibial tract

Tendon of rectus femoris

Patella

Patellar ligament

Psoas major and minor

Pectineus

Adductor longus

Gracilis

Adductor magnus

Sartorius

Vastus medialis

Creek

Anterior inferior iliac spine

Greater trochanter

Intertrochanteric line

Rectus femoris

Vastus lateralis

Patella

Patellar ligament

Tibia

Anterior superior iliac spine

Vastus intermedius

Sartorius

Vastus medialis

Patella

Patellar ligament

Tibia

UNIT IV

Table 9.15	**Medial Muscles That Move the Thigh**				
Muscle	**Derivation**	**Origin**	**Insertion**	**Action**	**Innervation**
Gracilis	*gracilis* = slender	Inferior edge of symphysis pubis	Proximal medial surface of tibia	Adducts thigh at hip joint; flexes and rotates leg at knee joint	Obturator n.
Pectineus	*pecten* = comb	Pectineal line of pubis	Distal to lesser trochanter of femur	Adducts and flexes thigh at hip joint	Femoral n.
Adductor longus	*ad* = toward; *longus* = long	Pubis—below pubic crest	Linea aspera of femur	Adducts, flexes, and laterally rotates thigh at hip joint	Obturator n.
Adductor brevis	*brevis* = short	Inferior ramus of pubis	Linea aspera of femur	Adducts, flexes, and laterally rotates thigh at hip joint	Obturator n.
Adductor magnus	*magnus* = large	Inferior ramus of ischium and pubis	Linea aspera and medial epicondyle of femur	Adducts, flexes, and laterally rotates thigh at hip joint	Obturator n.

Figure 9.40   Muscles of the right posterior thigh.

Gluteus medius

Gluteus maximus

Tensor fasciae latae

Iliotibial tract

Gracilis

Adductor magnus

Vastus lateralis

Long head of biceps femoris

Semitendinosus

Short head of biceps femoris

Semimem-branosus

Sartorius

Tibial nerve

Popliteal artery

Popliteal vein

Common fibular nerve

Gastrocnemius

Ischial tuberosity

Long head of biceps femoris

Semi-tendinosus

Semi-membranosus

Head of fibula

Linea aspera

Short head of biceps femoris

Adductor minimus (part of adductor magnus)

Adductor magnus

Vastus lateralis

Creek

insertion on the patella via the tendon of the rectus femoris muscle. The tendon of the rectus femoris muscle is continuous over the patella and becomes the *patellar ligament* as it attaches to the head of the tibial tuberosity (fig. 9.39). These muscles function synergistically to extend the leg, as in kicking a football. The four muscles of the quadriceps femoris muscle are the rectus femoris, vastus lateralis, vastus medialis, and vastus intermedius.

The **rectus femoris muscle** occupies a superficial position and is the only one of the four quadriceps that functions in both the hip and knee joints. The laterally positioned **vastus lateralis muscle** is the largest muscle of the quadriceps femoris. It is a common intramuscular injection site in infants who have small, underdeveloped buttock and shoulder muscles. The

vastus medialis muscle occupies a medial position along the thigh. The **vastus intermedius muscle** lies deep to the rectus femoris muscle.

The anterior thigh muscles that move the leg at the hip joint are summarized in table 9.16.

Posterior, or Flexor, Muscles   There are three posterior thigh muscles antagonistic to the quadriceps femoris muscles in flexing the knee. These muscles are known as the **hamstrings** (fig. 9.40). Interestingly, this name derives from the old butchers' practice of using the tendons of these muscles to hang the hams of pork for smoke curing.

Figure 9.41 Muscles of the right medial thigh.

Psoas minor
Psoas major
Iliacus
Obturator internus
Pubic bone
Levator ani (cut)
Pectineus
Adductor longus
Rectus femoris
Adductor magnus
Sartorius
Vastus medialis
Patella

5th lumbar vertebra
Piriformis
Sacrospinous lig.
Coccygeus
Gluteus maximus
Adductor magnus
Gracilis
Semimembranosus
Semitendinosus
Biceps femoris
Medial head of gastrocnemius

Creek

The **biceps femoris muscle** occupies the posterior lateral aspect of the thigh. It has a superficial long head and a deep short head and causes movement at both the hip and knee joints. The superficial **semitendinosus muscle** is fusiform and is located on the posterior medial aspect of the thigh. It also works over two joints. The **semimembranosus muscle** is a flat muscle that lies deep to the semitendinosus muscle on the posterior medial aspect of the thigh.

The posterior thigh muscles that move the leg at the hip joint are summarized in table 9.17. The relationship of the muscles of the thigh is illustrated in figure 9.42.

Hamstring injuries are a common occurrence in some sports. The injury usually occurs when sudden lateral or medial stress to the knee joint tears the muscles or tendons. Because of its structure and the stress applied to it in competition, the knee joint is highly susceptible to injury. To reduce the incidence of knee injury, the rules in contact sports should be altered, or at least additional support and protection provided for this vulnerable joint.

Muscles of the Leg That Move the Ankle, Foot, and Toes

The muscles of the leg, the **crural muscles,** are responsible for the movements of the foot. There are three groups of crural muscles: anterior, lateral, and posterior. The anteromedial aspect of the leg along the shaft of the tibia lacks muscle attachment.

Anterior Crural Muscles The anterior crural muscles include the tibialis anterior, extensor digitorum longus, extensor hallucis longus, and peroneus tertius muscles (figs. 9.43, 9.44, and 9.45).

UNIT IV

Table 9.16	**Anterior Thigh Muscles That Move the Leg**				
Muscle	**Derivation**	**Origin**	**Insertion**	**Action**	**Innervation**
Sartorius	*sartor* = tailor	Anterior superior iliac spine	Medial surface of tibia	Flexes leg and thigh, abducts thigh, rotates thigh laterally, and rotates leg medially at hip joint	Femoral n.
Quadriceps femoris	*quad* = four; *cephalic* = head; *femoris* = femur		Patella by common tendon, which continues as patellar ligament to tibial tuberosity	Extends leg at knee joint	Femoral n.
Rectus femoris	*rectus* = straight fibers	Anterior inferior iliac spine and lip of acetabulum			
Vastus lateralis	*vastus* = large; *lateralis* = lateral	Greater trochanter and linea aspera of femur			
Vastus medialis	*medialis* = medial	Medial surface and linea aspera of femur			
Vastus intermedius	*intermedius* = middle	Anterior and lateral surfaces of femur			

Table 9.17	Posterior Thigh Muscles That Move the Leg				
Muscle	**Derivation**	**Origin**	**Insertion**	**Action**	**Innervation**
Biceps femoris	*bi* = two; *cephal* = head; *femoris* = femur	Long head—ischial tuberosity; short head—linea aspera of femur	Head of fibula and lateral epicondyle of tibia	Flexes leg at knee joint; extends and laterally rotates thigh at hip joint	Tibial n.
Semitendinosus	*semi* = half; *tendo* = tendon	Ischial tuberosity	Proximal portion of medial surface of shaft of tibia	Flexes leg at knee joint; extends and medially rotates thigh at hip joint	Tibial n.
Semimembranosus	*membran* = membrane	Ischial tuberosity	Medial epicondyle of tibia	Flexes leg at knee joint; extends and medially rotates thigh at hip joint	Tibial n.

Figure 9.42   A transverse section of the right thigh as seen from above. (Note the position of the vessels and nerves.)

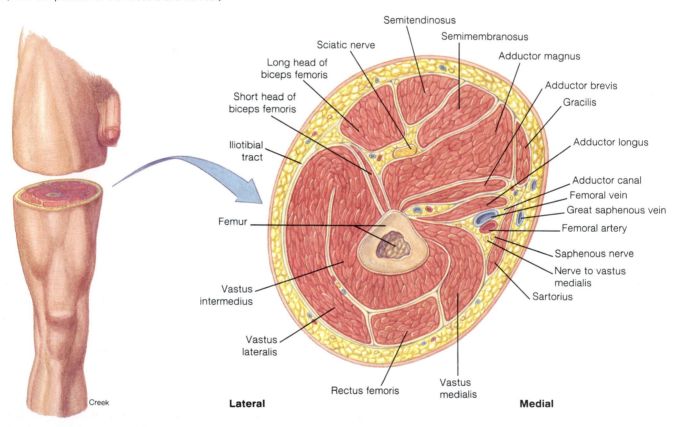

The large, superficial **tibialis anterior muscle** can be easily palpated on the anterior lateral portion of the tibia (fig. 9.43). It parallels the prominent anterior crest of the tibia. The **extensor digitorum longus muscle** lies lateral to the tibialis anterior on the anterolateral surface of the leg. The **extensor hallucis longus muscle** is positioned deep between the tibialis anterior muscle and the extensor digitorum longus muscle. The **peroneus tertius muscle** is a small muscle that is continuous with the distal portion of the extensor digitorum longus muscle.

Lateral Crural Muscles   The lateral crural muscles are the peroneus longus and peroneus brevis (figs. 9.43 and 9.44). The long, flat **peroneus longus muscle** is a superficial lateral muscle that overlies the fibula. The **peroneus brevis muscle** lies deep to the peroneus longus muscle and is positioned closer to the foot. These two muscles are synergistic in flexion and eversion of the foot (see table 9.18).

Figure 9.43 Anterior crural muscles.

Patella
Patellar ligament
Peroneus longus
Tibialis anterior
Extensor digitorum longus
Peroneus brevis
Extensor hallucis longus
Inferior extensor retinaculum
Tuberosity of tibia
Medial head of gastrocnemius
Tibia
Soleus
Superior extensor retinaculum
Creek

Tibia
Tibialis anterior
Medial cuneiform and first metatarsal bones
Fibula
Extensor hallucis longus
Peroneus tertius
Fifth metatarsal bone
Distal phalanx of hallux

Posterior Crural Muscles   The seven posterior crural muscles can be grouped into a superficial and a deep group. The superficial group is composed of the gastrocnemius *(gas''trok-ne'me-us)*, soleus, and plantaris muscles (fig. 9.46). The four deep posterior crural muscles are the popliteus, flexor hallucis longus, flexor digitorum longus, and tibialis posterior muscles (fig. 9.47).

The **gastrocnemius muscle** is a large superficial muscle that forms the major portion of the calf of the leg. It consists of two distinct heads, which arise from the posterior surfaces of the medial and lateral condyles of the femur. This muscle and the deeper soleus muscle insert onto the calcaneus bone via the common *tendo calcaneus (tendon of Achilles)*. The gastrocnemius acts over two joints to cause flexion of the knee joint and plantar flexion of the foot at the ankle joint.

The **soleus muscle** lies deep to the gastrocnemius. These two muscles are frequently referred to as a single muscle, the **triceps surae muscle.** The soleus muscle has a common insertion with the gastrocnemius muscle but acts on only one joint, the ankle, in plantar flexing the foot.

The small **plantaris muscle** arises just above the origin of the lateral head of the gastrocnemius muscle on the lateral condyle of the femur. It has a very long, slender tendon of insertion onto the calcaneus bone. The tendon of this muscle is frequently mistaken for a nerve by those dissecting it for the first time. The plantaris is a weak muscle, with limited ability to flex the leg and plantar flex the foot at the ankle joint.

The thin, triangular **popliteus muscle** is situated deep to the heads of the gastrocnemius muscle, where it forms part of the floor of the popliteal fossa. The *popliteal fossa* is the depression on the back side of the knee joint (fig. 9.48). The popliteus muscle is a medial rotator of the tibia on the femur. The bipennate **flexor hallucis longus muscle** lies deep to the soleus muscle on the posterolateral side of the leg. It flexes the big toe (hallux) and assists in plantar flexing and inverting the foot.

The **flexor digitorum longus muscle** also lies deep to the soleus muscle, and it parallels the flexor hallucis longus muscle on the medial side of the leg. Its distal tendon passes behind the medial malleolus and continues along the plantar

Figure 9.44 Lateral crural muscles.

surface of the foot, where it branches into four tendinous slips that attach to the bases of the terminal phalanges of the second, third, fourth, and fifth toes (fig. 9.49). The flexor digitorum longus muscle works over several joints, flexing four of the digits and assisting in plantar flexing and inverting the foot.

The **tibialis posterior muscle** is located deep to the soleus muscle, directly on the interosseous membrane on the posterior surface of the leg. The distal tendon of the tibialis posterior passes behind the medial malleolus and inserts on the plantar surfaces of the navicular, the cuneiforms, the cuboid, and the second, third, and fourth metatarsal bones (fig. 9.49). The tibialis posterior plantar flexes and inverts the foot and gives support to the arches of the foot.

The crural muscles are summarized in table 9.18.

Muscles of the Foot

With the exception of one additional intrinsic muscle, the **extensor digitorum brevis muscle,** the muscles of the foot are similar in number and nomenclature to those of the hand. The functions of the muscles of the foot are different, however, because the foot is adapted to provide support while bearing body weight rather than to grasp objects.

The muscles of the foot are topographically arranged into four layers (fig. 9.49) that are difficult to dissociate, even in dissection. The muscles function either to move the toes or to support the arches of the foot through their contraction. Because of their complexity, the muscles of the foot will be presented only in illustrations (see figs. 9.49 and 9.50).

1. List all the muscles that either originate from or insert on the scapula.
2. On the basis of function, categorize the muscles of the upper extremity as flexors, extensors, abductors, adductors, or rotators. (Each muscle may fit into two or more categories.)
3. Which muscles of the lower extremity span two joints, and therefore have two different actions?

Figure 9.45 A medial view of crural muscles.

Sartorius

Tendon of gracilis

Patella

Tendon of semimembranosus

Tendon of semitendinosus

Medial head of gastrocnemius

Tibia

Tibialis anterior

Soleus

Tendo calcaneus

Tendon of plantaris

Tibialis posterior

Flexor digitorum longus

Inferior extensor retinaculum

Flexor hallucis longus

Flexor retinaculum

Creek

Abductor hallucis (cut)

UNIT IV

Clinical Considerations

Compared to the other systems of the body, the muscular system is extremely durable. If properly conditioned, the muscles of the body can adequately serve a person for a lifetime. Muscles are capable of doing incredible amounts of work; through exercise, they can become even stronger.

Clinical considerations include the diagnosis of muscle condition, functional conditions in muscles, and diseases of muscles.

Diagnosis of Muscle Condition

The clinical symptoms of muscle diseases are usually weakness, loss of muscle mass (atrophy), and pain. The diagnosis and extent of the disease depends on the severity of the symptoms.

Several standard procedures are used to assess skeletal muscle dysfunction and disease. The most obvious is a clinical examination of the patient. Following this, it may be necessary to test muscle function using **electromyography (EMG)** to measure conduction rates and motor unit activity within a muscle. Laboratory tests may include serum enzyme assays or muscle biopsies. Muscle biopsy is perhaps the definitive source of diagnostic information. Progressive atrophy, polymyositis, and metabolic diseases of muscles can be determined through a biopsy.

Table 9.18	Muscles of the Lower Leg That Move the Ankle, Foot, and Toes				
Muscle	**Derivation**	**Origin**	**Insertion**	**Action**	**Innervation**
Tibialis anterior	*tibialis* = tibia; *ante* = in front	Lateral condyle and body of tibia	First metatarsal and first cuneiform	Dorsiflexes ankle and inverts foot at ankle	Deep fibular n.
Extensor digitorum longus	*extensor* = increase joint angle; *digit* = finger or toe; *longus* = long	Lateral condyle of tibia and anterior surface of fibula	Extensor expansions of digits II–V	Extends digits II–V and dorsiflexes foot at ankle	Deep fibular n.
Extensor hallucis longus	*hallux* = great toe	Anterior surface of fibula and interosseous membrane	Distal phalanx of digit I	Extends joints of big toe and assists dorsiflexion of foot at ankle	Deep fibular n.
Peroneus tertius	*perone* = fibula; *tertius* = third	Anterior surface of fibula and interosseous membrane	Dorsal surface of fifth metatarsal	Dorsiflexes and everts foot at ankle	Deep fibular n.
Peroneus longus	*longus* = long	Lateral condyle of tibia and head and shaft of fibula	First cuneiform and metatarsal I	Plantar flexes and everts foot at ankle	Superficial fibular n.
Peroneus brevis	*brevis* = short	Lower aspect of fibula	Metatarsal V	Plantar flexes and everts foot at ankle	Superficial fibular n.
Gastrocnemius	*gaster* = belly; *kneme* = leg	Lateral and medial condyle of femur	Posterior surface of calcaneus	Plantar flexes foot at ankle; flexes knee joint	Tibial n.
Soleus	*soleus* = sole of foot	Posterior aspect of fibula and tibia	Calcaneus	Plantar flexes foot at ankle	Tibial n.
Plantaris	*planta* = plantar surface of foot	Lateral supracondylar ridge of femur	Calcaneus	Plantar flexes foot at ankle	Tibial n.
Popliteus	*poples* = ham of the knee	Lateral condyle of femur	Upper posterior aspect of tibia	Flexes and medially rotates leg at knee joint	Tibial n.
Flexor hallucis longus	*flexor* = decrease angle	Posterior aspect of fibula	Distal phalanx of big toe	Flexes joints of distal phalanx of big toe	Tibial n.
Flexor digitorum longus		Posterior surface of tibia	Distal phalanges of digits II–V	Flexes joints of distal phalanges of digits II–V	Tibial n.
Tibialis posterior	*posterior* = back	Tibia and fibula and interosseous membrane	Navicular, cuneiforms, cuboid, and metatarsal bones II–IV	Plantar flexes and inverts foot at ankle; supports arches	Tibial n.

Functional Conditions in Muscles

Muscles depend on systematic, periodic contraction to maintain optimal health. Obviously, overuse or disease will cause a change in muscle tissue. The immediate effect of overexertion of muscle tissue is the accumulation of lactic acid, resulting in fatigue and soreness. Excessive contraction of a muscle can also damage the fibers or associated connective tissue, resulting in a **strained muscle.**

When skeletal muscles are not contracted, either because the motor nerve supply is blocked or because the limb is immobilized (as when a broken bone is in a cast), the muscle fibers **atrophy,** or diminish in size. Atrophy is reversible if exercise is resumed, as after a healed fracture, but tissue death is inevitable if the nerves cannot be stimulated.

Figure 9.46 Posterior crural muscles and popliteal region.

The fibers in healthy muscle tissue increase in size, or **hypertrophy,** if a muscle is systematically exercised. This increase in muscle size and strength is due not to an increase in the number of muscle cells but rather to the increased production of myofibrils, accompanied by a strengthening of the associated connective tissue.

A **cramp** within a muscle is an involuntary, painful, and prolonged contraction. Cramps can occur while muscles are in use or at rest. The precise cause of cramps is unknown, but evidence indicates that they may be related to conditions within the muscle (e.g., calcium or oxygen deficiencies) or to stimulation of the motor neurons.

A condition called **rigor mortis** (rigidity of death) affects skeletal muscle tissue several hours after death, as depletion of ATP within the fibers causes a state of muscle contracture and stiffness of the joint. After a few days, however, as the muscle proteins decompose, the rigidity of the corpse disappears.

UNIT IV

Figure 9.47 Deep posterior crural muscles.

Popliteal artery and vein
Plantaris (cut)
Tibial nerve
Lateral head of gastrocnemius (cut)
Popliteus
Common fibular nerve (cut)
Posterior tibial artery
Soleus (cut)
Anterior tibial artery
Tibial nerve
Peroneal artery
Flexor digitorum longus
Peroneus longus
Tibialis posterior
Flexor hallucis longus
Posterior tibial artery
Tibial nerve
Peroneus brevis
Tendon of tibialis posterior
Tendo calcaneus
Peroneal artery
Creek

Tibia
Tibialis posterior
Fibula
Interosseous ligament
Tarsal and metatarsal bones
Flexor digitorum longus
Phalanges of lesser toes

Lateral condyle of femur
Popliteus
Fibula
Flexor hallucis longus
Distal phalanx of hallux

Figure 9.48 Muscles that surround the popliteal fossa.

Semitendinosus
Semimembranosus
Gracilis
Great saphenous vein
Medial superior genicular artery
Popliteal artery and vein
Sartorius
Saphenous nerve
Medial head of gastrocnemius
Small saphenous vein

Long head of biceps femoris
Short head of biceps femoris
Vastus lateralis
Tibial nerve
Lateral superior genicular artery
Common fibular nerve
Lateral sural cutaneous nerve
Plantaris
Medial sural cutaneous nerve
Lateral head of gastrocnemius
Communicating branch of fibular nerve
Creek

UNIT IV

Figure 9.49 The four musculotendinous layers of the plantar aspect of the foot. (*a*) Superficial layer, (*b*) second layer, (*c*) third layer, and (*d*) deep layer.

Fibrous digital sheaths

Lumbricales

Flexor digiti minimi brevis

Abductor digiti minimi

Plantar interosseous

Tendon of flexor hallucis longus

Flexor hallucis brevis

Flexor digitorum brevis

Abductor hallucis

Plantar aponeurosis (cut)

Calcaneal tuberosity

(a)

Tendons of flexor digitorum brevis (cut)

Tendon of flexor hallucis longus

Flexor hallucis brevis

Lumbricales

Tendon of flexor digitorum longus

Quadratus plantae

Flexor digitorum brevis (cut)

Abductor hallucis (cut)

Flexor digiti minimi brevis

Abductor digiti minimi

(b)

Tendon of flexor digitorum brevis (cut)

Plantar interossei

Quadratus plantae (cut)

Tendon of lumbrical (cut)

Tendons of flexor digitorum longus (cut)

Transverse head and

Oblique head of adductor hallucis

Flexor hallucis brevis

Tendon of flexor hallucis longus (cut)

Tendon of flexor digitorum longus (cut)

(c)

Plantar ligaments

Opponens digiti minimi

Tendon of peroneus longus

Peroneus brevis tendon

Articular capsules

Sesamoid bones

Dorsal interossei

Plantar interossei

Tendon of tibialis posterior

Long plantar ligament

Creek

(d)

UNIT IV

Developmental Exposition

The Muscular System

Explanation

The formation of skeletal muscle tissue begins during the fourth week of embryonic development as specialized mesodermal cells called *myoblasts* begin rapid mitotic division (exhibit I). The proliferation of new cells continues while the myoblast cells migrate and fuse together into *syncytial (sin-sish' al) myotubes.* A syncytium is a multinucleated protoplasmic mass formed by the union of originally separate cells. At 9 weeks, primitive myofilaments course through the myotubes, and the nuclei of the contributing myoblasts are centrally located. Growth in length continues through the addition of myoblasts.

The process of muscle fiber development occurs within specialized mesodermal masses called *myotomes* in the embryonic trunk area and from loosely organized masses of mesoderm in the head and appendage areas. At 6 weeks, the trunk of an embryo is segmented into distinct myotomes (exhibit II) that are associated dorsally with specific sclerotomes. Sclerotomes give rise to vertebrae. As will be explained in chapter 12, spinal nerves arise from the spinal cord and exit between vertebrae to innervate developing muscles in the adjacent myotomes. As myotomes develop, they elongate ventrally, toward the midline of the body, or distally, into the developing limbs. The muscles of the entire muscular system have been differentiated and correctly positioned by the eighth week. The orientation of the developing muscles is preceded and influenced by cartilaginous models of bones.

It is not certain when skeletal muscle is sufficiently developed to sustain contractions, but by week 17 the fetal movements known as *quickening* are strong enough to be recognized by the mother. The individual muscle fibers have now thickened, the nuclei have moved peripherally, and the filaments (myofilaments) can be recognized as alternating dark and light bands. Growth in length still continues through addition of myoblasts. Shortly before a baby is born, the formation of myoblast cells ceases, and all of the muscle cells have been determined. Differences in strength, endurance, and coordination are somewhat genetically determined but are primarily the result of individual body conditioning. Muscle coordination is an ongoing process of achieving a fine neural control of muscle fibers. Mastery of muscle movement is comparatively slow in humans. It is several months before a human infant has the coordination to crawl, and about a year before it can stand or walk. By contrast, most mammals can walk and run within a few hours after they are born.

myoblast: Gk. *myos*, muscle; *blastos*, germ
syncytial: Gk. *syn*, with; *cyto*, cell

Exhibit I The development of skeletal muscle fibers. (*a*) At 5 weeks, the myotube is formed as individual cell membranes are broken down. Myotubes grow in length by incorporating additional myoblasts; each adds an additional nucleus. (*b*) Muscle fibers are distinct at 9 weeks, but the nuclei are still centrally located, and growth in length continues through the addition of myoblasts. (*c*) At 5 months, thin (actin) and thick (myosin) myofilaments are present and moderate growth in length still continues. (*d*) By birth, the striated myofilaments have aggregated into bundles, the fiber has thickened, and the nuclei have shifted to the periphery. Myoblast activity ceases and all the muscle fibers a person will have are formed. (*e*) The appearance of a mature muscle fiber.

Waldrop

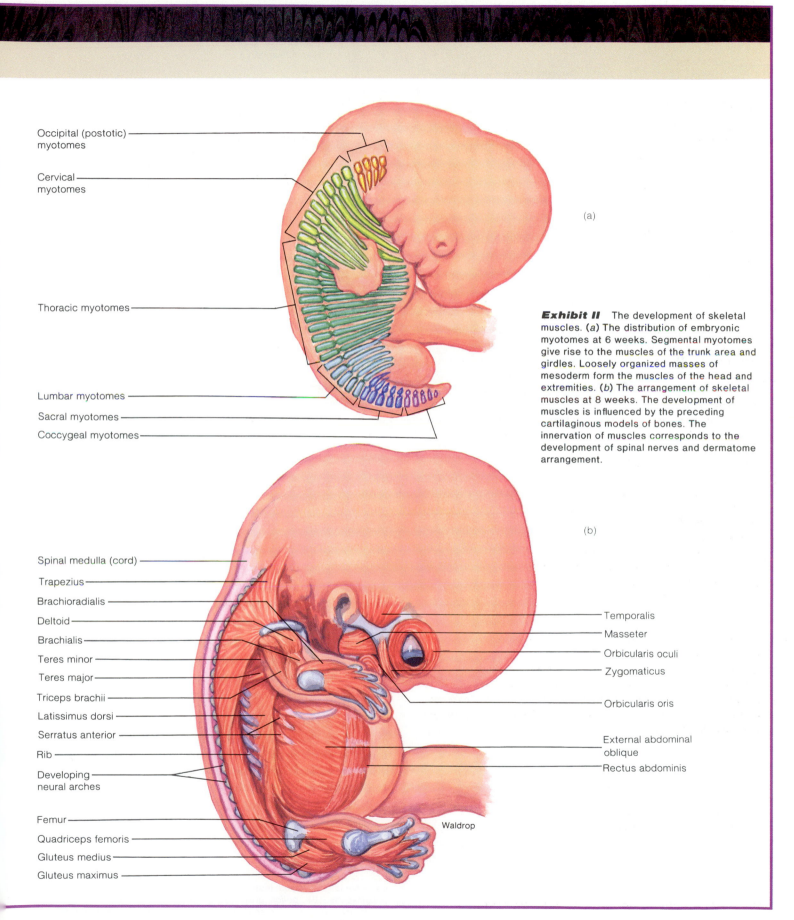

Occipital (postotic) myotomes

Cervical myotomes

Thoracic myotomes

Lumbar myotomes

Sacral myotomes

Coccygeal myotomes

(a)

Exhibit II   The development of skeletal muscles. (*a*) The distribution of embryonic myotomes at 6 weeks. Segmental myotomes give rise to the muscles of the trunk area and girdles. Loosely organized masses of mesoderm form the muscles of the head and extremities. (*b*) The arrangement of skeletal muscles at 8 weeks. The development of muscles is influenced by the preceding cartilaginous models of bones. The innervation of muscles corresponds to the development of spinal nerves and dermatome arrangement.

UNIT IV

(b)

Spinal medulla (cord)

Trapezius

Brachioradialis

Deltoid

Brachialis

Teres minor

Teres major

Triceps brachii

Latissimus dorsi

Serratus anterior

Rib

Developing neural arches

Femur

Quadriceps femoris

Gluteus medius

Gluteus maximus

Temporalis

Masseter

Orbicularis oculi

Zygomaticus

Orbicularis oris

External abdominal oblique

Rectus abdominis

Waldrop

Figure 9.50 An anterior view of the dorsum of the foot.

Superior extensor retinaculum

Peroneus tertius

Tendon of extensor digitorum longus

Lateral malleolus

Inferior extensor retinaculum

Extensor digitorum brevis

Tendon of peroneus brevis

Tuberosity of fifth metatarsal bone

Tendon of peroneus tertius

Abductor digiti minimi

Tendons of extensor digitorum brevis

Tendons of extensor digitorum longus

Anterior tibial artery and deep fibular nerve

Tendon of tibialis anterior

Medial malleolus

Anterior medial malleolar artery

Tendon of extensor hallucis longus

Dorsalis pedis artery

Deep peroneal nerve

Extensor hallucis brevis

Arcuate artery

Abductor hallucis

First dorsal interosseous

Tendon of extensor hallucis brevis

Extensor expansions

Dorsal digital arteries

Dorsal digital branches of superficial peroneal nerve

Creek

Diseases of Muscles

Fibromyositis is an inflammation of both skeletal muscular tissue and the associated connective tissue. Its causes are not fully understood. Fibromyositis frequently occurs in the extensor muscles of the lumbar region of the vertebral column, where there are extensive aponeuroses. Fibromyositis of this region is called **lumbago,** or **rheumatism.**

Muscular dystrophy is a genetic disease characterized by a gradual atrophy and weakening of muscle tissue. There are several kinds of muscular dystrophy, none of whose etiology is completely understood. The most frequent type affects children and is sex-linked to the male child. As muscular dystrophy progresses, the muscle fibers atrophy and are replaced by adipose tissue. Most children who have muscular dystrophy die before the age of 20.

The disease **myasthenia gravis** is characterized by extreme muscle weakness and low endurance. Transmission of impulses at the neuromuscular junction is defective. Myasthenia gravis is believed to be an autoimmune disease, and it typically affects women between the ages of 20 and 40.

Poliomyelitis (polio) is actually a viral disease of the nervous system that causes a paralysis of muscles. The viruses are usually localized in the anterior (ventral) horn of the spinal cord, where they affect the motor nerve impulses to skeletal muscles.

Neoplasms of muscle are rare, but when they do occur they are usually malignant. **Rhabdomyosarcoma** *(rab''do-*

lumbago: L. *lumbus,* loin

myasthenia: Gk. *myos,* muscle; *astheneia,* weakness
poliomyelitis: Gk. *polios,* gray; *myolos,* marrow
rhabdomyosarcoma: Gk. *rhabdos,* rod; *myos,* muscle; *oma,* a growth

Figure 9.51 A gradual diminishing of muscle strength occurs after the age of 35, as shown with a graph of hand-grip strength.

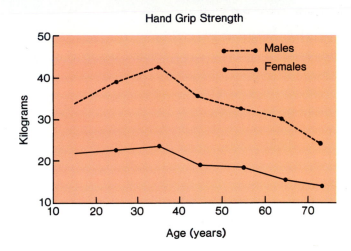

Hand Grip Strength

Figure 9.52 Competitive runners in late adulthood, recognizing the benefits of exercise.

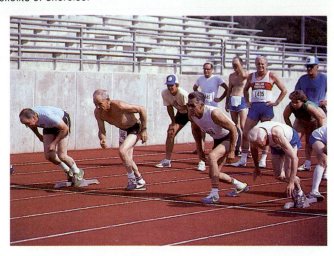

UNIT IV

mi''ō-sar-ko'mă) is a malignant tumor of skeletal muscle. It can arise in any skeletal muscle, and most often afflicts young children and elderly people.

Aging of Muscles

Although the aged experience a general decrease in the strength, endurance, and agility of skeletal muscle (fig. 9.51), the extent of senescent changes varies considerably among individuals. Apparently the muscular system is one of the body systems in which a person may actively slow senescent changes. A decrease in muscle mass, in part, is due to changes in connective and circulatory tissues. Atrophy of the muscles of the appendages causes the arms and legs to appear thin and bony. Degenerative changes in the nervous system decrease the effectiveness of motor activity. Muscle reflexes become less efficient, causing a marked reduction in physical capabilities.

Diminished muscular capabilities may affect the functioning of other body systems. A decrease in the strength of the respiratory muscles may limit the ability of the lungs to ventilate. Reduced muscularity of the urinary bladder causes difficult micturition and may cause urinary infections.

Exercise is important at all stages of life but is especially beneficial as one approaches old age (fig. 9.52). Exercise not only strengthens bones and muscles, but it also contributes to a healthy circulatory system and thus ensures an adequate blood supply to all body tissues. If an older person does not maintain muscular strength through exercise, he or she will be more prone to debility. Learning to use a cane, crutches, or a walker would be extremely difficult for such a person.

CLINICAL CASE STUDY ANSWER

When cancer of the head or neck involves lymph nodes in the neck, a number of structures on the affected side are removed surgically. This procedure usually includes the sternocleidomastoid muscle. This muscle, which originates on the sternum and clavicle, inserts on the mastoid process of the temporal bone. When it contracts, the mastoid process, which is located posteriorly at the base of the skull, is pulled forward, causing the chin to rotate away from the contracting muscle. This explains why the patient who had his left sternocleidomastoid muscle removed would have a disability in turning his head to the right.

Internal Affairs

The Relational Workings of the Body Systems

Muscular System

• Facial musculature attached to skin produces facial expression when contracted

---

• Enables body movement and stabilizes joints
• Muscle contractions maintain the health and strength of bone

---

• Sensory receptors monitor body position via autonomic nervous system
• Muscles permit eye movement and speech

• Provides protection to certain endocrine glands
• Epinephrine and norepinephrine stimulate muscle contraction

---

• Tonus and voluntary muscle contractions assist blood movement, particularly within veins

---

• Supports and protects superficial lymph nodes

• Respiratory muscles enable ventilation of lungs; sound production

---

• Supports and protects organs of GI tract
• Muscular sphincters located at openings of GI tract

• Muscles of urinary tract surround urinary bladder and form urethral sphincter

---

• Supports pelvic viscera
• Enables movement of gametes in male and female reproductive tracts

Integumentary System
• Covers and protects the body musculature
• Removes excessive body heat
• Synthesizes vitamin D

Skeletal System
• Source of calcium and phosphate
• Provides attachment sites for muscles
• Joints of skeleton provide levers for movement

Nervous System
• Coordinates muscle contraction
• Increases cardiac output and respiratory rates during periods of muscle activity

Endocrine System
• Sex hormones promote muscle development and maintenance
• Specific hormones regulate calcium and phosphate concentrations

Circulatory System
• Transports O_2 and CO_2, nutrients, and fluids to and from muscles; removes lactic acid and heat

Lymphatic System
• Maintains balanced amount of interstitial fluid within muscle tissue
• Lymphocytes provide defense against infection

Respiratory System
• Provides O_2 and eliminates CO_2 to and from muscles

Digestive System
• Provides nutrients for growth, maintenance, and repair of muscles
• Regulates blood glucose

Urinary System
• Eliminates metabolic wastes from muscles
• Assists regulation of calcium and phosphate concentrations

Reproductive System
• Gonads produce sex hormones that promote muscle development and maintenance

UNIT IV

IMPORTANT CLINICAL TERMINOLOGY

convulsion An involuntary, spasmodic contraction of skeletal muscle.

fibrillation (*fib-ri-la'shun*) A series of rapid, uncoordinated, and spontaneous contractions of individual motor units within a muscle.

hernia The rupture or protrusion of a portion of the underlying viscera through muscle tissue. The most common hernias occur in the normally weak places of the abdominal wall. Hernias are of four basic types:

1. **femoral**—viscera passing through the femoral ring
2. **hiatal**—the superior portion of the stomach protruding through the esophageal opening of the diaphragm
3. **inguinal**—viscera protruding through the inguinal canal
4. **umbilical**—a hernia occurring at the navel

intramuscular injection A hypodermic injection at certain heavily muscled areas to avoid damaging nerves. The most common site is the buttock.

myalgia Pain within a muscle resulting from any muscular disorder or disease.

myokymia (*mi''o-kim'e-ah*) Continual quivering of a muscle.

myoma A tumor of muscle tissue.

myopathy Any muscular disease.

myotomy (*mi-ot'ŏ-me*) Surgical cutting of muscle tissue.

myotonia A prolonged muscular spasm.

paralysis The loss of nervous control of a muscle.

shinsplints Tenderness and pain on the anterior surface of the leg caused by straining the tibialis anterior or extensor digitorium longus muscles.

torticollis (wryneck) A persistent contraction of a sternocleidomastoid muscle, drawing the head to one side and distorting the face. Torticollis may be acquired or congenital.

CHAPTER SUMMARY

Organization and General Functions (pp. 224–225)

1. The contraction of skeletal muscle fibers results in body motion, heat production, and the maintenance of posture and body support.
2. The four basic properties characteristic of all muscle tissue are irritability, contractility, extensibility, and elasticity.
3. Axial muscles include facial muscles, neck muscles, and trunk muscles. Appendicular muscles include those that act on the girdles and those that move the segments of the appendages.

Structure of Skeletal Muscles (pp. 226–230)

1. The origin of a muscle is the more stationary attachment. The insertion is the more movable attachment.
2. Individual muscle fibers are covered by endomysium. Muscle bundles, called fasciculi, are covered by perimysium. The entire muscle is covered by epimysium.
3. Synergistic muscles contract together. Antagonistic muscles perform in opposition to a particular group of muscles.
4. Muscles may be classified according to fiber arrangement as parallel, convergent, pennate, or circular.
5. Motor neurons conduct nerve impulses to the muscle fiber, stimulating it to contract. Sensory neurons conduct nerve impulses away from the muscle fiber to the central nervous system.

Skeletal Muscle Fibers and Mode of Contraction (pp. 230–235)

1. Each skeletal muscle fiber is a multinucleated, striated cell, composed of myofibrils and enclosed by a sarcolemma.
   (a) Myofibrils are characterized by alternating A and I bands; each I band is bisected by a Z line, and the subunit between two Z lines is the sarcomere.
   (b) Extending through the sarcoplasm is a network of membranous channels called the sarcoplasmic reticulum and a system of transverse tubules (T-tubules).
2. The activity of the crossbridges causes sliding of the myofilaments.
   (a) The actin on each side of the A bands is pulled toward the center.
   (b) The H bands thus appear to be shorter as more actin overlaps the myosin.
   (c) The I bands also appear to be shorter as adjacent A bands are pulled closer together.
   (d) The A bands stay the same length because the myofilaments (both thick and thin) do not shorten during muscle contraction.
3. When a muscle exerts tension without shortening, the contraction is termed isometric; when shortening does occur, the contraction is isotonic.
4. The neuromuscular junction is the area consisting of the motor end plate and the sarcolemma of a muscle fiber. In response to a nerve impulse, the synaptic vesicles of the terminal end of an axon secrete a neurotransmitter, which diffuses across the neuromuscular cleft of the neuromuscular junction and stimulates the muscle fiber.
5. The motor neuron and the muscle fibers innervated by the motor neuron are called a motor unit.
   (a) When a muscle is composed of many motor units (such as in the hand), there is a fine control of muscle contraction.
   (b) The large muscles of the thigh have relatively few units, which are correspondingly large.
   (c) Graded contractions are produced by the asynchronous stimulation of different motor units.

Naming of Muscles (pp. 235–236)

1. Skeletal muscles are named on the basis of shape, location, attachment, orientation of fibers, relative position, and function.
2. Most muscles are paired; that is, the right side of the body is a mirror image of the left.

Muscles of the Axial Skeleton (pp. 236–249)

The muscles of the axial skeleton include those of facial expression, mastication, eye movement, tongue movement, neck movement, respiration, the abdominal wall, the pelvic outlet, and the vertebral column.

Muscles of the Appendicular Skeleton (pp. 249–271)

The muscles of the appendicular skeleton include those of the pectoral girdle, humerus, forearm, wrist, hand, and fingers, and those of the pelvic girdle, thigh, leg, ankle, foot, and toes.

REVIEW ACTIVITIES

Objective Questions

1. The site at which a nerve impulse is transmitted from the motor nerve ending to the skeletal muscle cell membrane is
   (a) the sarcomere.
   (b) the neuromuscular junction.
   (c) the myofilament.
   (d) the Z line.
2. Muscles capable of highly dexterous movements contain
   (a) one motor unit per muscle fiber.
   (b) many muscle fibers per motor unit.
   (c) few muscle fibers per motor unit.
   (d) many motor units per muscle fiber.

UNIT IV

3. Which of the following is *not* used as a means of naming muscles?
   (a) location
   (b) action
   (c) shape
   (d) attachment
   (e) strength of contraction

4. Neurotransmitters are stored in synaptic vesicles within
   (a) the sarcolemma.
   (b) the motor units.
   (c) the myofibrils.
   (d) the axon terminals.

5. Which of the following muscles have motor units with the lowest innervation ratio?
   (a) brachial muscles
   (b) muscles of the forearm
   (c) thigh muscles
   (d) abdominal muscles

6. An eyebrow is drawn toward the midline of the face through contraction of which muscle?
   (a) the corrugator   (c) the nasalis
   (b) the risorius   (d) the frontalis

7. A flexor of the shoulder joint is
   (a) the pectoralis major.
   (b) the supraspinatus.
   (c) the teres major.
   (d) the trapezius.
   (e) the latissimus dorsi.

8. Which of the following muscles does *not* have either an origin or insertion upon the humerus?
   (a) the teres minor
   (b) the biceps brachii
   (c) the supraspinatus
   (d) the brachialis
   (e) the pectoralis major

9. Of the four quadriceps femoris muscles, which one may act on the hip and knee joints?
   (a) the vastus medialis
   (b) the vastus intermedius
   (c) the rectus femoris
   (d) the vastus lateralis

10. Which of the following muscles plantar flexes and inverts the foot as it supports the arches?
    (a) the flexor digitorum longus
    (b) the tibialis posterior
    (c) the flexor hallucis longus
    (d) the gastrocnemius

Essay Questions

1. Describe how muscle fibers are formed and explain why the fibers are multinucleated.
2. List the general functions of muscles and discuss functions that are passive, involuntary events.
3. Distinguish between fascia, aponeurosis, and retinaculum.
4. Describe the structural arrangement of the muscle fibers and fasciculi within muscle.
5. What are the advantages and disadvantages of pennate-fibered muscles?
6. Why is it necessary to have dual (sensory and motor) innervation to a muscle?
7. List the major components of a skeletal muscle fiber and describe the function of each part.
8. What is a motor unit, and what is its role in muscle contraction?
9. Give three examples of synergistic muscle groups within the upper extremity and identify the antagonistic muscle group for each.
10. Which muscles of the neck either originate or insert on the hyoid bone?
11. List all the muscles that either originate or insert on the scapula.
12. Discuss the position of flexor and extensor muscles relative to the shoulder, elbow, and wrist joints.

13. Based on function, describe exercises that would strengthen the following muscles: (a) the pectoralis major; (b) the deltoid; (c) the triceps; (d) the pronator teres; (e) the rhomboideus major; (f) the trapezius; (g) the serratus anterior; and (h) the latissimus dorsi.
14. Give three examples of synergistic muscle groups within the lower extremity and identify the antagonistic muscle group for each.
15. Describe the flexor and extensor compartments of muscles within the forearm.
16. List the muscles that border the popliteal fossa. Describe the structures that are located within this region.
17. Describe the actions of the muscles of inspiration. Which muscles participate in forced expiration?
18. Which muscles of the pelvic outlet support the floor of the pelvic cavity and which are associated with the genitalia?
19. Firmly press your fingers on the front, sides, and back of your ankle as you move your foot. The tendons of which muscles can be palpated anteriorly, laterally, and posteriorly?
20. Compare muscular dystrophy and myasthenia gravis as to causes, symptoms, and the effect they have on muscle tissue.

Gundy/Weber Software

The tutorial software accompanying Chapter 9, Muscular System, is Volume 4, Articulations and the Muscular System.

Surface and Regional Anatomy

Chapter Outline

Introduction to Surface Anatomy

Surface anatomy, a branch of gross anatomy, is concerned with identifying body structures through observation and palpation. The study of surface anatomy has tremendous application to maintaining physical fitness and diagnosing and treating diseases or dysfunctions of the body.

Objective 1. Describe the value of surface anatomy in learning about internal anatomical structures.

Objective 2. Discuss how surface anatomy is important in the diagnosis and treatment of various diseases or dysfunctions of the body.

Objective 3. Describe subcutaneous differences in adult males and females that may have a bearing on surface anatomy.

It is amazing how much anatomical information can be acquired by studying the surface anatomy of one's own body. Surface anatomy is the study of the structure and markings of the surface of the body through observation or palpation. Surface features can be readily identified through *visual observation,* and anatomical features beneath the skin can be located by *palpation* (feeling with firm pressure or perceiving by the sense of touch). Knowledge of surface anatomy is clinically important in locating precise sites for *percussion* (tapping sharply to detect resonating vibrations) and *auscultation* (listening to sounds emitted from organs).

With the exception of certain cranial bones, the bones of the entire skeleton can be palpated. Once the position, shape, and processes of these bones are identified, these skeletal features can serve as landmarks for other anatomical structures. One can observe the location of many skeletal muscles and their tendinous attachments as they are contracted and caused to move. The location and range of movement of each of the joints of the body can be determined as the articulating bones are moved by muscle contractions. On some individuals, the positions of superficial veins can be located and their courses traced. Even the location and function of the valves within the veins can be demonstrated on the surface of the skin (fig. 10.1). The locations of some of the arteries can be seen as they pulsate beneath the skin. Knowing where the arterial pressure points are is an important clinical aspect of surface anatomy. Other structures can be identified from the surface, including certain nerves, lymph nodes, glands, and other internal organs.

Surface anatomy is an essential part of the study of anatomy. Knowing the location of muscles and muscle groups as observed in surface anatomy can be extremely important in maintaining physical fitness and body conditioning. In many medical and paramedical professions, the surface anatomy of a patient is of immeasurable value in diagnosis and treatment. Knowing where to record a pulse, insert needles and tubes, listen to the functioning of internal organs, take radiographs, and perform physical therapy requires a knowledge of surface body landmarks.

Figure 10.1 A demonstration of the presence and function of valves within the veins of the forearm, conducted by the great English anatomist William Harvey. (After William Harvey, *On the Motion of the Heart and Blood in Animals,* 1628.) Surface anatomy was extremely important to understanding the concept of a closed circulatory system (e.g., blood contained within vessels).

The effectiveness of observation and palpation in studying a person's surface anatomy is dependent on the amount of subcutaneous adipose tissue present (fig. 10.2). In examining an obese person, it may be extremely difficult to observe or palpate certain internal structures that are readily discernible in a thin person. The subcutaneous tissue of females is normally thicker than that of males (fig. 10.3). This tends to smooth the surface contours of females and obscure the internal features, such as muscles, veins, and bony prominences, that are apparent in males.

This chapter will be of great value in reviewing the bones of the skeleton, articulations, and muscles you have already studied. Photographs of dissected cadavers will permit you to observe the relationship between the various body organs and systems in specific regions. Refer back to this chapter to learn the anatomy of the remaining body systems. Reviewing in this way will reinforce an external perspective of where the various organs and structures are located.

By using yourself as a model from which to learn and review, anatomy as a science will take on a new meaning. As you learn about a bone or a process on a bone, palpate that part of your body. Contract the muscles you are studying so that you better understand their locations, attachments, and actions. In this way, you will become better acquainted with your body, and anatomy will become more enjoyable and easier to learn. Your body is one crib sheet you can take with you to exams.

1. Define the terms *observation* and *palpation* and explain the value of surface anatomy in learning the location of internal structures.
2. Discuss the medical importance of surface anatomy.
3. How does the hypodermis layer of the skin differ in males and females? What are the clinical implications of this difference?

UNIT IV

Figure 10.2 Subcutaneous adipose tissue. (*a*) An anterior view of the left brachial region and (*b*) a posterior view of the left leg.

Clavicle bone

Pectoralis major muscle

Deltoid muscle

Cephalic vein

Basilic vein

Median nerve

Biceps brachii muscle

Median cubital vein

Basilic vein

Median vein of forearm

Cephalic vein

(a)

Peroneal nerve

Great saphenous vein

Small saphenous vein

Tibial nerve

Sural nerve

Gastrocnemius muscle

Tendo calcaneus

Nelson

(b)

UNIT IV

Surface Anatomy of the Newborn

The surface anatomy of a neonatal infant represents an early stage of human development; therefore, it differs from that of an adult. Certain aspects of the surface anatomy of a neonate are of clinical importance in ascertaining the degree of physical development, general health, and possible congenital abnormalities.

Objective 4. Describe the surface anatomy of a normal, full-term neonate.

Objective 5. List some internal anatomical structures that can be palpated in a neonate.

The birth of a baby is a dramatic culmination of a 9-month gestation, during which the miraculous development of the fetus prepares it for extrauterine life. The normal, full-term neonate

is physiologically prepared for life but totally dependent on parental care. The physical assessment of the neonate is extremely important to ensure its survival. Much of the assessment is performed through inspection and palpation of its surface anatomy. The surface anatomy of a neonate obviously differs from that of an adult because of the transitional stage of development from fetus to infant.

Although the surface anatomy of a neonate is discussed at this point in the text, prenatal development and body growth with its accompanying physiological changes are discussed in chapter 22. A summary of the surface anatomy of the neonate is presented in table 10.1.

General Appearance

As a result of in utero position, the posture of the full-term neonate is one of flexion (fig. 10.4). The neonate born vertex (head first) keeps the neck and vertebral column flexed, with the chin resting on the upper chest. The hands are clenched into fists,

neonate: Gk. *neos*, new; L. *natalis*, birth

Figure 10.3 Principal areas of adipose deposition of a female. (*a*) An anterior view and (*b*) a lateral view. The outline of a male is superposed in both views. There is significantly more adipose tissue interlaced in the fascia covering the muscles, vessels, and nerves in a female than in a male. The hypodermis layer of the skin is also approximately 8% thicker in a female than in a male.

(a)

(b)   Sims

Table 10.1	Surface Anatomy of the Neonate	
Body Structure	**Normal Conditions**	**Common Variations**
General posture	Joints of vertebral column and extremities flexed	Extended legs and neck; abducted and rotated thighs (breech birth)
Skin	Red or pink with vernix caseosa and lanugo; edematous face, extremities, and genitalia	Neonatal jaundice; integumentary blisters; mongolian spots
Skull	Fontanels large, flat, and firm, but soft to touch	Molded skull, bulging fontanels; cephalhematoma
Eyes	Lids edematous; color—gray, dark blue, or brown; absence of tears; corneal, pupillary, and blink reflexes	Conjunctivitis, subconjunctival hemorrhage
Ears	Auricle flexible, cartilage present; top of auricle positioned on horizontal line with outer canthus of eye	Auricle flat against head
Neck	Short and thick, surrounded by neck folds	Torticollis
Chest	Equal anteroposterior and lateral dimensions; xiphoid process evident; breast enlargement	Funnel or pigeon chest; additional nipples (polythelia); secretions from breast (witch's milk)
Abdomen	Cylindric in shape; liver and kidneys palpable	Umbilical hernia
Genitalia	(♂ and ♀) Edematous and darkly pigmented; (♂) testes palpable in scrotum; periodic erection of penis	(♀) Blood-tinged discharge (pseudomenstruation); hymenal tag; (♂) testes palpable in inguinal canal; inability to retract prepuce; inguinal hernia
Extremities	Symmetrical; 10 fingers and toes; soles flat with moderate to deep creases	Partial syndactyly; asymmetric length of toes

UNIT IV

Figure 10.4 The flexion position of a neonate (*a,b*) is an indication of a healthy gestation and a normal delivery.

(a)

(b)

Figure 10.5 Sole creases at different ages of gestation as seen from footprints of two premature babies (*a,b*) and a full-term baby (*c*). (*a*) At 26 weeks, only an anterior transverse crease is present. (*b*) By 33 weeks, creases have developed along the medial instep. (*c*) The entire sole has developed creases by 38 weeks.

(a)   (b)   (c)

Abnormal skin color is clinically important in the physical assessment of the neonate. *Cyanosis* (bluish discoloration) is usually due to a pulmonary disease like atelectasis or pneumonia or to congenital heart disease. Although *jaundice* (yellowish discoloration) is common in infants and is usually of no concern, it may indicate liver or bone marrow problems. *Pallor* (paleness) may indicate anemia, edema, or shock.

The appearance of the nails and nail beds is especially valuable in determining body dysfunctions, certain genetic conditions, and even normal gestations. Cyanosis, pallor, and capillary pulsations are best observed at the nails. Jaundice is common in postmature neonates and can be visibly detected by yellow nails.

the elbow joints are flexed, and the arms are held toward the chest. The knee and hip joints are flexed, bringing the thighs toward the abdomen. The ankle joints are dorsiflexed.

The skin is the one organ of the neonate that is completely visible and is therefore a source of considerable information concerning state of the development and the clinical condition of the newborn. At birth, the skin is covered with a grayish, cheeselike substance called **vernix caseosa.** If it is not washed away during bathing, it will dry and disappear within a couple of days. Fine, silklike hair called **lanugo** *(lă-noo′go)* may be present on the forehead, cheeks, shoulders, and back. Distended sebaceous glands called **milia** *(mil′e-ă)* may appear as tiny white papules on the nose, cheeks, and chin. Skin color depends on genetic background, although certain areas, such as the genitalia, areola, and linea alba may appear darker because of a response to the maternal and placental hormones that enter the fetal circulation. **Mongolian spots** occur in about 90% of newborn blacks, Asians, and American Indians. These blue-gray pigmented areas vary in size and are usually located in the lumbosacral region. Mongolian spots generally fade within the first year or two.

Local edema (swelling) is not uncommon, particularly in the skin of the face, legs, hands, feet, and genitalia. Creases on the palms of the hands and soles of the feet should be prominent; the absence of creases accompanies prematurity (fig. 10.5). The nose is usually flattened after birth and there may be bruises there, or on other areas of the face. The auricle of the ear is flexible, with the top edge positioned on a horizontal line with the outer canthus (corner) of the eye.

The neck of a neonate is short and thick and is surrounded by neck folds. The chest is rounded in cross section, and the abdomen is cylindrical. The abdomen may bulge in the upper right quadrant due to the large liver.

If the newborn is thin, peristaltic intestinal waves may be observed. At birth, the **umbilical cord** appears bluish white and moist. After clamping, it begins to dry and appears yellowish brown. It progressively shrivels and becomes greenish black prior to its falling off by the second week.

The genitalia of both sexes may appear darkly pigmented due to maternal hormonal influences. In a female neonate, a **hymenal** *(hi′men-al)* **tag** is frequently present and is visible from the posterior opening of the vagina. It is composed of tissue from the hymen and labia minora. The hymenal tag usually disappears by the end of the first month.

vernix caseosa: L. *vernix*, varnish; *caseus*, cheese
milia: L. *miliarius*, relating to millet

Palpable Structures

The six **fontanels** can be lightly palpated as the "soft spots" on the infant's head. The liver is palpable 2–3 cm (1 in.) below the right costal margin. A physician doing a physical examination of a neonate will palpate both kidneys soon after delivery before the intestines fill with air. The suprapubic area is also palpated for an abnormal distended urinary bladder. The newborn should void urine within the first 24 hours after birth.

The testes of the male neonate should always be palpated in the scrotum. If the neonate is small or premature, the testes may be palpable in the inguinal canals. An examination for inguinal hernias is facilitated by the crying of an infant, which creates abdominal pressure.

1. Describe the appearance of each of the following in a normal neonate: skin, head, thorax, abdomen, genitalia, and extremities. What is meant by the normal flexion position of a neonate?
2. Which internal body organs are palpable in a neonate?

Head

The head is the most highly integrated region of the body as it communicates with and controls all of the body systems. The head is of clinical concern because it contains important sense organs and provides openings into the respiratory and digestive systems. Of social importance is the aesthetics (pleasing appearance) of the head, which in some cases is also of clinical concern.

Objective 6. Identify various surface features of the cranial and facial regions by observation or palpation.

Objective 7. Describe the basic internal anatomy of the head.

Surface Anatomy

The head contains the brain and sense organs—the eyes, ears, nose, and taste buds. It also provides openings into the respiratory and digestive systems. The head is structurally and developmentally divided into the cranium and the face.

Cranium The **cranium,** also known as the **braincase,** is covered by the *scalp.* The scalp is attached anteriorly, at the level of the eyebrows, to the **supraorbital ridges.** The scalp continues posteriorly over the area commonly called the forehead and across the crown (vertex) of the top of the head to the **superior nuchal** (*noo'kal*) **line,** a ridge on the back of the skull. Both the supraorbital ridge above the **orbit,** or socket of the

cranium: Gk. *kranion,* skull

eye, and the superior nuchal line at the back of the skull can be easily palpated. Laterally, the scalp covers the **temporal region** and terminates at the fleshy portion of the ear called the **auricle,** or **pinna.** The temporal region is the attachment for the **temporalis muscle,** which can be palpated when the jaw is repeatedly clenched. This region is clinically important since it is a point of entrance to the cranial cavity in many surgical procedures.

Only a portion of the scalp is covered with hair, and the variable **hairline** is genetically determined. The scalp is clinically important because of the dense connective tissue layer that supports nerves and vessels beneath the skin. When the scalp is cut, the wound is held together by the connective tissue, but at the same time the vessels are held open, resulting in profuse bleeding.

Face The **face** (fig. 10.6) is composed of four regions: the **ocular region,** which includes the eye and associated structures; the **auricular region,** which includes the ear; the **nasal region,** which includes the external and internal structures of the nose; and the **oral region,** which includes the mouth and associated structures.

The skin of the face is relatively thin and contains many sensory receptors, particularly in the oral region. Certain facial regions also have numerous **sweat glands** and **sebaceous** (*seba'shus*) (oil-secreting) **glands.** Acne is a serious facial dermatological problem for many teenagers. Facial hair appears over most of the facial region in males after they go through puberty; unwanted facial hair may occur sparsely on some females and can be a social problem.

The muscles of facial expression are important in their effect on surface features. Various emotions are conveyed when these muscles are contracted. These muscles originate on the different facial bones and insert into the dermis (second major layer) of the skin. Repeated contraction of these muscles may eventually cause permanent crease lines in the skin.

Because the organs of the facial region are so complex and specialized, there are professional fields of specialty associated with the various regions. *Optometry* and *ophthalmology* are concerned with the structure and function of the eye. *Dentistry* is entirely devoted to the health and functional and cosmetic problems of the oral region, particularly the teeth. An *otorhinolaryngologist* is an ear, nose, and throat specialist.

The *ocular region* includes the eyeball and associated structures. Most of the surface features of the ocular region protect the eye. **Eyebrows** protect against potentially damaging sunlight and mechanical blows, **eyelids** reflexly close to protect against objects or visual stimuli, **eyelashes** prevent airborne objects from contacting the eyeball, and **lacrimal** (*lak'ri-mal*) **secretions** (tears) wash away chemical or foreign materials and prevent the surface of the eyeball from drying. Many of the surface features of the ocular region are shown in figure 10.7 and described in table 10.2.

Figure 10.6   The surface anatomy of the facial region. (*a*) An anterior view and (*b*) a lateral view.

Forehead
Root of nose
Bridge of nose
Superior palpebral sulcus
Lacrimal caruncle
Inferior palpebral sulcus
Auricle
Nasofacial angle
Alar nasal sulcus
Nostril
Philtrum
Corner of mouth
Mentolabial sulcus
Chin (mentalis)

Hair line
Superciliary ridge
Eyebrow
Eyelashes
Zygomatic arch
Apex of nose
Ala nasi
Lips
Angle of mandible
Body of mandible

(a)   (b)

UNIT IV

Figure 10.7   The surface anatomy of the ocular region.

Eyebrow
Sclera
Palpebral fissure
Lateral commissure
Conjunctiva
Lower eyelid

Pupil
Iris
Upper eyelid
Lacrimal caruncle
Medial commissure
Eyelashes

Table 10.2   Surface Anatomy of the Ocular Region

Structure	Comments	Structure	Comments
Eyebrow	Ridge of hair that superiorly arches the eye. It protects the eye against sunlight and is important in facial expression.	Iris	Circular, colored, muscular portion of the eyeball that surrounds the pupil. It reflexly regulates the amount of incoming light.
Eyelids	Movable folds of skin and muscle that cover the eyeball anteriorly. They assist in lubricating the anterior surface of the eyeball and reflexly close to protect the eyeball.	Pupil	Opening in the center of the iris through which light enters the eyeball
Eyelashes	Rows of hairs on the margin of the eyelids. They prevent airborne substances from contacting the eyeball.	Palpebral fissure	Space between the eyelids when they are open
		Subtarsal sulcus	Groove beneath the eyelid that parallels the margin of the lid. It traps small foreign particles that contact the conjunctiva.
Conjunctiva	Thin mucous membrane that covers the anterior surface of the eyeball and lines the undersurface of the eyelids. It aids in reducing friction during blinking.	Medial commissure	Medial junction of the upper and lower eyelids
Sclera	Outer fibrous layer of the eyeball; the "white" of the eye that gives form to the eyeball	Lateral commissure	Lateral junction of the upper and lower eyelids
Cornea	Transparent anterior portion of the eyeball. It is slightly convex to refract incoming light waves.	Lacrimal caruncle	Fleshy, pinkish elevation at the medial commissure. It contains sebaceous and sweat glands.

Figure 10.8 The surface anatomy of the auricular region.

Helix

Triangular fossa

Antihelix

Concha

Tragus

External auditory canal

Antitragus

Earlobe

Table 10.3   Surface Anatomy of the Auricular Region

Structure	Comments
Auricle (pinna)	Expanded, fleshy portion of the ear that projects from the side of the head. It funnels sound waves into the external auditory canal.
Helix	Outer rim of the auricle. It gives form and shape to the pinna.
Earlobe	Inferior, fleshy portion of the auricle
Tragus	Small projection of the auricle, just anterior to the external auditory canal.
Antitragus	Small, cartilaginous anterior projection opposite the tragus
Antihelix	Semicircular ridge anterior to the greater portion of the helix
Concha	Depressed hollow of the auricle. It funnels sound waves.
External auditory canal	Tube extending inward to the tympanic membrane. It is slightly S-shaped and contains glands that secrete earwax for protection.
Triangular fossa	Triangular depression in the superior part of the antihelix

The *auricular region* includes the visible surface structures and the internal organs that function in hearing and in maintaining equilibrium. The fleshy **auricle** and the tubular opening into the middle ear, called the **external auditory canal,** are the only observable surface features of the auricular region. The rim of the auricle is called the **helix** and the inferior portion is referred to as the **earlobe.** The earlobe is composed primarily of connective and fatty tissue and therefore can be easily pierced. For this reason, it is sometimes used when obtaining blood for a blood count. The **tragus** (tra'gus) is a small, posteriorly directed projection partially covering and protecting the external auditory canal. Further protection is provided by the many fine hairs that surround the opening into this canal. The condyle of the mandible can be palpated at the opening of the external

auditory canal by placing the little finger in the opening and then vigorously moving the jaw. Refer to figure 10.8 and table 10.3 for an illustration and description of other surface features of the auricular region.

The inspection of some of the internal structures of the ear is part of a routine physical examination and is performed using an otoscope. Earwax may accumulate in the canal, but this is a protective substance. It waterproofs the tympanic membrane (eardrum) and because of its bitter taste is thought to be an insect repellent.

Figure 10.9 The surface anatomy of the nasal and oral regions. (*a*) The nose and lips and (*b*) the teeth.

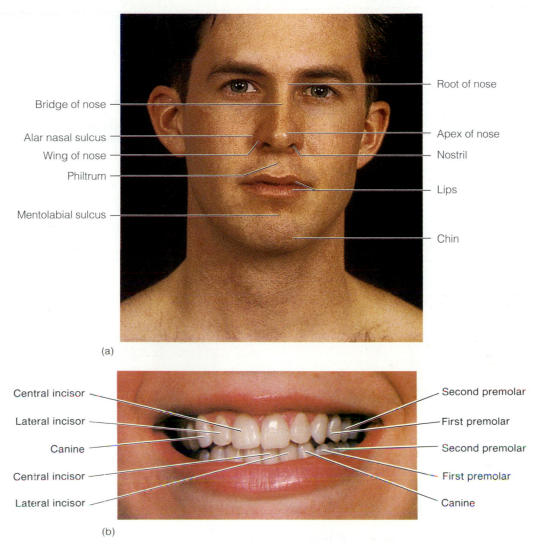

Bridge of nose
Alar nasal sulcus
Wing of nose
Philtrum
Mentolabial sulcus

Root of nose
Apex of nose
Nostril
Lips
Chin

(a)

Central incisor
Lateral incisor
Canine
Central incisor
Lateral incisor

Second premolar
First premolar
Second premolar
First premolar
Canine

(b)

A few structural features of the *nasal region* are apparent from its surface anatomy (fig. 10.9 and table 10.4). The principal function of the nose is associated with the respiratory system, and the need for a permanent body opening to permit gaseous ventilation accounts for its surface features. The **root** (nasion) of the nose is located at about the level of the eyebrows where the nose begins to protrude from the forehead. The firm, narrow part between the eyes is the **bridge** of the nose and is formed by the union of the nasal bones. The nose below this level has a pliable cartilaginous framework that maintains an opening. The tip of the nose is called the **apex.** The **nostrils,** or **nares,** are the paired openings into the nose. The **wing** (ala) of the nose forms the flaired outer margin of each nostril.

Structures of the *oral region* that are important in surface anatomy include the fleshy upper and lower **lips** (labia), the **chin** (mentum), and the structures of the *oral cavity* that can be observed when the mouth is open. The lips and chin are shown in figure 10.9 and the structures of the oral cavity, in figure 10.10.

Table 10.4	**Surface Anatomy of the Nasal and Oral Regions**
Structure	**Comments**
Root of nose (nasion)	Superior attachment of the nose to the cranium
Bridge of nose	Anterior bony framework of the nose formed by the union of nasal bones
Alar nasal sulcus	Lateral depression where the ala of the nose contacts the tissues of the face
Apex of nose	Tip of the nose
Nostril (naris)	External opening into the nasal cavity
Wing of nose (ala)	Laterally expanded border of the nostril
Philtrum	Vertical depression in the medial part of the upper lip
Lip (labium)	Upper and lower anterior borders of the mouth
Chin (mentum)	Anterior portion of the lower jaw

UNIT IV

Figure 10.10 Surface structures of the oral cavity (a) with the mouth open and (b) with the mouth open and the tongue elevated.

Soft palate
Palatoglossal arch
Palatopharyngeal arch
Uvula
Palatine tonsil
Posterior wall of oral pharynx

(a)

Lingual frenulum
Opening of submandibular duct

(b)

Figure 10.11 An anterior view of the muscles of the head (m. = muscle, mm. = muscles, a. = artery).

1 Frontalis m.	7 Zygomaticus mm.
2 Supratrochlear a.	8 Facial a.
3 Corrugator m.	9 Orbicularis oris m.
4 Orbicularis oculi m.	10 Risorius m.
5 Levator labii superioris m.	11 Depressor angularis oris m.
6 Alar cartilage	12 Mentalis m.

The color of the lips and other mucous membranes of the oral cavity are diagnostic of certain body dysfunctions. The lips may appear pale in patients with severe anemia or bluish or cyanotic in people lacking in oxygen. A lemon yellow tint to the lips may indicate pernicious anemia or jaundice.

Internal Anatomy

The internal anatomy of the head from cadaver dissections is shown in figures 10.11, 10.12, and 10.13. Figure 10.13 depicts the brain in sagittal section within the cranium. Additional details of the brain are discussed and illustrated in chapter 11. The sensory organs of the head (eyes, ears, taste buds, and olfactory receptors) are discussed and illustrated in chapter 15.

1. What are the boundaries of the cranial region and why is this region clinically important?
2. Why do scalp wounds bleed so freely? How might this relate to infections?
3. What are the subdivisions of the facial region?

Neck

Major organs are located within the flexible neck, and several structures (the spinal cord, vessels, trachea, and esophagus) that are essential for body sustenance pass through the neck.

Objective 8. Discuss the functions of the neck.

Objective 9. Name and locate the triangles of the neck and list the structures contained within these triangles.

The **neck** is a complex region of the body that connects the head to the thorax. The spinal cord, digestive and respiratory tracts, and major vessels traverse this highly flexible area. Several organs and glands are located here as well. Remarkable musculature in the neck produces an array of movements. Because of this complexity, the neck is a clinically important area. The surface features provide landmarks for determining the location of internal structures.

UNIT IV

Figure 10.12 A lateral view of the deep muscles of the head (m. = muscle, a. = artery, v. = vein, n. = nerve).

1 Temporalis m.
2 Superficial temporal a.
3 Supraorbital margin
4 Temporomandibular joint
5 Orbital fat body
6 Greater alar cartilage
7 Lateral alar cartilage
8 Facial v.
9 Parotid duct
10 Buccinator m.
11 Retromandibular v.
12 Mental a., v., n.
13 Facial a.
14 Submandibular gland

UNIT IV

Figure 10.13 A sagittal section of the head and neck.

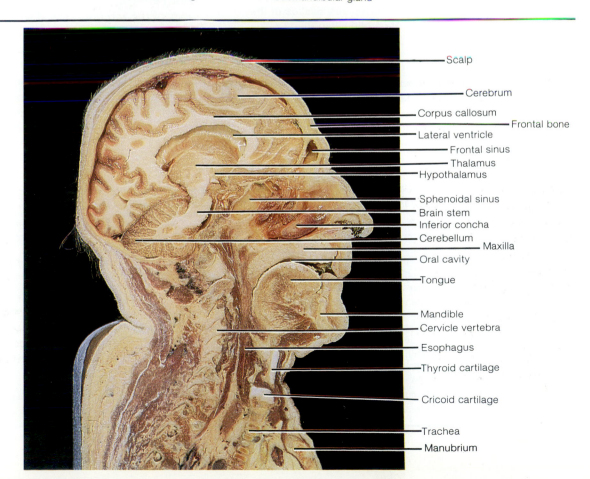

Scalp
Cerebrum
Corpus callosum
Frontal bone
Lateral ventricle
Frontal sinus
Thalamus
Hypothalamus
Sphenoidal sinus
Brain stem
Inferior concha
Cerebellum
Maxilla
Oral cavity
Tongue
Mandible
Cervicle vertebra
Esophagus
Thyroid cartilage
Cricoid cartilage
Trachea
Manubrium

Surface Anatomy

The neck is divided into four regions: (1) an *anterior region* called the **cervix** that contains the digestive and respiratory tracts, the **larynx** *(lar'ingks)* (voice box), vessels to and from the head, nerves, and the **thyroid** and **parathyroid glands;** (2) right and (3) left *lateral regions,* each composed of major neck muscles and **cervical** *(ser'vi-kal)* **lymph nodes;** and (4) a *posterior region,* referred to as the **nucha** *(noo'kă)* which includes the spinal cord, cervical vertebrae, and associated structures.

The most prominent structure of the cervix of the neck is the **thyroid cartilage** (Adam's apple) of the larynx (fig. 10.14). The thyroid cartilage supports the vocal cords and is larger in sexually mature males than in females, which accounts for the male's deeper voice. The **hyoid bone** can be palpated just above the larynx. Both of these structures are elevated during swallowing, which is one of the actions that directs food and fluid into the esophagus. Note this action on yourself by gently cupping your fingers on the larynx, then swallowing. Below the thyroid cartilage is the **cricoid** *(kri'koid)* **cartilage** followed by the **trachea** *(tra'ke-ă)* (windpipe), both of which can be palpated. The trachea is clinically important in the event that a respiratory tube must be inserted during a *tracheotomy.* The thyroid gland can be palpated on either side, just below the level of the larynx. Pulsations of the **common carotid artery** can be observed and palpated on either side of the neck, just lateral and a bit superior to the level of the larynx.

> The arteries of the head and neck are rarely damaged because of their elasticity. In a severe lateral blow to the head, however, the internal carotid artery may rupture, resulting in the perception of a roaring sound as blood rushes into the cavernous sinuses of the temporal bone. Containment of carotid hemorrhage within the sinuses may actually be lifesaving.

The **jugular notch** is the depression in the midline of the cervix, just superior to the sternum. The two **clavicles** are obvious in all people because they are subcutaneous (just under the skin).

cervix: L. *cervix,* neck
larynx: Gk. *larynx,* upper windpipe
hyoid: Gk. *hyoeides,* U-shaped

The **sternocleidomastoid** and **trapezius muscles** are the prominent structures of each lateral region (figs. 10.14 and 10.15). The sternocleidomastoid muscle can be palpated its entire length when the head is turned to the side. The tendon of this muscle is especially prominent to the side of the jugular notch. The trapezius muscle can be felt when the shoulders are shrugged. An inflammation of the trapezius muscle causes a "stiff neck." If a person is angry or if a shirt collar is too tight, the **external jugular vein** can be seen as it courses obliquely across the sternocleidomastoid muscle. **Cervical lymph nodes** of the lateral neck region may become swollen and painful from infectious diseases of the oral or pharyngeal regions.

Most of the structures of the nucha are too deep to be of importance in surface anatomy. The spines of the lower cervical vertebrae (especially C7), however, can be observed and palpated when the neck is flexed. In this same position, a firm ridge called the **ligamentum nuchae** (not shown) is raised, extending superiorly from C7 to the external occipital protuberance of the skull. Clinically, the nucha is extremely important because of the debilitating damage it sustains from whiplash injury or a broken neck.

Triangles of the Neck   The triangles of the neck, created by the arrangement of specific muscles and bones, are clinically important because of the specific structures included in each. The structures of the neck that are important in surface anatomy have already been identified, however, so only a summary of the two major and six minor triangles will be depicted in figure 10.15 and presented in table 10.5. The sternocleidomastoid muscle obliquely transects the neck, dividing it into an **anterior cervical triangle** and a **posterior cervical triangle.** The apex of the anterior cervical triangle is directed inferiorly. The median line of the neck forms the anterior boundary of the anterior cervical triangle, the inferior border of the mandible forms its superior boundary, and the sternocleidomastoid muscle forms its posterior boundary. The posterior cervical triangle is formed by the sternocleidomastoid muscle anteriorly and the trapezius muscle posteriorly, and the clavicle forms its base inferiorly.

Table 10.5	Location and Contents of the Triangles of the Neck	
Triangle	**Boundaries**	**Contents**
Anterior cervical	Sternocleidomastoid muscle, median line of neck, inferior border of mandible	Four lesser triangles, salivary glands, larynx, trachea, thyroid glands, various vessels and nerves
Carotid	Sternocleidomastoid, posterior digastric, and omohyoid muscles	Carotid arteries, internal jugular vein, vagus nerve
Submandibular	Digastric muscle (both heads), inferior border of mandible	Salivary glands
Submental	Digastric muscle, hyoid bone (This is the only unpaired triangle of the neck.)	Muscles of the floor of the mouth, salivary glands and ducts
Omotracheal	Sternocleidomastoid and omohyoid muscles, midline of neck	Larynx, trachea, thyroid gland, carotid sheath
Posterior cervical	Sternocleidomastoid and trapezius muscles, clavicle	Nerves and vessels
Occipital	Sternocleidomastoid, trapezius, and omohyoid muscles	Cervical plexus, accessory nerve
Omoclavicular	Sternocleidomastoid and omohyoid muscles; clavicle	Brachial plexus, subclavian artery

UNIT IV

Figure 10.14   An anterolateral view of the neck.

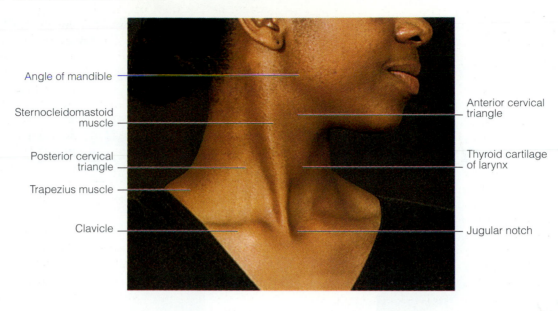

Angle of mandible

Sternocleidomastoid muscle

Posterior cervical triangle

Trapezius muscle

Clavicle

Anterior cervical triangle

Thyroid cartilage of larynx

Jugular notch

Figure 10.15   Triangles of the neck. (*a*) The two large triangular divisions, (*b*) the six lesser triangular subdivisions, and (*c*) the detailed muscular anatomy of the neck.

Anterior cervical triangle

Posterior cervical triangle

(a)

Occipital

Submandibular

Submental

Carotid

Omotracheal

Omoclavicular

(b)

Mastoid process of temporal bone

Splenius capitis

Levator scapulae

Trapezius

Scalenus

Omohyoid

Clavicle

Mandible

Mylohyoid

Digastric

Thyrohyoid

Omohyoid

Sternocleidomastoid

Sternohyoid

Pectoralis major

(c)

UNIT IV

Figure 10.16 An anterior view of the right cervical region.

1 Accessory nerve	9 Digastric m.
2 Trapezius m.	10 Submandibular gland
3 Supraclavicular nerve	11 Hyoid bone
4 Omohyoid m.	12 Omohyoid m.
5 Brachial plexus	13 Transverse cervical nerve
6 Clavicle	14 Sternohyoid m.
7 Facial artery	15 Sternocleidomastoid m.
8 Mylohyoid m.	16 External jugular vein

Three structures traversing the neck are extremely important and potentially vulnerable. These structures are the common carotid artery, which carries blood to the head; the internal jugular vein, which drains blood from the head; and the vagus nerve, which conducts nerve impulses to visceral organs. These structures are protected in the neck by their deep position behind the sternocleidomastoid muscle and by being enclosed in a tough connective tissue sheath called the *carotid sheath.*

Internal Anatomy

The internal anatomy of the neck from cadaver dissections is shown in figures 10.16 and 10.17. The organs of the neck are highly integrated and compacted into a relatively small area. The neck has to provide support to the head while permitting flexibility.

1. List four functions of the neck and state which body systems are located, in part, within the neck.
2. What are the structural regions of the neck? What structures are included in each region?
3. Using your knowledge of the triangles of the neck, describe where to palpate to feel a pulse, the trachea, cervical lymph nodes, and the thyroid gland.

Figure 10.17 A posterior view of the deep cervical muscles.

1 Occipital bone	6 Longissimus cervicis m.
2 Greater occipital nerve	7 Serratus posterior m.
3 Ligamentum nuchae	8 Occipital artery
4 Semispinalis capitis m.	9 Levator scapulae m.
5 Longissimus capitis m.	

Trunk

The locations of vital visceral organs in the cavities of the trunk make the surface anatomy of this body region especially important.

Objective 10. Identify various surface features of the trunk by observation or palpation.

Objective 11. List the auscultation sites of the thorax and abdomen.

Surface Anatomy

The **trunk,** or *torso,* is divided into the **back, thorax** (chest), **abdomen** (venter), and **pelvis.** A region called the **perineum** (*per''i-ne'um*) forms the floor of the pelvis and includes the external genitalia. The surface anatomy of these regions is particularly important in determining the location and condition of the visceral organs. The usability of some of these surface features varies with age, sex, and body weight.

Back No matter how obese a person is, a **median furrow** can be seen on the back, along with some of the spinous process of the vertebrae (fig. 10.18). The entire series of vertebral spines can be observed if the vertebral column is flexed. This position is important in determining vertebral-column defects (see clinical comments in chapters 8 and 11). The back of the **scapula**

Figure 10.18 The surface anatomy of the back. (*a*) Adduction of the limbs at the shoulder joints and (*b*) abduction of the shoulder joints and flexion of the elbow joints.

Spinous process of seventh cervical vertebra

Acromion of scapula

Triangle of auscultation

Trapezius muscle

Teres major muscle

Inferior angle of scapula

Median furrow over spinous processes

Latissimus dorsi muscle

Erector spinae muscle

(a)

Anterior head

Medial head

Posterior head

Deltoid muscle

Trapezius muscle

Infraspinatus muscle

Triangle of auscultation

Inferior angle of scapula

Latissimus dorsi muscle

Erector spinae muscle

(b)

UNIT IV

Figure 10.19   An anterolateral view of the trunk and axilla.

Deltoid muscle
Axilla
Pectoralis major muscle
Latissimus dorsi muscle
Nipple
Serratus anterior muscle
External intercostal muscle
Costal arch
Tendinous inscription
Rectus abdominis muscle

presents other important surface landmarks. The base of the spine of the scapula is level with the third thoracic vertebra, and the inferior angle of the scapula is even with the seventh thoracic vertebra. Several muscles of the scapula can be observed on a lean, muscular person and are identified in figure 10.18. Many of the ribs and muscles that attach to the ribs can be seen in a lateral view (fig. 10.19).

Two pairs of triangles on the back are clinically important. The **triangle of auscultation** (fig. 10.18) is bounded by the trapezius muscle, the latissimus dorsi muscle, and the medial border of the scapula. Because there is a space between the superficial back muscles in this area, heart and respiratory sounds are not muffled by the muscles when a stethoscope is placed here.

Thorax   The leading causes of death in the United States are associated with disease or dysfunction of the thoracic organs. With the exception of the breasts and surrounding lymph nodes, the organs of the thorax are within the rib cage. Because of the location of the thoracic visceral organs and their clinical importance, the surface anatomy of the thorax is extremely important.

The flexible rib cage presents several bony landmarks that can be observed or palpated (fig. 10.20). The paired clavicles and the jugular notch have already been identified as being important surface features of the neck. These structures are also important in the thoracic region as reference points for counting the ribs. Many of the ribs can be seen on a thin person. All of the ribs except the first, and at times the twelfth, can be palpated. The sternum is composed of three separate bones (manubrium, body, xiphoid process), each of which can be palpated. The **sternal angle** is felt as an elevation between the manubrium and body of the sternum. The sternal angle is important

Figure 10.20   The surface anatomy of the anterior thoracic region of the male.

Jugular notch
Supraclavicular fossa
Acromion
Clavopectoral triangle
Deltoid muscle
Nipple
Rib
Trapezius muscle
Clavicle
Body of sternum
Axilla
Xiphoid process
Linea alba

because it is located at the level of the second rib. The articulation between the body of the sternum and the xiphoid process, called the **xiphisternal** *(zif"i-ster'nal)* **joint,** is positioned over the lower border of the heart and the diaphragm. The **costal arch** of the rib cage is the lower oblique boundary and can be easily identified when a person inhales and holds his or her breath.

The nipples in the male are located at the fourth intercostal spaces (the area between the fourth and fifth ribs) and about 10 cm (4 in.) from the midline. They vary in position in sexually mature women according to age, size, and the pendulousness of the breasts (fig. 10.21). The position of the left nipple in males is an important landmark for knowing where to listen to various heart sounds and for determining whether the heart is enlarged. For diagnostic purposes, an imaginary line, the **midclavicular line,** can be extended vertically from the middle of the clavicle through the nipple. Several superficial chest muscles can be observed or palpated and are therefore important surface features. These muscles and the structures described above are depicted in figures 10.20 and 10.21.

In addition to helping one know where to listen with a stethoscope to heart sounds, surface features of the thorax are important for auscultations of the lungs, radiographs, tissue biopsies, sternal taps for bone marrow studies, and thoracic surgery. Although the ana-

tomical features of the rib cage are quite consistent, slight deformities and asymmetries do occur. These generally cause no disability and require no treatment. Most of the abnormalities are congenital and include such conditions as a projecting sternum (pigeon breast) or a receding sternum (funnel chest).

Abdomen The abdomen is the portion of the body between the diaphragm and the pelvis. The abdomen does not have a bony framework as does the thorax, so the surface anatomical features are not as well defined. Bony landmarks of both the thorax and pelvis are used when referring to abdominal structures (fig. 10.22). The **right costal arch** of the rib cage is located over the liver on the right side and the **left costal arch** is positioned over the stomach on the left. The xiphoid process is important because from this point a tendinous, midventral raphe, the **linea alba,** extends the length of the abdomen to attach to the **symphysis pubis.** The symphysis pubis can be palpated at the anterior union of the two halves of the pelvic girdle. The **navel,** or **umbilicus,** is the former site of attachment of the fetal umbilical cord and is located along the linea

linea alba: L. *linea*, line; *alba*, white
navel: O. E. *nafela*, umbilicus
umbilicus: L. *umbilicus*, navel

Figure 10.21 The surface anatomy of the female breast.

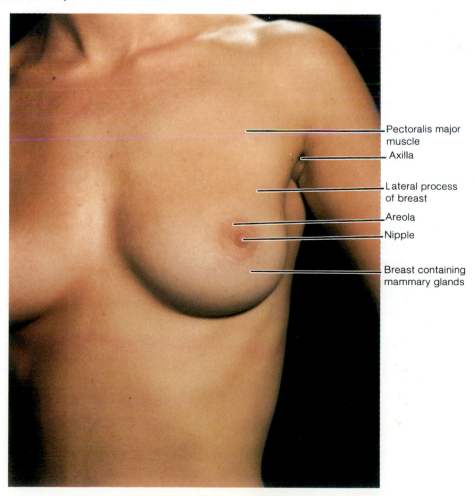

Pectoralis major muscle

Axilla

Lateral process of breast

Areola

Nipple

Breast containing mammary glands

Figure 10.22 The surface anatomy of the anterior abdominal region.

Serratus anterior muscle

Rectus abdominis muscle

Linea semilunaris

McBurney's point

Anterior superior iliac spine

External abdominal oblique muscle

Groove over inguinal ligament

Xiphoid process

Linea alba

Tendinous inscription across rectus abdominis muscle

Umbilicus

alba. The linea alba separates the paired, straplike **rectus abdominis muscles,** which can be seen when a person flexes the abdomen (as when doing sit-ups).

Clinically, the linea alba is a favored site for abdominal surgery because an incision made along this line severs no muscles and few vessels or nerves. Moreover, the linea alba heals readily. It has been said that only a zipper would provide a more convenient entry to the abdominal cavity.

The lateral margin of the rectus abdominis muscle can be observed on some persons, and the surface line it produces is called the **linea semilunaris.** The **external abdominal oblique muscle** is the superficial layer of the muscular abdominal wall. The **iliac crest** is subcutaneous and can be palpated along its entire length. The highest point of the crest lies opposite the body of the fourth lumbar vertebra, an important level in spinal anesthesia. Another important landmark is **McBurney's point,** located one-third of the distance from the right anterior-superior iliac crest to the navel (see fig. 10.22). This point overlies the appendix of the intestinal tract.

The abdominal region is frequently divided into nine regions or four quadrants in order to describe the location of internal organs and to clinically identify the sites of various pains or conditions. These regions have been adequately described in chapter 2 (see figs. 2.15 and 2.16).

Although the position of the umbilicus is relatively consistent in all people, its shape and health is not. Embryological remains of the umbilicus may cause clinical problems. For example, there may be an opening to the outside, called a *fistula,* or herniation of some of the abdominal contents. Acquired umbilical hernias may develop in children who have a weak abdominal wall in this area, or they may develop in pregnant women because of the extra pressure exerted at this time. The umbilicus is a common site for an incision into the abdominal cavity in a procedure called a *laparotomy.* A laparotomy is frequently done to examine or perform surgery on the internal female reproductive organs. A depressed umbilicus on an obese person is difficult to keep clean, and so various types of infections may occur there.

Internal Anatomy

Thorax Included in the internal anatomy of the thorax (figs. 10.23 through 10.29) is the rib cage and its contents, the thoracic musculature, and the mammary glands and breasts of a female. The rib cage is formed by the sternum, the costal cartilages, and the ribs attached to the thoracic vertebrae. It gives protection to the lungs, several large vessels, and the heart. It also affords a site of attachment for the muscles of the thorax, upper extremities, back, and diaphragm. The principal organs

McBurney's point: from Charles McBurney, U.S. surgeon, 1845–1914

laparotomy: Gk. *lapara,* flank; *tome,* incision

UNIT IV

Figure 10.23 An anterior view of the right thorax, shoulder, and brachium.

Figure 10.24 An anterior view of the deep muscles of the right thorax, shoulder, and brachium.

1 Deltoid m.	6 Pectoralis major m.
2 Cephalic vein	7 Serratus anterior m.
3 Latissimus dorsi m.	8 External abdominal oblique m.
4 Biceps brachii m.	9 Rectus sheath
5 Brachioradialis m.	

1 Deltoid m.	7 External abdominal oblique m.
2 Coracobrachialis m.	8 Subclavius m.
3 Cephalic vein	9 Brachial plexus
4 Biceps brachii m.:	10 Internal intercostal m.
4a Short head	11 Pectoralis minor m.
4b Long head	12 External intercostal m.
5 Latissimus dorsi m.	
6 Serratus anterior m.	

of the respiratory and circulatory systems are located within the thorax, and the esophagus of the digestive system passes through the thorax. Because the viscera of the thoracic cavity are vital organs, the thorax is of immense clinical importance.

Abdomen The abdomen is that portion of the trunk positioned between the diaphragm and the pelvis. Because of the domed shape of the diaphragm, some of the abdominal viscera lie within the protection of the rib cage. The cavity of the abdomen contains the stomach and intestines, the liver and gallbladder, the kidneys and adrenal glands, the spleen, the internal genitalia, and major vessels and nerves. The abdominal region is shown in photographs of cadavers in figures 10.30, 10.31, and 10.32.

1. Which structures of the torso can be readily observed? Which can be palpated?
2. Where are the common auscultation sites of the trunk located?
3. Why are the linea alba, costal margins, linea semilunaris, and McBurney's point important landmarks?

Pelvis and Perineum

The surface features of the pelvic region are important primarily to identify reproductive organs and clinical problems of these organs.

Objective 12. Describe the location of the perineum and list the organs of the pelvic and perineal regions.

The important bony structures of the **pelvis** include the crest of the ilium and the symphysis pubis, located anteriorly, and the ischium and os coccyx, which are palpable posteriorly. An **inguinal** *(ing'wi-nal)* **ligament** extends from the crest of the ilium to the symphysis pubis and is clinically important because hernias occur along it. Although the inguinal ligament cannot be seen, an oblique groove overlying the ligament is an apparent surface feature.

The **perineum** *(per''i-ne'um)* (see fig. 2.17) is the region that contains the external sex organs and the anal opening. The surface features of this region are further discussed in chapters 20 and 21. The surface anatomy of the perineum of a female becomes particularly important during parturition.

1. Define the term *perineum*. What structures are located within the perineum?
2. List three body systems that have openings within the pelvic region.

inguinal: L. *inguinalis*, groin

UNIT IV

Figure 10.25 Viscera of the thorax. The heart has been coronally sectioned to expose the chambers.

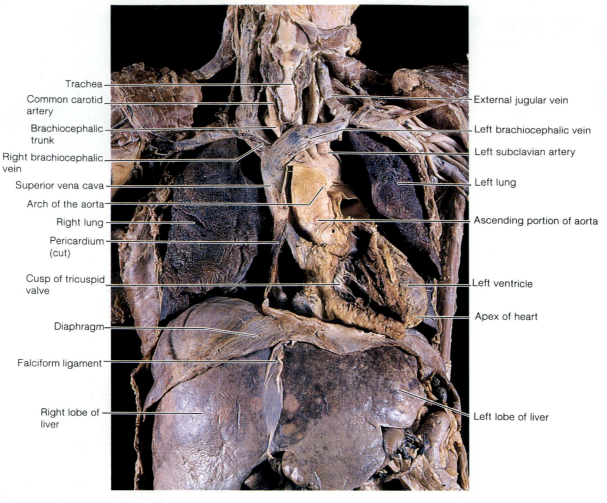

Trachea

Common carotid artery

Brachiocephalic trunk

Right brachiocephalic vein

Superior vena cava

Arch of the aorta

Right lung

Pericardium (cut)

Cusp of tricuspid valve

Diaphragm

Falciform ligament

Right lobe of liver

External jugular vein

Left brachiocephalic vein

Left subclavian artery

Left lung

Ascending portion of aorta

Left ventricle

Apex of heart

Left lobe of liver

UNIT IV

Figure 10.26
The thoracic cavity with the heart and lungs removed.

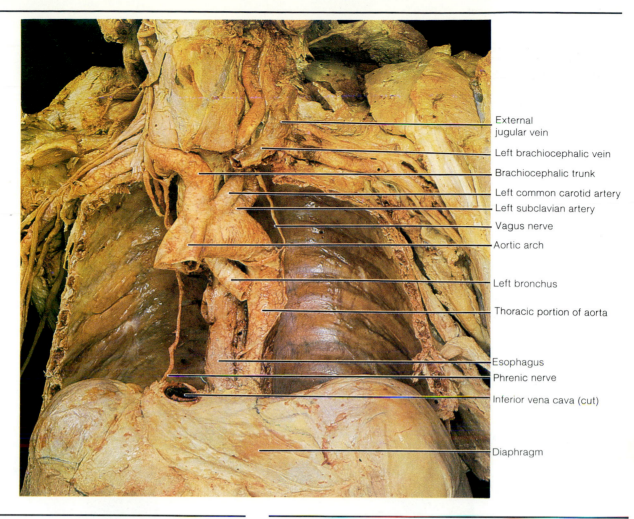

External jugular vein

Left brachiocephalic vein

Brachiocephalic trunk

Left common carotid artery

Left subclavian artery

Vagus nerve

Aortic arch

Left bronchus

Thoracic portion of aorta

Esophagus

Phrenic nerve

Inferior vena cava (cut)

Diaphragm

UNIT IV

Figure 10.27 A posterior view of the right thorax and neck.

1 External occipital protuberance
2 Trapezius m.
3 Triangle of auscultation
4 Occipital artery
5 Greater occipital n.
6 Lesser occipital nerve
7 Sternocleidomastoid m.
8 Great auricular nerve
9 Infraspinatus m.
10 Rhomboideus major m.
11 Latissimus dorsi m.

Figure 10.28 A posterior view of the deep structures of the right thorax and neck.

1 External occipital protuberance
2 Splenius capitis m.
3 Rhomboideus minor m.
4 Rhomboideus major m.
5 Occipital artery
6 Great occipital nerve
7 Sternocleidomastoid m. (cut)
8 Levator scapulae m.
9 Supraspinatus m.
10 Spine of scapula
11 Infraspinatus m.
12 Latissimus dorsi m.

Figure 10.29 A posterior view of the trunk with deep muscles exposed on the left.

Trapezius muscle (cut)

Rhomboideus major muscle

Infraspinatus muscle

Longissimus thoracis muscle

Deltoid muscle

Supraspinatus muscle

Trapezius muscle

Teres minor muscle

Infraspinatus muscle

Teres major muscle

Triceps brachii muscle

Latissimus dorsi muscle

External abdominal oblique muscle

Gluteus maximus muscle

UNIT IV

Figure 10.30 An anterior view of the structures of the abdominal wall.

1 Rectus abdominis m.
2 Rectus sheath
3 Umbilicus
4 Linea alba
5 Pyramidalis m.

6 Transverse abdominis m.
7 Inferior epigastric artery
8 Inguinal ligament
9 Spermatic cord

UNIT IV

Figure 10.31 Viscera of the abdomen.

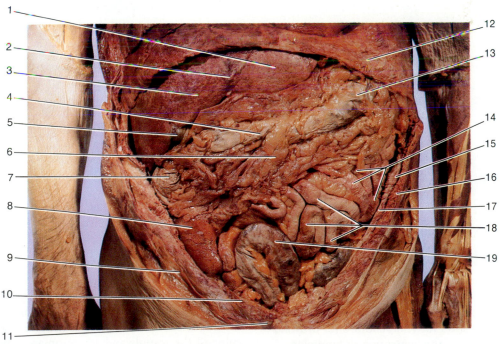

1 Left lobe of liver
2 Falciform ligament
3 Right lobe of liver
4 Transverse colon
5 Gallbladder
6 Greater omentum
7 Hepatic flexure of colon

8 Fat deposit within greater omentum
9 Aponeurosis of internal abdominal oblique m.
10 Rectus abdominis m. (cut)
11 Rectus sheath (cut)
12 Diaphragm

13 Splenic flexure of colon
14 Jejunum
15 Transversus abdominis m. (cut)
16 Internal and external abdominal oblique mm. (cut)
17 Parietal peritoneum (cut)
18 Ileum
19 Sigmoid colon

Figure 10.32 Viscera of the abdomen with the greater omentum removed and the small intestine displaced to the left.

UNIT IV

Diaphragm

Liver

Transverse colon

Superior mesenteric artery

Superior mesenteric vein

Ascending colon

Mesentery

Small intestine

Sigmoid colon

Shoulder and Upper Extremity

The anatomy of the shoulder and upper extremity is of clinical importance because of frequent trauma to these body regions. In addition, vessels of the upper extremity are used as pressure points and as sites for venipuncture for drawing blood, intravenous feeding, or administration of medicine.

Objective 13. Identify various surface features of the shoulder and upper extremity by observation or palpation.

Objective 14. Discuss the clinical importance of the axilla, cubital fossa, and wrist.

Surface Anatomy

Shoulder The scapula, clavicle, and proximal portion of the humerus form the shoulder, and portions of each of these bones are important surface landmarks in this region. Posteriorly, the spine of the scapula and acromion are subcutaneous and easily located.

The acromion and clavicle, as well as several large shoulder muscles, can be seen anteriorly (fig. 10.33). The rounded curve of the shoulder is formed by the deltoid muscle covering the

acromion: Gk. *akros*, extreme, tip; *omion*, small shoulder greater tuberosity of the humerus. The deltoid muscle frequently serves as a site for intramuscular injections. The large pectoralis major muscle is prominent as it crosses the shoulder joint and attaches to the humerus. A small depression, the **clavipectoral triangle** (fig. 10.33), is situated below the outer third of the clavicle and is bounded on either side by the deltoid and pectoralis major muscles.

Axilla The axilla is commonly called the armpit. This depressed region of the shoulder supports axillary hair in sexually mature individuals. The axilla is clinically important because of the subcutaneous position of vessels, nerves, and lymph nodes in this region. Two muscles form the anterior and posterior borders (fig. 10.34). The anterior axillary fold is formed by the pectoralis major muscle and the posterior axillary fold consists primarily of the latissimus dorsi muscle extending from the lumbar vertebrae to the humerus. Axillary lymph nodes are palpable in some individuals.

> In sexually mature females, the lateral process of the mammary gland, which is positioned on the pectoralis major muscle (see figs. 9.22 and 10.21), extends partially into the axilla. In doing a *breast self-examination* (see fig. 21.22), a woman should palpate the axillary area as well as the entire breast because the lymphatic drainage pathway is toward the axilla.

Brachium Several muscles are clearly visible in the brachium (figs. 10.34 and 10.35). The belly of the biceps brachii muscle becomes prominent when the elbow is flexed. While the arm is in this position, the deltoid muscle can be traced as it inserts upon the humerus. The triceps brachii muscle forms the bulk of the posterior surface of the brachium. A groove forms on the medial side of the brachium between the biceps brachii and triceps brachii muscles, where pulsations of the brachial artery may be felt as it carries blood toward the forearm. This region is clinically important because it is where arterial blood pressure is taken with a sphygmomanometer. It is also the place to apply pressure in case of severe arterial hemorrhage in the forearm or hand.

Three bony prominences can be located in the region of the elbow (fig. 10.36). The medial and lateral epicondyles are processes on the humerus, whereas the olecranon is a proximal process of the ulna. When the elbow is extended, these prominences lie on the same straight plane; when the elbow is flexed, these points form a triangle. The ulnar nerve can be palpated in the groove behind the medial epicondyle. This area is commonly known as the funny bone, or crazy bone.

Figure 10.33 An anterior view of the shoulder region.

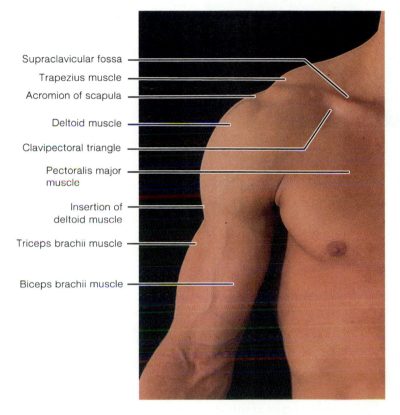

Supraclavicular fossa
Trapezius muscle
Acromion of scapula
Deltoid muscle
Clavipectoral triangle
Pectoralis major muscle
Insertion of deltoid muscle
Triceps brachii muscle
Biceps brachii muscle

Figure 10.34 An anterior view of the upper extremity.

Deltoid muscle
Brachioradialis muscle
Biceps brachii muscle
Olecranon of ulna
Medial epicondyle of humerus
Sulcus of ulnar nerve
Triceps brachii muscle

Pectoralis major muscle (anterior axillary fold)
Axilla
Latissimus dorsi muscle (posterior axillary fold)

UNIT IV

Figure 10.35 A lateral view of the upper extremity.

- Acromion of scapula
- Deltoid muscle
- Biceps brachii muscle
- Brachioradialis muscle
- Extensor carpi radialis muscle
- Long head of triceps brachii muscle
- Lateral head of triceps brachii muscle
- Lateral epicondyle of humerus
- Olecranon of ulna

Figure 10.36 A posterior view of the elbow.

- Triceps brachii muscle:
  - Lateral head
  - Long head
  - Medial head
- Lateral epicondyle of humerus
- Brachioradialis muscle
- Olecranon of ulna
- Medial epicondyle of humerus
- Extensor carpi radialis longus muscle
- Extensor digitorum muscle
- Extensor carpi ulnaris muscle
- Basilic vein

The **cubital fossa** is the triangular depression on the anterior surface of the elbow region where the median cubital vein links the cephalic and basilic veins. These veins are subcutaneous and become more conspicuous when a proximal compression is applied. For this reason, these veins (particularly the median cubital) are an important location for the removal of venous blood for analyses and transfusions or for intravenous therapy (fig. 10.37).

Antebrachium Contained within the antebrachium (forearm) are two parallel bones (the ulna and radius) and the muscles that control the movements of the hand. The muscles of the forearm taper distally over the wrist, where their tendons attach to various bones of the hand. Several muscles of the forearm can be identified as surface features and are depicted in figures 10.37 and 10.38.

Because of the frequency of fractures involving the forearm, bony landmarks are clinically important when setting broken bones. The ulna can be palpated its entire length from the olecranon to the styloid process. The distal half of the radius is palpable as the forearm is rotated, and its styloid process can be located.

Nerves, tendons, and vessels are close to the surface at the wrist, making cuts to this area potentially dangerous. Tendons from four flexor muscles can be observed as surface features if the anterior forearm muscles are strongly contracted while making a fist. The tendons that can be observed along this surface, from lateral to medial, are from the following muscles: flexor carpi radialis, palmaris longus, superficial digital flexor, and flexor carpi ulnaris. The median nerve going to the hand is located under the tendon of the palmaris longus muscle (see fig. 10.37), and the ulnar nerve is lateral to the tendon of the flexor carpi ulnaris muscle. The radial artery lies along the surface of the radius, immediately lateral to the tendon of the flexor carpi radialis muscle. This is the artery commonly used when monitoring the pulse. By careful palpation, pulsations can also be detected in the ulnar artery, lateral to the tendon of the flexor carpi ulnaris.

Two tendons that attach to the thumb can be seen on the posterior surface of the wrist as the thumb is extended backward. The tendon of the extensor pollicis brevis muscle is positioned anterolaterally along the thumb, and the tendon of the

Figure 10.37   An anterior view of the forearm and hand.

Site for palpation of
brachial artery

Cephalic vein

Basilic vein

Cubital fossa

Median cubital vein

Brachioradialis muscle

Median vein of forearm

Ulnar vein

Radial vein

Tendon of palmaris longus muscle

Tendon of flexor carpi radialis longus muscle

Tendon of superficial digital flexor muscle

Site for palpation of radial artery

Styloid process of ulna

Thenar eminence

Hypothenar eminence

Figure 10.38   A posterior view of the forearm and hand.

Lateral head of triceps brachii muscle

Long head of triceps brachii muscle

Brachioradialis muscle

Extensor carpi radialis muscle

Olecranon of ulna

Anconeus muscle

Extensor digitorum muscle

Extensor carpi ulnaris muscle

Abductor pollicus longus muscle

Styloid process of radius

First dorsal interosseous
muscle

extensor pollicis longus muscle lies posteromedially (fig. 10.39). The depression created between these two tendons as they are pulled taut is referred to as the anatomical snuffbox. Pulsations of the radial artery can be detected in this depression.

> The median nerve, which serves the thumb, is the nerve most commonly injured by stab wounds or the penetration of glass into the wrist or hand. If this nerve is severed, the adductor muscles of the thumb are paralyzed and waste away, resulting in an inability to oppose the thumb in grasping.

Hand   Much of the surface anatomy of the hand, such as flexion creases, fingerprints, and fingernails, includes features of the skin that are discussed in chapter 5. Other surface features are the extensor tendons from the extensor digitorum muscle, which can be seen going to each of the fingers on the back side of the hand as the hand is extended (fig. 10.40). The knuckles of the hand are the distal ends of the second through the fifth metacarpal bones. Each of the joints of the fingers and the individual phalanges can be palpated. The **thenar** *(the′nar)* **eminence** is the thickened, muscular portion of the hand forming the base of the thumb (fig. 10.41).

Figure 10.39   The anatomical snuffbox.

Tendon of extensor
pollicus brevis
muscle

Styloid process
of ulna

Anatomical
snuff box

Tendon of extensor
pollicus longus
muscle

Tendon of extensor
digiti minimi
muscle

Tendons of extensor
digitorum muscle

Figure 10.40 A posterior view of the hand.

Nail

Distal interphalangeal joints

Proximal interphalangeal joints

Metacarpophalangeal joints

Tendons of extensor digitorum muscle

Tendon of extensor pollicus longus muscle

Styloid process of radius

Figure 10.41 An anterior view of the wrist and hand (*a*) with the hand open and (*b*) with a clenched fist.

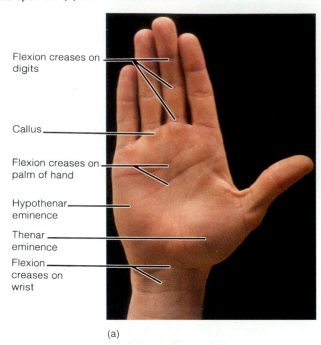

Flexion creases on digits

Callus

Flexion creases on palm of hand

Hypothenar eminence

Thenar eminence

Flexion creases on wrist

(a)

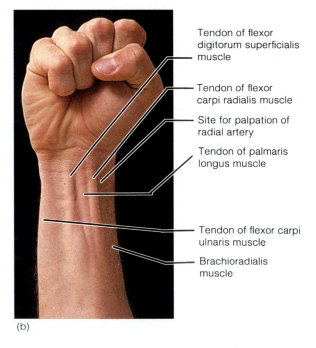

Tendon of flexor digitorum superficialis muscle

Tendon of flexor carpi radialis muscle

Site for palpation of radial artery

Tendon of palmaris longus muscle

Tendon of flexor carpi ulnaris muscle

Brachioradialis muscle

(b)

Internal Anatomy

The internal anatomy of the shoulder and upper extremity includes the structures of the shoulder, brachium, cubitus (elbow) antebrachium, and the hand. The principal structures of these regions are shown in the cadaver dissections in figures 10.42 through 10.46.

1. List the clinically important structures that can be observed or palpated in the shoulder and upper extremity.
2. Describe the locations of the axilla, brachium, cubital fossa, and wrist.
3. Bumping the ulnar nerve causes a tingling sensation along the medial part of the forearm to the little finger of the hand. What does this tell you about its distribution?
4. Which of the two bones of the forearm is the more stationary as the arm is rotated?

Figure 10.42   A posterior view of the right shoulder and brachium.

1 Trapezius m.
2 Infraspinatus m.
3 Rhomboideus major m.
4 Triangle of auscultation
5 Latissimus dorsi m.
6 External abdominal oblique m.

7 Deltoid m.
8 Teres minor m.
9 Teres major m.
10 Lateral head of triceps brachii m.
11 Long head of triceps brachii m.

Figure 10.43   A posterior view of the deep muslces of the right shoulder and brachium.

1 Supraspinatus m.
2 Spine of scapula
3 Infraspinatus m.
4 Teres minor m.
5 Teres major m.
6 Latissimus dorsi m.
7 External abdominal oblique m.

8 Deltoid m.
9 Axillary nerve
10 Radial nerve
11 Triceps brachii m.:
    11a long head
    11b medial head
    11c lateral head

UNIT IV

Figure 10.44   An anterolateral view of the head, neck, and thorax.

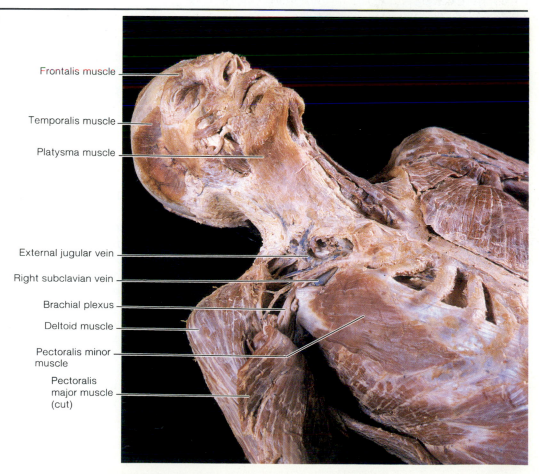

Frontalis muscle

Temporalis muscle

Platysma muscle

External jugular vein

Right subclavian vein

Brachial plexus

Deltoid muscle

Pectoralis minor muscle

Pectoralis major muscle (cut)

Figure 10.45 A posterior view of the left forearm and hand.

1 Brachioradialis m.
2 Extensor carpi radialis longus tendon
3 Extensor carpi radialis brevis m.
4 Extensor digitorum communis m.
5 Abductor pollicis longus m.
6 Extensor pollicis brevis m.
7 Extensor pollicis longus m.
8 Radius
9 Extensor retinaculum
10 Tendon of extensor carpi radialis longus m.
11 Tendon of extensor pollicis longus m.
12 Tendon of extensor pollicis brevis m.
13 First dorsal interosseous m.
14 Extensor carpi ulnaris m.
15 Extensor digiti minimi m.
16 Ulna
17 Tendon of extensor carpi radialis brevis m.
18 Tendon of extensor indicis m.
19 Tendon of extensor digiti minimi m.
20 Tendons of extensor digitorum m.
21 Intertendinous connections

Figure 10.46 An anterior view of the left forearm and hand.

1 Flexor carpi ulnaris m.
2 Extensor carpi ulnaris m.
3 Flexor digitorum profundis m.
4 Pisiform bone
5 Abductor digiti minimi m.
6 Flexor digiti minimi m.
7 Opponens digiti minimi m.
8 Lumbrical m.
9 Tendon of superficial digital flexor m.
10 Tendon of deep digital flexor m.
11 Fibrous digital sheath
12 Tendon of palmaris longus m.
13 Tendon of flexor carpi radialis m.
14 Pronator quadratus m.
15 Tendon of extensor pollicis brevis m.
16 Tendon of extensor pollicis longus m.
17 Abductor pollicis brevis m.
18 Flexor pollicis brevis m.
19 Adductor pollicis m. (oblique head)
20 Adductor pollicis m. (transverse head)
21 Opponens pollicis m.

Buttock and Lower Extremity

The massive bones and muscles of the buttock and lower extremity are important as weight bearers and locomotors. Many of the surface features of these regions are important with respect to locomotion or locomotor dysfunction.

Objective 15. Identify various surface features of the buttock and lower extremity by observation or palpation.

Objective 16. Discuss the clinical importance of the buttock, femoral triangle, popliteal fossa, ankle, and arches of the foot.

Surface Anatomy

Buttock The superior borders of the buttocks, or *gluteal region,* are formed by the iliac crests (fig. 10.47). Each crest can be palpated medially to the level of the second sacral vertebra. From this point, the **natal cleft** extends vertically to separate the buttocks into two prominences, each formed by pads of fat and by the massive gluteal muscles. An ischial tuberosity can be palpated in the lower portion of each buttock. In a sitting position, the ischial tuberosities support the weight of the body. When standing, these processes are covered by the gluteal muscles. The sciatic nerve, which is the major nerve to the lower extremity, lies under the gluteus maximus muscle. The inferior border of the gluteus maximus muscle forms the **fold of the buttock.**

buttock: O. E. *buttuc,* end or rump

UNIT IV

Figure 10.47 The buttocks and the posterior aspect of the thigh. (Note the relation of the angle of the elbow joint to the pelvic region, which is characteristic of females.)

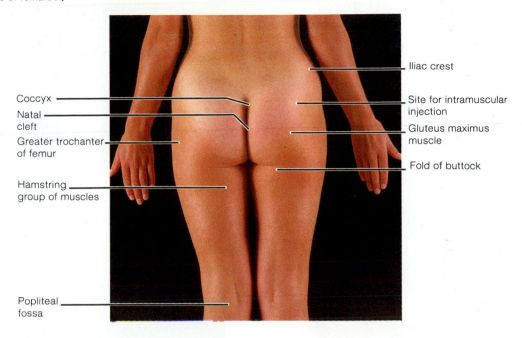

Coccyx

Natal cleft

Greater trochanter of femur

Hamstring group of muscles

Popliteal fossa

Iliac crest

Site for intramuscular injection

Gluteus maximus muscle

Fold of buttock

Because of the thickness of the gluteal muscles and the rich blood supply, the buttock is a preferred site for intramuscular injections. Care must be taken, however, not to inject into the sciatic nerve. For this reason, the surface landmark of the iliac crest is important. The injection is usually administered 5–7 cm (2–3 in.) below the iliac crest in what is known as the upper lateral quadrant of the buttock.

Thigh   The femur is the only bone of the thigh, but there are three groups of thigh muscles. The anterior group of muscles, referred to as the **quadriceps femoris muscles,** extend the leg when they are contracted (fig. 10.48). The medial muscles are the adductors, and when contracted they draw the thigh medially. The *"hamstrings"* are positioned on the posterior aspect of the thigh (see fig. 10.47) and serve to extend the hip joint, as well as flex the knee joint, when they are contracted. The tendinous attachments of the hamstrings can be palpated along the posterior aspect of the knee joint when it is flexed. The hamstrings or their attachments are often injured in athletic competition.

The **femoral triangle** is an extremely important part of the surface anatomy of the thigh. It can be seen as a depression inferior to the location of the inguinal ligament on the anterior surface in the upper part of the thigh (see fig. 16.30). The major vessels of the lower extremity, as well as the femoral nerve, traverse through this region. Hernias are frequent in this area. More importantly, the femoral triangle is an arterial pressure point where it is vital to apply pressure in the case of uncontrolled hemorrhage of the lower extremity.

The greater trochanter of the femur can be palpated on the lateral, upper surface of the thigh (fig. 10.47). At the knee, the lateral and medial condyles of the femur and tibia can be identified (fig. 10.48). The patella (kneecap) can be easily located

Figure 10.48 An anterior view of the right thigh and knee.

Femoral triangle

Quadriceps femoris muscles

Adductor group of muscles

Rectus femoris muscle

Vastus lateralis muscle

Sartorius muscle

Vastus medialis muscle

Tendon of rectus femoris muscle

Patella

Lateral epicondyle

Medial epicondyle

Patellar ligament

Tibial tuberosity

Figure 10.49   The (a) lateral, (b) posterior, and (c) medial surfaces of the leg.

Tensor fasciae latae muscle

Vastus lateralis muscle

Iliotibial tract

Biceps femoris muscle

Patella

Biceps femoris tendon of insertion

Lateral malleolus

Head of fibula

Tibialis anterior muscle

(a)

Adductor magnus muscle

Semitendinosus muscle

Vastus lateralis muscle

Long head of biceps femoris muscle

Short head of biceps femoris muscle

Semimembranosus muscle

Popliteal fossa

Lateral epicondyle

Medial epicondyle

Medial head of gastrocnemius muscle

Lateral head of gastrocnemius muscle

(b)

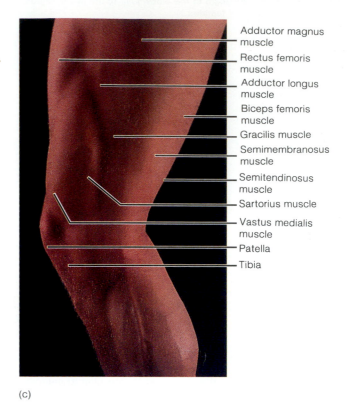

Adductor magnus muscle

Rectus femoris muscle

Adductor longus muscle

Biceps femoris muscle

Gracilis muscle

Semimembranosus muscle

Semitendinosus muscle

Sartorius muscle

Vastus medialis muscle

Patella

Tibia

(c)

within the patellar tendon anterior to the knee joint. Stress or injury to this joint may cause swelling, commonly called water on the knee.

The depression on the posterior aspect of the knee joint is referred to as the **popliteal** (*pop''li-te'al*) **fossa** (fig. 10.49). This area becomes more clinically important in elderly people who suffer degenerative conditions. Aneurysms of the popliteal artery are common, as are popliteal abscesses due to infected lymph nodes. The small saphenous vein as it traverses the popliteal fossa may become varicose in the elderly.

Leg   Portions of the tibia and fibula, the bones of the leg, can be observed as surface features. The medial surface and anterior border (commonly called shin) of the tibia are subcutaneous and

are palpable throughout their length. At the ankle, the medial malleolus of the tibia and the lateral malleolus of the fibula are easy to observe as prominent eminences (fig. 10.50). Of clinical importance in setting fractures of the leg is knowing that the top of the medial malleolus lies about 1.3 cm (0.6 in.) proximal to the level of the tip of the lateral malleolus.

The heel is not part of the leg but the posterior part of the calcaneus bone. It needs to be mentioned with the leg, however, because of its functional relationship to it. The **tendo calcaneus** (Achilles tendon) is the strong, cordlike tendon that attaches to the calcaneus from the calf of the leg. The muscle forming the belly of the calf is the gastrocnemius. Pulsations from the posterior tibial artery can be detected by palpating between the medial malleolus and the calcaneus.

The superficial veins of the leg can be observed on many individuals (see fig. 10.2). The great saphenous vein can be seen subcutaneously along the medial aspect of the leg. The less conspicuous small saphenous vein drains the lateral surface of the leg. If these veins become excessively enlarged, they are called varicose veins.

Leg injuries are common among athletes. *Shin splints*, probably the result of a stress fracture or periosteum damage of the tibia, is a common condition in runners. A fracture of one or both malleoli is caused by a severe twisting of the ankle region. Skiing fractures are generally caused by strong torsion forces on the shafts of the tibia or fibula.

UNIT IV

Figure 10.50   The leg and foot. (*a*) A lateral view, (*b*) a medial view, (*c*) an anterior view, and (*d*) a posterior view.

Lateral head of gastrocnemius muscle

Tibialis anterior muscle

Soleus muscle

Tendon of tibialis anterior muscle

Tendo calcaneus

Tendon of peroneus longus muscle

Lateral malleolus

Calcaneus

Extensor digitorum brevis muscle

Tendons of extensor digitorum longus muscle

(a)

Medial head of gastrocnemius muscle

Great saphenous vein

Soleus muscle

Tibia

Tendo calcaneus

Medial malleolus

Calcaneus

Abductor hallucis muscle

Longitudinal arch

Head of first metatarsal bone

(b)

Lateral malleolus of fibula

Medial malleolus of tibia

Site for palpation of dorsal pedis artery

Tendons of extensor digitorum longus muscle

Tendon of extensor hallucis longus muscle

(c)

Medial head of the gastrocnemius muscle

Soleus muscle

Tendo calcaneus

Medial malleolus of the tibia

Lateral malleolus of the fibula

Site for palpation of the posterior tibial artery

Tendon of peroneus longus muscle

Calcaneus

(d)

UNIT IV

Figure 10.51 Common clinical conditions of the foot and toes. (*a*) Ingrown toenail, (*b*) hammer toe, and (*c*) corn.

(a)

(b)

(c)

Figure 10.52 A posterior view of the right abdominal and gluteal regions.

1 Trapezius m.	4 External abdominal oblique m.
2 Lumbar aponeurosis	5 Gluteus maximus m.
3 Latissimus dorsi m.	6 Fascia lata

Foot   The feet are adapted to support the weight of the body, to maintain balance, and to function mechanically during locomotion. The structural features and surface anatomy of the **foot** are indicative of these functions. The **arch of the foot** is located on the medial portion of the plantar surface (fig. 10.50) and provides a spring effect when locomoting. The head of the first metatarsal bone forms the medial ball of the foot, just proximal to the hallux (big toe).

> The feet and toes are adapted to endure tremendous compression forces during locomotion. Although appropriate shoes help to minimize trauma to the feet and toes, there is still an array of common clinical conditions (fig. 10.51) that may impede walking or running. An *ingrown toenail* occurs as the sharp edge of a toenail embeds into and injures the skin fold, resulting in inflammation and infection. *Hammer toe* is a condition resulting from a forceful hyperextension at the metatarsophalangeal joint with flexion at the proximal interphalangeal joint. A *corn* is a thickening of skin (callus) resulting from recurrent pressure on the skin over a bony prominence.

The fifth metatarsal bone forms much of the lateral border of the plantar surface of the foot. The tendons of the extensor digitorum longus muscle can be seen along the dorsal surface of the foot, especially if the toes are elevated. Pulsations of the dorsal pedal artery can be palpated on the dorsal surface of foot between the first and second metatarsal bones. The individual phalanges of the toes, the joints between these bones, and the toenails are obvious surface landmarks.

Internal Anatomy

The internal anatomy of the buttock and lower extremity include the structures of the hip, thigh, knee, leg, and the foot. The principal structures of these regions are shown in the cadaver dissections in figures 10.52 through 10.58.

1. What surface features form the boundaries of a buttock?
2. List the clinically important structures that can be observed or palpated in the buttock and lower extremity.
3. Describe the anatomical location where each of the following could be observed or palpated: the distal tendinous attachments of the hamstring muscles; the greater trochanter; the great and small saphenous veins; the femoral, posterior tibial, and dorsal pedal arteries; and the medial malleolus.

Clinical Considerations

Head and Neck Regions

The highly specialized head and neck regions are extremely vulnerable to trauma and disease. Furthermore, because of the incredible complexity of these body regions, they are susceptible to numerous congenital conditions occurring during prenatal development.

The aggressive nature of humans, reflected in part by a penchant for contact sports and fast-moving vehicles, puts the

UNIT IV

Figure 10.53 A posterior view of the deep muscles of the right abdominal and gluteal regions.

1 Superior gluteal vessels	9 Internal abdominal oblique m.
2 Inferior gluteal vessels	10 Lumbar aponeurosis
3 Sacrotuberous ligament	11 Gluteus medius m.
4 Levator ani m.	12 Piriformis m.
5 Serratus anterior m.	13 Obturator internus m.
6 Erector spinae m.	14 Quadratus femoris m.
7 Serratus posterior m.	15 Sciatic nerve
8 External intercostal m.	

Figure 10.54 An anterior view of the right thigh.

- Iliacus muscle
- Femoral nerve
- Femoral artery
- Femoral vein
- Tensor fasciae latae muscle
- Vastus lateralis muscle
- Adductor longus muscle
- Rectus femoris muscle
- Sartorius muscle
- Gracilis muscle
- Vastus lateralis muscle
- Vastus medialis muscle
- Tendon of rectus femoris muscle

UNIT IV

human head and neck in constant danger of injury. Pathogens readily gain access to the internal structures of the head and neck through the several openings into the head. Also, the risk for contracting certain diseases is greatly increased by the social nature of humans.

Developmental Conditions Congenital malformations of the head and neck regions result from genetic or environmental (teratogenic) causes and are generally very serious. The less severe malformations may result in functional disability, whereas the more severe malformations usually make life impossible.

Anencephaly *(an″en-sef′ă-le),* a severe underdevelopment of the brain and surrounding cranial bones, is always fatal. The cause of anencephaly is unknown, but there does seem to be some hereditary and geographic influences. South Wales, for example, reports incidences of anencephaly as high as 1 in every 105 births. It occurs more frequently in females than males, and the damage to the developing embryo occurs between the 16th and 26th day after conception.

Figure 10.55 A posterior view of the right hip and thigh.

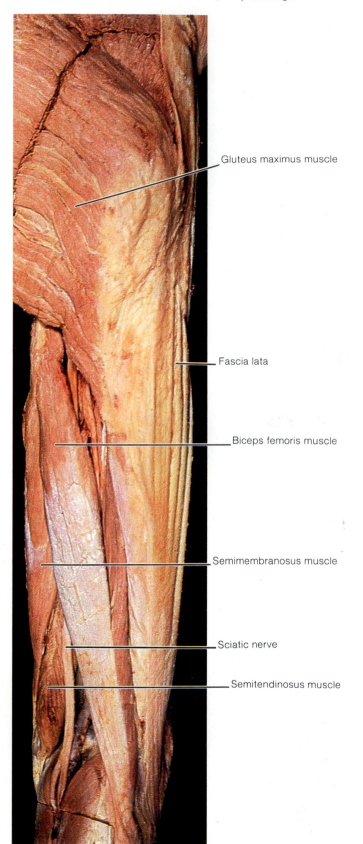

Gluteus maximus muscle

Fascia lata

Biceps femoris muscle

Semimembranosus muscle

Sciatic nerve

Semitendinosus muscle

Figure 10.56 An anterior view of the right leg.

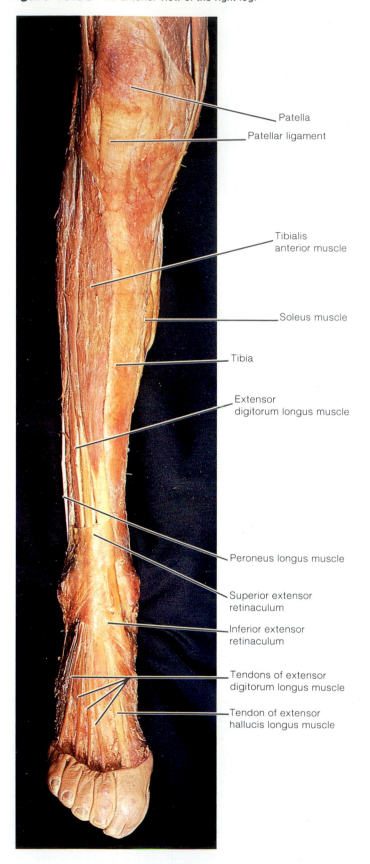

Patella

Patellar ligament

Tibialis anterior muscle

Soleus muscle

Tibia

Extensor digitorum longus muscle

Peroneus longus muscle

Superior extensor retinaculum

Inferior extensor retinaculum

Tendons of extensor digitorum longus muscle

Tendon of extensor hallucis longus muscle

UNIT IV

Figure 10.57 A lateral view of the right leg.

Patella

Patellar ligament

Tibialis anterior muscle

Extensor digitorum
longus muscle

Gastrocnemius muscle

Peroneus longus muscle

Soleus muscle

Peroneus tertius muscle

Tendo calcaneus

Superior extensor
retinaculum

Tendons of extensor
digitorum longus muscle
and peroneus tertius muscle

Figure 10.58 A posterior view of the right leg.

Sciatic nerve

Biceps femoris muscle

Semitendinosus muscle

Common peroneal nerve

Tibial nerve

Gastrocnemius muscle

Soleus muscle

Peroneus longus muscle

Peroneus brevis muscle

Tendo calcaneus

UNIT IV

Altered cranial bones and sutures, resulting in pressure on the brain, accompany several kinds of congenital conditions. **Microcephaly** is characterized by premature closure of the sutures of the skull. If the child is untreated, there will be underdevelopment of the brain and mental retardation. **Cranial encephalocele** is a condition in which the skull does not develop properly and portions of the brain often protrude through it. In **hydrocephalus** there is an excessive accumulation of cerebrospinal fluid dilating the ventricles of the brain and causing a separation of the cranial bones.

A **cleft palate** and **lip** is a common congenital condition of varying degrees of severity. A vertical split on one side, where the maxillary and median nasal processes fail to unite, is referred to as a *unilateral cleft*. A *bilateral*, or *double, cleft* occurs when the maxillary and median nasal processes on both sides fail to unite.

Normally by the age of 30 the sutures of the skull synostose and cranial bone growth ceases. **Premature synostosis (microcephaly)** is an early union of the cranial sutures before the brain has reached its normal size. **Scaphocephaly** is a malformation in which the sagittal suture prematurely closes. **Oxycephaly** is a condition in which the coronal suture closes prematurely. The skull will be noticeably crooked in a condition called **plagiocephaly.**

Torticollis (wryneck) is a congenital or acquired condition of the neck in which the head is drawn to one side and rotated so that the face is directed upward. It is not known what causes the congenital condition, but trauma to the neck, or certain diseases like rheumatism may cause its occurrence in newborns. The muscles that are involved in the spasmodic contractions are those innervated by the accessory cranial nerve, particularly the sternocleidomastoid muscle.

Trauma to the Head and Neck The head and neck are extremely susceptible to trauma and blows, which are frequently physically debilitating if not fatal. Striking the head from the front or back often causes **subdural hemorrhage,** resulting from the tearing of the superior cerebral veins at their points of entrance to the superior sagittal sinuses. Blows to the side of the head tend to be less severe because the falx cerebelli and tentorium cerebelli restrict the sideways displacement of the brain. With a sudden violent lateral movement of the head, such as in a serious automobile accident, the serrated edge of the lesser wing of the sphenoid may severely damage the brain and frequently tear cranial nerves.

The arteries of the head and neck are rarely damaged because of their elasticity. In a severe lateral blow to the head, however, the internal carotid artery may rupture, and a roaring sound will be perceived by the injured person as blood rushes into the cavernous sinuses of the temporal bone.

Skull fractures are fairly common in adults but much less common in children. The cranial bones of a child are resilient, and the sutures are not ossified. The cranium of an adult, however, is much like an eggshell in that it has limited resilience and tends to splinter. A hard blow to the head frequently breaks the bone on the opposite side of the skull in what is called a **contrecoup fracture.** Actually, the sphenoid, with its numerous foramina, is the weakest bone of the cranium and frequently sustains a contrecoup fracture as a result of a hard blow to the top of the head.

The two most commonly fractured bones of the face are the nasal and mandible bones. Trauma to either of them generally results in simple fracture, which is not usually serious. If the nasal septum or cribriform plate is fractured, however, careful treatment is required. If the cribriform plate is severely fractured, there may be a tear in the meninges, causing a sudden loss of cerebrospinal fluid and death.

Whiplash is a common injury to the neck due to a sudden and forceful displacement of the head (see fig. 11.46). The muscle, bone, or ligaments may be injured, in addition to the spinal cord and cervical nerves. A whiplash is usually extremely painful and difficult to treat because of the complexity in diagnosing the extent of the injury.

The sensory organs within the head are unfortunately very prone to trauma. The eyes and middle ears are extremely vulnerable to protruding objects. Sudden bright light and loud noise can also damage structures in each of these respective organs. A nonpenetrating blow to the eye may result in a herniation of the orbital contents through a fracture created in the floor of the orbit. The nerves that control the eye may also be damaged.

Diseases of the Head and Neck The head and neck are extremely susceptible to infection, especially along the mucous membranes lining body openings. Sinusitis, tonsillitis, laryngitis, pharyngitis, esophagitis, and colds are common, periodically recurring ailments of the mucous-lined digestive and respiratory tracts of the head and neck.

The cutaneous area of the head most susceptible to infections extends from the upper lip to the midportion of the scalp. An infection of the scalp may spread via the circulatory system to the bones of the skull, causing **osteomyelitis.** The infection may even spread into the sagittal venous sinus, causing **venous sinus thrombosis.** A **boil** in the facial region may secondarily cause thrombosis of the facial vein or the spread of the infection to the sinuses of the skull. Before antibiotics, such sinus infections had a mortality rate of 90%.

Close observation of the head by a physician can be helpful in diagnosing several diseases and body conditions. The nose becomes greatly enlarged in a person with **acromegaly** and very wide in a person with **hypothyroidism.** The bridge of the nose is depressed in a person with **congenital syphilis.**

The color of the mucous membranes of the mouth may be important in diagnoses. Pale lips generally indicate *anemia,* yellow lips indicate *pernicious anemia,* and blue lips are characteristic of *cyanosis,* or cardiovascular problems. In *Addison's disease,* the normally pinkish mucous membranes of the cheeks have brownish areas of pigmentation.

---

synostose: Gk. *syn*, together; *osteon*, bone
scaphocephalus: Gk. *skaphe*, boat; *kephale*, head
oxycephalus: Gk. *oxys*, pointed; *kephale*, head
plagiocephaly: Gk. *plagios*, oblique; *kephale*, head
torticollis: L. *tortus*, twisted; *collum*, neck

UNIT IV

Thoracic Region

Developmental Conditions   Although the rib cage varies widely among individuals, its development is usually complete. When serious deformities of the chest do occur, they are almost always due to an overgrowth of the ribs. In **pigeon breast** (*pectus carinatum*), the sternum is pushed forward and downward like the keel of a boat. In **funnel chest** (*pectus excavatum*), the sternum is pushed posteriorly, causing an anterior concavity in the thorax. Rarely, there may be a congenital absence of a pair, or pairs, of ribs. The absence of ribs is due to incomplete development of the thoracic vertebrae, a condition termed **hemivertebrae,** and may result in impaired respiratory functions. There is a 0.5% occurrence of **cervical rib,** and half the time it is bilateral. A cervical rib is attached to the transverse process of the seventh cervical vertebra, and it either has a free anterior portion or is attached to the first (thoracic) rib. Pressure of a cervical rib onto the brachial plexus may produce paresthesia (burning, prickling) along the ulnar border of the forearm and atrophy of the medial (hypothenar) muscles of the hand.

The pectoralis major muscle may be congenitally absent, either partially or wholly. A person with this anomaly will have a sunken appearance in the chest area and will have to rely on the contractions of synergistic muscles to the pectoralis major muscle for flexion at the shoulder joint.

The rapid and complex development of the heart and major thoracic vessels accounts for the numerous congenital abnormalities that may affect these organs (see chapter 16). Congenital heart problems occur in approximately 3 of every 100 births and account for about 50% of early childhood deaths. Cardiac malformations usually arise from developmental defects in the heart valves, septa (atrial and/or ventricular), or both. A **patent foramen ovale** is an example of a septal defect. Such a malformation permits venous blood from the right atrium to mix freely with the oxygenated blood in the left atrium. A **ventricular septal defect** usually occurs in the upper portion of the interventricular septum and is generally more serious than an atrial septal defect because of the greater fluid pressures in the ventricles and the greater chance of heart failure.

Congenital valvular problems are classified as either an **incompetence** (leakage) or a **stenosis** (constriction) of valves. Improper closure of a valve permits some backflow of blood, causing an abnormal sound referred to as a **murmur.** Murmurs are common and generally do not affect a person's health.

A **tetralogy of Fallot** is a combination of four defects within a newborn and immediately causes a cyanotic condition (blue baby). The four characteristics of this anomaly are (1) a ventricular septal defect, (2) an overriding aorta, (3) pulmonary stenosis, and (4) right ventricular hypertrophy. Although tetralogy of Fallot is one of the most common cardiac defects, it is one of the simplest to correct surgically.

Abnormal development of the primitive aortic arches may result in a functional right aortic arch and occasionally both a right and a left aortic arch. When these conditions occur, there are generally anomalies of other vessels as well.

Trauma to the Thorax   Because of its resilience, the rib cage generally provides considerable protection to the thoracic viscera. The ribs of children are highly elastic and fractures are rare. By contrast, the ribs of adults, although strong, are frequently fractured by direct trauma or, indirectly, by crushing injuries. Ribs 3 through 8 are the most commonly fractured. The first and second ribs are somewhat protected by the clavicle and the last four ribs are more flexible to blows. The costal cartilages in elderly people may undergo some ossification, reducing the flexibility of the rib cage and causing some confusion when examining a chest radiograph.

A common complication of a rib fracture is a puncture of the lung or the protrusion of a bone fragment through the skin. In either case, pleural membranes will likely be ruptured resulting in a **pneumothorax** (air into the pleural cavity) or a **hemothorax** (blood into the pleural cavity).

Any puncture wound to the thorax—from a bullet or a knife, for example—may cause a pneumothorax. A **collapsed lung** generally results from a pneumothorax, making respiration extremely difficult. Because each lung is contained within its own pleural cavity, trauma to one lung does not usually directly affect the other lung.

The heart, the ascending aorta, and the pulmonary trunk are enclosed by the fibrous pericardium. A severe blow to the chest, such as hitting the steering wheel in an automobile accident, may cause a sudden surge of blood from the ventricles sufficient to rupture the ascending aorta. Such an injury will flood the pericardial sac with blood, causing **cardiac tamponade** (fluid compression) and almost immediate death.

Diseases of the Thorax   The leading causes of death in the United States are due to disease or dysfunction of the thoracic organs. Consequently, the surface features of the thorax are extremely important to a physician in providing reference locations for performing *inspections* (visual observation), *palpations* (feeling with firm pressure), *percussions* (detecting densities through tapping), and *auscultations* (listening with a stethoscope). Many of the ribs can be seen on a thin person, and all of the ribs, except the first and at times the twelfth, can be palpated. The sternum, clavicles, and scapulae also provide important bony landmarks for doing a physical examination. The nipples in the male and prepubescent female are located at the fourth intercostal spaces. The position of the left nipple in males is important in knowing where to listen for various heart sounds and for determining whether the heart is enlarged.

The clinical importance of the female breasts lies in their periodic phases of activity (during pregnancy and lactation) and their susceptibility to neoplastic change. The breasts and mammary glands are highly susceptible to infections, cysts, and tumors. The superficial position of the breasts permits a chance for effective treatment by surgery and radiotherapy if tumors are recognized early. The importance of *breast self-examination (BSE),* therefore, cannot be overemphasized (see chapter 21). Untreated breast cancer is eventually fatal (fig. 10.59). Breast cancer, or **carcinoma of the breast,** is the most common malignancy in women. One in nine women will develop breast cancer, and one-third of these will die from the disease.

UNIT IV

UNIT IV

Figure 10.59   Advanced stages of untreated breast cancer that has metastasized to afflict both breasts.

The lungs can be examined through percussion, auscultation, or observation. *Bronchoscopy* enables a physician to examine the trachea, carina, primary bronchi, and secondary bronchi. A bronchoscope also enables a physician to remove foreign objects from these passageways. Swallowed objects that are aspirated beyond the glottis usually lodge (90% of the time) in the right primary bronchus because of its near vertical position to the trachea. *Bronchiograms,* or chest radiographs, are perhaps the best means of inspecting the lungs.

The lungs are a common site for cancer. Fortunately, each lung is divided into different segments allowing a surgeon to remove a diseased portion and leave the remainder of the lung intact. Pneumonia, tuberculosis, asthma, pleurisy, and emphysema are other common diseases that directly or indirectly afflict the lungs.

Cardiovascular diseases are the leading cause of death in the United States. Included among these diseases are heart attacks, which are caused by an insufficient blood supply to the myocardium (called **myocardial ischemia**). Poor cardiac circulation is due to an accumulation of artherosclerotic plaques or the presence of a thrombus (clot). A heart attack is generally accompanied by severe chest pains (**angina pectoris**) and usually by referred pain, perceived as arising from the left arm and shoulder. If the coronary deficiency is continuous, local tissue

necrosis results, causing a permanent loss of cardiac muscle fibers (**myocardial infarction**). Extensive cardiac necrosis results in cardiac arrest.

Ventricular fibrillation follows a functional breakdown of the conduction system of the heart. Such a loss of coordinated ventricular contraction impairs coronary circulation, resulting in low blood pressure, or **hypotension.** If untreated, continuous ventricular fibrillation results in death.

Various other heart diseases include infections of the serous membrane (**pericarditis**), infection of the lining of the heart chambers (**endocarditis**), and infection of the valves (a complication of **rheumatic fever**). Valvular disease may cause the cusps to function poorly and result in an enlarged heart.

Abdominal Region

Developmental Conditions   The development of the abdominal viscera from endoderm and mesoderm is a highly integrated, complex, and rapidly occurring process, and is therefore susceptible to a wide range of congenital malformations.

The diaphragm develops in four directions simultaneously as skeletal muscular tissues coalesce toward the posterior center at the hiatus, where the esophagus, abdominal aorta, and inferior vena cava traverse. Failure of these four sections to fuse results in a **congenital diaphragmatic hernia.** This anomalous opening may permit certain abdominal viscera to project into the thoracic cavity.

The role of the umbilicus in the development of the fetal urinary, circulatory, and digestive systems may present some interesting congenital defects. A **patent urachus** is an opening from the urinary bladder to the outside through the umbilicus. For a short time during development, this opening is natural as urine passes from the fetus into the allantois. Closure of the urachus occurs in most fetuses during further development of the urinary system. A patent urachus will generally go undetected unless there is extreme difficulty with urination, such as that caused by an enlarged prostate. In such a case, some urine may be forced through the patent urachus and out the umbilicus.

Meckel's diverticulum is the most common anomaly of the small intestine. It is the result of failure of the embryonic yolk sac to completely atrophy. Present in about 3% of the population, a Meckel's diverticulum consists of a pouch approximately 6.5 cm (2.5 in.) in length, resembling the appendix and arising near the center of the ileum. It may terminate freely or be attached to the anterior abdominal wall near the umbilicus. Like the appendix, a Meckel's diverticulum is prone to infections and may become inflamed, producing symptoms similar to appendicitis. Because of its clinical importance, it is usually removed as a precautionary measure when discovered during abdominal surgery.

The connection from the ileum of the small intestine to the outside sometimes is patent at the time of birth; this condition is called a *fecal fistula*. This anomalous opening permits the passage of fecal material through the umbilicus and must be surgically corrected in a newborn.

Other parts of the GI tract are also common sites for congenital problems. **Pyloric stenosis** is a condition in which there is a narrowing of the pyloric orifice of the stomach due to hypertrophy of the muscular layer of the pyloric sphincter. This anomaly is more common in males than in females, and the symptoms usually appear early in infancy. The constricted opening interferes with the passage of food into the duodenum and therefore causes dilation of the stomach, vomiting, and loss of weight. Treatment involves a surgical incision of the pyloric sphincter.

Congenital megacolon (Hirschsprung's disease) is a condition in which there is failure of ganglia to develop in the submucosal (Meissner's) and the myenteric (Auerbach's) plexuses in a portion of the colon. The absence of these ganglia results in enlargement of the affected portion of the colon due to a lack of innervation and muscle tone. With the loss of peristalsis there is severe constipation. Treatment involves surgical resection of the affected portion of the colon.

Congenital malformations may develop in any of the abdominal viscera, most of which are inconsequential. Accessory spleens, for example, occur in about 10% of the population. Located near the hilum of the spleen, these anomalous organs are small (about 1 cm. in diameter), number from two to five, and are only moderately functional. Generally they atrophy within a few years after birth.

Tremendous variation can occur in the formation of the kidneys. Frequently they are multilobed, fused, or malpositioned. They may also differ in number from the normal two (see fig. 19.17). When there is an anomalous kidney, there is usually an accompanying variation in the vascular supply. It is common to have multiple renal arteries serving a kidney. Most renal anomalies do not pose serious problems.

Because of the effects of sex hormones on the developing genitalia, there may be considerable malformation of these organs. These anomalies may range from being of cosmetic concern to rendering the organ nonfunctional. Some anomalies may be so severe as to preclude determination of an individual's sex based on external appearance. Most of these conditions can be surgically corrected. Also of clinical concern and treatable are the various problems that may occur during descent of the testes into the scrotum. In the normal development of the male fetus, the testes will be in scrotal position the 28th week of gestation. If they are undescended at birth (**cryptorchidism**), medical intervention may be necessary.

Trauma to the Abdomen The rib cage, the omentum (see fig. 18.3), and the pendant support of the abdominal viscera offer some protection from trauma. Puncture wounds, compression, and severe blows to the abdomen do result, however, in serious abdominal injury.

The large and dense liver, located in the upper right quadrant of the abdomen, is quite vulnerable to blunt trauma, stab

wounds, or puncture wounds by fractured ribs. A lacerated liver is extremely serious because of the possibility of internal hemorrhage from such a vascular organ.

The spleen is another highly vascular organ that is frequently injured, especially from blunt abdominal trauma. A ruptured spleen causes severe internal hemorrhage and shock. A prompt splenectomy is necessary to keep the patient from bleeding to death. The spleen may also rupture spontaneously due to infectious diseases that cause it to hypertrophy.

Rupture of the pancreas is not nearly as common as rupture of the spleen, but it could occur if a strong compression of the upper abdomen were to force the pancreas against the vertebral column. The danger of a ruptured pancreas is the flow of pancreatic juice into the peritoneal cavity, the subsequent digestive action, and peritonitis.

The kidneys are vulnerable to trauma in the lumbar region, such as from a kidney punch. Because the kidney is fluid-filled, a blow to one side propagates through the kidney and may possibly rupture the renal pelvis or the proximal portion of the ureter. Blood in the urine is symptomatic of kidney damage. The medical treatment of a traumatized kidney is dependent upon the severity of the injury.

Trauma to the external genitalia of both the male and female is a relatively common occurrence. The pendant position of the penis and scrotum makes them vulnerable to compression forces—such as would be experienced by a construction worker slipping and straddling a steel beam, which would compress the external genitalia between the beam and his pubic bone. In such an injury, the penis (including the urethra) may split open, and one or both testes may be crushed.

Trauma to the female genitalia usually results from sexual abuse. Vaginal tearing and a displaced uterus are common in molested girls. The physical and mental consequences are generally severe.

Diseases of the Abdomen Each of the abdominal organs may be afflicted by an array of diseases. It is beyond the scope of this text to discuss any of these diseases in depth, so only an overview of some general conditions will be presented.

Knowledge of the clinical regions of the abdomen (see figs. 2.15 and 2.16) and the organs within these regions (see table 2.3) is fundamental to the physician in performing a physical examination. Also important are the locations of the linea alba, extending from the xiphoid process to the symphysis pubis; the umbilicus; the inguinal ligament; the bones and processes that can be palpated on the rib cage; and the pelvic girdle.

Peritonitis is of major clinical concern. The peritoneum is the serous membrane of the abdominal cavity that lines the abdominal wall, as the parietal peritoneum, and covers the visceral organs, as the visceral peritoneum. The peritoneal cavity is the moistened space between the parietal and visceral portions of the peritoneum. Peritonitis results from any type of contamination of the peritoneal cavity, such as from a puncture wound, blood-borne diseases, or a ruptured visceral organ. In females, peritonitis is frequently a complication of infections of the re-

---

Hirschsprung's disease: from Harold Hirschsprung, Danish physician, 1830–1916

UNIT IV

productive tract that have entered the peritoneal cavity via the uterine tubes. Without medical treatment, peritonitis is generally fatal.

Ulcers may occur throughout the GI tract. *Peptic ulcers* are erosions of the mucous membranes of the stomach or duodenum, produced by the action of hydrochloric acid (HCl) contained in gastric juice. Agents that weaken the mucosal lining of the stomach, such as alcohol and aspirin, and abnormally high secretions of gastric juice increase the likelihood of developing peptic ulcers. Many people subject to chronic stress produce too much gastric juice and develop a peptic ulcer as a result.

Enteritis, or inflammation of the intestinal mucosa, is frequently referred to as intestinal flu. Causes of enteritis include bacterial or viral infections, irritating foods or fluids, and emotional stress. The symptoms are abdominal pain, nausea, and diarrhea. *Diarrhea* is symptomatic of inflammation, stress, and other body dysfunctions. Diarrhea in children is of immense clinical importance because of the rapid loss of body fluids.

Shoulder and Upper Extremity

Developmental Conditions Twenty-eight days after conception, a limb bud appears on the upper lateral side of the embryo, which eventually becomes a shoulder and an upper extremity. Three weeks later (7 weeks after conception) the shoulder and upper extremity are present in the form of mesenchymal primordium of bone and muscle. It is during this crucial 3 weeks of development that malformations of the extremities can occur due to genetic expression or to the effect of teratogens.

If a pregnant woman uses certain teratogenic drugs or is exposed to certain diseases (*Rubella* virus, for example) during development of the embryo, there is a strong likelihood that the appendage will be incompletely developed. A large number of limb deformities occurred between 1957 and 1962 as a result of women ingesting thalidomide during early pregnancy to relieve morning sickness. It is estimated that 7,000 infants were malformed by thalidomide. The malformations ranged from *micromelia* (short limbs) to *amelia* (the absence of limbs).

Although the genetic deformities of the shoulder and upper extremity are many, only a few are relatively common. **Sprengel's deformity** affects the development of one or both scapulae. In this condition, the scapula is smaller than normal and is positioned at an elevated level. As a result, abduction of the arm is not possible beyond a right angle to the plane of the body.

Minor defects of the extremities are relatively common malformations. Extra digits, a condition called **polydactyly,** is the most common limb deformity. Usually an extra digit is incompletely formed and does not function. **Syndactyly,** or webbed digits, is likewise a relatively common limb malformation. Polydactyly is inherited as a dominant trait, whereas syndactyly is a recessive trait.

Trauma to the Shoulder and Upper Extremity The wide variety of injuries to the shoulder and upper extremity range from damaged bones and surrounding muscles, tendons, vessels, and nerves to damaged joints in the form of sprains or dislocations.

It is not uncommon to traumatize the shoulder and upper extremity of a newborn during a difficult delivery. **Upper arm birth palsy** (*Erb–Duchenne palsy*) is the most common type of birthing injury, caused by a forcible widening of the angle between the head and shoulder. Using forceps to rotate the fetus *in utero* or pulling on the head during delivery may cause this injury. The site of injury is at the junction of C5 and C6 (*Erb's point*), as they form the upper trunk of the brachial plexus. The expression of the injury is paralysis of the abductors and lateral rotators of the shoulder and the flexors of the elbow. The arm will permanently hang at the side in medial rotation.

The stability of the shoulder is largely dependent on the support of the clavicle and acromion of the scapula superiorly and the tendons forming the rotator cuff anteriorly. Because support is weak along the inferior aspect of the shoulder joint, dislocations are frequent in this direction. Injuries of this sort are common in athletes engaging in contact sports. Sudden jerks of the arm are also likely to dislocate the shoulder, especially in children who have weak muscles spanning this area.

Fractures are common in any location of the shoulder and upper extremity. Many fractures are the result of extending the arm to break a fall. The clavicle is the most frequently broken bone in the body. Fractures of the humerus are also common and often serious because of injury to the nerves and vessels that parallel the bone. The surgical neck of the humerus is a common fracture site. At this point, the axillary nerve is often damaged, thus limiting abduction of the arm. A fracture in the middle third of the humerus may damage the radial nerve, causing paralysis of the extensor muscles of the hand (wristdrop). A fracture of the olecranon of the ulna often damages the ulnar nerve, resulting in paralysis of the flexor muscles of the hand and the adductor muscles of the thumb. The distal part of the radius is frequently fractured (*Colles' fracture*) when one falls on an outstretched arm. In this fracture, the hand is displaced backward and upward.

Sports injuries frequently involve the upper extremity. Repeated extension of the wrist against a force, such as occurs during the backhand stroke in tennis, may cause **lateral epicondylitis** (tennis elbow). Wearing an elbow brace or a compression band may help reduce the pain, but only abstaining from tennis for a while will heal the area.

Athletes frequently jam a finger when a ball forcefully strikes a distal phalanx as the fingers are extended, causing a sharp flexion at the joint between the middle and distal phalanges. Splinting the finger for a period of time may be curative; however, surgery may be required to avoid a permanent crook in the finger.

Sprengel's deformity: from Otto G. K. Sprengel, German surgeon, 1852–1915

Erb-Duchenne palsy: from Wilhelm H. Erb, German neurologist, 1840–1921, and Guillaume G. A. Duchenne, French neurologist, 1806–75
Colles' fracture: from Abraham Colles, Irish surgeon, 1773–1843

Diseases of the Shoulder and Upper Extremity Infections in specific locations of the shoulder or upper extremity are the only common diseases that are endemic to these regions. **Bursitis,** for example, may specifically afflict any of the numerous bursae of the shoulder, elbow, or wrist joints. There are several types of **arthritis,** but generally they involve synovial joints throughout the body rather than just in the hands and fingers.

Carpal tunnel syndrome is caused by compression of the median nerve by the transverse carpal ligament that forms the palmar aspect of the carpal tunnel. The nerve compression results in a painful burning sensation or numbness of the first three fingers and some muscle atrophy. The compression is due to an inflammation of the transverse carpal ligament. Surgery eases the pressure of the ligament and provides the entry for antibiotic treatment.

Tenosynovitis is an infection of the synovial tendon sheath in the wrist or hand. Digital sheath infections are quite common following a puncture wound in which pathogens enter the closed synovial sheath. The increased pressure from the swollen, infected sheath may cause severe pain and eventually result in necrosis of the flexor tendons. The loss of hand function can be prevented by draining the synovial sheath and providing antibiotic treatment.

Hip and Lower Extremity

Developmental Conditions The embryonic development of the hip and lower extremity follows the developmental pattern of the shoulder and upper extremity: the appearance of the limb bud is followed by the formation of the mesenchymal primordium of bone and muscle in the shape of an appendage. Development of the lower extremity, however, lags behind that of the upper extremity by 3 or 4 days.

The likelihood of congenital deformities of the hips and lower extremities in a newborn is slim if the pregnant mother has been healthy and well nourished, especially prior to and during embryonic development. The few congenital malformations that exist generally have a genetic basis.

In **congenital dislocation of the hip,** the acetabulum fails to develop adequately, and the head of the femur slides out of the acetabulum onto the gluteal surface of the ilium. If untreated, the infant would never be able to walk normally.

Polydactyly and syndactyly occur in the feet as well as in the hands. Treatment of the feet is the same as treatment of the hands.

Talipes, or **clubfoot,** is a congenital malformation in which the sole of the foot is twisted medially. It is uncertain if abnormal positioning or restricted movement in utero causes this anomaly, but both genetical and environmental conditions are involved in most cases.

Trauma to the Hip and Lower Extremity As with the shoulder and upper extremity, a variety of traumatic conditions are associated with the hip and lower extremity. These range from injury to the bones and surrounding muscles, tendons, vessels, and nerves to damage of the joints in the form of sprains or dislocations.

Dislocation of the hip is a common and severe result of an automobile accident when a seat belt is not worn. When the hip is in the flexed position, as in sitting in a seat, a sudden force onto the distal end of the femur will drive the head of the femur out of the acetabular socket, fracturing the posterior acetabular lip. In this kind of injury, there is usually damage to the sciatic nerve.

Trauma to the nerve roots that form the sciatic nerve may also occur from a herniated disc or pressure from the uterus during pregnancy. An improperly administered injection into the buttock may damage the sciatic nerve itself. Sciatic nerve damage is usually very painful and is expressed throughout the posterior length of the lower extremity.

Fractures are common in any location of the hip and lower extremity. Athletes (such as skiers) and elderly people seem to be most vulnerable. *Osteoporosis* markedly weakens the bones of the hip and thigh regions, making them vulnerable to fracture. A common fracture site, especially in elderly women, is across the femoral neck. A fracture of this kind may be complicated with vascular and nerve interruption. A direct blow to the knee will frequently fracture the patella. A potentially more serious knee trauma, however, is a clipping injury, caused by a blow to the lateral side. In this type of injury, there is generally damage to the cruciate ligament and menisci. Serious complications arise if the common peroneal nerve, traversing the popliteal fossa, is damaged. Damage to this nerve results in paralysis of the ankle and foot extensors (footdrop) and inversion of the foot.

Stress fractures of the long bones of the lower extremity are common to athletes. *Shinsplints,* which are trauma to the anterior muscles of the leg and their periosteal attachments, are often accompanied by tibial stress fractures. Stress fractures of the metatarsal bones may be very painful and yet not even show up on radiographs. Frequently, the only way to heal stress fractures is to abstain from exercise.

Sprains are common in the joints of the lower extremity. Ligaments and tendons are torn to varying degrees in sprains. Sprains are usually accompanied by *synovitis,* an inflammation of the joint capsule.

Diseases of the Hip and Lower Extremity As in the shoulder and upper extremity, infections in the hip and lower extremity—such as bursitis and tendonitis—can be localized in any part of the hip or lower extremity. Likewise, several types of arthritis may affect joints in these regions.

A variety of skin diseases afflict the foot, including athlete's foot, plantar warts, and dyshidrosis. Most of the diseases of the feet can be prevented, or if they do occur, they can be treated effectively.

Because arterial occlusive disease is common in elderly people, palpation of the posterior tibial artery is clinically important in general physical assessment. This can be accomplished by gently palpating between the medial malleolus and the tendo calcaneus.

Many neuromuscular diseases have a direct effect on the functional capabilities of the lower extremities. Muscular dystrophy and poliomyelitis are both serious immobilizing diseases because of muscle paralysis.

UNIT IV

UNIT IV

CHAPTER SUMMARY

Introduction to Surface Anatomy (p. 284)

1. Surface anatomy is concerned with identifying body structures through observation and palpation. Surface anatomy has tremendous application to the maintenance of physical fitness and to medical diagnosis and treatment.
2. Most of the bones of the skeleton are palpable and provide landmarks for other anatomical structures.
3. The effectiveness of observation and palpation in studying a person's surface anatomy depends on the thickness of the hypodermis, which varies in accordance with the amount of subcutaneous adipose tissue present.

Surface Anatomy of the Newborn (pp. 285–288)

1. Certain aspects of the surface anatomy of a neonate are of clinical importance in ascertaining the degree of physical development, general health, and possible congenital abnormalities.
2. The posture of a full-term, normal neonate is one of flexion.
3. Portions of the skin and subcutaneous tissues are edematous. Vernix caseosa covers the body of a neonate, and lanugo may be present on the head, neck, and back.
4. The fontanels, liver, kidneys, and testes of a male should be palpable.

Head (pp. 288–292)

1. Surface features of the cranium include the forehead, the crown, the temporalis muscles, and the hair and hairline.
2. The face is composed of the ocular region that surrounds the eye, the auricular region of the ear, the nasal region serving the respiratory system, and the oral region of the mouth.

Neck (pp. 292–296)

1. Major organs are located within the flexible neck, and structures that are essential for body sustenance pass through the neck to the trunk.
2. The neck consists of an anterior cervix, right and left lateral regions, and a posterior nucha.

3. Two major and six minor triangles, each of which contains specific structures, are located on both sides of the neck.
   (a) The anterior cervical triangle contains the carotid, submandibular, submental, and muscular triangles.
   (b) The posterior cervical triangle contains the occipital and omoclavicular triangles.

Trunk (pp. 296–301)

1. Vital visceral organs in the trunk make the surface anatomy of this region especially important.
2. The median furrow is observable and the vertebral spines and scapulae are palpable on the back.
3. Palpable structures of the thorax include the sternum, the ribs, and the costal arch.
4. The important surface anatomy of the abdomen includes the linea alba, umbilicus, costal arch, iliac crests, and the pubis.

Pelvis and Perineum (p. 301)

1. The crest of the ilium, the symphysis pubis, and the inguinal ligament are important pelvic landmarks.
2. The perineum is the region that contains the external genitalia and the anal opening.

Shoulder and Upper Extremity (pp. 306–312)

1. The surface anatomy of the shoulder and upper extremity is important because of frequent trauma to these regions. Vessels of the upper extremity are also used as pressure points and for intravenous injections or blood withdrawal.
2. The scapula, clavicle, and humerus are palpable in the shoulder.
3. The axilla is clinically important because of the vessels, nerves, and lymph nodes located there.
4. The brachial artery is an important pressure point in the brachium. The median cubital vein is important for the removal of blood or for intravenous therapy.
5. The ulna, radius, and their processes are palpable landmarks of the forearm.

6. The knuckles, fingernails, and tendons for the extensor muscles of the forearm can be observed on the posterior aspect of the hand.
7. Flexion creases and the thenar eminence are important surface features on the anterior surface of the hand.

Buttock and Lower Extremity (pp. 312–316)

1. The massive bones and muscles in the buttock and lower extremity serve as weight bearers and locomotors. Many of the surface features of these regions are important with respect to locomotion or locomotor dysfunctions.
2. The prominences of the buttocks are formed by the gluteal muscles and are separated by the natal cleft.
3. The thigh has three muscle groups: anterior (quadriceps), medial (adductors), and posterior (hamstrings).
4. The femoral triangle and popliteal fossa are clinically important surface landmarks.
5. The structures of the leg include the tibia and fibula, the muscles of the calf, and the saphenous veins.
6. The surface anatomy of the foot includes structures adapted to support the weight of the body, maintain balance, and function during locomotion.

REVIEW ACTIVITIES

Objective Questions

1. Eyebrows are located on
   (a) the palpebral fissure.
   (b) the subtarsal sulcus.
   (c) the scalp.
   (d) the supraorbital ridges.
   (e) both (c) and (d).
2. Which of the following structures is *not* part of the auricle (pinna) of the ear?
   (a) the tragus   (c) the earlobe
   (b) the ala   (d) the helix
3. Which of the following clinical–structural word pairs is *incorrectly* matched?
   (a) cleft lip/philtrum
   (b) broken nose/nasion
   (c) pierced ear/earlobe
   (d) black eye/concha

4. The conjunctiva
   (a) covers the entire eyeball.
   (b) is a thick nonmucous membrane.
   (c) secretes tears.
   (d) none of the above.
5. Which of the following could *not* be palpated within the cervix of the neck?
   (a) the larynx
   (b) the hyoid bone
   (c) the trachea
   (d) the cervical vertebrae
6. Palpation of a pulse to the head is best accomplished within
   (a) the carotid triangle.
   (b) the occipital triangle.
   (c) the submental triangle.
   (d) the submandibular triangle.
   (e) the omotracheal triangle.
7. Which nerve traverses behind the medial epicondyle of the humerus?
   (a) ulnar   (d) brachial
   (b) median   (e) cephalic
   (c) radial
8. Which of the following surface features could *not* be observed on obese people?
   (a) the jugular notch
   (b) scapular muscles
   (c) clavicles
   (d) vertebral spines
   (e) the natal cleft

9. Which pair of muscles forms the anterior and posterior borders of the axilla?
   (a) deltoid/pectoralis minor
   (b) biceps brachii/triceps brachii
   (c) latissimus dorsi/pectoralis major
   (d) triceps brachii/pectoralis major
   (e) latissimus dorsi/deltoid
10. Varicose veins occur when which of the following become(s) excessively enlarged?
   (a) saphenous veins
   (b) tibial veins
   (c) the external iliac vein
   (d) the popliteal vein
   (e) all of the above

Essay Questions

1. List four surface features of the cranium and explain how the cranium relates to the scalp.
2. Identify the four regions of the face and list at least two surface features of each region.
3. Which surface features can be observed on the trunk of any person, regardless of how obese that person might be?
4. Identify the two major triangles of the neck and list the associated six minor triangles. Discuss the importance of knowing the boundaries of these triangles and the specific structures included in each.
5. Name four structures that are palpable along the anterior midline of the neck.
6. What three bones are found in the shoulder region? List the surface features that can be either observed or palpated on each of these bones.
7. Identify the tendons or vessels that can be observed or palpated along the anterior surface of the wrist. Which nerves pass through this region?
8. Describe the locations of the arteries that can be palpated as they pulsate in the following regions: neck, brachium, antebrachium, thigh, and ankle. Which of these are considered clinical pressure points?
9. Describe the locations and clinical importance of the following regions: cubital fossa, femoral triangle, axilla, perineum, and popliteal fossa.
10. Explain why an understanding of surface anatomy is of immense importance in clinical application and in learning the basic gross anatomy of internal organs.

UNIT IV

Nervous Tissue and the Central Nervous System

Chapter Outline

CLINICAL CASE STUDY

A 56-year-old woman visits her family doctor for evaluation of a headache that has persisted for nearly a month. Upon questioning the patient, the doctor learns that her left arm, as she puts it, "has been a bit unwieldly, hard to control, and weak." Through examination, the doctor determines that the entire left upper extremity is generally weak. He additionally finds weakness, although less significant, of the left lower extremity. Sensation in the limbs seems to be normal, although mild rigidity and hyperactive reflexes are present. Expressing concern, the doctor tells the patient that she needs to have a CT scan of her head, explaining that there could be a problem within the brain, possibly a tumor or other lesion. The doctor then picks up the phone and contacts a radiologist, explaining the patient's case and parenthetically adding that he thinks he knows specifically where the problem is located.

Why does the doctor suggest to the patient that there may be a problem within her brain when the diagnosed symptoms are weakness of the extremities, and then just on one side of her body?

Hints: Remember the controlling and integrating function of the brain. How does the doctor know the location of the suspected tumor within the brain? In which side of the brain and in which lobe would it be? Explain the muscle weakness in terms of neuronal pathways from the brain to the periphery. Carefully study the information and accompanying figures for the structures and functions of the brain and the neuronal tracts.

Table 11.1	Anatomical Terms Used in Describing the Nervous System
Term	**Definition**
Central nervous system (CNS)	Brain and spinal cord
Peripheral nervous system (PNS)	Nerves and ganglia
Sensory nerve fiber (afferent fiber)	Neuron that transmits impulses from a sensory receptor to the CNS
Motor nerve fiber (efferent fiber)	Neuron that transmits impulses from the CNS to an effector organ; for example, muscle
Nerve	Cablelike collection of nerve fibers; may be "mixed" (contain both sensory and motor fibers)
Plexus	Network of interlaced nerves
Somatic motor nerve	Nerve that stimulates contraction of skeletal muscles
Autonomic motor nerve	Nerve that stimulates contraction (or inhibits contraction) of smooth muscle and cardiac muscles and that stimulates secretion of glands
Ganglion	Grouping of neuron cell bodies located outside the CNS
Nucleus	Grouping of neuron cell bodies within the CNS
Tract	Grouping of nerve fibers that interconnect regions of the CNS

From Stuart Ira Fox, *Human Physiology*, 4th ed. Copyright © 1993 Wm. C. Brown Communications, Inc., Dubuque, Iowa. All Rights Reserved. Reprinted by permission.

UNIT V

Organization and Functions of the Nervous System

The central nervous system and the peripheral nervous system are structural components of the nervous system, whereas the autonomic nervous system is a functional component. Together they orient the body, coordinate body activities, permit the assimilation of experiences, and program instinctual behavior.

Objective 1. Describe the divisions of the nervous system.

Objective 2. Define *neurology;* define *neuron.*

Objective 3. List the functions of the nervous system.

The immensely complex brain and its myriad of connecting pathways constitute the nervous system. The nervous system, along with the endocrine system, regulates the functions of other body systems. The brain, however, does much more than that, and its potential is perhaps greatly underestimated. It is incomprehensible that one's personality, thoughts, and aspirations result from the functioning of a body organ. Plato referred to the brain as "the divinest part of us." The thought processes of this organ have devised the technology for launching rockets into space, curing diseases, and splitting atoms. But with all of these achievements, the brain still remains amazingly ignorant of its own workings.

Neurology, or the study of the nervous system, has been referred to as the last frontier of functional anatomy. Basic questions concerning the functioning of the nervous system remain unanswered: How do nerve cells store and retrieve memory? What are the roles of the many chemical compounds within the brain? What causes mental illness or senility? Scientists are still developing the technology and skills necessary to understand the functional complexity of the nervous system. The next few decades should be very interesting as scientists come to a better understanding of this system.

Organization of the Nervous System

The nervous system is divided into the **central nervous system (CNS),** which includes the **brain** and **spinal cord,** and the **peripheral nervous system (PNS),** which includes the **cranial nerves** arising from the brain and the **spinal nerves** arising from the spinal cord (fig. 11.1 and table 11.1).

The **autonomic nervous system (ANS)** is a functional subdivision of the nervous system. The controlling centers of

neurology: Gk. *neuron,* nerve; L. *logos,* study of

Figure 11.1 The nervous system. The central nervous system (CNS) consists of the brain and spinal cord. The peripheral nervous system (PNS) consists of cranial nerves and spinal nerves. Also part of the PNS are the plexuses and additional nerves that arise from the cranial and spinal nerves. The autonomic nervous system (ANS) is a functional division of the nervous system.

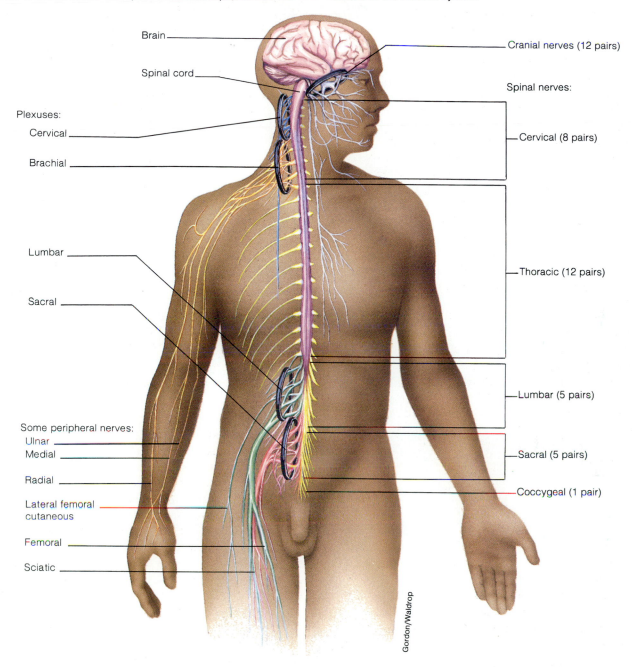

the ANS are within the brain and are considered part of the CNS; the peripheral portions of the ANS are subdivided into the **sympathetic** and **parasympathetic** divisions.

Functions of the Nervous System

The nervous system is specialized for perceiving and responding to events in our internal and external environments. An awareness of one's environment is made possible by **neurons** (nerve cells), which are highly specialized in the properties of excitability and conductivity. The nervous system functions throughout the body in conjunction with the endocrine system (see chapter 14) to closely coordinate the activities of the other body systems. In addition to integrating body activities, the nervous system has the ability to store experiences *(memory)* and to establish patterns of response based on prior experiences *(learning)*.

The functions of the nervous system include

1. orientation of the body to internal and external environments;
2. coordination and control of body activities;
3. assimilation of experiences requisite to memory, learning, and intelligence; and
4. programming of instinctual behavior (apparently more important in vertebrates other than humans).

Developmental Exposition

The Brain

Exhibit I The early development of the nervous system from embryonic ectoderm. (*a*) A dorsal view of an 18-day-old embryo showing the formation of the neural plate and the position of a transverse cut indicated in (*a₁*). (*b*) A dorsal view of a 22-day-old embryo showing cranial and caudal neuropores and the positions of three transverse cuts indicated in (*b₁*–*b₃*). (Note the amount of fusion of the neural tube at the various levels of the 22-day-old embryo. Note also the relationship of the notochord to the neural tube.)

Waldrop

An instinct may also be called a fixed action pattern and is typically genetically specified with little environmental modification. It is triggered only by a specific stimulus. Some of the basic instincts in humans include survival, feeding, drinking, voiding, and specific vocalization. Some ethologists (scientists who study animal behavior) believe that reproduction becomes an instinctive behavior following puberty.

1. State the meaning of ANS, PNS, and CNS. Which of these is a functional component of the nervous system and is further subdivided into sympathetic and parasympathetic divisions?
2. Explain why neurology is considered a dynamic science.

Exhibit II The developmental sequence of the brain. During the 4th week, the three principal regions of the brain are formed. During the 5th week, a five-regioned brain develops and specific structures begin to form.

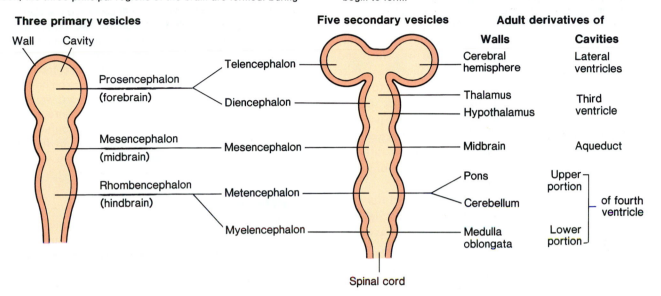

Explanation

The first indication of nervous tissue development occurs about 17 days following conception, when a thickening appears along the entire dorsal length of the embryo. This thickening, called the **neural plate** (exhibit I), differentiates and eventually gives rise to all of the **neurons** and to most of the **neuroglial cells,** which are the supporting cells of the nervous system. As development progresses, the midline of the neural plate invaginates to become the **neural groove.** At the same time, there is a proliferation of cells along the lateral margins of the neural plate, which become the thickened **neural folds.** The neural groove continues to deepen as the neural folds elevate. By day 20, the neural folds have met and fused at the midline and the neural groove has become a **neural tube.** For a short time, the neural tube is open both cranially and caudally. These openings, called **neuropores,** close during the fourth week. Once formed, the neural tube separates from the surface ectoderm and eventually develops into the central nervous system (brain and spinal cord). The **neural crest** forms from the neural folds as they fuse longitudinally along the dorsal midline. Most of the peripheral nervous system (cranial and spinal nerves) forms from the neural crest. Some neural crest cells break away from the main tissue mass and migrate to other locations, where they differentiate into motor nerve cells of the sympathetic nervous system or into *neurolemmocytes* (Schwann cells), which are a type of neuroglial cell important in the peripheral nervous system.

The brain begins its embryonic development as the cephalic end of the neural tube starts to grow rapidly and differentiate (exhibit II). By the middle of the fourth week, three distinct swellings are evident: the **prosencephalon** (*pros''en-sef'ă-lon*) (forebrain), the **mesencephalon** (midbrain), and the **rhombencephalon** (hindbrain). Further development during the fifth week results in the formation of five specific regions. The **telencephalon** and the **diencephalon** (*di''en-sef'-ă-lon*) derive from the forebrain, the mesencephalon remains unchanged, and the **metencephalon** and **myelencephalon** form from the hindbrain. The caudal portion of the myelencephalon is continuous with and resembles the spinal cord.

Mitotic activity within nervous tissue is completed during prenatal development. Thus, a person is born with all the neurons he is capable of producing. However, nervous tissue continues to grow and to specialize after a person is born, particularly in the first several years of postnatal life.

Neurons and Neuroglia

Neurons have many forms, but all contain dendrites for reception and an axon for the conduction of nerve impulses. Neurons contain axons that may be sensory and/or motor in function. Different types of neuroglial cells support the neurons, both structurally and functionally.

UNIT V

Objective 4. Describe the microscopic structure of a neuron.

Objective 5. Describe how a neurolemmal sheath and a myelin sheath are formed.

Objective 6. List the types of neuroglial cells and describe their functions.

Objective 7. Describe the functions and locations of sensory and motor nerve fibers.

The highly specialized and complex nervous system is composed of only two principal categories of cells—neurons and neuroglia. **Neurons** are the basic structural and functional units of the nervous system. They are specialized to respond to physical and chemical stimuli, conduct impulses, and release specific chemical regulators. Through these activities, neurons perform such functions as storing memory, thinking, and regulating other organs and glands. Neurons cannot divide mitotically, although some neurons can regenerate a severed portion or sprout small new branches under some conditions.

Neuroglia, or **glial cells,** are supportive cells in the nervous system that aid the function of neurons. Neuroglia cells are about five times more abundant than neurons and have limited mitotic abilities.

Neurons

Although **neurons** vary considerably in size and shape, they all have three principal components: (1) a cell body, (2) dendrites, and (3) an axon (figs. 11.2 and 11.3).

The **cell body** is the enlarged portion of the neuron that contains the nucleus and nucleolus surrounded by cytoplasm. Besides containing organelles typically found in cells, the cytoplasm of neurons is characterized by the presence of **chromatophilic substances** (Nissl bodies) and filamentous strands of protein called **neurofibrils** *(noor'ō-fi'brilz)*. Chromatophilic substances are specialized layers of granular (rough) endoplasmic reticulum, whose function is protein synthesis, and minute **microtubules,** which appear to be involved in transporting material within the cell. The cell bodies within the CNS are frequently clustered into regions called **nuclei** (not to be confused with the nucleus of a cell). Cell bodies in the PNS generally occur in clusters called **ganglia.**

Dendrites are branched processes that extend from the cytoplasm of the cell body. Dendrites function to receive stimuli and conduct impulses to the cell body. Some dendrites are covered with minute **dendritic spinules** that greatly increase their surface area and provide contact points for other neurons. The area occupied by dendrites is referred to as the **dendritic zone** of a neuron.

The **axon** is the second type of cytoplasmic extension from the cell body. The term *nerve fiber* is commonly used in reference to either an axon or an elongated dendrite. An axon is a relatively long, cylindrical process that conducts impulses away from the cell body. Axons vary in length from a few millimeters in the CNS to over a meter between the distal portions of the extremities and the spinal cord. Side branches called **collateral branches** extend a short distance from the axon. The cytoplasm of an axon contains many mitochondria, microtubules, and neurofibrils.

Proteins and other molecules are transported rapidly through the axon by two different mechanisms: axoplasmic flow and axonal transport. *Axoplasmic flow,* the slower of the two, results from rhythmic waves of contraction that push cytoplasmic contents from the axon hillock, where the axon originates, to the nerve fiber endings. *Axonal transport,* which is more rapid and more selective, may occur in a retrograde as well as a forward direction. Indeed, such retrograde transport may be responsible for the movement of herpes virus, rabies virus, and tetanus toxin from nerve terminals into cell bodies.

Neuroglia

Unlike other organs that are packaged in connective tissue derived from mesoderm, all but one type of supporting **neuroglial (glial) cells** (fig. 11.4) are derived from the same ectoderm that produces neurons. There are six categories of neuroglial cells: (1) **neurolemmocytes** (Schwann cells), which form myelin layers around axons in the PNS; (2) **oligodendrocytes** *(ol''i-go-den'drō-sītz),* which form myelin layers around axons in the CNS; (3) **microglia** *(mi-krog'le-ă),* which are derived from mesoderm and are phagocytic cells that migrate through the CNS and remove foreign and degenerated material; (4) **astrocytes,** which help regulate the passage of molecules from the blood to the brain; (5) **ependymal** *(ē-pen'di-mal)* **cells,** which line the ventricles of the brain and the central canal of the spinal cord; and (6) **ganglionic gliocytes,** which support neuron cell bodies within the ganglia of the PNS (table 11.2).

Myelination Neurons are either *myelinated* or *unmyelinated.* **Myelination** is the process in which a neuroglial cell surrounds a portion of the axon or dendrite to provide support and aid in the conduction of impulses (figs. 11.5 and 11.6). The neuroglial cells that participate in the myelination process contain a white lipid-protein substance called **myelin.** As several neuroglia are positioned in sequence, a **myelin layer** that encloses the axon or dendrite is formed. Myelinated neurons occur both in the CNS and the PNS. Myelin is responsible for the color of the *white matter* of the brain and spinal cord and the white coloration of nerves.

Myelination in the PNS occurs as neurolemmocytes grow and wrap around an axon or dendrite (figs. 11.2, 11.5, and

---

Nissl body: from Franz Nissl, German neuroanatomist, 1860–1919
ganglion: Gk. *ganglion,* swelling
dendrite: Gk. *dendron,* tree branch
axon: Gk. *axon,* axis

neuroglia: Gk. *neuron,* nerve; *glia,* glue
Schwann cell: from Theodor Schwann, German histologist, 1810–82
oligodendrocyte: Gk. *oligos,* few; L. *dens,* tooth; Gk. *kytos,* hollow (cell)
microglia: Gk. *mikros,* small; *glia,* glue
astrocyte: Gk. *aster,* star; *kytos,* hollow (cell)
ependyma: Gk. *ependyma,* upper garment
myelin: Gk. *myelos,* marrow

UNIT V

Figure 11.2 (a) A motor neuron conducts impulses away from the CNS and (b) a sensory neuron conducts nerve impulses toward the CNS.

Figure 11.3 A photomicrograph of neurons from the anterior column of gray matter of the spinal cord (120×).

Dendrites

Chromatophilic substance

Cell body

Nucleolus

Nucleus

Neurofibrils

Nucleus of neurolemmocyte

Axon

Myelin

Collateral branch

Direction of conduction

Neurofibril nodes

Axon terminals

(a)

Cell body

Nucleus

Nucleolus

Axon

Myelin

Nucleus of neurolemmocyte

Chromatophilic substance

Neurofibril nodes

Direction of conduction

Marshburn

Dendrites

(b)

UNIT V

Figure 11.4 The four types of neuroglia found within the central nervous system.

Oligodendrocyte

Microglial cell

Ependymal cell

Neuron

Capillary

Astrocyte

Moon

Table 11.2	Structure and Function of Neuroglia				
Type	**Structure**	**Function**	**Type**	**Structure**	**Function**
Astrocytes	Stellate with numerous processes	Form structural support between capillaries and neurons within the CNS; blood-brain barrier	Ependymal cells	Columnar cells that may have ciliated free surfaces	Line ventricles and the central canal within the CNS where cerebrospinal fluid is circulated by ciliary motion
Oligodendrocytes	Similar to astrocytes, but with shorter and fewer processes	Form myelin in the CNS; guide development of neurons within the CNS	Ganglionic gliocytes	Small, flattened cells	Support ganglia within the PNS
Microglia	Minute cells with few short processes	Phagocytize pathogens and cellular debris within the CNS	Neurolemmocytes (Schwann cells)	Flattened cells arranged in series around axons or dendrites	Form myelin within the PNS

Figure 11.5 A myelinated neuron. A myelin layer is formed by the wrapping of neurolemmocytes around the axon of a neuron.

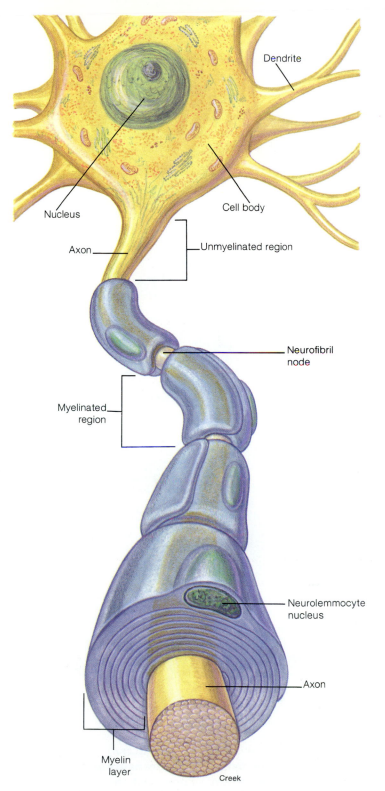

Nucleus

Dendrite

Cell body

Axon

Unmyelinated region

Neurofibril node

Myelinated region

Neurolemmocyte nucleus

Axon

Myelin layer

Creek

11.6). The outer surface of the myelin sheath is encased in a glycoprotein **neurolemmal sheath** that promotes neuron regeneration should the neuron become injured. Each neurolemmocyte wraps only about 1 mm of axon, leaving gaps of exposed axon between the adjacent neurolemmocyte. These gaps in the myelin sheath and neurolemmal sheath are known as the **neurofibril nodes** (nodes of Ranvier). It is at the neurofibril nodes that a nerve impulse is propagated along a neuron.

The myelin sheaths of the CNS are formed by **oligodendrocytes.** Unlike a neurolemmocyte, which forms a myelin sheath around only one axon, each oligodendrocyte has extensions that form myelin sheaths around several axons.

Regeneration of a Cut Axon   When an axon in a peripheral nerve is cut, the distal region of the axon that was severed from the cell body degenerates and is phagocytosed by neurolemmocytes. The neurolemmocytes then form a *regeneration tube* (fig. 11.7), as the part of the axon that is connected to the cell body begins to grow and exhibit ameboid movement. The neurolemmocytes of the regeneration tube are believed to secrete chemicals that attract the growing axon tip, and the tube helps to guide the regenerating axon to its proper destination. Even a severed major nerve may be surgically reconnected and the function of the nerve reestablished if the surgery is performed before tissue death. The CNS lacks neurolemmocytes, and central axons are generally believed to have a much more limited ability to regenerate than peripheral axons.

Astrocytes and the Blood-Brain Barrier   Astrocytes are large, stellate cells with numerous cytoplasmic processes that radiate outward. These are the most abundant neuroglia in the CNS, and in some locations in the brain they constitute 90% of the nervous tissue.

Capillaries in the brain, unlike those of most other organs, do not have pores between adjacent endothelial cells. Molecules within these capillaries must thus be moved through the endothelial cells by active transport, endocytosis, and exocytosis. Astrocytes within the brain have numerous extensions called **vascular processes** that surround most of the outer surface of the brain capillaries (fig. 11.8). Before molecules in the blood can enter neurons in the CNS, they may have to pass through both the endothelial cells and the astrocytes. Astrocytes, therefore, contribute to the **blood-brain barrier,** which is highly selective; some molecules are permitted to pass, whereas closely related molecules may not be allowed to cross the barrier.

The blood-brain barrier presents difficulties in the chemotherapy of brain diseases because drugs that could enter other organs may not be able to enter the brain. In the treatment of *Parkinson's disease*, for example, patients who need a chemical called dopamine in the brain must be given a precursor molecule called levodopa (L-dopa). This is because dopamine cannot cross the blood-brain barrier, whereas L-dopa can enter neurons and be changed to dopamine in the brain.

nodes of Ranvier: from Louis A. Ranvier, French pathologist, 1835–1922

UNIT V

Figure 11.6 The process of myelination. (*a*) An axon of a peripheral neuron becomes myelinated as a myelin layer is formed by the wrapping of neurolemmocytes around the axon. (*b*) A photomicrograph shows a cross section of a myelinated axon.

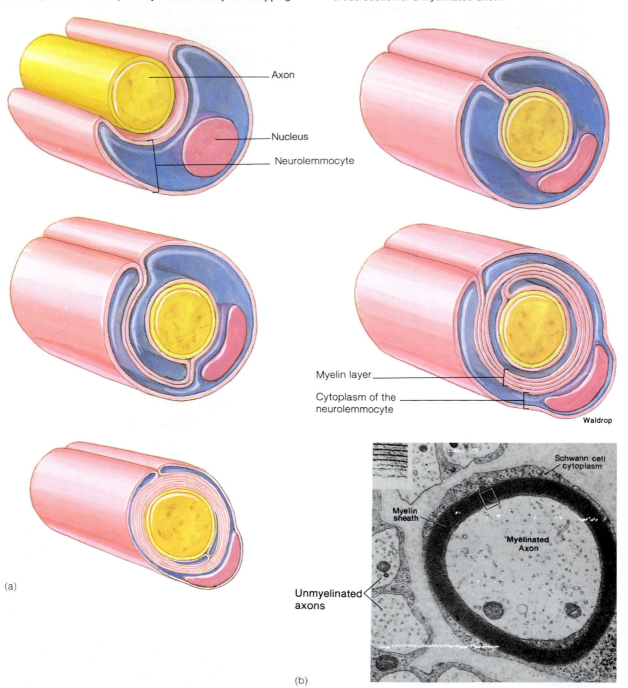

Axon

Nucleus

Neurolemmocyte

Myelin layer

Cytoplasm of the neurolemmocyte

Waldrop

(a)

Schwann cell cytoplasm

Myelin sheath

Myelinated Axon

Unmyelinated axons

(b)

Classification of Neurons and Nerves

Neurons may be classified according to structure or function. The functional classification is based on the direction of conducted impulses. Sensory impulses originate in sensory receptors and are conducted by **sensory,** or **afferent, neurons** to the CNS. Motor impulses originate in the CNS and are conducted by **motor,** or **efferent, neurons** to a muscle or gland (fig. 11.9).

Motor neurons may be *somatic* (nonvisceral) or *autonomic* (visceral). **Association neurons,** or **interneurons,** are located between sensory and motor neurons and are found within the spinal cord and brain.

The structural classification of neurons is based on the number of processes that extend from the cell body of the neuron (fig. 11.10). The spindle-shaped **bipolar neuron** has a process at both ends; this type occurs in the retina of the eye. **Multipolar neurons** are the most common type and are char-

Figure 11.7 Neuron regeneration. (*a*) If a neuron is severed through a myelinated axon, the proximal portion may survive, but (*b*) the distal portion degenerates through phagocytosis. (*c–d*) The myelin layer provides a pathway for the regeneration of an axon and (*e*) innervation is restored.

acterized by several dendrites and one axon extending from the cell body. Motor neurons are a good example of this type. A **pseudounipolar neuron** has a single process that divides into two. Most sensory neurons are pseudounipolar and have their cell bodies located in sensory ganglia of spinal and cranial nerves.

A **nerve** is a collection of nerve fibers outside the CNS. Fibers within a nerve are held together and strengthened by loose fibrous connective tissue (fig. 11.11). Each individual nerve fiber is enclosed in a connective tissue sheath called the **endoneurium** (*en"do-nyoo're-um*). A group of fibers, called a **fasciculus** (*fă-sik'yŭ-lus*), is surrounded by a connective tissue sheath called a **perineurium.** The entire nerve is surrounded and supported by connective tissue called the **epineurium,** which contains tiny blood vessels and often adipose cells. Perhaps less than a quarter of the bulk of a nerve consists of nerve fibers. More than half is associated connective tissue, and approximately a quarter is the myelin that surrounds the nerve fibers.

Figure 11.8 A photomicrograph showing the vascular processes of astrocytes, which cover most of the surface area of brain capillaries and participate in the blood-brain barrier.

fasciculus: L. diminutive of *fascis*, bundle

Figure 11.9 The relationship between sensory and motor fibers of the peripheral nervous system (PNS) and the central nervous system (CNS).

Figure 11.10 The three morphological types of neurons.

Figure 11.11 The structure of a nerve.

Figure 11.12 The classification of nerve fibers by origin and function.

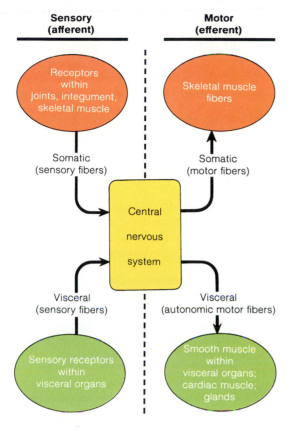

Most nerves are composed of both motor and sensory fibers, and thus are called **mixed nerves.** Some of the cranial nerves, however, are composed either of sensory neurons only **(sensory nerves)** or of motor neurons only **(motor nerves).** Sensory nerves serve the special senses, such as taste, smell, sight, and hearing. Motor nerves conduct impulses to muscles, causing them to contract, or to glands, causing them to secrete.

Neurons and their fibers within nerves may be classified according to the area of innervation into the following scheme (fig. 11.12):

1. **Somatic sensory.** Sensory receptors within the skin, bones, muscles, and joints receive stimuli and convey nerve impulses through somatic sensory (afferent) fibers to the CNS for interpretation. Sensory receptors within the eyes and ears are also somatic sensory.

2. **Somatic motor.** Impulses from the CNS through somatic motor (efferent) fibers cause the contraction of skeletal muscles.

3. **Visceral sensory.** Visceral sensory (afferent) fibers convey impulses from visceral organs and blood vessels to the CNS for interpretation. Sensory receptors within the tongue (taste) and nasal epithelium (smell) are also visceral sensory.

4. **Visceral motor.** Visceral motor (efferent) fibers, also called **autonomic motor fibers,** are part of the autonomic nervous system. They originate in the CNS and innervate cardiac muscle, glands, and smooth muscle within the visceral organs.

1. Briefly describe the functions of dendrites, cell bodies, and axons and describe the differences between bipolar, pseudounipolar, and multipolar neurons.
2. Distinguish between myelinated and unmyelinated axons.
3. Discuss specific ways in which neuroglia aid neurons.
4. Explain the nature of the blood-brain barrier and describe its structure.
5. Explain the differences between a neuron, nerve fiber, and nerve. Describe how nerves are classified.

Transmission of Impulses

Movements of sodium and potassium ions result in an impulse traveling through a neuron toward a synapse. Synaptic transmission is facilitated by the secretion of a neurotransmitter chemical.

Objective 8. Explain how a nerve fiber first becomes depolarized and then repolarized.

Objective 9. Describe the structure of a presynaptic nerve fiber ending and explain how neurotransmitters are released.

Nerve Impulse

Two properties for which neurons are specialized are irritability and conductivity, both of which are involved in the transmission of a nerve impulse. *Irritability* is the ability of dendrites and cell bodies to respond to a stimulus and convert it into an impulse. *Conductivity* is the transmission of an impulse along the axon or a dendrite of a neuron. A *nerve impulse* is the actual movement, or exchange, of sodium (Na^+) and potassium (K^+) ions progressively along a nerve fiber, resulting in the creation of a stimulus to another neuron or another tissue.

Before a nerve fiber can respond to a stimulus, it must be *polarized.* A polarized nerve fiber has an abundance of sodium ions on the outside of the axon membrane, which produces a difference in electrical charge called the *resting potential.* When a stimulus of sufficient strength arrives at the receptor portion of the neuron, the polarized nerve fiber becomes *depolarized* and a nerve impulse is initiated (fig. 11.13). Once depolarization has started, a sequence of ionic exchange occurs along the axon, and the nerve impulse is transmitted (fig. 11.14). *Repolarization* occurs after an impulse has passed through a nerve fiber to reestablish the original concentrations of sodium and potassium ions. As a result, the resting potential returns and the nerve fiber is again ready to send another impulse.

A nerve impulse travels in one direction only and is an *all-or-none* response. This means that if an impulse is initiated, it will travel the length of the nerve fiber and proceed without a loss in voltage. The speed of a nerve impulse is determined by the diameter of the nerve fiber, its type (myelinated or unmyelinated), and the general physiological condition of the neuron. For example, unmyelinated nerve fibers with small diameters conduct impulses at the rate of about 0.5 m/sec; myelinated neurons conduct impulses at speeds up to 130 m/sec.

UNIT V

Figure 11.13 A nerve impulse is initiated if a sufficient stimulus occurs at the receptor site, causing movement of ions across the membrane of the nerve fiber.

Figure 11.14 (a–c) Depolarization occurs in a sequential pattern along a nerve fiber as sodium ions diffuse inward and a wave of action potentials, or a nerve impulse, moves in one direction along the fiber.

Synapse

A **synapse** is a minute space of 200 to 300 Å between the axon terminal of a **presynaptic neuron** and a dendrite of a **postsynaptic neuron** (fig. 11.15). The **axon terminal,** or **synaptic knob,** is the distal portion of the presynaptic neuron at the end of the axon; it is characterized by the presence of numerous mitochondria and **synaptic vesicles.** Synaptic vesicles contain a neurotransmitter chemical, the most common of which is *acetylcholine.* When a nerve impulse reaches the axon terminal, some of the vesicles respond by releasing their neurotransmitter. If a sufficient number of nerve impulses occur in a short interval of time, enough neurotransmitter accumulates within the synaptic cleft to stimulate the postsynaptic neuron to depolarize.

Neurotransmitters are decomposed by enzymes present in the synaptic cleft. *Cholinesterase* is the enzyme that decomposes acetylcholine and prepares the synapse to receive additional neurotransmitter with the next impulse.

Synaptic transmission may be affected by various drugs or diseases. Caffeine is a stimulant that increases the rate of transmission across the synapse. Aspirin causes a moderate decrease in the transmission rate. The drug strychnine profoundly reduces synaptic transmission. It affects breathing and paralyzes respiratory structures, causing death. *Parkinson's disease* is caused by a deterioration of neurons within the brain that synthesize a neurotransmitter called *dopamine.*

1. Define the terms *depolarization* and *repolarization* and illustrate these processes graphically.
2. How does a neurotransmitter chemical affect impulse conduction across a synapse?

General Features of the Brain

The brain is enclosed by the skull and meninges and is bathed in cerebrospinal fluid. The brain has a tremendous metabolic rate and is susceptible to oxygen deprivation and certain toxins.

Objective 10. Describe the general features of the brain.

Objective 11. Comment on the importance of neuro-peptides.

Objective 12. List the metabolic needs of the brain.

The entire delicate CNS is protected by a bony encasement—the cranium (figs. 11.16 and 11.17) surrounding the brain and the vertebral column surrounding the spinal cord. The **meninges** (*mĕ-nin'jēz*) are connective tissue encasements that form a protective membrane between the bone and the soft tissue of the CNS. The CNS is bathed in **cerebrospinal fluid** that circulates within the hollow **ventricles** of the brain, the **central canal** of the spinal cord, and the **subarachnoid** (*sub''ă-rak'noid*) **space** surrounding the entire CNS (see figs. 11.34 and 11.35).

synapse: Gk. *syn,* together; *haptein,* to fasten

Figure 11.15   Synaptic transmission. When a nerve impulse reaches the synaptic knob, neurotransmitter chemicals are released into the synaptic cleft. Synaptic transmission occurs if sufficient amounts of the neurotransmitter chemicals are released.

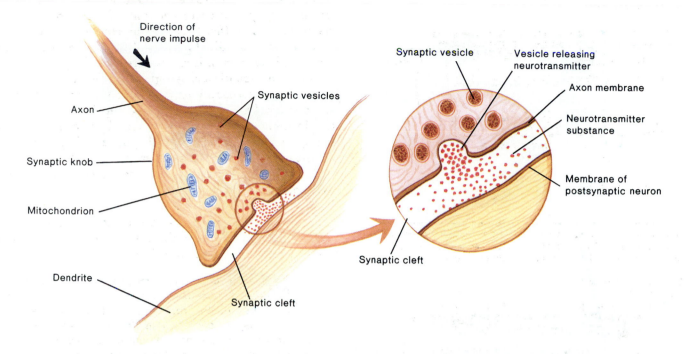

Figure 11.16   The CNS consists of the brain and the spinal cord, both of which are covered with meninges and bathed in cerebrospinal fluid.

Figure 11.17 The brain. (*a*) A lateral view, (*b*) an inferior view, (*c*) a sagittal view, and (*d*) a sagittal view of a magnetic resonance image (MRI) of the skull, brain, and cervical portion of the spinal cord.

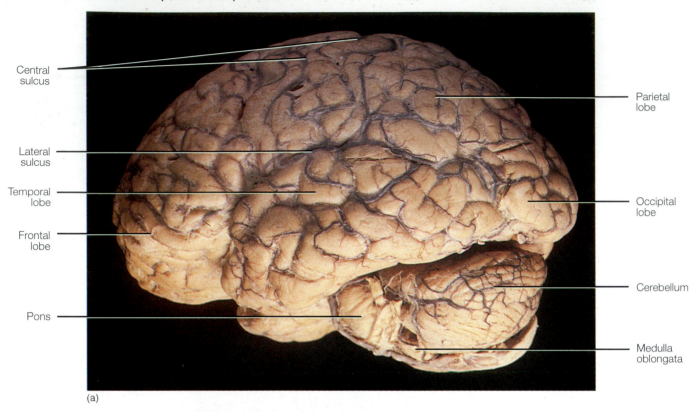

Central sulcus

Lateral sulcus

Temporal lobe

Frontal lobe

Pons

Parietal lobe

Occipital lobe

Cerebellum

Medulla oblongata

(a)

Olfactory tract

Internal carotid artery
Pons

Abducens nerve

Trigeminal nerve

Facial nerve

Vestibulocochlear nerve

Dura mater

Attached eyeball

Extrinsic eye muscles

Optic nerve

Pituitary gland

Temporal lobe

Basilar artery

Vertebral arteries

Medulla oblongata

Cerebellum

Spinal cord

(b)

UNIT V

Figure 11.17   continued

Parietal lobe

Splenium of
corpus callosum

Occipital lobe

Colliculi of
midbrain

Mesencephalic
aqueduct

Fourth ventricle

Cerebellum

Medulla
oblongata

Spinal cord

Body of
corpus callosum
Choroid plexus

Intermediate mass

Genu of corpus
callosum

Frontal lobe

Optic nerve

Pons

Temporal lobe

(c)

Cerebrum

Corpus
callosum

Ventricle

Cerebellum

Pons

Medulla
oblongata

Spinal
cord

006/011

(d)

UNIT V

The CNS is composed of gray and white matter. **Gray matter** consists of either nerve cell bodies and dendrites or bundles of unmyelinated axons and neuroglia. The gray matter of the brain exists as the outer convoluted **cortex layer** of the cerebrum and cerebellum. In addition, specialized gray matter clusters of nerve cells called **nuclei** are found deep within the white matter. **White matter** forms the tracts within the CNS and consists of aggregations of dendrites and myelinated axons and associated neuroglia.

The brain of an adult weighs nearly 1.5 kg (3–3.5 lbs) and is composed of an estimated 100 billion (10^{11}) neurons. Neurons communicate with one another by means of innumerable synapses between the axons and dendrites within the brain. Neurotransmission within the brain is regulated by specialized neurotransmitter chemicals called *neuropeptides*. These specialized chemical protein messengers are thought to account for specific mental functions.

> Over 200 neuropeptides have been identified within the brain, yet most of their functions are little understood. Two groups of neuropeptides that have received considerable attention are *enkephalins* and *endorphins*. These classes of substances numb the brain to pain, functioning in a manner similar to morphine. They are released in response to stress or pain in a traumatized person.

The brain has a tremendous metabolic rate and needs a continuous supply of oxygen and nutrients. The brain accounts for only 2% of a person's body weight, and yet it receives approximately 20% of the total resting cardiac output. This amounts to a flow of about 750 mℓ of blood per minute. The volume remains relatively constant even during physical or mental activity. This continuous flow is so crucial that a failure of cerebral circulation for as short an interval as 10 seconds causes unconsciousness. The brain is composed of perhaps the most sensitive tissue of the body. Due to its high metabolic rate, not only does the brain require continuous oxygen, but it also requires a continuous nutrient supply and rapid removal of wastes. The brain is also very sensitive to certain toxins and drugs. The cerebrospinal fluid aids the metabolic needs of the brain through the distribution of nutrients and the removal of wastes. Cerebrospinal fluid also maintains a protective homeostatic environment within the brain. The blood-brain barrier and the secretory activities of neural tissue also help to maintain homeostasis. The brain has an extensive vascular supply through the paired internal carotid and vertebral arteries that unite at the cerebral arterial circle (circle of Willis) (see chapter 16 and fig. 16.24).

Figure 11.18 Positron emission tomographic (PET) brain scan (in transverse section) of an unmedicated schizophrenic patient. Red areas indicate high glucose use (uptake of 18–F-deoxyglucose). The scan shows highest glucose uptake in the posterior region, where the brain's visual center is located.

Anterior

Posterior

> The brain of a newborn is especially sensitive to oxygen deprivation or to excessive oxygen. If complications arise during childbirth and the oxygen supply from the mother's blood to the baby is interrupted while it is still in the birth canal, the infant may be stillborn or suffer brain damage that can result in cerebral palsy, epilepsy, paralysis, or mental retardation. Excessive oxygen administered to a newborn may cause blindness.

> Measurable increases within the brain in regional blood flow and in glucose and oxygen metabolism accompany mental functions, including perception and emotion. These metabolic changes can be assessed through the use of *positron emission tomography (PET)*. The technique of a PET scan (fig. 11.18) is based on injecting radioactive tracer molecules labeled with carbon-11, fluorine-18, and oxygen-15 into the bloodstream and photographing the gamma rays that are emitted from the patient's brain through the skull. PET scans are of value in studying neurotransmitters and neuroreceptors, as well as the substrate metabolism of the brain.

UNIT V

Table 11.3	Derivation and Functions of the Major Brain Structures		
	Region	**Structure**	**Function**
Prosencephalon (forebrain)	Telencephalon	Cerebrum	Control of most sensory and motor activities; reasoning, memory, intelligence, etc.; instinctual and limbic functions
	Diencephalon	Thalamus	Relay center: all impulses (except olfactory) going into cerebrum synapse here; some sensory interpretation; initial autonomic response to pain
		Hypothalamus	Regulation of renal water flow, body temperature, hunger, heartbeat, etc.; control of secretory activity in anterior pituitary gland; instinctual and limbic functions
		Pituitary gland	Regulation of other endocrine glands
Mesencephalon (midbrain)	Mesencephalon	Superior colliculi	Visual reflexes (eye-hand coordination)
		Inferior colliculi	Auditory reflexes
		Cerebral peduncles	Reflex coordination; contain many motor fibers
Rhombencephalon (hindbrain)	Metencephalon	Cerebellum	Balance and motor coordination
		Pons	Relay center; contains nuclei (pontine nuclei)
	Myelencephalon	Medulla oblongata	Relay center; contains many nuclei; visceral autonomic center (e.g., respiration, heart rate, vasoconstriction)

From the five basic regions of the brain, the development of which was described earlier in this chapter, distinct functional structures are formed (table 11.3). These structures will be discussed in greater detail in the following sections.

1. What characteristics do the brain and spinal cord have in common? Describe the general features of each.
2. Discuss how the study of neuropeptides will enhance understanding of brain function.
3. Using specific examples, describe the metabolic requirements of the brain.

Cerebrum

The cerebrum, consisting of five paired lobes within two convoluted hemispheres, is concerned with higher brain functions, such as the perception of sensory impulses, the instigation of voluntary movement, the storage of memory, thought processes, and reasoning ability. The cerebrum is also concerned with instinctual and limbic (emotional) functions.

Objective 13. Describe the structure of the cerebrum and list the functions of the cerebral lobes.

Objective 14. Define the term *electroencephalogram* and discuss its clinical importance.

Objective 15. Describe the fiber tracts within the cerebrum.

Structure of the Cerebrum

The **cerebrum,** located in the region of the telencephalon, is the largest and most obvious portion of the brain. It accounts for about 80% of the mass of the brain and is responsible for the higher mental functions, including memory and reason. The cerebrum consists of the **right** and **left hemispheres,** which are incompletely separated by a **longitudinal cerebral fissure** (fig. 11.19). Portions of the two hemispheres are connected internally by the **corpus callosum** *(ka-lo'sum),* a large tract of white matter (see fig. 11.16). A portion of the meninges called the **falx** *(falks)* **cerebri** extends into the longitudinal fissure. Each cerebral hemisphere contains a central cavity called the **lateral ventricle** (fig. 11.20), which is lined with ependymal cells and filled with cerebrospinal fluid.

The two cerebral hemispheres carry out different functions. In most people, the left hemisphere controls analytical and verbal skills, such as reading, writing, and mathematics. The right hemisphere is the source of spatial and artistic kinds of intelligence. The corpus callosum unifies attention and awareness between the two hemispheres and permits a sharing of learning and memory. Severing the corpus callosum is a radical treatment to control severe epileptic seizures. Although this surgery is successful, it results in the cerebral hemispheres functioning as separate structures, each with its own memories and thoughts, competing for control. A more recent and effective technique of controlling epileptic seizures is a precise laser treatment of the corpus callosum.

cerebrum: L. *cerebrum,* brain

Figure 11.19 The cerebrum. (*a*) A lateral view and (*b*) a superior view.

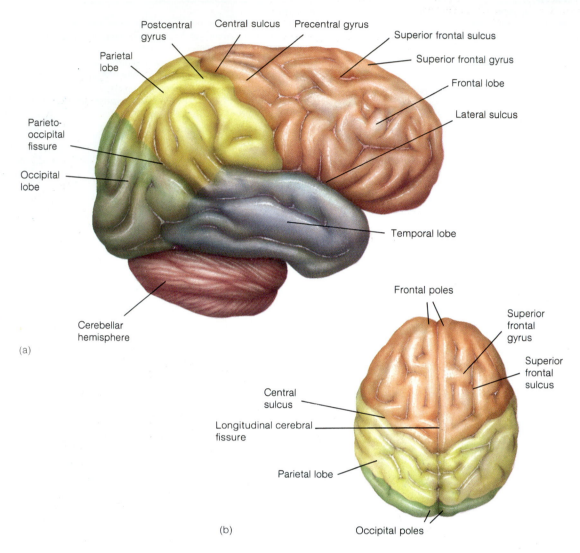

(a)

Postcentral gyrus
Central sulcus
Precentral gyrus
Superior frontal sulcus
Superior frontal gyrus
Parietal lobe
Frontal lobe
Lateral sulcus
Parieto-occipital fissure
Occipital lobe
Temporal lobe
Cerebellar hemisphere

Frontal poles
Superior frontal gyrus
Superior frontal sulcus
Central sulcus
Longitudinal cerebral fissure
Parietal lobe
Occipital poles

(b)

The cerebrum consists of two layers. The surface layer, referred to as the **cerebral cortex,** is composed of gray matter that is 2–4 mm (0.08–0.16 in.) thick (fig. 11.20). Beneath the cerebral cortex is the thick **white matter** of the cerebrum, which constitutes the second layer. The cerebral cortex is characterized by numerous folds and grooves called **convolutions.** Convolutions form during early fetal development, when brain size increases rapidly and the cortex enlarges out of proportion to the underlying white matter. The elevated folds of the convolutions are the **cerebral gyri** (singular, *gyrus*), and the depressed grooves are the **cerebral sulci** (*sul′si*) (singular, *sulcus*). The convolutions greatly increase the area of the gray matter, which is composed of nerve cell bodies.

Recent studies indicate that with increased learning, there is an increase in the number of synapses between neurons within the cerebrum. Although the number of neurons is established during prenatal development, the number of synapses is variable depending upon the learning process. The number of cytoplasmic extensions from the cell body of a neuron determines the extent of nerve impulse conduction and the associations that can be made to cerebral areas already containing stored information.

Lobes of the Cerebrum

Each cerebral hemisphere is subdivided into five lobes by especially deep sulci called fissures. Four of these lobes appear on the surface of the cerebrum and are named according to the overlying cranial bones (fig. 11.21). The reasons for the separate cerebral lobes, as well as two cerebral hemispheres, have to do with specificity of function (table 11.4).

Frontal Lobe The frontal lobe forms the anterior portion of each cerebral hemisphere (fig. 11.21). A prominent deep furrow

gyrus: Gk. *gyros,* circle
sulcus: L. *sulcus,* a furrow or ditch

UNIT V

Figure 11.20 Sections through the cerebrum and diencephalon. (*a*) A coronal section and (*b*) a transverse section.

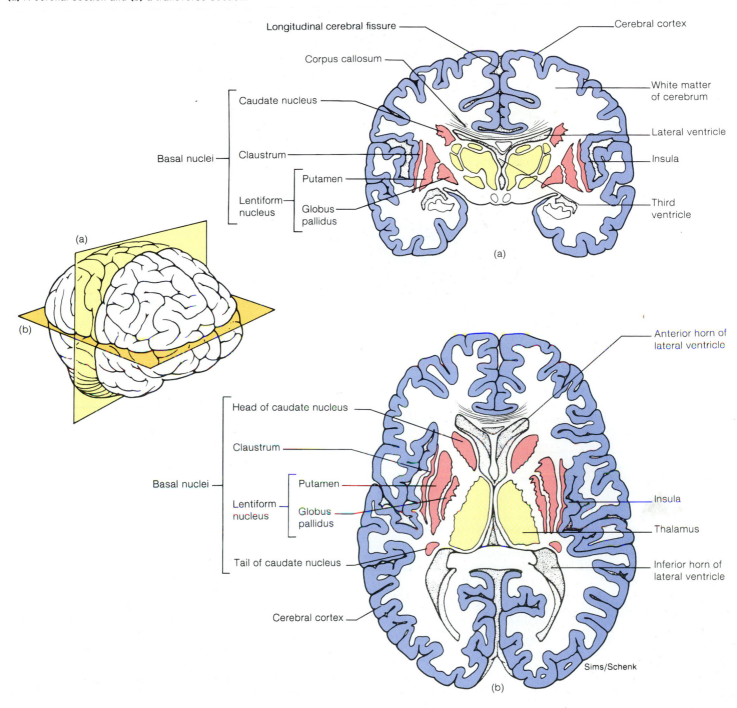

called the **central sulcus** (fissure of Rolando) separates the frontal lobe from the parietal lobe. The central sulcus extends at right angles from the longitudinal fissure to the lateral sulcus. The **lateral sulcus** (fissure of Sylvius) extends laterally from the inferior surface of the cerebrum to separate the frontal and temporal lobes. The **precentral gyrus** (see figs. 11.19 and

11.21), an important motor area, is positioned immediately in front of the central sulcus. The frontal lobe's functions include initiating voluntary motor impulses for the movement of skeletal muscles, analyzing sensory experiences, and providing responses relating to personality. The frontal lobes also involve responses related to memory, emotions, reasoning, judgment, planning, and verbal communication.

Parietal Lobe The parietal lobe is posterior to the central sulcus of the frontal lobe. An important sensory area called the **postcentral gyrus** (see figs. 11.19 and 11.21) is positioned

fissure of Rolando: from Luigi Rolando, Italian anatomist, 1773–1831
fissure of Sylvius: from Franciscus Sylvius de la Boe, Dutch anatomist, 1614–72

Figure 11.21 The lobes of the left cerebral hemisphere showing the principal motor and sensory areas of the cerebral cortex.

Table 11.4	Functions of the Cerebral Lobes		
Lobe	**Functions**	**Lobe**	**Functions**
Frontal	Voluntary motor control of skeletal muscles; personality; higher intellectual processes (e.g., concentration, planning, and decision making); verbal communication	Temporal	Interpretation of auditory sensations; storage (memory) of auditory and visual experiences
		Occipital	Integration of movements in focusing the eye; correlation of visual images with previous visual experiences and other sensory stimuli; conscious perception of vision
Parietal	Somatesthetic interpretation (e.g., cutaneous and muscular sensations); understanding speech and formulating words to express thoughts and emotions; interpretation of textures and shapes	Insula	Memory; integration of other cerebral activities

immediately behind the central sulcus. The postcentral gyrus is designated as a somatesthetic area because it responds to stimuli from cutaneous and muscular receptors throughout the body.

The size of the portions of the precentral gyrus responsible for motor movement and the size of the portions of the postcentral gyrus that respond to sensory stimuli do not correspond to the size of the part of the body being served but to the number of motor units activated or the density of receptors (fig. 11.22). For example, because the hand has many motor units and sensory receptors, larger portions of the precentral and postcentral gyri serve it than serve the thorax, even though the thorax is much larger.

In addition to responding to somatesthetic stimuli, the parietal lobe functions in the understanding of speech and in verbal articulation of thoughts and emotions. The parietal lobe also interprets the textures and shapes of objects as they are handled.

Temporal Lobe The temporal lobe is located below the parietal lobe and the posterior portion of the frontal lobe. It is separated from both by the lateral sulcus (see fig. 11.21). The temporal lobe contains auditory centers that receive sensory fibers from the cochlea of the ear. This lobe also interprets some sensory experiences and stores memories of both auditory and visual experiences.

Occipital Lobe The occipital lobe forms the posterior portion of the cerebrum and is not distinctly separated from the temporal and parietal lobes (see fig. 11.21). The occipital lobe is superior to the cerebellum and is separated from it by an infolding of the meningeal layer called the **tentorium cerebelli** (see fig. 11.16). The principal functions of the occipital lobe concern vision. The occipital lobe integrates eye movements by directing and focusing the eye. It is also responsible for visual

Figure 11.22 Motor and sensory areas of the cerebral cortex. Motor areas control skeletal muscles and sensory areas receive somatesthetic sensations.

association, correlating visual images with previous visual experiences and other sensory stimuli.

Insula The insula is a deep lobe of the cerebrum that cannot be viewed on the surface (see fig. 11.20). It is deep to the lateral sulcus and is covered by portions of the frontal, parietal, and temporal lobes. Little is known of the function of the insula, though it primarily integrates other cerebral activities and may have some function in memory.

> Because of its size and position, portions of the cerebrum frequently suffer brain trauma. A concussion to the brain may cause a temporary or permanent impairment of cerebral functions; a stroke usually affects cerebral function. Much of what is known about cerebral function comes from observing body dysfunctions when specific regions of the cerebrum are traumatized.

Brain Waves Neurons within the cerebral cortex continuously generate electrical activity that can be recorded by electrodes attached to precise locations on the scalp, producing an **electroencephalogram (EEG).** An EEG pattern, commonly called *brain waves,* is the collective expression of millions of action potentials from neurons.

Brain waves are first emitted from a developing brain during early fetal development and continue throughout a person's life. The cessation of brain-wave patterns may be a decisive factor in the legal determination of death.

Certain distinct EEG patterns signify healthy mental functions. Deviations from these patterns are of clinical significance in diagnosing trauma, mental depression, hematomas, and various diseases, such as tumors, infections, and epilepsy. Normally, there are four kinds of EEG patterns (fig. 11.23).

Figure 11.23 Brain waves. (*a*) A technician using an electroencephalograph to take the EEG of a patient. (*b*) Types of electroencephalograms.

(a)

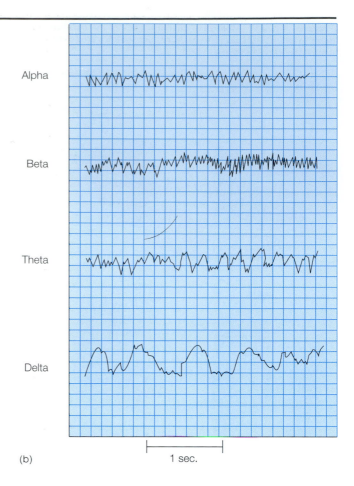

Alpha

Beta

Theta

Delta

(b)

|← 1 sec. →|

1. **Alpha waves** are best recorded from the parietal and occipital regions while a person is awake and relaxed but with the eyes closed. These waves are rhythmic oscillations of about 10–12 cycles/second. The alpha rhythm of a child under the age of 8 occurs at a slightly lower frequency of 4–7 cycles/second.
2. **Beta waves** are strongest from the frontal lobes, especially the area near the precentral gyrus. These waves are sensory-evoked and respond to visual and mental activity. Because they respond to stimuli from receptors and are superimposed on the continuous activity patterns of the alpha waves, they constitute *evoked activity.* The frequency of beta waves is 13–25 cycles/second.
3. **Theta waves** are emitted from the temporal and occipital lobes. They have a frequency of 5–8 cycles/second and are common in newborn infants. The recording of theta waves in adults generally indicates severe emotional stress and can be a forewarning of a nervous breakdown.
4. **Delta waves** are seemingly emitted in a general pattern from the cerebral cortex. These waves have a frequency of 1–5 cycles/second and are common during sleep and in an awake infant. The presence of delta waves in an awake adult indicates brain damage.

White Matter of the Cerebrum

The thick white matter of the cerebrum is deep to the cerebral cortex (see fig. 11.20) and consists of dendrites, myelinated axons, and associated neuroglia. These fibers form the billions of connections within the brain by which information in the form of electrical impulses is transmitted to the appropriate places. The three types of fiber tracts within the white matter are named according to location and the direction in which they conduct impulses (fig. 11.24).

1. **Association fibers** are confined to a given cerebral hemisphere and conduct impulses between neurons within that hemisphere.
2. **Commissural** *(kah-mis'yoo-ral)* **fibers** connect the neurons and gyri of one hemisphere with those of the other. The **corpus callosum** and **anterior commissure** (fig. 11.25) are composed of commissural fibers.
3. **Projection fibers** form the ascending and descending tracts that transmit impulses from the cerebrum to other parts of the brain and spinal cord and from the spinal cord and other parts of the brain to the cerebrum.

Basal Nuclei

The **basal nuclei** are specialized paired masses of gray matter located deep within the white matter of the cerebrum (see figs. 11.20 and 11.26). The most prominent of the basal nuclei is the **corpus striatum,** so named because of its striped appearance. The corpus striatum is composed of several masses of nuclei. The **caudate nucleus** is the upper mass. A thick band of white matter lies between the caudate nucleus and the next lower two masses, collectively called the **lentiform nucleus.** The lentiform nucleus consists of a lateral portion, called the

corpus striatum: L. *corpus,* body; *striare,* striped
lentiform: L. *lentis,* elongated

Figure 11.24 Types of fiber tracts within the white matter associated with the cerebrum. (*a*) Association fibers of a given hemisphere and (*b*) commissural fibers connecting the hemispheres and projection fibers connecting the hemispheres with other structures of the CNS. (Note the decussation [crossing-over] of projection fibers within the medulla oblongata.)

(a)

(b)

putamen (*pyoo-ta'men*), and a medial portion, called the **globus pallidus** (fig. 11.26). The claustrum is another portion of the basal nuclei. It is a thin layer of gray matter just deep to the cerebral cortex of the insula.

The basal nuclei are associated with other structures of the brain, particularly within the mesencephalon. The caudate nucleus and putamen of the basal nuclei control unconscious contractions of certain skeletal muscles, such as those of the upper extremities involved in involuntary arm movements during walking. The globus pallidus regulates the muscle tone necessary for specific, intentional body movements. Neural diseases or physical trauma to the basal nuclei generally cause a variety of motor movement dysfunctions, including rigidity, tremor, and rapid and aimless movements.

Language

Knowledge of the brain regions involved in language has been gained primarily by the study of *aphasias*—speech and language disorders caused by damage to specific language areas of the brain. These areas (fig. 11.27) are generally located in the cerebral cortex of the left hemisphere in both right-handed and left-handed people.

The **motor speech area** (Broca's area) is located in the left inferior gyrus of the frontal lobe. Neural activity in the motor speech area causes selective stimulation of motor impulses in motor centers elsewhere in the frontal lobe, which in turn causes coordinated skeletal muscle movement in the pharynx and larynx. At the same time, motor impulses are sent to the respiratory muscles to regulate air movement across the vocal cords. The combined muscular stimulation translates thought patterns into speech.

Wernicke's area is located in the superior gyrus of the temporal lobe and is directly connected to the motor speech area by a fiber tract called the **arcuate fasciculus.** People with *Wernicke's aphasia* produce speech that has been described as a "word salad." The words used may be real words that are chaotically mixed together, or they may be made-up words.

---

putamen: L. *putare*, to cut, prune
globus pallidus: L. *globus*, sphere; *pallidus*, pale
aphasia: L. *a*, without; Gk. *phasis*, speech

Broca's area: from Pierre P. Broca, French neurologist, 1824–80
Wernicke's area: from Karl Wernicke, German neurologist, 1848–1905

UNIT V

Figure 11.25 A midsagittal section through the brain. (*a*) A diagram and (*b*) a photograph.

Corpus callosum

Septum pellucidum

Intermediate commissure

Genu of corpus callosum

Choroid plexus of third ventricle

Splenum of corpus callosum

Pineal body

Anterior commissure

Thalamus

Optic chiasma

Infundibulum

Pituitary gland

Hypothalamus

Mammillary body

Pons

Medulla oblongata

Corpora quadrigemina

Cortex of cerebellum

Arbor vitae of cerebellum

(a)

(b)

Language comprehension has been destroyed in people with Wernicke's aphasia; they cannot understand either spoken or written language.

It appears that the concept of words to be spoken originates in Wernicke's area and is then communicated to the motor speech area through the arcuate fasciculus. Damage to the arcuate fasciculus produces *conduction aphasia,* which is fluent but nonsensical speech as in Wernicke's aphasia, even though both the motor speech area and Wernicke's area are intact.

The **angular gyrus,** located at the junction of the parietal, temporal, and occipital lobes, is believed to be a center for the integration of auditory, visual, and somatesthetic information. Damage to the angular gyrus produces aphasias, which suggests

that this area projects to Wernicke's area. Some patients with damage to the left angular gyrus can speak and understand spoken language but cannot read or write. Other patients can write a sentence but cannot read it, presumably due to damage to the projections from the occipital lobe (involved in vision) to the angular gyrus.

Recovery of language ability, by transfer to the right hemisphere after damage to the left hemisphere, is very good in children but decreases after adolescence. Recovery is reported to be faster in left-handed people, possibly because language ability is more evenly divided between the two hemispheres in left-handed people. Some recovery usually occurs after damage to the motor speech area, but damage to Wernicke's area produces more severe and permanent aphasias.

1. Diagram a lateral view of the cerebrum and label the four superficial lobes and the fissures that separate them.
2. List the functions of each of the paired cerebral lobes.
3. What is a brain-wave pattern? How are these patterns monitored clinically?
4. Describe the arrangement of the fiber tracts within the cerebrum.
5. Define *basal nuclei* and list their functions.
6. Describe the aphasias produced by damage to the motor speech and Wernicke's areas, by damage to the arcuate fasciculus, and by damage to the angular gyrus. Explain how these areas may interact in the production of speech.

Figure 11.26   Structures of the cerebrum containing neurons involved in the control of skeletal muscles (higher motor neurons). The thalamus is a relay center between the motor cerebral cortex and other brain areas.

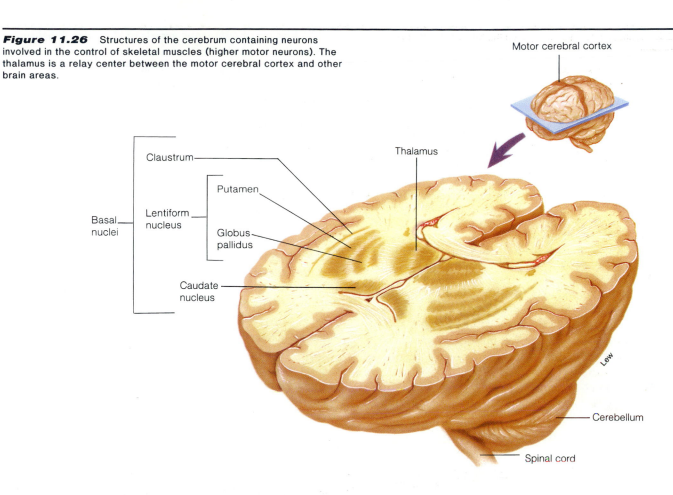

Figure 11.27   Brain areas involved in the control of speech. Arrows indicate the direct communication between these areas.

Diencephalon

The diencephalon is a major autonomic region of the brain that consists of such vital structures as the thalamus, hypothalamus, epithalamus, and pituitary gland.

Objective 16.   List the autonomic functions of the thalamus and the hypothalamus.

Objective 17.   Describe the location and structure of the pituitary gland.

The **diencephalon** (*di″en-sef′ă-lon*) is the second subdivision of the forebrain and is almost completely surrounded by the cerebral hemispheres of the telencephalon. The third ventricle (see fig. 11.36) forms a cavity on the median plane within the diencephalon. The most important structures of the diencephalon are the thalamus, hypothalamus, epithalamus, and pituitary gland.

Thalamus

The **thalamus** *(thal'ă-mus)* is a large ovoid mass of gray matter, constituting nearly four-fifths of the diencephalon. It is actually a paired organ, with each portion positioned immediately below the lateral ventricle of its respective cerebral hemisphere (see figs. 11.25 and 11.26). The principal function of the thalamus is to act as a relay center for all sensory impulses, except smell, to the cerebral cortex. Specialized masses of nuclei relay the incoming impulses to precise locations within the cerebral lobes for interpretation.

The thalamus also performs some sensory interpretation. The cerebral cortex discriminates pain and other tactile stimuli, but the thalamus responds to general sensory stimuli and provides crude awareness. The thalamus probably plays a role in the initial autonomic response of the body to intense pain and is, therefore, partially responsible for the physiological shock that frequently follows serious trauma.

Hypothalamus

The **hypothalamus** *(hi''po-thal'ă-mus)* is a small portion of the diencephalon located below the thalamus, where it forms the floor and part of the lateral walls of the third ventricle (figs. 11.25 and 11.28). The hypothalamus consists of several masses of nuclei interconnected with other parts of the nervous system. Despite its small size, the hypothalamus performs numerous vital functions, most of which relate directly or indirectly to the regulation of visceral activities. It also has emotional and instinctual functions (see also fig. 14.13).

The hypothalamus is an autonomic nervous center because of its role in accelerating or decreasing certain body functions. It secretes eight hormones, including two released from the posterior pituitary gland. These hormones and their functions are discussed in chapter 14. The principal autonomic and limbic (emotional) functions of the hypothalamus are as follows:

1. **Cardiovascular regulation.** Although the heart has an innate pattern of contraction, impulses from the hypothalamus cause autonomic acceleration or deceleration of the heart. Impulses from the posterior hypothalamus produce a rise in arterial blood pressure and an increase of the heart rate. Impulses from the anterior portion have the opposite effect. The impulses from these regions do not travel directly to the heart but pass first to the cardiovascular centers of the medulla oblongata.
2. **Body-temperature regulation.** Specialized nuclei within the anterior portion of the hypothalamus are sensitive to changes in body temperature. If the arterial blood flowing through this portion of the hypothalamus is above normal temperature, the hypothalamus initiates impulses that cause heat loss through sweating and

vasodilation of cutaneous vessels of the skin. A below-normal blood temperature causes the hypothalamus to relay impulses that result in heat production and retention through the initiation of shivering, the contraction of cutaneous blood vessels, and the cessation of sweating.

3. **Regulation of water and electrolyte balance.** Specialized *osmoreceptors* in the hypothalamus continuously monitor the osmotic concentration of the blood. An increased osmotic concentration due to lack of water causes antidiuretic hormone (ADH) to be produced by the hypothalamus and released from the posterior pituitary gland. At the same time, a *thirst center* within the hypothalamus produces feelings of thirst.
4. **Regulation of hunger and control of gastrointestinal activity.** The *feeding center* is a specialized portion of the lateral hypothalamus that monitors the blood glucose, fatty acid, and amino acid levels. Low levels of these substances in the blood are partially responsible for a sensation of hunger elicited from the hypothalamus. When sufficient amounts of food have been ingested, the *satiety (sah-ti'ĭ-te) center* in the midportion of the hypothalamus inhibits the feeding center. The hypothalamus also receives sensory impulses from the abdominal viscera and regulates glandular secretions and the peristaltic movements of the GI tract.
5. **Regulation of sleeping and wakefulness.** The hypothalamus has both a *sleep center* and a *wakefulness center* that function with other parts of the brain to determine the level of conscious alertness.
6. **Sexual response.** Specialized *sexual center* nuclei within the dorsal portion of the hypothalamus respond to sexual stimulation of the tactile receptors within the genital organs. The experience of orgasm involves neural activity within the sexual center of the hypothalamus.
7. **Emotions.** Found within the hypothalamus are a number of nuclei associated with specific emotional responses, such as anger, fear, pain, and pleasure (see fig. 14.13).
8. **Control of endocrine functions.** The hypothalamus produces neurosecretory chemicals that stimulate the anterior and posterior pituitary to release various hormones.

Obviously, the hypothalamus is one of the most vital structures of the body. Dysfunction of the hypothalamus may seriously affect the autonomic, somatic, or psychic body functions. Not surprisingly, the hypothalamus is implicated as a principal factor in *psychosomatic illness*. Insomnia, peptic ulcers, palpitation of the heart, diarrhea, and constipation are a few symptoms of psychosomatic problems.

Epithalamus

The **epithalamus** is the dorsal portion of the diencephalon that includes a thin roof over the third ventricle. The inside lining of the roof consists of a vascular **choroid plexus** where cere-

---

thalamus: L. *thalamus,* inner room

Figure 11.28 The pituitary gland is positioned within the sella turcica of the sphenoid bone and is attached to the brain by the infundibulum.

Anterior cerebral artery

Cerebral cortex

Optic chiasma

Infundibulum

Adenohypophysis

Sphenoidal sinus

Sphenoid bone

Hypothalamus

Neurohypophysis

Basilar artery

Sella turcica

Waldrop

brospinal fluid is produced (see fig. 11.25). A small cone-shaped mass called the **pineal gland** (epiphysis), which has a neuroendocrine function, extends outward from the posterior end of the epithalamus (see fig. 11.25). The posterior commissure, located inferior to the pineal gland, is a tract of commissural fibers that connects the superior colliculi (see fig. 11.31).

Pituitary Gland

The **pituitary gland,** or **cerebral hypophysis** (hi-pof'ĭ-sis), is positioned on the inferior aspect of the diencephalon and is attached to the hypothalamus by a stalklike structure called the **infundibulum** (in"fun-dib'yoo-lum) (see fig. 11.25). The pituitary is a rounded, pea-shaped gland measuring about 1.3 cm (0.5 in.) in diameter. It is covered by the dura mater and is supported by the sella turcica of the sphenoid bone (fig. 11.28). The cerebral arterial circle surrounds the highly vascular pituitary gland, providing it with a rich blood exchange (see fig. 16.24). The pituitary, which has an endocrine function, is structurally and functionally divided into an anterior portion, called the **adenohypophysis** (ad"ĕ-no-hi-pof'ĭ-sis), and a posterior portion, called the **neurohypophysis** (see chapter 14).

1. Discuss what is meant by the statement that the thalamus is the pain center of the brain.
2. List the body systems and the functions over which the hypothalamus has some control.
3. Describe the location of the pituitary gland relative to the rest of the diencephalon and the rest of the brain. What are the two portions of the pituitary gland and how are they supported?

pineal: L. *pinea*, pine cone
pituitary: L. *pituita*, phlegm (this gland was originally thought to secrete mucus into nasal cavity)
infundibulum: L. *infundibulum*, funnel

Mesencephalon

The mesencephalon contains the corpora quadrigemina, concerned with visual and auditory reflexes, and the cerebral peduncles, composed of fiber tracts. It also contains specialized nuclei that help to control posture and movement.

Objective 18. List the structures of the mesencephalon and explain their functions.

The *brain stem* contains nuclei for autonomic functions of the body and their connecting tracts. It is that portion of the brain that attaches to the spinal cord and includes the midbrain, pons, and medulla oblongata. The **mesencephalon** (mes"en-sef'ă-lon), or midbrain, is a short section of the brain stem between the diencephalon and the pons (see fig. 11.31). Within the midbrain is the **mesencephalic aqueduct** (aqueduct of Sylvius) (see fig. 11.36), which connects the third and fourth ventricle. The midbrain also contains the corpora quadrigemina (see fig. 11.25), cerebral peduncles (see fig. 11.24), red nucleus, and substantia nigra.

The **corpora quadrigemina** are the four rounded elevations on the dorsal portion of the midbrain. The two upper eminences, the **superior colliculi,** are concerned with visual reflexes. The two posterior eminences, the **inferior colliculi,** are responsible for auditory reflexes. The **cerebral peduncles** are a pair of cylindrical structures composed of ascending and descending projection fiber tracts that support and connect the cerebrum to the other regions of the brain.

The **red nucleus** is deep within the midbrain between the cerebral peduncle and the cerebral aqueduct. The red nucleus

aqueduct of Sylvius: from Jacobus Sylvius, French anatomist, 1478–1555
corpora quadrigemina: L. *corpus*, body; *quadri*, four; *geminus*, twin
colliculus: L. *colliculus*, small mound

is gray matter that connects the cerebral hemispheres and the cerebellum and functions in reflexes concerned with motor coordination and maintenance of posture. Another nucleus called the **substantia nigra** is inferior to the red nucleus. The substantia nigra is thought to inhibit forced involuntary movements.

---

1. Discuss symptoms that might be apparent as a result of a tumor in the midbrain.
2. Which structures of the midbrain function with the cerebellum in controlling posture and movement?

Metencephalon

The metencephalon contains the pons, which relays impulses, and the cerebellum, which coordinates skeletal muscle contractions.

Objective 19. Describe the location and structure of the pons and cerebellum and list their functions.

The **metencephalon** *(met''en-sef'ă-lon)* is the most superior portion of the hindbrain. Two vital structures of the metencephalon are the pons and cerebellum. The mesencephalic aqueduct of the mesencephalon enlarges to become the **fourth ventricle** (see fig. 11.36) within the metencephalon and myelencephalon.

Pons

The **pons** can be observed as a rounded bulge on the inferior surface of the brain, between the midbrain and the medulla oblongata (fig. 11.29). The pons consists of white fiber tracts that course in two principal directions. The surface fibers extend transversely to connect with the cerebellum through the middle cerebellar peduncles. The deeper longitudinal fibers are part of the motor and sensory tracts that connect the medulla oblongata with the tracts of the midbrain.

Scattered throughout the pons are several nuclei associated with specific cranial nerves. The cranial nerves that have nuclei within the pons include the trigeminal (V), which transmits impulses for chewing and sensory sensations from the head; the abducens (VI), which controls certain movements of the eyeball; the facial (VII), which transmits impulses for facial movements and sensory sensations from the taste buds; and the vestibular branches of the vestibulocochlear (VIII), which maintain equilibrium.

Other nuclei of the pons function with nuclei of the medulla oblongata to regulate the rate and depth of breathing. The two respiratory centers of the pons are called the **apneustic** and the **pneumotaxic areas** (fig. 11.29).

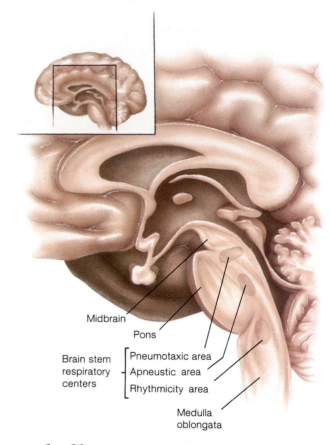

Figure 11.29 Nuclei within the pons and medulla oblongata constitute the respiratory center.

Midbrain

Pons

Brain stem respiratory centers
- Pneumotaxic area
- Apneustic area
- Rhythmicity area

Medulla oblongata

Cerebellum

The **cerebellum** is the second largest structure of the brain. It is located in the metencephalon and occupies the inferior and posterior aspect of the cranial cavity. The cerebellum is separated from the overlying cerebrum by a **transverse fissure** (see fig. 11.16). A portion of the meninges called the **tentorium cerebelli** extends into the transverse fissure. The cerebellum consists of two **hemispheres** and a central constricted area called the **vermis** (fig. 11.30). The **falx cerebelli** is the portion of the meninges that partially extends between the hemispheres.

Like the cerebrum, the cerebellum has a thin, outer layer of gray matter, the **cerebellar cortex,** and a thick, deeper layer of white matter. The cerebellum is convoluted into a series of slender, parallel **folia.** The tracts of white matter within the cerebellum have a distinctive branching pattern called the **arbor vitae** that can be seen in a sagittal view (fig. 11.30).

Three paired bundles of nerve fibers called **cerebellar peduncles** support the cerebellum and provide it with tracts for communicating with the rest of the brain (fig. 11.31). Following is a description of the cerebellar peduncles.

---

pons: L. *pons*, bridge

cerebellum: L. *cerebellum*, diminutive of *cerebrum*, brain
vermis: L. *vermis*, worm
arbor vitae: L. *arbor*, tree; *vitae*, life
peduncle: L. *peduncle*, diminutive of *pes*, foot

Figure 11.30   The structure of the cerebellum. (a) A superior view, (b) an inferior view, and (c) a sagittal view.

1. **Superior cerebellar peduncles** connect the cerebellum with the midbrain. The fibers within these peduncles originate primarily from specialized **dentate nuclei** within the cerebellum and pass through the red nucleus to the thalamus and then to the motor areas of the cerebral cortex. Impulses through the fibers of these peduncles provide feedback to the cerebrum.
2. **Middle cerebellar peduncles** convey impulses of voluntary movement from the cerebrum through the pons and to the cerebellum.
3. **Inferior cerebellar peduncles** connect the cerebellum with the medulla oblongata and the spinal cord. They contain both incoming vestibular and proprioceptive fibers and outgoing motor fibers.

The principal function of the cerebellum is coordinating skeletal muscle contractions by recruiting precise motor units within the muscles. Impulses for voluntary muscular movement originate in the cerebral cortex and are coordinated by the cerebellum. The cerebellum constantly initiates impulses to selective motor units for maintaining posture and muscle tone. The cerebellum also adjusts to incoming impulses from **proprioceptors** (*pro''pre-o-sep'torz*) within muscles, tendons, joints, and special sense organs. A proprioceptor is a sensory nerve ending that is sensitive to changes in the tension of a muscle or tendon.

Figure 11.31   The cerebellar peduncles can be seen when the cerebellum has been removed from its attachment to the brain stem.

Trauma or diseases of the cerebellum, such as a *stroke* or *cerebral palsy*, frequently cause an impairment of skeletal muscle function. Movements become jerky and uncoordinated in a condition known as *ataxia*. There is also a loss of equilibrium, resulting in a disturbance of gait. *Alcohol intoxication* causes similar uncoordinated body movements.

1. Describe the locations and relative sizes of the pons and the cerebellum.
2. Which cranial nerves have nuclei located within the pons?
3. Define the terms *tentorium cerebelli, vermis, arbor vitae,* and *cerebellar peduncles.*
4. List the functions of the pons and the cerebellum.

Myelencephalon

The medulla oblongata, contained within the myelencephalon, connects to the spinal cord and contains nuclei for the cranial nerves and vital autonomic functions.

Objective 20. Describe the location and structure of the medulla oblongata and list its functions.

Objective 21. Define the term *reticular formation* and explain its function.

Medulla Oblongata

The **medulla oblongata** is a bulbous structure about 3 cm (1 in.) long that is continuous with the pons anteriorly and the spinal cord posteriorly at the level of the foramen magnum (see figs. 11.24 and 11.25). Externally, the medulla oblongata resembles the spinal cord, except for the two triangular, elevated structures called **pyramids** on the inferior side and an oval enlargement called the **olive** (see fig. 11.24) on each lateral surface. The **fourth ventricle,** the space within the medulla oblongata, is continuous posteriorly with the central canal of the spinal cord and anteriorly with the mesencephalic aqueduct (see fig. 11.36).

The medulla oblongata is composed of vital nuclei and white matter that form all of the descending and ascending tracts communicating between the spinal cord and various parts of the brain. Most of the fibers within these tracts cross over to the opposite side through the pyramidal region of the medulla oblongata, permitting one side of the brain to receive information from and send information to the opposite side of the body (see fig. 11.24).

The gray matter of the medulla oblongata consists of several important nuclei for the cranial nerves, sensory relay, and for autonomic functions (fig. 11.32). The **nucleus ambiguus** and the **hypoglossal nucleus** are the centers from which arise the vestibulocochlear (VIII), glossopharyngeal (IX), accessory (XI), and hypoglossal (XII) nerves. The vagus nerves (X) arise from **vagus nuclei,** one on each lateral side of the medulla oblongata, adjacent to the fourth ventricle. The **nucleus gracilis** and the **nucleus cuneatus** relay sensory information to the thalamus, and then the impulses are relayed to the cerebral

Figure 11.32 A sagittal section of the medulla oblongata and pons showing the cranial nerve nuclei of gray matter.

Motor nuclei:
Accessory oculomotor
Oculomotor
Trochlear
Trigeminal
Abducens
Facial
Superior and inferior salivatory
Hypoglossal
Vagus
Nucleus ambiguus
Accessory

Sensory nuclei:
Mesencephalic nucleus of trigeminal
Trigeminal
Vestibular
Dorsal and ventral cochlear
Nucleus of tractus solitarius
Nucleus of spinal tract of trigeminal
Cerebellum

Krabach

cortex via the thalamic nuclei (not illustrated). The **inferior olivary nuclei** and the **accessory olivary nuclei** of the olive mediate impulses passing from the forebrain and midbrain through the inferior cerebellar peduncles to the cerebellum.

Three other nuclei within the medulla oblongata function as autonomic centers for controlling vital visceral functions.

1. **Cardiac center.** Both *inhibitory* and *accelerator fibers* arise from nuclei of the cardiac center. Inhibitory impulses constantly travel through the vagus nerves to slow the heartbeat. Accelerator impulses travel through the spinal cord and eventually innervate the heart through fibers within spinal nerves T1–T5.
2. **Vasomotor center.** Nuclei of the vasomotor center send impulses via the spinal cord and spinal nerves to the smooth muscles of arteriole walls, causing them to constrict and elevate arterial blood pressure.
3. **Respiratory center.** The respiratory center of the medulla oblongata controls the rate and depth of breathing and functions in conjunction with the respiratory nuclei of the pons (see fig. 11.29) to produce rhythmic breathing.

Other nuclei of the medulla oblongata function as centers for nonvital respiratory movements, such as sneezing, coughing, swallowing, and vomiting. Some of these activities (swallowing, for example) may be initiated voluntarily, but once they progress to a certain point they become involuntary and cannot be stopped.

Reticular Formation

The **reticular formation** is a complex network of nuclei and nerve fibers within the brain stem that functions as the *reticular activating system (RAS)* in arousing the cerebrum. Portions of the reticular formation are located in the spinal cord, pons, midbrain, and parts of the thalamus and hypothalamus (fig. 11.33). The reticular formation contains ascending and descending fibers from most of the structures within the brain.

Nuclei within the reticular formation generate a continuous flow of impulses unless they are inhibited by other parts of the brain. The principal functions of the RAS are to keep the cerebrum in a state of alert consciousness and to selectively monitor the sensory impulses perceived by the cerebrum. The RAS also helps the cerebellum activate selected motor units to maintain muscle tonus and produce smooth, coordinated contractions of skeletal muscles.

The RAS is sensitive to changes in and trauma to the brain. The sleep response is thought to occur because of a decrease in activity within the RAS, perhaps due to the secretion of specific neurotransmitters. A blow to the head or certain drugs and diseases may damage the RAS, causing unconsciousness. A *coma* is a state of unconsciousness and inactivity of the RAS that even the most powerful external stimuli cannot disturb.

1. When a nurse or physician prepares a case study, the location, structure, function, and possible dysfunction of an afflicted structure or organ is reported. Prepare a brief case study of a patient who has suffered severe trauma to the medulla oblongata from a blow to the back of the skull.
2. Which cranial nerves arise from the medulla oblongata?
3. Describe the location of the reticular formation. What are two functions of the RAS?

Meninges of the Central Nervous System

The CNS is covered by protective meninges; namely of a dura mater, an arachnoid mater, and a pia mater.

Objective 22. Describe the position of the meninges as they protect the CNS.

As mentioned previously in the discussion of the brain's general features, the entire delicate CNS is protected by a bony encasement—the cranium surrounding the brain and the vertebral column surrounding the spinal cord. It is also protected by three connective tissue membranous coverings called the **meninges** (figs. 11.34 and 11.35). Individually, from the outside in, they are known as the dura mater, the arachnoid mater, and the pia mater.

Dura Mater

The **dura mater** is in contact with the bone and is composed primarily of dense fibrous connective tissue. The **cranial dura mater** is a double-layered structure. The thicker, outer **perios-**

Figure 11.33 The reticular activating system. The arrows indicate the direction of impulses along nerve pathways that connect with the RAS.

Thalamus
Midbrain
Pons
Reticular formation
Medulla oblongata

Krabach

teal (*per''e-os'te-al*) **layer** adheres lightly to the cranium where it is the periosteum (fig. 11.34a). The thinner, inner **meningeal layer** follows the general contour of the brain. The **spinal dura mater** is not double layered but is similar to the meningeal layer of the cranial dura mater.

The two layers of the cranial dura mater are fused and cover most of the brain. In certain regions, however, the layers are separated, enclosing **dural sinuses** (see fig. 11.34) that collect venous blood and drain it to the internal jugular veins of the neck.

In four locations, the meningeal layer of the cranial dura mater forms distinct septa to partition major structures on the surface of the brain and anchor the brain to the inside of the cranial case. These septa have been previously identified in this chapter and are reviewed in table 11.5.

The spinal dura mater forms a tough, tubular **dural sheath** that continues into the vertebral canal and surrounds the spinal cord. There is no connection between the dural sheath and the vertebrae forming the vertebral canal, but instead there is a potential cavity called the **epidural space** (see fig. 11.35). The epidural space is highly vascular and contains loose fibrous and adipose connective tissues that form a protective pad around the spinal cord.

Arachnoid Mater

The **arachnoid mater** is the middle of the three meninges. This delicate, netlike membrane spreads over the CNS but generally does not extend into the sulci or fissures of the brain. The **subarachnoid space,** located between the arachnoid mater and the deepest meninx, the pia mater, contains cerebrospinal fluid.

---

meninges: L. plural form of *meninx*, membrane
dura mater: L. *dura*, hard; *mater*, mother

arachnoid: L. *arachnoides*, like a cobweb

UNIT V

Figure 11.34 Meninges and associated structures (a) surrounding the brain and (b) surrounding the spinal cord. The epidural space in the lower lumbar region is of clinical importance as a site for an epidural block that may be administered to facilitate parturition (childbirth).

(a)

Dura mater:
Periosteal layer
Meningeal layer
Arachnoid mater
Subarachnoid space
Pia mater

Skin of scalp
Galea aponeurotica
Skull
Arachnoid villi

Cerebral cortex

Dural sinus
Partition of meninges

(b)

Epidural space
Dura mater
Arachnoid mater
Subarachnoid space
Pia mater

Spinal cord

Spinal nerve

Body of vertebra

Waldrop

The subarachnoid space is maintained by delicate, weblike strands that connect the arachnoid mater and pia mater (see fig. 11.34).

Pia Mater

The thin **pia mater** is attached to the surfaces of the CNS and follows the irregular contours of the brain and spinal cord. The pia mater is composed of modified loose fibrous connective tissue. It is highly vascular and functions to support the vessels that nourish the underlying cells of the brain and spinal cord. The pia mater is specialized over the roofs of the ventricles, where it contributes to the formation of the choroid plexuses along with the arachnoid mater. Lateral extensions of the pia mater along the spinal cord form the **ligamentum denticulatum,** which attaches the spinal cord to the dura mater (see fig. 11.35).

pia mater: L. *pia*, soft or tender; *mater*, mother

Figure 11.35 The spinal cord and spinal meninges. (The meninges are labeled in boldface.)

Spinal cord
Denticulate ligament
Subarachnoid space
Dura mater
Pia mater
Anterior root
Spinal nerve
Arachnoid mater
Posterior root
Vertebral foramen
Lumbar vertebra

Krabach

Meningitis is an inflammation of the meninges, usually caused by certain bacteria or viruses. The arachnoid mater and the pia mater are the two meninges most frequently affected. Meningitis is accompanied by high fever and severe headache. Complications may cause sensory impairment, paralysis, or mental retardation. Untreated meningitis generally results in coma and death.

1. Contrast cranial and spinal dura mater.
2. Explain how the meninges support and protect the CNS.
3. Describe the location of the dural sinuses and the epidural space.

Ventricles and Cerebrospinal Fluid

The ventricles, central canal, and subarachnoid space contain cerebrospinal fluid, formed by the active transport of substances from blood plasma in the choroid plexuses.

Objective 23. Discuss the formation, function, and flow of cerebrospinal fluid.

Objective 24. Explain the importance of the blood-brain barrier in maintaining homeostasis within the brain.

Table 11.5	Septa of the Cranial Dura Mater
Septa	**Location**
Falx cerebri	Extends downward into the longitudinal fissure to partition the right and left cerebral hemispheres; anchored anteriorly to the crista galli of the ethmoid bone and posteriorly to the tentorium
Tentorium cerebelli	Separates the occipital and temporal lobes of the cerebrum from the cerebellum; anchored to the tentorium, petrous bones, and occipital bone
Falx cerebelli	Partitions the right and left cerebellar hemispheres; anchored to the occipital crest
Diaphragma sellae	Forms the roof of the sella turcica

Cerebrospinal fluid (CSF) is a clear, lymphlike fluid that forms a protective cushion around and within the CNS. The fluid also buoys the brain. CSF circulates through the various **ventricles** of the brain, the **central canal** of the spinal cord, and the **subarachnoid space** around the entire CNS. The cerebrospinal fluid returns to the circulatory system by draining through the walls of the **arachnoid villi,** which are venous capillaries.

Ventricles of the Brain

The ventricles of the brain are connected to one another and to the central canal of the spinal cord (figs. 11.36 and 11.37). Each of the two **lateral ventricles** (first and second ventricles) is located in one of the hemispheres of the cerebrum, inferior to the corpus callosum. The **third ventricle** is located in the diencephalon between the thalami. Each lateral ventricle is connected to the third ventricle by a narrow, oval opening called the **interventricular foramen.** The **fourth ventricle** is located in the brain stem, within the pons, cerebellum, and medulla oblongata. The **mesencephalic aqueduct** passes through the midbrain to link the third and fourth ventricles. The fourth ventricle also communicates posteriorly with the central canal. Cerebrospinal fluid exits from the fourth ventricle into the subarachnoid space (fig. 11.38) through three foramina: the **median aperture,** a medial opening, and two **lateral apertures** (not illustrated). Cerebrospinal fluid returns to the venous blood through the arachnoid villi.

Internal hydrocephalus is a condition in which cerebrospinal fluid builds up within the ventricles of the brain (fig. 11.37b). It is more common in infants whose cranial sutures have not yet strengthened or ossified. If the pressure is excessive, the condition may have to be treated surgically.

External hydrocephalus, an accumulation of fluid within the subarachnoid space, usually results from an obstruction of drainage at the arachnoid villi.

UNIT V

Figure 11.36 The ventricles of the brain. (*a*) An anterior view and (*b*) a lateral view.

Interventricular foramen

Lateral ventricle

Third ventricle

Mesencephalic aqueduct

Fourth ventricle

Central canal of spinal cord

(a)

Lateral ventricle

Mesencephalic aqueduct

Interventricular foramen

Fourth ventricle

Third ventricle

Central canal of spinal cord

Lew

(b)

Figure 11.37 CT scans of the brain showing (*a*) a normal configuration of the ventricles and (*b*) an abnormal configuration of the ventricles as a result of hydrocephalism.

Ventricles

Ventricles

(a)

(b)

UNIT V

Figure 11.38 The flow of cerebrospinal fluid. Cerebrospinal fluid is secreted by choroid plexuses in the ventricular walls. The fluid circulates through the ventricles and central canal, enters the subarachnoid space, and is reabsorbed into the blood of the dural sinuses through the arachnoid villi.

Cerebrospinal Fluid

CSF buoys the CNS and protects it from mechanical injury. CSF has a specific gravity of 1.007, which is a density close to that of brain tissue. The brain weight is about 1,500 g, but suspended in CSF its buoyed weight is about 50 g. This means that the brain has a near neutral buoyancy and can therefore function effectively as a relatively heavy organ. At a true neutral buoyancy, an object does not float or sink but is suspended in its fluid environment.

In addition to buoying the CNS, CSF reduces the damaging effect of an impact to the head by spreading the force over a larger area. It also helps to remove metabolic wastes from nervous tissue. Since the CNS lacks lymphatic circulation, the CSF moves cellular wastes into the venous return at its places of drainage.

The clear, watery CSF is continuously produced from materials within the blood by masses of specialized capillaries called **choroid plexuses** and, to a lesser extent, by secretions of the ependymal cells. The ciliated ependymal cells cover the choroid plexuses, as well as line the central canal, and presumably aid the movement of the CSF.

CSF is formed mainly by the active transport and ultrafiltration of substances within the blood plasma. CSF has more sodium, chloride, magnesium, and hydrogen ions than blood plasma and less calcium, potassium, and glucose. In addition, CSF contains some proteins, urea, and white blood cells.

Up to 800 mℓ of cerebrospinal fluid are produced each day, although only 140–200 mℓ are bathing the CNS at any given moment. A person lying in a horizontal position has a slow but continuous circulation of cerebrospinal fluid, with a fluid pressure of about 10 mmHg.

> The homeostatic consistency of the CSF composition is critical, and a chemical imbalance may have marked effects on CNS functions. An increase in amino acid glycine concentration, for example, produces hypothermia and hypotension as temperature and blood pressure regulatory mechanisms are disrupted. A slight change in pH may affect the respiratory rate and depth.

Blood-Brain Barrier

The **blood-brain barrier (BBB)** is a structural arrangement of capillaries, surrounding connective tissue, and specialized neuroglial cells called astrocytes (see fig. 11.4) that selectively determine which substances can move from the plasma of the blood to the extracellular fluid of the brain. Certain substances such as H_2O, O_2, CO_2, glucose, and lipid-soluble compounds (alcohol, for example) pass readily through the BBB. Certain inorganic ions such as Ca^+ and K^+ pass more slowly, so that the concentrations of these ions in the brain differ from those in the blood plasma. Other substances, such as proteins, lipids, creatine, urea, inulin, certain toxins, and most antibiotics, are restricted in passage. The BBB is an important factor to consider when planning drug therapy for neurological disorders.

> 1. Describe the location of the ventricles within the brain.
> 2. What are the functions of cerebrospinal fluid?
> 3. Where is cerebrospinal fluid produced and where does it drain?
> 4. Construct a flow chart depicting a substance that can flow freely across the BBB from the blood plasma to the nervous tissue. Also indicate a substance that passes more slowly and one that is restricted in its passage. Why is the BBB functionally important?

Spinal Cord

The spinal cord consists of centrally located gray matter, involved in reflexes, and peripherally located ascending and descending tracts of white matter, which conduct impulses to and from the brain.

Objective 25. Describe the structure of the spinal cord.

Objective 26. Describe the arrangement of ascending and descending tracts within the spinal cord.

The **spinal cord** is the portion of the CNS that extends through the vertebral canal of the vertebral column (fig. 11.39). It is continuous with the brain through the foramen magnum of the skull. The spinal cord has two principal functions.

1. **Impulse conduction.** It provides a means of neural communication to and from the brain through tracts of white matter. **Ascending tracts** conduct impulses from the peripheral sensory receptors of the body to the brain. **Descending tracts** conduct motor impulses from the brain to the muscles and glands.
2. **Reflex integration.** It serves as a center for spinal reflexes. Specific nerve pathways enable some movements to be reflexive rather than initiated voluntarily by the brain. Movements of this type are not confined to skeletal muscles; reflexive movements of cardiac and smooth muscles control heart rate, breathing rate, blood pressure, and digestive activities. Spinal nerve pathways are also involved in swallowing, coughing, sneezing, and vomiting.

Structure of the Spinal Cord

The spinal cord extends inferiorly from the position of the foramen magnum of the occipital bone to the level of the first lumbar vertebra (L1). The spinal cord is somewhat flattened posteroventrally, making it oval in cross section. Two prominent enlargements can be seen in a posterior view (see fig. 11.39). The **cervical enlargement** is located between the third cervical and the second thoracic vertebrae. Nerves emerging from this region serve the upper extremities. The **lumbar enlargement** lies between the ninth and twelfth thoracic vertebrae. Nerves from the lumbar enlargement supply the lower extremities.

The embryonic spinal cord develops more slowly than the associated vertebral column; thus, in the adult, the cord does not extend beyond L1. The tapering, terminal portion of the spinal cord is called the **conus medullaris.** The **filum terminale,** a fibrous strand composed mostly of pia mater, extends inferiorly from the conus medullaris at the level of L1 to the coccyx. Nerve roots also radiate inferiorly from the conus medullaris through the vertebral canal. These nerves are collectively referred to as the **cauda equina** because they resemble a horse's tail.

The spinal cord develops as 31 segments, each of which gives rise to a pair of **spinal nerves** that emerge from the spinal cord through the intervertebral foramina. Two grooves, an **anterior median fissure** and a **posterior median sulcus,** extend the length of the spinal cord and partially divide it into right

---

filum terminalis: L. *filum*, filament; *terminus*, end
cauda equina: L. *cauda*, tail; *equus*, horse

UNIT V

Figure 11.39 The spinal cord and plexuses.

Cervical plexus

Brachial plexus

Cervical enlargement

Thoracic spinal nerve

Dura mater

Lumbar enlargement

Conus medullaris

Filum terminale

Cauda equina

Lumbar plexus

Sacral plexus

and left portions. The spinal cord, like the brain, is protected by three distinct meninges and is cushioned by cerebrospinal fluid. The pia mater contains an extensive vascular network.

The **gray matter** of the spinal cord is centrally located and surrounded by white matter. It is composed of nerve cell bodies, neuroglia, and unmyelinated association neurons (interneurons). The **white matter** consists of bundles, or tracts, of myelinated fibers of sensory and motor neurons.

The relative size and shape of the gray and white matter varies throughout the spinal cord. The amount of white matter increases toward the brain as the nerve tracts become thicker. More gray matter is found in the cervical and lumbar enlargements where, respectively, innervations from the upper and lower extremities make connections.

The core of gray matter roughly resembles the letter H (fig. 11.40). Projections of the gray matter within the spinal cord are called horns, and are named according to the direction in which they project. The paired **posterior horns** extend posteriorly and the paired **anterior horns** project anteriorly. A pair of short **lateral horns** extend to the sides and are located between the other two pairs. Lateral horns are prominent only in the thoracic and upper lumbar regions. The transverse bar of gray matter that connects the paired horns across the center of the spinal cord is called the **gray commissure.** Within the gray commissure is the **central canal.** It is continuous with the ventricles of the brain and is filled with cerebrospinal fluid.

Spinal Cord Tracts

Impulses are conducted through the ascending and descending tracts of the spinal cord within the columns of white matter. The spinal cord has six columns of white matter called **funiculi** (*fyoo-nik-yu'li*), which are named according to their relative position within the spinal cord. The two **anterior funiculi** are located between the two anterior horns of gray matter to either lateral side of the anterior median fissure (fig. 11.40). The two **posterior funiculi** are located between the two posterior horns of gray matter to either lateral side of the posterior median sulcus. Two **lateral funiculi** are located between the anterior and posterior horns of gray matter.

Each funiculus consists of both ascending and descending tracts. The nerve fibers within the tracts are generally myelinated and have specific sites of origin and termination. In fact, the names of the various tracts reflect their origin and termination. The fibers of the tracts either remain on the same side of the brain and spinal cord or cross over within the medulla oblongata or the spinal cord. The crossing over of nerve tracts is referred to as *decussation (dek''uh-sa'shun).* Illustrated in figure 11.41 is a descending tract that decussates within the medulla oblongata, and in figure 11.42 is an ascending tract that decussates within the medulla oblongata.

The principal ascending and descending tracts within the funiculi are presented with their functions in table 11.6 and are illustrated in figure 11.43.

UNIT V

commissure: L. *commissura,* a joining
funiculus: L. diminutive of *funis,* cord, rope

Figure 11.40 The spinal cord in cross section. (a) A diagram and (b) a photomicrograph.

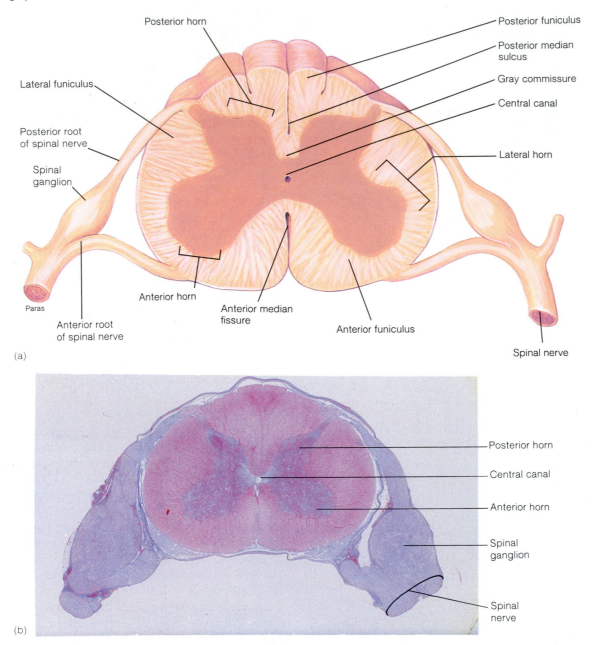

(a)

Posterior horn

Posterior funiculus

Posterior median sulcus

Gray commissure

Central canal

Lateral funiculus

Lateral horn

Posterior root of spinal nerve

Spinal ganglion

Paras

Anterior root of spinal nerve

Anterior horn

Anterior median fissure

Anterior funiculus

Spinal nerve

(b)

Posterior horn

Central canal

Anterior horn

Spinal ganglion

Spinal nerve

Descending tracts are grouped according to place of origin as either corticospinal or extrapyramidal. **Corticospinal (pyramidal) tracts** descend directly, without synaptic interruption, from the cerebral cortex to the lower motor neurons. The cell bodies of the neurons that contribute fibers to these tracts are located primarily in the precentral gyrus of the frontal lobe. Most (about 85%) of the corticospinal fibers decussate in the pyramids of the medulla oblongata (see fig. 11.24). The remaining 15% do not cross from one side to the other. The fibers that cross compose the **lateral corticospinal tracts,** and the remaining uncrossed fibers compose the **anterior corticospinal tracts.** Because of the crossing of fibers from higher motor neurons in the pyramids, the right hemisphere primarily controls the musculature on the left side of the body, whereas the left hemisphere controls the right musculature.

The corticospinal tracts appear to be particularly important in voluntary movements that require complex interactions between the motor cortex and sensory input. Speech, for example, is impaired when the corticospinal tracts are damaged in the thoracic region of the spinal cord, whereas involuntary breathing continues. Damage to the pyramidal motor system can be detected clinically by a positive *Babinski's reflex,* in which stimulation of the sole of the foot causes extension (upward movement) of the toes. This positive Babinski reflex is normal in infants because neural control is not yet fully developed.

UNIT V

Figure 11.41 Descending tracts composed of motor fibers that cross over within the medulla oblongata.

Sensory area of cerebral cortex

Primary motor area of cerebral cortex

Internal capsule

Thalamus

Mesencephalon

Red nucleus

Cerebral peduncle

Fasciculus cuneatus tract

Cerebellum

Pons

Lateral corticospinal tract

Rubrospinal tract

Anterior corticospinal tract

Sensory and motor fibers cross-over

Medulla oblongata

Lateral corticospinal tract

Pyramid

Anterior corticospinal tract

Fasciculus cuneatus

Cervical region

Cervical region

Sensory impulse from skin

Skeletal muscle

Motor impulse to skeletal muscle

Lumbar region

Krabach

Figure 11.42 Ascending tracts composed of sensory fibers that cross over within the medulla oblongata.

Postcentral gyrus

Postcentral gyrus

Axons of third-order neurons (not part of tracts)

Thalamus

Cerebral cortex

Mesencephalon

Cerebellum

Pons

Medial lemniscal tract (axons of second-order neurons)

Posterior spinocerebellar tract (+axons of second-order neurons)

Lateral spinothalamic tract (+axons of second-order neurons)

Nucleus gracilis

Nucleus cuneatus

Medulla oblongata

Muscle spindle (proprioceptor)

Fasciculus cuneatus (axon of first-order sensory neurons)

Joint stretch receptor (proprioceptor)

Pain receptor

Axons of first-order neurons (not part of spinocerebellar tract)

Cervical region

Axons of first-order neurons (not part of spinothalamic tract)

Fasciculus gracilis (axons of first-order sensory neuron)

Touch receptor

Krabach

Lumbar region

Temperature receptor

Table 11.6	Principal Ascending and Descending Tracts of the Spinal Cord			
Tract	**Funiculus**	**Origin**	**Termination**	**Function**
Ascending Tracts				
Anterior spinothalamic	Anterior	Posterior horn on one side of spinal cord but crosses to opposite side	Thalamus, then cerebral cortex	Conducts sensory impulses for crude touch and pressure
Lateral spinothalamic	Lateral	Posterior horn on one side of spinal cord but crosses to opposite side	Thalamus, then cerebral cortex	Conducts pain and temperature impulses that are interpreted within cerebral cortex
Fasciculus gracilis and fasciculus cuneatus	Posterior	Peripheral sensory neurons; does not cross over	Nucleus gracilis and nucleus cuneatus of medulla oblongata; crosses to opposite side; eventually thalamus, then cerebral cortex	Conducts sensory impulses from skin, muscles, tendons, and joints, which are interpreted as sensations of fine touch, precise pressures, and body movements
Posterior spinocerebellar	Lateral	Posterior horn; does not cross over	Cerebellum	Conducts sensory impulses from one side of body to same side of cerebellum for subconscious proprioception necessary for coordinated muscular contractions
Anterior spinocerebellar	Lateral	Posterior horn; some fibers cross, others do not	Cerebellum	Conducts sensory impulses from both sides of body to cerebellum for subconscious proprioception necessary for coordinated muscular contractions
Descending Tracts				
Anterior corticospinal	Anterior	Cerebral cortex on one side of brain; crosses to opposite side of spinal cord	Anterior horn	Conducts motor impulses from cerebrum to spinal nerves and outward to cells of anterior horns for coordinated, precise voluntary movements of skeletal muscle
Lateral corticospinal	Lateral	Cerebral cortex on one side of brain; crosses in base of medulla oblongata to opposite side of spinal cord	Anterior horn	Conducts motor impulses from cerebrum to spinal nerves and outward to cells of anterior horns for coordinated, precise voluntary movements
Tectospinal	Anterior	Mesencephalon; crosses to opposite side of spinal cord	Anterior horn	Conducts motor impulses to cells of anterior horns and eventually to muscles that move the head in response to visual, auditory, or cutaneous stimuli
Rubrospinal	Lateral	Mesencephalon (red nucleus); crosses to opposite side of spinal cord	Anterior horn	Conducts motor impulses concerned with muscle tone and posture
Vestibulospinal	Anterior	Medulla oblongata; does not cross over	Anterior horn	Conducts motor impulses that regulate body tone and posture (equilibrium) in response to movements of head
Anterior and medial reticulospinal	Anterior	Reticular formation of brain stem; does not cross over	Anterior horn	Conducts motor impulses that control muscle tone and sweat gland activity
Bulboreticulospinal	Lateral	Reticular formation of brain stem; does not cross over	Anterior horn	Conducts motor impulses that control muscle tone and sweat gland activity

UNIT V

Figure 11.43 A cross section showing the principal ascending and descending tracts within the spinal cord.

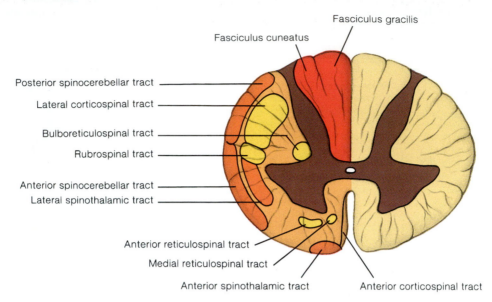

Fasciculus gracilis

Fasciculus cuneatus

Posterior spinocerebellar tract

Lateral corticospinal tract

Bulboreticulospinal tract

Rubrospinal tract

Anterior spinocerebellar tract

Lateral spinothalamic tract

Anterior reticulospinal tract

Medial reticulospinal tract

Anterior spinothalamic tract

Anterior corticospinal tract

The remaining descending tracts are **extrapyramidal tracts** that originate in the brain stem region (see fig. 11.26). Electrical stimulation of the cerebral cortex, the cerebellum, and the basal nuclei indirectly evokes movements because of synaptic connections within extrapyramidal tracts.

The **reticulospinal** *(rĕ-tik″yoo-lo-spi′nal)* **tracts** are the major descending pathways of the extrapyramidal system. These tracts originate in the reticular formation of the brain stem. Neurostimulation of the reticular formation by the cerebrum or cerebellum either facilitates or inhibits the activity of lower motor neurons (depending on the area stimulated) (fig. 11.44).

There are no descending tracts from the cerebellum. The cerebellum can influence motor activity only indirectly, through the vestibular nuclei, red nucleus, and basal nuclei. These structures, in turn, affect lower motor neurons via the **vestibulospinal tracts, rubrospinal tracts,** and reticulospinal tracts. Damage to the cerebellum interferes with the coordination of movements with spatial judgment. Underreaching or overreaching for an object may occur, followed by *intention tremor,* in which the limb moves back and forth in a pendulumlike motion.

The basal nuclei, acting through synapses in the reticular formation particularly, appear normally to exert an inhibitory influence on the activity of lower motor neurons. Damage to the basal nuclei thus results in decreased muscle tone. People with such damage display *akinesia (a″kĭ-ne′zhă)* (lack of desire to use the affected limb) and *chorea* (sudden and uncontrolled random movements).

Paralysis agitans, better known as *Parkinson's disease,* is a disorder of the basal nuclei involving the degeneration of fibers from the substantia nigra. These fibers, which use dopamine as a neurotrans-

Figure 11.44 Pathways involved in the higher motor neuron control of skeletal muscles. (The pyramidal [corticospinal] tracts are shown in red and the extrapyramidal tracts are shown in black.)

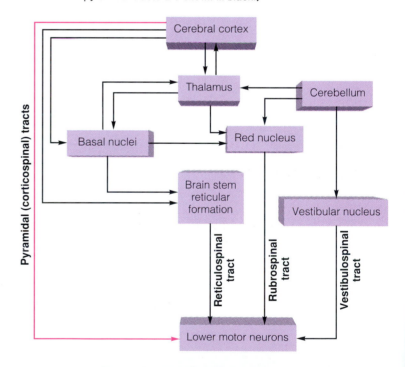

Cerebral cortex

Thalamus

Cerebellum

Basal nuclei

Red nucleus

Brain stem reticular formation

Vestibular nucleus

Pyramidal (corticospinal) tracts

Reticulospinal tract

Rubrospinal tract

Vestibulospinal tract

Lower motor neurons

mitter, are required to antagonize the effects of other fibers that use acetylcholine (ACh) as a transmitter. The relative deficiency of dopamine compared to ACh is believed to produce the symptoms of Parkinson's disease, including *resting tremor.* This shaking of the limbs tends to disappear during voluntary movements and then reappear when the limb is again at rest.

Parkinson's disease is treated with drugs that block the effects of ACh and by the administration of L-dopa, which can be converted to dopamine in the brain. (Dopamine cannot be given directly because it does not cross the blood-brain barrier.)

akinesia: Gk. *a*, without; *kinesis*, movement
chorea: Fr. *choros*, a dance

Developmental Exposition

The Spinal Cord

Exhibit III   The development of the spinal cord. (*a*) A dorsal view of an embryo at 23 days with the position of a transverse cut indicated in (*b*). (*c*) The formation of the alar and basal plates is evident in a transverse section through the spinal cord at 6 weeks. (*d*) The central canal has reduced in size, and functional posterior and anterior horns have formed at 9 weeks.

Explanation

The spinal cord, like the brain, develops as the neural tube undergoes differentiation and specialization. Throughout the developmental process, the hollow central canal persists while the specialized white and gray matter forms (exhibit III). Changes in the neural tube become apparent during the sixth week as the lateral walls thicken to form a groove called the **sulcus limitans** along each lateral wall of the central canal. A pair of **alar plates** forms dorsal to the sulcus limitans and a pair of **basal plates** forms ventrally. By the ninth week, the alar plates have specialized to become the **posterior horns,** containing fibers of the sensory cell bodies, and the basal plates have specialized to form the **anterior** and **lateral horns,** containing motor cell bodies. Sensory neurons of spinal nerves conduct impulses toward the spinal cord, whereas motor neurons conduct impulses away from the spinal cord.

1. Diagram a cross section of the spinal cord and label the structures of the gray matter and the white matter. Describe the location of the spinal cord.
2. List the structures and function of the corticospinal tracts. Make a similar list for the extrapyramidal tracts.
3. Explain why damage to the right side of the brain primarily affects motor activities on the left side of the body.

Clinical Considerations

The clinical aspects of the central nervous system are extensive and usually complex. Numerous diseases and developmental problems directly involve the nervous system, and the nervous system is indirectly involved with most of the diseases that afflict the body because of the location and activity of sensory pain receptors. Pain receptors are free nerve endings that are present throughout living tissue. The pain sensations elicited by disease or trauma are important for localizing and diagnosing specific diseases or dysfunctions.

Only a few of the many clinical considerations of the central nervous system will be discussed here. These include neurological assessment and drugs, developmental problems, injuries, infections and diseases, and degenerative disorders.

Neurological Assessment and Drugs

Neurological assessment has become exceedingly sophisticated and accurate in the past few years. In most physical examinations, only the basic aspects such as reflexes and sensory functions are assessed. But if the physician suspects abnormalities involving the nervous system, further neurological tests are done, employing the following techniques.

A **lumbar puncture** is performed by inserting a fine needle between the third and fourth lumbar vertebrae and withdrawing a sample of CFS from the subarachnoid space (fig. 11.45). A **cisternal puncture** is similar to a lumbar puncture except that the CFS is withdrawn from the cisterna magna, near the foramen magnum of the skull. The pressure of the CFS, which is normally about 10 mm of mercury, is measured with a *manometer*. Samples of CFS may also be examined for abnormal constituents. Also, excessive fluids, accumulated as a result of disease or trauma, may be drained.

The condition of the arteries of the brain can be determined through a **cerebral angiogram.** In this technique, a radiopaque substance is injected into the common carotid arteries and allowed to disperse through the cerebral vessels. Aneurysms and vascular constrictions or displacements by tumors may then be revealed on radiographs.

The development of the **CT scanner,** or **computerized axial tomographic scanner,** has revolutionized the diagnosis of brain disorders. The CT scanner projects a sharply focused, detailed tomogram, or cross section, of a patient's brain onto a television screen. The versatile CT scanner allows quick and accurate diagnoses of tumors, aneurysms, blood clots, and hemorrhage. The CT scanner may also be used to detect certain types of birth defects, brain damage, scar tissue, and old or recent strokes.

A machine with even greater potential than the CT scanner is the **DSR,** or **dynamic spatial reconstructor.** Like the CT scanner, the DSR is computerized to transform radiographs into composite video images. The advantage of the DSR over the CT is the three-dimensional view it presents and the speed with which the image is portrayed. DSR can produce 75,000 cross-sectional images in 5 seconds, whereas CT can produce only one. At that speed, body functions as well as structures may be studied. Blood flow through blood vessels of the brain can be observed. This type of data is important in detecting early symptoms of a stroke or other disorders.

Certain disorders of the brain may be diagnosed more simply by examining brain-wave patterns using an **electroencephalogram** (see fig. 11.23). Sensitive electrodes placed on the scalp record particular EEG patterns being emitted from evoked cerebral activity. EEG recordings are used to monitor epileptic patients to predict seizures and determine proper drug therapy, and also to monitor comatose patients.

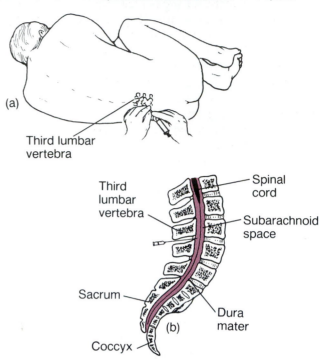

Figure 11.45 (a) A lumbar puncture is performed by inserting a needle between the third and fourth lumbar vertebrae (L3–L4) and (b) withdrawing cerebrospinal fluid from the subarachnoid space.

(a)

Third lumbar vertebra

Third lumbar vertebra

Spinal cord

Subarachnoid space

Sacrum

Dura mater

(b)

Coccyx

The fact that the nervous system is extremely sensitive to various drugs is fortunate, but it can also be potentially disastrous to a person. *Drug abuse* is a major clinical concern because of the addictive and devastating effect that certain drugs have on the nervous system. Much has been written on drug abuse, and it is beyond the scope of this text to elaborate on the effects of drugs. A positive aspect of drugs is their administration in medicine to temporarily interrupt the passage or perception of sensory impulses. Injecting an anesthetic drug near a nerve, as in dentistry, desensitizes a specific area and causes a *nerve block.* Nerve blocks of a limited extent occur if an appendage is cooled or if a nerve is compressed for a period of time. Before the discovery of pharmacological drugs, physicians would frequently cool an affected appendage with ice or snow before performing surgery. **General anesthetics** affect the brain and render a person unconscious. A **local anesthetic** causes a nerve block by desensitizing a specific area.

Developmental Problems

Congenital malformations of the CNS are common and frequently involve overlying bone, muscle, and connective tissue. The more severe abnormalities make life impossible, and the less severe malformations frequently result in functional disability.

Most congenital malformations of the nervous system occur during the sensitive embryonic period. Neurological malformations are generally caused by genetic abnormalities but may result from environmental factors, such as anoxia, infectious agents, drugs, and ionizing radiation.

Spina bifida is a defective fusion of the vertebral elements and may or may not involve the spinal cord. **Spina bifida oc-**

culta is the most common and least serious type of spina bifida. This defect usually involves few vertebrae, is not externally apparent except for perhaps a pigmented spot with a tuft of hair, and usually does not cause neurological disturbances. **Spina bifida cystica,** a severe type of spina bifida, is a saclike protrusion of skin and underlying meninges that may contain portions of the spinal cord and nerve roots. Spina bifida cystica is most common in the lower thoracic, lumbar, and sacral regions (see fig. 6.41). The position and extent of the defect determines the degree of neurological impairment.

Anencephaly *(an''en-sef'ă-le)* is a markedly defective development of the brain and the surrounding cranial bones. Anencephaly occurs once per thousand births and makes sustained extrauterine life impossible. This congenital defect apparently results from the failure of the neural folds at the cranial portion of the neural plate to fuse and form the prosencephalon.

Microcephaly is an uncommon condition in which brain development is not completed. If enough neurological tissue is present, the infant will survive but will be severely mentally retarded.

Defective skull development frequently causes **cranial encephalocele** *(en-sef'ă-lo-sēl)*. This condition occurs approximately once per two thousand births. Cranial encephalocele is characterized by a herniated portion of the brain and meninges through a defect in the skull, usually in the occipital region. Occasionally the herniation contains fluid with the meninges and not the brain tissue. In this case, it is referred to as a **cranial meningocele** *(mĕ-ning'go-sēl)*.

Hydrocephalus is the abnormal accumulation of cerebrospinal fluid in the ventricles and subarachnoid or subdural space. Hydrocephalus may be caused by the excessive production of or blocked flow of cerebrospinal fluid. It may also be associated with other congenital problems, such as spina bifida cystica or encephalocele. Hydrocephalus frequently causes the cranial bones to thin and the cerebral cortex to atrophy.

Many congenital disorders cause an impairment of intelligence, known as **mental retardation.** Chromosomal abnormalities, maternal and fetal infections like syphilis and German measles, and excessive irradiation of the fetus are all commonly associated with mental retardation.

Injuries

Although the brain and spinal cord seem to be well protected within a bony encasement, they are sensitive organs, highly susceptible to injury.

Certain symptomatic terms are used when determining possible trauma within the CNS. **Headaches** are the most common ailment of the CNS. Most headaches are due to dilated blood vessels within the meninges of the brain. Headaches are generally asymptomatic of brain disorders, associated rather with physiological stress, eyestrain, or fatigue. Persistent and intense headaches may indicate a more serious problem, such as a brain tumor. A **migraine** is a specific type of headache commonly preceded or accompanied by visual impairments and GI unrest. It is not known why only 5%–10% of the population periodically suffer from migraines or why they are more common in women. Fatigue, allergy, or emotional stress tends to trigger migraines.

Fainting is a brief loss of consciousness that may result from a rapid pooling of blood in the lower extremities. It may occur when a person rapidly arises from a reclined position, receives a blow to the head, or experiences an intense psychologic stimulus, such as viewing a cadaver for the first time. Fainting is of more concern when it is symptomatic of a particular disease.

A **concussion** is a sudden movement of the brain caused by a violent blow to the head, which may or may not fracture bones of the skull. A concussion usually results in a brief period of unconsciousness followed by mild **delirium,** in which the patient is in a state of confusion. **Amnesia** is a more intense disorientation in which the patient suffers various degrees of memory loss.

A person who survives a severe head injury may be **comatose** for a short or an extended period of time. A coma is a state of unconsciousness from which the patient cannot be aroused, even by the most intense external stimuli. The area of the brain most likely to cause a coma from trauma is the reticular activating system. Although a head injury is the most common cause of coma, chemical imbalances associated with certain diseases (e.g., diabetes) or the ingestion of drugs or poisons may also be responsible.

The flexibility of the vertebral column is essential for body movements, but because of this flexibility the spinal cord and spinal nerves are somewhat vulnerable to trauma. Falls or severe blows to the back are a common cause of injury. A skeletal injury, such as a fracture, dislocation, or compression of the vertebrae, usually traumatizes nervous tissue as well. Other frequent causes of trauma to the spinal cord include gunshot wounds, stabbings, herniated discs, and birth injuries. The consequences of the trauma depend on the location and severity of the injury and the medical treatment the patient receives. If nerve fibers of the spinal cord are severed, motor or sensory functions will be permanently lost.

Paralysis is a permanent loss of motor control, usually resulting from disease or a lesion of the spinal cord or specific nerves. Paralysis of both lower extremities is called **paraplegia.** Paralysis of both the upper and lower extremity on the same side is called **hemiplegia,** and paralysis of all four extremities is **quadriplegia.** Paralysis may be flaccid *(flak'sid)* or spastic. **Flaccid paralysis** generally results from a lesion of the anterior horn cells and is characterized by noncontractible muscles that atrophy. **Spastic paralysis** results from lesions of the corticospinal tracts of the spinal cord and is characterized by hypertonicity of the skeletal muscles.

Whiplash is a sudden hyperextension and flexion of the cervical vertebrae (fig. 11.46) such as may occur during a rear-end automobile collision. Recovery of a minor whiplash (muscle and ligament strains) is generally complete, albeit slow. Severe whiplash (spinal cord compression) may cause permanent paralysis to the structures below the level of injury.

amnesia: L. *amnesia*, forgetfulness
comatose: Gk. *koma*, deep sleep
paralysis: Gk. *paralysis*, loosening
paraplegia: Gk. *para*, beside; *plessein*,

Figure 11.46 Whiplash varies in severity from muscle and ligament strains to dislocation of the vertebrae and compression of the spinal cord. Injuries such as this may cause permanent loss of some or all of the spinal cord functions.

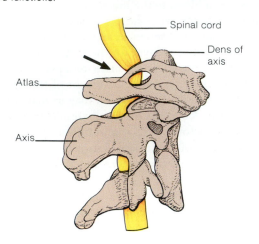

- Spinal cord
- Dens of axis
- Atlas
- Axis

Disorders of the Nervous System

Mental Illness   Mental illness is a major clinical consideration of the nervous system and is perhaps the least understood. The two principal categories of mental disorders are neurosis and psychosis. In *neurosis,* a maladjustment to certain aspects of life interferes with normal functioning, but contact with reality is maintained. An irrational fear *(phobia)* is an example of neurosis. Neurosis frequently causes intense anxiety or abnormal distress that brings about increased sympathetic stimulation. *Psychosis,* a more serious mental condition, is typified by a withdrawal from reality and is usually socially unacceptable. The more common forms of psychosis include *schizophrenia,* in which a person withdraws into a world of fantasy; *paranoia,* in which a person has systematized delusions, often of a persecutory nature; and *manic-depressive psychosis,* in which a person's moods swing widely from intense elation to deepest despair.

Epilepsy   Epilepsy is a relatively common brain disorder with a strong hereditary basis, but it also can be caused by head injuries, tumors, or childhood infectious diseases. It is sometimes idiopathic (without demonstrable cause). A person with epilepsy may periodically suffer from an *epileptic seizure,* which has various symptoms depending on the type of epilepsy.

The most common kinds of epilepsy are petit mal, psychomotor epilepsy, and grand mal. **Petit mal** *(pĕ-te'mahl')* occurs almost exclusively in children between the ages of 3 and 12. A child experiencing a petit mal seizure loses contact with reality for 5–30 seconds but does not lose consciousness or display convulsions. There may, however, be slight uncontrollable facial gestures or eye movements, and the child will stare, as if in a daydream. During a petit mal seizure, the thalamus and hypothalamus produce an extremely slow EEG pattern of

3 waves per second. Children with petit mal usually outgrow the condition by age 9 or 10 and generally require no medication.

Psychomotor epilepsy is often confused with mental illness because of the symptoms characteristic of the seizure. During such a seizure, EEG activity accelerates in the temporal lobes, causing a person to become disoriented and lose contact with reality. Occasionally during a seizure, specific cerebral motor areas will cause involuntary lip smacking or hand clapping. If motor areas in the brain are not stimulated, a person having a psychomotor epileptic seizure may wander aimlessly until the seizure subsides.

Grand mal is a more serious form of epilepsy characterized by periodic convulsive seizures that generally render a person unconscious. Grand mal epileptic seizures are accompanied by rapid EEG patterns of 25–30 waves per second. This sudden increase of EEG patterns from the normal of about 10 waves per second may cause an extensive stimulation of motor units and, therefore, uncontrollable urinary muscle activity. During a grand mal seizure, a person loses consciousness, convulses, and may lose bladder and bowel control. The unconsciousness and convulsions usually last a few minutes, after which the muscles relax and the person awakes but remains disoriented for a short time.

Epilepsy almost never affects intelligence and can be effectively treated with drugs in about 85% of the patients.

Cerebral Palsy   Cerebral palsy is a condition of motor disorders characterized by paresis (partial paralysis) and lack of muscular contraction. It is caused by damage to the motor areas of the brain during prenatal development, birth, or infancy. During neural development within an embryo, radiation or bacterial toxins (such as from German measles), transferred through the placenta of a pregnant mother, may cause cerebral palsy. Oxygen deprivation due to complications at birth and hydrocephalus in a newborn may also cause cerebral palsy. The three areas of the brain most severely affected by this disease are the cerebral cortex, the basal nuclei, and the cerebellum. The type of cerebral palsy is determined by the particular region of the brain that is affected.

Some degree of mental retardation occurs in 60%–70% of cerebral palsy victims. Partial blindness, deafness, and speech problems frequently accompany this disease. Cerebral palsy is nonprogressive (i.e., these impairments do not worsen as a person ages), but neither are there organic improvements.

Neoplasms of the CNS   Neoplasms of the CNS are either **intracranial tumors,** which affect cells within or associated with the brain, or they are **intravertebral (intraspinal) tumors,** which affect cells within or near the spinal cord. **Primary neoplasms** develop within the CNS. Approximately half of these are benign, but they may become lethal because of the pressure they exert upon vital centers as they grow. Patients with **secondary,** or **metastatic, neoplasms** within the brain have a poor prognosis because the cancer has already established itself in another body organ—frequently the liver, lung, or breast—and has only secondarily spread to the brain. The symptoms of a brain tumor include headache, convulsions, pain, paralysis, or a change in behavior.

epilepsy: Gk. *epi,* upon; *lepsis,* seize
petit mal: L. *pitinnus,* small child; *malus,* bad

Neoplasms of the CNS are classified according to the tissues in which the cancer occurs. Tumors arising in neuroglial cells are called **gliomas** (gle-o'maz) and account for about one-half of all primary neoplasms within the brain. Gliomas are frequently spread throughout cerebral tissue, develop rapidly, and usually cause death within a year after diagnosis. **Astrocytoma** (as''tro-si-to'mă), **oligodendroglioma** (ol''ĭ-go-den''drog-le-o'mă), and **ependymoma** (e-pen''dĭ-mo'mă) are common types of gliomas.

Meningiomas arise from meningeal coverings of the brain and account for about 15% of primary intracranial tumors. Meningiomas are usually benign if they can be treated readily.

Intravertebral tumors are classified as **extramedullary** when they develop on the outside of the spinal cord and as **intramedullary** when they develop within the substance of the spinal cord. Extramedullary neoplasms may cause pain and numbness in body structures distant from the tumor as the growing tumor compresses the spinal cord. An intramedullary neoplasm causes a gradual loss of sensory, temperature, and motor function below the spinal-segmental level of the affliction.

Methods of detecting and treating cancers within the CNS have greatly improved in the last few years. Early detection and competent treatment have lessened the likelihood of death from this disease and reduced the probability of physical impairment.

Dyslexia   Dyslexia is a defect in the language center within the brain. In dyslexia, otherwise intelligent people reverse the order of letters in syllables, of syllables in words, and of words in sentences. The sentence: "The man saw a red dog," for example might be read by the dyslexic as "A red god was the man." Dyslexia is believed to result from the failure of one cerebral hemisphere to respond to written language, perhaps due to structural defects. Dyslexia can usually be overcome by intense remedial instruction in reading and writing.

Meningitis   The nervous system is vulnerable to a variety of organisms and viruses that may cause abscesses or infections. Meningitis is an infection of the meninges. It may be confined to the spinal cord, in which case it is referred to as **spinal meningitis,** or it may involve the brain and associated meninges, in which case it is known as **encephalitis** or **encephalomyelitis** (en-sef''ă-lo-mi''ĕ-li'tis), respectively. The microorganisms that most commonly cause meningitis are meningococci, streptococci, pneumococci, and tubercle bacilli. Viral meningitis is more serious than bacterial meningitis. Nearly 20% of viral encephalitides are fatal. The organisms that cause meningitis probably enter the body through respiratory passageways.

Poliomyelitis   Poliomyelitis, or infantile paralysis, is primarily a childhood disease caused by a virus that destroys nerve cell bodies within the anterior horn of the spinal cord, especially those within the cervical and lumbar enlargements. This degenerative disease is characterized by fever, severe headache, stiffness and pain, and the loss of certain somatic reflexes. Muscle paralysis follows within several weeks, and eventually the muscles atrophy. Death results if the virus invades the vasomotor

and respiratory nuclei within the medulla oblongata or anterior horn cells controlling respiratory muscles. Poliomyelitis has been effectively controlled with immunization.

Syphilis   Syphilis is a sexually transmitted disease that, if untreated, progressively destroys body organs. When syphilis causes organ degeneration, it is said to be in the *tertiary stage* (10 to 20 years after the primary infection). The organs of the nervous system are frequently infected, causing a condition called **neurosyphilis.** Neurosyphilis is classified according to the tissue involved, and the symptoms vary correspondingly. If the meninges are infected, the condition is termed **chronic meningitis. Tabes dorsalis** is a form of neurosyphilis in which there is a progressive degeneration of the posterior funiculi of the spinal cord and posterior roots of spinal nerves. Motor control is gradually lost and patients eventually become bedridden, unable even to feed themselves.

Senescence of the Nervous System

With respect to the nervous system, little is known about the extent of senescent changes, or changes related to normal aging. It has been estimated that perhaps 100,000 neurons die each day of our adult life. Empirical studies, however, show that such claims are unfounded. It is believed that relatively few neural cells are lost during the normal aging process. Neurons are, however, extremely sensitive and susceptible to various drugs or interruptions of vascular supply such as caused by strokes or other cardiovascular diseases.

There is evidence that senescence alters neurotransmitters. Age-related conditions like depression or specific diseases like Parkinson's disease may be caused by an imbalance of neurotransmitter chemicals. Changes in sleeping patterns in elderly people also probably result from neurotransmitter problems.

The slowing of the nervous system with age is most apparent in tests of reaction time. It is not certain whether this is a result of a slowing in the transmission of impulses along neurons or in neurotransmitter relays at the synapses.

Although the nervous system does deteriorate with age (in ways that are not well understood), in most people it functions effectively throughout life. Brain dysfunction is not a common characteristic of senescence.

Degenerative Diseases of the Nervous System

Degenerative diseases of the CNS are characterized by a progressive, symmetrical deterioration of vital structures of the brain or spinal cord. The etiologies of these diseases are poorly understood, but it is thought that most of them are genetic.

Cerebrovascular Accident (CVA)   Cerebrovascular accident is the most common disease of the nervous system. It is the third highest cause of death in the United States and perhaps the major cause of disability. The term **stroke** is frequently used synonymously with CVA, but actually a stroke refers to the sudden and dramatic appearance of a neurological defect. Cerebral

UNIT V

thrombosis, in which a thrombus, or clot, forms in an artery of the brain, is the most common cause of CVA. Other causes of CVA include intracerebral hemorrhages, aneurysms, atherosclerosis, and arteriosclerosis of the cerebral arteries.

Patients who recover from CVA frequently suffer partial paralysis and mental disorders, such as loss of language skills. The dysfunction depends upon the severity of the CVA and the regions of the brain that were injured. Patients surviving a CVA can often be rehabilitated, but approximately two-thirds die within 3 years of the initial damage.

Multiple Sclerosis Multiple sclerosis (MS) is a relatively common neurological disease in people between the ages of 20 and 40. MS is a chronic, degenerating, remitting, and relapsing disease that progressively destroys the myelin layers of neurons in multiple areas of the CNS. Initially, lesions form on the myelin layers and soon develop into hardened *scleroses* or scars (hence the name). The destruction of myelin layers prohibits the normal conduction of impulses, resulting in a progressive loss of functions. Because myelin degeneration is widely distributed, MS has a greater variety of symptoms than any other neurologic disease. This characteristic, coupled with remission, frequently causes the disease to be misdiagnosed.

During the early stages of MS, many patients are believed to be neurotic because of the wide variety and temporary nature of their symptoms. As the disease progresses, the symptoms may include double vision (diplopia), spots in the visual field, blindness, tremor, numbness of appendages, and locomotor difficulty. Eventually the patient is bedridden, and death may occur anytime from 7 to 30 years after the first symptoms appear.

Syringomyelia Syringomyelia (*sĭ-ring''go-mi-e'le-ă*) is a relatively uncommon condition characterized by the appearance of cystlike cavities, called *syringes,* within the gray matter of the spinal cord. These syringes progressively destroy the cord from the inside out. Syringomyelia is a chronic, slow progressing disease of unknown cause. As the spinal cord deteriorates, the patient experiences muscular weakness and atrophy and sensory loss, particularly of the senses of pain and temperature.

Tay–Sachs Disease In Tay–Sachs disease, the myelin sheaths are destroyed by the excessive accumulation of one of the lipid components of the myelin. This results from an enzyme defect caused by the inheritance of genes carried by the parents in a recessive state. Tay–Sachs disease, which is inherited primarily by individuals of Eastern European Jewish descent, appears before the infant is a year old. It causes blindness, loss of mental and motor ability, and ultimately death by the age of 3. Potential parents can tell if they are carriers for this condition by the use of a special blood test for the defective enzyme.

Diseases Involving Neurotransmitters **Parkinson's disease,** or **paralysis agitans,** is a major cause of neurological disability in people over 60 years of age. It is a progressively degenerative disease of unknown cause. Nerve cells within the substantia nigra, an area within the basal nuclei of the brain, are destroyed. This causes muscle tremors, muscular rigidity, speech defects, and other severe problems. The symptoms of this disease can be partially treated by altering the neurotransmitter status of the brain. These patients are given L-dopa to increase the production of dopamine in the brain, and may also be given anticholinergic drugs to decrease the production of acetylcholine.

Alzheimer's disease is the most common cause of dementia, often beginning in middle age and producing progressive mental deterioration. The cause of Alzheimer's disease is unknown, but evidence suggests that it is associated with the decreased ability of the brain to produce acetylcholine. Attempts to increase ACh production by increased ingestion of precursor molecules (choline or lecithin) have thus far not been successful. Drugs that block acetylcholinesterase, an enzyme that inactivates ACh, offer promise but are so far still in the experimental stage.

At least some psychiatric disorders may be produced by dysfunction of neurotransmitters. There is evidence that **schizophrenia** (*skit''sŏ-fre-ne-ă*) is associated with hyperactivity of the neurons that use dopamine as a neurotransmitter. Drugs that are effective in the treatment of schizophrenia (e.g., chlorpromazine) act by blocking dopamine receptor proteins. **Depression** is associated with decreased activity of the neurons that use monoamines—norepinephrine and serotonin—as neurotransmitters. Antidepressant drugs enhance the action of these neurotransmitters. In a similar way, barbiturates and benzodiazepine (e.g., Valium), which decrease **anxiety,** act by enhancing the action of the neurotransmitter GABA in the CNS.

CLINICAL CASE STUDY ANSWER

The affected upper motor neurons have cell bodies that reside in the right cerebral hemisphere. They give rise to fibers that, as they course downward, cross in the medulla oblongata to the left side of the brain stem and spinal cord. They continue on the left side, eventually synapsing at the appropriate level with lower motor neurons in the anterior horn of the spinal cord. The lower motor neurons then give rise to fibers that travel to the periphery where they innervate end organs; in other words, muscle cells in the left side of the body. Since the patient's neurological deficits are all motor as opposed to sensory, the tumor is most likely located in the right frontal lobe. The parietal lobe contains sensory neurons. The persistent headache is due to the pressure of the tumorous mass on the meninges, which are heavily sensory innervated.

---

multiple sclerosis: L. *multiplus,* many parts; Gk. *skleros,* hardened
Tay–Sachs disease: from Warren Tay, English physician, 1843–1927, and Bernard Sachs, American neurologist, 1858–1944

Alzheimer's disease: from Alois Alzheimer, German neurologist, 1864–1915

Internal Affairs

The Relational Workings of the Body Systems

Nervous System

- Influences secretions from integumentary glands and contractions of arrector pili muscles

- Innervates bones and monitors movements within joints

- Innervates muscles for autonomic and voluntary muscle contractions
- Monitors include activities

- Innervates endocrine glands causing rapid, autonomic secretion of hormones

- Innervates the heart and blood vessels to modify heart rate and depth of respiration

- Innervates lymphoid organs

- Respiratory centers within brain stem control respiratory rates and depth of respiration

- Innervates digestive organs and autonomically regulates GI tract movements and secretions

- Innervates organs of urinary system to control urination
- Modifies renal blood pressure

- Innervates reproductive organs to control sexual function

Integumentary System
- Supports and protects peripheral receptors

Skeletal System
- Supports and protects the brain and spinal cord
- Stores calcium needed for neural function

Muscular System
- Generates body heat to maintain constant temperature for neural function
- Proprioceptors transmit impulses from muscles to brain

Endocrine System
- Hormones augment and sustain autonomic stimuli to body organs

Circulatory System
- Transports O_2 and CO_2, nutrients, and fluids to and from the brain and spinal cord

Lymphatic System
- Protects against infections within the brain and spinal cord

Respiratory System
- Provides O_2 and eliminates CO_2

Digestive System
- Provides nutrients for growth, maintenance, and repair of nervous system

Urinary System
- Eliminates metabolic wastes
- Regulates pH, body fluids, and electrolyte concentrations

Reproductive System
- Gonads produce sex hormones that influence brain development and sexual behavior

CHAPTER SUMMARY

Organization and Functions of the Nervous System (pp. 330–333)

1. The central nervous system (CNS) consists of the brain and spinal cord and contains gray and white matter. It is covered with meninges and is bathed in cerebrospinal fluid.
2. The functions of the nervous system are orientation, coordination, assimilation, and instinctual behavior.

Neurons and Neuroglia (pp. 333–341)

1. The dendrites of a neuron receive stimulation and the axon conducts nerve impulses. Neuroglial cells provide structural and functional support for the activities of neurons.
2. A neuron contains dendrites, a cell body, and an axon.
   (a) The cell body contains the nucleus, chromatophilic substances, neurofibrils, and other organelles.
   (b) Dendrites receive stimuli and the axon conducts nerve impulses away from the cell body.
3. Neuroglial cells are of six types: neurolemmocytes form myelin layers around axons in the PNS; oligodendrocytes form myelin layers around axons in the CNS; microglia perform a phagocytic function in the CNS; astrocytes regulate passage of substances from the blood into the CNS; ependymal cells assist the movement of cerebrospinal fluid in the CNS; and ganglionic gliocytes support neuron cell bodies in the PNS.
   (a) The neuroglial cells that surround an axon form a covering called a myelin layer.
   (b) Myelinated neurons have limited capabilities for regeneration following trauma.
4. A nerve is a collection of dendrites and axons in the PNS.
   (a) Sensory (afferent) neurons are pseudounipolar.
   (b) Motor (efferent) neurons are multipolar.
   (c) Association (interneurons) are located entirely within the CNS.
   (d) Somatic motor nerves innervate skeletal muscle; visceral motor (autonomic) nerves innervate smooth muscle, cardiac muscle, and glands.

Transmission of Impulses (pp. 341–342)

1. Irritability and conductivity are specialized properties of neurons that permit nerve impulse transmission.
2. Neurotransmitters facilitate synaptic impulse transmission.

General Features of the Brain (pp. 342–347)

1. The brain, composed of gray matter and white matter, is protected by meninges and is bathed in cerebrospinal fluid.
2. About 750 mℓ of blood flows to the brain each minute.

Cerebrum (pp. 347–355)

1. The cerebrum, consisting of two convoluted hemispheres, is concerned with higher brain functions, such as the perception of sensory impulses, the instigation of voluntary movement, the storage of memory, thought processes, and reasoning ability.
2. The cerebral cortex of the cerebral hemispheres is convoluted with gyri and sulci.
3. Each cerebral hemisphere contains frontal, parietal, temporal, occipital, and insula lobes.
4. Brain waves generated by the cerebral cortex are recorded as an electroencephalogram and may provide valuable diagnostic information.
5. The white matter of the cerebrum consists of association, commissural, and projection fibers.
6. Basal nuclei are specialized masses of gray matter located within the white matter of the cerebrum.

Diencephalon (pp. 355–357)

1. The diencephalon is a major autonomic region of the brain.
2. The thalamus is an ovoid mass of gray matter that functions as a relay center for sensory impulses and responds to pain.
3. The hypothalamus is an aggregation of specialized nuclei that regulate many visceral activities. It also performs emotional and instinctual functions.
4. The epithalamus contains the pineal gland and the vascular choroid plexus over the roof of the third ventricle.

Mesencephalon (pp. 357–358)

1. The mesencephalon contains the corpora quadrigemina, the cerebral peduncles, and specialized nuclei that help to control posture and movement.
2. The superior colliculi of the corpora quadrigemina are concerned with visual reflexes and the inferior colliculi are concerned with auditory reflexes.
3. The red nucleus and the substantia nigra are concerned with motor activities.

Metencephalon (pp. 358–360)

1. The pons consists of fiber tracts connecting the cerebellum and medulla oblongata to other structures of the brain. The pons also contains nuclei for certain cranial nerves and the regulation of respiration.
2. The cerebellum consists of two hemispheres connected by the vermis and supported by three paired cerebellar peduncles.
   (a) The cerebellum is composed of a white matter tract called the arbor vitae, surrounded by a thin convoluted cortex of gray matter.
   (b) The cerebellum is concerned with coordinated contractions of skeletal muscle.

Myelencephalon (pp. 360–361)

1. The medulla oblongata is composed of the ascending and descending tracts of the spinal cord and contains nuclei for several autonomic functions.
2. The reticular formation functions as the reticular activating system in arousing the cerebrum.

Meninges of the Central Nervous System (pp. 361–363)

1. The cranial dura mater consists of an outer periosteal layer and an inner meningeal layer. The spinal dura mater is a single layer surrounded by the vascular epidural space.
2. The arachnoid mater is a netlike meninx surrounding the subarachnoid space, which contains cerebrospinal fluid.
3. The thin pia mater adheres to the contour of the CNS.

UNIT V

Ventricles and Cerebrospinal Fluid (pp. 363–366)

1. The lateral (first and second), third, and fourth ventricles are interconnected chambers within the brain that are continuous with the central canal of the spinal cord.
2. These chambers are filled with cerebrospinal fluid, which also flows throughout the subarachnoid space.
3. Cerebrospinal fluid is continuously secreted by the choroid plexuses and is absorbed into the blood at the arachnoid villi.
4. The blood-brain barrier determines which substances within blood plasma can enter the extracellular fluid of the brain.

Spinal Cord (pp. 366–373)

1. The spinal cord is composed of 31 segments, each of which gives rise to a pair of spinal nerves.
   (a) It is characterized by a cervical enlargement, a lumbar enlargement, and two longitudinal grooves that partially divide it into right and left halves.
   (b) The conus medullaris is the terminal portion of the spinal cord and the cauda equina are nerve roots that radiate inferiorly from that point.
2. Ascending and descending spinal cord tracts are referred to as funiculi.
   (a) Descending tracts are grouped as either corticospinal (pyramidal) or extrapyramidal.
   (b) Many of the fibers in the funiculi decussate (cross over) in the spinal cord or in the medulla oblongata.

REVIEW ACTIVITIES

Objective Questions

Match the following structures of the brain to the region in which they are located:

1. cerebellum   (a) telencephalon
2. cerebral cortex   (b) diencephalon
3. medulla oblongata (c) mesencephalon
                        (d) metencephalon
                        (e) myelencephalon
4. The neuroglial cells that form myelin sheaths in the peripheral nervous system are
   (a) oligodendrocytes.
   (b) ganglionic gliocytes.
   (c) neurolemmocytes.
   (d) astrocytes.
   (e) microglia.

5. A collection of neuron cell bodies located outside the CNS is called
   (a) a tract.   (c) a nucleus.
   (b) a nerve.   (d) a ganglion.
6. Which of the following types of neurons is pseudounipolar?
   (a) sensory neurons
   (b) somatic motor neurons
   (c) neurons in the retina
   (d) autonomic motor neurons
7. Depolarization of an axon is produced by the movement of
   (a) Na^+ into the axon and K^+ out of the axon.
   (b) Na^+ into the axon to bond with K^+.
   (c) K^+ into the axon and Na^+ out of the axon.
   (d) Na^+ and K^+ within the axon toward the axon terminal.
8. The principal connection between the cerebral hemispheres is
   (a) the corpus callosum.
   (b) the pons.
   (c) the intermediate mass.
   (d) the vermis.
   (e) the precentral gyrus.
9. The structure of the brain that is most directly involved in the autonomic response to pain is
   (a) the pons.
   (b) the hypothalamus.
   (c) the medulla oblongata.
   (d) the thalamus.
10. Which statement is *false* concerning the basal nuclei?
    (a) They are located within the cerebrum.
    (b) They regulate the basal metabolic rate.
    (c) They consist of the caudate nucleus, lentiform nucleus, putamen and globus pallidus.
    (d) They indirectly exert an inhibitory influence on lower motor neurons.
11. The corpora quadrigemina, red nucleus, and substantia nigra are structures of
    (a) the diencephalon.
    (b) the metencephalon.
    (c) the mesencephalon.
    (d) the myelencephalon.
12. The fourth ventricle is contained within
    (a) the cerebrum.
    (b) the cerebellum.
    (c) the midbrain.
    (d) the metencephalon.

Essay Questions

1. Describe the formation of the neural crest and explain how derivatives of neural crest tissue can be located in different parts of the body.
2. Diagram and label a neuron. Beside each label, list the function of the identified structure. Why are neurons considered the basic functional units of the nervous system?
3. What is meant by a myelinated neuron? Describe how an injured nerve fiber may regenerate.
4. List the six principal types of neuroglia and discuss the location, structure, and function of each.
5. What is a nerve impulse and how does it relate to polarization, depolarization, and action potential?
6. What is a synapse and what is the role of a neurotransmitter in relaying a nerve impulse?
7. List the types of brain waves recorded on an electroencephalogram and explain the diagnostic value of each.
8. List the functions of the hypothalamus. Why is the hypothalamus considered a major part of the autonomic nervous system?
9. What structures are within the midbrain? List the nuclei located in the midbrain and state the function of each.
10. Describe the location and structure of the medulla oblongata. List the nuclei found within this structure. What are the functions of the medulla oblongata?
11. What is cerebrospinal fluid? Where is it produced and what is its path of circulation?
12. What do the following abbreviations stand for? EEG, ANS, CSF, PNS, RAS, CT scan, MS, DSR, and CVA.
13. Define the various psychological terms used to describe mental illness.
14. What is epilepsy? What causes it and how is it controlled?
15. What do meningitis, poliomyelitis and neurosyphilis have in common? How do these conditions differ?

Gundy/Weber Software

The tutorial software accompanying Chapter 11, Nervous Tissue and the Central Nervous System, is Volume 5, Nervous System.

UNIT V

Peripheral Nervous System

Chapter Outline

CLINICAL CASE STUDY

Following an auto accident, a 23-year-old male is brought to the emergency room for treatment of a fractured right humerus. Although the skin is not broken, there is obvious deformity caused by an angulated fracture at the midshaft. While conducting an examination on the patient's injured arm, the attending orthopedist notices that the patient is unable to extend the joints of his hand.

What structure has been injured in the brachial region of this patient that accounts for his inability to extend his hand? List the muscles that are unable to function and describe the movements that would be diminished. Would you suspect any other neurological deficits? Explain.

Hints: Because the nervous system functions to coordinate body movement, nerve trauma may be expressed in structures far removed from the site of injury. Carefully read the section dealing with the brachial plexus.

Introduction to the Peripheral Nervous System

The peripheral nervous system consists of receptors that respond to stimuli and of nerves that convey impulses to and from the central nervous system.

Objective 1. Define *peripheral nervous system* and distinguish between sensory, motor, and mixed nerves.

The **peripheral nervous system (PNS)** is that portion of the nervous system outside the central nervous system. The PNS functions to convey impulses to and from the brain or spinal cord. Sensory receptors within the sensory organs, neurons, nerves, ganglia, and plexuses are all part of the PNS, which serves virtually every part of the body (fig. 12.1). The sensory receptors are discussed in chapter 15.

Figure 12.1 The peripheral nervous system consists of cranial nerves, spinal nerves, plexuses, ganglia (not shown), and the peripheral nerves.

Plexuses:
Cervical
Brachial
Lumbar
Sacral

Some peripheral nerves:
Ulnar
Median
Radial
Lateral femoral cutaneous
Femoral
Sciatic

Cranial nerves (12 pairs)
Spinal nerves:
Cervical (8 pairs)
Thoracic (12 pairs)
Lumbar (5 pairs)
Sacral (5 pairs)
Coccygeal (1 pair)

Gordon/Waldrop

UNIT V

The nerves of the PNS are classified as cranial nerves or spinal nerves, depending on whether they arise from the brain or the spinal cord. The terms *sensory nerve, motor nerve,* and *mixed nerve* relate to the direction in which the nerve impulses are being conducted. **Sensory nerves** consist of sensory (afferent) neurons that convey impulses toward the CNS. **Motor nerves** consist of motor (efferent) neurons that convey impulses away from the CNS. **Mixed nerves** are composed of both sensory and motor neurons, and therefore convey impulses in both directions. The reflexes considered in this chapter involve sensory and motor nerves of the PNS and specific portions of the CNS. A cross section of a spinal nerve is depicted in figure 12.2.

1. The tongue responds to tastes and pain and moves to manipulate food. Make a quick sketch of the brain and the tongue to depict the relationship between the CNS and the PNS. Using lines and arrows indicate the sensory and motor innervation of the tongue. Define *mixed nerve.* What kinds of sensory stimulation arise from the tongue? What type of response is caused by motor stimulation to the tongue?
2. List the structures of the nervous system that are considered a part of the PNS.

Figure 12.2 A scanning electron micrograph of a spinal nerve seen in cross section (about 1,000×). (From: R. G. Kessel and R. H. Kardon, *Tissues and Organs: A Text-Atlas of Scanning Electron Microscopy,* © 1979 W. H. Freeman and Company.)

Epineurium

Perineurium

Endoneurium

Nerve fiber

Blood vessel

Fascicle

Cranial Nerves

Twelve pairs of cranial nerves emerge from the inferior surface of the brain and pass through the foramina of the skull to innervate structures in the head, neck, and visceral organs of the trunk.

Objective 2. List the 12 pairs of cranial nerves and describe the location and function of each.

Objective 3. Describe the clinical methods for determining cranial nerve dysfunction.

Structure and Function of the Cranial Nerves

Of the 12 pairs of cranial nerves, 2 pairs arise from the forebrain and 10 pairs arise from the midbrain and brain stem (fig. 12.3).

UNIT V

Figure 12.3 Cranial nerves. With the exception of the olfactory nerves, each cranial nerve is composed of a bundle of nerve fibers. The olfactory nerves are minute and diffuse strands of nerve fibers that attach to the olfactory bulb (see fig. 12.4).

Olfactory bulb

Olfactory tract

Optic chiasma

Optic tract

Abducens (VI)

Facial (VII)

Hypoglossal (XII)

Accessory (XI)

Olfactory (I)

Optic (II)

Oculomotor (III)

Trochlear (IV)

Trigeminal (V)

Vestibulocochlear (VIII)

Glossopharyngeal (IX)

Vagus (X)

Krabach

Figure 12.4 The olfactory nerve.

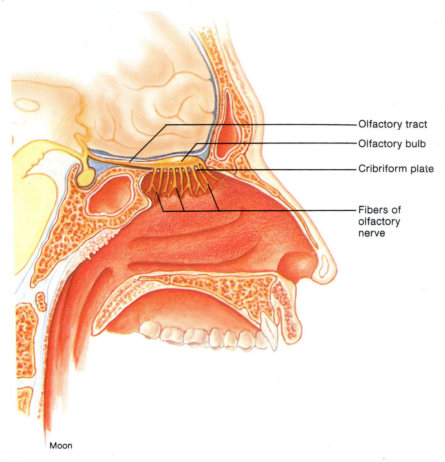

Olfactory tract
Olfactory bulb
Cribriform plate
Fibers of olfactory nerve

Moon

UNIT V

The cranial nerves are designated by Roman numerals and names. The Roman numerals refer to the order in which the nerves are positioned from the front of the brain to the back. The names indicate the structures innervated or the principal functions of the nerves. A summary of the cranial nerves is presented in table 12.1.

Although most cranial nerves are mixed, some are associated with special senses and consist of sensory neurons only. The cell bodies of sensory neurons are located in ganglia outside the brain.

Generations of anatomy students have used a mnemonic device to help them remember the order in which the cranial nerves emerge from the brain: "On old Olympus's towering top, a Finn and German viewed a hop." The initial letter of each word in this jingle corresponds to the initial letter of each pair of cranial nerves. A problem with this classic verse is that the eighth cranial nerve represented by *and* in the jingle, which used to be referred to as auditory, is currently recognized as the vestibulocochlear cranial nerve. Hence, the following topical mnemonic: "On old Olympus's towering top, a fat vicious goat vandalized a hat."

I Olfactory Nerve   Actually, numerous olfactory nerves relay sensory impulses of smell from the mucous membranes of the nasal cavity (fig. 12.4). Olfactory nerves are composed of bipolar neurons that function as *chemoreceptors,* responding to volatile chemical particles breathed into the nasal cavity. The dendrites and cell bodies of olfactory neurons are positioned within the mucosa, primarily that which covers the superior nasal conchae and adjacent nasal septum. The axons of these neurons pass through the cribriform plate of the ethmoid bone to the **olfactory bulb** where synapses are made, and the sensory impulses are passed through the **olfactory tract** to the primary olfactory area in the cerebral cortex.

II Optic Nerve   The optic nerve, another sensory nerve, conducts impulses from the *photoreceptors* (rods and cones) in the retina of the eye. Each optic nerve is composed of an estimated 1.25 million nerve fibers that converge at the back of the eyeball and enter the cranial cavity through the optic canal. The two optic nerves unite on the floor of the diencephalon to form the **optic chiasma** *(ki-as'ma)* (fig. 12.5). Nerve fibers that arise

olfactory: L. *olfacere,* smell out
optic: L. *optica,* see
chiasma: Gk. *chiasma,* an X-shaped arrangement

Figure 12.5 The optic nerve.

Eyeball
Retina
Optic nerve
Optic chiasma
Optic tract
Lateral geniculate nucleus of thalamus
Visual cortex

Sims

Figure 12.6 The oculomotor, trochlear, and abducens cranial nerves.

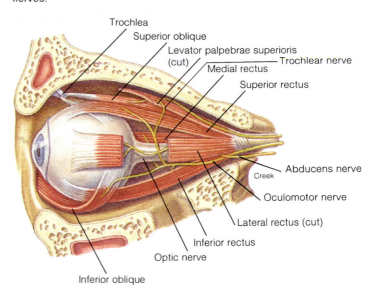

Trochlea
Superior oblique
Levator palpebrae superioris (cut)
Trochlear nerve
Medial rectus
Superior rectus
Abducens nerve
Creek
Oculomotor nerve
Lateral rectus (cut)
Inferior rectus
Optic nerve
Inferior oblique

from the medial half of each retina cross at the optic chiasma to the opposite side of the brain, whereas fibers arising from the lateral half remain on the same side of the brain. The optic nerve fibers pass posteriorly from the optic chiasma in the **optic tracts.** The optic tracts lead to the thalamus, where a majority of the fibers terminate within certain thalamic nuclei. A few of the ganglion-cell axons that reach the thalamic nuclei have collaterals that convey impulses to the superior colliculi. Synapses within the thalamic nuclei, however, permit impulses to pass through neurons to the **visual cortex** within the occipital lobes. Other synapses permit impulses to reach the nuclei for the oculomotor, trochlear, and abducens nerves, which regulate intrinsic (internal) and extrinsic (from orbit to eyeball) eye muscles. The visual pathway into the eyeball functions reflexively to produce motor responses to light stimuli. If an optic nerve is damaged, the eyeball served by that nerve is blinded.

III Oculomotor Nerve impulses through the oculomotor nerve produce certain extrinsic and intrinsic movements of the eyeball. The oculomotor is primarily a motor nerve that arises from nuclei within the midbrain. It divides into superior and inferior branches as it passes through the superior orbital fissure in the orbit (fig. 12.6). The superior branch innervates the **superior rectus muscle,** which moves the eyeball superiorly, and the **levator palpebrae** (le-va'tor pal'pĕ-bre) **superioris muscle,** which raises the upper eyelid. The inferior branch innervates the **medial rectus, inferior rectus,** and **inferior oblique eye muscles** for medial, inferior, and superior and lateral movement of the eyeball, respectively. In addition, fibers from the inferior branch of the oculomotor nerve enter the eyeball to supply autonomic motor innervation to the intrinsic smooth muscles of the iris for pupil constriction and to the muscles within the ciliary body for lens accommodation.

A few sensory fibers of the oculomotor nerve originate from proprioceptors within the intrinsic muscles of the eyeball. These fibers convey impulses that affect the position and activity of the muscles they serve. A person whose oculomotor nerve is damaged may have a drooping upper eyelid or a dilated pupil or be unable to move the eyeball in the directions permitted by the four extrinsic muscles innervated by this nerve.

IV Trochlear The trochlear nerve is a very small mixed nerve that emerges from a nucleus within the midbrain and passes from the cranium through the superior orbital fissure of the orbit. The trochlear nerve innervates the **superior oblique muscle** of the eyeball with both motor and sensory fibers (fig. 12.6). Motor impulses to the superior oblique muscle cause the eyeball to rotate downward and away from the midline. Sensory impulses originate in proprioceptors of the superior oblique muscle and provide information about its position and activity. Damage to the trochlear nerve impairs movement in the direction permitted by the superior oblique eye muscle.

V Trigeminal The large trigeminal nerve is a mixed nerve with motor functions originating from the nuclei within the pons and sensory functions terminating in nuclei within the midbrain, pons, and medulla oblongata. Two roots of the trigeminal nerve are apparent as they emerge from the ventrolateral side of the pons (see fig. 12.3). The larger, **sensory root** immediately enlarges into a swelling called the **trigeminal (semilunar) ganglion,** located in a bony depression on the inner surface of the petrous portion of the temporal bone. Three large nerves arise from the trigeminal ganglion (fig. 12.7): the **ophthalmic nerve** enters the orbit through the superior or-

---

trochlear: Gk. *trochos,* a wheel
trigeminal: L. *trigeminus,* three born together
ophthalmic: L. *ophthalmia,* region of the eye

UNIT V

Figure 12.7 The trigeminal nerve and its branches.

bital fissure, the **maxillary nerve** extends through the foramen rotundum, and the **mandibular nerve** passes through the foramen ovale. The smaller **motor root** consists of motor fibers of the trigeminal nerve that accompany the mandibular nerve through the foramen ovale and innervate the muscles of mastication and certain muscles in the floor of the mouth. Impulses through the motor portion of the mandibular nerve of the trigeminal ganglion stimulate contraction of the muscles involved in chewing, including the medial and lateral pterygoids, masseter, temporalis, mylohyoid, and the anterior belly of the digastric muscle.

Although the trigeminal is a mixed nerve, its sensory functions are much more extensive than its motor functions. The three sensory nerves of the trigeminal ganglion respond to touch, temperature, and pain sensations from the face. More specifically, the ophthalmic nerve consists of sensory fibers from the anterior half of the scalp, skin of the forehead, upper eyelid, surface of the eyeball, lacrimal (tear) gland, side of the nose, and upper mucosa of the nasal cavity. The maxillary nerve is composed of sensory fibers from the lower eyelid, lateral and inferior mucosa of the nasal cavity, palate and portions of the pharynx, teeth and gums of the upper jaw, upper lip, and skin of the cheek. Sensory fibers of the mandibular nerve transmit impulses from the teeth and gums of the lower jaw, anterior two-thirds of the tongue (not taste), mucosa of the mouth, auricle of the ear, and lower part of the face. Trauma to the trigeminal nerve results in a lack of sensation from specific facial structures. Damage to the mandibular nerve impairs chewing.

The trigeminal nerve is the principal nerve relating to the practice of dentistry. Before teeth are filled or extracted, anesthetic is injected into the appropriate nerve to block sensation. A *maxillary*, or *second-division, nerve block*, performed by injecting near the sphenopalatine ganglion (see fig. 12.7), desensitizes the teeth in the upper jaw. A *mandibular*, or *third-division, nerve block* desensitizes the lower teeth. This is performed by injecting anesthetic into the inferior alveolar nerve, a branch of the mandibular nerve where it enters the mandible through the mandibular foramen.

VI Abducens   The small abducens nerve originates from a nucleus within the pons and emerges from the lower portion of the pons and the anterior border of the medulla oblongata. It is a mixed nerve that traverses the superior orbital fissure of the orbit to innervate the **lateral rectus eye muscle** (see fig. 12.6). Impulses through the motor fibers of the abducens nerve cause the lateral rectus eye muscle to contract and the eyeball to move laterally away from the midline. Sensory impulses through the abducens nerve originate in proprioceptors in the lateral rectus muscle and are conveyed to the pons, where muscle contraction is mediated. If the abducens nerve is damaged, not only will the patient be unable to turn the eyeball laterally, but because of the lack of muscle tonus to the lateral rectus muscle, the eyeball will be pulled medially.

VII Facial   The facial nerve arises from nuclei within the lower portion of the pons, traverses the petrous part of the temporal bone (see fig. 12.9), and emerges on the side of the face near the parotid (salivary) gland. The facial nerve is mixed. Impulses through the motor fibers cause contraction of the posterior belly of the digastric muscle and the muscles of facial expression, including the scalp and platysma muscles (fig. 12.8).

Figure 12.8 The facial nerve and its superficial branches.

Temporal

Zygomatic

Posterior auricular

Facial nerve

Digastric and stylohyoid

Parotid gland

Buccal

Mandibular

Cervical

Creek

The submandibular and sublingual (salivary) glands also receive some autonomic motor innervation from the facial nerve, as does the lacrimal gland.

Sensory fibers of the facial nerve arise from taste buds on the anterior two-thirds of the tongue. Taste buds function as *chemoreceptors* because they respond to specific chemical stimuli. The **geniculate ganglion** is the enlargement of the facial nerve just before the sensory portion enters into the pons. Sensory sensations of taste are conveyed into nuclei within the medulla oblongata, through the thalamus, and finally to the gustatory (taste) area of the cerebral cortex of the insula.

Trauma to the facial nerve results in inability to contract facial muscles on the affected side of the face and distorts taste perception, particularly of sweets. The affected side of the face tends to sag because muscle tonus is lost. *Bell's palsy* is a functional disorder (probably of viral origin) of the facial nerve.

VIII Vestibulocochlear The vestibulocochlear nerve, also referred to as the **auditory, acoustic,** or **statoacoustic nerve,** serves structures contained within the skull. It is the only cranial nerve that does not exit the cranium through a foramen. A purely sensory nerve, the vestibulocochlear is composed of two nerves that arise within the inner ear (fig. 12.9). The **vestibular nerve** arises from the **vestibular organs** associated with equilibrium and balance. Bipolar neurons from the vestibular organs (saccule, utricle, and semicircular ducts) extend to the **vestibular ganglion,** where cell bodies are contained. From there, fibers convey impulses to the **vestibular nuclei** within the pons and medulla oblongata. Fibers from there extend to the thalamus and the cerebellum.

Bell's palsy: from Sir Charles Bell, Scottish physician, 1774–1842
vestibulocochlear: L. *vestibulum*, chamber; *cochlea*, snail shell

UNIT V

Figure 12.9 The vestibulocochlear nerve. The structures of the inner ear are served by this nerve. The semicircular ducts are the membranous structures contained within the semicircular canals and are concerned with balance and equilibrium. The cochlea contains the structures concerned with hearing.

Semicircular canals

Facial nerve

Vestibular nerve

Cochlear nerve

Vestibulocochlear nerve

Cochlea

The **cochlear nerve** arises from the **spiral organ** (organ of Corti) within the cochlea and is associated with hearing. The cochlear nerve is composed of bipolar neurons that convey impulses through the **spiral ganglion** to the **cochlear nuclei** within the medulla oblongata. From there, fibers extend to the thalamus and synapse with neurons that convey the impulses to the auditory areas of the cerebral cortex.

Injury to the cochlear nerve results in perception deafness, whereas damage to the vestibular nerve causes dizziness and an inability to maintain balance.

IX Glossopharyngeal

The glossopharyngeal nerve is a mixed nerve that innervates part of the tongue and pharynx (fig. 12.10). The motor fibers of this nerve originate in a nucleus within the medulla oblongata and pass through the jugular foramen. The motor fibers innervate the muscles of the pharynx and the parotid gland to stimulate the swallowing reflex and the secretion of saliva.

The sensory fibers of the glossopharyngeal nerve arise from the pharyngeal region, the parotid gland, the middle-ear cavity, and the taste buds on the posterior one-third of the tongue. These taste buds, like those innervated by the facial nerve, are *chemoreceptors.* Some sensory fibers also arise from sensory receptors within the carotid sinus of the neck and help regulate blood pressure. Impulses from the glossopharyngeal nerve travel through the medulla oblongata and into the thalamus, where they synapse with fibers that convey the impulses to the gustatory area of the cerebral cortex.

Damage to the glossopharyngeal nerve results in the loss of perception of bitter and sour taste from taste buds on the posterior portion of the tongue. If the motor portion of this nerve is damaged, swallowing becomes difficult.

X Vagus

The vagus nerve has motor and sensory fibers that innervate visceral organs of the thoracic and abdominal cavities (fig. 12.11). The motor portion arises from the **nucleus ambiguus** and **dorsal motor nucleus** of the vagus within the medulla oblongata and passes through the jugular foramen. The vagus is the longest of the cranial nerves, and through various branches it innervates the muscles of the pharynx, larynx, respiratory tract, lungs, heart, esophagus, and abdominal viscera, with the exception of the lower portion of the large intestine. One motor branch of the vagus nerve, the **recurrent laryngeal nerve,** innervates the larynx, enabling speech.

Sensory fibers of the vagus nerve convey impulses from essentially the same organs served by motor fibers. Impulses through the sensory fibers relay specific sensations, such as hunger pangs, distension, intestinal discomfort, or laryngeal movements. Sensory fibers also arise from proprioceptors in the muscles innervated by the motor fibers of this nerve.

If both vagus nerves are seriously damaged, death ensues rapidly because vital autonomic functions stop. The injury of

Figure 12.10 The glossopharyngeal nerve.

Glossopharyngeal nerve

Parotid gland

Tongue

Carotid sinus

Common carotid artery

one nerve causes vocal impairment, difficulty in swallowing, or other visceral disturbances.

XI Accessory

The accessory nerve is principally a motor nerve, but it does contain some sensory fibers from proprioceptors within the muscles it innervates. The accessory nerve is unique in that it arises from both the brain and the spinal cord (fig. 12.12). The **cranial root** arises from nuclei within the medulla oblongata (ambiguus and accessory), passes through the jugular foramen with the vagus nerve, and innervates the skeletal muscles of the soft palate, pharynx, and larynx, which contract reflexively during swallowing. The **spinal root** arises from the first five segments of the cervical portion of the spinal cord, passes cranially through the foramen magnum to join with the cranial root, and passes through the jugular foramen. The spinal root of the accessory nerve innervates the sternocleidomastoid and the trapezius muscles that move the head, neck, and shoulders. Damage to an accessory nerve makes it difficult to move the head or shrug the shoulders.

XII Hypoglossal

The hypoglossal nerve is a mixed nerve. The motor fibers arise from the hypoglossal nucleus within the medulla oblongata and pass through the hypoglossal canal of the skull to innervate both the extrinsic and intrinsic muscles of the tongue (fig. 12.12). Motor impulses along these fibers account for the coordinated contraction of the tongue muscles necessary for such activities as food manipulation, swallowing, and speech.

The sensory portion of the hypoglossal nerve arises from proprioceptors within the same tongue muscles and conveys impulses to the medulla oblongata regarding the position and function of the muscles.

If a hypoglossal nerve is damaged, a person will have difficulty in speaking, swallowing, and protruding the tongue.

organ of Corti: from Alfonso Corti, Italian anatomist, 1822–88
glossopharyngeal: L. *glossa,* tongue; Gk. *pharynx,* throat
vagus: L. *vagus,* wandering

hypoglossal: Gk. *hypo,* under; L. *glossa,* tongue

UNIT V

Figure 12.11 The visceral components of the vagus nerve.

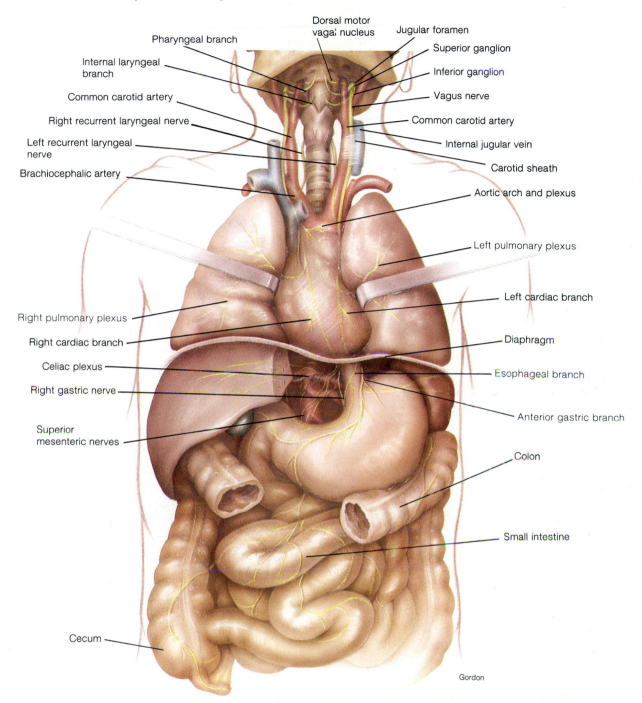

Dorsal motor
vagal nucleus

Jugular foramen

Pharyngeal branch

Superior ganglion

Internal laryngeal
branch

Inferior ganglion

Common carotid artery

Vagus nerve

Right recurrent laryngeal nerve

Common carotid artery

Left recurrent laryngeal
nerve

Internal jugular vein

Brachiocephalic artery

Carotid sheath

Aortic arch and plexus

Left pulmonary plexus

Right pulmonary plexus

Left cardiac branch

Right cardiac branch

Diaphragm

Celiac plexus

Esophageal branch

Right gastric nerve

Anterior gastric branch

Superior
mesenteric nerves

Colon

Small intestine

Cecum

Gordon

UNIT V

Figure 12.12 The accessory and hypoglossal nerves.

Table 12.1	Summary of Cranial Nerves			
Number and Name	**Foramen Transmitting**	**Composition**	**Location of Cell Bodies**	**Function**
I Olfactory	Foramina in cribriform plate of ethmoid bone	Sensory	Bipolar cells in nasal mucosa	Olfaction
II Optic	Optic canal	Sensory	Ganglion cells of retina	Vision
III Oculomotor	Superior orbital fissure	Motor	Oculomotor nucleus	Motor impulses to levator palpebrae superioris and extrinsic eye muscles except superior oblique and lateral rectus
		Motor: parasympathetic		Innervation to muscles that regulate amount of light entering eye and that focus the lens
		Sensory: proprioception		Proprioception from muscles innervated with motor fibers
IV Trochlear	Superior orbital fissure	Motor	Trochlear nucleus	Motor impulses to superior oblique muscle of eyeball
		Sensory: proprioception		Proprioception from superior oblique muscle of eyeball
V Trigeminal Ophthalmic nerve	Superior orbital fissure	Sensory	Trigeminal ganglion	Sensory impulses from cornea, skin of nose, forehead, and scalp
Maxillary nerve	Foramen rotundum	Sensory	Trigeminal ganglion	Sensory impulses from nasal mucosa, upper teeth and gums, palate, upper lip, and skin of cheek
Mandibular nerve	Foramen ovale	Sensory	Trigeminal ganglion	Sensory impulses from temporal region, tongue, lower teeth and gum, and skin of chin and lower jaw
		Sensory: proprioception		Proprioception from muscles of mastication
		Motor	Motor trigeminal nucleus	Motor impulses to muscles of mastication and muscle that tenses tympanum
VI Abducens	Superior orbital fissure	Motor	Abducens nucleus	Motor impulses to lateral rectus muscle of eyeball
		Sensory: proprioception		Proprioception from lateral rectus muscle of eyeball

Table 12.1	Continued			
Number and Name	**Foramen Transmitting**	**Composition**	**Location of Cell Bodies**	**Function**
VII Facial	Stylomastoid foramen	Motor	Motor facial nucleus	Motor impulses to muscles of facial expression and muscle that tenses the stapes
		Motor: parasympathetic	Superior salivatory nucleus	Secretion of tears from lacrimal gland and salivation from sublingual and submandibular glands
		Sensory	Geniculate ganglion	Sensory impulses from taste buds on anterior two-thirds of tongue; nasal and palatal sensation
		Sensory: proprioception		Proprioception from muscles of facial expression
VIII Vestibulocochlear	Internal acoustic meatus	Sensory	Vestibular ganglion	Sensory impulses associated with equilibrium
			Spiral ganglion	Sensory impulses associated with hearing
IX Glossopharyngeal	Jugular foramen	Motor	Nucleus ambiguus	Motor impulses to muscles of pharynx used in swallowing
		Sensory: proprioception	Petrosal ganglion	Proprioception from muscles of pharynx
		Sensory	Petrosal ganglion	Sensory impulses from taste buds on posterior one-third of tongue, pharynx, middle-ear cavity, and carotid sinus
		Parasympathetic	Inferior salivatory nucleus	Salivation from parotid gland
X Vagus	Jugular foramen	Motor	Nucleus ambiguus	Contraction of muscles of pharynx (swallowing) and larynx (phonation)
		Sensory: proprioception		Proprioception from visceral muscles
		Sensory	Nodose ganglion	Sensory impulses from taste buds on rear of tongue; sensations from auricle of ear; general visceral sensations
		Motor: parasympathetic	Dorsal motor nucleus	Regulate visceral motility
XI Accessory	Jugular foramen	Motor	Nucleus ambiguus	Laryngeal movement; soft palate
			Accessory nucleus	Motor impulses to trapezius and sternocleidomastoid muscles for movement of head, neck, and shoulders
		Sensory: proprioception		Proprioception from muscles that move head, neck, and shoulders
XII Hypoglossal	Hypoglossal canal	Motor	Hypoglossal nucleus	Motor impulses to intrinsic and extrinsic muscles of tongue and infrahyoid muscles
		Sensory: proprioception		Proprioception from muscles of tongue

UNIT V

Neurological Assessment of the Cranial Nerves

Head injuries and brain concussions are common occurrences in automobile accidents. The cranial nerves would seem to be well protected on the inferior side of the brain. But the brain, immersed in and filled with cerebrospinal fluid, is like a water-sodden log; a blow to the top of the head can cause a serious rebound of the brain from the floor of the cranium. Routine neurological examinations involve testing for cranial-nerve dysfunction.

Commonly used clinical methods for determining cranial-nerve dysfunction are presented in table 12.2.

1. Which cranial nerves consist of sensory fibers only?
2. Which cranial nerves pass through the superior orbital fissure? through the jugular foramen?
3. Which cranial nerves are involved in tasting, chewing and manipulating food, and swallowing?
4. Which cranial nerves have to do with the structure, function, or movement of the eyeball?
5. List the cranial nerves and indicate how each would be tested (both motor and sensory) for possible dysfunction.

Table 12.2	Methods of Determining Cranial-Nerve Dysfunction	
Nerve	**Techniques of Examination**	**Comments**
Olfactory	Patient asked to differentiate odors (tobacco, coffee, soap, etc.) with eyes closed.	Nasal passages must be patent and tested separately by occluding the opposite side.
Optic	Retina examined with ophthalmoscope; visual acuity tested with eye charts.	Visual acuity must be determined with lenses on, if patient wears them.
Oculomotor	Patient follows examiner's finger movement with eyes—especially cross-eyed movement; pupillary change observed by shining light into each eye separately.	Examiner should note rate of pupillary change and coordinated constriction of pupils. Light in one eye should cause a similar pupillary change in other eye, but to a lesser degree.
Trochlear	Patient follows examiner's finger movement with eyes—especially lateral and downward movement.	
Trigeminal	Motor portion: Temporalis and masseter muscles palpated as patient clenches teeth; patient asked to open mouth against resistance applied by examiner.	Muscles of both sides of the jaw should show equal contractile strength.
	Sensory portion: Tactile and pain receptors tested by lightly touching patient's entire face with cotton and then with pin stimulus.	Patient's eyes should be closed and innervation areas for all three nerves branching from the trigeminal nerve should be tested.
Abducens	Patient follows examiner's finger movement—especially lateral movement.	Motor functioning of cranial nerves III, IV, and VI may be tested simultaneously through selective movements of eyeball.
Facial	Motor portion: Patient asked to raise eyebrows, frown, tightly constrict eyelids, smile, puff out cheeks, and whistle.	Examiner should note lack of tonus expressed by sagging regions of face.
	Sensory portion: Sugar placed on each side of tip of patient's tongue.	Not reliable test for specific facial-nerve dysfunction because of tendency to stimulate taste buds on both sides of tip of tongue.
Vestibulocochlear	Vestibular portion: Patient asked to walk a straight line.	Not usually tested unless patient complains of dizziness or balance problems.
	Cochlear portion: Tested with tuning fork.	Examiner should note ability to discriminate sounds.
Glossopharyngeal and vagus	Motor: Examiner notes disturbances in swallowing, talking, and movement of soft palate; gag reflex tested.	Visceral innervation of vagus cannot be examined except for innervation to larynx, which is also served by glossopharyngeal.
Accessory	Patient asked to shrug shoulders against resistance of examiner's hand and to rotate head against resistance.	Sides should show uniformity of strength.
Hypoglossal	Patient requested to protrude tongue; tongue thrust may be resisted with tongue blade.	Tongue should protrude straight; deviation to side indicates ipsilateral-nerve dysfunction; asymmetry, atrophy, or lack of strength should be noted.

Spinal Nerves

Each of the 31 pairs of spinal nerves is formed by the union of a posterior and an anterior spinal root that emerges from the spinal cord through an intervertebral foramen to innervate a body dermatome.

Objective 4. Discuss how the spinal nerves are grouped.

Objective 5. Describe the general distribution of a spinal nerve.

The 31 pairs of **spinal nerves** (see fig. 12.1) are grouped as follows: 8 cervical, 12 thoracic, 5 lumbar, 5 sacral, and 1 coccygeal. With the exception of the first cervical nerve, the spinal nerves leave the spinal cord and vertebral canal through intervertebral foramina. The first pair of cervical nerves emerges between the occipital bone of the skull and the atlas vertebra. The second through the seventh pairs of cervical nerves emerge above the vertebrae for which they are named, whereas the eighth pair of cervical nerves passes between the seventh cervical and first thoracic vertebrae. The remaining pairs of spinal nerves emerge below the vertebrae for which they are named.

A spinal nerve is a mixed nerve attached to the spinal cord by a **posterior (dorsal) root** composed of sensory fibers and an **anterior (ventral) root** composed of motor fibers (fig. 12.13). The dorsal root contains an enlargement called the **spinal (sensory) ganglion,** where the cell bodies of sensory neurons are located. The axons of sensory neurons convey sensory impulses through the posterior root and into the spinal cord where synapses occur with dendrites of other neurons. The anterior root consists of axons of motor neurons that convey motor impulses away from the CNS. A spinal nerve is formed as the fibers from the posterior and anterior roots converge and emerge through an intervertebral foramen.

The disease *herpes zoster*, also known as *shingles*, is a viral infection of the spinal ganglia. Herpes zoster causes painful, often unilateral, clusters of fluid-filled vesicles in the skin along the paths of the affected peripheral sensory neurons. This disease requires no special treatment, as the lesions gradually heal, and is usually not serious except in elderly, debilitated patients, who may die from exhaustion.

Figure 12.13   A section of the spinal cord and thoracic spinal nerves.

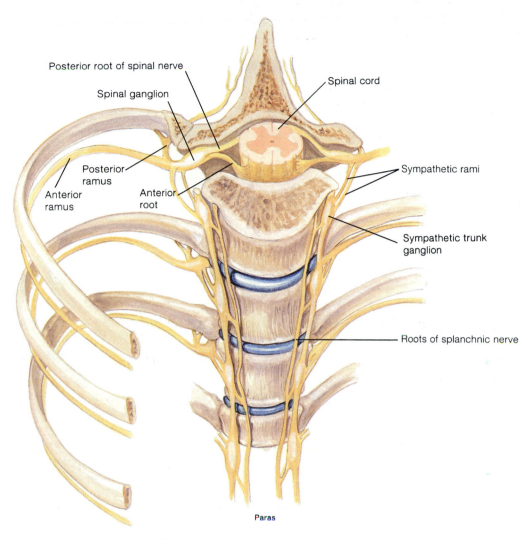

Posterior root of spinal nerve

Spinal ganglion

Spinal cord

Posterior ramus

Anterior ramus

Anterior root

Sympathetic rami

Sympathetic trunk ganglion

Roots of splanchnic nerve

Paras

UNIT V

A spinal nerve divides into several branches immediately after it emerges through the intervertebral foramen. The small **meningeal branch** reenters the vertebral canal to innervate the meninges, vertebrae, and vertebral ligaments. A larger branch, called the **posterior ramus,** innervates the muscles, joints, and skin of the back along the vertebral column (fig. 12.13). An **anterior ramus** of a spinal nerve innervates the muscles and skin on the lateral and anterior side of the trunk. Combinations of anterior rami innervate the limbs.

The **rami communicantes** are two branches from each spinal nerve that connect to a **sympathetic trunk ganglion,** which is part of the autonomic nervous system. The rami communicantes are composed of a **gray ramus,** containing unmyelinated fibers, and a **white ramus,** containing myelinated fibers.

---

1. List the number of nerves in each of the five regions of the vertebral column.
2. What are the four principal branches, or rami, from a spinal nerve, and what does each innervate?

Nerve Plexuses

Except in the thoracic nerves T2–T12, the ventral rami of the spinal nerves combine and then split again as networks of nerves referred to as plexuses. There are four plexuses of spinal nerves: the cervical, the brachial, the lumbar, and the sacral. Nerves emerge from the plexuses and are named according to the structures they innervate or the general course they take.

Objective 6.   List the spinal nerve composition of each of the plexuses arising from the spinal cord.

Objective 7.   List the principal nerves that emerge from the plexuses and describe their general innervation.

Cervical Plexus   The cervical plexus *(plek'sus)* is positioned deep on the side of the neck, lateral to the first four cervical vertebrae (fig. 12.14). It is formed by the anterior rami of the first four cervical nerves (C1–C4) and a portion of C5. Branches

Figure 12.14 The cervical plexus.

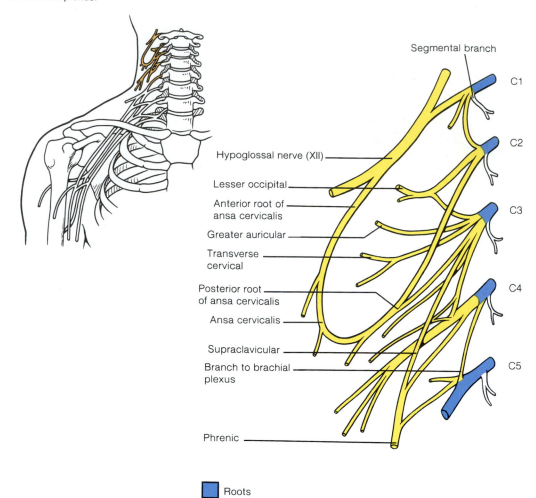

Segmental branch

C1

C2

Hypoglossal nerve (XII)

Lesser occipital

Anterior root of
ansa cervicalis

C3

Greater auricular

Transverse
cervical

Posterior root
of ansa cervicalis

C4

Ansa cervicalis

Supraclavicular

Branch to brachial
plexus

C5

Phrenic

☐ Roots

of the cervical plexus innervate the skin and muscles of the neck and portions of the head and shoulders. Branches of the cervical plexus also combine with the accessory and hypoglossal cranial nerves to supply dual innervation to some specific neck and pharyngeal muscles. Fibers from the third, fourth, and fifth cervical nerves unite to become the **phrenic** *(fren'ik)* **nerve,** which innervates the diaphragm. Motor impulses through the paired phrenic nerves cause the diaphragm to contract, inspiring air into the lungs.

The nerves of the cervical plexus are summarized in table 12.3.

Brachial Plexus The brachial plexus is positioned to the side of the last four cervical and first thoracic vertebrae. It is formed by the anterior rami of C5 through T1 with occasional contributions from C4 and T2. From its emergence, the brachial plexus extends downward and laterally, passes over the first rib behind the clavicle, and enters the axilla. Each brachial plexus innervates the entire upper extremity of one side, as well as a number of shoulder and neck muscles.

Structurally, the brachial plexus is divided into *roots, trunks, divisions,* and *cords* (fig. 12.15). The roots of the brachial plexus are simply continuations of the anterior rami of the cervical nerves. The anterior rami of C5 and C6 converge to become

Table 12.3	Branches of the Cervical Plexus	
Nerve	**Spinal Component**	**Innervation**
Superficial Cutaneous Branches		
Lesser occipital	C2,C3	Skin of scalp above and behind ear
Greater auricular	C2,C3	Skin in front of, above, and below ear
Transverse cervical	C2,C3	Skin of anterior aspect of neck
Supraclavicular	C3,C4	Skin of upper portion of chest and shoulder
Deep Motor Branches		
Ansa cervicalis		
Anterior root	C1,C2	Geniohyoid, thyrohyoid, and infrahyoid muscles of neck
Posterior root	C3,C4	Omohyoid, sternohyoid, and sternothyroid muscles of neck
Phrenic	C3,C4,C5	Diaphragm
Segmental branches	C1–C5	Deep muscles of neck (levator scapulae ventralis, trapezius, scalenus, and sternocleidomastoid)

Figure 12.15 The brachial plexus.

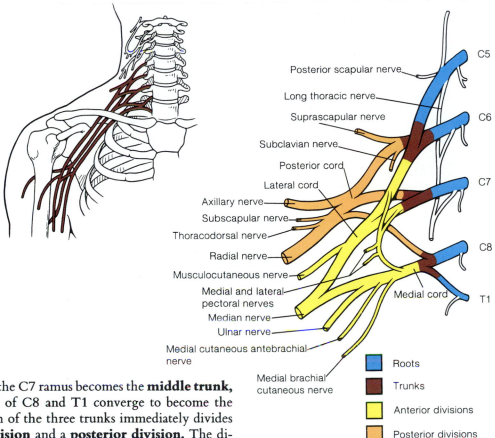

Posterior scapular nerve
Long thoracic nerve
Suprascapular nerve
Subclavian nerve
Posterior cord
Lateral cord
Axillary nerve
Subscapular nerve
Thoracodorsal nerve
Radial nerve
Musculocutaneous nerve
Medial and lateral pectoral nerves
Median nerve
Ulnar nerve
Medial cutaneous antebrachial nerve
Medial brachial cutaneous nerve
Medial cord

C5
C6
C7
C8
T1

Roots
Trunks
Anterior divisions
Posterior divisions

the **superior trunk,** the C7 ramus becomes the **middle trunk,** and the ventral rami of C8 and T1 converge to become the **inferior trunk.** Each of the three trunks immediately divides into an **anterior division** and a **posterior division.** The divisions then converge to form three cords. The **posterior cord** is formed by the convergence of the posterior divisions of the upper, middle, and lower trunks, and hence contains fibers from C5 through C8. The **medial cord** is a continuation of the anterior division of the lower trunk and primarily contains fibers from C8 and T1. The **lateral cord** is formed by the convergence of the anterior division of the upper and middle trunk and consists of fibers from C5 through C7.

In summary, the brachial plexus is composed of nerve fibers from spinal nerves C5 through T1 and a few fibers from C4 and T2. Roots are continuations of the anterior rami. The roots converge to form trunks, and the trunks branch into divisions. The divisions in turn form cords, and the nerves of the upper extremity arise from the cords.

Trauma to the brachial plexus sometimes occurs, especially if the clavicle, upper ribs, or lower cervical vertebrae are seriously fractured. Occasionally, the brachial plexus of a newborn is severely strained during a difficult delivery when the baby is pulled through the birth canal. In such cases, the arm of the injured side is paralyzed and eventually withers as the muscles atrophy in relation to the extent of the injury.

The entire upper extremity can be anesthetized in a procedure called a *brachial block* or *brachial anesthesia.* The site for injection of the anesthetic is midway between the base of the neck and the shoulder, posterior to the clavicle. At this point, the anesthetic can be injected in close proximity to the brachial plexus.

Five major nerves—the axillary, radial, musculocutaneous, ulnar, and median—arise from the three cords of the brachial plexus to supply cutaneous and muscular innervation to the upper extremity (table 12.4). The **axillary nerve** arises from the posterior cord and provides sensory innervation to the skin of the shoulder and to the shoulder joint and motor innervation to the deltoid and teres minor muscles (fig. 12.16). The **radial nerve** arises from the posterior cord and extends along the posterior aspect of the brachial region to the radial side of the forearm. The radial nerve provides sensory innervation to the skin of the posterior lateral surface of the upper extremity, including the posterior surface of the hand (fig. 12.17), and motor innervation to all of the extensor muscles of the upper extremity, the supinator muscle, and two muscles that flex the elbow joint.

The radial nerve is vulnerable to several types of trauma. *Crutch paralysis* may result when a person improperly supports the weight of the body for an extended period of time with a crutch pushed tightly into the axilla. Compression of the radial nerve between the top of the crutch and the humerus may result in radial nerve damage. Likewise, dislocation of the shoulder frequently traumatizes the radial nerve. Children are particularly at risk as adults yank on their arms. A fracture to the shaft of the humerus may damage the radial nerve that parallels the bone at this point. The principal symptom of radial nerve damage is *wristdrop,* in which the extensor muscles of the fingers and wrist fail to function, and as a result the joints of the fingers, wrist, and elbow are in a constant state of flexion.

Table 12.4	Selected Branches of the Brachial Plexus	
Nerve	**Cord and Spinal Components**	**Innervation**
Axillary	Posterior cord (C5,C6)	Skin of shoulder; shoulder joint, deltoid and teres minor muscles
Radial	Posterior cord (C5–C8,T1)	Skin of posterior lateral surface of arm, forearm, and hand; posterior muscles of brachium and antebrachium (triceps brachii, supinator, anconeus, brachioradialis, extensor carpi radialis brevis, extensor carpi radialis longus, extensor carpi ulnaris)
Musculocutaneous	Lateral cord (C5,C6,C7)	Skin of lateral surface of forearm; anterior muscles of brachium (coracobrachialis, biceps brachii, brachialis)
Ulnar	Medial cord (C8,T1)	Skin of medial third of hand; flexor muscles of anterior forearm (flexor carpi ulnaris, flexor digitorum), medial palm, and intrinsic flexor muscles of hand (profundus, third and fourth lumbricales)
Median	Medial cord (C6,C7,C8,T1)	Skin of lateral two-thirds of hand; flexor muscles of anterior forearm, lateral palm, and first and second lumbricales

Figure 12.16 Muscular and cutaneous distribution of the axillary nerve.

Figure 12.17 Muscular and cutaneous distribution of the radial nerve.

Figure 12.18 Muscular and cutaneous distribution of the musculocutaneous nerve.

Figure 12.19 Muscular and cutaneous distribution of the ulnar nerve.

The **musculocutaneous nerve** arises from the lateral cord and provides sensory innervation to the skin of the posterior lateral surface of the arm and motor innervation to the anterior muscles of the brachium (fig. 12.18). The **ulnar nerve** arises from the medial cord and provides sensory innervation to the skin on the medial (ulnar side) third of the hand (fig. 12.19). The motor innervation of the ulnar nerve is to two muscles of the forearm and the intrinsic muscles of the hand (except some that serve the thumb).

> The ulnar nerve can be palpated in the groove behind the medial epicondyle. This area is commonly known as the "funny bone" or "crazy bone." Damage to the ulnar nerve may occur as the medial side of the elbow is banged against a hard object. The immediate perception of this trauma is a painful tingling that extends down the ulnar side of the forearm and into the hand and medial two digits. Although common, ulnar nerve damage is generally not serious.

The **median nerve** arises from the medial cord and provides sensory innervation to the skin on the radial portion of the palm of the hand (fig. 12.20) and motor innervation to all but one of the flexor muscles of the forearm and most of the hand muscles of the thumb (thenar muscles).

> The median nerve, which serves the thumb, is the nerve in the forearm most commonly injured by stab wounds or the penetration of glass. If this nerve is severed, the muscles of the thumb are paralyzed and waste away, resulting in an inability to oppose the thumb in grasping.

Lumbar Plexus   The lumbar plexus is positioned to the side of the first four lumbar vertebrae. It is formed by the interior rami of spinal nerves L1–L4 and some fibers from T12 (fig. 12.21). The nerves that arise from the lumbar plexus innervate structures of the lower abdomen and anterior and medial portions of the lower extremity. The lumbar plexus is not as complex as the brachial plexus, having only roots and divisions rather than the roots, trunks, divisions, and cords as in the brachial plexus.

UNIT V

UNIT V

Figure 12.20 Muscular and cutaneous distribution of the median nerve.

Posterior cord

Lateral cord

Medial cord

Creek

Medial nerve

Pronator teres

Flexor carpi radialis

Palmaris longus

Flexor digitorum superficialis

Flexor digitorum profundus

Flexor pollicis longus

Pronator quadratus

Thenar muscles

Lateral lumbricales

Structurally, the **posterior division** of the lumbar plexus passes obliquely outward, deep to the psoas major muscle, whereas the **anterior division** is superficial to the quadratus lumborum muscle. The nerves that arise from the lumbar plexus are summarized in table 12.5. Because of their extensive innervation, the femoral nerve and the obturator nerve are illustrated in figures 12.22 and 12.23, respectively.

The **femoral nerve** arises from the posterior division of the lumbar plexus and provides cutaneous innervation to the anterior and lateral thigh and the medial leg and foot (fig. 12.22). The motor innervation of the femoral nerve is to the anterior muscles of the thigh, including the iliopsoas and sartorius muscles and the quadriceps femoris group.

The **obturator nerve** arises from the anterior division of the lumbar plexus and provides cutaneous innervation to the medial thigh and motor innervation to the adductor muscles of the thigh (fig. 12.23).

Sacral Plexus The sacral plexus is positioned immediately caudal to the lumbar plexus. It is formed by the anterior rami of spinal nerves L4, L5, and S1–S4 (fig. 12.24). The nerves arising from the sacral plexus innervate the lower back, pelvis, perineum, posterior surface of the thigh and leg, and dorsal and plantar surfaces of the foot (table 12.6). Like the lumbar plexus, the sacral plexus consists of **roots** and **anterior** and **posterior divisions** from which nerves arise. Because some of the nerves of the sacral plexus also contain fibers from the nerves of the lumbar plexus through the **lumbosacral trunk,** these two plexuses are frequently described collectively as the **lumbosacral plexus.**

sacral: L. *sacris,* sacred

Figure 12.21 The lumbar plexus.

Iliohypogastric nerve

Ilioinguinal nerve

Genitofemoral nerve

Lateral femoral cutaneous nerve

Femoral nerve

L1

L2

L3

L4

L5

Roots

Anterior divisions

Posterior divisions

Saphenous nerve

Obturator nerve

Lumbosacral trunk

Table 12.5	Branches of the Lumbar Plexus	
Nerve	**Spinal Components**	**Innervation**
Iliohypogastric	T12–L1	Skin of lower abdomen and buttock; muscles of anterolateral abdominal wall (external abdominal oblique, internal abdominal oblique, transversus abdominis)
Ilioinguinal	L1	Skin of upper median thigh, scrotum and root of penis in male and labia majora in female; muscles of anterolateral abdominal wall with iliohypogastric nerve
Genitofemoral	L1,L2	Skin of middle anterior surface of thigh, scrotum in male and labia majora in female; cremaster muscle in male
Lateral cutaneous femoral	L2,L3	Skin of anterior, lateral, and posterior aspects of thigh
Femoral	L2–L4	Skin of anterior and medial aspect of thigh and medial aspect of leg and foot; anterior muscles of thigh (iliacus, psoas major, pectineus, rectus femoris, sartorius) and extensor muscles of leg (rectus femoris, vastus lateralis, vastus medialis, vastus intermedius)
Obturator	L2–L4	Skin of medial aspect of thigh; adductor muscles of lower extremity (external obturator, pectineus, adductor longus, adductor brevis, adductor magnus, gracilis)
Saphenous	L2–L4	Skin of medial aspect of lower extremity

Figure 12.22 Muscular and cutaneous distribution of the femoral nerve.

Figure 12.23 Muscular and cutaneous distribution of the obturator nerve.

UNIT V

Figure 12.24 The sacral plexus.

Lumbosacral trunk
L5
S1
S2
S3
S4
S5
Co1

Superior gluteal nerve
Inferior gluteal nerve
Common fibular nerve
Tibial nerve
Sciatic nerve
Posterior cutaneous femoral nerve
Pudendal nerve

■ Roots
■ Anterior divisions
■ Posterior divisions

Table 12.6	Branches of the Sacral Plexus	
Nerve	**Spinal Components**	**Innervation**
Superior gluteal	L4,L5,S1	Abductor muscles of thigh (gluteus minimus, gluteus medius, tensor fasciae latae)
Inferior gluteal	L5–S2	Extensor muscle of hip joint (gluteus maximus)
Nerve to piriformis	S1,S2	Abductor and rotator of thigh (piriformis)
Nerve to quadratus femoris	L4,L5,S1	Rotators of thigh (gemellus inferior, quadratus femoris)
Nerve to internal obturator	L5–S2	Rotators of thigh (gemellus superior, internal obturator)
Perforating cutaneous	S2,S3	Skin over lower medial surface of buttock
Posterior cutaneous femoral	S1–S3	Skin over lower lateral surface of buttock, anal region, upper posterior surface of thigh, upper aspect of calf, scrotum in male and labia majora in female
Sciatic	L4–S3	Composed of two nerves (tibial and common fibular); splits into two portions at popliteal fossa; branches from sciatic in thigh region to "hamstring muscles" (biceps femoris, semitendinosus, semimembranosus) and adductor magnus muscle
Tibial (sural, medial and lateral plantar)	L4–S3	Skin of posterior surface of leg and sole of foot; muscle innervation includes gastrocnemius, soleus, flexor digitorum longus, flexor hallucis longus, tibialis posterior, popliteus, and intrinsic muscles of the foot
Common fibular (superficial and deep fibular)	L4–S2	Skin of anterior surface of the leg and dorsum of foot; muscle innervation includes peroneus tertius, peroneus brevis, peroneus longus, tibialis anterior, extensor hallucis longus, extensor digitorum longus, extensor digitorum brevis
Pudendal	S2–S4	Skin of penis and scrotum in male and skin of clitoris, labia majora, labia minora, and lower vagina in female; muscles of perineum

UNIT V

Figure 12.25 Muscular and cutaneous distribution of the tibial nerve.

- L4
- L5
- S1
- S2
- S3
- Tibial nerve
- Adductor magnus
- Long head of biceps femoris
- Semitendinosus
- Semimembranosus
- Plantaris
- Gastrocnemius
- Popliteus
- Soleus
- Flexor digitorum longus
- Tibialis posterior
- Flexor hallucis longus
- Creek
- Medial plantar nerve to plantar muscles
- Lateral plantar nerve to plantar muscles

Figure 12.26 Muscular and cutaneous distribution of the common fibular nerve.

- L4
- L5
- S1
- S2
- Common fibular nerve
- Short head of biceps femoris
- Tibialis anterior
- Peroneus longus
- Superficial fibular nerve
- Peroneus brevis
- Extensor digitorum longus
- Deep fibular nerve
- Peroneus tertius
- Extensor hallucis longus
- Extensor digitorum brevis
- Extensor hallucis brevis
- Creek

The **sciatic** *(si-at'ik)* **nerve** is the largest nerve arising from the sacral plexus and is the largest nerve in the body. The sciatic nerve passes from the pelvis through the greater sciatic notch of the os coxa and extends down the posterior aspect of the thigh. The sciatic nerve is actually composed of two nerves—the tibial and common fibular nerves—wrapped in a connective tissue sheath.

The **tibial nerve** arises from the anterior division of the sacral plexus, extends through the posterior regions of the thigh and leg, and branches in the foot to form the **medial** and **lateral plantar nerves** (fig. 12.25). The cutaneous innervation of the tibial nerve is to the calf of the leg and the plantar surface of the foot. The motor innervation of the tibial nerve is to most of the posterior thigh and leg muscles and to many of the intrinsic muscles of the foot.

The **common fibular nerve** (peroneal nerve) arises from the posterior division of the sacral plexus, extends through the posterior region of the thigh, and branches in the upper portion of the leg into the **deep** and **superficial fibular nerves** (fig. 12.26). The cutaneous innervation of the common fibular nerve and its branches is to the anterior and lateral leg and to the dorsum of the foot. The motor innervation is to the anterior and lateral muscles of the leg and foot.

The sciatic nerve in the buttock lies deep to the gluteus maximus muscle, midway between the greater trochanter and the ischial tuberosity. Because of its position, the sciatic nerve is of tremendous clinical importance. A posterior dislocation of the hip joint will generally injure the sciatic nerve. A *herniated disc* (fig. 12.27) or pressure from the uterus during pregnancy may

sciatic: L. *sciaticus*, hip joint

Figure 12.27 An MR scan of a herniated disc (arrow) in the lumbar region.

damage the roots of the nerves resulting in a condition called *sciatica (si-at'i-ka).* Sciatica is characterized by a sharp pain in the gluteal region and by pains that extend down the posterior side of the thigh. An improperly administered injection into the buttock may injure the sciatic nerve itself. Even a temporary compression of the sciatic nerve as a person sits on a hard surface for a period of time may result in the perception of tingling throughout the limb as the person stands up. The limb is said to have "gone to sleep."

1. Define *nerve plexus.* What are the four spinal nerve plexuses and which spinal nerves contribute to each?
2. Which of the spinal nerves are not involved in a plexus?
3. Distinguish between a posterior ramus and an anterior ramus. Which is involved in the formation of plexuses?
4. Construct a table that lists the plexus of origin and the general region of innervation for the following nerves: pudendal, phrenic, femoral, ulnar, median, sciatic, saphenous, axillary, radial, ansa cervicalis, tibial, and common fibular.

Reflex Arc and Reflexes

The conduction pathway of a reflex arc consists of a receptor, a sensory neuron, a motor neuron and its innervation in the PNS, and an association neuron in the CNS. The reflex arc provides the mechanism for a rapid, automatic response to a potentially threatening stimulus.

Objective 8. Define *reflex arc* and list its five components.

Objective 9. Distinguish between the various kinds of reflexes.

Specific **nerve pathways** provide routes by which impulses travel through the nervous system. Frequently, a nerve pathway begins with impulses being conducted to the CNS through sensory receptors and sensory neurons of the PNS. Once within the CNS, impulses may immediately travel back through motor portions of the PNS to activate specific skeletal muscles, glands, or smooth muscles and also be sent simultaneously to other parts of the CNS through ascending tracts within the spinal cord.

Components of the Reflex Arc

The simplest type of nerve pathway is a **reflex arc** (fig. 12.28). A reflex arc implies an automatic, unconscious, protective response to a situation in an attempt to maintain body homeostasis. A reflex arc leads by a short route from sensory to motor neurons and includes only two or three neurons. The five components of a reflex arc are the receptor, sensory neuron, center, motor neuron, and effector. The **receptor** includes the dendrite of a sensory neuron and the place where the electrical impulse is initiated. The **sensory neuron** relays the impulse through the posterior root to the CNS. The **center** is within the CNS and usually involves an association neuron (interneuron). It is here that the arc is made and other impulses are sent through synapses to other parts of the body. The **motor neuron** conducts the impulse to an **effector organ** (generally a skeletal muscle). The response of the effector is called a *reflex action* or, simply, a *reflex.*

Kinds of Reflexes

Visceral Reflexes   Reflexes that cause smooth or cardiac muscle to contract or glands to secrete are **visceral (autonomic) reflexes.** Visceral reflexes help control the body's many involuntary processes such as heart rate, respiratory rate, blood flow, and digestion. Swallowing, sneezing, coughing, and vomiting may also be reflexive, although they involve the involuntary action of skeletal muscles.

Somatic Reflexes   Somatic reflexes are those that result in the contraction of skeletal muscles. The three principal kinds of somatic reflexes are named according to the response they produce.

The **stretch reflex** involves only two neurons and one synapse in the pathway and is therefore called a *monosynaptic reflex arc.* Slight stretching of neuromuscular spindle receptors (described in chapter 15) within a muscle initiates an impulse along a sensory neuron to the spinal cord. A synapse with a motor neuron occurs in the anterior gray column and a motor unit is activated, causing specific muscle fibers to contract. Since the receptor and effector organs of the stretch reflex involve structures on the same side of the spinal cord, the reflex arc is an *ipsilateral reflex arc.* The knee-jerk reflex is an ipsilateral reflex (fig. 12.29), as are all monosynaptic reflex arcs.

A **flexor reflex,** or **withdrawal reflex,** consists of a *polysynaptic reflex arc* (fig. 12.30). Flexor reflexes involve association

Figure 12.28   The reflex arc.

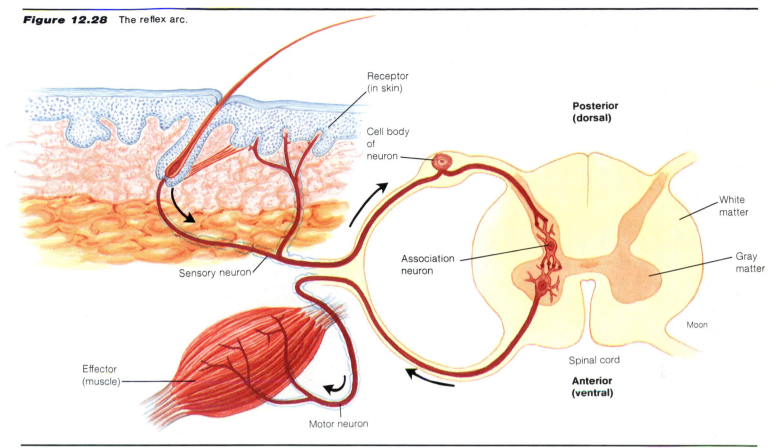

Receptor
(in skin)

Cell body
of
neuron

**Posterior
(dorsal)**

White
matter

Gray
matter

Association
neuron

Moon

Sensory neuron

Spinal cord

Effector
(muscle)

**Anterior
(ventral)**

Motor neuron

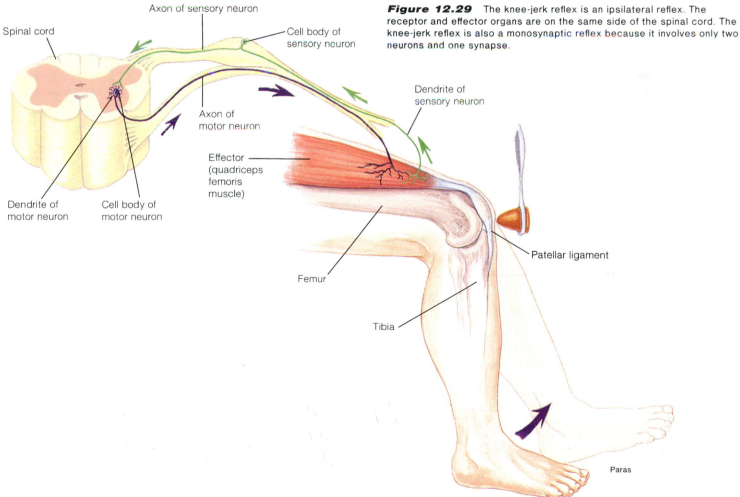

Axon of sensory neuron

Spinal cord

Cell body of
sensory neuron

Figure 12.29   The knee-jerk reflex is an ipsilateral reflex. The receptor and effector organs are on the same side of the spinal cord. The knee-jerk reflex is also a monosynaptic reflex because it involves only two neurons and one synapse.

Dendrite of
sensory neuron

Axon of
motor neuron

Effector
(quadriceps
femoris
muscle)

Dendrite of
motor neuron

Cell body of
motor neuron

Patellar ligament

Femur

Tibia

Paras

UNIT V

Figure 12.30 The flexor, or withdrawal, reflex is a polysynaptic (more than one synapse) reflex and involves association neurons in addition to sensory and motor neurons.

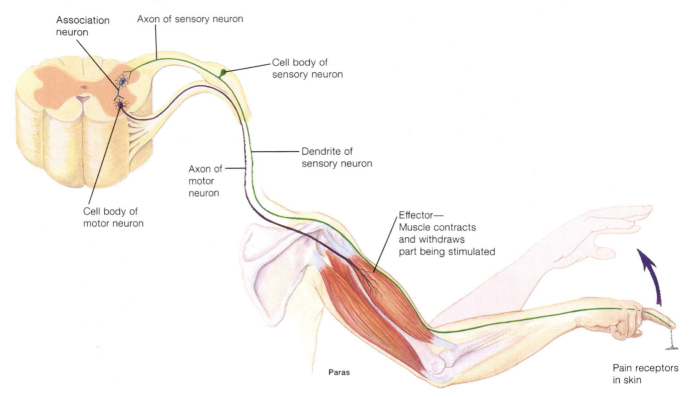

Association neuron

Axon of sensory neuron

Cell body of sensory neuron

Dendrite of sensory neuron

Axon of motor neuron

Cell body of motor neuron

Effector— Muscle contracts and withdraws part being stimulated

Pain receptors in skin

Paras

neurons in addition to the sensory and motor neurons. A flexor reflex is initiated as a person encounters a painful stimulus, such as a hot or sharp object. As a receptor organ is stimulated, sensory neurons transmit the impulse to the spinal cord where association neurons are activated. There, the impulses are directed through motor neurons to flexor muscles, which contract in response. Simultaneously, antagonistic muscles are inhibited (relaxed) so that the traumatized extremity can be quickly withdrawn from the harmful source of stimulation.

Several additional reflexes may be activated while a flexor reflex is in progress. In an *intersegmental reflex arc,* motor units from several segments of the spinal cord are activated by impulses coming in from the receptor organ. An intersegmental reflex arc produces stimulation of more than one effector organ. Frequently, sensory impulses from a receptor organ cross over through the spinal cord to activate effector organs within the opposite (*contralateral*) limb. This type of reflex is called a **crossed extensor reflex** (fig. 12.31) and is important for maintaining body balance while a flexor reflex is in progress. The reflexive inhibition of certain muscles to contract, called *reciprocal inhibition,* also helps maintain balance while either flexor or crossed extensor reflexes are in progress.

Certain reflexes are important for physiological functions, while others are important for avoiding injury. Some common reflexes are described in table 12.7 and illustrated in figure 12.32.

Part of a routine physical examination involves testing a person's reflexes. The condition of the nervous system, particularly the functioning of the synapses, may be determined by examining reflexes. In case of injury to some portion of the nervous system, testing certain reflexes may indicate the location and extent of the injury. Also, an anesthesiologist may try to initiate a reflex to ascertain the effect of an anesthetic.

1. How are reflexes important in maintaining body homeostasis?
2. List the five components of a reflex arc.
3. Define the following terms: *visceral reflex, somatic reflex, stretch reflex, flexor reflex, crossed extensor reflex, ipsilateral reflex,* and *contralateral reflex.*

anesthesia: Gk. *an,* without; *aisthesis,* sensation

Figure 12.31 A crossed extensor reflex causes a reciprocal inhibition of muscles of the opposite appendage. This type of reflex inhibition is important in maintaining balance.

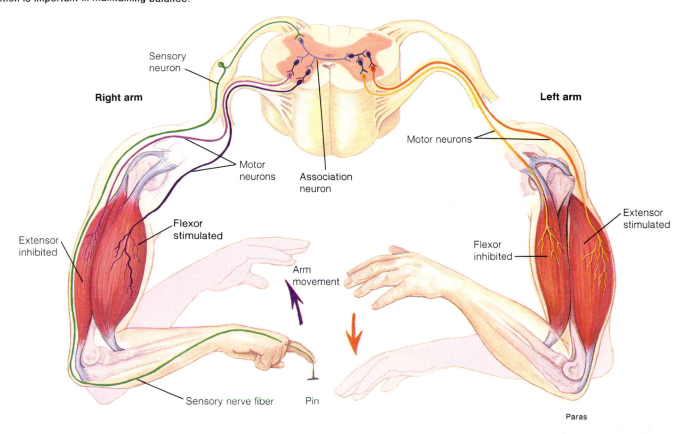

Table 12.7	Selected Reflexes of Clinical Importance		
Reflex	**Spinal Segment**	**Site of Receptor Stimulation**	**Effector Action**
Biceps reflex	C5, C6	Tendon of biceps brachii muscle, near attachment on radial tuberosity	Contracts biceps brachii muscle to flex elbow
Triceps reflex	C7, C8	Tendon of triceps brachii muscle, near attachment on olecranon	Contracts triceps brachii muscle to extend elbow
Supinator or brachioradialis reflex	C5, C6	Radial attachment of supinator and brachioradialis muscles	Supinates forearm and hand
Knee reflex	L2, L3, L4	Patellar ligament, just below patella	Contracts quadriceps muscle to extend the knee
Ankle reflex	S1, S2	Tendo calcaneus, near attachment on calcaneus	Plantar flexes ankle
Plantar reflex	L4, L5, S1, S2	Lateral aspect of sole, from heel to ball of foot	Plantar flexes foot and flexes toes
Babinski reflex *	L4, L5, S1, S2	Lateral aspect of sole, from heel to ball of foot	Dorsiflexes great toe and fans other toes
Abdominal reflexes	T8, T9, T10 above umbilicus and T10, T11, T12 below umbilicus	Sides of abdomen, above and below level of umbilicus	Contracts abdominal muscles and deviates umbilicus toward stimulus
Cremasteric reflex	L1, L2	Upper inside of thigh in males	Contracts cremasteric muscle and elevates testis on same side of stimulation

*If the Babinski reflex rather than the plantar reflex occurs as the sole of the foot is stimulated, it may indicate damage to the corticospinal tract within the spinal cord. However, the Babinski reflex is present in infants up to 12 months of age because of the immaturity of their corticospinal tracts.

Babinski reflex: from Joseph F. Babinski, French neurologist, 1857–1932

Figure 12.32 Some reflexes of clinical importance.

Glabellar reflex

Ankle (Achilles) reflex

Babinski reflex

Abdominal reflex

Plantar reflex

Knee (patellar) reflex

Biceps reflex

Supinator (brachioradialis) reflex

Triceps reflex

Developmental Exposition

The Peripheral Nervous System

Exhibit I The pattern of dermatomes within the body and the peripheral distribution of spinal nerves. (*a*) An anterior view and (*b*) a posterior view.

(a)

(b)

Explanation

Development of the peripheral nervous system produces the pattern of dermatomes within the body (exhibit I). A **dermatome** (*der'mă'tōm*) is an area of the skin innervated by all the cutaneous neurons of a certain spinal or cranial nerve. Most of the scalp and face is innervated by sensory neurons from the trigeminal nerve. With the exception of the first cervical nerve (C1), all of the spinal nerves are associated with specific dermatomes. Dermatomes are consecutive in the neck and trunk regions. In the appendages, however, adjacent dermatome innervations overlap. The apparently uneven dermatome arrangement in the appendages is due to the uneven rate of nerve growth

dermatome: Gk. *derma*, skin; *tomia*, a cutting

into the limb buds. Actually, the limbs are segmented, and dermatomes overlap only slightly.

The pattern of dermatome innervation is of clinical importance when a physician wants to anesthetize a particular portion of the body. Because adjacent dermatomes overlap in the appendages, at least three spinal nerves must be blocked to produce complete anesthesia in these regions. Abnormally functioning dermatomes provide clues about injury to the spinal cord or specific spinal nerves. If a dermatome is stimulated but no sensation is perceived, the physician can infer that the injury involves the innervation to that dermatome.

UNIT V

CLINICAL CASE STUDY ANSWER

The radial nerve lies in the radial groove of the humerus as it extends through the brachial region toward the arm and hand. In this position, the radial nerve is susceptible to injury in the case of a fracture to the midshaft of the humerus. The muscles innervated by the radial nerve include those of wrist extension (extensor carpi radialis longus and brevis, extensor carpi ulnaris); wrist adduction (extensor carpi ulnaris); wrist abduction (extensor carpi radialis brevis and longus); supination (supinator); and finger extension (finger extensors). A detectable weakness in arm flexion could also result from impairment of the brachioradialis muscle. The triceps brachii muscle would not be affected, however, because of the more proximal branching point of its motor nerve supply. The radial nerve is a mixed nerve, carrying both motor and sensory fibers; therefore, a sensory deficit would be present. Decreased sensation would be detectable on the posterolateral aspect of the hand (see fig. 12.17).

CHAPTER SUMMARY

Introduction to the Peripheral Nervous System (pp. 384–385)

1. The peripheral nervous system consists of nerves that convey impulses to and from the central nervous system.
2. The cranial nerves arise from the brain and the spinal nerves arise from the spinal cord.
3. Sensory (afferent) nerves convey impulses toward the CNS, whereas motor (efferent) nerves convey impulses away from the CNS. Mixed nerves are composed of both sensory and motor fibers.

Cranial Nerves (pp. 385–394)

1. Twelve pairs of cranial nerves emerge from the inferior surface of the brain and, with the exception of the vestibulocochlear nerve, pass through foramina of the skull to innervate structures in the head, neck, and visceral organs of the torso.
2. The names of the cranial nerves indicate their primary function or the general distribution of their fibers.
3. The olfactory, optic, and vestibulocochlear cranial nerves are sensory only; the trigeminal, glossopharyngeal, and vagus are mixed; and the others are primarily motor, with the only sensory fibers of these being proprioceptive.
4. Some of the cranial nerve fibers are somatic; others are visceral.
5. A test for cranial-nerve dysfunction is clinically important in a neurological examination.

Spinal Nerves (pp. 394–395)

1. Each of the 31 pairs of spinal nerves is formed by the union of an anterior (ventral) and posterior (dorsal) spinal root that emerges from the spinal cord through an intervertebral foramen to innervate a body dermatome.
2. The spinal nerves are grouped according to the levels of the spinal column from which they arise, and they are numbered in sequence.
3. Each spinal nerve is a mixed nerve consisting of a posterior root of sensory fibers and an anterior root of motor fibers.
4. Just beyond its intervertebral foramen, each spinal nerve divides into several branches.

Nerve Plexuses (pp. 395–404)

1. Except in the thoracic nerves T2–T12, the anterior rami of the spinal nerves combine and then split again as networks of nerves referred to as plexuses.
   (a) There are four plexuses of spinal nerves: the cervical, the brachial, the lumbar, and the sacral.
   (b) Nerves that emerge from the plexuses are named according to the structures they innervate or the general course they take.
2. The cervical plexus is formed by the anterior rami of C1–C4 and by a portion of C5.
3. The brachial plexus is formed by the anterior rami of C5–T1 and occasionally by some fibers from C4 and T2.
   (a) The brachial plexus is divided into roots, trunks, divisions, and cords.
   (b) The axillary, radial, musculocutaneous, ulnar, and median are the five largest nerves arising from the brachial plexus.

4. The lumbar plexus is formed by the anterior rami of L1–L4 and by some fibers from T12.
   (a) The lumbar plexus is divided into roots and divisions.
   (b) The femoral and obturator are two important nerves arising from the lumbar plexus.
5. The sacral plexus is formed by the anterior rami of L4, L5, and S1–S4.
   (a) The sacral plexus is divided into roots and divisions.
   (b) The sciatic nerve, composed of the common fibular and tibial nerves, arises from the sacral plexus.
   (c) Frequently, the lumbar plexus and the sacral plexus are collectively referred to as the lumbosacral plexus.

Reflex Arc and Reflexes (pp. 404–408)

1. The conduction pathway of a reflex arc consists of a receptor, a sensory neuron, a motor neuron and its innervation in the PNS, and an association neuron in the CNS. The reflex arc enables a rapid, automatic response to a potentially threatening stimulus.
2. A reflex arc is the simplest type of nerve pathway.
3. Visceral reflexes cause smooth or cardiac muscle to contract or glands to secrete.
4. Somatic reflexes cause skeletal muscles to contract.
   (a) The stretch reflex is a monosynaptic reflex arc.
   (b) The flexor reflex is a polysynaptic reflex arc.

REVIEW ACTIVITIES

Objective Questions

1. Which of the following is a *false* statement concerning the peripheral nervous system?
   - (a) It consists of cranial and spinal nerves only.
   - (b) It contains components of the autonomic nervous system.
   - (c) Sensory receptors, neurons, nerves, ganglia, and plexuses are all part of the PNS.
2. An inability to cross the eyes would most likely indicate a problem with which cranial nerve?
   - (a) the optic nerve
   - (b) the oculomotor nerve
   - (c) the abducens nerve
   - (d) the facial nerve
3. Which cranial nerve innervates the muscle that raises the upper eyelid?
   - (a) the trochlear nerve
   - (b) the oculomotor nerve
   - (c) the abducens nerve
   - (d) the facial nerve
4. The inability to walk a straight line may indicate damage to which cranial nerve?
   - (a) the trigeminal nerve
   - (b) the facial nerve
   - (c) the vestibulocochlear nerve
   - (d) the vagus nerve

5. Which cranial nerve passes through the stylomastoid foramen?
   - (a) the facial nerve
   - (b) the glossopharyngeal nerve
   - (c) the vagus nerve
   - (d) the hypoglossal nerve
6. Which of the following cranial nerves does not contain parasympathetic fibers?
   - (a) the oculomotor nerve
   - (b) the accessory nerve
   - (c) the vagus nerve
   - (d) the facial nerve
7. Which of the following is not a spinal nerve plexus?
   - (a) the cervical plexus
   - (b) the brachial plexus
   - (c) the thoracic plexus
   - (d) the lumbar plexus
   - (e) the sacral plexus
8. Roots, trunks, divisions, and cords are characteristic of
   - (a) the sacral plexus.
   - (b) the thoracic plexus.
   - (c) the lumbar plexus.
   - (d) the brachial plexus.
9. Which of the following nerve–plexus associations is incorrect?
   - (a) median/sacral
   - (b) phrenic/cervical
   - (c) axillary/brachial
   - (d) femoral/lumbar

10. Extending the leg when the patellar ligament is tapped is an example of
   - (a) a visceral reflex.
   - (b) a flexor reflex.
   - (c) an ipsilateral reflex.
   - (d) a crossed extensor reflex.

Essay Questions

1. Explain the structural and functional relationship between the central nervous system, the autonomic nervous system, and the peripheral nervous system.
2. List the cranial nerves, and describe the major function(s) of each. How is each cranial nerve tested for dysfunction?
3. Describe the structure of a spinal nerve.
4. List the roots of each of the spinal plexuses. Describe where each is located and state the nerves that originate from them.
5. What is a reflex arc? Explain how reflexes are important in maintaining body homeostasis.
6. Distinguish between monosynaptic, polysynaptic, ipsilateral, stretch, and flexor reflexes.

Gundy/Weber Software

The tutorial software accompanying Chapter 12, Peripheral Nervous System, is Volume 5, Nervous System.

UNIT V

Autonomic Nervous System

Chapter Outline

UNIT V

CLINICAL CASE STUDY

A 67-year-old female with unresectable lung cancer is seen for a routine checkup at a tumor clinic. Her only trouble the 6 months following her diagnosis has been occasional coughing, slight shortness of breath, and a mild but nagging pain in the right side of her chest. She tells her doctor that the right side of her face has felt funny recently. These symptoms, upon closer questioning, include lack of perspiration on the right side of her face during her regular morning walks. Examination of the patient reveals a right pupil that is much smaller than the left, along with a drooping of the right eyelid. Recognizing the patient's peculiar face and eye findings as Horner's syndrome, the doctor turns to the chest CT scan in her file to look for the cause. He knowingly nods his head as he examines the copy of the scan, noting that the lung tumor, located in the medial aspect of the right upper lobe, has invaded the mediastinum and tracts upward almost into the base of the neck.

What area or general structure(s) within the autonomic nervous system has the tumor invaded and compromised? Explain, using autonomic nervous system terminology, the reason for each of the three findings (lack of right facial sweating, pupillary constriction, and eyelid droop) and include a description of neural pathways and end organs.

Introduction to the Autonomic Nervous System

The action of effectors, muscles, and glands is controlled to a large extent by motor neuron impulses. Skeletal muscles, which are the voluntary effectors, are regulated by somatic motor impulses. The involuntary effectors, including smooth muscle, cardiac muscle, and glands, are regulated by autonomic motor impulses through the autonomic nervous system.

Objective 1. Define the terms *preganglionic neuron* and *postganglionic neuron* and explain how the motor pathways of the somatic motor and autonomic motor systems differ.

Objective 2. Describe how the neural regulation of visceral effector organs can differ from the neural regulation of skeletal muscles.

Objective 3. Compare the structure and regulation of single-unit smooth muscle with that of multiunit smooth muscle.

Organization of the Autonomic Nervous System

The autonomic portion of the nervous system is concerned with maintaining homeostasis within the body by increasing or decreasing the activity of various organs in response to changing

Table 13.1 — Comparison of the Somatic Motor and Autonomic Motor Systems

Feature	Somatic Motor	Autonomic Motor
Effector organs	Skeletal muscles	Cardiac muscle, smooth muscle, and glands
Presence of ganglia	No ganglia	Cell bodies of postganglionic autonomic fibers located in paravertebral, prevertebral (collateral), and terminal ganglia
Number of neurons from CNS to effector	One	Two
Type of neuromuscular junction	Specialized motor end plate	No specialization of postsynaptic membrane; all areas of smooth muscle cells contain receptor proteins for neurotransmitters
Effect of nerve impulse on muscle	Excitatory only	Either excitatory or inhibitory
Type of nerve fibers	Fast-conducting, thick (9–13 μm), and myelinated	Slow-conducting; preganglionic fibers lightly myelinated but thin (3 μm); postganglionic fibers unmyelinated and very thin (about 1.0 μm)
Effect of denervation	Flaccid paralysis and atrophy	Minimal effect on muscle tone and function; target cells show denervation hypersensitivity

From Stuart Ira Fox, *Human Physiology*, 4th ed. Copyright © 1993 Wm. C. Brown Communications, Inc., Dubuque, Iowa. All Rights Reserved. Reprinted by permission.

physiological conditions. Although the autonomic nervous system (ANS) is composed of portions of both the central nervous system and peripheral nervous system, it functions independently and without a person's conscious control.

Autonomic motor nerves innervate organs whose functions are not usually under voluntary control. The effectors that respond to autonomic regulation include **cardiac muscle** (the heart), **smooth** (visceral) **muscles,** and **glands.** These effectors are part of the organs of the *viscera,* of blood vessels, and of specialized structures within other organs. The involuntary effects of autonomic innervation contrast with the voluntary control of skeletal muscles by way of somatic motor innervation.

Unlike the somatic motor system, in which impulses are conducted along a single axon from the spinal cord to the neuromuscular junction, the autonomic motor pathway involves two neurons in the motor transmission of impulses (table 13.1).

---

viscus, pl. viscera: L. *viscera,* internal organs

Figure 13.1 A comparison of a somatic motor reflex (*a*) and an autonomic motor reflex (*b*).

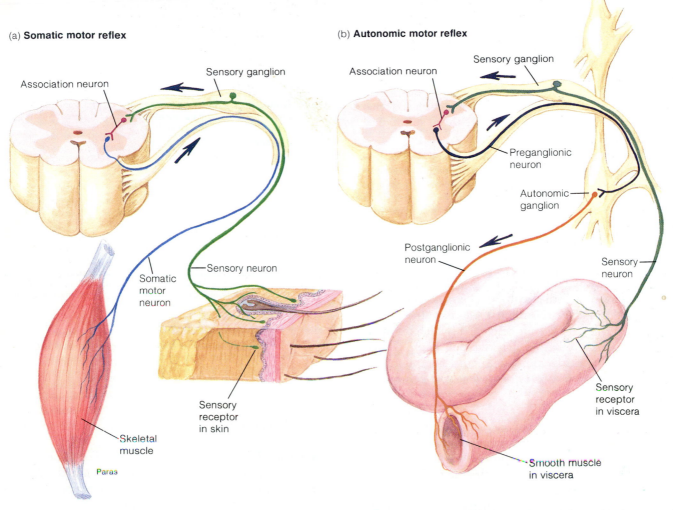

(a) **Somatic motor reflex**

Association neuron

Sensory ganglion

Somatic motor neuron

Sensory neuron

Sensory receptor in skin

Skeletal muscle

Paras

(b) **Autonomic motor reflex**

Association neuron

Sensory ganglion

Preganglionic neuron

Autonomic ganglion

Postganglionic neuron

Sensory neuron

Sensory receptor in viscera

Smooth muscle in viscera

UNIT V

The first of these autonomic motor neurons has its cell body in the gray matter of the brain or spinal cord. The axon of this neuron does not directly innervate the effector organ but instead synapses with a second neuron within an *autonomic ganglion*. (A ganglion is a collection of neuron cell bodies outside the CNS.) The first neuron is thus called a **preganglionic,** or presynaptic, **neuron.** The second neuron in this pathway, called a **postganglionic,** or postsynaptic, **neuron,** has an axon that leaves the autonomic ganglion and synapses with the cells of an effector organ (fig. 13.1).

Preganglionic autonomic neurons originate in the midbrain and hindbrain and from the upper thoracic to the fourth sacral portions of the spinal cord, with the exceptions of the area between L3 and S1. Autonomic ganglia are located in the head, neck, and abdomen. Chains of autonomic ganglia also parallel the right and left sides of the spinal cord. The origin of the preganglionic neurons and the location of the autonomic ganglia help to differentiate the **sympathetic** and **parasympathetic** divisions of the autonomic system, which will be discussed in later sections of this chapter.

autonomic: Gk. *auto*, self; *nomos*, law
ganglion: Gk. *ganglion*, a swelling or knot

Visceral Effector Organs

Unlike skeletal muscles, which enter a state of flaccid paralysis when their motor nerves are severed, the involuntary effectors are somewhat independent of their innervation. Smooth muscles maintain a resting tone (tension) in the absence of nerve stimulation. Damage to an autonomic nerve, in fact, makes its target muscle more sensitive than normal to stimulating agents.

In addition to their intrinsic (built-in) muscle tone, many smooth muscles and the cardiac muscle contract rhythmically, even in the absence of nerve stimulation, in response to electrical waves of depolarization initiated by the muscles themselves. Autonomic nerves also maintain a resting tone in the sense that they maintain a baseline firing rate that can be either increased or decreased. Changes in tonic neural activity produce changes in the intrinsic activity of the effector organ. A decrease in the excitatory input to the heart, for example, will slow its rate of beat.

Cardiac Muscle Like skeletal muscle fibers, cardiac muscle fibers are striated. The long, fibrous skeletal muscle fibers, however, are structurally and functionally separated from each other, whereas the cardiac fibers are short, branched, and interconnected by **intercalated** (*in-ter′kă-lāt-ed*) **discs.**

Electrical impulses that originate at any point in the mass of cardiac fibers called the **myocardium** (*mi''o-kar'de-um*) can spread to all cells in the mass that are joined by intercalated discs. Because all of the cells in the myocardium are electrically joined, the myocardium behaves as a single functional unit, or a *functional syncytium* (*sin-sish'e-um*). Unlike skeletal muscles, which can produce graded contractions with a strength that depends on the number of cells stimulated, the heart contracts with an *all-or-none contraction.*

Furthermore, whereas skeletal muscles require external stimulation by somatic motor nerves before they can produce action potentials and contract, cardiac muscle is able to produce action potentials automatically. Cardiac action potentials normally originate in a specialized group of cells called the *pacemaker* (see fig. 16.12). However, the rate of this spontaneous depolarization and, thus, the rate of the heartbeat is regulated by autonomic innervation.

Smooth Muscles Smooth (visceral) muscles are arranged in circular layers around the walls of blood vessels, bronchioles (small air passages in the lungs), and in the sphincter muscles of the GI tract. However, both circular and longitudinal smooth muscle layers are found in the tubular GI tract, the ureters (which transport urine), the ductus deferentia (which transport sperm), and the uterine tubes (which transport ova). The alternate contraction of circular and longitudinal smooth muscle layers produces **peristaltic waves** that propel the contents of these tubes in one direction.

Smooth muscle fibers do not contain sarcomeres (which produce striations in skeletal and cardiac muscle). Smooth muscle fibers do, however, contain a great amount of actin and some myosin, which produces a ratio of thin-to-thick myofilaments of about 16:1 (in striated muscles the ratio is 2:1).

The long length of myosin myofilaments and the fact that they are not organized into sarcomeres helps the smooth muscles to function optimally. Smooth muscles must be able to exert tension even when greatly stretched—in the urinary bladder, for example, the smooth muscle cells may be stretched up to two and a half times their resting length. Skeletal muscles, by contrast, lose their ability to contract when the sarcomeres are stretched to the point where actin and myosin no longer overlap.

Single-Unit and Multiunit Smooth Muscles Smooth muscles are often grouped into two functional categories: **single-unit** and **multiunit.** Single-unit smooth muscles have numerous gap junctions (electrical synapses) between adjacent cells that weld them together electrically; thus, they behave as a single unit. Multiunit smooth muscles have few, if any, gap junctions; thus, the individual cells must be stimulated separately by impulses through nerve endings. This is similar to the numerous motor units required for the control of skeletal muscles.

Single-unit smooth muscles display *pacemaker activity* in which certain cells stimulate others in the mass. Single-unit smooth muscles also display intrinsic, or *myogenic* (*mi''o-jen'ik*), electrical activity and contraction in response to stretch. For ex-

myogenic: Gk. *mys*, muscle; *genesis*, origin

Table 13.2	Comparison of Single-Unit and Multiunit Smooth Muscles	
	Single-Unit Muscle	**Multiunit Muscle**
Location	Gastrointestinal tract; uterus; ureter; small arteries (arterioles)	Arrector pili muscles of hair follicles; ciliary muscle (controls shape of lens); iris; ductus deferens; large arteries
Origin of electrical activity	Spontaneous activity by pacemakers (myogenic)	Not spontaneously active; potentials are neurogenic
Type of potentials	Action potentials	Graded depolarizations
Response to stretch	Responds to stretch by contraction; not dependent on nerve stimulation	No inherent response to stretch
Presence of gap junctions	Numerous gap junctions join all cells together electrically	Few (if any) gap junctions
Type of contraction	Slow, sustained contractions	Slow, sustained contractions

From Stuart Ira Fox, *Human Physiology*, 4th ed. Copyright © 1993 Wm. C. Brown Communications, Inc., Dubuque, Iowa. All Rights Reserved. Reprinted by permission.

ample, the stretch induced by an increase in the luminal contents of a small artery or a section of the GI tract can stimulate myogenic contraction. Such contraction does not require stimulation by autonomic nerves. By contrast, contraction of multiunit smooth muscles requires nerve stimulation. Single-unit and multiunit smooth muscles are compared in table 13.2.

Autonomic Innervation of Smooth Muscles The neural control of skeletal muscles and that of smooth muscles differ markedly. A skeletal muscle fiber has only one junction with a somatic nerve fiber, and the receptors for the neurotransmitter are localized at the neuromuscular junction in the membrane of the skeletal muscle fiber. By contrast, the entire surface of smooth muscle fibers contains transmitter receptor proteins. Neurotransmitter molecules are released along a stretch of an autonomic nerve fiber that is located some distance from the smooth muscle fibers. The regions of the autonomic fiber that release transmitters appear as bulges, or *varicosities,* and the neurotransmitters released from these varicosities stimulate a number of smooth muscle fibers.

1. Describe how the regulation of the contraction of cardiac and smooth muscle fibers differs from that of skeletal muscle fibers and how these muscles are affected by the experimental removal of their innervation.
2. Define the terms *preganglionic* and *postganglionic neurons* in the ANS and use a diagram to illustrate how motor innervation differs in somatic and autonomic nerves.
3. Compare single-unit and multiunit smooth muscles in terms of structure and describe the differences in how these two types are regulated by autonomic nerves.

Figure 13.2   The sympathetic trunk of ganglia relative to the vertebral column and the spinal cord.

Spinal cord

Posterior root

Anterior root

Spinal nerve

Rib

Ramus communicantes — Gray ramus

White ramus

Sensory ganglion

Sympathetic trunk

Ganglion of sympathetic trunk

Body of vertebra

Moon

UNIT V

Structure of the Autonomic Nervous System

The sympathetic and parasympathetic divisions of the autonomic nervous system consist of preganglionic neurons that originate in the CNS and postganglionic neurons that originate outside of the CNS in ganglia. The specific origin of the preganglionic neurons and the location of the ganglia, however, are different in the two subdivisions of the autonomic nervous system.

Objective 4. Describe the origin of preganglionic sympathetic neurons and the location of sympathetic ganglia.

Objective 5. Explain the relationship between the sympathetic division and the adrenal medulla.

Objective 6. Describe the origin of the preganglionic parasympathetic neurons and the location of the parasympathetic ganglia.

Objective 7. Describe the distribution of the vagus nerve and comment on its significance within the parasympathetic division.

Sympathetic (Thoracolumbar) Division

The **sympathetic division** is also called the *thoracolumbar division* of the ANS because its preganglionic neurons exit the vertebral column from the first thoracic (T1) to the second lumbar (L2) levels. Most sympathetic neurons, however, separate from the somatic motor neurons and synapse with postganglionic neurons within a chain of sympathetic trunk ganglia located on both lateral sides of the vertebral column (fig. 13.2).

Since the preganglionic sympathetic neurons are myelinated and thus appear white, they are called **white rami** (fig. 13.3). Some of these preganglionic sympathetic neurons synapse with postganglionic neurons located at their same level in the chain of sympathetic ganglia. Other preganglionic neurons travel up or down within the sympathetic chain before synapsing with postganglionic neurons. Since the postganglionic sympathetic neurons are unmyelinated and thus appear gray, they form the **gray rami**. The white ramus and the gray ramus constitute the **ramus communicantes.** Postganglionic axons

ramus: L. *ramus*, a branch

Figure 13.3 Sympathetic trunk ganglia, the sympathetic trunk, and rami communicantes of the sympathetic division of the ANS. The solid lines indicate preganglionic neurons and the dashed lines indicate postganglionic neurons.

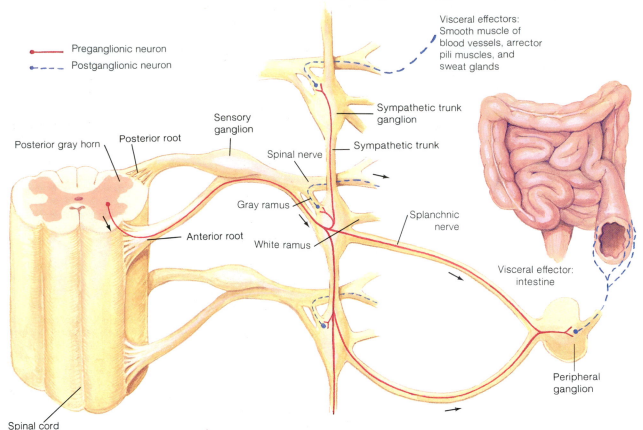

in the gray rami extend directly back to the anterior roots of the spinal nerves and travel distally within the spinal nerves to innervate their effector organs.

Within the sympathetic trunk ganglia, *divergence* is evident when preganglionic neurons branch to synapse with many postganglionic neurons located at different levels in the chain. *Convergence* is evident when a given postganglionic neuron receives synaptic input from a large number of preganglionic neurons. The divergence of impulses from the spinal cord to the ganglia and the convergence of impulses within the ganglia usually results in the *mass activation* of almost all of the postganglionic neurons. This explains why the sympathetic division is usually activated as a unit and affects all of its effector organs at the same time.

Many preganglionic neurons that exit the spinal cord in the upper thoracic level travel through the sympathetic chain into the neck, where they synapse in cervical sympathetic ganglia (fig. 13.4). Postganglionic neurons from here innervate the smooth muscles and glands of the head and neck.

Peripheral Ganglia   Many preganglionic neurons that exit the spinal cord below the level of the diaphragm pass through the sympathetic trunk of ganglia without synapsing. Beyond the sympathetic trunk, these preganglionic neurons form **splanchnic** *(splank'nik)* **nerves** (fig. 13.3). Preganglionic neu-

rons in the splanchnic nerves synapse in peripheral ganglia, which include the **celiac** *(se'le-ak)*, **superior mesenteric** *(mes"en-ter'ik)* and **inferior mesenteric ganglia** (figs. 13.5 and 13.6).

The *greater splanchnic nerve* arises from preganglionic sympathetic neurons T4–T9 and synapses in the celiac ganglion. These neurons contribute to the *celiac (solar) plexus.* Postganglionic neurons from the celiac ganglion innervate the stomach, spleen, liver, small intestine, and kidneys. The *lesser splanchnic nerve* terminates in the superior mesenteric ganglion. Postganglionic neurons from here innervate the small intestine and colon. The *lumbar splanchnic nerve* synapses in the inferior mesenteric ganglion, and the postganglionic neurons innervate the distal colon and rectum, urinary bladder, and genital organs.

Adrenal Glands   The paired adrenal glands are located above each kidney (see fig. 13.5). Each adrenal is composed of two parts: an outer **adrenal cortex** and an inner **adrenal medulla.** These two parts are really two functionally different glands with different embryonic origins, different hormones, and different regulatory mechanisms. The adrenal cortex secretes steroid hormones; the adrenal medulla secretes the hormone **epinephrine** (adrenaline) and, to a lesser degree, **norepinephrine** when it is stimulated by the sympathetic system.

---

splanchnic: Gk. *splanchno*, relating to viscera

adrenal: L. *ad*, to; *renes*, kidney
cortex: L. *cortex*, bark
medulla: L. *medulla*, marrow

Figure 13.4 The cervical sympathetic ganglia.

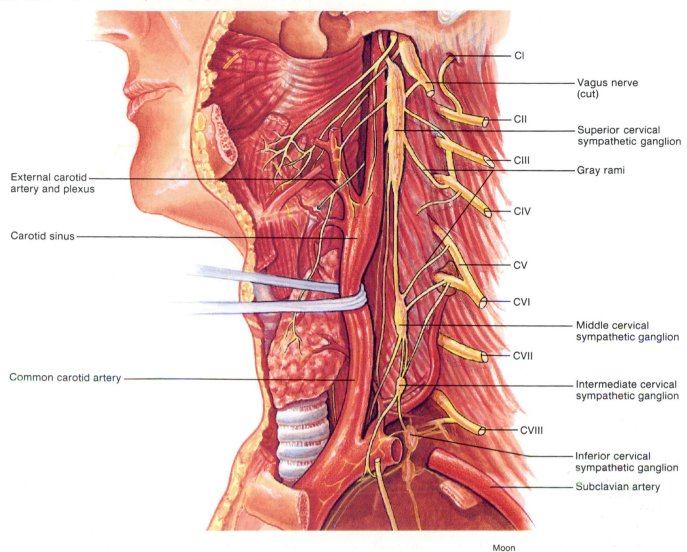

External carotid artery and plexus

Carotid sinus

Common carotid artery

CI

Vagus nerve (cut)

CII

Superior cervical sympathetic ganglion

CIII

Gray rami

CIV

CV

CVI

Middle cervical sympathetic ganglion

CVII

Intermediate cervical sympathetic ganglion

CVIII

Inferior cervical sympathetic ganglion

Subclavian artery

Moon

UNIT V

The adrenal medulla is a modified sympathetic ganglion whose cells are derived from postganglionic sympathetic neurons. The cells of the adrenal medulla are innervated by preganglionic sympathetic neurons originating in the thoracic level of the spinal cord; they secrete epinephrine into the blood in response to sympathetic stimulation. The effects of epinephrine are complementary to those of the neurotransmitter norepinephrine, which is released from postganglionic sympathetic nerve endings.

Parasympathetic (Craniosacral) Division

The **parasympathetic division** is also known as the *craniosacral division* of the autonomic system. This is because its preganglionic neurons originate in the brain (specifically, the midbrain and the medulla oblongata of the brain stem) and in the second through fourth sacral levels of the spinal cord. These preganglionic parasympathetic neurons synapse in ganglia that are located next to (or actually within) the organs innervated.

These parasympathetic ganglia, which are called **terminal ganglia,** supply the postganglionic neurons that synapse with the effector cells.

Tables 13.3 and 13.4 show the comparative structures of the sympathetic and parasympathetic divisions. It should be noted that, unlike sympathetic neurons, most parasympathetic neurons do not travel within spinal nerves. Cutaneous effectors (blood vessels, sweat glands, and arrector pili muscles) and blood vessels in skeletal muscles thus receive sympathetic but not parasympathetic innervation.

Four of the 12 pairs of cranial nerves contain preganglionic parasympathetic neurons. These are the oculomotor (III), facial (VII), glossopharyngeal (IX), and vagus (X) nerves. Parasympathetic neurons within the first three of these cranial nerves synapse in ganglia located in the head; neurons in the vagus nerve synapse in terminal ganglia located in many regions of the body.

Figure 13.5 Peripheral sympathetic plexuses and ganglia of abdominal viscera.

Adrenal gland

Superior mesenteric ganglion

First lumbar sympathetic ganglion

Celiac ganglion

Left kidney

Renal plexus

Plexus of abdominal aorta

Inferior mesenteric ganglion and plexus

Pelvic sympathetic trunk

Moon

The oculomotor cranial nerve contains somatic motor and parasympathetic neurons that originate in the oculomotor nuclei of the midbrain. These parasympathetic neurons synapse in the **ciliary ganglion,** whose postganglionic neurons innervate the ciliary muscle and constrictor neurons in the iris of the eye. Preganglionic neurons that originate in the medulla oblongata travel in the facial nerve to the **pterygopalatine** (ter''ĭ-go-pal'ă-tēn) **ganglion,** which sends postganglionic neurons to the nasal mucosa, pharynx, palate, and lacrimal glands. Another group of neurons in the facial nerve terminate in the **submandibular ganglion,** which sends postganglionic neurons to the submandibular and sublingual glands. Preganglionic neurons of the glossopharyngeal nerve synapse in the **otic ganglion,** which sends postganglionic neurons to innervate the parotid gland.

Other nuclei in the medulla oblongata contribute preganglionic neurons to the very long vagus cranial nerves, which provide the most extensive parasympathetic innervation in the

body. As the paired vagus nerves pass through the thorax, they contribute to the *pulmonary plexuses* within the mediastinum. Branches of the pulmonary plexuses accompany blood vessels and bronchi into the lungs. Below the pulmonary plexuses, the vagus nerves merge to form the *esophageal plexuses*.

At the lower end of the esophagus, vagal neurons collect to form an **anterior** and **posterior vagal trunk,** each composed of neurons from both vagus nerves. The vagal trunks enter the abdominal cavity through the esophageal hiatus (opening) in the diaphragm. Neurons from the vagal trunks innervate the stomach on the anterior and posterior sides. Branches of the vagus nerves within the abdominal cavity also contribute to the *celiac plexus* and *plexuses of the abdominal aorta*.

The preganglionic neurons in the vagus synapse with postganglionic neurons that are actually located *within* the innervated organs. These preganglionic neurons are thus quite long. They provide parasympathetic innervation to the heart, lungs,

vagus: L. *vagus,* wandering

UNIT V

Figure 13.6 The autonomic nervous system. The sympathetic division is shown in red; the parasympathetic, in blue. The solid lines indicate preganglionic neurons and the dashed lines indicate postganglionic neurons.

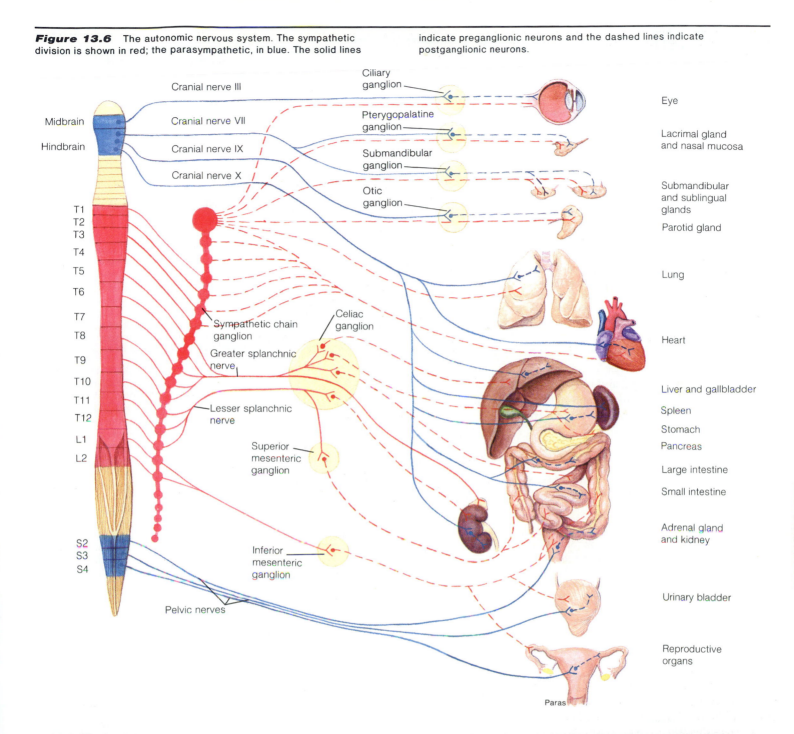

Table 13.3	The Sympathetic (Thoracolumbar) Division	
Parts of Body Innervated	**Spinal Origin of Preganglionic Neurons**	**Origin of Postganglionic Neurons**
Eye	C8, T1	Cervical ganglia
Head and neck	T1–T4	Cervical ganglia
Heart and lungs	T1–T5	Upper thoracic ganglia
Upper extremities	T2–T9	Lower cervical and upper thoracic ganglia
Upper abdominal viscera	T4–T9	Celiac and superior mesenteric ganglia
Adrenal glands	T10, T11	Adrenal medulla
Urinary and reproductive systems	T12–L2	Celiac and inferior mesenteric ganglia
Lower extremities	T9–L2	Lumbar and upper sacral ganglia

From Stuart Ira Fox, *Human Physiology*, 4th ed. Copyright © 1993 Wm. C. Brown Communications, Inc., Dubuque, Iowa. All Rights Reserved. Reprinted by permission.

UNIT V

Table 13.4	The Parasympathetic (Craniosacral) Division		
Effector Organs	**Origin of Preganglionic Neurons**	**Nerve**	**Origin of Postganglionic Neurons**
Eye (smooth muscle in iris and ciliary body)	Midbrain (cranial)	Oculomotor (third cranial) nerve	Ciliary ganglion
Lacrimal, mucous, and salivary glands in head	Medulla oblongata (cranial)	Facial (seventh cranial) nerve	Pterygopalatine and submandibular ganglia
Parotid gland	Medulla oblongata (cranial)	Glossopharyngeal (ninth cranial) nerve	Otic ganglion
Heart, lungs, GI tract, liver, pancreas	Medulla oblongata (cranial)	Vagus (tenth cranial) nerve	Terminal ganglia in or near organ
Lower half of large intestine, rectum, urinary bladder, and reproductive organs	S2–S4 (sacral)	Through pelvic spinal nerves	Terminal ganglia near organs

From Stuart Ira Fox, *Human Physiology*, 4th ed. Copyright © 1993 Wm. C. Brown Communications, Inc., Dubuque, Iowa. All Rights Reserved. Reprinted by permission.

esophagus, stomach, pancreas, liver, small intestine, and upper half of the large intestine. Postganglionic parasympathetic neurons arise from terminal ganglia within these organs and synapse with effector cells.

Preganglionic neurons from the sacral levels of the spinal cord provide parasympathetic innervation to the lower half of the large intestine, rectum, and to the urinary and reproductive systems. These neurons, like those of the vagus, synapse with terminal ganglia located within the effector organs. Parasympathetic nerves to the visceral organs thus consist of preganglionic neurons, whereas sympathetic nerves to these organs contain postganglionic neurons.

A composite view of the sympathetic and parasympathetic divisions is provided in figure 13.6, and the comparisons are summarized in table 13.5.

1. Compare the origins of preganglionic sympathetic and parasympathetic neurons and the locations of sympathetic and parasympathetic ganglia.
2. Using a simple line drawing, illustrate the sympathetic pathway from the spinal cord to the heart. Label the preganglionic neuron, postganglionic neuron, and the ganglion.
3. Use a simple line diagram to show the parasympathetic innervation of the heart. Label the preganglionic and postganglionic neurons, the nerve involved, and the terminal ganglion.
4. Describe the distribution of the vagus nerve and its functional significance.
5. Define the terms *white rami* and *gray rami* and explain why blood vessels in the skin and skeletal muscles receive sympathetic but not parasympathetic innervation.
6. Describe the structure of the adrenal gland and explain its relationship to the sympathetic division of the ANS.

Functions of the Autonomic Nervous System

The actions of the autonomic nervous system, together with the effects of hormones, help to maintain a state of dynamic constancy in the internal environment. The sympathetic di-

Table 13.5	Comparison of the Sympathetic and Parasympathetic Divisions	
Feature	**Sympathetic**	**Parasympathetic**
Origin of preganglionic outflow	Thoracolumbar portion of spinal cord	Midbrain, hindbrain, and sacral portion of spinal cord
Location of ganglia	Chain of paravertebral ganglia and prevertebral ganglia	Terminal ganglia in or near effector organs
Distribution of postganglionic neurons	Throughout the body	Mainly limited to the head and viscera
Divergence of impulses from pre- to postganglionic neurons	Great divergence (one preganglionic may activate 20 postganglionic neurons)	Little divergence (one preganglionic only activates a few postganglionic neurons)
Mass discharge of system as a whole	Usually	Not normally

From Stuart Ira Fox, *Human Physiology*, 4th ed. Copyright © 1993 Wm. C. Brown Communications, Inc., Dubuque, Iowa. All Rights Reserved. Reprinted by permission.

vision activates the body to "fight or flight" through adrenergic effects; the parasympathetic division often manifests antagonistic actions through cholinergic effects. Homeostasis thus depends, in large part, on the complementary and often antagonistic effects of sympathetic and parasympathetic innervation.

Objective 8. List the neurotransmitters of the preganglionic and postganglionic neurons of the sympathetic and parasympathetic divisions.

Objective 9. Describe the effects of acetylcholine released by postganglionic parasympathetic neurons.

Objective 10. Explain the antagonistic, complementary, and cooperative effects of sympathetic and parasympathetic innervation.

The sympathetic and parasympathetic divisions of the ANS (fig. 13.6) affect the visceral organs in different ways. Mass activation of the sympathetic division prepares the body for intense physical activity in emergencies; the heart rate increases, blood glucose rises, and blood is diverted to the skeletal muscles (away from the visceral organs and skin). These and other effects are listed in table 13.6. The theme of the sympathetic division is aptly summarized in the phrase **fight or flight.**

The effects of parasympathetic nerve stimulation are in many ways opposite to the effects of sympathetic stimulation. The parasympathetic division, however, is not normally activated as a whole. Stimulation of separate parasympathetic nerves can result in slowing of the heart, dilation of visceral blood vessels, and increased activity of the GI tract (table 13.6). The different responses of visceral organs to sympathetic and parasympathetic nerve activity is due to the fact that the postganglionic neurons of these two divisions release different neurotransmitters.

Neurotransmitters of the Autonomic System

The neurotransmitter released by most postganglionic sympathetic neurons is **norepinephrine** (noradrenaline). Transmission at these synapses is thus said to be **adrenergic** *(ad″rĕ-ner′jik).* There are a few exceptions to this rule: some sympathetic neurons that innervate blood vessels in skeletal muscles, as well as sympathetic neurons to sweat glands, release acetylcholine (are cholinergic).

Acetylcholine *(ă-sēt″l-ko′lēn)* **(ACh)** is the neurotransmitter of all preganglionic neurons (both sympathetic and parasympathetic). Acetylcholine is also the transmitter released by all parasympathetic postganglionic neurons at their synapses with effector cells (fig. 13.7). Transmission at the autonomic ganglia and at synapses of postganglionic neurons is thus said to be **cholinergic** *(ko″li-ner′jik).* In other words, a *cholinergic fiber* is a neuron that secretes ACh at the terminal end of its axon.

Responses to Adrenergic Stimulation

Adrenergic stimulation—by epinephrine in the blood and by norepinephrine released from sympathetic nerve endings—has both excitatory and inhibitory effects. The heart, dilatory muscles of the iris, and the smooth muscles of many blood vessels are stimulated to contract. The smooth muscles of the bronchioles and of some blood vessels, however, are inhibited from contracting; adrenergic chemicals, therefore, cause these structures to dilate.

Responses to Cholinergic Stimulation

Somatic motor neurons, all preganglionic autonomic neurons, and all postganglionic parasympathetic neurons are cholinergic—they use acetylcholine as a neurotransmitter. The

cholinergic: Gk. *chole,* bile; *ergon,* work

Table 13.6	**Effects of Autonomic Nerve Stimulation on Various Visceral Effector Organs**	
Effector Organ	**Sympathetic Effect**	**Parasympathetic Effect**
Eye		
Iris (pupillary dilator muscle)	Dilation of pupil	—
Iris (pupillary sphincter muscle)	—	Constriction of pupil
Ciliary muscle	Relaxation (for far vision)	Contraction (for near vision)
Glands		
Lacrimal (tear)	—	Stimulation of secretion
Sweat	Stimulation of secretion	—
Salivary	Decreased secretion; saliva becomes thick	Increased secretion; saliva becomes thin
Stomach	—	Stimulation of secretion
Intestine	—	Stimulation of secretion
Adrenal medulla	Stimulation of hormone secretion	—
Heart		
Rate	Increased	Decreased
Conduction	Increased rate	Decreased rate
Strength	Increased	—
Blood vessels	Mostly constriction; affects all organs	Dilation in a few organs (e.g., penis)
Lungs		
Bronchioles	Dilation	Constriction
Mucous glands	Inhibition of secretion	Stimulation of secretion
GI tract		
Motility	Inhibition of movement	Stimulation of movement
Sphincters	Closing stimulated	Closing inhibited
Liver	Stimulation of glycogen hydrolysis	—
Adipocytes (fat cells)	Stimulation of fat hydrolysis	—
Pancreas	Inhibition of exocrine secretions	Stimulation of exocrine secretions
Spleen	Stimulation of contraction	—
Urinary bladder	Muscle tone aided	Stimulation of contraction
Arrector pili muscles	Stimulation of hair erection, causing goose bumps	—
Uterus	If pregnant, contraction; if not pregnant, relaxation	—
Penis	Ejaculation	Erection (due to vasodilation)

From Stuart Ira Fox, *Human Physiology,* 4th ed. Copyright © 1993 Wm. C. Brown Communications, Inc., Dubuque, Iowa. All Rights Reserved. Reprinted by permission.

UNIT V

Figure 13.7 Neurotransmitters of the autonomic motor system. ACh = acetylcholine, NE = norepinephrine, E = epinephrine. Those nerves that release ACh are called cholinergic; those nerves that release NE are called adrenergic. The adrenal medulla secretes both epinephrine (85%) and norepinephrine (15%) as hormones into the blood.

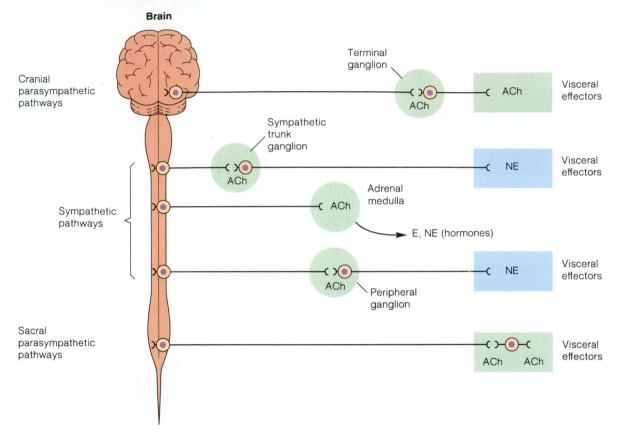

cholinergic effects of somatic motor neurons and preganglionic autonomic neurons are always excitatory. The cholinergic effects of postganglionic parasympathetic neurons are usually excitatory, with some notable exceptions; the parasympathetic neurons innervating the heart, for example, cause slowing of the heart rate.

The drug *muscarine (mus'kă-rēn)*, derived from some poisonous mushrooms, blocks the cholinergic effects of parasympathetic nerves in the heart, smooth muscles, and glands. This drug, however, does not block the effects of somatic motor nerves or the cholinergic transmission in autonomic ganglia. The acetylcholine receptors in visceral organs are, therefore, said to promote the *muscarinic effects* of ACh. These effects are identical to those produced by parasympathetic nerve stimulation and are believed to be mediated by different receptors than those that mediate other cholinergic effects.

 The muscarinic effects of ACh are specifically inhibited by the drug *atropine*, derived from the deadly nightshade plant (*Atropa belladonna*). Indeed, extracts of this plant were used by women during the Middle Ages to dilate their pupils (atropine inhibits parasympathetic stimulation of the iris). This was done to enhance their beauty (belladonna—beautiful woman). Atropine is used clinically today to dilate pupils during eye examinations, to dry mucous membranes of the respiratory tract prior to general anesthesia, and to inhibit spasmodic contractions of the lower GI tract.

Organs with Dual Innervation

Many organs receive a dual innervation—they are innervated by both sympathetic and parasympathetic neurons. When this occurs, the effects of these two divisions may be antagonistic, complementary, or cooperative.

Antagonistic Effects The effects of sympathetic and parasympathetic innervation on the pacemaker region of the heart is the best example of the antagonism of these two systems. In this case, sympathetic and parasympathetic neurons innervate the same cells. Adrenergic stimulation from sympathetic neurons increases the heart rate and cholinergic stimulation from parasympathetic neurons inhibits the pacemaker cells and, thus, decreases the heart rate. Antagonism is also seen in the GI tract, where impulses through sympathetic nerves inhibit intestinal movements and secretions and impulses through parasympathetic nerves stimulate these movements and secretions.

The effects of sympathetic and parasympathetic stimulation on the diameter of the pupil of the eye is analogous to the reciprocal innervation of flexor and extensor skeletal muscles by somatic motor neurons. This is because the iris contains antagonistic muscle layers. Contraction of the pupillary dilator muscle, which is stimulated by impulses through sympathetic nerve endings, causes dilation; contraction of the pupillary sphincter muscle, which is innervated by parasympathetic nerve endings, causes constriction of the pupil (fig. 13.8).

UNIT V

Figure 13.8 Reciprocal sympathetic and parasympathetic innervation of smooth muscles within the iris of the eye. Stimulation through sympathetic nerves produces contraction of the dilator muscles, which enlarges the size of the pupil. Stimulation through the parasympathetic nerves produces contraction of the sphincter muscles, which constricts the size of the pupil.

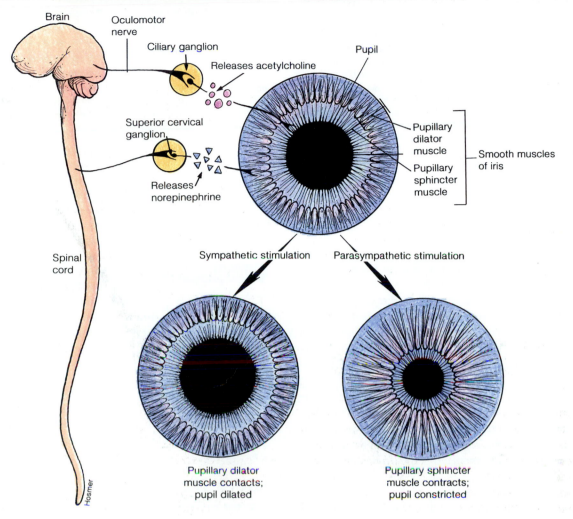

Complementary Effects The effects of sympathetic and parasympathetic stimulation on salivary gland secretion are complementary. The secretion of watery saliva is stimulated through parasympathetic nerves, which also stimulate the secretion of other exocrine glands in the GI tract. Impulses through sympathetic nerves stimulate the constriction of blood vessels throughout the GI tract. The resultant decrease in blood flow to the salivary glands causes the production of a thicker, more viscous saliva.

Cooperative Effects The effects of sympathetic and parasympathetic stimulation on the urinary and reproductive systems are cooperative. Erection of the penis, for example, is due to vasodilation resulting from parasympathetic nerve stimulation; ejaculation is due to stimulation through sympathetic nerves. Although the contraction of the urinary bladder is myogenic (independent of nerve stimulation), it is promoted in part by the action of parasympathetic nerves. This *micturition (mik"tŭ-rish'un)*, or urination, urge and reflex is also enhanced by sympathetic nerve activity, which increases the tone of the bladder muscles. Emotional states that are accompanied by high sym-

pathetic nerve activity may thus result in reflex urination at bladder volumes that are normally too low to trigger this reflex.

Organs without Dual Innervation

Although most organs are innervated by both sympathetic and parasympathetic nerves, some—including the adrenal medulla, arrector pili muscles, sweat glands, and most blood vessels—receive only sympathetic innervation. In these cases, regulation is achieved by increases or decreases in the "tone" (firing rate) of the sympathetic neurons. Constriction of blood vessels, for example, is produced by increased sympathetic activity, which stimulates adrenergic receptors, and vasodilation results from decreased sympathetic nerve stimulation.

Sympathetic activity is required for proper thermoregulatory responses to heat. In a hot room, for example, decreased sympathetic stimulation produces dilation of the blood vessels in the surface of the skin, which increases cutaneous blood flow and provides better heat radiation. During exercise, on the other hand, there is increased sympathetic activity, which causes

UNIT V

Table 13.7		Some Reflexes Stimulated by Input from Sensory Neurons in the Vagus Nerves That Are Transmitted to Nuclei in the Medulla Oblongata			
Organs	Type of Receptors	Reflex Effects	Organs	Type of Receptors	Reflex Effects
Lungs	Stretch receptors	Further inhalation inhibited; increase in cardiac rate and vasodilation stimulated	Aorta (cont.)	Baroreceptors	Stimulated by increased blood pressure—produces a reflex decrease in heart rate
	Type J receptors	Stimulated by pulmonary congestion—produces feelings of breathlessness and causes a reflex fall in cardiac rate and blood pressure	Heart	Atrial stretch receptors	Antidiuretic hormone secretion inhibited, thus increasing the volume of urine excreted
				Stretch receptors in ventricles	Produces a reflex decrease in heart rate and vasodilation
Aorta	Chemoreceptors	Stimulated by rise in CO_2 and fall in O_2—produces increased rate of breathing, fall in heart rate, and vasoconstriction	GI tract	Stretch receptors	Feelings of satiety, discomfort, and pain

From Stuart Ira Fox, *Human Physiology*, 4th ed. Copyright © 1993 Wm. C. Brown Communications, Inc., Dubuque, Iowa. All Rights Reserved. Reprinted by permission.

constriction of the blood vessels in the skin of the limbs and stimulation of sweat glands in the trunk.

The eccrine sweat glands in the trunk secrete a watery fluid in response to sympathetic stimulation. Evaporation of this dilute sweat helps to cool the body. The eccrine sweat glands also secrete a chemical called **bradykinin** (*brad"i-ki'nin*) in response to sympathetic stimulation. Bradykinin stimulates dilation of the surface blood vessels near the sweat glands, helping to radiate heat. At the conclusion of exercise, sympathetic stimulation is reduced and blood flow to the surface of the limbs is increased, which aids in the elimination of metabolic heat. Notice that all of these thermoregulatory responses are achieved without the direct involvement of the parasympathetic division.

1. Define the meaning of the terms *adrenergic* and *cholinergic* and use these terms to describe the neurotransmitters of different autonomic neurons.
2. Describe the effects of the drug *atropine* and explain these effects in terms of the actions of the parasympathetic division.
3. Explain how the sympathetic and parasympathetic divisions can have antagonistic, cooperative, and complementary effects. Give examples.

Control of the Autonomic Nervous System by Higher Brain Centers

Visceral functions are mainly regulated by autonomic reflexes. In most autonomic reflexes, sensory input is directed to brain centers, which in turn regulate the activity of descending pathways to preganglionic autonomic neurons. The neural centers that directly control the activity of autonomic nerves are influenced by higher brain areas, as well as by sensory input.

Objective 11. Describe the area of the brain that most directly controls the activity of autonomic nerves. Also describe the higher brain areas that influence autonomic activity.

Objective 12. In terms of the structures involved, explain how the activity of the autonomic system and the activity of the endocrine system can be coordinated.

Objective 13. In terms of the structures involved, explain how autonomic functions can be affected by emotions.

Medulla Oblongata

The **medulla oblongata** of the brain stem is the structure that most directly controls the activity of the ANS. Almost all autonomic responses can be elicited by experimental stimulation of the medulla oblongata, which contains centers for the control of the circulatory, respiratory, urinary, reproductive, and digestive systems. Much of the sensory input to these centers travels through the sensory neurons of the vagus nerves. These reflexes are listed in table 13.7.

Hypothalamus

The **hypothalamus** (fig. 13.9) is an extremely important autonomic structure within the brain located just above (superior to) the pituitary gland. By means of motor fibers to the brain stem and to the posterior pituitary, and also by means of hormones that regulate the anterior pituitary, the hypothalamus serves to orchestrate somatic, autonomic, and endocrine responses during various behavioral states.

Experimental stimulation of different areas of the hypothalamus can evoke the autonomic responses characteristic of aggression, sexual behavior, eating, or satiety. Chronic stimulation of the lateral hypothalamus, for example, can make an animal eat and become obese, whereas stimulation of the medial

UNIT V

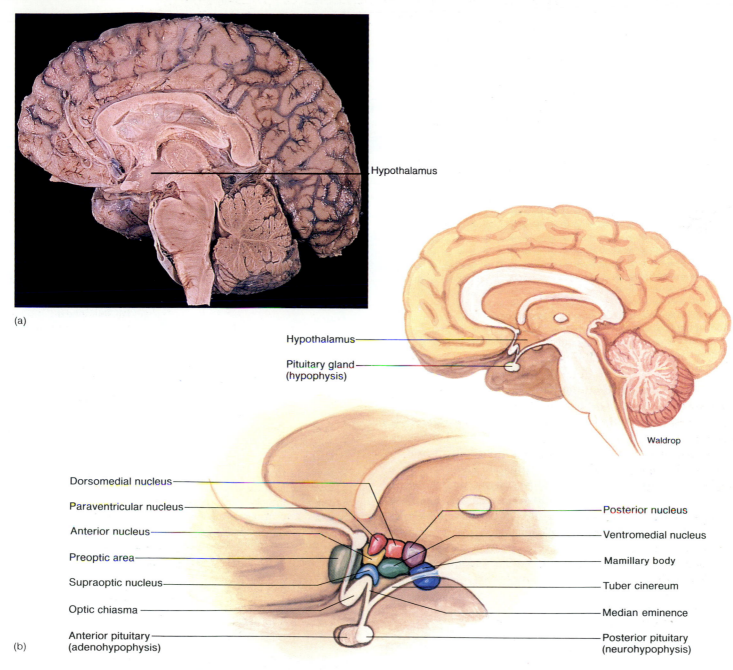

Figure 13.9 (*a*) A photograph of a sagittal section of the brain and (*b*) an illustration of the area of the hypothalamus.

(a)

Hypothalamus

Hypothalamus

Pituitary gland (hypophysis)

Waldrop

UNIT V

Dorsomedial nucleus

Paraventricular nucleus

Anterior nucleus

Preoptic area

Supraoptic nucleus

Optic chiasma

Anterior pituitary (adenohypophysis)

Posterior nucleus

Ventromedial nucleus

Mamillary body

Tuber cinereum

Median eminence

Posterior pituitary (neurohypophysis)

(b)

hypothalamus inhibits eating. Other areas contain osmoreceptors that stimulate thirst and the secretion of antidiuretic hormone (ADH) from the posterior pituitary.

The hypothalamus is also where the body's thermostat is located. Experimental cooling of the preoptic-anterior hypothalamus causes shivering (a somatic response) and nonshivering thermogenesis (sympathetic response). Experimental heating of this hypothalamic area results in hyperventilation (stimulated by somatic motor nerves), vasodilation, salivation, and sweat gland secretion (stimulated by autonomic nerves).

The coordination of sympathetic and parasympathetic reflexes by the medulla oblongata is thus integrated with the con-

trol of somatic and endocrine responses by the hypothalamus. The activities of the hypothalamus are in turn influenced by higher brain centers.

Limbic System, Cerebellum, and Cerebrum

The **limbic system** is a group of fiber tracts and nuclei that form a ring (limbus) around the brain stem. It includes the cingulate gyrus of the cerebral cortex, the hypothalamus, the fornix

limbic: L. *limbus,* edge or border

Figure 13.10 The limbic system and the pathways that interconnect the structures of the limbic system. (Note that the left temporal lobe of the cerebral cortex has been removed.)

Corpus callosum

Cingulate gyrus

Septal area

Preoptic area

Olfactory bulb

Olfactory tract

Hypothalamus

Amygdaloid nucleus

Cortex of
right hemisphere

Fornix

Thalamic nucleus

Mammilothalamic
tract

Mammillary body

Hippocampus

Cortex of
left hemisphere

Waldrop

(a fiber tract), the hippocampus, and the amygdaloid nucleus (fig. 13.10). These structures, which were derived early in the course of vertebrate evolution, were once called the *rhinencephalon (ri''nen-sef'ă-lon),* or "smell brain," because of their importance in the central processing of olfactory information.

In higher vertebrates, these structures are now recognized as centers involved in basic emotional drives—such as anger, fear, sex, and hunger—and in short-term memory. Complex circuits between the hypothalamus and other parts of the limbic system (illustrated in figure 13.10) contribute visceral responses to emotions.

The autonomic correlates of motion sickness—nausea, sweating, and cardiovascular changes—are eliminated by cutting the motor tracts of the cerebellum. This demonstrates that impulses from the cerebellum to the medulla oblongata influence activity of the ANS. Experimental and clinical observations have also demonstrated that the frontal and temporal lobes of the cerebral cortex influence lower brain areas as part of their involvement in emotion and personality.

One of the most dramatic examples of the role of higher brain areas in personality and emotion is the famous crowbar accident of 1848. A 25-year-old railroad foreman, Phineas P. Gage, was tamping gunpowder into a hole in a rock when it exploded. The rod—three feet, seven inches long and one and one-fourth inches thick—was driven through his left eye, passed through his brain, and emerged through the back of his skull.

After a few minutes of convulsions, Gage got up, rode a horse three-quarters of a mile into town, and walked up a long flight of stairs to see a doctor. He recovered well,

with no noticeable sensory or motor deficits. His associates, however, noted striking personality changes. Before the accident Gage was a responsible, capable, and financially prudent man. Afterward, he was much less inhibited socially, engaging, for example, in gross profanity (which he had never done previously). He also seemed to be tossed about by chance whims. Eventually Gage was fired from his job, and his old friends remarked that he was "no longer Gage."

1. Describe the role of the medulla oblongata in the regulation of the ANS.
2. Describe the role of the hypothalamus in the regulation of the autonomic and endocrine systems.
3. In terms of the neural pathways involved, explain how emotional states can be associated with particular autonomic responses (such as blushing).

Clinical Considerations

Autonomic Dysreflexia

Autonomic dysreflexia, a serious condition producing rapid elevations in blood pressure that can lead to stroke (cerebrovascular accident), occurs in 85% of people with quadriplegia and others with spinal cord lesions above the sixth thoracic level. Lesions to the spinal cord first produce the symptoms of spinal shock, characterized by the loss of both skeletal muscle and autonomic reflexes. After a period of time, both types of reflexes return in an exaggerated state; the skeletal muscles may become spastic due to the absence of higher inhibitory influ-

UNIT V

ences and the visceral organs experience denervation hypersensitivity. Patients in this state have difficulty emptying their urinary bladders and must often be catheterized.

Noxious stimuli, such as overdistension of the urinary bladder, can result in reflex activation of the sympathetic nerves below the spinal cord lesion. This produces goose bumps, cold skin, and vasoconstriction in the regions served by the spinal cord below the level of the lesion. The rise in blood pressure resulting from this vasoconstriction activates pressure receptors that transmit impulses along sensory neurons to the medulla oblongata. In response to this sensory input, the medulla oblongata directs a reflex slowing of the heart and vasodilation. Since descending impulses are blocked by the spinal lesion, however, the skin is warm and moist (due to vasodilation and sweat gland secretion) above the lesion but cold below the level of spinal cord damage.

CLINICAL CASE STUDY ANSWER

The tumor has invaded the sympathetic ganglia and connecting pathways within the base of the neck, known as the cervical ganglia and sympathetic trunk. Preganglionic sympathetic neurons originating in the upper thoracic spinal cord travel upward through the sympathetic trunk, where they synapse on a postganglionic neuron in one of the cervical ganglia. From that point, postganglionic neurons travel through various routes (mostly paralleling arteries) to innervate the end organs involved. In the case of our patient, these include the pupillary dilator muscle of the iris, the smooth muscle portion of the levator palpebrae superioris, and sweat glands on the right side of the face.

CHAPTER SUMMARY

Introduction to the Autonomic Nervous System (pp. 414–416)

1. The autonomic nervous system (ANS) is a functional division of the nervous system and is composed of portions of the central nervous system (CNS) and portions of the peripheral nervous system (PNS).
2. Preganglionic autonomic neurons originate from the brain or spinal cord; postganglionic neurons originate from ganglia outside the CNS.
3. Smooth muscle, cardiac muscle, and glands receive autonomic innervation.
   (a) The involuntary effectors are somewhat independent of their innervation and become hypersensitive when their innervation is removed.
   (b) Myocardial cells are interconnected by electrical synapses, or gap junctions, to form a functional syncytium with independent pacemaker activity.
   (c) Single-unit smooth muscles are characterized by gap junctions and pacemaker activity; multiunit smooth muscles are characterized by a lack of gap junctions and pacemaker activity.

Structure of the Autonomic Nervous System (pp. 417–422)

1. Preganglionic neurons of the sympathetic (thoracolumbar) division originate in the spinal cord, between the thoracic and lumbar levels.
   (a) Many of these neurons synapse with postganglionic neurons, whose cell bodies are located in a trunk of sympathetic ganglia outside the spinal cord.
   (b) Some preganglionic neurons synapse in peripheral ganglia; these are the celiac, superior mesenteric, and the inferior mesenteric ganglia.
   (c) Some preganglionic neurons innervate the adrenal medulla, which secretes epinephrine (and some norepinephrine) into the blood in response to this stimulation.
2. Preganglionic parasympathetic neurons originate in the brain and in the sacral levels of the spinal cord.
   (a) Preganglionic parasympathetic neurons contribute to cranial nerves III, VII, IX, and X.
   (b) Preganglionic neurons of the vagus (X) nerve are very long and synapse in terminal ganglia located next to or within the innervated organ; short postganglionic neurons then innervate the effector cells.
   (c) The vagus nerves provide parasympathetic innervation to the heart, lungs, esophagus, stomach, liver, small intestine, and upper half of the large intestine.
   (d) Parasympathetic outflow from the sacral levels of the spinal cord innervates terminal ganglia in the lower half of the large intestine, the rectum, and the urinary and reproductive systems.

Functions of the Autonomic Nervous System (pp. 422–426)

1. The sympathetic and parasympathetic divisions of the ANS may act in a complementary or antagonistic way to help maintain homeostasis. The sympathetic division activates the body to "fight or flight" through adrenergic effects; the parasympathetic division often manifests antagonistic actions through cholinergic effects.
2. All preganglionic autonomic neurons are cholinergic (use acetylcholine as a neurotransmitter).
   (a) All postganglionic parasympathetic neurons are cholinergic.
   (b) Most postganglionic sympathetic neurons are adrenergic (use norepinephrine at their synapses).
   (c) Sympathetic neurons that innervate sweat glands and those that innervate blood vessels in skeletal muscles are cholinergic.
3. Adrenergic effects include stimulation of the heart, vasoconstriction in the viscera and skin, bronchodilation, and glycogenolysis in the liver.
4. Cholinergic effects of parasympathetic nerves are promoted by the drug muscarine and inhibited by atropine.
5. In organs with dual innervation, the effects of the sympathetic and parasympathetic divisions can be antagonistic, complementary, or cooperative.
   (a) The effects are antagonistic in the heart and pupils.
   (b) The effects are complementary in the regulation of salivary gland secretion and are cooperative in the regulation of the reproductive and urinary system.

UNIT V

6. In organs without dual innervation (such as most blood vessels), regulation is achieved by variations in sympathetic nerve activity.

Control of the Autonomic System by Higher Brain Centers (pp. 426–428)

1. Visceral sensory input to the brain may result in the activity of the descending pathways to the preganglionic autonomic neurons. The centers in the brain that control autonomic activity are influenced by higher brain areas, as well as by sensory input.
2. The medulla oblongata of the brain stem is the structure that most directly controls the activity of the ANS.
   (a) The medulla oblongata is in turn influenced by sensory input and by input from the hypothalamus.
   (b) The hypothalamus orchestrates somatic, autonomic, and endocrine responses during various behavioral states.
3. The activity of the hypothalamus is influenced by input from the limbic system, cerebellum, and cerebrum; these interconnections provide an autonomic component to changes in body position, emotion, and various expressions of personality.

REVIEW ACTIVITIES

Objective Questions

1. Which of the following statements about the superior mesenteric ganglion is *true*?
   (a) It is a parasympathetic ganglion.
   (b) It is a paravertebral sympathetic ganglion.
   (c) It is located in the head.
   (d) It contains postganglionic sympathetic neurons.

2. The pterygopalatine, ciliary, submandibular, and otic ganglia are
   (a) collateral sympathetic ganglia.
   (b) cervical sympathetic ganglia.
   (c) parasympathetic ganglia that receive neurons from the vagus.
   (d) parasympathetic ganglia that receive neurons from the third, seventh, and ninth cranial nerves.
3. Parasympathetic ganglia are located
   (a) in a trunk parallel to the spinal cord.
   (b) in the posterior roots of spinal nerves.
   (c) next to or within the organs innervated.
   (d) in the brain.
4. The neurotransmitter of preganglionic sympathetic neurons is
   (a) norepinephrine.
   (b) epinephrine.
   (c) acetylcholine.
   (d) dopamine.
5. The preganglionic neurons of the sympathetic division of the autonomic nervous system originate in
   (a) the midbrain and the medulla oblongata.
   (b) the entire spinal nerve complex.
   (c) the first cervical (C1) to the first lumbar (L1) vertebrae.
   (d) the first thoracic (T1) to the second lumbar (L2) vertebrae.
6. Which of the following neurons release norepinephrine?
   (a) preganglionic parasympathetic neurons
   (b) postganglionic parasympathetic neurons
   (c) postganglionic sympathetic neurons in the heart
   (d) postganglionic parasympathetic neurons in sweat glands
   (e) all of these
7. The actions of sympathetic and parasympathetic neurons are cooperative in
   (a) the heart.
   (b) the reproductive system.
   (c) the digestive system.
   (d) the eyes.

8. Which of the following is *not* a result of parasympathetic nerve stimulation?
   (a) increased movement of the GI tract
   (b) increased mucus secretion
   (c) constriction of the pupils
   (d) constriction of visceral blood vessels
9. Atropine blocks parasympathetic nerve effects. It would therefore result in
   (a) dilation of the pupils.
   (b) a decrease in mucus secretion.
   (c) a decrease in GI tract movement.
   (d) an increase in heart rate.
   (e) all of these.
10. The area of the brain that is most directly involved in the reflex control of the autonomic system is
    (a) the hypothalamus.
    (b) the cerebral cortex.
    (c) the medulla oblongata.
    (d) the cerebellum.

Essay Questions

1. Compare the sympathetic and parasympathetic divisions in terms of ganglia location and nerve distribution.
2. Explain the structural and functional relationship between the sympathetic nervous system and the adrenal glands.
3. Compare the effects of adrenergic and cholinergic stimulation on the cardiovascular and digestive systems.
4. Explain how effectors that receive only sympathetic innervation are regulated by the ANS.
5. Explain why a person may sweat more profusely immediately after exercise than during exercise.

Gundy/Weber Software

The tutorial software accompanying Chapter 13, Autonomic Nervous System, is Volume 5, Nervous System.

UNIT V

Endocrine System

Chapter Outline

CLINICAL CASE STUDY

A 38-year-old female visits her family doctor complaining of chronic fatigue and weakness, especially in her legs. Upon greeting the patient, the doctor notes that although she is mildly obese, there is an unusually round contour to her face. During questioning, he learns that at her recent 20-year high school reunion, nobody recognized her because her face looked so different. He also learns that she has a chronic unrelenting headache. Physical examination yields, in addition to the facial findings, an unusual fat distribution consisting of a hump on the upper back and marked truncal obesity. Additional findings include hypertension and blindness in the lateral visual fields of both eyes (bitemporal hemianopsia). Laboratory findings are remarkable for elevated blood glucose and other evidence that glucocorticoid levels are high.

Can a correlation be drawn between the patient's visual findings and the rest of her clinical picture? Explain. In which endocrine organ does the primary disease process reside? Does that same organ produce the hormones that directly cause the patient's signs and symptoms? Explain. Do you suppose that similar findings could occur due to a tumor elsewhere in the endocrine system? *Hints:* Study figure 14.8 and note the position of the endocrine glands with respect to the optic nerve and optic chiasma. Also, study table 14.4 to review the actions of the hormones released by the endocrine glands within the brain.

Introduction to the Endocrine System

Hormones are regulatory chemicals secreted by the endocrine glands into the blood, which transports them to their target tissues. Feedback mechanisms in the target cells control the secretion (production) of the hormones.

Objective 1. Distinguish between endocrine and exocrine glands.

Objective 2. Compare and contrast the nervous system and the endocrine system in regulating body functions and maintaining homeostasis.

Objective 3. Define *mixed gland* and identify the endocrine glands that are mixed.

Objective 4. Describe the action of a hormone on its target cell and explain how negative feedback regulates hormonal secretion.

Objective 5. Differentiate between the three principal kinds of hormones: steroids, proteins, and amines.

Figure 14.1 Comparison of (*a*) an endocrine gland (e.g., the thyroid) with (*b*) an exocrine gland (e.g., a sweat gland in the skin). An endocrine gland is a ductless gland that releases hormones into the blood. By contrast, exocrine glands secrete their products into ducts that lead to body surfaces.

Thyroid gland cells

Hormone molecules within capillary

(a)

Blood flow

Duct

Sweat gland

(b)

The numerous glands of the body can be classified into two types based on structure and function: *exocrine* and *endocrine.* Exocrine glands, such as sweat, salivary, or mucous glands, produce secretions that are transported through ducts to their respective destinations. Each of the exocrine glands functions within a particular system of the body. The endocrine glands constitute their own system, the endocrine system, and have several distinctions. Endocrine glands are ductless glands that secrete specific chemicals called *hormones* directly into the blood (fig. 14.1). The blood then transports these hormones to specific sites called *target cells,* where they perform precise functions.

The endocrine system functions closely with the nervous system in regulating and integrating body processes and maintaining homeostasis. The nervous system regulates body activities through the action of electrochemical impulses that are transmitted by means of neurons, resulting in rapid, but usually brief, responses. By contrast, the glands of the endocrine system secrete chemical regulators that travel through the bloodstream to their intended sites where they have a slow but relatively long-lasting effect. Neurological responses are measured in milliseconds, but hormonal action requires seconds or days to elicit a response. Some hormones may have an effect that lasts for minutes; for others, the effect may last for weeks or months.

endocrine: Gk. *endon*, within; *krinein*, to separate
exocrine: Gk. *exo*, outside, *krinein*, to separate
hormone: Gk. *hormon*, to set in motion

UNIT V

Figure 14.2 (*a*) The location of some of the endocrine glands. (*b*) The pancreatic islets (islets of Langerhans) within the pancreas.

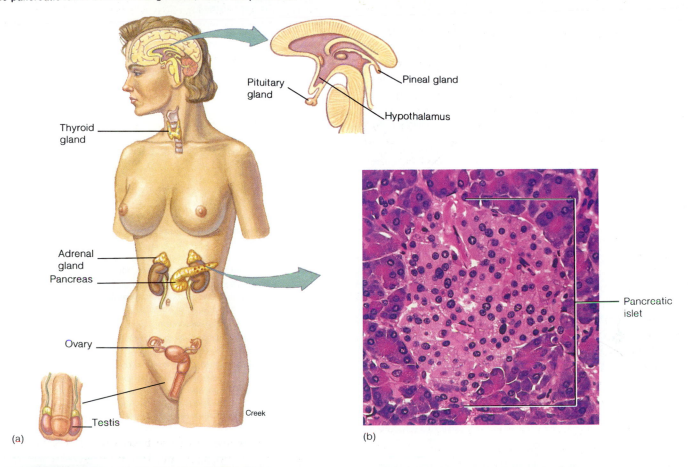

Pituitary gland

Pineal gland

Hypothalamus

Thyroid gland

Adrenal gland

Pancreas

Ovary

Testis

Creek

(a)

Pancreatic islet

(b)

UNIT V

Table 14.1	Basic Functions of the Endocrine and Nervous Systems	
Endocrine System	**Nervous System**
Secretes hormones that are transported to target cells via the blood	Transmits neurochemical impulses via nerve fibers
Causes changes in the metabolic activities in specific cells	Causes muscles to contract or glands to secrete
Exerts effects relatively slowly (seconds or even days)	Exerts effects relatively rapidly (milliseconds)
Effects are relatively prolonged	Effects are relatively brief

The nervous and endocrine systems are closely coordinated in autonomically controlling the functions of the body. Three endocrine glands are located within the cranial cavity, where certain structures of the brain routinely stimulate or inhibit the release of hormones. Likewise, certain hormones may stimulate or inhibit the activities of the nervous system. The functions of the two body control systems are compared in table 14.1.

Glands of the Endocrine System

The endocrine glands are located throughout the body (fig. 14.2). The **pituitary,** the **hypothalamus,** and the **pineal**

gland are associated with the brain within the cranial cavity. The **thyroid gland** and **parathyroid glands** are located in the neck. The **adrenal glands** and **pancreas** are located within the abdominal region. The **gonads** (ovaries) of the female are within the pelvic cavity, whereas the **gonads** (testes) of the male are in the scrotum. The pancreas and gonads are frequently classified as *mixed glands* because they have exocrine as well as endocrine functions.

In addition to the glands just mentioned, the *thymus* in the lower median neck region, the *stomach,* the *kidneys,* the *mucosal cells of the duodenum,* and the *placenta,* associated with the fetus, may be considered part of the endocrine system because each has endocrine functions. The principal endocrine glands of the body, their hormones, and the effects of the hormones are listed in table 14.2.

Hormones and Their Actions

Hormones are specific organic substances that act as the chemical messengers of the endocrine system. The three basic kinds of hormones (proteins, steroids, and amines) are derived either from amino acids or cholesterol (fig. 14.3).

Proteins are composed of amino acids bonded together in peptide chains (fig. 14.3a). Most of the hormones of the body are proteins, including calcitonin from the thyroid gland and hormones secreted by the pituitary gland, the pancreas, and the

Table 14.2	The Principal Endocrine Glands		
Endocrine Gland	**Major Hormones**	**Primary Target Organs**	**Primary Effects**
Adrenal cortex	Glucocorticoids Aldosterone	Liver, muscles Kidneys	Glucocorticoids influence glucose metabolism; aldosterone promotes Na^+ retention, K^+ excretion
Adrenal medulla	Epinephrine	Heart, bronchioles, blood vessels	Causes adrenergic stimulation
Hypothalamus	Releasing and inhibiting hormones	Anterior pituitary	Regulate secretion of anterior pituitary hormones
Intestine	Secretin and cholecystokinin	Stomach, liver, and pancreas	Inhibit gastric motility and stimulate bile and pancreatic juice secretion
Pancreatic islets	Insulin Glucagon	Many organs Liver and adipose tissue	Insulin promotes cellular uptake of glucose and formation of glycogen and fat; glucagon stimulates hydrolysis of glycogen and fat
Ovaries	Estradiol-17β and progesterone	Female reproductive tract and mammary glands	Maintain structure of reproductive tract and promote secondary sexual characteristics
Parathyroids	Parathyroid hormone	Bone, intestine, and kidneys	Increases Ca^{++} concentration in blood
Pineal	Melatonin	Hypothalamus and anterior pituitary	Affects secretion of gonadotrophic hormones
Pituitary, anterior	Trophic hormones	Endocrine glands and other organs	Stimulate growth and development of target organs; stimulate secretion of other hormones
Pituitary, posterior	Antidiuretic hormone Oxytocin	Kidneys, blood vessels Uterus, mammary glands	Antidiuretic hormone promotes water retention and vasoconstriction; oxytocin stimulates contraction of uterus and mammary secretory units
Stomach	Gastrin	Stomach	Stimulates acid secretion
Testes	Testosterone	Prostate, seminal vesicles, other organs	Stimulates secondary sexual development
Thymus	Thymosin	Lymph nodes	Stimulates white blood cell production
Thyroid	Thyroxine (T_4) and triiodothyronine (T_3); Calcitonin	Most organs; blood	Thyroxine and triiodothyronine promote growth and development and stimulate basal rate of cell respiration (basal metabolic rate or BMR); calcitonin regulates Ca^{++} levels within blood by inhibiting bone decalcification
Kidneys	Erythropoietin	Bone marrow	Stimulates red blood cell production

From Stuart Ira Fox, *Human Physiology*, 4th ed. Copyright © 1993 Wm. C. Brown Communications, Inc., Dubuque, Iowa. All Rights Reserved. Reprinted by permission.

UNIT V

Figure 14.3 Structural formulas of the three basic kinds of hormones: (*a*) protein hormone, (*b*) steroid hormone, and (*c*) amine hormone.

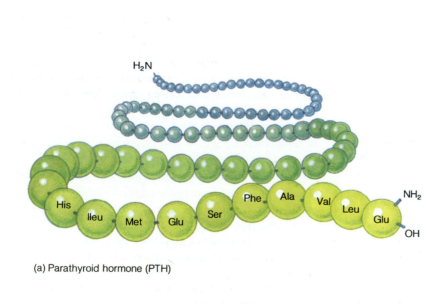

(a) Parathyroid hormone (PTH)

(b) Cortisol

(c) Norepinephrine

parathyroid glands. Protein hormones cannot be administered orally because the peptide bonds would be split during the hydrolytic reaction of digestion; thus, they must be injected intravenously, intramuscularly, or subcutaneously.

A *steroid* is a lipid synthesized from cholesterol. Steroids exist as complex rings of carbon and hydrogen atoms (fig. 14.3b). The types of atoms attached to the rings determine the specific kind of steroid. There are more than 20 steroid hormones in the body, including such common ones as *cortisol, cortisone, estrogen, progesterone,* and *testosterone.* The sex hormones produced by the gonads and the hormones produced by the adrenal cortex are all steroids. Steroids can be taken orally or intravenously to regulate body activity if glandular dysfunction prevents normal amounts from being produced naturally.

Amines are produced from amino acids but do not contain peptide bonds. Amine molecules contain atoms of carbon, hydrogen, and nitrogen and always have an amine group ($-NH_2$) associated with the molecule (fig. 14.3c). Examples of amines include thyroxine produced in the thyroid gland, epinephrine (adrenaline) and norepinephrine produced in the adrenal gland, and melatonin produced in the pineal gland.

Thyroxine is usually administered orally, whereas epinephrine is administered intravenously to produce a fast response. Epinephrine can also be administered as an inhalant when needed to enlarge the air-conducting passageways within the lungs.

A summary of the basic types of hormones is presented in table 14.3.

The usual action of hormones is to speed up or slow down cellular processes in the target tissue. Hormones are extremely specific as to which cells they affect and the cellular changes they elicit. The ability of a hormone to affect a particular cell depends on the presence of receptor molecules in the cell or specific receptor sites on its cell membrane.

Steroid hormones are soluble in lipids and can readily pass through a cell membrane. Once in a cell, the steroid combines with a protein to form a steroid-protein complex (fig. 14.4) that is necessary for the synthesis of specific kinds of messenger RNA molecules.

Protein and amine hormones are insoluble in lipids and therefore must attach to specific receptor sites on the cell membrane. These hormones do not enter the cell, but their attachment increases the activity of an enzyme in the cell membrane called *adenyl cyclase.* In the presence of adenyl cyclase, ATP molecules in the cell are converted to *cyclic AMP* (adenosine monophosphate), which in turn disperses throughout the cell to cause changes in cellular processes (fig. 14.5). These changes may include increasing protein synthesis, altering membrane permeability, or activating certain cellular enzymes.

Control of Hormone Secretion

The rate of secretion of a particular hormone and the rate of usage by the target cells are closely balanced. The stability of hormone levels is maintained by a negative feedback system and autonomic neural impulses.

Negative feedback (fig. 14.6) is a homeostatic mechanism that maintains the status quo of supply and demand between the

Table 14.3	Classes of Hormones		
Type of Hormone	Composition	Examples	Method of Administration
Protein	Amino acids bonded by peptide chains	Pituitary hormones, pancreatic hormones, parathyroid hormones, and calcitonin from thyroid gland	Intravenously, intramuscularly, or subcutaneously
Steroid	Lipid with a cholesterol-type nucleus	Sex hormones and hormones from adrenal cortex	Orally, intravenously, or intramuscularly
Amine	Amino acids with no peptide bonds; molecule contains $-NH_2$ group	Thyroxine from thyroid gland, adrenaline from adrenal gland, and melatonin from the pineal gland	Orally, subcutaneously, intravenously, or as an inhalant

normal level and the needs of the target cells. An endocrine gland will continue to secrete hormones that affect target cells until messages come back from the cells reporting that sufficient hormones are present. These messages are generally in the form of hormones secreted by the target cells. This chemical feedback information signals the endocrine gland to inhibit secretion (fig. 14.6). The primary endocrine gland will resume secretions when the levels of inhibiting chemicals in the blood become low again.

Neural impulses through the autonomic nervous system cause certain endocrine glands, such as the adrenal medulla, to secrete hormones (fig. 14.7). One specialized kind of neural impulse involves the hypothalamus and the pituitary gland (fig. 14.7). In this system, chemical secretions from neurosecretory cells in the hypothalamus called *releasing factors* influence specific target cells in the pituitary gland. Stimulation of the pituitary by a releasing factor causes the secretion of specific hormones. The hypothalamus continues to secrete the releasing factor until a given level of hormones is present in the body fluids and is detected by the hypothalamus through a negative feedback mechanism.

1. What are some of the ways in which endocrine and exocrine glands differ structurally?
2. How are the nervous and endocrine systems functionally related?
3. List the body organs that are exclusively endocrine in function. Also list the organs that secrete hormones and serve additional body functions.
4. Using diagrams, describe the mechanism of steroid hormone action and the mechanism of protein hormone action within cells.
5. List the three kinds of hormones, give examples of each, and describe their chemical compositions.

Figure 14.4 The mechanism of the action of a steroid hormone (H) on the target cells. A steroid hormone passes through a cell membrane and unites with a receptor protein in the cytoplasm. The steroid-protein complex then enters the cell nucleus and activates the synthesis of messenger RNA. The messenger RNA then disperses into the cytoplasm where it activates the synthesis of proteins by means of ribosomes.

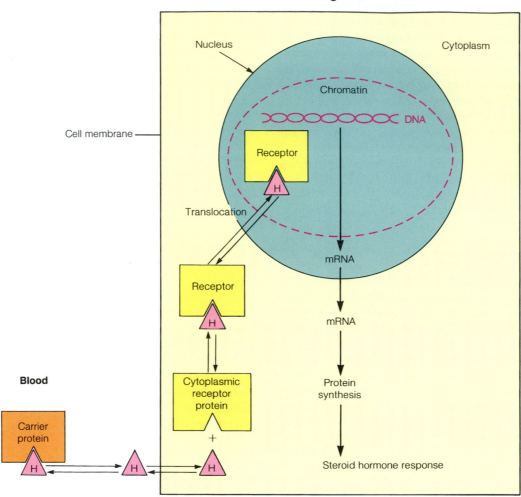

Pituitary Gland

The neurohypophysis of the pituitary gland releases hormones that are produced by the hypothalamus, whereas the adenohypophysis of the pituitary gland secretes its own hormones in response to regulation from hypothalamic hormones. The secretions of the pituitary gland are thus controlled by the hypothalamus, as well as by negative feedback influences from the target glands.

Objective 6. List the hormones secreted by the adenohypophysis and neurohypophysis.

Objective 7. Describe, in a general way, the actions of anterior pituitary hormones.

Objective 8. Describe how the secretions of anterior and posterior pituitary hormones are controlled by the hypothalamus.

Objective 9. Explain how the secretion of anterior pituitary hormones is regulated by negative feedback.

Description of the Pituitary Gland

The **pituitary gland,** or **cerebral hypophysis,** is located on the inferior aspect of the brain in the region of the diencephalon and is attached to the brain by a stalklike structure called the *pituitary stalk* (fig. 14.8). The *infundibulum* is the portion of the pituitary stalk that connects the hypothalamus to the posterior lobe of the pituitary gland. The pituitary is a rounded, pea-shaped gland measuring about 1.3 cm (0.5 in.) in diameter. It is covered by the dura mater and is supported by the sella turcica of the sphenoid bone. The cerebral arterial circle surrounds the highly vascular pituitary gland, providing it with a rich blood supply (see fig. 16.24).

The pituitary gland is structurally and functionally divided into an anterior lobe, or **adenohypophysis,** and a posterior lobe

pituitary: L. *pituita*, phlegm (this gland was originally thought to secrete mucus into the nasal cavity)
adenohypophysis: Gk. *adeno*, gland; *hypo*, under; *physis*, a growing

Figure 14.5 The relationship between a protein hormone, the target cells, and cyclic AMP. Hormones arrive at target cells through body fluids and attach to specific receptor sites on the cell membrane. This causes molecules of adenyl cyclase to diffuse into the cytoplasm and promote the change of ATP into cyclic AMP. Cyclic AMP in turn causes changes in cellular processes.

Figure 14.6 In a negative feedback system, gland A secretes hormone A (first step) that stimulates gland B to release its hormone; the hormone from gland B (second step) inhibits the secretion of hormone A by gland A.

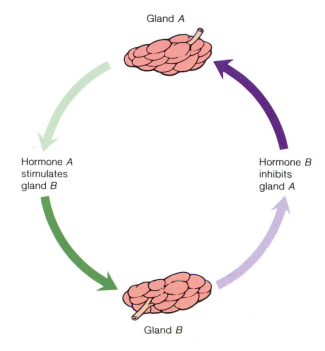

Figure 14.7 Neural control of endocrine secretion. (a) The adrenal medulla secretes hormones in response to autonomic stimulation. (b) The hypothalamus has a neurosecretory role in stimulating target cells in the pituitary gland to release hormones.

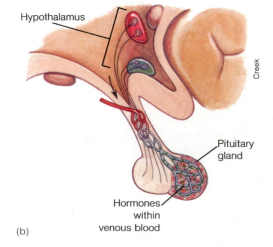

UNIT V

Figure 14.8 The pituitary gland. (*a*) The pituitary gland is attached to the hypothalamus and lies in the sella turcica of the sphenoid bone. (*b*) A diagram of the pituitary gland showing the various portions.

(a)

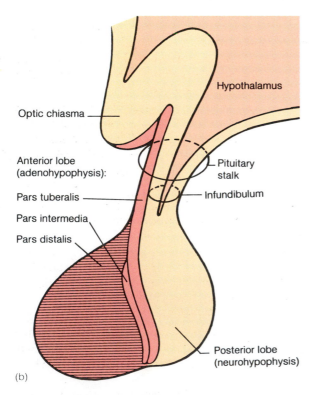

(b)

called the **neurohypophysis.** The adenohypophysis consists of three parts: (1) the **pars distalis** is the bulbar portion; (2) the **pars tuberalis** is the thin extension in contact with the infundibulum; and (3) the **pars intermedia** is located between the anterior and posterior parts of the pituitary. These parts are illustrated in figure 14.8. Refer to figure 14.9 for the histological structure of the adenohypophysis.

The neurohypophysis is the neural part of the pituitary gland. It consists of the bulbar **lobus nervosa,** which is in contact with the pars intermedia and the pars distalis of the adenohypophysis, and the **infundibulum,** which is the connecting stalk to the hypothalamus. Nerve fibers extend through the infundibulum along with minute neuroglia-like cells called **pituicytes** (*pĭ-too'ĭ-sītz*).

The pituitary is the structure of the brain perhaps most subject to neoplasms. A tumor of the pituitary is generally detected easily as it begins to grow and interfere with hormonal activity. If a tumor of the pituitary grows superiorly, it may exert sufficient pressure on the optic chiasma to cause *bitemporal hemianopia,* which is blindness in the temporal field of vision of both eyes. Surgical removal of a neoplasm of the pituitary gland (*hypophysectomy*) may be performed transcranially through the frontal bone or through the nasal cavity and sphenoidal sinus.

Pituitary Hormones

The pituitary gland releases nine important hormones. The first six in the following list are secreted by the anterior pituitary (the pars distalis of the adenohypophysis). The seventh hormone in the list is secreted by the intermediate lobe (pars intermedia) of the pituitary gland. The last two hormones in the list are produced in the hypothalamus, transported through axons in the infundibulum to the posterior pituitary, and released from the posterior pituitary (the pars nervosa of the neurohypophysis) into the blood.

infundibulum: L. *infundibulum,* a funnel

Figure 14.9 The histology of the adenohypophysis.

Chromophobes   Chromophils

Figure 14.10 Acromegaly is caused by a hypersecretion of growth hormone during adulthood.

The hormones secreted by the pars distalis are called **trophic** (*trof'ik*) **hormones.** The term *trophic* means food. This term is used because high amounts of the anterior pituitary hormones make their target glands hypertrophy.

1. **Growth hormone (GH).** Growth hormone, or *somatotropin,* regulates the rate of growth of all body cells and promotes mitotic activity. The secretion of GH is regulated by growth hormone releasing factor (GRF) and growth hormone release-inhibiting factor (GR-IH) from the hypothalamus. The precise mechanism of GH is not understood, but it seems to promote the movement of amino acids through cell membranes and the utilization of these substances in protein synthesis. Pathological hyposecretion of GH during adolescence limits body growth, causing a type of *dwarfism.* Hypersecretion of GH during adolescence may result in *gigantism.* In both of these conditions, body proportions are greatly distorted. Hypersecretion of GH within an adult, after the epiphyses (see chapter 6) are fused, causes *acromegaly.* In this condition, the soft tissues rapidly proliferate and certain body features, such as the hands, feet, nose, jaw, and tongue, are greatly distorted (fig. 14.10).
2. **Thyroid-stimulating hormone (TSH).** TSH, frequently called *thyrotropin,* regulates the hormonal activity of the thyroid gland. The secretion of TSH, however, is partly regulated by the hypothalamus through the secretion of *thyrotropin releasing factor (TRF).* External factors may influence the release of TSH as well. Exposure to cold, certain illnesses, and emotional stress may trigger an increased output of TSH.
3. **Adrenocorticotropic hormone (ACTH).** ACTH promotes normal functioning of the adrenal cortex. It also acts on all body cells by assisting in the breakdown of fats. The release of ACTH is controlled by a *corticotropin releasing factor (CRF)* produced in the hypothalamus. As with TSH, stress further influences the release of ACTH.
4. **Follicle-stimulating hormone (FSH).** In the male, FSH stimulates the testes to produce sperm. In the female, FSH regulates the monthly development of the follicle and egg. It also stimulates the secretion of the female sex hormone *estrogen.*
5. **Luteinizing hormone (LH).** LH works with FSH, and together they are referred to as *gonadotrophins,* which means their target cells are within the gonads or reproductive organs. In the female, LH works with FSH in bringing about ovulation. It also stimulates the formation of the corpus luteum and the production of another female sex hormone, *progesterone* (see chapter 21).

    In the male, the luteinizing hormone is called *interstitial-cell-stimulating hormone (ICSH)* and stimulates the interstitial cells of the testes to develop and secrete the male sex hormone testosterone (see chapter 20).

    The mechanism that controls the production and release of gonadotrophic hormones is not well understood. It is known, however, that following puberty (see chapters 20 and 21) the hypothalamus releases a hormone called *gonadotrophin-releasing hormone (GnRH)* that regulates the secretion of both LH and FSH.
6. **Prolactin.** Prolactin is secreted in both males and females, but it functions primarily in females after parturition. Prolactin assists other hormones in initiating and sustaining milk production by the mammary glands. The hypothalamus plays an important role in the release of this hormone through the production of *prolactin release-inhibiting factor (PR-IF).* When PR-IF is secreted, the secretion of prolactin is inhibited; when PR-IF is not secreted, prolactin is released.

UNIT V

7. **Melanocyte-stimulating hormone (MSH).** MSH increases skin pigmentation by stimulating the dispersal of melanin granules within melanocytes. Secretion of MSH is regulated by *melanocyte-stimulating hormone releasing factor (MRF)* and *melanocyte-stimulating hormone inhibiting factor (MIF),* which both come from the hypothalamus.

8. **Oxytocin.** Oxytocin is produced by specialized cells in the hypothalamus. It then travels through axons in the infundibulum to the lobus nervosa, where it is stored and released in response to neural impulses from the hypothalamus. Oxytocin influences physiological activity in the female reproductive system but has no known function in males. Oxytocin is released near the end of gestation and causes uterine contractions during labor. It also has an effect on the mammary glands, causing contractions in the tissue surrounding the alveoli and lactiferous ducts (see chapter 21). This results in the letdown of milk in the mammary glands and is necessary before milk can be removed by an infant's sucking.

Injections of oxytocin may be given to a woman during labor if she is having difficulties in parturition. Increased amounts of oxytocin assist uterine contractions and generally speed up delivery. Oxytocin administration after parturition causes the uterus to regress in size and squeezes the blood vessels, thus minimizing the danger of hemorrhage.

9. **Antidiuretic hormone.** Antidiuretic hormone (ADH) is similar to oxytocin in its site of production and release. Like oxytocin, it is a polypeptide. The major function of ADH is to act on the kidneys to inhibit the formation of urine, or more specifically, to reduce the amount of water excreted from the kidneys. This hormone also acts as a *vasopressin* to constrict blood vessels, which increases blood pressure.

Diabetes insipidus results from a marked decrease in ADH output caused by trauma or disease to the hypothalamus or neurohypophysis. The symptoms of this disease are polyuria (voiding excessive dilute urine), concentrated body fluids with dehydration, and a particularly heightened sensation of thirst.

Oxytocin and antidiuretic hormone are released by the posterior pituitary (pars nervosa of the neurohypophysis). These two hormones, however, are actually produced in neuron cell bodies of the **supraoptic nuclei** and **paraventricular nuclei** of the hypothalamus. These nuclei within the hypothalamus are thus endocrine glands; the hormones they produce are transported along axons of the **hypothalamo-hypophyseal tract** (fig. 14.11) to the posterior pituitary, which stores and later releases these hormones. The posterior pituitary is thus more of a storage organ than a true gland.

The secretion of oxytocin and ADH from the posterior pituitary is controlled by **neuroendocrine reflexes.** In nursing mothers, for example, the stimulus of sucking acts via sensory nerve impulses to the hypothalamus to stimulate the reflex se-

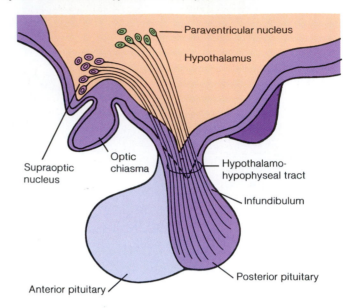

Figure 14.11 The posterior pituitary, or neurohypophysis, stores and secretes oxytocin and antidiuretic hormone produced in neuron cell bodies within the supraoptic and paraventricular nuclei of the hypothalamus. These hormones are transported to the posterior pituitary by nerve fibers of the hypothalamo-hypophyseal tract.

cretion of oxytocin. The secretion of ADH is stimulated by osmoreceptor neurons in the hypothalamus in response to a rise in blood osmotic pressure; its secretion is inhibited by sensory impulses from stretch receptors in the left atrium of the heart in response to a rise in blood volume.

At one time, the anterior pituitary was called the master gland because it secretes hormones that regulate some other endocrine glands (see table 14.4 and fig. 14.12). Adrenocorticotrophic hormone (ACTH), thyroid-stimulating hormone (TSH), and the gonadotrophic hormones (FSH and LH) stimulate the adrenal cortex, thyroid, and gonads, respectively, to secrete their hormones. The anterior pituitary hormones also have a trophic effect on their target glands in that the structure and health of the target glands depend on adequate stimulation by anterior pituitary hormones. The anterior pituitary, however, is not really the master gland, since secretion of its hormones is in turn controlled by hormones secreted by the hypothalamus.

Releasing and inhibiting hormones from the hypothalamus travel through the **hypothalamo-hypophyseal portal system** to control the secretion of hormones from the anterior pituitary. Neurons in the hypothalamus secrete hormones into the region of the **median eminence,** where they enter a network of primary capillaries (fig. 14.13). Venous drainage through the pituitary stalk transports the hypothalamic hormones to a network of secondary capillaries within the anterior pituitary. This system is considered a *portal system* because there are two sets of capillaries in a series (fig. 14.13). (This is analogous to the hepatic portal system that delivers venous blood from the pancreas, spleen, and GI tract to the liver, as described in chapter 16.)

Table 14.4	**Hormones Released by the Pituitary Gland**	
Hormone	**Action**	**Source of Regulation**
Adenohypophysis		
Growth hormone (GH), or somatotropin	Regulates mitotic activity and growth of body cells; promotes movement of amino acids through cell membranes	Growth hormone releasing factor (GRF) and growth hormone release-inhibiting factor (GR-IH) from the hypothalamus
Thyroid-stimulating hormone (TSH), or thyrotropin	Regulates hormonal activity of thyroid gland	Thyrotropin releasing factor (TRF) from the hypothalamus
Adrenocorticotropic hormone (ACTH)	Controls secretion of certain hormones from the adrenal cortex; assists in breakdown of fats	Corticotropin releasing factor (CRF) from the hypothalamus
Follicle-stimulating hormone (FSH)	In males, stimulates production of sperm cells; in females, regulates follicle development in ovary and stimulates secretion of estrogen	Gonadotrophin-releasing hormone (GnRH) from the hypothalamus
Luteinizing hormone (LH), or ICSH in males	Promotes secretion of sex hormones; in females, plays role in release of ovum and stimulates formation of corpus luteum and production of progesterone; in males, stimulates testosterone secretion	Gonadotrophin-releasing hormone (GnRH) from the hypothalamus
Prolactin	Promotes secretion of milk from mammary glands (lactation)	Hypothalamus through production of prolactin release-inhibiting factor (PR-IF)
Melanocyte-stimulating hormone (MSH)	Stimulates pigmentation within the melanocytes of the skin	Melanocyte-stimulating hormone releasing factor (MRF) and melanocyte-stimulating hormone inhibiting factor (MIF), both from the hypothalamus
Neurohypophysis		
Oxytocin	Stimulates contractions of muscles in uterine wall; causes contraction of muscles in mammary glands	Hypothalamus in response to stretch in uterine walls and stimulation of breasts
Antidiuretic hormone	Reduces water loss from kidneys; elevates blood pressure	Hypothalamus in response to changes in water concentration in the blood

UNIT V

1. List the hormones secreted by the anterior pituitary and describe how the hypothalamus controls the secretion of each hormone.
2. Which hormone secreted by the anterior pituitary does not affect another endocrine gland?
3. List the hormones released by the posterior pituitary. Describe the site of origin of these hormones and the mechanisms by which their secretions are regulated.
4. Which of the hormones secreted by the pituitary gland affect only females?
5. How does the hypothalamus regulate hormonal secretion from the pituitary gland?

Thyroid and Parathyroid Glands

The thyroid gland secretes thyroxine and triiodothyronine, which function in the regulation of energy metabolism. These hormones are critically important for proper growth and development. The thyroid also secretes calcitonin, which may antagonize the action of parathyroid hormone in the regulation of calcium and phosphate balance.

Objective 10. Describe the location and structure of the thyroid gland and list the actions of the thyroid hormones.

Objective 11. Describe the location and structure of the parathyroid glands and list the actions of parathyroid hormone.

Description of the Thyroid Gland

The **thyroid gland** is positioned just below the larynx (fig. 14.14). This gland consists of two lobes positioned on either lateral side of the trachea and connected anteriorly by a broad **isthmus.** The thyroid is the largest of the endocrine glands, weighing between 20 and 25 g. It receives an abundant blood supply (80–120 mℓ/min) through the paired superior thyroid branches of the external carotid arteries and the paired inferior thyroid branches of the subclavian arteries. The venous return is through the paired superior and middle thyroid veins that pass into the internal jugular veins and through the inferior thyroid veins that empty into the brachiocephalic veins.

thyroid: Gk. *thyreos*, oblong shield
isthmus: L. *isthmus*, narrow portion

Figure 14.12 The hormones secreted by the anterior pituitary and the target organs for those hormones.

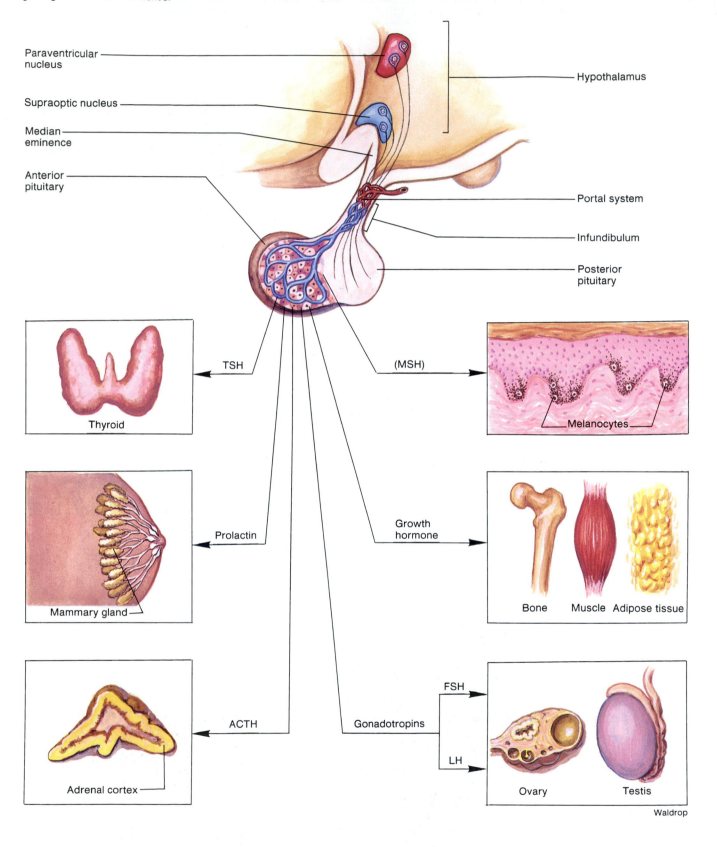

Paraventricular nucleus

Supraoptic nucleus

Median eminence

Anterior pituitary

Hypothalamus

Portal system

Infundibulum

Posterior pituitary

TSH

Thyroid

(MSH)

Melanocytes

Prolactin

Mammary gland

Growth hormone

Bone   Muscle   Adipose tissue

ACTH

Adrenal cortex

Gonadotropins

FSH

LH

Ovary   Testis

Waldrop

Figure 14.13   The hypothalmo-hypophyseal portal system. Hypothalmic hormones (shown as dots) enter this system in the first set of capillaries of the median eminence and are transported through the portal veins of the pituitary stalk to a second set of capillaries of the anterior pituitary.

Nuclei of the hypothalamus

Axon to first set of capillaries

Primary capillaries

Superior hypophyseal artery

Portal venules

Anterior pituitary

Second set of capillaries

Median eminence

Pituitary stalk

Posterior pituitary

Waldrop

Figure 14.14   The thyroid gland. (*a*) Its relationship to the larynx and trachea and (*b*) a scan of the thyroid gland 24 hours after the intake of radioactive iodine.

Thyroid cartilage of larynx

Cricoid cartilage of larynx

Thyroid gland

Trachea

Gordon

(a)

(b)

UNIT V

Figure 14.15 The histology of the thyroid gland showing numerous thyroid follicles. Each follicle consists of follicular cells surrounding the fluid known as colloid.

Follicular cells

Colloid

Table 14.5	Hormones of the Thyroid Gland	
Hormone	**Action**	**Source of Regulation**
Thyroxine (T_4)	Increases rate of protein synthesis and rate of energy release from carbohydrates; regulates rate of growth; stimulates maturity of nervous system	Hypothalamus and release of TSH from adenohypophysis of the pituitary gland
Triiodothyronine (T_3)	Same as above	Same as above
Calcitonin (thyrocalcitonin)	Lowers blood calcium by inhibiting the release of calcium from bone tissue	Calcium levels in blood

On a microscopic level, the thyroid gland consists of many spherical hollow sacs called **thyroid follicles** (fig. 14.15). These follicles are lined with a simple cuboidal epithelium composed of **follicular cells** that synthesize the two principal thyroid hormones (see table 14.5). The interior of the follicles contains **colloid** *(kol'oid)*, a protein-rich fluid. Between the follicles are epithelial cells called **parafollicular cells** that produce a hormone called *calcitonin* (or *thyrocalcitonin*).

The thyroid is innervated by postganglionic neurons from the superior and middle cervical sympathetic ganglia and preganglionic neurons from ganglia derived from the second through the seventh thoracic segment of the spinal cord.

Functions of the Thyroid Gland

The thyroid gland produces two major hormones, *thyroxine (T_4)* and *triiodothyronine (T_3)*, and the minor hormone *calcitonin (thyrocalcitonin)*. The release of thyroxine and triiodothyronine is controlled by the hypothalamus and by the TSH secreted from the adenohypophysis of the pituitary gland. Thyroxine and triiodothyronine are stored in the thyroid follicles and released as needed to control the metabolic rate of the body. More specifically, they act to increase the rate of protein synthesis and the rate of energy release from the carbohydrates. They also regulate the rate of growth in young persons and are associated with sexual maturity and early maturation of the nervous system.

Iodine is the most common component of thyroxine and triiodothyronine, and a continual intake of iodine is essential for normal thyroid function. Seafood contains adequate amounts of iodine, and commercial salt generally has iodine as an additive. Absorbed iodine is transported through the blood to the thyroid gland, where an active transport mechanism called an iodine pump moves the iodides into the follicle cells. Here, they combine with amino acids in the synthesis of thyroid hormones.

colloid: Gk. *kolla*, glue

Most of the thyroxine in blood is attached to carrier proteins. Only the very small percentage of thyroxine that is free in blood plasma can enter the target cells. In the target cells, it is converted to triiodothyronine and attached to nuclear receptor proteins. Through the activation of genes, thyroid hormones stimulate protein synthesis, promote maturation of the nervous system, and increase the rate of energy utilization by the body.

Thyroid-stimulating hormone (TSH) from the anterior pituitary stimulates the thyroid to secrete thyroxine and exerts a trophic effect on the thyroid gland. This trophic effect shows up dramatically in people who develop an *iodine-deficiency (endemic) goiter*. In the absence of sufficient dietary iodine, the thyroid cannot produce adequate amounts of T_4 and T_3. The resulting lack of negative feedback inhibition causes abnormally high levels of TSH secretion, which in turn stimulate the abnormal growth of the thyroid (a goiter). These events are summarized in figure 14.16.

Calcitonin is a polypeptide hormone produced by the parafollicular cells. It functions to regulate calcium levels in the blood by inhibiting the rate at which calcium leaves bone tissue.

Parathyroid Glands

The small, flattened **parathyroid glands** are embedded in the posterior surfaces of the lateral lobes of the thyroid gland (fig. 14.17). There are usually four parathyroid glands: a **superior** and an **inferior pair.** Each parathyroid gland is a small yellowish brown body measuring 3–8 mm (0.1–0.3 in.) in length, 2–5 mm (.07–0.2 in.) in width, and about 1.5 mm (0.05 in.) in depth.

On a microscopic level, the parathyroids are composed of two types of epithelial cells (fig. 14.18). The cells that synthesize parathyroid hormone are called **principal cells** and are scattered among **oxyphil** *(ok'sĕ-fil)* **cells.** Oxyphil cells support the principal cells and are believed to produce reserve quantities of parathyroid hormone.

The blood supply and drainage of the parathyroid glands is similar to that of the thyroid gland. The innervation of the

Figure 14.16   The mechanism of goiter formation in iodine deficiency. Low negative feedback inhibition results in excessive TSH secretion, which stimulates abnormal growth of the thyroid.

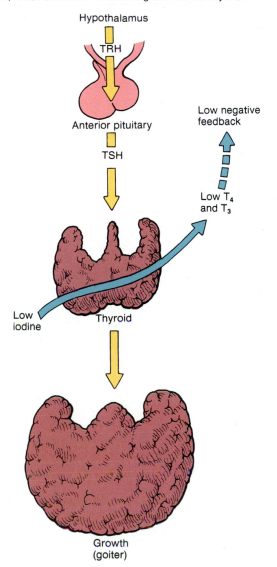

Figure 14.17   A posterior view of the parathyroid glands.

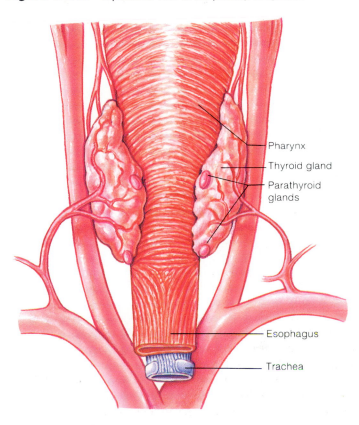

Figure 14.18   The histology of the parathyroid gland.

parathyroids, however, is a bit different. The parathyroids receive neurons from the pharyngeal branches of the vagus nerves in addition to neurons arising from the cervical sympathetic ganglia.

The parathyroid glands secrete one hormone called **parathyroid hormone (PTH).** This hormone promotes a rise in blood calcium levels by acting on the bones, kidneys, and large intestine (fig. 14.19).

1. Describe the location and structure of the thyroid gland and list the effects of thyroid hormones.
2. What may be the metabolic consequences of an overactive thyroid gland?
3. Why is a continual supply of iodine important for body metabolism?
4. Describe the location and structure of the parathyroid glands and identify the target organs of parathyroid hormone.

Pancreas

The pancreatic islets in the pancreas secrete two hormones, insulin and glucagon, which are critically involved in the regulation of blood sugar levels in the body.

Objective 12.   Describe the structure of the endocrine portion of the pancreas and the origin of insulin and glucagon.

Objective 13.   Describe the actions of insulin and glucagon.

UNIT V

Figure 14.19 Actions of parathyroid hormone. An increased level of parathyroid hormone causes the bones to release calcium, the kidneys to conserve calcium that is lost through the urine, and the small intestine to absorb calcium. Negative feedback of increased calcium levels in the blood inhibits the secretion of this hormone.

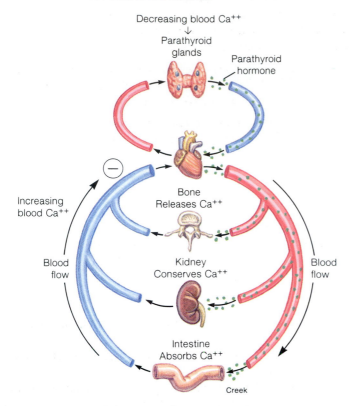

Decreasing blood Ca++
↓
Parathyroid glands

Parathyroid hormone

Increasing blood Ca++

Blood flow

Bone Releases Ca++

Kidney Conserves Ca++

Blood flow

Intestine Absorbs Ca++

Creek

Description of the Pancreas

The **pancreas** is both an endocrine and an exocrine gland. The gross structure of this gland and its exocrine functions in digestion are described in chapter 18. The endocrine portion of the pancreas consists of scattered clusters of cells called **pancreatic islets** (islets of Langerhans). These endocrine structures are most common in the body and tail of the pancreas (fig. 14.20) and principally consist of two types of secretory cells called **alpha cells** and **beta cells** (fig. 14.21).

Functions of the Pancreas

The endocrine function of the pancreas is to produce and secrete the hormones *glucagon* and *insulin.* The alpha cells of the pancreatic islets secrete glucagon and the beta cells secrete insulin.

Glucagon stimulates the liver to convert glycogen into glucose, which causes the blood glucose level to rise. Apparently, the alpha cells themselves regulate the glucagon output by monitoring the blood glucose level through a feedback mechanism. If for some reason alpha cells secrete glucagon continuously, high blood sugar, or *hyperglycemia,* may result.

Insulin has a physiological function opposite to that of glucagon: it decreases the level of blood sugar. Insulin promotes

pancreas: Gk. *pan,* all; *kreas,* flesh
islets of Langerhans: from Paul Langerhans, German anatomist, 1847–88
insulin: L. *insula,* island

Figure 14.20 The pancreas and the associated pancreatic islets.

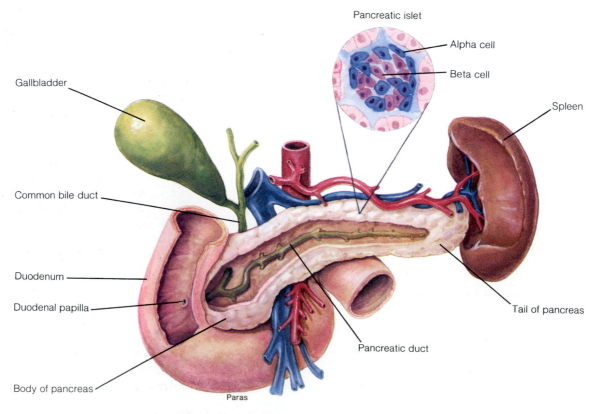

Pancreatic islet
Alpha cell
Beta cell
Spleen
Gallbladder
Common bile duct
Duodenum
Duodenal papilla
Body of pancreas
Tail of pancreas
Pancreatic duct
Paras

Figure 14.21   The histology of the pancreatic islets.

Table 14.6	Hormones of the Pancreas	
Hormone	**Action**	**Source of Regulation**
Glucagon	Stimulates the liver to convert glycogen to glucose, causing the blood glucose level to rise	Blood glucose level through negative feedback in pancreas
Insulin	Promotes movement of glucose through cell membranes; stimulates the liver to convert glucose to glycogen; promotes the transport of amino acids to cells; assists in synthesis of proteins and fats	Blood glucose level through negative feedback in pancreas

the movement of glucose through cell membranes, especially in muscle and adipose cells. As the glucose enters the cells, the sugar level of the blood decreases. Other functions of insulin include stimulating muscle and liver cells to convert glucose to glycogen, helping amino acids to enter cells, and assisting the synthesis of proteins and fats. Failure of beta cells to produce insulin causes the common hereditary disease *diabetes mellitus.*

The action of the hormones from the pancreatic islets is summarized in table 14.6.

1. Describe specific sites of glucagon and insulin production.
2. What is the function of glucagon? of insulin?
3. What disease is caused by the insufficient production of insulin?

Adrenal Glands

The adrenal cortex and adrenal medulla are structurally and functionally different. The adrenal medulla secretes catecholamine hormones that complement the action of the sympathetic division of the ANS. The adrenal cortex secretes corticosteroids that function in the regulation of mineral balance, energy balance, and reproductive activity.

Objective 14.   Describe the location of the adrenal glands and distinguish between the adrenal cortex and the adrenal medulla.

Objective 15.   List the hormones secreted by the adrenal glands and discuss their effects in the body.

Description of the Adrenal Glands

The **adrenal glands** (also called **suprarenal glands**) are paired organs that cap the superior borders of the kidneys (fig. 14.22). The adrenal glands, along with the kidneys, are retroperitoneal and are embedded against the muscles of the back in a protective pad of fat.

Each adrenal gland is about 50 mm (2 in.) in length, 30 mm (1.1 in.) in width, and 10 mm (0.4 in.) in depth and is generally pyramidal in appearance. Each consists of an outer adrenal cortex and inner adrenal medulla (figs. 14.22 and 14.23) that function as separate glands.

The **adrenal cortex** makes up the bulk of the gland and is histologically subdivided into three zones: an outer **zona glomerulosa** (glo-mer''yoo-lo'să), an intermediate **zona fasciculata** (fă-sik''yoo-lă'tă), and an inner **zona reticularis.** The **adrenal medulla** is composed of tightly packed clusters of **chromaffin** (kro-maf'in) **cells,** which are arranged around blood vessels. Each cluster of chromaffin cells receives direct autonomic innervation.

Like other endocrine glands, the adrenal glands are highly vascular. Three separate suprarenal arteries supply blood to each adrenal gland. One arises from the inferior phrenic artery, another from the aorta, and a third is a branch of the renal artery. The venous drainage goes through a suprarenal vein into the inferior vena cava for the right adrenal gland and through the suprarenal vein into the left renal vein for the left adrenal gland.

The adrenal glands are innervated by preganglionic neurons of the splanchnic nerves and by fibers of the celiac and associated sympathetic plexuses.

Functions of the Adrenal Glands

Over 30 hormones have been identified as being produced by the adrenal cortex. These hormones are called **corticosteroids** (kor''ti-ko-ste'roids), or **corticoids,** for short. The adrenal corticoids are grouped into three types: mineralocorticoids, glucocorticoids, and gonadocorticoids.

The *mineralocorticoids* are produced by the zona glomerulosa of the adrenal cortex and regulate the concentrations of extracellular electrolytes. Of the three hormones secreted by

adrenal: L. *ad,* to; *renes,* kidney

UNIT V

Figure 14.22 The structure of the adrenal gland. (*a*) The gross structure and (*b*) the histological structure showing the three zones of the adrenal cortex.

(a)

Surface of adrenal gland

Adrenal cortex
- Zona glomerulosa
- Zona fasciculata
- Zona reticularis

Adrenal medulla

Loechel

(b)

Figure 14.23 The histology of the adrenal gland (100×).

ADRENAL CORTEX

Capsule
Adrenal cortex
Adrenal medulla
Zona glomerulosa
Zona fasciculata
Zona reticularis

Cortisol and related anti-inflammatory compounds are commonly used to treat patients suffering from arthritis and various allergies. They are also frequently used in sports medicine to treat traumatized joints and in organ transplants and tissue grafts to help keep the host from rejecting donor tissues. Unfortunately, cortisol inhibits the regeneration of connective tissues, and therefore should be used sparingly and only when necessary.

Gonadocorticoids are the sex hormones that are secreted by the zona reticularis of the adrenal cortex. The majority of these hormones are adrenal *androgens,* but small quantities of adrenal *estrogens* and *progesterones* are produced as well. It is thought that these hormones supplement the sex hormones produced in the gonads. There is also evidence that the concentration of androgens in both the male and female plays a role in determining the sex drive.

A summary of the hormones produced by the adrenal cortex is presented in table 14.7.

The chromaffin cells of the adrenal medulla produce two closely related hormones: *epinephrine* and *norepinephrine.* Both of these hormones are classified as amines (more specifically, as catecholamines) because they contain amine groups (see fig. 14.3).

The effects of these hormones are similar to those caused by stimulation of the sympathetic division of the ANS, except that the hormonal effect lasts about 10 times longer. The hormones from the adrenal medulla increase cardiac output and heart rate, dilate coronary blood vessels, increase mental alertness, increase the respiratory rate, and elevate metabolic rate. A comparison of the effects of epinephrine and norepinephrine is presented in table 14.8.

this layer, *aldosterone* is the most important. Aldosterone affects the kidneys, causing them to reabsorb sodium and increase potassium excretion. At the same time, it promotes water reabsorption and reduces urine output.

The *glucocorticoids* are produced primarily by the zona fasciculata of the adrenal gland and influence the metabolism of carbohydrates, proteins, and fats. The glucocorticoids also help the body resist stress, promote vasoconstriction, and act as anti-inflammatory compounds. The most abundant and physiologically important glucocorticoid is *cortisol (hydrocortisone).*

Table 14.7	**Hormones of the Adrenal Cortex**	
Hormone	**Action**	**Source of Regulation**
Mineralocorticoids	Regulate the concentration of extracellular electrolytes, especially sodium and potassium	Electrolyte concentration in blood
Glucocorticoids	Influence the metabolism of carbohydrates, proteins, and fats; promote vasoconstriction; act as anti-inflammatories	ACTH from the adenohypophysis of the pituitary gland in response to stress
Gonadocorticoids	Supplement the sex hormones from the gonads	

Table 14.8	**Comparison of the Hormones From the Adrenal Medulla**
Epinephrine	**Norepinephrine**
Elevates blood pressure because of increased cardiac output and peripheral vasoconstriction	Elevates blood pressure because of generalized vasoconstriction
Accelerates respiratory rate and dilates respiratory passageways	Similar effect but to a lesser degree
Increases efficiency of muscular contraction	Similar effect but to a lesser degree
Increases rate of glycogen breakdown into glucose, so level of blood glucose rises	Similar effect but to a lesser degree
Increases rate of fatty acid released from fat, so level of blood fatty acids rises	Similar effect but to a lesser degree
Increases release of ACTH and TSH from the adenohypophysis of the pituitary gland	No effect

The adrenal medulla is innervated by sympathetic neurons. The impulses are initiated from the hypothalamus via the spinal cord in response to stress. Stress therefore activates the adrenal medulla, as well as the adrenal cortex. Activation of the adrenal medulla prepares the body for greater physical performance—the *fight-or-flight* response (fig. 14.24).

Excessive stimulation of the adrenal medulla can result in depletion of the body's energy reserves, and high levels of corticosteroid secretion from the adrenal cortex can significantly impair the immune system. It is reasonable to expect, therefore, that prolonged stress can result in increased susceptibility to disease. Indeed, many studies show that prolonged stress results in an increased incidence of cancer and other diseases.

1. Describe the location and appearance of the adrenal gland. Diagram and label the layers of the adrenal gland that can be seen in a sagittal section as, for example, in a histological slide.
2. List the hormones of the adrenal medulla and describe their effects.
3. List the categories of corticosteroids and the zones of the adrenal cortex that secrete these hormones.

Gonads and Other Endocrine Glands

The gonads produce sex hormones that control the development and function of the male and female reproductive systems. Additionally, many other organs secrete hormones that help regulate digestion, metabolism, growth, and immunity.

Objective 16. Discuss the endocrine functions of the gonads.

Objective 17. Describe the structure and anatomical location of the pineal and thymus glands and indicate their endocrine functions.

Gonads

The **gonads** are the male and female primary sex organs. The male gonads are called **testes** and the female gonads are called **ovaries.** They are mixed glands in that they produce both sex hormones and sex cells, or **gametes** (see chapters 20 and 21).

Testes   The **interstitial cells** of the testes produce and secrete the male sex hormone *testosterone.* Testosterone controls the development and function of the male secondary sex organs, such as the penis and accessory glands. It also promotes the male secondary sex characteristics (see chapter 20) and somewhat determines the sex drive.

Ovaries   The endocrine function of the ovaries is the production of the female sex hormones, *estrogens* and *progesterone.* Estrogens are produced in the **ovarian (graafian) follicles** and **corpus luteum** of the ovaries. They are also produced in the placenta, adrenal cortex, and even the testes of the male. Estrogens are responsible for (1) development and function of the secondary sex organs, (2) menstrual changes of the uterus, (3) development of the female secondary sex characteristics (see chapter 21), and (4) regulation of the sex drive.

Progesterone is produced by the corpus luteum and is primarily associated with pregnancy in preparing the uterus for implantation and preventing abortion of the fetus.

UNIT V

Figure 14.24 Stress is the physiological change in the body that prepares it for fight or flight. These changes are elicited by neural impulses from the hypothalamus that directly influence body tissue and by impulses to the adrenal medulla that secondarily influence the same body tissues through the release of epinephrine.

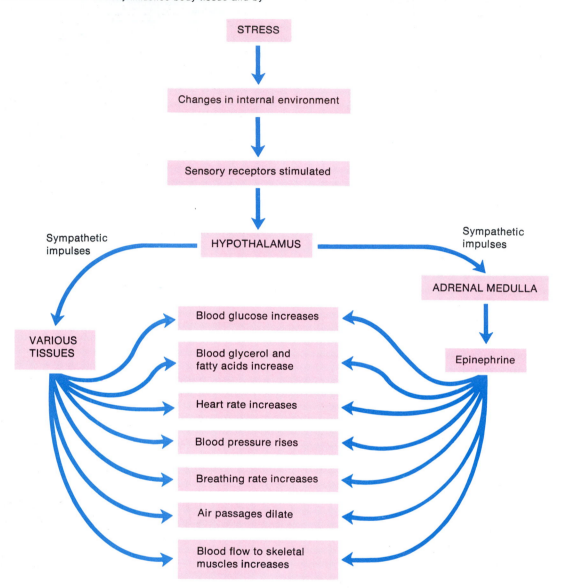

Most cultures of the world practice birth control, or contraception, in one form or another. It has a long history, dating back to the ancient Egyptians who used various substances to inhibit sperm survival and motility. In the age of hormonal biochemistry, birth control techniques have become increasingly sophisticated. The female, rather than the male, has been the target of hormonal birth control techniques for the following reasons: (1) ovulation is cyclic; (2) the genetic structure of each ovum is established by the time of the female's birth, whereas sperm production is a continuous process in the male and is therefore more vulnerable to genetic damage; (3) the female system has more potential sites for hormonal interference than does the male; and (4) the female is usually more conscientious about birth control because she has far more invested in pregnancy than does the male.

Pineal Gland

The small, cone-shaped **pineal gland,** or **pineal body** (see fig. 14.2), is located in the roof of the third ventricle, near the corpora quadrigemina, where it is encapsulated by the meninges covering the brain. The pineal gland of a child weighs about 0.2 g and is 5–8 mm (0.2–0.3 in.) long and 9 mm wide. The gland begins to regress in size at about the age of 7, and in the adult appears as a thickened strand of fibrous tissue. Histologically speaking, the pineal gland consists of specialized parenchymal and neuroglial cells. Although the pineal gland lacks direct nervous connection to the rest of the brain, it is highly innervated by the sympathetic division of the ANS from the superior cervical ganglion.

pineal: L. *pinea,* pine cone

Table 14.9 — Summary of the Physiological Effects of Gastrointestinal Hormones

Hormone	Secreted By	Effects
Gastrin	Stomach	Stimulates parietal cells to secrete HCl Stimulates chief cells to secrete pepsinogen Maintains structure of gastric mucosa
Secretin	Small intestine	Stimulates water and bicarbonate secretion in pancreatic juice Potentiates actions of cholecystokinin on pancreas
Cholecystokinin (CCK)	Small intestine	Stimulates contraction of the gallbladder Stimulates secretion of pancreatic juice enzymes Potentiates action of secretin on pancreas Maintains structure of exocrine pancreas (acini)
Gastric inhibitory peptide (GIP)	Small intestine	Inhibits gastric emptying Inhibits gastric acid secretion Stimulates secretion of insulin from pancreatic islets

From Stuart Ira Fox, *A Laboratory Guide to Human Physiology: Concepts and Clinical Applications*, 6th ed. Copyright © 1993 Wm. C. Brown Communications, Inc., Dubuque, Iowa. All Rights Reserved. Reprinted by permission.

The function of the pineal gland in some other vertebrates is well known but is not well understood in humans. There is evidence that it secretes a hormone called *melatonin*. Melatonin is thought to affect the hypothalamus by stimulating the secretion of certain releasing factors. These factors in turn affect the secretion of gonadotrophin and the ACTH from the adenohypophysis of the pituitary gland.

Thymus

The **thymus** is a bilobed organ positioned in the upper mediastinum, in front of the aorta and behind the manubrium of the sternum (fig. 14.25). Although the size of the thymus varies considerably from person to person, it is relatively large in newborns and children and then sharply regresses in size after puberty. Besides decreasing in size, the thymus of adults becomes infiltrated with strands of fibrous and fatty connective tissue.

The principal function of the thymus is associated with the lymphatic system (see chapter 16) in maintaining body immunity through the maturation and discharge of a specialized group of lymphocytes called *T-cells (thymus-dependent cells)*. The thymus also secretes a hormone called *thymosin*, which is believed to stimulate the T-cells after they leave the thymus.

Stomach and Small Intestine

Certain cells of the mucosal linings of the stomach and small intestine secrete hormones that promote digestive activities. These hormones are summarized in table 14.9.

Placenta

The **placenta** is the organ responsible for nutrient and waste exchange between the fetus and the mother (see chapter 22). The placenta is also an endocrine gland; it secretes large amounts of estrogens and progesterone, as well as a number of polypeptide and protein hormones that are similar to some hormones secreted by the anterior pituitary gland. These latter hormones include *human chorionic gonadotrophin (hCG)*, which is similar to LH, and *somatomammotrophin*, which is similar in action to

Figure 14.25 The thymus is a bilobed organ within the mediastinum of the thorax.

- Larynx
- Thyroid gland
- Trachea
- Thymus
- Lung
- Pericardium
- Heart

UNIT V

growth hormone and prolactin. Detection of hCG in urine is an indication of pregnancy and is the basis of a home pregnancy test.

1. Where is testosterone produced? What are its functions?
2. What are the functions of estrogens and progesterone? Where are they produced?
3. Describe the location of the pineal gland and discuss the possible function of melatonin.
4. Describe the location and function of the thymus.

thymus: Gk. *thymos*, a warty excrescence
placenta: L. *placenta*, flat cake

Developmental Exposition

The Endocrine System

Exhibit I   The development of the pituitary gland. (*a*) The head end of an embryo at 4 weeks showing the position of a midsagittal cut seen in the developmental sequence (*b–e*). The pituitary gland arises from a specific portion of the neuroectoderm, called the neurohypophyseal bud, which evaginates downward during the 4th and 5th weeks respectively in (*b*) and (*c*), and from a specific portion of the oral ectoderm, called the hypophyseal (Rathke's) pouch, which evaginates upward from a specific portion of the primitive oral cavity. At 8 weeks (*d*), the hypophyseal pouch is no longer connected to the pharyngeal roof of the oral cavity. During the fetal stage (*e*), the development of the pituitary gland is completed.

Waldrop

Explanation

The endocrine system is the only anatomical body system whose organs are not structurally connected. Because these organs are isolated from each other and are distributed throughout the body, each endocrine organ has a separate and independent development. All three embryonic germ layers (endoderm, mesoderm, and ectoderm) contribute to the development of the endocrine system.

The following sections describe, in turn, the development of the pituitary, thyroid, pancreas, and adrenal glands. The development of the testes and ovaries is discussed in chapters 20 and 21, respectively.

Pituitary Gland

Although the **pituitary gland** is a single organ, it is actually composed of two distinct types of tissues that have different embry-

onic origins. These two types of tissues release different hormones and are under different control systems. The anterior portion of the hypophysis, called the **adenohypophysis** (*ad''ĕ-no-hi-pof'ĭ-sis*), develops from ectoderm that lines the primitive oral cavity. The posterior portion, called the **neurohypophysis,** develops from neuroectoderm of the developing brain (exhibit I).

The adenohypophysis begins to develop during the third week as a diverticulum, a pouchlike extension, called the **hypophyseal** (Rathke's) **pouch.** It arises from the roof of the primitive oral cavity and grows toward the brain. At the same time, another diverticulum called the **infundibulum** forms from the diencephalon on the inferior aspect of the brain. As the two diverticula come in contact, the hypophyseal pouch loses its connection with the oral cavity and the primordial tissue of the

Rathke's pouch: from Martin H. Rathke, German anatomist, 1793–1860

Exhibit II The embryonic development of the thyroid gland. (*a*) At 4 weeks, a thyroid diverticulum begins to form in the floor of the pharynx at the level of the second branchial arch. (*b,c*) The thyroid diverticulum extends downward during the 5th and 6th weeks. (*d*) A sagittal section through the head and neck of an adult shows the path of development and final position of the thyroid gland.

(a)

Pharyngeal pouches

Thyroid diverticulum

Branchial arches

Laryngotracheal diverticulum

Esophagus

(c)

Foramen cecum

Tongue

Thyroglossal duct

Thyroid gland

Trachea

Lung buds

(b)

Foramen cecum

Primitive oral cavity

Thyroglossal duct

Thyroid diverticulum

(d)

Tongue

Larynx

Thyroid gland

Trachea

Foramen cecum

Developmental pathway

Esophagus

Waldrop

UNIT V

adenohypophysis is formed. The fully developed adenohypophysis includes the **pars distalis, pars tuberalis,** and **pars intermedia.**

The neurohypophysis develops as the infundibulum extends inferiorly from the diencephalon to come in contact with the developing adenohypophysis. The fully formed neurohypophysis consists of the infundibulum and the **lobus nervosa.** Specialized nerve fibers that connect the hypothalamus with the lobus nervosa develop within the infundibulum.

Notice that the neurohypophysis is essentially an extension of the brain, and indeed the entire pituitary gland, like the brain, is surrounded by the meninges. The adenohypophysis, by contrast, is derived from nonneural tissue (the same embryonic tissue that will form the epithelium over the roof of the mouth). These different embryologies have important consequences with regard to the functions of the two parts of the pituitary gland.

Thyroid Gland

The **thyroid gland** is derived from endoderm and begins its development during the third week as a thickening in the floor of the primitive pharynx. The thickening soon evaginates (out-

pouches) downward as the **thyroid diverticulum** (exhibit II). A narrow **thyroglossal duct** connects the descending primordial thyroid tissue to the pharynx. As the descent continues, the tongue starts to develop, and the opening into the thyroglossal duct, called the **foramen cecum** pierces through the base of the tongue. By the seventh week, the thyroid gland occupies a position immediately inferior to the larynx and around the front and lateral sides of the trachea. At this time, the thyroglossal duct disappears and the foramen cecum regresses in size to a vestigial pit that persists throughout life.

Pancreas

The **pancreas** begins development during the fifth week, as dorsal and ventral pancreatic buds of endoderm arise from the caudal portion of the foregut. These primordial pancreatic tissues continue to grow independently until the ventral bud is carried dorsally and fuses with the dorsal bud as the duodenum rotates to the right. The fusion of the two portions of the pancreas occurs during the seventh week.

foramen cecum: L. *foramen,* open passage; *caecum,* blind tube

Continued on next page

Developmental Exposition

Continued

The pancreas develops from and maintains connections with the small intestine (via the pancreatic duct). The exocrine products of the pancreas, contained in *pancreatic juice*, are channeled through the pancreatic duct to the small intestine. The endocrine parts of the pancreas—the **pancreatic islets**—however, secrete their products (insulin and glucagon) into the blood. The precise origin of the endocrine cells of the pancreas remains uncertain. It is believed that some of the endocrine cells develop as buds from pancreatic ductules and that others arise from neuroectodermal cells that migrate to the pancreas to form these endocrine cells and the autonomic innervation of the pancreas.

Adrenal Glands

The **adrenal glands** begin development during the fifth week from two different germ layers. Each adrenal gland has an outer part, or **adrenal cortex**, which develops from mesoderm, and an inner part, or **adrenal medulla**, which develops from neuroectoderm. The mesodermal ridge that forms the adrenal cortex is in the same region from which the gonads develop.

The neuroectodermal cells that form the adrenal medulla are derived from the neural crest of the neural tube. The developing adrenal medulla is gradually encapsulated by the adrenal cortex, a process that continues into the fetal stage. The formation of the adrenal gland is not completed until the end of the third year of age.

Notice that the adrenal gland, like the pituitary, has a dual origin; part is neural and part is not. Like the pituitary, the adrenal cortex and adrenal medulla are really two different endocrine tissues. They are located in the same organ but secrete different hormones and are regulated by different control systems.

Clinical Considerations

Endocrinology is the science concerned with the structure and function of the glands of the endocrine system and the consequences of glandular dysfunctions. The most frequent endocrine disorders result from an unusual increase (hypersecretion) or decrease (hyposecretion) of hormones. Hypersecretion of an endocrine gland is generally caused by hyperplasia (increase in size) of the gland, whereas hyposecretion of hormones is the consequence of a damaged or atrophied gland.

The diagnosis and treatment of endocrine problems can be difficult because of three complex physiological effects of hormones. (1) Hormones are extremely specific, and their target tissues are sensitive to minute changes in hormonal levels. (2) The responses of hormones are usually a type of chain reaction; that is, the hormones secreted from one gland cause the secretion of hormones by another gland. Thus, the clinical symptoms obscure the source of the problem. (3) Endocrine problems frequently manifest as metabolic problems because the primary action of hormones is to regulate metabolism.

Common diagnostic methods will be discussed in this section, along with the more important endocrine disorders that affect the major endocrine glands.

Diagnostic Tests for Endocrine Disorders

Certain endocrine disorders affect the physical appearance and behavior of the patient, so observation is very important in diagnosis. The patient's clinical history is also important in evaluating the speed of progression and the stage of development of an endocrine disorder.

The confirmation of endocrine disorders requires laboratory tests, particularly of blood and urine samples. These samples are important because hormones are distributed via the blood, and urine is produced from the metabolic wastes filtered from the blood. A **radioimmunoassay (RIA)** is a laboratory test to determine the concentration of hormones in blood and urine. Other blood tests include the **protein-bound iodine (PBI) test** to determine the iodine level within the blood and hence thyroid problems; measurement of blood cholesterol content for thyroid problems; measurement of sodium-potassium ratios to detect Addison's disease; and blood sugar tests, including testing glucose tolerance in a fasting patient to examine for diabetes mellitus.

A urinalysis can be important in the diagnosis of several endocrine disorders. A high level of glucose in a fasting patient indicates diabetes mellitus. A patient who has diabetes insipidus will have frequent urinations that are dilute and of low specific gravity. Certain diseases of the adrenal glands can also be detected by examining for changes in urine samples collected over a 24-hour period.

Basal metabolism rate (BMR) and **thyroid scans** (see fig. 14.14b) are tests for thyroid disorders. A visual-field examination may be important in detecting a tumor of the pituitary gland. Radiographs and electrocardiograms also may be helpful in diagnosing endocrine disorders.

Disorders of the Pituitary Gland

The pituitary is a remarkable gland. It simultaneously carries out several functions, yet more than 90% of the gland must be destroyed before pituitary function becomes severely impaired.

Panhypopituitarism   A reduction in the activity of the pituitary gland is called **hypopituitarism** *(hi″po-pi-too′i-tă-rizm″)* and can result from intracranial hemorrhage, a blood clot, prolonged steroid treatments, or a tumor. Total pituitary impairment, termed **panhypopituitarism,** brings about a progressive and general loss of hormonal activity. For example, the gonads stop functioning and the person suffers from amenorrhea (lack of menstruation) or aspermia (no sperm production) and loss of pubic and axillary hair. The thyroid and adrenals also eventually stop functioning. People with this condition and those who have had their pituitary surgically removed (a procedure called **hypophysectomy**) receive thyroxine, cortisone, growth hormone, and gonadal hormones throughout life to maintain normal body function.

Abnormal Growth Hormone Secretion   Inadequate growth hormone secretion during childhood causes **pituitary dwarfism.** Hyposecretion of growth hormone in an adult produces a rare condition called **pituitary cachexia** *(kă-kek′se-ă)* **(Simmonds' disease).** One of the symptoms of this disease is premature aging caused by tissue atrophy. By contrast, oversecretion of growth hormone during childhood causes **gigantism.** Excessive growth hormone secretion in an adult does not cause further growth in length because the epiphyseal discs have ossified. Hypersecretion of growth hormone in an adult causes **acromegaly** *(ak″ro-meg′ă-le),* in which the person's appearance gradually changes as a result of thickening of bones and growth of soft tissues, particularly in the face, hands, and feet.

Inadequate ADH Secretion   A dysfunction of the neurohypophysis results in a deficiency in ADH secretion, causing a condition called **diabetes insipidus.** Symptoms of this disease include polyuria (excessive urination), polydipsia (consumption of large amounts of water), and severe ionic imbalances. Diabetes insipidus is treated by injections of ADH.

Disorders of the Thyroid and Parathyroids

Hypothyroidism   The infantile form of hypothyroidism is known as **cretinism** *(kre′ti-nizm).* An affected child usually appears normal at birth because thyroxine is received from the mother through the placenta. The clinical symptoms of cretinism are stunted growth, thickened facial features, abnormal bone development, mental retardation, low body temperature, and general lethargy. If cretinism is diagnosed early, it can be successfully treated by administering thyroxine.

Simmonds' disease: from Morris Simmonds, German physician,
   1855–1925
acromegaly: Gk. *akron,* extremity; *megas,* large
diabetes: Gk. *diabetes,* to pass through a siphon

Myxedema   Hypothyroidism in an adult causes myxedema *(mik″sĕ-de′mă).* This disorder affects body fluids, causing edema and increasing blood volume, hence increasing blood pressure. A person with myxedema has a low metabolic rate, lethargy, and a tendency to gain weight. This condition is treated with thyroxine or with triiodothyronine, which is taken orally (as pills).

Endemic Goiter   A goiter is an abnormal growth of the thyroid gland. When this is a result of inadequate dietary intake of iodine, the condition is called endemic goiter (fig. 14.26). In this case, growth of the thyroid is due to excessive TSH secretion, which results from low levels of thyroxine secretion. Endemic goiter is thus associated with hypothyroidism.

---

myxedema: Gk. *myxa,* mucus; *oidema,* swelling

Figure 14.27 Hyperthyroidism is characterized by an increased metabolic rate, weight loss, muscular weakness, and nervousness. The eyes may also protrude.

Graves' Disease   Graves' disease, also called **toxic goiter,** involves growth of the thyroid associated with hypersecretion of thyroxine. This hyperthyroidism is produced by antibodies that act like TSH and stimulate the thyroid; it is an autoimmune disease. As a consequence of high levels of thyroxine secretion, the metabolic rate and heart rate increase, there is loss of weight, and the autonomic nervous system induces excessive sweating. In about half of the cases, **exophthalmos** *(ek''sof-thal'mos)* (bulging of the eyes) also develops (fig. 14.27) because of edema in the tissues of the eye sockets and swelling of the extrinsic eye muscles.

Disorders of the Parathyroid Glands   Surgical removal of the parathyroids sometimes unintentionally occurs when the thyroid is removed because of a tumor or the presence of Graves' disease. The resulting fall in parathyroid hormone (PTH) causes a decrease in plasma calcium levels, which can lead to severe muscle tetany. Hyperparathyroidism is usually caused by a tumor that secretes excessive amounts of PTH. This stimulates demineralization of bone, which makes the bones soft and raises the blood levels of calcium and phosphate. As a result of these

Graves' disease: from Robert James Graves, Irish physician, 1796–1853
exophthalmos: Gk. *ex,* out; *opthalmos,* eyeball

changes, bones are subject to deformity and fracture, and stones composed of calcium phosphate are likely to develop in the urinary tract.

Disorders of the Pancreatic Islets

Diabetes Mellitus   Diabetes mellitus is characterized by fasting hyperglycemia and the presence of glucose in the urine. There are two forms of this disease. **Type I,** or **juvenile-onset diabetes mellitus,** is caused by destruction of the beta cells and the resulting lack of insulin secretion. **Type II,** or **maturity-onset diabetes mellitus** (which is the more common form), is caused by decreased tissue sensitivity to the effects of insulin, so that increasingly large amounts of insulin are required to produce a normal effect. Both types of diabetes mellitus are also associated with abnormally high levels of glucagon secretion.

Treatment of diabetes mellitus requires administering the necessary amounts of insulin to maintain a balanced carbohydrate metabolism. If excessive amounts of insulin are given, the person will experience extreme nervousness and tremors, perhaps followed by convulsion and loss of consciousness. This condition is commonly called insulin shock and can be treated by administering glucose intravenously.

Reactive Hypoglycemia   People who have a genetic predisposition for type II diabetes mellitus often first develop reactive hypoglycemia. In this condition, the rise in blood glucose that follows the ingestion of carbohydrates stimulates an excessive secretion of insulin, which in turn causes the blood glucose levels to fall below the normal range. This can result in weakness, changes in personality, and mental disorientation.

Disorders of the Adrenal Glands

Tumors of the Adrenal Medulla   Tumors of the chromaffin cells of the adrenal medulla are referred to as **pheochromocytomas** *(fe-o-kro''mo-si-to'maz).* These tumors cause hypersecretion of epinephrine and norepinephrine, which produce an effect similar to continuous sympathetic nervous stimulation. The symptoms of this condition are hypertension, elevated metabolism, hyperglycemia and sugar in the urine, nervousness, digestive problems, and sweating. It does not take long for the body to become totally fatigued under these conditions, making the patient susceptible to other diseases.

pheochromocytomas: Gk. *phaios,* dusky; *chroma,* color; *oma,* tumor

Addison's Disease   This disease is caused by an inadequate secretion of both glucocorticoids and mineralocorticoids, which results in hypoglycemia, sodium and potassium imbalance, dehydration, hypotension, rapid weight loss, and generalized weakness. A person with this condition who is not treated with corticosteroids will die within a few days because of the severe electrolyte imbalance and dehydration. Another symptom of this disease is darkening of the skin. This is caused by high secretion of ACTH and possibly MSH (because MSH is derived from the same parent molecule as ACTH), which is a result of lack of negative feedback inhibition of the pituitary by corticosteroids.

Cushing's Syndrome   Hypersecretion of corticosteroids results in Cushing's syndrome. This is generally caused by a tumor of the adrenal cortex or by oversecretion of ACTH from the adenohypophysis. Cushing's syndrome is characterized by changes in carbohydrate and protein metabolism, hyperglycemia, hypertension, and muscular weakness. Metabolic problems give the body a puffy appearance and can cause structural changes characterized as buffalo hump and moon face. Similar effects are also seen when people with chronic inflammatory diseases receive prolonged treatment with corticosteroids, which are given to reduce inflammation and inhibit the immune response.

Adrenogenital Syndrome   Usually associated with Cushing's syndrome, this condition is caused by hypersecretion of adrenal sex hormones, particularly the androgens. Adrenogenital syndrome in young children causes premature puberty and enlarged genitals, especially the penis in a male and the clitoris in a female. An increase in body hair and a deeper voice are other characteristics. This condition in a mature woman can cause growth of a beard.

As with most other endocrine disorders, problems of the adrenal cortex are treated either through surgery (if there is a tumor causing hypersecretion) or by the administration of the necessary levels of appropriate hormones (in the case of tissue dysfunction with hyposecretion).

Addison's disease: from Thomas Addison, English physician, 1793–1860
Cushing's syndrome: from Harvey Cushing, American physician, 1869–1939

CLINICAL CASE STUDY ANSWER

The visual findings are likely caused by a pituitary tumor (adenoma) eroding and pushing upward on the optic chiasma. The optic chiasma contains nerve tracts that eminate from the medial portions of each retina that receive light from the lateral (temporal) visual fields. The pituitary tumor is secreting either increased or unregulated amounts of ACTH (adrenocorticotropic hormone), which in turn are causing overproduction of adrenal cortical hormones. These include glucocorticoids, mineralocorticoids, and various androgens, which in combination are producing the patient's constellation of findings.

This specific condition, hypersecretion of adrenal cortical hormones secondary to an ACTH-producing pituitary adenoma, is known as *Cushing's disease*. This condition is usually treated by transsphenoidal excision of the tumor. In some cases, adrenal-suppressing drugs are used. Elevated adrenal corticoid levels can be a result of endocrine tumors other than pituitary. An example would be an adrenal tumor(s), either cancerous or benign, that produces adrenal cortical hormones. The physical effects of this condition and all others that involve high adrenal corticoid levels are known collectively as *Cushing's syndrome*, which is not to be confused with *Cushing's disease*, a term specifically reserved for hyperadrenalism caused by a pituitary adenoma.

UNIT V

Internal Affairs

The Relational Workings of the Body Systems

UNIT V

Endocrine System

- Sex hormones maintain texture of adult skin through activation of sebaceous glands

Integumentary System
- Covers and protects the body

- Growth hormone and sex hormones promote bone development
- Stores minerals needed by endocrine glands

Skeletal System
- Provides protection and support of certain endocrine glands (e.g., the pituitary gland)

- Provides growth hormone and sex hormones necessary for muscle development and maintenance

Muscular System
- Skeletal muscles provide protection for certain endocrine organs

- Specific hormones affect neural metabolism
- Sex hormones affect brain development and behaviors

Nervous System
- Autonomic nervous system works in conjunction with certain adrenal hormones
- Hypothalamic hormones directly control hormonal release from the pituitary

- Several hormones elevate blood pressure
- Epinephrine elevates heart rate
- Erythropoietin regulates blood cell production

Circulatory System
- Transports O_2 and CO_2, nutrients, and fluids to and from endocrine glands
- Transports hormones to target cells

- Several hormones affect immune function
- Glucocorticoids have anti-inflammatory effects

Lymphatic System
- Maintains balanced amount of interstitial fluid within endocrine organs
- Lymphocytes provide defense against infection

- Epinephrine and norepinephrine stimulate respiratory activity and expand respiratory passageways

Respiratory System
- Provides O_2 and eliminates CO_2 to and from endocrine organs

- GI tract hormones coordinate secretory activities that affect digestion

Digestive System
- Provides nutrients for growth, maintenance, and repair of endocrine glands

- ADH and aldosterone regulate rates of fluid and electrolyte reabsorption in kidneys

Urinary System
- Eliminates metabolic wastes
- Releases renin and erythropoietin when local blood pressure declines

- Several hormones regulate reproductive activities and even determine sex drives

Reproductive System
- Gonads produce sex hormones that suppress secretory acitivities in hypothalamus and pituitary gland

CHAPTER SUMMARY

Introduction to the Endocrine System (pp. 432–436)

1. Hormones are regulatory molecules released into the blood by endocrine glands. The action of a hormone on target cells is dependent on its concentration and the specific receptor sites on cell membranes.
2. Hormones are classified chemically as steroids, proteins, and amines.
3. Negative feedback occurs when information concerning an imbalance in hormone concentration is fed back to an organ that acts to correct the imbalance.

Pituitary Gland (pp. 436–441)

1. The pituitary gland, or hypophysis, is divided into an anterior adenohypophysis and a posterior neurohypophysis.
   (a) The adenohypophysis has a glandular pars distalis, a thin proximal extension called the pars tuberalis, and a narrow pars intermedia.
   (b) The neurohypophysis consists of the lobus nervosa and the infundibulum.
2. The anterior pituitary produces and secretes growth hormone, thyroid-stimulating hormone, adrenocorticotropic hormone, follicle-stimulating hormone, luteinizing hormone, prolactin, and melanocyte-stimulating hormone.
3. The posterior pituitary releases antidiuretic hormone and oxytocin.
4. Secretions of the anterior pituitary are controlled by hypothalamic hormones and regulated by the feedback of hormones from the target cells. Release of hormones from the posterior pituitary are controlled by the hypothalamo-hypophyseal nerve tract.

Thyroid and Parathyroid Glands (pp. 441–445)

1. The bilobed thyroid gland is located in the neck, just below the larynx. Four small parathyroid glands are embedded on its posterior surface.
2. Thyroid follicles secrete thyroxine and triiodothyronine, which increase the rate of protein synthesis and the rate of energy release from carbohydrates. They also regulate the rate of growth and the rate of maturity of the nervous system.

3. Parafollicular cells of the thyroid secrete the hormone calcitonin, which lowers blood calcium by inhibiting the release of calcium from bone tissue.
4. Parathyroid hormone causes an increase in blood calcium and a decrease in blood phosphate levels. It acts on the large intestine, kidneys, and bones.

Pancreas (pp. 445–447)

1. The pancreas is a mixed endocrine and exocrine gland, located in the abdominal cavity.
2. The pancreatic islets within the pancreas contain beta cells that secrete insulin and alpha cells that secrete glucagon.
   (a) Insulin lowers blood glucose and stimulates the production of glycogen, fat, and protein.
   (b) Glucagon raises blood glucose by stimulating the breakdown of liver glycogen.

Adrenal Glands (pp. 447–449)

1. Each adrenal gland consists of an adrenal cortex and an adrenal medulla and is positioned along the superior border of a kidney.
2. Hormones of the adrenal cortex include mineralocorticoids, which regulate sodium reabsorption and potassium excretion; glucocorticoids, which influence metabolism by promoting vasoconstriction and resistance to stress, and gonadocorticoids, which supplement gonadal hormones.
3. Epinephrine and norepinephrine, secreted from the adrenal medulla, produce effects similar to those of the sympathetic division of the ANS.

Gonads and Other Endocrine Glands (pp. 449–451)

1. Testes are the male gonads that produce the male sex hormone testosterone within the interstitial cells.
2. Ovaries are the female gonads that produce estrogens within the ovarian follicles and corpus luteum.
3. The pineal gland, located in the roof of the third ventricle of the brain, secretes melatonin, which seems to have an effect on the hypothalamus in the release of gonadotrophin.
4. The thymus lies behind the sternum within the mediastinum, where it produces T-cells that are important in maintaining body immunity.
5. Certain gastrointestinal cells secrete hormones that aid digestion.
6. The maternal placenta secretes hormones that maintain pregnancy.

REVIEW ACTIVITIES

Objective Questions

Match the gland to its embryonic origin.
1. adenohypophysis
2. neurohypophysis
3. adrenal medulla
4. pancreas
5. thyroid gland

   (a) endoderm of pharynx
   (b) diverticulum from brain
   (c) endoderm of foregut
   (d) neural crest ectoderm
   (e) hypophyseal pouch

6. The sella turcica that supports the pituitary gland is located in which bone?
   (a) the ethmoid bone
   (b) the frontal bone
   (c) the sphenoid bone
   (d) the occipital bone
7. The hormone primarily responsible for setting the basal metabolic rate and for promoting the maturation of the brain is
   (a) cortisol.   (c) TSH.
   (b) ACTH.   (d) thyroxine.
8. Which of the following statements about the adrenal medulla is *true*?
   (a) It is developed from mesoderm.
   (b) It secretes some androgens.
   (c) Its secretion prepares the body for the fight-or-flight response.
   (d) The zona fasciculata is stimulated by ACTH.
   (e) All of the above are true.
9. The hormone insulin
   (a) is secreted by alpha cells in the pancreatic islets.
   (b) is secreted in response to a rise in blood glucose.
   (c) stimulates the production of glycogen and fat.
   (d) both (a) and (b)
   (e) both (b) and (c)

Match the hormone with the primary agent that stimulates its secretion.
10. epinephrine
11. thyroxine
12. corticosteroids
13. ACTH

   (a) TSH
   (b) ACTH
   (c) growth hormone
   (d) sympathetic nerve impulse
   (e) CRF

14. Steroid hormones are secreted by
   (a) the adrenal cortex.
   (b) the gonads.
   (c) the thyroid.
   (d) both (a) and (b).
   (e) both (b) and (c).

UNIT V

15. The secretion of which of the following hormones would be *increased* in a person with endemic goiter?
    (a) TSH
    (b) thyroxine
    (c) triiodothyronine
    (d) all of the above

Essay Questions

1. Compare and contrast the roles of the nervous system and the endocrine system in maintaining body homeostasis.
2. List the glands of the endocrine system and describe their general locations.
3. Give examples of steroids, proteins, and amines and describe the various ways in which they are usually administered.
4. Define negative feedback and explain why this is a reliable mechanism for controlling hormonal secretion.
5. Which endocrine glands, or portions of endocrine glands, develop as a process of germ layer invagination and which develop as an outgrowth or budding of germ layer tissue?
6. Why is the pituitary frequently considered two separate glands?
7. List the hormones secreted by the adenohypophysis and describe the general functions of each.
8. Describe the gross structure and position of the neurohypophysis. What hormones are released from this portion of the pituitary gland and how is their release regulated?
9. List the hormones secreted by the thyroid gland and discuss the general function of each.
10. Describe the location of the parathyroid glands. What specialized cells constitute these glands and what hormone do they secrete?
11. Describe the structure and position of the pancreatic islets, list the hormones they produce, and discuss the general function of each.
12. Distinguish between the adrenal cortex and adrenal medulla in structure, function, and value to the body.
13. Define *chromaffin cells, interstitial cells, alpha* and *beta cells, oxyphil cells, follicular cells,* and *pituicytes.*
14. Explain the causes of each of the following endocrine diseases: gigantism, acromegaly, pituitary cachexia, and pituitary dwarfism.
15. Distinguish between cretinism, myxedema, and Graves' disease.

Gundy/Weber Software

The tutorial software accompanying Chapter 14, Endocrine System, is Volume 7, Endocrine System.

Sensory Organs

Chapter Outline

UNIT V

CLINICAL CASE STUDY

A 50-year-old man visits his family doctor and complains of progressive hearing loss in his right ear. In order to rule out visible abnormalities or wax buildup, the physician performs an otoscopic examination, which reveals no abnormalities. The physician then strikes a tuning fork and places it on the patient's right mastoid process of the temporal bone, which causes the patient to remark: "I can really hear that well in my bad ear." After a moment or so the patient notes that the tone has nearly died out. The doctor then moves the barely vibrating instrument 2 centimeters away from the same ear. Now the patient is unable to hear any tone. The doctor explains that a person with normal hearing can hear the vibrating fork when it is held against the mastoid process but can hear it better when it is held just outside the external auditory canal.

What components of the patient's hearing mechanism are being bypassed when the handle of the tuning fork is placed on his mastoid process?

Hints: The hearing organs of the inner ear can receive and effectively process sound waves directly from the bone in which they are encased. List in order the structures bypassed by sound waves being processed through the mastoid process. Describe the hearing problem that exists in this patient. Is it a conduction problem or a perception problem? Carefully read the sections on the structure of the ear and sound waves and neural pathways for hearing. Further explanations may be obtained in the clinical section on functional impairments of the ear.

Overview of Sensory Perception

Sensory organs are highly specialized extensions of the nervous system in that they contain sensory neurons adapted to respond to specific stimuli and conduct nerve impulses to the brain for interpretation.

Objective 1. State the conditions necessary for perceiving a sensation.

Objective 2. Discuss the selectivity of sensory receptors for specific stimuli.

The sense organs are actually extensions of the nervous system that allow us to perceive our internal and external environments. These sense organs have been described as windows for the brain because it is through them that we achieve awareness of the environment. A stimulus must first be received before the sensation can be interpreted and the necessary body adjustments dictated by the central nervous system can be made. Not only do we depend on our sense organs to experience pleasure, they also ensure our very survival. For example, they enable us to hear warning sounds, see dangers, avoid ingesting toxic substances through taste, and perceive sensations of pain, hunger, and thirst.

A **sensation** is the arrival of a sensory impulse to the brain. The interpretation of a sensation is referred to as **perception.** In other words, one feels, sees, hears, tastes, and smells with the brain. In order to perceive a sensation, the following four conditions are necessary.

1. A **stimulus** sufficient to initiate a response in the nervous system must be present.
2. A **receptor** must convert the stimulus to a nerve impulse. A receptor is a specialized, peripheral dendritic ending of a sensory nerve fiber or the specialized receptor cell associated with it.
3. The **conduction of the nerve impulse** must occur from the receptor to the brain along a nervous pathway.
4. The **interpretation of the impulse** in the form of a perception must occur within a specific portion of the brain.

Only impulses reaching the cerebral cortex of the brain are consciously interpreted as sensations. If impulses terminate in the spinal cord or brain stem, they may initiate a reflexive motor activity response rather than a conscious sensation. Impulses reaching the higher brain centers travel through nerve fibers composing *sensory,* or *ascending, tracts.* Clusters of neuron cell bodies, called **nuclei,** are synaptic sites along sensory tracts within the CNS. The nuclei that sensory impulses pass through before reaching the cerebral cortex are located in the spinal cord, medulla oblongata, pons, and thalamus.

Through the use of scientific instruments, it is known that the senses act as energy filters that allow perception of only a narrow range of energy. Vision, for example, is limited to light waves in the visible spectrum. Other types of waves of the same type of energy as visible light, such as X rays, radio waves, and ultraviolet and infrared light, cannot normally excite the sensory receptors in the eyes. Although filtered and distorted by the limitations of sensory functions, perceptions allow us to interact effectively with our environment and are of obvious survival value.

1. Distinguish between sensation and perception.
2. List the four conditions necessary for perception and identify which of the four must always involve consciousness in order for perception to occur.
3. Using examples, explain the statement that each of the senses acts as a filter, allowing the perception of only a narrow range of energy.

Classification of the Senses

The senses are classified as general or special according to the degree of complexity of the receptors and neural pathways. They are also classified as somatic or visceral according to the location of the receptors.

Objective 3. Compare and contrast somatic, visceral, and special senses.

Objective 4. Define the terms *exteroceptors, visceroceptors,* and *proprioceptors* and give examples of each.

Structurally, the sensory receptor can be the dendrites of sensory neurons, which are either free (such as those in the skin that mediate pain and temperature) or encapsuled within nonneural structures (such as lamellated corpuscles or pressure receptors in the skin) (see table 15.1). Other receptors, including taste buds, photoreceptors in the eyes, and hair cells in the inner ears, derive from epithelial cells that synapse with sensory dendrites.

The senses of the body can be classified as general or special according to the degree of complexity of the receptors and neural pathways involved. **General senses** are widespread through the body and are simple in structure. Examples are touch, pressure, cold-heat, and pain. **Special senses** are localized in complex receptor organs and have extensive neural pathways. Among the special senses are taste, smell, sight, hearing, and balance.

The senses can also be classified as somatic or visceral according to the location of the receptors. **Somatic senses** are those in which the receptors are localized within the body wall. These include the cutaneous receptors and those within muscles, tendons, and joints. **Visceral senses** are those in which the receptors are located within visceral organs. These classes of senses intersect in certain organs. Hearing, for example, is a special somatic sense, whereas pain from the GI tract is a general visceral sense.

Senses are also functionally classified according to the location of the receptors and the types of stimuli to which they respond. The three basic kinds of receptors are exteroceptors, visceroceptors, and proprioceptors.

Exteroceptors   Exteroceptors (*ek"stē-ro-sep'torz*) are located near the surface of the body where they respond to stimuli from the external environment. They include the following:

1. cones and rods in the retina of the eye—*photoreceptors;*
2. hair cells in the spiral organ (organ of Corti) within the inner ear—*mechanoreceptors;*
3. olfactory receptors in the nasal epithelium of the nasal cavity—*chemoreceptors;*
4. taste receptors on the tongue—*chemoreceptors;* and
5. skin receptors within the dermis—*tactile receptors* for touch, *mechanoreceptors* for pressure, *thermoreceptors* for temperature, and *nocireceptors (no-ci"re-sep'torz)* for pain.

Pain receptors are stimulated by chemicals released from damaged tissue cells, and thus are a type of chemoreceptor. Although there are specific pain receptors, nearly all types of receptors when intensely stimulated transmit impulses that are perceived as pain. Pain receptors exist throughout the body, but only those located within the skin are classified as exteroceptors.

Visceroceptors (Enteroceptors)   As the name implies, visceroceptors (*vis"er-o-sep'torz*) produce sensations arising from the viscera, such as internal pain, hunger, thirst, fatigue, or nausea.

somatic: Gk. *somatikos,* body
visceral: L. *viscera,* body organs
nocireceptor: L. *nocco,* to injure; *ceptus,* taken

Specialized visceroceptors located within the circulatory system are sensitive to changes in blood pressure (baroreceptors).

Proprioceptors   Proprioceptors (*pro"pre-o-sep'torz*) relay information about body position, equilibrium, and movement. They are located in the inner ear, joints, tendons, and muscles.

The three types of receptors may be further classified on the basis of sensory adaptation (accommodation). Some receptors respond with a burst of activity when a stimulus is first applied but quickly decrease their firing rate as they adapt to the stimulus, even though the stimulus remains constant. Receptors with this response pattern are called **phasic receptors.** Receptors whose firing rate remains relatively constant as long as the stimulus is maintained are known as **tonic receptors.**

 Phasic receptors alert us to changes in sensory stimuli and are in part responsible for the fact that we can cease paying attention to constant stimuli. This ability is called *sensory adaptation.* Odor and touch, for example, adapt rapidly; bathwater feels hotter when we first enter it. Sensations of pain, in contrast, adapt little if at all.

1. Using examples, explain how sensory receptors can be classified according to complexity, location, structure, and the type of stimuli to which they respond.
2. Describe sensory adaptation in olfactory and pain receptors.

Somatic Senses

Somatic senses include cutaneous receptors and proprioceptors. The perception of somatic sensations is determined by the density of the receptors in the stimulated receptive field and the intensity of the sensation.

Objective 5.   Describe the structure, function, and location of the various tactile receptors and the neural pathway for somatic sensation.

Objective 6.   Explain the purpose of pain, the receptors that respond to pain, and the impulse pathway to the brain.

Objective 7.   Define and give examples of *referred pain* and *phantom pain.*

The **somatic senses,** or **somesthetic senses,** include the receptors located in the skin that cause cutaneous sensations, such as touch, tickling, pressure, cold, heat, and pain. The somatic senses also include proprioceptors located in the inner ear, joints, tendons, and muscles that relay information about body position, equilibrium, and movement.

Tactile and Pressure Receptors

Both tactile receptors and pressure receptors are sensitive to mechanical forces that distort or displace the tissue in which they are located. **Tactile receptors** respond to *fine,* or *light, touch*

proprioceptor: L. *proprius,* one's own; *ceptus,* taken

UNIT V

Table 15.1	**Cutaneous Receptors**		
Type	**Location**	**Function**	**Sensation**
Corpuscles of touch (Meissner's corpuscles) (mechanoreceptors)	Papillae of dermis; numerous in hairless portions of body (eyelids, fingertips, lips, nipples, external genitalia)	Detect light motion against surface of skin	Fine touch; texture
Free nerve endings (pressure receptors; pain receptors)	Lower layers of epidermis	Detect changes in pressure; detect tissue damage	Touch, pressure; pain
Root hair plexuses (tactile receptors)	Around hair follicles	Detect movement of hair	Touch
Lamellated (pacinian) corpuscles (mechanoreceptors)	Hypodermis; synovial membranes; perimysium; certain visceral organs	Detect changes in pressure	Deep pressure; vibrations
Organs of Ruffini (thermoreceptors)	Lower layers of dermis	Detect changes in temperature	Heat
Bulbs of Krause (thermoreceptors)	Dermis	Detect changes in temperature	Cold

Bulb of Krause

Root hair plexus

Lamellated corpuscle

Corpuscle of touch

Free nerve ending

Organ of Ruffini

Creek

and are located primarily in the dermis and hypodermis of the skin. **Pressure receptors** respond to *deep pressure* and are commonly found in the hypodermis of the skin and in the tendons and ligaments of joints. The three kinds of tactile receptors and one kind of pressure receptor are summarized in table 15.1.

Corpuscles of Touch A corpuscle of touch *(Meissner's corpuscle)* is an oval receptor composed of a mass of dendritic endings from two or three nerve fibers enclosed by connective tissue sheaths. These corpuscles are numerous in the hairless portions of the body, such as the eyelids, lips, tip of the tongue, fingertips, palms of the hands, soles of the feet, nipples, and external

genital organs. Corpuscles of touch lie within the papillary layer of the dermis, where they are especially sensitive to the motion of objects that barely contact the skin (see chapter 5). Sensations of fine or light touch are perceived as these receptors are stimulated. They also function when a person touches an object to determine its texture.

The high tactile acuity of the fingertips is exploited in the reading of Braille. Braille symbols consist of dots that are raised 1 mm up from the page and separated from each other by 2.5 mm. Experienced Braille readers can scan words at about the same speed that a sighted person can read aloud—a rate of about 100 words per minute.

Meissner's corpuscle: from George Meissner, German histologist, 1829–1905

corpuscle: L. *corpusculum*, diminutive of *corpus*, body

Braille: from Louis Braille, French teacher of the blind, 1809–52

Free Nerve Endings   Free nerve endings are the least modified and the most superficial of the tactile receptors. These receptors extend into the lower layers of the epidermis, where they end as knobs between the epithelial cells. Free nerve endings are most important as pain receptors, although they also respond to objects that are in continuous contact with the skin, such as clothing.

Hair Root Plexuses   Hair root plexuses are a specialized type of free nerve ending. They are coiled around hair follicles, where they respond to movement of the hair.

Lamellated Corpuscles   Lamellated (pacinian) corpuscles are large, onion-shaped receptors composed of the dendritic endings of several sensory nerve fibers enclosed by connective tissue layers. They are commonly found within the synovial membranes of synovial joints, in the perimysium of skeletal muscle tissue, and in certain visceral organs. Lamellated corpuscles are also abundant in the skin of the palms and fingers of the hand, soles of the feet, external genital organs, and breasts.

Lamellated corpuscles respond to heavy pressures, generally those that are constantly applied. They can also detect vibrations in tissues and organs.

Thermoreceptors

Thermoreceptors are widely distributed throughout the dermis of the skin but are especially abundant in the lips and the mucous membranes of the mouth and anal regions. There are two kinds of thermoreceptors—one that responds to heat and the other to cooler temperatures (see table 15.1).

Organs of Ruffini   The organs of Ruffini are heat receptors located deep within the dermis. Heat receptors are elongated, oval structures that are most sensitive to temperatures above 25° C (77° F). Temperatures above 45° C (113° F) elicit impulses through the organs of Ruffini that are perceived as painful, burning sensations.

Bulbs of Krause   The bulbs of Krause are receptors for the sensation of cold. They are more abundant than heat receptors and are closer to the surface of the skin. The bulbs of Krause are most sensitive to temperatures below 20° C (68° F). Temperatures below 10° C elicit responses through the bulbs of Krause that are perceived as painful, freezing sensations.

Pain Receptors

The principal receptors for pain are the **free nerve endings.** Several million free nerve endings are distributed throughout the skin and internal tissues. Pain receptors are sparse in certain visceral organs and absent within the nervous tissue of the brain. Although the free nerve endings are specialized to respond to tissue damage, all of the cutaneous receptors will relay impulses that are interpreted as pain if stimulated excessively.

The protective value of pain receptors is obvious. Unlike other cutaneous receptors, free nerve endings have little accommodation (few phasic receptors), so impulses are relayed continuously to the CNS as long as the irritating stimulus is present. Although pain receptors can be activated by all types of stimuli, they are particularly sensitive to chemical stimulation. Muscle spasms, muscle fatigue, or an inadequate supply of blood to an organ may also cause pain.

Impulses for pain are conducted to the spinal cord through sensory neurons. The pain sensations are then conducted to the thalamus along the *lateral spinothalamic tract* of the spinal cord, and from there to the somatesthetic area of the cerebral cortex. Although an awareness of pain occurs in the thalamus, the type and intensity of pain is interpreted in the specialized areas of the cerebral cortex.

The sensation of pain can be clinically classified as **somatic pain** or **visceral pain.** Stimulation of the cutaneous pain receptors results in the perception of superficial somatic pain. Deep somatic pain comes from stimulation of receptors in skeletal muscles, joints, and tendons.

Stimulation of the receptors within the viscera causes the perception of visceral pain. Through precise neural pathways, the brain is able to perceive the area of stimulation and project the pain sensation back to that area. The sensation of pain from certain visceral organs, however, may not be perceived as arising from those organs but from other somatic locations. This phenomenon is known as **referred pain** (fig. 15.1). The sensation of referred pain is consistent from one person to another and is clinically important in diagnosing organ dysfunctions. The pain of a heart attack, for example, may be perceived subcutaneously over the heart and down the medial side of the left arm. Ulcers of the stomach may cause pain that is perceived as coming from the upper central (epigastric) region of the torso. Pain from problems of the liver or gallbladder may be perceived as localized visceral pain or as referred pain arising from the right neck and shoulder regions.

Referred pain is not totally understood but seems to be related to the development of the tracts within the spinal cord. It is thought that there are some *common nerve pathways* used by sensory impulses coming from both the cutaneous areas and from visceral organs (fig. 15.2). Consequently, impulses along these pathways may be incorrectly interpreted as arising cutaneously rather than from within a visceral organ.

The perception of pain is of survival value because it alerts the body to an injury, disease, or organ dysfunction. *Acute pain* is sudden, usually short-term, and can generally be endured and attributed to a known cause. *Chronic pain,* however, is long-term and tends to weaken a person as it interferes with the ability to function effectively. Certain diseases, such as arthritis, are characterized by chronic pain. In these patients, relief of pain is of paramount concern. Treatment of chronic pain often requires the use of moderate pain-reducing drugs (analgesics) or intense narcotic drugs. Treatment in severely tormented chronic pain patients may include severing sensory nerves or implanting stimulating electrodes in appropriate nerve tracts.

---

pacinian corpuscle: from Filippo Pacini, Italian anatomist, 1812–83
organs of Ruffini: from Angelo Ruffini, Italian anatomist, 1864–1929
bulbs of Krause: from Wilhelm J. F. Krause, German anatomist, 1833–1910

Figure 15.1 Sites of referred pain are perceived cutaneously but actually originate from specific visceral organs.

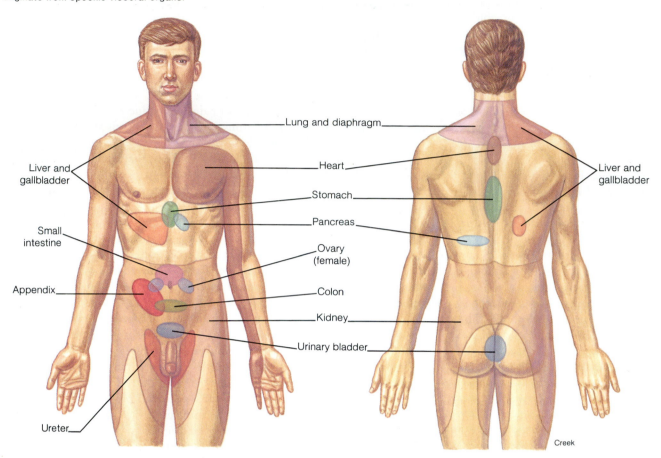

Lung and diaphragm

Liver and gallbladder

Heart

Stomach

Liver and gallbladder

Small intestine

Pancreas

Ovary (female)

Appendix

Colon

Kidney

Urinary bladder

Ureter

Creek

Phantom pain is frequently experienced by an amputee who continues to feel pain from the body part that was amputated, as if it were still there. After amputation, the severed sensory neurons heal and function in the remaining portion of the appendage. Although it is not known why impulses that are interpreted as pain are sent periodically through these neurons, the sensations evoked in the brain are projected to the region of the amputation, resulting in phantom pain.

Proprioceptors

Proprioceptors monitor internal conditions, frequently those under voluntary control. They are located within skeletal muscle tissue, tendons, and the synovial membranes and connective tissue surrounding joints. Proprioceptors are sensitive to changes in stretch and tension. Some of the sensory impulses from proprioceptors reach the level of consciousness as the **kinesthetic sense,** by which the position of the body parts is perceived. Other proprioceptor information is not consciously interpreted and is used to adjust the intensity and timing of muscle contractions to provide coordinated movements. With the kinesthetic sense, the position and movement of the limbs can be determined without visual sensations, such as when dressing or walking in the dark. The kinesthetic sense, along with hearing, becomes keenly developed in a blind person.

High-speed transmission is a vital characteristic of the kinesthetic sense because rapid feedback to various body parts is essential for quick, smooth, coordinated body movements.

Proprioceptors are of three types: joint kinesthetic receptors, neuromuscular spindles, and neurotendinous spindles.

Joint Kinesthetic Receptors   Joint kinesthetic receptors are located in the connective tissue capsule in synovial joints where they are stimulated by changes in position caused by movement at the joints.

Neuromuscular Spindles   Neuromuscular spindles are located in skeletal muscle, particularly in the muscles of the extremities. They consist of the endings of sensory neurons spiraled around specialized individual muscle fibers (fig. 15.3). Neuromuscular spindles are stimulated by an increase in muscle tension due to the lengthening or stretching of the individual spindles and thus provide information about the length of the muscle and the rate of change of length (i.e., velocity of contraction).

Neurotendinous Spindles   Neurotendinous spindles, also called *Golgi (gol'je) tendon organs,* are located where a muscle attaches to a tendon (fig. 15.3). They are stimulated by the tension produced in a tendon when the attached muscle is either stretched or contracted.

Figure 15.2 An explanation of referred pain. Pain originating from the myocardium of the heart may be perceived as coming from the skin of the left arm because sensory impulses from these two organs are conducted through common nerve pathways to the brain.

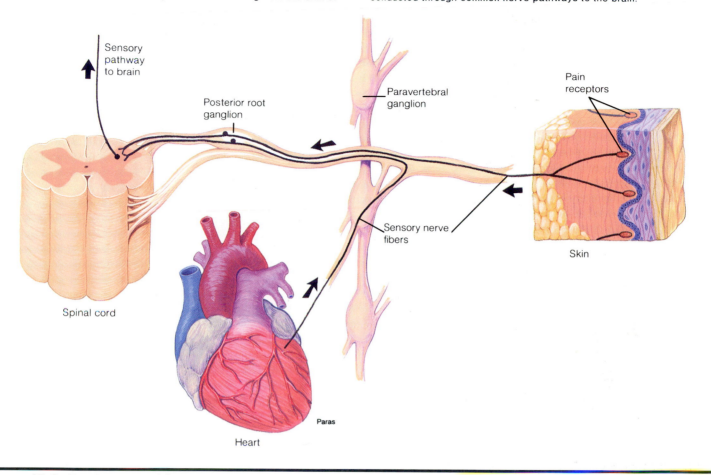

Figure 15.3 Proprioreceptors. (*a*) A neuromuscular spindle and (*b*) a neurotendinous spindle.

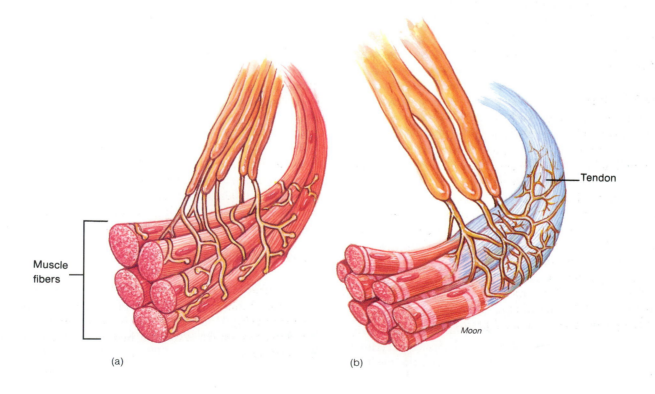

Figure 15.4 Pathways that lead from the cutaneous receptors and proprioreceptors into the postcentral gyrus in the cerebral cortex.

Postcentral gyrus

Ventrolateral nucleus of the thalamus

Midbrain

Pain, hot, and cold

Touch and pressure

Medulla oblongata

Lateral spinothalamic tract

Ventral spinothalamic tract

Proprioception

Spinal cord

Waldrop

Neural Pathways for Somatic Sensation

The conduction pathways for the somatic senses are shown in figure 15.4. Sensations of proprioception and touch and pressure are carried by large, myelinated nerve fibers that ascend in the posterior columns of the spinal cord on the ipsilateral (same) side. These fibers do not synapse until they reach the medulla oblongata of the brain stem; hence, fibers that carry these sensations from the feet are incredibly long. After synapsing in the medulla oblongata with other, second-order sensory neurons, information in the latter neurons crosses over to the contralateral (opposite) side as it ascends via a fiber tract called the **medial lemniscus** to the thalamus. Third-order sensory neurons in the thalamus that receive this input in turn project to the **postcentral gyrus.**

Figure 15.5   The olfactory receptive area of the nasal cavity as seen in sagittal view.

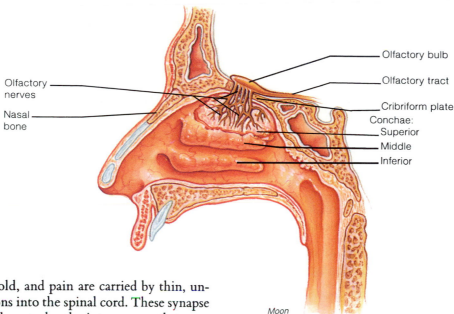

Olfactory nerves

Nasal bone

Olfactory bulb

Olfactory tract

Cribriform plate

Conchae:
Superior
Middle
Inferior

Moon

Sensations of heat, cold, and pain are carried by thin, unmyelinated sensory neurons into the spinal cord. These synapse within the spinal cord with second-order interneurons that cross over to the contralateral side and ascend to the brain in the **lateral spinothalamic tract.** Fibers that mediate touch and pressure ascend in the **ventral spinothalamic tract.** Fibers of both spinothalamic tracts synapse in the thalamus with third-order neurons, which in turn project to the postcentral gyrus of the parietal cerebral lobe. Note that, in all cases, somatic information is carried to the postcentral gyrus in third-order neurons. Also, because of crossing-over, somatic information from each side of the body is projected to the postcentral gyrus of the contralateral cerebral hemisphere.

All somatic information from the same area of the body projects to the same area of the postcentral gyrus, so that a *sensory homunculus* map of the body can be drawn on the postcentral gyrus to represent sensory projection points (see fig. 11.22). This map is greatly distorted, however, because it shows larger areas of cerebral cortex devoted to sensation in the face and hands than in other areas of the body. This disproportionately larger area of the cerebral cortex devoted to the face and hands reflects the fact that there is a higher density of sensory receptors in these regions.

1. List the types of tactile receptors and state where they are located. What portion of the brain interprets tactile sensations?
2. Discuss the importance of pain. List the receptors that respond to pain and the structures of the brain that are particularly important in the perception of pain sensation.
3. Using examples, distinguish between referred pain and phantom pain. Discuss why it is important for a physician to know the referred pain sites.
4. Using a flow diagram, describe the neural pathways leading from cutaneous pain and pressure receptors to the postcentral gyrus. Indicate where crossing-over occurs.

Olfactory Sense

Olfactory receptors are the dendritic endings of the olfactory (first cranial) nerve that respond to chemical stimuli and transmit the sensation of olfaction directly to the olfactory portion of the cerebral cortex.

Objective 8.   Describe the sensory pathway of olfaction.

Olfactory reception in humans is not highly developed compared to that of certain other vertebrates. Because humans do not rely on smell for communicating or for finding food, the olfactory sense is probably the least important of the senses; it is more important in detecting the presence of an odor rather than its intensity. Accommodation occurs relatively rapidly with this sense. Olfaction functions closely with gustation (taste) in that the receptors of both are chemoreceptors and require dissolved substances for stimuli.

Olfactory receptor cells are located in the nasal epithelium within the roof of the nasal cavity on both sides of the nasal septum (fig. 15.5). Olfactory cells are moistened and stabilized by the surrounding glandular goblet cells and the supporting columnar epithelial cells. The olfactory cells are actually bipolar neurons whose cell bodies lie between the supporting columnar cells. The free end of each olfactory cell contains several dendritic endings called **olfactory hairs** that constitute the sensitive portion of the receptor cell.

The sensory pathway for olfaction consists of several neural segments. The unmyelinated axons of the olfactory cells unite to form the **olfactory nerves,** which traverse the foramina of the cribriform plate and terminate in the paired masses of gray and white matter called the **olfactory bulbs.** The olfactory bulbs lie on both sides of the crista galli of the ethmoid bone, beneath the frontal lobes of the cerebrum. The neurons of the

UNIT V

olfactory nerves synapse within the olfactory bulb with dendrites of the **olfactory tract.** Sensory impulses are conveyed along the olfactory tract and into the olfactory portion of the cerebral cortex, where they are interpreted as odor and cause the perception of smell.

Unlike taste, which is divisible into only four modalities, thousands of distinct odors can be distinguished by people who are trained in this capacity (as in the perfume industry). The molecular basis of olfaction is not understood, but it is known that a single odorant molecule is sufficient to excite an olfactory receptor.

Only about 2% of inhaled air comes in contact with the olfactory receptors, which are positioned above the mainstream of airflow. Olfactory sensitivity can be increased by forceful sniffing, which draws the air into contact with the receptors. Certain chemicals activate the trigeminal (fifth) as well as the olfactory (first) cranial nerves and cause reactions. Pepper, for example, may cause sneezing; onions cause the eyes to water; and smelling salts (ammonium salts) initiate respiratory reflexes and are used to revive unconscious persons.

1. What are olfactory hairs? Where are they located?
2. Trace the pathway of an olfactory stimulation from the olfactory hairs to the cerebral cortex, where interpretation occurs.

Gustatory Sense

Taste receptors are specialized epithelial cells, clustered together into taste buds, that respond to chemical stimuli and transmit the sense of taste through the glossopharyngeal (ninth cranial) nerve or the facial (seventh cranial) nerve to the cortex of the parietal cerebral lobe for interpretation.

Objective 9. List the three types of papillae and explain how they function in the perception of taste.

Objective 10. Identify the cranial nerves and the sensory pathways of gustation.

The receptors of taste are located in the **taste buds.** Taste buds are specialized sensory organs that are most numerous on the dorsum of the tongue but also present on the soft palate and on the walls of the oropharynx. The cylindrical taste bud is composed of many sensory **gustatory cells** that are encapsulated by **supporting cells** (fig. 15.6). Each gustatory cell contains a dendritic ending called a **gustatory microvillus** that projects to the surface through an opening in the taste bud called the **taste pore.** The gustatory microvilli are the sensitive portion of the receptor cells. Saliva provides the moistened environment necessary for a chemical stimulus to activate the gustatory microvilli.

Taste buds are elevated by surrounding connective tissue and epithelium to form *papillae (pă-pil'e)* (fig. 15.6). Three types of papillae can be identified:

1. **Vallate papillae.** The largest but least numerous are the vallate papillae, which are arranged in an inverted V-shape pattern on the back of the tongue.
2. **Fungiform papillae.** Knoblike fungiform papillae are present on the tip and sides of the tongue.
3. **Filiform papillae.** Short, thickened, threadlike filiform papillae are located on the anterior two-thirds of the tongue.

Taste buds are found only in the vallate and fungiform papillae. Filiform papillae contain gustatory microvilli, which are not clustered into taste buds.

The four basic modalities of taste, or *gustation (gus-ta'shun),* are sensed most acutely on particular regions of the tongue (fig. 15.7). These are *sweet* (tip of tongue), *sour* (sides of tongue), *bitter* (back of tongue), and *salty* (over most of the tongue, but concentrated on the sides).

Sour taste is produced by hydrogen ions (H^+); all acids therefore taste sour. Most organic molecules, particularly sugars, taste sweet to varying degrees. Only pure table salt (NaCl) has a pure salty taste. Other salts, such as KCl (commonly used in place of NaCl by people with hypertension), taste salty but have bitter overtones. Bitter taste is evoked by quinine and seemingly unrelated molecules.

The sensory pathway of taste receptors to the brain mainly involves two cranial nerves (fig. 15.8). Taste buds on the posterior third of the tongue have a sensory pathway through the glossopharyngeal nerve, whereas the anterior two-thirds of the tongue is served by the chorda tympani branch of the facial nerve.

Taste sensations passing through the nerves just mentioned are conveyed through the medulla oblongata and thalamus to the parietal lobe of the cerebral cortex, where they are interpreted.

Because taste and smell are both chemoreceptors, they complement each other. We often confuse a substance's smell with its taste; and if we have a head cold or hold our nose while eating, food seems to lose its flavor.

1. Distinguish between papillae, taste buds, and gustatory microvilli. Discuss the function of each as related to taste.
2. List the three types of papillae.
3. Which cranial nerves have sensory innervation associated with taste? What are the sensory pathways to the brain where the perception of taste occurs?

gustatory: L. *gustare,* to taste

papilla: L. *papilla,* nipple

Figure 15.6 Papillae of the tongue and associated taste buds. (*a*) Numerous taste buds are positioned within each papilla. (*b*) Each gustatory cell and its associated gustatory microvillus are encapsuled by supporting cells. (*c*) A photomicrograph of some taste buds (indicated with arrows).

(a)

Vallate papillae

Taste buds

Sensory nerve fiber

Connective tissue

(b)

(c)

Gustatory cell

Supporting cell

Taste pore

Gustatory microvillus

Epithelium of tongue

Visual Sense

Rods and cones are the photoreceptors within the eyeball that are sensitive to light energy and are stimulated to transmit nerve impulses through the optic nerve and optic tract to the visual cortex of the occipital lobes, where the interpretation of vision occurs. Formation of the sensory components of the eye is complete at 20 weeks, and the accessory structures have been formed by 32 weeks.

Objective 11. Describe the associated structures of the eye and the structure of the eyeball.

Objective 12. Trace the path of light waves through the eye and explain how they are focused on distant and near objects.

Objective 13. Describe the neural pathway of a visual impulse and discuss the neural processing of visual information.

The eyes are organs that refract (bend) and focus the incoming light waves onto the sensitive photoreceptors at the back of each eye. Nerve impulses from the stimulated photoreceptors are conveyed through visual pathways within the brain to the occipital lobes of the cerebrum, where the sense of vision is perceived. The specialized photoreceptor cells can respond to an incredible one billion different stimuli each second. Further, these cells are sensitive to about 10 million gradations of light intensity and 7 million different shades of color.

The eyes are anteriorly positioned on the skull and set just far enough apart to achieve *binocular (stereoscopic) vision* when

Figure 15.7 Patterns of taste receptor distribution on the dorsum of the tongue. (*a*) Sweet receptors, (*b*) sour receptors, (*c*) salt receptors, and (*d*) bitter receptors.

(a)   (b)   (c)   (d)

Figure 15.8 The principal viscerosensory innervation (taste) is provided by the facial (VII) and the glossopharyngeal (IX) cranial nerves. The chorda tympani nerve is the branch of the facial nerve innervating the tongue. Branches from the vagus and the trigeminal nerves also provide some sensory innervation. The hypoglossal (XII) cranial nerve (not shown) provides motor innervation to the tongue. The lingual nerve transmits general sensory information from the tongue (hot, cold, pressure, and pain).

Lingual nerve

Glossopharyngeal nerve

Uvula

Tongue

Trigeminal ganglion

Trigeminal nerve

Geniculate ganglion

Facial nerve

Chorda tympani nerve

Vagus nerve

Laryngeal branch of vagus nerve

Moon

UNIT V

focusing on an object. This three-dimensional perspective allows a person to assess depth. Often likened to a camera (table 15.2), the eyes are responsible for approximately 80% of all knowledge that is assimilated.

The eyes of other vertebrates are basically similar to ours. Certain species, however, have adaptive modifications. Consider, for example, the extremely keen eyesight of a hawk, which soars high in the sky searching for food, or the eyesight of a nocturnal animal like an owl, which feeds only at night. Note how the location of the eyes on the head corresponds to behavior. Predatory species, such as cats, have eyes that are directed forward, allowing depth perception. Prey species, such as deer, have eyes positioned high on the sides of their heads, allowing panoramic vision to detect distant threatening movements, even while grazing.

Table 15.2	Eye Structures and Their Camera Analogies
Eye Structure	**Camera Structure**
Cornea and lens	Lens system
Iris and pupil	Variable aperture system
Eyelid	Lens cap
Sclera	Camera frame
Pigment epithelium and choroid	Black interior of camera
Retina *	Film

*Since neural processing begins in the retina, this eye structure may also be considered analogous (in part) to the photographer.
From Stuart Ira Fox, *Human Physiology*. Copyright © 1984 Wm. C. Brown Communications, Inc., Dubuque, Iowa. All Rights Reserved. Reprinted by permission.

Associated Structures of the Eye

Associated structures of the eye either protect the eyeball or provide eye movement. Protective structures include the bony orbit, eyebrow, facial muscles, eyelids, eyelashes, conjunctiva, and the lacrimal apparatus that produces tears. Eyeball movements are made possible by the actions of the extrinsic ocular eye muscles that arise from the orbit and insert on the outer layer of the eyeball.

Orbit   Each eyeball is positioned in a bony depression in the skull called the orbit (see fig. 6.21 and table 6.5). Seven bones of the skull (frontal, lacrimal, ethmoid, zygomatic, maxilla, sphenoid, and palatine) form the walls of the orbit that support and protect the eye.

Eyebrow   Eyebrows consist of short, thick hair positioned transversely above both eyes along the superior orbital ridges of the skull (figs. 15.9 and 15.10). Located here they effectively

Figure 15.9   Accessory structures of the eyeball as seen in a sagittal section of the eyelids and an anterior portion of the eyeball within the orbit.

Levator palpebrae superioris muscle
Superior rectus muscle
Retina
Anterior chamber
Suspensory ligament of lens
Posterior chamber
Lens
Iris
Ciliary body

Orbicularis oculi muscle
Eyebrow
Cornea
Palpebral conjunctiva
Superior tarsal plate
Bulbar conjunctiva
Eyelashes
Conjunctival sac
Orbicularis oculi muscle
Fornix of conjunctival sac

UNIT V

Figure 15.10 The surface anatomy of the eye.

shade the eyes from the sun and prevent perspiration or falling particles from getting into the eyes. Underneath the skin of each eyebrow is the orbital portion of the orbicularis oculi muscle and a portion of the corrugator muscle (see fig. 9.13). Contraction of either of these muscles causes the eyebrow to move, often reflexively, to protect the eye.

Eyelids and Eyelashes Eyelids (palpebrae) develop as reinforced folds of skin with attached skeletal muscle so that they are movable. In addition to the orbicularis oculi muscle attached to the skin that surrounds the front of the eye, the **levator palpebrae** *(le-va'tor pal'pĕ-bre)* **superioris muscle** attaches along the upper eyelid and provides it with greater movability than the lower eyelid. Contraction of the orbicularis oculi muscle closes the eyelids over the eye, and contraction of the levator palpebrae superioris muscle elevates the upper eyelid to expose the eye. The eyelids protect the eyeball from desiccation by reflexively blinking about every 7 seconds and moving fluid across the anterior surface of the eyeball. To avoid a blurred image, the eyelid will generally blink when the eyeball moves to a new position of fixation.

The **palpebral fissure** (fig. 15.10) is the interval between the upper and lower eyelids. The **commissures (canthi)** of the eye are the medial and lateral angles where the eyelids come together. The **medial commissure** is broader than the **lateral commissure** and is characterized by a small, reddish, fleshy elevation called the **lacrimal caruncle** *(kar'ung-kl)* (fig. 15.11), which contains sebaceous and sudoriferous glands. The shape of the palpebral fissure is elliptical when the eyes are open.

Each eyelid supports a row of numerous **eyelashes,** which protect the eye from airborne particles. The shaft of each eyelash is surrounded by a root hair plexus that provides the hair with the sensitivity necessary to elicit a reflexive closure of the lids. Eyelashes of the upper lid are long and turn upward, whereas those of the lower lid are short and turn downward.

Figure 15.11 The lacrimal apparatus consists of the lacrimal gland and the drainage pathway of lacrimal fluid into the nasal cavity. The lacrimal gland produces lacrimal fluid (tears), which moistens and cleanses the conjunctiva that lines the underside of the eyelids and covers the anterior surface of the eyeball.

In addition to the layers of the skin and the underlying connective tissue and orbicularis oculi muscle fibers, each eyelid contains a tarsal plate, tarsal glands, and conjunctiva. The **tarsal plates,** composed of dense regular connective tissue, are important in maintaining the shape of the eyelids (fig. 15.9). Specialized sebaceous glands called **tarsal glands** are embedded within the tarsal plates along the exposed inner surfaces of the eyelids. The ducts of the tarsal glands open onto the edges of the eyelids, and their oily secretions help keep the eyelids from adhering to each other. A *chalazion (kă-la'ze-on)* is a tumor or

---

palpebra: L. *palpebra*, eyelid (related to *palpare*, to pat gently)
commissure: L. *commissura*, a joining
caruncle: L. *caruncula*, diminutive of *caro*, flesh

tarsal: Gk. *tarsos*, flat basket
chalazion: Gk. *chalazion*, hail; a small tubercle

Figure 15.12 The extrinsic ocular muscles of the right eyeball.

Medial rectus muscle

Levator palpebrae superioris muscle

Optic nerve

Superior rectus muscle

Superior oblique muscle

Trochlea

Lateral rectus muscle (cut)

Inferior rectus muscle

Inferior oblique muscle

UNIT V

cyst on the eyelid that results from an infection of the tarsal glands. Modified sweat glands called **ciliary glands** are also located within the eyelids, along with additional sebaceous glands at the bases of the hair follicles of the eyelashes. An infection of these sebaceous glands is referred to as a *sty*.

Conjunctiva   The conjunctiva (con''jungk-ti'vă) is a thin mucus-secreting epithelial membrane that lines the interior surface of each eyelid and exposed anterior surface of the eyeball (fig. 15.9). It consists of stratified squamous epithelium that varies in thickness in different regions. The **palpebral conjunctiva** is thick and adheres to the tarsal plates of the eyelids. As the conjunctiva reflects onto the anterior surface of the eyeball, it is known as the **bulbar conjunctiva.** This portion is transparent and especially thin where it covers the cornea. Because the conjunctiva is continuous and reflects from the eyelids to the anterior surface of the eyeball, a space called the **conjunctival sac** exists when the eyelids are closed. The conjunctival sac protects the eyeball by preventing objects (including a contact lens) from passing beyond the confines of the sac. The conjunctiva can rapidly repair itself if it is scratched.

Lacrimal Apparatus   The lacrimal apparatus consists of the lacrimal gland, which secretes the *lacrimal fluid* (tears), and a series of ducts that drain the secretion into the nasal cavity (fig. 15.11). The **lacrimal gland,** which is about the size and shape of an almond, is located in the superolateral portion of the orbit. It is a compound tubuloacinar gland that secretes lacrimal fluid through several excretory lacrimal ductules into the conjunc-

tival sac of the upper eyelid. With each blink of the eyelids, lacrimal fluid passes medially and downward and drains into two small openings, called **lacrimal puncta,** on both sides of the lacrimal caruncle. From here, the lacrimal fluid drains through the lacrimal duct into the lacrimal sac and continues through the nasolacrimal duct to the inferior meatus of the nasal cavity.

Lacrimal fluid is an aqueous, mucous secretion that contains a bactericidal enzyme called *lysozyme.* Lacrimal fluid not only moistens and lubricates the conjunctival sac but also reduces the chance of eye infections. Normally about one milliliter of lacrimal fluid is produced each day by the lacrimal gland of each eye. If irritating substances, such as particles of sand or chemicals from onions, make contact with the conjunctiva, the lacrimal glands are stimulated to oversecrete. The extra lacrimal fluid protects the eye by diluting and washing away the irritating substance.

Humans are the only animals known to weep in response to emotional stress. While crying, the volume of lacrimal secretion is so great that the tears may spill over the edges of the eyelids and the nasal cavity fill with fluid. The crying response is an effective means of communicating one's emotions and results from stimulation of the lacrimal glands by parasympathetic motor neurons.

Extrinsic Eye Muscles   The movements of the eyeball are controlled by six extrinsic eye muscles called the **extrinsic ocular muscles** (figs. 15.12 and 15.13). Each extrinsic ocular muscle

Figure 15.13 The positions of the eyes as the ocular muscles are contracted. (*a*) Right eye, inferior oblique muscle; left eye, superior and medial recti muscles. (*b*) Both eyes, superior recti and inferior oblique muscles. (*c*) Right eye, superior and medial recti muscles; left eye, inferior oblique muscle. (*d*) Right eye, lateral rectus muscle; left eye, medial rectus muscle. (*e*) Primary position with the eyes fixed on a distant fixation point. (*f*) Right eye, medial rectus muscle; left eye, lateral rectus muscle. (*g*) Right eye, superior oblique muscle; left eye; inferior and medial recti muscles. (*h*) Both eyes, inferior recti and superior oblique muscles. (*i*) Right eye, inferior and medial recti muscles; left eye, superior oblique muscle.

originates from the bony orbit and inserts by a tendinous attachment to the tough outer tunic of the eyeball. Four **recti muscles** maneuver the eyeball in the direction indicated by their names **(superior, inferior, lateral,** and **medial),** and two **oblique muscles (superior** and **inferior)** rotate the eyeball on its axis (see also fig. 9.17). One of the extrinsic ocular muscles, the superior oblique, passes through a pulleylike cartilaginous loop, called the **trochlea** *(trok'le-ă),* before attaching to the eyeball. Although stimulation of each muscle causes a precise movement of the eyeball, most of the movements involve the combined contraction of typically two muscles.

The motor units of the extrinsic ocular muscles are the smallest in the body. This means that a single motor neuron serves about 10 muscle fibers, resulting in precision movements. The eyes move in synchrony by contracting synergistic muscles while relaxing antagonistic muscles.

The extrinsic ocular muscles are innervated by three cranial nerves (table 15.3). Innervation of the other skeletal and smooth muscles that serve the eye is also indicated in table 15.3.

A physical examination may include an eye movement test. As the patient's eyes follow the movement of a physician's finger, the physician can assess weaknesses in specific muscles or dysfunctions of specific cranial nerves. The patient experiencing *double vision (diplopia)* when moving his eyes may be suffering from muscle weakness. Looking laterally tests the abducens cranial nerve, looking inferiorly and laterally tests the trochlear cranial nerve, and looking cross-eyed tests the oculomotor and trochlear cranial nerves of both eyes.

Structure of the Eyeball

The eyeball of an adult is essentially spherical and is approximately 25 mm (1 in.) in diameter. About four-fifths of the eyeball is positioned within the orbit of the skull. The eyeball consists of three basic layers: the fibrous tunic, the vascular tunic, and the internal tunic (fig. 15.14).

Fibrous Tunic The fibrous tunic is the outer layer of the eyeball. It is divided into two regions: the posterior five-sixths is the opaque sclera and the anterior one-sixth is the transparent cornea (fig. 15.15).

The toughened **sclera** is the white of the eye. It is composed of tightly bound elastic and collagenous fibers that give

trochlea: Gk. *trochos,* a wheel

sclera: Gk. *skleros,* hard

Table 15.3	Muscles of the Eye	
Extrinsic Muscles (Skeletal)		
Superior rectus	Oculomotor nerve (III)	Rotates eye upward and toward midline
Inferior rectus	Oculomotor nerve (III)	Rotates eye downward and toward midline
Medial rectus	Oculomotor nerve (III)	Rotates eye toward midline
Lateral rectus	Abducens nerve (VI)	Rotates eye away from midline
Superior oblique	Trochlear nerve (IV)	Rotates eye downward and away from midline
Inferior oblique	Oculomotor nerve (III)	Rotates eye upward and away from midline
Intrinsic Muscles (Smooth)		
Ciliary muscles	Oculomotor nerve (III) parasympathetic fibers	Causes suspensory ligament to relax
Iris, circular muscles	Oculomotor nerve (III) parasympathetic fibers	Causes size of pupil to decrease
Iris, radial muscles	Sympathetic fibers	Causes size of pupil to increase

From Stuart Ira Fox, *Human Physiology*, 4th ed. Copyright © 1993 Wm. C. Brown Communications, Inc., Dubuque, Iowa. All Rights Reserved. Reprinted by permission.

shape to the eyeball and protect its inner structures. The sclera is avascular but does contain sensory receptors for pain. The large **optic nerve** exits through the sclera at the posterior part of the eyeball.

The **cornea** is transparent and convex to permit the passage and cause the refraction (bending) of incoming light waves. The transparency of the cornea is due to tightly packed, avascular, dense connective tissue. Also, the relatively few cells that are present in the cornea are arranged in unusually regular patterns. The circumferential edge of the cornea is continuous structurally with the sclera. The outer surface of the cornea is covered with a thin, nonkeratinized, stratified squamous epithelial layer called the **anterior epithelium,** which is actually a continuation of the bulbar conjunctiva of the sclera (fig. 15.9).

A defective cornea can be replaced with a donor cornea in a surgical procedure called a *corneal transplant (keratoplasty)*. A defective cornea is one that does not transmit or refract light effectively due to its shape, scars, or disease. During a corneal transplant, the defective cornea is excised and replaced with a transplanted cornea that is sutured into place. It is considered to be the most successful type of homotransplant (between individuals of the same species).

optic: L. *optica*, see
cornea: L. *cornu*, horn

Figure 15.14 Sagittal sections of the eyeball. (*a*) A posterior portion showing the tunics of the eye, the optic disc, and the optic nerve and (*b*) an anterior portion showing the cornea and the lens.

Tunicas:
- Internal
- Vascular
- Fibrous

Optic disc

Optic nerve

(a)

Cornea

Lens

Anterior chamber

Vitreous chamber

(b)

UNIT V

Figure 15.15 The internal anatomy of the eyeball.

Figure 15.16 The structure of the anterior portion of the eyeball. (a) A sagittal view, (b) an anterior view of the lens and supporting structures, and (c) a scanning electron micrograph in anterior view showing the relationship between the lens, zonular fibers, and ciliary muscle of the eye. (Parts a and b adapted from "How the Eye Focuses" by James F. Koretz and George Handleman. Copyright © 1988 by Scientific American, Inc. All rights reserved.)

Vascular Tunic   The vascular tunic of the eyeball consists of the choroid, the ciliary body, and the iris (fig. 15.15).

The **choroid** is a thin, highly vascular layer that lines most of the internal surface of the sclera. The choroid contains nu-

merous pigment-producing melanocytes, which give it a dark brownish color that prevents light waves from being reflected out of the eyeball. There is an opening in the choroid at the back of the eyeball where the optic nerve is located.

The **ciliary body** is the thickened, anterior portion of the vascular tunic that forms an internal muscular ring toward the front of the eyeball (fig. 15.16). Three distinct planes of smooth muscle fibers called **ciliary muscles** are found within the ciliary body.

choroid: Gk. *chorion,* membrane

Figure 15.17 Changes in the shape of the lens during accommodation. (*a*) The lens is flattened for distant vision when the ciliary muscle fibers are relaxed and the suspensory ligament is taut. (*b*) The lens is more spherical for close-up vision when the ciliary muscle fibers are contracted and the suspensory ligament is relaxed.

Figure 15.18 Dilation and constriction of the pupil. In dim light, the radially arranged smooth muscle fibers are stimulated to contract by sympathetic stimulation, dilating the pupil. In bright light, the circularly arranged smooth muscle fibers are stimulated to contract by parasympathetic stimulation, constricting the pupil.

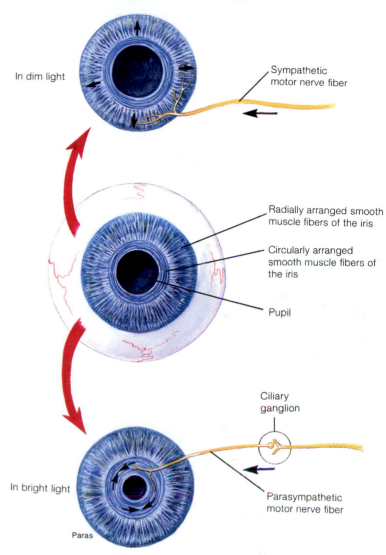

Numerous extensions of the ciliary body called **ciliary processes** attach to the **zonular fibers,** which in turn attach to the lens. Collectively, the zonular fibers constitute the **suspensory ligament.** The transparent **lens** consists of tight layers of protein fibers arranged like the layers of an onion. A thin, clear **lens capsule** encloses the lens and provides attachment for the suspensory ligament (see figs. 15.16 and 15.24).

The shape of the lens determines the extent to which the light waves that pass through will be refracted. Constant tension of the suspensory ligament, when the ciliary muscles are relaxed, flattens the lens somewhat (fig. 15.17). Contraction of the ciliary muscles relaxes the suspensory ligament and makes the lens more spherical. The constant tension within the lens capsule results in the surface of the lens becoming more convex when the suspensory ligament is not taut. A flattened lens facilitates viewing a distant object, whereas a rounded lens facilitates viewing a close object.

The **iris** is the anterior portion of the vascular tunic and is continuous with the choroid. The iris is viewed from the outside as the colored portion of the eyeball (figs. 15.15 and 15.16). It consists of smooth muscle fibers arranged in a circular and a radial pattern. The contraction of the smooth muscle fibers regulates the diameter of the **pupil** (table 15.4 and fig. 15.18), an opening in the center of the iris. Contraction of the circularly arranged fibers of the iris, stimulated by bright light, constricts the pupil and diminishes the amount of light entering the eyeball (see fig. 13.8). Contraction of the radially arranged fibers, in response to dim light, dilates the pupil and permits more light to enter.

---

zonular fibers: L. *zona,* a girdle

iris: Gk. *irid,* rainbow

Table 15.4	Location and Functions of Structures of the Eyeball		
Tunic and Structure	**Location**	**Composition**	**Function**
Fibrous tunic	Outer layer of eyeball	Avascular connective tissue	Gives shape to eyeball
Sclera	Posterior, outer layer; white of the eye	Tightly bound elastic and collagen fibers	Supports and protects eyeball
Cornea	Anterior surface of eyeball	Tightly packed dense connective tissue—transparent and convex	Transmits and refracts light
Vascular tunic (uvea)	Middle layer of eyeball	Highly vascular pigmented tissue	Supplies blood; prevents reflection
Choroid	Middle layer in posterior portion of eyeball	Vascular layer	Supplies blood to eyeball
Ciliary body	Anterior portion of vascular tunic	Smooth muscle fibers and glandular epithelium	Supports the lens through suspensory ligament and determines its thickness; secretes aqueous humor
Iris	Anterior portion of vascular tunic continuous with ciliary body	Pigment cells and smooth muscle fibers	Regulates the diameter of the pupil, and hence the amount of light entering the vitreous chamber
Internal tunic	Inner layer of eyeball	Tightly packed photoreceptors, neurons, blood vessels, and connective tissue	Provides location and support for rods and cones
Retina	Principal portion of internal tunica	Photoreceptor neurons (rods and cones), bipolar neurons, and ganglion neurons	Photoreception; transmits impulses
Lens (not part of any tunic)	Between posterior and vitreous chambers; supported by suspensory ligament of ciliary body	Tightly arranged protein fibers; transparent	Refracts light and focuses onto fovea centralis

Internal Tunic The internal tunic is the innermost layer of the eyeball. The **retina** is the structure of the internal tunic that covers the choroid. The retina consists of a thin, outer **pigmented layer** in contact with the choroid and a thick, inner **nervous layer,** or visual portion. The retina is principally in the posterior part of the eyeball (fig. 15.15). The visual layer of the retina terminates in a jagged margin near the ciliary body called the **ora serrata.** The pigmented layer extends anteriorly over the back of the ciliary body and iris.

> The pigmented layer and nervous layer of the retina are not attached to each other, except surrounding the optic nerve and at the ora serrata. Because of this loose connection, the two layers may become separated as a *detached retina.* Such a separation can be corrected by fusing the layers with a laser.

The nervous layer of the retina is composed of three principal layers of neurons. Listing them in the order in which they conduct impulses, they are rod and cone cells, bipolar neurons, and ganglion neurons (fig. 15.19). In terms of the passage of light, however, the order is reversed. Light must first pass through the layer of ganglion cells and then bipolar cells before reaching the photoreceptors.

Rods and **cones** are photoreceptors. Rods number over 100 million per eye and are more slender and elongated than cones (fig. 15.20). Rods are positioned on the peripheral parts of the retina, where they respond to dim light for black-and-white vision. They also respond to form and movement but provide poor visual acuity. Cones number about 7 million per eye. They provide daylight color vision and are responsible for visual acuity. The photoreceptors synapse with **bipolar neurons,** which in turn synapse with the **ganglionic neurons.** The axons of ganglionic neurons leave the eye as the optic nerve.

Cones are concentrated in a depression near the center of the retina called the **fovea centralis,** which is the area of keenest vision (figs. 15.15 and 15.21). Surrounding the fovea centralis is the yellowish **macula lutea,** which also has an abundance of cones (fig. 15.21). There are no photoreceptors at the point where the optic nerve is attached. This area is a blind spot and is referred to as the **optic disc** (figs. 15.21 and 15.22). Normally a person is unaware of the blind spot because the eyes continually move about and because an object is viewed from a different angle with each eye. The blind spot can easily be demonstrated as described in figure 15.23.

Blood Supply to the Eyeball Both the choroid and the retina are richly supplied with blood. Two **ciliary arteries** pierce the sclera at the posterior aspect of the eyeball and traverse the choroid to the ciliary body and base of the iris. Although the ciliary arteries enter the eyeball independently, they anastomose (connect) extensively throughout the choroid.

The **central artery** (central retinal artery) branches from the ophthalmic artery and enters the eyeball in contact with the optic nerve. As the central artery passes through the optic disc, it divides into superior and inferior branches, each of which then divides into temporal and nasal branches to serve the inner layers of the retina (see fig. 15.15). The **central vein** drains blood from the eyeball through the optic disc. The branches of the central artery can be observed within the eyeball through an ophthalmoscope (fig. 15.21).

ora serrata: L. *ora*, margin; *serra*, saw

fovea: L. *fovea*, small pit
macula lutea: L. *macula*, spot; *luteus*, yellow

Figure 15.19 The layers of the retina. The retina is inverted, so that light must pass through various layers of nerve cells before reaching the photoreceptors (rods and cones). (a) A schematic diagram and (b) a light micrograph.

(a)

(b)

An examination of the internal eyeball with an ophthalmoscope is frequently part of a routine physical examination. Capillary vessels can be seen within the eyeball. If they appear abnormal (for example, constricted, dilated, or hemorrhaged), they may be symptomatic of certain diseases or body dysfunctions. Diseases such as arteriosclerosis, diabetes, cataracts, or glaucoma can be detected by examining the internal eyeball.

Figure 15.20 Photoreceptor cells of the retina.

Cavities and Chambers of the Eyeball The interior of the eyeball is separated by the lens into an **anterior cavity** and a **posterior cavity** (see fig. 15.15). The anterior cavity is subdivided by the iris into an **anterior chamber** and a **posterior chamber** (see fig. 15.15). The anterior chamber is located in front of the iris and behind the cornea. The posterior chamber is located between the iris and the suspensory ligament and the lens. The anterior and posterior chambers connect through the pupil and are filled with a watery fluid called **aqueous humor.** The constant production of aqueous humor maintains an *intraocular pressure* of about 12 mm of mercury within the anterior and posterior chambers. Aqueous humor also provides nutrients and oxygen to the avascular lens and cornea. An estimated 5–6 mℓ of aqueous humor is secreted each day from the vascular epithelium of the ciliary body (fig. 15.24). From its site of secretion within the posterior chamber, the aqueous humor passes through the pupil into the anterior chamber. From here, it drains from the eyeball through the **scleral venous sinus** (canal of Schlemm) into the bloodstream. The scleral venous sinus is located at the junction of the cornea and iris.

Between the lens and retina is the large **vitreous chamber,** which is filled with a transparent jellylike **vitreous humor.**

canal of Schlemm: from Friedrich S. Schlemm, German anatomist, 1795–1858
vitreous: L. *vitreus,* glassy

Figure 15.21 (a) A view of the retina as seen with an ophthalmoscope and (b) a diagrammatic view. Optic nerve fibers leave the eyeball at the optic disc to form the optic nerve. (Note the blood vessels that can be seen entering the eyeball at the optic disc.)

(a)

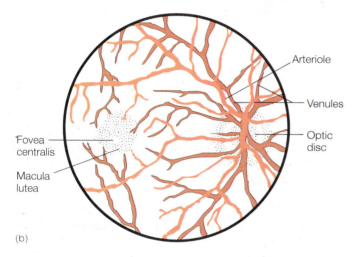

Arteriole

Venules

Optic disc

Fovea centralis

Macula lutea

(b)

Figure 15.22 The optic disc is a small area of the retina where the fibers of the ganglion neurons emerge forming the optic nerve. The optic disc is frequently called the blind spot because it is devoid of rods and cones.

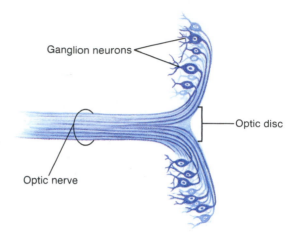

Ganglion neurons

Optic disc

Optic nerve

Figure 15.23 The blind spot. Hold the drawing about twenty inches from your face with your left eye closed and your right eye focused on the circle. Slowly move the drawing closer to your face until the cross disappears. This occurs because the image of the cross is focused on the optic disc, where photoreceptors are absent.

Figure 15.24 Aqueous humor maintains the intraocular pressure within the anterior and posterior chambers of the eyeball. It is secreted into the posterior chamber, flows through the pupil into the anterior chamber, and drains from the eyeball through the scleral venous sinus.

Vitreous humor also contributes to the intraocular pressure to maintain the shape of the eyeball and to hold the retina against the choroid. Unlike aqueous humor, vitreous humor is not continuously produced but formed prenatally. Additional vitreous humor forms as a person's eyes become larger through normal body growth.

The location and functions of the structures within the eyeball are summarized in table 15.4.

Puncture wounds to the eyeball are especially dangerous and frequently cause blindness. Protective equipment such as goggles, shields, and shatterproof lenses should be used in hazardous occupations and certain sports. If the eye is punctured, the main thing to remember is to *not remove the object* if it is still impaling the eyeball. Removal may allow the fluids to drain from the eyeball and cause the loss of intraocular pressure. This is particularly serious if the nonreplaceable vitreous humor drains from the eyeball.

Aqueous humor

Cornea

Iris

Scleral venous sinus

Anterior chamber

Ciliary body

Lens

Lens capsule

Posterior chamber

Vitreous humor

Lew

Figure 15.25   The refraction of light waves within the eyeball causes the image of an object to be inverted on the retina.

Function of the Eyeball

The focusing of light waves and stimulation of photoreceptors of the retina require five basic processes:

1. the transmission of light waves through transparent media of the eyeball;
2. the refraction of light waves through media of different densities;
3. accommodation of the lens to focus the light waves;
4. constriction of the pupil by the iris to regulate the amount of light entering the vitreous chamber; and
5. convergence of the eyeballs so that visual acuity is maintained.

Visual impairment may result if one or more of these processes does not function properly (see "Clinical Considerations").

Transmission of Light Waves   Light waves entering the eyeball pass through four transparent media before they stimulate the photoreceptors. In sequence, the media through which light waves pass are the cornea, aqueous humor, lens, and vitreous humor. The cornea and lens are solid media composed of tightly packed, avascular protein fibers. An additional thin, transparent membranous continuation of the conjunctiva covers the outer surface of the cornea. The aqueous humor is a low-viscosity fluid and the vitreous humor is jellylike in consistency.

Refraction of Light Waves   Refraction is the bending of light waves. Refraction occurs as light waves pass at an oblique angle from a medium of one optical density to a medium of another optical density. The convex cornea is the principal refractive medium; the fluids within the various chambers produce minimal refraction. The lens is particularly important for refining

and altering refraction. Of the refractive media, only the lens can be altered in shape to achieve precise refraction.

The refraction of light waves is so extensive that the visual image is formed upside down on the retina (fig. 15.25). Nerve impulses of the image in this position are relayed to the visual cortex of the occipital lobe, where the inverted image is interpreted as right side up.

Accommodation of the Lens   Accommodation is the automatic adjustment of the curvature of the lens by contraction of ciliary muscles to bring light waves into sharp focus on the retina. The lens of the eyeball is biconvex. When an object is viewed from a distance of less than about 20 feet, the lens must make an adjustment, or accommodation, for clear focus on the retina. Contraction of the smooth muscle fibers of the ciliary body causes the suspensory ligament to relax and the lens to become thicker (see fig. 15.17). A thicker, more convex lens causes a greater convergence of the refracted light waves necessary for viewing close objects.

Constriction of the Pupil   Constriction of the pupil through parasympathetic stimulation that causes contraction of the circular muscle fibers of the iris (see fig. 15.18) is important for two reasons. One is the reduction in the amount of light that enters the vitreous chamber. A reflexive pupil constriction protects the retina from sudden or intense bright light. Another more important function is that a reduced pupil diameter prevents light waves from entering the vitreous chamber through the periphery of the lens. Light waves refracted from the periphery would not be brought into focus on the retina and would cause blurred vision. Autonomic constriction of the pupil and accommodation of the lens occur simultaneously.

UNIT V

Figure 15.26 The electromagnetic spectrum (*top*) is shown in Angstrom units (1Å = 10⁻¹⁰ meter). The visible spectrum (*bottom*) comprises only a small range of this spectrum and is shown in nanometer units (1 nm = 10⁻⁹ meter).

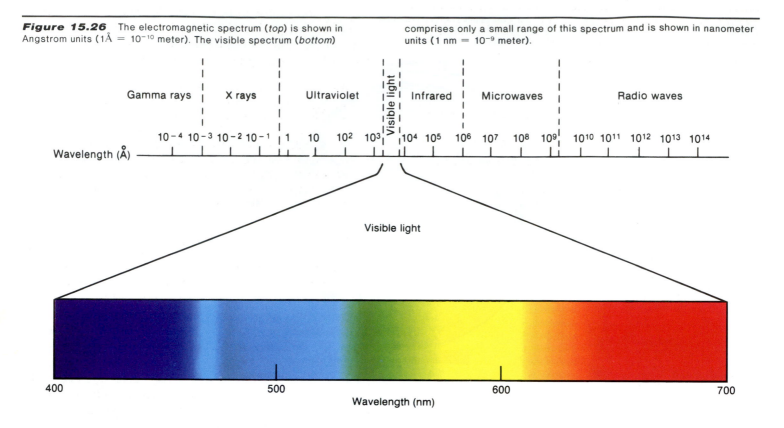

Visible light

Convergence of the Eyeballs Convergence of the eyeballs means that the eyes rotate more medially when viewing closer objects. In fact, focusing on an object close to the tip of the nose causes a person to appear cross-eyed. The eyeballs must converge when viewing close objects because only then can the light waves focus on the same portions in both retinas.

> *Amblyopia exanopsia*, commonly called lazy eye, is a condition of ocular muscle weakness causing a deviation of one eye so that there is not a concurrent convergence of both eyeballs. Because of disconjugate fixation, two images are received by the optic portion of the cerebral cortex—one of which is suppressed to avoid *diplopia (double vision)* or images of unequal clarity. A person who has amblyopia will experience dimness of vision and partial loss of sight. Amblyopia is frequently tested for in young children because if untreated before age 6, little can be done to strengthen the afflicted muscle.

Visual Spectrum The eyes transduce energy in the *electromagnetic spectrum* (fig. 15.26) into nerve impulses. Only a limited part of this spectrum can excite the photoreceptors. Electromagnetic energy with wavelengths between 400 and 700 nanometers (nm) constitute *visible light*. Light of longer wavelengths, which are in the infrared regions of the spectrum, do not have sufficient energy to excite photoreceptors but are perceived as heat. Ultraviolet light, which has shorter wavelengths and more energy than visible light, is filtered out by the yellow color of the eye's lens. Certain insects, such as honeybees, and people who have had their lenses removed, can see light in the ultraviolet range.

> Three distinct and specialized types of cones within the retina permit color vision (fig. 15.26). Different photosensitive pigments enable each type to absorb light waves primarily in the blue, green, or red portion of the color spectrum. Blue cones are stimulated by wavelengths between 400 and 550 nm; green cones are stimulated by wavelengths between 450 and 550 nm; and red cones are stimulated by wavelengths between 500 and 700 nm. *Color blindness* is the inability to distinguish colors, particularly reds and greens. It affects about 5% of the U.S. population. Color blindness for the majority of these people is a misnomer, however, since in nearly all cases there is a deficiency in the number of specific cones, not a total lack of them. Hence, there is some ability to distinguish color.

Neural Pathways of Vision, Eye Movements, and Processing of Visual Information

The photoreceptor neurons, rods and cones, are the functional units of sight in that they respond to light waves and produce nerve impulses. Nerve impulses from the rods and cones pass through bipolar neurons to ganglion neurons (see fig. 15.19). The optic nerve consists of axons of aggregated ganglion neurons that emerge through the posterior aspect of the eyeball. The two optic nerves (one from each eyeball) converge at the **optic chiasma** (fig. 15.27). The fibers partially cross at the optic chiasma so that all the fibers arising from the medial (nasal) half of each retina cross to the opposite side. The fibers of the optic

UNIT V

Figure 15.27 Visual fields of the eyes and neural pathways for vision. An overlapping of the visual field of each eye provides binocular vision, which is the ability to perceive depth.

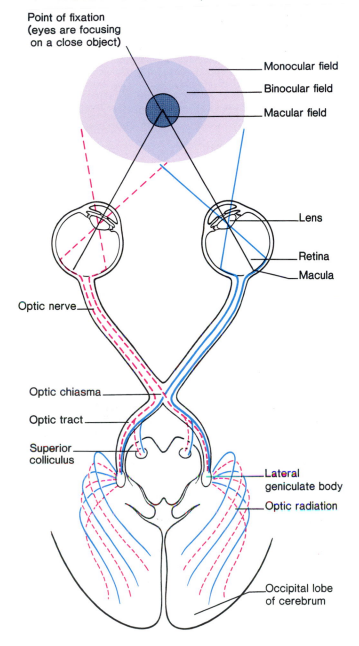

Point of fixation (eyes are focusing on a close object)

Monocular field

Binocular field

Macular field

Lens

Retina

Macula

Optic nerve

Optic chiasma

Optic tract

Superior colliculus

Lateral geniculate body

Optic radiation

Occipital lobe of cerebrum

optic radiation transmits impulses to the **striate** *(stri'āt)* **cortex** area of the occipital cerebral lobe. This entire arrangement of visual fibers is known as the **geniculostriate system,** which is responsible for perception of the visual field.

> The nerve fibers that cross at the optic chiasma arise from the retinas in the medial portions of the eyeballs. The photoreceptors of these fibers are stimulated by light entering the eyeball from the periphery. If the optic chiasma were severed longitudinally, peripheral vision would be lost and leave only "tunnel vision." If an optic tract were severed, both eyes would be partially blind, the lateral field of vision lost for one eye, and the medial field of vision lost for the other.

Superior Colliculi and Eye Movements Neural pathways from the superior colliculi to motor neurons in the spinal cord help mediate the startle response to the sight, for example, of an unexpected intruder. Other nerve fibers from the superior colliculi stimulate the **extrinsic eye muscles** (table 15.3), which are the skeletal muscles that move the eyes.

There are two types of eye movements coordinated by the superior colliculi. *Smooth pursuit movements* track moving objects and keep the image focused on the fovea centralis. *Saccadic (să-kad-ik) eye movements* are short (lasting 20–50 msec), jerky movements that occur while the eyes appear to be still. These saccadic movements are believed to be important in maintaining visual acuity.

The tectal system is also involved in the control of the **intrinsic eye muscles**—the iris and the muscles of the ciliary body. Shining a light into one eye stimulates the *pupillary reflex* in which both pupils constrict. This is caused by activation of parasympathetic neurons by fibers from the superior colliculi. Postganglionic neurons in the ciliary ganglia behind the eyes, in turn, stimulate constrictor fibers in the iris. Contraction of the ciliary body during *accommodation* also involves stimulation of the superior colliculi.

Processing of Visual Information In order for visual information to have meaning, it must be associated with past experience and integrated with information from other senses. Some of this higher processing occurs in the inferior temporal lobes of the cortex. Experimental removal of these areas from monkeys impairs their ability to remember visual tasks that they previously learned and hinders their ability to associate visual images with the significance of the object. Monkeys with their inferior temporal lobes removed, for example, will handle a snake without fear. The symptoms produced by loss of the inferior temporal lobes are known as the *Kluver–Bucy syndrome.*

In an attempt to reduce the symptoms of severe epilepsy, surgeons at one time cut the corpus callosum in some patients. This fiber tract, as previously described, transmits impulses between the right and left cerebral hemispheres. The right cerebral hemisphere of patients with such *split brains,* therefore, receives sensory information from only the left half of the external world. The left hemisphere, similarly cut off from communication with the right hemisphere, receives sensory information from only the right half of the external world.

nerve that arise from the lateral (temporal) half of the retina do not cross. The **optic tract** is a continuation of optic nerve fibers from the optic chiasma and is composed of fibers arising from the retinas of both eyeballs.

As the optic tracts enter the brain, some of the fibers in the tracts terminate in the **superior colliculi** *(ko-lik'yoo-li)*. These fibers and the motor pathways they activate constitute the **tectal system,** which is responsible for body-eye coordination.

Approximately 70%–80% of the fibers in the optic tract pass to the **lateral geniculate** *(jĕ-nik'yŭ-lit)* **body** of the thalamus (fig. 15.27). Here the fibers synapse with neurons whose axons constitute a pathway called the **optic radiation.** The

UNIT V

Developmental Exposition

The Eye

Exhibit I The development of the eye. (*a*) An anterior view of the developing head of a 22-day-old embryo and the formation of the optic vesicle from the neuroectoderm of the prosencephalon (forebrain). (*b*) The development of the optic cup. The lens vesicle is formed (*c*) as the ectodermal lens placode invaginates during the 4th week.

The hyaloid vessels become enclosed within the optic nerve (c_1 and e_1) as there is fusion of the optic fissure. (*d*) The basic shape of the eyeball and the position of its internal structures are established during the 5th week. The successive development of the eye is shown at 6 weeks (*e*) and at 20 weeks (*f*), respectively. (*g*) The eye of the newborn.

(a)

Mesoderm

Surface ectoderm

Optic vesicle

Prosencephalon

Neuroectoderm

(b)

Optic cup:

Outer layer

Inner layer

Lens fovea

Optic stalk

(c)

Lens vesicle

(c_1)

Optic fissure

Hyaloid vessels

(e_1)

Optic nerve

(d)

Undifferentiated mesoderm

Lens

Optic nerve

Hyaloid vessels

(e)

Developing cornea

Developing sclera

(f)

Retina:

Sensory layer

Pigmented layer

Vitreous body

Choroid

Sclera

Ciliary body

Iris

Eyelids fused

Cornea

Conjunctival sac

Lens capsule

(g)

Suspensory ligament

Eyelids open

Conjunctiva

Cornea

Hyaloid canal

Explanation

The development of the eye is a complex, rapid process starting early in the fourth week and involving the precise interaction of neuroectoderm, surface ectoderm, and mesoderm. The initial differentiation involves neuroectoderm forming a lateral diverticulum on each side of the prosencephalon (forebrain). As the diverticulum increases in size, the distal portion dilates to become the **optic vesicle** and the proximal portion constricts to become the **optic stalk** (exhibit I). Once the optic vesicle is formed, the overlying surface ectoderm thickens and invaginates. The thickened portion is the **lens placode** and the invagination is the **lens fovea.**

During the fifth week, the lens placode is depressed and eventually cut off from the surface ectoderm, causing the formation of the **lens vesicle.** Simultaneously, the optic vesicle invaginates and differentiates into the two-layered **optic cup.** A groove called the **optic fissure** appears along the inferior surface of the optic cup in which the **hyaloid artery** and **hyaloid vein** traverse to serve the developing eyeball. The walls of the optic fissure eventually close so that the hyaloid vessels are within the tissue of the optic stalk. They become the **central vessels of the retina** of the mature eye. The optic stalk eventually becomes the **optic nerve,** composed of sensory axons from the retina.

By the early part of the seventh week, the optic cup has differentiated into two sheets of epithelial tissue that become the sensory and pigmented layers of the **retina.** Both of these layers also line the entire vascular coat, including the **ciliary body, iris,** and the **choroid.** A proliferation of cells in the lens vesicle leads to the formation of the lens. The **lens capsule** forms from the mesoderm surrounding the lens, as does the **vitreous body.** Mesoderm surrounding the optic cup differentiates into two distinct layers of the developing eyeball. The inner layer of mesoderm becomes the vascular **choroid** and the outer layer becomes the toughened **sclera** posteriorly and the transparent **cornea** anteriorly. Once the cornea has formed, additional surface ectoderm gives rise to the thin **conjunctiva** covering the anterior surface of the eyeball. Epithelium of the **eyelids** and the **lacrimal glands** and **duct** develop from surface ectoderm, whereas the **extrinsic eye muscles** and all connective tissues associated with the eye develop from mesoderm. These accessory structures of the eye gradually develop during the embryonic period and into the fetal period as late as the fifth month.

---

hyaloid: Gk. *hyalos,* glass; *eiodos,* form

Experiments with split-brain patients have revealed that the functions of the two hemispheres are different. This is true even though each hemisphere would normally receive input from both halves of the external world through the corpus callosum. If the sensory image of an object, such as a key, is delivered to only the left hemisphere (by showing it to only the right visual field), the object can be named. If the object is presented to the right cerebral cortex, the person knows what the object is but cannot name it. Experiments such as this suggest that (in right-handed people) the left hemisphere is needed for language and the right hemisphere is responsible for pattern recognition.

1. List the associated structures of the eye that either cause the eye to move or protect it within the orbit.
2. Diagram the structure of the eye and label the following: sclera, cornea, choroid, retina, fovea centralis, iris, pupil, lens, and ciliary body. What are the principal cells or tissues in each of the three layers of the eye?
3. Trace the path of light through the three chambers of the eye and explain the mechanism of light refraction. Describe how the eye is focused for viewing distant and near objects.
4. List the different layers of the retina and describe the path of light and of nerve activity through these layers. Continue tracing the path of a visual impulse to the cerebral cortex and list in order the structures traversed.

Senses of Hearing and Balance

Structures of the outer, middle, and inner ear are involved in the sense of hearing. The inner ear also contains structures that provide a sense of balance, or equilibrium. The development of the ear begins during the fourth week and is complete by week 32.

Objective 14. Describe the structures of the ear that relate to hearing and list their locations and functions.

Objective 15. Trace the path of sound waves through the ear and describe how they are transmitted and converted to nerve impulses.

Objective 16. Explain the mechanisms by which equilibrium is maintained.

Structure of the Ear

The ear is the organ of hearing and equilibrium. It contains receptors that respond to movements of the head and receptors that convert sound waves into nerve impulses. Impulses from both receptor types are transmitted through the vestibulocochlear (eighth cranial) nerve to the brain for interpretation. The ear consists of three principal regions: the outer ear, the middle ear, and the inner ear (fig. 15.28).

UNIT V

Figure 15.28 The ear. (Note the outer, middle, and inner regions indicated by dashed lines.)

Outer Ear The outer ear consists of the auricle, or pinna, and the external auditory canal. The external auditory canal is the fleshy tube that is fitted into the bony tube called the *external acoustic meatus.* The **auricle** *(or'ĭ-kul)* is the visible fleshy appendage attached to the side of the head. It consists of a cartilaginous framework of elastic connective tissue covered with skin. The rim of the auricle is called the **helix,** and the inferior fleshy portion is the **earlobe** (fig. 15.29). The earlobe is the only portion of the auricle that is not supported with cartilage. The auricle has a ligamentous attachment to the skull and poorly developed auricular muscles inserting anteriorly, superiorly, and posteriorly within it. The blood supply to the auricle is from the posterior auricular and occipital arteries, which branch from the external carotid and superficial temporal arteries, respectively. The structure of the auricle directs sound waves into the external auditory canal.

The **external auditory canal** is a slightly S-shaped tube about 2.5 cm (1 in.) in length, extending slightly upward from the auricle to the tympanic *(tim-pan'ik)* membrane (fig. 15.28). The skin that lines the canal contains fine hairs and sebaceous glands near the entrance. Specialized wax-secreting glands, called **ceruminous** *(sĕ-roo'mi-nus)* **glands,** are located in the skin, deep within the canal. *Cerumen* (earwax) secreted from ceruminous glands keeps the tympanic membrane soft and waterproof. Cerumen and the hairs also help to prevent small foreign objects from reaching the tympanic membrane. The bitter cerumen is probably an insect repellent as well.

The **tympanic membrane** (eardrum) is a thin, double-layered, epithelial partition between the external auditory canal

and the middle ear. It is approximately 1 cm in diameter and is composed of an outer concave layer of stratified squamous epithelium and an inner convex layer of low columnar epithelium. The tympanic membrane is extremely sensitive to pain and is innervated by the auriculotemporal nerve (a branch of the mandibular nerve of the trigeminal [V] cranial nerve) and the auricular nerve (a branch of the vagus [X] cranial nerve).

Inspecting the tympanic membrane with an otoscope during a physical examination yields significant information about the condition of the middle ear. A diagram of the normal tympanic membrane is presented in figure 15.30. The color, curvature, presence of lesions, and position of the malleus of the middle ear are features of particular importance. If ruptured, the tympanic membrane can generally regenerate and readily heal itself.

Middle Ear The laterally compressed middle ear is an air-filled chamber called the **tympanic cavity** in the petrous part of the temporal bone (see figs. 15.28 and 15.31). The tympanic membrane separates the middle ear from the external auditory canal of the outer ear. A bony partition containing the **vestibular window** (oval window) and the **cochlear window** (round window) separates the middle ear from the inner ear.

There are two openings into the tympanic cavity. The **epitympanic recess** in the posterior wall connects the tympanic cavity to the **mastoidal air cells** within the mastoid

otoscope: Gk. *otikos,* ear; *skopein,* to examine

Figure 15.29 The surface anatomy of the auricle of the ear.

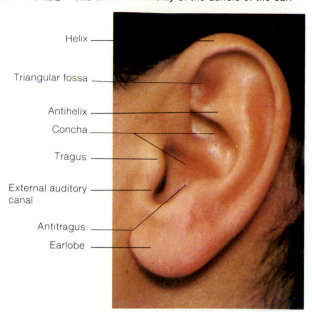

Helix

Triangular fossa

Antihelix

Concha

Tragus

External auditory canal

Antitragus

Earlobe

Figure 15.30 The right tympanic membrane as seen with an auriscope (a type of otoscope). When the tympanic membrane is illuminated with an auriscope, a slight concavity at the manubrium of the malleus produces a cone of light that radiates anteriorly and inferiorly.

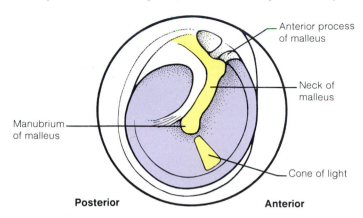

Anterior process of malleus

Neck of malleus

Manubrium of malleus

Cone of light

Posterior   **Anterior**

process of the temporal bone. The **auditory** (eustachian) **tube** connects the tympanic cavity anteriorly with the nasopharynx and equalizes air pressure on both sides of the tympanic membrane.

A series of three **auditory ossicles** crosses the tympanic cavity from the tympanic membrane to the vestibular window (fig. 15.31). These tiny bones, from outer to inner, are the **malleus** *(mal'e-us)* (hammer), **incus** (anvil), and **stapes** (stirrup). The auditory ossicles are attached to the wall of the tympanic cavity by ligaments. Vibrations of the tympanic membrane cause the auditory ossicles to move and transmit sound waves across the tympanic cavity to the vestibular window. Vibration of the vestibular window moves a fluid within the inner ear and stimulates the receptors of hearing.

As the auditory ossicles transmit vibrations, they act as a lever system to increase the force of vibration. In addition, the force of vibration is intensified as it is transmitted from the relatively large surface of the tympanic membrane to the smaller surface area of the vestibular window. The combined effect increases the force of the vibrations about 20 times.

Two small skeletal muscles, the **tensor tympani** and the **stapedius** *(sta-pi'de-us)* (fig. 15.31), attach to the malleus and stapes, respectively, and contract reflexively to protect the inner ear against loud noises. When contracted, the tensor tympani muscle pulls the malleus inward, and the stapedius muscle pulls the stapes outward. This combined action reduces the force of vibration of the auditory ossicles.

eustachian tube: from Bartolommeo E. Eustachio, Italian anatomist, 1520–74
malleus: L. *malleus*, hammer
incus: L. *incus*, anvil
stapes: L. *stapes*, stirrup

The mucous membranes that line the tympanic cavity, the mastoidal air cells, and the auditory tube are continuous with those of the nasopharynx. For this reason, infections of the nose or throat may spread to the tympanic cavity and cause a middle-ear infection, or even to the mastoidal air cells and cause *mastoiditis*. Forcefully blowing the nose advances the spread of the infection.

An equalization of air pressure on both sides of the tympanic membrane is of functional importance to hearing. When atmospheric pressure is increased, as occurs when descending from higher altitudes, the tympanic membrane bulges inward in response to the greater air pressure on the outside of the tympanic cavity. The bulging is painful and may impair hearing by reducing flexibility. The auditory tube, which is collapsed most of the time in adults, opens during swallowing or yawning and allows the air pressures to equalize.

Inner Ear   The inner ear, or **labyrinth** *(lab'i-rinth)*, contains the functional organs for hearing and equilibrium. The labyrinth consists of two parts: an outer **bony labyrinth** and a **membranous labyrinth** contained within the bony labyrinth (fig. 15.32). The space between the bony labyrinth and the membranous labyrinth is filled with a fluid called *perilymph*, secreted by cells lining the bony canals. Within the tubular chambers of the membranous labyrinth is another fluid called *endolymph*. These two fluids provide a liquid-conducting medium for the vibrations involved in hearing and the maintenance of equilibrium.

The bony labyrinth is structurally and functionally divided into three areas: vestibule, semicircular canals, and cochlea.

Vestibule   The vestibule is the central portion of the bony labyrinth that contains the vestibular window, into which the stapes fits, and the cochlear window on the opposite end (fig. 15.32).

The membranous labyrinth within the vestibule consists of two connected sacs called the **utricle** *(yoo'tri-kul)* and the **saccule** *(sak'yool)*. The utricle is larger than the saccule and lies in the upper back portion of the vestibule. Both the utricle and saccule contain receptors that are sensitive to gravity and linear movement of the head.

UNIT V

Figure 15.31 (*a*) The auditory ossicles and associated structures within the tympanic cavity. (*b*) The stapedius muscle arises from a bony protrusion called the pyramid.

Figure 15.32 The labyrinths of the inner ear. The membranous labyrinth is contained within the bony labyrinth. The principal structures of the inner ear are the cochlea for hearing and the vestibular organs (semicircular canals, utricle, and saccule) for balance and equilibrium.

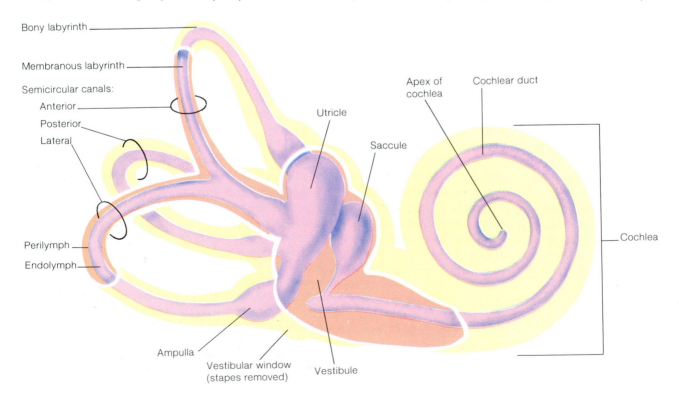

Figure 15.33 A cross section of the cochlea showing its three turns (indicated in boldface) and its three compartments—scala vestibuli, cochlear duct, and scala tympani.

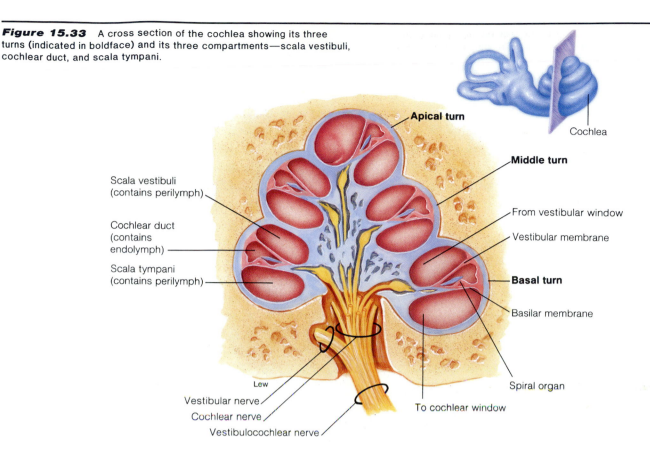

Scala vestibuli (contains perilymph)

Cochlear duct (contains endolymph)

Scala tympani (contains perilymph)

Apical turn

Middle turn

From vestibular window

Vestibular membrane

Basal turn

Basilar membrane

Spiral organ

To cochlear window

Lew

Vestibular nerve

Cochlear nerve

Vestibulocochlear nerve

Cochlea

Semicircular Canals   The three bony semicircular canals of each ear are at right angles to each other and are positioned posteriorly to the vestibule. The thinner **semicircular ducts** form the membranous labyrinth within the semicircular canals (fig. 15.32). Each of the three semicircular ducts has a **membranous ampulla** *(am-pool'ă)* at one end and connects with the upper back part of the utricle. Receptors within the semicircular ducts are sensitive to angular acceleration and deceleration of the head, as in rotational movement.

Cochlea   The snail-shaped cochlea *(kok'le-ă)* is coiled two and a half times around a central axis of bone (fig. 15.33). There are three chambers in the cochlea (fig. 15.34). The upper chamber, the **scala** *(ska'lă)* **vestibuli,** begins at the vestibular window and is continuous with the vestibule. The lower chamber, the **scala tympani,** terminates at the cochlear window. The **cochlear duct** is the middle chamber. The roof of the cochlear duct is called the **vestibular membrane** and the floor is called the **basilar membrane.** The scala vestibuli and the scala tympani contain perilymph and are completely separated, except at the narrow apex of the cochlea called the **helicotrema** *(hel''ĭ-kŏ-tre'mă),* where they are continuous (see fig. 15.36). The cochlear duct is filled with endolymph and ends at the helicotrema.

The **spiral organ** (organ of Corti) is contained within the cochlear duct of the cochlea. The sound receptors that trans-

cochlea: L. *cochlea,* snail shell
scala: Gk. *scala,* staircase
organ of Corti: from Alfonso Corti, Italian anatomist, 1822–88

form mechanical vibrations into nerve impulses are located along the basilar membrane of this structure, making it the functional unit of hearing. The epithelium of the spiral organ consists of supporting cells and hair cells (figs. 15.34 and 15.35). The bases of the hair cells are anchored in the basilar membrane, and their tips are embedded in the tectorial membrane that forms a gelatinous canopy over them.

Sound Waves and Neural Pathways for Hearing

Sound Waves   Sound waves travel in all directions from their source, like ripples in a pond after a stone is dropped. These waves are characterized by their frequency and intensity. The **frequency,** or distances between crests of the sound waves, is measured in *hertz (Hz),* which is the modern designation for *cycles per second (cps).* The *pitch* of a sound is directly related to its frequency—the higher the frequency of a sound, the higher its pitch.

The **intensity,** or loudness of a sound, is directly related to the amplitude of the sound waves. This is measured in units known as *decibels (db).* A sound that is barely audible—at the threshold of hearing—has an intensity of zero decibels. Every 10 decibels indicates a tenfold increase in sound intensity: a sound is 10 times higher than threshold at 10 db, 100 times higher at 20 db, a million times higher at 60 db and 10 billion times higher at 100 db.

The ear of a trained, young individual can hear sound over a frequency range of 20,000–30,000 Hz, yet can distinguish

UNIT V

Figure 15.34 A section of the cochlea showing (*a*) the spiral organ within the cochlear duct. (*b*) The spiral organ in greater detail.

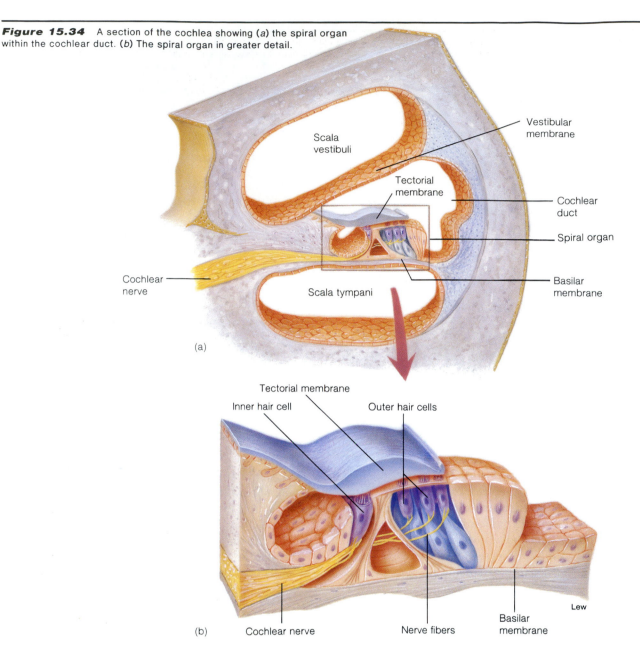

Scala vestibuli

Vestibular membrane

Tectorial membrane

Cochlear duct

Spiral organ

Cochlear nerve

Scala tympani

Basilar membrane

(a)

Tectorial membrane

Inner hair cell

Outer hair cells

Lew

(b)   Cochlear nerve

Nerve fibers

Basilar membrane

Figure 15.35 A scanning electron micrograph of the hair cells of the spiral organ.

Outer hair cells

Inner hair cells

between two pitches that have only a 0.3% difference in frequency. The human ear can detect differences in sound intensities of only 0.1 to 0.5 db, whereas the range of audible intensities covers 12 orders of magnitude (10^{12}), from the barely audible to the limits of painful loudness.

Sound waves funneled through the external auditory canal produce extremely small vibrations of the tympanic membrane. Movements of the tympanum during speech (average intensity of 60 db) are estimated to be equal to the diameter of a molecule of hydrogen.

Sound waves passing through the solid medium of the auditory ossicles are amplified about 20 times as they reach the footplate of the stapes against the vestibular window. As the vestibular window is displaced, pressure waves pass through the fluid medium of the scala vestibuli (fig. 15.36), around the hel-

Figure 15.36 An illustration of the cochlea, straightened to show the mechanism of sound-wave generation at specific sites along the basilar membrane. The scala vestibuli and the scala tympani, which contain perilymph, are continuous at the helicotrema. The cochlear duct, which contains endolymph, separates the scala vestibuli and the scala tympani. Sounds of low frequency (blue arrow) cause pressure waves of perilymph to pass through the helicotrema and displace the basilar membrane near its apex. Sounds of medium frequency (green arrow) cause pressure waves to displace the basilar membrane near its center. Sounds of high frequency (red arrow) cause pressure waves to displace the basilar membrane near its base. (The frequency of sound waves is measured in hertz [Hz].)

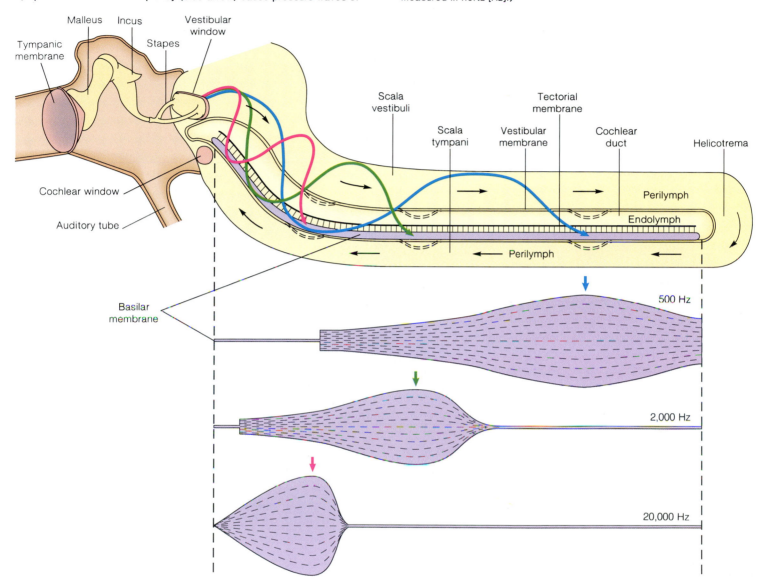

icotrema to the scala tympani. Movements of perilymph within the scala tympani, in turn, displace the cochlear window into the tympanic cavity.

When the sound frequency (pitch) is sufficiently low, there is adequate time for the pressure waves of perilymph within the upper scala vestibuli to travel around the helicotrema to the scala tympani. As the sound frequency increases, however, pressure waves of perilymph within the scala vestibuli do not have time to travel all the way to the apex of the cochlea. Instead, they are transmitted through the *vestibular membrane,* which sepa-

rates the scala vestibuli from the cochlear duct, and through the *basilar membrane,* which separates the cochlear duct from the scala tympani, to the perilymph of the scala tympani. The distance that these pressure waves travel, therefore, decreases as the sound frequency increases.

Sounds of low pitch (with frequencies below about 50 Hz) cause movements of the entire length of the basilar membrane—from the base to the apex. Higher sound frequencies result in maximum displacement of the basilar membrane closer to its base, as illustrated in figure 15.36.

Displacement of the basilar membrane and hair cells by movements of perilymph causes the microvilli that are

helicotrema: Gk. *helix,* a spiral; *trema,* a hole

Figure 15.37 Neural pathways for the sense of hearing.

Thalamus

Auditory cortex (temporal lobe)

Medial geniculate body of thalamus

Inferior colliculus

Midbrain

Cochlear nuclei

Spiral organ

Medulla oblongata

Cochlear nerve

Figure 15.38 The neural processing involved in maintaining equilibrium and balance.

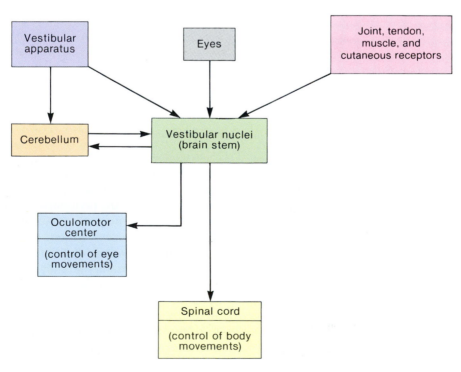

Vestibular apparatus

Eyes

Joint, tendon, muscle, and cutaneous receptors

Cerebellum

Vestibular nuclei (brain stem)

Oculomotor center

(control of eye movements)

Spinal cord

(control of body movements)

embedded in the tectorial membrane to bend. This stimulation excites the sensory cells, which causes the release of an unknown neurotransmitter that excites sensory endings of the cochlear nerve.

Neural Pathways for Hearing   Cochlear sensory neurons in the eighth cranial nerve synapse with neurons in the medulla oblongata (fig. 15.37) that project to the inferior colliculi of the midbrain. Neurons in this area in turn project to the thalamus, which sends axons to the auditory cortex of the temporal lobe.

Mechanics of Equilibrium

Maintaining equilibrium is a complex process that depends on constant input from the vestibular organs of both inner ears. Although the vestibular organs are the principal source of sensory information for equilibrium, visual input from the eyes, tactile receptors within the skin, and proprioceptors of tendons, muscles, and joints also provide sensory input that is vital for maintaining equilibrium (fig. 15.38).

The vestibular organs provide the CNS with two kinds of receptor information. One kind is provided by receptors within the saccule and utricle, which are sensitive to gravity and to linear acceleration and deceleration of the head, such as in riding in a car. The other is provided by receptors within the semicircular canals, which are sensitive to angular acceleration and deceleration, such as turning the head, spinning, or tumbling.

The receptor hair cells of the vestibular organs contain 20 to 50 microvilli and one cilium, called a **kinocilium** *(ki″no-sil′e-um)* (fig. 15.39). When the hair cells are displaced in the direction of the kinocilium, the cell membrane is depressed and becomes depolarized. When the hair cells are displaced in the opposite direction, the membrane becomes hyperpolarized.

Figure 15.39   Sensory hair cells of a vestibular organ. (*a*) A scanning electron photomicrograph of a kinocilium. (*b*) Sensory hairs (microvilli) and one kinocilium. (*c*) When hair cells are bent in the direction of the kinocilium, the cell membrane is depressed (see arrow) and the sensory neuron innervating the hair cell is stimulated. (*d*) When the hairs are bent in the opposite direction, the sensory neuron is inhibited.

(a)

Kinocilium

(b)   **At rest**

(c)   **Stimulated**

(d)   **Inhibited**

UNIT V

Developmental Exposition

The Ear

Exhibit II   The development of the inner ear. (*a*) A lateral view of a 22-day-old embryo showing the position of a transverse cut through the otic placode. (*a₁*) The otic placode of surface ectoderm begins to invaginate at 22 days. (*a₂*) By 24 days, a distinct otic fovea has formed and the neural ectoderm is positioned to give rise to the brain. (*b*) A lateral view of a 28-day-old embryo showing the position of a transverse cut through the otocyst. (*b₁*) By 28 days, the otic fovea has become a distinct otocyst. (*b₂*) The otocyst is in position in the 30-day-old embryo, where it gives rise to the structures of the inner ear. (*c–e*) Lateral views of the differentiating otocyst into the cochlea and semicircular canals from the 5th to the 8th week.

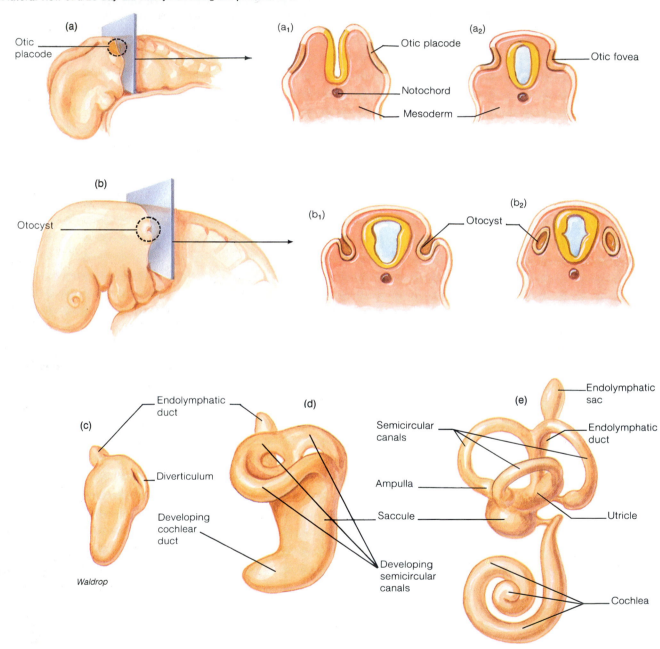

UNIT V

Exhibit III The formation of the cochlea and the spiral organ from the otic capsule. (*a–d*) Successive stages of development of the perilymphatic space and the spiral organ from the 8th to the 20th week.

Cartilaginous otic capsule

Cochlear duct

(a)

Perilymph vacuoles

(b)

Perilymphatic spaces

Developing scala vestibuli

Developing spiral organ

Developing scala tympani

Developing cochlear nerve

(c)

Bony labyrinth

Scala vestibuli

Cochlear duct

Spiral organ

Scala tympani

Cochlear nerve

(d)

Waldrop

Explanation

The ear begins to develop at the same time as the eye, early during the fourth week. All three embryonic germ layers—ectoderm, mesoderm, and endoderm—are involved in the formation of the ear. Both types of ectoderm (neuroectoderm and surface ectoderm) play a role.

The ear of an adult is structurally and functionally divided into an **outer ear**, a **middle ear**, and an **inner ear**, each of which has a separate embryonic origin. The inner ear does not develop from deep embryonic tissue as one might expect, but rather begins to form early in the fourth week when a plate of surface ectoderm called the **otic** (*o'tik*) **placode** appears lateral to the developing hindbrain (exhibit II). The otic placode soon invaginates and forms an **otic fovea.** Toward the end of the fourth week, the outer edges of the invaginated otic fovea come together and fuse to form an **otocyst.** The otocyst soon pinches off and separates from the surface ectoderm. The otocyst further differentiates to form a dorsal **utricular portion** and a ventral **saccular portion.** Three separate diverticula extend outward from the utricular portion and develop into the **semicircular canals,** which later function in balance and equilibrium. A tubular diverticulum called the **cochlear duct** extends in a coiled fashion from the saccular portion and forms the membranous portion of the **cochlea** of the ear (exhibit II). The **spiral organ,** which is the functional portion of the cochlea, differentiates from cells along the wall of the cochlear duct (exhibit III). The sensory nerves that innervate the inner ear are derived from neuroectoderm from the developing brain.

The differentiating otocyst is surrounded by mesodermal tissue that soon forms a cartilaginous **otic capsule** (exhibit IV). As the otocyst and surrounding otic capsule grow, vacuoles containing the fluid **perilymph** form within the otic capsule. The vacuoles soon enlarge and coalesce to form the **perilymphatic space,** which divides into the **scala tympani** and the **scala vestibuli.** Eventually, the cartilaginous otic capsule ossifies to form the **bony (osseous) labyrinth** of the inner ear. The middle-ear chamber is referred to as the **tympanic cavity** and derives from the first pharyngeal pouch (exhibit IV). The **auditory ossicles,** which amplify incoming sound waves, derive from the first and second pharyngeal arch cartilages. As the tympanic cavity enlarges, it surrounds and encloses the developing ossicles (exhibit IV). The connection of the tympanic cavity to the pharynx gradually elongates to develop into the **auditory (eustachian) tube,** which remains patent throughout life and is important in maintaining an equilibrium of air pressure between the pharyngeal and tympanic cavities.

UNIT V

Continued on next page

Developmental Exposition

Continued

Exhibit IV The development of the outer- and middle-ear regions and the auditory ossicles. (*a*) A lateral view of a 4-week-old embryo showing the position of the cut depicted in the sequential development (*b–e*). (*b*) The embryo at 4 weeks illustrating the invagination of the surface ectoderm and the evagination of the endoderm at the level of the first pharyngeal pouch. (*c*) During the 5th week, mesenchymal condensations are apparent, from which the auditory ossicles will be derived. (*d*) Further invagination and evagination at 6 weeks correctly position the structures of the outer- and middle-ear regions. (*e*) By the end of the 8th week, the auditory ossicles, tympanic membrane, auditory tube, and external auditory canal have formed.

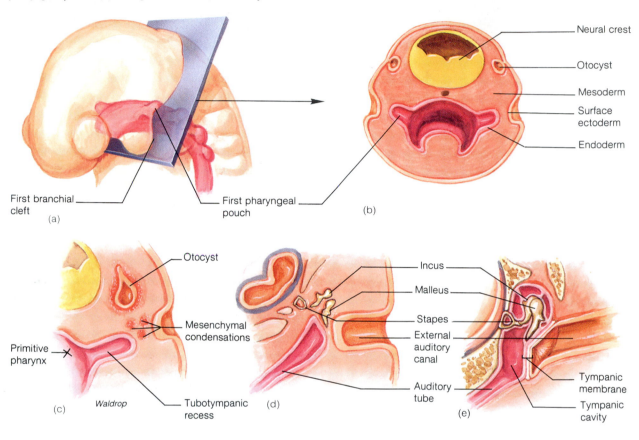

The outer ear includes the fleshy **auricle** attached to the side of the head and the tubular **external acoustic meatus** that extends into the temporal bone of the skull. The external acoustic meatus forms from the surface ectoderm that covers the dorsal end of the first branchial groove (exhibit IV). A solid epithelial plate called the **meatal plug** soon develops at the bottom of the funnel-shaped branchial groove. The meatal plug is involved in the formation of the inner wall of the external acoustic meatus and contributes to the **tympanic membrane** (eardrum). The tympanic membrane has a dual origin from surface ectoderm and the endoderm lining the first pharyngeal pouch (exhibit IV).

Saccule and Utricle   Receptor hair cells of the saccule and utricle are located in a small, thickened area of the walls of these organs called the **macula** (fig. 15.40). Cytoplasmic extensions of the hair cells project into a gelatinous material that supports microscopic crystals of calcium carbonate called **statoconia.** The statoconia increase the weight of the gelatinous mass, which results in a higher *inertia* (resistance to change in movement).

macula: L. *macula*, spot

When a person is upright, the hairs of the utricle are oriented vertically, whereas those of the saccule are oriented horizontally into the statoconial membrane. During forward acceleration, the statoconial membrane lags behind the hair cells, so the hairs are pushed backward. This is similar to the backward thrust of the body when a car accelerates rapidly forward. The inertia of the statoconial membrane similarly causes the hairs of the utricle to be pushed upward when a person jumps from a raised platform. These effects, and the opposite ones that

Figure 15.40 The macula of the utricle. (*a*) When the head is in an upright position, the weight of the statoconia applies direct pressure to the sensitive cytoplasmic extensions of the hair cells. (*b*) As the head is tilted forward, the extensions of the hair cells bend in response to gravitational force and cause the hair cells to be stimulated.

occur when a person accelerates backward or upward, allow us to maintain our equilibrium with respect to gravity during linear acceleration.

Sensory impulses from the vestibular organs are conveyed to the brain by way of the vestibular portion of the vestibulocochlear nerve.

Semicircular Canals Receptors of the semicircular canals are contained within the ampulla at the base of each semicircular duct. An elevated area of the ampulla called the **crista ampullaris** (*am"poo-lar'is*) contains numerous hair cells and supporting cells (fig. 15.41). Like the saccule and utricle, the hair cells have cytoplasmic extensions that project into a dome-shaped gelatinous mass called the **cupula** (*kyoop'yoo-la*). When the hair cells within the cupula are bent by rapid displacement of the fluid within the semicircular ducts, as in spinning around, sensory impulses travel to the brain by way of the vestibular portion of the vestibulocochlear nerve.

Neural Pathways Stimulation of the hair cells in the vestibular apparatus activates the sensory neurons of the *vestibulocochlear* (*eighth cranial*) *nerve*. These fibers transmit impulses to the cerebellum and to the vestibular nuclei of the medulla oblongata. The vestibular nuclei, in turn, send fibers to the oculomotor center of the brain stem and to the spinal cord. Neurons in the oculomotor center control eye movements and neurons in the spinal cord stimulate movements of the head, neck, and limbs. Movements of the eyes and body produced by these pathways serve to maintain balance and track the visual field during rotation.

1. List the structures of the outer ear, middle ear, and inner ear and explain the function of each as related to hearing.
2. Use a flow chart to describe how sound waves in air within the external auditory canal are transduced into movements of the basilar membrane.
3. Explain how movements of the basilar membrane of the spiral organ can code for different sound frequencies (pitches).
4. Explain how the vestibular organs maintain a sense of balance and equilibrium.

Clinical Considerations

Numerous disorders and diseases afflict the sensory organs. Some of these occur during the sensitive period of prenatal development and have a congenital effect. Other sensory impairments, some of which are avoidable, can occur at any time of life. Still others are the result of changes during the natural aging process. The loss of a sense is frequently a traumatic adjustment. Fortunately, however, when a sensory function is impaired or lost, the other senses seem to become keener to lessen the extent of the handicap. A blind person, for example, compensates somewhat for the loss of sight by developing a remarkable hearing ability.

Entire specialities within medicine are devoted to treating the disorders of specific sensory organs. It is beyond the scope

cupula: L. *cupula*, cup-shaped

UNIT V

Figure 15.41 (a) A crista ampullaris within an ampulla. (b) Movement of the endolymph during rotation causes the cupula to displace, thus stimulating the hair cells.

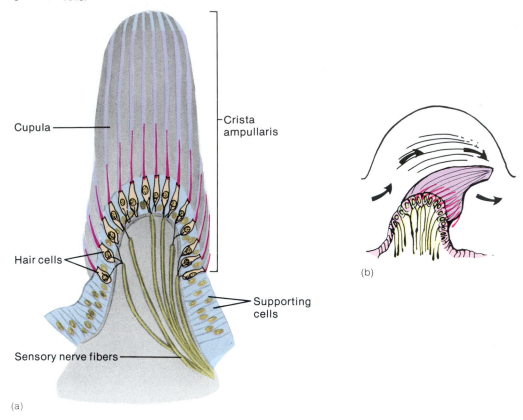

Cupula

Crista ampullaris

Hair cells

Supporting cells

Sensory nerve fibers

(a)

(b)

of this text to attempt a comprehensive discussion of the numerous diseases and dysfunctions of these organs. Some general comments will be made, however, on the diagnosis of sensory disorders and on developmental problems that can affect the eyes or ears. In addition, the more common diseases and dysfunctions of these organs are noted.

Diagnosis of Sensory Organs

Eye   There are two distinct professional specialities concerned with the structure and function of the eye. *Optometry* is the paramedical profession concerned with assessing vision and treating visual problems. An *optometrist* prescribes corrective lenses or visual training but is not a medical doctor and does not treat eye diseases. *Ophthalmology (of''thal-mol'ŏ-je)* is the speciality of medicine concerned with diagnosing and treating eye diseases.

Although the eyeball is an extremely complex organ, it is quite accessible to examination. Various techniques and instruments are used during an eye examination, but the following are used most frequently: (1) a *cycloplegic drug,* which is instilled into the eyes to dilate the pupils and temporarily inactivate the ciliary muscles; (2) a *Snellen's chart,* which is used to determine the visual acuity of a person standing 20 feet from the chart (a reading of 20/20 is considered normal for the test); (3) an *ophthalmoscope,* which contains a light, mirrors, and lenses to illuminate and magnify the interior of the eyeball so that the

structures within may be examined; and (4) a *tonometer,* used to measure ocular tension, which is important in detecting glaucoma.

Ear   Otorhinolaryngology (o''to-ri''no-lar''in-gol'ŏ-je) is the speciality of medicine dealing with the diagnosis and treatment of diseases or conditions of the ear, nose, and throat. *Audiology* is the study of hearing, particularly assessment of the ear and its functioning.

There are three common instruments or techniques used to examine the ears to determine auditory function: (1) an *otoscope* is an instrument used to examine the tympanic membrane of the ear; abnormalities of this membrane are informative when diagnosing specific auditory problems, including middle-ear infections; (2) *tuning fork tests* are useful in determining hearing acuity and especially for discriminating the various kinds of hearing loss; and (3) *audiometry* is a functional examination for hearing sensitivity and speech discrimination.

Developmental Problems of the Eyes and Ears

Although there are many congenital abnormalities of the eyes and ears, most of them are rare. The sensitive period of development of these organs is from about 24 to 45 days after conception. Indeed, 85% of newborns suffer anomalies if infected during this developmental time period. Most congenital dis-

UNIT V

orders of the eyes and ears are caused by genetic factors or intrauterine infections such as *rubella virus*.

If a pregnant woman contracts rubella (German measles), there is a 90% probability that the embryo or fetus will contract it also. An embryo afflicted with rubella is 30% more likely to be aborted, stillborn, or congenitally deformed than one that is not afflicted. Rubella interferes with the mitotic process and thus causes underdeveloped organs. An embryo with rubella may suffer from a number of physical deformities, *cataracts* and *glaucoma* being common deformities of the eye.

Eye Most **congenital cataracts** are hereditary, but they may also be caused by maternal rubella infection during the critical fourth to sixth week of eye development. In this condition, the lens is opaque and frequently appears grayish white.

Cyclopia is a rare condition in which the eyes are partially fused into a median eye enclosed by a single orbit. Other severe malformations, which are incompatible with life, are generally expressed with this condition.

Ear **Congenital deafness** is generally caused by an autosomal recessive gene but may also be caused by a maternal rubella infection. The actual functional impairment is generally either a defective set of auditory ossicles or improper development of the neurosensory structures of the inner ear.

Although the shape of the auricle varies widely, **auricular abnormalities** are not uncommon, especially in infants with chromosomal syndromes causing mental deficiencies. In addition, the external auditory canal frequently does not develop in these children, producing a condition called **atresia** (ă-tre'zhă) of the external auditory canal.

Functional Impairments of the Eye

Few people have perfect vision. Slight variations in the shape of the eyeball or curvature of the cornea or lens cause an imperfect focal point of light waves onto the retina. Most variations are slight, however, and the error of refraction goes unnoticed. Severe deviations that are not corrected may cause blurred vision, fatigue, chronic headaches, and depression.

The primary clinical considerations associated with defects in the refractory structures or general shape of the eyeball are myopia, hyperopia, presbyopia, and astigmatism. **Myopia** (nearsightedness) is an elongation of the eyeball that causes light waves to focus at a point in the vitreous humor in front of the retina (fig. 15.42). Only light waves from close objects can be focused clearly on the retina; distant objects appear blurred. **Hyperopia** (farsightedness) is a condition in which the eyeball is too short; consequently, light waves have a focal point behind the retina. Although visual accommodation aids a hyperopic person, it generally does not help enough for the person to clearly see very close or distant objects. **Presbyopia** is a condition in which the lens tends to lose its elasticity and ability

myopia: Gk. *myein*, to shut; *ops*, eye
hyperopia: Gk. *hyper*, over; *ops*, eye
presbyopia: Gk. *presbys*, old man; *ops*, eye

Figure 15.42 In a normal eye (*a*), parallel rays of light are brought to a focus on the retina by refraction in the cornea and lens. If the eye is too long, as in myopia (*b*), the focus is in front of the retina. This can be corrected by a concave lens. If the eye is too short, as in hyperopia (*c*), the focus is behind the retina. This is corrected by a convex lens. In astigmatism (*d*), light refraction is uneven due to an abnormal shape of the cornea or lens.

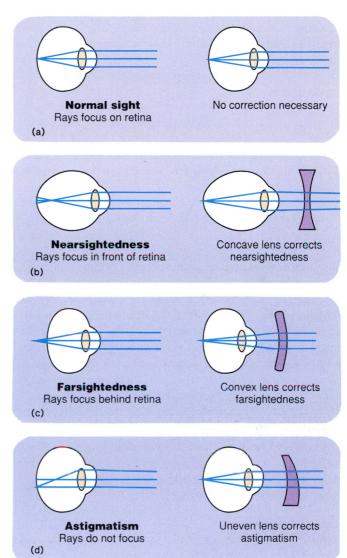

Normal sight
Rays focus on retina
(a)

No correction necessary

Nearsightedness
Rays focus in front of retina
(b)

Concave lens corrects nearsightedness

Farsightedness
Rays focus behind retina
(c)

Convex lens corrects farsightedness

Astigmatism
Rays do not focus
(d)

Uneven lens corrects astigmatism

to accommodate. It is relatively common in elderly people. In order to read print on a page, a person with presbyopia must hold the page farther from the eyes. **Astigmatism** is a condition in which an irregular curvature of the cornea or lens of the eye distorts the refraction of light waves. If a person with astigmatism views a circle, the image will not appear clear in all 360 degrees; the parts of the circle that appear blurred can be used to map the astigmatism.

Various glass or plastic lenses are generally prescribed for people with the visual impairments just described. Myopia may be corrected with a biconcave lens; hyperopia with a biconvex

astigmatism: Gk. *a*, without; *stigma*, point

lens; and presbyopia with bifocals, or a combination of two lenses adjusted for near and distant vision. Correction for astigmatism requires a careful assessment of the irregularities and a prescription of a specially ground corrective lens.

The condition of myopia may also be treated by a surgical procedure called **radial keratotomy.** In this technique, 8 to 16 microscopic slashes, like the spokes of a wheel, are made in the cornea from the center to the edge. The ocular pressure inside the eyeball bulges the weakened cornea and flattens its center, changing the focal length of the eyeball. Side effects may include vision fluctuations, sensitivity to glare, and incorrect corneal alteration.

Cataracts   A cataract is a clouding of the lens that leads to a gradual blurring of vision and the eventual loss of sight. A cataract is not a growth within or upon the eye but a chemical change in the protein of the lens caused by injury, poisons, infections, or age degeneration.

Cataracts are the leading cause of blindness. A cataract can be removed surgically, however, and vision restored by implanting a tiny intraocular lens that either clips to the iris or is secured into the vacant lens capsule. Special contact lenses or thick lenses for glasses are other options.

Detachment of the Retina   Retinal detachment is a separation of the nervous or visual layer of the retina from the underlying pigment epithelium. It generally begins as a minute tear in the retina that gradually extends as vitreous fluid accumulates between the layers. Retinal detachment may result from hemorrhage, a tumor, degeneration, or trauma from a violent blow to the eye. A detached retina may be repaired by using laser beams, cryoprobes, or intense heat to destroy the tissue beneath the tear and rejoin the layers.

Strabismus   Strabismus is a condition in which both eyes do not focus upon the same axis of vision. This prevents stereoscopic vision, and the persons afflicted will have varied visual impairments. Strabismus is usually caused by a weakened extrinsic eye muscle.

Strabismus is assessed while the patient attempts to look straight ahead. If the afflicted eye is turned toward the nose, the condition is called convergent strabismus **(esotropia).** If the eye is turned outward, it is called divergent strabismus **(exotropia).** Disuse of the afflicted eye causes a visual impairment called **amblyopia.** Visual input from the normal eye and the eye with strabismus results in **diplopia,** or double vision. A normal, healthy person who has excessively consumed alcohol may experience diplopia.

Infections and Diseases of the Eye

Infections   Infections and inflammation can occur in any of the associated structures of the eye and in structures within or on the eyeball itself. The causes of infections are usually microorganisms, mechanical irritation, or sensitivity to particular substances.

Conjunctivitis (inflammation of the conjunctiva) may result from sunburn or an infection by organisms such as staphylococci, viruses, or streptococci. Bacterial conjunctivitis is commonly called "pinkeye."

Keratitis (inflammation of the cornea) may develop secondarily from conjunctivitis or be caused by such diseases as tuberculosis, syphilis, mumps, or measles. Keratitis is painful and may cause blindness if untreated.

A **chalazion** is a cyst caused by an infection and a subsequent blockage in the ducts of the sebaceous glands along the free edge of the eyelids.

Styes (hordeola) are relatively common, but mild, infections of the follicle of an eyelash or the sebaceous gland of the follicle. Styes may spread readily from one eyelash to another if untreated. Poor hygiene and the excessive use of cosmetics may contribute to the development of styes.

Diseases   **Trachoma** (*trǎ-ko′mǎ*) is a highly contagious bacterial disease of the conjunctiva and cornea. Although rare in the United States, it is estimated that over 500 million people are afflicted by this disease. Trachoma may be treated readily with sulfonamides and some antibiotics, but if untreated it will spread progressively until it covers the cornea. At this stage, vision is lost and the eye undergoes degenerative changes.

Glaucoma is the second leading cause of blindness and is particularly common in underdeveloped countries. Although it can afflict individuals of any age, 95% of the cases involve people over the age of 40. Glaucoma is an abnormal increase in the intraocular pressure of the eyeball. Aqueous humor does not drain through the scleral venous sinus as quickly as it is produced. Accumulation of fluid causes compression of the blood vessels in the eyeball and compression of the optic nerve. Retinal cells die and the optic nerve may atrophy, producing blindness.

Infections, Diseases, and Functional Impairments of the Ear

Disorders of the ear are common and may affect both hearing and the vestibular functions. The ear is afflicted by numerous infections and diseases—some of which can be prevented.

Infections and Diseases   **External otitis** is a general term for infections of the outer ear. The causes of external otitis range from dermatitis to fungal and bacterial infections.

Acute purulent otitis media is a middle-ear infection. Pathogens of this disease usually enter through the auditory tube, most often following a cold or tonsillitis. Children frequently have middle-ear infections because of their susceptibility to infections and their short and straight auditory tubes. As a middle-ear infection progresses to the inflammatory stage,

---

cataract: Gk. *katarrhegnynai*, to break down
amblyopia: Gk. *amblys*, dull; *ops*, vision

---

glaucoma: Gk. *glaukos*, gray

the auditory tube is closed and drainage prohibited. An intense earache is a common symptom of a middle-ear infection. The pressure from the inflammation may eventually rupture the tympanic membrane to permit drainage.

Repeated middle-ear infections, particularly in children, usually call for an incision of the tympanic membrane known as a **myringotomy** *(mir″in-got′ŏ-me)* and the implantation of a tiny tube within the tympanic membrane (fig. 15.43) to assist the patency of the auditory tube. The tubes, which are eventually sloughed out of the ear, permit the infection to heal and help prohibit further infections by keeping the auditory tube open.

Perforation of the tympanic membrane may occur as the result of infections or trauma. Sudden, intense noise might rupture the membrane. Spontaneous perforation of the membrane usually heals rapidly, but scar tissue may form and lessen the sensitivity to sound vibrations.

Otosclerosis is a progressive deterioration of the normal bone of the bony labyrinth and its replacement with vascular spongy bone. This frequently causes hearing loss as the auditory ossicles are immobilized. Surgical scraping of the bone growth and replacing the stapes with a prosthesis is the most frequent treatment for otosclerosis.

Ménière's disease afflicts the inner ear and may cause hearing loss as well as equilibrium disturbance. The causes of Ménière's disease vary and are not completely understood, but they are thought to be related to a dysfunction of the autonomic nervous system that causes a vasoconstriction within the inner ear. The disease is characterized by recurrent periods of **vertigo** (dizziness and a sensation of rotation), **tinnitus** *(ti-ni′tus)* (ringing in the ear), and progressive deafness in the affected ear. Ménière's disease is chronic and affects both sexes equally. It is more common in middle-aged and elderly people.

Nystagmus and Vertigo   When a person first begins to spin, the inertia of the endolymph within the semicircular canals causes the cupula to bend in the opposite direction. As the spin continues, however, the inertia of the endolymph is overcome and the cupula straightens. At this time the endolymph and the cupula are moving in the same direction and at the same speed. If movement is suddenly stopped, the greater inertia of the endolymph causes it to continue moving in the previous direction of spin and to bend the cupula in that direction.

Bending of the cupula after movement has stopped affects muscular control of the eyes and body. The eyes slowly drift in the direction of the previous spin and are then rapidly jerked back to the midline position, producing involuntary oscillations. These movements are called postrotatory **vestibular nystagmus** *(nis-tag′mus)*. People experiencing this may feel that they, or the room, are spinning. The loss of equilibrium that results is called **vertigo.** If the vertigo is sufficiently severe or the person particularly susceptible, the autonomic nervous system may become involved. This can produce dizziness, pallor, sweating, and nausea.

Figure 15.43   An implanted ventilation tube in the tympanic membrane following a *myringotomy*.

Auditory Impairment   Loss of hearing results from disease, trauma, or developmental problems involving any portion of the auditory apparatus, cochlear nerve and auditory pathway, or areas of auditory perception within the brain. Hearing impairment varies from slight disablement, which may or may not worsen, to total deafness. Some types of hearing impairment, including deafness, can be mitigated through hearing aids or surgery.

Auditory impairments fall into two broad categories. **Conduction deafness** is caused by an interference with the sound waves through the outer or middle ear. Conduction problems include impacted cerumen (wax), a ruptured tympanic membrane, a severe middle-ear infection, or adhesions of one or more auditory ossicles *(otosclerosis)*. Conductive deafness can be successfully treated.

Perceptive (sensorineural) deafness results from disorders that affect the inner ear, the cochlear nerve or nerve pathway, or auditory centers within the brain. Perceptive impairment ranges in severity from the inability to hear certain frequencies to deafness. Such deafness may be caused by a number of factors, including diseases, trauma, and genetic or developmental problems. Hearing impairments of this type, which are common in the elderly, are related to age. The ability to perceive high-frequency sounds is generally affected first. Hearing aids may help patients with perceptive deafness. This type of deafness is permanent, however, because it involves sensory structures that cannot heal themselves through regeneration or replication.

UNIT V

Ménière's disease: from Prosper Ménière, French physician, 1799–1862
vertigo: L. *vertigo*, dizziness
tinnitus: L. *tinnitus*, ring or tingle

CLINICAL CASE STUDY ANSWER

When the handle of the vibrating tuning fork is placed on the mastoid process, the temporal bone transmits sound waves directly to the inner ear. This bypasses the conductive components of the middle ear, which are (beginning with the first to receive sound waves) the tympanic membrane, malleus, incus, and stapes. Since the patient can hear well when the sound waves are transmitted through the temporal bone, the inner ear organs, as well as the neural pathways, can be assumed to be functioning. It can, therefore, be concluded that the problem lies in the conductive components. This is often the result of otosclerosis (a spongy proliferation of bone) or other conditions affecting the auditory ossicles and/or tympanic membrane, many of which are treatable surgically.

CHAPTER SUMMARY

Overview of Sensory Perception (p. 462)

1. Sensory organs are specialized extensions of the nervous system that respond to specific stimuli and conduct nerve impulses.
2. A stimulus to a receptor that conducts an impulse to the brain is necessary for perception.
3. Senses act as energy filters that permit perception of only a narrow range of energy.

Classification of the Senses (pp. 462–463)

1. The senses are classified according to receptors or on the basis of the stimulus energy they transduce.
2. General senses are widespread throughout the body and are simple in structure. Special senses are localized in complex receptor organs and have extensive neural pathways.
3. Somatic senses are localized within the body wall; visceral senses are located within the visceral organs.
4. Phasic receptors respond quickly to a stimulus but then adapt and decrease firing rate. Tonic receptors produce a constant rate of firing.

Somatic Senses (pp. 463–469)

1. Corpuscles of touch, free nerve endings, and hair root plexuses are tactile receptors, responding to light touch.
2. Lamellated corpuscles are pressure receptors in the dermis.
3. Thermoreceptors include the organs of Ruffini (heat) and bulbs of Krause (cold).
4. Free nerve endings are the principal pain receptors, serving to protect the body from injury.

5. Joint kinesthetic receptors, neuromuscular spindles, and neurotendinous spindles are proprioceptors that are sensitive to changes in stretch and tension.
6. The acuity of sensation depends on receptor density and lateral inhibition.

Olfactory Sense (pp. 469–470)

1. Olfactory receptors of the olfactory nerve respond to chemical stimuli and transmit the sensation of olfaction to the cerebral cortex.
2. Olfaction functions closely with gustation in that the receptors of both are chemoreceptors, requiring dissolved substances for stimuli.

Gustatory Sense (pp. 470–471)

1. Taste receptors in taste buds are chemoreceptors and transmit the sensation of gustation to the cerebral cortex.
2. Gustatory cells compose the taste buds located in vallate, fungiform, and filiform papillae.
3. The four modalities of taste sensation are sweet, salty, sour, and bitter.

Visual Sense (pp. 471–487)

1. Accessory structures of the eye (the orbit, eyebrow, eyelids, eyelashes, conjunctiva, lacrimal gland, and ocular muscles) either protect the eyeball or provide for eye movement.
2. The eyeball consists of the fibrous tunic, which is divided into the sclera and cornea; the vascular tunic, which consists of the choroid, the ciliary body, and the iris; and the internal tunic, which consists of the retina. The retina has an outer pigmented layer and an inner nervous layer. The transparent lens is not part of any tunic.
3. Rods and cones, which are the photoreceptors in the nervous layer of the retina, respond to dim and colored light, respectively. Cones are concentrated in the fovea centralis, the area of keenest vision.

4. Rods and cones contain specific pigments that provide sensitivity to different light waves.
5. The visual process includes the transmission and refraction of light waves, accommodation of the lens, constriction of the pupil, and convergence of the eyes.
   (a) Refraction is achieved as incoming light waves pass through the cornea, aqueous humor, lens, and vitreous humor.
   (b) A sharp focus is accomplished as the curvature of the lens is changed by autonomic contraction of smooth muscles within the ciliary body.
6. Neural pathways from the retina to the superior colliculus help to regulate eye and body movements. Most fibers from the retina project to the lateral geniculate body and from there to the striate cortex.
7. The sensory components of the eye have been formed by 20 weeks; the accessory structures have been formed by 32 weeks.

Senses of Hearing and Balance (pp. 487–499)

1. The outer ear consists of the auricle and the external auditory canal.
2. The middle ear, bounded by the tympanic membrane and the vestibular and cochlear windows, contains the auditory ossicles (malleus, incus, and stapes) and the auditory muscles (tensor tympani and stapedius).
3. The vestibule of the middle ear connects to the pharynx through the auditory tube.
4. The inner ear contains the functional organs for hearing and equilibrium: the spiral organ, within the cochlear duct, and the semicircular canals, saccule, and utricle.
5. The development of the ear begins during the fourth week and is complete by the thirty-second week.

REVIEW ACTIVITIES

Objective Questions

1. Which of the following conditions is/ are necessary for the perception of a sensation to take place?
   (a) presence of a stimulus
   (b) nerve impulse conduction
   (c) activation of a receptor
   (d) all of the above
2. The cutaneous receptor sensitized for the detection of heavy pressure is
   (a) a hair root plexus.
   (b) a lamellated corpuscle.
   (c) the organ of Ruffini.
   (d) a free nerve ending.
3. Proprioceptors that are located within the connective tissue capsule in synovial joints are called
   (a) neuromuscular spindles.
   (b) Golgi tendon organs.
   (c) neurotendinous spindles.
   (d) joint kinesthetic receptors.
4. The sensation of visceral pain perceived as arising from another somatic location is known as
   (a) related pain.
   (b) phantom pain.
   (c) referred pain.
   (d) parietal pain.
5. When a person with normal vision views an object from a distance of at least 20 feet,
   (a) the ciliary muscles are relaxed.
   (b) the suspensory ligament is taut.
   (c) the lens is flat, having the least convex shape.
   (d) all of the above.
6. Which of the following is an avascular ocular tissue?
   (a) the sclera
   (b) the choroid layer
   (c) the ciliary body
   (d) the iris

7. Pupils are dilated in response to contraction of
   (a) ciliary muscles.
   (b) radial iris muscles.
   (c) circular iris muscles.
   (d) orbicularis oculi muscles.
8. The stimulation of hair cells in the semicircular canals results from the movement of
   (a) endolymph.
   (b) perilymph.
   (c) the statoconial membrane.
9. The middle ear is separated from the inner ear by
   (a) the cochlear window.
   (b) the tympanic membrane.
   (c) the vestibular window.
   (d) both (a) and (c).
10. Glasses with concave lenses help to correct
    (a) presbyopia.
    (b) myopia.
    (c) hyperopia.
    (d) astigmatism.

Essay Questions

1. What four events are necessary for perception of a sensation? What is considered a sensation and how does it differ from a reflex?
2. Outline the major events in the development of the eye and ear. When are congenital deformities most likely to occur?
3. Name the senses of the body and differentiate between the special and somatic senses. How are these two senses similar?
4. List the functions of proprioceptors and differentiate between the three types. What role do proprioceptors play in the kinesthetic sense?

5. Compare and contrast olfaction and gustation. Identify the cranial nerves that serve both of these senses and trace the sensory pathways of each.
6. Identify and describe the accessory structures of the eye and list their functions.
7. Diagram the structure of the eyeball and label the sclera, cornea, choroid, macula lutea, ciliary body, suspensory ligament, lens, iris, pupil, retina, optic disc, and fovea centralis.
8. Outline and explain the process of focusing light waves onto the fovea centralis.
9. Diagram the ear and label the structures of the outer, middle, and inner ear.
10. Trace a sound wave through the structures of the ear and explain how the mechanism of hearing functions.
11. Explain the mechanism by which equilibrium is maintained and the role played by the two kinds of receptor information.
12. List some techniques used to examine the eye and the ear. Give examples of congenital abnormalities of the eye and ear.

Gundy/Weber Software

The tutorial software accompanying Chapter 15, Sensory Organs, is Volume 6, Special Senses.

UNIT V

Circulatory System

Chapter Outline

CLINICAL CASE STUDY

A 65-year-old woman recently discharged from the hospital following a myocardial infarction (heart attack) returns to the hospital emergency room complaining of the sudden onset of pain in her right lower extremity. The attending physician notices that the patient's leg is also pale and cool from the knee distally. Furthermore, he is unable to detect a pulse over the posterior dorsal pedal and popliteal arteries. A good femoral pulse, however, is palpable in the inguinal region. Upon questioning, the patient states that she was discharged from the hospital with blood-thinning medication because tests had revealed that the heart attack had caused a blood clot to form within her heart. She added that the doctors were worried that "a piece of the clot could break off and go to other parts of the body."

Explain how the patient's heart attack several weeks earlier could have led to her current leg complaints. In which side of her heart (right or left) did the previously diagnosed blood clot most likely reside? Explain anatomically where the cause of her new problem is located and describe how it arrived there.

Functions and Major Components of the Circulatory System

The functions of the components that make up the circulatory system are essential to the life of complex, multicellular organisms.

Objective 1. Describe the functions of the circulatory system.

Objective 2. Describe the major components of the circulatory system.

A unicellular organism can provide for its own maintenance and continuity by performing the wide variety of functions needed for life. By contrast, the complex human body is composed of trillions of specialized cells that demonstrate a division of labor. Cells of a multicellular organism depend on one another for the very basics of their existence. The majority of the cells of the body are firmly implanted in tissues and are incapable of procuring food and oxygen on their own, or even moving from their own wastes. Therefore, a highly specialized and effective means of transporting materials within the body is needed.

The blood serves this transportation function. An estimated 60,000 miles of vessels throughout the body of an adult ensure that continued sustenance reaches each of the trillions of living cells. However, the blood can also serve to transport disease-causing viruses, bacteria, and their toxins. To guard against this, the circulatory system has protective mechanisms—the white blood cells and lymphatic system. In order to perform its various functions, the circulatory system works together with the respiratory, urinary, digestive, endocrine, and integumentary systems in maintaining homeostasis (fig. 16.1).

Functions of the Circulatory System

The many functions of the circulatory system can be grouped into two areas: transportation and protection.

1. **Transportation.** All of the substances involved in cellular metabolism are transported by the circulatory system. These substances can be categorized as follows:
   a. *Respiratory.* Red blood cells called **erythrocytes** *(ĕ-rith'rŏ-sītz)* transport oxygen to the tissue cells. In the lungs, oxygen from the inhaled air attaches to hemoglobin molecules within the erythrocytes and is transported to the cells for aerobic respiration. Carbon dioxide produced by cellular respiration is carried by the blood to the lungs for elimination in the exhaled air (fig. 16.2).
   b. *Nutritive.* The digestive system is responsible for the mechanical and chemical breakdown of food so that it can be absorbed through the intestinal wall and into the blood vessels of the circulatory system. The blood then carries these absorbed products of digestion through the liver and to the cells of the body.
   c. *Excretory.* Metabolic wastes, excessive water and ions, as well as other molecules in plasma (the fluid portion of blood) are filtered through the capillaries of the kidneys into kidney tubules and excreted in urine.
   d. *Regulatory.* The blood carries hormones and other regulatory molecules from their site of origin to distant target tissues.
2. **Protection.** The circulatory system protects against injury and foreign microbes or toxins introduced into the body. The clotting mechanism protects against blood loss when vessels are damaged, and white blood cells called **leukocytes** *(loo'kŏ-sītz)* render the body immune to many disease-causing agents. Leukocytes may also protect the body through phagocytosis (see fig. 3.12).

Major Components of the Circulatory System

The circulatory system is frequently divided into the **cardiovascular system,** which consists of the heart and blood vessels, and the **lymphatic system,** which consists of lymph vessels and lymph nodes.

The **heart** is a four-chambered double pump. Its pumping action creates the pressure needed to push blood in the vessels to the lungs and body cells. At rest, the heart of an adult pumps about 5.0 liters of blood/min. It takes about one minute for blood to be circulated to the most distal extremity and back to the heart.

Blood vessels form a tubular network that permits blood to flow from the heart to all living cells of the body and then

UNIT VI

Figure 16.1 The relationship of the circulatory system to the other body systems in maintaining homeostasis.

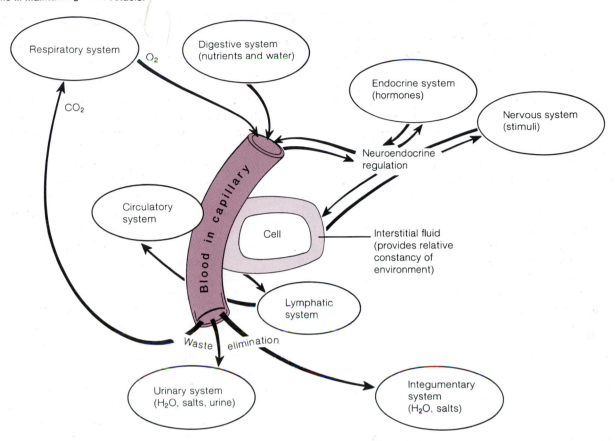

back to the heart. *Arteries* carry blood away from the heart, while *veins* return blood to the heart. Arteries and veins are continuous with each other through smaller blood vessels.

Arteries branch extensively to form a network of progressively smaller vessels. Those that are microscopic in diameter are called *arterioles*. Conversely, microscopic-sized veins called *venules (ven'yoolz)* deliver blood into progressively larger vessels that empty into the large veins. Blood passes from the arterial to the venous system in *capillaries (kap'i-lar''ēz)*, which are the thinnest and most numerous blood vessels. All exchanges of fluid, nutrients, and wastes between the blood and tissue cells occur across the walls of capillaries.

Fluid derived from plasma passes out of capillary walls into the surrounding tissues, where it is called *tissue fluid,* or *interstitial (in''ter-stish'al) fluid.* Some of this fluid returns directly to capillaries and some enters into **lymph vessels** located in the connective tissues around the blood vessels. Fluid in lymph vessels is called *lymph (limf).* This fluid is returned to the venous blood at particular sites. **Lymph nodes,** positioned along the way, cleanse the lymph prior to its return to the venous blood.

Endothermic (warm-blooded) animals, including humans, need an efficient circulatory system to transport oxygen-rich blood rapidly to all parts of the body. Of the vertebrates, only birds and mammals with their consistently warm body temperatures are considered endothermic, and only birds, mammals, and a few reptiles (crocodiles and alligators) have a four-chambered heart.

1. Name the components of the circulatory system that function in oxygen transport, in the transport of nutrients from the digestive system, and in protection.
2. Define the terms *artery, vein,* and *capillary* and describe the function of each.
3. Define the terms *interstitial fluid* and *lymph.* Describe their relationships to blood plasma.

Blood

Blood, a highly specialized connective tissue, consists of formed elements (erythrocytes, leukocytes, and thrombocytes) that are suspended and carried in the plasma. The constituents of blood function in transport, immunity, and blood-clotting mechanisms.

Objective 3. List the different types of formed elements of blood and describe their appearance.

Objective 4. Describe the origin of erythrocytes, leukocytes, and thrombocytes.

Objective 5. List the different types of substances found in plasma.

Objective 6. Describe the origin and function of the different categories of plasma proteins.

Figure 16.2 A schematic diagram of the circulatory system.

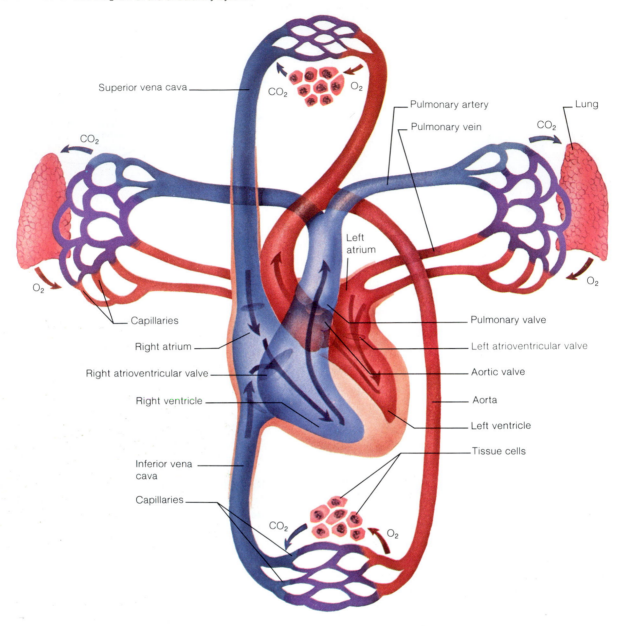

Superior vena cava

CO_2 O_2

CO_2

O_2

Pulmonary artery

Pulmonary vein

Lung

CO_2

O_2

Capillaries

Left atrium

Right atrium

Right atrioventricular valve

Right ventricle

Pulmonary valve

Left atrioventricular valve

Aortic valve

Aorta

Left ventricle

Tissue cells

Inferior vena cava

Capillaries

CO_2 O_2

The average-sized adult has about 5 liters of blood, which constitute about 7.7% of total body weight. Blood leaving the heart is referred to as *arterial blood.* Arterial blood, with the exception of that going to the lungs, is bright red in color due to a high concentration of oxyhemoglobin (the combination of oxygen with hemoglobin) in the erythrocytes. *Venous blood* is blood returning to the heart and, except for the venous blood from the lungs, has a darker color (due to increased amounts of hemoglobin that are no longer combined with oxygen). Blood has a viscosity that ranges between 4.5 and 5.5. This means that it is thicker than water, which has a viscosity of 1. Blood has a pH range of 7.35 to 7.45 and a temperature within the thorax of the body of about 38° C (100.4° F).

Blood is composed of two portions: a cellular portion called **formed elements** and a fluid portion called **plasma.** When a

blood sample is centrifuged, the heavier formed elements are packed into the bottom of the tube, leaving plasma at the top (fig. 16.3). The formed elements constitute approximately 45% of the total blood volume (the *hematocrit*); the plasma accounts for the remaining 55%.

Formed Elements of Blood

The formed elements of blood include **erythrocytes,** or red blood cells; **leukocytes,** or white blood cells; and **thrombocytes,** or platelets. Erythrocytes are by far the most numerous of these three types: a cubic millimeter of blood contains 5.1 million to 5.8 million erythrocytes in males and 4.3 million to

plasma: Gk. *plasma,* to form or mold

hematocrit: Gk. *haima,* blood; *krino,* to separate
erythrocytes: Gk. *erythros,* red; *kytos,* hollow (cell)
leukocytes: Gk. *leukos,* white; *kytos,* hollow (cell)

Figure 16.3 Blood cells become packed at the bottom of the test tube when whole blood is centrifuged, leaving the fluid plasma at the top of the tube. Red blood cells are the most abundant of the blood cells—white blood cells and platelets form only a thin, light-colored buffy coat at the interface of the packed red blood cells and the plasma.

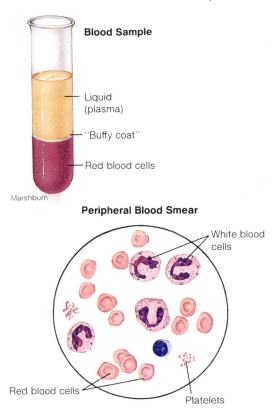

Blood Sample

Liquid
(plasma)

"Buffy coat"

Red blood cells

Marshburn

Peripheral Blood Smear

White blood
cells

Red blood cells

Platelets

5.2 million erythrocytes in females. By contrast, the same volume of blood contains only 5,000 to 9,000 leukocytes and 130,000 to 360,000 thrombocytes.

Erythrocytes Erythrocytes are flattened, biconcave discs, about 7.5 µm in diameter and 2.5 µm thick. Their unique shape relates to their function of transporting oxygen and provides an increased surface area through which gas can diffuse (fig. 16.4). Mature erythrocytes lack a nucleus and mitochondria (they get energy from anaerobic respiration). Because of these deficiencies, erythrocytes have a circulating life span of only about 120 days before they are destroyed by phagocytic cells in the liver, spleen, and bone marrow.

Oxygen molecules attached to *hemoglobin molecules* within erythrocytes give blood its red color. A hemoglobin molecule consists of a protein called *globin* and an iron-containing pigment called *heme.* Each iron portion of heme is able to combine with four molecules of oxygen. Considering that each erythrocyte contains approximately 280 million hemoglobin molecules, a single erythrocyte can transport over a billion molecules of oxygen. It is within the lungs that the oxygen molecules contained in inhaled air attach to the hemoglobin molecules and are transported via erythrocytes to the trillions of body cells.

hemoglobin: Gk. *haima,* blood; *globus,* globe

Figure 16.4 Erythrocytes. (*a*) A diagram and (*b*) a scanning electron micrograph.

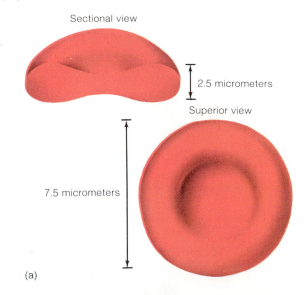

Sectional view

2.5 micrometers

Superior view

7.5 micrometers

(a)

(b)

Leukocytes Leukocytes differ from erythrocytes in several ways. Leukocytes contain nuclei and mitochondria and can move in an amoeboid fashion (erythrocytes are not able to move independently). Because of their amoeboid ability, leukocytes can squeeze through pores in capillary walls and get to a site of infection, whereas erythrocytes usually remain confined within blood vessels. The movement of leukocytes through capillary walls is called *diapedesis.*

Leukocytes, which are almost invisible under the microscope unless they are stained, are classified according to their stained appearance. Those leukocytes that have granules in their cytoplasm are called **granular leukocytes;** those that do not are **agranular** (or nongranular) **leukocytes.** The granular leukocytes are also identified by their oddly shaped nuclei, which in some cases are contorted into lobes separated by thin strands. The granular leukocytes are therefore known as **polymorphonuclear** (*pol"e-mor"fo-noo'kle-ar*) **leukocytes (PMN).**

diapedesis: Gk. *dia,* through; *pedester,* on foot

Figure 16.5 Types of formed elements in blood.

Neutrophils   Eosinophils   Basophils

Lymphocytes   Monocytes   Platelets   Erythrocytes

The stain used to identify leukocytes is usually a mixture of a pink-to-red stain called eosin and a blue-to-purple (hematoxylin) stain called a basic stain. Granular leukocytes with pink-staining granules are therefore called **eosinophils** *(e''ō-sin'ō-filz)* (fig. 16.5), and those with blue-staining granules are called **basophils.** Those with granules that have little affinity for either stain are **neutrophils.** Neutrophils are the most abundant type of leukocyte, constituting 54%–62% of the leukocytes in the blood.

The leukocytes of the agranular group include **monocytes** and **lymphocytes.** Monocytes are the largest cells found in the blood, and their large nuclei may vary considerably in shape. Lymphocytes have large nuclei surrounded by a relatively thin layer of cytoplasm.

Thrombocytes Thrombocytes are the smallest of the formed elements and are actually fragments of large cells called **megakaryocytes** *(meg''ă-kar'e-ō-sītz)* found in bone marrow. (This is why the term *formed elements* is used rather than *blood cells* to describe erythrocytes, leukocytes, and thrombocytes.) The fragments that enter the circulation as platelets lack nuclei but, like leukocytes, are capable of amoeboid movement. The platelet count per cubic millimeter of blood is 130,000 to 360,000. Platelets survive for about 5 to 9 days and then are destroyed by the spleen and liver.

Platelets play an important role in blood clotting. They constitute the major portion of the mass of the clot, and phospholipids in their cell membranes activate the clotting factors in blood plasma that result in threads of fibrin, which reinforce the platelet plug. Platelets that attach together in a blood clot also release a chemical called *serotonin,* which stimulates constriction of blood vessels, reducing the flow of blood to the injured area.

The appearance of the formed elements of the blood is shown in figures 16.5 and 16.6, and a summary of the characteristics of these formed elements is presented in table 16.1.

> Blood cell counts are an important source of information in assessing the health of a person. An abnormal increase in erythrocytes, for example, is termed *polycythemia (pol''e-si-the'me-ă)* and is indicative of several dysfunctions. An abnormally low red blood cell count is *anemia.* An elevated leukocyte count, called *leukocytosis,* is often associated with localized infection. A large number of immature leukocytes within a blood sample is diagnostic of the disease *leukemia.*

Hemopoiesis

Blood cells are constantly formed through a process called *hemopoiesis* (fig. 16.6). The term **erythropoiesis** *(e-rith''ro-poi-e'sis)* refers to the formation of erythrocytes; **leukopoiesis** refers to

hemopoiesis: Gk. *haima*, blood; *poiesis*, production
erythropoiesis: Gk. *erythros*, red; *poiesis*, production
leukopoiesis: Gk. *leukos*, white; *poiesis*, production

Table 16.1	Formed Elements of Blood		
Component	**Description**	**Number Present**	**Function**
Erythrocyte (red blood cell)	Biconcave disc without nucleus; contains hemoglobin; survives 100–120 days	4,000,000 to 6,000,000/mm³	Transports oxygen and carbon dioxide
Leukocytes (white blood cells)		5,000 to 10,000/mm³	Aid in defense against infections by microorganisms
Granulocytes	About twice the size of red blood cells; cytoplasmic granules present; survive 12 hours to 3 days		
1. Neutrophil	Nucleus with 2–5 lobes; cytoplasmic granules stain slightly pink	54%–62% of white cells present	Phagocytic
2. Eosinophil	Nucleus bilobed; cytoplasmic granules stain red in eosin stain	1%–3% of white cells present	Helps to detoxify foreign substances; secretes enzymes that break down clots
3. Basophil	Nucleus lobed; cytoplasmic granules stain blue in hematoxylin stain	Less than 1% of white cells present	Releases anticoagulant heparin
Agranulocytes	Cytoplasmic granules absent; survive 100–300 days		
1. Monocyte	2–3 times larger than red blood cell; nuclear shape varies from round to lobed	3%–9% of white cells present	Phagocytic
2. Lymphocyte	Only slightly larger than red blood cell; nucleus nearly fills cell	25%–33% of white cells present	Provides specific immune response (including antibodies)
Thrombocyte (platelet)	Cytoplasmic fragment; survives 5–9 days	130,000 to 360,000/mm³	Enables clotting

the formation of leukocytes. These processes occur in two classes of tissues. **Myeloid tissue** is the red bone marrow of the humeri, femora, ribs, sternum, pelvis, and portions of the skull that produces erythrocytes, granular leukocytes, and platelets.

Lymphoid tissue—including the lymph nodes, tonsils, spleen, and thymus—produces the agranular leukocytes (monocytes and lymphocytes). During embryonic and fetal development, the hemopoietic centers are located in the yolk sac, liver, and spleen. After birth, the liver and spleen are sites of blood cell destruction.

Erythropoiesis is an extremely active process. It is estimated that about 2.5 million erythrocytes are produced every second in order to replace the number that are continuously destroyed by the spleen and liver. (Recall that the life span of an erythrocyte is approximately 120 days.) During the destruction of erythrocytes, iron is salvaged and returned to the red bone marrow where it is used again in the formation of erythrocytes. Agranular leukocytes remain functional for 100–300 days under normal body conditions. Granular leukocytes, by contrast, have an extremely short life span of 12 hours to 3 days.

Hemopoiesis begins the same way in both myeloid and lymphoid tissues (fig. 16.6). Undifferentiated mesenchymal-like cells develop into stem cells called **hemocytoblasts.** These stem cells are able to divide rapidly. Some of the daughter cells become new stem cells (so the stem cell population is never depleted), whereas other daughter cells become specialized along different paths of blood cell formation. Hemocytoblasts, for example, may develop into **proerythroblasts** *(pro''ē-rith'rō-blasts),* which form erythrocytes; **myeloblasts,** which form granular leukocytes (neutrophils, eosinophils, and basophils);

lymphoblasts, which form lymphocytes; **monoblasts,** which form monocytes; or **megakaryoblasts** *(meg''ă-kar'e-ō-blasts),* which form thrombocytes.

Plasma

Plasma is the fluid portion of the blood, or blood with the formed elements removed. It constitutes approximately 55% of a given volume of blood. Plasma is a straw-colored liquid that is about 90% water. The remainder of the plasma consists of proteins, inorganic salts, carbohydrates, lipids, amino acids, vitamins, and hormones. The functions of plasma include transporting nutrients, gases, and vitamins; regulating electrolyte and fluid balances; and maintaining a consistent blood pH.

Plasma proteins constitute 7%–9% of the plasma. These proteins remain within the blood and interstitial fluid and assist in maintaining body homeostasis. The three types of plasma proteins are albumins, globulins, and fibrinogen. **Albumins** *(al-byoo'minz)* account for about 60% of the plasma proteins and are the smallest of the three types. They are produced by the liver and provide the blood with the necessary viscosity that is directly related to maintaining and regulating blood pressure. **Globulins** *(glob'yoo-linz)* make up about 36% of the plasma proteins. The three types of globulins are **alpha globulins, beta globulins,** and **gamma globulins.** The alpha and beta globulins are synthesized in the liver and function in transporting lipids and fat-soluble vitamins. Gamma globulins are produced

megakaryoblasts: Gk. *megas,* great; *karyon,* nut; *blastos,* germ
albumin: L. *albumen,* white
globulin: L. *globulus,* small globe

Figure 16.6 The processes of hemopoiesis. Formed elements begin as hemocytoblasts and differentiate into the various kinds of blood cells, depending on the needs of the body.

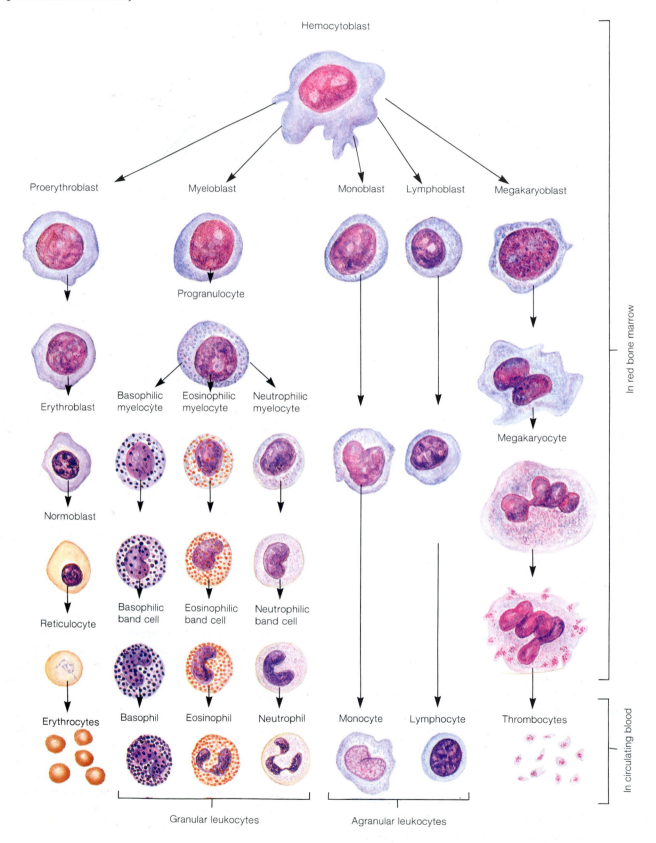

Table 16.2	Plasma Proteins		
Protein	**Percentage of Total**	**Origin**	**Function**
Albumin	60	Liver	Gives viscosity to blood and assists in maintaining blood osmotic pressure
Globulin Alpha globulins	36	Liver	Transport lipids and fat-soluble vitamins
Beta globulins		Liver	Transport lipids and fat-soluble vitamins
Gamma globulins		Lymphoid tissue	Constitute antibodies of immunity
Fibrinogen	4	Liver	Assists thrombocytes in the formation of clots

by lymphoid tissues and are antibodies of immunity. The third type of plasma protein, **fibrinogen** (*fi-brin'o-jen*), accounts for only about 4% of the protein content. Fibrinogens are large molecules that are synthesized in the liver and play an important role with thrombocytes in clotting the blood. When fibrinogen is removed, the remaining fluid is referred to as *serum*.

The plasma proteins of blood are summarized in table 16.2.

1. Describe erythropoiesis in terms of its location, rate, and the specific steps that lead to the production of mature erythrocytes. What does a mature erythrocyte look like?
2. Describe the different types of leukocytes and explain where and how these different types are produced.
3. Explain how platelets differ from other formed elements of blood and describe how platelets are produced.
4. List the different classes of plasma proteins and describe their origin and functions.
5. Explain how leukocytes get to the site of an infection.

Heart

The structure of the heart muscle and the action of its valves allow it to function as a double pump. In this capacity, the heart pumps blood low in oxygen to the lungs and pumps oxygen-rich blood to the rest of the body.

Objective 7. Describe the location of the heart in relation to other organs of the thoracic cavity and their associated serous membranes.

Objective 8. Describe the structure and functions of the three layers of the heart wall.

Objective 9. Identify the chambers, valves, and associated vessels of the heart.

---

fibrinogen: L. *fibra*, fibrous

Figure 16.7 A radiograph of the heart situated within the mediastinum between the lungs.

Clavicle
Mediastinum
Ribs
Shadow of breast
Heart
Cardiac impression
Diaphragm

Objective 10. Trace the flow of blood through the heart and distinguish between the pulmonary and systemic circulations.

Objective 11. Describe the location of the parts of the conduction system of the heart and trace the pathway of impulse initiation and conduction.

Location and General Description

The **heart** is a hollow, four-chambered muscular organ, roughly the size of the clenched fist. It averages 255 grams in adult females and 310 grams in adult males. It is estimated that the heart contracts some 42 million times and ejects 700,000 gallons of blood during a year.

The heart is located within the thoracic cavity between the lungs in the mediastinum (fig. 16.7). About two-thirds of the heart is located left of the midline, with its *apex*, or cone-shaped end, pointing downward in contact with the diaphragm. The *base* of the heart is the broad superior end, where the large vessels attach.

The heart is enclosed and protected by a loose-fitting, serous sac of dense fibrous connective tissue called the **parietal pericardium** (*per''i-kar'de-um*) (fig. 16.8). The parietal pericardium separates the heart from the other thoracic organs and forms the wall of the *pericardial cavity*, which contains a watery, lubricating *pericardial fluid*. The parietal pericardium is actually composed of an outer *fibrous layer* and an inner *serous layer*. The serous layer secretes the pericardial fluid.

---

pericardium: Gk. *peri*, around; *kardia*, heart

Figure 16.8 The position of the heart and associated serous membranes within the thoracic cavity.

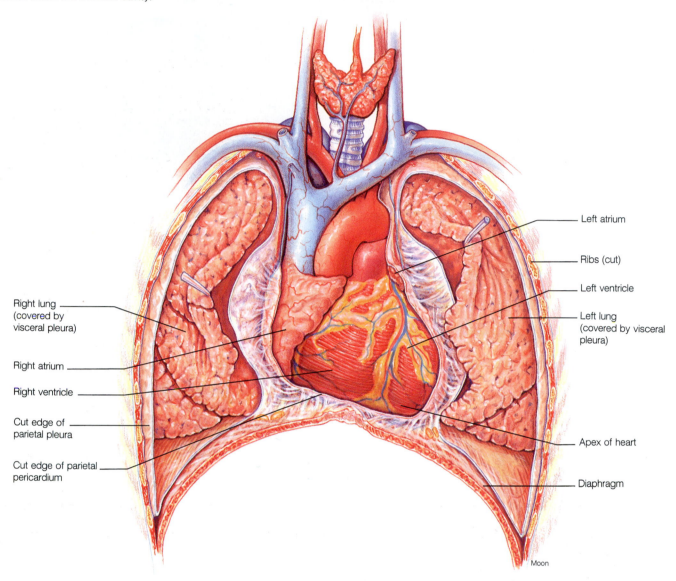

Right lung (covered by visceral pleura)

Right atrium

Right ventricle

Cut edge of parietal pleura

Cut edge of parietal pericardium

Left atrium

Ribs (cut)

Left ventricle

Left lung (covered by visceral pleura)

Apex of heart

Diaphragm

Moon

Pericarditis is an inflammation of the parietal pericardium that results in an increased secretion of fluid into the pericardial cavity. Because the tough, fibrous portion of the parietal pericardium is inelastic, an increase in fluid pressure impairs the movement of blood into and out of the chambers of the heart. Some of the pericardial fluid may be withdrawn for analysis by injecting a needle to the left of the xiphoid process to pierce the parietal pericardium.

Heart Wall

The wall of the heart is composed of three distinct layers (table 16.3). The outer layer is the **epicardium,** also called the *visceral pericardium.* The space between this layer and the parietal pericardium is the pericardial cavity, just described. The thick middle layer of the heart is called the **myocardium.** It is composed of cardiac muscle tissue (see chapter 4) and arranged in such a way that the contraction of the muscle bundles results in squeezing or wringing of the heart chambers. The thickness

of the myocardium varies in accordance with the force needed to eject blood from the particular chambers. The thickest portion of the myocardium, therefore, surrounds the left ventricle. The atrial walls are relatively thin. The inner layer, called the **endocardium,** is continuous with the endothelium of blood vessels. The endocardium also forms part of the valves of the heart. Inflammation of the endocardium is called *endocarditis.*

Chambers and Valves

The interior of the heart is divided into four chambers: upper right and left **atria** and lower right and left **ventricles.** The atria contract and empty simultaneously into the ventricles (fig. 16.9), which also contract in unison. Each atrium has an ear-shaped appendage called an **auricle.** The atria are separated from each other by the thin, muscular **interatrial septum;** the ven-

atrium: L. *atrium,* chamber
ventricle: L. *ventriculus,* diminutive of *venter,* belly
auricle: L. *auricula,* a little ear

UNIT VI

Table 16.3 Layers of the Heart Wall

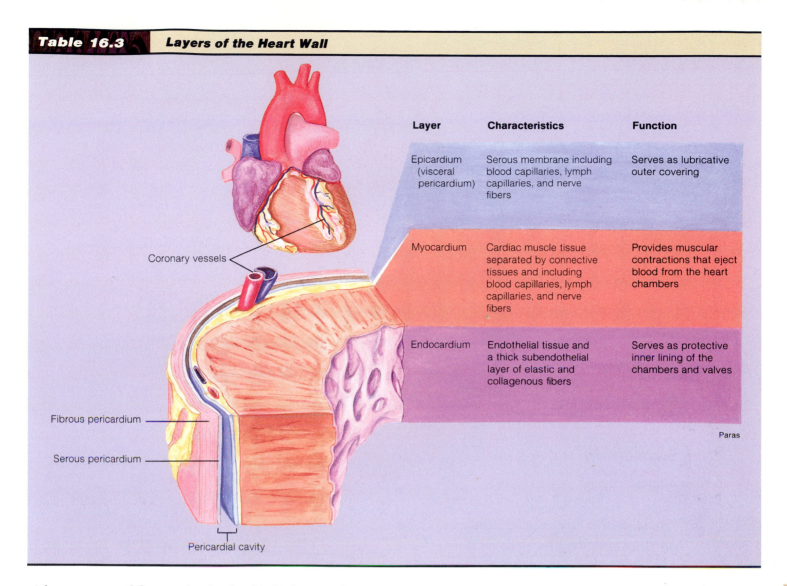

Layer	Characteristics	Function
Epicardium (visceral pericardium)	Serous membrane including blood capillaries, lymph capillaries, and nerve fibers	Serves as lubricative outer covering
Myocardium	Cardiac muscle tissue separated by connective tissues and including blood capillaries, lymph capillaries, and nerve fibers	Provides muscular contractions that eject blood from the heart chambers
Endocardium	Endothelial tissue and a thick subendothelial layer of elastic and collagenous fibers	Serves as protective inner lining of the chambers and valves

Coronary vessels

Fibrous pericardium

Serous pericardium

Pericardial cavity

Paras

tricles are separated from each other by the thick, muscular **interventricular septum.** Left and right *atrioventricular valves* are located between the atria and ventricles and *semilunar valves* are located at the bases of the two large vessels leaving the heart. These valves maintain the flow of blood in one direction.

Grooved depressions on the surface of the heart indicate the partitions between the chambers and also contain **cardiac vessels** that supply blood to the muscular wall of the heart. The most prominent groove is the *coronary sulcus* that encircles the heart and marks the division between the atria and ventricles. The partition between right and left ventricles is denoted by two (anterior and posterior) *interventricular sulci.*

The following discussion follows the sequence in which blood flows through the structures of the atria, ventricles, and valves.

Right Atrium The right atrium (*a'tre-um*) receives systemic venous blood from the **superior vena cava,** which drains the upper portion of the body, and from the **inferior vena cava,** which drains the lower portion (fig. 16.9). There is an addi-

tional venous opening into the right atrium from the *coronary sinus,* which returns venous blood from the myocardium of the heart itself.

Right Ventricle Blood from the right atrium passes through the **right atrioventricular valve** (also called the *tricuspid valve*) to fill the right ventricle. The right atrioventricular (AV) valve is characterized by three valve leaflets, or *cusps.* Each cusp is held in position by strong tendinous cords called **chordae tendineae** that are secured to the ventricular wall by cone-shaped **papillary muscles.** These structures prevent the valves from everting, like an umbrella in a strong wind, when the ventricles contract.

Ventricular contraction causes the right atrioventricular valve to close and the blood to leave the right ventricle through the **pulmonary trunk** and to enter the capillaries of the lungs via the right and left **pulmonary arteries.** The **pulmonary valve** (also called the *pulmonary semilunar valve*) lies at the base of the pulmonary trunk, where it prevents the backflow of ejected blood into the right ventricle.

vena cava: L. *vena,* vein; *cava,* empty

chordae tendineae: L. *chorda,* string; *tendere,* to stretch

UNIT VI

Figure 16.9 The structure of the heart. (*a*) An anterior view, (*b*) a posterior view, and (*c*) an internal view.

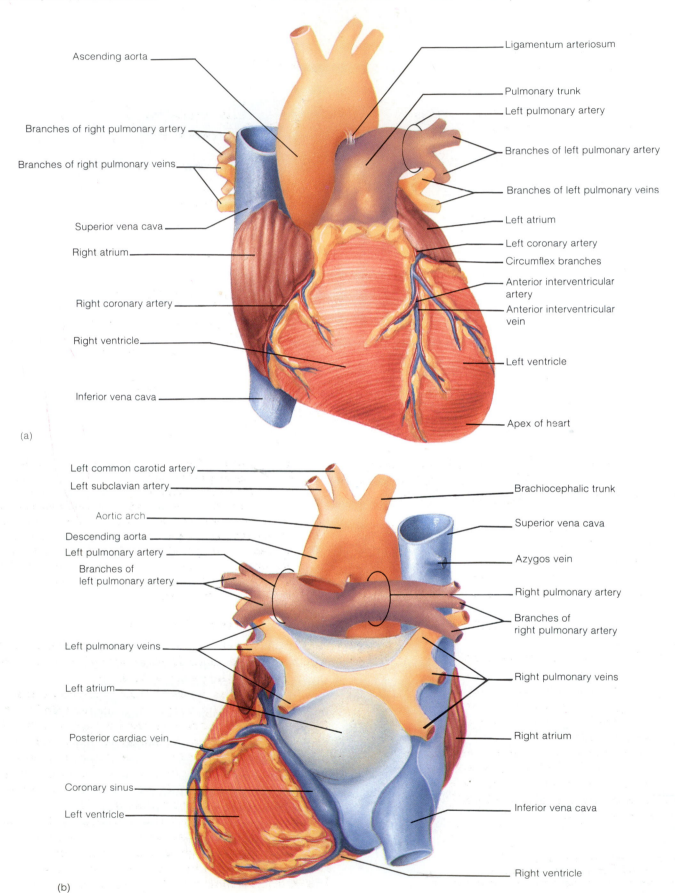

Ascending aorta

Branches of right pulmonary artery

Branches of right pulmonary veins

Superior vena cava

Right atrium

Right coronary artery

Right ventricle

Inferior vena cava

(a)

Ligamentum arteriosum

Pulmonary trunk

Left pulmonary artery

Branches of left pulmonary artery

Branches of left pulmonary veins

Left atrium

Left coronary artery

Circumflex branches

Anterior interventricular artery

Anterior interventricular vein

Left ventricle

Apex of heart

Left common carotid artery

Left subclavian artery

Aortic arch

Descending aorta

Left pulmonary artery

Branches of left pulmonary artery

Left pulmonary veins

Left atrium

Posterior cardiac vein

Coronary sinus

Left ventricle

(b)

Brachiocephalic trunk

Superior vena cava

Azygos vein

Right pulmonary artery

Branches of right pulmonary artery

Right pulmonary veins

Right atrium

Inferior vena cava

Right ventricle

UNIT VI

Figure 16.9—Continued

Aortic arch

Superior vena cava

Left pulmonary artery

Pulmonary trunk

Right pulmonary veins

Left pulmonary veins

Pulmonary valve

Left atrium

Aortic valve

Right atrium

Left atrioventricular valve

Right atrioventricular valve

Papillary muscle

Chordae tendineae

Interventricular septum

Left ventricle

Inferior vena cava

Trabeculae carneae

Right ventricle

Left Atrium   After gas exchange has occurred within the capillaries of the lungs, oxygenated blood is transported to the left atrium through four branches of the **pulmonary veins.**

Left Ventricle   The left ventricle receives blood from the left atrium. These two chambers are separated by the **left atrioventricular valve** (also called the *mitral valve* or *bicuspid valve*). When the left ventricle is relaxed, the valve is open and allows blood to flow from the atrium to the ventricle; when the left ventricle contracts, the valve closes. Closing of the valve during ventricular contraction prevents the backflow of blood into the atrium.

The walls of the left ventricle are thicker than those of the right ventricle. The endocardium of both chambers is characterized by distinct ridges called **trabeculae** *(trǎ-bek'yǔ-lē)* **carneae** (fig. 16.9c). Oxygenated blood leaves the left ventricle through the **ascending portion of the aorta** *(a-or'tǎ)*. The **aortic valve** (also called the *aortic semilunar valve*), located at the base of the ascending portion of the aorta, closes as a result of the pressure of the blood when the left ventricle relaxes, and thus prevents the backflow of blood into the relaxed ventricle.

The valves are shown in figure 16.10, and their actions are summarized in table 16.4.

mitral: L. *mitra*, like a bishop's mitre
bicuspid: L. *bi*, two; *cuspis*, tooth point or spike
trabeculae carneae: L. *trabecula*, small beams; *carneus*, flesh

Cardiac catheterization is a procedure used in the diagnosis of certain heart disorders. A tube, or catheter, is inserted into a blood vessel in the leg or the arm and threaded through the lumen of the vessel until it enters the heart. The location of the catheter can be monitored by means of a fluoroscope. When the catheter reaches the desired location within the heart, a radiopaque dye is released. The patient may then be radiographed, blood samples removed for analyses, or blood pressures monitored.

Circulatory Routes

The circulatory routes of the blood are illustrated in figure 16.2. The principal divisions of the circulatory blood flow are the pulmonary and systemic circulations.

The *pulmonary circulation* includes blood vessels that transport blood to the lungs for gas exchange and then back to the heart. It consists of the right ventricle that ejects the blood, the pulmonary trunk with its pulmonary valve, the pulmonary arteries that transport blood to the lungs, the pulmonary capillaries within each lung, the pulmonary veins that transport oxygenated blood back to the heart, and the left atrium that receives the blood from the pulmonary veins.

The *systemic circulation* is composed of all of the remaining vessels of the body that are not part of the pulmonary circulation. It includes the right atrium, the left ventricle, the aorta with its aortic valve, all of the branches of the aorta, all capillaries other than those in the lungs, and all veins other than the

UNIT VI

Figure 16.10 A superior view of the heart valves.

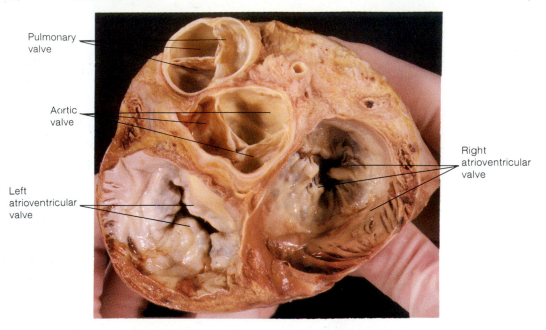

Table 16.4	Valves of the Heart	
Valve	**Location**	**Comments**
Right atrioventricular valve	Between right atrium and right ventricle	Composed of three cusps that prevent a backflow of blood from right ventricle into right atrium during ventricular contraction
Pulmonary valve	Entrance to pulmonary trunk	Composed of three half-moon-shaped flaps that prevent a backflow of blood from pulmonary trunk into right ventricle during ventricular relaxation
Left atrioventricular (mitral) valve	Between left atrium and left ventricle	Composed of two cusps that prevent a backflow of blood from left ventricle to left atrium during ventricular contraction
Aortic valve	Entrance to ascending aorta	Composed of three half-moon-shaped flaps that prevent a backflow of blood from aorta into left ventricle during ventricular relaxation

UNIT VI

pulmonary veins. The right atrium receives all of the venous return of oxygen-depleted blood from the systemic veins.

Coronary Circulation   The myocardium is supplied with blood by the right and left **coronary arteries** (fig. 16.15). These two vessels arise from the aorta immediately beyond the aortic semilunar valve. The coronaries encircle the heart within the atrioventricular sulcus, the depression between the atria and ventricles. Two branches arise from both the right and left coronaries to serve the atrial and ventricular walls. The left coronary artery gives rise to the **anterior interventricular branch,** which courses within the anterior interventricular sulcus to serve both ventricles, and the **circumflex branch,** which supplies oxygenated blood to the walls of the left atrium

circumflex: L. *circum,* around; *flectere,* to bend

and left ventricle. The right coronary artery gives rise to the **right marginal branch,** which serves the walls of the right atrium and right ventricle, and the **posterior interventricular branch,** which courses through the posterior interventricular sulcus to serve the two ventricles. The main trunks of the right and left coronaries anastomose (join together) on the posterior surface of the heart.

From the capillaries in the myocardium, the blood enters the **cardiac veins.** The course of these vessels parallels that of the coronary arteries. The cardiac veins, however, have thinner walls and are more superficial than the arteries. The two principal cardiac veins are the **anterior interventricular vein,** which returns blood from the anterior aspect of the heart, and the **posterior cardiac vein,** which drains the posterior aspect of the heart. These cardiac veins converge into a large venous channel on the posterior surface of the heart called the **coronary sinus** (figs. 16.9 and 16.11). The coronary venous blood then enters the heart through an opening into the right atrium.

> **Heart attacks** are the most common cause of death in the United States. The most common type of heart attack involves an occlusion of a coronary artery, which reduces the delivery of oxygen to the myocardium. Several strategies are recommended to minimize the risk of heart attack: (1) reduce excess weight; (2) avoid hypertension through proper diet, stress reduction, or medication if necessary; (3) maintain a low cholesterol level through proper diet; (4) avoid smoking; and (5) exercise regularly.

Conduction System of the Heart

Cardiac muscle has an intrinsic rhythmicity that allows the heartbeat to originate in and be conducted through the heart without extrinsic stimulation. Specialized strands of cardiac muscle tissue that coordinate cardiac contraction constitute the **conduction system.** The conduction system enables the **cardiac cycle,** which is defined as the events surrounding the filling and emptying of the chambers of the heart. The conduction system consists of specialized tissues that generate and distribute electrical impulses through the heart. The components of the conduction system are the **sinoatrial node** (SA node), **atrioventricular node** (AV node), **atrioventricular bundle** (bundle of His), and **conduction myofibers** (Purkinje fibers). None of these tissues can be seen grossly, but their locations can be noted (fig. 16.12). The SA node, or pacemaker, is located in the posterior wall of the right atrium. The SA node initiates the cardiac cycle by producing an electrical impulse that spreads over both atria, causing them to contract simultaneously and force blood into the ventricles. The basic depolarization rate of the SA node is 70–80 times per minute. The impulse then passes to the AV node, located in the inferior portion of the interatrial septum. From here, the impulse continues through the atrioventricular bundle, located at the top of the interventricular septum, where it divides into right and left branches. These

bundle of His: from Wilhelm His Jr., Swiss physician, 1863–1934
Purkinje fibers: from Johannes E. von Purkinje, Bohemian anatomist, 1787–1869

Figure 16.11 Coronary circulation. (*a*) A plastic cast of the coronary vessels and their major branches. (*b*) An anterior view of the arterial supply to the heart. (*c*) An anterior view of the venous drainage.

(a)

(b)

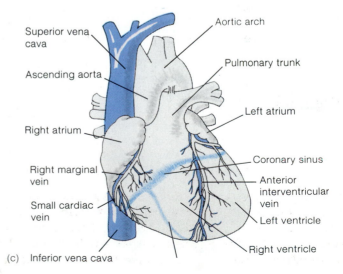

(c)

UNIT VI

Figure 16.12 The conduction system of the heart.

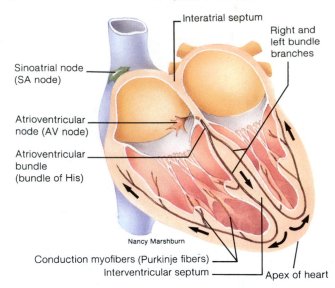

Interatrial septum

Right and left bundle branches

Sinoatrial node (SA node)

Atrioventricular node (AV node)

Atrioventricular bundle (bundle of His)

Nancy Marshburn

Conduction myofibers (Purkinje fibers)

Interventricular septum

Apex of heart

branches are continuous with the conduction myofibers within the ventricular walls. Stimulation of these fibers causes the ventricles to contract simultaneously.

Contraction of the ventricles is referred to as **systole** *(sis'tŏ-le)*. Systole, together with the tension of the elastic fibers and contraction of smooth muscles within the systemic arteries, accounts for the systolic pressure within arteries. Ventricular relaxation is called **diastole** *(di-as'tŏ-le)*. During diastole, the diastolic pressure within arteries can be recorded.

> Variation in functioning within vital organs should be regarded as normal for a healthy adult. The heartbeat can vary by as much as 20 or 30 beats per minute in 24 hours, but the heart maintains an average of about 70 beats per minute during a day. Blood pressure recorded at 120 over 80 in the morning can rise to 140 over 100 by evening. Normal body temperature does not stay at 98.6° F, but varies by 1½ to 2 degrees during the day—from 97 to 99° F.

Although the heart does have an innate contraction pattern, it is also innervated from the autonomic nervous system in order to respond to the ever-changing physiological needs of the body. The SA and AV nodes have both sympathetic and parasympathetic innervations. Sympathetic stimulation accelerates the heart rate and dilates the coronary arteries, enabling the heart to meet its own increased metabolic demands as well as those of the rest of the body. Parasympathetic stimulation has the opposite effect. Sympathetic innervation is through fibers from the cervical and upper thoracic ganglia. Parasympathetic innervation is through branches of the vagus nerves. Branches from the right vagus innervate the SA node and branches from the left vagus innervate the AV node.

systole: Gk. *systole*, contraction
diastole: Gk. *diastole*, prolonged or expansion

> *Cardiac output* is the volume of blood ejected from the heart into systemic circulation over the course of 1 minute. It is determined by the stroke rate, or heart rate, times the stroke volume of blood. The stroke volume is the amount of blood pumped from the heart into systemic circulation with each ventricular contraction, which amounts to about 70 mℓ of blood. A normal resting cardiac output ranges between 4.2 and 5.6 liters per minute. Exercise increases the heart rate, as do the hormones epinephrine and thyroxine. Atropine, caffeine, and camphor are drugs that have a stimulatory effect. Blood pressure and body temperature also have a profound effect on the heart rate.

Electrocardiogram

The electrical impulses that pass through the conduction system of the heart during the cardiac cycle can be recorded as an *electrocardiogram* (ECG or EKG). The electrical changes result from depolarization and repolarization of cardiac muscle fibers and can be detected on the surface of the skin using an instrument called the *electrocardiograph*.

An ECG of a normal cardiac cycle is illustrated in figure 16.13. The several waves, or deflections, are each produced as specific events of the cardiac cycle occur. The wave deflections have been designated as P, Q, R, S, and T. Any pathologic condition that disturbs the electrical activity of the heart will produce characteristic changes in one or more of the waves, so understanding the normal wave-deflection patterns is clinically important.

P Wave   The P wave is a small upward deflection that accompanies depolarization of the atrial fibers through stimulation of the SA node. The actual contraction of the atria follows the P wave by a fraction of a second. The ventricles of the heart are in diastole during the expression of the P wave. A missing or otherwise abnormal P wave may indicate a dysfunction of the SA node.

P-R Interval   The period of time from the start of the P wave to the beginning of the QRS complex is known as the P-R interval. This period indicates the time required for the SA depolarization to reach the ventricles. A prolonged P-R interval suggests a conduction delay in the AV node.

QRS Complex   The QRS wave begins as a short downward deflection followed by a sharp upward spike and ends as a downward deflection at the base. The QRS complex represents the depolarization of the ventricles. During this interval, the ventricles of the heart are in systole and blood is being ejected from the heart. An abnormal QRS complex generally indicates cardiac problems involving the ventricles. An enlarged R spike, for example, generally indicates enlarged ventricles.

S-T Segment   The time duration known as the S-T segment represents the period between the completion of ventricular depolarization and initiation of repolarization. The S-T segment is depressed when the heart receives insufficient oxygen and is elevated in acute myocardial infarction.

UNIT VI

Figure 16.13 The electrocardiogram indicates the conduction of electrical impulses through the heart (a) and measures and records both the intensity of this electrical activity (in millivolts) and the time intervals involved (b).

(a)

Lew

(b)

T Wave The T wave is produced by ventricular repolarization. Various heart diseases, such as an arteriosclerotic heart, will produce altered T waves.

Heart Sounds

Closing of the AV and semilunar valves produces sounds that can be heard at the surface of the chest with a stethoscope. These sounds are often described onomatopoetically as *lub-dub*. The "lub," or **first sound,** is produced by the closing of the AV valves. The "dub," or **second sound,** is produced by the closing of the aortic and pulmonary valves. The first sound is thus heard when the ventricles contract at systole and the second sound is heard when the ventricles relax at the beginning of diastole.

Heart sounds are of clinical importance because they provide information about the condition of the heart valves and other heart problems. Abnormal sounds are referred to as *heart murmurs* and are caused by valvular leakage or turbulence of the blood as it passes through the heart. In general, three basic conditions cause murmurs: (1) *valvular insufficiency*, in which the cusps of the valves do not form a tight seal; (2) *stenosis*, in which the walls surrounding a valve are roughened or constricted; and (3) a *functional murmur*, which is frequent in children and is caused by turbulence of the blood moving through the heart during heavy exercise. Functional murmurs are not pathological and are considered normal.

stenosis: Gk. *stenosis*, a narrowing

The valves of the heart are positioned directly deep to the sternum, which tends to obscure and dissipate valvular sounds. For this reason, a physician will listen with a stethoscope for the heart sounds at locations designated as **valvular auscultatory areas,** which are named according to the valve that can be detected (fig. 16.14). The **aortic area** is at the right second intercostal space near the sternum. The **pulmonic area** is directly across from the aortic area at the left second intercostal space near the sternum. The **right atrioventricular** and **left atrioventricular (mitral) areas** are both at the fifth intercostal space, with the left atrioventricular area farther out. Surface landmarks are extremely important in identifying auscultatory areas.

1. Distinguish between the pericardial and the thoracic cavities. Describe the cardiac serous membranes.
2. List the three layers of the heart wall and describe the structures associated with each layer.
3. List the valves that control blood flow through the heart and describe their locations and functions.
4. Trace the flow of electrical impulses through the conduction system of the heart.

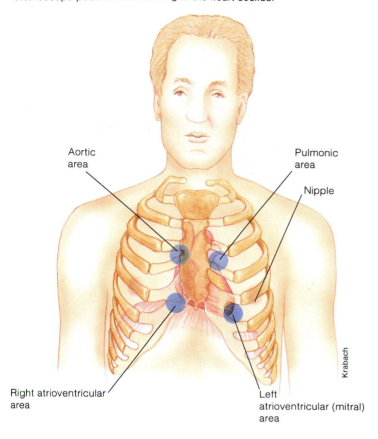

Aortic area

Pulmonic area

Nipple

Right atrioventricular area

Left atrioventricular (mitral) area

Krabach

Blood Vessels

The structure of arteries and veins allows them to function in the transport of blood from the heart to the capillaries and from the capillaries back to the heart. The structure of capillaries provides the mechanism for the exchange of plasma fluid and dissolved molecules between the blood and the surrounding tissues.

Objective 12. Describe the structure, size, and function of arteries, capillaries, and veins.

Objective 13. Explain why capillaries are considered the functional units of the circulatory system.

Blood vessels form a tubular network throughout the body that permits blood to flow from the heart to all the living cells of the body and then back to the heart. Blood leaving the heart passes through vessels of progressively smaller diameters referred to as **arteries, arterioles,** and **capillaries.** Capillaries are microscopic vessels that join the arterial flow to the venous flow. Blood returning to the heart from the capillaries passes through vessels of progressively larger diameters called **venules** and **veins.**

The walls of arteries and veins are composed of three coats, or tunics (fig. 16.15). The outermost layer is the **tunica** (_too'ni-_

tunic: L. _tunica,_ covering or coat

_kă_) **externa,** the middle layer is the **tunica media,** and the inner layer is the **tunica intima.** The tunica externa is composed of connective tissue. The tunica media is composed primarily of smooth muscle. The tunica intima consists of two parts: (1) an innermost simple squamous epithelium, the _endothelium (en''do-the'le-um),_ which lines the lumina of all blood vessels, and (2) a layer of elastic fibers, or _elastin._

Although both arteries and veins have the same basic structure, there are some important differences between the two types. Arteries have more muscle in proportion to their diameter than do comparably sized veins. Also, arteries appear rounder than veins in cross section. Veins are usually partially collapsed due to the fact that they are not usually filled to their capacity. They can stretch when they receive more blood, and thus function as reservoirs or capacitance vessels. In addition, many veins have valves, which are absent in arteries.

Arteries

The aorta and other large arteries contain numerous layers of elastin fibers between smooth muscle cells in the tunica media. These large arteries expand when the pressure of the blood rises as a result of the heart's contraction; they recoil, like a stretched rubber band, when blood pressure falls during relaxation of the heart. This elastic recoil helps to produce a smoother, less pulsatile flow of blood through the smaller arteries and arterioles.

The small arteries and arterioles are less elastic than the larger arteries and have a thicker layer of smooth muscle in proportion to their diameter. Unlike the larger _elastic arteries,_ therefore, the smaller _muscular arteries_ retain almost the same diameter as the pressure of the blood rises and falls during the heart's pumping activity. Since small muscular arteries and arterioles have narrow lumina, they provide the greatest resistance to blood flow through the arterial system.

Small muscular arteries that are 100 μm or less in diameter branch to form smaller arterioles (20–30 μm in diameter). In some tissues, blood from the arterioles can enter the venules directly through **arteriovenous anastomoses** called **metarterioles** (fig. 16.16). In most cases, however, blood from arterioles passes into capillaries. Capillaries are the narrowest of blood vessels (7–10 μm in diameter), and serve as the functional unit of the circulatory system, in which exchanges of gases and nutrients between the blood and the tissues occur.

Capillaries

The arterial system branches extensively to deliver blood to over 40 billion capillaries in the body. The extensiveness of these branchings is indicated by the fact that all tissue cells are located within a distance of only 60–80 μm of a capillary and by the fact that capillaries provide a total surface area of 1,000 square miles for exchanges between blood and tissue fluid.

Despite their large number, capillaries contain only about 250 mℓ of blood at any time, out of a total blood volume of about 5,000 mℓ (most is contained in the venous system). The amount of blood flowing through a particular capillary bed is determined in part by the action of the **precapillary sphincter**

lumina: L. _lumen,_ opening

UNIT VI

Figure 16.15 The structure of a medium-sized artery and vein showing the relative thickness and composition of the tunics.

Artery

Vein

Tunica intima
(endothelial cells)

(elastin)

Valve

Tunica media
(smooth muscle)

Tunica externa
(loose fibrous
connective tissue)

Serosa

Figure 16.16 Microcirculation at the capillary level. Metarterioles (arteriovenous anastomoses) provide a path of least resistance between arterioles and venules. Precapillary sphincter muscles regulate the flow of blood through the capillaries.

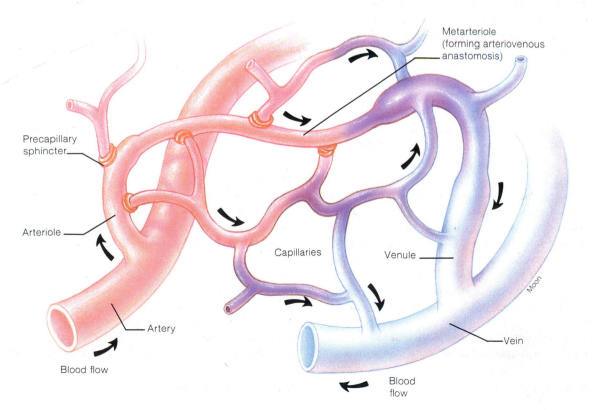

Metarteriole
(forming arteriovenous
anastomosis)

Precapillary
sphincter

Arteriole

Capillaries

Venule

Artery

Vein

Blood flow

Blood
flow

Moon

Figure 16.17 Diagrams of continuous, fenestrated, and discontinuous capillaries as they appear in the electron microscope. This classification is based on the continuity of the endothelial layer. (Dark circles in the cytoplasm indicate pinocytotic vesicles.)

Continuous

Fenestrated

Discontinuous

Creek

muscles (fig. 16.16). These muscles allow only 5%–10% of the capillary beds in skeletal muscles, for example, to be open at rest. Blood flow to an organ is regulated by the action of these precapillary sphincters and by the degree of resistance to blood flow (due to constriction or dilation) provided by the small arteries and arterioles in the organ.

Unlike the vessels of the arterial and venous systems, the walls of capillaries are composed of only one cell layer—a simple squamous epithelium, or endothelium. The absence of smooth muscle and connective tissue layers allows for a more rapid rate of transport of materials between the blood and the tissues.

Types of Capillaries   Different organs have different types of capillaries, which are distinguished by significant differences in structure. In terms of their endothelial lining, these capillary types include those that are *continuous,* those that are *discontinuous,* and those that are *fenestrated (fen'es-trāt''ed)* (fig. 16.17).

Continuous capillaries are those in which adjacent endothelial cells are closely joined together. These are found in muscles, lungs, adipose tissue, and in the central nervous system. The fact that continuous capillaries in the CNS lack intercellular channels contributes to the blood-brain barrier. Continuous capillaries in other organs have narrow intercellular channels (about 40–45 Å in width) that permit the passage of molecules other than protein between the capillary blood and tissue fluid.

The examination of endothelial cells with an electron microscope has revealed the presence of pinocytotic vesicles, which suggests that the intracellular transport of material may occur across the capillary walls. This type of transport appears to be the only mechanism of capillary exchange available within the central nervous system and may account, in part, for the selective nature of the blood-brain barrier.

The kidneys, endocrine glands, and intestines have **fenestrated capillaries,** characterized by wide intercellular pores (800–1,000 Å) that are covered by a layer of mucoprotein, which may serve as a diaphragm. In the bone marrow, liver, and spleen, the distance between endothelial cells is so great that these **discontinuous capillaries** appear as little cavities (*sinusoids*) in the organ.

Veins

The average pressure in the veins is only 2 mm Hg (millimeters of mercury), compared to a much higher average arterial pressure of about 100 mm Hg. These pressures represent the hydrostatic pressure that the blood exerts on the walls of the vessels, and the numbers indicate the differences from atmospheric pressure.

The low venous pressure is insufficient to return blood to the heart, particularly from the lower limbs. Veins, however, pass between skeletal muscle groups that produce a massaging action as they contract (fig. 16.18). As the veins are squeezed by contracting skeletal muscles, a one-way flow of blood to the heart is ensured by the presence of **venous valves.**

The effect of the massaging action of skeletal muscles on venous blood flow is often described as the *skeletal muscle pump.* The rate of venous return to the heart is dependent, in large part, on the action of skeletal muscle pumps. When these pumps are less active, as when a person stands still or is bedridden, blood accumulates in the veins and causes them to bulge. When a person is more active, blood returns to the heart at a faster rate and less is left in the venous system.

The accumulation of blood in the veins of the legs over a long period of time, as may occur in people with occupations that require standing still all day, can cause the veins to stretch to the point where the venous valves are no longer efficient. This can produce *varicose veins*. During walking, the movements of the foot activate the soleus muscle pump. This effect can be produced in bedridden people by upward and downward manipulations of the feet.

Blood Pressure

Blood pressure exists throughout the vascular system but is greatest within arteries, where it is commonly measured and used as an indication of health. Blood pressure is created primarily by the action of the heart and is the force exerted by the blood against the vascular walls.

Arterial blood pressure can be measured by using a device called a *sphygmomanometer* (fig. 16.19). This instrument has an inflatable cuff that is used to constrict an artery at a pressure point (see fig. 16.31). The most common site of application is the brachial artery compressed against the humerus. As the blood pulsates through the constricted brachial artery, the pressure produced is indicated by the sphygmomanometer.

The normal blood pressure of an adult is about 120/80. This is an expression of the **systolic pressure** (120 mm Hg) over the **diastolic pressure** (80 mm Hg). The systolic pressure is created as blood is ejected from the heart during ventricular systole. When the ventricles relax during ventricular diastole,

UNIT VI

Figure 16.18   The action of the one-way venous valves. The contraction of skeletal muscles helps to pump blood toward the heart, but it is prevented from pushing blood away from the heart by closure of the venous valves.

Figure 16.19   The use of a pressure cuff and a sphygmomanometer to measure blood pressure.

Figure 16.20   Blood pressure as recorded at various vascular sites.

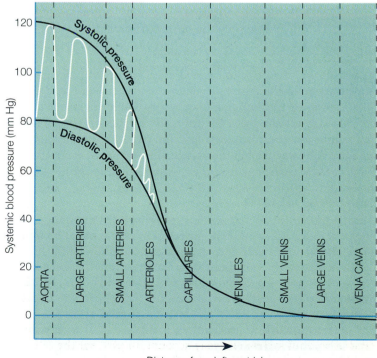

the arterial pressure drops and the diastolic pressure is recorded. For the most part, diastolic pressure results from contraction of the smooth muscles within the arterial walls. The difference between the systolic and diastolic pressure is called the **pulse pressure** and is generally about 40 mm Hg.

Several factors can influence the arterial blood pressure. Blood pressure decreases as the distance from the heart increases (fig. 16.20). An increased cardiac output (stroke volume × heart rate) increases blood pressure. Blood volume and blood viscosity both influence blood pressure, as do various drugs. Changes in the diameters of the vascular lumina through autonomic vasoconstriction or vasodilation have a direct bearing on blood pressure. Blood pressures are also greatly influenced by

UNIT VI

the general health of the cardiovascular system. Elevated blood pressure is referred to as *hypertension* and is potentially dangerous because of the extra strain it puts on the heart.

One of the harmful effects of nicotine inhaled from cigarette smoking is that it causes vasoconstriction of arterioles. It also is a stimulant to the heart and increases cardiac output. Both of these responses raise the blood pressure and increase the strain on the heart.

1. Describe the basic structural pattern of arteries and veins. Describe how arteries and veins differ in structure and how these differences contribute to the resistance function of arteries and the capacitance function of veins.
2. Describe the functional significance of the skeletal muscle pump and explain the action of venous valves.
3. Discuss the functions of capillaries and describe the structural differences between capillaries in different organs.
4. Describe the cardiovascular events that determine systolic and diastolic blood pressures.

Principal Arteries of the Body

The aorta ascends from the left ventricle to a position just above the heart, where it arches to the left and then descends through the thorax and abdomen. Branches of the aorta carry oxygenated blood to all of the cells of the body.

Objective 14. In the form of a flow chart, list the arterial branches of the ascending aorta and aortic arch.

Objective 15. Describe the arterial supply to the brain.

Objective 16. Describe the arterial pathways that supply the upper extremity.

Objective 17. Describe the major arteries to the thorax, abdomen, and lower extremity.

Contraction of the left ventricle forces oxygenated blood into the arteries of the systemic circulation. The principal arteries of the body are shown in figure 16.21. They will be described by region and identified in order from largest to smallest, or as the blood flows through the system. The major systemic artery is the **aorta,** from which all the primary systemic arteries arise.

Aortic Arch

The systemic vessel that ascends from the left ventricle of the heart is called the **ascending portion of the aorta.** The right and left **coronary arteries** serving the myocardium of the heart with blood are the only branches that arise from the ascending aorta. The aorta arches to the left and posteriorly over the pulmonary arteries as the **aortic arch** (fig. 16.22). Three vessels arise from the aortic arch: the **brachiocephalic** *(brak''e-o-sĕ-fal'ik)* **trunk,** the **left common carotid** *(ka-rot'id)* **artery,** and the **left subclavian artery.**

The brachiocephalic trunk is the first vessel to branch from the aortic arch and, as its name suggests, supplies blood to the

tissues of the arm and head on the right side of the body. It is a short vessel rising superiorly through the mediastinum to a point near the junction of the sternum and the right clavicle. There it bifurcates into the **right common carotid artery,** which extends to the right side of the neck and head, and the **right subclavian artery,** which carries blood to the right upper extremity.

The remaining two branches from the arch of the aorta are the left common carotid and the left subclavian arteries. The left common carotid artery transports blood to the left side of the neck and head, while the left subclavian artery supplies the left upper extremity.

Arteries of the Neck and Head

The common carotid arteries course upward in the neck along the lateral sides of the trachea (fig. 16.23). Each common carotid artery bifurcates into the **internal** and **external carotid arteries** slightly below the angle of the mandible. By pressing gently in this area, a pulse can be detected (see fig. 16.31). At the base of the internal carotid artery, near the bifurcation, is a slight dilation called the **carotid sinus.** The carotid sinus contains *baroreceptors,* which monitor blood pressure, and *chemoreceptors* within the **carotid body,** which respond to chemical changes in the blood.

Blood Supply to the Brain The brain is supplied with arterial blood that arrives through four vessels that eventually unite on the inferior aspect of the brain surrounding the pituitary gland (fig. 16.24). The four vessels are the paired internal carotid arteries and the paired vertebral arteries. The value of having four separate vessels that anastomose at one location is that if one becomes occluded, the three alternate routes may allow for an adequate blood supply to the brain.

The **vertebral arteries** arise from the subclavian arteries at the base of the neck (fig. 16.23). They pass superiorly through the transverse foramina of the cervical vertebrae and enter the skull through the foramen magnum. Within the braincase, the two vertebral arteries unite to form the **basilar artery** at the level of the pons. The basilar ascends along the inferior surface of the brain stem and terminates by forming two **posterior cerebral arteries,** which supply the posterior portion of the cerebrum (fig. 16.24). The **posterior communicating arteries** are branches that arise from the posterior cerebral arteries and participate in forming the **cerebral arterial circle** around the pituitary gland, also called the *circle of Willis.*

The **internal carotid artery** branches from the common carotid artery and ascends in the neck until it reaches the base of the skull, where it enters the carotid canal of the temporal bone. Several branches arise from the internal carotid artery once it is on the inferior surface of the brain. Three of the more important ones are the **ophthalmic** *(of'thal'mik)* **artery** (fig. 16.23a), which supplies the eye and associated structures, and the **anterior** and **middle cerebral arteries,** which provide blood to the cerebrum. The internal carotid arteries are connected to the posterior cerebral arteries at the cerebral arterial circle.

circle of Willis: from Thomas Willis, English physician, 1621–75

Figure 16.21 Principal arteries of the body (a. = artery, aa. = arteries).

External carotid a.

Internal carotid a.

Lt. common carotid a.

Lt. subclavian a.

Brachiocephalic trunk

Aortic arch

Coronary aa.

Celiac trunk

Intercostal a.

Superior mesenteric a.

Inferior mesenteric a.

Radial a.

Dorsal pedal a.

Carotid sinus

Rt. vertebral a.

Rt. subclavian a.

Axillary a.

Internal thoracic a.

Brachial a.

Descending portion of aorta

Ulnar a.

Radial a.

Common iliac a.

Internal iliac a.

External iliac a.

Testicular a.

Deep femoral a.

Femoral a.

Popliteal a.

Posterior tibial a.

Anterior tibial a.

Margulies/Waldrop

UNIT VI

Figure 16.22 The structural relationship between the major arteries and veins to and from the heart (v. = vein, vv. = veins).

Right common carotid a.
Right internal jugular v.
Right subclavian a.
Brachiocephalic trunk
Right brachiocephalic v.
Superior vena cava
Ascending portion of aorta
Right pulmonary a.
Right pulmonary vv.
Right auricle

Left common carotid a.
Left internal jugular v.
Left subclavian a.
Left brachiocephalic v.
Aortic arch
Ligamentum arteriosum
Left pulmonary a.
Left pulmonary v.
Left auricle
Pulmonary trunk

Moon

Capillaries within the pituitary gland receive both arterial and venous blood. The venous blood arrives from venules immediately superior to the pituitary gland, which drain capillaries in the hypothalamus of the brain. This arrangement of two capillary beds in series—whereby the second capillary bed receives venous blood from the first—is called a *portal system*. The venous blood that travels from the hypothalamus to the pituitary contains hormones from the hypothalamus that help regulate pituitary gland hormone secretion.

External Carotid Artery The external carotid artery gives off several branches as it extends upward along the side of the neck and head (see fig. 16.23). The names of these branches are determined by the areas or structures that they serve. The principal vessels that arise from the external carotid artery are the following:

1. **superior thyroid artery,** which serves the muscles of the hyoid region, the larynx and vocal cords, and the thyroid gland;
2. **ascending pharyngeal artery** (not seen in illustration), which serves the pharyngeal area and various lymph nodes;
3. **lingual artery,** which provides extensive vascularization of the tongue and sublingual gland;
4. **facial artery,** which traverses a notch on the inferior margin of the mandible to serve the pharyngeal area,

palate, chin, lips, and nasal region—an important vessel to apply pressure to when controlling bleeding from the face;
5. **occipital artery,** which serves the posterior portion of the scalp, the meninges over the brain, the mastoid process, and certain posterior neck muscles; and
6. **posterior auricular artery,** which serves the auricle of the ear and scalp over the auricle.

The external carotid artery terminates at a level near the mandibular condyle by dividing into **maxillary** and **superficial temporal arteries.** The maxillary artery gives off branches to the teeth and gums, the muscles of mastication, nasal cavity, eyelids, and meninges. The superficial temporal artery supplies blood to the parotid gland and to the superficial structures on the side of the head. Pulsations through the temporal artery can be easily detected by placing the fingertips immediately in front of the ear at the level of the eye (see fig. 16.31). This vessel is frequently used by anesthesiologists to check a patient's pulse rate during surgery.

Headaches are usually caused by vascular pressure on the sensitive meninges covering the brain. The two principal vessels serving the meninges are the meningeal branches of the occipital and maxillary arteries. Vasodilation of these vessels creates excessive pressure on the sensory receptors within the meninges, resulting in a headache.

UNIT VI

Figure 16.23   Arteries of the neck and head. (a) Major branches of the right common carotid and right subclavian arteries. (b) A radiograph of the head following a radiopaque injection of the arteries.

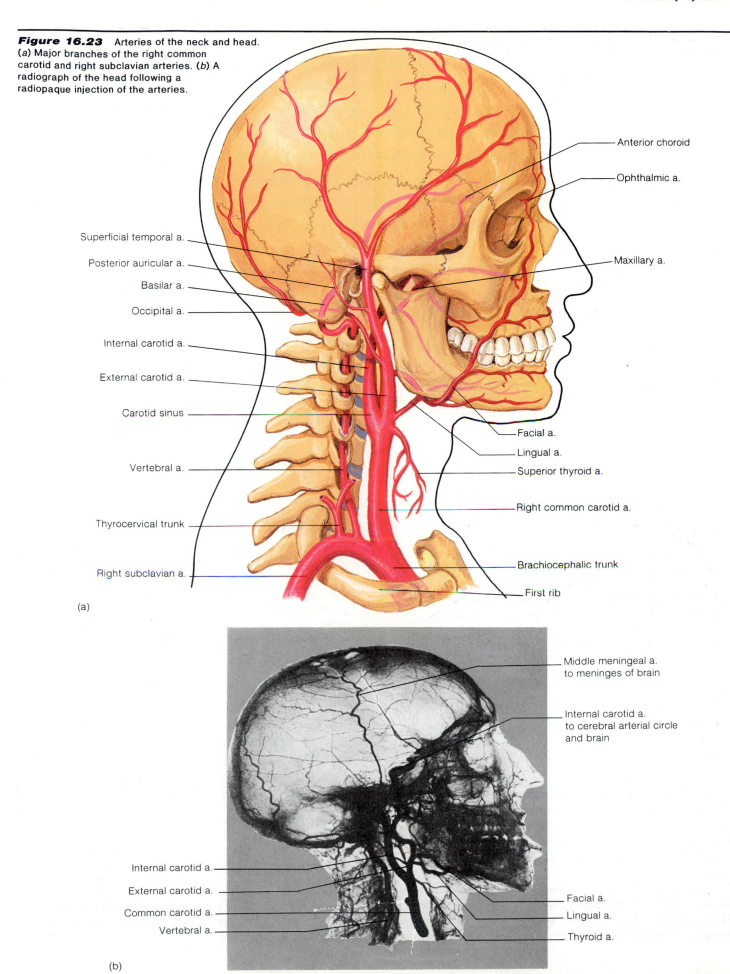

Anterior choroid

Ophthalmic a.

Maxillary a.

Superficial temporal a.

Posterior auricular a.

Basilar a.

Occipital a.

Internal carotid a.

External carotid a.

Carotid sinus

Facial a.

Lingual a.

Superior thyroid a.

Vertebral a.

Right common carotid a.

Thyrocervical trunk

Brachiocephalic trunk

Right subclavian a.

First rib

(a)

Middle meningeal a. to meninges of brain

Internal carotid a. to cerebral arterial circle and brain

Internal carotid a.

External carotid a.

Common carotid a.

Vertebral a.

Facial a.

Lingual a.

Thyroid a.

(b)

UNIT VI

Figure 16.24 Arteries that supply blood to the brain. (*a*) An inferior view of the brain and (*b*) a close-up view of the region of the pituitary gland.

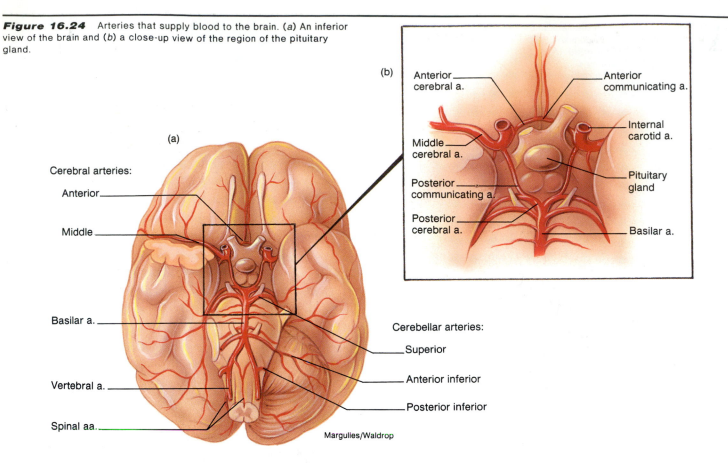

(a)

Cerebral arteries:

Anterior

Middle

Basilar a.

Vertebral a.

Spinal aa.

(b)

Anterior cerebral a.

Middle cerebral a.

Posterior communicating a.

Posterior cerebral a.

Anterior communicating a.

Internal carotid a.

Pituitary gland

Basilar a.

Cerebellar arteries:

Superior

Anterior inferior

Posterior inferior

Margulies/Waldrop

Arteries of the Upper Extremity

The right subclavian artery branches off the brachiocephalic, and the left subclavian artery arises directly from the aortic arch (see fig. 16.22). The **subclavian artery** passes laterally deep to the clavicle, carrying blood toward the arm (fig. 16.25). The pulsations of the subclavian artery can be detected by pressing firmly on the tissue just above the medial portion of the clavicle (see fig. 16.31). From each subclavian artery arises a **vertebral artery** that carries blood to the brain (already described), a short **thyrocervical trunk** that serves the thyroid gland, trachea, and larynx, and an **internal thoracic artery** that descends into the thorax to serve the thoracic wall, thymus, and pericardium. The **costocervical trunk** branches to serve the upper intercostal muscles, posterior neck muscles, and the spinal cord and its meninges. A branch of the internal thoracic artery supplies blood to the muscles and tissues (mammary glands) of the anterior thorax.

The subclavian artery becomes the **axillary artery** as it passes into the axillary region. The axillary artery is that portion of the major artery of the upper extremity between the outer border of the first rib and the lower border of the teres major muscle. Several small branches arise from the axillary artery and supply blood to the tissues of the upper thorax and shoulder region.

The **brachial artery** is the continuation of the axillary artery through the brachial region. The brachial artery courses on the medial side of the humerus, where it is a major pressure point and the most common site for determining blood pres-

sure. A **deep brachial artery** branches from the brachial artery and curves posteriorly near the radial nerve to supply the triceps brachii muscle. Two additional branches from the brachial, the **anterior** and **posterior humeral circumflex arteries,** form a continuous ring of vessels around the proximal portion of the humerus.

Just proximal to the cubital fossa, the brachial bifurcates into the **radial** and **ulnar arteries,** which supply blood to the forearm and a portion of the hand and digits. The radial artery courses down the lateral, or radial, side of the arm, where it sends numerous small branches to the muscles of the forearm. The **radial recurrent artery** serves the region of the elbow and is the first and largest branch of the radial artery. The radial artery is important as a site to record the pulse near the wrist.

The ulnar artery is a bit larger in caliber than the radial artery. It extends down the ulnar side of the forearm and gives off many small branches to the muscles on the medial side of the forearm. It too has an initial large branch, the **ulnar recurrent artery,** which arises from the proximal portion near the elbow. At the wrist, the ulnar and radial arteries anastomose to form the **palmar arches** in the hand, from which **digital arteries** extend into the fingers.

Branches of the Thoracic Portion of the Aorta

The **thoracic portion of the aorta** is a continuation of the aortic arch as it descends through the thoracic cavity to the diaphragm. This large vessel gives off branches to the organs and

UNIT VI

Figure 16.25 Arteries of the upper extremity. (*a*) A radiograph of the forearm and hand following a radiopaque injection of the arteries. (*b*) An anterior view of major arteries.

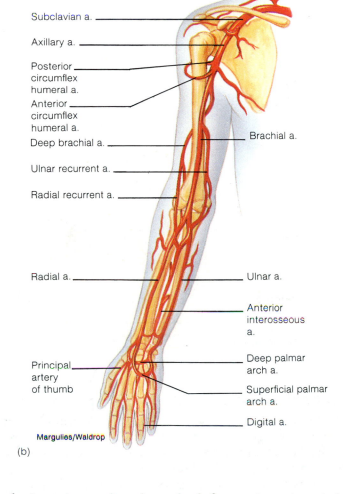

muscles of the thoracic region. These branches include **pericardial arteries,** going to the pericardium of the heart; **bronchial arteries** for systemic circulation to the lungs; **esophageal arteries,** going to the esophagus as it passes through the mediastinum; segmental **posterior intercostal arteries,** serving the intercostal muscles and structures of the wall of the thorax (fig. 16.26); and **superior phrenic** *(fren'ik)* **arteries,** supplying blood to the diaphragm.

These vessels are summarized according to their location and function in table 16.5.

Branches of the Abdominal Portion of the Aorta

The **abdominal portion of the aorta** is the segment of the aorta between the diaphragm and the level of the fourth lumbar vertebra, where it divides into the right and left **common iliac arteries.** Small **inferior phrenic arteries** that serve the diaphragm are the first vessels to arise from the abdominal portion of the aorta. The next vessel is the short and thick **celiac** *(se'le-ak)* **trunk** that divides immediately into three arteries: the

splenic, going to the spleen; the **left gastric,** going to the stomach; and the **common hepatic,** going to the liver (fig. 16.27).

The **superior mesenteric artery** is another unpaired vessel and arises anteriorly from the abdominal portion of the aorta, just below the celiac trunk. The superior mesenteric supplies blood to the small intestine (except for a portion of the duodenum), the cecum, the appendix, the ascending colon, and the transverse colon.

The next major vessels to arise from the abdominal portion of the aorta are the paired **renal arteries,** which carry blood to the kidneys. Smaller **suprarenal arteries** are located just above the renal arteries and serve the adrenal (suprarenal) glands. The **testicular arteries** in the male and the **ovarian arteries** in the female are small paired vessels that arise from the abdominal portion of the aorta, just below the renal arteries. These vessels serve the gonads.

The **inferior mesenteric artery** is the last major branch of the abdominal portion of the aorta. It is an unpaired anterior vessel that arises just before the iliac bifurcation. The inferior mesenteric supplies blood to the descending colon, sigmoid colon, and rectum.

Figure 16.26 Arteries that serve the thoracic wall.

Body of thoracic vertebra

Posterior intercostal a.

Internal intercostal m.

Rib

Thoracic portion of aorta

External intercostal m.

Internal thoracic a.

Sternum

Anterior intercostal a.

Costal cartilage

Table 16.5	Segments and Branches of the Aorta				
Segment of Aorta	Arterial Branch	General Region or Organ Served	Segment of Aorta	Arterial Branch	General Region or Organ Served
Ascending portion of aorta	Right and left coronary aa.	Heart	Abdominal portion of aorta	Inferior phrenic aa.	Inferior surface of diaphragm
Aortic arch	Brachiocephalic trunk Right common carotid a.	Right side of head and neck		Celiac trunk Common hepatic a.	Liver, upper pancreas, duodenum
	Right subclavian a.	Right shoulder and right upper extremity		Left gastric a. Splenic a.	Stomach, esophagus Spleen, pancreas, stomach
	Left common carotid a.	Left side of head and neck		Superior mesenteric a.	Small intestine, pancreas, cecum, appendix, ascending and transverse colons
	Left subclavian a.	Left shoulder and left upper extremity			
Thoracic portion of aorta	Pericardial aa.	Pericardium of heart		Suprarenal aa.	Adrenal (suprarenal)
	Posterior intercostal aa.	Intercostal and thoracic muscles, pleurae		Renal aa. Gonadal aa. Testicular aa.	Kidneys Testes
	Bronchial aa.	Bronchi of lungs		Ovarian aa.	Ovaries
	Superior phrenic aa.	Superior surface of diaphragm		Inferior mesenteric a.	Transverse, descending, and sigmoid colons; rectum
	Esophageal aa.	Esophagus			
				Common iliac aa. External iliac aa.	Lower extremities
				Internal iliac aa.	Genital organs, gluteal muscles

UNIT VI

Figure 16.27   An anterior view of the abdominal aorta and its principal branches. In (*a*) the abdominal viscera have been removed and in (*b*) they are intact.

Margulies/Waldrop

Several **lumbar arteries** branch posteriorly from the abdominal portion of the aorta throughout its length and serve the muscles and the spinal cord in the lumbar region. In addition, an unpaired **middle sacral artery** (fig. 16.28) arises from the posterior terminal portion of the abdominal portion of the aorta to supply the sacrum and coccyx.

Arteries of the Pelvis and Lower Extremity

The aorta terminates in the posterior pelvic area as it bifurcates into the right and left common iliac arteries. These vessels pass downward approximately 5 cm on their respective sides and terminate by dividing into the **internal** and **external iliac arteries.**

The internal iliac artery has extensive branches to supply arterial blood to the gluteal muscles and the organs of the pelvic region (fig. 16.28). The wall of the pelvis is served by the **iliolumbar** and **lateral sacral arteries.** The internal visceral organs of the pelvis are served by the **middle rectal** and the **superior, middle,** and **inferior vesicular arteries** to the urinary bladder. In addition, **uterine** and **vaginal arteries** branch from the internal iliac arteries to serve the reproductive organs of the female. The muscles of the buttock are served by the

superior and **inferior gluteal arteries.** Some of the upper medial thigh muscles are supplied with blood from the **obturator artery.** The **internal pudendal artery** of the internal iliac artery serves the musculature of the perineum and the external genitalia of males and females. The internal pudendal artery is an important vessel in sexual activity. Erection of the penis in the male and corresponding vascular engorgement of the genitalia of the female are vascular phenomena controlled by the autonomic nervous system (see chapter 13).

The external iliac artery passes out of the pelvic cavity beneath the inguinal ligament (fig. 16.29) and becomes the **femoral artery.** Two branches arise from the external iliac artery, however, before it passes beneath the inguinal ligament. An **inferior epigastric artery** branches from the external iliac artery and passes superiorly to supply the skin and muscles of the abdominal wall. The **deep circumflex iliac artery** is a small branch that extends laterally to supply the muscles attached to the iliac fossa.

The femoral artery passes through an area called the **femoral triangle** on the upper medial portion of the thigh (figs. 16.29 and 16.30). At this point, the femoral artery is close to the surface and not only can be palpated but is an important pressure point. Several vessels arise from the femoral artery to serve the thigh region. The largest of these, the **deep femoral**

UNIT VI

Figure 16.28 Arteries of the pelvic region.

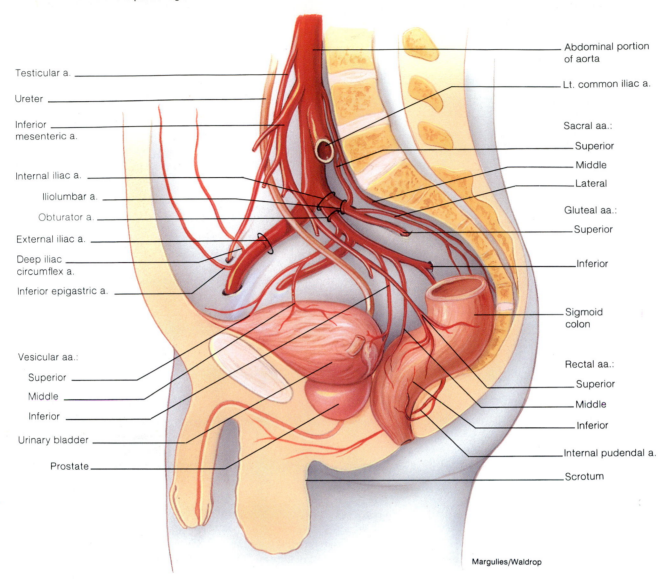

Testicular a.

Ureter

Inferior mesenteric a.

Internal iliac a.

Iliolumbar a.

Obturator a.

External iliac a.

Deep iliac circumflex a.

Inferior epigastric a.

Vesicular aa.:

Superior

Middle

Inferior

Urinary bladder

Prostate

Abdominal portion of aorta

Lt. common iliac a.

Sacral aa.:

Superior

Middle

Lateral

Gluteal aa.:

Superior

Inferior

Sigmoid colon

Rectal aa.:

Superior

Middle

Inferior

Internal pudendal a.

Scrotum

Margulies/Waldrop

artery, passes posteriorly to serve the hamstring muscles. The **lateral** and **medial femoral circumflex arteries** encircle the proximal end of the femur and serve muscles in this region. The femoral artery becomes the **popliteal** *(pop''li-te'al)* **artery** as it passes across the posterior aspect of the knee.

Hemorrhage can be a serious problem in many accidents. Therefore, one should know the pressure points where the arterial blood flow can be curtailed to prevent a victim from bleeding to death (see fig. 16.31). The pressure points for the appendages are the brachial artery on the medial side of the arm and the femoral artery in the groin. Firmly applied pressure to these regions greatly diminishes the flow of blood to traumatized areas below. A tourniquet may have to be applied if bleeding is severe enough to endanger life.

The popliteal artery supplies small branches to the knee joint and then divides into an **anterior tibial artery** and **posterior tibial artery** (fig. 16.29). These vessels traverse the anterior and posterior aspects of the leg, respectively, providing blood to the muscles of these regions and to the foot.

At the ankle, the anterior tibial artery becomes the **dorsal pedal artery** that serves the ankle and dorsum of the foot and then contributes to the formation of the **plantar arch** of the foot. Clinically, palpation of the dorsal pedal artery can provide information about circulation to the foot; but more important, it can provide information about the circulation in general because these pulses are taken at the most distal portion of the body.

The posterior tibial artery sends a large **fibular artery** to serve the peroneal muscles of the leg. At the ankle, the poste-

Figure 16.29   Arteries of the right lower extremity. (*a*) An anterior view and (*b*) a posterior view.

Right common iliac a.

Right external iliac a.

Inguinal ligament

Lateral femoral circumflex a.

Descending branch of lateral femoral circumflex a.

Lateral genicular aa.

Anterior tibial a.

Dorsal pedal a.

Internal iliac a.

Obturator a.

Femoral a.

Deep femoral a.

Medial genicular aa.

Posterior tibial a.

Medial plantar a.

Medial femoral circumflex a.

Lateral femoral circumflex a.

Popliteal a.

Fibular a.

Lateral plantar a.

Digital a.

(a)

(b)

Margulies/Waldrop

UNIT VI

rior tibial bifurcates into the **lateral** and **medial plantar arteries** that supply the muscles and structures on the sole of the foot. The lateral plantar artery anastomoses with the dorsal pedal artery to form the plantar arch, similar to the arterial arrangement in the hand. **Digital arteries** arise from the plantar arch to supply the toes with blood.

1. Describe the blood supply to the brain. Where is the cerebral arterial circle located and how is it formed?
2. Describe the clinical significance of the brachial and radial arteries.
3. Describe the arterial pathway from the subclavian artery to the digital arteries.
4. List the arteries that supply blood to the lower abdominal wall, the external genitalia, the hamstring muscles, the knee joint, and the dorsum of the foot.

Figure 16.30 The femoral triangle. The structures within the femoral triangle are shown in (a) and the boundries of the triangle are shown in (b).

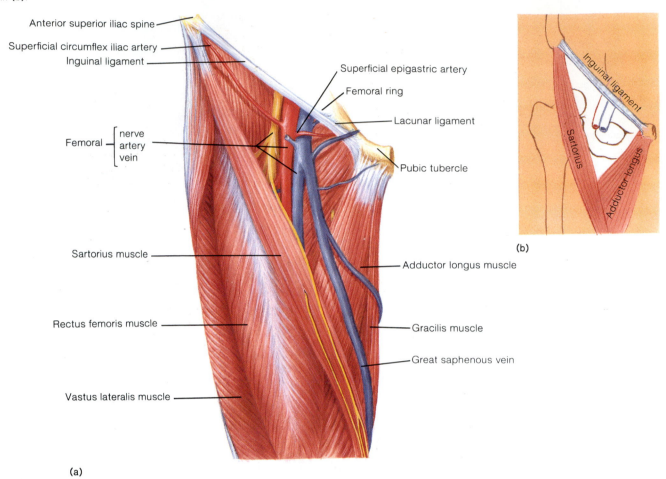

(a)

(b)

Principal Veins of the Body

After systemic blood has passed through the tissues and become depleted of oxygen, it returns through veins of progressively larger diameters to the right atrium of the heart.

Objective 18. Describe the venous drainage of the head, neck, and upper extremity.

Objective 19. Describe venous drainage of the thorax, lower extremities, and abdominal region.

Objective 20. Describe the vessels involved in the hepatic portal system.

In the venous portion of the systemic circulation, blood flows from smaller vessels into larger ones, so that a vein receives smaller tributaries instead of giving off branches as an artery does. The veins from all parts of the body converge into two major vessels that empty into the right atrium: the **superior** and **inferior vena cavae.** Veins are more numerous than arteries and are both superficial and deep. Superficial veins can generally be seen just beneath the skin and are clinically important in drawing blood and giving injections. Deep veins are close to the principal arteries and are usually similarly named. As with arteries, veins are named according to the region in

which they are found or the organ that they serve (when a vein serves an organ, it drains blood away from it).

The principal systemic veins of the body are illustrated in figure 16.32.

Veins Draining the Head and Neck

Blood from the scalp, portions of the face, and the superficial neck regions is drained by the **external jugular veins** (fig. 16.33). These vessels descend on the lateral sides of the neck, superficial to the sternocleidomastoid muscle and deep to the platysma muscle. They drain into the right and left **subclavian veins,** located just behind the clavicles.

The paired **internal jugular veins** drain blood from the brain, meninges, and deep regions of the face and neck. The internal jugular veins are larger and deeper than the external jugular veins. They arise from numerous cranial **venous sinuses,** which are a series of both paired and unpaired channels positioned between the two layers of dura mater. The venous sinuses, in turn, receive venous blood from the **cerebral,** the

jugular: L. *jugulum,* throat or neck

Figure 16.31 Important arterial pressure points and the locations at which arterial pulsations can best be detected.

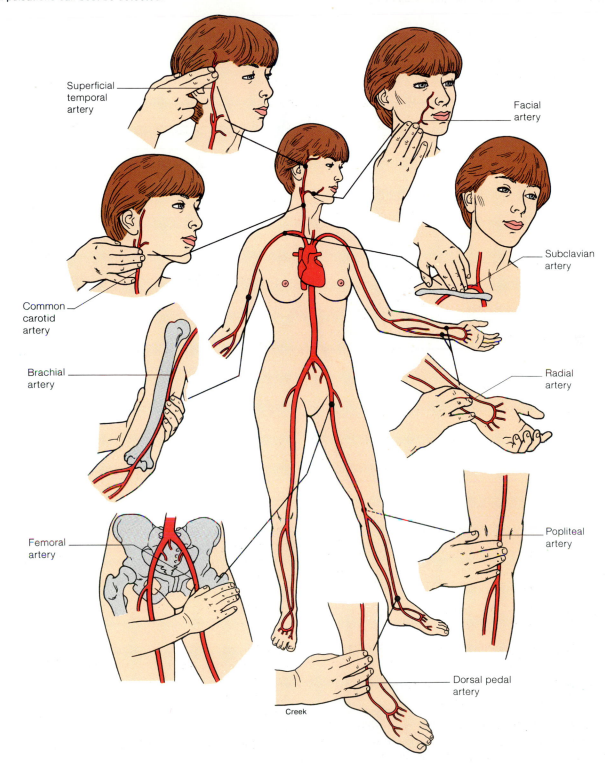

Superficial temporal artery

Facial artery

Common carotid artery

Subclavian artery

Brachial artery

Radial artery

Femoral artery

Popliteal artery

Dorsal pedal artery

Creek

Figure 16.32 Principal veins of the body. Superficial veins are depicted in the left extremities and deep veins in the right extremities.

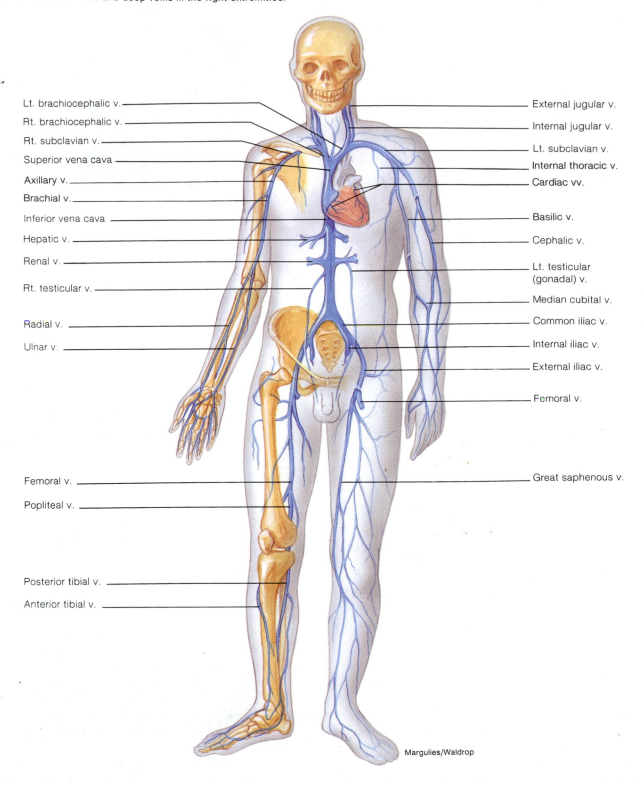

Lt. brachiocephalic v.
Rt. brachiocephalic v.
Rt. subclavian v.
Superior vena cava
Axillary v.
Brachial v.
Inferior vena cava
Hepatic v.
Renal v.
Rt. testicular v.
Radial v.
Ulnar v.
Femoral v.
Popliteal v.
Posterior tibial v.
Anterior tibial v.

External jugular v.
Internal jugular v.
Lt. subclavian v.
Internal thoracic v.
Cardiac vv.
Basilic v.
Cephalic v.
Lt. testicular (gonadal) v.
Median cubital v.
Common iliac v.
Internal iliac v.
External iliac v.
Femoral v.
Great saphenous v.

Margulies/Waldrop

UNIT VI

cerebellar, the **ophthalmic** and the **meningeal veins.** The principal cranial venous sinuses and the formation of the internal jugular vein are illustrated in figure 16.34.

The internal jugular vein passes inferiorly down the neck adjacent to the common carotid artery and the vagus nerve. All three are surrounded by the protective *carotid sheath* and are po-

sitioned beneath the sternocleidomastoid muscle. The internal jugular empties into the subclavian vein, and the union of these two vessels forms the large **brachiocephalic vein** on each side. The two brachiocephalic veins then merge to form the **superior vena cava,** which drains into the right atrium of the heart (see fig. 16.32).

Figure 16.33 Veins that drain the head and neck.

Figure 16.34 Cranial venous sinuses and the internal jugular vein.

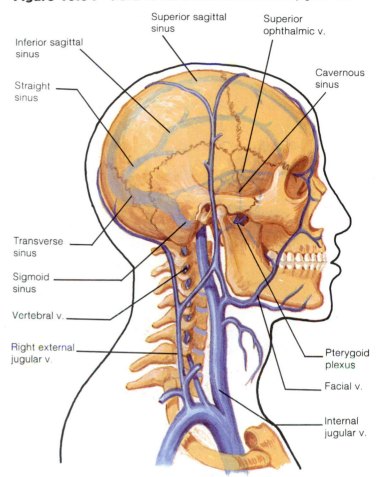

The external jugular vein is the vessel that is seen on the side of the neck when a person is angry or wearing a tight collar. You can voluntarily distend this vein by performing the *Valsalva maneuver.* To do this, take a deep breath and hold it while you forcibly contract your abdominal muscles as in a forced exhalation. This procedure is automatically performed when lifting a heavy object or defecating. The increased thoracic pressure that results compresses the vena cavae and interferes with the return of blood to the right atrium. Thus, the Valsalva maneuver can be dangerous if performed by an individual with cardiovascular disease.

Veins of the Upper Extremity

The upper extremity has both superficial and deep venous drainage (fig. 16.35). The superficial veins are highly variable and form an extensive network just below the skin. The deep veins accompany the arteries of the same region and bear similar names. The deep veins of the upper extremity will be described first. Both the **radial vein** on the lateral side of the forearm and the **ulnar vein** on the medial side drain blood from the **deep** and **superficial palmar** (volar) **arches** of the hand.

Valsalva maneuver: from Antonio Valsalva, Italian anatomist, 1666–1723

The radial and ulnar veins join in the cubital fossa to form the **brachial vein,** which continues up the medial side of the brachium.

The main superficial vessels of the upper extremity are the **basilic vein** and the **cephalic vein.** The basilic vein passes on the ulnar side of the forearm and the medial side of the arm. Near the head of the humerus, the basilic vein merges with the brachial vein and forms the **axillary vein.**

The cephalic vein drains the superficial portion of the hand and forearm on the radial side and then continues up the lateral side of the arm. In the shoulder region, the cephalic vein pierces the fascia and joins the axillary vein. The axillary vein then passes the first rib to form the subclavian vein, which unites with the internal jugular to form the brachiocephalic vein of that side.

Superficially, in the cubital fossa of the elbow, the **median cubital vein** ascends from the cephalic vein on the lateral side to connect with the basilic vein on the medial side. The median cubital vein is a frequent site for venipuncture in order to remove a sample of blood or add fluids to the blood.

Veins of the Thorax

The **superior vena cava,** formed by the union of the two brachiocephalic veins, empties venous blood from the head, neck,

UNIT VI

Figure 16.35 An anterior view of the veins that drain the upper right extremity. (a) Superficial veins and (b) deep veins.

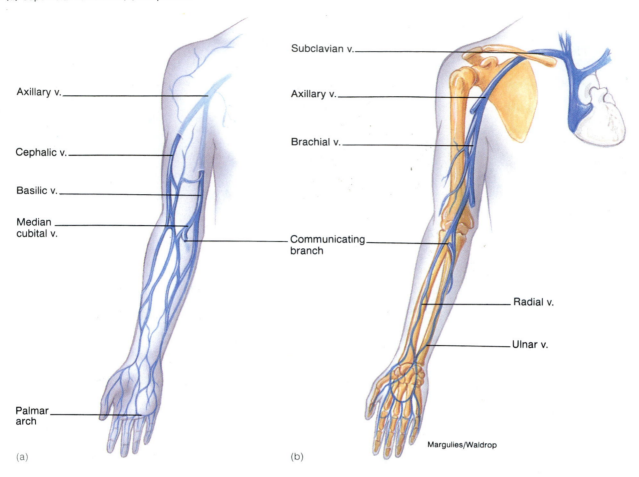

Subclavian v.

Axillary v.

Cephalic v.

Basilic v.

Median cubital v.

Palmar arch

(a)

Subclavian v.

Axillary v.

Brachial v.

Communicating branch

Radial v.

Ulnar v.

Margulies/Waldrop

(b)

and upper extremities directly into the right atrium of the heart. These large vessels lack the valves that are characteristic of most other veins in the body.

In addition to receiving blood from the brachiocephalic veins, the superior vena cava collects blood from the azygos (*az'ĭ-gos*) system of veins arising from the posterior thoracic wall (fig. 16.36). The **azygos vein** extends superiorly along the dorsal abdominal and thoracic walls on the right side of the vertebral column. It ascends through the mediastinum to join the superior vena cava at the level of the fourth thoracic vertebra. Tributaries of the azygos vein include the **ascending lumbar veins** that drain the lumbar and sacral regions, **intercostal veins** draining from the intercostal muscles, and the **accessory hemiazygos** and **hemiazygos veins,** which form the major tributaries to the left of the vertebral column.

Veins of the Lower Extremity

The lower extremities, like the upper extremities, have both a deep and a superficial group of veins (fig. 16.37). The deep veins accompany corresponding arteries and have more valves than do the superficial veins. The deep veins will be described first.

The **posterior** and **anterior tibial veins** originate in the foot and course upward behind and in front of the tibia to the

back of the knee, where they merge to form the **popliteal vein.** The popliteal vein receives blood from the knee region. Just above the knee, this vessel becomes the **femoral vein.** The femoral vein in turn continues up the thigh and receives blood from the **deep femoral vein** near the groin. Just above this, the femoral vein receives blood from the **great saphenous (*să-fe'nus*) vein** and then becomes the **external iliac vein** as it passes under the inguinal ligament. The external iliac curves upward to the level of the sacroiliac joint, where it merges with the **internal iliac vein** at the pelvic and genital regions to form the **common iliac vein.** At the level of the fifth lumbar vertebra, the right and left common iliacs unite to form the large **inferior vena cava** (fig. 16.37).

The superficial veins of the lower extremity are the **small** and **great saphenous veins.** The small saphenous vein arises from the lateral side of the foot, courses posteriorly along the surface of the calf of the leg, and descends deep to enter the popliteal vein behind the knee. The great saphenous vein is the longest vessel in the body. It originates at the arch of the foot and ascends superiorly along the medial aspect of the leg and thigh before draining into the femoral vein.

---

azygos: Gk. *a*, without; *zygon*, yoke

saphenous: L. *saphena*, the hidden one

UNIT VI

Figure 16.36 Veins of the thoracic region. (The lungs and heart have been removed.)

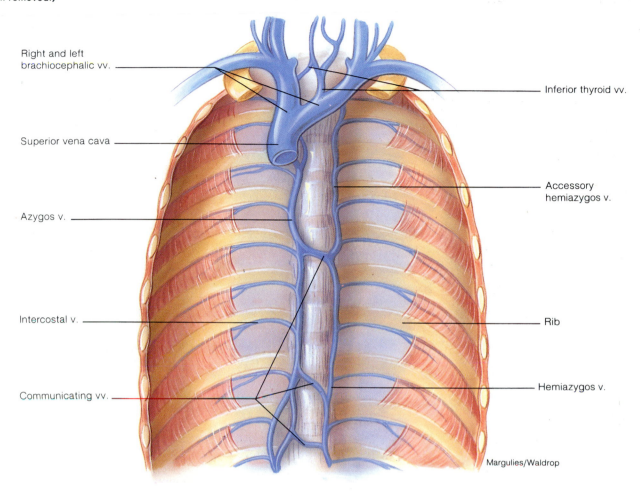

Right and left brachiocephalic vv.

Superior vena cava

Azygos v.

Intercostal v.

Communicating vv.

Inferior thyroid vv.

Accessory hemiazygos v.

Rib

Hemiazygos v.

Margulies/Waldrop

Veins of the Abdominal Region

The **inferior vena cava** parallels the abdominal aorta on the right side as it ascends through the abdominal cavity to penetrate the diaphragm and enter the right atrium (see fig. 16.32). It is the largest in diameter of the vessels in the body and is formed by the union of the two common iliac veins draining the lower extremities. As the inferior vena cava ascends through the abdominal cavity, it receives tributaries from veins that correspond in name and position to arteries previously described.

Four paired **lumbar veins** (not shown) drain the posterior abdominal wall, the vertebral column, and the spinal cord. The **renal veins** drain blood from the kidneys and ureters into the inferior vena cava. The **right testicular vein** or the **right ovarian vein,** draining the corresponding gonads, and the **right suprarenal vein,** draining the right adrenal gland, each empty into the inferior vena cava. The **left testicular vein** or **left ovarian vein,** and the **left suprarenal vein,** by contrast, each drain into the left renal vein. The **inferior phrenic veins** receive blood from the inferior side of the diaphragm and drain into the inferior vena cava. **Right** and **left hepatic veins** originate from the capillary sinusoids of the liver and empty into the inferior vena cava immediately below the diaphragm.

Note that the inferior vena cava does not receive blood directly from the GI tract, pancreas, or spleen. Instead, the venous outflow from these organs first passes through capillaries in the liver.

Hepatic Portal System

A *portal system* is one in which the veins that drain one group of capillaries deliver blood to another group of capillaries, which in turn are drained by more usual systemic veins that carry blood to the vena cavae and the right atrium of the heart. There are thus two capillary beds in series. The hepatic portal system is composed of veins that drain blood from capillaries in the intestines, pancreas, spleen, stomach, and gallbladder into capillaries in the liver (called *sinusoids*) and of the **right** and **left hepatic veins** that drain the liver and empty into the inferior vena cava (fig. 16.38). As a consequence of the hepatic portal system, the absorbed products of digestion must first pass through the liver before entering the general circulation.

The **hepatic portal vein** is the large vessel that receives blood from the digestive organs. It is formed by a union of the **superior mesenteric vein,** which drains nutrient-rich blood from the small intestine, and the **splenic vein.** The splenic vein

UNIT VI

Figure 16.37 Veins of the lower extremity. (*a*) Superficial veins, medial and posterior aspects, and (*b*) deep veins, medial view.

Superficial epigastric v.

Femoral v.

Great saphenous v.

Popliteal v.

Small saphenous v.

Krabach

(a)

Inferior vena cava

Right common iliac v.

Internal iliac v.

External iliac v.

Inguinal ligament

Femoral v.

Great saphenous v. (cut)

Femoral circumflex vv.

Deep femoral v.

Femoral v.

Popliteal v.

Small saphenous v. (cut)

Anterior tibial v.

Posterior tibial v.

Dorsal pedal v.

Lateral plantar v.

Medial plantar v.

Margulies/Waldrop

(b)

Figure 16.38   The hepatic portal system.

Hepatic vv.

Inferior vena cava

Left gastric v.

Stomach

Left gastroepiploic v.

Spleen

Right gastric v.

Right gastroepiploic v.

Splenic v.

Liver

Hepatic sinusoids

Gallbladder

Hepatic portal v.

Pancreas

Superior
mesenteric v.

Small intestine

Ascending colon

Inferior
mesenteric v.

Descending colon

Lew

Rectum

drains the spleen but is enlarged because of a convergence of the following three tributaries: (1) the **inferior mesenteric vein** from the large intestine, (2) the **pancreatic vein** from the pancreas, and (3) the **left gastroepiploic vein** from the stomach. The **right gastroepiploic vein,** also from the stomach, drains directly into the superior mesenteric vein.

Three additional veins empty into the portal vein. The **right** and **left gastric veins** drain the lesser curvature of the stomach and the **cystic vein** drains blood from the gallbladder.

One of the functions of the liver is to detoxify harmful substances, such as alcohol, that are absorbed into the blood from the small intestine. However, excessive quantities of alcohol cannot be processed during a single pass through the liver, and so a person becomes intoxicated. Eventually, the liver is able to process the alcohol as the circulating blood is repeatedly exposed to the liver sinusoids via the hepatic artery. Alcoholics may eventually suffer from *cirrhosis* of the liver as the normal liver tissue is destroyed.

In summary, it is important to note that the sinusoids of the liver receive blood from two sources. The hepatic artery supplies oxygen-rich blood to the liver, whereas the hepatic

portal vein transports nutrient-rich blood from the intestine for processing. These two blood sources become mixed in the liver sinusoids. Liver cells exposed to this blood obtain nourishment from it and are uniquely qualified (because of their anatomical position and enzymatic ability) to modify the chemical nature of the venous blood that enters the general circulation from the GI tract.

1. Using a flow chart, indicate the venous drainage from the head and neck to the superior vena cava. Point out which vein may bulge in the side of the neck when a person performs the Valsalva maneuver and which vein is commonly used as a site for venipuncture.
2. Describe the positions, sources, and drainages of the small and great saphenous veins.
3. Describe the hepatic portal system and comment on the functional importance of this system.

Fetal Circulation

The fetal circulation is adapted to the fact that the fetal lungs are nonfunctional and that oxygen and nutrients are obtained from the placenta.

Objective 21.   Describe the fetal circulation to and from the placenta.

gastroepiploic: Gk. *gastros,* stomach; *epiplein,* to float on (referring to greater omentum)

Figure 16.39   Fetal circulation. (Arrows indicate the direction of blood flow.)

Arch of the aorta
Superior vena cava
Foramen ovale
Rt. atrium
Rt. ventricle
Inferior vena cava
Ductus venosus
Liver
Hepatic a.
Hepatic portal vein
Umbilical vein
Umbilicus (navel)
Umbilical arteries

Ductus arteriosus
Pulmonary trunk
Pulmonary artery
Pulmonary vein
Lt. atrium
Lt. ventricle
Diaphragm
Renal vein
Renal artery
Kidney
Abdominal aorta
Common iliac artery
External iliac artery
Internal iliac artery

Margulies/Waldrop

Objective 22.   Describe the structure and function of the foramen ovale, ductus venosus, and ductus arteriosus.

The circulation of blood through a fetus is by necessity different from that of a newborn (fig. 16.39). Respiration, the procurement of nutrients, and the elimination of metabolic wastes occur through the maternal blood instead of through the organs of the fetus. The capillary exchange between the maternal and fetal circulation occurs within the **placenta.** This remarkable structure includes maternal and fetal capillary beds and is discharged following delivery as the afterbirth.

The **umbilical cord** is the connection between the placenta and the fetal umbilicus. It includes one **umbilical vein** and two **umbilical arteries** surrounded by a gelatinous substance. Oxygenated and nutrient-rich blood flows through the umbilical vein toward the inferior surface of the liver. At this point, the umbilical vein bifurcates into one branch that joins with the portal vein, while the other branch, called the **ductus venosus,** enters the inferior vena cava. Thus, oxygenated blood is mixed with venous blood returning from the lower extremities of the fetus before it enters the heart. The umbilical vein is the only vessel of the fetus that carries fully oxygenated blood.

The inferior vena cava empties into the right atrium of the fetal heart. From here, most of the blood is directed through the **foramen ovale,** an opening between the two atria. Here it mixes with a small quantity of blood returning through the pulmonary circulation. The blood then passes into the left ventricle, from which it is pumped into the aorta and through the body of the fetus. Some blood entering the right atrium passes into the right ventricle and out of the heart via the pulmonary trunk. Since the lungs of the fetus are not functional, only a small portion of blood continues through the pulmonary circulation (the resistance to blood flow is very high in the collapsed fetal lungs). Most of the blood in the pulmonary trunk passes through the **ductus arteriosus** into the aortic arch, where it mixes with blood coming from the left ventricle. Blood is returned to the placenta by the two umbilical arteries that arise from the internal iliac arteries.

Notice that, in the fetus, oxygen-rich blood is transported by the inferior vena cava to the heart and via the foramen ovale and ductus arteriosus to the systemic circulation.

Important changes occur in the cardiovascular system at birth. The foramen ovale, ductus arteriosus, ductus venosus, and the umbilical vessels are no longer necessary. The foramen ovale abruptly closes with the first breath of air because the reduced

UNIT VI

Table 16.6	Cardiovascular Structures of the Fetus and Changes in the Neonate		
Structure	**Location**	**Function**	**Neonate Transformation**
Umbilical vein	Connects the placenta to the liver; forms a major portion of umbilical cord	Transports nutrient-rich, oxygenated blood from the placenta to the fetus	Forms the round ligament of the liver
Ductus venosus	Venous shunt within the liver to connect with the inferior vena cava	Transports oxygenated blood directly into the inferior vena cava	Forms the ligamentum venosum, a fibrous cord in the liver
Foramen ovale	Opening between the right and left atria	Acts as a shunt to bypass the pulmonary circulation	Closes at birth and becomes the fossa ovalis, a depression in the interatrial septum
Ductus arteriosus	Between pulmonary trunk and arch of the aorta	Acts as a shunt to bypass the pulmonary circulation	Closes shortly after birth, atrophies, and becomes the ligamentum arteriosum
Umbilical arteries	Arise from internal iliac arteries and associated with umbilical cord	Transport blood from the fetus to the placenta	Atrophies to become the lateral umbilical ligaments

pressure in the right side of the heart causes a flap to cover the opening. This reduction in pressure when the lungs fill with air occurs because the vascular resistance to blood flow in the pulmonary circulation falls far below that of the systemic circulation. The pressure in the inferior vena cava and right atrium falls as a result of the loss of the placental circulation.

The constriction of the ductus arteriosus occurs gradually over a period of about 6 weeks after birth as the vascular smooth muscle fibers constrict in response to the higher oxygen concentration in the postnatal blood. The remaining structure of the ductus arteriosus gradually atrophies and becomes nonfunctional as a blood vessel. Transformation of the unique fetal cardiovascular system is summarized in table 16.6.

1. Describe the path of blood from the fetal heart through the placenta and back to the fetal heart.
2. Trace the pathway of oxygenated blood through the fetal circulation.
3. Describe the foramen ovale and ductus arteriosus and explain why blood flows the way it does through these structures in the fetal circulation.

Lymphatic System

The lymphatic system consists of lymphatic vessels and lymph nodes. It functions to return tissue fluid to the venous system and to help protect the body from diseases.

Objective 23. Describe the pattern of lymph flow from the lymphatic capillaries to the venous system.

Objective 24. Describe the structure, location, and function of the lymph nodes and the location and function of the spleen and thymus.

The lymphatic system is closely related anatomically and physiologically to the circulatory system (fig. 16.40). The functions of the lymphatic system are basically threefold: (1) it transports excess interstitial (tissue) fluid, which was initially formed as a blood filtrate, back to the bloodstream; (2) it serves as the route by which absorbed fat from the intestine is transported to the blood; and (3) it helps provide immunological defenses against disease-causing agents.

Lymph and Lymph Capillaries

The smallest vessels of the lymphatic system are the **lymph capillaries** (fig. 16.41). Lymph capillaries are microscopic, closed-ended tubes that form vast networks in the intercellular spaces within most tissues. Within the villi of the small intestine, for example, lymph capillaries, called *lacteals (lak'te-alz)*, transport the products of fat absorption away from the GI tract.

Because the walls of lymph capillaries are composed of a highly permeable simple squamous epithelium, tissue fluid can readily enter. This fluid is formed as a filtrate of plasma through blood capillaries and is identical in composition to plasma except for a lower protein concentration (proteins are largely held back by the filtration process). Adequate lymphatic drainage is needed to prevent the accumulation of tissue fluid, a condition called *edema (ě-de'mä)*. Once fluid enters the lymphatic capillaries, it is referred to as **lymph.**

Lymph Vessels

From merging lymph capillaries, the lymph is carried into larger **lymph ducts.** The walls of lymph ducts are similar to those of veins in that they have the same three layers and contain valves to prevent backflow (fig. 16.42). Interconnecting lymph ducts eventually empty into one of the two principal vessels: the **thoracic duct** and the **right lymphatic duct** (fig. 16.43).

The larger thoracic duct drains lymph from the lower extremities, abdomen, left thoracic region, left upper extremity, and left side of the head and neck. The main trunk of this vessel ascends along the spinal column and drains into the left subclavian vein. In the abdominal area, there is a saclike

lacteal: L. *lacteus,* milk
lymph: L. *lympha,* clear water
edema: Gk. *edema,* a swelling

Figure 16.40 The schematic relationship of the circulatory and lymphatic systems. Lymphatic vessels transport lymph fluid from interstitial spaces to the venous bloodstream.

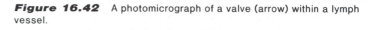

Figure 16.41 A diagram showing the structural relationship of a capillary bed, lymph capillaries, and lymph ducts.

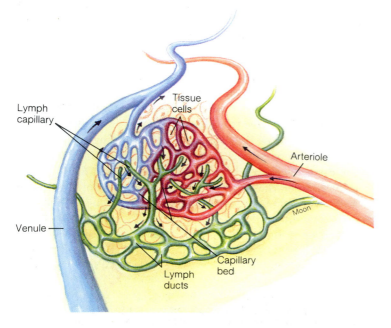

Figure 16.42 A photomicrograph of a valve (arrow) within a lymph vessel.

Figure 16.43 Lymphatic vessels. (*a*) The major lymph drainage of the body and (*b*) a magnified view of the upper right quadrant showing the lymph drainage of the right breast and axilla.

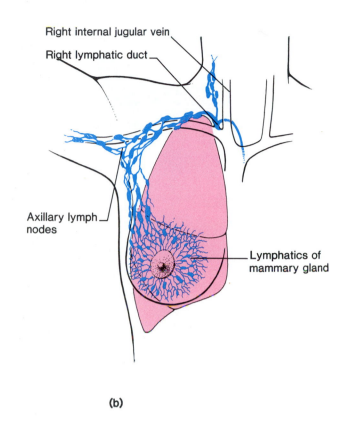

(a)

(b)

enlargement of the thoracic duct called the **cisterna chyli** *(sis-ter'nǎ ki'le)*. The shorter right lymphatic duct drains lymph vessels from the right upper extremity, right thoracic region, and right side of the head and neck. The right lymphatic duct empties into the right subclavian vein near the internal jugular vein.

The pressure that pushes lymph through the lymph ducts is the result of massaging actions from skeletal muscle contractions, intestinal movements, and other body movements. Valves in the walls of lymph vessels keep lymph moving in one direction.

Lymph Nodes

Lymph filters through the reticular tissue of **lymph nodes** (fig. 16.44). The reticular tissue contains phagocytic cells that help purify the fluid. Lymph nodes are small oval bodies enclosed within fibrous connective tissue *capsules*. Specialized connective tissue bands called *trabeculae* divide the node. *Afferent lymphatic vessels* carry lymph into the node, where it is circulated through sinuses in the *cortical tissue*. Lymph leaves the node through the *efferent lymphatic vessel,* which emerges from the *hilum*—the depression on the concave side. **Lymphatic nodules** within the node are the sites of lymphocyte production.

Lymph nodes usually occur in clusters in specific regions of the body (fig. 16.45). Some of the principal groups of lymph

nodes are the **popliteal** (not illustrated) and **inguinal nodes** of the lower extremity, the **lumbar nodes** of the pelvic region, the **cubital** and **axillary nodes** of the upper extremity, the **thoracic nodes** of the chest, and the **cervical nodes** of the neck. The submucosa of the small intestine contains numerous scattered lymphocytes and lymphatic nodules and larger clusters of lymphatic tissue called **mesenteric** (Peyer's) **patches.**

> Migrating cancer cells (metastases) are especially dangerous if they enter the lymphatic system, which can disperse them widely. On entering the lymph nodes, the cancer cells can multiply and establish secondary tumors in organs that are distal to the lymphatic drainage of the site of the primary tumor.

Lymphoid Organs

The **spleen** and the **thymus** are lymphoid organs. The spleen is located posterior and lateral to the stomach, from which it is suspended (fig. 16.46). The spleen is not a vital organ in an adult, but it does assist other body organs in producing lymphocytes, filtering the blood, and destroying old erythrocytes. In an infant, it is an important site for erythrocyte production.

Peyer's patches: from Johann K. Peyer, Swiss anatomist, 1653–1712
spleen: L. *splen*, low spirits (thought to cause melancholy)
thymus: Gk. *thymos*, thyme (compared to the flower of this plant by Galen)

cisterna chyli: L. *cisterna*, box; Gk. *chylos*, juice

Figure 16.44 The structure of a lymph node. (a) A schematic diagram of a sectioned lymph node and associated vessels. (b) A photomicrograph of lymphatic tissue. (c) A photograph of a lymph node positioned near a blood vessel.

(a)

(b)

(c)

Figure 16.45 Major locations of lymph nodes.

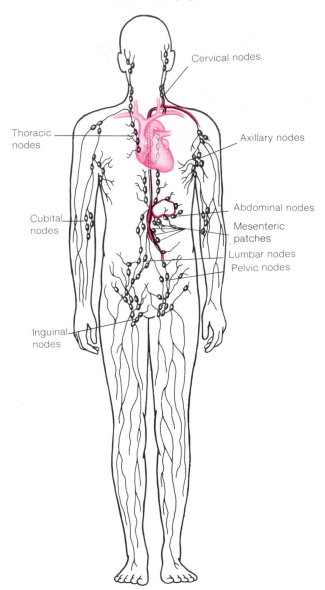

The thymus is located in the anterior thorax, deep to the manubrium of the sternum (fig. 16.47). It is much larger in a fetus and child than in an adult, because it regresses in size during puberty. The thymus plays a key role in the immune system.

The lymphatic organs are summarized in table 16.7.

The tonsils, of which there are three pairs, are actually lymphatic organs of the pharyngeal region (see chapter 18). The function of the tonsils is to combat infection of the ear, nose, and throat regions. Because of the persistent infections that some children suffer, the tonsils may become so overrun with infections that they actually become a source of infections that spread to other parts of the body. A *tonsillectomy* may then have to be performed. This operation is not as common as in the past because of the availability of powerful antibiotics and because the functional value of the tonsils is appreciated to a greater extent.

Figure 16.46   The structure of the spleen.

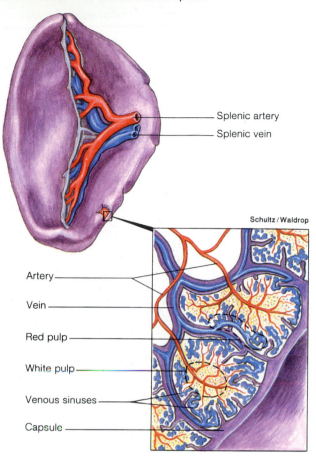

Splenic artery
Splenic vein

Schultz / Waldrop

Artery
Vein
Red pulp
White pulp
Venous sinuses
Capsule

1. Describe the relationship between lymph and blood in terms of their origin, composition, and fate.
2. List the two major lymph vessels of the body and describe their relationships to the vascular system.
3. Describe the structure, location, and function of the lymph nodes and list the lymphoid organs.

Table 16.7	**Lymphatic Organs**	
Organ	**Location**	**Function**
Lymph nodes	In clusters or chains along the paths of larger lymphatic vessels	Sites of lymphocyte production; house T-lymphocytes and B-lymphocytes that are responsible for immunity; phagocytes filter foreign particles and cellular debris from lymph
Spleen	In upper left portion of abdominal cavity, beneath the diaphragm and suspended from the stomach	Serves as blood reservoir; phagocytes filter foreign particles, cellular debris, and worn erythrocytes from the blood; houses lymphocytes
Thymus	Within the mediastinum, behind the manubrium	Important site of immunity in a child; houses lymphocytes; changes undifferentiated lymphocytes into T-lymphocytes

Figure 16.47   The location and structure of the thymus. (*a*) The thymus is located within the mediastinum between the lungs. (*b*) A photomicrograph of thymus tissue.

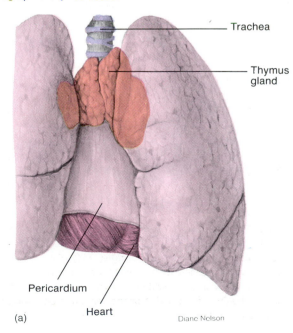

Trachea
Thymus gland
Pericardium
Heart

Diane Nelson

(a)

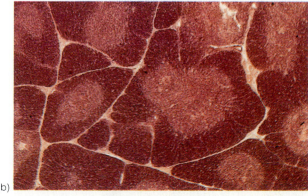

(b)

Developmental Exposition

The Circulatory System

Exhibit I The early development of the heart from embryonic mesoderm. (*a*) A dorsal view of an embryo at day 20 showing the position of a transverse cut depicted in (*a₁*), (*b*), and (*c*). (*a₁*) At day 20, the heart tubes are formed from the heart cords. (*b*) By day 21, the medial migration of the heart tubes has brought them together within the forming pericardial cavity. (*c*) A single heart tube is completed at day 22 and is suspended in the pericardial cavity by the dorsal mesocardium.

Waldrop

Explanation

The cardiovascular system is one of the first systems to form in the embryo, delivering nutrients to the mitotically active cells and disposing of waste products through its association with the maternal blood vessels in the placenta. Blood is formed and is circulated through the vessels by the pumping action of the heart approximately 25 days after conception.

Throughout pregnancy, the fetus is dependent on the mother's circulatory system for nutrient, gaseous, and waste exchange. Therefore, some unique structures of the fetal circulatory system develop to accommodate the placental arrangement. The discussion that follows focuses on the development of the fetal heart and associated major vessels. The circulation of blood through the fetal cardiovascular system has already been discussed in the section "Fetal Circulation."

Development of the Heart

The development of the heart from two separate segments of mesoderm requires only 6 to 7 days. Heart development is first apparent at day 18 or 19 in the *cardiogenic (kar′′de-o-jen′ik) area* of the mesoderm layer (exhibit I). A small paired mass of spe-

cardiogenic: Gk. *kardia*, heart; *genesis*, be born (origin)

Exhibit II Formation of the heart chambers. (*a*) The heart tubes fuse during days 21 and 22. (*b*) The developmental chambers are formed during day 23. (*c*) Differential growth causes folding between the chambers during day 24, and vessels are developed to transport blood to and from the heart. The embryonic heart generally has begun rhythmic contractions and pumping blood by day 25.

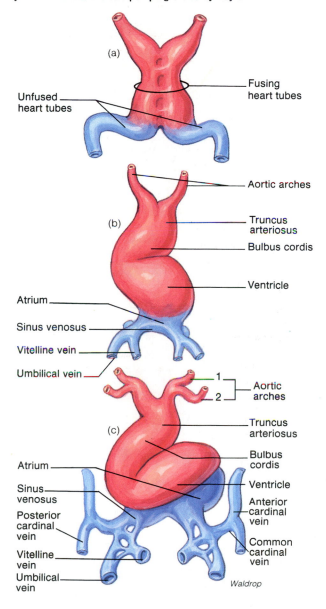

Waldrop

Table 16.8	Structural Changes in the Development of the Heart
Primitive Cardiac Dilation	**Developmental Fate**
Truncus arteriosus	Differentiates to form the aorticopulmonary septum, which partitions the ascending aorta and the pulmonary trunk
Bulbus cordis	Incorporated into the walls of the ventricles
Sinus venosus	Differentiates to form the coronary sinus and a portion of the wall of the right atrium
Ventricle	Divided into right and left chambers by growth of the interventricular septum
Atrium	Partially partitioned into right and left chambers by the septum secundum. The opening in the partition, called the foramen ovale, persists throughout development.

After the fusion of the heart tubes and the formation of distinct dilations, the heart begins to pump blood. Partitioning of the heart chambers begins during the middle of the fourth week and is complete by the end of the fifth week. During this crucial time, many congenital heart problems develop.

Major changes occur in each of the five primitive dilations of the developing heart during the week-and-a-half embryonic period beginning in the middle of the fourth week. The truncus arteriosus differentiates to form a partition between the aorta and the pulmonary trunk. The bulbus cordis is incorporated in the formation of the walls of the ventricles. The sinus venosus forms the **coronary sinus** and a portion of the wall of the right atrium. The ventricle is divided into the right and left chambers by the growth of the **interventricular septum.** The atrium is partially partitioned into right and left chambers by the **septum secundum.** An opening between the two atria called the **foramen ovale** persists throughout fetal development. This opening is covered by a flexible valve that permits blood to pass from the right to the left side of the heart.

The development of the heart is summarized in table 16.8.

Congenital defects of the cardiac septa are a relatively common form of birth abnormality. Approximately 0.7% of live births and 2.7% of stillbirths show cardiac abnormalities. Ventricular septal defects are the most common of the cardiac defects. An infant with a congenital cardiac defect may suffer from inadequate oxygenation of blood and thus be termed a ''blue baby.''

Development of the Major Arteries

The formation of the major arteries occurs simultaneously with the development of the heart. The most complex and fascinating vascular formation is the development of the **aortic arches** associated with the pharyngeal pouches and branchial arches in

cialized cells called **heart cords** form here. Shortly after, a hollow center develops in each heart cord, and each structure is then referred to as a **heart tube.** The heart tubes begin to migrate toward each other during day 21 and soon fuse to form a single median heart tube. During this time, the heart tube undergoes dilations and constrictions so that when fusion is completed during the fourth week, five distinct regions of the heart can be identified. These are the **truncus arteriosus, bulbus cordis, ventricle, atrium,** and **sinus venosus** (exhibit II).

Continued on next page

UNIT VI

Developmental Exposition

Continued

Exhibit III Formation of the aortic arch and major arteries of the thoracic region. (*a*) Schematic lateral view of the pharyngeal pouches and aortic arches within the neck region of a 28-day embryo.

(*b*) Aortic arches at 6 weeks. (Note that by this time arches one, two, and five have largely disappeared.) (*c*) Aortic arches at 7 weeks. (*d*) Arterial configuration at 8 weeks.

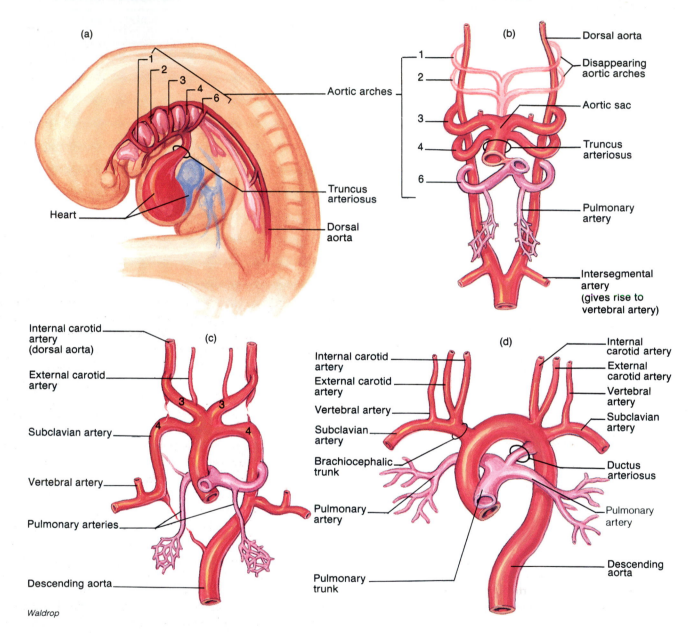

Waldrop

the neck region (exhibit III). These aortic arches arise from the truncus arteriosus. Although six pairs of aortic arches develop, they are not all present at the same time, and none of them persists in entirety through fetal development.

The transformation of the six aortic arches into the basic adult arterial arrangement occurs during the sixth to the eighth week of embryonic development. The first and second pairs of

aortic arches disappear before the formation of the sixth. The third pair of aortic arches form the common carotid arteries and the external carotid arteries. The right fourth aortic arch forms the base of the right subclavian artery, and the left fourth arch contributes to part of the aortic arch. The first, second, and fifth pairs of aortic arches have no derivatives and soon atrophy. The right sixth aortic arch forms the proximal portion of the right pul-

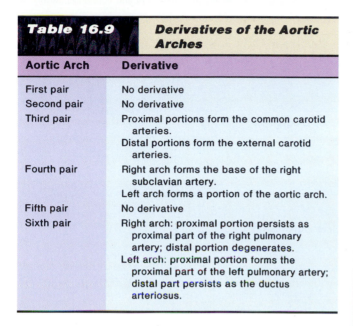

Table 16.9	Derivatives of the Aortic Arches
Aortic Arch	**Derivative**
First pair	No derivative
Second pair	No derivative
Third pair	Proximal portions form the common carotid arteries.
	Distal portions form the external carotid arteries.
Fourth pair	Right arch forms the base of the right subclavian artery.
	Left arch forms a portion of the aortic arch.
Fifth pair	No derivative
Sixth pair	Right arch: proximal portion persists as proximal part of the right pulmonary artery; distal portion degenerates.
	Left arch: proximal portion forms the proximal part of the left pulmonary artery; distal part persists as the ductus arteriosus.

monary artery. The left sixth arch forms the proximal portion of the left pulmonary artery, and the distal portion of this arch persists as an embryonic shunt between the pulmonary trunk and the aorta called the **ductus arteriosus.**

The derivatives of the aortic arches are summarized in table 16.9.

Although the embryonic vessels that develop from the truncus arteriosus are called aortic arches, they should not be confused with the aortic arch that is the major systemic artery leaving the heart in the adult. Only the left fourth arch participates in the formation of the adult aortic arch.

An examination of the persisting aortic arches in the adults of different classes of vertebrates reveals interesting evolutionary relationships. Fish have six aortic arches persisting in the gill region. Reptiles have only one pair, a branch to the left and one to the right. All birds have a single right aortic arch, and all mammals have a single left aortic arch.

Clinical Considerations

It has been said that as the circulatory system goes, so goes the rest of the body. The circulatory system is the lifeline to the other organs of the body. If the vital organs are deprived of the essentials for life, they will falter in their functions, ultimately resulting in body death. Some gerontologists believe that the first system of the body to show signs of aging is the circulatory system, and that this accelerates the aging process in the other systems.

Clinically speaking, the circulatory system is extremely important because of the frequent irregularities and diseases that afflict it and the effect of its dysfunction on maintaining homeostasis. Given the confines of this text, it would not be practical to attempt a comprehensive discussion of the numerous clinical aspects of the cardiovascular system. However, some general comments will be made on cardiovascular diagnosis; on dysfunctions of the blood, lymphatics, vessels, and heart; and on the prevention and treatment of heart problems. In addition, first-aid treatment for victims of hemorrhage and shock will be discussed.

Cardiovascular Assessment

Cardiovascular assessment is an extremely important aspect of a physical examination for identifying existing and potential health problems. Several techniques are routinely used in gathering information. Auscultation of heart sounds with a **stethoscope** is not only important for detecting murmurs, but may help the physician evaluate general cardiac functioning or the progress of a patient with a cardiovascular problem.

An **electrocardiograph** (ECG) is an important way of assessing electrical activity within the heart. The most important diagnostic use of the resting ECG is identifying abnormal cardiac rhythms. Injury to cardiac muscle, as in myocardial infarction, elevates or depresses the S-T segment while inverting and shortening the T wave.

Several diagnostic techniques are employed for determining the size and position of the heart, the condition of the pericardial sac, and even the internal structure and condition of the vessels, chambers, and valves. Radiograms, CT scans, and ultrasound are important for gaining a perspective of how the heart functions and relates to other thoracic structures. The accumulation of fluids is easy to detect using these techniques. **Catheterization** of the heart in conjunction with a **fluoroscopic analysis** permits observation of the vessels, chambers, and valves of the heart. In addition, internal blood pressures can be recorded and samples of blood taken. Selective **arteriography** enables a physician to study the wall of a particular vessel. An **angiocardiogram** is a radiograph of the chambers of the heart and its vessels following an intravenous injection of a radiopaque substance.

Arterial blood pressure can be measured using a **sphygmomanometer** (see fig. 16.19). Blood pressure is indicative

stethoscope: Gk. *stethos*, breast; *skopos*, watch, look at
sphygmomanometer: Gk. *sphygmos*, throbbing; *manos*, at intervals;
  L. *metrum*, measure

UNIT VI

of the general health of the cardiovascular system. **Hypertension** is sustained high blood pressure and can result in mechanical damage to the cardiovascular system. Hypertension can be caused by resistance in the vessels from physiological actions (vasoconstriction) or by various vascular diseases (arteriosclerosis). Certain characteristics of the blood itself may cause hypertension.

Cardiac Arrhythmia   Arrhythmias, or abnormal heart rhythms, can be detected and described by the abnormal ECG patterns they produce. Since a heartbeat occurs whenever a normal QRS complex (see fig. 16.13) is seen, and since the ECG chart paper moves at a known speed, its x-axis indicates time, and the cardiac rate (beats per minute) can easily be obtained from the ECG recording. A cardiac rate slower than 60 beats per minute indicates **bradycardia** *(brad″i-kar′de-ă);* a rate faster than 100 beats per minute is described as **tachycardia** *(tak″i-kar′de-ă).*

Both bradycardia and tachycardia can occur normally (fig. 16.48a). Endurance-trained athletes, for example, commonly have a slower heart rate than the general population. This *athlete's bradycardia* occurs as a result of higher levels of parasympathetic inhibition of the SA node and is a beneficial adaptation. Activation of the sympathetic division of the ANS during exercise or emergencies causes a normal tachycardia to occur.

An abnormal bradycardia may be caused by a heart block, various drugs, shock, or increased intracranial pressure. An abnormal tachycardia occurs if the rate increases when a person is at rest. This may result from abnormally fast pacing by the atria due to drugs or to the development of abnormally fast *ectopic pacemakers*—cells located outside the SA node that assume a pacemaker function. This abnormal atrial tachycardia thus differs from normal "sinus" (SA node) tachycardia. *Ventricular tachycardia* results when abnormally fast ectopic pacemakers in the ventricles cause them to beat rapidly and independently of the atria (fig. 16.48b). This is very dangerous because it can quickly degenerate into a lethal condition known as *ventricular fibrillation.*

Ventricular Fibrillation   Fibrillation is caused by a continuous recycling of electrical waves through the myocardium. Because the myocardium enters a refractory period simultaneously at all regions, this recycling is normally prevented. If some cells emerge from their refractory periods before others, however, electrical waves can be continuously regenerated and conducted. The recycling of electrical waves along continuously changing pathways produces uncoordinated contraction and an impotent pumping action. These effects can be the result of damage to the myocardium.

Fibrillation can sometimes be stopped by a strong electric shock delivered to the chest—a procedure called **electrical defibrillation.** The electric shock depolarizes all the myocardial cells at the same time, causing them to enter a refractory state. The conduction of random, recirculating impulses thus stops, and—within a short time—the SA node can begin to stimulate

bradycardia: Gk. *bradys,* slow; *kardia,* heart
tachycardia: Gk. *tachys,* rapid; *kardia,* heart

Figure 16.48   In (*a*) the heartbeat is paced by the normal pacemaker—the SA node (hence the name sinus rhythm). This can be abnormally slow (bradycardia—46 beats per minute in this example) or fast (tachycardia—136 beats per minute in this example). Compare the pattern of tachycardia in (*a*) with the tachycardia in (*b*). Ventricular tachycardia is produced by an ectopic pacemaker in the ventricles. This dangerous condition can quickly lead to ventricular fibrillation, also shown in (*b*).

Sinus Bradycardia

Sinus Tachycardia

(a)

Ventricular tachycardia

Ventricular fibrillation

(b)

contraction in a normal fashion. Although this does not correct the initial problem that caused the abnormal electrical patterns, it can keep a person alive long enough to take other corrective measures.

Blood Disorders

Since blood is the functional component of the circulatory system and since the circulatory system works in such close association with other body systems, blood analysis is perhaps the most informative part of a physical exam. Peripheral arterial pulsations, usually obtained at the radial artery, are informative of blood flow. Capillary filling, following **blanching,** is an indicator of peripheral arterial circulation and is generally tested at the nail bed. This is done by firmly pressing the thumbnail against the patient's toenail or fingernail and then quickly releasing the pressure. If the pinkish color returns quickly to the whitened, or blanched, area, circulation is considered normal. Lack of peripheral coloration indicates vascular insufficiency.

Figure 16.49 Abnormal patterns of blood flow due to septal defects. Left-to-right shunting of blood is shown (shaded circles) because the left side is at a higher pressure than the right side. However, under other abnormal conditions (not illustrated here), the pressure in the right atrium may exceed that in the left, causing right-to-left shunting of blood through a septal defect in the atria.

Septal defect
in atria

Septal defect
in ventricles   BECK

Certain cardiovascular and blood abnormalities are expressed through the skin. **Cyanosis** is a condition characterized by a bluish coloration of the skin due to a decreased oxygen concentration. **Anemia** is a condition characterized by a pallor of the skin due to a deficiency of erythrocytes or hemoglobin. **Jaundice** is a condition in which the skin is yellowed because of excessive bile pigment (bilirubin) in the blood. **Edema** is an excessive accumulation of fluid within the tissue spaces, causing a swelling of a portion of the body. **Erythema** is a redness of the skin usually due to an infection, inflammation, toxic reaction, sunburn, or a lesion. None of these conditions are themselves diseases, but each is symptomatic of problems that may involve the cardiovascular system.

Actual examination of the blood is very informative and is a part of any thorough physical examination. Blood cell counts are used to determine the percentage of formed elements in the blood. An excess of red blood cells, called **polycythemia** (more than 6 million/mm³), may indicate certain bone diseases. Excessive leukocyte production, called **leukocytosis,** is generally diagnostic of infections or diseases within the body. **Leukopenia** is a decrease in leukocyte count. The disease **leukemia** causes the unrestricted reproduction of immature leukocytes, which depresses erythrocyte and platelet formation and causes anemia and a tendency to bleed. Coagulation time, blood sedimentation rates, prothrombin time, and various serum analyses are other blood tests that provide specific information about body function or dysfunction.

Several blood diseases are distinguished by their rate of occurrence. **Sickle-cell anemia** is an autosomal recessive disease that is more prevalent in blacks than in other human races. Although about 10% of American blacks have the sickle-cell trait, fortunately fewer than 1% have sickle-cell disease. The disfigured shape of the diseased cells reduces their capacity to transport oxygen, resulting in an abnormally high destruction of erythrocytes. With the decrease in erythrocytes, the patient becomes anemic. **Mononucleosis** is an infectious disease that is transmitted by a virus in saliva and is therefore commonly called the kissing disease. Mononucleosis is characterized by atypical lymphocytes. It affects primarily adolescents, causing fever, sore throat, enlarged lymph glands, and fatigue.

Heart Diseases

Heart diseases can be classified as congenital or acquired. **Congenital heart problems** result from abnormalities in embryonic development and may be attributed to heredity, nutritional problems (poor diet) of the pregnant mother, or viral infections, such as rubella. Congenital heart diseases occur in approximately 3 of every 100 births and account for about 50% of early childhood deaths. Many congenital heart defects can be corrected surgically, however, and others are not of a serious nature.

Heart murmurs are both congenital and acquired but are generally of no clinical significance. Nearly 10% of all people have heart murmurs, ranging from slight to severe. A **septal defect** is the most common type of congenital heart problem (fig. 16.49). An **atrial septal defect,** or **patent foramen ovale,** is a failure of the fetal foramen ovale to close at the time of birth. A **ventricular septal defect** is caused by an abnormal development of the interventricular septum. This condition may interfere with closure of the atrioventricular valves and may be indicated by cyanosis and abnormal heart sounds. **Pulmonary stenosis** is a narrowing of the opening into the pulmonary trunk from the right ventricle. It may lead to a pulmonary embolism and is usually recognized by extreme lung congestion.

UNIT VI

Figure 16.50 The tetralogy of Fallot. The four defects of this anomaly are (1) a ventricular septal defect, (2) an overriding aorta, (3) pulmonary stenosis, and (4) right ventricular hypertrophy.

Figure 16.51 An arteriogram of the left coronary artery in a patient (a) when the ECG was normal and (b) when the ECG showed evidence of myocardial ischemia. Notice that a coronary artery spasm—see arrow in (b)—appears to accompany the ischemia.

(a)   (b)

The **tetralogy of Fallot** is a combination of four defects within a newborn and immediately causes a cyanotic condition, leading to the newborn being termed a blue baby. The four characteristics of this anomaly are (1) a ventricular septal defect, (2) an overriding aorta, (3) pulmonary stenosis, and (4) right ventricular hypertrophy (fig. 16.50). The pulmonary stenosis obstructs blood flow to the lungs and causes hypertrophy of the right ventricle. In an overriding aorta, the ascending portion arises midway between the right and left ventricles. Open-heart surgery is necessary to correct tetralogy of Fallot, and the overall mortality rate is about 5%.

Acquired heart disease may develop suddenly or gradually. Heart attacks are included in this category and are of immense clinical importance because they are the leading cause of death in the United States. It is estimated that one in five individuals over the age of 60 will succumb to a heart attack. The immediate cause of a heart attack is generally one of the following: inadequate coronary blood supply, an anatomical disorder, or conduction disturbances.

Other types of acquired heart diseases affect the layers of the heart. **Bacterial endocarditis** is a disease of the lining of the heart, especially cusps of the valves. It is caused by infectious organisms that enter the bloodstream. **Myocardial disease** is an inflammation of the heart muscle followed by cardiac enlargement and congestive heart failure. **Pericarditis** causes an inflammation of the pericardium—the covering membrane of the heart. Its distinctive feature is pericardial friction rub, a transitory scratchy sound heard during auscultation.

A tissue is said to be **ischemic** (is-kem'ik) when it receives an inadequate supply of oxygen because of an inadequate blood

flow. The most common cause of myocardial ischemia is atherosclerosis of the coronary arteries. The adequacy of blood flow is relative—it depends on the metabolic requirements of the tissue for oxygen. An obstruction in a coronary artery, for example, may allow sufficient blood flow at rest but may produce ischemia when the heart is stressed by exercise or emotional factors (fig. 16.51). In patients with this condition, coronary artery bypass surgery may be performed (fig. 16.52).

Myocardial ischemia is associated with increased concentrations of blood lactic acid produced by anaerobic respiration of the ischemic tissue. This condition often causes substernal pain, which may also be referred to the left shoulder and arm, as well as to other areas. This referred pain is called **angina pectoris**. People with angina frequently take nitroglycerin or related drugs that help relieve the ischemia and pain. These drugs are effective because they stimulate vasodilation, which improves circulation to the heart and decreases the work that the heart must perform to eject blood into the arteries.

Myocardial cells are adapted to respire aerobically and cannot respire anaerobically for more than a few minutes. If ischemia and anaerobic respiration continue for more than a few minutes, *necrosis* (cellular death) may occur in the areas most deprived of oxygen. A sudden, irreversible injury of this kind is called a **myocardial infarction,** or **MI.**

Myocardial ischemia may be detected by characteristic changes in the ECG (fig. 16.53). The diagnosis of myocardial infarction is aided by examining the blood concentration of various enzymes that are released from the damaged cells.

Vascular Disorders

Hypertension, or high blood pressure, is the most common type of vascular disorder. In hypertension, the resting systolic blood pressure exceeds 140 mm Hg. An estimated 22 million adult Americans are afflicted by hypertension. About 15% of the cases are the result of other body problems, such as kidney diseases, adrenal hypersecretion, or arteriosclerosis, and are di-

tetralogy of Fallot: from Etienne-Louis A. Fallot, French physician, 1850–1911

UNIT VI

Figure 16.52 An illustration of double coronary artery bypass surgery. Several vessels may be used in the autotransplant, including the internal thoracic artery and the great saphenous vein.

Nelson

Figure 16.53 Depression of the S–T segment of the electrocardiogram as a result of myocardial ischemia.

Normal

Ischemia

agnosed as secondary hypertension. Primary hypertension is more common and cannot be attributed to any particular body dysfunction. If hypertension is not controlled by diet, exercise, or drugs that reduce the blood pressure, it can damage various vital body organs, such as the heart or kidneys.

Arteriosclerosis, or hardening of the arteries, is a generalized degenerative vascular disorder that results in loss of elasticity and thickening of the arteries. **Atherosclerosis** is a type of arteriosclerosis in which plaque material called **atheroma** forms on the tunica intima, narrowing the lumina of the arteries and prohibiting the normal flow of blood (fig. 16.54). Furthermore, an atheroma often creates a rough surface that can initiate the formation of a blood clot called a **thrombus.** An **embolism,** or **embolus,** is a thrombus that has dislodged from the wall of a vessel and is moving through the bloodstream. Both a thrombus and an embolism can occlude blood flow. An embolism lodged in a coronary artery is called a coronary thrombus; in a vessel of the lung it is a pulmonary thrombus; and in the brain it is a cerebral thrombus, which could cause a **stroke.**

The causes of atherosclerosis are not well understood, but the disease seems to be associated with improper diet, smoking, hypertension, obesity, lack of exercise, and heredity.

Aneurysms, coarctations, and varicose veins are all types of vascular disfigurations. An **aneurysm** *(an'yŭ-riz-ĕm)* is an expansion or bulging of the heart, aorta, or any other artery (fig. 16.55). Aneurysms are caused by weakening of the tunicas and may rupture or lead to embolisms. A **coarctation** is a constriction of a segment of a vessel, usually the aorta, and is frequently caused by a remnant of the ductus arteriosus tightening around the vessel. **Varicose veins** are weakened veins that become stretched and swollen. They are most common in the legs because the force of gravity tends to weaken the valves and overload the veins. Varicose veins can also occur in the rectum, in which case they are called **hemorrhoids.** *Vein stripping* is the surgical removal of superficial weakened veins. **Phlebitis** *(flĕ-bi'tis)* is inflammation of a vein. It may develop as a result of trauma or as an aftermath of surgery. Frequently, however, it appears for no apparent reason. Phlebitis interferes with normal venous circulation.

Disorders of the Lymphatic System

Infections of the body are generally accompanied by a swelling and tenderness of lymph nodes near the infection. An inflammation of lymph nodes is referred to as **lymphadenitis.** In prolonged lymphadenitis, an **abscess** usually forms in the nodal tissue. An abscess is a localized pocket of pus formed by tissue destruction. If an infection is not contained by localized lymph nodes, **lymphangitis** may ensue. In this condition, narrow red streaks can be seen through the skin extending proximally from the infected area. Lymphangitis is potentially dangerous because the uncontained infection may cause **septicemia** (blood poisoning).

The term **lymphoma** is used to describe primary malignancies within lymphoid tissues. Lymphomas are generally classified as **Hodgkin's disease lymphomas** or **non-Hodgkin's lymphomas.** Hodgkin's disease manifests itself as

---

arteriosclerosis: Gk. *arterio,* artery; *skleros,* hard
atheroma: Gk. *athere,* mush; *oma,* tumor
thrombus: Gk. *thrombos,* a clot
embolism: Gk. *embolos,* a plug

septicemia: Gk. *septikos,* septic; *haima,* blood
Hodgkin's disease: from Thomas Hodgkin, English physician, 1798–1866

UNIT VI

Figure 16.54 Atherosclerosis. (*a*) The lumen of a human coronary artery is almost completely occluded by an atheroma. (*b*) A close-up view of the cleared left anterior descending coronary artery containing calcified atherosclerotic plaques. The heart is of an 85-year-old female. (*c*) The structure of an atheroma is diagrammed.

swollen lymph nodes in the neck and then progresses to involve the spleen, liver, and bone marrow. The prognosis for Hodgkin's disease is good if it is detected early. Non-Hodgkin's lymphomas include an array of specific and more obscure lymphatic cancers.

The lymphatic system is also frequently infected by metastasizing carcinomas. Fragmented cells from the original tumor may enter the lymphatic ducts with the lymph and travel to the lymph nodes, where they may cause secondary cancerous growths. Breast cancer will typically do this. The surgical treatment involves the removal of the infected nodes as well as some of the healthy nodes downstream to ensure that the cancer is eliminated.

Trauma

Hemorrhage and shock are two clinical considerations that directly involve the circulatory system. Knowledge of how to treat a victim experiencing these conditions is of paramount impor-

Figure 16.55 An aneurysm is an outpouching of the vascular wall. (*a*) A diagram of two types of aortic aneurysms and (*b*) an angiograph of a saccular aneurysm (see arrow) of the middle cerebral artery.

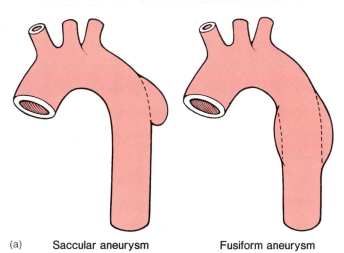

(a)   Saccular aneurysm   Fusiform aneurysm

(b)

tance in administering first aid. Techniques to control bleeding and to administer first aid to victims experiencing shock are presented in the next two sections. Also review the important arterial pressure points presented in figure 16.31.

Control of Bleeding Because serious bleeding is life threatening, the principal first-aid concern is to stop the loss of blood. The following are recommended steps in treating a victim who is hemorrhaging.

1. To reduce the chance that the victim will faint, lay the person down and slightly elevate his or her legs. If possible, elevate the site of bleeding above the level of the trunk. To minimize the chance of shock, cover the victim with a blanket.
2. Without causing further trauma, carefully remove any dirt or debris from the wound. *Do not remove any impaling objects.* This should be done at the hospital by trained personnel.
3. Apply direct pressure to the wound with a sterile bandage, clean cloth, or an article of clothing (fig. 16.56).
4. Maintain direct pressure until the bleeding stops. Dress the wound with clean bandages or cloth lightly bound in place.
5. If the bleeding does not stop and continues to seep through the dressing, *do not remove the dressing.* Rather, place additional absorbent material on top of it and continue to apply direct pressure.
6. If direct pressure does not stop the bleeding, the pressure point to the wound site may need to be compressed. In the case of a severe wound to the hand, for example, compress the brachial artery against the humerus. This should be done while pressure continues to be applied to the wound itself.

7. Once the bleeding has stopped, leave the bandage in place and immobilize the injured body part. Get the victim to the hospital or medical treatment center at once.

Recognizing and Treating Victims of Shock **Shock** is defined as the condition of the body as it responds to trauma or severe pain by dilation or relaxation of blood vessels, producing circulatory failure. Symptoms of patients experiencing shock include the following:

1. **Skin.** The skin is pale or gray, cool, and clammy.
2. **Pulse.** The heartbeat is weak and rapid. Blood pressure is reduced, frequently to below measurable values.
3. **Respiration.** The respiratory rate is hurried, shallow, and irregular.
4. **Eyes.** The eyes are staring and lusterless. The pupils may be dilated.
5. **State of being.** The victim may be conscious or unconscious. If conscious, he or she is likely to feel faint, weak, and confused. Frequently, the victim is anxious and excited.

Most trauma victims will experience some degree of shock, especially if there has been considerable blood loss. Immediate first-aid treatment for shock is essential and includes the following steps:

1. Get the victim to lie down (fig. 16.57). Lay the person on his or her back with the feet elevated. This position maintains blood flow to the brain and may relieve faintness and mental confusion. Keep movement to a minimum. If the victim has sustained an injury in which raising the legs causes additional pain, leave the person flat on his or her back.

UNIT VI

Figure 16.56 Control of bleeding. (*a*) Direct pressure and (*b*) direct pressure and compression to arterial pressure point.

(a)

(b)

Figure 16.57 Treatment of a victim experiencing shock.

4. Treat injuries appropriately. If the victim is bleeding, treat accordingly (see arterial pressure points and control of bleeding). Immobilize fractures and sprains. Always be aware of the chance of spinal injuries and take the necessary precautions.
5. Get hospital care as soon as possible.

CLINICAL CASE STUDY ANSWER

Mural thrombus (a blood clot adherent to the inner surface of one of the heart's chambers) is a fairly common complication of myocardial infarction. Once a thrombus forms within the heart, a piece may break off and travel throughout the body. This is the most likely cause of the symptoms in the right leg of our patient. Because the embolus traveled and lodged in the systemic circulation (as opposed to the pulmonary circulation), the mural thrombus was probably located in the left side of the heart. The embolus traveled to a point in the femoral artery where it lodged, thus occluding blood flow to the popliteal artery and its distal branches. The route of travel was as follows: left side of heart → ascending aorta → descending thoracic aorta → abdominal aorta → right common iliac artery → right external iliac artery → femoral artery.

The standard treatment for this problem is emergency surgery to extract the clot from the leg. Anticoagulation (blood-thinning) therapy is then continued or instituted.

2. Keep the victim warm and comfortable. If the weather is cold, place a blanket under and over the person. If the weather is hot, position the person in the shade on top of a blanket. Loosen tight collars, belts, or other restrictive clothing. *Do not give the person anything to drink,* even if he or she complains of thirst.
3. Take precautions for internal bleeding or vomiting. If the victim has blood coming from the mouth, or if there is indication that the victim may vomit, position the person on his or her side to prevent choking or inhaling the blood or vomitus.

Internal Affairs

The Relational Workings of the Body Systems

Circulatory System

- Transports O_2 and nutrients; eliminates CO_2 and metabolic wastes; provides clotting mechanism to heal breaks in skin

- Transports O_2 and nutrients; eliminates CO_2 and metabolic wastes

- Transports O_2 and nutrients; eliminates CO_2 and metabolic wastes, including lactic acid

- Transports O_2 and nutrients; eliminates CO_2 and metabolic wastes; certain capillaries participate in blood-brain barrier

- Transports O_2, hormones, and nutrients; eliminates CO_2 and metabolic wastes

- Transports O_2 and nutrients; eliminates CO_2 and metabolic wastes

- Transports O_2 and nutrients; eliminates CO_2 and metabolic wastes

- Transports O_2 and nutrients; eliminates CO_2 and metabolic wastes

- Transports O_2 and nutrients; eliminates CO_2 and metabolic wastes

- Transports O_2, sex hormones, and nutrients; eliminates CO_2 and metabolic wastes

Integumentary System
- Provides site for thermoregulation

Skeletal System
- Provides site (bone marrow) for hemopoiesis; protects heart and thoracic vessels

Muscular System
- Muscle contraction assists blood movement

Nervous System
- Provides autonomic regulation of cardiac rate and force
- Maintains blood pressure and controls blood distribution

Endocrine System
- Various hormones affect blood pressure

Lymphatic System
- Maintains balanced amount of interstitial fluid within body tissues

Respiratory System
- Provides O_2 to bloodstream and eliminates CO_2 at alveoli of lungs

Digestive System
- Provides nutrients for blood formation, including iron and B vitamins

Urinary System
- Eliminates metabolic wastes from blood
- Helps regulate blood volume and pressure

Reproductive System
- Gonads produce sex hormones for maintenance of blood and blood vessels

CHAPTER SUMMARY

Functions and Major Components of the Circulatory System (pp. 508–509)

1. Blood transports oxygen and nutritive molecules to the tissue cells and carbon dioxide and other wastes away from tissue cells; it also carries hormones and other regulatory molecules to their target organs.
2. Leukocytes and their products serve to protect the body from infection.

Blood (pp. 509–515)

1. Blood, a highly specialized connective tissue, consists of formed elements (erythrocytes, leukocytes, and thrombocytes) that are suspended and carried in the plasma. The constituents of blood function in transport, immunity, and blood-clotting mechanisms.
2. Erythrocytes are disc-shaped cells that lack nuclei but do contain hemoglobin. There are between 4.3 and 5.8 million erythrocytes per cubic millimeter of blood and they live for about 120 days.
3. Leukocytes have nuclei and are classified as granular (eosinophils, basophils, and neutrophils) or agranular (monocytes and lymphocytes). Leukocytes function to defend the body against infections by microorganisms.
4. Thrombocytes, or platelets, are cytoplasmic fragments that assist in the formation of clots to prevent blood loss.
5. Erythrocytes are formed through a process called erythropoiesis; leukocytes are formed through leukopoiesis.
6. Prenatal hemopoietic centers are the yolk sac, liver, and spleen. In the adult, bone marrow and lymphoid tissues perform this function.

Heart (pp. 515–524)

1. The heart is enclosed within a pericardial sac. The wall of the heart consists of the epicardium, myocardium, and endocardium.
   (a) The right atrium receives blood from the superior and inferior venae cavae and the right ventricle pumps blood into the pulmonary trunk to the pulmonary arteries.
   (b) The left atrium receives blood from the pulmonary veins and pumps blood into the aorta.
   (c) The heart contains right and left atrioventricular valves, a pulmonary valve, and an aortic (mitral) valve.

2. The two principal circulatory divisions are the pulmonary and the systemic; in addition, the coronary system serves the heart.
   (a) The pulmonary circulation includes the vessels that carry blood from the right ventricle through the lungs and from there to the left atrium.
   (b) The systemic circulation includes all other arteries, capillaries, and veins in the body. These vessels carry blood from the left ventricle through the body and return blood to the right atrium.
3. Contraction of the atria and ventricles is produced by action potentials that originate in the sinoatrial (SA) node.
   (a) These electrical waves spread over the atria and then enter the atrioventricular (AV) node.
   (b) From here, the impulses are conducted by the atrioventricular bundle and conduction myofibers into the ventricular walls.
4. During contraction of the ventricles, the intraventricular pressure rises and causes the AV valves to close; during relaxation, the pulmonary and aortic valves close because the pressure is greater in the arteries than in the ventricles.
5. Closing of the AV valves causes the first sound (lub); closing of the pulmonary and aortic valves causes the second sound (dub). Heart murmurs are commonly caused by abnormal valves or by septal defects.
6. A recording of the pattern of electrical conduction is called an electrocardiogram (ECG or EKG).

Blood Vessels (pp. 524–528)

1. Arteries and veins have a tunica externa, tunica media, and tunica intima.
   (a) Arteries have thicker muscle layers in proportion to their diameters than do veins because arteries must withstand a higher blood pressure.
   (b) Veins have venous valves that direct blood to the heart when the veins are compressed by the skeletal muscle pumps.
2. Capillaries are composed of endothelial cells only, and may be continuous or discontinuous.

Principal Arteries of the Body (pp. 528–538)

1. Three arteries arise from the aortic arch: the brachiocephalic trunk, the left common carotid artery, and the left subclavian artery. The brachiocephalic trunk divides into the right common carotid artery and the right subclavian artery.
2. The head and neck receive an arterial supply from branches of the internal and external carotid arteries and the vertebral arteries.
   (a) The brain receives blood from the paired internal carotid arteries and the paired vertebral arteries, which form the cerebral arterial circle around the pituitary gland.
   (b) The external carotid artery gives off many branches that supply the head and neck.
3. The upper extremity is served by the subclavian artery and its derivatives.
   (a) The subclavian artery becomes first the axillary artery and then the brachial artery as it enters the arm.
   (b) The brachial artery bifurcates to form the radial and ulnar arteries, which supply blood to the forearm and a portion of the hand and digits.
4. The abdominal portion of the aorta produces the following branches: the inferior phrenic, celiac trunk, superior mesenteric, renal, suprarenal, testicular (or ovarian), and inferior mesenteric arteries.
5. The common iliac arteries divide into the internal and external iliac arteries, which supply branches to the pelvis and lower extremities.

Principal Veins of the Body (pp. 538–545)

1. Blood from the head and neck is drained by the external and internal jugular veins; blood from the brain is drained by the internal jugular veins.
2. The upper extremity is drained by superficial and deep veins.
3. In the thorax, the superior vena cava is formed by the union of the two brachiocephalic veins and also collects blood from the azygos vein.
4. The lower extremity is drained by both superficial and deep veins. At the level of the fifth lumbar vertebra, the right and left common iliac veins unite to form the inferior vena cava.

UNIT VI

5. Blood from capillaries in the GI tract is drained via the hepatic portal vein to the liver.
   (a) This venous blood then passes through hepatic sinusoids and is drained from the liver in the hepatic veins.
   (b) The pattern of circulation in which two capillary beds are in a series is called a portal system.

Fetal Circulation (pp. 545–547)

1. Structural adaptations in the fetal cardiovascular system reflect the fact that oxygen and nutrients are obtained from the placenta rather than from the fetal lungs and GI tract.
2. Fully oxygenated blood is carried only in the umbilical vein, which drains the placenta. This blood is carried via the ductus venosus to the inferior vena cava of the fetus.
3. Partially oxygenated blood is shunted from the right to the left atrium via the foramen ovale and from the pulmonary trunk to the aorta via the ductus arteriosus.

Lymphatic System (pp. 547–551)

1. The lymphatic system functions to return tissue fluid to the venous system and to help protect the body from disease.
2. Lymph capillaries drain tissue fluid, which is formed from blood plasma; when this fluid enters lymph capillaries, it is called lymph.
3. Lymph is returned to the venous system via two large lymph ducts—the thoracic duct and the right lymphatic duct.
4. Lymph filters through lymph nodes, which contain phagocytic cells and lymphatic nodules that produce lymphocytes.
5. The spleen and thymus are lymphoid organs because they produce lymphocytes.

REVIEW ACTIVITIES

Objective Questions

1. Which of the following is *not* a formed element of blood cell?
   (a) a leukocyte
   (b) an eosinophil
   (c) a fibrinogen
   (d) a thrombocyte
2. An elevated white blood cell count is referred to as
   (a) leukocytosis.
   (b) polycythemia.
   (c) leukopoiesis.
   (d) leukemia.
3. All arteries in the body contain oxygen-rich blood with the exception of
   (a) the aorta.
   (b) the pulmonary arteries.
   (c) the renal arteries.
   (d) the coronary arteries.
4. Blood from the coronary circulation directly enters
   (a) the inferior vena cava.
   (b) the superior vena cava.
   (c) the right atrium.
   (d) the left atrium.
5. The "lub," or first heart sound, is produced by the closing of
   (a) the aortic valve.
   (b) the pulmonary valve.
   (c) the right atrioventricular valve.
   (d) the left atrioventricular valve.
   (e) both (a) and (b).
   (f) both (c) and (d).
6. The QRS wave of an ECG is produced by
   (a) depolarization of the atria.
   (b) repolarization of the atria.
   (c) depolarization of the ventricles.
   (d) repolarization of the ventricles.
7. Which of the following does/do *not* arise from the aortic arch?
   (a) the brachiocephalic trunk.
   (b) the coronary arteries.
   (c) the left common carotid artery.
   (d) the left subclavian artery.
8. Which of the following does *not* supply blood to the brain?
   (a) the external carotid artery.
   (b) the internal carotid artery.
   (c) the vertebral artery.
   (d) the basilar artery.
9. The maxillary and superficial temporal arteries are derived from
   (a) the external carotid artery.
   (b) the internal carotid artery.
   (c) the vertebral artery.
   (d) the facial artery.
10. Which of the following organs is/are served by a portal vein?
    (a) the liver
    (b) the pituitary
    (c) both (a) and (b)
    (d) neither (a) nor (b)
11. The heart derives from mesoderm and is complete in embryonic form by the
    (a) third week.
    (b) fifth week.
    (c) tenth week.
    (d) seventeenth week.
12. Which of the following fetal blood vessels carries fully oxygenated blood?
    (a) the ductus arteriosus
    (b) the ductus venosus
    (c) the umbilical vein
    (d) the aorta

Essay Questions

1. What are the functions of the circulatory system? Identify the body systems with which the circulatory system closely functions in maintaining homeostasis.
2. Explain how the development of the aortic arches contributes to the formation of the major vessels associated with the heart.
3. Distinguish between granulocytes and agranulocytes. What are the functions of leukocytes? How do they differ from erythrocytes and thrombocytes?
4. Describe how the heart is compartmentalized within the thoracic cavity. What is the function of the pericardium?
5. Describe the cardiac cycle. Why is the sinoatrial node called the pacemaker of the heart?
6. Diagram and label a normal electrocardiogram. What is the significance of the deflection waves of an ECG?
7. Compare the structure, size, and function of arteries, capillaries, and veins.
8. What is the difference between blood pressure and pulse pressure? What cardiac event determines the systolic blood pressure? What accounts for the diastolic blood pressure?
9. Name the vessels of the thoracic and shoulder regions that are not symmetrical (do not have a counterpart on the opposite side of the body).
10. Trace the path to the brain of a glucose injection into the median cubital vein of the arm and list in sequence all the blood vessels and chambers of the heart through which it passes.
11. What is significant about the hepatic portal system? What do we mean when we say that the liver has two blood supplies?
12. List the functions of the lymphatic system and describe the relationship between the lymphatic system and the circulatory system.
13. Briefly describe the development of the heart.
14. What are five fetal circulatory structures that cease to function in a newborn?
15. Distinguish between congenital and acquired heart diseases and identify several kinds of each.

Gundy/Weber Software

The tutorial software accompanying Chapter 16, Circulatory System, are Volume 8, Blood and the Lymphatic System, and Volume 9, Cardiovascular System.

17

Respiratory System

Chapter Outline

CLINICAL CASE STUDY

A 37-year-old male comes to the emergency room after having been stabbed above the left clavicle with an ice pick. The patient's main complaint is pain in the left side of the chest. Initial evaluation reveals a minute puncture wound superior to the left clavicle, just lateral to the sternocleidomastoid muscle. The vital signs are normal except for a moderately high respiratory rate. The chest radiograph shows the left lung to be collapsed to half its normal size and surrounded by air and blood. (This condition is known as *hemopneumothorax*, meaning blood and air in the thoracic cavity.)

Explain how the air got inside the thoracic cavity (assuming it did not enter through the skin puncture wound). What is the name of the space where the air and blood are located? What membranes contain the space?

Introduction to the Respiratory System

The respiratory system can be divided into upper and lower divisions on the basis of function and embryological development.

Objective 1. Describe the functions associated with the term *respiration.*

Objective 2. Identify the organs of the respiratory system and describe their locations.

Objective 3. List the functions of the respiratory system.

The term *respiration* refers to three separate but related functions: (1) **ventilation** (breathing), (2) **gas exchange** between the air and blood in the lungs and between the blood and tissues, and (3) **oxygen utilization** by the tissues in the energy-liberating reactions of cell respiration. The exchange of gases (oxygen and carbon dioxide) between the air and blood is called *external respiration.* Gas exchange between the blood and tissues is known as *internal respiration.*

It is estimated that a relaxed adult breathes about 9 to 20 times a minute, ventilating about 5 to 6 liters of air during this period. Strenuous exercise increases the demand for oxygen and increases the respiratory rate fifteenfold to twentyfold, so that about 100 liters of air are breathed each minute. If breathing stops, a person will lose consciousness after 4 or 5 minutes. Brain damage may occur after 7 to 8 minutes, and the person will die after 10 minutes. Knowledge of the structure and function of the respiratory system, therefore, is of the utmost clinical importance.

respiration: L. *re*, back; *spirare*, to breathe

Physical Requirements of the Respiratory System

The respiratory system includes those organs and structures that function together to bring gases in contact with the blood of the circulatory system. In order to be effective, the respiratory system must comply with certain physical requirements.

1. The membrane through which gas is exchanged between the circulatory and respiratory system must be thin-walled and selectively permeable so that diffusion can occur easily.
2. This membrane must be kept moist so that oxygen and carbon dioxide can be dissolved in water to facilitate diffusion.
3. The system must be nourished by a rich blood supply.
4. The surface for gas exchange must be located deep in the body to allow the incoming air to be sufficiently warmed, moistened, and cleansed of airborne particles so as not to damage this delicate membrane.
5. The system must include an effective ventilation mechanism to constantly replenish the air.

The respiratory system adequately meets all of these requirements to ensure that all of the trillions of cells of the body can carry on energy metabolism.

Functions of the Respiratory System

The four basic functions of the respiratory system, not all of which are associated with breathing, are as follows:

1. It provides oxygen to the bloodstream and removes carbon dioxide.
2. It enables sound production or vocalization as expired air passes over the vocal cords.
3. It assists in abdominal compression during micturition (urination), defecation (passing of feces), and parturition (childbirth). The abdominal muscles become more effective during a deep breath when the air is held in the lungs by closing the glottis and fixing the diaphragm. This same technique is used when lifting a heavy object, in which case the diaphragm indirectly assists the back muscles.
4. It includes protective and reflexive nonbreathing air movements, such as coughing and sneezing, to keep the air passageways clean.

Basic Structure of the Respiratory System

The major passages and structures of the respiratory system are the **nasal cavity, pharynx, larynx** and **trachea,** and the **bronchi, bronchioles,** and **alveoli** within the **lungs** (fig. 17.1). The respiratory system is frequently divided into the *conducting division* and the *respiratory division.* The conducting portion includes all of the cavities and structures that transport gases to and from alveoli. The alveoli, the functional units of the system, constitute the respiratory portion.

Figure 17.1   The basic anatomy of the respiratory system.

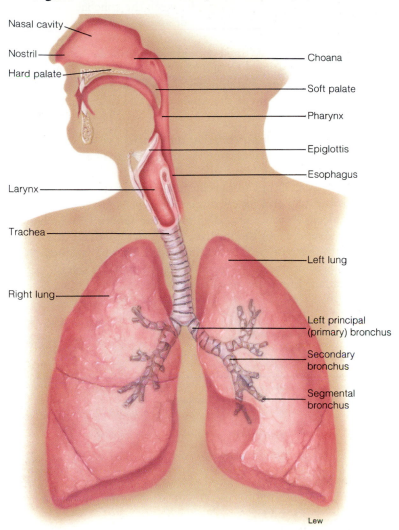

Nasal cavity
Nostril
Hard palate
Choana
Soft palate
Pharynx
Epiglottis
Esophagus
Larynx
Trachea
Left lung
Right lung
Left principal (primary) bronchus
Secondary bronchus
Segmental bronchus

Lew

Early Greek and Roman scientists placed great emphasis on the invisible material that was breathed in. They knew nothing about oxygen or the role of the blood in transporting this vital substance to cells. For that matter, they knew nothing about microscopic structures like cells because the microscope had not yet been invented. They did know, however, that respiration was essential for life. Early Greeks referred to air as an intangible, divine spirit called *pneuma*. In Latin, the term for breath, *spiritus*, meant life-force.

1. Define the terms *external respiration* and *internal respiration*.
2. Differentiate between the physical requirements and the functions of the respiratory system.
3. List in order the major passages and structures through which inspired air would pass from the nostrils to the alveoli of the lungs.

Conducting Passages

Air is conducted through the oral and nasal cavities to the pharynx and then through the larynx to the trachea and bronchial tree. These structures deliver warmed and humidified air to the respiratory division in the lungs.

Objective 4.   List the types of epithelial tissue that characterize each region of the respiratory tract and comment on the significance of the special attributes of each type.

Objective 5.   Identify the boundaries of the nasal cavity and discuss the relationship of the paranasal sinuses to the rest of the respiratory system.

Objective 6.   Describe the three regions of the pharynx and identify the structures located in each.

Objective 7.   Discuss the role of the laryngeal region in digestion and respiration.

Objective 8.   Identify the anatomical features of the larynx associated with sound production and respiration.

The conducting passages serve to transport air to the respiratory structures of the lungs. The passageways are lined with various types of epithelia to prepare the air properly for utilization. The majority of the conducting passages are held permanently open by muscle or a bony or cartilaginous framework.

Nose

The **nose** includes an external portion that juts out from the face and an internal **nasal cavity** for the passage of air. The external portion of the nose is covered with skin and supported by paired **nasal bones,** forming the bridge, and pliable cartilage, forming the distal portions (fig. 17.2). The **septal cartilage** forms the anterior portion of the **nasal septum** and the paired **lateral cartilages** and **alar cartilages** form the framework around the **nostrils.**

The perpendicular plate of the **ethmoid bone** and the **vomer bone** (see fig. 6.17), together with the septal cartilage, constitute the supporting framework of the nasal septum, which divides the nasal cavity into two lateral halves. Each half is referred to as a **nasal fossa.** The **vestibule** is the anterior expanded portion of the nasal fossa (fig. 17.3). Each nasal fossa opens anteriorly through the nostril, or **naris,** and communicates posteriorly with the **nasopharynx** through the **choana** (ko-a-nă). The roof of the nasal cavity is formed anteriorly by the frontal bone and paired nasal bones, medially by the cribriform plate of the ethmoid bone, and posteriorly by the sphenoid bone (see figs. 6.17 and 6.20). The palatine and maxillary bones form the floor of the cavity. On the lateral walls are the

nose: O.E. *nosu,* nose
nostril: O.E. *nosu,* nose; *thyrel,* hole
choana: Gk. *choane,* funnel

UNIT VI

Figure 17.2 The supporting framework of the nose.

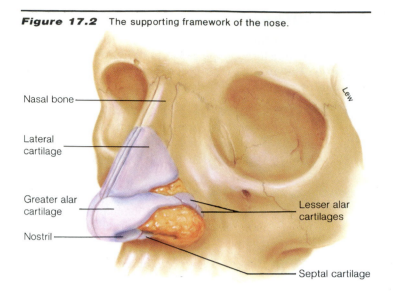

Nasal bone

Lateral cartilage

Greater alar cartilage

Nostril

Lesser alar cartilages

Septal cartilage

superior, middle, and **inferior conchae** *(kong'ke)* (turbinates) (see fig. 6.26). Air passages between the conchae are referred to as **meatuses** (fig. 17.3). The anterior openings of the nasal cavity are lined with stratified squamous epithelium, whereas the conchae are lined with pseudostratified ciliated columnar epithelium (figs. 17.4 and 17.5). Mucus-secreting goblet cells are present in great abundance throughout both regions.

The three functions of the nasal cavity and its contents are as follows:

1. The nasal epithelium covering the conchae serves to warm, moisten, and cleanse the inspired air. The nasal epithelium is highly vascular and covers an extensive surface area. This is important for warming the air but unfortunately also makes humans susceptible to nosebleeds. Nasal hairs called **vibrissae** *(vi-bris'e),* which often extend from the nostrils, filter macroparticles that might otherwise be inhaled. Fine particles, such as dust, pollen, or smoke, are trapped along the moist mucous membrane lining the nasal cavity.

concha: L. a shell
vibrissa: L. *vibrare,* to vibrate

Figure 17.3 A sagittal section of the head showing the structures of the upper respiratory tract. There are several openings into the lateral wall of the nasal cavity, including the openings of the various paranasal sinuses, the nasolacrimal ducts that drain from the eyes, and the auditory tubes that drain from the middle ear chambers.

Frontal sinus

Middle concha (cut)

Opening of frontal sinus

Inferior concha (cut)

Vestibule

Opening of nasolacrimal duct

Nostril

Hard palate

Upper lip

Vermillion

Oral cavity

Tongue

Vestibule

Hyoid bone

Trachea

Superior meatus

Opening of ethmoidal sinus

Opening of maxillary sinus

Opening of sphenoidal sinus

Sphenoidal sinus

Meatus

Choana

Pharyngeal tonsil

Opening of auditory tube

Soft palate

Uvula

Palatine tonsil

Lingual tonsil

Epiglottis

Esophagus

UNIT VI

Figure 17.4 The various types of epithelial tissues found throughout the respiratory system.

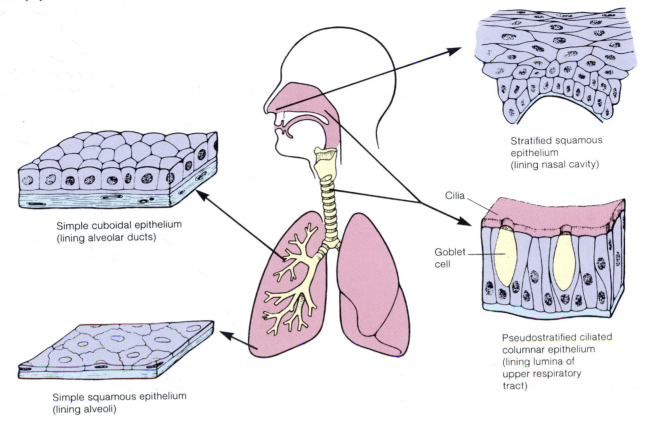

Stratified squamous epithelium (lining nasal cavity)

Simple cuboidal epithelium (lining alveolar ducts)

Cilia

Goblet cell

Pseudostratified ciliated columnar epithelium (lining lumina of upper respiratory tract)

Simple squamous epithelium (lining alveoli)

2. Olfactory epithelium in the upper medial portion of the nasal cavity is concerned with the sense of smell.

3. The nasal cavity is associated with voice phonetics by functioning as a resonating chamber.

There are several drainage openings into the nasal cavity (see fig. 17.3). The paranasal ducts drain mucus from the paranasal (the frontal, ethmoidal, sphenoidal, and maxillary) sinuses and the nasolacrimal ducts drain tears from the eyes. An excessive secretion of tears causes the nose to run as the tears drain into the nasal cavity. The auditory tube from the middle ear enters the upper respiratory tract posterior to the nasal cavity in the nasopharynx. With all these accessory connections, it is no wonder that infections can spread so easily from one chamber to another throughout the facial area. To avoid causing damage or spreading infections to other areas, one must be careful not to blow the nose too forcefully.

Paranasal Sinuses

Paired air spaces in certain bones of the skull are called **paranasal sinuses.** These sinuses are named according to the bones in which they are found; thus, there are the **maxillary, frontal, sphenoidal,** and **ethmoidal sinuses** (fig. 17.6). Each sinus communicates via drainage ducts within the nasal cavity on its own side (see fig. 17.3). Paranasal sinuses may help to warm

sinus: L. *sinus,* bend or curve

Figure 17.5 A color-enhanced scanning electron micrograph of a bronchial wall showing cilia. In the trachea and bronchi, there are about 300 cilia per cell. The cilia move mucus-dust packages toward the pharynx, where they can either be swallowed or expectorated.

and moisten the inspired air. These sinuses are responsible for some sound resonance, but most importantly, they function to decrease the weight of the skull while giving structural strength.

You can observe your own paranasal sinuses. Face a mirror in a darkened room and shine a small flashlight into your face. The frontal sinuses will be illuminated by directing the light just below the eyebrow. The maxillary sinuses are illuminated by shining the light into the oral cavity and closing your mouth around the flashlight.

UNIT VI

Figure 17.6 The paranasal sinuses.

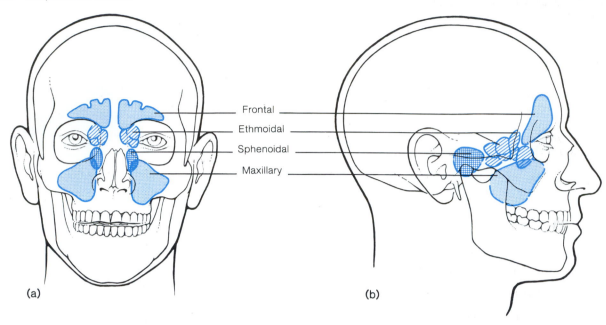

Frontal
Ethmoidal
Sphenoidal
Maxillary

(a)　　　　　　　　　　　　(b)

Pharynx

The **pharynx** *(far'ingks)* is a funnel-shaped passageway approximately 13 cm (5 in.) long that connects the nasal and oral cavities to the larynx at the base of the skull. The supporting walls of the pharynx are composed of skeletal muscle and the lumen is lined with a mucous membrane. Within the pharynx are several paired lymphoid organs called **tonsils** (fig. 17.7). Commonly referred to as the throat or gullet, the pharynx has both respiratory and digestive functions. It also provides a resonating chamber for certain speech sounds. The pharynx is divided on the basis of location and function into three regions: nasal, oral, and laryngeal.

The **nasopharynx** *(na''zo-far'ingks)* has only a respiratory function. It is the uppermost portion of the pharynx, directly posterior to the nasal cavity and above the soft palate. A pendulous **uvula** hangs from the middle lower border of the soft palate. The paired **auditory,** or eustachian, **tubes** connect the nasopharynx with the middle-ear cavities. **Pharyngeal tonsils,** or **adenoids,** are situated in the posterior wall of this cavity. During the act of swallowing, the soft palate and uvula are elevated to block the nasal cavity and prevent food from entering. Occasionally a person may suddenly exhale air (as with a laugh) while in the process of swallowing fluid. If this occurs before the uvula effectively blocks the nasopharynx, fluid will be discharged through the nasal cavity.

The **oropharynx** *(o''ro-far'ingks)* is the middle portion of the pharynx between the soft palate and the level of the hyoid bone. The base of the tongue forms the anterior wall of the oropharynx. Paired **palatine tonsils** are located along the pos-

terior lateral wall of the oropharynx, and the **lingual tonsils** are found on the base of the tongue. This portion of the pharynx has both a respiratory and a digestive function.

The **laryngopharynx** *(la-ring''go-far'ingks)* is the lowermost portion of the pharynx. It extends posteriorly from the level of the hyoid bone to the larynx and opens into the esophagus and larynx. It is at the lower laryngopharynx that the respiratory and digestive systems become distinct. Food is directed into the esophagus, whereas air is moved anteriorly into the larynx.

During a physical examination, a physician commonly depresses the patient's tongue and examines the condition of the palatine tonsils. Tonsils are pharyngeal lymphoid organs and tend to become swollen and inflamed after persistent infections. As mentioned in chapter 16, tonsils may have to be surgically removed when they become so overrun with pathogens after repeated infections that they themselves become the source of infection. The removal of the palatine tonsils is called a *tonsillectomy,* whereas the removal of the pharyngeal tonsils is called an *adenoidectomy.*

Larynx

The **larynx** *(lar'ingks),* or voice box, forms the entrance into the lower respiratory system as it connects the laryngopharynx with the trachea. It is positioned in the anterior midline of the neck at the level of the fourth through sixth cervical vertebrae. The larynx has two functions. Primarily, it serves to prevent food or fluid from entering the trachea and lungs during swallowing and to permit passage of air while breathing. A secondary function is to produce sounds.

pharynx: L. *pharynx,* throat
tonsil: L. *toles,* goiter or swelling
uvula: L. *uvula,* small grape
adenoid: Gk. *adenoeides,* glandlike

larynx: Gk. *larynx,* upper windpipe

UNIT VI

Figure 17.7 A sagittal section of the head showing the structures of the upper respiratory tract and the three pharyngeal regions.

Frontal sinus
Middle concha (cut)
Opening of frontal sinus
Inferior concha (cut)
Opening of nasolacrimal duct
Nostril
Hard palate
Upper lip
Vermillion
Oral cavity
Tongue
Vestibule
Hyoid bone

Superior meatus
Opening of ethmoidal sinus
Opening of maxillary sinus
Opening of sphenoidal sinus
Sphenoidal sinus
Nasal cavity
Choana
Pharyngeal tonsil
Opening of auditory tube
Nasopharynx
Soft palate
Oropharynx
Uvula
Palatine tonsil
Lingual tonsil
Epiglottis
Laryngopharynx
Esophagus
Lew
Trachea

Laryngitis is the inflammation of the mucosal epithelium of the larynx and vocal cords, which causes a hoarseness of a person's voice or an inability to speak above a whisper. Laryngitis may result from overuse of the voice, inhalation of an irritating chemical, or a bacterial or viral infection. Mild cases are temporary and seldom of major concern.

The larynx is shaped like a triangular box (fig. 17.8). It is composed of a framework involving nine cartilages: three are large single structures and six are smaller paired structures. The largest of the unpaired cartilages is the anterior **thyroid cartilage.** The **laryngeal prominence** of the thyroid cartilage, commonly called the Adam's apple, forms an anterior vertical ridge along the larynx that can be palpated on the midline of the neck. The thyroid cartilage is typically larger and more prominent in males than in females because of the effect of male sex hormones on the development of the larynx during puberty.

The spoon-shaped **epiglottis** has a cartilaginous framework. It is behind the root of the tongue and aids in closing the **glottis,** or laryngeal opening, during swallowing.

The lower end of the larynx is formed by the ring-shaped **cricoid** *(kri'koid)* **cartilage.** This third unpaired cartilage connects the thyroid cartilage above and the trachea below. The paired **arytenoid** *(ar''i-te'noid)* **cartilages,** located above the cricoid and behind the thyroid, furnish the posterior attachment of the **vocal cords.** The other paired **cuneiform cartilages** and **corniculate** *(kor-nik'yŭ-lāt)* **cartilages** are small accessory cartilages that are closely associated with the arytenoid cartilages (fig. 17.9).

Two pairs of strong connective tissue bands are stretched across the upper opening of the larynx from the thyroid cartilage anteriorly to the paired arytenoid cartilages posteriorly. These are the **true vocal cords** and the **false vocal cords** (fig. 17.9). The false vocal cords, or **ventricular folds,** support the true vocal cords and, as their name implies, are not used in sound production. Stratified squamous epithelium lines the true vocal cords, whereas the rest of the larynx is lined with pseudostratified ciliated columnar epithelium. This is an important

thyroid: Gk. *thyreos,* shieldlike

cricoid: Gk. *krikos,* ring; *eidos,* form
arytenoid: Gk. *arytaina,* ladle- or cup-shaped
cuneiform: L. *cuneus,* wedge-shaped
corniculate: L. *corniculum,* diminutive of *cornu,* horn

UNIT VI

Figure 17.8 The structure of the larynx. (*a*) An anterior view, (*b*) a lateral view, and (*c*) a sagittal view.

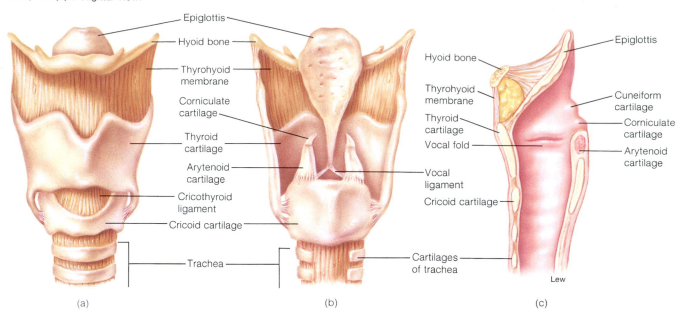

Figure 17.9 A superior view of the vocal cords. In (*a*) the vocal cords are taut; in (*b*) they are relaxed and the glottis is opened. (*c*) A photograph through a laryngoscope showing the true and false vocal cords and the glottis.

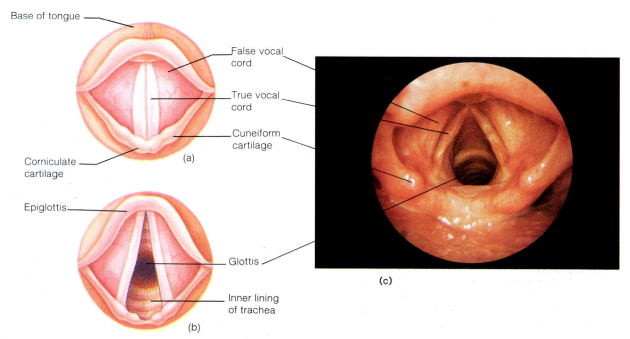

anatomical modification considering the tremendous vibratory action of the true vocal cords in the production of sound.

The **laryngeal muscles** are extremely important in closing the glottis during swallowing and in speech. There are two groups of laryngeal muscles: **extrinsic muscles,** responsible for elevating the larynx during swallowing, and **intrinsic muscles** that, when contracted, change the length, position, and tension of the vocal cords. Various pitches are produced as air passes over the altered vocal cords. If the vocal cords are taut, vibration is more rapid and causes a higher pitch. Less tension

on the cords produces lower sounds. Mature males generally have thicker and longer vocal cords than females; therefore, the vocal cords of males vibrate more slowly and produce lower pitches. The loudness of vocal sound is determined by the force of the air passed over the vocal cords and the amount of vibration. The vocal cords do not vibrate when whispering.

Sounds originate in the larynx, but other structures are necessary to convert sound into recognizable speech. Vowel sounds, for example, are produced by constriction of the walls of the pharynx. The pharynx, paranasal sinuses, and oral and nasal

Figure 17.10 An anterior view of the larynx, trachea, and bronchi.

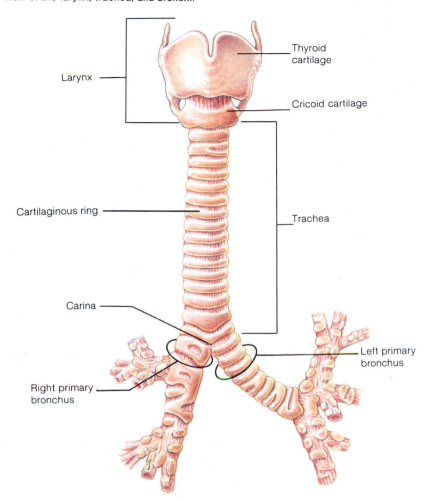

Larynx

Thyroid cartilage

Cricoid cartilage

Cartilaginous ring

Trachea

Carina

Left primary bronchus

Right primary bronchus

UNIT VI

cavities act as resonating chambers. The final enunciation of words is accomplished through movements of the lips and tongue.

> Swallowing is a complex action involving several neurological pathways and various groups of muscles. One of the swallowing responses is the elevation of the larynx to close the glottis against the epiglottis. This movement can be noted by cupping the fingers lightly over the larynx and then swallowing. Food may become lodged within the glottis if it is not closed as it should be. In this case, the *abdominal thrust (Heimlich) maneuver* can be used to prevent suffocation.

Trachea

The **trachea** *(tra′ke-ă)*, commonly called the windpipe, is a rigid tube, approximately 12 cm (4½ in.) long and 2.5 cm (1 in.) in diameter, connecting the larynx to the primary bronchi (fig. 17.10). It is positioned anterior to the esophagus as it extends into the thoracic cavity. A series of 16 to 20 C-shaped rings of hyaline cartilage form the walls of the trachea (fig. 17.11). The open part of the C is positioned posteriorly and is

covered by fibrous connective tissue and smooth muscle. These cartilages provide a rigid but flexible tube in which the lumen, or airway, is permanently open. The mucosa (surface lining of the lumen) consists of pseudostratified ciliated columnar epithelium containing many mucus-secreting **goblet cells** (see figs. 17.4 and 17.5). Inhaled dust particles stick to the mucus, and the cilia sweep it upward to the pharynx, where it is removed through a cough reflex. The tracheal bifurcation forming the **right** and **left primary bronchi** is reinforced by the **carina** *(kă-ri′nă)*, a keel-like cartilage plate (fig. 17.10).

> If the trachea becomes occluded through inflammation, excessive secretion, trauma, or aspiration of a foreign object, it may be necessary to create an emergency opening into this tube so that ventilation can still occur. A *tracheotomy* is the procedure of surgically opening the trachea and a *tracheostomy* involves inserting a tube into the trachea to permit breathing and to keep the passageway open (fig. 17.12). A tracheotomy should be performed only by a competent physician as there is a great risk of cutting a recurrent laryngeal nerve or the common carotid artery, which is also in this area.

---

trachea: L. *trachia*, rough air vessel

carina: L. *carina*, keel

Figure 17.11 The histology of the trachea. (*a*) A photomicrograph showing the relationship of the trachea to the esophagus (3×) and (*b*) a photomicrograph of tracheal cartilage (63×).

(a)

Lumen of esophagus

Trachealis muscle

Lumen of trachea

Tracheal epithelium

Thyroid gland

Tracheal cartilage

(b)

Tracheal epithelium

Tracheal cartilage

Tracheal cartilage

Adventitia

Figure 17.12 The site for a surgical tracheostomy.

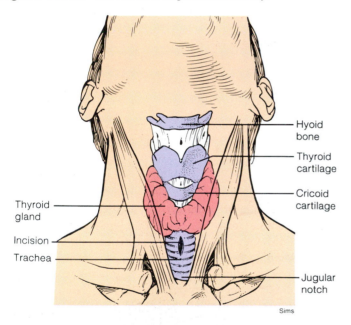

Hyoid bone

Thyroid cartilage

Cricoid cartilage

Thyroid gland

Incision

Trachea

Jugular notch

Sims

Figure 17.13 A photograph of a plastic cast of the conducting airways from the trachea to the terminal bronchioles.

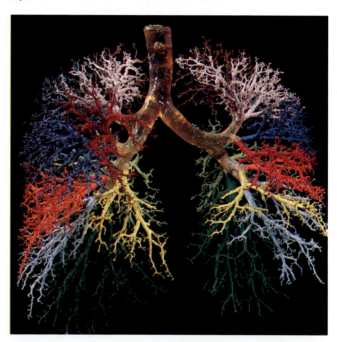

Bronchial Tree

The **bronchial tree** is so named because it is composed of a series of respiratory tubes that branch into progressively narrower tubes as they extend into the lung (fig. 17.13). The trachea bifurcates into a **right** and **left primary bronchus** (*brong'kus*) at the level of the sternal angle behind the manubrium. Each primary bronchus has hyaline cartilage wedges within its wall surrounding the lumen to keep it open as it extends into the lung. Because of the more vertical position of the right primary bronchus, foreign particles are more likely to lodge here than in the left primary bronchus.

bronchus: L. *bronchus*, windpipe

The primary bronchus divides deeper in the lungs to form **secondary bronchi** and **segmental (tertiary) bronchi** (fig. 17.14). The bronchial tree continues to branch into yet smaller tubules called **bronchioles** (fig. 17.15). There is little cartilage in the bronchioles, which contain thick smooth muscle that can constrict or dilate these airways. Bronchioles provide the greatest resistance to air flow in the conducting passages, and thus their function is analogous to that of arterioles in the vascular system. A simple cuboidal epithelium lines the bronchioles rather than the pseudostratified columnar epithelium that lines the bronchi (see fig. 17.4). Numerous **terminal bronchioles** connect to

Figure 17.14 An anterior view of the lower respiratory system showing the bronchial tree, alveoli, and lungs.

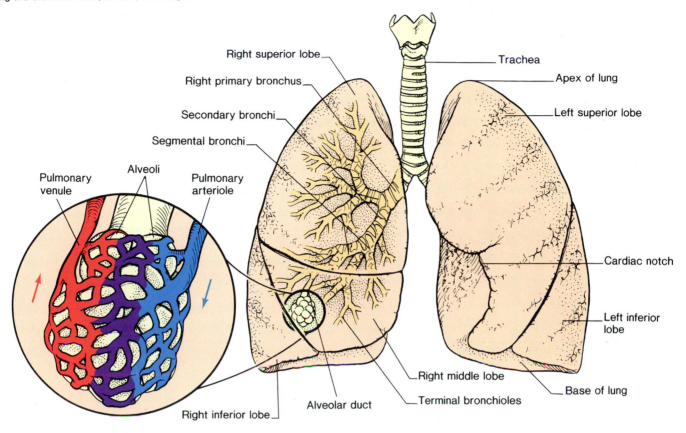

respiratory bronchioles that lead into **alveolar ducts,** and then into the **alveolar sacs** (see fig. 17.17). The conduction portion of the respiratory system ends at the terminal bronchioles and the respiratory portion begins at the respiratory bronchioles.

Asthma is an infectious or allergenic condition that involves the bronchi. During an asthmatic attack, there is a spasm of the smooth muscles in the respiratory bronchioles. Because of an absence of cartilage at this level, the air passageways constrict.

A fluoroscopic examination of the bronchi with a radiopaque medium is called *bronchography.* This technique enables the physician to visualize the bronchial tree on a bronchogram. A bronchogram of the lungs is shown in figure 17.15 and the change in the bronchial tree during respiration is diagrammed in figure 17.16.

1. List in order the types of epithelia through which inspired air would pass in going through the nasal cavity to the alveoli of the lungs. What is the function of each of these epithelia?
2. What are the functions of the nasal cavity?
3. Identify the structures that make up the nasal septum.
4. Describe the location of the nasopharynx and list the structures found in this cavity.
5. List the paired and unpaired cartilages of the larynx and describe the functions of the larynx.
6. Describe the structure of the conducting airways from the trachea to the terminal bronchioles.

Figure 17.15 An anteroposterior bronchogram of the lungs.

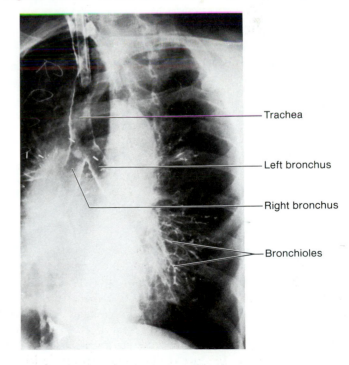

Figure 17.16 A diagram of the change in the bronchial tree during respiration. During inspiration, the bronchi lengthen and increase in diameter. During expiration, they return to their original length and diameter.

Expiration   Inspiration

Alveoli, Lungs, and Pleurae

Alveoli are the functional units of the lungs, where gas exchange occurs. Right and left lungs are separately contained in pleural membranes.

Objective 9. Describe the structure and function of the alveoli.

Objective 10. Describe the surface anatomy of the lungs in relation to the thorax.

Objective 11. Discuss the structural arrangement of the thoracic serous membranes and explain their functions.

Respiratory Sacs

The respiratory bronchioles branch into many **alveolar ducts** that lead directly into **alveolar sacs** and **alveoli** (fig. 17.17). The last three structures make up the *respiratory division* of the lungs. Gas exchange occurs across the walls of the tiny (0.25–0.50 mm in diameter) alveoli, the functional units of the respiratory system. The enormous number of these structures (about 350 million per lung) provides a very large surface area (60–80 square meters, or 760 square feet) for the diffusion of

alveolus: L. diminutive of *alveus*, cavity

Figure 17.17 The respiratory division of the respiratory system. The respiratory tubes end in minute alveoli, each of which is surrounded by an extensive capillary network.

Waldrop

Figure 17.18 A diagram showing the relationship between lung alveoli and pulmonary capillaries.

Figure 17.19 A scanning electron micrograph of lung tissue showing alveolar sacs and alveoli.

Alveolar sacs   Alveoli

gases. The diffusion rate is further increased by the fact that each alveolus is only one cell layer thick, so that the total air-blood barrier is only one alveolar cell with its basement membrane and one blood capillary cell across, or about 2 micrometers. This is an average distance because the type II alveolar cells are thicker than the type I alveolar cells (fig. 17.18). Type I alveolar cells permit diffusion and type II alveolar cells (septal cells) secrete *surfactant.*

Alveoli are polyhedral in shape and are usually clustered together, like the units of a honeycomb, in groups called alveolar sacs (fig. 17.19). These clusters usually occur at the ends of alveolar ducts. Individual alveoli also occur as separate outpouchings along the length of the alveolar ducts. Although the distance between each alveolar duct and its terminal alveoli is only about 0.5 mm, these units together compose most of the mass of the lungs.

Lungs

The **lungs** are large, spongy, paired organs within the thoracic cavity. They lie against the rib cage anteriorly and posteriorly and extend from the diaphragm to a point just above the clavicles. The lungs are separated from one another by the heart and other structures of the **mediastinum** *(me''de-ă-sti'num),* as shown in figure 17.20. The mediastinum is the area between the lungs. All structures of the respiratory system beyond the primary bronchi, including the bronchial tree and alveoli, are contained in the lungs.

Each lung presents four borders or surfaces that match the contour of the thoracic cavity. The **mediastinal (medial)**

surface of each lung is slightly concave and contains a vertical slit, the **hilum** *(hi'lum),* through which pulmonary vessels, nerves, and bronchi traverse. The inferior border of the lung, or **base,** is concave as it fits over the convex dome of the diaphragm. Anteriorly, the portion of the lungs that extends above the level of the clavicle is called the **apex,** or **cupola.** Finally, the broad, rounded surface in contact with the membranes covering the ribs is called the **costal surface.**

Although the right and left lungs are basically similar, there are some distinct differences. The left lung is somewhat smaller than the right and has a **cardiac notch** on its medial surface to accommodate the heart. The left lung is subdivided into a **superior lobe** and an **inferior lobe** by a single fissure. The right lung is subdivided by two fissures into three lobes: **superior, middle,** and **inferior lobes** (see fig. 17.14). Each lobe of the lung is divided into many small lobules, which in turn contain the alveoli. Lobular divisions of the lungs make up specific **bronchial segments.** The right lung contains 10 bronchial segments and the left lung contains 8 (fig. 17.21).

The lungs of a newborn are pink but may become discolored in an adult as a result of smoking or air pollution. Smoking not only discolors the lungs, it may also cause deterioration of the alveoli. *Emphysema (em''fĭ-se'mă)* and *lung cancer* are diseases that have been linked to smoking. If a person moves to a less polluted environment or gives up smoking, the lungs will get pinker and function more efficiently, unless they have been permanently damaged by disease.

Pleurae

Pleurae are serous membranes investing the lungs (figs. 17.20 and 17.22). The pleura *(ploo-r'ā)* of each lung is composed of both a visceral and parietal portion. The **visceral pleura** ad-

mediastinum: L. *mediastinus,* intermediate

hilum: L. *hilum,* a trifle (little significance)
cupola: L. *cupula,* diminutive of *cupa,* dome, tub
pleura: Gk. *pleura,* side or rib

UNIT VI

Figure 17.20 A cross section of the thoracic cavity showing the mediastinum and pleural membranes.

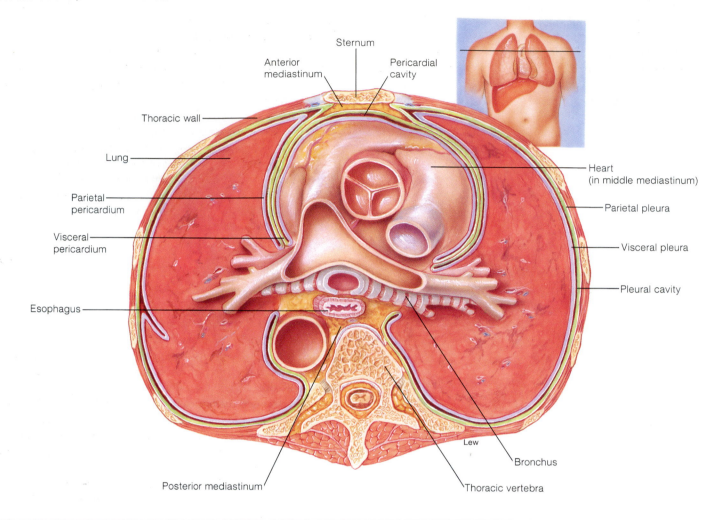

Sternum

Anterior mediastinum

Pericardial cavity

Thoracic wall

Lung

Parietal pericardium

Visceral pericardium

Esophagus

Heart (in middle mediastinum)

Parietal pleura

Visceral pleura

Pleural cavity

Lew

Bronchus

Posterior mediastinum

Thoracic vertebra

Figure 17.21 Lobes, lobules, and bronchopulmonary segments of the lungs.

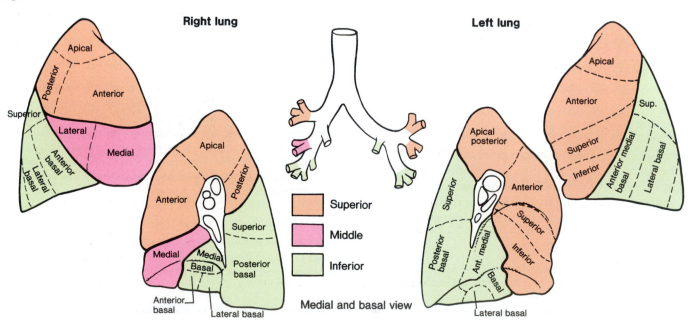

Right lung

Left lung

Apical

Posterior

Anterior

Superior

Lateral

Medial

Anterior basal

Lateral basal

Apical

Anterior

Posterior

Superior

Medial

Medial Basal

Posterior basal

Anterior basal

Lateral basal

Apical

Anterior

Sup.

Superior

Inferior

Anterior medial basal

Lateral basal

Apical posterior

Superior

Anterior

Posterior basal

Ant. medial

Superior

Inferior

Basal

Lateral basal

Superior

Middle

Inferior

Medial and basal view

heres to the outer surface of the lung and extends into each of the interlobar fissures. The **parietal pleura** lines the thoracic walls and the thoracic surface of the diaphragm. A continuation of the parietal pleura around the heart and between the lungs forms the boundary of the mediastinum. Between the visceral and parietal pleurae is a moistened space called the **pleural cavity.** An inferiorly extending reflection of the pleural layers around the roots of each lung is called the **pulmonary ligament.** The pulmonary ligaments help to support the lungs.

---

pulmonary: Gk. *pleumon*, lung

Figure 17.22   The position of the lungs and associated pleurae.

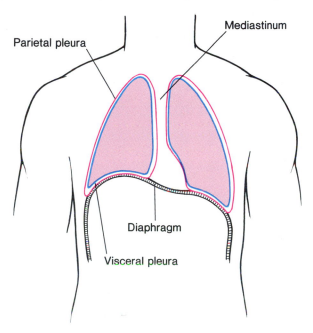

Mediastinum

Parietal pleura

Diaphragm

Visceral pleura

The arrangement of the pleurae serves three functions. First, a small amount of fluid within the pleural cavities acts as a lubricant to allow the lungs to slide along the chest wall. Second, the pressure in the pleural cavity is lower than the pressure in the lungs, which is needed for ventilation. The third function is the effective separation of the thoracic organs. Although the serous membrane is composed of simple squamous epithelium and fibrous connective tissue, it provides tight compartments for each major thoracic organ.

The thoracic cavity has four distinct compartments. A pleural cavity surrounds each lung, the heart is situated in the pericardial cavity, and the esophagus, thoracic duct, major vessels, various nerves, and portions of the respiratory tract are located in the mediastinum. This *compartmentalization* has protective value in that infections are usually confined to one compartment and damage to one organ usually will not involve another. *Pleurisy,* for example, which is an inflamed pleura, is generally confined to one side. A penetrating injury to one side, such as a knife wound, might cause one lung to collapse but not the other.

The moistened serous membranes of the visceral and parietal pleurae are normally flush against each other, so that the lungs are stuck to the thoracic wall in the same manner that two wet pieces of glass stick to each other. The pleural cavity (intrapleural space) between the two moistened membranes contains only a thin layer of fluid secreted by the pleural membranes. The pleural cavity in a healthy, living person is thus potential rather than real; it can become real only in abnormal situations when air enters the intrapleural space. Since the lungs normally remain in contact with the thoracic wall, they get larger and smaller along with the thoracic cavity during respiratory movements.

The functions of the major organs of the respiratory system along with a description of each is presented in table 17.1.

Table 17.1	Major Structures of the Respiratory System	
Structure	**Description**	**Function**
Nose	Jutting external portion that is part of the face, plus internal nasal cavity	Warms, moistens, and filters inhaled air as it is conducted to the pharynx
Paranasal sinuses	Air spaces in certain facial bones	Produce mucus; provide sound resonance; lighten the skull
Pharynx	Chamber connecting oral and nasal cavities to the larynx	Serves as passageway for air into the larynx and for food into the esophagus
Larynx	Voice box; short passageway that connects the pharynx to the trachea	Serves as passageway for air; produces sound; prevents foreign materials from entering the trachea
Trachea	Flexible, tubular connection between the larynx and bronchial tree	Serves as passageway for air; pseudostratified ciliated columnar epithelium cleanses the air
Bronchial tree	Bronchi and branching bronchioles in the lung; tubular connection between the trachea and alveoli	Serves as passageway for air; continued cleansing of air
Alveoli	Microscopic, membranous air sacs within the lungs	Functional units of respiration; site of gaseous exchange between the respiratory and circulatory systems
Lungs	Major organs of the respiratory system; located in pleural cavities in the thorax	Contain bronchial trees, alveoli, and associated pulmonary vessels
Pleurae	Serous membranes covering the lungs and lining the thoracic cavity	Compartmentalize, protect, and lubricate the lungs

UNIT VI

Figure 17.23 The mechanics of pulmonary ventilation.

At rest

Atmospheric pressure (760 mmHg)

Inspiration

Intrapulmonic pressure (760 mmHg)

Intrapleural pressure (756 mmHg)

(a)

Diaphragm

758 mmHg

754 mmHg

(b)

Expiration

763 mmHg

756 mmHg

(c)

Creek

---

1. Describe the structure of the respiratory division of the lungs and explain how this structure aids in gas exchange.
2. Compare the structure of the right and left lung.
3. Describe the structure and location of the mediastinum and list the organs it contains.
4. Describe the location of the visceral pleura, parietal pleura, and pleural cavity. Comment on the functional significance of the pleural membranes.

Mechanics of Breathing

Normal, quiet inspiration results from muscle contraction and normal expiration from muscle relaxation and elastic recoil. These actions can be forced by contractions of the accessory respiratory muscles. The amount of air inspired and expired can be measured in a number of ways to test pulmonary function.

Objective 12. Identify and describe the actions of the muscles involved in both quiet and forced inspiration.

Objective 13. Describe how quiet expiration occurs and identify and describe the actions of the muscles involved in forced expiration.

Objective 14. List the various lung volumes and capacities and describe how pulmonary function tests help diagnose lung disorders.

Breathing, or **pulmonary ventilation**, entails the movement of air through the respiratory passageways into the alveoli of the lungs, followed by a reversal of air movement out of the lungs. It consists of two phases, inspiration and expiration. Inspiration (inhaling) and expiration (exhaling) are accomplished by alternately increasing and decreasing the volume of the thoracic cavity (fig. 17.23). Pulmonary ventilation is accomplished because pressure gradients are created. Breathing in takes place when the air pressure within the lungs is lower than the atmospheric pressure; breathing out takes place when the air pressure within the lungs is greater than the atmospheric pressure.

Pressure gradients change as the size of the thoracic cavity changes. The thorax, therefore, must be sufficiently rigid to protect the vital organs it contains, provide extensive attachment surface for many short, powerful muscles, and be flexible enough to function as a bellows during the ventilatory cycle. The rigidity and the extensive surfaces for muscle attachment are provided by the bony composition of the rib cage. The rib cage has pliability because the ribs are separated from one another and because most ribs (upper 10 of the 12 pairs) are at-

UNIT VI

Figure 17.24 A change in lung volume, as shown by radiographs, during expiration (*a*) and inspiration (*b*). The increase in lung volume during full inspiration is shown by comparison with the lung volume in full expiration (dashed lines).

tached to the sternum by resilient costal cartilage. The vertebral attachment likewise allows for considerable mobility. The structure of the rib cage and associated cartilage provides continuous elastic tension, so that an expanded thorax will return passively to its resting position when relaxed.

Inspiration

The size of the thoracic cavity increases with inspiration (fig. 17.24). This increase in dimension normally takes place in three directions: anteroposteriorly, laterally, and vertically.

During normal inspiration, the muscles of importance are the *diaphragm,* the *external intercostal muscles,* and the interchondral portion of the *internal intercostal muscles* (fig. 17.25). Downward contraction of the dome-shaped diaphragm vertically increases thoracic volume. A simultaneous contraction of the external intercostal muscles and interchondral portion of the internal intercostal muscles increases the lateral and anteroposterior dimensions of the thorax.

The *scalenus* and *sternocleidomastoid muscles* are involved in deep inspiration or forced breathing. When these muscles are contracted, the ribs are elevated in an anteroposterior direction. At the same time, the upper rib cage is stabilized so that the intercostal muscles become more effective. The expanded thoracic cavity decreases the air pressure within the pleural spaces to below that of the atmosphere. It is this pressure difference that causes the lungs to become inflated.

Expiration

For the most part, expiration is a passive process that occurs as the muscles of inspiration are relaxed and the rib cage rebounds to its original position. Even the lungs recoil during expiration as the alveoli draw together to eliminate gaseous wastes. Lowering of the surface tension in the alveoli, which brings on elastic recoiling, is due to a lipoprotein substance called *surfactant* that lines the walls of the alveoli. Surfactant is extremely important not only in causing the alveoli to recoil during expiration, but

also in equalizing the overall surface tension as the alveoli expand and contract. A deficiency in surfactant in premature infants can cause *respiratory distress syndrome* (RDS) or, as it is commonly called, *hyaline membrane disease.*

Even under normal conditions, the first breath of life is a difficult one because the newborn must overcome large surface tension forces in order to inflate its partially collapsed alveoli. The transpulmonary pressure required for the first breath is 15 to 20 times that required for subsequent breaths, and an infant with *respiratory distress syndrome* must duplicate this effort with every breath. Fortunately, many babies with this condition can be saved by mechanical ventilators that keep them alive long enough for their lungs to mature and manufacture sufficient surfactant.

During forced expiration, such as coughing or sneezing, contraction of the interosseous portion of the internal intercostal muscles causes the rib cage to be depressed. The *abdominal muscles* may also aid expiration because, when contracted, they force abdominal organs up against the diaphragm and further decrease the volume of the thorax. Thus, intrapulmonary pressure can rise to 20 or 30 mmHg above the atmospheric pressure. The events that occur during inspiration and expiration are summarized in table 17.2.

Respiratory Air Volumes

The respiratory system is somewhat inefficient because the air enters and exits at the same place, through either the nose or the mouth. Consequently, there is an incomplete exchange of gas during each ventilatory cycle, and approximately five-sixths of the air present in the lungs still remains when the next inspiration begins.

Clinically speaking, it is important to know the amount of air that is breathed in a given time as well as the degree of difficulty in breathing. The amount of air exchanged during pulmonary ventilation varies from person to person according to age, gender, activity level, and general health. Respiratory air

diaphragm: Gk. *dia,* across; *phragma,* fence

Figure 17.25 The muscles of respiration. The principal muscles of inspiration are shown on the right side of the trunk and the principal muscles of forced expiration are shown on the left side. For the most part, expiration is passive.

Muscles of inspiration

Muscles of expiration

Sternocleidomastoid
Scalenes
External intercostals
Internal intercostals (interchondral part)
Diaphragm
Creek
Internal intercostals (excluding interchondral part)
External abdominal oblique
Internal abdominal oblique
Transversus abdominis
Rectus abdominis

Table 17.2	Events of Inspiration and Expiration During a Pulmonary Ventilatory Cycle*
Nerve Stimulus	**Event**
Inspiration Phrenic nerves	The diaphragm contracts inferiorly, increasing the volume of the thorax. The diaphragm is the principal muscle involved in quiet inspiration.
Intercostal nerves	Contraction of the external intercostal muscles and the interchondral portion of the internal intercostal muscles elevates the ribs, thus increasing the capacity of the thoracic cavity.
Accessory, cervical, and thoracic nerves	Forced inspiration is accomplished through contraction of the sternocleidomastoid and scalenus muscles, which increases the dimension of the thoracic cavity anteroposteriorly. Pulmonary ventilation during forced inspiration usually occurs through the mouth rather than the nose.
	As the dimension of the thoracic cavity is increased, the pressure within the pleural cavities decreases and the lungs inflate because of the greater atmospheric pressure.
Expiration	Nerve stimuli to the inspiratory muscles cease and the muscles relax.
	The rib cage and lungs recoil as air is forced out of the lungs due to the increased pressure.
Intercostal and lower spinal nerves	Forced expiration occurs when the interosseus portion of the internal intercostal and abdominal muscles are contracted.

*Some of the events during inspiration and expiration may occur simultaneously.

volumes are measured with a *spirometer* (fig. 17.26). Any ventilatory abnormalities can then be compared to what is accepted as normal. The normal adult respiratory air volumes are presented in table 17.3 and figure 17.27.

People with pulmonary disorders frequently complain of *dyspnea (disp'ne-ă)*, which is a subjective feeling of shortness of breath. Dyspnea may occur even when ventilation is normal, however, and may not occur even when the total volume of air movement is very high as in exercise. Some of the terms used to define ventilation are defined in table 17.4.

Nonrespiratory Air Movements

Air movements through the respiratory system other than for pulmonary ventilation are termed *nonrespiratory movements*. Such movements are part of emotional displays such as laughing, sighing, crying, or yawning, or they may function to expel foreign matter from the respiratory tract, as in coughing and sneezing. Generally, nonrespiratory movements are reflexive. Some of them, however, can be voluntarily initiated. These types of air movements and the reflexive mechanisms involved are summarized in table 17.5.

dyspnea: Gk. *dys*, bad; *pnoe*, breathing

UNIT VI

Figure 17.26 A spirometer. With the exception of the residual volume, which is measured using special techniques, this instrument can determine respiratory air volumes.

Table 17.3	Respiratory Air Volumes for a Healthy Adult Male	
Volume	**Quantity of Air**	**Description**
Tidal volume (TV)	500 cc	Volume moved in or out of the lungs during quiet breathing
Inspiratory reserve volume (IRV)	3,000 cc	Volume that can be inhaled during forced breathing in addition to tidal volume
Expiratory reserve volume (ERV)	1,000 cc	Volume that can be exhaled during forced breathing in addition to tidal volume
Vital capacity (VC)	4,500 cc	Maximum volume of air that can be exhaled after taking the deepest breath possible: VC = TV + IRV + ERV
Residual volume (RV)	1,500 cc	Volume that remains in the lungs at all times
Total lung capacity (TLC)	6,000 cc	Total volume of air that the lungs can hold: TLC = VC + RV

From John W. Hole, Jr., *Human Anatomy and Physiology*, 6th ed. Copyright © 1993 Wm. C. Brown Communications, Inc., Dubuque, Iowa. All Rights Reserved. Reprinted by permission.

Figure 17.27 Respiratory air volumes.

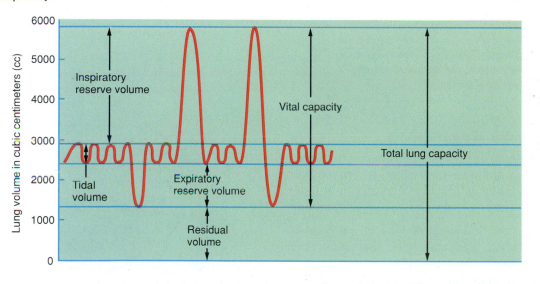

Table 17.4	Ventilation Terminology		
Term	**Definition**	**Term**	**Definition**
Air spaces	Alveolar ducts, alveolar sacs, and alveoli	Hyperventilation	An alveolar ventilation that is excessive in relation to metabolic rate; results in abnormally low alveolar CO_2
Airways	Structures that conduct air from the mouth and nose to the respiratory bronchioles	Hypoventilation	An alveolar ventilation that is low in relation to metabolic rate; results in abnormally high alveolar CO_2
Alveolar ventilation	Removal and replacement of gas in alveoli; equal to the tidal volume minus the volume of dead space	Orthopnea	Inability to breathe horizontally
Anatomical dead space	Volume of the conducting airways to the zone where gas exchange occurs	Physiological dead space	Combination of anatomical dead space and underventilated alveoli that do not contribute normally to blood-gas exchange
Apnea	Cessation of breathing		
Dyspnea	Unpleasant subjective feeling of difficult or labored breathing	Pneumothorax	Presence of gas in the pleural space (the space between the visceral and parietal pleural membranes) causing lung collapse
Eupnea	Normal, comfortable breathing at rest	Torr	Unit of pressure very nearly equal to the millimeter of mercury (760 mmHg = 760 torr)

From Stuart Ira Fox, *Human Physiology*, 4th ed. Copyright © 1993 Wm. C. Brown Communications, Inc., Dubuque, Iowa. All Rights Reserved. Reprinted by permission.

UNIT VI

Table 17.5   *Nonrespiratory Air Movements*

Air Movement	Mechanism	Comments
Coughing	Deep inspiration followed by a closure of the glottis. Suddenly the glottis is opened, and a forceful expiration of air is sent through the upper respiratory tract.	Reflexive or voluntary. Stimulus may be foreign material irritating larynx or trachea.
Sneezing	Similar to a cough, except that the forceful expired air is directed primarily through the nasal cavity. The eyelids are reflexively closed during a sneeze.	Reflexive response to irritating stimulus of the nasal mucosa. Sneezing clears the upper respiratory passages.
Sighing	Deep, prolonged inspiration followed by a rapid, forceful expiration.	Reflexive or voluntary, usually in response to boredom or sadness.
Yawning	Deep inspiration through a widely opened mouth. The inspired air is usually held for a short period before sudden expiration.	Usually reflexive in response to drowsiness, fatigue, or boredom, but exact stimulus-receptor cause is unknown.
Laughing	Deep inspiration followed by a rapid convulsive expiration. Air movements are accompanied by expressive facial distortions.	Reflective or voluntary to express emotions.
Crying	Similar to laughing, but glottis remains open during entire expiration and different facial muscles are involved.	Somewhat reflexive but under voluntary control.
Hiccuping	Spasmodic contraction of the diaphragm while the glottis is closed, producing a sharp inspiratory sound.	Reflexive and serves no known function.

---

1. Describe the actions of the diaphragm and intercostal muscles during relaxed inspiration.
2. Describe how forced inspiration and forced expiration are produced.
3. Define the terms *tidal volume* and *vital capacity*.
4. Indicate the respiratory volumes being used during a sneeze, a deep inspiration prior to jumping into a swimming pool, maximum ventilation while running, and quiet breathing while sleeping.

Regulation of Breathing

The rhythm of breathing is controlled by centers in the brain stem. These centers are influenced by higher brain function and regulated by sensory information that makes breathing responsive to the changing respiratory needs of the body.

Objective 15. Describe the functions of the pneumotaxic, apneustic, and rhythmicity centers in the brain stem.

Objective 16. Identify the chemoreceptors and describe their pathway of innervation.

Pulmonary ventilation is primarily an involuntary, rhythmic action so effective that it continues to function even when a person is unconscious. In order for the neural control center of respiration to function effectively, it must possess monitoring, stimulating, and inhibiting properties so that the body can respond appropriately to increased or decreased metabolic needs. In addition, the center must be connected to the cerebrum to receive the voluntary impulses from a person who wants to change the rate of respiration.

The three **respiratory centers** of the brain are the rhythmicity, the apneustic, and the pneumotaxic areas (fig. 17.28). The **rhythmicity area** is located in the medulla oblongata and contains two aggregations of nerve cell bodies that form the **inspiratory** and **expiratory portions.** Nerve impulses from

Figure 17.28 Approximate locations of the brain stem respiratory centers.

Midbrain

Pons

Brain stem respiratory centers — Pneumotaxic area / Apneustic area / Rhythmicity area

Medulla oblongata

the inspiratory portion travel through the phrenic and intercostal nerves to stimulate the diaphragm and intercostal muscles. Impulses from the expiratory portion stimulate the muscles of expiration. These two portions act reciprocally; that is, when one is stimulated, the other is inhibited. Stretch receptors in the visceral pleura of the lungs provide feedback through the vagus nerves to stimulate the expiratory portion.

The **apneustic** and **pneumotaxic areas** are located in the pons of the brain. These areas function to modify and control

Figure 17.29 The peripheral chemoreceptors (aortic and carotid bodies) regulate the brain stem respiratory centers by means of sensory nerve stimulation.

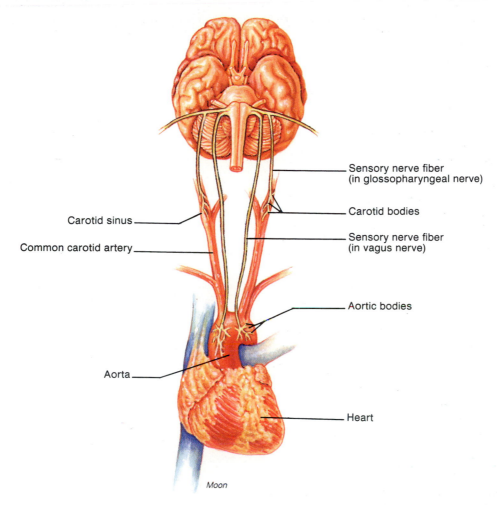

Sensory nerve fiber
(in glossopharyngeal nerve)

Carotid bodies

Sensory nerve fiber
(in vagus nerve)

Aortic bodies

Carotid sinus

Common carotid artery

Aorta

Heart

Moon

the activity of the rhythmicity area. The apneustic center promotes inspiration and the pneumotaxic center inhibits the activity of inspiratory neurons.

The brain stem respiratory centers produce rhythmic breathing even in the absence of other neural input. This intrinsic respiratory pattern, however, is modified by input from higher brain centers and by input from receptors sensitive to the chemical composition of the blood. The influence of higher brain centers is evidenced by the fact that you can voluntarily hyperventilate and hypoventilate (as in breath holding). The inability to hold your breath for more than a short period is due to reflex breathing in response to input from the chemoreceptors *(ke''mo-re-cep'tors)*.

There are two groups of chemoreceptors that respond to changes in blood chemistry. These are the *central chemoreceptors* in the medulla oblongata and the *peripheral chemoreceptors*. The peripheral chemoreceptors include the **aortic bodies,** located

in the aortic arch, and the **carotid bodies,** located at the junctions of the internal and external carotid arteries (fig. 17.29). These peripheral chemoreceptors control breathing indirectly via sensory neurons to the medulla oblongata. The aortic bodies send sensory information to the medulla oblongata in the vagus (tenth cranial) nerves; the carotid bodies stimulate sensory fibers in the glossopharyngeal (ninth cranial) nerve.

1. Identify the two structures within the brain where the three respiratory areas are located. Which of these three areas is responsible for autonomic rhythmic breathing?
2. Identify the locations of the peripheral chemoreceptors and the two paired cranial nerves that carry sensory impulses from these sites to the respiratory centers within the brain stem.

Developmental Exposition

The Respiratory System

Exhibit I The development of the upper respiratory system. (*a*) An anterior view of the developing head of an embryo at 4 weeks showing the position of a transverse cut depicted in (*a₁*), (*a₂*), and (*a₃*). (*a₂*) Development at 5 weeks and (*a₃*) at 5½ weeks. (*b*) An anterior view of the developing head of an embryo at 6 weeks showing the position of a sagittal cut depicted in (*b₁*), (*b₂*), (*b₃*), and (*b₄*) at 14 weeks.

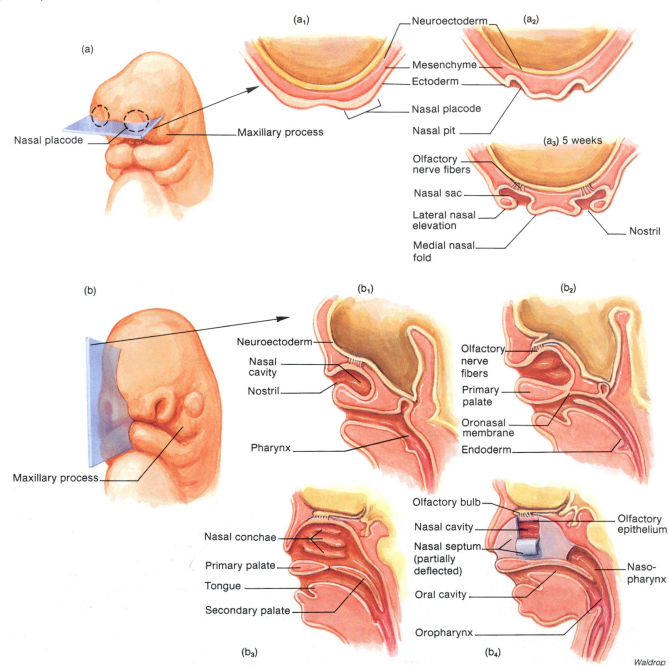

Waldrop

Explanation

The development of the respiratory system is initiated early in embryonic development and involves both ectoderm and endoderm. Although the structures of the upper respiratory system (nose and pharynx) and those of the lower (trachea and lungs) develop simultaneously, we will consider each system separately because of the different germ layers involved.

Development of the Upper Respiratory System

Cephalization is the evolutionary tendency toward structural and functional differentiation of the cephalic, or head, end of an organism from the rest of the body. In humans, cephalization is apparent early in development. One of the initial events is the

cephalization: Gk. *kephale*, head

Exhibit II The development of the lower respiratory tract. Developments in (a) and (b) occur during the first month. Those in (c), (d), and (e) occur during the second month.

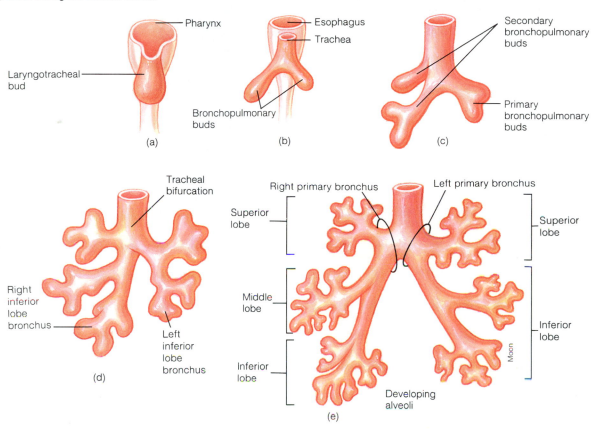

(a) — Pharynx, Laryngotracheal bud

(b) — Esophagus, Trachea, Bronchopulmonary buds

(c) — Secondary bronchopulmonary buds, Primary bronchopulmonary buds

(d) — Tracheal bifurcation, Right inferior lobe bronchus, Left inferior lobe bronchus

(e) — Right primary bronchus, Left primary bronchus, Superior lobe, Middle lobe, Inferior lobe, Superior lobe, Inferior lobe, Developing alveoli

formation of the nasal cavity at 3½ to 4 weeks of embryonic life. A region of thickened ectoderm called the **olfactory** (nasal) **placode** *(plak′od)* appears on the front and inferior part of the head (exhibit I). The placode invaginates to form the **olfactory pit,** which extends posteriorly to connect with the **foregut.** The foregut, derived of endoderm, later develops into the pharynx.

The mouth, or oral cavity, develops at the same time as the nasal cavity, and for a short time there is a thin **oronasal** *(or″o-na′zal)* **membrane** separating the two cavities. This membrane ruptures during the seventh week, and a single large **oronasal cavity** forms. Shortly after, tissue plates of mesoderm begin to grow horizontally across the cavity. At approximately the same time, a vertical plate develops inferiorly from the roof of the nasal cavity. These plates have completed their formation by 3 months of development. The vertical plate forms the nasal septum, and the horizontal plates form the hard palate.

 A *cleft palate* forms when the horizontal plates fail to meet in the midline. This condition can be corrected surgically. The more immediate and serious problem facing an infant with a cleft palate is that it may be unable to create enough suction to nurse properly.

Development of the Lower Respiratory System

The lower respiratory system begins as a diverticulum *(di″ver-tik′yoo-lum),* or outpouching, from the ventral surface of endoderm along the lower pharyngeal region (exhibit II). This diverticulum forms during the fourth week of development and is referred to as the **laryngotracheal** *(lă-ring″go-tra′ke-al)* **bud.** As the bud grows, the proximal portion forms the trachea and the distal portion bifurcates (splits) into a right and left bronchus.

The buds continue to elongate and split until all the tubular network within the lower respiratory tract is formed (exhibit II). As the terminal portion forms air sacs, called **alveoli,** at about 8 weeks of development, the supporting lung tissue begins to form. The complete structure of the lungs, however, is not fully developed until about 26 weeks of fetal development, so premature infants born prior to this time require special artificial respiratory equipment to live.

UNIT VI

Clinical Considerations

The respiratory system is particularly vulnerable to inflammatory and infectious diseases simply because many pathogens are airborne, humans are highly social, and the warm, moistened environment along the respiratory tract allows pathogens to thrive. Injury and trauma are likewise frequent problems. Protruding noses are subject to fractures, the large, spongy lungs are easily penetrated by broken ribs, and portions of the respiratory tract are liable to become occluded because they also have a digestive function.

Developmental Problems

Birth defects, inherited disorders, and premature births commonly cause problems in the respiratory system of infants. A **cleft palate** is a developmental deformity of the hard palate of the mouth in which an opening persists between the oral and nasal cavities, making it difficult, if not impossible, for an infant to nurse. A cleft palate may be hereditary or a complication of some disease (e.g., German measles) contracted by the mother during pregnancy. A **cleft lip** is a genetically based developmental disorder in which the two sides of the upper lip fail to fuse. Cleft palates and cleft lips can be treated very effectively with cosmetic surgery.

As mentioned earlier in this chapter, **hyaline membrane disease** is a fairly common respiratory disorder that accounts for about one-third of neonatal deaths. This condition results from the deficient production of surfactant. **Cystic fibrosis** (sis'tik fi-bro'sis) is a genetic disorder that affects the respiratory system, as well as other systems of the body, and accounts for approximately one childhood death in twenty. The effect of this disease on the respiratory system is usually a persistent inflammation and infection of the respiratory tract.

Alveoli of the lungs are not developed sufficiently to sustain life until after week 20 of gestation. Thus, extrauterine life prior to that time is extremely difficult even with life-supporting devices.

Trauma or Injury

Humans are especially susceptible to **epistaxes** (ep''i-stak'sēz) (nosebleeds) because the nose is located where it can be easily bumped and because the nasal mucosa has extensive vascularity for warming the inspired air. Such conditions or diseases as high blood pressure or leukemia may cause epistaxes.

When air enters the pleural cavity surrounding either lung and causes the lung to collapse, the condition is referred to as a **pneumothorax** (fig. 17.30). A pneumothorax can result from an external injury, such as a stabbing, bullet wound, or penetrating fractured rib, or it can be the result of internal conditions. A severely diseased lung, as in emphysema, can create a pneumothorax as the wall of the lung deteriorates along with the visceral pleura and permits air to enter the pleural cavity.

Choking on a foreign object such as aspirated food is a common serious trauma to the respiratory system. More than eight Americans choke to death each day on food lodged in their trachea. A simple process termed the **abdominal thrust**

Figure 17.30 A pneumothorax of the right lung. The right side of the thorax appears uniformly dark because it is filled with air; the spaces between the ribs are also greater than those on the left due to release from the elastic tension of the lungs. The left lung appears denser (less dark) because of shunting of blood from the right to the left lung.

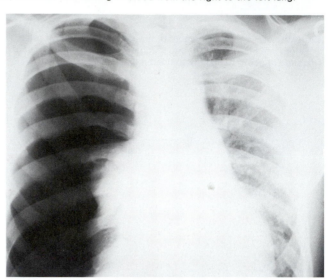

(Heimlich) **maneuver** can save the life of a person who is choking. The abdominal thrust maneuver is performed as follows:

A. If the victim is standing or sitting:
   1. Stand behind the victim or the victim's chair and wrap your arms around his or her waist.
   2. Grasp your fist with your other hand and place the fist against the victim's abdomen, slightly above the navel and below the rib cage.
   3. Press your fist into the victim's abdomen with a quick upward thrust.
   4. Repeat several times if necessary.
B. If the victim is lying down:
   1. Position the victim on his or her back.
   2. Face the victim, and kneel on his or her hips.
   3. With one of your hands on top of the other, place the heel of your bottom hand on the abdomen, slightly above the navel and below the rib cage.
   4. Press into the victim's abdomen with a quick upward thrust.
   5. Repeat several times if necessary.

If you are alone and choking, use whatever is available to apply force just below your diaphragm. Press into a table or a sink, or use your own fist.

People saved from drowning and victims of shock frequently experience apnea (cessation of breathing) and will soon die if not revived by artificial respiration. The accepted treatment for reviving a person who has stopped breathing is illustrated in figure 17.31.

UNIT VI

Figure 17.31 Artificial respiration.

Mouth-to-Mouth Method

1. **Check for unresponsiveness.**
Gently shake the victim and shout, "Are you okay?" If no response, get the attention of someone who can phone for help. Make sure that the victim is on his or her back.

2. **Open the airway.**
Tilt the victim's head back by pushing on his or her forehead with your hand and lifting the chin with your fingers under his or her jaw. This will open the airway by moving the tongue away from the back of the victim's throat.

3. **Check for breathing.**
Put your ear close to the victim's face to listen and feel for any return of air. At the same time, look to see if there is chest movement. Check for breathing for about 5 seconds.

4. **If no breathing, give two full breaths.**
While maintaining the victim in the head-tilt position, pinch his or her nose to close off the nasal passageway. Take a deep breath, then seal your mouth around the victim's mouth and give two full breaths. (After the first breath, raise your head slightly to inhale quickly and then give the second breath.)

5. **Check for pulse.**
While maintaining head tilt, feel for a carotid pulse for 5 to 10 seconds on the side of the victim's neck.

6. **Continue rescue breathing.**
With the victim in the head-tilt position and his or her nostrils pinched, give one breath every 5 seconds. Observe for signs of breathing between breaths. For an infant, give one gentle puff every 3 seconds.

7. **Recheck for pulse.**
Feel for a carotid pulse at 1-minute intervals. If the victim has a pulse but is not breathing, continue rescue breathing.

Mouth-to-Nose Method

1. **Open the airway.**
Place the victim in the head-tilt position as described above.

2. **Blow into the victim's nose.**
Using the same sequence described above, blow into the victim's nose while holding his or her mouth closed.

3. **Feel and observe for breathing.**
With the victim's mouth held open, detect for breathing between giving forced breaths.

Common Respiratory Disorders

A cough is the most common symptom of respiratory disorders. Acute problems may be accompanied by dyspnea or wheezing. Respiratory or circulatory problems may cause **cyanosis,** a blue discoloration of the skin due to blood with a low oxygen content.

Although the **common cold** is the most widespread of all respiratory diseases, there is still no cure for this ailment—only medications that offer symptomatic relief. Colds occur repeatedly because acquired immunity for one virus does not protect against other viruses. Cold viruses generally incite acute inflammation in the respiratory mucosa, causing flow of mucus, sometimes accompanied by fever and/or headache.

Nearly all of the structures and regions of the respiratory passageways can become infected and inflamed. **Influenza** is a viral disease that causes an inflammatory condition of the upper respiratory tract. Influenza can be epidemic, but fortunately vaccines are available. **Sinusitis** is an inflammation of the paranasal sinuses. Sinusitis can be quite painful if the drainage ducts from the sinuses into the nasal cavity become blocked. **Tonsillitis** may involve any or all of the tonsils and frequently follows other lingering diseases of the oral or pharyngeal regions. Chronic tonsillitis generally requires a tonsillectomy. **Laryngitis** is inflammation of the larynx, which often produces a hoarse voice and limits the ability to talk. **Tracheobronchitis** and **bronchitis** are infections of the regions for which they are named. Severe inflammation in these areas can cause smaller respiratory tubules to collapse, blocking the passage of air.

Diseases of the lungs are likewise common and usually serious. **Pneumonia** is an acute infection and inflammation of lung tissue accompanied by exudation (accumulation of fluid). It is usually caused by bacteria, most commonly the pneumococcus bacterium. Viral pneumonia is caused by a number of different viruses. **Tuberculosis** is an inflammatory disease of the lungs contracted by inhaling air sneezed or coughed by someone who is carrying active tuberculosis bacteria. Tuberculosis softens lung tissue, which eventually becomes ulcerated. **Asthma** is a disease that affects persons who are allergic to certain inhaled antigens, such as pollen. Asthma causes a swelling and blocking of lower respiratory tubes, often accompanied by the formation of mucous plugs. **Pleurisy** is an inflammation of the pleura and is usually secondary to some other respiratory disease. Inspiration may become painful, and fluid may collect within the pleural space. **Emphysema** is a disease that causes the breakdown of the alveoli, thus increasing the size of air spaces and decreasing the surface area (fig. 17.32). It is a frequent cause of death among heavy cigarette smokers.

Cancer in the respiratory system is known to be caused by the repeated inhalation of irritating substances, such as cigarette smoke. Cancers of the lip, larynx, and lungs are especially common in smokers over the age of 50.

cyanosis: Gk. *kyanosis,* dark-blue color
influenza: L. *influentia,* a flowing in
tuberculosis: L. *tuberculum,* diminutive of *tuber,* swelling
asthma: Gk. *asthma,* panting
emphysema: Gk. *emphysan,* blow up, inflate

Figure 17.32 Photomicrographs of tissue (a) from a normal lung and (b) from the lung of a person with emphysema. In emphysema, lung tissue is destroyed, resulting in the presence of fewer and larger alveoli.

(a)

(b)

Disorders of Respiratory Control

A variety of disease processes can result in cessation of breathing during sleep, or *sleep apnea.* **Sudden infant death syndrome (SIDS)** is an especially tragic form of sleep apnea that claims the lives of about 10,000 babies annually in the United States. Victims of this condition are apparently healthy 2-to-5-month-old babies who die in their sleep without apparent reason—hence, the layperson's term, *crib death.* These deaths seem to be caused by failure of the respiratory control mechanisms in the brain stem and/or by failure of the carotid bodies to be stimulated by reduced arterial oxygen.

Abnormal breathing patterns often appear prior to death from brain damage or heart disease. The most common of these abnormal patterns is **Cheyne–Stokes breathing,** in which the depth of breathing progressively increases and then progressively decreases. These cycles of increasing and decreasing tidal volumes may be followed by periods of apnea of varying durations. Cheyne–Stokes breathing may be caused by neurological damage or by insufficient oxygen delivery to the brain. The latter may result from heart disease or from a brain tumor that diverts a large part of the vascular supply from the respiratory centers.

Cheyne–Stokes breathing: from John Cheyne, British physician, 1777–1836, and William Stokes, Irish physician, 1804–78.

UNIT VI

Internal Affairs

The Relational Workings of the Body Systems

Respiratory System

Integumentary System
- Nasal hairs and mucus protect the entry of respiratory tract
- Provides O_2 and eliminates CO_2

Skeletal System
- Protects the lungs within the rib cage and the upper respiratory tract within the skull
- Provides O_2 and eliminates CO_2

Muscular System
- Provides the respiratory muscles necessary for ventilation of the lungs
- Provides for sound production through activity of vocal cords, tongue, and lips
- Provides O_2 and eliminates CO_2

Nervous System
- Monitors and regulates respiratory rates
- Provides O_2 and eliminates CO_2

Endocrine System
- Dilates the bronchioles through the production of epinephrine
- Provides O_2 and eliminates CO_2

Circulatory System
- Transports O_2 and CO_2 within the blood
- Provides O_2 and eliminates CO_2

Lymphatic System
- Protects against pathogen invasion
- Prevents edema (retention of interstitial fluid)
- Provides O_2 and eliminates CO_2
- Contains immune cells within tonsils

Digestive System
- Provides nutrients for growth and maintenance of respiratory system
- Provides O_2 and eliminates CO_2

Urinary System
- Excretes metabolic wastes
- Provides O_2 and eliminates CO_2

Reproductive System
- Promotes changes in respiratory rate and depth during sexual arousal
- Provides O_2 and eliminates CO_2

UNIT VI

CLINICAL CASE STUDY ANSWER

The ice pick traversed the parietal pleura, the visceral pleura, and then entered the airway, at least at the alveolar-terminal bronchiole level, but possibly at larger airways. This allowed inspired air to escape from the airway into the pleural space. The air would have taken the following route: pharynx → larynx → trachea → left primary bronchus → secondary bronchus (of left upper lobe) → apical seg- mental (tertiary) bronchus → bronchioles → air escapes through laceration into pleural space. This condition is treated by inserting a tube (tube thoracostomy) into the pleural space to allow suction evacuation of air and blood resulting in re-expansion of the lung. The laceration usually seals over within one to two days. Persistent bleeding may necessitate thoracotomy for repair.

CHAPTER SUMMARY

Introduction to the Respiratory System (pp. 568–569)

1. Respiration refers not only to the act of breathing, but also to the exchange of gases between the atmosphere, the blood, and individual cells.
2. In order for the respiratory system to be functional, the respiratory membranes must be moist, thin-walled, highly vascular, and differentially permeable.
3. The functions of the respiratory system are gaseous exchange, sound production, assistance in abdominal compression, and coughing and sneezing.

Conducting Passages (pp. 569–578)

1. The nose is supported by nasal bones and cartilages.
   (a) The nasal epithelium warms, moistens, and cleanses the inspired air.
   (b) Olfactory epithelium is associated with the sense of smell and the nasal cavity acts as a resonating chamber for the voice.
2. The paranasal sinuses are found in the maxillary, frontal, sphenoid, and ethmoid bones.
   (a) These sinuses lighten the skull and are lined with mucus-secreting goblet cells.
   (b) Sinusitis is an inflammation of one or more of these sinuses.
3. The pharynx is a funnel-shaped passageway that connects the oral and nasal cavities with the larynx.
   (a) The nasopharynx, connected by the auditory tubes to the middle-ear cavities, contains the pharyngeal tonsils, or adenoids.
   (b) The oropharynx is the middle portion, extending from the soft palate to the level of the hyoid bone, which contains the palatine and lingual tonsils.
   (c) The laryngopharynx extends from the hyoid bone to the larynx.

4. The larynx is composed of a number of cartilages that keep the passageway to the trachea open during breathing and close the respiratory passageway during swallowing.
   (a) The epiglottis is a spoon-shaped structure that aids in closing the laryngeal opening, or glottis, during swallowing.
   (b) The vocal cords in the larynx are controlled by intrinsic muscles and are used in sound production.
5. The trachea is a rigid tube, supported by rings of cartilage, that leads from the larynx to the bronchial tree.
6. The bronchial tree includes a right and left primary bronchus, which divides to produce secondary bronchi, tertiary bronchi, and bronchioles; the conducting division ends with the respiratory bronchioles, which connect to the alveoli.

Alveoli, Lungs, and Pleurae (pp. 578–582)

1. Alveoli are the functional units of the lungs, where gas exchange occurs; they are numerous, small, thin-walled air sacs.
2. The right and left lungs are separated by the mediastinum. Each lung is divided into lobes and lobules.
   (a) The right lung is subdivided by two fissures into superior, middle, and inferior lobes.
   (b) The left lung is subdivided into a superior lobe and an inferior lobe by a single fissure.
3. The lungs are covered by visceral pleura and the thoracic cavity is lined by a parietal pleura.
   (a) A potential space between these two pleural membranes is called the pleural cavity.
   (b) The pleural membranes package each lung separately and exclude the structures located in the mediastinum.

Mechanics of Breathing (pp. 582–586)

1. Quiet (unforced) inspiration is due to contraction of the diaphragm and certain intercostal muscles. Forced inspiration is aided by the scalenus, pectoralis minor, and sternocleidomastoid muscles.
2. Quiet expiration is produced by relaxation of the respiratory muscles and elastic recoil of the lungs and thorax. Forced expiration is aided by certain intercostal muscles and the abdominal muscles.
3. Among the air volumes exchanged in ventilation are tidal, inspiratory reserve, and expiratory reserve volumes.
4. Any air movement for purposes other than respiration is nonrespiratory, including coughing, sneezing, sighing, yawning, laughing, crying, and hiccuping.

Regulation of Breathing (pp. 586–587)

1. Ventilation is directly controlled by the rhythmicity center in the medulla oblongata, which in turn is influenced by the pneumotaxic and apneustic centers in the pons.
2. These brain stem areas are affected by higher brain function and by sensory input from chemoreceptors.
3. Central chemoreceptors are located in the medulla oblongata; peripheral chemoreceptors are located in the aortic and carotid bodies.

UNIT VI

REVIEW ACTIVITIES

Objective Questions

1. Which of the following is a *false* statement?
   - (a) The term *respiration* can be used in reference to ventilation (breathing) or oxygen utilization by body cells.
   - (b) The incoming (inhaled) air that contacts the alveoli of the lung is unchanged from that which surrounds the body.
   - (c) As a resonating chamber, the nasal cavity is important in voice phonetics.
   - (d) It is only through the walls of the alveoli within the respiratory system that gaseous exchange occurs.

2. Which is *not* a component of the nasal septum?
   - (a) the palatine bone
   - (b) the vomer bone
   - (c) the ethmoid bone
   - (d) septal cartilage

3. An adenoidectomy is the removal of
   - (a) the uvula.
   - (b) the pharyngeal tonsils.
   - (c) the palatine tonsils.
   - (d) the lingual tonsils.

4. Which is *not* a paranasal sinus?
   - (a) the palatine sinus
   - (b) the ethmoidal sinus
   - (c) the sphenoidal sinus
   - (d) the frontal sinus
   - (e) the maxillary sinus

5. Which of the following is *not* characteristic of the left lung?
   - (a) a cardiac notch
   - (b) a superior lobe
   - (c) a single fissure
   - (d) an inferior lobe
   - (e) a middle lobe

6. The epithelial lining of the wall of the thorax is called
   - (a) the parietal pleura.
   - (b) the pleural peritoneum.
   - (c) the mediastinal pleura.
   - (d) the visceral pleura.
   - (e) the costal pleura.

7. Which muscle group combination permits inspiration?
   - (a) diaphragm, abdominal complex
   - (b) internal intercostals, diaphragm
   - (c) external intercostals, internal intercostals
   - (d) external intercostals, diaphragm, internal intercostals (interchondral part)

8. The vocal cords are attached to
   - (a) the cricoid and thyroid cartilages.
   - (b) the cuneiform and cricoid cartilages.
   - (c) the corniculate and thyroid cartilages.
   - (d) the arytenoid and thyroid cartilages.

9. The maximum amount of air that can be expired after a maximum inspiration is
   - (a) the tidal volume.
   - (b) the forced expiratory volume.
   - (c) the vital capacity.
   - (d) the maximum expiratory flow rate.
   - (e) the residual volume.

10. The rhythmic control of breathing is produced by the activity of inspiratory and expiratory neurons in
    - (a) the medulla oblongata.
    - (b) the apneustic center of the pons.
    - (c) the pneumotaxic center of the pons.
    - (d) the cerebral cortex.
    - (e) the hypothalamus.

Essay Questions

1. Why is warming, moistening, and filtering the air so important for a healthy, functioning respiratory system?
2. Explain the structural and functional differences between the three regions of the pharynx.
3. What is a bronchial tree? Why are alveoli rather than bronchial trees considered the functional units of the respiratory system?
4. Define each of the following terms relating to the structure of the lung: *base, hilum, apex, costal surface, fissure, lobe, lobule, pulmonary ligament,* and *bronchial segment.*
5. Diagram the location of the lungs in the thoracic cavity with respect to the heart. Identify the various thoracic serous membranes.
6. List the kinds of epithelial tissues found within the respiratory system and describe the location of each.
7. Identify two places in the respiratory system where large surface areas of capillary networks are found. What is the function of each of these areas?
8. What protective devices of the respiratory system guard against pollutants, the spread of infections, and dual collapse of the lungs?
9. What are the functions of the larynx? List some of the clinical conditions that could involve this organ.
10. What are the advantages of compartmentalization of the thoracic organs?
11. Explain the sequence of pulmonary ventilation. Discuss the mechanisms of inspiration and expiration. How are air pressures related to ventilation?
12. Distinguish between tidal volume, vital capacity, and total lung capacity.
13. What is meant by a rhythmicity respiratory area? How are the apneustic and pneumotaxic areas related to the rhythmicity area?
14. Describe how the abdominal thrust maneuver is performed. When should it be used? On what principle is this maneuver based?
15. When would mouth-to-mouth respiration rather than the abdominal thrust maneuver be used to revive a person?

Gundy/Weber Software

The tutorial software accompanying Chapter 17, Respiratory System, is Volume 10, Respiratory System.

C H A P T E R

18

Digestive System

Chapter Outline

CLINICAL CASE STUDY

A 25-year-old male construction worker is admitted to the emergency room after suffering a blow to the upper abdomen by a swinging beam. Initial assessment is significant for marked tenderness in the right upper quadrant of the abdomen and for vital signs and examination findings consistent with mild hemorrhagic shock. Intravenous fluids are administered, causing stabilization of vital signs. A chest radiograph reveals no abnormalities. Peritoneal lavage (see chapter 2, Clinical Case Study) is likewise negative for blood. There are no externally detectable signs of hemorrhage. A CT scan demonstrates a significant hematoma (collection of blood) deep within the substance of the liver as well as a notable amount of blood in the small intestine. The decision is made to operate. Initial exploration reveals no trauma to the stomach or small intestine.

What is the likely source of the bleeding? Explain anatomically how the blood found its way into the small intestine, noting each step of its path. Given that the hepatic arterial system and hepatic venous system are possible sources of the bleeding, is there another system of blood vessels relative to the liver that could also be a source of hemorrhage?

Hint: As you study the digestive system, pay close attention to the anatomical position of each accessory digestive organ and note how each connects to the lumen of the gastrointestinal (GI) tract.

Introduction to the Digestive System

The organs of the digestive system are specialized for the digestion and absorption of food. The digestive system consists of a tubular gastrointestinal tract and accessory digestive organs.

Objective 1. List the activities of the digestive system and distinguish between digestion and absorption.

Objective 2. Identify the major structures and regions of the digestive system.

Objective 3. Define the terms *viscera* and *gut*.

It is apparent that food is necessary to sustain life and that many of our daily endeavors involve procuring and eating food. The food is utilized at the cellular level, where nutrients are required for chemical reactions involving synthesis of enzymes, cellular division and growth, repair, and the production of heat energy. Most of the food eaten, however, is not suitable for cellular utilization until it is mechanically and chemically reduced to forms that can be absorbed through the intestinal wall and transported to the cells by the blood. Ingested food is not technically in the body until it is absorbed; and, in fact, a large portion of consumed food is not digested and passes through as waste material.

The principal function of the digestive system is to prepare food for cellular utilization. This involves the following functional activities:

1. **Ingestion**—the taking of food into the digestive system by way of the mouth
2. **Mastication**—chewing movements to pulverize food and mix it with saliva
3. **Deglutition**—swallowing of food to move it from the mouth to the stomach
4. **Digestion**—mechanical and chemical breakdown of food material
5. **Absorption**—passage of molecules of food through the mucous membrane of the small intestine and into the circulatory or lymphatic systems for distribution to cells
6. **Peristalsis**—rhythmic, wavelike intestinal contractions that move food through the digestive tract
7. **Defecation**—the discharge of indigestible wastes, called feces, from the body

Anatomically and functionally, the digestive system can be divided into a tubular **digestive tract,** or **gastrointestinal tract** (GI tract), and **accessory organs.** The GI tract is approximately 9 m (30 ft) long and extends from the mouth to the anus. It traverses the thoracic cavity and enters the abdominal cavity at the level of the diaphragm. The anus is located at the inferior portion of the pelvic cavity. The organs of the GI tract include the **oral (buccal) cavity, pharynx, esophagus, stomach, small intestine,** and **large intestine** (fig. 18.1). The accessory digestive organs include the **teeth, tongue, salivary glands, liver, gallbladder,** and **pancreas.** The term **viscera** is frequently used to refer to the abdominal organs of digestion, but actually viscera can be any of the organs (lungs, stomach, spleen, etc.) of the thoracic and abdominal cavities. **Gut** is an anatomical term that generally refers to the developing stomach and intestines in the embryo.

It usually takes about 24 hours for food to travel the length of the GI tract. Food ingested through the mouth passes in assembly-line fashion through the GI tract where complex molecules are progressively broken down. Each region of the GI tract has specific functions in preparing food for utilization (table 18.1).

ingestion: L. *ingerere*, carry in
mastication: Gk. *mastichan*, gnash the teeth
deglutition: L. *deglutire*, swallow down
peristalsis: Gk. *peri*, around; *stellein*, compress
defecation: L. *de*, from, away; *faecare*, cleanse

Figure 18.1 The digestive system, including the gastrointestinal tract and the accessory digestive organs.

Table 18.1	Regions of the GI Tract and Basic Functions
Region	**Function**
Oral cavity	Ingests food; receives saliva; grinds food and mixes it with saliva (mastication); initiates digestion of carbohydrates; forms and swallows bolus (deglutition)
Pharynx	Receives bolus from oral cavity; autonomically continues deglutition of bolus to esophagus
Esophagus	Transports bolus to stomach by peristalsis; lower esophageal sphincter restricts backflow of food
Stomach	Receives bolus from esophagus; churns bolus with gastric juice; initiates digestion of proteins; carries out limited absorption; moves chyme into duodenum and prohibits backflow of chyme; regurgitates when necessary
Small intestine	Receives chyme from stomach and secretions from liver and pancreas; chemically and mechanically breaks down chyme; absorbs nutrients; transports wastes through peristalsis to large intestine; prohibits backflow of intestinal wastes from large intestine
Large intestine	Receives undigested wastes from small intestine; absorbs water and electrolytes; forms, stores, and expels feces through defecation reflex

1. Which functional activities of the digestive system break down food? Define *absorption*. Which functional activities move the food through the GI tract?
2. List in order the regions of the GI tract through which ingested food passes from the mouth to the anus.
3. List the abdominal visceral organs of the digestive system.
4. Write a sentence in which the term *gut* is used correctly.

Although there is an abundance of food in the United States, too many people eat the wrong kinds of foods, eat irregularly, or overeat, to the point that eating patterns have become a critical public health concern. Obesity is a major health problem. People who are grossly overweight are at greater risk for cardiovascular disease, hypertension, osteoarthritis, and diabetes mellitus. Those who practice good nutritional habits are better able to withstand trauma, are less likely to get sick, and are usually less seriously ill when they do become sick.

Serous Membranes and Tunics of the Gastrointestinal Tract

Serous membranes line the abdominal cavity and cover the visceral organs. The wall of the GI tract is composed of four tunics.

Objective 4. Describe the arrangement of the serous membranes within the abdominal cavity.

Objective 5. Describe the generalized structure of the four tunics composing the wall of the GI tract.

Serous Membranes

Most of the GI tract and abdominal accessory digestive organs are positioned within the **abdominal cavity.** These organs are not firmly embedded in solid tissue but are supported and covered by **serous membranes.** A serous membrane is an epithelial membrane that lines the thoracic and abdominal cavities

UNIT VI

and covers the organs that lie within these cavities. A serous membrane has a *parietal portion* lining the body wall and a *visceral portion* covering the internal organs. The serous membranes associated with the lungs are called pleurae. The serous membranes of the abdominal cavity are called **peritoneal membranes,** or **peritoneum** *(per''i-to-ne'um).* The peritoneum is the largest serous membrane of the body. It is composed of simple squamous epithelium with portions reinforced with connective tissue.

The **parietal peritoneum** lines the wall of the abdominal cavity (fig. 18.2). Along the posterior aspect of the abdominal cavity, the parietal peritoneum comes together to form a double-layered peritoneal fold called the **mesentery** that supports the GI tract. The mesentery gives the pendulous small intestine freedom for peristaltic movement and provides a structure through which intestinal nerves and vessels traverse. The **mesocolon** is a specific portion of the mesentery that supports the large intestine (fig. 18.3c, d). The peritoneal covering continues around the intestinal viscera as the **visceral peritoneum.** The **peritoneal cavity** is the space between the parietal and visceral portions of the peritoneum.

Extensions of the parietal peritoneum, located in the peritoneal cavity, serve specific functions (fig. 18.3). The **falciform ligament,** a serous membrane reinforced with connective tissue, attaches the liver to the diaphragm and anterior abdominal wall. The **lesser omentum** passes from the lesser curvature of the stomach and the upper duodenum to the inferior surface of the liver. The **greater omentum** extends from the greater curvature of the stomach to the transverse colon, forming an apronlike structure over most of the small intestine. Functions of the greater omentum include storing fat, cushioning visceral organs, supporting lymph nodes, and protecting against the spread of infections. In cases of localized inflammation, such as appendicitis, the greater omentum may compartmentalize the inflamed area, sealing it off from the rest of the peritoneal cavity.

Certain abdominal organs are not within the peritoneal cavity and are therefore not covered by the parietal peritoneum. These organs are said to be *retroperitoneal* and include most of the pancreas, kidneys, a portion of the duodenum, and the abdominal aorta.

> *Peritonitis* is an inflammation of the peritoneum. Peritonitis may be caused by trauma, rupture of a visceral organ, ectopic pregnancy, or postoperative complications. Peritonitis is usually extremely painful and serious. Treatment usually involves the injection of massive doses of antibiotics and perhaps peritoneal intubation to permit drainage.

Layers of the Gastrointestinal Tract

The GI tract from the esophagus to the anal canal is composed of four layers, or **tunics.** Each tunic contains a dominant tissue type that performs specific functions in the digestive process.

peritoneum: Gk. *peritonaion,* stretched over
mesentery: Gk. *mesos,* middle; *enteron,* intestine
omentum: L. *omentum,* apron
tunica: L. *tunica,* covering or coat

Figure 18.2 A diagrammatic representation of the serous membranes.

The four tunics of the GI tract, from the inside out, are the **mucosa, submucosa, muscularis,** and **serosa** (fig. 18.4a).

Mucosa The mucosa surrounds the lumen of the GI tract and is the absorptive and major secretory layer. It consists of a simple columnar epithelium supported by the **lamina propria,** which is a thin layer of connective tissue. The lamina propria contains numerous lymph nodules, which are important in protecting against disease (fig. 18.4b). External to the lamina propria are thin layers of smooth muscle called the **muscularis mucosa.** This is the muscle layer responsible for the numerous folds, called *plicae circulares* (see fig. 18.20) that greatly increase the absorptive surface area of the small intestine portion of the GI tract. Specialized **goblet cells** in the mucosa secrete mucus throughout most of the GI tract.

Submucosa The relatively thick submucosa is a highly vascular layer of connective tissue serving the mucosa. Absorbed molecules that pass through the columnar epithelial cells of the mucosa enter into blood vessels or lymph ductules of the submucosa. In addition to blood vessels, the submucosa contains glands and nerve plexuses. The *submucosal plexus (Meissner's plexus)* (fig. 18.4b) provides autonomic innervation to the muscularis mucosa.

Muscularis The muscularis (also called the muscularis externa) is responsible for segmental contractions and peristaltic movement through the GI tract. The muscularis has an inner circular and an outer longitudinal layer of smooth muscle. Contractions of these layers move the food peristaltically through the tract and physically pulverize and churn the food with digestive enzymes. The *myenteric plexus (Auerbach's plexus)* located between the two muscle layers provides the major nerve supply to the GI tract and includes neurons and ganglia from both the sympathetic and parasympathetic divisions of the ANS.

Meissner's plexus: from Georg Meissner, German histologist, 1829–1905
Auerbach's plexus: from Leopold Auerbach, German anatomist, 1828–97

Figure 18.3 The structural arrangement of the abdominal organs and peritoneal membranes within the peritoneal cavity. (*a*) The greater omentum, (*b*) the lesser omentum with the liver lifted, (*c*) the mesentery with the greater omentum lifted, and (*d*) the relationship of the peritoneal membranes to the visceral organs as shown in a sagittal view.

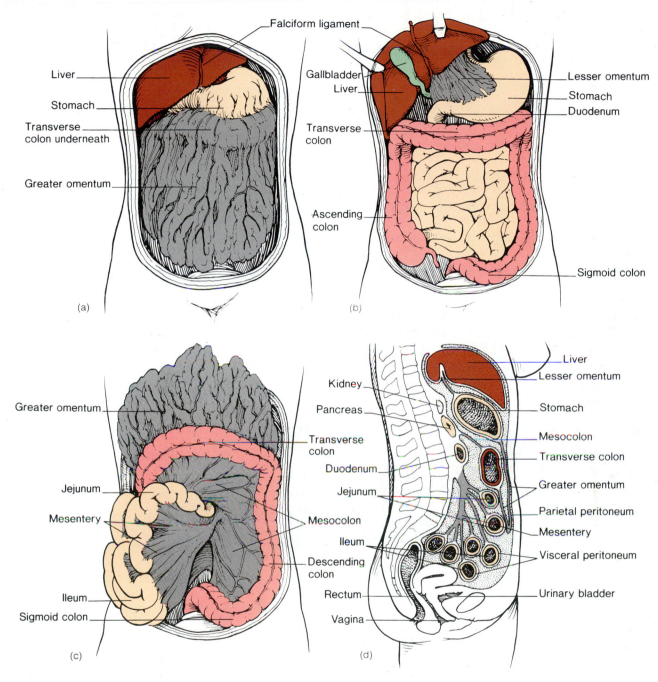

Serosa The outer serosa layer completes the wall of the GI tract. It is a binding and protective layer consisting of loose fibrous connective tissue covered with a layer of simple squamous epithelium and subjacent connective tissue. The simple squamous epithelium is actually the visceral peritoneum.

> The body has several defense mechanisms to protect against ingested material that may be harmful if absorbed. The acidic environment of the stomach and the lymphatic system kill many harmful bacteria. A mucous lining throughout the GI tract serves as a protective layer. Vomiting and, in certain cases, diarrhea are reactions to substances that irritate the GI tract. Vomiting is a reflexive response to many toxic chemicals and as such can be beneficial, even though unpleasant.

Innervation The GI tract is innervated by the sympathetic and parasympathetic divisions of the ANS. The vagus nerve is the source of parasympathetic activity in the esophagus, stomach, pancreas, gallbladder, small intestine, and upper portion of the large intestine. The lower portion of the large intestine receives parasympathetic innervation from spinal nerves in the sacral region. The submucosal plexus and myenteric plexus are the

UNIT VI

Figure 18.4 The tunics (layers) of the GI tract. (*a*) A section of the small intestine with each of the four tunics exposed and (*b*) a section showing the detailed structure and relative thickness of each tunic. (Note the location of the exocrine gland and the innervation of the small intestine.)

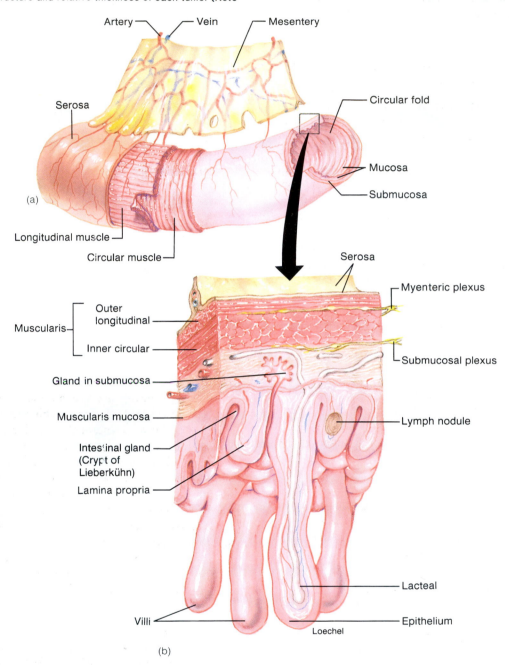

sites where preganglionic neurons synapse with postganglionic neurons that innervate the smooth muscle of the GI tract. Stimulation of the parasympathetic neurons increases peristalsis and the secretions of the GI tract.

Postganglionic sympathetic fibers pass through the submucosal and myenteric plexuses and innervate the GI tract. Sympathetic nerve stimulation acts antagonistically to the effects of parasympathetic nerves by reducing peristalsis and secretions and stimulating the contraction of sphincter muscles along the GI tract.

1. Describe the functions of the mesentery and greater omentum. Identify the location of the peritoneum and list the organs that are retroperitoneal.
2. List the four tunics of the GI tract and identify their major tissue types. Explain the functions of these four tunics.

Mouth, Pharynx, and Associated Structures

Ingested food is changed by the mechanical action of teeth and by the chemical activity of saliva into a bolus, which is swallowed in the process of deglutition.

The functions of the **mouth** and associated structures are to form a receptacle for food, to initiate digestion through mastication, to swallow food, and to form words in speech. The mouth can also assist the respiratory system in the passage of air. The **pharynx,** which is posterior to the mouth, serves as a common passageway for both the respiratory and digestive systems. Both the mouth and pharynx are lined with nonkeratinized stratified squamous epithelium, which is constantly moistened by the secretion of saliva. The mouth is also known as the **oral,** or **buccal** *(buk'al),* **cavity** (fig. 18.5). It is formed by the **cheeks, lips, hard** and **soft palates,** and **tongue.** The **vestibule** of the oral cavity is the depression between the cheeks and lips externally and the gums and teeth internally (fig. 18.6). The opening of the oral cavity is referred to as the **oral orifice** and the opening between the oral cavity and the pharynx is called the **fauces.**

Cheeks and Lips

The **cheeks** form the lateral walls of the oral cavity and consist of outer layers of skin, subcutaneous fat, facial muscles that assist in manipulating food in the oral cavity, and inner linings of moistened stratified squamous epithelium. The anterior portion of the cheeks terminates in the superior and inferior lips that surround the oral orifice.

The **lips** are fleshy, highly mobile organs whose principal function in humans is associated with speech. Lips also serve for suckling, manipulating food, and keeping food between the upper and lower teeth. Each lip is attached from its inner surface to the gum by a midline fold of mucous membrane called the **labial frenulum** *(fren'yŭ-lum)* (fig. 18.5). The lips are formed from the orbicularis oris muscle and associated connective tissue and covered with soft, pliable skin. Between the outer skin and the mucous membrane of the oral cavity is a transition zone called the **vermilion.** Lips are red to reddish-brown in color because of blood vessels close to the surface. The many sensory receptors in the lips aid in determining the temperature and texture of food.

  Suckling is an innate characteristic of newborns. The lips are well formed for this activity and even contain blisterlike milk pads that aid in suckling. The wide nostrils and receding lower jaw of infants also facilitate suckling.

pharynx: L. *pharynx,* throat
buccal: L. *bucca,* cheek
fauces: L. *fauces,* throat
vermilion: O.E. *vermeylion,* red colored

Figure 18.5 The superficial structures of the oral cavity.

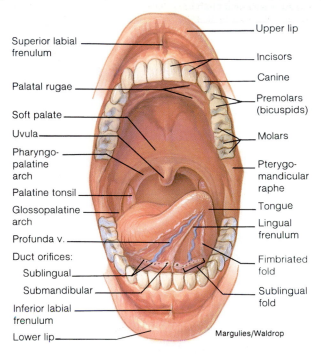

Superior labial frenulum — Upper lip
Palatal rugae — Incisors
Soft palate — Canine
Uvula — Premolars (bicuspids)
Pharyngo-palatine arch — Molars
Palatine tonsil — Pterygo-mandicular raphe
Glossopalatine arch — Tongue
Profunda v. — Lingual frenulum
Duct orifices: — Fimbriated fold
Sublingual — Sublingual fold
Submandibular
Inferior labial frenulum
Lower lip — Margulies/Waldrop

Tongue

As a digestive organ, the **tongue** functions to move food around in the mouth during mastication and to assist in swallowing food. It contains **taste buds** through which various food tastes are sensed and it is also essential in producing articulated speech. The tongue is a mass of skeletal muscle covered with a mucous membrane. Extrinsic tongue muscles (those that insert upon the tongue) move the tongue from side to side and in and out. Only the anterior two-thirds of the tongue lies in the oral cavity; the remaining one-third lies in the pharynx (fig. 18.6) and is attached to the hyoid bone. Rounded masses of **lingual tonsils** are located on the superior surface of the base of the tongue (fig. 18.7). The inferior of the tongue is connected along the midline anteriorly to the floor of the mouth by the vertically positioned **lingual frenulum** (fig. 18.5).

When a short lingual frenulum restricts tongue movements, the person is said to be *tongue-tied.* If this developmental problem is severe, the infant may have difficulty suckling. Older children with this problem may have faulty speech. These functional problems can be easily corrected through surgery.

On the superior surface of the tongue are numerous small elevations called **papillae** that give the tongue a distinct roughened surface, which aids the handling of food. These papillae also contain taste buds that can distinguish sweet, salty, sour, and bitter sensations. The three types of papillae on the superior surface of the tongue are **filiform, fungiform,** and

papilla: L. *papula,* little nipple
filiform: L. *filum,* thread; *forma,* form
fungiform: L. *fungus,* fungus; *forma,* form

Figure 18.6 A sagittal section of the facial region showing the oral cavity, nasal cavity, and pharynx.

Figure 18.7 The dorsum of the tongue.

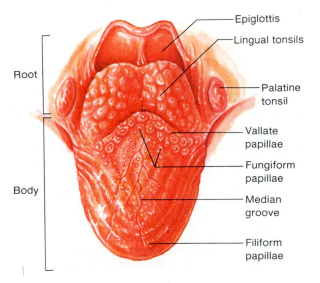

vallate (circumvallate) (fig. 18.7). Filiform papillae, by far the most numerous, are sensitive to touch and have tapered tips. The larger and rounded fungiform papillae are scattered among the filiform type. The few vallate papillae are arranged in a **V** shape on the posterior surface of the tongue.

Palate

The **palate** is the roof of the oral cavity and consists of the bony hard palate anteriorly and the soft palate posteriorly (figs. 18.5 and 18.6). The hard palate is formed by the palatine processes of the maxillae and the horizontal plates of the palatine bones and is covered with a mucous membrane. Transverse ridges called **palatal rugae** *(roo'je)* are located along the mucous membrane of the hard palate. These structures serve as friction ridges against which the tongue is placed during swallowing. The soft palate is a muscular arch covered with mucous membrane and is anteriorly continuous with the hard palate. Suspended from the middle lower border of the soft palate is a cone-

vallate: L. *vallatus,* surrounded with a rampart

UNIT VI

Figure 18.8 Dentitions and the sequence of eruptions. (*a*) Deciduous teeth and (*b*) permanent teeth.

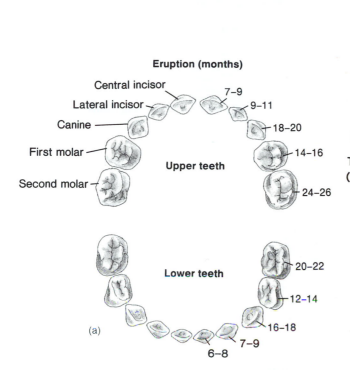

Eruption (months)

Central incisor — 7–9
Lateral incisor — 9–11
Canine — 18–20
First molar — 14–16
Second molar — 24–26

Upper teeth

Lower teeth

20–22
12–14
16–18
7–9
6–8

(a)

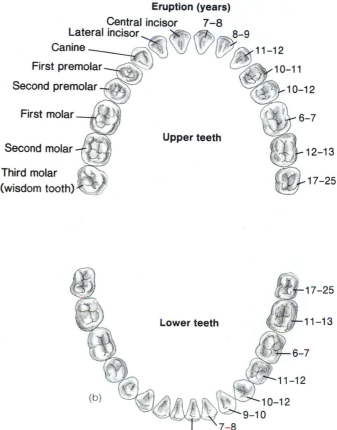

Eruption (years)

Central incisor — 7–8
Lateral incisor — 8–9
Canine — 11–12
First premolar — 10–11
Second premolar — 10–12
First molar — 6–7

Upper teeth

Second molar — 12–13
Third molar (wisdom tooth) — 17–25

Lower teeth

17–25
11–13
6–7
11–12
10–12
9–10
7–8
6–7

(b)

shaped projection called the **uvula.** As one swallows, the soft palate and uvula are drawn upward, closing the nasopharynx and preventing food and fluid from entering the nasal cavity.

Two muscular folds extend downward from both sides of the base of the uvula (fig. 18.5). The anterior fold is called the **glossopalatine arch** and the posterior fold is the **pharyngopalatine** (*fā-ring″go-pal′ā-tīn*) **arch.** Between these two arches, toward the posterior lateral portion of the oral cavity, is the **palatine tonsil.**

Teeth

Humans, being mammals, have *heterodont dentition*. This means that the **teeth** vary structurally and are adapted to handle food in different ways (fig. 18.8). The four pairs (upper and lower jaws) of anteriormost teeth are the **incisors** (*in-si′zorz*). The chisel-shaped incisors are adapted for cutting and shearing food. The two pairs of cone-shaped **canines,** or **cuspids,** are located at the anterior corners of the mouth and are adapted for holding and tearing. Incisors and canines are further characterized by a single root on each tooth. Located behind the canines are the **premolars,** or **bicuspids,** and **molars.** These teeth have two or three roots and have somewhat rounded, irregular surfaces called **cusps** for crushing and grinding food.

Humans are *diphyodont;* that is, normally two sets of teeth develop in a person's lifetime. Twenty **deciduous (milk) teeth**

uvula: L. *uvula*, small grapes
heterodont: Gk. *heteros*, other; *odous*, tooth
incisor: L. *incidere*, to cut
canine: L. *canis*, dog
molar: L. *mola*, millstone
deciduous: L. *deciduus*, to fall away

begin to erupt at about 6 months of age (fig. 18.9 and tables 18.2 and 18.3), beginning with the incisors. All of the deciduous teeth have erupted by the age of 2½. Thirty-two **permanent teeth** replace the deciduous teeth in a predictable sequence. This process begins at about age 6 and continues until about age 17. The **third molars,** or **wisdom teeth,** are the last to erupt. Wisdom teeth are less predictable, and if they do erupt, it is between the ages of 17 and 25. Because the jaws have formed by this time and other teeth are in place, the eruption of wisdom teeth may cause serious problems of crowding or impaction.

A **dental formula** is a graphic representation of the types, number, and position of teeth in the oral cavity. A dental formula can be written for each species of mammal that has heterodontia. Following are the deciduous and permanent dental formulae for humans:

Formula for deciduous dentition:
I 2/2, **C** 1/1, **DM** 2/2 = 10 × 2 = 20 teeth

Formula for permanent dentition:
I 2/2, **C** 1/1, **P** 2/2, **M** 3/3 = 16 × 2 = 32 teeth

(**I** = incisor; **C** = canine; **P** = premolar;
DM = deciduous molar; **M** = molar)

The cusps of the upper and lower premolar and molar teeth occlude for chewing food (mastication), whereas the upper incisors normally form an overbite with the incisors of the lower jaw. An overbite of the upper incisors creates a shearing action

UNIT VI

Figure 18.9 The skull of a youth (9 to 12 years old) showing the eruption of teeth.

- Permanent second molar
- Permanent first molar
- Permanent canine
- Permanent second molar
- Permanent first premolar
- Deciduous canine
- Deciduous canine
- Deciduous first molar
- Permanent canine
- Permanent first premolar
- Deciduous second molar
- Permanent second premolar
- Permanent first molar
- Permanent second molar
- Permanent third molar

as these teeth slide past one another. Masticated food is mixed with saliva, which initiates chemical digestion and facilitates swallowing. The soft, flexible mass of food that is swallowed is called a *bolus*.

A tooth consists of an exposed **crown** supported by a **neck** that is anchored firmly into the jaw by one or more **roots** (fig. 18.10). The roots of teeth fit into sockets, called **alveoli**, in the alveolar processes of the mandible and maxillae. Each socket is lined with a connective tissue periosteum, specifically called the **periodontal membrane.** The root of a tooth is covered with a bonelike material called the **cementum;** fibers in the periodontal membrane insert into the cementum and fasten the tooth in its socket. The **gingiva** *(jin-ji'vă),* or **gum,** is the mucous membrane surrounding the alveolar processes in the oral cavity.

The bulk of a tooth consists of **dentin,** a substance similar to bone, but harder. Covering the dentin on the outside and forming the crown is a tough, durable layer of **enamel.** Enamel is composed primarily of calcium phosphate and is the hardest substance in the body. The central region of the tooth contains the **pulp cavity.** The pulp cavity contains the **pulp,** which is composed of connective tissue with blood vessels, lymph vessels, and nerves. A **root canal** is continuous with the pulp cavity through the root to an opening at the base called the **apical foramen.** The tooth receives nourishment through vessels traversing the apical foramen. Proper nourishment is particularly important during embryonic development. The diet of the mother should contain an abundance of calcium and vitamin D during pregnancy to ensure the proper development of the baby's teeth.

Enamel is the hardest substance in the body, but it can be dissolved by bacterial activity that results in *dental caries (cavities).* These caries must be artificially filled because new enamel is not produced after a tooth erupts. The rate of tooth decay decreases after age 35, but then periodontal diseases may develop. *Periodontal diseases* result from plaque or tartar buildup at the gum line. This buildup wedges the gum away from the teeth, allowing bacterial infections to develop.

bolus: Gk. *bolos,* lump
gingiva: L. *gingiva,* gum
dentin: L. *dens,* tooth

UNIT VI

Table 18.2	Eruption Sequence and Loss of Deciduous Teeth		
	Average Age of Eruption		
Type of Tooth	**Lower**	**Upper**	**Average Age of Loss**
Central incisors	6–8 mos	7–9 mos	7 yrs
Lateral incisors	7–9 mos	9–11 mos	8 yrs
First molars	12–14 mos	14–16 mos	10 yrs
Canines (cuspids)	16–18 mos	18–20 mos	10 yrs
Second molars	20–22 mos	24–26 mos	11–12 yrs

Table 18.3	Eruption Sequence of Permanent Teeth	
	Average Age of Eruption	
Type of Tooth	**Lower**	**Upper**
First molars	6–7 yrs	6–7 yrs
Central incisors	6–7 yrs	7–8 yrs
Lateral incisors	7–8 yrs	8–9 yrs
Canines (cuspids)	9–10 yrs	11–12 yrs
First premolars (bicuspids)	10–12 yrs	10–11 yrs
Second premolars (bicuspids)	11–12 yrs	10–12 yrs
Second molars	11–13 yrs	12–13 yrs
Third molars (wisdom)	17–25 yrs	17–25 yrs

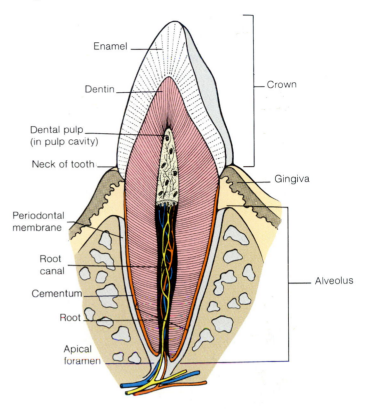

Figure 18.10 The structure of a tooth shown in a vertical section through a canine tooth.

Enamel — Crown
Dentin
Dental pulp (in pulp cavity)
Neck of tooth
Gingiva
Periodontal membrane
Root canal
Cementum
Root
Alveolus
Apical foramen

Salivary Glands

Salivary glands are accessory digestive glands that produce a fluid secretion called *saliva.* Saliva functions as a solvent in cleansing the teeth and dissolving food chemicals so they can be tasted. Saliva also contains enzymes, which digest starch, and mucus, which lubricates the pharynx to facilitate swallowing. Saliva is secreted continuously but usually only in sufficient amounts to keep the mucous membranes of the oral cavity moist. Numerous minor salivary glands called **buccal glands** are located in the mucous membranes of the palatal region of the oral cavity. Most of the saliva, however, is produced by three pairs of salivary glands outside of the oral cavity and is transported to the mouth via **salivary ducts.** The three major pairs of salivary glands are the parotid *(pă-rot'id),* submandibular, and sublingual (fig. 18.11).

The **parotid gland** is the largest of the salivary glands and is positioned below and in front of the auricle of the ear, between the skin and the masseter muscle. The **parotid** (Stensen's) **duct** parallels the zygomatic arch across the masseter muscle, pierces the buccinator muscle, and drains into the oral cavity opposite the second upper molar. It is the parotid gland that becomes infected and swollen with the mumps.

The **submandibular gland** is inferior to the body of the mandible about midway along the inner side of the jaw. This gland is covered by the more superficial mylohyoid muscle. The **submandibular** (Wharton's) **duct** empties into the floor of the mouth on both lateral sides of the lingual frenulum.

The **sublingual gland** lies under the mucosa in the floor of the mouth on the side of the tongue. Each sublingual gland possesses several small ducts that empty into the floor of the mouth in an area posterior to the papilla of the submandibular duct.

Two types of secretory cells, **serous** and **mucous cells,** are found in all salivary glands in various proportions (fig. 18.12). Serous cells produce a watery fluid containing digestive enzymes; mucous cells secrete a thick, stringy mucus. Cuboidal epithelial cells line the lumina of the salivary ducts.

The salivary glands are innervated by both divisions of the autonomic nervous system. Sympathetic impulses stimulate the secretion of small amounts of viscous saliva. Parasympathetic stimulation causes the secretion of large volumes of watery saliva. Physiological responses of this type occur whenever a person sees, smells, tastes, or even thinks about desirable food. The amount of saliva secreted daily ranges from 1,000 to 1,500 ml. Information about the salivary glands is summarized in table 18.4.

parotid: Gk. *para,* beside; *otos,* ear
Stensen's duct: from Nicholaus Stensen, Danish anatomist, 1638–86

Wharton's duct: from Thomas Wharton, English physician, 1614–73

UNIT VI

Figure 18.11 The salivary glands.

Figure 18.12 The histology of the salivary glands. (a) The parotid gland, (b) the submandibular gland, and (c) the sublingual gland.

Table 18.4	Major Salivary Glands			
Gland	**Location**	**Duct**	**Entry into Oral Cavity**	**Type of Secretion**
Parotid gland	Anterior and inferior to auricle; subcutaneous over masseter muscle	Parotid (Stensen's duct)	Lateral to upper second molar	Watery serous fluid, salts, and enzyme
Submandibular gland	Inferior to the base of the tongue	Submandibular (Wharton's duct)	Papilla lateral to lingual frenulum	Watery serous fluid with some mucus
Sublingual gland	Anterior to submandibular; under tongue	Several small ducts (Rivinus's ducts)	Ducts along the base of the tongue	Mostly thick, stringy mucus, salts, and enzyme

Figure 18.13 A posterior view of the constrictor muscles of the pharynx. The right side has been cut away to illustrate the interior structures in the pharynx.

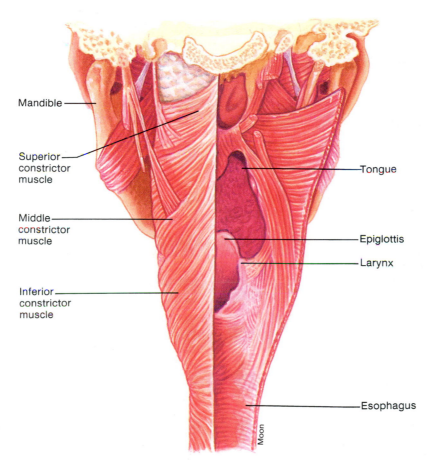

UNIT VI

Pharynx

The funnel-shaped **pharynx** *(far'ingks)* is a passageway approximately 13 cm (5 in.) long connecting the oral and nasal cavities to the esophagus and trachea. The pharynx has both digestive and respiratory functions. The supporting walls of the pharynx are composed of skeletal muscle and the lumen is lined with a mucous membrane of stratified squamous epithelium.

The external *circular layer* of pharyngeal muscles, called **constrictors** (fig. 18.13), serves to compress the lumen of the pharynx involuntarily during swallowing. The **superior constrictor muscle** attaches to bony processes of the skull and mandible and encircles the upper portion of the pharynx. The

middle constrictor muscle arises from the hyoid bone and stylohyoid ligament and encircles the middle portion of the pharynx. The **inferior constrictor muscle** arises from the cartilages of the larynx and encircles the lower portion of the pharynx. During breathing, the lower portion of the inferior constrictor muscle is contracted, preventing air from entering the esophagus.

The motor and most of the sensory innervation to the pharynx is via the **pharyngeal plexus,** situated chiefly on the middle constrictor muscle. It is formed by the pharyngeal branches of the glossopharyngeal (ninth cranial) and the vagus (tenth cranial) nerves, together with a deep sympathetic branch from the superior cervical ganglion.

The pharynx is served by the ascending pharyngeal and inferior thyroid arteries, both of which are branches of the external carotid arteries. Venous return is via the internal jugular veins.

---

1. Define the terms *diphyodont* and *heterodont*. Which of the four kinds of teeth is not found in deciduous dentition?
2. Identify the locations of the enamel, dentin, cementum, and pulp of a tooth. Explain how a tooth is anchored into its socket.
3. Identify the location of the parotid and the submandibular ducts and describe where they empty in the oral cavity.
4. Describe the effects of autonomic stimulation on salivary secretion.

Esophagus and Stomach

A bolus of food is passed from the esophagus to the stomach, where it is churned and mixed with gastric secretions. The chyme thus produced is sent past the pyloric sphincter of the stomach to the duodenum.

Objective 9. Describe the steps in deglutition.

Objective 10. Describe the location, gross structure, and functions of the stomach.

Objective 11. Describe the histological structure of the esophagus and stomach. List the cell types in the gastric mucosa and their secretory products.

Esophagus

The **esophagus** (*ē-sof'ă-gus*) is that portion of the GI tract that connects the pharynx to the stomach. It is a collapsible muscular tube approximately 25 cm (10 in.) long, originating at the larynx and located posterior to the trachea.

The esophagus is located within the mediastinum of the thorax and passes through the diaphragm just above the opening into the stomach. The opening through the diaphragm is called the **esophageal hiatus** (*ē-sof''ă-je'al hi-a'tus*). The esophagus is lined with a nonkeratinized stratified squamous epithelium (fig. 18.14); its walls contain either skeletal or smooth muscle, depending on the location. The upper third of the esophagus contains skeletal muscle; the middle third contains both skeletal and smooth muscle, and the terminal portion contains only smooth muscle.

The lumen of the terminal portion of the esophagus is slightly narrowed because of a thickening of the circular muscle fibers in its wall. This portion is referred to as the **lower esophageal (gastroesophageal) sphincter.** After food passes into the stomach, the muscle fibers of this region constrict to help prevent the stomach contents from regurgitating into the

esophagus: Gk. *oisein*, to carry; *phagema*, food

Figure 18.14 The histology of the esophagus.

- Serosa
- Muscularis
- Submucosa
- Mucosa
- Lumen

esophagus. Regurgitation would occur because the pressure in the abdominal cavity is greater than the pressure in the thoracic cavity as a result of respiratory movements. The lower esophageal sphincter must remain closed, therefore, until food is pushed through it by peristalsis into the stomach.

The lower esophageal sphincter is not a true sphincter muscle that can be identified histologically, and it does at times permit the acidic contents of the stomach to enter the esophagus. This can create a burning sensation commonly called *heartburn,* although the heart is not involved. In infants under a year of age, the lower esophageal sphincter may function erratically, causing them to "spit up" following meals. Certain mammals, such as rodents, have a true lower esophageal sphincter and cannot regurgitate, which is why poison grains are effective in killing mice and rats.

Swallowing Mechanisms

Swallowing, or **deglutition** (*de''gloo-tish'un*), is the complex mechanical and physiological act of moving food or fluid from the oral cavity to the stomach. For descriptive purposes, deglutition is divided into three stages.

The first deglutitory stage is voluntary and follows mastication, if food is involved. During this stage, the mouth is closed and breathing is temporarily interrupted. A bolus is formed as the tongue is elevated against the palate through contraction of the mylohyoid and styloglossus muscles and the intrinsic muscles of the tongue.

The second stage of deglutition is the passage of the bolus through the pharynx. The events of this stage are involuntary and are elicited by stimulation of sensory receptors located at the opening of the oropharynx. Pressure of the tongue against the rugae of the hard palate seals the nasopharynx from the oral cavity, creates a pressure, and forces the bolus into the oropharynx. The soft palate and pendulant uvula are elevated to close the nasopharynx as the bolus passes. The hyoid bone and the larynx are also elevated. Elevation of the larynx against the epiglottis seals the glottis so that food or fluid is less likely to enter the trachea. Sequential contraction of the constrictor muscles of the pharynx moves the bolus through the pharynx to the esophagus. This stage is completed in 1 second or less.

Figure 18.15 An anteroposterior radiograph of the esophagus showing peristaltic contraction and movement of a bolus into the stomach.

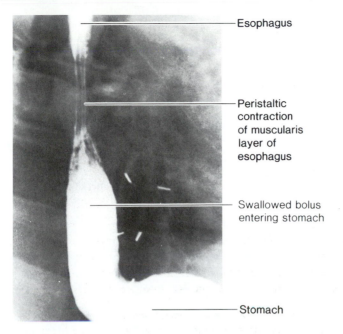

Esophagus

Peristaltic contraction of muscularis layer of esophagus

Swallowed bolus entering stomach

Stomach

Figure 18.16 The stomach. (a) As seen from a cadaver, the stomach is a J-shaped organ. (b) A radiograph of the stomach is of clinical value for detecting ulcers, constrictions, or ingested objects. A patient swallows radiopaque barium, which coats the lining of the stomach and duodenum. These structures and any abnormalities will then show up clearly in the radiograph.

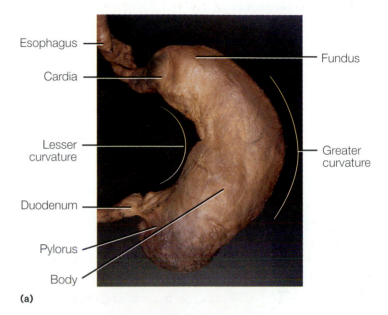

Esophagus

Cardia

Lesser curvature

Duodenum

Pylorus

Body

Fundus

Greater curvature

(a)

Fundus

Body

Pylorus

Duodenum

(b)

The third stage, the entry and passage of food through the esophagus, is also involuntary. The bolus is moved through the esophagus by peristalsis (fig. 18.15). The entire time for deglutition varies, but it is slightly more than 1 second in the case of fluids and 5 to 8 seconds with solid food material.

Stomach

The **stomach** is the most distensible part of the GI tract and is positioned in the upper left quadrant of the peritoneal cavity immediately below the diaphragm. It is a J-shaped pouch that is continuous with the esophagus superiorly and empties into the duodenal portion of the small intestine inferiorly (fig. 18.16). The functions of the stomach are to store food as it is mechanically churned with gastric secretions; to initiate the digestion of proteins; to carry out limited absorption; and to move food into the small intestine as a pasty material called *chyme (kīm)*.

The stomach is divided into four regions: the cardia, fundus, body, and pylorus (fig. 18.17). The **cardia** is the narrow upper region immediately below the lower esophageal sphincter. The **fundus** is the dome-shaped portion to the left and in direct contact with the diaphragm. The **body** is the large central portion, and the **pylorus** is the funnel-shaped terminal portion. The pylorus communicates with the duodenal portion of the small intestine through a **pyloric sphincter.** *Pylorus* is a Greek word meaning gatekeeper, and this junction is just that, regulating the movement of chyme into the small intestine and prohibiting backflow.

chyme: L. *chymus,* juice
cardia: Gk. *kardia,* heart (upper portion, nearer the heart)
fundus: L. *fundus,* bottom
pylorus: Gk. *pyloros,* gatekeeper

The stomach has two surfaces and two borders. The broadly rounded surfaces are referred to as the **anterior** and **posterior surfaces.** The medial concave border is the **lesser curvature** (fig. 18.17), and the lateral convex border is the **greater curvature.** The lesser omentum extends between the lesser curvature and the liver and the greater omentum is attached to the greater curvature.

The wall of the stomach consists of the same four layers found in other regions of the GI tract, with certain modifications. The muscularis is composed of three layers of smooth muscle named according to the direction of fiber arrangement: an outer **longitudinal layer,** a middle **circular layer,** and an

UNIT VI

Cardia

Esophagus

Adventitia

Longitudinal muscle

Fundus

Body

Lesser curvature

Duodenum

Circular muscle

Gastric rugae

Oblique muscle

Pyloric sphincter

Greater curvature

Pylorus

Loechel

Submucosa

Mucosa

inner **oblique layer** (fig. 18.17). The circular muscle layer is further thickened at the gastroduodenal junction to form the pyloric sphincter.

The mucosa is shaped into numerous longitudinal folds called **gastric rugae** that permit stomach distension. The gastric rugae gradually smooth out as the stomach fills. The mucosal layer contains a simple columnar epithelium. The entire mucosal layer is folded to form **gastric pits** and **gastric glands** (figs. 18.18 and 18.19).

Gastric glands have several kinds of secreting cells: *chief,* or *zymogenic cells, parietal cells, mucous cells, argentaffin cells,* and *G cells.* The chief cells secrete various digestive enzymes and the parietal cells produce hydrochloric acid. The product of these two cell types is called *gastric juice.* Mucous cells secrete large quantities of protective mucus. Argentaffin cells secrete serotonin and histamine, and G cells secrete the hormone gastrin. In addition to these products, the gastric mucosa (probably the parietal cells) secretes a polypeptide called *intrinsic factor,* which is required for absorption of vitamin B_{12} in the intestine.

Figure 18.18 The histological structure of the mucosa of the stomach.

Columnar epithelium of mucosal ridge

Gastric pit

Gastric glands with chief and parietal cells

Lamina propria

argentaffin cells: L. *argentum,* silver; *affinis,* attraction (become colored with silver stain)

Figure 18.19 Gastric pits and gastric glands of the mucosa. (a) Gastric pits are the openings of the gastric glands. (b) Gastric glands consist of mucous cells, chief cells, and parietal cells, each of which produces a specific secretion.

The stomach is sensitive to emotional stress. Mucus, secreted by mucous cells of the stomach, is important in preventing hydrochloric acid and the digestive enzyme pepsin from eroding the stomach wall. *Peptic ulcers* may be caused by an increase in cellular secretion or by not enough protective mucus being secreted. Another protective feature is the rapid mitotic activity of the columnar epithelium of the stomach. The entire lining of the stomach is usually replaced every few days. Nevertheless, in the United States approximately 10% of the male population and 4% of the female population develop peptic ulcers.

Table 18.5	Phases of Gastric Secretion
Phase	**Response**
Cephalic phase	Parasympathetic response via vagus nerves; reflexes respond to sight, taste, smell, or mental stimuli; 50–150 mℓ of gastric juice is secreted
Gastric phase	Food in the stomach stretches the mucosa and chemical breakdown of protein stimulates the release of gastrin; gastrin stimulates the production of 600–750 mℓ of gastric juice
Intestinal phase	Chyme entering the duodenum stimulates intestinal cells to release intestinal gastrin; intestinal gastrin stimulates the production of additional small quantities of gastric juice

Regulation of gastric activity is autonomic. The sympathetic neurons arise from the celiac plexus and the parasympathetic neurons, from the vagus (tenth cranial) nerves. Parasympathetic neurons synapse in the myenteric plexus between the muscular layers and in the submucosal plexus in the submucosa. Parasympathetic impulses promote gastric activity, the phases of which are presented in table 18.5.

Vomiting is a reflexive response of emptying the stomach through the esophagus, pharynx, and oral cavity. This action is controlled by the **vomiting center** of the medulla oblongata. Stimuli within the GI tract, especially the duodenum, may activate the vomiting center, as may nauseating odors or sights, motion sickness, or body stress. Various drugs called *emetics* can also stimulate a vomiting reflex. The mechanics of vomiting are as follows: (1) strong, sustained contractions of the upper small intestine, followed by a contraction of the pyloric sphincter; (2) relaxation of the lower esophageal sphincter and contraction of the pyloric portion of the stomach; (3) a shallow inspiration and closure of the glottis; and (4) compression of the stomach against the liver by contraction of the diaphragm and the abdominal muscles. This reflexive sequence causes a forceful ejection of vomit.

The feeling of *nausea* is caused by stimuli in the vomiting center and may or may not cause vomiting.

UNIT VI

Figure 18.20 The small intestine. (*a*) The regions and the mesenteric attachment. (*b*) A section of the intestinal wall showing the folds of the mucosa and submucosa into structures called plicae circulares.

(a)

(b)

Gordon

The only function of the stomach that appears to be essential for life is the secretion of *intrinsic factor*. This polypeptide is needed for the intestinal absorption of vitamin B_{12}, which is required for maturation of red blood cells in the bone marrow. A patient with a gastrectomy has to receive B_{12} orally (together with intrinsic factor) or through injections so that he or she will not develop *pernicious anemia*.

1. Describe the three stages in deglutition, including the structures involved.
2. Describe the structure and function of the lower esophageal sphincter.
3. List the functions of the stomach. What is the function of the gastric rugae?
4. Identify the modifications of the stomach that aid in mechanical and chemical digestion.

Small Intestine

The small intestine, consisting of the duodenum, jejunum, and ileum, is the site where digestion is completed and nutrients are absorbed. The surface area of the intestinal wall is increased by plicae circulares, villi, and microvilli.

Objective 12. Describe the location, support, and regions of the small intestine.

Objective 13. List the functions of the small intestine and describe the anatomical specializations that accomplish these functions.

Objective 14. Describe the movements that occur within the small intestine.

The **small intestine** is that portion of the GI tract between the pyloric sphincter of the stomach and the ileocecal valve opening into the large intestine. It is positioned in the central and lower portions of the abdominal cavity and is supported, except for the first portion, by the **mesentery** (fig. 18.20). The attachment of the fan-shaped mesentery to the small intestine maintains mobility but leaves little chance for the intestine to become twisted or kinked. Enclosed within the mesentery are blood vessels, nerves, and lymphatic vessels that supply the intestinal wall.

The small intestine is approximately 3 m (10 ft) long and 2.5 cm (1 in.) wide in a living person, but it will measure nearly twice this length in a cadaver when the muscular wall is relaxed. It is called the small intestine because of its relatively small diameter compared to that of the large intestine. The functions of the small intestine are the reception of the secre-

UNIT VI

Table 18.6	Digestive Enzymes	
Enzyme	**Source**	**Digestive Action**
Salivary enzyme		
Amylase	Salivary glands	Begins carbohydrate digestion by converting starch and glycogen to disaccharides
Gastric juice enzyme		
Pepsin	Gastric glands	Begins the digestion of nearly all types of proteins
Intestinal juice enzymes		
Peptidase	Intestinal glands	Converts proteins into amino acids
Sucrase	Intestinal glands	Converts disaccharides into monosaccharides
Maltase		
Lactase		
Lipase	Intestinal glands	Converts fats into fatty acids and glycerol
Amylase	Intestinal glands	Converts starch and glycogen into disaccharides
Nuclease	Intestinal glands	Converts nucleic acids into nucleotides
Enterokinase	Intestinal glands	Activates trypsin
Pancreatic juice enzymes		
Amylase	Pancreas	Converts starch and glycogen into disaccharides
Lipase	Pancreas	Converts fats into fatty acids and glycerol
Peptidases	Pancreas	Converts proteins or partially digested proteins into amino acids
Trypsin		
Chymotrypsin		
Carboxypeptidase		
Nuclease	Pancreas	Converts nucleic acids into nucleotides

From John W. Hole, Jr., *Human Anatomy and Physiology*, 6th ed. Copyright © 1993 Wm. C. Brown Communications, Inc., Dubuque, Iowa. All Rights Reserved. Reprinted by permission.

tions from the liver and pancreas, chemical and mechanical breakdown of chyme, absorption of nutrients, and transportation of the remaining undigested material to the large intestine. A summary of the digestive enzymes of the GI tract is presented in table 18.6.

The small intestine is innervated by the **superior mesenteric** *(mes″en-ter′ik)* **plexus.** The branches of the plexus contain sensory neurons, postganglionic sympathetic neurons, and preganglionic parasympathetic neurons. The small intestine's arterial blood supply comes through the superior mesenteric artery and small branches from the celiac trunk and the inferior mesenteric artery. Venous drainage is through the superior mesenteric vein, which unites with the splenic vein to form the hepatic portal vein carrying nutrient-rich blood to the liver.

Regions of the Small Intestine

The small intestine is divided into three regions on the basis of function and histological structure. These three regions are the duodenum *(doo″ŏ-de′num* or *doo-od′ĕ-num)*, jejunum *(jĕ-joo′num)*, and ileum *(il′e-um)*.

The **duodenum** is a relatively fixed, C-shaped tube, measuring approximately 25 cm (10 in.) from the pyloric sphincter of the stomach to the **duodenojejunal** *(du-od″ĕ-no″jĕ-joo′nal)* **flexure.** The concave surface of the duodenum faces to the left, where it receives bile secretions through the **common bile duct** from the liver and gallbladder and pancreatic secretions through the **pancreatic duct** of the pancreas (fig. 18.21). These two ducts unite to form a common entry into the duodenum called the **hepatopancreatic ampulla** *(hep″ă-to-pan″kre-at′ik am-pool′ă)* (or ampulla of Vater), which pierces the duodenal

wall and drains into the duodenum from an elevation called the **duodenal papilla.** The duodenum is retroperitoneal, except for a short portion near the stomach. It differs histologically from the rest of the small intestine by the presence of **duodenal** (Brunner's) **glands** in the submucosa (fig. 18.22). These compound tubuloalveolar glands secrete mucus and are most numerous near the superior end of the duodenum.

The **jejunum** is approximately 1 m (3 ft) long and extends from the duodenum to the ileum. The jejunum has a slightly larger lumen and more internal folds than does the ileum but lacks unique histological structures.

The **ileum** makes up the remaining 2 m (6–7 ft) of the small intestine. The terminal portion of the ileum empties into the medial side of the cecum through the **ileocecal** *(il″e-ŏ-se′kal)* **valve.** The walls of the ileum have an abundance of lymphatic tissue. Aggregates of lymph nodules, called **mesenteric** (Peyer's) **patches,** are characteristic of the ileum.

Structural Modifications of the Small Intestinal Wall

The products of digestion are absorbed across the epithelial lining of the intestinal mucosa. Absorption occurs primarily in the jejunum, although some also occurs in the duodenum and ileum. The many folds in the small intestine and modifications in the structure of its walls provide a large mucosal surface area; hence, absorption occurs at a rapid rate. The mucosa and submucosa form large folds called the **plicae** *(pli′se)* **circulares,** which can be observed with the unaided eye. The surface area is further increased by the microscopic folds of mucosa called

---

duodenum: L. *duodeni,* twelve each (length of twelve fingers' breadth)
ampulla of Vater: from Abraham Vater, German anatomist, 1684–1751

Brunner's glands: from Johann C. Brunner, Swiss anatomist, 1653–1727
Peyer's patches: from Johann K. Peyer, Swiss anatomist, 1653–1712
plica: L. *plicatus,* folded

Figure 18.21 The duodenum and associated structures.

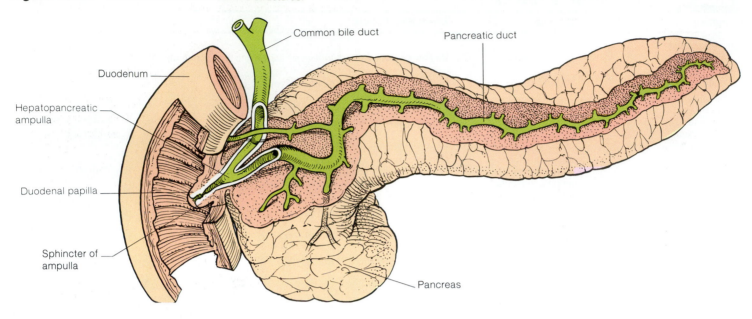

Common bile duct

Pancreatic duct

Duodenum

Hepatopancreatic ampulla

Duodenal papilla

Sphincter of ampulla

Pancreas

Figure 18.22 The histology of the duodenum.

Villus
(lined with simple columnar epithelium)

Lamina propria

Muscularis mucosa

Duodenal glands

villi and by the foldings of the apical cell membrane of epithelial cells (which can only be seen with an electron microscope), called *microvilli.*

Each **villus** is a fingerlike fold of mucosa that projects into the intestinal lumen. The villi are covered with columnar epithelial cells, and interspersed among these are the mucus-secreting goblet cells. The lamina propria forms the connective tissue core of each villus and contains numerous lymphocytes,

villus: L. *villosus*, shaggy

blood capillaries, and a lymphatic vessel called the **lacteal** (fig. 18.23). Absorbed monosaccharides and amino acids enter the blood capillaries; absorbed fatty acids enter the lacteals.

Exposed epithelial cells of the villi are continuously exfoliated and are replaced by deeper cells that are pushed up from the germinal layer. The epithelium at the base of the villi invaginates downward at various points to form tubular **intestinal glands** (crypts of Lieberkühn), which extend downward into the mucous membrane (fig. 18.24). The intestinal glands secrete digestive enzymes and are also the site for the production of new epithelial cells. In a light microscope, the **microvilli** produce a somewhat vague **brush border** on the edges of the columnar epithelial cells. The term *brush border* is thus often used synonymously with microvilli in descriptions of the small intestine (fig. 18.25).

Mechanical Activities of the Small Intestine

Contractions of the longitudinal and circular muscles of the small intestine produce three distinct types of movement: rhythmic segmentation, pendular movements, and peristalsis.

Rhythmic segmentations are local, ringlike contractions of the circular muscular layer that occur at the rate of about 12 to 16 per minute in regions containing chyme. This action churns chyme with digestive juices and brings it in contact with the mucosa. During these contractions, the vigorous motion of the villi stirs the chyme and facilitates absorption.

Pendular movements primarily occur in the longitudinal muscle layer. In this motion, a constrictive wave moves

lacteal: L. *lacteus,* milk
crypts of Lieberkühn: from Johann N. Lieberkühn, German anatomist,
1711–56

Figure 18.23 A diagram of the structure of an intestinal villus.

Villus

Simple columnar epithelium

Lacteal

Capillary network

Goblet cells

Intestinal crypt

Lymph vessel

Arteriole

Venule

Waldrop

Figure 18.24 Villi and intestinal crypts are specializations of the mucosa of the small intestine.

Villi

Paras

Lamina propria

Muscularis mucosa

Opening of crypt

Intestinal crypt

Figure 18.25 An electron micrograph of microvilli (arrow) at the exposed surface of a columnar epithelial cell in the small intestine.

along a segment of the small intestine and then reverses and moves in the opposite direction, moving the chyme back and forth. Pendular movements also mix the chyme but do not seem to have a particular frequency.

Peristalsis is responsible for the propulsive movement of the chyme through the small intestine. These wavelike contractions are usually weak and relatively short, occurring at a frequency of about 15 to 18 per minute. Chyme requires 3 to 10 hours to travel the length of the small intestine. Both muscle layers are involved in peristalsis.

 The sounds of digestive peristalsis can be easily heard through a stethoscope placed at various abdominal locations. These sounds can be detected even through clothing. The sounds, mostly clicks and gurgles, occur at a frequency of 5 to 30 per minute.

1. Describe the small intestine. Where is it located? How is it subdivided? How is it supported?
2. What are the functions of the small intestine?
3. List three structural modifications of the small intestine that increase its absorptive surface area.
4. Describe the movements of the small intestine. Which movements are produced by the circular layer of the tunica muscularis?
5. Which region of the small intestine is the longest? Which is the shortest? How long does it take a portion of chyme to move through the small intestine?

Large Intestine

The large intestine absorbs water and electrolytes from the chyme it receives from the small intestine and, through the action of sphincter muscles, passes waste products out of the body through the rectum and anal canal.

Figure 18.26 The large intestine.

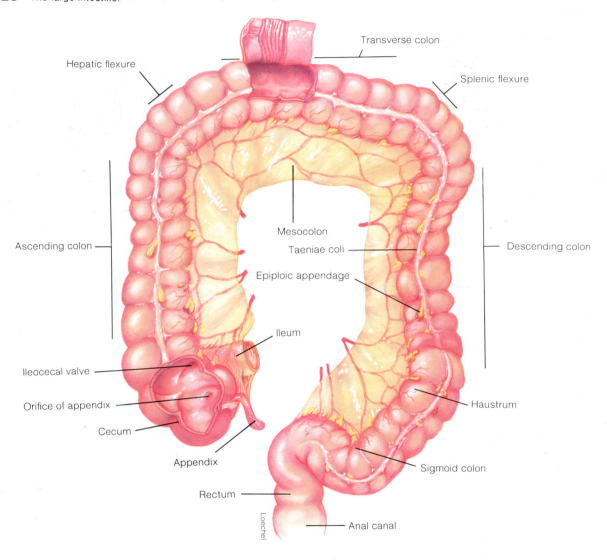

UNIT VI

| **Objective 15.** | Identify the regions of the large intestine and describe its gross and histological structure. |

| **Objective 16.** | Describe the functions of the large intestine and explain how defecation is accomplished. |

The **large intestine** is about 1.5 m (5 ft) long and 6.5 cm (2.5 in.) in diameter. It is called the large intestine because its diameter is larger than that of the small intestine. The large intestine begins at the terminal end of the ileum in the lower right quadrant of the abdominal cavity. From there, it leads superiorly on the right side to just below the liver, then crosses to the left, descends into the pelvis, and terminates at the anus. A specialized portion of the mesentery, the **mesocolon,** supports the transverse portion of the large intestine along the posterior abdominal wall.

The large intestine has little or no digestive function, but it does absorb water and electrolytes from the remaining chyme.

In addition, the large intestine functions to form, store, and expel *feces* from the body.

Regions and Structures of the Large Intestine

The large intestine is structurally divided into the cecum (*se′kum*), colon, rectum, and anal canal (figs. 18.26 and 18.27).

The **cecum** is a dilated pouch that hangs inferiorly, slightly below the ileocecal valve. The ileocecal valve is a fold of mucous membrane at the junction of the small and large intestine that prohibits the backflow of chyme. A fingerlike projection called the **appendix** is attached to the inferior medial margin of the cecum. The 8 cm (3 in.) appendix has an abundance of lymphatic tissue (fig. 18.28), which may serve to resist infection.

cecum: L. *caecum*, blind pouch
appendix: L. *appendix*, attachment

Figure 18.27 A radiograph after a barium enema showing the regions, flexures, and the haustra of the large intestine.

Figure 18.28 The histology of the appendix.

Center of lymphatic nodule

Intestinal glands

Lumen

Muscularis

Submucosa

Serosa

A common disorder of the large intestine is inflammation of the appendix, or *appendicitis*. Wastes that accumulate in the appendix cannot be moved easily by peristalsis since the appendix has only one opening. The symptoms of appendicitis include muscular rigidity, localized pain in the lower right quadrant, loss of appetite, and occasionally vomiting. The chief danger of appendicitis is that the appendix might rupture, resulting in *peritonitis*.

The open, superior portion of the cecum is continuous with the colon. The colon consists of ascending, transverse, descending, and sigmoid portions (fig. 18.26). The **ascending colon** extends superiorly from the cecum along the right abdominal wall to the inferior surface of the liver. Here the colon bends sharply to the left at the **hepatic flexure (right colic flexure)** and transversely crosses the upper abdominal cavity as the **transverse colon.** At the left abdominal wall, another right-angle bend called the **splenic flexure (left colic flexure)** marks the beginning of the **descending colon.** The descending colon traverses inferiorly along the left abdominal wall to the pelvic region. The colon then angles medially from the brim of the pelvis to form an S-shaped bend, known as the **sigmoid colon.**

The terminal 20 cm (7.5 in.) of the GI tract is the **rectum,** and the last 2 to 3 cm of the rectum is referred to as the **anal canal** (fig. 18.29). The rectum lies anterior to the sacrum where it is firmly attached by peritoneum. The **anus** is the external opening of the anal canal. Two sphincter muscles guard the

colon: Gk. *kolon,* member of the whole
sigmoid: Gk. *sigmoeides,* shaped like a sigma, Σ
rectum: L. *rectum,* straight tube
anus: L. *anus,* ring

Figure 18.29 The anal canal.

Rectum

Levator ani muscle

Anal canal

Hemorrhoidal vessels

Anal columns

External anal sphincter

Internal anal sphincter

Anus

Loechel

anal opening: the **internal anal sphincter,** which is composed of smooth muscle fibers, and the **external anal sphincter,** composed of skeletal muscle. The mucous membrane of the anal canal is arranged into highly vascular, longitudinal folds called **anal columns.**

A *hemorrhoid (hem'o-roid)* is a mass of varicose veins in the anal area caused, in part, by difficulty in defecating. Hemorrhoids, in reference to the condition in which such masses occur, are also called *piles.* A first-degree hemorrhoid is internally contained within the anal canal. A second-degree hemorrhoid prolapses, or

UNIT VI

Figure 18.30 The histology of the large intestine. (*a*) A photomicrograph of the tunics. (*b*) A scanning electron micrograph of the mucosa. The arrow indicates the opening of a goblet cell into the intestinal lumen. (From: R. G. Kessel and R. H. Kardon, *Tissues and Organs: A Text-Atlas of Scanning Electron Microscopy*, © 1979 W. H. Freeman and Company.)

Serosa

Muscularis

Submucosa

Mucosa

Columnar epithelium with goblet cells

(a)

(b)

extends outward during defecation. A third-degree hemorrhoid remains prolapsed through the anal orifice. Rubber band constriction is a common medical treatment for a prolapsed hemorrhoid. In this technique, a rubber band is tied around the hemorrhoid, constricting its blood supply so that the tissue dries and falls off. In a relatively new treatment, infrared photocoagulation, a high-energy light beam coagulates the hemorrhoid.

Although the large intestine consists of the same tunics as the small intestine, there are some structural differences. The large intestine lacks villi but does have numerous goblet cells in the mucosal layer (fig. 18.30). The longitudinal muscle layer of the muscularis forms three distinct muscle bands called **taeniae coli** *(te'ne-e co'li)* that run the length of the large intestine. A series of bulges in the walls of the large intestine form sacculations, or **haustra** *(hows'tra),* along its entire length (see figs. 18.26 and 18.27). Finally, the large intestine has small but numerous fat-filled pouches called **epiploic appendages** (see fig. 18.26) that are attached superficially to the taeniae coli in the serous layer.

The large intestine has both types of autonomic innervation. The sympathetic innervation arises from superior and inferior mesenteric plexuses, as well as from the celiac plexus. The parasympathetic innervation arises from the paired pelvic splanchnic and vagus nerves. Sensory fibers from the large intestine respond to bowel pressure and signal the need to defecate.

Branches from the superior mesenteric and inferior mesenteric arteries supply blood to the large intestine. Venous blood is returned through the superior and inferior mesenteric veins, which are tributaries to the hepatic portal vein that drains the liver.

taenia: L. *tainia,* a ribbon
haustrum: L. *haustrum,* bucket or scoop
epiploic: Gk. *epiplein,* to float on

Mechanical Activities of the Large Intestine

Chyme enters the large intestine through the ileocecal valve. About 15 mℓ of pasty material enters the cecum with each rhythmic opening of the valve. The ingestion of food intensifies peristalsis of the ileum and increases the frequency with which the ileocecal valve opens; this is called the **gastroileal reflex.** Material entering the large intestine accumulates in the cecum and ascending colon.

Three types of movements occur throughout the large intestine: peristalsis, haustral churning, and mass movement. **Peristaltic movements** of the colon are similar to those of the small intestine, though they are usually more sluggish. In **haustral churning,** a relaxed haustrum is filled until it reaches a certain point of distension, and then the muscularis layer is stimulated to contract. This contraction not only moves the material to the next haustrum but churns the contents and exposes it to the mucosa where water and electrolytes are absorbed. **Mass movement** is a very strong peristaltic wave, involving the action of the taeniae coli, which moves the colonic contents toward the rectum. Mass movements generally occur only two or three times a day, usually during or shortly after a meal. This response to eating is called the **gastrocolic reflex** and can best be observed in infants who have a bowel movement during or shortly after feeding.

As material passes through the large intestine, Na$^+$, K$^+$, and water are absorbed. It has been estimated that an average of 850 mℓ of water per day is absorbed across the mucosa of the colon. The waste material that is left then passes to the rectum, leading to an increase in rectal pressure and the urge to defecate. If the urge to defecate is denied, feces are prevented from entering the anal canal by the internal anal sphincter. In this case, the feces remain in the rectum and may even back up into the sigmoid colon. The **defecation reflex** normally occurs when the rectal pressure rises to a particular level that is determined, to

Table 18.7	Mechanical Activity in the GI Tract			
Region	**Type of Motility**	**Frequency**	**Stimulus**	**Result**
Oral cavity	Mastication	Variable	Initiated voluntarily, proceeds reflexively	Subdivision, mixing with saliva
Oral cavity and pharynx	Deglutition	Maximum of 20 per min	Initiated voluntarily, reflexively controlled by swallowing center	Clears oral cavity of food
Esophagus	Peristalsis	Depends on frequency of swallowing	Initiated by swallowing	Movement through the esophagus
Stomach	Receptive relaxation	Matches frequency of swallowing	Unknown	Permits filling of stomach
	Tonic contraction	15–20 per min	Autonomic plexuses	Mixing and churning
	Peristalsis	1–2 per min	Autonomic plexuses	Evacuation of stomach
	Hunger contractions	3 per min	Low blood sugar level	Feeding
Small intestine	Peristalsis	15–18 per min	Autonomic plexuses	Transfer through intestine
	Rhythmic segmentation	12–16 per min	Autonomic plexuses	Mixing
	Pendular movements	Variable	Autonomic plexuses	Mixing
Large intestine	Peristalsis	3–12 per min	Autonomic plexuses	Transport
	Mass movements	2–3 per day	Stretch	Fills sigmoid colon
	Haustral churning	3–12 per min	Autonomic plexuses	Mixing
	Defecation	Variable: 1 per day to 3 per week	Reflex triggered by rectal distension	Defecation

a large degree, by habit. At this point, the internal anal sphincter relaxes to admit feces into the anal canal.

During the act of defecation the longitudinal rectal muscles contract to increase rectal pressure and the internal and external anal sphincter muscles relax. Excretion is aided by contractions of abdominal and pelvic skeletal muscles, which raise the intra-abdominal pressure and help push the feces from the rectum through the anal canal and out the anus.

The various mechanical activities of the GI tract are summarized in table 18.7.

Constipation occurs when fecal material accumulates because of longer-than-normal periods between defecations. The slower rate of elimination allows more time for water absorption, so that the waste products become harder. Although uncomfortable and sometimes painful, this condition is usually not serious. *Diarrhea* occurs when waste material passes too quickly through the colon, so that insufficient time is allowed for water absorption. Excessive diarrhea can result in dangerous levels of dehydration and electrolyte imbalance, particularly in infants because of their small body size.

1. Identify the four principal regions of the large intestine and describe the functions of the colon.
2. Describe the haustra and the taeniae coli and explain their role in the movements of the large intestine.
3. Describe the location of the rectum, anal canal, and anal sphincter muscles and explain how defecation is accomplished.

Liver, Gallbladder, and Pancreas

The liver, consisting of four lobes, processes nutrients and secretes bile, which is stored and concentrated in the gallbladder prior to discharge into the duodenum. The pancreas, consisting of endocrine (islet) cells and exocrine (acini) cells, secretes important hormones into the blood and essential digestive enzymes into the duodenum.

Objective 17. Describe the location, structure, and functions of the liver.

Objective 18. Describe the location of the gallbladder and trace the flow of bile through the systems of ducts into the duodenum.

Objective 19. Describe the location, structure, and functions of the pancreas.

Three accessory digestive organs in the abdominal cavity aid in the chemical breakdown of food. These are the liver, gallbladder, and pancreas. The liver and pancreas function as exocrine glands in this process because their secretions are transported to the lumen of the GI tract via ducts.

Liver

The **liver** is the largest internal organ of the body, weighing about 1.3 kg (3.5–4.0 lbs) in an adult. It is positioned immediately beneath the diaphragm in the epigastric and right hypochondriac regions (see fig. 2.15 and table 2.4) of the abdominal cavity. The liver is reddish-brown in color because of its great vascularity.

The liver has two major lobes and two minor lobes. Anteriorly, the **right lobe** is separated from the smaller **left lobe**

Figure 18.31 The liver. (*a*) An anterior view of the gross structure. (*b*) An inferior view showing the lobes of the liver, gallbladder, hepatic vessels, and ducts. (*c*) A cross section of the abdomen showing the relative position of the liver to other abdominal organs.

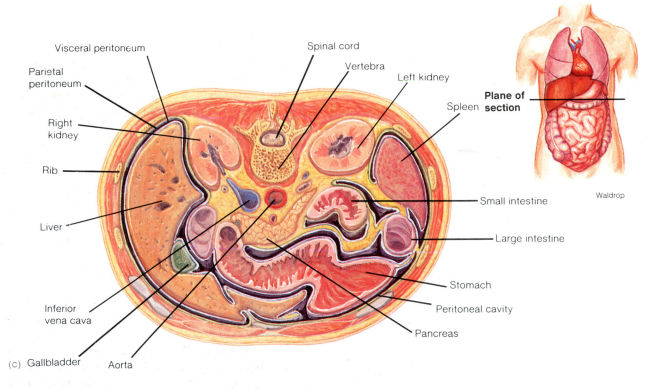

by the **falciform ligament** (fig. 18.31). Inferiorly, the **caudate** *(kaw'dāt)* **lobe** is near the inferior vena cava and the **quadrate lobe** is adjacent to the gallbladder. The falciform ligament attaches the liver to the anterior abdominal wall and the diaphragm. A **ligamentum teres** (round ligament) is continuous along the free border of the falciform ligament to the umbilicus. The ligamentum teres is the remnant of the umbilical vein of the fetus. The **porta** of the liver is where the hepatic artery, portal vein, lymphatics, and nerves enter the liver and where the **hepatic ducts** exit.

Although the liver is the largest internal organ, it is, in a sense, only one to two cells thick. This is because the liver cells, or **hepatocytes,** form **hepatic plates** that are one to two cells thick and separated from each other by large capillary spaces called **sinusoids** (fig. 18.32). The sinusoids are lined with phagocytic **Kupffer** *(koop'fer)* **cells,** but the large intercellular

---

falciform: L. *falcis,* sickle; *forma,* form
porta: L. *porta,* gate

hepatic: Gk. *hepatos,* liver
Kupffer cells: from Karl Wilhelm von Kupffer, Bavarian anatomist and embryologist, 1829–1902

Figure 18.32 The structure of the liver. (*a*) A scanning electron micrograph of the liver. Hepatocytes are arranged in plates so that blood that passes through sinusoids (*b*) will be in contact with each liver cell.

(Photo from: R. G. Kessel and R. H. Kardon, *Tissues and Organs: A Text-Atlas of Scanning Electron Microscopy,* © 1979 W. H. Freeman and Company.)

- Portal vein
- Hepatic artery
- Bile duct

(a)

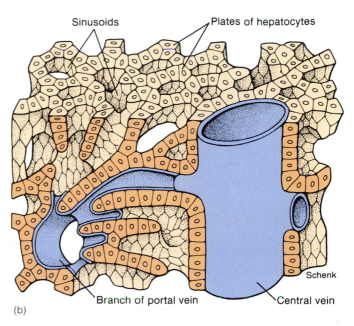

Sinusoids — Plates of hepatocytes

Branch of portal vein — Central vein — Schenk

(b)

gaps between adjacent Kupffer cells make these sinusoids more highly permeable than other capillaries. The plate structure of the liver and the high permeability of the sinusoids allow each hepatocyte to be in direct contact with the blood.

The hepatic plates are arranged into functional units called **liver lobules** (fig. 18.33). In the middle of each lobule is a **central vein,** and at the periphery of each lobule are branches of the hepatic portal vein and of the hepatic artery, which open into the spaces *between* hepatic plates. Arterial blood and portal venous blood, containing molecules absorbed in the GI tract, thus mix as the blood flows within the sinusoids from the periphery of the lobule to the central vein (fig. 18.34). The central veins of different liver lobules converge to form the hepatic vein, which carries blood from the liver to the inferior vena cava.

The liver lobules carry out numerous functions, including synthesis, storage, and release of vitamins; synthesis, storage, and release of glycogen; synthesis of blood proteins; phagocytosis of old red blood cells and certain bacteria; removal of toxic substances; and production of bile. Bile is stored in the gallbladder and is eventually secreted into the duodenum for the emulsification (breaking down into smaller particles) and absorption of fats.

Bile is produced by the hepatocytes and secreted into thin channels called **bile canaliculi** *(kan″ă-lik′yoo-li)* located *within* each hepatic plate (fig. 18.34). These bile canaliculi are drained at the periphery of each lobule by **bile ducts,** which in turn drain into **hepatic ducts** that carry bile away from the liver. Since blood travels in the sinusoids and bile travels in the opposite direction within the hepatic plates, blood and bile do not mix in the liver lobules.

The liver receives parasympathetic innervation from the vagus nerves and sympathetic innervation from thoracolumbar nerves through the celiac ganglia.

Gallbladder

The **gallbladder** is a saclike organ attached to the inferior surface of the liver (see figs. 18.21 and 18.31). This organ stores and concentrates bile, which drains to it from the liver by way of the bile ducts, hepatic duct, and **cystic duct,** respectively. A sphincter valve at the neck of the gallbladder allows a storage capacity of about 35 to 50 mℓ. The inner mucosal layer of the gallbladder is arranged in rugae similar to those of the stomach. When the gallbladder fills with bile, it expands to the size and shape of a small pear. Bile is a yellowish-green fluid containing bile salts, bilirubin, (a product resulting from a breakdown of blood), cholesterol, and other compounds. Contraction of the muscularis ejects bile from the cystic duct into the **common bile duct,** which conveys bile into the duodenum (fig. 18.35).

Bile is continuously produced by the liver and drains through the hepatic and common bile ducts to the duodenum. When the small intestine is empty of food, the **sphincter of ampulla** (Oddi) constricts, and bile is forced up the cystic duct to the gallbladder for storage.

The gallbladder is supplied with blood from the cystic artery, which branches from the right hepatic artery. Venous blood is returned through the cystic vein, which empties into the hepatic portal vein. Autonomic innervation of the gallbladder is similar to that of the liver; both receive parasympathetic innervation from the vagus nerves and sympathetic innervation from thoracolumbar nerves through the celiac ganglia.

UNIT VI

cystic: Gk. *kystis,* pouch
sphincter of Oddi: from Ruggero Oddi, Italian physician, nineteenth century

Figure 18.33 A liver lobule and the histology of the liver. (*a*) A cross section of a liver lobule and (*b*) a longitudinal section. Blood enters a liver lobule through the vessels in a hepatic triad, passes through hepatic sinusoids, and leaves the lobule through a central vein. The central veins converge to form hepatic veins that transport venous blood from the liver. (*c*) A photomicrograph of a liver lobule in cross section.

Figure 18.34 The flow of blood and bile in a liver lobule. Blood flows within sinusoids from a portal vein to the central vein (from the periphery to the center of a lobule). Bile flows within hepatic plates from the center to bile ducts at the periphery of a lobule.

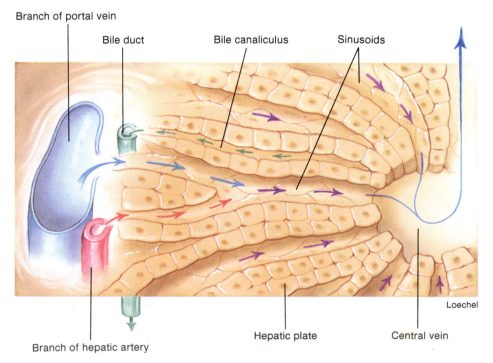

UNIT VI

Figure 18.35 The pancreatic duct joins the common bile duct to empty its secretions through the duodenal papilla into the duodenum. The release of bile and pancreatic juice into the duodenum is controlled by the sphincter of ampulla (Oddi).

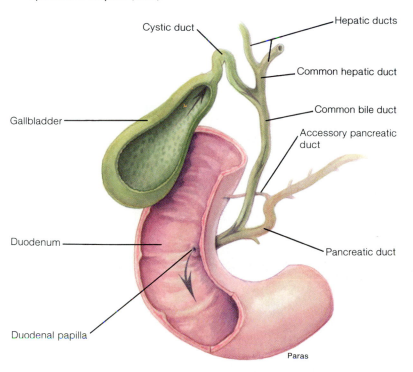

Figure 18.36 (a) A radiograph of a gallbladder that contains gallstones. (b) A posterior view of a gallbladder that has been removed (cholecystectomy) and cut open to reveal its gallstones (biliary calculi). (A dime is placed in the photo to show relative size.)

(a)

(b)

A common clinical problem of the gallbladder is the development of *gallstones*. Bile is composed of various salts, pigments, and cholesterols that become concentrated as water is removed. Normally cholesterols remain in solution, but under certain conditions they precipitate to form solid crystals. Large crystals may block the bile duct and have to be surgically removed. The radiograph in figure 18.36 shows gallstones in position and the photograph shows removed gallstones.

Pancreas

The **pancreas** is a soft, lobulated, glandular organ that has both exocrine and endocrine functions. The endocrine function is performed by clusters of cells called the **pancreatic islets** (islets of Langerhans) that secrete the hormones insulin and glucagon into the blood. As an exocrine gland, the pancreas secretes pancreatic juice through the pancreatic duct (fig. 18.37) into the duodenum. The pancreas is horizontally positioned along the posterior abdominal wall, adjacent to the greater curvature of the stomach. The pancreas is about 12.5 cm (6 in.) long and 2.5 cm (1 in.) thick. It has an expanded **head** near the duodenum, a centrally located **body,** and a tapering **tail** near the

spleen. All but a portion of the head is positioned retroperitoneally. Within the lobules of the pancreas are the exocrine secretory units, called **acini** *(as'i-ni),* and the endocrine secretory units, called *islet cells.* Each acinus consists of a single layer of epithelial cells surrounding a lumen into which the constituents of pancreatic juice are secreted.

The pancreas is innervated by branches of the celiac plexus. The glandular portion of the pancreas receives parasympathetic innervation, whereas the pancreatic blood vessels receive sympathetic innervation. The pancreas is supplied with blood by the pancreatic branch of the splenic artery arising from the celiac

---

pancreas: Gk. *pan,* all; *kreas,* flesh
islets of Langerhans: from Paul Langerhans, German anatomist, 1847–88

acinus: L. *acinus,* grape

Figure 18.37 The pancreas is both an exocrine and an endocrine gland. Pancreatic juice—the exocrine product—is secreted by acinar cells into the pancreatic duct. Scattered islands of cells, called pancreatic islets (islets of Langerhans), secrete the hormones insulin and glucagon into the blood.

artery and by the pancreatoduodenal branches arising from the superior mesenteric artery. Venous blood is returned through the splenic and superior mesenteric veins into the hepatic portal vein.

Pancreatic cancer has the worst prognosis of all types of cancer. This is probably because of the spongy, vascular nature of this organ and its vital exocrine and endocrine functions. Pancreatic surgery is a problem because the soft, spongy tissue is difficult to suture.

1. Describe the liver. Where is it located? List the lobes of the liver and the supporting ligaments.
2. List the principal functions of the liver.
3. Describe the structure of liver lobules and trace the flow of blood and bile in the lobules.
4. Explain how the liver receives a double blood supply.
5. Discuss the way in which the gallbladder is filled with bile secretions and the mechanism through which bile and pancreatic secretion enter into the duodenum.
6. Identify the exocrine and endocrine functions of the pancreas. What are the various cellular secretory units of the pancreas?

Clinical Considerations

Developmental Disorders

Most of the congenital disorders of the digestive system have unknown causes and occur during the fourth or fifth week of embryonic development. A **cleft palate** is a congenital opening between the oral and nasal cavities and therefore involves both the digestive and respiratory systems. It occurs in about 1 in 1,000 live births and has immediate importance because it interferes with suckling and swallowing mechanisms. **Esophageal atresia,** or failure to develop the normal structure of the esophageal-stomach area, is another disorder of the upper GI tract that requires surgery to correct. **Pyloric stenosis** is a common abnormality in which the pyloric sphincter muscle is hypertrophied, reducing the size of the lumen. This condition affects approximately 1 in 200 newborn males and 1 in 1,000 newborn females. Stenoses, atresias, and malrotations of various portions of the GI tract may occur as the gut develops. Umbilical problems involving the GI tract are fairly common, as is some form of *imperforate anus,* which occurs in about 1 in 5,000 births.

UNIT VI

Developmental Exposition

The Digestive System

Exhibit I A sagittal section of a 5-week-old embryo showing the development of the digestive system and its association with the extraembryonic membranes and organs.

Pharynx

Heart

Aorta

Foregut

Liver

Stomach

Celiac artery

Superior mesenteric artery

Proximal limb — **Midgut**

Distal limb

Inferior mesenteric artery

Hindgut

Limb bud

Umbilical cord

Septum transversum

Vitelline duct

Allantois

Proctodeum

Cloaca

Amnion

Yolk sac

Chorion

WALDROP

UNIT VI

Explanation

The entire digestive system develops from modifications of an elongated tubular structure called the **primitive gut.** These modifications are initiated during the fourth week of embryonic de-velopment. The primitive gut is composed solely of endoderm and for descriptive purposes can be divided into three regions: the foregut, midgut, and hindgut (exhibit I).

Continued on next page

Developmental Exposition

Continued

Exhibit II Progressive stages of development of the foregut to form the stomach, duodenum, liver, gallbladder, and pancreas: (*a*) 4 weeks, (*b*) 5 weeks, (*c*) 6 weeks, and (*d*) 7 weeks.

Septum transversum
Hepatic buds
Dorsal mesentery
Ventral mesentery
Foregut
Vitelline duct
Midgut
Hindgut
Peritoneal cavity

(a)

Esophagus
Stomach
Hepatic cords
Dorsal pancreatic bud
Ventral pancreatic bud
Gallbladder

(b)

Gallbladder
Ventral pancreatic bud
Dorsal pancreatic bud
Liver
Direction of rotation
Gallbladder
Duodenum

(c)

Stomach
Hepatic ducts
Cystic duct
Common bile duct
Dorsal pancreas
Ventral pancreas
Pancreatic ducts

(d)

Waldrop

Foregut

The **stomodeum** (*sto''mŏ-de'um*), or **oral pit,** is not part of the foregut but an invagination of ectoderm that breaks through a thin **oral membrane** to become continuous with the foregut and form part of the oral cavity, or mouth. Structures in the mouth, therefore, are ectodermal in origin. The esophagus, pharynx, stomach, a portion of the duodenum, the pancreas, liver, and gallbladder are the organs that develop from the foregut (exhibit II). Along the GI tract, only the inside epithelial lining of the lumen is derived from the endoderm of the primitive gut. The vascular portion and smooth muscle layers are formed from mesoderm that develops from the surrounding splanchnic mesenchyme. The

stomach first appears as an elongated dilation of the foregut. The dorsal border of the stomach undergoes more rapid growth than the ventral border, forming a distinct curvature. The caudal portion of the foregut and the cranial portion of the midgut form the duodenum. The liver and pancreas arise as small **hepatic** and **pancreatic buds,** respectively, from the wall of the duodenum. The hepatic bud experiences incredible growth to form the gallbladder, associated ducts, and the various lobes of the liver (exhibit II). By the sixth week, the liver is carrying out hemopoiesis (the formation of blood cells). By the ninth week, the liver has developed to the point where it represents 10% of the total weight of the fetus.

The pancreas develops from dorsal and ventral pancreatic buds of endodermal cells. As the duodenum grows, it rotates clockwise and the two pancreatic buds fuse (exhibit II).

stomodeum: Gk. *stoma,* mouth; *hodaios,* on the way to

Exhibit III Progressive stages of the development of the midgut to form the distal portion of the small intestine and the proximal portion of the large intestine: (*a*) 5 weeks, (*b*) 6 weeks, (*c*) 7 weeks, (*d*) 10 weeks, and (*e*) 18 weeks.

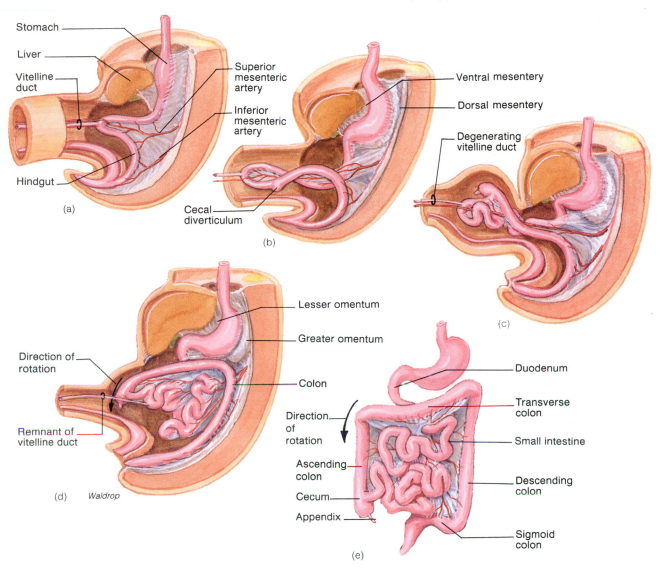

Midgut

During the fourth week of the embryonic stage (exhibit I), the midgut is continuous with the yolk sac. By the fifth week, the midgut has formed a ventral U-shaped **midgut loop,** which projects into the umbilical cord (exhibit III). As development continues, the anterior limb of the midgut loop coils to form most of the small intestine. The posterior limb of the midgut loop expands to form a portion of the small and large intestines. A **cecal diverticulum** (*di''ver-tik'yoo-lum*) appears during the fifth week.

During the tenth week, the intestines are drawn up into the abdominal cavity and further differentiation and rotation occur. The cecal diverticulum continues to develop, forming the cecum and appendix. The remainder of the midgut gives rise to the ascending colon and hepatic flexure (exhibit III).

Hindgut

The hindgut extends from the midgut to the **cloacal membrane** (exhibit IV). The **proctodeum** (*prok''tŏ-de'um*), or **anal pit,** is a depression in the anal region formed from an invagination of ectoderm that contributes to the cloacal membrane. The **allantois** (ă-lan'to-is), which receives urinary wastes from the fetus, connects to the hindgut at a region called the **cloaca** (klo-a'kă), as seen in exhibit IV. A band of mesenchymal cells called the **urorectal septum** grows caudally during the fourth through seventh

---

proctodeum: Gk. *proktos,* anus; *hodaios,* on the way to
cloaca: L. *cloaca,* sewer

Continued on next page

Developmental Exposition

Continued

Exhibit IV The progressive development of the hindgut illustrating the developmental separation of the digestive system from the urogenital system. (*a*) An anterolateral view of an embryo at 4 weeks showing the position of a sagittal cut depicted in (*a₁*), (*b*), and (*c*). (*a₁*) At 4 weeks, the hindgut, cloaca, and allantois are connected.

(*b*) At 6 weeks, the connections between the gut and extraembryonic structures are greatly diminished. (*c*) By 7 weeks, structural and functional separation between the digestive system and the urogenital system has almost been completed.

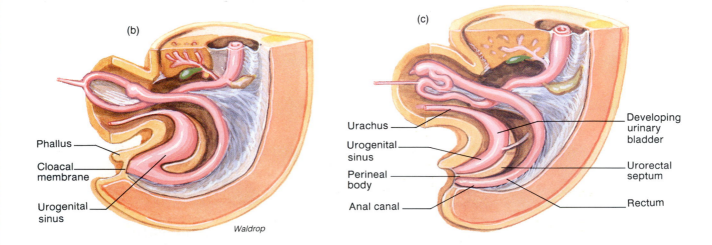

Waldrop

week until a complete partition separates the cloaca into a dorsal **anal canal** and a ventral **urogenital sinus**. With the completion of the urorectal septum, the cloacal membrane is divided into an anterior **urogenital membrane** and a posterior **anal membrane**. Toward the end of the seventh week, the anal membrane per-

forates and forms the anal opening, which is lined with ectodermal cells. About this time, the urogenital membrane ruptures to provide further development of the urinary and reproductive systems.

Pathogens and Poisons

The GI tract is a suitable environment for an array of parasitic helminths (worms) and microorganisms (fig. 18.38). Many of the microorganisms are beneficial, but some bacteria and protozoa can cause diseases. Only a few examples of the pathogenic microorganisms will be discussed here.

Dysentery *(dis'en-ter''e)* is an inflammation of the intestinal mucosa, characterized by the discharge of loose stools that contain mucus, pus, and blood. The most common dysentery is **amoebic dysentery,** which is caused by the protozoan *Entamoeba histolytica.* Cysts from this organism are ingested in contaminated food, and after the protective coat is removed by HCl in the stomach, the vegetative form invades the mucosal walls of the ileum and colon.

Food poisoning comes from ingesting pathogenic bacteria or their toxins. *Salmonella* is a bacterium that commonly causes food poisoning. **Botulism** is the most serious type of food poisoning and is caused by ingesting food contaminated with the bacterium *Clostridium botulinum.* This organism is widely distributed in nature, and the spores from it are frequently on food being processed by canning. For this reason food must be heated to 120°C (248°F) before it is canned. It is the toxins produced by the bacterium growing in the food that are pathogenic rather than the organisms themselves. The poison is a neurotoxin that is readily absorbed in the blood, where it affects the nervous system.

Clinical Problems of the Teeth and Salivary Glands

Dental caries, or tooth decay, is the gradual decalcification of tooth enamel (fig. 18.39) and underlying dentin, produced by

dysentery: Gk. *dys*, bad; *entera*, intestine

Figure 18.38   The gastrointestinal tract and the accessory digestive organs are susceptible to an array of pathogenic (disease-causing) agents. This woman is afflicted with hydatid cysts in her liver as a result of ingesting an egg from the tapeworm *Echinococcus granulosus.* If untreated, conditions such as this are eventually fatal.

Figure 18.39   Clinical problems of the teeth. (*a*) Trench mouth and dental caries, (*b*) severe alveolar bone destruction from periodontitis, (*c*) pyogenic granuloma and dental caries, and (*d*) malposition of teeth.

(a)

(b)

(c)

(d)

the acid products of bacteria. These bacteria thrive between teeth where food particles accumulate and form part of the thin layer of bacteria, proteins, and other debris called *plaque* that covers teeth. The development of dental caries can be reduced by brushing at least once a day and by flossing between teeth at regular intervals.

People over the age of 35 are particularly susceptible to **periodontal diseases.** There are many types, but basically they all cause inflammation and deterioration of the gingiva, alveolar sockets, periodontal membranes, and teeth cementum. Some of the symptoms are loosening of the teeth, bad breath, bleeding gums when brushing, and some edema. Some of the causes of periodontal disease are impacted plaque, cigarette smoking, malocclusion, and poor diet.

Mumps is a viral disease of the parotid glands and in advanced stages may involve the pancreas and testes. In children, it is generally not serious, but in adults it may cause deafness and destroy the pancreatic islet tissue or testicular cells.

Disorders of the Liver

The liver is a remarkable organ that has the ability to regenerate itself even if up to 80% has been removed. The most serious diseases of the liver (hepatitis, cirrhosis, and hepatomas) affect the liver throughout so that it cannot repair itself. **Hepatitis** is inflammation of the liver. Certain chemicals may cause hepatitis, but generally it is caused by infectious viral agents. **Hepatitis A** (infectious hepatitis) is a viral disease transmitted through contaminated foods and liquids. **Hepatitis B** (serum hepatitis) is also caused by a virus and is transmitted in serum or plasma during transfusions or by improperly sterilized needles and syringes. Other types of viral hepatitis are designated as hepatitis C, D, and E.

In **cirrhosis,** large numbers of liver lobules are destroyed and replaced with permanent connective tissue and regenerative nodules of hepatocytes. These nodules do not have the platelike structure of normal liver tissue and are therefore less functional. Cirrhosis may be caused by chronic alcohol abuse, viral hepatitis, or by other agents that attack liver cells.

Jaundice is a yellow staining of the tissue produced by high blood concentrations of either free or conjugated bilirubin. Since free bilirubin is derived from heme, abnormally high concentrations of this pigment may result from an unusually high rate of red blood cell destruction. This can occur, for example, as a result of Rh disease (erythroblastosis fetalis) in an Rh positive baby born to a sensitized Rh negative mother. Jaundice may also occur in healthy infants because red blood cells are normally destroyed at about the time of birth (hemoglobin con-

centrations decrease from 19 g per 100 mℓ to 14 g per 100 mℓ near the time of birth). This condition is called *physiological jaundice of the newborn* and is not indicative of disease. Premature infants may also develop jaundice due to inadequate amounts of hepatic enzymes necessary to conjugate bilirubin and excrete it in the bile. Jaundice due to high levels of conjugated bilirubin in the blood is commonly exhibited in adults when the excretion of bile is blocked by gallstones.

Hepatomas *(hep''ă-to'maz)* are malignant tumors that originate in or secondarily invade the liver. Those that originate in the liver (primary hepatomas) are relatively rare, but those that secondarily metastasize to the liver from other organs (secondary hepatomas) are common. Carcinoma of the liver is usually fatal.

Disorders of the GI Tract

Peptic ulcers are erosions of the mucous membranes of the stomach or duodenum produced by the action of HCl. Agents that weaken the mucosal lining of the stomach, such as alcohol and aspirin, and abnormally high secretions of HCl thus increase one's likelihood of developing peptic ulcers. Many people subject to chronic stress produce too much gastric acid and develop a peptic ulcer as a result (fig. 18.40).

Enteritis is inflammation of the intestinal mucosa and is frequently referred to as intestinal flu. Causes of enteritis include bacterial or viral infections, irritating foods or fluids (including alcohol), and emotional stress. The symptoms are abdominal pain, nausea, and diarrhea. **Diarrhea** is the passage of watery, unformed stools. This condition is symptomatic of inflammation, stress, and other body dysfunctions.

A **hernia** is a protrusion of a portion of a visceral organ, usually the small intestine, through a weakened portion of the abdominal wall. Inguinal, femoral, umbilical, and hiatal hernias are the most common. With a **hiatal hernia,** a portion of the stomach pushes superiorly through the esophageal hiatus in the diaphragm and into the thorax. The potential dangers of a hernia are strangulation of the blood supply followed by gangrene, blockage of chyme, or rupture—each of which can threaten life.

Diverticulosis is a condition in which the intestinal wall weakens and an outpouching occurs. Studies suggest that suppressing the passage of flatus (intestinal gas) may contribute to diverticulosis, especially in the sigmoid colon. **Diverticulitis,** or inflammation of a diverticulum, can develop if fecal material becomes impacted in these pockets.

Peritonitis is inflammation of the peritoneum lining the abdominal cavity and covering the viscera. The causes of peritonitis include bacterial contamination of the abdominal cavity through accidental or surgical wounds in the abdominal wall or perforation of the intestinal wall (as with ulcers or a ruptured appendix).

cirrhosis: Gk. *kirrhos,* yellow-orange
jaundice: L. *galbus,* yellow

UNIT VI

Figure 18.40 Common sites of upper GI disorders.

Esophagus
Esophagitis
Tumor
Varices
Stomach
Duodenum
Peptic ulcer
Laceration
Ulcer
Gastritis
Tumor
Acute or chronic peptic ulcer
Polyposis

Loechel

CLINICAL CASE STUDY ANSWER

The likely source of bleeding in such a case is a rupture of the internal portion of the liver involving significant blood vessels, either of the hepatic arterial or venous circulation, but also possibly of the portal venous circulation. Nearby lacerated bile vessels then receive blood into the biliary tree. From there, the course of the blood is as follows: large hepatic ducts to common hepatic duct, to common bile duct through duodenal papilla and into the duodenum of the small intestine.

This type of bleeding in some instances may be stopped by placing radiographic-guided catheters into the arterial system (angiographic embolization). However, surgery is sometimes required.

Internal Affairs

The Relational Workings of the Body Systems

Digestive System

- Provides lipids for storage in adipocytes (adipose cells)

Integumentary System
- Covers and protects the body
- Synthesizes vitamin D
- Provides site for lipid storage

- Absorbs calcium and phosphate ions for formation and maintenance of bone tissue

Skeletal System
- Supports and protects certain digestive organs
- Stores calcium and phosphate ions

- Provides nutrients for muscle contraction and maintenance

Muscular System
- Supports and protects certain digestive organs
- Assists the processing of food—preparing, chewing, swallowing

- Provides nutrients for neural function and maintenance

Nervous System
- Provides autonomic innervation to GI tract

- Provides nutrients for neural function and maintenance
- Liver eliminates hormones from blood

Endocrine System
- Regulates digestive activity

- Provides nutrients for circulatory function and maintenance
- Absorbs iron for blood synthesis

Circulatory System
- Transports O_2 and CO_2, nutrients, and fluids to and from the digestive organs

- Provides nutrients for lymphatic function and maintenance

Lymphatic System
- Maintains balanced amount of intersitial fluid within the digestive organs
- Protects the GI tract against infection

- Provides nutrients for function and maintenance of respiratory organs

Respiratory System
- Provides O_2 and eliminates CO_2

- Provides nutrients for function and maintenance of urinary organs

Urinary System
- Eliminates metabolic wastes

- Provides nutrients for function and maintenance of reproductive organs
- Provides nutrients in a pregnant woman for development of embryo and fetus

Reproductive System
- Influences metabolic rates through effects of steriods

UNIT VI

IMPORTANT CLINICAL TERMINOLOGY

chilitis Inflammation of the lips.

colitis Inflammation of the colon.

colostomy The formation of an abdominal exit from the GI tract by bringing a loop of the colon to the surface of the abdomen. If the rectum is removed because of cancer, the colostomy provides a permanent outlet for the feces.

cystic fibrosis An inherited disease of the exocrine glands, particularly the pancreas. Pancreatic secretions are too thick to drain easily, causing the ducts to become inflamed and promoting connective tissue formation that occludes drainage from the ducts still further.

gingivitis Inflammation of the gums. It may result from improper hygiene, poorly fitted dentures, improper diet, or certain infections.

halitosis Offensive breath odor. It may result from dental caries, certain diseases, or eating particular foods.

heartburn A burning sensation of the esophagus and stomach. It may result from the regurgitation of gastric juice into the lower portion of the esophagus.

hemorrhoids Varicose veins of the rectum and anus.

nausea Gastric discomfort and sensations of illness with a tendency to vomit. This feeling is symptomatic of motion sickness and other diseases and may occur during pregnancy.

pyorrhea The discharge of pus at the base of the teeth at the gum line.

regurgitation (vomiting) The forceful expulsion of gastric contents into the mouth. Nausea and vomiting are common symptoms of almost any dysfunction of the digestive system.

trench mouth A contagious bacterial infection that causes inflammation, ulceration, and painful swelling of the floor of the mouth. Generally it is contracted through direct contact by kissing an infected person. Trench mouth can be treated with penicillin and other medications.

vagotomy The surgical removal of a section of the vagus nerve where it enters the stomach in order to eliminate nerve impulses that stimulate gastric secretion. This procedure may be used to treat chronic ulcers.

CHAPTER SUMMARY

Introduction to the Digestive System (pp. 598–599)

1. The digestive system goes through a series of mechanical and chemical processes in which food is reduced to forms that can be absorbed through the intestinal wall and transported by the blood and lymph for use at the cellular level.
2. The digestive system consists of a gastrointestinal (GI) tract and accessory digestive organs.

Serous Membranes and Tunics of the Gastrointestinal Tract (pp. 599–602)

1. Peritoneal membranes line the abdominal wall and cover the visceral organs. The GI tract is supported by a double layer of peritoneum called the mesentery.
    (a) The lesser omentum and greater omentum are folds of peritoneum that extend from the stomach.
    (b) Visceral organs not within the peritoneal cavity and therefore not covered with peritoneum are retroperitoneal.
2. The layers (tunics) of the abdominal GI tract are, from the inside outward, the mucosa, submucosa, muscularis, and serosa.
    (a) The mucosa consists of a simple columnar epithelium, a thin layer of connective tissue called the lamina propria, and a thin layer of smooth muscle called the muscularis mucosa.
    (b) The submucosa is composed of connective tissue, the muscularis consists of layers of smooth muscles, and the serosa is connective tissue covered with the visceral peritoneum.
    (c) The submucosa contains the submucosal plexus and the muscularis contains the myenteric plexus of autonomic nerves.

Mouth, Pharynx, and Associated Structures (pp. 602–610)

1. The oral cavity is formed by the cheeks, lips, hard and soft palates, and tongue.
    (a) Located on the tongue are lingual tonsils and papillae with taste buds.
    (b) Structures of the palate include palatal rugae, a cone-shaped projection called the uvula, and palatine tonsils.

2. The four most anterior pairs of teeth are the incisors and canines, which have one root each; the bicuspids and molars have two or three roots.
    (a) Humans are diphyodont; they have deciduous and permanent sets of teeth.
    (b) The tooth is anchored to its bony alveolar socket by a periodontal membrane, which contains fibers that insert into the cementum of the tooth root.
    (c) Enamel forms the outer layer of the tooth crown; beneath the enamel is dentin.
    (d) The interior of a tooth contains a pulp cavity, which is continuous through the apical foramen of the root with the connective tissue around the tooth.
3. The major salivary glands are the parotid glands and the submandibular glands.
4. The muscular pharynx is a passageway connecting the oral and nasal cavities to the esophagus and trachea.

Esophagus and Stomach (pp. 610–614)

1. Peristaltic waves of contraction push food through the lower esophageal sphincter into the stomach.
2. Swallowing, or deglutition, occurs in three phases and involves structures of the buccal cavity, pharynx, and esophagus.
3. The stomach consists of the cardia, fundus, body, and pylorus; the pylorus terminates at the pyloric sphincter.
    (a) The lining of the stomach is characterized by distensible folds, or gastric rugae, and the mucosa is formed into gastric pits and gastric glands.
    (b) The parietal cells of the gastric glands secrete HCl and the chief cells secrete pepsinogen.

Small Intestine (pp. 614–617)

1. Regions of the small intestine include the duodenum, jejunum, and ileum; the common bile duct and pancreatic duct empty into the duodenum.
2. Fingerlike extensions of mucosa, called villi, project into the lumen, and at the bases of the villi the mucosa forms intestinal glands.
    (a) New epithelial cells are formed in the crypts.
    (b) The membrane of intestinal epithelial cells is folded to form microvilli; this brush border of the mucosa increases the absorptive surface area.

UNIT VI

3. Movements of the small intestine include rhythmic segmentation, pendular movement, and peristalsis.

Large Intestine (pp. 617–621)

1. The large intestine absorbs water and electrolytes from the chyme and passes waste products out of the body through the rectum and anal canal.
2. The large intestine is divided into the cecum, colon, rectum, and anal canal.
   (a) The appendix is attached to the inferior medial margin of the cecum.
   (b) The colon consists of ascending, transverse, descending, and sigmoid portions.
   (c) Haustra are bulges in the walls of the large intestine.
3. Movements of the large intestine include peristalsis, haustral churning, and mass movement.

Liver, Gallbladder, and Pancreas (pp. 621–626)

1. The liver is divided into right, left, quadrate, and caudate lobes. Each lobe contains hepatic lobules, the functional units of the liver.
   (a) Liver lobules consist of plates of hepatic cells separated by capillary sinusoids.
   (b) Blood flows from the periphery of each lobule, where the hepatic artery and portal vein empty, through the sinusoids and out the central vein.
   (c) Bile flows within the hepatic plates, in canaliculi, to the bile ducts at the periphery of each lobule.
2. The gallbladder stores and concentrates the bile, and releases it through the cystic duct and common bile duct to the duodenum.
3. The pancreas is both an exocrine and an endocrine gland.
   (a) The endocrine portion is known as the pancreatic islets and secretes the hormones insulin and glucagon.
   (b) The exocrine acini of the pancreas produce pancreatic juice, which contains various digestive enzymes.

UNIT VI

REVIEW ACTIVITIES

Objective Questions

1. Viscera are the only body organs that are
   (a) concerned with digestion.
   (b) located in the abdominal cavity.
   (c) covered with peritoneal membranes.
   (d) located within the thoracic and abdominal cavities.
2. Which of the following types of teeth are found in the permanent but not in the deciduous dentition?
   (a) incisors   (c) premolars
   (b) canines   (d) molars
3. The double layer of peritoneum that supports the GI tract is called
   (a) the visceral peritoneum.
   (b) the mesentery.
   (c) the greater omentum.
   (d) the lesser omentum.
4. Which of the following tissue layers in the small intestine contains the lacteals?
   (a) submucosa
   (b) muscularis mucosa
   (c) lamina propria
   (d) muscularis
5. Which of the following organs is *not* considered a part of the digestive system?
   (a) the pancreas
   (b) the spleen
   (c) the tongue
   (d) the gallbladder
6. The numerous small elevations on the dorsal surface of the tongue that support taste buds and aid in handling food are called
   (a) cilia.   (c) villi.
   (b) rugae.   (d) papillae.
7. Most digestion occurs in
   (a) the mouth.
   (b) the stomach.
   (c) the small intestine.
   (d) the large intestine.
8. Stenosis (constriction) of the sphincter of ampulla (Oddi) would interfere with
   (a) transport of bile and pancreatic juice.
   (b) secretion of mucus.
   (c) passage of chyme into the small intestine.
   (d) peristalsis.
9. The first organ to receive the blood-borne products of digestion is
   (a) the liver.   (c) the heart.
   (b) the pancreas. (d) the brain.
10. Which of the following statements about hepatic portal blood is true?
   (a) It contains absorbed fat.
   (b) It contains ingested proteins.
   (c) It is mixed with bile in the liver.
   (d) It is mixed with blood from the hepatic artery in the liver.

Essay Questions

1. Define *digestion*. Differentiate between the mechanical and chemical aspects of digestion.
2. Distinguish between the gastrointestinal tract, viscera, accessory digestive organs, and gut.
3. List the specific portions or structures of the digestive system formed by each of the three embryonic germ layers.
4. Define *serous membrane*. How are the serous membranes of the abdominal cavity classified and what are their functions?
5. Describe the structures of the four tunics in the wall of the GI tract.
6. Why are there two autonomic innervations to the GI tract? Identify the specific sites of autonomic stimulation in the tunic layers.
7. Define the following: *dental formula, diphyodont, deciduous teeth, permanent teeth,* and *wisdom teeth.*
8. Outline the stages of deglutition. What biomechanical roles do the tongue, hard and soft palates, pharynx, and hyoid bone perform in deglutition?
9. What are the specializations of the mucosa of the stomach that permit distension and secretion?
10. Describe the structural modifications of the small intestine that increase the surface area for absorption.
11. Diagram a villus and explain why villi are considered the functional units of the digestive system.
12. What are the regions of the large intestine? In what portion of the abdominal cavity and pelvic cavity is each region located?
13. Describe the location and the gross structure of the liver. What structures are associated with the porta of the liver?
14. Describe how the gallbladder is filled with and emptied of bile fluid. What is the function of bile?
15. List the functions of the large intestine. What are the biomechanical movements of the large intestine that make these functions possible?
16. Define *cirrhosis* and explain why it is so devastating to the liver. What are some of the causes of cirrhosis?

Gundy/Weber Software

The tutorial software accompanying Chapter 18, Digestive System, is Volume 11, Digestive System.

19

Urinary System

Chapter Outline

CLINICAL CASE STUDY

A 17-year-old male is involved in a knife fight in which he sustains two stab wounds to the anterior abdomen. He is brought to the emergency room complaining of mild abdominal pain and an urgent need to urinate. Although neither of the stab wounds is externally bleeding, examination by the surgeon reveals signs of moderate hemorrhagic shock. One wound is 3 cm below the right costal arch at the midclavicular line; the other is just medial to the right anterior superior iliac spine. The surgeon immediately orders preparations for an emergency exploratory laparotomy. She parenthetically adds that the patient's urinary bladder is quite full and chides the intern for not having placed a urinary catheter. Placement of the catheter yields a brisk flow of bright red blood.

How would you explain the phenomenon of hematuria (blood in the urine) in this case? Which of the two stab wounds would most likely be associated with the hematuria? Regarding the blood draining into the catheter, trace and explain its course of flow. Begin at a point in the abdominal aorta, ending the course with drainage into the catheter. Assuming that the surgeon is prompted to remove the kidney in order to quickly control hemorrhage, what possible anatomical variant should she keep in mind? *Hints:* Study the positions of the kidneys within the abdominal cavity and note the location of the supportive and protective serous membranes. Carefully examine the specific location of urine production and the route of urine passage through the urinary system.

Introduction to the Urinary System

The urinary system maintains the composition and properties of the body fluid that establishes the internal environment of the body cells. The end product of the urinary system is urine, which is voided from the body during micturition.

Objective 1. List the functions of the urinary system.

Objective 2. Identify the arteries that transport blood to the urinary system for filtration.

The urinary system, along with the respiratory, digestive, and integumentary systems, excretes substances from the body. For this reason, these systems are occasionally referred to as excretory systems. In the process of cellular metabolism, nutrients taken in by the digestive system and oxygen from inhaled air are utilized to synthesize a variety of substances and provide the energy needed for body maintenance. Metabolic processes, however, produce cellular wastes that must be eliminated if homeostasis is to be maintained. Just as the essential elements are transported to the cells by the blood, the cellular wastes are removed through the circulatory system to the appropriate ex-

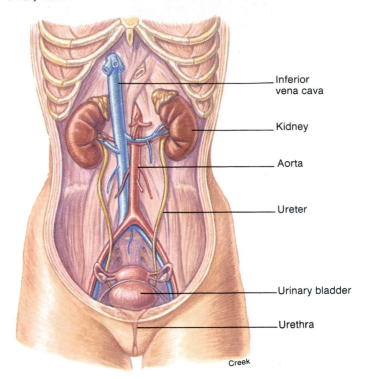

Figure 19.1 The anatomical locations of the kidneys, ureters, and urinary bladder.

Inferior vena cava

Kidney

Aorta

Ureter

Urinary bladder

Urethra

Creek

cretory system. Carbon dioxide is eliminated through the respiratory system; excessive water, salts, nitrogenous wastes, and even excessive metabolic heat are removed through the integumentary system; and various bile digestive wastes are eliminated through the digestive system.

The urinary system is the principal system responsible for water-and-electrolyte balance. Electrolytes are compounds that separate into ions when dissolved in water, and a balance is achieved when the number of electrolytes entering the body equals the number leaving. Hydrogen ions, for example, are maintained in precise concentration so that an acid-base, or pH, balance exists in the body.

A second major function of the urinary system is the excretion of toxic metabolic products that are nitrogenous compounds—specifically, urea and creatinine. Other functions of the urinary system include the elimination of toxic wastes that may result from bacterial action and the removal of various drugs that have been taken into the body. The end product of all of these processes is **urine.**

The urinary system consists of two kidneys, two ureters, the urinary bladder, and the urethra (fig. 19.1). Tubules in the kidneys are intertwined with vascular networks of the circulatory system to enable the production of urine. After the urine is formed, it is moved through the ureters to the urinary bladder for storage. **Micturition** *(mik″tŭ-rish′un),* or voiding of urine from the urinary bladder, occurs through the urethra.

Blood that is to be processed by a kidney enters through the large **renal artery,** and after the filtration process (see chapter 3) exits through the **renal vein.** The importance of filtration of blood is demonstrated by the fact that during normal resting conditions the kidneys receive approximately 20%–25% of the entire cardiac output. Every minute the kidneys process approximately 1,200 ml of blood.

---

1. Drawing on your knowledge of the functions of the urinary system, list the basic substances that compose normal urine.
2. Explain the role of the renal vessels in maintaining homeostasis. What percentage of the body blood is filtered in the kidneys each minute during normal resting conditions?

Kidneys

The kidney consists of an outer renal cortex and an inner renal medulla containing the renal pyramids. Urine is formed as a filtrate from the blood at the nephrons and collects in the calyces and renal pelvis before flowing from the kidney via the ureter.

Objective 3. Describe the gross structure of the kidney.

Objective 4. Describe the tubular and vascular components of a nephron.

Objective 5. Describe the position of cortical and juxtamedullary nephrons in the gross structure of the kidney.

Position and Appearance

The reddish-brown **kidneys** lie on each lateral side of the vertebral column, high in the abdominal cavity, between the levels of the twelfth thoracic and the third lumbar vertebrae (fig. 19.2). The right kidney is usually 1.5–2.0 cm lower than the left because of the large area occupied by the liver.

The kidneys are retroperitoneal, which means that they are positioned behind the peritoneum (fig. 19.3). Thus, strictly speaking, they are not within the peritoneal cavity. Each adult kidney is an anteroposteriorly compressed lima-bean-shaped organ about 11.25 cm (4 in.) long, 5.5–7.7 cm (2–3 in.) wide, and 2.5 cm (1 in.) thick. The lateral border of each kidney is convex, whereas the medial border is strongly concave (figs. 19.2 and 19.3). The **hilum** of the kidney is the depression along the medial border through which the **renal artery** enters and the **renal vein** and **ureter** exit. Innervation to the kidney is also at the hilum.

Each kidney is embedded in a fatty fibrous pouch consisting of three layers. The **renal capsule** or **fibrous capsule,** is the innermost layer that forms a strong, transparent fibrous attachment to the kidney. The renal capsule protects the kidney from trauma and the spread of infections. The second layer is formed by a firm protective layer of adipose tissue called the

Figure 19.2 A color-enhanced radiograph of the calyces and renal pelvises of the kidneys, the ureters, and the urinary bladder. (Notice the position of the kidneys relative to the vertebral column and ribs.)

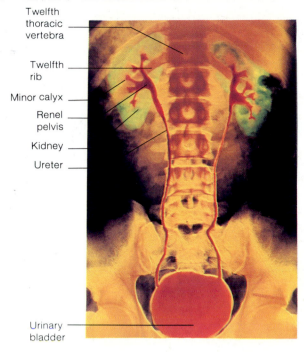

Twelfth thoracic vertebra

Twelfth rib

Minor calyx

Renel pelvis

Kidney

Ureter

Urinary bladder

renal adipose capsule (see fig. 19.3). The outermost layer, called the **renal fascia,** is a supportive layer that anchors the kidney to the peritoneum and the abdominal wall.

Although the kidneys are firmly supported by the renal adipose capsule, renal fascia, and even the renal vessels, under certain conditions these structures may give in to the force of gravity and the kidneys may drop a bit in position. This condition is called *renal ptosis (to' sis)* and generally occurs in elderly people who are extremely thin and have insufficient amounts of supportive fat in the adipose capsular layer. It may also affect victims of *anorexia nervosa* who suffer from extreme weight loss. The potential danger of renal ptosis is that the ureter may kink, blocking the flow of urine from the affected kidney.

Gross Structure

A coronal section of the kidney shows two distinct regions and a major cavity (figs. 19.4 and 19.5). The outer **renal cortex,** in contact with the capsule, is reddish brown and appears granular because of its many capillaries. The deeper region, or **renal medulla,** is darker and striped because of the presence of microscopic tubules and blood vessels. The renal medulla is composed of 8 to 15 conical **renal pyramids** separated by **renal columns.** The apexes of the renal pyramids are all directed into the renal sinus and are known as the renal papillae.

UNIT VI

---

hilum: L. *hilum,* a trifle

ptosis: Gk. *ptosis,* a falling

Figure 19.3 The positions of the kidneys as seen in cross section through the upper abdominal cavity. The kidneys are embedded in adipose tissue behind the parietal peritoneum.

Figure 19.4 The kidney. (*a*) The anterior surface of a human kidney and (*b*) a coronal section of a human kidney.

(a)

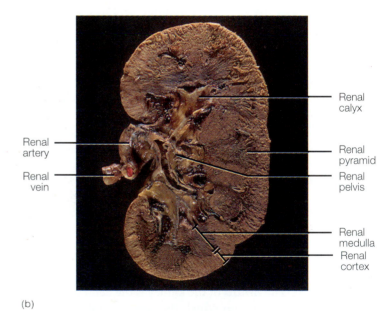

(b)

The cavity of the kidney collects and transports urine from the kidney to the ureter. It is divided into several portions. Each papilla of a renal pyramid projects into a small depression called the **minor calyx** *(ka'liks)*—in the plural, *calyces*. Several minor calyces unite to form a **major calyx.** In turn, the major calyces join to form the funnel-shaped **renal pelvis.** The renal pelvis is actually an expanded portion of the ureter in the kidney and serves to collect urine from the calyces and transport it to the ureter.

Microscopic Structure

The **nephron** *(nef'ron)* is the functional unit of the kidney that is responsible for the formation of urine. Each kidney contains more than a million nephrons. A nephron consists of a **urinary tubule** and associated small blood vessels. Fluid formed by capillary filtration enters the tubule and is subsequently modified by transport processes. The resulting fluid that leaves the tubule is urine.

Figure 19.5 The internal structures of a kidney. (*a*) A coronal section showing the structure of the renal cortex, renal medulla, and renal pelvis. (*b*) A diagrammatic magnification of a renal pyramid and renal cortex to depict the tubules. (*c*) A diagrammatic view of a single nephron and a collecting duct.

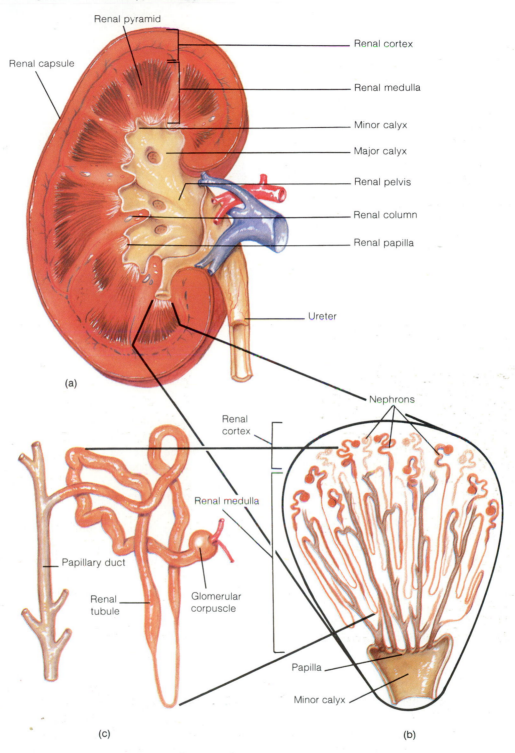

Renal pyramid

Renal capsule

Renal cortex

Renal medulla

Minor calyx

Major calyx

Renal pelvis

Renal column

Renal papilla

Ureter

(a)

Nephrons

Renal cortex

Renal medulla

Papillary duct

Renal tubule

Glomerular corpuscle

Papilla

Minor calyx

(c)

(b)

UNIT VI

Figure 19.6 The vascular structure of the kidneys. (a) An illustration of the major arterial supply and (b) a scanning electron micrograph of a glomerulus.

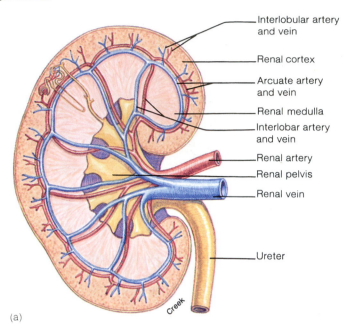

Interlobular artery and vein
Renal cortex
Arcuate artery and vein
Renal medulla
Interlobar artery and vein
Renal artery
Renal pelvis
Renal vein
Ureter

(a)

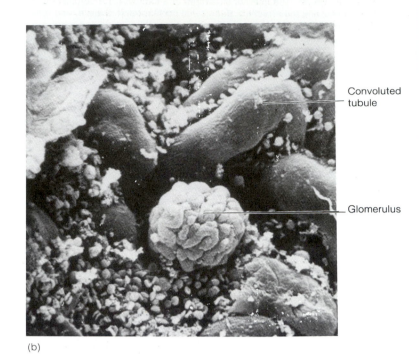

Convoluted tubule
Glomerulus

(b)

Renal Blood Vessels Arterial blood enters the kidney at the hilum through the **renal artery,** which divides into **interlobar** *(in"ter-lo'bar)* **arteries** (fig. 19.6) that pass between the renal pyramids through the renal columns. **Arcuate** *(ar'kyoo-āt)* **arteries** branch from the interlobar arteries at the boundary of the renal cortex and renal medulla. Many **interlobular arteries** radiate from the arcuate arteries and subdivide into numerous **afferent glomerular arterioles,** which are microscopic in size. The afferent arterioles deliver blood into capillary networks called **glomeruli,** which produce a blood filtrate that enters the urinary tubules. The blood remaining in the glomerulus leaves through an **efferent glomerular arteriole,** which delivers the blood into either the **peritubular capillaries** surrounding the convoluted tubules or into the **vasa recta** surrounding the ascending and descending tubules (fig. 19.7). From these capillary networks, the blood is drained into veins that parallel the course of the arteries in the kidney. These are the **interlobular veins, arcuate veins,** and **interlobar veins.** The interlobar veins descend between the renal pyramids, converge, and leave the kidney as a single **renal vein** that empties into the inferior vena cava.

In summary, vessels are at a capillary level between the arterioles in each glomerulus of the kidney. The blood pressure at this level is strong enough to force considerable water, including the wastes, from the blood into the urinary tubular portion of the nephron. A secondary capillary network of peritubular capillaries surrounds various tubular portions of the nephron. This capillary plexus, however, is between arterioles and venules and functions in the reabsorption of water and other substances that should not be excreted with the urine.

> Although the kidneys are generally well protected by being encapsulated retroperitoneally, they may be injured by a hard blow to the lumbar region. The immense vascularity of the kidney makes it highly susceptible to hemorrhage. As a result, such an injury can produce blood in the urine.

Nephron The tubular nephron consists of a glomerular capsule, proximal convoluted tubule, descending limb of the nephron loop (loop of Henle), ascending limb of the nephron loop, and distal convoluted tubule (fig. 19.7).

The **glomerular** (Bowman's) **capsule** surrounds the glomerulus. The glomerular capsule and its associated glomerulus are located in the renal cortex of the kidney and together constitute the **renal corpuscle** (fig. 19.8). The glomerular capsule contains an inner visceral layer of epithelium around the glomerular capillaries and an outer parietal layer. The space between these two layers, called the **capsular space,** receives the glomerular filtrate.

The glomerular epithelium has pores called **fenestrae** *(fĕ-nes'tre)* that permit the filtrate to pass from the blood into the glomerular capsule (fig. 19.8). Although the fenestrae are large, they are still small enough to prevent the passage of blood cells, platelets, and most plasma proteins into the filtrate. The inner

UNIT VI

---

arcuate: L. *arcuare,* to bend
glomerulus: L. diminutive of *glomus,* ball

loop of Henle: from Friedrich Gustav Jacob Henle, German physician, anatomist, and pathologist, 1809–85
Bowman's capsule: from Sir William Bowman, English anatomist, 1816–92

Figure 19.7 A simplified illustration of blood flow from a glomerulus to an efferent arteriole, to the peritubular capillaries, to the venous drainage of the kidneys.

layer of the glomerular capsule is composed of unique cells called **podocytes** (pod'ō-sītz) with numerous cytoplasmic extensions known as **pedicels** (ped'ĭ-selz). Pedicels interdigitate, like the fingers of clasped hands, as they wrap around the glomerular capillaries (figs. 19.8 and 19.9). The narrow **slit pores** between adjacent pedicels provide the passageways through which filtered molecules must pass to enter the interior of the glomerular capsule.

Filtrate in the glomerular capsule passes into the lumen of the **proximal convoluted tubule.** The wall of the proximal convoluted tubule consists of a single layer of cuboidal cells containing millions of microvilli; these serve to increase the surface area for reabsorption. In the process of reabsorption, salt, water, and other molecules needed by the body are transported from the lumen, through the tubular cells, and into the surrounding peritubular capillaries.

The glomerulus, glomerular capsule, and proximal convoluted tubule are located in the renal cortex. Fluid passes from the proximal convoluted tubule to the **nephron loop** (loop of Henle). This fluid is carried into the renal medulla in the **descending limb** of the loop and returns to the renal cortex in the **ascending limb** of the loop. Back in the renal cortex, the tubule becomes coiled again and is called the **distal convoluted tubule.** The distal convoluted tubule is shorter than the

proximal tubule and has fewer microvilli. It is the last segment of the nephron and terminates as it empties into a collecting duct.

The two principal types of nephrons are classified according to their positions in the kidney and the lengths of their nephron loops. Nephrons that originate in the inner one-third of the cortex—called **juxtamedullary nephrons**—have longer loops than the **cortical nephrons** that originate in the outer two-thirds of the renal cortex (fig. 19.10).

The distal convoluted tubules of several nephrons drain into a **papillary duct** (collecting duct). Fluid is then drained by the papillary duct from the renal cortex into the renal medulla as the collecting duct passes through a renal pyramid. This fluid, now called urine, passes out of a renal papilla into a minor calyx. Urine is then funneled through the renal pelvis and out of the kidney into the ureter.

The kidneys have an autonomic nerve supply derived from the **renal plexus** of the tenth, eleventh, and twelfth thoracic nerves. Sympathetic stimulation of the renal plexus produces a vasomotor vascular network response in the kidney. This response determines the circulation of the blood by regulating the diameters of arterioles.

The physiological functions of the nephron are summarized in table 19.1.

Figure 19.8 The structure of a glomerulus. (a) A renal corpuscle is composed of a glomerulus and a glomerular (Bowman's) capsule. Note that the diameter of the efferent arteriole carrying blood away from the glomerulus is smaller than that of the afferent arteriole transporting blood into the glomerulus. This is one factor contributing to the maintenance of high blood pressure within the glomerulus. The first step of urine formation is the filtration through the glomerular membrane into the glomerular space of the glomerular capsule. (b) A scanning electron micrograph of a glomerulus (8,000×).

(a)

(b)

Pedicels   Primary process of podocyte

Urine from a healthy individual is virtually bacteria-free but easily becomes contaminated after voiding because its organic components serve as a nutrient source for contaminating microbes. The breakdown of urine by bacterial action produces ammonia. A urinalysis is a common procedure in any routine physical examination. The appearance and pH of the urine are noted, as well as the presence of such abnormal constituents as albumin, blood, glucose, and acetone. Abnormal urine is symptomatic of a variety of diseases or problems in the urinary system.

1. Describe the gross appearance of the renal cortex and renal medulla and indicate the microscopic structures found in each layer.
2. Trace the course of blood through the kidney from the renal artery to the renal vein.
3. Trace the course of tubular fluid from the glomerular capsules to the ureter.
4. Draw a diagram of a nephron. Label the segments and indicate which parts are in the renal cortex and which are in the renal medulla.

Figure 19.9 Glomerular filtration. (*a*) A transmission electron micrograph of a glomerular capillary and the glomerular membranes and (*b*) accompanying line art. Substances in the blood are filtered through capillary fenestrae. The filtrate then passes across the basement membrane and through slit pores between the pedicels and enters the capsular space. From here, the filtrate is transported to the lumen of the proximal convoluted tubule (see fig. 19.8).

(a)

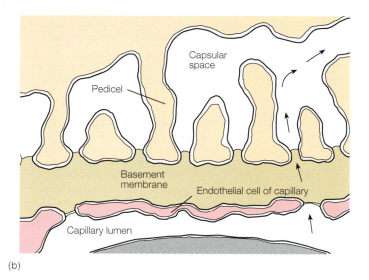

(b)

Figure 19.10 Cortical nephrons are located almost exclusively within the renal cortex. Juxtamedullary nephrons are located, for the most part, within the outer portion of the renal medulla.

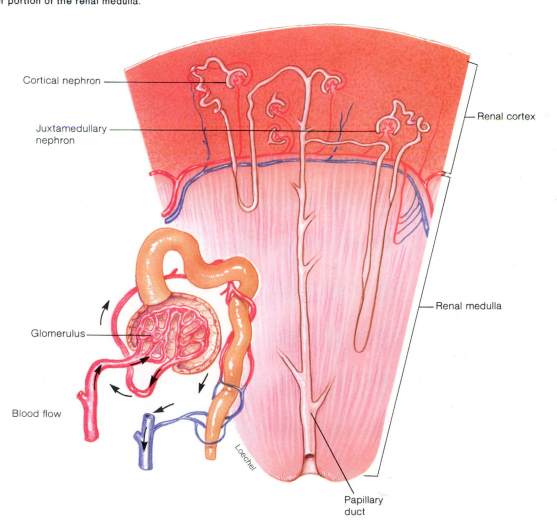

Table 19.1	Physiological Functions of the Nephron in Urine Formation
Structure	**Function**
Renal Corpuscle	
Glomerulus	Filtration of water and dissolved substances from blood plasma
Glomerular capsule	Receives glomerular filtrate
Nephron	
Proximal convoluted tubule	Reabsorption of water by osmosis
	Reabsorption of glucose, amino acids, creatine, lactic acid, citric acid, uric acid, ascorbic acid, phosphate ions, sulfate ions, calcium ions, potassium ions, and sodium ions by active transport
	Reabsorption of proteins by pinocytosis
	Reabsorption of chloride ions and other negatively charged ions by electrochemical attraction
	Active secretion of substances like penicillin, histamine, and hydrogen ions
Descending limb of the nephron loop	Reabsorption of water by osmosis
Ascending limb of the nephron loop	Reabsorption of chloride ions by active transport and passive reabsorption of sodium ions
Distal convoluted tubule	Reabsorption of water by osmosis
	Reabsorption of sodium ions by active transport
	Active secretion of hydrogen ions
	Passive secretion of potassium ions by electrochemical attraction
Papillary duct	Passive resorption of water

Ureters, Urinary Bladder, and Urethra

Urine is channeled from the kidneys to the urinary bladder by the ureters and expelled from the body through the urethra. The mucosa of the urinary bladder permits distension, and the muscles of the urinary bladder and urethra are used in the control of micturition.

Objective 6. Describe the location, structure, and function of the ureters.

Objective 7. Describe the gross and histological structure of the urinary bladder and its innervation.

Objective 8. Explain the process of micturition.

Objective 9. Compare and contrast the structure of the male and female urethra.

Ureters

The **ureters,** like the kidneys, are retroperitoneal. Each ureter is a tubular organ about 25 cm (10 in.) long that begins at the renal pelvis and courses inferiorly to enter the urinary bladder at the posterolateral angle of its base. The thickest portion of the ureter is near where it enters the urinary bladder and is approximately 1.0 cm (0.4 in.) in diameter.

The wall of the ureter consists of three layers, or tunics. The inner **mucosa** is continuous with the linings of the renal tubules and the urinary bladder. The mucosa consists of transitional epithelium (fig. 19.11). The cells of this layer secrete a mucus that coats the walls of the ureter with a protective film.

Figure 19.11 A photomicrograph of a ureter in transverse section. (Note the transitional epithelium of the mucosal layer.)

Lumen

Transitional epithelium

Mucosa

Muscularis

Adventitia (serosa)

UNIT VI

The middle layer of the ureter is called the **muscularis.** It consists of an inner longitudinal and an outer circular layer of smooth muscle. In addition, the proximal one-third of the ureter contains another longitudinal layer outside of the circular layer. Muscular peristaltic waves move the urine through the ureter. The peristaltic waves are initiated by the presence of urine in the renal pelvis, and their frequency is determined by the volume of urine. The waves, which occur from every few seconds to every few minutes, force urine through the ureter and cause it

to spurt into the urinary bladder. The outer layer of the ureter is called the **adventitia.** The adventitia is composed of loose fibrous connective tissue that not only covers the ureter but has extensions that anchor the ureter in place.

The arterial supply of the ureter comes from several sources. Branches from the renal artery serve the superior portion. The testicular (or ovarian) artery (also called gonadal artery) supplies the middle portion, and the superior vesicular artery serves the pelvic region. The venous return is through corresponding veins.

A *calculus,* or *renal stone* (fig. 19.12), may obstruct the ureter and greatly increase the frequency of peristaltic waves in an attempt to pass the stone. The pain from a lodged calculus is extreme and extends throughout the pelvic area. A lodged calculus also causes a sympathetic ureterorenal reflex that results in constriction of renal arterioles, thus reducing the production of urine in the kidney on the affected side.

Urinary Bladder

The **urinary bladder** is a storage sac for urine. It is located posterior to the symphysis pubis and anterior to the rectum (fig. 19.13). In females, the urinary bladder is in contact with the uterus and vagina. In males, the prostate is positioned below the urinary bladder.

The shape of the urinary bladder is determined by the volume of urine it contains. An empty urinary bladder is pyramidal in shape, with an apex, body, fundus, and neck. The apex of the urinary bladder is superior to the symphysis pubis and is secured to the **median umbilical ligament,** a fibrous remnant of the embryonic urachus (see Developmental Exposition on page 650). The body of the urinary bladder receives

calculus: L. *calculus,* small stone

Figure 19.12 A calculus, or renal stone, positioned with a dime for size comparison. Factors contributing to stone formation may include the ingestion of excessive mineral salts, a decrease in water intake, and overactivity of the parathyroid glands. A renal stone generally consists of calcium oxalate, calcium phosphate, and uric acid crystals.

Figure 19.13 The position of the urinary bladder within the pelvic cavity. In the male (*a*), the urinary bladder is located between the symphysis pubis and the rectum. In the female (*b*), the urinary bladder is located between the symphysis pubis and the uterus and upper portion of the vagina. In a female, the volume capacity of the urinary bladder is diminished during the last trimester of pregnancy, when the greatly enlarged uterus puts constant pressure on the urinary bladder.

Figure 19.14 The urinary bladder and urethra. (a) The urethra of a male transports both urine and seminal fluid. It consists of prostatic, membranous, and spongy portions. The prostatic portion is surrounded by the prostate. (b) The urethra of a female is considerably shorter and transports only urine.

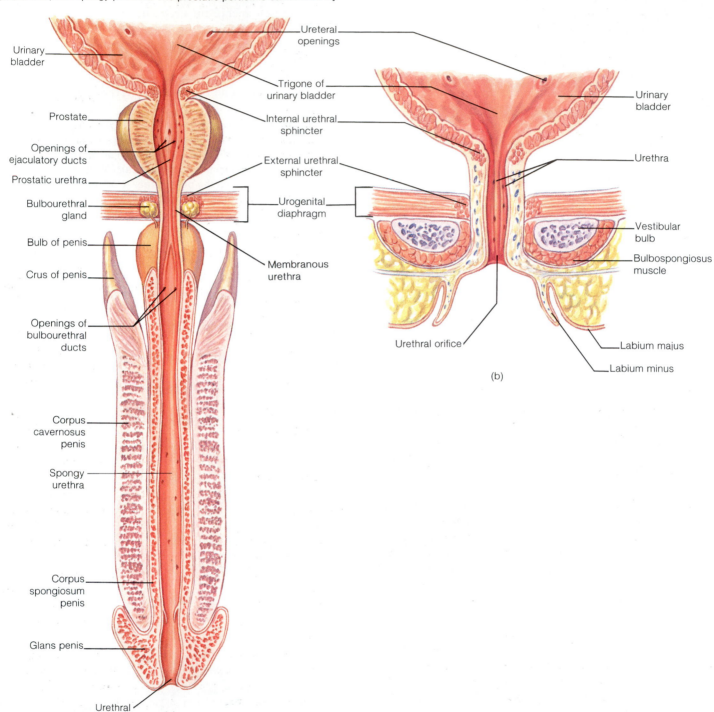

the ureters along the superolateral angles, and the urethra exits at the neck (fig. 19.14). The urethra is a tubular continuation of the neck of the urinary bladder.

The wall of the urinary bladder consists of four layers: the mucosa, submucosa, muscularis, and adventitia (serosa). The **mucosa** (fig. 19.15) is the innermost layer. It is composed of transitional epithelium that decreases in thickness as the urinary bladder distends and the cells are stretched. Further distension is permitted by folds of the mucosa, called **rugae,** which can be seen when the urinary bladder is empty (fig. 19.14). Fleshy flaps of mucosa located where the ureters pierce into the urinary bladder act as valves over the openings of the ureters to

Figure 19.15 The histology of the urinary bladder (50×).

Lumen of urinary bladder

Mucosa of transitional epithelium

Lamina propria

Smooth muscle fibers of detrusor muscle

second, third, and fourth sacral nerves to serve the detrusor muscle. The sensory receptors of the urinary bladder respond to distension and relay impulses to the central nervous system via the pelvic spinal nerves.

> The urinary bladder becomes infected easily, and because a woman's urethra is so much shorter than a man's, women are particularly susceptible to urinary bladder infections. A urinary bladder infection, called *cystitis*, may easily ascend from the urinary bladder to the ureters since the mucous linings are continuous. An infection that involves the renal pelvis is called *pyelitis*; if it continues into the nephrons, it is known as *pyelonephritis*.

Urethra

The tubular **urethra** conveys urine from the urinary bladder to the outside of the body (see fig. 19.14). The urethral wall has an inside lining of mucous membrane surrounded by a relatively thick layer of smooth muscle, the fibers of which are directed longitudinally. Specialized **urethral glands,** embedded in the urethral wall, secrete mucus into the urethral canal.

Two muscular sphincters surround the urethra. The upper involuntary smooth muscle sphincter is the **internal urethral sphincter,** which is formed from the detrusor muscle of the urinary bladder. The lower sphincter is composed of voluntary skeletal muscle fibers and is called the **external urethral sphincter** (fig. 19.14).

The urethra of the female is a simple tube about 4 cm (1.5 in.) long that empties urine through the **urethral orifice** into the vestibule between the labia minora. The urethral orifice is positioned anterior to the vaginal orifice and about 2.5 cm posterior to the clitoris.

The urethra of the male serves both the urinary and reproductive systems. It is about 20 cm (8 in.) long and S-shaped because of the shape of the penis. Three regions can be identified in the male urethra: the prostatic urethra, membranous urethra, and spongy urethra (fig. 19.14).

The **prostatic urethra** is the proximal portion, about 2.5 cm long, that passes through the **prostate** located near the neck of the urinary bladder. The prostatic urethra receives drainage from small ducts of the prostate and two **ejaculatory ducts** of the reproductive system.

The **membranous urethra** is the short (0.5 cm) portion of the urethra that passes through the urogenital diaphragm. The external urethral sphincter muscle is located in this region.

The **spongy urethra** is the longest portion (15 cm), extending from the outer edge of the urogenital diaphragm to the external urethral orifice on the glans penis. This portion is surrounded by erectile tissue as it passes through the corpus spongiosum of the penis. The paired ducts of the **bulbourethral glands** (Cowper's glands) of the reproductive system attach to the spongy urethra near the urogenital diaphragm.

prevent a reverse flow of urine toward the kidneys as the bladder fills. A triangular area known as the **trigone** *(tri'gōn)* is formed on the mucosa between the two ureter openings and the single urethral opening. The internal trigone lacks rugae and is therefore smooth in appearance and remains relatively fixed in position as the urinary bladder changes shape during distension and contraction.

The second layer of the urinary bladder, the **submucosa,** functions to support the mucosa. The **muscularis** consists of three interlaced smooth muscle layers and is referred to as the **detrusor muscle.** At the neck of the urinary bladder, the detrusor muscle is modified to form the upper (the internal) of two muscular sphincters surrounding the urethra. The outer covering of the urinary bladder is the **adventitia.** It appears only on the superior surface of the urinary bladder and is actually a continuation of the peritoneum.

The arterial supply to the urinary bladder comes from the **superior** and **inferior vesicular arteries,** which arise from the internal iliac arteries. Blood draining the urinary bladder enters from a **vesicular venous plexus** and empties into the internal iliac veins.

The autonomic nerves serving the urinary bladder are derived from pelvic plexuses. Sympathetic innervation arises from the last thoracic and first and second lumbar spinal nerves to serve the trigone, ureteral openings, and blood vessels of the urinary bladder. Parasympathetic innervation arises from the

trigone: L. *trigonum,* triangle
detrusor: L. *detrudere,* thrust or forced down

Cowper's glands: from William Cowper, English surgeon, 1666–1709

Developmental Exposition

The Urinary System

Exhibit I The embryonic development of the kidney. (*a*) An anterolateral view of an embryo at 5 weeks showing the position of a transverse cut depicted in (*a₁*). During the sixth week (*b* and *b₁*), the kidney is forming in the pelvic region and the mesonephric duct and gonadal ridge are prominent. The kidney begins migrating during the seventh week (*b₂*), and the urinary bladder and gonad are formed.

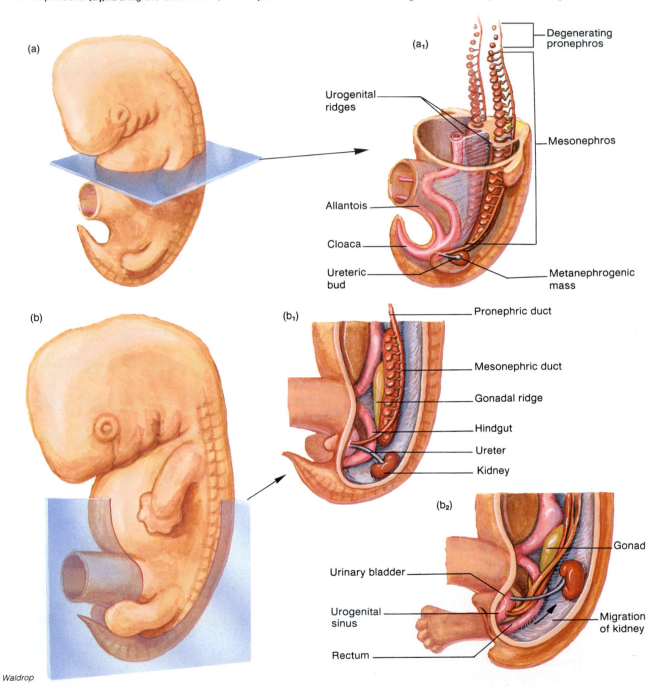

Waldrop

Explanation

The urinary and reproductive systems originate from a specialized elevation of mesodermal tissue called the **urogenital ridge.** The two systems share common structures for part of the developmental period, but by the time of birth two separate systems have formed. The separation in the male is not totally complete, however, since the urethra serves to transport both urine and semen. The development of both systems is initiated during the embryonic stage, but the development of the urinary system starts and ends sooner than that of the reproductive system.

Three successive types of kidneys develop in the human embryo: the pronephros, mesonephros, and metanephros (exhibit I). The third type, or metanephric kidney, remains as the permanent kidney.

The **pronephros** (*pro-nef'ros*) develops during the fourth week after conception and persists only through the sixth week. It is the most superior in position on the urogenital ridge of the three kidneys and is connected to the embryonic **cloaca** by the **pronephric duct.** Although the pronephros is nonfunctional and degenerates in humans, most of its duct is used by the mesonephric kidney (exhibit I), and a portion of it is important in the formation of the metanephros.

The **mesonephros** (*mez''ō-nef'ros*) develops toward the end of the fourth week as the pronephros becomes vestigial. The mesonephros forms from an intermediate portion of the urogenital ridge and functions throughout the embryonic period of development.

Although the **metanephros** (*met''ă-nef'ros*) begins its formation during the fifth week, it does not become functional until immediately before the start of the fetal stage of development at the end of the eighth week. The paired metanephric kidneys continue to form urine throughout fetal development. The urine is expelled through the urinary system into the amniotic fluid.

The metanephros develops from two mesodermal sources (exhibit II). The meaty (glomerular) part of the kidney forms from a specialized caudal portion of the urogenital ridge called the **metanephrogenic mass.** The tubular drainage portion of the kidneys forms as a diverticulum emerging from the wall of the mesonephric duct near the cloaca. This diverticulum, known as the **ureteric** (*yoo''rē-ter'ik*) **bud,** expands into the metanephrogenic mass to form the drainage pathway for urine. The stalk of the ureteric bud develops into the ureter, whereas the expanded terminal portion forms the renal pelvis, calyces, and papillary ducts. A combination of the ureteric bud and metanephrogenic mass forms the other tubular channels within the kidney.

Once the metanephric kidneys are formed, they begin to migrate from the pelvis to the upper posterior portion of the abdomen. The renal blood supply develops as the kidneys become positioned in the posterior body wall.

The development of the kidneys illustrates the concept that ontogeny (embryonic development) recapitulates phylogeny (evolution). This means that the development of the three kidney types follows an evolutionary pattern. The larvae of a few of the more primitive vertebrates have functional pronephric kidneys. Adult fishes and amphibians have mesonephric kidneys, whereas adult reptiles, birds, and mammals have metanephric kidneys.

The urinary bladder develops from the urogenital sinus, which is connected to the embryonic umbilical cord by the fetal membrane called the allantois (exhibit I). By the twelfth week, the two ureters are emptying into the urinary bladder, the urethra is draining, and the connection of the urinary bladder to the allantois has been reduced to a supporting structure called the **urachus.**

Exhibit II   The development of the metanephric kidney at (*a*) 4 weeks, (*b*) 5 weeks, (*c*) 7 weeks, and (*d*) birth. (*d₁*) A magnified view of the arrangement of papillary ducts within a papilla.

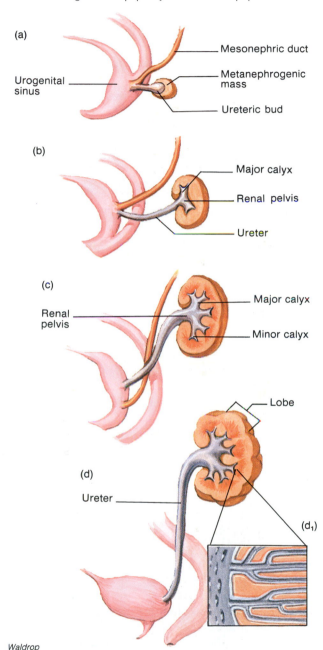

(a)

Urogenital sinus — Mesonephric duct — Metanephrogenic mass — Ureteric bud

(b)

Major calyx — Renal pelvis — Ureter

(c)

Renal pelvis — Major calyx — Minor calyx

Lobe

(d)

Ureter

(d₁)

Waldrop

Occasionally a *patent urachus* is present in a newborn and is discovered when urine is passed through the umbilicus, especially if there is a urethral obstruction. Usually, however, it remains undiscovered until old age, when an enlarged prostate obstructs the urethra and forces urine through the patent urachus and out the umbilicus (navel).

UNIT VI

Figure 19.16 Innervation of the ureter, urinary bladder, and urethra.

Micturition

Micturition *(mik"tŭ-rish'un)*, commonly called urination or voiding, is a reflex action that expels urine from the urinary bladder. It is a complex function that requires a stimulus from the urinary bladder and a combination of involuntary and voluntary nerve impulses to the appropriate muscular structures of the urinary bladder and urethra.

In young children, micturition is a simple reflex action that occurs when the urinary bladder becomes sufficiently distended. Voluntary control of micturition is normally developed by the time a child is 2 or 3 years old. Voluntary control requires the development of an inhibitory ability by the cerebral cortex and a maturing of various portions of the spinal cord.

The volume of urine produced by an adult averages about 1,200 ml per day, but it can vary from 600–2,500 ml. The average capacity of the urinary bladder is 700–800 ml. A volume of 200–300 ml will distend the urinary bladder enough to stimulate stretch receptors and trigger the micturition reflex, creating a desire to urinate.

The micturition reflex center is located in the second, third, and fourth sacral segments of the spinal cord. Following stimulation of this center by impulses arising from stretch receptors in the urinary bladder, parasympathetic nerves that stimulate the detrusor muscle and the internal urethral sphincter are activated. Stimulation of these muscles causes a rhythmic contraction of the urinary bladder wall and a relaxation of the internal urethral sphincter. At this point, a sensation of urgency is perceived in the brain, but there is still voluntary control over the external urethral sphincter. At the appropriate time, the conscious activity of the brain activates the motor nerve fibers (S4) to the external urethral sphincter via the pudendal nerve (S2, S3, and S4), causing the sphincter to relax and urination to occur.

The innervation of the ureter, urinary bladder, and urethra is shown in figure 19.16, and a summary of the process of micturition is presented in table 19.2.

Table 19.2	**Events of Micturition**

1. The urinary bladder becomes distended as it fills with urine.
2. Stretch receptors in the urinary bladder wall are stimulated, and impulses are sent to the micturition center in the spinal cord.
3. Parasympathetic nerve impulses travel to the detrusor muscle and the internal urethral sphincter.
4. The detrusor muscle contracts rhythmically, and the internal urethral sphincter relaxes.
5. The need to urinate is sensed as urgent.
6. Urination is prevented by voluntary contraction of the external urethral sphincter and by inhibition of the micturition reflex by impulses from the midbrain and cerebral cortex.
7. Following the decision to urinate, the external urethral sphincter is relaxed, and the micturition reflex is facilitated by impulses from the pons and the hypothalamus.
8. The detrusor muscle contracts, and urine is expelled through the urethra.
9. Neurons of the micturition reflex center are inactivated, the detrusor muscle relaxes, and the urinary bladder begins to fill with urine.

Urinary retention, or the inability to void, may occur postoperatively, especially following surgery of the rectum, colon, or internal reproductive organs. The difficulty may be due to nervous tension, the effects of anesthetics, or pain and edema at the site of the operation. If urine is retained beyond 6 to 8 hours, *catheterization* may become necessary. In this procedure, a tube or catheter is passed through the urethra into the urinary bladder so that urine can flow freely.

1. Describe the location and the structure of a ureter and indicate the function of its muscularis layer.
2. Describe the structure of the urinary bladder and indicate the anatomical structures that permit the organ to be distended.
3. Compare the male urinary system with that of the female.
4. Explain the structures and processes involved in the control of micturition.

UNIT VI

Clinical Considerations

The importance of kidney function in maintaining homeostasis and the ease with which urine can be collected and used as a mirror of the plasma's chemical composition make the clinical study of renal function and urine composition particularly significant. **Urology** is the medical specialty concerned with dysfunctions of the urinary system. Urinary dysfunctions can be congenital or acquired; they may result from physical trauma or from conditions that secondarily involve the urinary organs.

Congenital Malformations of the Urinary Organs

Abnormalities of the organs of the urinary system occur in about 12% of newborn babies. These deviations range from insignificant anomalies to those that are incompatible with life.

Kidneys   Common malformations of the kidneys are illustrated in figure 19.17. One common deformity is **renal agenesis** *(ă-jen′ĕ-sis)*, the unilateral absence of a kidney as a result of failure of a uretic bulb to develop. **Renal ectopia** is the condition in which one or both kidneys are in an abnormal position. Generally, this condition occurs when a kidney remains in the pelvic area. **Horseshoe kidney** refers to a fusion of the kidneys across the midline. The incidence of this asymptomatic condition is about 1 in 600.

Ureters   **Duplication of the ureters** and the associated renal pelvis is a frequent anomaly of the urinary tract. Unilateral duplication occurs in 1 in 200 births, whereas bilateral duplication occurs in 1 in 1,200 births. Occasionally there is partial duplication, which may become very important because of the attendant propensity for urinary infections. A completely duplicated ureter frequently opens into areas other than the urinary bladder and requires surgical correction.

Urinary Bladder   Protrusion of the posterior wall of the urinary bladder is called **exstrophy of the bladder** and occurs when the wall of the perineum fails to close. Associated with this condition, which occurs in about 1 in 40,000 births, are defective urethral sphincter muscles.

Urethra   The most common anomaly of the urethra is a condition called **hypospadias** *(hi″po-spād′e-as)*, where the urethra of the male opens on the ventral surface of the penis instead of at the tip of the glans penis (fig. 19.18). There is a similar defect in the female in which the urethra opens into the vagina. **Epispadias** is a failure of closure on the dorsal surface of the penis. Hypospadias and epispadias can be corrected surgically.

Symptoms and Diagnosis of Urinary Disorders

Normal micturition is painless. **Dysuria** *(dis-yur′e-ă)*, or painful urination, is a sign of a urinary tract infection or obstruction of the urethra—as in an enlarged prostate in a male. **Hematuria** means blood in the urine and is usually associated with trauma.

Figure 19.17   Various congenital anomalies involving the kidneys.

Pelvic kidney

Unilateral double kidney

Rosette kidney

Horseshoe kidney

Aberrant renal artery causing urinary obstruction

Aberrant renal arteries

Polycystic kidney

Bacteriuria means bacteria in the urine, and **pyuria** is the term for pus in the urine, which may result from a prolonged infection. **Oliguria** is a scanty output of urine, whereas **polyuria** is an excessive output. Low blood pressure and kidney failure are two causes of oliguria. **Uremia** is a condition in which substances ordinarily excreted in the urine accumulate in the blood. **Enuresis** *(en″yŭ-re-sis)*, or **incontinence**, is the inability to control micturition. It may be caused by psychological factors or by structural impairment.

The palpation and inspection of urinary organs is an important aspect of physical assessment. The right kidney is palpable in the supine position; the left kidney usually is not. The distended urinary bladder is palpable along the superior pelvic rim.

UNIT VI

Figure 19.18 Epispadias and hypospadias of the male urethra.

Epispadias   Hypospadias

Figure 19.19 Cystoscopic examination of a male.

The urinary system may be examined using radiographic techniques. An **intravenous pyelogram** (IVP) permits radiographic examination of the kidneys following the injection of radiopaque dye. In this procedure, the dye that has been injected intravenously is excreted by the kidneys so that the renal pelvises and the outlines of the ureters and urinary bladder can be observed in a radiograph.

Cystoscopy *(si-stos'kŏ-pe)* is the inspection of the inside of the urinary bladder using an instrument called a cystoscope (fig. 19.19). By means of this technique, tissue samples can be obtained, as well as urine samples from each kidney prior to mixing in the urinary bladder. Once the cystoscope is in the urinary bladder, the ureters and pelvis can be viewed through urethral catheterization and inspected for obstructions.

A **renal biopsy** is a diagnostic test for evaluating certain types and stages of kidney diseases. The biopsy is performed either through a skin puncture (closed biopsy) or through a surgical incision (open biopsy).

In a **urinalysis,** the voided urine specimen is tested for color, specific gravity, chemical composition, and for the presence of microscopic bacteria, crystals, and *casts*. Casts are accumulations of proteins that have leaked through the glomeruli and have been pushed through the tubules like toothpaste through a tube.

Infections of the Urinary Organs

Urinary tract infections are a significant cause of illness and are also a major factor in the development of chronic renal failure. Females are more predisposed to urinary tract infections than are males, and the incidence of infection increases directly with sexual activity and aging. The higher infection rate in females has been attributed to a shorter urethra, which is close to the rectum, and to the lack of protection provided by prostatic secretions in males. To reduce the risk of urinary infections, a young girl should be taught to wipe her anal region in a posterior direction, away from the urethral orifice, after a bowel movement.

Infections of the urinary tract are named according to the infected organ. An infection of the urethra is called **urethritis** and involvement of the urinary bladder is **cystitis.** Cystitis is frequently a secondary infection from some other part of the urinary tract.

Nephritis is inflammation of the kidney tissue. **Glomerulonephritis** *(glo-mer''yu-lo-nĕ-fri'tis)* is inflammation of the glomeruli. Glomerulonephritis may occur following an upper respiratory tract infection because antibodies produced against streptococci bacteria can produce an autoimmune inflammation in the glomeruli. This inflammation may permanently change the glomeruli and figure significantly in the development of chronic renal disease and renal failure.

Any interference with the normal flow of urine, such as from a renal stone or an enlarged prostate in a male, causes stagnation of urine in the renal pelvis and the development of pyelitis. **Pyelitis** is an inflammation of the renal pelvis and its calyces. **Pyelonephritis** is inflammation involving the renal pelvis, the calyces, and the tubules of the nephron within one or both kidneys. Bacterial invasion from the blood or from the lower urinary tract is another cause of both pyelitis and pyelonephritis.

Trauma to the Urinary Organs

A sharp blow to a lumbar region of the back may cause a contusion or rupture of a kidney. Symptoms of kidney trauma include hematuria and pain in the upper abdominal quadrant and flank on the injured side.

Pelvic fractures from accidents may result in perforation of the urinary bladder and urethral tearing. When driving an automobile, it is advisable to stop periodically to urinate because

UNIT VI

Figure 19.20 A diagram of a hemodialysis machine.

an attached seat belt over the region of a full urinary bladder can cause it to rupture in even a relatively minor accident. Urethral injuries are more common in men than in women because of the position of the urethra in the penis. In a straddle injury, for example, a man walking along a raised beam may slip and compress his urethra and penis between the hard surface and his pubic arch, rupturing the urethra.

Hemodialysis Hemodialysis equipment is designed to filter the wastes from the blood of a patient who has chronic renal failure. During hemodialysis, the blood of a patient is pumped via a tube from an artery through the machine where it is cleansed and then returned to the body via a vein (fig. 19.20). The cleaning process involves pumping the blood past a semipermeable cellophane membrane that separates the blood from an isotonic solution containing molecules needed by the body (such as glucose). In a process called **dialysis,** waste products diffuse out of the blood through the membrane, while glucose and other molecules needed by the body remain in the blood.

Obstruction The urinary system can become obstructed anywhere along the tract. Calculi (stones) are the most common cause, but blockage can also come from trauma, strictures, tumors or cysts, spasms or kinks of the ureters, or congenital anomalies. If not corrected, an obstruction causes urine to col-

lect behind the blockage and generate pressure that may cause permanent functional and anatomic damage to one or both kidneys. As a result of pressure buildup in a ureter, a distended ureter, or **hydroureter,** develops. Dilation in the renal pelvis is called **hydronephrosis.**

Calculi, or renal stones, are generally the result of infections or metabolic disorders that cause the excretion of large amounts of organic and inorganic substances (see fig. 19.12). As the urine becomes concentrated, these substances may crystalize and form granules. The granules then serve as cores for further precipitation and development of larger calculi. This becomes dangerous when a calculus grows large enough to cause an obstruction. It also causes intense pain when it passes through the urinary tract.

Renal Failure An output of 50–60 cc of urine per hour is considered normal. If the output drops to less than 30 cc per hour, it may indicate renal failure. Renal failure is the loss of the kidney's ability to maintain fluid and electrolyte balance and to excrete waste products.

Renal failure can be either acute or chronic. **Acute renal failure** is the sudden loss of kidney function caused by shock and hemorrhage, thrombosis, or other physical trauma to the kidneys. The kidneys may sustain a 90% loss of their nephrons through tissue death and still continue to function without

UNIT VI

apparent difficulty. If a patient suffering acute renal failure is stabilized, the nephrons have an excellent capacity to regenerate.

A person with **chronic renal failure** cannot sustain life independently. Chronic renal failure is the end result of kidney disease in which the kidney tissue is progressively destroyed. As renal tissue continues to deteriorate, the options for sustaining life are hemodialysis or kidney transplantation.

Urinary Incontinence   The inability to voluntarily retain urine in the urinary bladder is known as urinary incontinence. It has a number of temporary or permanent causes. Emotional stress is a cause of temporary incontinence in adults. Causes of permanent incontinence include neurological trauma, various urinary diseases, and tissue damage within the urinary bladder or urethra. Remarkable advances have been made in treating permanent urinary incontinence through the implantation of an artificial urethral sphincter (fig. 19.21).

CLINICAL CASE STUDY ANSWER

The hematuria is likely the result of a right renal laceration caused by the upper abdominal knife wound. The right lower quadrant stab most likely did not violate the urinary tract. The course of blood seen in the catheter begins and proceeds as follows: Abdominal aorta → right renal artery → smaller parenchymal artery → through the lacerated vessel(s) → into the lacerated urinary collecting system either at the calyx or the renal pelvis or proximal right ureter → urinary bladder → into catheter. During the operation, the surgeon should keep in mind that the presence of only one kidney is a relatively common occurrence (2%–4% of the population). If, therefore, she is prompted to remove the damaged kidney, she should first confirm the presence of a second functioning kidney. If a second kidney is not present, every effort should be made to correct the problem without performing a nephrectomy, which would consign the patient to chronic hemodialysis or kidney transplant.

Figure 19.21   An artificial urethral sphincter. The entire device is implanted internally and consists of an inflation bulb that is manually squeezed to inflate the urethral cuff from the fluid stored in a reservoir (pressure-regulating balloon). Manual pumping of the deflation bulb (implanted in the opposite side of the scrotum) releases the pressure of the urethral cuff and allows urine to flow from the urinary bladder. The device for a female is similar except that the inflation and deflation bulbs are implanted in the tissue of the labia minora.

Internal Affairs

The Relational Workings of the Body Systems

Urinary System

- Eliminates metabolic wastes
- Maintains normal acid-base (pH), fluid, and electrolyte levels

Integumentary System
- Covers the body and protects it from excessive fluid loss
- Provides site for evaporative water loss

- Eliminates metabolic wastes
- Maintains normal acid-base (pH), fluid, and electrolyte levels

Skeletal System
- Supports and protects certain organs of the urinary system
- Stores calcium and phosphate ions

- Eliminates metabolic wastes
- Maintains normal acid-base (pH), fluid, and electrolyte levels

Muscular System
- Supports and protects certain organs of the urinary system
- Assists the storage and voiding of urine

- Eliminates metabolic wastes
- Maintains normal acid-base (pH), fluid, and electrolyte levels

Nervous System
- Provides autonomic innervation to the urinary system
- Provides motor control of micturition

- Eliminates metabolic wastes
- Maintains normal acid-base (pH), fluid, and electrolyte levels

Endocrine System
- Provides site for the production of the hormone erythropoietin within the kidney

- Eliminates metabolic wastes
- Maintains normal acid-base (pH), fluid, and electrolyte levels

Circulatory System
- Transports O_2 and CO_2, nutrients, and fluids to and from the organs of the urinary system

- Eliminates metabolic wastes
- Maintains normal acid-base (pH), fluid, and electrolyte levels

Lymphatic System
- Maintains a balanced amount of interstitial fluid within the organs of the urinary system
- Protects urinary tract against infection

- Eliminates metabolic wastes
- Maintains normal acid-base (pH), fluid, and electrolyte levels

Respiratory System
- Provides O_2 and eliminates CO_2

- Eliminates metabolic wastes
- Maintains normal acid-base (pH), fluid, and electrolyte levels

Digestive System
- Provides nutrients for growth, maintenance, and repair of the urinary system

- Eliminates metabolic wastes
- Maintains normal acid-base (pH), fluid, and electrolyte levels

Reproductive System
- Provides organ (penis) in male for passage of urethra

CHAPTER SUMMARY

Introduction to the Urinary System (pp. 638–639)

1. The urinary system consists of two kidneys, two ureters, the urinary bladder, and the urethra.
2. The urinary system maintains the composition and properties of the body fluid, which establishes the internal environment of the body cells. The end product of the urinary system is urine, which is voided from the body during micturition.

Kidneys (pp. 639–646)

1. The kidneys are retroperitoneal, embedded in a renal adipose capsule.
2. Each kidney is contained by a renal capsule and divided into an outer renal cortex and an inner renal medulla.
   (a) The renal medulla is composed of renal pyramids separated by renal columns.
   (b) The renal papillae empty urine into the minor calyces and then into the major calyces, which drain into the renal pelvis. From there, urine flows through the ureter.
3. Each kidney contains more than a million microscopic functional units called nephrons.
   (a) Filtration occurs in the glomerulus, which receives blood from an afferent glomerular arteriole.
   (b) Glomerular blood is drained by an efferent glomerular arteriole that delivers blood to peritubular capillaries that surround the nephron tubules.
   (c) The glomerular capsule and distal convoluted tubules are located in the renal cortex.
   (d) The nephron loop is located in the renal medulla.
   (e) Filtrate from the distal convoluted tubule is drained into papillary ducts that plunge through the renal medulla to empty urine into the calyces.

Ureters, Urinary Bladder, and Urethra (pp. 646–652)

1. Urine is channeled from the kidneys to the urinary bladder by the ureters and expelled from the bladder through the urethra. The detrusor muscle of the urinary bladder and the sphincter muscles of the urethra are used in the control of micturition.
   (a) Each ureter contains three layers: the mucosa, muscularis, and adventitia.
   (b) The lumen of the urinary bladder is lined by transitional epithelium, which is folded into rugae. These structures enhance the ability of the urinary bladder to distend.
   (c) The urethra has an internal sphincter of smooth muscle and an external sphincter of skeletal muscle.
   (d) The male urethra conducts urine during urination and seminal fluid during ejaculation. The female urethra is much shorter than that of a male and conducts only urine.
   (e) The male urethra is composed of prostatic, membranous, and spongy portions.
2. Micturition is controlled by reflex centers in the second, third, and fourth sacral segments of the spinal cord.

REVIEW ACTIVITIES

Objective Questions

1. Which of the following statements about the renal pyramids is *false?*
   (a) They are located in the renal medulla.
   (b) They contain glomeruli.
   (c) They contain papillary ducts.
   (d) They open through the renal papillae into the renal calyces.
2. Renal vessels and the ureter attach at the concave medial border of the kidney called
   (a) the renal pelvis.
   (b) the urachus.
   (c) the calyx.
   (d) the hilum.
3. The renal medulla of the kidney contains
   (a) glomerular capsules.
   (b) glomeruli.
   (c) renal pyramids.
   (d) adipose capsules.

4. Urine flowing from the papillary ducts enters directly into
   (a) the renal calyces.
   (b) the ureter.
   (c) the renal pelvis.
   (d) the distal convoluted tubules.
5. Which of the following statements concerning the kidneys is *false?*
   (a) They are retroperitoneal.
   (b) They each contain 8 to 15 renal pyramids.
   (c) They each have distinct renal cortex and renal medulla regions.
   (d) They are positioned between the third and fifth lumbar vertebrae.
6. A calculus, or renal stone, would most likely cause stagnation of the urine in which portion of the urinary system?
   (a) the urinary bladder
   (b) the renal column
   (c) the ureter
   (d) the renal pelvis
   (e) the urethra
7. Distention of the urinary bladder is possible because of the presence of
   (a) the rugae.
   (b) the trigone.
   (c) the adventitia.
   (d) the transitional epithelium.
   (e) both (a) and (d).
8. The detrusor muscle is located in
   (a) the kidneys.
   (b) the ureters.
   (c) the urinary bladder.
   (d) the urethra.
9. The internal urethral sphincter is innervated by
   (a) sympathetic neurons.
   (b) parasympathetic neurons.
   (c) somatic motor neurons.
   (d) all of the above.
10. Which of the following statements about metanephric kidneys is *true?*
    (a) They become functional at the end of the eighth week.
    (b) They are active throughout fetal development.
    (c) They are the third pair of kidneys to develop.
    (d) All of the above are true.

UNIT VI

Essay Questions

1. Describe the location of the kidneys in relation to the abdominal cavity and the peritoneal membranes.
2. Diagram the kidney structures that can be identified in a coronal section.
3. Describe how the kidney is supported against the posterior abdominal wall. How is this support related to the condition called renal ptosis?
4. Trace a drop of blood from an interlobular artery through a glomerulus and to an interlobular vein. List in order all the vessels through which the blood passes. How do structural differences in afferent and efferent arterioles ensure the necessary high blood pressure for filtrate formation?
5. In a male, trace the pathway of urine from the site of filtration at the renal corpuscle, through the urinary tubules of the nephron, and to the outside of the body. List in order all the structures through which the urine passes.
6. What is a nephron? Describe the two types of nephrons found in a kidney. Why are nephrons considered the functional units of the urinary system?
7. Describe the mechanism involved in the passage of urine from the renal pelvis to the urinary bladder.
8. Describe the urinary bladder with regard to position, histological structure, blood supply, and innervation.
9. Compare and contrast the male and female urethra.
10. What is the micturition reflex? Discuss the physiological and functional events of a voluntary micturition response.
11. What is a metanephrogenic mass and a ureteric bud? Discuss the sequential development of these embryonic structures to form the urinary system. How can a greater knowledge of this development process lead to a better understanding of congenital abnormalities?
12. Define *urology, cystoscopy, renal calculi,* and *urinalysis.*
13. List four common congenital malformations of the urinary system. Which of these require surgical correction?
14. Define each of the following terms: *dysuria, hematuria, bacteriuria, pyuria, oliguria, polyuria, uremia,* and *enuresis.*

Gundy/Weber Software

The tutorial software accompanying Chapter 19, Urinary System, is Volume 12, Urinary System.

20

Male Reproductive System

Chapter Outline

CLINICAL CASE STUDY

A 27-year-old man visits his family doctor, stating that he and his wife have been unable to conceive a child after nearly 2 years of trying. Furthermore, he says that his wife has undergone a thorough gynecological evaluation revealing no female cause for the infertility. Upon palpating the patient's testes, the doctor finds nothing unusual. When he examines the scrotal sac above the testes, however, he appears to be perplexed. He states that there is a tubular structure, one for each testis, for which he is palpating but cannot find. He tells his patient that these structures probably have been missing from birth. During a follow-up visit, the doctor informs the patient that examination of his ejaculate revealed azoospermia (no viable sperm).

Explain how the result of the semenalysis relates to the patient's physical exam findings. List the structures involved. Does it seem peculiar that the patient is capable of producing an ejaculate? Explain.

Introduction to the Male Reproductive System

The organs of the male and female reproductive systems are adapted to produce and allow the union of gametes that contain specific genes. A random combination of the genes during sexual reproduction results in the propagation of individuals with genetic differences.

Objective 1. Explain why sexual reproduction is biologically advantageous.

Objective 2. List the functions of the male reproductive system and compare them with those of the female.

Objective 3. Distinguish between primary and secondary sex organs.

Sexual reproduction is common in organisms. Through this process, individuals of a species are propagated and have a genetic makeup inherited from both parents. Sexual reproduction provides the important advantage of genetic recombination through which genetic changes become part of the population's gene pool. These genetic changes help some of the offspring adapt to environmental changes that might otherwise cause the population to become extinct.

Because sexual reproduction requires the production of two types of **gametes** *(gam'ētz)*, or sex cells, the species has a male and female form, each with its own unique reproductive system. The functions of the female reproductive system are more complex than those of the male. The functions of the male reproductive system are to produce the male gametes, **spermatozoa**

gamete: Gk. *gameta*, husband or wife
spermatozoa: Gk. *sperma*, seed; *zoon*, animal

(sperm), and to transfer them to the female through the process of *coitus (sexual intercourse),* or *copulation.* The female not only produces her own gametes, called **ova,** and receives the sperm from the male, but her reproductive organs are specialized to provide a site for fertilization, implantation of the blastocyst, development of the embryo and fetus, and delivery of a baby. The reproductive system of the female also provides a means of nourishing the baby through the secretions of the mammary glands in the breast.

The reproductive system is a unique body system in three respects. First, whereas all of the other body systems function to sustain the individual, the reproductive system is specialized to perpetuate the species and pass genetic material from generation to generation. Second, the anatomy and physiology of the reproductive organs constitute the major differences between the male and female of the species. Other systems, such as the integumentary, skeletal, and urinary, have minor gender-based differences, but none to the extent of the reproductive system. The third unique characteristic of the reproductive system is its latent development under hormonal control. The other body systems are totally functional at birth or shortly thereafter, whereas the reproductive system does not become functional until it is acted on by hormones during puberty. *Puberty* is the period of human development during which the reproductive organs become functional.

The structures of the male reproductive system can be categorized on a functional basis as follows:

1. **Primary sex organs.** The primary sex organs are called **gonads;** specifically, the **testes** in the male. Gonads produce the gametes, or spermatozoa, and produce and secrete sex hormones. The appropriate amounts and timing of the production and secretion of male sex hormones cause the development of secondary sex organs and the expression of secondary sex characteristics.
2. **Secondary sex organs.** Secondary sex organs are those structures that are essential in caring for and transporting spermatozoa. The three categories of secondary sex organs are the sperm-transporting ducts, the accessory glands, and the copulatory organ. The ducts that transport sperm include the **epididymides, ductus deferentia, ejaculatory ducts,** and **urethra.** The accessory glands are the **seminal vesicles,** the **prostate,** and the **bulbourethral** *(bul"bo-yoo-re'thral)* **glands.** The **penis,** which contains erectile tissue, is the copulatory organ. The **scrotum** is a pouch of skin that encloses the testes.
3. **Secondary sex characteristics.** Secondary sex characteristics are features that, although not essential for the reproductive process, are generally considered to be sexual attractants. Physique, body hair pattern, and voice pitch are examples.

The organs of the male reproductive system are depicted in figure 20.1 and their functions are summarized in table 20.1.

puberty: L. *puberty*, grown up

UNIT VII

Figure 20.1 Organs of the male reproductive system. (*a*) A sagittal view and (*b*) a posterior view.

(a)

Urinary bladder

Symphysis pubis

Ductus deferens

Urethra

Penis

Glans penis

Prepuce

Ampulla of ductus deferens

Seminal vesicle

Ejaculatory duct of ductus deferens

Prostate

Bulbourethral gland

Anus

Ductus deferens

Epididymis

Testis

Scrotum

(b)

Ampulla

Seminal vesicle

Ejaculatory duct

Bulbourethral gland

Epididymis

Testis

Penis

Urethra

Ureter

Urinary bladder

Prostate

Ductus deferens

Testis (sectioned to show lobules)

Glans penis

Table 20.1 | **Functions of the Male Reproductive Organs**

Organ(s)	Function(s)
Testes	
Seminiferous tubules	Produce spermatozoa
Interstitial cells	Secrete male sex hormones
Epididymides	Site of sperm maturation; store and convey spermatozoa to ductus deferentia
Ductus deferentia	Store spermatozoa; convey spermatozoa to ejaculatory ducts
Ejaculatory ducts	Receive spermatozoa and additives to produce seminal fluid
Seminal vesicles	Secrete alkaline fluid containing nutrients and prostaglandins
Prostate	Secretes alkaline fluid that helps neutralize acidic environment of the vagina and enhances motility of spermatozoa
Bulbourethral glands	Secrete fluid that lubricates urethra and end of penis
Scrotum	Encloses and protects testes
Penis	Conveys urine and seminal fluid to outside of body; organ of coitus

1. What is the primary importance of sexual reproduction?
2. What are the functions of the male reproductive system?
3. List the organs of the male reproductive system and indicate whether they are primary or secondary sex organs.
4. With respect to the other body systems, discuss the latent (delayed) development of the reproductive system.

Perineum and Scrotum

The perineum is the specific portion of the pelvic region that contains the external genitalia and the anal opening. The perineum is subdivided into the urogenital triangle and the anal triangle.

Objective 4. Describe the location of the perineum and distinguish between the urogenital and anal triangles.

Objective 5. Discuss the structure and function of the scrotum.

Perineum

The **perineum** (per''i-ne'um) is a diamond-shaped region between the posterior portion of the symphysis pubis and the coccyx (fig. 20.2). It is the muscular region of the outlet of the pelvis (see fig. 9.24). The perineum is divided into a *urogenital triangle* in front and an *anal triangle* in back. In the male perineum, the penis and scrotum are attached at the anterior portion of the urogenital triangle and the anus is located within the posterior portion of the anal triangle.

Scrotum

The saclike **scrotum** is suspended immediately behind the base of the penis (see figs. 20.1 and 20.2). The functions of the scrotum are to support and protect the testes and to regulate their position relative to the pelvic region of the body. The soft textured skin of the scrotum is covered with sparse hair in mature males and is darker in color than most of the other skin of the body. It also contains numerous sebaceous glands. The external appearance of the scrotum varies at different times in the same individual, depending on environmental conditions and actions of the *dartos (dar'tos)* and *cremaster (kre-mas'ter) muscles*. The **dartos** is a layer of smooth muscle fibers in the subcutaneous tissue of the scrotum and the **cremaster** is a thin strand of skeletal muscle extending through the spermatic cord (see fig. 20.15). The **spermatic cord** is the organ attached to the testis that contains the nerves, genital tube, and vessels that serve the testis. Both the dartos and cremasteric muscles involuntarily contract in response to cold temperatures to move the testes closer to the heat of the body in the pelvic region. The cremaster muscle is a continuation of the internal abdominal oblique muscle of the abdominal wall, which is derived as the testes descend into the scrotum. Because it is a skeletal muscle, it can be contracted voluntarily as well. When these muscles are contracted, the scrotum is tightly wrinkled. Warm temperatures cause the dartos and cremaster muscles to relax and the testes to become pendent in the flaccid scrotum. The temperature of the testes is maintained at about 35° C (95° F, or about 3.6 degrees below normal body temperature) by the contraction or relaxation of the scrotal muscles. This temperature has been determined to be an optimum temperature for the production and storage of spermatozoa.

The scrotum is subdivided into two longitudinal compartments by a fibrous **scrotal septum** (fig. 20.3). The purpose of the scrotal septum is to compartmentalize each testis; thus, the trauma or infection of one will generally not affect the other. In addition, the left testis is generally suspended lower in the scrotum than the right so that the two are less likely to be compressed forcefully together. The site of the scrotal septum is apparent on the surface of the scrotum along a median longitudinal ridge called the **scrotal raphe** (ra'fe). The scrotal raphe extends forward to the undersurface of the penis and backward to the anal opening.

The scrotum has an extensive blood and nerve supply. The arteries that serve the scrotum are the internal pudendal branch of the internal iliac artery, the external pudendal branch from the femoral artery, and the cremasteric branch of the inferior epigastric artery. The venous drainage follows a pattern similar to that of the arteries. The scrotal nerves are primarily sensory and include the pudendal nerves, ilioinguinal nerves, and posterior cutaneous nerves of the thigh.

dartos: Gk. *dartos*, skinned or flayed
cremaster: Gk. *cremaster*, a suspender, to hang
septum: L. *septum*, a partition
pudendal: L. *pudeo*, to feel ashamed

Figure 20.2 A superficial view of (*a*) the male perineum, and (*b*) the female perineum (see chapter 21). Divided into urogenital and anal triangles, the perineum is the region between the symphysis pubis and coccyx.

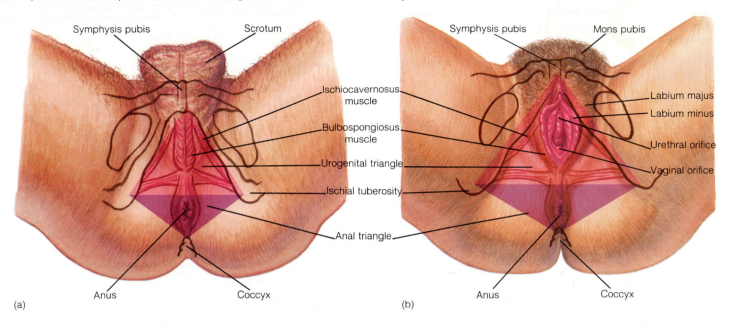

(a)

(b)

Figure 20.3 Structural features of the testis and epididymis. (*a*) A longitudinal view and (*b*) a transverse view.

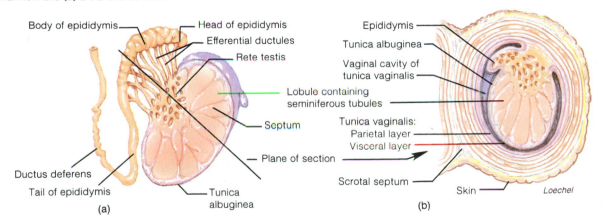

(a)

(b)

Loechel

Although uncommon, male infertility may result from excessively high temperatures of the testes over an extended period of time. Tight clothing that constricts the testes close to the body or frequent hot baths or saunas may destroy sperm to the extent that the sperm count of discharged semen will be below that necessary for fertilization to occur.

1. Describe the location of the perineum.
2. Explain how the temperature of the testes is maintained. Why is it important to maintain a particular testicular temperature?
3. Describe the innervation and the blood supply to the testes.

Testes

The testes, located within the scrotum, produce spermatozoa and androgens. Androgens regulate spermatogenesis and the development and functioning of the secondary sex organs.

Objective 6. Describe the location, structure, and functions of the testes.

Objective 7. Explain how hormones control the activities of the male reproductive organs and how hormones are related to the development of male secondary sex characteristics.

UNIT VII

Objective 8. List the events of spermatogenesis and distinguish between spermatogenesis and spermiogenesis.

Objective 9. Diagram the structure of a sperm cell and explain the function of each of its parts.

Structure of the Testis

The testes are paired, whitish, ovoid organs, about 4 cm (1.5 in.) long and 2.5 cm (1 in.) in diameter. Each testis weighs between 10 and 14 g.

Two tissue layers, or tunics, cover the testis. The outer **tunica vaginalis** is a thin serous sac derived from the peritoneum during the descent of the testis. The **tunica albuginea** (al''byoo-jin'e-ă) is a tough fibrous membrane that directly encapsules the testis (fig. 20.3). Fibrous inward extensions of the tunica albuginea partition the testis into 250 to 300 wedge-shaped **lobules.**

Each lobule of the testis contains one to three tightly convoluted **seminiferous tubules** (fig. 20.4) that may exceed 70 cm (28 in.) in length if uncoiled. The seminiferous tubules are the functional units of the testis because they are the structures that produce spermatozoa. Spermatozoa are produced at the rate of thousands per second throughout the life of a healthy, sexually mature male in a process called *spermatogenesis.*

Various stages of meiosis can be observed in a section of seminiferous tubules (see fig. 20.7). The germinal cells, called **spermatogonia** *(sper-mat''e-go'ne'ă),* are in contact with the basement membrane. Spermatogonia are in a constant state of mitotic division to produce, in order of advancing maturity, the primary spermatocytes, secondary spermatocytes, and spermatids (see fig. 20.6). Forming the walls of the seminiferous tubules are **sustentacular cells** (Sertoli cells or nurse cells) that produce and secrete nutrients to the developing spermatozoa embedded between them (see fig. 20.7). The spermatozoa are formed, but not fully mature, by the time they reach the lumina of the seminiferous tubules.

Between the seminiferous tubules are specialized endocrine cells called **interstitial cells** (cells of Leydig). The function of these cells is to produce and secrete the male sex hormones. The testes are considered mixed exocrine and endocrine glands because they produce both spermatozoa and androgens.

Newly produced spermatozoa move through the seminiferous tubules and enter the **rete testis** for further maturation (fig. 20.4). Cilia are located on some of the cells of the rete testis, presumably for moving the spermatozoa. The spermatozoa are transported out of the testis and into the epididymis through a series of **efferential ductules.**

The testes receive blood through the **testicular arteries** that arise from the abdominal aorta immediately below the origin of the renal arteries. The **testicular veins** drain the testes. The testicular vein of the right side enters directly into the inferior vena cava, whereas the testicular vein of the left side drains into the left renal vein (see fig. 16.32).

Testicular nerves innervate the testes with both motor and sensory neurons arising from the tenth thoracic segment of the spinal cord. Innervation is primarily through sympathetic neurons, but there is limited parasympathetic stimulation as well.

The primary cause of male infertility is a condition called *varicocele (var'ĭ-ko-sēl'').* Varicocele occurs when one or both of the testicular veins draining from the testes becomes swollen, resulting in poor vascular circulation in the testes. A varicocele generally occurs on the left side because the left testicular vein drains into the renal vein where the blood pressure is higher than it is in the inferior vena cava, into which the right testicular vein empties.

Endocrine Functions of the Testis

Testosterone is clearly the major androgen secreted by the adult testis. This hormone is responsible for the initiation and maintenance of the body changes of puberty in males. Androgens are sometimes called *anabolic steroids* because they stimulate the growth of muscles and other structures (table 20.2). Increased testosterone secretion during puberty is also required for the growth of the accessory sex organs, primarily the seminal vesicles and prostate. The removal of androgens by castration results in atrophy of these organs.

Androgens stimulate growth of the larynx (resulting in lowering of the voice), hemoglobin synthesis (males have higher

---

Figure 20.4 The testis and epididymis seen in sagittal section.

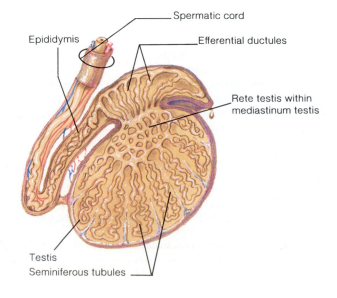

Spermatic cord

Epididymis

Efferential ductules

Rete testis within mediastinum testis

Testis

Seminiferous tubules

---

tunica: L. *tunica,* a coat
vaginalis: L. *vagina,* a sheath
albuginea: L. *albus,* white
spermatogonia: Gk. *sperma,* seed; *gone,* generation
Sertoli cells: from Enrico Sertoli, Italian histologist, 1842–1910
cells of Leydig: from Franz von Leydig, German anatomist, 1821–1908
androgen: Gk. *andros,* male producing
efferential ductules: L. *efferre,* to bring out; *ducere,* to lead

UNIT VII

hemoglobin levels than females), and bone growth. The effect of androgens on bone growth is self-limiting, however, because androgens are ultimately responsible for the conversion of cartilage to bone in the epiphyseal plates, thus sealing the plates and preventing further lengthening of the bones.

Spermatogenesis

The germ cells that migrate from the yolk sac to the testes during early embryonic development become stem cells called **spermatogonia** within the outer region of the seminiferous tubules. Spermatogonia are diploid ($2n$) cells (with 46 chromosomes) that ultimately give rise to mature haploid (n) gametes by a process of cell division called *meiosis (mi-o'sis)*.

Meiosis occurs within the testes of males who have gone through puberty and involves two nuclear divisions, as summarized in table 20.3. During the first part of this process, the DNA duplicates (prophase I) and homologous chromosomes are separated (during anaphase I), producing two daughter cells (at telophase I). Since each daughter cell contains only one of each homologous pair of chromosomes, the cells formed at the end of this *first meiotic division* contain 23 chromosomes each and are haploid (fig. 20.5). Each of the 23 chromosomes at this stage, however, consists of two strands (called *chromatids*) of identical DNA. During the *second meiotic division,* these duplicate chromatids are separated (at anaphase II), producing daughter cells (at telophase II). Meiosis of one diploid spermatogonia cell therefore produces four haploid daughter cells.

Actually, only about 1,000–2,000 stem cells migrate from the yolk sac into the embryonic testes. In order to produce many millions of spermatozoa throughout adult life, these spermatogonia cells duplicate themselves by mitotic division, and only one of the two cells—now called a **primary spermatocyte**— undergoes meiotic division (fig. 20.6). In this way, spermatogenesis can occur continuously without exhausting the number of spermatogonia.

When a diploid primary spermatocyte completes the first meiotic division (at telophase I), the two haploid daughter cells thus produced are called **secondary spermatocytes.** At the end of the second meiotic division, each of the two secondary spermatocytes produce two haploid **spermatids.** One primary spermatocyte therefore produces four spermatids.

The stages of spermatogenesis follow a sequential arrangement in the wall of the seminiferous tubule. The epithelial wall of the tubule, called the **germinal epithelium,** is indeed composed of germ cells in different stages of spermatogenesis. The spermatogonia and primary spermatocytes are located toward the outer side of the tubule, whereas the spermatids and mature spermatozoa are located toward the lumen (fig. 20.7).

At the end of the second meiotic division, the four spermatids produced by the meiosis of two secondary spermatocytes are interconnected with each other—their cytoplasm does not completely pinch off at the end of each division. The development of these interconnected spermatids into separate, mature

Table 20.2	Actions of Androgens in the Male
Category	**Action**
Sex determination	Growth and development of mesonephric ducts into epididymides, ductus deferentia, seminal vesicles, and ejaculatory ducts
	Development of the urogenital sinus and tubercle into prostate
	Development of male external genitalia
Spermatogenesis	At puberty: meiotic division and early maturation of spermatids
	After puberty: maintenance of spermatogenesis
Secondary sex characteristics	Growth and maintenance of accessory sex organs
	Growth of the penis
	Growth of facial and axillary hair
	Body growth
Anabolic effects	Protein synthesis and muscle growth
	Growth of bones
	Growth of other organs (including the larynx)
	Erythropoiesis (red blood cell formation)

From Stuart Ira Fox, *Human Physiology*, 4th ed. Copyright © 1993 Wm. C. Brown Communications, Inc., Dubuque, Iowa. All Rights Reserved. Reprinted by permission.

UNIT VII

Table 20.3	Stages of Meiosis		
Stage	**Events**	**Stage**	**Events**
First Meiotic Division		**Second Meiotic Division**	
Prophase I	Chromosomes appear double-stranded. Each strand, called a chromatid, contains duplicate DNA joined together by a structure known as a centromere.	Prophase II	Chromosomes appear, each containing two chromatids.
	Homologous chromosomes pair up side by side.	Metaphase II	Chromosomes line up single file along equator as spindle formation is completed.
Metaphase I	Homologous chromosome pairs line up at equator; spindle apparatus is completed.	Anaphase II	Centromeres split and chromatids move to opposite poles.
Anaphase I	Homologous chromosomes separate; each member of a homologous pair moves to opposite poles.	Telophase II	Cytoplasm divides to produce two haploid cells from each of the haploid cells formed at telophase I.
Telophase I	Cytoplasm divides to produce two haploid cells.		

From Stuart Ira Fox, *Human Physiology*, 4th ed. Copyright © 1993 Wm. C. Brown Communications, Inc., Dubuque, Iowa. All Rights Reserved. Reprinted by permission.

Figure 20.5 Meiosis, or reduction division. In the first meiotic division, the homologous chromosomes of a diploid parent cell are separated into two haploid daughter cells. Each of these chromosomes contains duplicate strands, or chromatids. In the second meiotic division, these chromatids are distributed to two new haploid daughter cells.

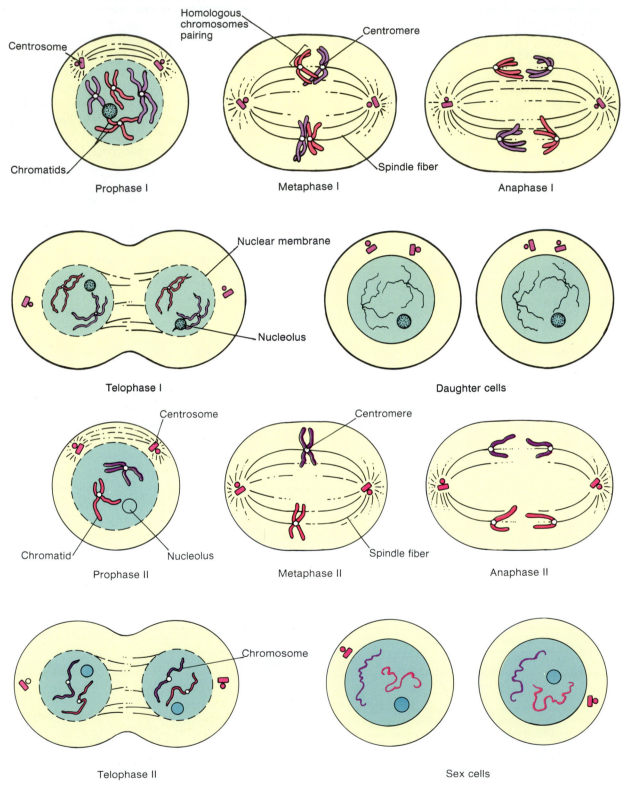

Centrosome

Homologous
chromosomes
pairing

Centromere

Chromatids

Prophase I

Metaphase I

Spindle fiber

Anaphase I

Nuclear membrane

Nucleolus

Telophase I

Daughter cells

Centrosome

Centromere

Chromatid

Nucleolus

Prophase II

Metaphase II

Spindle fiber

Anaphase II

Chromosome

Telophase II

Sex cells

Twomey

Figure 20.6 Spermatogonia undergo mitotic division to replace themselves and produce a daughter cell that will undergo meiotic division. This cell is called a primary spermatocyte. Upon completion of the first meiotic division, the daughter cells are called secondary spermatocytes. Each of these completes a second meiotic division to form spermatids. Notice that the four spermatids produced by the meiosis of a primary spermatocyte are interconnected. Each spermatid forms a mature spermatozoon.

Figure 20.7 Seminiferous tubules. (a) A cross section with surrounding interstitial tissue and (b) the stages of spermatogenesis within the germinal epithelium of a seminiferous tubule in which the relationship between sustentacular cells and developing spermatozoa is shown.

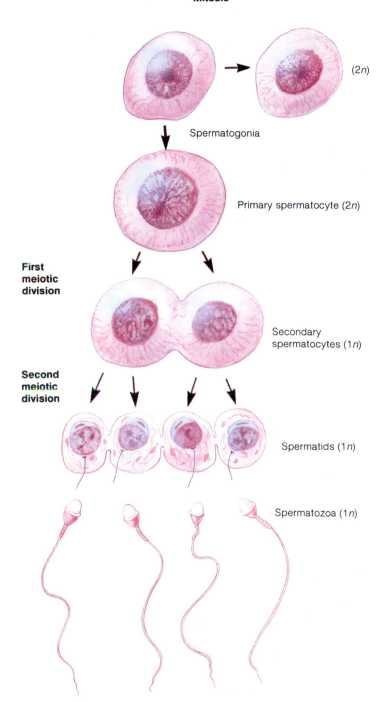

Mitosis

Spermatogonia

Primary spermatocyte (2n)

First meiotic division

Secondary spermatocytes (1n)

Second meiotic division

Spermatids (1n)

Spermatozoa (1n)

Interstitial tissue

Lumen of seminiferous tubule

(a)

Spermatozoa (23 chromosomes)

Lumen of seminiferous tubule

Sustentacular cell

Spermatids (23 chromosomes)

Secondary spermatocytes (23 chromosomes)

Primary spermatocytes (46 chromosomes)

Wall of seminiferous tubule

Spermatogonia (46 chromosomes)

(b)

Lew

UNIT VII

Figure 20.8 The processing of spermatids into spermatozoa (spermiogenesis). As the spermatids develop into spermatozoa, most of their cytoplasm is pinched off as residual bodies and ingested by the surrounding sustentacular cells.

Lew

spermatozoa (a process called *spermiogenesis*) requires the participation of another type of cell in the tubules—the sustentacular cells.

The **sustentacular** (Sertoli) **cells** are the only nongerminal cell type in the tubules. They form a continuous layer, connected by tight junctions, around the circumference of each tubule. In this way, the sustentacular cells create a *blood-testis barrier* because molecules from the blood must pass through the cytoplasm of these cells before entering germinal cells. The cytoplasm of the sustentacular cells extends through the width of the tubule and envelops the developing germ cells, so that it is often difficult to tell where the cytoplasm of sustentacular cells and that of germ cells is separated.

In the process of *spermiogenesis* (the conversion of spermatids to spermatozoa), most of the spermatid cytoplasm is eliminated (fig. 20.8). This occurs through phagocytosis by the sustentacular cells of the residual bodies of cytoplasm from the spermatids. It is believed that phagocytosis of the residual bodies may transmit informational molecules from the germ cells to the sustentacular cells. The sustentacular cells, in turn, may provide many molecules needed by the germ cells. It is known, for example, that the X chromosome of the germ cells is inactive during meiosis. Since this chromosome contains the genes needed to produce many essential molecules, it is believed that these molecules are provided by the sustentacular cells while meiosis is taking place.

Structure of Spermatozoa

A mature sperm cell, or **spermatozoon,** is a microscopic, tadpole-shaped structure about 0.06 mm long (fig. 20.9). It consists of an oval-shaped **head** and an elongated **flagellum.** The head of a spermatozoon contains a nucleus with 23 chromosomes. The tip of the head, called the *acrosome,* contains enzymes that help the sperm penetrate the ovum. The flagellum contains numerous mitochondria spiraled around a filamentous core. The mitochondria provide the energy necessary for locomotion. The flagellum propels the spermatozoon with a lashing movement. The maximum unassisted rate of spermatozoa movement is about 3 mm per hour.

The life expectancy of ejaculated spermatozoa is between 48 and 72 hours at body temperature. Many of the ejaculated spermatozoa, however, are defective and are of no value. It is not uncommon for spermatozoa to have enlarged heads, dwarfed and misshapen heads, two flagella, or a flagellum that is bent. Spermatozoa such as these are unable to propel themselves adequately.

1. Describe the location and structure of the testes.
2. What is the function of the seminiferous tubules, the germinal epithelial cells, and the interstitial cells?
3. Discuss the effect of testosterone on the production of sperm and the development of the secondary sex characteristics.
4. Using a diagram, describe the stages of spermatogenesis. Explain why spermatogenesis can continue throughout life without using up all of the spermatogonia.
5. Diagram a spermatozoon and its adjacent sustentacular cell and explain the functions of sustentacular cells in the seminiferous tubules.

UNIT VII

Figure 20.9 A human spermatozoon. (*a*) A diagrammatic representation and (*b*) a scanning electron micrograph of a spermatozoon in contact with an egg.

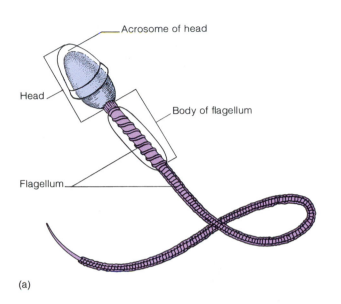

Acrosome of head

Head

Body of flagellum

Flagellum

(a)

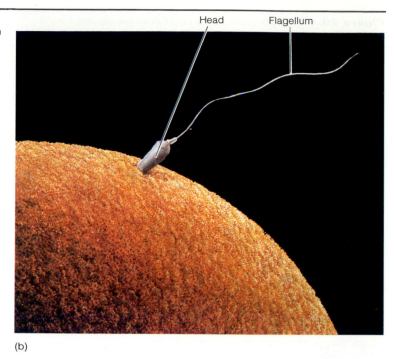

Head   Flagellum

(b)

Spermatic Ducts, Accessory Glands, and the Urethra

The spermatic ducts store spermatozoa and transport them from the testes to the urethra. The accessory reproductive glands provide additives to the spermatozoa to form semen, which is discharged from the erect penis during ejaculation.

Objective 10. List the various spermatic ducts and describe the location and structure of each segment.

Objective 11. Describe the structure and contents of the spermatic cord.

Objective 12. Describe the location, structure, and function of the ejaculatory ducts, seminal vesicles, prostate, and bulbourethral glands.

Spermatic Ducts

The duct system, which stores and transports spermatozoa from the testes to the urethra, includes the epididymides, the ductus deferentia, and the ejaculatory ducts.

Epididymis The **epididymis** (*ep″ĭ-did′ĭ-mis*)—in the plural, *epididymides*—is a long, flattened organ attached to the posterior surface of the testis (see figs. 20.1 and 20.3). The duct of the epididymis is highly coiled and contains millions of spermatozoa in their final stages of maturation. It is estimated that if the epididymis were uncoiled, it would measure 5–6 m (about 17 ft). The upper expanded portion of the epididymis is the **head,** the tapering middle section is the **body,** and the lower portion is the **tail.** The tail of the epididymis is continuous with

the ductus deferens. Both the tail of the epididymis and the ductus deferens store the spermatozoa that is to be discharged during ejaculation (fig. 20.10). The time required to produce mature spermatozoa—from meiosis in the seminiferous tubules to storage in the ductus deferens—is approximately 2 months.

Ductus Deferens The **ductus deferens** (*duk′tus def′er-enz*)—in the plural, *ductus deferentia*—is a fibromuscular tube about 45 cm (18 in.) long and 2.5 mm thick (see fig. 20.1) that conveys spermatozoa from the epididymis to the ejaculatory duct. The ductus deferens originates where the tail of the epididymis becomes less convoluted and is no longer attached to the testis. This first portion of the ductus deferens is important for the storage of spermatozoa. The ductus deferens exits from the scrotum as it ascends along the posterior border of the testis. From here, it penetrates the inguinal canal, enters the pelvic cavity, and passes to the side of the urinary bladder on the medial side of the ureter. The **ampulla** of the ductus deferens is the terminal portion that joins the ejaculatory duct.

The histological structure of the ductus deferens includes a layer of pseudostratified ciliated columnar epithelium in contact with the tubular lumen and surrounded by three layers of tightly packed smooth muscle (fig. 20.11). Sympathetic nerves from the pelvic plexus serve the ductus deferens. Stimulation through these nerves causes peristaltic contractions of the muscular layer, which forcefully ejects the stored spermatozoa toward the ejaculatory duct.

Much of the ductus deferens is located within a structure known as the **spermatic cord** (see figs. 20.4 and 20.15). The

deferens: L. *deferens,* conducting away
ampulla: L. *ampulla,* a two-handled bottle

UNIT VII

Figure 20.10 The histology of the epididymis showing sperm in the lumina (50×).

- Pseudostratified ciliated columnar epithelium
- Smooth muscle
- Sperm in lumen
- Connective tissue

Figure 20.11 The histology of the ductus deferens (250×).

- Sperm in lumen
- Pseudostratified ciliated columnar epithelium
- Inner longitudinal smooth muscle layer

Figure 20.12 The histology of the seminal vesicle.

- Seminal crypt
- Smooth muscle
- Glandular epithelium

spermatic cord extends from the testis to the inguinal ring and consists of the ductus deferens, spermatic vessels, nerves, cremaster muscle, lymph vessels, and connective tissue. The portion of the spermatic cord that is anterior to the pubic bone can be palpated on a male as it is compressed between the skin and the bone. The **inguinal canal** is a passageway through the abdominal wall for the spermatic cord to traverse the inguinal ligament; it is a potentially weak area and a common site for development of a hernia.

Ejaculatory Duct   The ejaculatory *(e-jak'yoo-lă-tor-e)* duct is 2 cm (1 in.) long and is formed by the union of the ampulla of the ductus deferens and the duct of the seminal vesicle. The ejaculatory duct then pierces the capsule of the prostate on its posterior surface and continues through this gland (fig. 20.1).

Both ejaculatory ducts receive secretions from the seminal vesicles and prostate and then eject the spermatozoa with its additives into the prostatic urethra.

Accessory Glands

Accessory reproductive glands include the seminal vesicles, the prostate, and the bulbourethral glands (see fig. 20.1). The contents of the seminal vesicles and the prostate are mixed with the spermatozoa during ejaculation to form *semen (seminal fluid)*. The fluid from the bulbourethral glands is released in response to sexual stimulation prior to ejaculation. Both the seminal vesicles and prostate are androgen-dependent organs. They atrophy if deprived of androgen as a result of castration.

Seminal Vesicles   The seminal vesicles, which are convoluted club-shaped glands about 5 cm (2 in.) long, are immediately posterior to and at the base of the urinary bladder. They secrete a sticky, slightly alkaline, yellowish substance that serves as a fluid medium to enhance spermatozoa movement and longevity. The secretion from the seminal vesicles contains a variety of nutrients, including fructose, a monosaccharide that provides spermatozoa with an energy source. It also contains citric acid, coagulation proteins, and prostaglandins. The discharge from the seminal vesicles makes up about 60% of the volume of semen.

Histologically, the seminal vesicle appears as a mass of cuts embedded in connective tissue (fig. 20.12). The extensively coiled mucosal layer breaks the lumen into numerous intercommunicating spaces that are lined by pseudostratified columnar and cuboidal secretory epithelia (referred to as glandular epithelium).

UNIT VII

Blood is supplied to the seminal vesicles by branches from the middle rectal arteries. The seminal vesicles are innervated by both sympathetic and parasympathetic neurons. Sympathetic stimulation causes the contents of the seminal vesicles to empty into the ejaculatory ducts of their respective sides.

Prostate   The firm prostate *(pros'tāt)* is the size and shape of a horse chestnut. It is about 4 cm (1.6 in.) across and 3 cm (1.2 in.) thick and positioned immediately below the urinary bladder, surrounding the beginning portion of the urethra (see fig. 20.1). It is enclosed by a fibrous capsule and divided into lobules formed by the urethra and the ejaculatory ducts extending through the gland. The ducts from the lobules open into the urethra. Extensive bands of smooth muscular tissue course throughout the prostate to form a meshwork that supports the glandular tissue (fig. 20.13). Contraction of the smooth muscle within the prostate empties the contents from the gland and provides part of the propulsive force needed to ejaculate the semen. The thin, milky-colored prostatic secretion assists spermatozoa motility as a liquefying agent, and its alkalinity protects sperm in their passage through the acidic environment of the female vagina. The prostate also secretes the enzyme *acid phosphatase,* which is often measured clinically to assess prostate function. The discharge from the prostate makes up about 40% of the volume of the semen.

Blood is supplied to the prostate from branches of the middle rectal and inferior vesical arteries. The venous return forms the prostatic venous plexus, along with blood draining from the penis. The prostatic venous plexus drains into the internal iliac veins. The prostate has both sympathetic and parasympathetic innervation arising from the pelvic plexuses.

A routine physical examination of the male includes rectal palpation of the prostate. Enlargement or overgrowth of the glandular substance of the prostate, called *benign prostatic hyperplasia,* is relatively common in older men. This may constrict the urethra and cause difficult micturition. An enlarged prostate sometimes requires surgery. If the obstruction is slight, the surgery may be accomplished through the urethral canal using a technique called a *transurethral prostatic resection,* in which excessive tissue is cut and cauterized.

Bulbourethral Glands   The paired, pea-sized, brownish-colored bulbourethral (Cowper's) glands are inferior to the prostate. Each bulbourethral gland is about 1 cm in diameter and drains by a 2.5 cm (1 in.) duct into the urethra (see fig. 20.1). Upon sexual excitement and prior to ejaculation, the bulbourethral glands are stimulated to secrete a mucoid substance that coats the lining of the urethra to neutralize the pH of the urine residue and lubricates the tip of the penis in preparation for sexual intercourse.

Urethra

The **urethra** of the male serves as a common tube for both the urinary and reproductive systems. However, urine and semen cannot simultaneously pass through the urethra because the

Figure 20.13   The histology of the prostate.

nervous reflex during ejaculation automatically inhibits micturition. The urethra of the male is about 20 cm (8 in.) long and S-shaped due to the shape of the penis. Three regions can be identified—the prostatic urethra, the membranous urethra, and the spongy urethra (see fig. 19.14).

The **prostatic urethra** is the proximal (2.5 cm) portion of the urethra that passes through the prostate. The prostatic urethra receives drainage from the small ducts of the prostate and the two ejaculatory ducts.

The **membranous urethra** is the short (0.5 cm) portion of the urethra that passes through the urogenital diaphragm. The external urethral sphincter muscle is located in this region.

The **spongy urethra** is the longest portion (15 cm), extending from the outer edge of the urogenital diaphragm to the external urethral orifice on the glans penis. This portion is surrounded by erectile tissue as it passes through the corpus spongiosum of the penis. The paired ducts from the bulbourethral glands attach to the spongy urethra near the urogenital diaphragm.

The wall of the urethra has an inside lining of mucous membrane, composed of transitional epithelium (fig. 20.14) and surrounded by a relatively thick layer of smooth muscle tissue called *tunica muscularis.* Specialized **urethral glands** are embedded in the urethral wall and function to secrete mucus into the urethral canal.

1. List in order the structures through which spermatozoa pass during ejaculation.
2. Differentiate between the ductus deferens and the spermatic cord.
3. Describe the structure and location of each of the accessory glands.
4. Why is the position of the prostate of clinical importance?
5. Distinguish between the segments of the urethra.

UNIT VII

Figure 20.14 The histology of the urethra (10×).

- Urethral glands
- Tunica muscularis
- Lumen of urethra
- Transitional epithelium
- Corpus spongiosum penis

Penis

The penis, containing the spongy urethra and covered with loose-fitting skin, is specialized with three columns of erectile tissue to become engorged with blood for insertion into the vagina during coitus.

Objective 13. Describe the structure and function of the penis.

The **penis** is composed mainly of erectile tissue and, when distended, serves as the copulatory organ of the male reproductive system. **Erectile tissue** contains numerous vascular spaces that become engorged with blood. The penis is a pendent structure anterior to the scrotum and attached to the pubic arch. It is divided into a proximal attached root, an elongated tubular body, and a distal cone-shaped glans penis (fig. 20.15).

The **root of the penis** expands posteriorly to form the **bulb of the penis** and the **crus** (*krus*) **of the penis.** The bulb is positioned in the urogenital triangle of the perineum, where it is attached to the undersurface of the urogenital diaphragm and enveloped by the bulbocavernosus muscle (see fig. 9.24). The crus, in turn, attaches the root of the penis to the pubic arch (ischiopubic ramus) and to the perineal membrane. The crus, which is superior to the bulb, is enveloped by the ischiocavernosus muscle.

The **body of the penis** is composed of three cylindrical columns of erectile tissue that are bound together by fibrous tissue and covered with skin (fig. 20.15). The paired dorsally positioned masses are named the **corpora cavernosa penis.** The fibrous tissue between the two corpora forms a **median septum.** The **corpus spongiosum penis** is ventral to the other two and surrounds the spongy urethra. The penis is flaccid and relaxed when the spongelike tissue is not engorged with blood but becomes firm and erect when the spaces are filled.

The **glans penis** is the cone-shaped terminal portion of the penis that is formed from the expanded corpus spongiosum. The opening of the urethra at the tip of the glans penis is called the **external urethral ostium.** The **corona glandis** is the prominent posterior ridge of the glans penis. On the undersurface of the glans penis, a vertical fold of tissue called the **frenulum** (*fren'yŭ-lum*) attaches the skin covering the penis to the glans penis.

The skin covering the penis is hairless, contains no fat cells, and generally has a darker pigment than the other body skin. The skin of the body of the penis is loosely attached and is continuous over the glans penis as a retractable sheath called the **prepuce** (*pre'pyoos*) or **foreskin.** The prepuce is commonly removed in male infants by a surgical procedure called *circumcision.*

The penis is supplied with blood through the superficial external pudendal branch of the femoral artery and the internal pudendal branch of the internal iliac artery. The venous return is through a superficial median dorsal vein that drains into the great saphenous vein in the thigh and through the deep median vein that drains into the prostatic plexus.

A *circumcision* is generally performed for hygienic purposes because the glans penis is easier to clean if exposed. A sebaceous secretion from the glans penis, called *smegma*, will accumulate along the border of the corona glandis if good hygiene is not practiced. Smegma can foster bacteria that may cause infections and therefore should be removed through washing. Cleaning the glans penis of an uncircumcised male requires retraction of the prepuce. Occasionally, a child is born with a prepuce that is too tight to permit retraction. This condition is called *phimosis* and necessitates circumcision.

crus: L. *crus*, leg, resembling a leg
cavernosa: L. *cavus*, hollow

glans: L. *glans*, acorn
corona: L. *corona*, garland, crown
frenulum: L. diminutive of *frenum*, a bridle
prepuce: L. *prae*, before; *putium*, penis
phimosis: Gk. *phimosis*, a muzzling

UNIT VII

Figure 20.15 The structure of the penis showing the attachment, blood and nerve supply, and the arrangement of the erectile tissue.

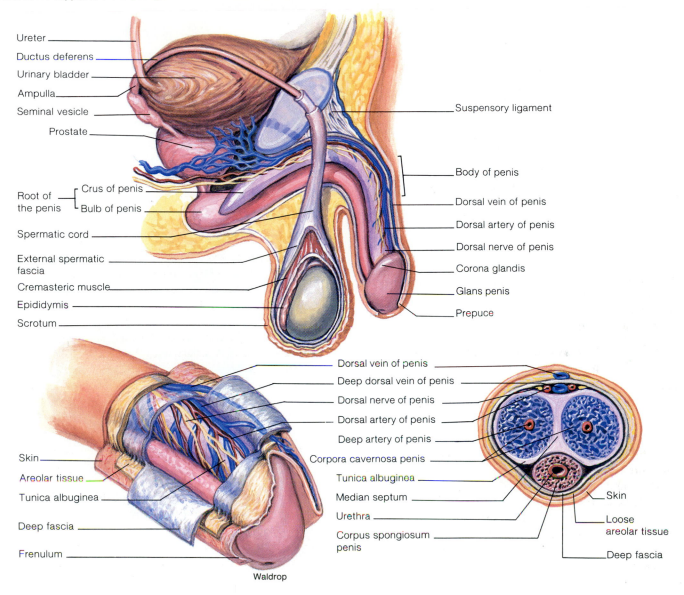

Ureter
Ductus deferens
Urinary bladder
Ampulla
Seminal vesicle
Prostate

Root of the penis { Crus of penis
Bulb of penis

Spermatic cord
External spermatic fascia
Cremasteric muscle
Epididymis
Scrotum

Suspensory ligament

Body of penis
Dorsal vein of penis
Dorsal artery of penis
Dorsal nerve of penis
Corona glandis
Glans penis
Prepuce

Skin
Areolar tissue
Tunica albuginea
Deep fascia
Frenulum

Dorsal vein of penis
Deep dorsal vein of penis
Dorsal nerve of penis
Dorsal artery of penis
Deep artery of penis
Corpora cavernosa penis
Tunica albuginea
Median septum
Urethra
Corpus spongiosum penis

Skin
Loose areolar tissue
Deep fascia

Waldrop

The penis has many sensory tactile receptors, especially in the glans penis, that make it a highly sensitive organ. In addition, the penis has extensive sympathetic and parasympathetic motor innervation.

1. Describe the position of the penis relative to the scrotum and accessory glands.
2. Describe the external structure of the penis and the internal arrangement of the erectile tissue within the penis.
3. Define *circumcision* and explain why this procedure is commonly performed.

Mechanisms of Erection, Emission, and Ejaculation

Erection of the penis results from parasympathetic induced vasodilation of arteries within the penis and a decrease in venous drainage. Emission and ejaculation are stimulated by sympathetic impulses, resulting in the forceful expulsion of semen from the penis.

Objective 14. Distinguish between erection, emission, and ejaculation.

Objective 15. Describe the events that result in erection of the penis.

Objective 16. Explain the physiological process of ejaculation.

Objective 17. Describe the characteristic properties of semen.

Erection, emission, and ejaculation are interrelated events that are necessary for the deposition of semen into the vagina by the natural means of coitus (sexual intercourse). *Erection* usually occurs as a male becomes sexually aroused; the erectile tissue of the penis becomes engorged with blood, causing the penis to become wider, longer, and firmer. *Emission* is the movement of spermatozoa from the epididymides to the ejaculatory ducts. *Ejaculation* is the forceful expulsion of the ejaculate, or *semen* (seminal fluid), from the ejaculatory ducts and urethra of the penis. Emission and ejaculation do not have to follow erection of the penis and occur only if there is sufficient stimulation of the sensory receptors in the penis to elicit the ejaculatory response.

Erection of the Penis

Erection of the penis depends on the volume of blood that enters the arteries of the penis as compared to the volume that exits through venous drainage (fig. 20.16). Normally, constant sympathetic stimuli to the arterioles of the penis maintain a partial constriction of smooth muscles within the arteriole walls so that there is an even flow of blood throughout the penis. During sexual excitement, however, parasympathetic impulses cause marked vasodilation within the arterioles of the penis, resulting in more blood entering than venous blood draining. Also, during parasympathetic stimulation there is inhibition of sympathetic impulses to arterioles of the penis. At the same time, there may be slight vasoconstriction of the dorsal vein of the penis and an increase in cardiac output. These combined events cause the spongy tissue of the corpora cavernosa and the corpus spongiosum to become distended with blood and the penis to become turgid. In this condition, the penis can be inserted into the vagina of the female and function as a copulatory organ to discharge semen.

Erection is controlled by two portions of the central nervous system—the hypothalamus in the brain and the sacral portion of the spinal cord. The hypothalamus controls conscious sexual thoughts that originate in the cerebral cortex. Nerve impulses from the hypothalamus elicit parasympathetic responses from the sacral region that cause vasodilation of the arterioles within the penis. Conscious thought is not required for an erection, however, and stimulation of the penis can cause an erection because of a reflex response in the spinal cord. This reflexive action makes possible an erection in a sleeping male or in an infant—perhaps from the stimulus of a diaper. The mechanism of erection of the penis is summarized in figure 20.17.

Ejaculation is the expulsion of semen through the urethra of the penis. In contrast to erection, ejaculation is a response involving the sympathetic innervation of the accessory reproductive organs. Ejaculation is preceded by continued sexual stimulation, usually through activated tactile receptors in the glans penis and the skin of the body. Rhythmic friction of these

Figure 20.16 Thermograms of the penis showing differential heat radiation during erection. Red, yellow, and purple indicate the warmest portions, whereas blue and green indicate the coolest.

structures during coitus causes sensory impulses to be transmitted to the thalamus and cerebral cortex. The first sympathetic response, which occurs prior to ejaculation, is the discharge from the bulbourethral glands. The fluids from these glands are usually discharged before penetration into the vagina and serve to lubricate the urethra and the glans penis.

Emission and Ejaculation of Semen

Emission Continued sexual stimulation following erection of the penis causes emission. Emission is the movement of sperm from the epididymides to the ejaculatory ducts and the secretions of the accessory glands into the ejaculatory ducts and urethra in the formation of semen. Emission occurs as sympathetic impulses from the pelvic plexus cause a rhythmic contraction of the smooth muscle of the epididymides, ductus deferentia, ejaculatory ducts, seminal vesicles, and prostate.

Ejaculation Ejaculation immediately follows emission and is accompanied by *orgasm,* which is considered the climax of the

UNIT VII

Figure 20.17   The mechanism of erection of the penis.

sex act. Ejaculation occurs in a series of spurts of semen from the urethra. This takes place as parasympathetic impulses traveling through the pudendal nerves stimulate the bulbospongiosus muscles (see fig. 9.24) at the base of the penis and cause them to contract rhythmically. There is also sympathetic stimulation of the smooth muscles in the urethral wall that peristaltically contract to help eject the semen.

Sexual function in the male thus requires the synergistic action (rather than antagonistic action) of the parasympathetic and sympathetic divisions of the ANS. The mechanism of emission and ejaculation is summarized in figure 20.18.

Immediately following ejaculation or a cessation of sexual stimulus, sympathetic impulses cause vasoconstriction of the arterioles within the penis, reducing the inflow of blood. At the same time, cardiac output returns to normal as does venous return of blood from the penis. With the normal flow of blood through the penis, it returns to its flaccid condition. Following an ejaculation of semen, another erection and ejaculation cannot be triggered for 10 to 30 minutes or longer.

Adolescent males may experience erection of the penis and spontaneous emission and ejaculation of semen during sleep. These *nocturnal emissions* are thought to be caused by changes in hormonal concentrations that accompany adolescent development.

Semen   Semen consists of spermatozoa plus the additives from the accessory reproductive glands. It is the substance discharged during ejaculation (table 20.4). Generally, between 1.5 and 5.0 mℓ of semen are ejected during ejaculation. The bulk of the fluid (about 60%) is produced by the seminal vesicles, and the rest (about 40%) is contributed by the prostate. Spermatozoa constitute less than 1% of the volume of an ejaculate. There are usually between 60 and 150 million sperm per milliliter of ejaculate. In the condition of *oligospermia,* the male ejaculates fewer

Table 20.4	The Clinical Examination of Semen
Characteristic	**Reference Value**
Volume of ejaculate	1.5–5.0 mℓ
Sperm count	40–250 million/mℓ
Sperm motility	
Percentage of motile forms:	
1 hour after ejaculation	70% or more
3 hours after ejaculation	60% or more
Leukocyte count	0–2,000/mℓ
pH	7.2–7.8
Fructose concentration	150–600 mg/100 mℓ

Modified from Glasser, L., "Seminal Fluid and Subfertility," *Diagnostic Medicine,* July/August 1981, p. 28. By permission.

than 10 million sperm per milliliter and is likely to have fertility problems.

Human semen can be frozen and stored in sperm banks for future artificial insemination. In this procedure, the semen is diluted with 10% glycerol, monosaccharide, and distilled water buffer and frozen in liquid nitrogen. The freezing process destroys defective and abnormal sperm. For some unknown reason, however, not all human sperm is suitable for freezing.

1. Define the terms *erection, emission,* and *ejaculation.*
2. Explain the statement that male sexual function is an autonomic synergistic action.
3. Compose a flow chart to explain the physiological and physical events of erection, emission, and ejaculation.
4. Describe the components of a normal ejaculate.

Developmental Exposition

The Reproductive System

Exhibit I Differentiation of the male and female gonads and genital ducts. (*a*) An embryo at 6 weeks showing the positions of a transverse cut depicted in (*a₁*), (*b*), and (*c*). (*a₁*) At 6 weeks, the developing gonads (primary sex cords) are still indifferent. By 4 months, the gonads have differentiated into male (*b*) or female (*c*). The oogonia have formed within the ovaries (*c₁*) by 6 months.

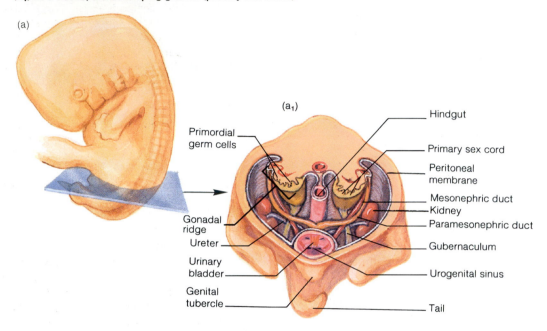

(a)

(a₁)

Primordial germ cells
Gonadal ridge
Ureter
Urinary bladder
Genital tubercle

Hindgut
Primary sex cord
Peritoneal membrane
Mesonephric duct
Kidney
Paramesonephric duct
Gubernaculum
Urogenital sinus
Tail

(b) **Male**

Testis
Rete testis
Mesonephric duct persists (developing ductus deferens)
Descent of testes at 8 months

Parietal peritoneum
Ureter
Gubernaculum
Urinary bladder

Waldrop

(c) **Female**

Ovary
Degenerating mesonephric duct
Primordial follicle
Paramesonephric duct persists (developing uterine tube)

Follicular cell
Oogonium

(c₁)

UNIT VII

Explanation

Sex Determination

Sexual identity is initiated at the moment of conception when the *genetic sex* of the zygote (fertilized egg) is determined. The ovum is fertilized by a spermatozoon containing either an X or a Y sex chromosome. If the spermatozoon contains an X chromosome, it will pair with the X chromosome of the ovum and a female child will develop. A spermatozoon carrying a Y chromosome results in an XY combination, and a male child will develop.

Genetic sex determines whether the gonads will be testes or ovaries. If testes develop, they will produce and secrete male sex hormones during late embryonic and early fetal development and cause the secondary sex organs of the male to develop.

Embryonic Development

The male and female reproductive systems follow a similar pattern of development, with sexual distinction resulting from the influence of hormones. A significant fact of embryonic development is that the sexual organs for both male and female are derived from the same developmental tissues and are considered *homologous structures*.

The first sign of development of either the male or the female reproductive organs occurs during the fifth week as the medial aspect of each mesonephros (see chapter 19) enlarges to form the **gonadal ridge** (exhibit I). The gonadal ridge continues to grow behind the developing peritoneal membrane. By the sixth week, stringlike masses called **primary sex cords** form within the enlarging gonadal ridge. The primary sex cords in the male will eventually mature to become the seminiferous tubules. In the female, the primary sex cords will contribute to nurturing tissue of developing ova. Each gonad develops near a **mesonephric duct** and a **paramesonephric duct.** In the male embryo, each testis connects through a series of tubules to the mesonephric duct. During further development, the connecting tubules become the seminiferous tubules, and the mesonephric duct becomes the *efferent ductules, epididymis, ductus deferens, ejaculatory duct,* and *seminal vesicle*. The paramesonephric duct in the male degenerates without contributing any functional structures to the reproductive system. In the female embryo, the mesonephric duct degenerates, and the paramesonephric duct contributes greatly to structures of the female reproductive system. The distal ends of the paired paramesonephric ducts fuse to form the *vagina* and *uterus*. The proximal unfused portions become the *uterine tubes*.

Externally, by the sixth week a swelling called the **genital tubercle** is apparent anterior to the small embryonic tail (future coccyx). The mesonephric and paramesonephric ducts open to the outside through the genital tubercle. The genital tubercle consists of a *glans*, a *urethral groove*, paired *urethral folds*, and paired *labioscrotal swellings* (see exhibit II). As the glans portion of the genital tubercle enlarges, it becomes known as the *phallus*. Early in fetal development (tenth through twelfth week), sexual distinction of the external genitalia becomes apparent. In the male, the phallus enlarges and develops into the *glans* of the penis. The urethral folds fuse around the urethra to form the *body* of the penis. The urethra opens at the end of the glans as the *urethral orifice*. The labioscrotal swellings fuse to form the *scrotum* into which the testes will descend. In the female, the phallus gives rise to the *clitoris*, the urethral folds remain apart as the *labia minora,* and the labioscrotal swellings become the *labia majora*. The urethral groove is retained as a longitudinal cleft known as the *vestibule*.

Descent of the Testes

The descent of the testes from the site of development begins between the sixth and tenth week. Descent into the scrotal sac, however, does not occur until about week 28, when paired inguinal canals form in the abdominal wall to provide openings from the pelvic cavity to the scrotal sac. The process by which a testis descends is not well understood, but it seems to be associated with the shortening and differential growth of the *gubernaculum,* which is attached to the testis and extends through the inguinal canal to the wall of the scrotum (exhibit III). As the testis descends, it passes to the side of the urinary bladder and anterior to the symphysis pubis. It carries with it the ductus deferens, the testicular vessels and nerve, a portion of the internal abdominal oblique muscle, and lymph vessels. All of these structures remain attached to the testis and form what is known as the **spermatic cord** (see fig. 20.15). By the time the testis has taken its position in the scrotal sac, the gubernaculum is no more than a remnant of scarlike tissue.

During the physical examination of a neonatal male, a physician will palpate the scrotum to determine if the testes are in position. If one or both are not in the scrotal sac, it may be possible to induce descent by administering certain hormones. If this procedure does not work, surgery is necessary. The surgery is generally performed before the age of 5. Failure to correct the situation may result in sterility and possibly the development of a tumorous testis.

UNIT VII

---

homologous: Gk. *homos,* the same

gubernaculum: L. *gubernaculum,* helm

Continued on next page

Developmental Exposition

Continued

Exhibit II (*a*) Differentiation of the external genitalia in the male and female. (*a₁*) A sagittal view. At 6 weeks, the urethral fold and labioscrotal swelling are differentiated from the genital tubercle. (*b*) At 8 weeks, a distinct phallus is present during the indifferent stage. By the twelfth week, the genitalia have become distinctly male (*c*) or female (*d*), being derived from homologous structures. (*e,f*) At 16 weeks, the genitalia are formed. (*g,h*) Photographs at week 10 of male and female genitalia, respectively.

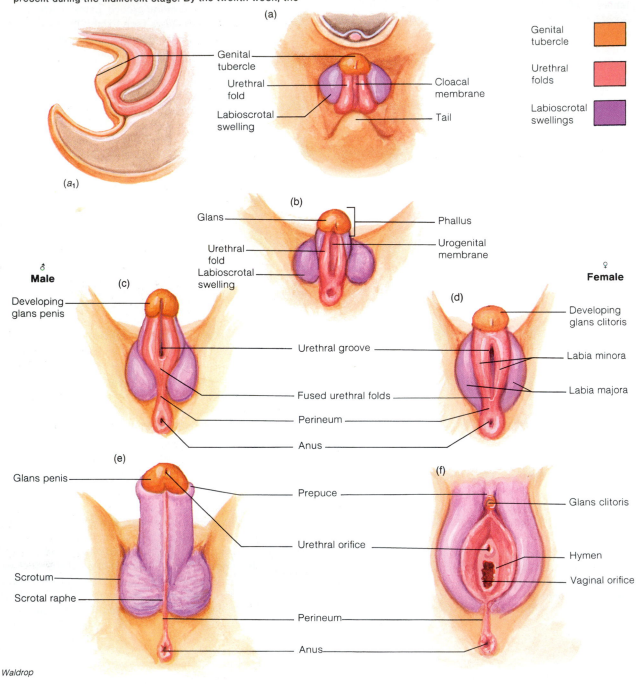

Waldrop

UNIT VII

Exhibit II   continued

(g)

(h)

Exhibit III   The descent of the testes. (*a*) At 10 weeks, (*b*) at 18 weeks, and (*c*) at 28 weeks. During development, each testis descends through an inguinal canal in front of the symphysis pubis and enters the scrotum.

Epididymis
Testis
Rectum
Gubernaculum
Symphysis pubis
Labioscrotal swelling
Abdominal wall
Pelvic cavity
Developing penis
(a)

Testis
Inguinal canal
Gubernaculum
(b)

Ductus deferens
Spermatic cord
Testis
Gubernaculum
Scrotum
(c)

Clinical Considerations

Sexual dysfunction is a broad area of medical concern that includes developmental and psychogenic problems as well as conditions resulting from various diseases. Psychogenic problems of the reproductive system are extremely complex, poorly understood, and beyond the scope of this book. Only a few of the principal developmental conditions, functional disorders, and diseases that affect the physical structure and function of the male reproductive system will be discussed here.

Developmental Abnormalities

The reproductive organs of both sexes develop from similar embryonic tissue that follows a consistent pattern of formation well into the fetal period. Because an embryo has the potential to differentiate into a male or a female, developmental errors can result in various degrees of intermediate sex, or **hermaphroditism** (*her-maf′rō-di-tiz″em*). A person with undifferentiated or ambiguous external genitalia is called a **hermaphrodite.**

True hermaphroditism—in which both male and female gonadal tissues are present, in either the same or opposite gonads—is a rare anomaly. True hermaphrodites usually have a 46 XX chromosome constitution. **Male pseudohermaphroditism** occurs more frequently and generally results from hormonal influences during early fetal development. This condition is caused either by inadequate amounts of androgenic hormones being secreted or by the delayed development of the reproductive organs after the period of tissue sensitivity has passed. These individuals have a 46 XY chromosome constitution and male gonads, but the genitalia are intersexual and variable. The treatment of hermaphroditism varies, depending on the extent of ambiguity of the reproductive organs. Although people with this condition are sterile, they may engage in normal sexual relations following hormonal therapy and plastic surgery.

Chromosomal anomalies result from the improper separation of the chromosomes during meiosis and are usually expressed in deviations of the reproductive organs. The two most frequent chromosomal anomalies cause Turner's syndrome and Klinefelter's syndrome. **Turner's syndrome** occurs when only one X chromosome is present. About 97% of embryos lacking an X chromosome die; the remaining 3% survive and appear to be females, but their gonads are rudimentary or absent, and they do not mature at puberty. A person with **Klinefelter's syndrome** has an XXY chromosome constitution, develops breasts and male genitalia, but has underdeveloped seminiferous tubules and is generally retarded.

A more common developmental problem than genetic abnormalities, and fortunately less serious, is cryptorchidism. **Cryptorchidism** (*krip-tor′ki-diz″em*) means hidden testis and

UNIT VII

---

hermaphrodite: Gk. (mythology) *Hermaphroditos*, son of Hermes (Mercury)
Turner's syndrome: from Henry H. Turner, American endocrinologist, 1892–1970
Klinefelter's syndrome: from Harry F. Klinefelter Jr., American physician, b. 1912
cryptorchidism: Gk. *crypto*, hidden; *orchis*, testis

Figure 20.18 The mechanism of emission and ejaculation.

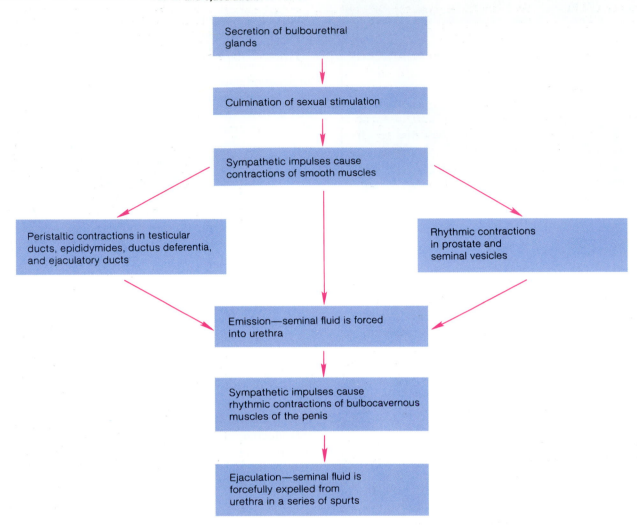

is characterized by the failure of one or both testes to descend into the scrotum. A cryptorchid testis is usually located along the path of descent but can be anywhere in the pelvic cavity (fig. 20.19). It occurs in about 3% of male infants and should be treated before the infant has reached the age of 5 to reduce the chance of infertility or other complications.

Functional Considerations

Functional disorders of the male reproductive system include impotence *(im'pŏ-tens),* infertility, and sterility. **Impotence** is the inability of a sexually mature male to achieve and maintain penile erection and/or the inability to achieve ejaculation. The causes of impotence may be physical, such as abnormalities of the penis, vascular irregularities, neurological disorders, or certain diseases. Generally, however, the cause of impotence is psychological, and the patient requires skilled counseling by a sex therapist.

Infertility is the inability of the sperm to fertilize the ovum. Infertility problems may be due to male or female factors, or both. The term *impotence* should not be confused with infertility. In males, the most common cause of infertility is the inadequate production of viable spermatozoa. This may result from alcoholism, dietary deficiencies, local injury, varicocele, excessive heat, or exposure to X rays. A hormonal imbalance may also contribute to infertility. Many of the causes of infertility can be treated through proper nutrition, gonadotrophic hormone treatment, or microsurgery. If corrective treatment is not possible, however, it may be possible to concentrate the spermatozoa obtained through *masturbation* (in males, self-stimulation to the point of ejaculation) and use this concentrate to artificially inseminate the woman.

Sterility is similar to infertility, except that it is a permanent condition. Sterility may be genetically caused, or it may be the result of degenerative changes in the seminiferous tu-

impotence: L. *im,* not; *potens,* potent

sterility: L. *sterilis,* barren

UNIT VII

bules (for example, mumps in a mature male may secondarily infect the testes and cause irreversible tissue damage).

Voluntary sterilization of the male in a procedure called a **vasectomy** is a common technique of birth control. In this procedure, a small section of each ductus deferens near the epididymis is surgically removed and the cut ends of the ducts are tied (fig. 20.20). A vasectomy interferes with sperm transport but does not directly affect the secretion of androgens from interstitial cells in the interstitial tissue. Since spermatogenesis continues, the spermatozoa cannot be drained from the testes and instead accumulate in the crypts that form in the seminiferous tubules and ductus deferens. These crypts present sites of inflammatory reactions in which spermatozoa are phagocytosed and destroyed by the immune system.

Diseases of the Male Reproductive System

Sexually Transmitted Diseases   Sexually transmitted diseases (STDs), frequently called venereal diseases (VDs), are contagious diseases that affect the reproductive systems of both the male and the female (table 20.5) and are transmitted during sexual activity. The frequency of sexually transmitted diseases in the United States is regarded by health authorities as epidemic. These diseases have not been eradicated mainly because humans cannot develop immunity to them and because increased sexual promiscuity increases the chances of infection and reinfection.

Gonorrhea *(gon''ŏ-re'ă),* commonly called clap, is caused by the bacterium gonococcus, or *Neisseria gonorrhoeae.* Males with this disease suffer inflammation of the urethra, accompanied by painful urination and frequently the discharge of pus. In females, the condition is usually asymptomatic, and therefore many women may be unsuspecting carriers of the disease. Advanced stages of gonorrhea in females may infect the uterus and the uterine tubes. A pregnant woman with gonorrhea that is not treated may transmit the bacteria to the eyes of her newborn during its passage through the birth canal, possibly causing blindness.

Syphilis *(sif'ĭ-lis)* is caused by the bacterium *Treponema pallidum.* Syphilis is less common than gonorrhea but is the more serious of the two diseases. During the *primary stage* of syphilis, a lesion called a *chancre* develops at the point where contact was made with the infectious syphilitic lesion of a carrier. The chancre—an ulcerated sore that has hard edges—endures for 10 days to 3 months. It is only during the primary stage that syphilis can be spread to another sexual partner. The chancre will heal with time, but if not treated it will be followed by secondary and tertiary stages of syphilis. During the initial contact, the bacteria enter the bloodstream and spread throughout the body. The *secondary stage* of syphilis is expressed by lesions

Figure 20.19   Cryptorchidism. (*a*) Incomplete descent of a testis may involve four separate regions: (1) in the pelvic cavity, (2) in the inguinal canal, (3) at the superficial inguinal ring, and (4) in the upper scrotum. (*b*) An ectopic testis may be (1) in the superficial fascia of the anterior pelvic wall, (2) at the root of the penis, or (3) in the perineum, in the thigh alongside the femoral vessels.

(a)

(b)

or a rash of the skin and mucous membranes, accompanied by fever (fig. 20.21). This stage lasts from 2 weeks to 6 months, and the symptoms disappear of their own accord. The *tertiary stage* occurs 10 to 20 years following the primary infection. The circulatory, integumentary, skeletal, and nervous systems are

---

vasectomy: L. *vas,* vessel; Gk. *ektome,* excision
venereal: L. (mythology) from *Venus,* the goddess of love
gonorrhea: L. *gonos,* seed; *rhoia,* a flow
chancre: Fr. *chancre,* indirectly from L. *cancer,* a crab

UNIT VII

Table 20.5	Sexually Transmitted diseases		
Name	**Organism**	**Resulting Condition**	**Treatment**
Gonorrhea	*Gonococcus* (bacterium)	Adult: sterility due to scarring of epididymides and tubes; rarely: septicemia; newborn: blindness	Penicillin injections; tetracycline tablets; eye drops (silver nitrate or penicillin) in newborns as preventative
Syphilis	*Treponema pallidum* (bacterium)	Adult: gummas, cardiovascular neurosyphilis; newborn: congenital syphilis (abnormalities, blindness)	Penicillin injections; tetracycline tablets
Chancroid (soft chancre)	*Hemophilus ducreyi* (bacterium)	Chancres, buboes	Tetracycline; sulfa drugs
Urethritis in men	Various microorganisms	Clear discharge	Tetracycline
Vaginitis	*Trichomonas* (protozoan)	Frothy white or yellow discharge	Metronidazole
	Candida albicans (yeast)	Thick, white, curdy discharge (moniliasis)	Nystatin
Acquired immune deficiency syndrome (AIDS)	Human immunodeficiency virus (HIV)	Early symptoms include extreme fatigue, weight loss, fever, diarrhea; extreme susceptibility to pneumonia, rare infections, and cancer	Azidothymidine (AZT, or Retrovir); no cure available
Chlamydia	*Chlamydia trachomatis* (bacterium)	Whitish discharge from penis or vagina; pain during urination	Tetracycline and sulfonamides
Lymphogranuloma venereum (LGV)	Microorganism	Ulcerating buboes; rectal stricture	Tetracycline; sulfa drugs
Granuloma venereum (inguinale)	*Donovania granulomatis*	Raw, open, extended sore	Tetracycline
Venereal warts	Virus	Warts	Podophyllin; cautery, cryosurgery, or laser treatment
Genital herpes	Herpes simplex virus	Sores	Palliative treatment
Crabs	Arthropod	Itching	Gamma benzene hexachloride

Figure 20.20 A simplified illustration of a vasectomy in which a segment of the ductus deferens is removed through an incision in the scrotum. The procedure is then repeated on the opposite side.

Figure 20.21   The secondary stages of syphilis as expressed by lesions of the skin of this young woman.

particularly vulnerable to the degenerative changes caused by this disease. The end result of untreated syphilis is blindness, insanity, and eventual death.

AIDS, or **acquired immune deficiency syndrome,** is a viral disease that is transmitted primarily through coitus and drug abuse (by sharing contaminated syringe needles). Additional information about this fatal disease, for which there is no known cure, is presented in table 20.5.

Disorders of the Prostate   The prostate is subject to several disorders, most of which are common in older men. The four most frequent prostatic problems are acute prostatitis, chronic prostatitis, benign prostatic hyperplasia, and carcinoma of the prostate.

Acute prostatitis is common in sexually active young men through infections acquired from a gonococcus bacterium. The symptoms of acute prostatitis are a swollen and tender prostate, painful urination, and in extreme conditions, pus dripping from the penis. It is treated with penicillin, bed rest, and increased fluid intake.

Chronic prostatitis is one of the most common afflictions of middle-aged and elderly men. The symptoms of this condition range from irritation and slight difficulty in urinating to

extreme pain and urine blockage, which commonly causes secondary renal infections. In this disease, several kinds of infectious microorganisms are believed to be harbored in the prostate and are responsible for inflammations elsewhere in the body, such as in the nerves (neuritis), the joints (arthritis), the muscles (myositis), and the iris (iritis).

Prostatic hyperplasia, or an enlarged prostate, occurs in approximately one-third of all males over the age of 60. In this condition, an overgrowth of granular material compresses the prostatic urethra. The cause of prostatic hyperplasia is not known. As the prostate enlarges, urination becomes painful and difficult. If the urinary bladder is not emptied completely, cystitis eventually occurs. People with cystitis may become incontinent and dribble urine continuously. Prostatic hyperplasia is usually treated by the surgical removal of portions of the gland through transurethral curetting (cutting and removal of a small section) or the removal of the entire prostate, called **prostatectomy.**

Prostatic carcinoma, or cancer of the prostate, is common in males over the age of 60; in fact, it is the second leading cause of death from cancer in males in the United States. When prostatic cancer is confined to the prostate, it is generally small and asymptomatic. But as the cancer grows and invades surrounding nerve plexuses, it becomes extremely painful and easily detected. The metastases of this cancer to the spinal column and brain are generally what kills the patient.

As prostatic carcinoma develops, it has symptoms nearly identical to prostatic hyperplasia—painful urination and cystitis. When examined by rectal palpation with a gloved finger, however, a hard cancerous mass can be detected in contrast to the enlarged, soft, and tender prostate diagnostic of prostatic hyperplasia. Prostatic carcinoma is treated by prostatectomy and frequently by the removal of the testes (called **orchiectomy**) as well. An orchiectomy inhibits metastases by eliminating testosterone secretion.

Disorders of the Testes and Scrotum   A **hydrocele** *(hi'dro-sēl)* is a benign fluid mass within the tunica vaginalis that causes swelling of the scrotum. It is a frequent, minor disorder in male infants as well as in adults. The cause is unknown.

An infection in the testes is called **orchitis.** Orchitis may develop from a primary infection from a tubercle bacterium or as a secondary complication of mumps contracted after puberty. If orchitis from mumps involves both testes, it usually causes sterility.

Trauma to the testes and scrotum is common because of their pendent position. The testes are extremely sensitive to pain, and a male responds reflexively to protect the groin area.

UNIT VII

Internal Affairs

The Relational Workings of the Body Systems

Reproductive System

- Sex hormones affect distribution of body hair and deposition of subcutaneous fat

Integumentary System
- Covers and protects reproductive organs
- Provides sites for sexual sensory receptors
- Mammary glands represent specialized integumentary structures

- Sex hormones stimulate bone growth and maintenance

Skeletal System
- Pelvis provides protection and support for reproductive organs

- Sex hormones stimulate muscle growth and maintenance

Muscular System
- Certain pelvic muscles aid coitus; involuntary action of smooth muscles aids movement of gametes

- Sex hormones influence brain development and sexual behavior

Nervous System
- Controls sexuality

- Certain hormones (gonadotropins and GnRH) regulate function of gonads

Endocrine System
- Sex hormones aid development and maintenance of sex organs; influence sex drives and functions

- Estrogens maintain healthy blood vessels (lower cholesterol levels)

Circulatory System
- Transports O_2, sex hormones, and nutrients; eliminates CO_2 and metabolic wastes

Lymphatic System
- Maintains a balanced amount of interstitial fluid within reproductive tissues

Respiratory System
- Provides O_2 to reproductive organs and eliminates CO_2

Digestive System
- Provides nutrients for organ function
- Provides nutrients for embryonic and fetal development in pregnant woman

Urinary System
- Eliminates metabolic wastes
- Male urethra transports semen during ejaculation

CLINICAL CASE STUDY ANSWER

The tubular structures that are apparently absent in our patient are the ductus deferentia. This condition, known as *congenital bilateral absence of the ductus deferentia*, prevents spermatozoa from being transported from the testes to the ejaculatory ducts. This explains the absence of spermatozoa in the patient's ejaculate since the accessory reproductive glands that contribute to the production of seminal fluid add their secretions at a point in the reproductive tract distal to the absent duct. The accessory reproductive glands include the seminal vesicles (which in many cases are also absent or nonfunctional in this deformity) and the prostate. Until recently, this condition would have categorically prevented our patient from becoming a father. Microsurgical extraction of spermatozoa from the epididymides is now possible, however, and has allowed many afflicted men to father children.

CHAPTER SUMMARY

Introduction to the Male Reproductive System (pp. 662–664)

1. The functions of the male reproductive system are to produce spermatozoa, secrete testosterone, and transfer spermatozoa to the reproductive system of the female during coitus (sexual intercourse).
2. The male reproductive system consists of primary sex organs (the testes), secondary sex organs (those that are essential for reproduction), and secondary sex characteristics (features that are sexual attractants, expressed after puberty).

Perineum and Scrotum (pp. 664–665)

1. The saclike scrotum within the urogenital portion of the perineum supports and protects the testes and regulates their position relative to the pelvic region of the body.
2. Each testis is contained within its own scrotal compartment and is separated from the other by the scrotal septum.

Testes (pp. 665–670)

1. The testes are partitioned into wedge-shaped lobules; the lobules are composed of seminiferous tubules that produce sperm and of interstitial tissue that produces androgens.
2. Spermatogenesis occurs by meiotic division of the cells that line the seminiferous tubules.
   (a) At the end of the first meiotic division, two secondary spermatocytes have been produced.
   (b) At the end of the second meiotic division, four haploid spermatids have been produced.
3. The conversion of spermatids to spermatozoa is called spermiogenesis.

4. A sperm consists of a head and a flagellum and matures in the epididymides prior to ejaculation.
   (a) The acrosome of the head contains digestive enzymes for penetrating an ovum.
   (b) The flagellum provides undulating movement of about 3 mm per hour.

Spermatic Ducts, Accessory Glands, and the Urethra (pp. 671–673)

1. The epididymides, ductus deferentia, and ejaculatory ducts are the components of the spermatic ducts.
   (a) The highly coiled epididymides are the tubular structures on the testes where spermatozoa mature and are stored.
   (b) The ductus deferentia convey spermatozoa from the epididymides to the ejaculatory ducts during emission. Each ductus deferens forms a component of a spermatic cord.
2. The seminal vesicles and prostate provide additives to the spermatozoa in the formation of semen.
   (a) The seminal vesicles are located posterior to the base of the urinary bladder and secrete about 60% of the additive fluid of semen.
   (b) The prostate surrounds the urethra just below the urinary bladder and secretes about 40% of the additive fluid of semen.
   (c) Spermatozoa constitute less than 1% of the volume of an ejaculate.
   (d) The small bulbourethral glands secrete fluid that serves as a lubricant for the erect penis in preparation for coitus.
3. The male urethra, which serves both the urinary and reproductive systems, is divided into prostatic, membranous, and spongy portions.

Penis (pp. 674–675)

1. The penis is specialized to become erect for insertion into the vagina during coitus.
2. The body of the penis consists of three columns of erectile tissue, the spongy urethra, and associated vessels and nerves.
3. The root of the penis is attached to the pubic arch and urogenital diaphragm.
4. The glans penis is the terminal end, which is covered with the prepuce in an uncircumcised male.

Mechanisms of Erection, Emission, and Ejaculation (pp. 675–677)

1. Erection of the penis occurs as the erectile tissue becomes engorged with blood. Emission is the movement of the spermatozoa from the epididymides to the ejaculatory ducts and ejaculation is the forceful expulsion of semen from the ejaculatory ducts and urethra of the penis.
2. Parasympathetic stimuli to arteries in the penis cause the erectile tissue to engorge with blood as arteriole flow increases and venous drainage decreases.
3. Ejaculation is the result of sympathetic reflexes in the smooth muscles of the reproductive organs.

UNIT VII

REVIEW ACTIVITIES

Objective Questions

1. The perineum is
   (a) a membranous covering over the testis.
   (b) a ligamentous attachment of the penis to the symphysis pubis.
   (c) a diamond-shaped region between the symphysis pubis and the coccyx.
   (d) a fibrous sheath supporting the scrotum.
2. The dartos is a layer of smooth muscle fibers found within
   (a) the scrotum.
   (b) the penis.
   (c) the epididymis.
   (d) the prostate.
3. The interstitial cells (cells of Leydig)
   (a) nourish spermatids.
   (b) produce testosterone.
   (c) produce spermatozoa.
   (d) secrete alkaline fluid.
   (e) both (b) and (d).
4. The bulb and crus of the penis are located at
   (a) the glans penis.
   (b) the corona glandis.
   (c) the body of the penis.
   (d) the prepuce.
   (e) the root of the penis.
5. Which of the following is *not* a spermatic duct?
   (a) the epididymis.
   (b) the spermatic cord.
   (c) the ejaculatory duct.
   (d) the ductus deferens.
6. Spermatozoa are stored prior to emission and ejaculation in
   (a) the epididymides.
   (b) the seminal vesicles.
   (c) the spongy urethra.
   (d) the prostate.
7. Urethral glands function to
   (a) secrete mucus.
   (b) produce nutrients.
   (c) secrete hormones.
   (d) regulate spermatozoa production.

8. Which statement is *false* regarding erection of the penis?
   (a) It is a parasympathetic response.
   (b) It may be both a voluntary and involuntary response.
   (c) It has to be followed by emission and ejaculation.
   (d) It is controlled by the hypothalamus of the brain and sacral portion of the spinal cord.
9. An embryo with the genotype XY develops male accessory sex organs because of
   (a) androgens.
   (b) estrogens.
   (c) the absence of androgens.
   (d) the absence of estrogens.
10. Which of the following does *not* arise from the embryonic mesonephric duct?
   (a) the epididymis
   (b) the ductus deferens
   (c) the seminal vesicle
   (d) the prostate

Essay Questions

1. What are the functions of the male reproductive system? How do these functions compare with those of the female reproductive system?
2. Discuss how and when the genetic sex of a zygote is determined.
3. Comment on the significance of the indifferent stage in reproductive development. When does the indifferent stage occur, and what are its implications in the potential development of abnormalities of the reproductive organs?
4. Define *perineum*. Contrast the appearance of the perineum of an embryo during the indifferent stage of genital development with that of an adult male.
5. Define what is meant by homologous structures. List the structures of the male reproductive system that form from the phallus, urethral folds, labioscrotal swellings, and the gonadal ridge.

6. Describe the location and structure of the scrotum. Explain how the scrotal muscles regulate the position of the testes in the scrotum. Why is this important?
7. Describe the internal structure of a testis. Discuss the function of the sustentacular cells, interstitial cells, seminiferous tubules, rete testis, and efferent ductules.
8. List the structures that constitute the spermatic cord. Where is the inguinal canal? Why is the inguinal canal clinically important?
9. Diagram a spermatozoon and list the function of each of its principal parts. Define *semen*. How much semen is ejected during ejaculation? What does it consist of and what are its properties?
10. Describe the position, function, and histological structure of the epididymis, ductus deferens, and ejaculatory duct.
11. Compare the seminal vesicles and the prostate in terms of location, structure, and function.
12. Describe the structure of the penis and explain the mechanisms that result in erection, emission, and ejaculation.
13. Distinguish between impotence, infertility, and sterility.
14. Distinguish between gonorrhea and syphilis. Describe the stages through which syphilis will progress if untreated.
15. Define *hydrocele, orchitis,* and *orchiectomy*. What conditions would warrant an orchiectomy?

Gundy/Weber Software

The tutorial software accompanying Chapter 20, Male Reproductive System, is Volume 13, Reproductive System.

21

Female Reproductive System

Chapter Outline

CLINICAL CASE STUDY

A 28-year-old female is brought to the emergency room with a 4-day history of moderate right-sided pelvic pain. On the morning of the fifth day the pain had become more severe, prompting her to seek medical attention. She complains of weakness and light-headedness and states that she hasn't had a period for about 8 weeks. A urine pregnancy test is positive. The consulting gynecologist says that a ruptured ectopic pregnancy is likely. He orders a blood test, the results of which suggest that the patient has suffered a slight amount of hemorrhage. A *culdocentesis* (needle sampling of the peritoneal cavity via the posterior vaginal wall to detect pooled blood) is positive. The patient is prepared for surgery.

What is an ectopic pregnancy? Where is it most likely to occur? Briefly explain the sequence of events leading up to the rupture of the ectopic pregnancy beginning with ovulation. Differentiate normal from abnormal events. Explain how blood from a ruptured ectopic pregnancy can be aspirated through the vagina.

Hints: Carefully study the anatomical position of the uterus and uterine tubes with respect to the ovaries within the peritoneal cavity. Read about ectopic pregnancies in the clinical sections of this chapter and in chapter 22 refer to figure 22.37.

Introduction to the Female Reproductive System

The female reproductive system produces ova, secretes sex hormones, receives spermatozoa from the male, and provides sites for fertilization and implantation. Parturition follows gestation, and secretion from the mammary glands provides nourishment for the baby.

Objective 1. Explain the functional differences between the male and female reproductive systems.

Objective 2. Define *puberty, menstruation, ovulation,* and *menopause.*

Objective 3. Identify the primary and secondary sex organs in a female, and describe the secondary sex characteristics.

The reproductive systems of the male and female have some basic similarities and some specialized differences. The two systems are similar in that (1) most of the reproductive organs of both sexes develop from similar embryonic tissues and are therefore *homologous;* (2) both systems have gonads that produce gametes and sex hormones; and (3) both systems experience latent development of the reproductive organs as they mature and become functional during puberty as a result of the influence of sex hormones secreted by the gonads.

The differences between the reproductive systems of the female and male are based on the specific functions of each in sexual reproduction and on the cyclic events that are characteristic of the female. The reproductive organs of the sexually mature, healthy male continuously produce male gametes, or spermatozoa, and transfer them to the female during *coitus (sexual intercourse).* A male does not produce any spermatozoa until puberty (at about age 13), but is then capable of producing viable spermatozoa throughout his life if he remains healthy. The gametes, or ova, of a female are completely formed, but not totally matured, during fetal development of the ovaries. The ova are generally discharged, or *ovulated,* one at a time in a cyclic pattern throughout the reproductive period of the female, which extends from puberty to menopause. *Menstruation* is the discharge of *menses* (blood and solid tissue) from the uterus at the end of each ovulation cycle. *Menopause* is the period marked by the termination of ovulation and menstruation. The reproductive period in females generally extends from about age 12 to about age 50. The cyclic reproductive pattern of ovulation and the age span of fertility are determined by hormonal action.

The functions of the female reproductive system are (1) to produce ova; (2) to secrete sex hormones; (3) to receive the spermatozoa from the male during coitus; (4) to provide sites for fertilization, implantation of the blastocyst (see chapter 22), and embryonic and fetal development; (5) to facilitate *parturition,* or delivery of the baby; and (6) to provide nourishment for the baby through the secretion of milk from the mammary glands in the breasts.

The organs of the female reproductive system, like those of the male, are categorized on a functional basis as follows:

1. **Primary sex organs.** The primary sex organs are called **gonads,** and in the female are known more specifically as the **ovaries.** Ovaries produce the gametes (ova), or eggs, and produce and secrete sex steroid hormones. Secretion of the female sex hormones at puberty contributes to the development of secondary sexual characteristics and causes cyclic changes in the secondary sex organs that are required for reproductive function.

2. **Secondary sex organs.** Secondary sex organs (fig. 21.1 and table 21.1) are those structures that are essential for successful fertilization of the ovum, implantation of the blastocyst, development of the embryo and fetus, and parturition. The secondary sex organs include the **vagina,** which receives the penis and ejaculated semen during coitus and through which the baby passes during delivery; the external genitalia, which protect the vaginal orifice (opening); the **uterine** (fallopian) **tubes,** through one of which an egg is transported toward the uterus

---

coitus: L. *coitio,* a coming together
menopause: Gk. *men,* month; *pausis,* cessation
vagina: L. *vagina,* sheath or scabbard

UNIT VII

Figure 21.1   Organs of the female reproductive system seen in sagittal section.

Uterine tube

Ovary

Uterus

Urinary bladder

Symphysis pubis

Urethra

Clitoris

Labium minor

Labium major

Vaginal orifice

Fimbriae of uterine tube

Posterior portion of vaginal fornix

Cervix of uterus

Rectum

Vagina

Anus

Waldrop

Table 21.1	Functions of the Female Reproductive Organs
Organ(s)	**Function(s)**
Ovaries	Produce ova and female sex hormones
Uterine tubes	Convey ova toward uterus; site of fertilization; convey developing blastocyst to uterus
Uterus	Site of implantation; protects and sustains life of embryo and fetus during pregnancy; plays active role in parturition
Vagina	Conveys uterine secretions to outside of body; receives erect penis and semen during coitus and ejaculation; serves as passageway for fetus during parturition
Labia majora	Form margins of pudendal cleft; enclose and protect other external reproductive organs
Labia minora	Form margins of vestibule; protect openings of vagina and urethra
Clitoris	Glans of the clitoris is richly supplied with sensory nerve endings associated with feeling of pleasure during sexual stimulation
Pudendal cleft	Cleft between labia majora within which labia minora and clitoris are located.
Vaginal vestibule	Cleft between labia minora within which vaginal and urethral openings are located
Vestibular glands	Secrete fluid that moistens and lubricates the vestibule and vaginal opening during coitus
Mammary glands	Produce and secrete milk for nourishment of an infant

after ovulation and where fertilization takes place; and the **uterus** (womb), where implantation and development occur. The muscular walls of the uterus play an active role in parturition. **Mammary glands** are also considered secondary sex organs because the milk they secrete after parturition provides nourishment to the child.

3. **Secondary sex characteristics.** Secondary sex characteristics are features that are not essential for the reproductive process but are generally considered to be sexual attractants. Distribution of fat to the breasts, abdomen, mons pubis, and hips; body hair pattern; and broad pelvis are examples of female secondary sex characteristics. Although the breasts contain the mammary glands, large breasts are not essential for nursing the young. In fact, all female mammals have mammary glands, but only human females have protruding breasts that function as a sexual attractant.

The onset of puberty in females generally occurs between the ages of 12 and 14 in accordance with the nutritional condition, genetic background, and even sexual exposure of the individual. Generally, girls attain puberty 6 months to 1 year earlier than boys, accompanied by an earlier growth spurt (fig. 21.2). Puberty in girls is heralded by the onset of menstruation, or *menarche (mĕ-nar′ke)*. Puberty results from the increased secretion of gonadotropins from the anterior pituitary, which stimulates the ovaries to begin their cyles of ova development and sex steroid secretion.

UNIT VII

Figure 21.2 A comparison of male and female rates of growth in height.

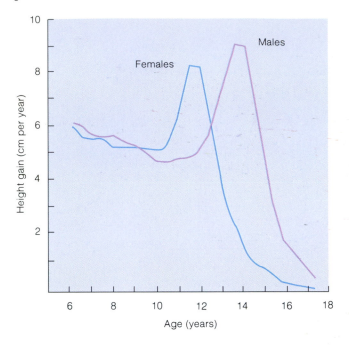

1. List the functions of the female reproductive system.
2. Define *puberty* and explain the process by which it occurs. Define *menstruation* and *ovulation*. What is the usual age span of female fertility? Define *menses* and *menopause*.
3. Distinguish between the primary sex organs, secondary sex organs, and secondary sex characteristics.

Structure and Function of the Ovaries

The ovary contains numerous primary follicles with primary oocytes. During an ovarian cycle, some of these become secondary follicles and one becomes a mature vesicular ovarian follicle that ruptures and releases its oocyte in ovulation. The ruptured follicle becomes a corpus luteum and regresses to become a corpus albicans. These cyclic changes in structure are supporting cyclic changes in hormone secretion.

Objective 4. Describe the position of the ovaries and the ligaments supporting the ovaries and genital ducts.

Objective 5. Describe the structural changes that occur in the ovaries leading to and following ovulation.

Objective 6. Describe oogenesis and explain why meiosis of one primary oocyte results in the formation of only one mature ovum.

Objective 7. Discuss the hormonal secretions of the ovaries during an ovarian cycle.

Position and Structure of the Ovaries

Ovaries are the paired primary sex organs of the female that produce *gametes*, or *ova*, and the sex hormones, *estrogens* and *progesterone*. The ovaries of a sexually mature female are solid, ovoid structures about 3.5 cm (1.4 in.) long, 2 cm (0.8 in.) wide, and 1 cm (0.4 in.) thick. The color and texture of the ovaries vary according to the age and reproductive stage of the female. The ovaries of a young girl are smooth and pinkish. Following puberty, the ovaries are pinkish-gray and have an irregular surface because of the scarring caused by ovulation. On the medial portion of each ovary is a **hilum,** which is the point of entrance for ovarian vessels and nerves. The lateral portion of the ovary is positioned near the open end of the uterine tube (fig. 21.3).

The ovaries are positioned in the upper pelvic cavity on both lateral sides of the uterus. Each ovary is situated in a shallow depression of the posterior body wall, the **ovarian fossa,** and secured by several membranous attachments. The principal supporting membrane of the female reproductive tract is the **broad ligament.** The broad ligament is the parietal peritoneum that supports the uterine tubes and uterus. The **mesovarium** (*mes''ō-va're-um*) is a specialized posterior extension of the broad ligament that attaches to an ovary. Each ovary is additionally supported by an **ovarian ligament** that is anchored to the uterus and a **suspensory ligament** that is attached to the pelvic wall (fig. 21.3).

Each ovary consists of four layers. The **superficial epithelium** is the thin outermost layer composed of cuboidal epithelial cells (see fig. 21.6). A collagenous connective tissue layer called the **tunica albuginea** (*al''byoo-jin'e-ă*) is located immediately below the superficial epithelium. The principal substance of the ovary is divided into an outer **ovarian cortex** and an inner, vascular **ovarian medulla,** although the boundary between these layers is not distinct. The **stroma**—the material of the ovary in which follicles and blood vessels are embedded—lies in both the cortical and medullary layers.

Blood Supply and Innervation Blood is supplied by ovarian arteries that arise from the lateral sides of the abdominal aorta, just below the origin of the renal arteries. An additional supply comes from the ovarian branches of the uterine arteries. Venous return is through the ovarian veins. The right ovarian vein empties into the inferior vena cava, whereas the left ovarian vein drains into the left renal vein.

The ovaries have both sympathetic and parasympathetic innervation from the ovarian plexus. Innervation to the ovaries, however, is only to the vascular networks and not to the follicular substance within the stroma. All of the vessels and nerves to an ovary enter by way of the hilum, which is supported by the ovarian ligament.

---

ovaries: L. *ovum*, egg
stroma: Gk. *stroma*, a couch or bed

Figure 21.3 An anterior view of the female reproductive organs showing the relationship of the ovaries, uterine tubes, uterus, cervix, and vagina.

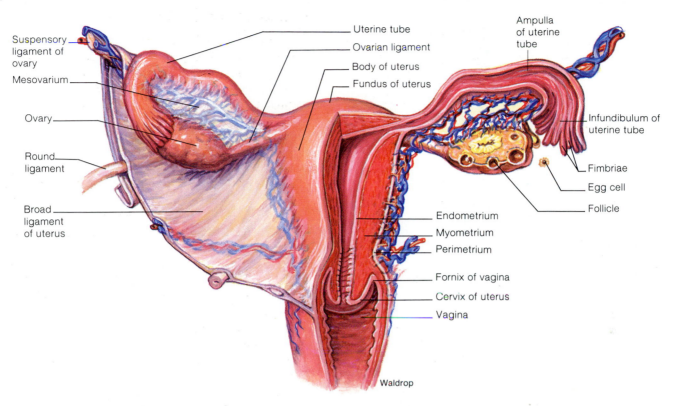

Suspensory ligament of ovary

Mesovarium

Ovary

Round ligament

Broad ligament of uterus

Uterine tube

Ovarian ligament

Body of uterus

Fundus of uterus

Ampulla of uterine tube

Infundibulum of uterine tube

Fimbriae

Egg cell

Follicle

Endometrium

Myometrium

Perimetrium

Fornix of vagina

Cervix of uterus

Vagina

Waldrop

Normal, healthy ovaries usually cannot be palpated either by vaginal or abdominal examination. If the ovaries become swollen or displaced, however, they are palpable through the vagina. There are many types of nonmalignant tumors of the ovaries, most of which cause swelling and some localized tenderness. The ovaries atrophy during menopause, and ovarian enlargement in postmenopausal women is usually cause for concern.

Ovarian Cycle

The germ cells that migrate into the ovaries during early embryonic development multiply, so that at about 5 months of gestation the ovaries contain approximately 6 to 7 million oogonia. Production of new oogonia stops at this point and never resumes again. Toward the end of gestation, the oogonia begin meiosis, at which time they are called **primary oocytes** (o'ō-sītz). Oogenesis progresses to prophase I of the first meiotic division, at which point it is arrested. The number of primary oocytes decreases throughout a woman's reproductive years. The ovaries of a newborn girl contain about 2 million oocytes—all she will ever have. By the time she reaches puberty, this number has been reduced to about 300,000–400,000. Oogenesis ceases entirely at menopause (when menstruation stops).

Primary oocytes that are not stimulated to complete the first meiotic division are contained within tiny follicles called

primordial follicles. In response to gonadotrophin stimulation, some of these oocytes and follicles get larger, and the follicular cells divide to produce the **follicular epithelium** that surrounds the oocyte and fills the follicle. A follicle at this stage in development is called a **primary follicle.**

Some primary follicles will be stimulated to grow still bigger and develop a fluid-filled cavity called an *antrum,* at which time they are called **secondary follicles** (fig. 21.4). The follicular epithelium of secondary follicles forms a ring around the circumference of the follicle and a mound that supports the oocyte. This mound is called the *cumulus oophorous (o-of'ō-rus).* Some follicular cells also encircle the oocyte, forming a *corona radiata.* Between the oocyte and the corona radiata is a thin gel-like layer of proteins and polysaccharides called the *zona pellucida (pĕ-loo'sĭ-dă).* Under stimulation of follicle-stimulating hormone (FSH) from the anterior pituitary, the follicular cells secrete increasing amounts of estrogen as the follicles grow. Interestingly, the follicular cells produce estrogen from its precursor testosterone, which is supplied by cells of the *theca interna layer,* immediately outside the follicle.

As the follicle develops, the primary oocyte completes its first meiotic division. This does not form two complete cells,

UNIT VII

---

oogonium: Gk. *oion,* egg; *gonos,* generation

cumulus oophorous: L. *cumulus,* a mound; G. *oophoros,* egg-bearing
corona radiata: Gk. *korone,* crown; L. *radiata,* radiate
zona pellucida: L. *zone,* girdle; L. *pellis,* skin
theca interna: Gk. *theke,* a box; L. *internus,* interior

Figure 21.4 Photomicrographs of (a) primordial and primary follicles and (b) a mature ovarian (graafian) follicle.

Primary follicles

Secondary follicle

Vesicle

(a)

Follicular cells

Antrum

Secondary oocyte

Corona radiata

Cumulus oophorus

Zona pellucida

Theca interna

(b)

however, because only one cell—the **secondary oocyte**—gets all the cytoplasm. The other cell formed at this time becomes a small *polar body* (fig. 21.5) that eventually fragments and disappears. The secondary oocyte enters the second meiotic division, but meiosis is arrested at metaphase II and is never completed unless fertilization occurs.

Ovulation By about the tenth to the fourteenth day following the first day of a menstrual period, usually just one follicle has matured fully to become a **vesicular ovarian** (graafian)

follicle (fig. 21.6). Other secondary follicles regress during that menstrual cycle and become *atretic (ă-tret′ik)*. The vesicular ovarian follicle is so large that it forms a bulge on the surface of the ovary. Under proper hormonal stimulation (a sudden burst of luteinizing hormone from the anterior pituitary triggered by a peak level of estrogen) this follicle will rupture—much like the popping of a blister—and extrude its secondary oocyte into the peritoneal cavity near the opening of the uterine tube in the process of *ovulation* (fig. 21.7).

The released secondary oocyte is surrounded by the zona pellucida and corona radiata. If it is not fertilized, it disinte-

vesicular ovarian (graafian) follicle: from Regnier de Graaf, Dutch
anatomist and physician, 1641–73

atretic: Gk. *atretos,* not perforated

Figure 21.5 (a) A primary oocyte at metaphase I of meiosis. (Note the alignment of chromosomes [arrow]). (b) A human secondary oocyte formed at the end of the first meiotic division and the first polar body (arrow).

(a)

(b)

Figure 21.6 A vesicular ovarian (graafian) follicle within the ovary of a monkey.

Superficial epithelium   Primordial follicle   Vesicular ovarian follicle   Atretic follicle   Blood vessel

Stroma

Corpus albicans   Primary follicle

UNIT VII

grates in a couple of days. If a spermatozoon passes through the corona radiata and zona pellucida and enters the cytoplasm of the secondary oocyte, the oocyte completes the second meiotic division, becoming a mature ovum. In this process, the cytoplasm is again not divided equally; most of the cytoplasm remains in the zygote (fertilized egg), leaving another polar body, which like the first, disintegrates (fig. 21.8).

Changes continue in the ovary following ovulation. The empty follicle, under the influence of luteinizing hormone, undergoes structural and biochemical changes to become a

corpus luteum. Unlike the ovarian follicles that secrete only estrogen, the corpus luteum secretes two sex steroid hormones: estrogen and progesterone. Toward the end of a nonfertile cycle, the corpus luteum regresses and is changed into a nonfunctional **corpus albicans.** These cyclic changes in the ovary are summarized in figure 21.9.

corpus luteum: L. *corpus*, body; *luteum*, yellow
albicans: L. *albicare*, to whiten

Figure 21.7 Ovulation from a human ovary. An ovulated oocyte is free in the peritoneal cavity until it enters the lumen of the uterine tube.

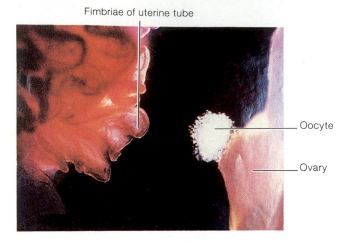

Fimbriae of uterine tube

Oocyte

Ovary

1. Describe the position of the ovaries relative to the uterine tubes and describe the position and functions of the broad ligament and mesovarium.
2. Compare the structure and contents of a primordial follicle, primary follicle, secondary follicle, and vesicular ovarian follicle.
3. Define *ovulation* and describe the changes that occur in the ovary following ovulation in a nonfertile cycle.
4. Describe oogenesis and explain why only one mature ovum is produced by this process.
5. Compare the hormonal secretions of the vesicular ovarian follicle with those of a corpus luteum.

Secondary Sex Organs

The uterine tube conducts the zygote to the uterus, where implantation into the endometrial layer typically occurs. The muscular layer of the uterus, or myometrium, is functional in labor and delivery. Sperm enter the female genital tract through the vagina, which also serves as the birth canal during parturition.

Figure 21.8 A schematic diagram of the process of oogenesis. During meiosis, each primary oocyte produces a single haploid gamete. If the secondary oocyte is fertilized, it forms a secondary polar body and becomes a zygote.

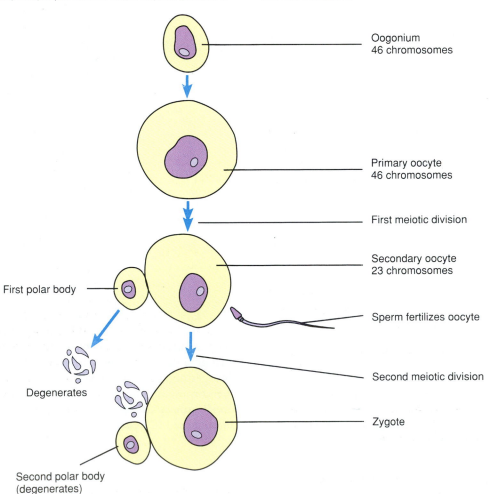

Oogonium
46 chromosomes

Primary oocyte
46 chromosomes

First meiotic division

Secondary oocyte
23 chromosomes

First polar body

Sperm fertilizes oocyte

Degenerates

Second meiotic division

Zygote

Second polar body
(degenerates)

UNIT VII

Figure 21.9 A schematic diagram of an ovary showing the various stages of ovum and follicle development.

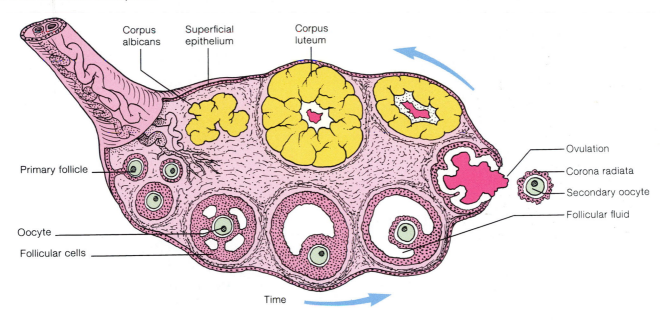

Objective 8. Describe how an ovum is moved through a uterine tube to the uterus.

Objective 9. Describe the structure and function of each of the three layers of the uterus.

Objective 10. Describe the structure and functions of the vulva and vagina.

Objective 11. Discuss the changes that occur in the female reproductive system during sexual excitement and coitus.

The secondary sex organs described in this section include the internal organs of the uterine tubes, uterus, and vagina, as well as the external genitalia (vulva). The mammary glands, which are also considered to be secondary sex organs, are described in a separate section.

Uterine Tubes

The paired **uterine tubes,** also known as the fallopian *(fă-lo'pe-an)* tubes, or oviducts, transport ova from the ovaries to the uterus. Fertilization normally occurs within the uterine tube. Each uterine tube is approximately 10 cm (4 in.) long and 0.7 cm (0.3 in.) in diameter and is positioned between the folds of the broad ligament of the uterus (see fig. 21.3).

> The term *salpinx* is occasionally used to refer to the uterine tubes. It is a Greek word meaning trumpet or tube and is the root of such clinical terms as *salpingitis (sal"pin-ji'tis),* or inflammation of the uterine tubes; *salpingography* (radiography of the uterine tubes); and *salpingolysis* (the breaking up of adhesions of the uterine tube to correct female infertility).

The funnel-shaped, open-ended portion of the uterine tube is called the **infundibulum** (see fig. 21.3). Although the infundibulum is close to the ovary, it is not attached. A number of fringed, fingerlike processes called **fimbriae** project from the margins of the infundibulum over the lateral surface of the ovary. Movement of the fimbriae directs oocytes into the lumen of the uterine tube. From the infundibulum, the uterine tube extends medially and inferiorly to open into the superolateral cavity of the uterus at the **uterine opening.** The **ampulla** *(am-pool'ă)* is the longest and widest portion of the uterine tube (see fig. 21.3).

The wall of the uterine tube consists of three histological layers. The internal **mucosa** lines the lumen and is composed of ciliated columnar epithelium (fig. 21.10) that is drawn into numerous folds. The **muscularis** is the middle layer, composed of a thick, circular layer of smooth muscle and a thin outer layer of smooth muscle. Peristaltic contractions of the muscularis and ciliary action of the mucosa move the oocyte through the lumen of the uterine tube. The outer **serous layer** of the uterine tube is part of the visceral peritoneum.

The oocyte takes 4 to 5 days to move through the uterine tube. If enough viable sperm are ejaculated into the vagina during coitus and if there is an oocyte in the uterine tube, fertilization will occur within hours after discharge of the semen. The zygote will move toward the uterus where implantation occurs. If the developing embryo (called a *blastocyst*) implants into the uterine tube instead of the uterus, the pregnancy is termed an *ectopic pregnancy,* meaning an implantation of a blastocyst in a site other than the uterus.

UNIT VII

---

fallopian tubes: from Gabriele Fallopius, Italian anatomist, 1523–62
oviduct: L. *ovum,* egg; *ductus,* a leading

ectopic: Gk. *ex,* out; *topos,* place

Figure 21.10 The histology of the uterine tube.

- Connective tissue layer
- Basement membrane
- Ciliated columnar epithelium
- Cytoplasm
- Cilia
- Nucleus
- Cell membrane

Since the infundibulum of the uterine tube is unattached, it provides a pathway for pathogens to enter the peritoneal cavity. The mucosa of the uterine tube is continuous with that of the uterus and vagina, and it is possible for infectious agents to enter the vagina and cause infections that may ultimately spread to the peritoneal linings resulting in *pelvic inflammatory disease* (PID). There is no opening into the peritoneal cavity other than through the uterine tubes. The peritoneal cavity of a male is totally sealed from external contamination.

The uterine tubes are supplied with blood through the ovarian and uterine arteries. Venous drainage is through uterine veins that parallel the arteries. Both the uterine artery and vein can be observed in the broad ligament that supports the uterine tube (see fig. 21.3).

The uterine tubes have both sympathetic and parasympathetic innervation from the hypogastric plexus and pelvic splanchnic nerves. The nerve supply to the uterine tubes regulates the activity of the smooth muscles and blood vessels.

Uterus

The **uterus** receives the blastocyst that develops from a fertilized oocyte and provides a site for implantation. Prenatal development continues within the uterus until gestation is completed, at which time the uterus plays an active role in the delivery of the baby.

Structure The uterus is a hollow, thick-walled, muscular organ with the shape of an inverted pear. Although the shape and position of the uterus changes immensely during pregnancy (fig. 21.11), in its nonpregnant state it is about 7 cm (2.8 in.) long, 5 cm (2 in.) wide (through its broadest region), and 2.5 cm (1 in.) in diameter. The anatomical regions of the uterus include the uppermost dome-shaped portion above the entrance

Figure 21.11 The size and position of the uterus in a full-term pregnant woman in sagittal section.

- Diaphragm
- Liver
- Stomach
- Pancreas
- Transverse colon
- Small intestine
- Inferior vena cava
- Uterus
- Cervix
- Rectum
- Symphysis pubis
- Urinary bladder
- Vagina

Gordon/Waldrop

of the uterine tubes called the **fundus,** the enlarged main portion called the **body,** and the inferior constricted portion opening into the vagina called the **cervix** (fig. 21.12). The uterus is located between the urinary bladder anteriorly and the rectum and sigmoid colon posteriorly. The fundus projects anteriorly and slightly superiorly over the urinary bladder. The cervix projects posteriorly and inferiorly, joining the vagina at nearly a right angle.

The **uterine cavity** is the space within the regions of the fundus and body. The lumina of the uterine tubes open into the uterine cavity on the superolateral portions. The uterine cavity is continuous inferiorly with the **cervical canal,** which extends through the cervix and opens into the lumen of the vagina. The junction of the uterine cavity with the cervical canal is called the **isthmus of uterus,** whereas the opening of the cervical canal into the cavity of the vagina is called the **uterine ostium** (fig. 21.12).

Support of the Uterus The uterus is maintained in position by muscular support and ligaments that extend from the pelvic girdle or body wall to the uterus. Muscles of the perineum, especially the levator ani muscle (see fig. 9.25), provide the prin-

fundus: L. *fundus*, bottom
cervix: L. *cervix*, neck

Figure 21.12 Layers of the uterine wall.

cipal muscular support. The ligaments that support the uterus undergo marked hypertrophy during pregnancy, regress in size after parturition, and atrophy after menopause.

Four paired ligaments support the uterus in position within the pelvic cavity. The paired **broad ligaments** are folds of the peritoneum that extend from the pelvic walls and floor to the lateral walls of the uterus (see fig. 21.3). The ovaries and uterine tubes are also supported by the broad ligaments. The paired **rectouterine folds** (not illustrated) are also continuations of peritoneum that curve along the lateral pelvic wall on both sides of the rectum to connect the uterus to the sacrum. The **cardinal** (lateral cervical) **ligaments** (not illustrated) are fibrous bands within the broad ligament that extend laterally from the cervix and vagina across the pelvic floor, where they attach to the wall of the pelvis. The cardinal ligaments contain some smooth muscle as well as vessels and nerves to the cervix and vagina. The fourth paired ligaments are the **round ligaments** (see fig. 21.3). The round ligaments are actually continuations of the ovarian ligaments that support the ovaries. Each round ligament extends from the lateral border of the uterus just below the point where the uterine tube attaches to the lateral pelvic wall. Similar to the course taken by the ductus deferens in the male, the round ligaments continue through the inguinal canal of the abdominal wall, where they attach to the deep tissues of the labium majus.

Although the uterus has extensive support, considerable movement is possible. The uterus tilts slightly posteriorly as the urinary bladder fills, and it moves anteriorly during defecation. In some women, the uterus may become displaced and interfere with the normal progress of pregnancy. A posterior tilting of the uterus is called *retroflexion,* whereas an anterior tilting is called *anteflexion.*

Uterine Wall   The wall of the uterus is composed of three layers: the perimetrium, myometrium, and endometrium (fig. 21.12). The **perimetrium** is the thin outer serosal covering and a part of the peritoneum. The lateral portion of the perimetrium is continuous with the broad ligament. A shallow pouch called the **vesicouterine** (*ves″i-ko-yoo′ter-in*) **pouch** is formed as the peritoneum is reflected over the urinary bladder. The **rectouterine** (*rek″to-yoo′ter-in*) **pouch** (pouch of Douglas) is formed as the peritoneum is reflected onto the rectum. The rectouterine pouch is the lowest point in the pelvic cavity and provides a site for surgical entry into the peritoneal cavity.

The thick **myometrium** is composed of three poorly defined layers of smooth muscle arranged in longitudinal, circular, and spiral patterns. The myometrium is thickest in the fundus and thinnest in the cervix. During parturition, the muscles of this layer are stimulated to contract forcefully.

The **endometrium** is the inner mucosal lining of the uterus. The endometrium has two distinct layers. The superficial **stratum functionale,** composed of columnar epithelium and containing secretory glands, is shed as *menses* during menstruation and built up again under the stimulation of ovarian steroid hormones. The deeper **stratum basale** is highly vascular and serves to regenerate the stratum functionale after each menstruation.

The uterus undergoes tremendous change during pregnancy. Its weight increases more than 16 times (from about 60 g to about 1,000 g), and its capacity increases from about 2.5 mℓ to over 5,000 mℓ. The principal change in the myometrium is a marked hypertrophy, or elongation, of the individual muscle cells to as much as 10 times their original length. There is some atrophy of the muscle cells after parturition, but the uterus never returns to its original size.

Uterine Blood Supply and Innervation   The uterus is supplied with blood through the **uterine arteries** that arise from the internal iliac arteries and by the **uterine branches** of the **ovarian arteries** (fig. 21.13). Each pair of these two vessels anastomose on the upper lateral margin of the uterus. The blood from the uterus returns through uterine veins that parallel the pattern of the arteries.

The uterus has both sympathetic and parasympathetic innervation from the pelvic and hypogastric plexuses. Both autonomic innervations serve the arteries of the uterus, whereas the smooth muscle of the myometrium receives only sympathetic innervation.

perimetrium: Gk. *peri,* around; *metra,* uterus
vesicouterine: L. *vesico,* bladder; *uterus,* womb
pouch of Douglas: from James Douglas, English anatomist and physician, 1675–1742
myometrium: Gk. *mys,* muscle; *metra,* uterus
endometrium: Gk. *endon,* within; *metra,* uterus
menses: L. *menses,* plural of *mensis,* monthly

UNIT VII

Figure 21.13 Vascular supply to the uterus.

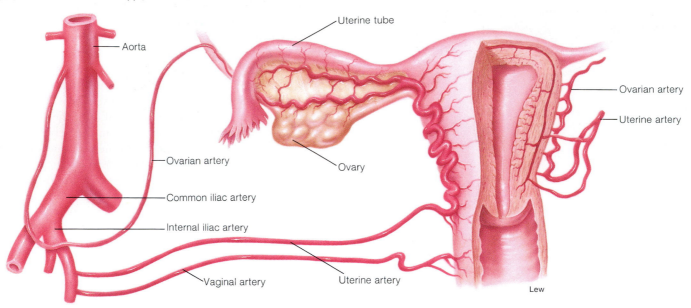

Vagina

The **vagina** is the organ that receives sperm through the ure-thra of the erect penis during coitus. It also serves as the birth canal during parturition and provides for the passage of menses to the outside. The musculomembranous vagina is a tubular organ about 9 cm (3.6 in.) long, passing from the cervix of the uterus to the vestibule. The vagina is positioned between the urinary bladder and urethra anteriorly and the rectum poste-riorly, where it is continuous with the cervical canal of the uterus. The cervix attaches to the vagina at a nearly 90 degree angle. The deep recess surrounding the protrusion of the cervix into the vagina is called the **fornix** (see fig. 21.3).

> The fornix is of clinical importance since it permits the cervix to be palpated during a gynecological examination. Occasionally, the deep posterior portion of the fornix provides surgical access to the pelvic cavity through the vagina. In addition, the fornix is impor-tant in the placement of two birth control devices—the cervical cap and the diaphragm (see fig. 21.26).

The exterior opening of the vagina, at its lower end, is called the **vaginal orifice.** A thin fold of mucous membrane called the **hymen** may partially cover the vaginal orifice.

The vaginal wall is composed of three layers: an inner mucosa, a middle muscularis, and an outer fibrous layer. The **mucosal layer** consists of nonkeratinized stratified squamous epithelium that forms a series of transverse folds called **vaginal rugae** (fig. 21.14). The vaginal rugae permit considerable dis-tension of the vagina for penetration of the erect penis. They also provide friction ridges for stimulation of the erect penis during coitus. The mucosal layer contains few glands; the acidic mucus that is present in the vagina comes primarily from glands within the uterus. The acidic environment of the vagina retards microbial growth. The additives within semen, however, tem-

Figure 21.14 The histology of a vaginal ruga.

porarily neutralize the acidity of the vagina to ensure the sur-vival of the ejaculated spermatozoa deposited within the vagina.

The **muscularis layer** consists of longitudinal and circular bands of smooth muscle interlaced with distensible connective tissue. The distension of this layer is especially important during parturition. Skeletal muscle strands, including the levator ani muscle, near the vaginal orifice partially constrict this opening.

The **fibrous layer** covers the vagina and attaches it to sur-rounding pelvic organs. This layer consists of dense regular connective tissue interlaced with strands of elastic fibers.

The blood supply to the highly vascular vagina is primarily from the vaginal branches of the internal iliac artery. Blood is also supplied to the vagina through branches of the uterine, middle rectal, and internal pudendal arteries. Blood draining from the vagina returns through vaginal veins that parallel the course of the arteries.

The vagina has sympathetic innervation from the hypogastric plexus and parasympathetic innervation from the second and third sacral nerves. Sensory innervation is through the pudendal plexus and is especially well developed near the vaginal orifice.

Vulva

The external genitalia of the female are referred to collectively as the **vulva,** which is shown in figure 21.15. The structures of the vulva surround the vaginal orifice and include the mons pubis, labia majora, labia minora, clitoris, vaginal vestibule, vestibular bulbs, and vestibular glands.

The **mons pubis** is the subcutaneous pad of adipose connective tissue covering the symphysis pubis. At puberty, the mons pubis becomes covered with a pattern of coarse pubic hair that is somewhat triangular, usually with a horizontal upper border. The elevated and padded mons pubis cushions the symphysis pubis and vulva during coitus.

The **labia majora** (labium majus, singular) are two thickened longitudinal folds of skin that contain loose fibrous connective tissue and adipose tissue, as well as some smooth muscle. The labia majora are continuous anteriorly with the mons pubis, are separated longitudinally by the **pudendal cleft,** and converge again posteriorly on the **perineum** (*per"i-ne'um*). The labia minora and clitoris are located within the pudendal cleft. The labia majora are covered with hair and contain numerous sebaceous and sweat glands. The labia majora are homologous to the scrotum of the male and function to enclose and protect the other organs of the vulva.

> An *episiotomy* (*ē-pe'ze-ot"o-me*) is a surgical incision, for obstetrical purposes, of the vaginal orifice that extends into the perineum. An episiotomy may be done during parturition to facilitate delivery and accommodate the head of an emerging fetus when laceration seems imminent. After delivery the cut is sutured.

Positioned close together between the labia majora are two smaller longitudinal folds called the **labia minora** (labium minus, singular). The labia minora are hairless but do contain sebaceous glands. On the anterior side, the labia minora split to form the **prepuce** (*pre'pyoos*) of the clitoris. These inner folds of skin further protect the vaginal and urethral openings.

The **clitoris** (*klit'or-is*) is a small rounded projection at the upper portion of the pudendal cleft. The clitoris corresponds in structure and origin to the penis in the male; it is, however, smaller and has no urethra. Although most of the clitoris is embedded, it does have an exposed **glans clitoris** of erectile tissue that is richly innervated with sensory endings. The clitoris is about 2 cm (0.8 in.) long and 0.5 cm (0.2 in.) in diameter. The unexposed portion of the clitoris is composed of two columns of erectile tissue called the **corpora cavernosa** that diverge posteriorly to form the **crura** and attach to the sides of the pubic arch.

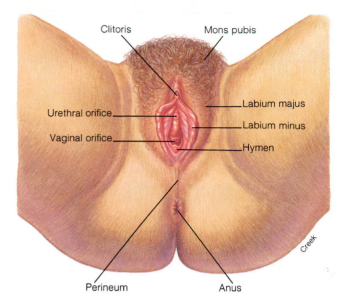

Figure 21.15 The external female genitalia.

The **vaginal vestibule** is the longitudinal cleft enclosed by the labia minora. The openings for the urethra and vagina are located in the vaginal vestibule. The external opening of the urethra is about 2.5 cm (1 in.) behind the glans of the clitoris and immediately in front of the vaginal orifice. The vaginal orifice is lubricated during sexual excitement by secretions from paired major and minor **vestibular glands** (Bartholin's glands) located within the wall of the region immediately inside the vaginal orifice. The ducts from these glands open into the vestibule near the lateral margins of the vaginal orifice. Bodies of vascular erectile tissue, called **vestibular bulbs,** are located immediately below the skin forming the lateral walls of the vestibule. The vestibular bulbs are separated from each other by the vagina and urethra and extend from the level of the vaginal orifice to the clitoris.

The vulva is highly vascular and is supplied with arterial blood from internal pudendal branches of the internal iliac arteries and external pudendal branches from the femoral arteries. Extensive vascular networks are found within most of the organs of the vulva. The venous return is through vessels that correspond in name and position to the arteries.

> During pregnancy, the vulva becomes swollen and bluish—especially the labia minora—due to increased vascularity and venous congestion. This discoloration is an important diagnostic indicator of pregnancy. It appears at about the eighth to the twelfth week and becomes more apparent as pregnancy progresses.

The vulva has both sympathetic and parasympathetic innervation, as well as extensive somatic neurons that respond to

---

vulva: L. *volvere*, to roll, wrapper
mons pubis: L. *mons*, mountain: *pubis*, genital area

vestibule: L. *vestibule*, an entrance, court
Bartholin's glands: from Casper Bartholin Jr., Danish anatomist, 1655–1738

UNIT VII

sensory stimulation. Parasympathetic stimulation causes a response similar to that of the male: dilation of the arterioles of the genital erectile tissue and compression of the venous return.

Mechanism of Erection and Orgasm

The homologous structures of the male and female reproductive systems respond to sexual stimulation in a similar fashion. The erectile tissues of a female, like those of a male, become engorged with blood and swollen during sexual arousal. During sexual excitement, the hypothalamus of the brain sends parasympathetic nerve impulses through the sacral segments of the spinal cord, which cause dilation of the arteries serving the clitoris and vestibular bulbs. This increased blood flow causes the erectile tissues to swell. In addition, the erectile tissues in the areola of the breasts become engorged.

Simultaneous with the erection of the clitoris and vestibular bulbs, the vagina expands and elongates to accommodate the erect penis of the male, and parasympathetic impulses cause the vestibular glands to secrete mucus near the vaginal orifice. The vestibular secretion moistens and lubricates the tissues of the vestibule, thus facilitating the penetration of the erect penis into the vagina during coitus. Mucus continues to be secreted during coitus so that the male and female genitalia do not become irritated as they would if the vagina became dry.

The position of the sensitive clitoris usually allows it to be stimulated during coitus. If stimulation of the clitoris is of sufficient intensity and duration, a woman will experience a culmination of pleasurable psychological and physiological release called *orgasm.*

Associated with orgasm is a rhythmic contraction of the muscles of the perineum and the muscular walls of the uterus and uterine tubes. These reflexive muscular actions are thought to aid the movement of spermatozoa through the female reproductive tract toward the upper end of a uterine tube where an ovum might be located.

Following an orgasm or completion of the sexual act, sympathetic impulses cause a reduction in arterial flow to the erectile tissues, and their size diminishes to that prior to sexual stimulation.

1. Describe the structure and position of the uterine tubes and explain how an ovum is transported through these tubes to the uterus.
2. Describe the structure and function of the three layers of the uterus and explain why the endometrium is subdivided into a stratum functionale and a stratum basale.
3. Describe the structures of the vagina and the vulva and explain how these structures change during sexual excitement and coitus.

---

orgasm: Gk. *orgasmos,* to swell; to become excited

Mammary Glands

Mammary glands are modified sweat glands composed of secretory alveoli and ducts. The glands develop in the female breasts at puberty and function in lactation.

Objective 12. Distinguish between the mammary glands and the breast and describe the structure of the mammary glands.

In structure, the **mammary glands,** located in the **breasts,** are modified sweat glands and are a part of the integumentary system. In function, however, these glands are associated with the reproductive system since they secrete milk for the nourishment of the young. The size and shape of the breasts vary considerably from person to person because of genetic differences, age, percentage of body fat, or pregnancy. At puberty, estrogen from the ovaries stimulates growth of the mammary glands and the deposition of adipose tissue within the breasts. Mammary glands hypertrophy in pregnant and lactating women and usually atrophy somewhat after menopause.

Structure of the Breast and the Mammary Glands

Each breast is positioned over ribs 2 through 6 and overlies the pectoralis major muscle, the pectoralis minor muscle, and portions of the serratus anterior and external abdominal oblique muscles (see figs. 9.22 and 21.16). The medial boundary of the breast is over the lateral margin of the sternum and the lateral margin of the breast is along the anterior border of the axilla. The **axillary process** of the breast extends upward and laterally toward the axilla where it comes into close relationship with the axillary vessels. This region of the breast is clinically significant because of the high incidence of breast cancer within the lymphatic drainage of the axillary process.

Each mammary gland is composed of 15 to 20 **lobes,** divided by adipose tissue. Each lobe has its own drainage pathway to the outside. The amount of adipose tissue determines the size and shape of the breast but has nothing to do with the ability of a woman to nurse. Each lobe is subdivided into **lobules** that contain the glandular **alveoli** (fig. 21.17). The alveoli secrete the milk of a lactating female. **Suspensory ligaments** between the lobules extend from the skin to the deep fascia overlying the pectoralis major muscle and support the breasts. The clustered alveoli secrete milk into a series of **mammary ducts,** which in turn converge to form **lactiferous** (*lak-tif′er-us*) **ducts** (fig. 21.16). The lumen of each lactiferous duct expands near the nipple to form a **lactiferous sinus,** where milk may be stored before it drains at the tip of the nipple.

The **nipple** is a cylindrical projection containing some erectile tissue. A circular pigmented **areola** (*ă-re′ŏ-lă*) surrounds the nipple. The surface of the areola may appear rough because of the sebaceous **areolar glands** close to the surface. The secretions of the areolar glands keep the nipple pliable. The color of the areola and nipple varies with the complexion of the woman and whether or not she is pregnant. During pregnancy, the areola becomes darker and enlarges somewhat, presumably to become more conspicuous to a nursing infant.

UNIT VII

Figure 21.16 The structure of the breast and mammary glands.
(*a*) A sagittal section and (*b*) an anterior view partially sectioned.

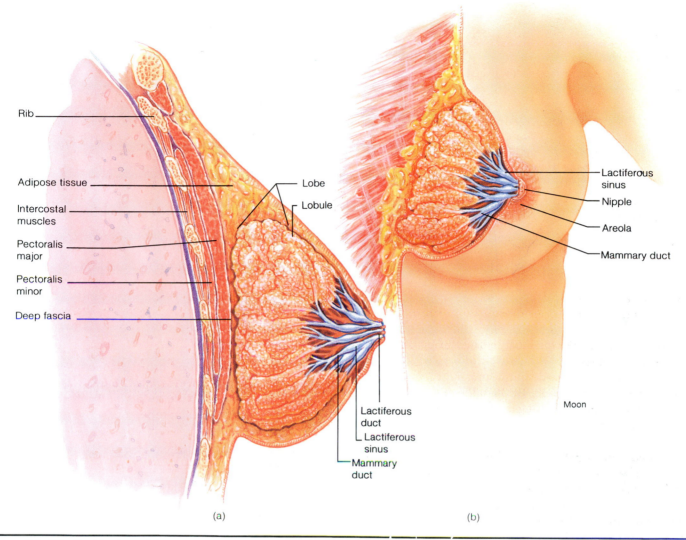

Rib

Adipose tissue

Intercostal muscles

Pectoralis major

Pectoralis minor

Deep fascia

Lobe

Lobule

Lactiferous duct

Lactiferous sinus

Mammary duct

Lactiferous sinus

Nipple

Areola

Mammary duct

Moon

(a)

(b)

Figure 21.17 The histology of the mammary gland. (*a*) Nonlactating (63×) and (*b*) lactating (63×).

Secretory tubules

Interlobular dense connective tissue

Alveoli with secretions

(a)

(b)

UNIT VII

Figure 21.18   Lymphatic drainage of the mammary gland.

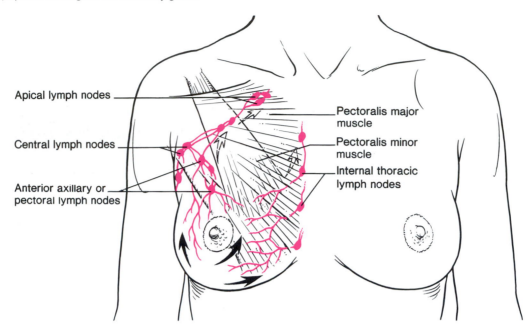

Apical lymph nodes

Central lymph nodes

Anterior axillary or
pectoral lymph nodes

Pectoralis major
muscle

Pectoralis minor
muscle

Internal thoracic
lymph nodes

Blood is supplied to the mammary gland through the perforating branches of the internal thoracic artery, which enter the breast through the second, third, and fourth intercostal spaces just lateral to the sternum, and through the more superficial mammary artery, branching from the lateral thoracic artery. Venous return is through a series of veins that parallel the pattern of the arteries. A superficial venous plexus may be apparent through the skin of the breast, especially during pregnancy and lactation.

The breast is innervated primarily through sensory somatic neurons that are derived from the anterior and lateral cutaneous branches of the fourth, fifth, and sixth thoracic nerves. Sensory nerve endings in the nipple and areola are especially important in stimulating the release of milk from the mammary glands to a suckling infant (see chapter 14).

Lymphatic drainage and the location of lymph nodes within the breast are of considerable clinical importance because of the frequency of *breast cancer* and the high incidence of metastases. About 75% of the lymph drains through the axillary process of the breast into the axillary lymph nodes (fig. 21.18). Some 20% of the lymph passes toward the sternum to the internal thoracic lymph nodes. The remaining 5% of the lymph is subcutaneous and follows the lymph drainage pathway in the skin toward the back, where it reaches the intercostal nodes near the neck of the ribs.

1. Describe the structure of the breasts and mammary glands and explain why variations in breast size do not affect the ability to lactate.
2. List in order the parts of the mammary glands through which milk passes during lactation.

Ovulation and Menstruation

Ovulation and menstruation are reproductive cyclic events that are regulated by follicle-stimulating hormone (FSH) and luteinizing hormone (LH), secreted by the anterior pituitary gland, and by estrogen and progesterone, secreted by structures of the ovaries.

Objective 13.   Describe the hormonal changes that occur during the cycles of ovulation and menstruation.

Objective 14.   Describe the structural changes that occur in the endometrium during the cycles of ovulation and menstruation and explain how these changes are controlled by hormones.

Both ovulation and menstruation are reproductive functions of sexually mature females and are largely regulated by hormones secreted by the anterior pituitary gland and the ovaries. Both occur approximately every 28 days, as menstruation follows ovulation and is regulated by the hormonal activity of the ovaries.

In ovulation, the follicular wall of the ovary ruptures, and a secondary oocyte is released and passes into the uterine tube. Ovulation generally occurs from alternate ovaries. If fertilization occurs in the uterine tube, mitotic divisions are initiated and the blastocyst implants on the uterine wall (see chapter 22).

If the egg is not fertilized, the menstrual cycle is initiated usually 14 days after ovulation. The cycle is divided into three phases (fig. 21.19):

1. **Menstrual phase (menstruation).** The menstrual phase is characterized by a bloody discharge of endometrial

Figure 21.19 The cycle of ovulation and menstruation.

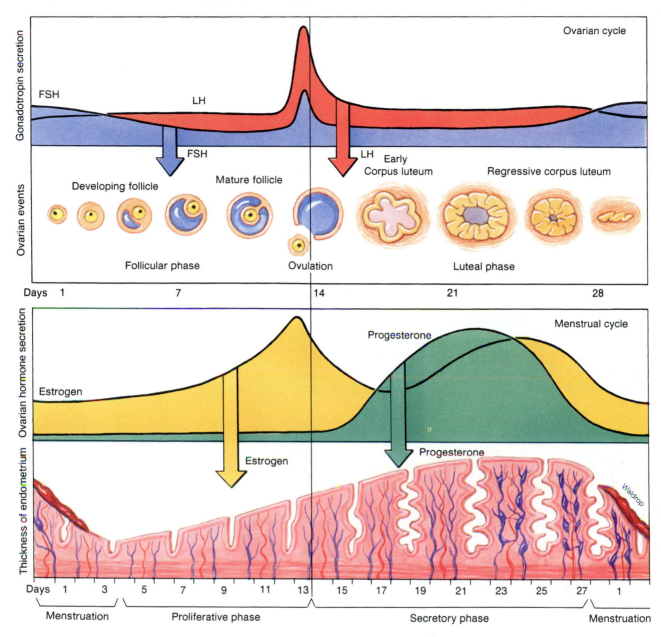

tissue from the uterus during the first 3 to 5 days of the cycle.

2. **Proliferative phase.** During the proliferative phase, days 5 to 14 of the cycle, endometrial tissue regrows.

3. **Secretory phase.** The secretory phase, lasting from day 14 to day 28, is characterized by an increase in glandular secretions and blood to the endometrium in preparation for nourishing a blastocyst. The last 2 or 3 days of the secretory phase may be characterized by cramping and external spotting of blood. This period of the cycle is sometimes referred to as the *premenstrual phase* (not shown in figure 21.19) and involves the initial breakdown of the endometrial lining.

The controlling center for ovulation and menstruation is the hypothalamus. On a regular cycle, the hypothalamus releases a *gonadotropin-releasing factor* (GRF), which in turn stimulates the anterior pituitary gland to secrete *follicle-stimulating hormone* (FSH) and *luteinizing hormone* (LH) at the appropriate time. The FSH stimulates the maturation of a follicle within an ovary. During the middle of the menstrual cycle, the anterior pituitary releases a relatively large quantity of LH and an increased amount of FSH. This increased concentration of hormones causes the mature follicle to swell rapidly and rupture. Ovulation is complete as the oocyte is discharged, along with its follicular fluid, toward the uterine tube.

UNIT VII

Although there are several different female sex hormones, they all belong to two major groups that are referred to as *estrogen* and *progesterone*. The principal source of estrogen (in a nonpregnant female) is the ovaries. Estrogen, as associated with the menstrual cycle, causes the endometrium to thicken. Estrogen also plays an important role in the development and maintenance of the secondary sex organs and secondary sex characteristics. Progesterone is secreted by the ovaries also (in a nonpregnant female) and helps estrogen maintain the endometrium.

The principal events of ovulation and menstruation are outlined in table 21.2.

---

1. Define *ovulation*. Define *menstruation*.
2. Explain the role of GRF in regulating female reproductive functions.
3. Diagram the relative thickness of the endometrium during the three phases of the menstrual cycle.
4. Summarize the major events in a menstrual cycle.

Clinical Considerations

Females are more prone to dysfunctions and diseases of the reproductive organs than are males because of cyclic changes in reproductive events, problems associated with pregnancy, and the susceptibility of the female breasts to infections and neoplasms. The termination of reproductive capabilities at menopause can also cause complications due to hormonal alterations. *Gynecology* is the specialty of medicine concerned with dysfunction and diseases of the female reproductive system, whereas *obstetrics* is the specialty dealing with pregnancy and childbirth. Frequently a physician will specialize in both obstetrics and gynecology (OBGYN).

A comprehensive discussion of the numerous clinical aspects of the female reproductive system is beyond the scope of this text; thus, only the most important conditions will be addressed in the following sections. Coverage includes diagnostic procedures, developmental abnormalities, problems involving the ovaries and uterine tubes, problems involving the uterus, diseases of the vagina and vulva, and diseases of the breasts and mammary glands. In addition, the more popular methods of birth control are described.

Diagnostic Procedures

A gynecological, or pelvic, examination is generally included in a thorough physical examination, especially prior to marriage, during pregnancy, or if problems involving the reproductive organs are suspected. In a gynecological examination, the physician inspects the vulva for irritations, lesions, or abnormal vaginal discharge and palpates the vulva and internal organs. Most of the internal organs can be palpated through the vagina, especially if they are enlarged or tender. Inserting a lubricated *speculum* into the vagina allows visual examination of the cervix and vaginal walls. A speculum is an instrument for opening or distending a body opening to permit visual inspection.

Table 21.2	**Principal Events in an Ovulation and Menstruation Cycle**

1. The hypothalamus releases GRF, which stimulates the anterior pituitary.
2. The anterior pituitary gland secretes FSH.
3. FSH stimulates the maturation of a follicle.
4. Follicular cells produce and secrete estrogen.
   (a) Estrogen maintains secondary sexual traits.
   (b) Estrogen causes the uterine lining to thicken.
5. The anterior pituitary gland secretes relatively large amounts of LH and FSH, which stimulate ovulation.
6. Follicular cells become corpus luteum cells, which secrete estrogen and progesterone.
   (a) Estrogen continues to stimulate the uterine wall development.
   (b) Progesterone stimulates the uterine lining to become more glandular and vascular.
   (c) Estrogen and progesterone inhibit the secretion of FSH and LH from the anterior pituitary gland.
7. If the oocyte is not fertilized, the corpus luteum degenerates.
8. As concentrations of estrogen and progesterone decline, blood vessels in the uterine lining constrict.
9. The uterine lining disintegrates and sloughs away, producing menstrual flow.
10. The anterior pituitary gland, which is no longer inhibited, again secretes FSH.
11. The cycle is repeated.

In special cases, it may be necessary to examine the cavities of the uterus and uterine tubes by *hysterosalpingography (his″ter-o-sal″ping-gog′rǎ-fe)* (fig. 21.20). This technique involves injecting a radiopaque dye into the reproductive tract. The patency of the uterine tubes, irregular pregnancies, and various types of tumors may be detected using this technique. A *laparoscopy (lap″ǎ-ros′kŏ-pe)* permits in vivo visualization of the internal reproductive organs. The entrance for the laparoscope may be the umbilicus, a small incision in the lower abdominal wall, or through the posterior fornix of the vagina into the rectouterine pouch. Although a laparoscope is used primarily in diagnosis, it can be used when performing a **tubal ligation** (fig. 21.21), which is a method of sterilizing a female by tying off the uterine tubes.

One diagnostic procedure that should be routinely performed by a woman is a *breast self-examination* (BSE). The importance of a BSE is not to prevent diseases of the breast but to detect any problems before they become life-threatening. One in nine women will develop breast cancer during her lifetime. Early detection of breast cancer and follow-up medical treatment minimizes the necessary surgical treatment and increases the patient's prognosis. Breast cancer is curable if it is caught early.

A woman should examine her breasts monthly. If she has not yet reached menopause, the ideal time for a BSE is 1 week after her period ends because the breasts are less likely to be swollen and tender at that time. A woman no longer menstruating should just pick any day of the month and do a BSE on that same day on a monthly basis. Visual inspection and

Developmental Exposition

The Female Reproductive System

Exhibit I   The development of the female genital tract. (*a*) A lateral view and (*b*) an anterior view.

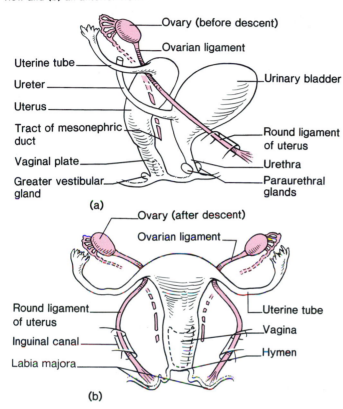

(a)

(b)

Table 21.3	**Homologous Reproductive Organs and the Undifferentiated Structures from Which They Develop**	
Indifferent Stage	**Male**	**Female**
Gonads	Testes	Ovaries
Urogenital groove	Membranous urethra	Vaginal vestibule
Genital tubercle	Glans penis	Clitoris
Urethral folds	Spongy urethra	Labia minora
Labioserotal folds	Scrotum	Labia majora
	Bulbourethral glands	Vestibular glands

Explanation

Although the genetic sex is determined at fertilization, both sexes develop similarly through the indifferent stage of the eighth week. The gonads of both sexes develop from the **gonadal ridges** medial to the mesonephros. **Primary sex cords** form within the gonadal ridges during the sixth week. The **genital tubercle** also develops during the sixth week as an external swelling cephalic to the cloacal membranes.

The ovaries develop more slowly than do the testes. Ovarian development begins at about the tenth week when **primordial follicles** begin to form within the ovarian medulla. Each of the primordial follicles consists of an **oogonium** (*o''ō-go'ne-um*) surrounded by a layer of **follicular cells.** Mitosis of the oogonia occurs during fetal development, so that thousands of germ cells are formed. Unlike the male reproductive system, in which spermatogonia are formed by mitosis throughout life, all oogonia are formed prenatally and their number continuously decreases after birth.

The **genital tract** includes the uterus and the uterine tubes. These organs develop from a pair of embryonic tubes called the

paramesonephric (müllerian) **ducts.** The paramesonephric ducts form to the sides of the mesonephric ducts that develop into the kidneys. As the mesonephric ducts regress, the paramesonephric ducts develop into the female genital tract (exhibit I). The lower portion of both ducts fuse to form the uterus. The upper portions remain unfused and give rise to the uterine tubes.

The epithelial lining of the vagina develops from the endoderm of the urogenital sinus. A thin membrane called the **hymen** forms to separate the lumen of the vagina from the urethral sinus. The hymen usually is perforated during later fetal development.

The external genitalia of both sexes appear the same during the indifferent stage of the eighth week (see page 680). A prominent **phallus** (*fal'us*) forms from the genital tubercle, and a **urethral groove** forms on the ventral side of the phallus. Paired **urogenital folds** surround the urethral groove on the lateral sides. In the male embryo these indifferent structures become masculinized by testosterone secreted by the testes. In the female embryo, in the absence of testosterone, feminization occurs.

In the process of feminization, the size of the phallus is inhibited and forms the relatively small **clitoris,** the urethral folds remain unfused to form an inner **labia minora,** and the labioscrotal swellings remain unfused to form the prominent **labia majora** (table 21.3). The external genitalia of a female are completely formed by the end of the twelfth week.

There is tremendous individual variation in the structure of the hymen. The hymen of a baby girl may be absent, or it may partially, or occasionally completely, cover the vaginal orifice, in which case it is called an *imperforate hymen.* An imperforate hymen is usually not detected until the first menstruation (menarche) when the discharge cannot be expelled. If the hymen is present, it may be ruptured during childhood in the course of normal exercise. On the other hand, a hymen may be so elastic that it persists even after coitus. Therefore, the presence of a hymen is not a reliable sign of virginity.

---

primordial: L. *prima,* first; *ordior,* to begin
follicle: L. diminutive of *follis,* bag

hymen: Gk. (mythology) Hymen was god of marriage; *hymen,* thin skin or membrane

UNIT VII

Figure 21.20 A hysterosalpingogram showing the cavity of the uterus and lumina of the uterine tubes.

Lumen of uterine tube

"Spill" of radiopaque medium into peritoneal cavity

Cavity of uterus

Application tube for hysterosalpingogram

Figure 21.21 Tubal ligation involves removal of a portion of each uterine tube.

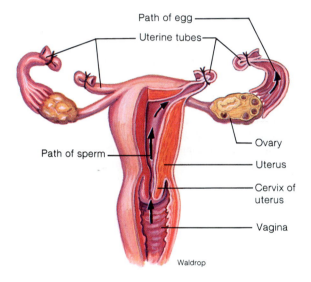

Path of egg

Uterine tubes

Path of sperm

Ovary

Uterus

Cervix of uterus

Vagina

Waldrop

palpation (fig. 21.22) are equally important in doing a BSE. The steps involved in this procedure are the following:

1. **Observation before a mirror.** Inspect the breasts with the arms at the sides. Next, raise the arms high overhead. Look for any changes in the contour of each breast—a swelling, dimpling of skin or changes in the nipple. The left and right breast will not match exactly—few women's breasts do. Finally, squeeze the nipple of each breast gently between the thumb and index finger to check for discharge. Any discharge from the nipple should be reported to a physician.

2. **Palpation during bathing.** Examine the breasts during a bath or shower when the hands will glide easily over wet skin. With the fingers flat, move gently over every part of each breast. Use the right hand to examine the left breast, and the left hand for the right breast. Palpate for any lump, hard knot, or thickening. If the breasts are

normally fibrous or lumpy (fibrocystic tissue), the locations of these lumps should be noted and checked each month for changes in size and locations.

3. **Palpation while lying down.** To examine the right breast, put a pillow or folded towel under the right shoulder. Place the right hand behind the head—this distributes the breast tissue more evenly on the rib cage. With the fingers of the left hand held flat, press gently in small circular motions around an imaginary clock face. Begin at outermost top of the right breast for 12 o'clock, then move to 1 o'clock, and so on around the circle back to 12 o'clock. A ridge of firm tissue in the lower curve of each breast is normal. Then move in an inch, toward the nipple, and keep circling to examine every part of the breast, including the nipple. Also examine the armpit carefully for enlarged lymph nodes. Repeat the procedure on the left breast.

Another important diagnostic procedure is a *Papanicolaou (Pap) smear.* The Pap smear permits a microscopic examination of cells obtained near the uterine ostium of the cervix. Samples of cells are obtained by using a vaginal pipette or a specially designed wooden spatula. Women should have routine Pap smears for the early detection of cervical cancer.

Developmental Abnormalities

Many of the developmental abnormalities of the female reproductive system also occur in the reproductive system of the male and were discussed in the previous chapter. Hermaphroditism and irregularities of the sex chromosomes, for example, are developmental conditions that cause a person to develop both male and female characteristics.

Other developmental abnormalities of the female reproductive system may occur during the formation of the uterus and vagina. Failure of the paramesonephric ducts to fuse normally can result in a **double uterus,** a **bicornuate** (bi-kor′nyoo-āt) **uterus,** or a **unicornuate uterus.** These types of uterine

Pap smear: from George N. Papanicolaou, American anatomist and
     physician, 1883–1962
bicornuate: L. *bi,* two; *cornu,* horn

Figure 21.22 The technique of conducting a breast self-examination (BSE) involves (*a*) visual inspection and (*b*) palpation. The sites and incidence of occurrence of breast cancer are shown in (*c*).

(a)

(b)

Creek

(c) Sites and incidence of occurrence of breast cancer.

abnormalities and others are diagrammed in figure 21.23. In about 1 in 4,000 females, the vagina is absent. This is usually accompanied by an absence of the uterus as well.

Problems Involving the Ovaries and Uterine Tubes

Most **ovarian neoplasms** are nonmalignant ovarian cysts lined by cuboidal epithelium and filled with a serous albuminous fluid. These tumors may frequently be palpated during a gynecological examination and may require surgical removal if they exceed about 4 cm in diameter. They are generally removed as a precaution because it is impossible to determine by palpation whether the mass is malignant or benign.

Malignant ovarian tumors, which are most likely to occur in women over the age of 60, may reach massive size. Ovarian tumors of 5 kg (14 lbs) are not uncommon and ovarian tumors of 110 kg (300 lbs) have been reported. Some ovarian

tumors produce estrogen and thus cause feminization in elderly women, including recommencement of menstrual periods. The prognosis for women with ovarian tumors varies depending on the type of tumor and whether or not metastasis has occurred.

Two frequent problems involving the uterine tubes are salpingitis and ectopic pregnancies. **Salpingitis** is an inflammation of one or both uterine tubes. Infection of the uterine tubes is generally caused by a sexually transmitted disease, although secondary bacterial infections from the vagina may also cause salpingitis. Salpingitis may cause sterility if the uterine tubes become occluded.

Ectopic pregnancy results from implantation of the blastocyst in a location other than the body or fundus of the uterus. The most frequent ectopic site is in the uterine tube, where an implanted blastocyst causes what is commonly called a **tubular pregnancy.** One danger of a tubular pregnancy is the enlargement, rupture, and subsequent hemorrhage of the uterine tube where implantation has occurred. A tubular pregnancy is frequently treated by removing the affected tube.

---

neoplasm: Gk. *neos*, new; *plasma*, something formed

UNIT VII

Figure 21.23 Congenital uterine abnormalities. (*a*) A double uterus and double vagina, (*b*) a double uterus with a single vagina, (*c*) a bicornuate uterus, (*d*) a bicornuate uterus with a rudimentary left horn, (*e*) a septate uterus, and (*f*) a unicornuate uterus.

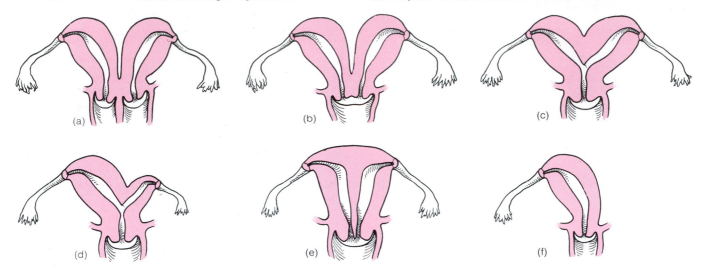

Infertility, or the inability to conceive, is a clinical problem that may involve the male or female reproductive system. Based on the number of people who seek help for this problem, it is estimated that 10% to 15% of couples have impaired fertility. Generally, when a male is infertile, it is because of inadequate sperm counts. Female infertility is frequently caused by an obstruction of the uterine tubes or abnormal ovulation.

Problems Involving the Uterus

Abnormal menstruations are among the most common disorders of the female reproductive system. Abnormal menstruations may be directly related to problems of the reproductive organs and pituitary gland or associated with emotional and psychological stress.

Amenorrhea *(a-men-o-re-ă)* is the absence of menstruation and can be categorized as normal, primary, or secondary. **Normal amenorrhea** follows menopause, occurs during pregnancy, and in some women may occur during lactation. **Primary amenorrhea** is the failure to have menstruated by the age when menstruation normally begins. Primary amenorrhea is generally accompanied by failure of the secondary sex characteristics to develop. Endocrine disorders may cause primary amenorrhea and abnormal development of the ovaries or uterus.

Secondary amenorrhea is the cessation of menstruation in women who previously have had normal menstrual periods and who are not pregnant and have not gone through menopause. Various endocrine disturbances and psychological factors may cause secondary amenorrhea. It is not uncommon, for example, for young women who are in the process of making major changes or adjustments in their lives to miss menstrual periods. Secondary amenorrhea is also frequent in women

athletes during periods of intense training. A low percentage of body fat may be a contributing factor. Sickness, fatigue, poor nutrition, or emotional stress also may cause secondary amenorrhea.

Dysmenorrhea is painful or difficult menstruation accompanied by severe menstrual cramps. The causes of dysmenorrhea are not totally understood but may include endocrine disturbances (inadequate progesterone levels), a faulty position of the uterus, emotional stress, or some type of obstruction that prohibits menstrual discharge.

Abnormal uterine bleeding includes **menorrhagia** *(men''ŏ-ra'je-ă),* or excessive bleeding during the menstrual period, and **metrorrhagia,** or spotting between menstrual periods. Other types of abnormal uterine bleeding are menstruations of excessive duration, too-frequent menstruations, and postmenopausal bleeding. These abnormalities may be caused by hormonal irregularities, emotional factors, or various diseases and physical conditions.

Uterine neoplasms are an extremely common problem of the female reproductive tract. Most of the neoplasms are benign and include cysts, polyps, and smooth muscle tumors (leiomyoma). Any of these conditions may provoke irregular menstruations and may cause infertility if the neoplasms are massive.

Cancer of the uterus is the most common malignancy of the female reproductive tract. The most common site of uterine cancer is in the cervix (fig. 21.24). Cervical cancer is second only to cancer of the breast in frequency of occurrence and is a disease of relatively young women (ages 30 through 50), especially those who have had frequent coitus with multiple partners during their teens and onward. If detected early through regular Pap smears, the disease can be cured before it metasta-

UNIT VII

Figure 21.24 Sites of various conditions and diseases of the female reproductive tract, each of which could cause an abnormal discharge of blood.

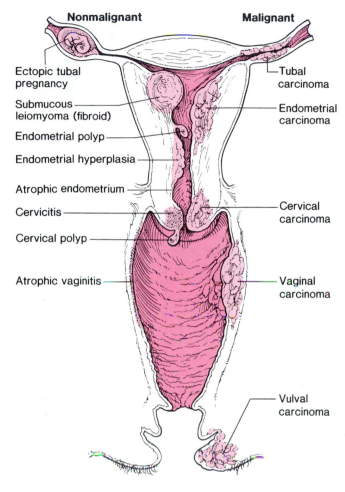

Nonmalignant

- Ectopic tubal pregnancy
- Submuous leiomyoma (fibroid)
- Endometrial polyp
- Endometrial hyperplasia
- Atrophic endometrium
- Cervicitis
- Cervical polyp
- Atrophic vaginitis

Malignant

- Tubal carcinoma
- Endometrial carcinoma
- Cervical carcinoma
- Vaginal carcinoma
- Vulval carcinoma

sizes. The treatment of cervical cancer depends on the stage of the malignancy and the age and health of the woman. In the case of women who are through having children, a **hysterectomy** (surgical removal of the uterus) is usually performed.

Endometriosis is a condition characterized by the presence of endometrial tissues at sites other than the inner lining of the uterus. Frequent sites of ectopic endometrial cells are on the ovaries, outer layer of the uterus, abdominal wall, and urinary bladder. Although it is not certain how endometrial cells become established outside the uterus, it is speculated that some discharged endometrial tissue might be flushed backward from the uterus and through the uterine tubes during menstruation. Women with endometriosis will bleed internally with each menstrual period because the ectopic endometrial cells are stimulated along with the normal endometrium by ovarian hormones. The most common symptoms of endometriosis are extreme dysmenorrhea and a feeling of fullness during each

menstruation period. Endometriosis can cause infertility. It is treated by suppressing the endometrial tissues with oral contraceptive pills or by surgery. An oophorectomy, or removal of the ovaries, may be necessary in extreme cases.

Uterine displacements are relatively common in elderly women. When uterine displacements occur in younger women, they are important because of dysmenorrhea, fertility, or parturition problems. **Retroversion,** or **retroflexion,** is a displacement backward; **anteversion,** or **anteflexion,** is displacement forward. **Prolapse of the uterus** is a marked downward displacement into the vagina.

An **abortion** is defined as the termination of a pregnancy before the twenty-eighth week of gestation. A **spontaneous abortion,** or **miscarriage,** occurs without mechanical aid or medicinal intervention and may happen in as many as 10% of all pregnancies. Spontaneous abortions usually occur when there is abnormal development of the fetus or disease of the maternal reproductive system. An **induced abortion** is removal of the fetus from the uterus by mechanical means or drugs. Induced abortions are a major issue of controversy because of questions regarding individual rights—those of the mother and those of the fetus—the definition of life, and morality.

Diseases of the Vagina and Vulva

Pelvic inflammatory disease (PID) is a general term for inflammation of the female reproductive organs within the pelvis. The infection may be confined to a single organ, or it may involve all of the internal reproductive organs. The pathogens generally enter through the vagina during coitus, induced abortion, childbirth, or postpartum. Inflammation of the ovaries is called **oophoritis** and inflammation of the uterine tube is **salpingitis** *(sal''pin-ji'tis)*.

The vagina and vulva are generally resistant to infection because of the acidity of the vaginal secretions. Occasionally, however, localized infections and inflammations do occur; these are termed **vaginitis,** if confined to the vagina, or **vulvovaginitis,** if both the vagina and external genitalia are affected. The symptoms of vaginitis are a discharge of pus *(leukorrhea)* and itching *(pruritus)*. The two most common organisms that cause vaginitis are the protozoan *Trichomonas vaginalis* and the fungus *Candida albicans.*

Diseases of the Breasts and Mammary Glands

The breasts and mammary glands of females are highly susceptible to infections, cysts, and tumors. Infections involving the mammary glands usually follow the development of a dry and cracked nipple during lactation. Bacteria enter the wound and establish an infection within the lobules of the gland. During an infection of the mammary gland, a blocked duct frequently causes a lobe to become engorged with milk. This localized swelling is usually accompanied by redness, pain, and

UNIT VII

Figure 21.25 (a) A mammogram of a patient with carcinoma of the breast. (Note the presence of a neoplasm indicated with an arrow.) (b) In mammography, the breasts are placed alternately on a metal plate and radiographed from the side and from above.

(a)

(b)

UNIT VII

an elevation of temperature. Administering specific antibiotics and applying heat are the usual treatments.

Nonmalignant cysts are the most frequent diseases of the breast. These masses are generally of two types, neither of which is life-threatening. **Dysplasia** (fibrocystic disease) is a broad condition involving several nonmalignant diseases of the breast. All dysplasias are benign neoplasms of various sizes that may become painful during or prior to menstruation. Most of the masses are small and remain undetected. Dysplasia affects nearly 50% of women over the age of 30 prior to menopause.

A **fibroadenoma** (fi″bro-ad″ĕ-no′mă) is a benign tumor of the breast that frequently occurs in women under the age of 35. Fibroadenomas are nontender, rubbery masses that are easily moved about in the mammary tissue. A fibroadenoma can be excised in a physician's office under local anesthetics.

Carcinoma of the breast is the most common malignancy in women (see fig. 10.59). One in nine women will develop breast cancer and one-third of these will die from the disease. Breast cancer is the leading cause of death in women between 40 and 50 years of age. Men are also susceptible to breast cancer, but it is 100 times more frequent in women. Breast cancer in men is usually fatal.

The causes of breast cancer are not known, but women who are most susceptible are those who are over age 35, who have a family history of breast cancer, and who are nulliparous (never having given birth). The early detection of breast cancer is important because the progressed state of the disease will determine the treatment and prognosis.

Confirming suspected breast cancer generally requires *mammography* (fig. 21.25). If the mammogram suggests breast cancer, surgery is performed so that a biopsy can be obtained and the tumor assessed. If the tumor is found to be malignant, extensive surgery is performed depending on the size of the tumor and whether metastasis has occurred. The surgical treatment for breast cancer is generally some degree of *mastectomy*. A *simple mastectomy* is removal of the entire breast but not the underlying lymph nodes. A *modified radical mastectomy* is the complete removal of the breast, the lymphatic drainage, and perhaps the pectoralis major muscle. A *radical mastectomy* is similar to a modified except that the pectoralis major muscle is always removed, as well as the axillary lymph nodes and adjacent connective tissue.

Methods of Contraception

The *rhythm method* of birth control is used by many people, but the popularity of this technique has declined as more successful methods of contraception have been introduced (fig. 21.26). In the rhythm method of birth control, the woman attempts to predict the day of her ovulation and restricts coitus to safe times of the cycle that allow no chance for fertilization to occur. The day of ovulation can be determined by a drop in basal body temperature or by a change in mucous discharge from the vagina. This technique has one of the highest failure rates of any of the widely used birth control methods, largely because women's cycles are often irregular.

More popular methods of birth control include (1) sterilization, (2) oral contraceptives, (3) intrauterine devices (IUDs), and (4) barrier methods—including condoms for the male and female and diaphragms for the female. All of these techniques are effective, but they vary with respect to safety, side effects, and efficacy.

Figure 21.26 Various types of birth control devices. (*a*) IUD, (*b*) contraceptive sponge, (*c*) diaphragm, (*d*) birth control pills, (*e*) vaginal spermicide, (*f*) condom, (*g*) female condom, and (*h*) Norplant.

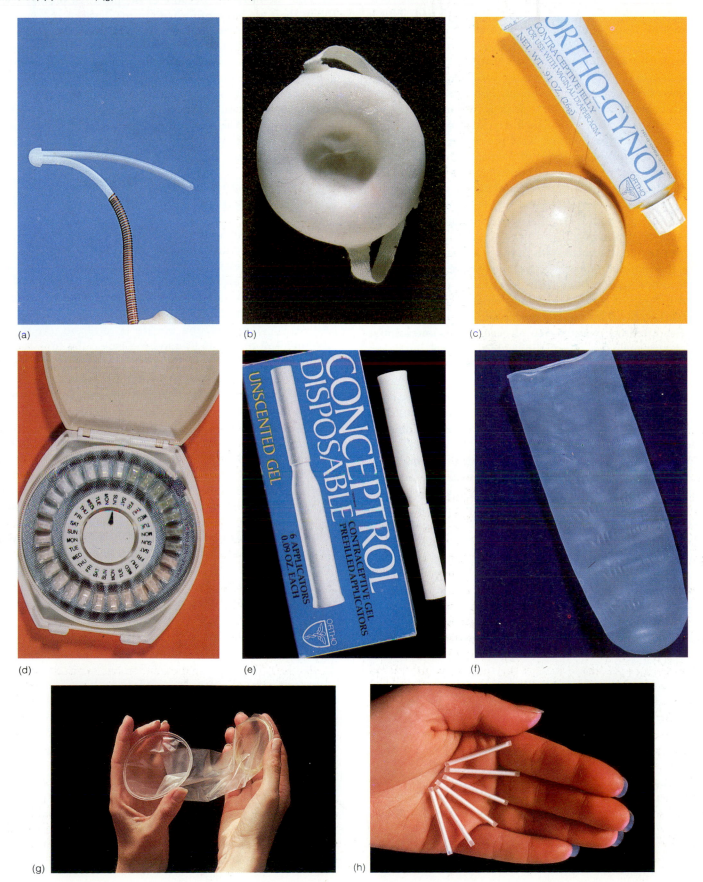

(a)

(b)

(c)

(d)

(e)

(f)

(g)

(h)

UNIT VII

Sterilization techniques include *vasectomy* for the male and *tubal ligation* for the female. In the latter technique (which accounts for over 60% of sterilization procedures performed in the United States), the uterine tubes are cut and tied. This is analogous to the procedure performed on the ductus deferentia in a vasectomy and prevents fertilization of the ovulated ovum.

About 10 million women in the United States and 60 million women in the world are currently using **oral steroid contraceptives.** These contraceptives usually consist of a synthetic estrogen combined with a synthetic progesterone in the form of pills, which are taken once each day for 3 weeks after the last day of a menstrual period. This procedure causes an immediate increase in blood levels of ovarian steroids (from the pill), which is maintained for the normal duration of a monthly cycle. As a result of *negative feedback inhibition* of gonadotrophin secretion, *ovulation never occurs.* The entire cycle is like a false luteal phase, with high levels of progesterone and estrogen and low levels of gonadotrophins.

Since the contraceptive pills contain ovarian steroid hormones, the endometrium proliferates and becomes secretory just as it does during a normal cycle. In order to prevent an abnormal growth of the endometrium, women stop taking the pill after 3 weeks. This causes estrogen and progesterone levels to fall, and permits menstruation to occur. The contraceptive pill is an extremely effective method of birth control, but it does have potentially serious side effects—including an increased incidence of thromboembolism and cardiovascular disorders. It has been pointed out, however, that the mortality risk of contraceptive pills is still much lower than the risk of death from the complications of pregnancy—or from automobile accidents.

Another way in which to deliver hormonal contraceptives to a woman's body is by means of a **subdermal implant.** Implants are 2-in. rods filled with a hormonal contraceptive drug and implanted just under the skin, usually on the upper arm, through a tiny incision. The contraceptive hormone gradually leaches out through the walls of the rod and enters the bloodstream, preventing pregnancy for at least 5 years.

Intrauterine devices (IUDs) don't prevent ovulation, but instead prevent implantation of the blastocyst into the uterus in the event that fertilization occurs. The mechanisms by which these contraceptive effects are produced are not well understood, but the efficiency of different IUDs appears to be related to their ability to cause inflammatory reactions in the uterus. Uterine perforations are the foremost complication associated with the use of IUDs. Because of the potential problems with IUDs, their use has diminished.

Barrier contraceptives—condoms, diaphragms, and cervical caps—are slightly less effective than the other methods of contraception, but they do not have serious side effects. Barrier contraceptives are quite effective, especially when they are used in conjunction with spermicidal foams, gels, and sponges. Many couples avoid them, however, because they detract from the spontaneity of sexual intercourse. Latex condoms offer an additional benefit; they provide some protection against sexually transmitted diseases, including AIDS.

CLINICAL CASE STUDY ANSWER

An ectopic pregnancy is any pregnancy that implants outside of the uterine cavity. This is most likely to occur in the uterine tube. The events leading up to our patient's problem are as follows: An oocyte is extruded from the ovary during ovulation and received into the uterine tube. Soon after arrival into the tube, the oocyte is fertilized, creating a zygote. The zygote is transported inside the tube toward the uterine cavity. Up to this point, the process is no different from that which occurs in a normal pregnancy. In the case of tubal pregnancy, however, the transporting ability of the uterine tube fails, causing the conceptus to be retained in the tube. Implantation then occurs within tissues that are not well-suited for that purpose. For example, the uterine tube does not expand well to accommodate a growing embryo, nor does it possess the necessary epithelium and glandular structures as does the endometrium. The result is overexpansion and erosion of the tubal wall, leading to rupture and hemorrhage. Because the uterine tube is basically exposed to the peritoneal cavity, blood from the site of rupture can flow into and collect in the area known as the *culdesae*, which is posterior to the dome of the uterus and vaginal fornix. There, it is easily aspirated by a needle placed through the vaginal wall.

CHAPTER SUMMARY

Introduction to the Female Reproductive System (pp. 690–692)

1. The reproductive period of a female is between puberty (about age 12) to menopause (about age 50). During this age span, cyclic ovulation and menstruation patterns occur when a female is not pregnant.
2. The functions of the female reproductive system are to produce ova; secrete sex hormones; receive sperm from the male; provide sites for fertilization implantation, and development of the embryo and fetus; facilitate parturition; and secrete milk from the mammary glands.
3. The female reproductive system consists of (1) primary sex organs—the ovaries; (2) secondary sex organs—those that are essential for sexual reproduction, characterized by latent development; and (3) secondary sex characteristics—features that are sexual attractants, expressed after puberty.

Structure and Function of the Ovaries (pp. 692–696)

1. The ovaries are supported by the mesovarium from the broad ligament, the ovarian ligament, and the suspensory ligament.
2. The ovarian follicles within the ovarian cortex of the ovary undergo cyclic changes.
   (a) Primary oocytes, arrested at prophase I of meiosis, are contained within primordial follicles.
   (b) Due to gonadotrophin stimulation, some of the primordial follicles enlarge to become primary follicles.
   (c) When a follicle develops a fluid-filled antrum, it is called a secondary follicle.
   (d) Generally only one follicle continues to grow to become a vesicular ovarian follicle.
   (e) The vesicular ovarian follicle contains a secondary oocyte arrested at metaphase II of meiosis.
   (f) In the process of ovulation, the vesicular ovarian follicle ruptures and releases its secondary oocyte, which becomes an ovum upon fertilization.
   (g) After ovulation, the empty follicle becomes a corpus luteum.

Secondary Sex Organs (pp. 696–702)

1. The uterine tube, which conveys ova from the ovary to the uterus, is the site for fertilization.
   (a) The open-ended portion of each uterine tube is expanded; its margin bears fimbriae that extend over the lateral surface of the ovary.
   (b) Movement of an ovum is aided by ciliated cells that line the lumen and by peristaltic contractions in the wall of the uterine tube.
2. The uterus is supported by the broad ligaments, uterosacral ligaments, cardinal ligaments, and round ligaments.
   (a) The endometrium consists of a stratum basale and a stratum functionale; the superficial stratum functionale is shed in menstruation.
   (b) The myometrium produces the muscular contractions needed for labor and parturition.
3. The vagina serves to receive the erect penis, to convey menses to the outside, and to transport the fetus during parturition.
   (a) The vaginal wall is composed of an inner mucosa, a middle muscularis, and an outer fibrous layer.
   (b) The vaginal orifice may be partially covered by a thin membranous hymen.
4. The external genitalia, or vulva, include the labia majora and minora, clitoris, vaginal vestibule, vestibular bulbs, and vestibular glands.
5. Impulses through parasympathetic nerves stimulate erectile tissues in the clitoris and vestibular bulbs; in orgasm, muscular contraction occurs in the perineum, uterus, and uterine tubes.

Mammary Glands (pp. 702–704)

1. Mammary glands are located within the breasts and are modified sweat glands.
   (a) Each mammary gland is composed of 15 to 20 lobes; the lobes are subdivided into lobules that contain alveoli.
   (b) During lactation, the alveoli secrete milk, which passes through mammary ducts, lactiferous ducts, and lactiferous sinuses and is discharged through the nipple.
2. The nipple is a cylindrical projection from the breast surrounded by the circular pigmented areola.

Ovulation and Menstruation (pp. 704–706)

1. Ovulation and menstruation are reproductive cyclic events that are regulated by hormones secreted by the hypothalamus, the anterior pituitary gland, and the ovaries.
2. Menstruation is divided into menstrual, proliferative, and secretory phases.
3. The principal hormones that regulate ovulation and menstruation are estrogen, progesterone, follicle-stimulating hormone (FSH), and luteinizing hormone (LH).

REVIEW ACTIVITIES

Objective Questions

1. The cervix is a portion of
   (a) the vulva.
   (b) the vagina.
   (c) the uterus.
   (d) the uterine tubes.
2. Fertilization normally occurs in
   (a) the ovary.
   (b) the uterine tube.
   (c) the uterus.
   (d) the vagina.
3. The secretory phase of the endometrium corresponds to which of the following ovarian phases?
   (a) the follicular phase
   (b) ovulation
   (c) the luteal phase
   (d) the menstrual phase
4. Which of the following statements about oogenesis is *true*?
   (a) Oogonia, like spermatogonia, form continuously during postnatal life.
   (b) Primary oocytes are haploid.
   (c) Meiosis is completed prior to ovulation.
   (d) In ovulation, secondary oocytes are released from vesicular ovarian follicles.
5. The function of the mesovarium is
   (a) movement of the sperm to the ova.
   (b) nourishment of the ovarian walls.
   (c) muscular contraction of the uterus.
   (d) suspension of the ovary.
6. Which of the following layers is shed as menses?
   (a) the perimetrial layer
   (b) the fibrous layer
   (c) the functionalis layer
   (d) the menstrual layer
7. The transverse folds in the mucosal layer of the vagina are called
   (a) perineal folds.
   (b) vaginal rugae.
   (c) fornices.
   (d) labia gyri.

UNIT VII

8. Which of the following is *not* a part of the vulva?
   (a) the mons pubis
   (b) the clitoris
   (c) the vaginal vestibule
   (d) the vagina
   (e) the labia minora
9. The paramesonephric (müllerian) ducts give rise to
   (a) the uterine tubes.
   (b) the uterus.
   (c) the pudendum.
   (d) both (a) and (b).
   (e) both (b) and (c).
10. In a female, the homologue of the male scrotum is/are
    (a) the labia majora.
    (b) the labia minora.
    (c) the clitoris.
    (d) the vestibule.

Essay Questions

1. Define *puberty, ovulation, menstruation,* and *menopause.*
2. List the homologous reproductive organs of the male and female reproductive systems and note the undifferentiated structures from which they develop.
3. Describe the follicular changes within the ovarian cortex of an ovary during the events that precede and follow ovulation.
4. Define *oogenesis.* When is the process initiated and when is it completed?
5. Describe the gross and histologic structure of the uterus and comment on the significance of the strata functionale and basale of the endometrium.
6. Identify the secondary sex organs and explain their functions.
7. Summarize the events in a menstrual cycle and explain the roles of estrogen and progesterone.
8. List the functions of the vagina and describe its structure.
9. Distinguish between the labia majora and labia minora, between the pudendal cleft and vaginal vestibule, and between the vestibular bulbs and vestibular glands.
10. List the events that cause an erection of the female genitalia, and explain the process of orgasm.
11. Describe the structure and position of the mammary glands in the breasts.
12. Define *gynecology, obstetrics, speculum, laparoscopy, breast self-examination,* and *Pap smear.*
13. Distinguish between normal, primary, and secondary amenorrhea. What causes each?
14. List the various kinds of uterine neoplasms. Why is it important to know about cancer of the uterus? How is it generally detected?
15. Distinguish between dysplasia, fibroadenoma, and carcinoma of the breast.

Gundy/Weber Software

The tutorial software accompanying Chapter 21, Female Reproductive System, is Volume 13, Reproductive System.

Developmental Anatomy, Postnatal Growth, and Inheritance

Chapter Outline

CLINICAL CASE STUDY

A 27-year-old woman gives birth to twin boys, followed by an apparently single placenta. After examining the two infants, the pediatrician informs the mother that one of her babies has a cleft palate but that the other is normal. She then adds that in some instances cleft palate is a genetic disorder and asks if any family members have had the problem. The mother says that she knows of none. Further examination of the placenta reveals two amnions and only one chorion.

Does the presence of two amnions and one chorion indicate monozygotic or dizygotic twins? How would you account for the fact that one baby has a cleft palate, while the other does not?

Hints: Carefully read the section on multiple pregnancy at the end of the chapter and study the examples presented in figure 22.40. Can any conclusions be drawn regarding genetic similarities between the two infants? Note that identical twins are always the same gender.

Fertilization

Upon fertilization of a secondary oocyte by a spermatozoon in the uterine tube, meiotic development is completed and a diploid zygote is formed.

Objective 1. Define *morphogenesis, capacitation,* and *fertilization.*

Objective 2. Describe the changes that occur in the spermatozoon and ovum prior to, during, and immediately following fertilization.

The structure of the human body forms before birth, or *prenatally,* through a process called **morphogenesis** (*mor''fo-jen'ĕ-sis*). Through morphogenic events, the organs and systems of the body are established in a functional relationship. A comprehensive view of the structure and function of each body system requires an understanding of morphogenesis. Associated with each organ and system are sensitive stages of morphogenesis during which genetic or environmental factors may affect the normal development of the baby. Many clinical problems are congenital in nature and originate during morphogenic development.

The prenatal development of a human is a fascinating and awesome event. It begins with a single fertilized egg and culminates some 38 weeks later with a complex organization composed of billions of cells. Prenatal development can be divided into a *pre-embryonic period,* initiated by the fertilization of a secondary oocyte; an *embryonic period,* during which the body's organ systems are formed; and a *fetal period,* culminating in parturition, or the birth of the baby.

During coitus (sexual intercourse), a male ejaculates between 100 million and 500 million sperm into the female's

vagina. This tremendous number is needed because of the high rate of sperm fatality—only about 100 survive to contact the secondary oocyte in the uterine tube. During passage through the acidic female tract, sperm gain the ability to fertilize a secondary oocyte through a process called **capacitation.** The changes occurring in sperm capacitation are not fully understood. Evidence indicates, however, that when placed in the acidic environment of the female reproductive tract, there is an enzymatic reaction resulting in a molecular alteration in the cell membrane of the sperm.

Experiments confirm that freshly ejaculated sperm are infertile and must be in the female reproductive tract for at least 7 hours before they can fertilize a secondary oocyte. During in vitro fertilization, sperm capacitation is induced artificially by treating the ejaculate with a solution of gamma globulin, free serum, follicular fluid, dextran, serum dialysate, and adrenal gland extract.

A woman usually ovulates one secondary oocyte a month, totaling approximately 400 during her reproductive years. Each ovulation releases a secondary oocyte arrested at metaphase of the second meiotic division. As the secondary oocyte enters the uterine tube, it is surrounded by a thin transparent layer of protein and polysaccharides, called the *zona pellucida,* and a layer of granulosa cells, called the *corona radiata* (fig. 22.1).

The head of each spermatozoon is capped by an organelle called an **acrosome** (figs. 22.1 and 22.2). The acrosome contains a trypsinlike protein-digesting enzyme and *hyaluronidase* (*hi''ă-loo-ron'ĭ-dās*), which digests hyaluronic acid, an important constituent of connective tissue. When a spermatozoon meets a secondary oocyte in the uterine tube, an *acrosomal reaction* occurs that exposes the acrosome's digestive enzymes and allows the spermatozoon to penetrate through the corona radiata and the zona pellucida.

As a spermatozoon penetrates the zona pellucida, a chemical change in the zona prevents other sperm from entering. Only one spermatozoon, therefore, is allowed to fertilize a secondary oocyte. When the cell membranes of the spermatozoon and the secondary oocyte merge, the secondary oocyte becomes an **ovum.** As fertilization occurs, the ovum is stimulated to complete its second meiotic division (fig. 22.3). Like the first meiotic division, the second produces one cell that contains all of the cytoplasm and one *polar body.* The healthy cell is the mature ovum, and the second polar body, like the first, ultimately fragments and disintegrates.

At fertilization, the entire spermatozoon enters the cytoplasm of the much larger ovum (fig. 22.4). Within 12 hours, the nuclear membrane in the ovum disappears, and the *haploid number* of chromosomes (23) in the ovum is joined by the haploid number of chromosomes from the spermatozoon. A fertilized egg, or zygote (*zi'gōt*), containing the *diploid number* of chromosomes (46) is thus formed.

morphogenesis: Gk. *morphe,* form; *genesis,* beginning

capacitation: L. *capacitas,* capable of
zona pellucida: L. *zone,* a girdle; *pellis,* skin
corona radiata: Gk. *korone,* crown; *radiata,* radiate
acrosome: Gk. *akron,* extremity; *soma,* body
haploid: Gk. *haplous,* single; L. *ploideus,* multiple in form
diploid: Gk. *diplous,* double; L. *ploideus,* multiple in form

Figure 22.1 The process of fertilization. (*a,b*) Diagrammatic representations. As the head of the sperm encounters the gelatinous corona radiata of the egg, which has progressed in meiotic development to the secondary oocyte stage (2), the acrosomal vesicle ruptures and the sperm digests a path for itself by the action of enzymes released from the acrosome (3, 4). When the cell membrane of the sperm contacts the cell membrane of the egg (5), they become continuous, and the sperm nucleus and other contents move into the egg cytoplasm. (*c*) A scanning electron micrograph of sperm bound to the egg surface.

First polar body
Perivitelline space
Cytoplasm of secondary oocyte
(a)

Corona radiata
Second meiotic spindle
Zona pellucida
Cell membrane of secondary oocyte

Nucleus containing chromosomes
Acrosome containing enzymes
Perforations in acrosome wall

(b)

A secondary oocyte that is ovulated but not fertilized does not complete its second meiotic division, but instead it disintegrates 12 to 24 hours after ovulation. Fertilization therefore cannot occur if coitus takes place beyond 1 day following ovulation. Sperm, by contrast, can survive up to 3 days in the female reproductive tract. Fertilization therefore can occur if coitus takes place within 3 days prior to the day of ovulation.

(c)

1. Explain why capacitation and the acrosomal reaction of spermatozoa are necessary to accomplish fertilization of a secondary oocyte.
2. Define *morphogenesis* and discuss its importance in the development of a healthy baby.
3. Discuss the changes that occur in a spermatozoon from the time of ejaculation to the time of fertilization. What changes occur in a secondary oocyte following ovulation to the time of fertilization?

Pre-embryonic Period

The events of the 2-week pre-embryonic period include fertilization, transportation of the zygote through the uterine tube, mitotic divisions, implantation, and the formation of primordial embryonic tissue.

Objective 3. Describe the events of pre-embryonic development that result in the formation of the blastocyst.

Objective 4. Discuss the role of the trophoblast in the implantation and development of the placenta.

Objective 5. Explain how the primary germ layers develop and list the structures produced by each layer.

Objective 6. Define *gestation* and explain how the parturition date is determined.

UNIT VII

Figure 22.2 A transmission electron micrograph showing the head of a human spermatozoon with its nucleus and acrosome.

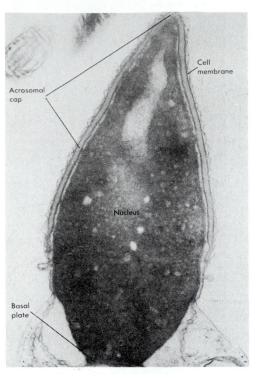

Cleavage and Formation of the Blastocyst

Fertilization occurs within the uterine tube, usually about 12 to 24 hours following ovulation. The fertilized egg (ovum) is referred to as a **zygote.** Within 30 hours, the **cleavage** process begins with a mitotic division that results in the formation of two identical daughter cells called *blastomeres* (fig. 22.5). Several more cleavages occur as the structure passes down the uterine tube and enters the uterus on about the third day. It is now composed of a ball of 16 or more cells called a **morula.** Although the morula has undergone several mitotic divisions, it is not much larger than the zygote because no additional nutrients necessary for growth have been entering the cells.

The developing structure remains unattached in the uterine cavity for about 3 days. During this time, the center of the morula fills with fluid passing in from the uterine cavity. As the fluid-filled space develops inside the morula, two distinct groups of cells form, and the structure becomes known as a **blastocyst.** The single outer layer forming the wall of the blastocyst is known as the **trophoblast,** whereas the small, inner aggregation of cells is called the **embryoblast,** or *internal cell mass.* With further development, the trophoblast differentiates into a structure called the **chorion,** which will become a portion of the placenta, and the embryoblast will become the embryo. The hollow, fluid-filled center of the blastocyst is called the **blastocyst cavity.** A diagrammatic summary of the ovarian cycle, fertilization, and the morphogenic events of the first week is presented in figure 22.6.

zygote: Gk. *zygotos*, yolked, joined

Figure 22.3 A secondary oocyte, arrested at metaphase II of meiosis, is released at ovulation. If this cell is fertilized, it becomes an ovum, completes its second meiotic division, and produces a second polar body.

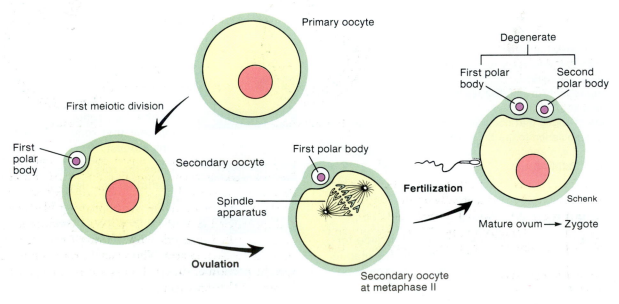

UNIT VII

Figure 22.4 Fertilization and the union of chromosomes from the sperm cell and ovum to form the zygote. (a) Sperm penetration, (b) the haploid number of chromosomes within the nucleus of each sex cell, (c) degeneration of nuclear membranes, and (d) matching and alignment of chromosomes.

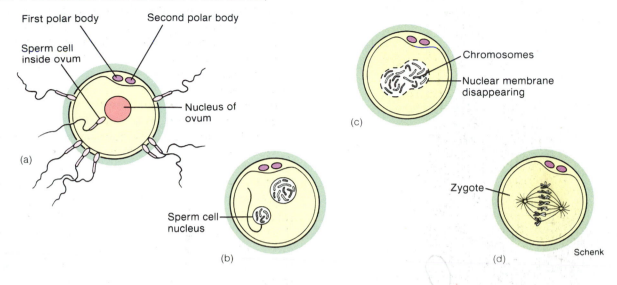

Figure 22.5 Sequential illustrations from the first cleavage of the zygote to the formation of the blastocyst. (Note the deterioration of the zona pellucida in the early blastocyst.)

Implantation

The process of **implantation,** or *nidation,* begins between the fifth and seventh day. Attachment is usually upon the posterior wall of the body of the uterus, with the side containing the embryoblast against the endometrium (fig. 22.7). Implantation is made possible by the secretion of *proteolytic enzymes* by the trophoblast, which digest a portion of the endometrium. The blastula sinks into the depression and endometrial cells move back to cover the defect in the wall. At the same time, the part of the uterine wall below the implanting blastocyst thickens, and specialized cells of the trophoblast produce fingerlike projections, called **syncytiotrophoblasts** *(sin-sit″e-ō-trof′ō-blasts),* into the thickened area. The syncytiotrophoblasts arise from a specific portion of the trophoblast called the **cytotrophoblast,** located at the embryonic pole.

implantation: L. *im,* in; *planto,* to plant
nidation: L. *nidus,* nest

Figure 22.6 A diagrammatic representation of the ovarian cycle, fertilization, and the morphogenic events of the 1st week. Implantation of the blastocyst begins between the 5th and 7th day and is generally completed by the 10th day.

The blastocyst saves itself from being aborted by secreting a hormone that indirectly prevents menstruation. Even before the sixth day when implantation begins, the syncytiotrophoblasts secrete **chorionic gonadotropin** *(kor''e-on-ik go-nad-o-tro'pin)*, or **hCG** (the *h* stands for *human*). This hormone is identical to LH in its effects, and therefore is able to maintain the corpus luteum past the time when it would otherwise regress. The secretion of estrogen and progesterone is maintained and menstruation is normally prevented (fig. 22.8).

The secretion of hCG declines by the tenth week of pregnancy. Actually, this hormone is required only for the first 5 to 6 weeks of pregnancy because the placenta itself becomes an active steroid-secreting gland by this time. At the fifth to sixth week, the mother's corpus luteum begins to regress (even in the presence of hCG), but the placenta secretes more than sufficient amounts of steroids to maintain the endometrium and prevent menstruation.

All *pregnancy tests* assay for the presence of hCG in the blood or urine because this hormone is secreted by the blastocyst but not by the mother's endocrine glands. Modern pregnancy tests detect the presence of hCG by use of antibodies against hCG or by the use of cellular receptor proteins for hCG.

Formation of Germ Layers

As the blastocyst completes implantation during the second week of development, the embryoblast undergoes marked differentiation. A slitlike space called the **amniotic cavity** forms between the embryoblast and the invading trophoblast (fig. 22.9). The embryoblast flattens into the **embryonic disc** (see fig. 22.11), which consists of two layers: an upper **ectoderm,** which is closer to the amniotic cavity, and a lower **endoderm,** which borders the blastocyst cavity. A short time later, a third layer called the **mesoderm** forms between the endoderm and ectoderm. These three layers constitute the **primary germ layers.** Once they are formed, at the end of the second week, the pre-embryonic period is completed and the embryonic period begins.

The primary germ layers are important because various cells and tissues of the body are derived from them. Ectodermal cells form the nervous system; the outer layer of skin (epidermis), including hair, nails, and skin glands; and portions of the sensory organs. Mesodermal cells form the skeleton, muscles, blood, reproductive organs, dermis of the skin, and connective tissue. Endodermal cells produce the lining of the GI tract, the digestive organs, the respiratory tract and lungs, and the urinary bladder and urethra.

The events of the pre-embryonic period are summarized in table 22.1 and the derivatives of the three primary germ layers are listed in table 22.2. Refer to figure 22.10 for an illustration of the organs and body systems that derive from each of the primary germ layers.

ectoderm: Gk. *ecto*, outside; *derm*, skin
endoderm: Gk. *endo*, within; *derm*, skin
mesoderm: Gk. *meso*, middle; *derm*, skin

UNIT VII

Figure 22.7 The blastocyst adheres to the endometrium on about the sixth day as seen in (*a*), a photomicrograph, and (*b*), a diagram of the specific cells of the blastocyst. By the seventh day (*c*), specialized syncytiotrophoblasts from the trophoblast have begun to invade the endometrium. The syncytiotrophoblasts secrete hCG to sustain pregnancy and will eventually participate in the formation of the placenta for embryonic and fetal sustenance.

(a)

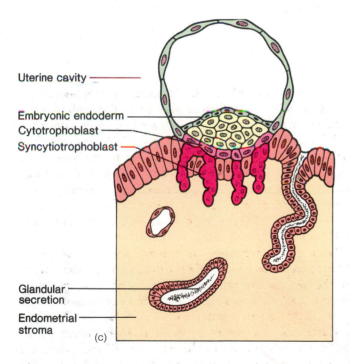

(c)

Figure 22.8 Human chorionic gonadotropin (hCG) is secreted by syncytiotrophoblasts during the first trimester of pregnancy. This hormone maintains the mother's corpus luteum for the first 5½ weeks. After that time the placenta becomes the major sex-hormone-producing gland, secreting increasing amounts of estrogen and progesterone throughout pregnancy.

Table 22.1	Morphogenic Stages and Principal Events of Pre-embryonic Development	
Stage	**Time Period**	**Principal Events**
Zygote	24 to 30 hours following ovulation	Egg is fertilized; zygote has 23 pairs of chromosomes (diploid) from haploid sperm and haploid egg and is genetically unique
Cleavage	30 hours to third day	Mitotic divisions produce increased number of cells
Morula	Third to fourth day	Hollow ball-like structure forms, a single layer thick
Blastocyst	Fifth day to end of second week	Embryoblast and trophoblast form; implantation occurs; embryonic disc forms, followed by primary germ layers

Figure 22.9 The completion of implantation. (*a*) A photomicrograph of an implanted blastocyst in the uterine wall. (*b*) The formation of the primary germ layers at the end of 2 weeks demarcates the end of the pre-embryonic period and the beginning of the embryonic period. (Note the formation of extraembryonic membranes at this early period of development.)

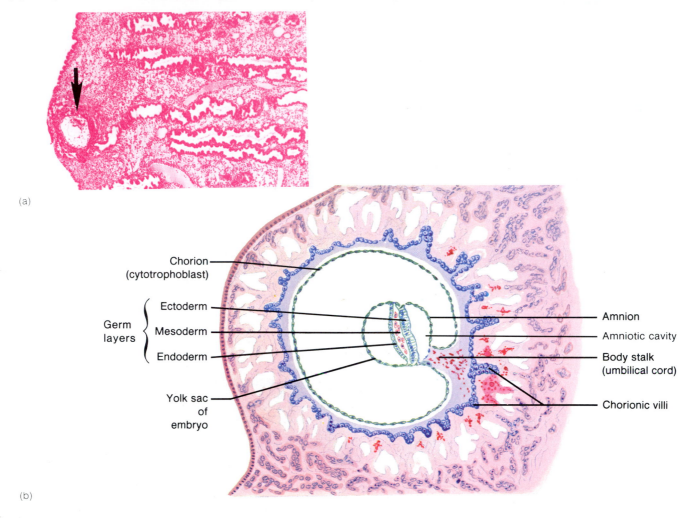

(a)

(b)

Table 22.2	Derivatives of Germ Layers	
Ectoderm	**Mesoderm**	**Endoderm**
Epidermis of skin and epidermal derivatives: hair, nails, glands of the skin; linings of oral, nasal, anal, and vaginal cavities	Muscle: smooth, cardiac, and skeletal	Epithelium of pharynx, auditory canal, tonsils, thyroid, parathyroid, thymus, larynx, trachea, lungs, GI tract, urinary bladder and urethra, and vagina
Nervous tissue; sense organs	Connective tissue: embryonic, connective tissue proper, cartilage, bone, blood	
Lens of eye; enamel of teeth	Dermis of skin; dentin of teeth	Liver and pancreas
Pituitary gland	Epithelium of blood vessels, lymphatic vessels, body cavities, joint cavities	
Adrenal medulla	Internal reproductive organs	
	Kidneys and ureters	
	Adrenal cortex	

UNIT VII

Figure 22.10   The body systems and the primary germ layers from which they develop.

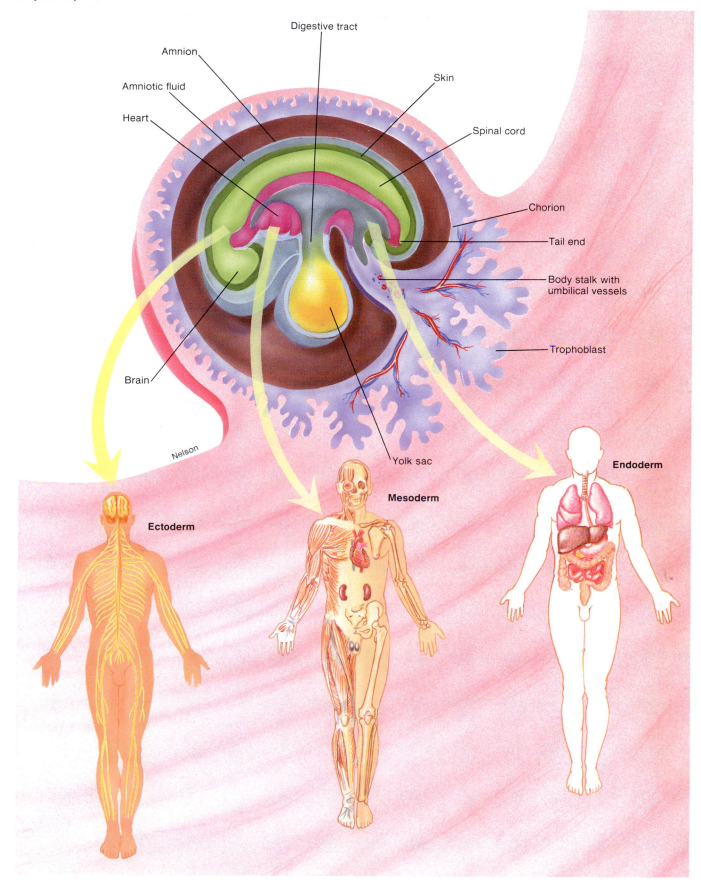

Digestive tract

Amnion

Amniotic fluid

Skin

Heart

Spinal cord

Chorion

Tail end

Body stalk with umbilical vessels

Trophoblast

Brain

Nelson

Yolk sac

Endoderm

Mesoderm

Ectoderm

The period of prenatal development is referred to as *gestation*. Normal gestation for humans is 9 months. Knowing this and the pattern of menstruation make it possible to determine the delivery date of a baby. In a typical reproductive cycle, a woman ovulates 14 days prior to the onset of the next menstruation and is fertile for approximately 20 to 24 hours following ovulation. Adding 9 months, or 38 weeks, to the time of ovulation gives one the estimated delivery date, or parturition date.

1. List the structural characteristics of a zygote, morula, and blastocyst. Approximately when do each of these stages of the pre-embryonic period of development occur?
2. Discuss the process of implantation and describe the physiological events that ensure pregnancy.
3. Describe the development of the placenta.
4. List the major structures or organs that derive from each germ layer.
5. What is the length of time of a typical pregnancy? Define *gestation*. How is the parturition date determined?

Embryonic Period

The events of the 6-week embryonic period include the differentiation of the germ layers into specific body organs and the formation of the placenta, the umbilical cord, and the extraembryonic membranes. Through these morphogenic events, the needs of the embryo are met.

Objective 7. Define *embryo* and describe the major events of the embryonic period of development.

Objective 8. List the embryonic needs that must be met to avoid a spontaneous abortion.

Objective 9. Describe the structure and function of each of the extraembryonic membranes.

Objective 10. Describe the development and function of the placenta and the umbilical cord.

The embryonic period lasts from the beginning of the third week to the end of the eighth week. At this period, the developing organism can correctly be called an **embryo.** During the embryonic period all of the body organs form, as well as the placenta, umbilical cord, and extraembryonic membranes. The term **conceptus** refers to the embryo, or fetus, and all of the extraembryonic structures—the products of conception. During parturition, the baby and the remaining portion of the conceptus are expelled from the uterus.

Embryology is the study of the sequential changes in an organism as the various tissues, organs, and systems develop. Chick embryos are frequently studied because of the easy access through the shell and their rapid development. Mice and pig embryos are also extensively studied as mammalian models. Genetic manipulation, induction of drugs, exposure to disease, radioactive tagging or dyeing of developing tissues, and X-ray treatments are some of the commonly conducted experiments that provide information that can be applied to human development and birth defects.

During the pre-embryonic period of cell division and differentiation, the developing structure is self-sustaining. The embryo, however, is not self-supporting and must derive sustenance from the mother. For morphogenesis to continue, certain immediate needs must be met. These needs include (1) formation of a vascular connection between the uterus of the mother and the embryo so that nutrients and oxygen can be provided and metabolic wastes and carbon dioxide can be removed; (2) establishment of a constant, protective environment around the embryo that is conducive to development; (3) establishment of a structural foundation for embryonic morphogenesis along a longitudinal axis; (4) provision for structural support for the embryo both internally and externally; and (5) coordination of the morphogenic events through genetic expression. If these needs are not met, a spontaneous abortion will generally occur.

The first and second of these needs are provided for by extraembryonic structures; the last three are provided for intraembryonically. The extraembryonic membranes, the placenta, and the umbilical cord will be considered separately, prior to a discussion of the development of the embryo.

 Most serious developmental defects cause the embryo to be naturally aborted. About 25% of early aborted embryos have chromosomal abnormalities. Other abortions may be caused by environmental factors, such as infectious agents or teratogenic drugs (drugs that cause birth defects). In addition, an implanting, developing embryo is regarded as foreign tissue by the immune system of the mother, and is rejected and aborted unless maternal immune responses are suppressed.

Extraembryonic Membranes

While the many intraembryonic events are forming the body organs, a complex system of extraembryonic membranes is developing as well (fig. 22.11). The **extraembryonic membranes** are the amnion *(am'ne-on)*, yolk sac, allantois, and chorion. These membranes are responsible for the protection, respiration, excretion, and nutrition of the embryo and subsequent fetus. At parturition, the placenta *(plah-sen'tă)*, umbilical cord, and extraembryonic membranes separate from the fetus and are expelled from the uterus as the *afterbirth*.

Amnion The amnion is a thin extraembryonic membrane derived from ectoderm and mesoderm. It loosely envelops the embryo, forming an **amniotic sac** that is filled with *amniotic fluid* (fig. 22.12). In later fetal development, the amnion expands to come in contact with the chorion. Amniotic development is initiated early in the embryonic period, at which time its margin is attached around the free edge of the embryonic disc (fig. 22.11). As the amniotic sac enlarges during the late embryonic period (at about 8 weeks), the amnion gradually sheaths the developing umbilical cord with an epithelial covering (fig. 22.13).

gestation: L. *gestatus*, to bear

UNIT VII

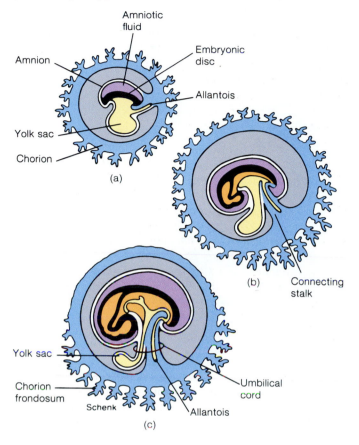

Figure 22.11 The formation of the extraembryonic membranes during a single week of rapid embryonic development: (*a*) at 3 weeks, (*b*) 3½ weeks, and (*c*) 4 weeks.

Amniotic fluid

Amnion

Embryonic disc

Allantois

Yolk sac

Chorion

(a)

(b) Connecting stalk

Yolk sac

Chorion frondosum

Schenk

Umbilical cord

Allantois

(c)

As a buoyant medium, amniotic fluid performs four functions for the embryo and subsequent fetus: (1) it permits symmetrical structural development and growth, (2) it cushions and protects by absorbing jolts that the mother may receive, (3) it helps to maintain consistent pressure and temperature, and (4) it allows the fetus to develop freely, which is important for musculoskeletal development and blood flow.

Amniotic fluid is formed initially as an isotonic fluid absorbed from the maternal blood in the endometrium surrounding the developing embryo. Later, the volume is increased and the concentration changed by urine excreted from the fetus into the amniotic sac. Amniotic fluid also contains cells that are sloughed off from the fetus, placenta, and amniotic sac. Since all of these cells are derived from the same fertilized egg, all have the same genetic composition. Many genetic abnormalities can be detected by aspirating this fluid and examining the cells obtained in a procedure called *amniocentesis (am″ne-o-sen-te′sis)*.

Amniotic fluid is normally swallowed by the fetus and absorbed in the GI tract. Prior to delivery, the amnion is naturally or surgically ruptured, and the amniotic fluid (bag of waters) is released.

Amniocentesis (fig. 22.14) is usually performed at the fourteenth or fifteenth week of pregnancy, when the amniotic sac contains 175–225 mℓ of fluid. Genetic diseases, such as *Down syndrome* (in which there are three instead of two number–21 chromosomes), can be detected by examining chromosomes. Diseases such as *Tay–Sachs disease*, in which there is a defective enzyme involved in formation of myelin sheaths, can be detected by biochemical techniques.

Down syndrome: from John L. H. Down, English physician, 1828–96

Figure 22.12 An implanted embryo at approximately 4½ weeks. (*a*) The interior of a uterus showing the implantation site and decidua capsularis, elevated by the expanded chorion. (*b*) The developing embryo, extraembryonic membranes, and the formation of the placenta.

Villi of chorion frondosum

Chorion

Amnion

Amniotic sac

Yolk sac

Decidua capsularis

Body of uterus

Umbilical blood vessels

(a)

(b)

UNIT VII

Figure 22.13 The embryo, extraembryonic membranes, and placenta at approximately 7 weeks of development. At this time, the amnion and chorion are adherent and are frequently referred to as the amniochorionic membrane. Blood from the embryo is carried to and from the chorion frondosum by the umbilical arteries and vein. The maternal tissue between the chorionic villi is known as the decidua basalis; this tissue, together with the villi, form the functioning placenta.

Labels on figure:
Decidua basalis
Chorion frondosum
Maternal vein
Maternal artery
Umbilical cord
Chorion
Amnion
Amniotic sac containing amniotic fluid
Placenta

Yolk Sac   The yolk sac is established during the end of the second week as cells from the trophoblast form a thin *exocoelomic (ek″so-sē-lo′mik) membrane.* Unlike that of many vertebrates, the human yolk sac contains no nutritive yolk but is an essential structure during early embryonic development. It is attached to the underside of the embryonic disc (figs. 22.11 and 22.12), where it produces blood for the embryo until the liver forms during the sixth week. The dorsal portion of the yolk sac is involved in the formation of the primitive gut. In addition, primordial germ cells form in the wall of the yolk sac and migrate during the fourth week to the developing gonads, where they become the primitive germ cells (spermatogonia or oogonia).

The stalk of the yolk sac usually detaches from the gut by the sixth week. Following this, the yolk sac gradually shrinks as pregnancy advances. Eventually it becomes very small and serves no additional developmental functions.

Allantois   The allantois forms during the third week as a small outpouching, or diverticulum, from the caudal wall of the yolk sac (see fig. 22.11). The allantois remains small but is involved in the formation of blood cells and gives rise to the fetal umbilical arteries and vein. It also contributes to the development of the urinary bladder.

The extraembryonic portion of the allantois degenerates during the second month. The intraembryonic portion involutes to form a thick urinary tube called the **urachus.** After birth, the urachus becomes a fibrous cord called the median umbilical ligament that attaches to the urinary bladder.

Chorion   The chorion is a highly specialized extraembryonic membrane that participates in the formation of the placenta (see fig. 22.11). It is the outermost membrane and originates from the trophoblast of the blastocyst. Numerous small, fingerlike extensions called **villi** (see fig. 22.12) form from the chorion and penetrate deeply into the uterine tissue. Initially, the entire surface of the chorion is covered with villi. But those villi on the surface toward the uterine cavity gradually degenerate and produce a smooth, bare area known as the **smooth chorion.** As this occurs, the villi associated with the uterine wall rapidly increase in number and branch out. This portion of the chorion

allantois: Gk. *allanto,* sausage; *iodos,* resemblance

chorion: Gk. *chorion,* external fetal membrane

UNIT VII

Figure 22.14 Amniocentesis. In this procedure, amniotic fluid containing suspended cells is withdrawn for examination. Various genetic diseases can be detected prenatally by this means.

Uterus

Amniotic sac

Placenta

is known as the **villous chorion.** The villous chorion becomes highly vascular, and as the embryonic heart begins to function, blood is pumped in close proximity to the uterine wall.

Chorionic villus biopsy is a technique used to detect genetic disorders much earlier than amniocentesis permits. In chorionic villus biopsy, a catheter is inserted through the cervix to the chorion, and a sample of chorionic villus is obtained by suction or cutting. Genetic tests can be performed directly on the villus sample, since this sample contains much larger numbers of fetal cells than does a sample of amniotic fluid. Chorionic villus biopsy can provide genetic information at 10 to 12 weeks' gestation.

Placenta

The **placenta** is a vascular structure by which an unborn child is attached to its mother's uterine wall and through which metabolic exchange occurs (fig. 22.15). The placenta is formed in part from maternal tissue and in part from embryonic tissue. The embryonic portion of the placenta consists of the villi of the **chorion frondosum,** whereas the maternal portion is composed of the area of the uterine wall called the **decidua basalis** (see fig. 22.13), into which the villi penetrate. Blood does not flow directly between these two portions, but because of the close membranous proximity, certain substances diffuse readily.

UNIT VII

villous: L. *villus*, tuft of hair

placenta: L. *placenta*, a flat cake

Figure 22.15 The circulation of blood within the placenta. Maternal blood is delivered to and drained from the spaces between the chorionic villi. Fetal blood is brought to blood vessels within the villi by branches of the umbilical artery and is drained by branches of the umbilical vein.

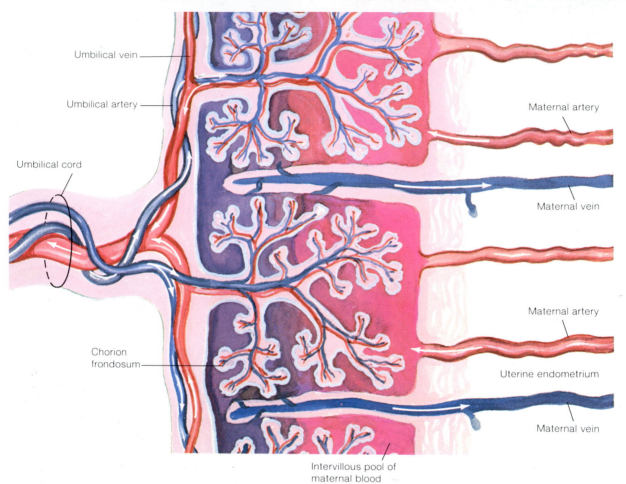

When fully formed, the placenta is a reddish brown oval disc with a diameter of 15–20 cm (8 in.) and a thickness of 2.5 cm (1 in.). It weighs 500–600 gm, about one-sixth the weight of the fetus.

Exchange of Molecules across the Placenta The two **umbilical arteries** deliver fetal blood to vessels within the villi of the chorion frondosum of the placenta. This blood circulates within the villi and returns to the fetus via the **umbilical vein** (see fig. 22.16). Maternal blood is delivered to and drained from the cavities within the decidua basalis, which are located between the chorionic villi. In this way, maternal and fetal blood are brought close together but never mix within the placenta.

The placenta serves as a site for the exchange of gases and other molecules between the maternal and fetal blood. Oxygen diffuses from mother to fetus, and carbon dioxide diffuses in the opposite direction. Nutrient molecules and waste products likewise pass between maternal and fetal blood.

The placenta is not merely a passive conduit for exchange between maternal and fetal blood, however. It has a very high metabolic rate, utilizing about one-third of all the oxygen and glucose supplied by the maternal blood. The rate of protein synthesis is, in fact, higher in the placenta than it is in the liver. Like the liver, the placenta produces a great variety of enzymes

capable of converting biologically active molecules (such as hormones and drugs) into less active, more water-soluble forms. In this way, potentially dangerous molecules in the maternal blood are often prevented from harming the fetus.

Most drugs ingested by a pregnant woman can readily pass through the placenta to the developing baby. Many drugs that are not particularly harmful to the mother may be deleterious to her child. For example, the nicotine taken in by a chain-smoking pregnant woman will stunt the growth of her fetus. Hard drugs, such as heroin, can lead to fetal drug addiction. Depressant drugs given to a mother during labor can readily cross the placenta and, if given in high dosages, can cause respiratory depression in the newborn infant.

Although the placenta is an effective barrier against diseases of bacterial origin, viruses like rubella and certain blood-borne diseases like syphilis can diffuse through the placenta and affect the fetus. During parturition, small amounts of fetal blood may pass across the placenta to the mother. If the fetus is Rh positive and the mother Rh negative, the antigens of the fetal red blood cells elicit an antibody response in the mother. In a subsequent pregnancy, the maternal antibodies then cross the placenta and cause a breakdown of fetal red blood cells—a condition called *erythroblastosis fetalis.*

UNIT VII

Table 22.3	Hormones Secreted by Placenta	
Hormones	**Effects**	
Pituitary-Like Hormones		
Chorionic gonadotropin (hCG)	Similar to LH; maintains mother's corpus luteum for first 5½ weeks of pregnancy; may be involved in suppressing immunological rejection of embryo; also exhibits TSH-like activity	
Chorionic somatomammotropin (hCS)	Similar to prolactin and growth hormone; in the mother, hCS acts to promote increased fat breakdown and fatty acid release from adipose tissue and decreased glucose use by maternal tissues (diabetic-like effects)	
Sex Steroids		
Progesterone	Helps maintain endometrium during pregnancy; helps suppress gonadotropin secretion; promotes uterine sensitivity to oxytocin; helps stimulate mammary gland development	
Estrogens	Help maintain endometrium during pregnancy; help suppress gonadotropin secretion; help stimulate mammary gland development; inhibit prolactin secretion	

Endocrine Functions of the Placenta   The placenta functions as an endocrine gland in secreting both steroid and glycoprotein hormones. The glycoprotein hormones perform actions similar to those of some anterior pituitary hormones. The importance of these hormones is to maintain pregnancy and to ensure optimal nutrition to the fetus.

The placenta converts androgens, secreted from the mother's adrenal glands, into estrogens to help protect the female embryo from becoming masculinized. In addition, the placenta secretes large amounts of estriol, a weak estrogen that helps to maintain the endometrium and stimulate the development of the mother's mammary glands. The production of estriol increases tenfold during pregnancy, so that by the third trimester estriol accounts for about 90% of the estrogens excreted in the mother's urine. Since almost all of this estriol comes from the placenta (rather than from maternal tissues), measurements of urinary estriol can be used clinically to assess the health of the placenta.

In the later stages of gestation, the steroid hormones stimulate the development of the mammary glands. The hormones secreted by the placenta and the effect they have upon their target tissues are summarized in table 22.3.

Umbilical Cord

The **umbilical cord** forms as the yolk sac shrinks and the amnion expands to envelop the tissues on the underside of the embryo. The formation of the umbilical cord is illustrated in figure 22.16.

The umbilical cord usually attaches near the center of the placenta. When fully formed, it is about 1–2 cm (0.5–1 in.) in diameter and approximately 55 cm (2 ft) long. The umbilical cord contains two **umbilical arteries,** which carry deoxygenated blood toward the placenta, and one **umbilical vein,** which carries oxygenated blood from the placenta to the embryo. These vessels are surrounded by embryonic connective tissue called **mucoid connective tissue** (Wharton's jelly).

The umbilical cord has natural twists because the umbilical vein is longer than the arteries. In about one-fifth of all deliveries, the cord is looped once around the baby's neck. If drawn tightly, the cord could cause death or serious perinatal problems.

Structural Changes of the Embryo by Weeks

Third Week   Early in the third week, a thick linear band called the **primitive line,** appears along the dorsal midline of the embryonic disc (see figs. 22.11 and 22.17). The primitive line is derived from the mesodermal cells, which are specialized from the ectodermal layer of the embryonic mass. The primitive line establishes a structural foundation for embryonic morphogenesis along a longitudinal axis. As the primitive line elongates, a prominent thickening called the **primitive node** appears at its cranial end (fig. 22.17). The primitive node later gives rise to the mesodermal structures of the head and the **notochord.** To give support to the embryo, the notochord forms a midline axis that is the basis of the embryonic skeleton. The primitive line also gives rise to loose embryonic connective tissue called **intraembryonic mesoderm** (mesenchyme), which differentiates into all the various kinds of connective tissue found in the adult. One of the earliest formed organs is the skin, which develops to support and maintain homeostasis within the embryo.

A tremendous amount of change and specialization occurs during the embryonic stage. The factors that cause precise, sequential change from one cell or tissue type to another are not fully understood. It is known, however, that the potential for change is programmed into the genetics of each cell and that under conducive environmental conditions these characteristics become expressed. The process of developmental change is referred to as **induction.** Induction occurs when one tissue, called the **inductor tissue,** has a marked effect on an adjacent tissue causing it to become **induced tissue** and stimulating it to differentiate.

UNIT VII

Wharton's jelly: from Thomas Wharton, English anatomist, 1614–73

induction: L. *inductus,* to lead in

Figure 22.16 The formation of the umbilical cord and other extraembryonic structures as seen in sagittal sections of the gravid uterus from week 4 to week 22. (*a*) A connecting stalk forms as the developing amnion expands around the embryo and finally meets ventrally. (*b*) The umbilical cord begins to take form as the amnion ensheathes the yolk sac. (*c*) A cross section of the umbilical cord showing the embryonic vessels, mucoid connective tissue, and the tubular connection to the yolk sac. (*d*) By week 22, the amnion and chorion have fused and the umbilical cord and placenta have become well-developed structures. (*e*) A 16-week-old fetus.

Figure 22.17 The appearance of the primitive line and primitive node along the embryonic disc. These progressive changes occur through the process of induction.

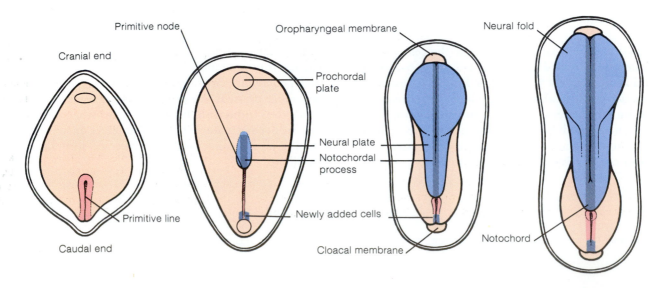

Figure 22.18   Progressive illustrations of 4-week-old embryos.

Neural groove

Somite

Neuropores — Rostral

Caudal

Neural tube

22 ± 1 day   2–3 mm

Optic sulcus

Neural folds fusing to form primary brain vesicles

Somites

Caudal neuropore

Actual length

23 ± 1 day   2.25–3.42 mm

Rostral neuropore closing

Heart prominence

Connecting stalk

Caudal neuropore

Mandibular arch

Forebrain prominence

Hyoid arch

Somites

Tail

24 ± 1 day   2.5–3.8 mm

Otic pit

Third branchial arch

Superior bud

Inferior bud

26 ± 1 day   3.0–4.5 mm

Fourth branchial arch

Lens placode

28 ± 1 day   4.5–6.0 mm

Fourth Week   During the fourth week of development, the embryo increases about 4 mm in length. A **connecting stalk,** which is later involved in the formation of the umbilical cord, is established from the body of the embryo to the developing placenta (fig. 22.18). By this time, the heart is beating firmly to pump blood to all parts of the embryo. The head and jaws are apparent, and the primordial tissue that will form the eyes, brain, spinal cord, lungs, and digestive organs has developed. The **superior** and **inferior limb buds** are recognizable as small swellings on the lateral body walls.

Fifth Week   Embryonic changes during the fifth week are not as extensive as those during the fourth week. The head enlarges, and the developing eyes, ears, and nasal pit are obvious (fig. 22.19). The appendages have formed from the limb buds, and paddle-shaped hand plates develop digital ridges called **finger,** or **digital, rays.**

Sixth Week   During the sixth week, the embryo is 16–24 mm long. The head is much larger than the trunk, and the brain has undergone marked differentiation. This is the most vulnerable period of development for many organs. An interruption at this critical time can easily cause congenital damage. The limbs undergo considerable change during this week. The fore-

limbs are lengthened and slightly flexed, and notches appear between the rays in the hand and foot plates.

Seventh and Eighth Weeks   During the last 2 weeks of the embryonic stage, the embryo, which is now 28–40 mm long, has distinct human characteristics (figs. 22.20 and 22.21). The body organs are formed, and the nervous system begins coordinating body activity. The neck region is apparent, and the abdomen is less protuberant. The eyes are well developed, but the lids are stuck together to protect against probing fingers during muscular movement. The nostrils are developed but plugged with mucus. The external genitalia are forming but are still undifferentiated. The body systems are developed by the end of the eighth week, and from this time on the embryo is called a **fetus.**

The most precarious time of prenatal development is during the embryonic period, when there is much tissue differentiation and organ formation. Frequently, however, a woman does not even realize that she is pregnant until she is well into this period of development. For this reason, a woman should consistently take good care of herself and abstain from taking certain drugs (including some antibiotics) if there is even a remote chance that she is pregnant or might become pregnant in the near future.

UNIT VII

Figure 22.19 Progressive illustrations of 5-week-old embryos and a photograph of a 5-week-old embryo. (Note the digital primordia and the developing eyes and ears.)

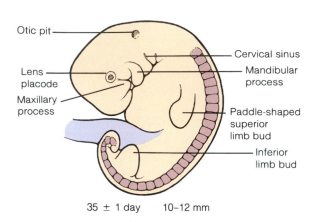

35 ± 1 day   10–12 mm

37 ± 1 day   12.5–15.75 mm

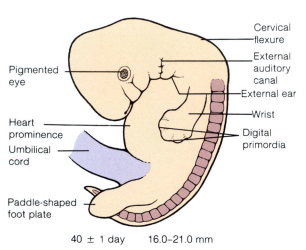

40 ± 1 day   16.0–21.0 mm

1. Distinguish between *embryo* and *fetus*. Briefly summarize the structural changes that an embryo undergoes between the fourth and eighth weeks of development.
2. Describe the origin of hCG and explain why it is needed to maintain pregnancy for the first 10 weeks.
3. What are the five embryonic needs that must be met in order to avoid spontaneous abortion? Which of these needs are met by the extraembryonic membranes?
4. Describe the formation and function of each of the extraembryonic membranes.
5. Identify the fetal and maternal components of the placenta and describe the circulation in these two components. Describe how fetal and maternal gas exchange occurs.

Fetal Period

The fetal period lasts from 9 weeks until birth and is characterized by tremendous growth and the specialization of body structures.

Objective 11. Define *fetus* and discuss the major events of the fetal period of development.

Objective 12. Describe the various techniques available for examining the fetus or monitoring fetal activity.

Since most of the tissues and organs of the body become apparent during the embryonic period, the **fetus** is recognizable as a human being at 9 weeks and is far less vulnerable than the

Figure 22.20 A photograph of a 6-week-old embryo and progressive illustrations of 6- and 7-week-old embryos.

Cervical flexure

External ear

Notches between digital primordia

Ear

Eyelid

Webbed fingers

Early umbilical herniation

Notches between toe primordia

Toe rays

(a) 45 ± 1 day   22–24 mm

(b) 49 ± 1 day   28–30 mm

Fingers separated

Fan-shaped webbed toes

Toes separated

(c) 52 ± 1 day   32–34 mm

(d) 56 ± 1 day   34–40 mm

embryo to deformation from viruses, drugs, and radiation. A small amount of tissue differentiation and organ development still occurs during the fetal period, but for the most part fetal development is primarily limited to body growth. Changes in external appearance of the fetus from the ninth through the thirty-eighth week are depicted in figure 22.22. The following is a discussion of the weekly structural changes of the fetus as it grows and matures.

Nine to Twelve Weeks   At the beginning of the ninth week, the head is as large as the rest of the body. The eyes are widely spaced, and the ears are set low. Head growth slows during the next 3 weeks, whereas lengthening of the body accelerates. Os-

sification centers appear in most bones during the ninth week. Differentiation of the external genitalia becomes apparent at the end of the ninth week, but the genitalia are not developed to the point of sex determination until the twelfth week. By the end of the twelfth week, the fetus is 87 mm (3.5 in.) long and weighs about 45 g (1.6 oz). It can swallow, digest the fluid that passes through its system, and defecate and urinate into the amniotic fluid. The nervous system and muscle coordination are developed enough so that the fetus will withdraw its leg if tickled. The fetus begins inhaling through its nose but can take in only amniotic fluid. The external appearance of fetuses at 10 and 12 weeks is depicted in figure 22.23.

UNIT VII

Figure 22.21 Photograph of an 8-week-old embryo. The body systems have developed by the end of the eighth week, and the embryo is recognizable as human.

Major structural abnormalities, which may not be predictable from genetic analysis, can often be detected by *ultrasonography* (fig. 22.24). Sound-wave vibrations are reflected from the interface of tissues with different densities—for example, from the interface between the fetus and amniotic fluid—and used to produce an image. This technique is so sensitive that it can be used to detect a fetal heartbeat several weeks before it can be detected by a stethoscope.

Thirteen to Sixteen Weeks By the thirteenth week, the facial features are well formed, and epidermal structures such as eye-lashes, eyebrows, hair on the head, fingernails, and nipples, begin to develop. The appendages lengthen, and by the six-teenth week the skeleton is sufficiently developed to show up clearly on radiographs. During the sixteenth week, the fetal heartbeat can be heard by applying a stethoscope to the mother's abdomen. By the end of the sixteenth week, the fetus is 140 mm long (5.5 in.) and weighs about 200 g (7 oz).

After the sixteenth week, fetal length can be determined from radiographs. The reported length of a fetus is generally derived from a straight line measurement from the crown of the head to the developing ischium (crown-rump length). Measurements made on an embryo prior to the fetal stage, however, are not reported as crown-rump measurements but as total length.

Seventeen to Twenty Weeks Between the seventeenth and twentieth weeks, the legs achieve their final relative proportions and fetal movements, known as **quickening,** are commonly felt by the mother. The skin is covered with a white, cheeselike material known as **vernix caseosa.** It consists of fatty

vernix caseosa: L. *vernix*, varnish; L. *caseus*, cheese

Figure 22.22 Changes in the external appearance of the fetus from week 9 through week 38.

Creek

9   12   16   20   25   29   38   Full term

UNIT VII

Figure 22.23 External appearances of fetuses (a) at 10 weeks and (b) at 12 weeks.

(a)

(b)

Figure 22.24 Ultrasonography. (a) Sound-wave vibrations are reflected from the internal tissues of a person's body. (b) Structures of the human fetus observed through an ultrasound scan.

(a)

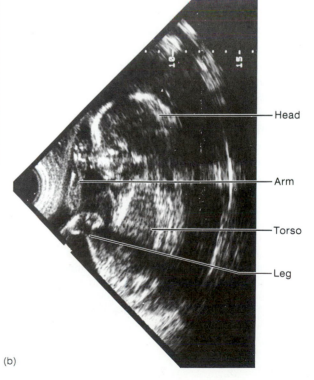

Head

Arm

Torso

Leg

(b)

UNIT VII

secretions from the sebaceous glands and dead epidermal cells. The function of vernix caseosa is to protect the fetus while it is bathed in amniotic fluid. Twenty-week-old fetuses usually have fine, silklike fetal hair called **lanugo** *(lă-noo'go)* covering the skin. Lanugo is thought to hold the vernix caseosa on the skin and produce a ciliarylike motion that moves amniotic fluid. The length of a 20-week-old fetus is about 190 mm (7.5 in.) and it weighs about 460 gm (16 oz). Because of cramped space, the fetus develops a marked spinal flexure and is in what is commonly called the fetal position, with the head bent down in contact with the flexed knees.

Twenty-One to Twenty-Five Weeks Between the twenty-first and twenty-fifth weeks, the fetus increases its weight substantially to about 900 gm (32 oz). Body length increases only moderately (240 mm), however, so the weight is evenly proportioned. The skin is quite wrinkled and is translucent. Because the blood flowing in the capillaries is now visible, the skin appears pinkish.

Twenty-Six to Twenty-Nine Weeks Toward the end of the 26-to-29-week period, the fetus will be about 275 mm (11 in.) long and will weigh about 1,300 gm (46 oz). A fetus might now survive if born prematurely, but the mortality rate is high. Its body metabolism cannot yet maintain a constant temperature, and the respiratory muscles have not matured enough to provide a regular respiratory rate. If, however, the premature infant is put in an incubator and a respirator is used to maintain its breathing, it may survive. The eyes open during this period, and the body is well covered with lanugo. If the fetus is a male, the testes should have begun descent into the scrotum (see exhibit III, chapter 20). As the time of birth approaches, the fetus rotates to a **vertex position** (fig. 22.25). The head repositions toward the cervix because of the shape of the uterus and because the head is the heaviest part of the body.

Thirty to Thirty-Eight Weeks At the end of 38 weeks, the fetus is considered full-term. It has reached a crown-rump length of 360 mm (14 in.) and weighs about 3,400 gm (7.5 lb). The average total length from crown to heel is 50 cm (20 in.). Most fetuses are plump with smooth skin because of the accumulation of subcutaneous fat. The skin is pinkish-blue even in fetuses of dark-skinned parents, because melanocytes do not produce melanin until the skin is exposed to sunlight. Lanugo hair is sparse and is generally found on the head and back. The chest is prominent, and the mammary area protrudes in both sexes. The external genitalia are somewhat swollen.

1. Explain why the ninth week is designated as the beginning of the fetal stage of development.
2. List the approximate fetal age at which each of the following occur: first detection of fetal heartbeat; presence of vernix caseosa and lanugo; fetal rotation into vertex position.
3. Compare ultrasound and radiographic techniques in determining fetal development and structure.

lanugo: L. *lana*, wool
vertex: L. *vertex*, summit

Figure 22.25 A fetus in vertex position. Toward the end of most pregnancies, the weight of the fetal head causes a rotation of its entire body such that the head is positioned in contact with the cervix of the uterus.

Amniotic fluid

Amniochorionic membrane

Umbilical cord

Placenta

Uterine wall

Cervix

Labor and Parturition

Labor and parturition are the culmination of gestation and require the action of oxytocin, secreted by the posterior pituitary, and prostaglandins, produced in the uterus.

Objective 13. Describe the hormonal action that controls labor and parturition.

Objective 14. Describe the three stages of labor.

The time of prenatal development, or the time of pregnancy, is called **gestation.** The human gestational period is usually 266 days or about 280 days from the beginning of the last menstrual period to **parturition,** or birth. Most fetuses are born within 10 to 15 days before or after this time. Parturition is accompanied by a sequence of physiological and physical events called **labor.**

The *onset of labor* is denoted by rhythmic and forceful contractions of the myometrial layer of the uterus. In *true labor,* the pains from uterine contractions occur at regular intervals and intensify as the interval between contractions shortens. A reliable indication of true labor is dilation of the cervix and a show, or discharge, of blood-containing mucus in the cervical canal and vagina. In *false labor,* abdominal pain is experienced at irregular intervals and cervical dilation and cervical show are absent.

The uterine contractions of labor are stimulated by two agents: (1) **oxytocin** *(ok"si-to'sin),* a polypeptide hormone produced in the hypothalamus and released from the posterior pituitary and (2) **prostaglandins** *(pros"tă-glan'dinz),* a class of fatty acids produced within the uterus itself. Labor can indeed be induced artificially by injections of oxytocin or by the insertion of prostaglandins into the vagina as a suppository.

gestation: L. *gestatus*, to bear

UNIT VII

Figure 22.26 The stages of labor and parturition. (*a*) The position of the fetus prior to labor. (*b*) The ruptured amniotic sac and early dilation of the cervix. (*c*) The expulsion stage, or period of parturition. (*d*) The placental stage.

The hormone *relaxin,* produced by the corpus luteum, may also be involved in labor and parturition. Relaxin is known to soften the symphysis pubis in preparation for parturition and is thought to also soften the cervix in preparation for dilation. It may be, however, that relaxin does not affect the uterus, but rather that progesterone and estradiol may be responsible for this effect. Further research is necessary to understand the total physiological effect of these hormones.

As illustrated in figure 22.26, labor is divided into three stages:

1. **Dilation stage.** In this period, the cervix dilates to a diameter of approximately 10 cm. Contractions are regular during this stage and the amniotic sac (bag of waters) generally ruptures. If the amniotic sac does not rupture spontaneously, it is done surgically. The dilation stage generally lasts from 8 to 24 hours.
2. **Expulsion stage.** This is the period of parturition, or actual childbirth. It consists of forceful uterine contractions and abdominal compressions to expel the fetus from the uterus and through the vagina. This stage

may require 30 minutes in a first pregnancy but only a few minutes in subsequent pregnancies.
3. **Placental stage.** Generally within 10 to 15 minutes after parturition, the placenta is separated from the uterine wall and expelled as the *afterbirth.* Forceful uterine contractions characterize this stage, constricting uterine blood vessels to prevent hemorrhage. In a normal delivery, blood loss does not exceed 350 mℓ.

A *pudendal nerve block* may be administered during the early part of the expulsion stage to ease the trauma of delivery for the mother and to allow for an episiotomy.

Five percent of newborns are born *breech.* In a breech birth, the fetus has not rotated and the buttocks are the presenting part. The principal concern of a breech birth is the increased time and difficulty of the expulsion stage of parturition. Attempts to rotate the fetus through the use of forceps may injure the infant. If an infant cannot be delivered breech, a *cesarean (si-zar'e-an) section* must be performed. A cesarean section is delivery of the fetus through an incision made into the abdominal wall and the uterus.

UNIT VII

1. Distinguish between labor and parturition.
2. Describe the hormonal mechanisms responsible for labor and explain two techniques for inducing labor.
3. Describe the three stages of labor and state how long each stage lasts.

Periods of Postnatal Growth

The course of human life after birth is seen in terms of physical and physiological changes and the attainment of maturity in the neonatal, infant, childhood, adolescent, and adulthood periods.

Objective 15. Describe the growth and developmental events of the neonatal, infant, childhood, and adolescent periods.

Objective 16. Define *puberty* and discuss what determines its onset in males and females.

Objective 17. Define the term *adulthood* and discuss sexual dimorphism in adult humans.

Neonatal Period

The **neonatal period** extends from birth to the end of four weeks. Although growth is rapid during this period, the most drastic changes are physiological. The body of a newborn must immediately adapt to major environmental changes, including thermal stress; rapid bacterial colonization of the skin, oral cavity, and GI tract; a barrage of sensory stimuli; and sudden demands on cardiorespiratory, gastrointestinal, and renal functions.

The most critical need of the newborn is the establishment of an adequate respiratory rate to ensure sufficient amounts of oxygen. The normal respiratory rate of a newborn is 30 to 40 respirations per minute. An adequate heart rate is also imperative. The heart of a newborn seems enlarged relative to the thoracic cavity (compared to the heart of an adult) and has a rapid rate that ranges from 120 to 160 beats per minute.

Most full-term newborn babies appear chubby because of the deposition of fat within adipose tissue during the last trimester of pregnancy. Dehydration is a serious threat because of the inability of the kidneys to excrete concentrated urine; large volumes of dilute urine are eliminated. Immunity is not well developed and is limited to that obtained from the mother through placental transfer. For this reason, newborns need to be guarded against exposure to infectious diseases.

Although virtually all of the neurons of the nervous system are present in a newborn, they are immature and the newborn has little coordination. Most infant behavior appears to be governed by lower cerebral centers and the spinal cord.

A newborn has many reflexes, some indicative of neuromuscular maturity and others essential for life itself. Four reflexes critical to survival are (1) the *suckling reflex*, triggered by anything that touches the newborn's lips; (2) the *rooting reflex*, which helps a baby find a nipple as it turns its head and starts to suckle in response to something that brushes its cheek; (3) the *crying reflex*, triggered by an empty stomach or other discomforts; and (4) the *breathing reflex*, apparent in a normal newborn even before the umbilical cord, with its supply of oxygen, is cut.

Babies born more than 3 weeks before the due date are generally considered *premature*, but because errors are commonly made in calculating the conception date, prematurity is defined by neonatal body weight rather than due date. Newborns weighing less than 2,500 grams (5.5 lbs) are considered premature. By this definition, approximately 8% of newborns in the United States are premature.

Postmature babies are those born 2 or more weeks after the due date. They frequently weigh less than they would have if they had been born at term because the placenta often becomes less efficient after a full-term pregnancy. Approximately 10% of newborns in the United States are postmature.

Infancy

The period of **infancy** extends from the end of the neonatal period at 4 weeks until 2 years of age. Infancy is characterized by tremendous growth, increased coordination, and mental development.

A full-term child will generally double its birth weight by 5 months and triple it in a year. The formation of subcutaneous adipose tissue reaches its peak at about 9 months, causing the infant to appear chubby. Growth decelerates during the second year, during which time the infant gains only about 2.5 kg (5–6 lbs). During the second year, the infant develops locomotor and manipulative control and gradually becomes more lean and muscular.

Body length increases during the first year by 25 to 30 cm (10 to 12 in.). There is an additional 12 cm (5 in.) of growth during the second year. The brain and circumference of the head also grow rapidly during the first year and only moderately during the second. Head circumference increases by approximately 12 cm (5 in.) during the first year and only by an additional 2 cm during the second. The anterior fontanel gradually diminishes in size after 6 months and becomes effectively closed at any time from 9 to 18 months. It is the last of the fontanels to close. The brain is two-thirds of its adult size at the end of the first year and four-fifths of its adult size by the end of the second year.

By 2 years, most infants weigh approximately four times their birth weight and are 81 to 91 cm (32 to 36 in.) long. The body proportions of a 2-year-old are certainly not the same as those of an adult (fig. 22.27). Growth is a differential process, resulting in gradual changes in body proportions.

Deciduous teeth begin to erupt in most infants between 5 and 9 months. By the time they are a year old, most infants have 6 to 8 teeth. Eight more teeth erupt during the second year, making a total of 14 to 16, including the first deciduous molars and canine teeth.

UNIT VII

neonatal: Gk. *neos*, new; *natus*, born

premature: L. *prae*, before; *maturus*, ripe
infancy: L. *in*, not; *fans*, speaking

Figure 22.27   The relative proportions of the body from embryo to adulthood. The head of a newborn accounts for a quarter of the total body length, and the lower appendages make up about one-third. In an adult, the head accounts for about 13% of the total body length, whereas the length of the lower appendages constitutes approximately one-half.

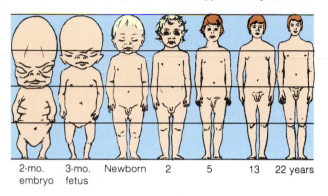

2-mo.   3-mo.   Newborn   2   5   13   22 years
embryo   fetus

The growth rates of children vary tremendously. Body lengths and weights are not always reliable indicators of normal growth and development. A more objective evaluation of a child's physical development is determined through radiographic analysis of skeletal ossification in the carpal region (fig. 22.28).

Childhood

Childhood is the period of growth and development extending from infancy to adolescence, at which time puberty begins. The duration of childhood varies because puberty begins at different ages for different people.

Childhood years are a period of relatively steady growth until preadolescence, which is characterized by a growth spurt. The average weight gain during childhood is about 3 to 3.5 kg (7 lb) per year. There is an average increase in height of 6 cm (2.5 in.) per year. The circumference of the head increases slightly—by about 3 to 4 cm (1.5 in.) during childhood—and by adolescence the head and brain are virtually adult size.

The facial bones continue to develop during childhood (fig. 22.29). Especially significant is the enlargement of the sinuses. The first permanent teeth generally erupt during the seventh year, and then the deciduous teeth are shed approximately in the same sequence as they were acquired. Deciduous teeth are replaced at a rate of about four per year over the next 7 years.

Although there is an average rate of growth during childhood due to genetics, there is a wide range in what is considered normal growth. If, for example, the 8-year-olds who were within the tallest and heaviest 10% of their age group were to stop growing for a year while their classmates grew normally, they would still be taller than half of their contemporaries and heavier than three-quarters of them.

During childhood, the average child becomes thinner and stronger each year as he or she grows taller. The average 10-year-old, for example, can throw a ball twice as far as the average 6-year-old. Visceral organs, particularly the heart and lungs, develop tremendously during this period, enabling a child to run faster and exercise longer.

Lymphatic tissue is at its peak of development during midchildhood and generally exceeds the amount of such tissue in the normal adult. Children need the extra lymphatic tissue to combat childhood diseases, especially in countries where nutrition is poor and health care is minimal.

Childhood *obesity* can become a serious physical and psychological problem if not corrected. Overweight children usually exercise less and run a greater risk of contracting serious illnesses. Frequently, they are teased and rejected by classmates, which causes psychological stress and learning impairments. At least 5% of children in the United States can be classified as obese. Childhood obesity and adult obesity usually go hand in hand. Obesity in adults is a major health problem considering that one of five adults is at least 30% over his or her ideal, healthy weight. A controlled diet and regular exercise are fundamental in correcting obesity.

Adolescence

Adolescence is the period of growth and development between childhood and adulthood. It begins around the age of 10 in girls and the age of 12 in boys. Adolescence is frequently said to end at 20 years of age, but it is not clearly delineated and varies with the developmental, physical, emotional, mental, or cultural criteria that define an adult.

Puberty *(pyoo'ber-te)* is the stage of early adolescence when the secondary sex characteristics become expressed and the sexual organs become functional. **Pubescence** *(pyoo-bes'ens)* refers to the continuum of physical changes during puberty, particularly in regard to body hair. Although puberty is under hormonal control, a complex interaction of other factors, including nutrition and socioeconomic forces, has a decisive influence on the onset and duration of puberty. Thus, for both sexes there is wide individual variation.

The end result of puberty is that the sexes are **sexually dimorphic** *(di-mor'fik);* that is, they are structurally different. The average adult male, for example, has a deeper voice and more body hair and is taller than the average adult female. Prior to puberty, male and female children have few major structural differences aside from the general appearance of the external genitalia.

Puberty actually begins before it is physically expressed. In most instances, significant amounts of sex hormones appear in the blood of females by the age of 10 and in males by the age of 11. Sexual changes are usually thought of as the only features of puberty, but major musculoskeletal changes take place as well. During late childhood, the body proportions of the sexes are similar, males being slightly taller. Under the influence of hormones, females experience a growth spurt in early adolescence that precedes that of males by nearly 2 years. During this time, females are temporarily taller. Once puberty begins in males, the heights of both sexes equalize, and at the culmination of puberty, males are approximately 10 cm (4 in.) taller than females on the average. By the time growth is completed at the end of adolescence, males are generally 13 cm (5 in.) taller than females.

adolescence: L. *adolescere,* to grow up
puberty: L. *pubertas,* adult form
dimorphic: Gk. *di,* two; *morphe,* form

UNIT VII

Figure 22.28 Radiographs of the right hand (a) of a child, (b) of an adolescent, and (c) of an adult.

(a)

(b)

(c)

Other dimorphic differences involving skeletal structures include a broadening of the pelvic girdle in females. The muscles of males become more massive and stronger than those of females. Females acquire a thicker subcutaneous layer of the skin during adolescence, which gives them a softer appearance.

Sexual maturation during adolescence includes not only the development of the reproductive organs but the appearance of secondary sex characters (see fig. 2.9). The sequence of average sexual maturation and the expression of secondary sex characteristics for both males and females is presented in table 22.4.

In females, the first physical indication of puberty is the appearance of **breast buds,** which are swellings of the breasts and slight enlargement and pigmentation of the areolar areas. Breast buds generally appear in healthy girls at about 11 years of age, but the age may range from 9 to 13 years. Approximately 3 years are required after the appearance of breast buds for the maturation of the breasts. Usually pubic hair begins to appear shortly after the breast buds, but in about one-third of all girls, sparse pubic hair appears before the breast buds. Axillary hair appears a year or two after pubic hair.

The first menstrual period, referred to as **menarche** (mĕ-nar'ke), generally occurs at age 13 but may occur as early as the age of 9 or as late as 17. During puberty, the vaginal secretions change from alkaline to acidic.

The onset of puberty in males varies just as much as in females but generally lags behind by about 1½ years. The first indication of puberty in boys is growth of the testes and the appearance of sparse pubic hair at the age of 12 on the average. This is followed by growth of the penis, which continues for about 2 years, and the appearance of axillary hair. Vocal changes generally begin during early puberty but are not completed until

UNIT VII

Table 22.4	Sequence of Adolescent Physical Development			
Females	**Age Span**		**Males**	
Growth spurt begins; breast buds and sparse pubic hair appear	10–11 yrs	11.5–13 yrs	Growth of testes and scrotum; sparse pubic hair appears; growth spurt begins; penis growth begins	
Appearance of straight, pigmented pubic hair; some deepening of voice; rapid growth of ovaries, uterus, and vagina; acidic vaginal secretion; menarche; further enlargement of breasts; kinky pubic hair; age of maximum growth	11–14 yrs	13–16 yrs	Appearance of straight, pigmented pubic hair; deepening of voice; maturation of penis, testes, scrotum, and accessory reproductive glands; ejaculation of semen; axillary hair; kinky pubic hair; sparse facial hair; age of maximum growth	
Appearance of axillary hair; breasts of adult size and shape; culmination of physical growth	14–16 yrs	16–18 yrs	Increased body hair; marked vocal change; culmination of physical growth	

Figure 22.29 Growth of the skull. The height of the cranial vault (distance between planes A and B) is drawn the same in both the infant and adult skulls. Growth of the skull occurs almost exclusively within the bones of the facial region.

midpuberty. Facial hair and chest hair (which may or may not be present) first appear toward the end of puberty. The mean age when semen can be ejaculated is 13.7 years, but sufficient mature sperm for fertility are generally not produced until 14 to 16 years of age.

Acne is an inflammatory disease of the integument that frequently occurs during adolescence. The increase in hormonal activity that is responsible for the physical changes taking place during puberty also affects the activity of the sebaceous glands and promotes the formation of inflamed superficial pustules and comedones. Tension and emotional stress may also promote acne. Acne in teenagers may cause serious psychological problems related to the need for peer acceptance.

Adulthood

Adulthood is the final stage of human physical change. It is the period of life beyond adolescence. An adult has reached maximum physical stature as determined by genetic, nutritional, and environmental factors. Although skeletal maturity is reached in early adulthood, anatomical and physiological changes continue throughout adulthood and are part of the aging process.

Sexual dimorphism in human adults goes beyond the obvious primary and secondary sex differences. The sexes differ anatomically, physiologically, metabolically, and behaviorally (psychologically and socially). Some of these differences manifest themselves prenatally and during childhood. Others are characteristics of adolescence and adulthood. It is uncertain to what extent specific dimorphisms of the sexes are genetically determined through hormonal action or influenced by environmental (including cultural) factors. It is also unclear how these governing factors are expressed in observed physical characteristics.

The shape of the adult body is determined primarily by the skeleton and attached muscles and also by the subcutaneous connective tissue (especially adipose tissue) and extracellular body fluids. Although the body proportions of adult males and females vary widely between individuals (fig. 22.30), in general, adult males have relatively longer appendages than females, their shoulders are relatively broader, and their pelvises are narrower. A male also has a relatively longer neck.

The general body composition of males and females can also be compared numerically. Mean data of body composition are summarized in table 22.5. These data show that total body fluid and skeletal weight is lower in adult females than in adult males. Females, however, have a higher percentage of body fat. Other differences not shown in table 22.5 are that adult females have lower blood pressures, erythrocyte counts (hematocrits), basal metabolic rates, and respiratory rates than adult males. Females, however, have higher heart rates and oral temperatures.

Some of the physical and behavioral characteristics of each stage of human growth and development are summarized in table 22.6.

Physical anthropologists have long been interested in the body proportions of racial groups. *Anthropometry* is the study of the physical differences, particularly skeletal, between racial groups. Proportions in anthropometric studies are expressed in indices; that is, one measurement reckoned as a percentage of another measurement. The cranial index, for example, is the breadth of the skull expressed as a percentage of its length.

anthropometry: Gk. *anthropos*, human; *metron*, measure

UNIT VII

Figure 22.30 Relative differences in the physiques of an adult male and female. A male has proportionately longer appendages and a longer neck than a female.

Male | Female

Anterior

Female | Male

Posterior

Table 22.5	Body Composition in Average Adults							
	Males				**Females**			
	Absolute		**Relative (% Body Weight)**		**Absolute**		**Relative (% Body Weight)**	
	Age 25	Age 65	Age 25	Age 65	Age 25	Age 65	Age 25	Age 65
Body weight	70.0 kg	70.0 kg			69.0 kg	60.0 kg		
Body fluids	41.0 liters	37.0 liters	58.9%	52.9%	30.8 liters	28.0 liters	51.3%	46.7%
Intracellular	24.0 liters	19.2 liters	34.3%	27.4%	16.6 liters	14.3 liters	27.7%	23.8%
Extracellular	17.2 liters	17.8 liters	24.6%	25.5%	14.2 liters	13.7 liters	23.6%	22.9%
Plasma volume	3,302 ml	2,940 ml	4.7%	4.2%	2,760 ml	2,462 ml	4.6%	4.1%
Lean body weight	56.3 kg	50.5 kg	80.4%	72.1%	42.0 kg	38.2 kg	70.2%	63.7%
Body fat	13.7 kg	19.5 kg	19.6%	27.9%	17.9 kg	21.8 kg	29.8%	36.3%
Skeletal weight	5.8 kg	5.7 kg	8.3%	8.1%	4.4 kg	4.2 kg	7.3%	7.0%

From K. H. Oleson, "Body Composition in Normal Adults" in *Human Body Composition*, Vol. 7:177–190. Copyright © 1965 Pergamon Press Ltd., Oxford, England. Reprinted by permission.

UNIT VII

Table 22.6	Summary of Stages of Postnatal Periods	
Stage	Time Period	Physical and Behavioral Characteristics
Neonatal period	Birth to end of fourth week	Stabilizing of body systems necessary to carry on respiration, obtain nutrients, digest nutrients, excrete wastes, regulate body temperature, and circulate blood
Infancy	End of fourth week to two years	Tremendous growth; teeth begin to erupt; muscular and nervous systems develop so that motor activities are possible; verbal communication begins
Childhood	One year to puberty	Consistent growth; deciduous teeth erupt and are replaced by permanent teeth; motor control improves; urinary bladder and bowel controls are established; intellect develops rapidly
Adolescence	Puberty to adulthood	Reproductive system matures; growth spurts in skeletal and muscular systems; intellect and emotional maturity continue to develop
Adulthood	Adolescence to old age	Maximum physical stature and strength obtained; anatomical and physiological degenerative changes begin
Senescence	Old age to death	Senescence continues; body becomes less able to cope with diseases and physical demands; death—usually from physical disturbances in the cardiovascular system or disease processes in vital organs

There are some racial differences in the physical stature of both males and females. A particularly standardized expression of racial differences are the indices of appendage lengths relative to sitting height. People of African ancestry, for example, have comparatively long appendages to sitting height; moreover, the forearm and leg are long relative to the brachium and thigh (fig. 22.31). Australian Aborigines have even longer legs proportionately than do Native Africans. People of African ancestry also have the narrowest pelvic girdle for a given shoulder width. People of Asian ancestry have relatively short appendage lengths to sitting heights. These differences provide distinct advantages and disadvantages in certain sports. Blacks have an advantage in many track events, particularly the sprints and high hurdles, whereas whites are generally adapted to distance running. Asians often excel in gymnastics and weight lifting.

1. Construct a table that lists the stages of postnatal growth from infancy through adolescence and indicate the events or characteristics of each stage.
2. List four reflexes in a newborn that are critical for survival.
3. Define the terms *puberty, pubescence, sexual dimorphism,* and *menarche.* What is the mean age of puberty and what causes its occurrence?
4. Describe the physical characteristics of adulthood. Compare the body structure of an adult female to that of an adult male.

Inheritance

Inheritance is the passage of hereditary traits carried by the genes on chromosomes from one generation to another.

Objective 18.   Define *genetics.*

Objective 19.   Discuss the variables that account for a person's phenotype.

Objective 20.   Explain how probability is involved in predicting inheritance and use a Punnett square to illustrate selected probabilities.

Figure 22.31   Comparison of Caucasian and Afro-American physiques. Photographs are of two Olympic 400-meter runners and have been scaled so that both have the same sitting height.

UNIT VII

Figure 22.32 A karyotype of homologous pairs of chromosomes obtained from a human diploid cell. The first 22 pairs of chromosomes are called the autosomal chromosomes. The sex chromosomes are (a) XY for a male and (b) XX for a female.

Figure 22.33 A pair of homologous chromosomes. Homologous chromosomes contain genes for the same characteristic at the same locus.

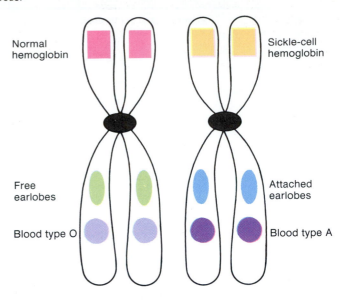

Genetics is the branch of biology that deals with inheritance. Genetics and inheritance are important in anatomy and physiology because of the numerous developmental and functional disorders that have a genetic basis. The knowledge of which disorders and diseases are inherited finds practical application in genetic counseling. The genetic inheritance of an individual begins with conception.

Each zygote inherits 23 chromosomes from the mother and 23 chromosomes from the father. This does not produce 46 different chromosomes; rather, it produces 23 pairs of *homologous chromosomes*. Each member of a homologous pair, with the important exception of the sex chromosomes, looks like the other and contains similar genes (such as those coding for eye color, height, and so on). These homologous pairs of chromosomes can be **karyotyped** (photographed or illustrated) and identified, as shown in figure 22.32. Each cell that contains 46 chromosomes (that is *diploid*) has two number-1 chromosomes, two number-2 chromosomes, and so on through chromosomes number 22. The first 22 pairs of chromosomes are called **autosomal** (*aw″tŏ-so′-mal*) **chromosomes.** The twenty-third pair of chromosomes are the **sex chromosomes,** which may look different and may carry different genes. In a female these consist of two X chromosomes, whereas in a male there is one X chromosome and one Y chromosome.

Genes and Alleles A **gene** is the portion of the DNA of a chromosome that contains the information needed to synthesize a particular protein molecule. Although each diploid cell has a pair of genes for each characteristic, these genes may be present in a number of alternate forms. Those alternative forms of a gene that affect the same characteristic, but produce different expressions of that characteristic, are called **alleles** (*a-lēlz*). One allele of each pair originates from the female parent and the other from the male. The shape of a person's ears, for example, is determined by the kind of allele received from each parent and how the alleles interact with one another. Alleles are always located on the same spot (called a **locus**) on homologous chromosomes (fig. 22.33).

For any particular pair of alleles in a person, the two alleles are either identical or not identical. If the alleles are identical, the person is said to be **homozygous** (*ho″mo-zi′gus*) for that particular characteristic. But if the two alleles are different, the person is **heterozygous** (*het″er-o-zi′gus*) for that particular trait.

Genotype and Phenotype A person's DNA contains a catalog of genes known as the **genotype** of that person. The expression of those genes results in certain observable characteristics referred to as the **phenotype.**

If the alleles for a particular trait are homozygous, the characteristic expresses itself in a specific manner (two alleles for attached earlobes, for example, results in a person with attached earlobes). If the alleles for a particular trait are heterozygous, however, the allele that expresses itself and the way in which the genes for that trait interact will determine the phenotype. Often one of the alleles expresses itself as the **dominant allele,** while the other does not and is the **recessive allele.** The combinations of dominant and recessive alleles are responsible for a person's hereditary traits (table 22.7).

UNIT VII

Table 22.7	Hereditary Traits in Humans Determined by Single Pairs of Dominant and Recessive Alleles		
Dominant	**Recessive**	**Dominant**	**Recessive**
Free earlobes	Attached earlobes	Color vision	Color blindness
Dark brown hair	All other colors	Broad lips	Thin lips
Curly hair	Straight hair	Ability to roll tongue	Lack of this ability
Pattern baldness (♂♂)	Baldness (♀♀)	Arched feet	Flat feet
Pigmented skin	Albinism	A or B blood factor	O blood factor
Brown eyes	Blue or green eyes	Rh blood factor	No Rh blood factor

Figure 22.34 Inheritance of ear shape. Two parents with free earlobes can have a child with attached earlobes.

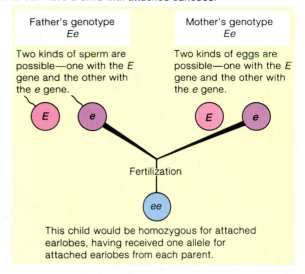

Figure 22.35 Inheritance of the shape of earlobes.

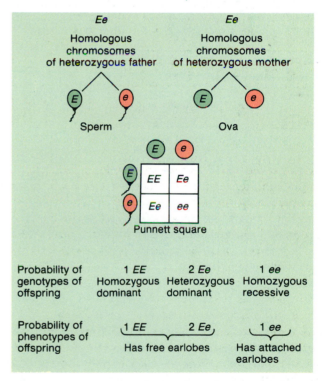

In describing genotypes, it is traditional to use letter symbols to refer to the alleles of an organism. The dominant alleles are symbolized by uppercase letters and the recessive alleles are symbolized by lowercase. Thus, the *genotype* of a person who is homozygous for free earlobes due to a dominant allele is symbolized *EE;* a heterozygous pair is symbolized *Ee.* In both of these instances, the *phenotypes* of the individuals would be free earlobes due to the presence of a dominant allele in each genotype. A person who inherited two recessive alleles for earlobes would have the genotype *ee* and would have attached earlobes.

Thus, three genotypes are possible when gene pairing involves dominant and recessive alleles. They are *homozygous dominant (EE), heterozygous (Ee),* and *homozygous recessive (ee).* Only two phenotypes are possible, however, since the dominant allele is expressed in both the homozygous dominant (*EE*) and the heterozygous (*Ee*) individuals. The recessive allele is expressed only in the homozygous recessive (*ee*) condition. Refer to figure 22.34 for an illustration of how a homozygous recessive trait may be expressed in a child of parents who are heterozygous dominant.

Probability A **Punnett square** is a convenient way to express the probabilities of allele combinations for a particular inheritable trait. In constructing a Punnett square, the male gametes (spermatozoa) carrying a particular trait are placed at the side of the chart, and the female gametes (ova) at the top, as in figure 22.35. The four spaces on the chart represent the possible combinations of male and female gametes that could form zygotes. The probability of an offspring having a particular genotype is 1 in 4 (.25) for homozygous dominant and homozygous recessive and 1 in 2 (.50) for heterozygous dominant.

A genetic study in which a single characteristic (e.g., ear shape) is followed from parents to offspring is referred to as a **monohybrid cross.** A genetic study in which two characteristics are followed from parents to offspring is referred to as a

dihybrid cross (fig. 22.36). The term **hybrid** refers to an offspring descended from parents who have different genotypes.

Sex-Linked Inheritance Certain inherited traits are located on a sex-determining chromosome and are called **sex-linked** characteristics. The allele for *red-green color blindness,* for example, is determined by a recessive allele (designated *c*) found

Figure 22.36 A dihybrid cross studies the probability of two characteristics at the same time. Any of the combinations of genes that have a *D* and an *E* (nine possibilities) will have free earlobes and dark hair. These are indicated with an asterisk (∗). Three of the possible combinations have two alleles for attached earlobes (*ee*) and at least one allele for dark hair. They are indicated with a dot (•). Three of the combinations have free earlobes and light hair. These are indicated with a square (■). The remaining possibility has the genotype *eedd* for attached earlobes and light hair.

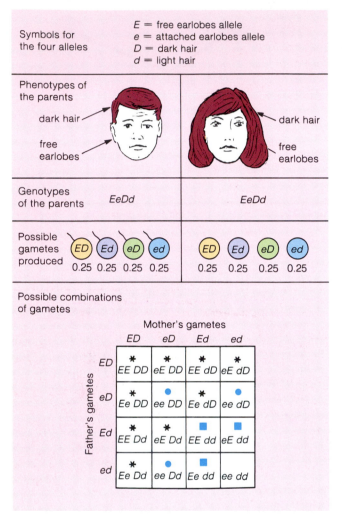

on the X chromosome but not on the Y chromosome. Normal color vision (designated *C*) dominates. The ability to discern red-green colors, therefore, depends entirely on the X chromosomes. The genotype possibilities are

$X^C Y$ Normal male
$X^c Y$ Color-blind male
$X^C X^C$ Normal female
$X^C X^c$ Normal female carrying the recessive allele
$X^c X^c$ Color-blind female

In order for a female to be red-green color-blind, she must have the recessive allele on both of her X chromosomes. Her father would have to be red-green color-blind and her mother would have to be a carrier for this condition. A male with only one such allele on his X chromosome, however, will show the

characteristic. Since a male receives his X chromosome from his mother, the inheritance of sex-linked characteristics usually passes from mother to son.

> *Hemophilia* is a sex-linked condition caused by a recessive allele. The blood in a person with hemophilia fails to clot or clots very slowly after an injury. If *H* represents normal clotting and *h* represents abnormal clotting, then males with $X^H Y$ will be normal and males with $X^h Y$ will be hemophiliac. Females with $X^h X^h$ will have the disorder.

1. Define *genetics, genotype, phenotype, allele, dominant, recessive, homozygous,* and *heterozygous.*
2. List several dominant and recessive traits inherited in humans. What are some variables that determine a person's phenotype?
3. Construct a Punnett square to show the possible genotypes for color blindness of an $X^C Y$ male and an $X^C X^c$ female.

Clinical Considerations

Pregnancy and childbirth are natural events in human biology and generally progress smoothly without complications. Prenatal development is amazingly precise, and although traumatic, childbirth for most women in the world takes place without the aid of a physician. Occasionally, however, serious complications arise, and the knowledge of an obstetrician is required. The physician's knowledge of what constitutes normal development and what factors are responsible for congenital malformations ensures the embryo and fetus every possible chance to develop normally. Many of the clinical aspects of prenatal development involve what might be referred to as applied developmental biology.

In clinical terms, gestation is frequently divided into three phases, or **trimesters,** each lasting three calendar months. By the end of the **first trimester** all of the major body systems are formed, the fetal heart can be detected, the external genitalia are developed, and the fetus is about the width of the palm of an adult's hand. During the **second trimester,** fetal quickening can be detected, epidermal features are formed, and the vital body systems are functioning. The fetus, however, still would be unlikely to survive if birth were to occur. At the end of the second trimester, fetal length is about equal to the length of an adult's hand. The fetus experiences a tremendous amount of growth and refinement in system functioning during the **third trimester.** A fetus of this age may survive if born prematurely, and of course, the chances of survival improve as the length of pregnancy approaches the natural delivery date.

Many clinical considerations are associated with prenatal development, some of which relate directly to the female reproductive system. Other developmental problems are genetically related and will be mentioned only briefly. Of clinical concern for developmental anatomy are such topics as ectopic pregnancies, so-called test-tube babies, multiple pregnancy, fetal monitoring, and congenital defects.

UNIT VII

Figure 22.37 The various sites of ectopic pregnancies. The normal site is indicated by an X and the abnormal sites are indicated by letters in order of frequency of occurrence.

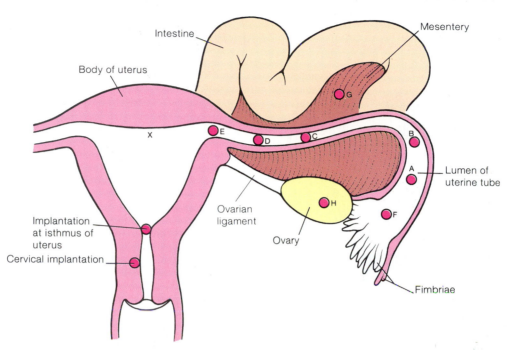

Abnormal Implantation Sites

In an **ectopic pregnancy** the blastocyst implants outside the uterus or in an abnormal site within the uterus (fig. 22.37). About 95% of the time, the ectopic location is within the uterine tube and is referred to as a **tubal pregnancy.** Occasionally, implantation occurs near the cervix, where development of the placenta blocks the cervical opening. This condition, called **placenta previa,** causes serious bleeding. Ectopic pregnancies will not develop normally in unfavorable locations, and the fetus seldom survives beyond the first trimester. Tubal pregnancies are terminated through medical intervention. If a tubal pregnancy is permitted to progress, however, the uterine tube generally ruptures, followed by hemorrhaging. Depending on the location and the stage of development (hence vascularity) of a tubal pregnancy, it may or may not be life-threatening to the woman.

In Vitro Fertilization and Artificial Implantation

Reproductive biologists have been able to fertilize a human oocyte in vitro (outside the body), culture it to the blastocyst stage, and then perform artificial implantation, leading to a full-term development and delivery. This is the so-called test-tube baby. To obtain the oocyte, a specialized laparoscope (fig. 22.38) is used to aspirate the preovulatory egg from a mature vesicular ovarian follicle. The oocyte is then placed in a suitable culture

Figure 22.38 A laparoscope, used for various abdominal operations including the extraction of a preovulatory ovum.

medium, where it is fertilized with spermatozoa. After the zygote forms, the sequential pre-embryonic development continues until the blastocyst stage, at which time implantation is performed. In vitro fertilization with artificial implantation is a means of overcoming infertility problems due to damaged, blocked, or missing uterine tubes in females or low spermatozoa counts in males.

previa: L. *previa,* appearing before or in front of

UNIT VII

Figure 22.39 The formation of dizygotic twins. Twins of this type are not identical but fraternal and may have (a) separate or (b) fused placentas. (c) A photograph of fraternal twins at 11 weeks.

Multiple Pregnancy

Twins occur about once in 85 pregnancies. They can develop in two ways. **Dizygotic** (fraternal) **twins** develop from two zygotes resulting from two spermatozoa fertilizing two oocytes in the same ovulatory cycle (fig. 22.39). **Monozygotic** (identical) **twins** form from a single zygote (fig. 22.40). Approximately one-third of twins are monozygotic.

Dizygotic twins may be of the same sex or different sexes and are not any more alike than brothers or sisters born at different times. Dizygotic twins always have two chorions and two amnions, but the chorions and the placentas may be fused.

Monozygotic twins are of the same sex and are genetically identical. Any physical differences in monozygotic twins are caused by environmental factors during morphogenic development (e.g., there might be a differential vascular supply that causes slight differences to be expressed). Monozygotic twinning is usually initiated toward the end of the first week when the embryoblast divides to form two embryonic primordia. Monozygotic twins have two amnions but only one chorion and a common placenta. If the embryoblast fails to completely divide, **conjoined twins** (Siamese twins) may form.

Triplets occur about once in 7,600 pregnancies and may be (1) all from the same ovum and identical, (2) two identical and the third from another ovum, or (3) three zygotes from three different ova. Similar combinations occur in quadruplets, quintuplets, and so on.

Fetal Monitoring

Obstetrics has benefited greatly from the advancements made in fetal monitoring in the last two decades. Before these techniques became available, physicians could determine the welfare of the unborn child only by auscultation of the fetal heart and palpation of the fetus. Currently, there are several tests that provide much information about the fetus during any stage of development. Fetal conditions that can now be diagnosed and evaluated include genetic disorders, hypoxia, blood disorders, growth retardation, placental functioning, prematurity, postmaturity, and intrauterine infections. These tests also help determine the advisability of an abortion.

Radiographs of the fetus were once commonly performed but were found harmful and have been replaced by other methods of evaluation that are safer and more informative. **Ultrasonography,** produced by a mechanical vibration of high frequency, produces a safe high resolution (sharp image) of fetal structure (fig. 22.41). Ultrasonic imaging is a reliable way to determine pregnancy as early as 6 weeks after ovulation. It can also be used to determine fetal weight, length, and position, as well as to diagnose multiple fetuses.

Amniocentesis is a technique used to obtain a small sample (5–10 mℓ) of amniotic fluid with a syringe so that the fluid can

amniocentesis: Gk. *amnion,* lamb (fetal membrane); *kentesis,* puncture

Figure 22.40   The formation of monozygotic twins. Twins of this type develop from a single zygote and are identical. Such twins have two amnions but one chorion and a common placenta.

One placenta

One chorionic sac

Two amniotic sacs

Implantation of blastocyst

Zygote

Two-cell stage

Blastocyst with two inner cell masses

(a)

(b)

Single placenta

Anastomosis of placental vessels

Figure 22.41   A color-enhanced ultrasonogram of a fetus during the 3rd trimester. The left hand is raised, as if waving to the viewer.

Amniotic fluid

Placenta

Left cerebral hemisphere

Orbit of eye

Left hand

Uterine wall

Thorax

Figure 22.42   Fetoscopy.

Syringe withdraws blood

Endoscope (guided to exact location by pulsed sound waves)

Placenta

Fiber optics

Needle punctures fetal vein on placenta

be assessed (see fig. 22.14). Amniocentesis is most often performed to determine fetal maturity, but it can also help predict serious disorders like *Down syndrome* and *Gaucher's disease* (a metabolic disorder).

Fetoscopy (fig. 22.42) goes beyond amniocentesis by allowing direct examination of the fetus. Using fetoscopy, physicians scan the uterus with pulsed sound waves to locate fetal structures, the umbilical cord, and the placenta. Skin samples are taken from the head of the fetus and blood samples extracted from the placenta. The principal advantage of fetoscopy is that external features of the fetus (such as fingers, eyes, ears, mouth, and genitals) can be carefully observed. Fetoscopy is also

fetoscopy: L. *fetus,* offspring; *skopein,* to view

Figure 22.43 Monitoring fetal heart rate and uterine contractions using an FHR-UC device.

used to determine several diseases, including hemophilia, thalassemia, and sickle-cell anemia cases, 40% of which are missed by amniocentesis.

Most hospitals are now equipped with instruments that monitor fetal heart rate and uterine contractions during labor and can detect any complication that may arise during the delivery. This procedure is called Electronic Monitoring of Fetal Heart Rate and Uterine Contractions (FHR-UC Monitoring). The extent of stress to the fetus from uterine contractions can be determined through FHR-UC monitoring (fig. 22.43). Long, arduous deliveries are taxing to both the mother and fetus. If the baby's health and vitality are diagnosed to be in danger because of a difficult delivery, the physician may decide to perform a cesarean section.

Congenital Defects

Major developmental problems called **congenital malformations** occur in approximately 2% of all newborn infants. The causes of congenital conditions include genetic inheritance, mutation (genetic change), and environmental factors. About 15% of neonatal deaths are attributed to congenital malformations. The branch of developmental biology concerned with abnormal development and congenital malformations is called *teratology*. Many congenital problems have been discussed in previous chapters, in connection with the body system in which they occur.

CLINICAL CASE STUDY ANSWER

Two amnions and one chorion in all but very unusual cases prove the twins to be monozygotic. The two infants, therefore, are genetically identical. Thus, when considering genetic disorders such as cleft palate, one would expect a high degree of concordance (both twins of a monozygotic pair exhibit a particular anomaly). Many such defects however can be present in only one twin—a consequence of nongenetic factors, such as intrauterine environment. An example would be inadequate blood supply to only one twin, resulting in a defect in that twin but not in the other.

congenital: L. *congenitus,* born with

teratology: Gk. *teras,* monster; *logos,* study of

GENETIC DISORDERS OF CLINICAL IMPORTANCE

cystic fibrosis An autosomal recessive disorder characterized by the formation of thick mucus in the lungs and pancreas that interferes with normal breathing and digestion.

familial cretinism An autosomal recessive disorder characterized by a lack of thyroid secretion due to a defect in the iodine transport mechanism. Untreated children are dwarfed, sterile, and may be mentally retarded.

galactosemia An autosomal recessive disorder characterized by an inability to metabolize galactose, a component of milk sugar. Patients with this disorder have cataracts, damaged livers, and mental retardation.

gout An autosomal dominant disorder characterized by an accumulation of uric acid in the blood and tissue due to an abnormal metabolism of purines.

hepatic porphyria An autosomal dominant disorder characterized by painful GI disorders and neurologic disturbances due to an abnormal metabolism of porphyrins.

hereditary hemochromatosis A sex-influenced, autosomal dominant disorder characterized by an accumulation of iron in the pancreas, liver, and heart, resulting in diabetes, cirrhosis, and heart failure.

hereditary leukomelanopathy An autosomal recessive disorder characterized by decreased pigmentation in the skin, hair, and eyes and abnormal white blood cells. People with this condition are generally susceptible to infections and early deaths.

Huntington's chorea An autosomal dominant disorder characterized by uncontrolled twitching of skeletal muscles and the deterioration of mental capacities. A latent expression of this disorder allows the mutant gene to be passed to children before symptoms develop.

Marfan's syndrome An autosomal dominant disorder characterized by tremendous growth of the extremities, extreme looseness of the joints, dislocation of the lenses, and congenital cardiovascular defects.

phenylketonuria (*fen''il-kēt''on-oor'e-ă*) **(PKU)** An autosomal recessive disorder characterized by an inability to metabolize the amino acid phenylalanine. This is accompanied by brain and nerve damage and mental retardation.

pseudohypertrophic muscular dystrophy A sex-linked recessive disorder characterized by progressive muscle atrophy. It usually begins during childhood and causes death in adolescence.

retinitis pigmentosa A sex-linked recessive disorder characterized by progressive atrophy of the retina and eventual blindness.

Tay–Sachs disease An autosomal recessive disorder characterized by a deterioration of physical and mental abilities, early blindness, and early death.

CHAPTER SUMMARY

Fertilization (pp. 718–719)

1. Upon fertilization of a secondary oocyte by a spermatozoon in the uterine tube, meiotic development is completed and a diploid zygote is formed.
2. Morphogenesis is the sequential formation of body structures during the prenatal period of human life. The prenatal period lasts 38 weeks and is divided into a pre-embryonic, an embryonic, and a fetal period.
3. A capacitated spermatozoon digests its way through the zona pellucida and corona radiata layers of the secondary oocyte to complete the fertilization process and formation of a zygote.

Pre-embryonic Period (pp. 719–726)

1. Cleavage of the zygote is initiated within 30 hours and continues until a morula forms; the morula enters the uterine cavity on about the third day.
2. A hollow, fluid-filled space forms within the morula, and it is then called a blastocyst.
3. Implantation begins between the fifth and seventh day and is enabled by the secretion of enzymes that digest a portion of the endometrium.
   (a) During implantation, the trophoblast cells secrete human chorionic gonadotropin (hCG), which prevents the breakdown of the endometrium and menstruation.
   (b) The secretion of hCG declines by the tenth week as the developed placenta secretes steroids that maintain the endometrium.

4. The embryoblast of the implanted blastocyst flattens into the embryonic disc, from which the primary germ layers of the embryo develop.
   (a) Ectoderm gives rise to the nervous system, the epidermis of the skin and epidermal derivatives, and portions of sensory organs.
   (b) Mesoderm gives rise to bones, muscles, blood, reproductive organs, the dermis of the skin, and connective tissue.
   (c) Endoderm gives rise to linings of the GI tract, digestive organs, the respiratory tract and lungs, and the urinary bladder and urethra.

Embryonic Period (pp. 726–734)

1. The events of the 6-week embryonic period include the differentiation of the germ layers into specific body organs and the formation of the placenta, the umbilical cord, and the extraembryonic membranes. These events make it possible for morphogenesis to continue.
2. The extraembryonic membranes include the amnion, yolk sac, allantois, and chorion.
   (a) The amnion is a thin membrane surrounding the embryo. It contains amniotic fluid that cushions and protects the embryo.
   (b) The yolk sac produces blood for the embryo.
   (c) The allantois also produces blood for the embryo and gives rise to the umbilical arteries and vein.
   (d) The chorion participates in the formation of the placenta.
3. The placenta, formed from both maternal and embryonic tissue, has a transport role in providing for the metabolic needs of the fetus and in removing its wastes.
   (a) The placenta produces steroid and polypeptide hormones.
   (b) Nicotine, drugs, alcohol, and viruses can cross the placenta to the fetus.
4. The umbilical cord, containing two umbilical arteries and one umbilical vein, is formed as the amnion envelops the tissues on the underside of the embryo.

Huntington's chorea: from George Huntington, American physician, 1850–1916

Marfan's syndrome: from Antoine Bernard-Jean Marfan, French physician, 1858–1942

Tay–Sachs disease: from Warren Tay, English physician, 1843–1927, and Bernard Sachs, American neurologist, 1858–1944

UNIT VII

5. From the third to the eighth week, the structure of all the body organs, except the genitalia, becomes apparent.
   (a) During the third week, the primitive node forms from the primitive line, which later gives rise to the notochord and intraembryonic mesoderm.
   (b) By the end of the fourth week, the heart is beating; the primordial tissues of the eyes, brain, spinal cord, lungs, and digestive organs are properly positioned; and the superior and inferior limb buds are recognizable.
   (c) At the end of the fifth week, the sense organs are formed in the enlarged head and the appendages have developed with digital primordia evident.
   (d) During the seventh and eighth weeks the body organs, except the genitalia, are formed and the embryo appears distinctly human.

Fetal Period (pp. 734–738)

1. A small amount of tissue differentiation and organ development occurs during the fetal period, but for the most part fetal development is primarily limited to body growth.
2. Between weeks 9 and 12, ossification centers appear, the genitalia are formed, and the digestive, urinary, respiratory, and muscle systems show functional activity.
3. Between weeks 13 and 16, facial features are formed and the fetal heartbeat can be detected with a stethoscope.
4. During the 17-to-20-week period, quickening can be felt by the mother, and vernix caseosa and lanugo cover the skin of the fetus.
5. During the 21-to-25-week period, substantial weight gain occurs and the fetal skin becomes wrinkled and pinkish.
6. Toward the end of the 26-to-29-week period, the eyes have opened, the gonads have descended in a male, and the fetus is developed to the extent that it might survive if born prematurely.
7. At 38 weeks, the fetus is full-term; the normal gestation is 266 days.

Labor and Parturition (pp. 738–740)

1. Labor and parturition are the culmination of gestation and require the action of oxytocin, secreted by the posterior pituitary, and prostaglandins, produced in the uterus.
2. Labor is divided into the dilation, expulsion, and placental stages.

Periods of Postnatal Growth (pp. 740–745)

1. The course of human life after birth is seen in terms of physical and physiological changes and the attainment of maturity in the neonatal, infant, childhood, adolescent, and adulthood periods.
2. The neonatal period, extending from birth to the end of the fourth week, is characterized by major physiological changes.
   (a) The most critical need of the newborn is to establish adequate respiratory and heart rates. A normal respiratory rate is 30 to 40 respirations per minute, and a normal heart rate ranges from 120 to 160 beats per minute.
   (b) The four reflexes in the newborn critical to survival are the suckling reflex, the rooting reflex, the crying reflex, and the breathing reflex.
3. Infancy, extending from 4 weeks until the age of 2 years, is characterized by tremendous growth, increased coordination, and mental development.
   (a) By 2 years, most infants weigh about four times their birth weight and average between 32 and 36 inches in length.
   (b) Growth is a differential process resulting in gradual changes from infant to adult body proportions.
4. Childhood, extending from infancy to adolescence, is characterized by steady growth until preadolescence, at which time there is a marked growth spurt.
   (a) During childhood, the average child becomes thinner and stronger each year as he or she grows taller.
   (b) The fact that disease and death are relatively rare during childhood may be because lymphatic tissue is then at its peak of development and is present in greater amounts in children than in adults.
5. Adolescence is the period of growth and development between childhood and adulthood.
   (a) Puberty is the stage of early adolescence when the secondary sex characteristics are expressed and the sex organs become functional.
   (b) The end result of puberty is the structural expression of gender, or sexual dimorphism.
   (c) Menarche generally occurs in adolescent girls at the age of 13, but ranges from 9 to 17 years. At this time, vaginal secretions change from alkaline to acid.
   (d) The first physical indications of puberty are the appearance of breast buds in females and the growth of the testes and the appearance of sparse pubic hair in males.
   (e) Although semen may be ejaculated at age 13, sufficient mature spermatozoa for fertility are not produced until 14 to 16 years of age.
6. Adulthood, the final period of human physical change, is characterized by gradual senescence as a person ages.
   (a) Although skeletal maturity is reached in early adulthood, anatomical and physiological changes continue throughout adulthood and are part of the aging process.
   (b) Sexual dimorphism in human adults is evident anatomically, physiologically, metabolically, and behaviorally.
   (c) Male and Female differences in the stature, proportions, and composition of the body may become more apparent with age.

Inheritance (pp. 745–748)

1. Inheritance is the passage of hereditary traits carried on the genes of chromosomes from one generation to another.
2. Each zygote contains 22 pairs of autosomal chromosomes and 1 pair of sex chromosomes—XX in a female and XY in a male.
3. A gene is the portion of a DNA molecule that contains information for the production of one kind of protein molecule. Alleles are different forms of genes that occupy corresponding positions on homologous chromosomes.
4. The combination of genes in an individual's cells constitutes his or her genotype; the appearance of a person is his or her phenotype.
   (a) Dominant alleles are symbolized by uppercase letters and recessive alleles are symbolized by lowercase letters.
   (b) The three possible genotypes are homozygous dominant, heterozygous, and homozygous recessive.
5. A Punnett square is a convenient means for expressing probability.
   (a) The probability of a particular genotype is 1 in 4 (0.25) for homozygous dominant and homozygous recessive and 1 in 2 (0.50) for heterozygous dominant.
   (b) A single trait is studied in a monohybrid cross; two traits are studied in a dihybrid cross.
6. Sex-linked traits like color blindness and hemophilia are carried on the sex-determining chromosome.

UNIT VII

REVIEW ACTIVITIES

Objective Questions

1. The pre-embryonic period is completed when
   - (a) the blastocyst implants.
   - (b) the placenta forms.
   - (c) the blastocyst reaches the uterus.
   - (d) the primary germ layers form.
2. The yolk sac produces blood for the embryo until
   - (a) the heart is functional.
   - (b) the kidneys are functional.
   - (c) the liver is functional.
   - (d) the baby is delivered.
3. Which of the following is a function of the placenta?
   - (a) production of steroids and hormones
   - (b) diffusion of nutrients and oxygen
   - (c) production of enzymes
   - (d) all of the above
4. The decidua basalis is
   - (a) a component of the umbilical cord.
   - (b) the embryonic portion of the villous chorion.
   - (c) the maternal portion of the placenta.
   - (d) a vascular membrane derived from the trophoblast.
5. Which of the following could diffuse across the placenta?
   - (a) nicotine
   - (b) alcohol
   - (c) heroin
   - (d) all of the above
6. During which week following conception does the embryonic heart begin pumping blood?
   - (a) the fourth week
   - (b) the fifth week
   - (c) the sixth week
   - (d) the eighth week
7. Which of the following is the period of growth from birth to the end of the fourth week?
   - (a) the neonatal period
   - (b) the fetal period
   - (c) infancy
   - (d) the suckling period
8. The normal newborn heart rate is
   - (a) 70–80 beats/min.
   - (b) 120–160 beats/min.
   - (c) 100–120 beats/min.
   - (d) 180–200 beats/min.
9. The continuum of physical change occurring in adolescence that regulates the growth of body hair is known as
   - (a) puberty.
   - (b) pubal progression.
   - (c) pubescence.
   - (d) dimorphism.
10. Which is *not* characteristic of the female as compared to the male?
    - (a) lower blood pressure
    - (b) higher basal metabolic rate
    - (c) lower red blood cell count
    - (d) faster heart rate
11. The first physical indication of puberty in females is generally
    - (a) alkaline vaginal secretions.
    - (b) a widening pelvis.
    - (c) breast buds.
    - (d) axillary hair.
12. Twins that develop from two zygotes resulting from the fertilization of two ova by two spermatozoa in the same ovulatory cycle are referred to as
    - (a) monozygotic.
    - (b) conjoined.
    - (c) dizygotic.
    - (d) identical.
13. Match the genotype descriptions in the left-hand column with the correct symbols in the right-hand column.
    homozygous recessive   *Bb*
    heterozygous   *bb*
    homozygous dominant   *BB*
14. An allele that is *not* expressed in a heterozygous genotype is called
    - (a) recessive.
    - (b) dominant.
    - (c) genotypic.
    - (d) phenotypic.
15. If the genotypes of both parents are *Aa* and *Aa*, the offspring probably will be
    - (a) ½ *AA* and ½ *aa*.
    - (b) all *Aa*.
    - (c) ¼ *AA*, ½ *Aa*, ¼ *aa*.
    - (d) ¾ *AA* and ¼ *aa*.

Essay Questions

1. Describe the implantation of the blastocyst into the uterine wall and the involvement of the trophoblast in the formation of the placenta.
2. Explain how the primary germ layers form. What major structures does each germ layer give rise to?
3. Explain why development during the embryonic period is so critical and list the embryonic needs that must be met during the embryonic period for morphogenesis to continue.
4. State the approximate time period (in weeks) for the following occurrences:
   - (a) appearance of the arm and leg buds.
   - (b) differentiation of the external genitalia.
   - (c) perception of quickening by the mother.
   - (d) functioning of the embryonic heart.
   - (e) initiation of bone ossification.
   - (f) appearance of lanugo and vernix caseosa.
   - (g) survival of fetus if born prematurely.
   - (h) formation of all major body organs completed.
5. Define the term *infancy* and discuss the growth and developmental events characteristic of this period of life.
6. Define the term *childhood* and discuss the growth and developmental events characteristic of this period of life.
7. Define the term *adolescence* and discuss the role of puberty in this period of life for both males and females.
8. Distinguish between puberty and pubescence.
9. Define the term *adulthood* and describe the characteristics of this period of life.
10. List the structural features of adults that exemplify sexual dimorphism.
11. Define *anthropometry* and give examples of anthropometric characteristics.
12. State the features of a genetic disorder that would lead one to believe that it was a form of sex-linked inheritance.

UNIT VII

Appendixes

Appendix A
Answers to Objective Questions with Explanations

Appendix B
Laboratory Demonstrations in Anatomy

Appendix C
Medical and Pharmacological Abbreviations

Appendix D
Units of Measurement and Their Equivalents

Appendix A

Answers to Objective Questions with Explanations

Chapter 1

1. (a) The word *anatomy* is derived from the Greek prefix *ana,* meaning "up" and the suffix *tome,* "a cutting." In ancient times, the Greek word *anatomize* was more commonly used than the word *dissect.*

2. (a) In William Harvey's time, there were many misconceptions regarding the flow of blood. Using the scientific method, Harvey demonstrated the flow of blood and described the function of valves within veins.

3. (c) The concept of body humors was widely accepted by physicians in Greek and Roman times as an explanation for a person's disposition and general health. It was not until the Renaissance that this concept was gradually discarded in the light of scientific research.

4. (d) The Greeks held the body in great esteem and were eager to learn about its structure and function.

5. (a) Hippocrates is regarded as the father of medicine because of the sound principles of medical practice that he established. While the Hippocratic oath cannot be directly credited to him, it undoubtedly represents his ideals and principles.

6. (c) Phlegm was thought to be the body humor associated with the lungs. It was believed that too much phlegm, or not enough, could result in respiratory disorders.

7. (c) Andreas Vesalius completed *De Humani Corporis Fabrica* in A.D. 1543 when he was 28 years old. This beautifully illustrated book corrected many anatomical errors that had been handed down as fact and provided a visual guide for studying anatomy.

8. (d) The development of movable type allowed for widespread access to printed material, and was therefore important in ushering in the Renaissance.

9. (b) Aristotle thought that the heart was the seat of intelligence and that the warm blood from the heart was cooled by the fluids surrounding the brain.

10. (a) Wilhelm Konrad Roentgen discovered X rays in 1895. His achievement was recognized in 1901, when he was awarded the Nobel Prize.

Chapter 2

1. (b) Although chordates have a distinct head, thorax, and abdomen, so do many invertebrate organisms.

2. (a) Because we have prehensile hands, digits modified for grasping, and large, well-developed brains, we are classified as primates along with monkeys and great apes.

3. (b) The tissue level of body organization is intermediate between the cell and organ levels.

4. (a) The urinary system maintains blood homeostasis. Basically, the kidneys are filtering organs of the blood.

5. (b) The cubital fossa is located on the anterior surface of the upper extremity, at the junction of the brachium and antebrachium. It is the anterior surface of the elbow.

6. (c) An organ is a body structure composed of two or more tissue types.

7. (c) By definition, both the coronal and sagittal planes extend vertically through the body, but there is no vertical plane designation.

8. (b) The external genitalia are located within the perineum, which is the area between the coccyx and the symphysis pubis.

9. (b) Within the mediastinum are the primary bronchi, esophagus, and heart, along with major vessels that transport blood toward and away from these organs.

10. (b) The epigastric region of the abdomen is centrally located, medial to the right and left hypochondriac regions.

11. (c) The mesenteries are modified serous membranes that support portions of the small and large intestines within the abdominal cavity.

12. (c) A coronal section divides the body into front (anterior) and back (posterior) portions.

13. (c) The anatomical postion provides a consistent frame of reference when describing structures of the body.

14. (d) Some listening can be accomplished without the aid of an instrument, but the technique of choice is to use a stethoscope.

Chapter 3

1. (c) Electrolytes are formed from the breakdown of inorganic compounds in water. All three kinds of electrolytes—acids, bases, and salts—are important for normal cellular function.

2. (d) The elements carbon, oxygen, nitrogen, and hydrogen compose over 95% of the body and are often linked to form molecules and compounds.

3. (a) The strong hydrolytic enzymes in lysosomes digest foreign molecules and worn and damaged cells.

4. (e) Ciliated cells promote movement of substances through the tubes or ductules of sections of the respiratory and reproductive systems. For example, mucus containing trapped particles is moved in the bronchioles and trachea, sperm are moved in the ductus deferentia and ova are moved in the uterine tubes.

5. (b) Osmosis is movement of water and solvent molecules through a semipermeable membrane as a result of concentration differences.

6. (c) Metaphase is the stage of mitosis immediately preceding the separation of the chromatids.

7. (d) Anaphase is the stage of mitosis immediately following metaphase; it precedes the formation of two identical cells.

8. (a) The Golgi apparatus synthesizes carbohydrate molecules in the production of glycoproteins.

9. (b) Compensatory hypertrophy is an increase in cell mass in response to greater demands, such as the enlargement of a muscle fiber as a result of exercise.

10. (b) Hyperplasia is a protective mechanism that ensures the availability of an adequate number of cells to perform a particular task.

Chapter 4

1. (b) The four principal types of tissues are epithelial tissues, connective tissues, muscular tissues, and nervous tissues. The integument, or skin, is an organ.

2. (c) Blood is a fluid tissue that flows through blood vessels.

3. (a) Many body structures derive from mesoderm, including cartilage, bone, and other connective tissue; smooth, cardiac, and skeletal muscle; and the dermis of the skin.

4. (b) Elastic and collagenous fibers are characteristic of certain kinds of connective tissues.

5. (d) The entire abdominal portion of the gastrointestinal (GI) tract contains simple columnar epithelium lining the lumen.

6. (a) Based on structural classification, the mammary glands are classified as compound acinar. Based on secretory classification, the mammary glands are classified as apocrine.

7. (c) Dense regular connective tissue is the principal component of tendons, which accounts for their great strength.

8. (a) Reticular tissue, which contains white blood cells, is phagocytic in cleansing body fluids.

9. (b) Cartilage tissue is slow to heal because it is avascular (without blood vessels). Cartilage generally derives its nutrients from surrounding fluids rather than from permeating blood vessels and capillaries.

10. (d) Like skeletal muscle, cardiac muscle is striated. As compared to skeletal muscle or smooth muscle tissues, however, cardiac muscle tissue is unique in containing intercalated discs and experiencing rhythmical involuntary contractions.

Chapter 5

1. (a) Ectoderm is the outermost of the three embryonic germ layers, and it is from this germ layer that hair, nails, integumentary glands, and the epidermis of the skin is formed.

2. (b) The appearance and general health of nails is an indicator of general body health and of certain dietary deficiencies, such as iron deficiency.

3. (b) The stratum lucidum is an epidermal layer that occurs only in the skin of the palms of the hands and the soles of the feet.

4. (a) The dermal papillae may contribute to surface features of the skin, such as print patterns, but does not contribute to skin coloration.

5. (c) Mitosis occurs in the stratum basale, and to a limited extent, in the stratum spinosum.

6. (a) The sebum from sebaceous glands is emptied into a hair follicle, where it is dissipated along the shaft of the hair to the surface of the skin.

7. (c) Lanugo, or fetal hair, is thought to be important in the development of the hair follicles.

8. (b) Melanoma is an aggressive malignant skin cancer that is life threatening if untreated.

9. (b) A second-degree burn is serious, but it generally does not require skin grafts.

10. (d) A comedo is a blackhead or whitehead resulting from a small, localized skin infection.

Chapter 6

1. (c) A bone, such as the femur, is considered to be an organ because it is composed of more than two tissues integrated to perform a particular function.

2. (a) Skin, rather than bone, is involved in the synthesis of vitamin D.

3. Matching: 1(c), 2(d), 3(e), 4(a), 5(b)

4. (c) The mandible articulates with the mandibular fossae of the temporal bones.

5. (a) The sella turcica of the sphenoid bone is located immediately superior to the sphenoidal sinus, where it supports the pituitary gland.

6. (d) Osteoclasts are important in releasing stored minerals within bone tissue and in the continuous remodeling of bone.

7. (a) Located in the squamous part of the temporal bone, the mandibular fossa is the depression for articulation of the condyloid process of the mandible.

8. (b) The crista galli is the portion of the ethmoid bone that contains numerous olfactory foramina for the passage of olfactory nerves from the nasal epithelium.

9. (d) Cervical vertebrae contain transverse foramina that permit the passage of the vertebral vessels to and from the brain.

10. (c) A person may gain some protection against osteoporosis by maintaining a healthy diet and a regular exercise program.

Chapter 7

1. (a) When in anatomical position, a person is erect and facing anteriorly. The subscapular fossa is found on the anterior side of the scapula. The supraspinatus and infraspinatus fossae are located on the posterior side.

APPENDIX

2. (c) The sternal extremity of the clavicle articulates with the manubrium of the sternum. The acromial extremity of the clavicle articulates with the acromion of the scapula.

3. (d) The clavicle has a conoid tubercle near its acromial extremity.

4. (b) The bony prominence of the elbow is the olecranon of the ulna.

5. (b) Sesamoid bones are formed in tendons, and none of the carpal bones are sesamoid bones. Sesamoid bones are fairly common, however, at specific joints of the digits.

6. (d) Pelvimetry measures the dimension of the lesser pelvis in a pregnant woman to determine whether a cesarean delivery might be necessary.

7. (e) The linea aspera is a vertical ridge on the posterior surface of the body of the femur, where the posterior gluteal muscles of the hip attach.

8. (b) The intertrochanteric line is located on the anterior side of the femur between the greater trochanter and the lesser trochanter.

9. (d) Sex-related structural differences in the pelvis reflect modifications for childbirth. For example, the female pelvis as compared to that of the male has a wider pelvic outlet, a shallower and shorter symphysis pubis, and a wider pubic arch.

10. (a) Talipes is a congenital malformation in which the sole of the foot is twisted medially.

Chapter 8

1. (b) The structural classification of joints includes fibrous, cartilaginous, and synovial types.

2. (c) Synchondroses are cartilaginous joints that have hyaline cartilage between the bone segments.

3. (d) Syndesmoses are fibrous joints found only in the antebrachium (forearm) and leg, where adjacent bones are held together by interosseous ligaments.

4. (d) Syndesmoses do not occur within the skull; rather, they are located only in the upper and lower extremities.

5. (c) The only saddle joint in the body is at the base of the thumb, where the trapezium articulates with the first metacarpal bone.

6. (d) Only a few of the synovial joints contain a meniscus.

7. (c) A pivotal joint is a synovial joint that permits rotational movement.

8. (b) The coxal (hip) joint has a wide range of movement, including hyperextension as the lower extremity is moved posteriorly beyond the vertical position of the body in anatomical position.

9. (b) The shoulder joint, with its relatively shallow socket and weak ligamentous support, is vulnerable to dislocation. In addition, we often place our arms in vulnerable positions as we engage in various activities.

10. (d) Rheumatoid arthritis is a chronic disease that frequently leaves the patient crippled. It occurs most commonly between the ages of 30 and 35.

Chapter 9

1. (b) The neuromuscular cleft is a slight gap within the neuromuscular junction where a motor nerve fiber and a skeletal muscle fiber meet.

2. (c) When there are many motor units present within a muscle, a person can be more selective at which ones are recruited, and thus have greater dexterity than when few motor units are present.

3. (e) Muscles are named on the basis of structural features, location, attachment, relative position, or function (action).

4. (d) When a nerve impulse reaches an axon terminal, a neurotransmitter chemical is released into the neuromuscular cleft at the neuromuscular junction.

5. (c) With respect to the muscles of the thigh, a single motor neuron serves a large number of muscle fibers and, therefore, has a low innervation ratio.

6. (a) The corrugator muscle is located beneath the medial portion of the eyebrow. When the muscle contracts, the eyebrow is drawn toward the midline.

7. (a) A description of a muscle's contraction is always made in reference to a person in anatomical position. In anatomical position, the shoulder joint is positioned vertically at 180°. When the pectoralis muscle is contracted while in this position, the angle of the shoulder joint is decreased; therefore, the joint is flexed.

8. (b) Although it is positioned on the anterior surface of the humerus, the biceps brachii muscle arises from the coracoid process of the scapula and from the tuberosity above the glenoid fossa of the scapula. Both heads insert on the radial tuberosity.

9. (c) Spanning both the hip and knee joints, the rectus femoris muscle can flex the hip joint and extend the knee joint.

10. (b) The tibialis posterior muscle is deep to the soleus muscle. Its tendinous insertion passes across the arches of the foot and inserts on the plantar surfaces of a number of foot bones. In this position, the tibialis posterior muscle supports the arches of the foot and plantar flexes and inverts the foot as it contracts.

Chapter 10

1. (e) The scalp is attached anteriorly to the supraorbital ridges. The eyebrows are attached to the scalp, just above the supraorbital ridges.

2. (b) The ala is a portion of the nose, lateral to the apex.

3. (d) The concha is a bony projection into the nasal cavity.

4. (d) The conjunctiva is a thin mucous membrane that covers the anterior surface of the eyeball and lines the undersurface of the eyelids.

5. (d) The cervix of the neck is the anterior portion. The cervical vertebrae can be palpated in the posterior portion, called the nucha.

6. (a) The carotid triangle is located on the lateral side of the neck and is bordered by the sternocleidomastoid, posterior digastric, and omohyoid muscles. Located within the carotid triangle are the common carotid artery, internal jugular vein, and vagus nerve.

7. (a) The ulnar nerve passes through the ulnar sulcus of the elbow. A fracture of the olecranon of the ulna often damages the ulnar nerve.

8. (b) The scapular muscles are usually obscure on an obese person.

9. (c) The axilla is the depression commonly known as the armpit. The anterior axillary fold is formed by the pectoralis major muscle and the posterior axillary fold consists of the latissimus dorsi muscle.

10. (a) The great saphenous vein and small saphenous vein are superficial veins of the leg that are frequently varicosed in elderly people.

Chapter 11

1. (d) The cerebellum is a structure within the metencephalon of the brain.

2. (a) The cerebral cortex is the outer portion of the cerebrum, and the cerebrum is a structure within the telencephalon of the brain.

3. (e) The medulla oblongata is a structure within the myelencephalon of the brain.

4. (c) As the nervous system matures in an infant, the neurolemmocytes wrap around the axons and some of the dendrites of neurons within the peripheral nervous system. Myelination is the process of forming myelin layers that protect neurons and aid conduction.

5. (d) Ganglia are collections of nerve cell bodies outside the CNS. Collections of nerve cell bodies within the CNS are called nuclei.

6. (a) A pseudounipolar neuron has a single process that divides into two, and its cell body is located in posterior root ganglia of the spinal and cranial nerves.

7. (a) Depolarization of an axon is produced by the movement of Na^+ into the axon and K^+ out of the axon. Once the impulse has completed its course to the axon terminal, repolarization of the axon is produced by the movement of K^+ into the axon and the movement of Na^+ out of the axon.

8. (a) The corpus collosum is composed of commissural fibers that connect the two cerebral hemispheres.

9. (d) The thalamus autonomically responds to pain by activating the sympathetic nervous system. It also relays pain sensations to the parietal lobes of the cerebrum for perception.

10. (b) The basal nuclei consist of cell bodies of motor neurons that regulate contraction of skeletal muscles. Basal metabolic rate is regulated, for the most part, in the hypothalamus and medulla oblongata.

11. (c) Located within the mesencephalon, the corpora quadrigemina is concerned with visual and hearing reflexes, the red nucleus is concerned with motor coordination and posture maintenance, and the substantia nigra is thought to inhibit forced involuntary movements.

12. (d) The fourth ventricle is located inferior to the cerebellum and contains cerebrospinal fluid.

Chapter 12

1. (a) Anatomically speaking, the PNS consists of all of the structures of the nervous system outside of the CNS. That means, then, that all nerves, sensory receptors, neurons, ganglia, and plexuses are part of the PNS.

2. (b) The oculomotor cranial nerve innervates the medial rectus eye muscles that, when simultaneously contracted, cause the eyes to be directed medially.

3. (b) The oculomotor cranial nerve innervates the levator palpebrae superioris muscle with motor fibers. This muscle elevates the upper eyelid when contracted.

4. (c) The vestibulocochlear nerve serves the vestibular organs of the inner ear with sensory fibers. These organs are associated with equilibrium and balance.

5. (a) Passing through the stylomastoid foramen, the facial nerve innervates the muscles of facial expression with motor fibers and the taste buds on the anterior two-thirds of the tongue with sensory fibers.

6. (b) The accessory cranial nerve innervates several muscles that move the head and neck with motor fibers.

7. (c) The four spinal nerve plexuses are the cervical, brachial, lumbar, and sacral.

8. (d) Only the brachial plexus consists of roots, trunks, divisions, and cords. The nerves of the upper extremity arise from the cords.

9. (a) The median nerve arises from the brachial plexus.

10. (c) The knee-jerk reflex is an ipsilateral reflex because the receptor and effector organs are on the same side of the spinal cord.

Chapter 13

1. (d) Postganglionic neurons from the superior mesenteric ganglion innervate the small intestine and colon.

2. (d) Parasympathetic neurons within the oculomotor, facial, and glossopharyngeal cranial nerves synapse in ganglia located in the head.

3. (c) Parasympathetic ganglia, also called terminal ganglia, synapse with the effector cells near or within the organs being served.

4. (c) Secreted from synaptic vesicles, the neurotransmitter chemical acetylcholine facilitates transmission across a synapse.

5. (d) Because the preganglionic neurons of the sympathetic division of the autonomic nervous system exit the vertebral column from the first thoracic to the second lumbar levels, the sympathetic division is also called the thoracolumbar division.

6. (c) Nerve impulses through the postganglionic sympathetic neurons in the heart release norepinephrine, which stimulates the heart to contract.

7. (b) The cooperative effect of sympathetic and parasympathetic stimulation is evident in the erection of the penis (parasympathetic response) and the ejaculation of semen (sympathetic response).

8. (d) Most blood vessels dilate in response to sympathetic stimulation. A few blood vessels (e.g., those serving the external genitalia) constrict in response to parasympathetic stimulation. Visceral blood vessels do not constrict in response to parasympathetic stimulation.

9. (e) Parasympathetic stimulation increases digestive activity, constricts pupils, and decreases the heart rate. Because atropine blocks parasympathetic stimulation, it results in decreased mucus secretion and GI tract movement and causes dilation of the pupils. It also results in an increased heart rate.

10. (c) The medulla oblongata is the structure of the brain that most directly controls the activity of the ANS. The medulla oblongata contains control centers for the circulatory, respiratory, urinary, reproductive, and digestive systems.

Chapter 14

1. (e) The adenohypophysis derives from the hypophyseal pouch in the roof of the oral cavity.

2. (b) The neurohypophysis derives from the neurohypophyseal bud of neuroectoderm within the developing brain.

3. (d) The adrenal medulla derives from neural crest ectoderm within the primitive celomic cavity.

4. (c) The pancreas derives from an outpouching of the endoderm along the developing foregut.

5. (a) The thyroid gland derives from the thyroid diverticulum of endoderm within the developing pharynx.

6. (c) The sella turcica is a depression within the sphenoid bone that supports the pituitary gland.

7. (d) Secreted from the thyroid gland, thyroxine determines the basic metabolic rate of most organs and promotes the maturation of the brain.

8. (e) Activation of the adrenal medulla causes the secretion of epinephrine and norepinephrine. Both of these hormones prepare the body for greater physical performance—the fight-or-flight response.

9. (e) Insulin is secreted from beta cells within the pancreatic islets in response to a rise in blood glucose. Insulin stimulates the production of glycogen and fat.

10. (d) The adrenal medulla releases epinephrine upon receiving sympathetic nerve impulse stimulation.

11. (a) The thyroid gland releases thyroxine upon receiving TSH stimulation from the adenohypophysis of the pituitary gland.

12. (b) The adrenal cortex releases corticosteroids upon receiving ACTH stimulation from the adenohypophysis of the pituitary gland in response to stress.

13. (e) The adenohypophysis of the pituitary gland releases ACTH in response to receiving CRF stimulation from the hypothalamus.

14. (d) Both the adrenal cortex and the gonads secrete steroid hormones.

15. (a) In the case of an endemic goiter, growth of the thyroid is due to excessive TSH secretion, which results from low levels of thyroxine secretion.

Chapter 15

1. (d) In order for perception to occur, there must be a stimulus at a receptor site that causes a nerve impulse to be conducted to the cerebrum of the brain.

2. (b) Located within the dermis of the skin, in certain visceral organs, and near synovial joints, lamellated corpuscles respond to heavy pressures.

3. (d) Located within connective tissue capsules in synovial joints, joint kinesthetic receptors are stimulated by changes in position caused by movements at the joints.

4. (c) Because sensations of referred pain are consistent from one person to another, an understanding of this phenomenon is of immense clinical importance in diagnosing organ dysfunction.

5. (d) When an object is viewed at a distance of at least 20 feet by someone with normal vision, the suspensory ligaments of the eyes are taut and the lenses are flat because the ciliary muscles are relaxed.

6. (a) Composed of tightly bound elastic and collagenous fibers, the toughened sclera is avascular but does contain sensory receptors for pain.

7. (b) In dim light, sympathetic motor impulses cause the radially arranged smooth muscle fibers within the iris to contract, permitting the pupil to become larger.

8. (a) The semicircular canals contain endolymph and hair cells that respond to movements of the head and convey sensations to the brain that are important for maintaining equilibrium and balance.

9. (d) The cochlear window is at the footplate of the stapes and the vestibular window borders between the scala tympani and the middle-ear cavity. Both windows separate the middle ear from the cochlea within the inner ear.

10. (b) A concave lens corrects myopia (nearsightedness) by causing the light waves to focus deeper within the vitreous chamber upon the fovea centralis.

Chapter 16

1. (c) Fibrinogen is a protein in blood plasma that aids in clotting.

2. (a) An excessive leukocyte count, called leukocytosis, is generally diagnostic of infections or diseases within the body.

3. (b) The pulmonary arteries transport deoxygenated blood from the right ventricle of the heart to the lungs.

4. (c) The right atrium receives venous blood from the superior and inferior vena cavae and the coronary sinus. The coronary sinus collects venous blood from coronary circulation prior to delivery into an opening of the right atrium.

5. (f) Closure of the atrioventricular valves causes the "lub" sound of the heart.

6. (c) The QRS complex represents the depolarization of the ventricles. During this interval, the ventricles of the heart are in systole and blood is being ejected from the heart.

7. (b) Arising from ascending aorta, the coronary arteries feed directly into the myocardium of the heart to ensure a rich blood supply to the cardiac muscles.

8. (a) The external carotid arteries supply blood to the entire head, excluding the brain. The paired internal carotid arteries and vertebral arteries (that unite to form the basilar artery) supply blood to the brain.

9. (a) Branching from the external carotid artery at the level of the mandibular condyle, the maxillary artery supplies blood to the teeth and gums of the upper jaw and the superficial temporal artery supplies blood to the parotid gland and to the superficial temporal region.

10. (a) The hepatic portal vein drains nutrient-rich blood into the liver, where the blood is processed by a venous portal system. An arteriole portal system provides the pituitary gland with blood.

11. (b) Although the heart begins pumping blood at 25 days following conception, its development is not complete until the end of the fifth week (at about 35 days).

12. (c) The umbilical vein receives oxygenated blood from the placenta and transports it to the fetal heart.

Chapter 17

1. (b) Inhaled air is cleansed, moistened, and warmed prior to its arrival at the alveoli.

2. (a) The paired palatine bones support the nasal septum but are not part of its structure.

3. (b) Adenoid is the common name of the pharyngeal tonsil. An adenoidectomy is the removal of both pharyngeal tonsils.

4. (a) There are four paranasal sinuses, each named according to the bone in which it is located. Hence, we have ethmoidal, sphenoidal, frontal, and maxillary sinuses.

5. (e) Unlike the right lung that has three lobes, the left lung has only a superior lobe and an inferior lobe.

6. (a) The parietal pleura lines the wall of the thoracic cavity and the visceral pleura covers the lung. The space between the parietal pleura and the visceral pleura is referred to as the pleural cavity.

7. (d) When contracted, the muscles of inspiration increase the dimensions of the thoracic cavity, causing a decrease in the air pressure surrounding the lungs. Air flows through the respiratory tract, inflating the lungs.

8. (d) The vocal cords are attached between the arytenoid and thyroid cartilages on either lateral side of the glottis.

9. (c) The vital capacity is the greatest amount of air that can be exhaled following a maximal inhalation—an approximate volume of 4,500 cc of air.

10. (a) The nuclei for normal breathing are located within the medulla oblongata of the brain.

Chapter 18

1. (d) Viscera are the organs located within the trunk of the body. In the thoracic cavity, the viscera include the heart, lungs, and esophagus. In the abdominopelvic cavity, the viscera include the stomach, small and large intestines, liver, gallbladder, spleen, pancreas, kidneys, and adrenal glands, along with major vessels.

2. (c) The deciduous dentition includes 8 incisors, 4 canines, and 8 molars. The permanent dentition includes 8 incisors, 4 canines, 8 premolars, and 12 molars. The third molars are called wisdom teeth.

3. (b) It is through the supporting mesentery that vessels and nerves supply the abdominal viscera.

4. (c) Lacteals are lymph ductules found within the lamina propria of villi.

5. (b) Composed of lymphoid tissue, the spleen stores red blood cells and is an organ of the circulatory system.

6. (d) The papillae of the dorsum of the tongue provide a roughened surface for physically handling food. The papillae also support taste buds for responding to the chemical stimuli of various foods.

7. (c) Following the formation of chyme within the stomach, the food entering into the small intestine is ready for additional digestion and absorption. Bile and pancreatic juice enter the lumen of the duodenum to continue the chemical breakdown of food. Intestinal movements aid in its mechanical breakdown. The nutrients from digested food enter the bloodstream as absorption occurs within the small intestine.

8. (a) As the sphincter of ampulla opens, bile and pancreatic juice enter the duodenum. Stenosis of the sphincter of ampulla would prohibit the entry of these products.

9. (a) The hepatic portal vein transports absorbed nutrients within the bloodstream to the liver, where they are processed.

APPENDIX

10. (d) Fats are absorbed into the lymphatic system. Proteins are broken down into amino acids before they are absorbed. Bile is secreted into the duodenum. The liver is served with blood from the hepatic portal vein and the hepatic artery; thus, it has a double blood supply, which becomes mixed at the capillary level.

Chapter 19

1. (b) The glomeruli are enclosed within the glomerular capsules in the renal cortex.
2. (d) The hilum of the kidney is the concave medial surface where the renal vessels enter and exit and where the ureter is located.
3. (c) Renal pyramids, containing the papillary ducts, are located within the renal medulla.
4. (a) The minor calyx receives urine directly from the papillary ducts. The urine then passes through the major calyx and into the renal pelvis.
5. (d) The kidneys are located high in the abdominal cavity between the levels of the twelfth thoracic and the third lumbar vertebrae.
6. (d) A renal calculus is most likely to cause a blockage of a ureter and backup of urine into the renal pelvis.
7. (e) Located only in the linings of the urinary bladder and ureters, transitional epithelium permits distension as urine is stored in the lumen of the urinary bladder.
8. (c) Collectively, the three layers of smooth muscle within the wall of the urinary bladder are referred to as the detrusor muscle. The detrusor muscle plays an active role in forcing urine from the urinary bladder during micturition.
9. (b) The internal urethral sphincter is actually a modified portion of the detrusor muscle and is innervated by parasympathetic neurons.
10. (d) Metanephros are the last type of developmental kidneys. They are continuously functioning throughout the fetal period, with the urine produced being expelled into the amniotic fluid.

Chapter 20

1. (c) The perineum is of functional and clinical importance because the external genitalia are located there.
2. (a) The dartos muscle is embedded within the wall of the scrotum. Along with the cremasteric muscle, it regulates the position of the testes within the scrotum through involuntary contraction in response to cold temperatures.

3. (b) Produced by interstitial cells, testosterone maintains male sexuality, including production of spermatozoa, activity of accessory sex organs, expression of secondary sex characteristics, and determination of sex drives.
4. (e) As components of the root of the penis, the bulb attaches to the undersurface of the urogenital diaphragm and the crus is the expanded proximal portion of the corpora cavernosa penis.
5. (b) The spermatic cord contains the ductus deferens, which is a spermatic duct. The spermatic cord, however, also contains spermatic vessels, a nerve, and the cremasteric muscle.
6. (a) The epididymides are long, flattened organs that store sperm in their last stage of maturation. Sperm are also stored in the ductus deferentia.
7. (a) Mucus secreted by the urethral glands lubricates the urethra and retards the entry of pathogens into the urinary bladder.
8. (c) Emission is movement of stored sperm from the epididymides and ductus deferentia to the ejaculatory ducts. Ejaculation is the forceful discharge of semen from the erect penis. Emission and ejaculation occur only if stimulation is sufficient, as in masturbation or coitus.
9. (a) The ovum from a female may be fertilized by a sperm cell containing either an X or a Y chromosome. If the sperm cell contains an X chromosome, it will pair with the X chromosome of the ovum and a female child will develop. A sperm cell carrying a Y chromosome results in an XY combination, and a male child will develop. The Y chromosome, therefore, is responsible for the subsequent production of androgens, which cause masculinization.
10. (d) The epididymides, ductus deferentia, and seminal vesicles derive from the mesonephric duct. The prostate arises from an endodermal outgrowth of the urogenital sinus.

Chapter 21

1. (c) The cervix of the uterus is the inferior constricted portion that opens into the vagina.
2. (b) The usual site of fertilization is within a uterine tube. From there, the fertilized egg continues to develop and enters into the cavity of the uterus about 3 days after conception.

3. (c) The secretory phase of the endometrium is characterized by an increase in glandular secretion and blood flow into the endometrium in preparation for implantation. During this time the ovary is in its luteal phase, characterized by a regressive corpus luteum.
4. (d) A secondary oocyte is discharged from an ovary during ovulation. If a spermatozoon passes through the corona radiata and zona pellucida and enters the cytoplasm of the secondary oocyte, the second meiotic division is completed and a mature ovum is formed.
5. (d) In addition to the mesovarium, each ovary is supported by an ovarian ligament and a suspensory ligament.
6. (c) The endometrium consists of a superficial stratum functionale layer that is shed as menses during menstruation and a deeper stratum basale layer that replenishes the stratum functionale layer after each menstruation.
7. (b) Vaginal rugae permit distension of the vagina during coitus and also during parturition.
8. (d) The vulva is the collective term for the female external genitalia. The vagina is an internal reproductive organ.
9. (d) The paramesonephric ducts give rise to the genital tract of the female reproductive system, which includes the uterus and the uterine tubes.
10. (a) Both the labia majora of the female and the scrotum of the male derive from the embryonic labioscrotal swellings.

Chapter 22

1. (d) The pre-embryonic period of developments lasts 14 days and is completed when the three primary germ layers have been formed and are in place to begin migration and differentiation.
2. (c) The yolk sac only produces blood for about a 2-week period (the third to the fifth week) prior to the formation of the liver. The liver then produces blood until the bone marrow is sufficiently developed to carry out the task.
3. (d) The placenta is an organ that serves many physiological functions, including the exchange of gases and other molecules between the maternal and fetal blood and the production of enzymes. In addition, it serves as a barrier against many harmful substances and as an endocrine gland in secreting steroid and glycoprotein hormones.

4. (c) The embryonic portion of the placenta is the chorion frondosum, and the maternal portion is the decidua basalis.
5. (d) Although the placenta is an effective barrier against diseases of bacterial origin, viruses and certain blood-borne diseases can diffuse through the vascular tissues. Furthermore, most drugs ingested by a pregnant woman can readily pass through the placenta, including nicotine, alcohol, and heroin.
6. (a) The embryonic heart begins pumping blood on about day 25, or during the fourth week of development.
7. (a) The neonatal period from birth to the end of the fourth week is characterized by major physiological changes, including the establishment of a stable heart rate and respiratory rate, and a consistent body temperature.
8. (b) A rapid heart rate of 120–160 beats/min. ensures an adequate oxygen supply to all of the cells and helps to maintain a constant body temperature.
9. (c) Pubescence accompanies puberty and refers to the continuum of physical changes that occur during this period of maturation, especially with regard to body hair.
10. (b) As compared to males, females have a lower basal metabolic rate. This may account, in part, for the longer life span of females.
11. (c) The average age at which breast buds appear in healthy girls in the United States is about 11 years, but the range is from 9 to 13 years.
12. (c) Dizygotic, or fraternal, twins may be of the same sex or of different sexes and are not any more alike than brothers or sisters born at different times.
13. homozygous recessive (bb)
    heterozygous (Bb)
    homozygous dominant (BB)
14. (a) A recessive allele is not expressed in a heterozygous genotype. This means that the particular recessive trait is not physically apparent.
15. (c) In a monohybrid cross, the probability of an offspring having a particular genotype is one in four for homozygous dominant and homozygous recessive and one in two for heterozygous dominant.

Appendix B

Laboratory Demonstrations in Anatomy

The study of human anatomy can be augmented by dissecting various vertebrate organs such as the brain, heart, eye, trachea, lung, and kidney.

Fresh organs obtained from a local slaughterhouse can be preserved by placing them in an embalming solution made according to the following formula:

Carbolic acid (melted crystals)	5 parts
Formalin (40%)	5 parts
Glycerin	5 parts
Water	85 parts

Carbolic acid is a disinfectant and helps maintain natural tissue color. Formalin is the main preservative. Glycerin prevents the anatomical structures from drying out too rapidly. Once an organ has been preserved, it can be stored for further, more detailed, dissection at a later date.

Using human cadavers is of obvious benefit in learning human anatomy. Cadavers have many advantages over anatomical charts, models, and illustrations when demonstrating size, position, and regional relationships within the human body. Furthermore, advanced anatomy students may be able to dissect the cadaver for use in an elementary human anatomy course.

Appropriate radiographs displayed on a viewing screen give an added dimension to the anatomy course. Outdated radiographs are usually available from doctors' offices, clinics, or hospitals.

Directions for preparing many interesting and valuable anatomical demonstrations can be found in the following books:

Hildebrand, M. 1968. *Anatomical preparations.* Berkeley: Univ. of California Press.

Tompsett, D. H. 1970. *Anatomical techniques.* 2d ed. Edinburgh: Livingstone.

Appendix C

Medical and Pharmacological Abbreviations

aa	of each	D.T.D.	give of such doses	Lat. dol.	to the painful side
a.c.	before meals	Dur. dolor.	while pain lasts		
A/G	albumin globulin ratio	Dx	diagnosis	M.	mix
ANS	autonomic nervous system			man.	in the morning
		e	out of, with	mEq.	milliequivalent
Bib.	drink	ECG, EKG	electrocardiogram	mg.	milligram
bid	twice a day	EEG	electroencephalogram	ml.	milliliter
bihor	during two hours	e.m.p.	in the manner prescribed	MRI	magnetic resonance imaging
B.M.R.	basal metabolic rate				
bp	blood pressure	feb	fever	Noct.	at night
BUN	blood urea nitrogen			Noct. maneq.	night and morning
b.v.	vapor bath	GI	gastrointestinal	N.P.O.	nothing by mouth
		gm.	gram	O.D.	right eye
c̄	with	gr.	grain	o.d.	every day
caps.	capsule	Grad.	gradually	Omn. hor.	every hour
c.b.c.	complete blood count	gtt.	drop(s)	Omn. man.	every morning
cc.	cubic centimeter(s)			Omn. noct.	every night
cm.	centimeter(s)	h.	hour	O.S.	left eye
CNS	central nervous system	Hct	hematocrit	O.U.	both eyes
Co., Comp.	compound	Hg.	mercury	oz.	ounce
cr	tomorrow	Hgb	hemoglobin		
CFS	cerebrospinal fluid	hs	at bedtime	P	pulse
CT	computed tomography	Hx	history	Part. aeq.	equal parts
CVP	central venous pressure			pc	after meals
CXR	chest x-ray	ibid.	in the same place	pCO₂	partial pressure of carbon dioxide
		IM	intramuscular	PNS	peripheral nervous system
d.	a day	incid.	cut	PO	by mouth
D & C	dilatation and curettage	in d.	in a day	pO₂	partial pressure of oxygen
D.C.	discontinue	inj.	an injection	p.p.a.	having first shaken the bottle
D, Det.	give	int. cib.	between meals	p.r.n.	as needed
de d. in d.	from day to day	int. noct.	during the night	pro. us. ext.	for external use
Dieb. secund	every second day	IV	intravenous	pt	patient
Dieb. tert.	every third day			pt.	let it be continued
dim.	one-half	kg.	kilogram		
d. in dup.	give twice as much			q.	each; every
D. in p. aeq.	divide into equal parts			qd	every day
dr.	dram				

qh	every hour	s̄	without	t.	three times
q. ____ h.	every ____ hours	Semih.	half an hour	tab.	tablet
qhs	every evening	Sig.	write, label	tid	three times a day
qid	four times a day	S.O.S.	if needed		
qod	every other day	sp. gr.	specific gravity	ung.	ointment
q.q.	also	ss., s̄s̄	one-half	Ut. dict.	as directed
q.s.	sufficient quantity	s.s.s.	layer on layer		
		stat.	immediately	vic.	times
RBC	red blood cell	sum.	take		
		s.v.r.	alcohol	WBC	white blood cell

Appendix D

Units of Measurement and Their Equivalents

Apothecaries' Weights and Their Metric Equivalents

1 grain (gr) =
0.05 scruple (s)
0.017 dram (dr)
0.002 ounce (oz)
0.0002 pound (lb)
0.065 gram (g)
65. milligrams (mg)

1 scruple (s) =
20. grains (gr)
0.33 dram (dr)
0.042 ounce (oz)
0.004 pound (lb)
1.3 grams (g)
1,300. milligrams (mg)

1 dram (dr) =
60. grains (gr)
3. scruples (s)
0.13 ounce (oz)
0.010 pound (lb)
3.9 grams (g)
3,900. milligrams (mg)

1 ounce (oz) =
480. grains (gr)
24. scruples (s)
8. drams (dr)
0.08 pound (lb)
31.1 grams (g)
31,100. milligrams (mg)

1 pound (lb) =
5,760. grains (gr)
288. scruples (s)
96. drams (dr)
12. ounces (oz)
373. grams (g)
373,000. milligrams (mg)

Apothecaries' Volumes and Their Metric Equivalents

1 minim (min) =
0.017 fluid dram (fl dr)
0.002 fluid ounce (fl oz)
0.0001 pint (pt)
0.06 milliliter (ml)
0.06 cubic centimeter (cc)

1 fluid dram (fl dr) =
60. minims (min)
0.13 fluid ounce (fl oz)
0.008 pint (pt)
3.70 milliliters (ml)
3.70 cubic centimeters (cc)

1 fluid ounce (fl oz) =
480. minims (min)
8. fluid drams (fl dr)
0.06 pint (pt)
29.6 milliliters (ml)
29.6 cubic centimeters (cc)

1 pint (pt) =
7,680. minims (min)
128. fluid drams (fl dr)
16. fluid ounces (fl oz)
473. milliliters (ml)
473. cubic centimeters (cc)

Metric Weights and Their Apothecaries' Equivalents

1 gram (g) =
0.001 kilogram (kg)
1,000. milligrams (mg)
1,000,000. micrograms (μg)
15.4 grains (gr)
0.032 ounce (oz)

1 kilogram (kg) =
1,000. grams (g)
1,000,000. milligrams (mg)
1,000,000,000. micrograms (μg)
32. ounces (oz)
2.7 pounds (lb)

1 milligram (mg) =
0.000001 kilogram (kg)
0.001 gram (g)
1,000. micrograms (μg)
0.0154 grains (gr)
0.000032 ounce (oz)

Metric Volumes and Their Apothecaries' Equivalents

1 liter (l) =
1,000. milliliters (ml)
1,000. cubic centimeters (cc)
2.1 pints (pt)
270. fluid drams (fl dr)
34. fluid ounces (fl oz)

1 milliliter (ml) =
0.001 liter (l)
1. cubic centimeter (cc)
16.2 minims (min)
0.27 fluid dram (fl dr)
0.034 fluid ounce (fl oz)

Approximate Equivalents of Household Measures

1 teaspoon (tsp) =
4. milliliters (ml)
4. cubic centimeters (cc)
1. fluid dram (fl dr)

1 tablespoon (tbsp) =
15. milliliters (ml)
15. cubic centimeters (cc)
0.5 fluid ounce (fl oz)
3.7 teaspoons (tsp)

1 cup (c) =
240. milliliters (ml)
240. cubic centimeters (cc)
8. fluid ounces (fl oz)
0.5 pint (pt)
16. tablespoons (tbsp)

1 quart (qt) =
960. milliliters (ml)
960. cubic centimeters (cc)
2. pints (pt)
4. cups (c)
32. fluid ounces (fl oz)

Conversion of Units from One Form to Another

Refer to the preceding equivalency lists when converting one unit to another equivalent unit.

To convert a unit shown in bold type to one of the equivalent units listed immediately below it, multiply the first number (boldtype unit) by the appropriate equivalent unit listed below it.

Sample Problems:

1. Convert 320 grains into scruples (1 gr = 0.05 s).

$$320 \text{ gr} \times \frac{0.05 \text{ s}}{1 \text{ gr}} = 16.0 \text{ s}$$

2. Convert 320 grains into drams (1 gr = 0.017 dr).

$$320 \text{ gr} \times \frac{0.017 \text{ dr}}{1 \text{ gr}} = 5.44 \text{ dr}$$

3. Convert 320 grains into grams (1 gr = 0.065 g).

$$320 \text{ gr} \times \frac{0.065 \text{ g}}{1 \text{ gr}} = 20.8 \text{ g}$$

Body Temperatures in ° Fahrenheit and ° Celsius

°F	°C	°F	°C
95.0	35.0	100.0	37.8
95.2	35.1	100.2	37.9
95.4	35.2	100.4	38.0
95.6	35.3	100.6	38.1
95.8	35.4	100.8	38.2
96.0	35.5	101.0	38.3
96.2	35.7	101.2	38.4
96.4	35.8	101.4	38.6
96.6	35.9	101.6	38.7
96.8	36.0	101.8	38.8
97.0	36.1	102.0	38.9
97.2	36.2	102.2	39.0
97.4	36.3	102.4	39.1
97.6	36.4	102.6	39.2
97.8	36.6	102.8	39.3
98.0	36.7	103.0	39.4
98.2	36.8	103.2	39.6
98.4	36.9	103.4	39.7
98.6	37.0	103.6	39.8
98.8	37.1	103.8	39.9
99.0	37.2	104.0	40.0
99.2	37.3	104.2	40.1
99.4	37.4	104.4	40.2
99.6	37.6	104.6	40.3
99.8	37.7	104.8	40.4
		105.0	40.6

To convert ° F to ° C:

Subtract 32 from ° F and multiply by 5/9.
——° F − 32 × 5/9 = ——° C

To convert ° C to ° F:

Multiply ° C by 9/5 and add 32.
——° C × 9/5 + 32 = ——° F

Glossary

The words in this glossary are followed by a phonetic guide to pronunciation. This is a simplified system that is standard in medical usage and terminology.

Any unmarked vowel that ends a syllable or stands alone as a syllable is long. Any unmarked vowel that is followed by a consonant has the short sound.

If a long vowel appears in the middle of a syllable (followed by a consonant), it is marked with a macron (–). Similarly, if a vowel stands alone or ends a syllable, but should have a short sound, it is marked with a breve (˘).

abdomen (ab′-dŏ-men, ab-do′men)   The portion of the trunk between the diaphragm and pelvis.

abduction (ab-duk′shun)   The movement of a body part away from the axis or midline of the body; movement of a digit away from the axis of the limb.

accessory organs (ak-ses′ŏ-re)   Organs that assist the functioning of other organs within a system.

acetabulum (as″ĕ-tab′yŭ-lum)   A socket in the lateral surface of the hipbone (os coxa) with which the head of the femur articulates.

Achilles tendon (ă-kil′ēz)   (see tendo calcaneus)

actin (ak′tin)   A protein in muscle fibers that together with myosin is responsible for contraction.

adduction (ad-duk′shun)   The movement of a body part toward the axis or midline of the body; movement of a digit toward the axis of the limb.

adrenal cortex (ă-dre′nal kor′teks)   The outer layer of the adrenal gland.

adrenal glands   The two endocrine glands superior to the kidneys.

adventitia (ad″ven-tish′ă)   The outermost epithelial layer of a visceral organ; also called *serosa*.

afferent (af′er-ent)   Conveying or transmitting to.

afferent arteriole (ar-tir′e-ōl)   A blood vessel within the kidney that supplies blood to the glomerulus.

afferent neuron (noor′on)   A sensory nerve cell that transmits an impulse toward the central nervous system.

agonist (ag′ŏ-nist)   The prime mover muscle, which is directly engaged in the contraction that produces the desired movement.

alimentary canal (al″ĭ-men′tre)   The tubular portion of the digestive tract; also called the *gastrointestinal (GI) tract.*

allantois (ă-lan′to-is)   An extraembryonic membranous sac involved in the formation of blood cells. It gives rise to the fetal umbilical arteries and vein and also contributes to the formation of the urinary bladder.

all-or-none principle   The statement of the fact that muscle fibers of a motor unit contract to their maximum extent when exposed to a stimulus of threshold strength.

alveolar sacs (al-ve′ŏ-lar)   A cluster of alveoli that share a common chamber or central atrium.

alveolus (al-ve′ŏ-lus)   **1.** An individual air capsule within the lung. The alveoli are the basic functional units of respiration. **2.** The socket that secures a tooth (tooth socket).

amniocentesis (am″ne-o-sen-te′sis)   A procedure in which a sample of amniotic fluid is aspirated to examine suspended cells for various genetic diseases.

amnion (am′ne-on)   A developmental membrane surrounding the fetus that contains amniotic fluid.

amphiarthrosis (am″fe-ar-thro′sis)   A slightly movable articulation in a functional classification of joints.

ampulla (am-pool′ă)   A saclike enlargement of a duct or tube.

ampulla of Vater (fă-ter)   (see hepatopancreatic ampulla)

anal canal (a′nal)   The terminal tubular portion of the large intestine that opens through the anus of the GI tract.

anal glands   Enlarged and modified sweat glands that empty into the anal opening.

anastomosis (ă-nas″tŏ-mo′sis)   An interconnecting aggregation of blood vessels or nerves that form a network plexus.

anatomical position (an″ă-tom′ĭ-kal)   An erect body stance with the eyes directed forward, the arms at the sides, the palms of the hands facing anterior, and the fingers pointed straight down.

anatomy (ah-nat′ŏ-me)   The branch of science concerned with the structure of the body and the relationship of its organs.

antagonist (an-tag′-ŏ-nist)   A muscle that acts in opposition to another muscle.

antebrachium (an″te-bra′ke-um)   The forearm.

anterior (ventral) (an-tir′e-or)   Toward the front; the opposite of *posterior (dorsal).*

anterior root   The anterior projection of the spinal cord composed of axons of motor, or efferent, fibers.

anus (a′nus)   The terminal opening of the GI tract.

aorta (a-or′tă)   The major systemic vessel of the arterial system of the body, emerging from the left ventricle.

aortic arch   The superior left bend of the aorta between the ascending and descending portions; also called *arch of aorta.*

apex (a′peks)   The tip or pointed end of a conical structure.

apocrine gland (ap′ŏ-krin)   A type of sweat gland that functions in evaporative cooling. It may respond during periods of emotional stress.

aponeurosis (ap″ŏ-noo-ro′sis)   A fibrous or membranous sheetlike tendon.

appendix (ă-pen′diks)   A short pouch that attaches to the cecum; also called the *vermiform appendix.*

aqueous humor (a′kwe-us)   The watery fluid that fills the anterior and posterior chambers of the eye.

arachnoid mater (ă-rak′noid)   The weblike middle covering (meninx) of the central nervous system.

arbor vitae (ar′bor vi′te)   The branching arrangement of white matter within the cerebellum.

arm (brachium)   The portion of the upper extremity from the shoulder to the elbow.

arrector pili muscle (ă-rek′tor pi′li)   The smooth muscle attached to a hair follicle that, upon contraction, pulls the hair into a more vertical position, resulting in goose bumps.

arteriole (ar-tir′e-ōl)   A minute arterial branch.

artery (ar′tĕ-re)   A blood vessel that carries blood away from the heart.

arthrology (ar-throl′ŏ-je)   The scientific study of the structure and function of joints.

articular cartilage (ar-tik′yŭ-lar kar′ti-lij)   A hyaline cartilaginous covering over the articulating surface of bones of synovial joints.

articulation (ar-tik″yŭ-la′shun)   A joint.

arytenoid cartilages (ar″ĕ-te′noid)   A pair of small cartilages located on the superior aspect of the larynx.

ascending colon (ko′lon)   The portion of the large intestine between the cecum and the hepatic (right colic) flexure.

association neuron (noor′on)   A nerve cell located completely within the central nervous system. It conveys impulses in an arc from sensory to motor neurons; also called *internuncial neuron.*

atrioventricular bundle (a″tre-o-ven-trik′yŭ-lar)   A group of specialized cardiac fibers that conducts impulses from the atrioventricular node to the ventricular muscles of the heart; also called the *bundle of His* or *AV bundle.*

atrioventricular node   A microscopic aggregation of specialized cardiac fibers located in the interatrial septum of the heart that are a part of the conduction system of the heart; *AV node.*

atrioventricular valve   A cardiac valve located between an atrium and a ventricle of the heart; *AV valve.*

atrium (a′tre-um)   Either of the two superior chambers of the heart that receive venous blood.

atrophy (at′rŏ-fe)   A gradual wasting away or decrease in the size of a tissue or organ.

auditory (aw′di-tor-e)   Pertaining to the structures of the ear associated with hearing.

auditory tube A narrow canal that connects the middle ear chamber to the pharynx; also called the *eustachian canal.*

auricle (or'i-k'l) **1.** The fleshy pinna of the ear. **2.** An ear-shaped appendage of each atrium of the heart.

autonomic nervous system (aw''to-nom'ik) The sympathetic and parasympathetic portions of the nervous system that function to control the actions of the visceral organs and skin.

axilla (ak-sil'ă) The depressed hollow commonly called the armpit.

axon (ak'son) The elongated process of a nerve cell that transmits an impulse away from the cell body of a neuron.

ball-and-socket joint The most freely movable type of synovial joint (e.g., the shoulder or hip joint).

baroreceptor (bar''o-re-sep'tor) A cluster of neuroreceptors stimulated by pressure changes. Baroreceptors monitor blood pressure.

basal ganglion (ba'sal gang'gle-on) A mass of nerve cell bodies located deep within a cerebral hemisphere of the brain.

basement membrane A thin sheet of extracellular substance to which the basal surfaces of membranous epithelial cells are attached; also called the *basal lamina.*

basophil (ba'so-fil) A granular leukocyte that readily stains with basophilic dye.

belly The thickest circumference of a skeletal muscle.

benign (bi-nīn') Not malignant.

bifurcation (bi''fur-ka'shun) Forked; divided into two branches.

bile (bīl) A liver secretion that is stored and concentrated in the gallbladder and released through the common bile duct into the duodenum. It is essential for the absorption of fats.

bipennate (bi-pen'āt) Denoting muscles that have a fiber architecture coursing obliquely on both sides of a tendon.

blastula (blas'tyoo-lă) An early stage of prenatal development between the morula and embryonic stages.

blood The fluid connective tissue that circulates through the cardiovascular system to transport substances throughout the body.

blood-brain barrier A specialized mechanism that inhibits the passage of certain materials from the blood into brain tissue and cerebrospinal fluid.

bolus (bo'lus) A moistened mass of food that is swallowed from the oral cavity into the pharynx.

bone A solid, rigid, ossified connective tissue forming the skeletal system.

bony labyrinth (lab'i-rinth) A series of chambers within the petrous part of the temporal bone associated with the vestibular organs and the cochlea.

Bowman's capsule (bo'manz kap'sul) (see glomerular capsule)

brachial plexus (bra'ke-al plek'sus) A network of nerve fibers that arise from spinal nerves C5–C8 and T1. Nerves arising from the brachial plexuses supply the upper extremity.

brain The enlarged superior portion of the central nervous system located in the cranial cavity of the skull.

brain stem The portion of the brain consisting of the medulla oblongata, pons, and midbrain.

bronchial tree (brong'ke-al) The bronchi and their branching bronchioles.

bronchiole (brong'ke-ōl) A small division of a bronchus within the lung.

bronchus (brong'kus) A branch of the trachea that leads to a lung.

buccal cavity (buk'al) The mouth, or oral cavity.

bulbourethral glands (bul''bo-yoo-re'thral) A pair of glands that secrete a viscous fluid into the male urethra during sexual excitement; also called *Cowper's glands.*

bundle of His (see atrioventricular bundle)

bursa (bur'să) A saclike structure filled with synovial fluid. Bursae are located at friction points, as around joints, over which tendons can slide without contacting bone.

buttock (but'ok) The rump or fleshy mass on the posterior aspect of the lower trunk, formed primarily by the gluteal muscles.

calyx (ka'liks) A cup-shaped portion of the renal pelvis that encircles a renal papilla.

canaliculus (kan''ă-lik'yū-lus) A microscopic channel in bone tissue that connects lacunae.

canal of Schlemm (shlem) (see scleral venous sinus)

cancellous bone (kan'sĕ-lus) Spongy bone; bone tissue with a latticelike structure.

capillary (kap'i-lar''e) A microscopic blood vessel that connects an arteriole and a venule; the functional unit of the circulatory system.

carotid sinus (kă-rot'id) An expanded portion of the internal carotid artery located immediately above the point of branching from the external carotid artery; contains baroreceptors that monitor blood pressure.

carpus (kar'pus) The proximal portion of the hand that contains the eight carpal bones.

cartilage (kar'ti-lij) A type of connective tissue with a solid elastic matrix.

cartilaginous joint (kar''ti-laj'i-nus) A joint that lacks a joint cavity, permitting little movement between the bones held together by cartilage.

cauda equina (kaw'dă e-kwi'nă) The lower extension off of the spinal cord, where the roots of spinal nerves have a tail-like appearance.

caudal (kaw'dal) Referring to a position more toward the tail.

cecum (se'kum) The pouchlike portion of the large intestine to which the ileum of the small intestine is attached.

cell The structural and functional unit of an organism; the smallest structure capable of performing all of the functions necessary for life.

cementum (se-men'tum) Bonelike material that binds the root of a tooth to the periodontal membrane of the bony socket.

central canal An elongated longitudinal channel in the center of an osteon in bone tissue that contains branches of the nutrient vessels and a nerve; also called a *haversian canal.*

central nervous system (CNS) The brain and the spinal cord.

centrosome (sen'tro-sōm) A dense body near the nucleus of a cell that contains a pair of centrioles.

cerebellar peduncle (ser''ĕ-bel'ar pĭ-dung'k'l) An aggregation of nerve fibers connecting the cerebellum with the brain stem.

cerebellum (ser''ĕ-bel'um) The portion of the brain concerned with the coordination of skeletal muscle contraction. Part of the metencephalon, it consists of two hemispheres and a central vermis.

cerebral aqueduct (ser'ĕ-bral ak'wĕ-dukt'') The channel that connects the third and fourth ventricles of the brain; also called the *aqueduct of Sylvius* or the *mesencephalic aqueduct.*

cerebral arterial circle (ar-tir'e-al) An arterial vessel located on the ventral surface of the brain around the pituitary gland; also called the *circle of Willis.*

cerebral peduncles A paired bundle of nerve fibers along the ventral surface of the midbrain, that conduct impulses between the pons and the cerebral hemispheres.

cerebrospinal fluid (ser''ĕ-bro-spi'nal) A fluid produced by the choroid plexus of the ventricles of the brain. It fills the ventricles and surrounds the central nervous system in association with the meninges.

cerebrum (ser'ĕ-brum) The largest portion of the brain, composed of the right and left hemispheres.

ceruminous gland (sĕ-roo'mĭ-nus) A specialized integumentary gland that secretes cerumen, or earwax, into the external auditory canal.

cervical (ser'vi-kal) Pertaining to the neck or a necklike portion of an organ.

cervical ganglion (gang'gle-on) A cluster of postganglionic sympathetic nerve cell bodies located in the neck, near the cervical vertebrae.

cervical plexus (plek'sus) A network of spinal nerves formed by the anterior branches of the first four cervical nerves.

cervix (ser'viks) **1.** The narrow necklike portion of an organ. **2.** The inferior end of the uterus that adjoins the vagina (cervix of the uterus).

chemoreceptor (ke''mo-re-sep'tor) A neuroreceptor that is stimulated by the presence of chemical molecules.

chiasma (ki-as'mă) A crossing of nerve tracts from one side of the CNS to the other; also called a *chiasm.*

choanae (ko-a'ne) The two posterior openings from the nasal cavity into the nasopharynx; also called the *internal nares.*

chondrocranium (kon''dro-kra'ne-um) The portion of the skull that supports the brain. It is derived from endochondral bone.

chondrocyte (kon'dro-sīt) A cartilage cell.

chordae tendineae (kor'de ten-din'e-e) Tendinous bands that connect papillary muscles to the leaflets of the atrioventricular valves within the ventricles of the heart.

chorion (ko're-on) An extraembryonic membrane that contributes to the formation of the placenta.

choroid (kor'oid) The vascular, pigmented middle layer of the wall of the eye.

choroid plexus A mass of vascular capillaries from which cerebrospinal fluid is secreted into the ventricles of the brain.

chromatophilic substances (kro''mă-to-fil'ik) Clumps of rough endoplasmic reticulum in the cell body of a neuron; also called *Nissl bodies.*

chromosome (kro′mŏ-sōm) Structures in the nucleus that contain the genes for genetic expression.

chyme (kīm) The mass of partially digested food that passes from the pylorus of the stomach into the duodenum of the small intestine.

cilia (cil′e-ă) Microscopic hairlike processes that move in a wavelike manner on the exposed surfaces of certain epithelial cells.

ciliary body (sil′e-er″e) A portion of the choroid layer of the eye that secretes aqueous humor. It contains the ciliary muscle.

circle of Willis (see cerebral arterial circle)

circumduction (ser″kum-duk′shun) A movement of a body part that outlines a cone, such that the distal end moves in a circle while the proximal portion remains relatively stable.

clitoris (klit′or-is, kli′tor-is) A small, erectile structure in the vulva of the female, homologous to the glans penis in the male.

coccygeal (kok-sij′e-al) Pertaining to the region of the coccyx; the caudal termination of the vertebral column.

cochlea (kok′le-ă) The spiral portion of the inner ear that contains the spiral organ (organ of Corti).

cochlear window A round, membrane-covered opening between the middle and inner ear, directly below the vestibular window; also called the *round window.*

coelom (se′lom) The abdominal cavity.

collateral (kŏ-lat′er-al) A small branch of a blood vessel or nerve fiber.

colon (ko′lon) The first portion of the large intestine.

common bile duct A tube formed by the union of the hepatic duct and cystic duct that transports bile to the duodenum.

compact (dense) bone Tightly packed bone, superficial to spongy bone and covered by the periosteum.

conduction myofibers Specialized cardiac muscle fibers that conduct electrical impulses from the AV bundle into the ventricular walls; also called *Purkinje fibers.*

condyle (kon′dīl) A rounded process at the end of a long bone that forms an articulation.

cone (kōn) A color receptor cell in the retina of the eye.

congenital (kon-jen′ĭ-tal) Present at the time of birth.

conjunctiva (con″jungk-ti′vă) The thin membrane covering the anterior surface of the eyeball and lining the eyelids.

connective tissue One of the four basic tissue types within the body. It is a binding and supportive tissue with abundant matrix.

conus medullaris (ko′nus med″yŭ-lar′is) The caudal, tapering portion of the spinal cord.

convolution (con-vŏ-loo′shun) An elevation on the surface of a structure and an infolding of the tissue upon itself.

cornea (kor′ne-ă) The transparent convex anterior portion of the outer layer of the eyeball.

coronary circulation (kor′ŏ-nar″e) The arterial and venous blood circulation to the wall of the heart.

coronary sinus A large venous channel on the posterior surface of the heart into which the cardiac veins drain.

corpora quadrigemina (kor′por-ă kwad″rĭ-jem′ĭ-nă) Four dorsal lobes of the midbrain concerned with visual and auditory functions.

corpus callosum (kor′pus kă-lo′sum) A large tract of white matter within the brain that connects the right and left cerebral hemispheres.

cortex (kor′teks) **1.** The outer layer of an internal organ or body structure, as of the kidney or adrenal gland. **2.** The convoluted layer of gray matter that covers the surface of each cerebral hemisphere.

costal cartilage (kos′tal) The cartilage that connects the ribs to the sternum.

cranial (kra′ne-al) Pertaining to the cranium.

cranial nerve One of 12 pairs of nerves that arise from the ventral surface of the brain.

cranium (kra′ne-um) The endochondral bones of the skull that enclose or support the brain and the organs of sight, hearing, and balance.

crest A thickened ridge of bone for the attachment of muscle.

cricoid cartilage (kri′koid) A ring-shaped cartilage that forms the inferior end of the larynx.

crista (kris′tă) A crest, such as the crista galli that extends superiorly from the cribriform plate.

cubital (kyoo′bĭ-tal) Pertaining to the elbow. The cubital fossa is the anterior aspect of the elbow joint.

cystic duct (sis′tik dukt) The tube that transports bile from the gallbladder to the common bile duct.

cytology (si-tol′ŏ-je) The science dealing with the study of cells.

cytoplasm (si′tŏ-plaz″em) In a cell, the protoplasm located outside of the nucleus.

deciduous (dĭ-sij′oo-us) Pertaining to something cast off or shed in a particular sequence. Deciduous teeth are shed and replaced by permanent teeth during development.

decussation (dek″uh-sa′shun) A crossing of nerve fibers from one side of the CNS to the other.

defecation (def″ĕ-ka′shun) The elimination of feces from the rectum through the anal canal and out the anus.

deglutition (de″gloo-tish′un) The act of swallowing.

dendrite (den′drīt) A nerve cell process that transmits impulses toward the cell body of a neuron.

dentin (den′tin) The main substance of a tooth, covered by enamel over the crown of the tooth and by cementum on the root.

dentition (den-tish′un) The number, arrangement, and shape of teeth.

dermal papilla (der′mal pă-pil′ă) A projection of the dermis into the epidermis.

dermis (der′mis) The second, or deep, layer of skin beneath the epidermis.

descending colon The segment of the large intestine that descends on the left side from the level of the spleen to the level of the left iliac crest.

diaphragm (di′ă-fram) A sheetlike dome of muscle and connective tissue that separates the thoracic and abdominal cavities.

diaphysis (di-af′ĭ-sis) The shaft of a long bone.

diarthrosis (di″ar-thro′sis) A type of functionally classified joint in which the articulating bones are freely movable; also called a *synovial joint.*

diastole (di-as′tŏ-le) The sequence of the cardiac cycle during which a heart chamber wall is relaxed.

diencephalon (di″en-sef-ă-lon) A major region of the brain that includes the third ventricle, epithalamus, thalamus, hypothalamus, and pituitary gland.

digestion The process by which larger molecules of food substance are broken down mechanically and chemically into smaller molecules that can be absorbed.

diploe (dip′lo-e) The spongy layer of bone positioned between the inner and outer layers of compact bone.

distal (dis′tal) Away from the midline or origin; the opposite of *proximal.*

dorsal (dor′sal) Pertaining to the back or posterior portion of a body part; the opposite of *ventral.*

dorsiflexion (dor″si-flek′shun) Movement at the ankle or wrist as the dorsum of the foot or hand is elevated.

ductus arteriosus (duk′tus ar-tir″e-o′sus) The blood vessel that connects the pulmonary trunk and the aorta in a fetus.

ductus deferens (def′er-enz) A tube that carries spermatozoa from the epididymis to the ejaculatory duct; also called the *vas deferens* or *seminal duct.*

ductus venosus (ven-o′sus) A fetal blood vessel that connects the umbilical vein and the inferior vena cava.

duodenum (doo″ŏ-de′num, doo-od′ĕ-num) The first portion of the small intestine that leads from the pylorus of the stomach to the jejunum.

dura mater (door′ă ma′ter) The outermost meninx.

eccrine gland (ek′rin) A sweat gland that functions in thermoregulation.

ECG Electrocardiogram; EKG.

ectoderm (ek′tŏ-derm) The outermost of the three primary germ layers of an embryo.

edema (ĕ-de′mă) An excessive accumulation of fluid in the body tissues.

effector (ĕ-fek′tor) An organ, such as a gland or muscle, that responds to a motor stimulation.

efferent (ef′er-ent) Conveying away from the center of an organ or structure.

efferent arteriole (ar-tir′e-ōl) An arteriole of the renal vascular system that conducts blood away from the glomerulus of a nephron.

efferential ductules (duk′toolz) A series of coiled tubules that convey spermatozoa from the rete testis to the epididymis.

efferent neuron (noor′on) A motor nerve cell that conducts impulses from the central nervous system to effector organs, such as muscles or glands.

ejaculation (ĕ-jak″yoo-la′shun) The discharge of semen from the male urethra during climax.

ejaculatory duct (ĕ-jak′yoo-lă-tor-e) A tube that transports spermatozoa from the ductus deferens to the prostatic urethra.

elastic fibers (ĕ-las′tik) Protein strands found in certain connective tissue that have contractile properties.

elbow The synovial joint between the brachium and the antebrachium.

electrocardiogram (ĕ-lek″tro-kar′de-ŏ-gram″) A recording of the electrical activity that accompanies the cardiac cycle; ECG or EKG.

electroencephalogram (ē-lek″tro-en-sef′ă-lŏ-gram″)   A recording of the brain wave patterns or electrical impulses of the brain; EEG.

electromyogram (ē-lek″tro-mi′ŏ-gram″)   A recording of the electrical impulses or activity of a muscle; EMG.

embryology (em′bre-ol″ŏ-je)   The study of prenatal development from conception through the eighth week in utero.

enamel (ē-nam′el)   The outer dense substance covering the crown of a tooth.

endocardium (en″do-kar′de-um)   The endothelial lining of the heart chambers and valves.

endochondral bone (en″dŏ-kon′dral)   Bones that develop as hyaline cartilage models first and then are ossified.

endocrine gland (en′dŏ-krin)   A ductless, hormone-producing gland that is part of the endocrine system.

endoderm (en′dŏ-derm)   The innermost of the three primary germ layers of an embryo.

endolymph (en′dŏ-limf)   A fluid within the membranous labyrinth and cochlear duct of the inner ear that aids in the conduction of vibrations involved in hearing and in the maintenance of equilibrium.

endometrium (en″do-me′tre-um)   The inner lining of the uterus.

endomysium (en″do-mis′e-um)   The connective tissue sheath that surrounds each skeletal muscle fiber, separating the muscle cells from one another.

endoneurium (en″do-nyoo′re-um)   The connective tissue sheath that surrounds each nerve fiber, separating the nerve fibers one from another within a nerve.

endoplasmic reticulum (en-do-plaz′mik rĕ-tik′yoo-lum)   A cytoplasmic organelle composed of a network of canals running through the cytoplasm of a cell.

endothelium (en″do-the′le-um)   The layer of epithelial tissue that forms the thin inner lining of blood vessels and heart chambers.

eosinophil (e″ŏ-sin′ŏ-fil)   A type of white blood cell characterized by the presence of cytoplasmic granules that become stained by acidic eosin dye. Eosinophils normally constitute about 2%–4% of the white blood cells.

epicardium (ep″i-kar′de-um)   A thin, outer layer of the heart; also called the *visceral pericardium.*

epicondyle (ep″i-kon′dīl)   A projection of bone above a condyle.

epidermis (ep″i-der′mis)   The outermost layer of the skin, composed of stratified squamous epithelium.

epididymis (ep″i-did′i-mis)   A highly coiled tube located along the posterior border of the testis. It stores spermatozoa and transports them from the seminiferous tubules of the testis to the ductus deferens.

epidural space (ep″i-door′al)   A space between the spinal dura mater and the bone of the vertebral canal.

epiglottis (ep″i-glot′is)   A leaflike structure positioned on top of the larynx. It covers the glottis during swallowing.

epimysium (ep″i-mis′e-um)   A fibrous outer sheath of connective tissue surrounding a skeletal muscle.

epinephrine (ep″i-nef′rin)   A hormone secreted from the adrenal medulla initating actions similar to those initiated by sympathetic nervous system stimulation; also called *adrenaline.*

epineurium (ep″i-nyoo′re-um)   A fibrous, outer sheath of connective tissue surrounding a nerve.

epiphyseal plate (ep″i-fiz′e-al)   A hyaline cartilaginous layer located between the epiphysis and diaphysis of a long bone. It functions as a longitudinal growing region.

epiphysis (ē-pif′i-sis)   The end segment of a long bone, separated from the diaphysis early in life by an epiphyseal plate but later becoming part of the larger bone.

episiotomy (ē-piz″e-ot′o-me)   An incision of the perineum at the end of the second stage of labor to facilitate delivery and avoid tearing the perineum.

epithelial tissue (ep″i-the′le-al)   One of the four basic tissue types; the type of tissue that covers or lines all exposed body surfaces.

eponychium (ep″o-nik′e-um)   The thin layer of stratum corneum of the epidermis of the skin that overlaps and protects the lunula of the nail.

erythrocyte (ē-rith′rŏ-sīt)   A red blood cell.

esophagus (ē-sof′ă-gus)   A tubular portion of the GI tract that leads from the pharynx to the stomach as it passes through the thoracic cavity.

estrogen (es′tro-jen)   Any of several female sex hormones secreted from the ovarian (Graafian) follicle.

etiology (e″te-ol′ŏ-je)   The study of cause, especially of disease, including the origin and what pathogens, if any, are involved.

eustachian canal (yoo-sta′ke-an)   (see auditory tube)

eversion (e-ver′zhun)   A movement of the foot in which the sole is turned outward.

exocrine gland (ek′sŏ-krin)   A gland that secretes its product to an epithelial surface, directly or through ducts.

expiration (ek″spi-ra′shun)   The process of expelling air from the lungs through breathing out; also called *exhalation.*

extension (ek-sten′shun)   A movement that increases the angle between parts of a joint.

external (superficial)   Located on or toward the surface.

external acoustic meatus (aw′di-tor″e me-a′tus)   An opening through the temporal bone that connects with the tympanum and the middle-ear chamber and through which sound vibrations pass; also called *external auditory meatus.*

external ear   The outer portion of the ear, consisting of the auricle (pinna) and the external auditory canal.

extrinsic (eks-trin′sik)   Pertaining to an outside or external origin.

face   **1.** The anterior aspect of the head not supporting or covering the brain. **2.** The exposed surface of a structure.

facet (fas′et)   A small, smooth surface of a bone where articulation occurs.

falciform ligament (fal′si-form lig′ă-ment)   An extension of parietal peritoneum that separates the two major lobes of the liver.

fallopian tube (fă-lo′pe-an)   (see uterine tube)

false vocal cords   (see vestibular folds)

falx cerebelli (falks)   A fold of the dura mater anchored to the occipital bone. It projects inward between the cerebellar hemispheres.

falx cerebri   A fold of dura mater anchored to the crista galli. It extends between the right and left cerebral hemispheres.

fascia (fash′e-ă)   A tough sheet of fibrous tissue binding the skin to underlying muscles or supporting and separating muscles.

fasciculus (fă-sik′yū-lus)   A small bundle of muscle or nerve fibers.

fauces (faw′sēz)   The passageway between the mouth and the oropharynx.

feces (fe′sēz)   Material expelled from the GI tract during defecation, composed of food residue, bacteria, and secretions; also called *stool.*

fetus (fe′tus)   A prenatal human after 8 weeks of development.

fibroblast (fi′bro-blast)   An elongated connective tissue cell with cytoplasmic extensions that is capable of forming collagenous fibers or elastic fibers.

fibrous joint (fi′brus)   A type of articulation bound by fibrous connective tissue that allows little or no movement (e.g., a syndesmosis).

filiform papillae (fil′i-form pă-pil′e)   Numerous small projections over the entire dorsal surface of the tongue in which taste buds are absent.

filum terminale (fi′lum ter-mi-nal′e)   A fibrous, threadlike continuation of the pia mater, extending inferiorly from the terminal end of the spinal cord to the coccyx.

fimbriae (fim′bre-e)   Fringelike extensions from the borders of the open end of the uterine tube.

fissure (fish′ur)   A groove or narrow cleft that separates two parts, such as the cerebral hemispheres of the brain.

flexion (flek′shun)   A movement that decreases the angle between parts of a joint.

fontanel (fon″tă-nel′)   A membranous-covered region on the skull of a fetus or baby where ossification has not yet occurred; commonly called a *soft spot.*

foot   The terminal portion of the lower extremity, consisting of the tarsus, metatarsus, and digits.

foramen (fŏ-ra′men), pl. *foramina*   An opening in an anatomical structure, usually in a bone, for the passage of a blood vessel or a nerve.

foramen ovale (o-val′e)   An opening through the interatrial septum of the fetal heart.

forearm   The portion of the upper extremity between the elbow and the wrist; also called the *antebrachium.*

fornix (for′niks)   **1.** A recess around the cervix of the uterus where it protrudes into the vagina. **2.** A tract within the brain connecting the hippocampus with the mammillary bodies.

fossa (fos′ă)   A depressed area, usually on a bone.

fourth ventricle (ven′tri-k′l)   A cavity within the brain, between the cerebellum and the medulla oblongata and pons, containing cerebrospinal fluid.

fovea centralis (fo′ve-ă sen-tra′lis)   A depression on the macula lutea of the eye where only cones are located; the area of keenest vision.

frenulum (fren′yu-lum)   A membranous structure that serves to anchor and limit the movement of a body part.

frontal   **1.** Pertaining to the region of the frontal bone. **2.** A plane through the body, dividing the body into anterior and posterior portions; also called the *coronal plane.*

fungiform papillae (fun′ji-form pă-pil′e)   Flattened, mushroom-shaped projections interspersed over the dorsal surface of the tongue in which taste buds are present.

gallbladder A pouchlike organ attached to the underside of the liver in which bile secreted by the liver is stored and concentrated.

gamete (gam′ēt) A haploid sex cell; either an egg cell or a sperm cell.

ganglion (gang′gle-on) An aggregation of nerve cell bodies occurring outside the central nervous system.

gastrointestinal tract (GI tract) (gas″tro-in-tes′ti-nal) The portion of the digestive tract that includes the stomach and the small and large intestines.

gingiva (jin-ji′vă) The fleshy covering over the mandible and maxilla through which the teeth protrude within the mouth; also called the *gum*.

gland An organ that produces a specific substance or secretion.

glans penis (glanz pe′nis) The enlarged, sensitive, distal end of the penis.

gliding joint A type of synovial joint in which the articular surfaces are flat, permitting only side-to-side and back-and-forth movements.

glomerular capsule (glo-mer′yŭ-lar) The double-walled proximal portion of a renal tubule that encloses the glomerulus of a nephron; also called *Bowman's capsule*.

glomerulus (glo-mer′yŭ-lus) A coiled tuft of capillaries surrounded by the glomerular capsule. It filters urine from the blood.

glottis (glot′is) A slitlike opening into the larynx, positioned between the vocal folds.

goblet cell A mucus-secreting unicellular gland associated with columnar epithelia.

Golgi apparatus (gol′je) A network of fibrils or a series of four to eight channels forming an organelle in the cytoplasm of a cell.

Golgi tendon organ (see neurotendinous spindle)

gonad (go′nad) A reproductive organ, testis or ovary, that produces gametes and sex hormones.

gray matter The region of the central nervous system composed of nonmyelinated nerve tissue.

greater omentum (o-men′tum) A double-layered peritoneal membrane that originates on the greater curvature of the stomach. It hangs inferiorly like an apron over the contents of the abdominal cavity.

gross anatomy The branch of anatomy concerned with structures of the body that can be studied without a microscope.

gustatory (gus′tă-tor″e) Pertaining to the sense of taste.

gut (gŭt) The GI tract or a portion thereof; generally used in reference to the embryonic digestive tube, consisting of the foregut, midgut, and hindgut.

gyrus (jī′rus) A convoluted elevation or ridge.

hair A threadlike appendage of the epidermis consisting of keratinized dead cells that have been pushed up from a dividing basal layer.

hair cells Specialized receptor nerve endings for detecting sensations, such as in the spiral organ.

hair follicle (fol′ĭ-k'l) A tubular depression in the dermis of the skin in which a hair develops.

hand The terminal portion of the upper extremity, consisting of the carpus, metacarpus, and digits.

hard palate (pal′it) The bony partition between the oral and nasal cavities, formed by the maxillae and palatine bones and lined by mucous membrane.

haustra (haws′trah) Sacculations or pouches of the colon.

haversian canal (ha-ver′shan) (see central canal)

haversian system (see osteon)

head The uppermost portion of a human that contains the brain and major sense organs.

heart A four-chambered, muscular pumping organ positioned in the thoracic cavity, slightly to the left of midline.

hemoglobin (he″mo-glo′bin) The pigment of red blood cells that constitutes about 33% of the cell volume and transports oxygen and carbon dioxide.

hemopoiesis (hem″ŏ-poi-e′sis) The production of red blood cells.

hepatic duct (hĕ-pat′ik) Tubules that drain bile from the liver and merge with the cystic duct from the gallbladder to form the common bile duct.

hepatic portal circulation The return of venous blood from the digestive organs and spleen through a capillary network within the liver before draining into the heart.

hepatopancreatic ampulla A short ductule formed by the combined pancreatic and common bile ducts within the duodenal papilla; also called the *ampulla of Vater*.

hiatus (hi-a′tus) An opening or fissure; a foramen.

hilum (hi′lum) A concave or depressed area where vessels or nerves enter or exit an organ.

hinge joint A type of synovial articulation characterized by the convex surface of one bone fitting into the concave surface of another so that movement is confined to one plane, as in the knee or interphalangeal joint.

histology (his-tol′ŏ-je) Microscopic study of the structure and function of tissues.

horizontal (transverse) A directional plane that divides the body, organ, or appendage into superior and inferior or proximal and distal portions.

hormone (hor′mōn) A chemical substance produced in an endocrine gland and secreted into the bloodstream to cause an effect in a specific target organ.

hyaline cartilage (hi′ă-lĭn) A cartilage with a homogeneous matrix. It is the most common type, occurring at the articular ends of bones, in the trachea, and within the nose. Most of the bones in the body are formed from hyaline cartilage.

hymen (hi′men) A developmental remnant (vestige) of membranous tissue that partially covers the vaginal opening.

hyperextension (hi″per-ek-sten′shun) Extension beyond the normal anatomical position or 180°.

hypertension (hi″per-ten′shun) Elevated or excessive blood pressure.

hyponychium (hi″pŏ-nik′e-um) A thickened supportive layer of the stratum corneum at the distal end of a digit under the free edge of the nail.

hypothalamus (hi″po-thal′ă-mus) A structure within the diencephalon that lies below the thalamus, where it functions as an autonomic center and regulates the pituitary gland.

ileocecal valve (il″e-ŏ-se′kal) A modification of the mucosa at the junction of the small and large intestine that forms a one-way passage and prevents the backflow of food materials.

ileum (il′e-um) The terminal portion of the small intestine between the jejunum and cecum.

incus (ing′kus) The middle of three auditory ossicles within the middle-ear chamber; commonly called the *anvil*.

inferior vena cava (ve′nă ka′vă) A large systemic vein that collects blood from the body regions inferior to the level of the heart and returns it to the right atrium.

infundibulum (in″fun-dib′yoo-lum) The stalk that attaches the pituitary gland to the hypothalamus of the brain.

ingestion (in-jes′chun) The process of taking food or liquid into the body by way of the oral cavity.

inguinal (ing′wi-nal) Pertaining to the groin region.

inguinal canal The passage in the abdominal wall through which a testis descends into the scrotum.

insertion The more movable attachment of a muscle, usually more distal.

inspiration (in″spi-ra′shun) The act of breathing air into the alveoli of the lungs; also called *inhalation*.

insula (in′sŭ-lă) A cerebral lobe deep to the lateral sulcus and covered by portions of the frontal, parietal, and temporal lobes.

integument (in-teg′yoo-ment) The skin; the largest organ of the body.

intercalated disc (in-ter′kă-lāt-ed) A thickened portion of the sarcolemma that extends across a cardiac muscle fiber and indicates the boundary between cells.

intercellular substance (in″ter-sel′yŭ-lar) The matrix or material between cells that largely determines tissue types.

internal (deep) Toward the center, away from the surface of the body.

internal ear The innermost portion or chamber of the ear, containing the cochlea and the vestibular organs.

intervertebral disc (in″ter-ver′tĕ-bral) A pad of fibrocartilage located between the bodies of adjacent vertebrae.

intestinal gland (in-tes′ti-nal) A simple tubular digestive gland that opens onto the surface of the intestinal mucosa and secretes digestive enzymes; also called the *crypt of Lieberkühn*.

intramembranous ossification (see membraneous bone)

intrinsic (in-trin′sik) Situated in or pertaining to internal origin.

inversion (in-ver′zhun) A movement of the foot in which the sole is turned inward.

iris (i′ris) The pigmented portion of the vascular tunic of the eye that surrounds the pupil and regulates its diameter.

islets of Langerhans (i′letz of lang′er-hanz″) (see pancreatic islets)

isthmus (is′mus) A narrow neck or portion of tissue connecting two structures.

jejunum (jĕ-joo′num) The middle portion of the small intestine, located between the duodenum and the ileum.

joint capsule The fibrous tissue that encloses the joint cavity of a synovial joint.

keratin (ker′ă-tin) An insoluble protein present in the epidermis and in epidermal derivatives, such as hair and nails.

kidney (kid′ne) One of the paired organs of the urinary system that contains nephrons and that filters wastes from the blood in the formation of urine.

kinesiology (ki-ne″se-ol′ŏ-je) The study of body movement.

knee A region in the lower extremity between the thigh and the leg that contains a synovial hinge joint.

labial frenulum (la′be-al fren′yŭ-lum) A longitudinal fold of mucous membrane that attaches the lips to the gum along the midline of both the upper and lower lip.

labia majora (la′be-ă mă-jor′ă) The portion of the external genitalia of a female consisting of two longitudinal folds of skin extending downward and backward from the mons pubis.

labia minora (mĭ-nor′ă) Two small folds of skin, devoid of hair and sweat glands, lying between the labia majora of the external genitalia of a female.

labyrinth (lab′ĭ-rinth) The complex system of interconnecting tubes within the inner ear that includes the semicircular ducts and the cochlear and vestibular labyrinths.

lacrimal canaliculus (lak′rĭ-mal) A drainage duct for tears, located at the medial corner of an eyelid. It conveys the tears medially into the nasolacrimal sac.

lacrimal gland A tear-secreting gland, located on the superior lateral portion of the eyeball underneath the upper eyelid.

lactation (lak-ta′shun) The production and secretion of milk by the mammary glands.

lacteal (lak′te-al) A small lymphatic duct associated with a villus of the small intestine.

lacuna (lă-kyoo′nă) A small, hollow chamber that houses an osteocyte in mature bone tissue or a chondrocyte in cartilage tissue.

lambdoidal suture (lam′doid-al soo′chur) The immovable joint in the skull between the parietal bones and the occipital bone.

lamella (lă-mel′ă) A concentric ring of matrix surrounding the central canal in an osteon of mature bone tissue.

lamellated corpuscle (lam′ĕ-la-ted kor′pus′l) A sensory receptor for pressure, found in tendons, around joints, and in visceral organs; also called a *pacinian corpuscle.*

lamina (lam′ĭ-nă) A thin plate of bone that extends superiorly from the body of a vertebra to form both sides of the arch of a vertebra.

lanugo (lă-noo′go) Short, silky fetal hair, which may be present for a short time on a premature infant.

large intestine The last major portion of the GI tract, consisting of the cecum, colon, rectum, and anal canal.

laryngopharynx (lă-ring″go-far′ingks) The inferior or lower portion of the pharynx in contact with the larynx.

larynx (lar′ingks) The structure located between the pharynx and trachea that houses the vocal folds (cords); commonly called the *voice box.*

lateral (lat′er-al) Pertaining to the side; farther from the median plane.

lateral ventricle (ven′tri-k′l) A cavity within the cerebral hemisphere of the brain that is filled with cerebrospinal fluid.

leg The portion of the lower extremity between the knee and ankle.

lens (lenz) A transparent refractive organ of the eye positioned posterior to the pupil and iris.

lesser omentum (o-men′tum) A peritoneal fold of tissue extending from the lesser curvature of the stomach to the liver.

leukocyte (loo′kŏ-sīt) A white blood cell; also spelled leucocyte.

ligament (lig′ă-ment) A tough cord or fibrous band of connective tissue that binds bone to bone to strengthen and provide flexibility to a joint. It also may support viscera.

limbic system (lim′bik) A portion of the brain concerned with emotions and autonomic activity.

linea alba (lin′e-ă al′bă) A vertical fibrous band extending down the anterior medial portion of the abdominal wall.

lingual frenulum (ling′gwal fren′yŭ-lum) A longitudinal fold of mucous membrane that attaches the tongue to the floor of the oral cavity.

lower extremity A lower appendage, including the hip, thigh, knee, leg, and foot.

lumbar (lum′bar) Pertaining to the region of the loins.

lumbar plexus (plek′sus) A network of nerves formed by the anterior branches of spinal nerves L1 through L4.

lumen (loo′men) The space within a tubular structure through which a substance passes.

lung One of the two major organs of respiration positioned within the thoracic cavity on either lateral side of the mediastinum.

lunula (loo′nyoo-lă) The half-moon-shaped whitish area at the proximal portion of a nail.

luteinizing hormone (LH) (loo′te-ĭ-ni″zing) A hormone secreted by the adenohypophysis (anterior lobe) of the pituitary gland that stimulates ovulation and progesterone secretion by the corpus luteum, influences the mammary glands of milk secretion in females, and stimulates testosterone secretion by the testes in males.

lymph (limf) A clear, plasmalike fluid that flows through lymphatic vessels.

lymph node A small ovoid mass of reticular tissue located along the course of lymph vessels.

lymphocyte (lim′fŏ-sīt) A type of white blood cell characterized by a granular cytoplasm. Lymphocytes usually constitute about 20%–25% of the white cell count.

macrophage (mak′rŏ-fāj) A wandering phagocytic cell.

macula lutea (mak′yŭ-lă loo′te-ă) A yellowish depression in the retina of the eye that contains the fovea centralis, the area of keenest vision.

malignant Threatening to life; virulent. Of a tumor, cancerous, tending to metastasize.

malleus (mal′e-us) The first of three auditory ossicles that attaches to the tympanum; commonly called the *hammer.*

mammary gland (mam′er-e) The gland of the female breast responsible for lactation and nourishment of the young.

marrow (mar′o) The soft connective tissue found within the inner cavity of certain bones that produces red blood cells.

mastication (mas″ti-ka′shun) The chewing of food.

matrix (ma′triks) The intercellular substance of a tissue.

meatus (me-a′tus) A passageway or opening into a structure.

mechanoreceptor (mek″ă-no-re-sep′tor) A sensory receptor that responds to a mechanical stimulus.

medial (me′de-al) Toward or closer to the midline of the body.

mediastinum (me″de-ă-sti′num) The partition in the center of the thorax between the two pleural cavities.

medulla (mĕ-dul′ă) The center portion of an organ.

medulla oblongata (ob″long-ga′tă) A portion of the brain stem, located between the spinal cord and the pons.

medullary (marrow) cavity (med′l-er″e) The hollow core of the diaphysis of a long bone in which marrow is found.

meiosis (mi-o′sis) A specialized type of cell division by which gametes, or haploid sex cells, are formed.

Meissner's corpuscle (mīs′nerz kor′pus′l) A touch sensory receptor found in the papillary layer of the dermis of the skin.

melanin (mel′ă-nin) A dark pigment found within the epidermis or epidermal derivatives of the skin.

melanocyte (mel′ă-no-sīt) A specialized melanin-producing cell found in the deepest layer of epidermis.

melanoma (mel″ă-no′mă) A dark malignant tumor of the skin; frequently forms in moles.

membranous bone (mem′bră-nus) Bone that forms from membranous connective tissue rather than from cartilage; also called *intramembranous bone.*

menarche (mĕ-nar′ke) The first menstrual discharge.

meninges (mĕ-nin′jēz), sing. *meninx* A group of three fibrous membranes covering the central nervous system, composed of the dura mater, arachnoid mater, and pia mater.

menisci (mĕ-nis′ke) Wedge-shaped fibrocartilages in certain synovial joints.

menopause (men′o-pawz) The period marked by the cessation of menstrual periods in the human female.

menstrual cycle (men′stru-al) The rhythmic female reproductive cycle, characterized by changes in hormone levels and physical changes in the uterine lining.

menstruation (men″stru-a′shun) The discharge of blood and tissue from the uterus at the end of the female reproductive cycle.

mesenchyme (mez′en-kīm) An embryonic connective tissue that can migrate, and from which all connective tissues arise.

mesenteric patches (mes″en-ter′ik) Clusters of lymph nodes on the walls of the small intestine; also called *Peyer's patches.*

mesentery (mes′en-ter″e) A fold of peritoneal membrane that attaches an abdominal organ to the abdominal wall.

mesoderm (mes′o-derm) The middle one of the three primary germ layers.

mesothelium (mes″o-the′leum) A simple squamous epithelial tissue that lines body cavities and covers visceral organs; also called *serosa.*

mesovarium (mes″o-va′re-um) The peritoneal fold that attaches an ovary to the broad ligament of the uterus.

metabolism (mĕ-tab′ŏ-liz-em) The sum total of the chemical changes that occur within a cell.

metacarpus (met″ă-kar′pus)   The region of the hand between the wrist and the digits, including the five bones that support the palm of the hand.

metarteriole (met″ar-tir′e-ōl)   A small blood vessel that emerges from an arteriole, passes through a capillary network, and empties into a venule.

metastasis (mě-tas′tă-sis)   The spread of a disease from one organ or body part to another.

metatarsus (met″ă-tar′sus)   The region of the foot between the ankle and the digits that includes five bones.

metencephalon (met″en-sef′ă-lon)   The most superior portion of the hindbrain that contains the cerebellum and the pons.

microglia (mi-krog′le-ă)   Small phagocytic cells found in the central nervous system.

microvilli (mi″kro-vil′i)   Microscopic hairlike projections of cell membranes on certain epithelial cells.

micturition (mik″tŭ-rish′un)   The process of voiding urine; also called *urination*.

midbrain   The portion of the brain between the pons and the forebrain.

middle ear   The middle of the three portions of the ear that contains the three auditory ossicles.

midsagittal (mid-saj′ĭ-tal)   A plane that divides the body or an organ into right and left halves; also called the *median plane*.

mitosis (mi-to′sis)   The process of cell division that results in two identical daughter cells, containing the same number of chromosomes.

mitral valve (mi′tral)   The left atrioventricular heart valve; also called the *bicuspid valve*.

mixed nerve   A nerve that contains both motor and sensory nerve fibers.

monocyte (mon′o-sīt)   A phagocytic type of white blood cell, normally constituting about 3%–8% of the white blood cell count.

mons pubis (monz pu′bis)   A fatty tissue pad covering the symphysis pubis and covered by pubic hair in the female.

morula (mor′yoo-lă)   An early stage of embryonic development characterized by a solid ball of cells.

motor area   A region of the cerebral cortex from which motor impulses to muscles or glands originate.

motor nerve   A nerve composed of motor nerve fibers.

motor neuron (noor′on)   A nerve cell that conducts action potentials away from the central nervous system and innervates effector organs (muscles and glands). Motor neurons form the ventral roots of the spinal nerves.

motor unit   A single motor neuron and the muscle fibers it innervates.

mucosa (myoo-ko′să)   A mucous membrane that lines cavities and tracts opening to the exterior.

mucous cell (myoo′kus)   A specialized unicellular gland that produces and secretes mucus; also called a *goblet cell*.

multipolar neuron (mul″tĭ-po′lar noor′on)   A nerve cell with many processes originating from the cell body.

muscle (mus′el)   A major type of tissue adapted to contract. The three kinds of muscle are cardiac, smooth, and skeletal.

muscularis (mus″kyŭ-la′ris)   A muscular layer or tunic of an organ, composed of smooth muscle tissue.

myelencephalon (mi″ĕ-len-sef′ă-lon)   The posterior portion of the hindbrain that contains the medulla oblongata.

myelin (mi′ĕ-lin)   A lipoprotein material that forms a sheathlike covering around nerve fibers.

myenteric plexus (mi″en-ter′ik plek′sus)   A network of sympathetic and parasympathetic nerve fibers located in the muscularis tunic of the small intestine; also called the *plexus of Auerbach*.

myocardium (mi″o-kar′de-um)   The cardiac muscle layer of the heart.

myofibril (mi″ŏ-fi′bril)   A bundle of contractile fibers within muscle cells.

myogram (mi′ŏ-gram)   A recording of electrical activity within a muscle.

myology (mi-ol′ŏ-je)   The science or study of muscle structure function.

myometrium (mi″o-me′tre-um)   The layer or tunic of smooth muscle within the uterine wall.

myoneural junction (mi″o-noor′al)   The site of contact between an axon of a motor neuron and a muscle fiber.

myopia (mi-o′pe-ă)   A visual defect in which objects may be seen distinctly only when very close to the eyes; also called *nearsightedness.*.

myosin (mi′ŏ-sin)   A thick filament protein that together with actin causes muscle contraction.

nail   A hardened, keratinized plate that develops from the epidermis and forms a protective covering of the distal phalanges of fingers and toes.

nares (na′rēz)   The openings into the nasal cavity; also called the *nostrils.*

nasal cavity (na′zal)   A mucosa-lined space above the oral cavity, divided by a nasal septum. It is the first chamber of the respiratory system.

nasal concha (kong′kă)   A scroll-like bone extending medially from the lateral wall of the nasal cavity; also called a *turbinate bone.*

nasal septum (sep′tum)   A bony and cartilaginous partition that separates the nasal cavity into two portions.

nasopharynx (na″zo-far′ingks)   The first or uppermost chamber of the pharynx, positioned posterior to the nasal cavity and extending down to the soft palate.

neck   **1.** Any constricted portion, such as the neck of an organ. **2.** The cervical region of the body between the head and thorax.

necrosis (nĕ-kro′sis)   Cell death or tissue death due to disease or trauma.

neonatal (ne″o-na′tal)   The stage of life from birth to the end of 4 weeks.

nephron (nef′ron)   The functional unit of the kidney, consisting of a glomerulus, glomerular capsule, convoluted tubules, and the loop of the nephron.

nerve   A bundle of nerve fibers outside the central nervous system.

neurilemma (noor″ĭ-lem′ă)   A thin, membranous covering surrounding the myelin sheath of a nerve fiber.

neurofibril (noor′ŏ-fi′bril)   One of many delicate threadlike structures within the cytoplasm of a cell body and the axon hillock of a neuron.

neurofibril node   A gap in the myelin sheath of a nerve fiber; also called the *node of Ranvier.*

neuroglia (noo′rog-le-ă)   Specialized supportive cells of the central nervous system.

neurohypophysis (noor″o-hi-pof′ĭ-sis)   The posterior lobe of the pituitary gland.

neurolemmocyte   A specialized neuroglia cell that surrounds an axon fiber of a peripheral neuron and forms the neurilemmal sheath.

neuron (noor′on)   The structural and functional cell of the nervous system, composed of a cell body, dendrites, and an axon; also called a *nerve cell.*

neurotendinous spindle   A sensory receptor found near the junction of tendons and muscles.

neutrophil (noo′tro-fil)   A type of phagocytic white blood cell, normally constituting about 60%–70% of the white blood cell count.

nipple   A dark-pigmented, rounded projection at the tip of the breast.

Nissl bodies (nis′l)   (see chromatophilic substances)

node of Ranvier (ran′ve-a)   (see neurofibril node)

notochord (no′to-kord)   A flexible rod of tissue that extends the length of the back of an embryo.

nucleoplasm (noo′kle-ŏ-plaz″em)   The protoplasmic contents of the nucleus of a cell.

nucleus (noo′kle-us)   A spheroid body within a cell that contains the genetic factors of the cell.

nucleus pulposus (pul-po′sus)   The soft, pulpy core of an intervertebral disc; a remnant of the notochord.

olfactory (ol-fak′to-re)   Pertaining to the sense of smell.

olfactory bulb   An aggregation of sensory neurons of an olfactory nerve, lying inferior to the frontal lobe of the cerebrum on both sides of the crista galli of the ethmoid bone.

olfactory tract   The olfactory sensory tract of axons that conveys impulses from the olfactory bulb to the olfactory portion of the cerebral cortex.

oligodendrocyte (ol″ĭ-go-den′dro-sīt)   A type of neuroglial cell concerned with the formation of the myelin of nerve fibers within the central nervous system.

oocyte (o′ŏ-sīt)   A developing egg cell.

oogenesis (o″ŏ-jen′ĕ-sis)   The process of female gamete formation.

optic (op′tik)   Pertaining to the eye.

optic chiasma (ki-az′mă)   An X-shaped structure on the inferior aspect of the brain, anterior to the pituitary gland, where there is a partial crossing over of fibers in the optic nerves; also called the *optic chiasm.*

optic disc   A small region of the retina where the fibers of the ganglion neurons exit from the eyeball to form the optic nerve; also called the *blind spot.*

optic tract   A bundle of sensory axons located between the optic chiasma and the thalamus that functions to convey visual impulses from the photoreceptors within the eye.

oral   Pertaining to the mouth.

ora serrata (o′ră)   The jagged peripheral margin of the retina.

organ   A structure consisting of two or more tissues that performs a specific function.

organ of Corti (kor′te)   (see spinal organ)

organelle (or″gă-nel′)   A minute living structure of a cell with a specific function.

organism An individual living creature.

orifice (or'ĭ-fis) An opening into a body cavity or tube.

origin The place of muscle attachment—usually the more stationary point or proximal bone; opposite the insertion.

oropharynx (o''ro-far'ingks) The second portion of the pharynx, located posterior to the oral cavity and extending from the soft palate to the hyoid bone.

osseous tissue (os'e-us) Bone tissue.

ossicle (os'si-k'l) One of the three bones of the middle ear; also called the *auditory ossicle.*

ossification (os''ĭ-fi-ka'shun) The process of bone tissue formation.

osteoblast (os'te-o-blast'') A bone-forming cell.

osteoclast (os'te-o-klast'') A cell that causes erosion and resorption of bone tissue.

osteocyte (os'te-o-sīt'') A mature bone cell.

osteology (os''te-ol'o-je) The study of the structure and function of bone and the entire skeleton.

osteon (os'te-on) A group of osteocytes and concentric lamellae surrounding a central canal, constituting the basic unit of structure in osseous tissue; also called a *haversian system.*

oval window (see vestibular window)

ovarian follicle (o-va're-an fol'li-k'l) A developing ovum and its surrounding epithelial cells.

ovarian ligament (lig'ă-ment) A cordlike connective tissue that attaches the ovary to the uterus.

ovary (o'vă-re) The female gonad in which ova and certain sexual hormones are produced.

oviduct (o'vĭ-dukt) The tube that transports ova from the ovary to the uterus; also called the *uterine tube* or *fallopian tube.*

ovulation (ov-yŭ-la'shun) The rupture of an ovarian follicle with the release of an ovum.

ovum (o'vum) A secondary oocyte capable of developing into a new individual when fertilized by a spermatozoon.

pacinian corpuscle (pă-sin'e-an kor'pus'l) (see lamellated corpuscle)

palate (pal'at) The roof of the oral cavity.

palatine (pal'ă-tin) Pertaining to the palate.

palmar (pal'mar) Pertaining to the palm of the hand.

palpebra (pal'pĕ-bră) An eyelid.

pancreas (pan'kre-as) A mixed organ in the abdominal cavity that secretes gastric juices into the GI tract and insulin and glucagon into the blood.

pancreatic duct (pan''kre-at'ik) A drainage tube that carries pancreatic juice from the pancreas into the duodenum of the hepatopancreatic ampulla.

pancreatic islets A cluster of cells within the pancreas that forms the endocrine portion and secretes insulin and glucagon; also called *islets of Langerhans.*

papillae (pă-pil'e) Small, nipplelike projections.

papillary muscle (pap'i-ler''e) Muscular projections from the ventricular walls of the heart to which the chordae tendineae are attached.

paranasal sinus (par''ă-na'zal si'nus) An air chamber lined with a mucous membrane that communicates with the nasal cavity.

parasympathetic (par''ă-sim''pă-thet'ik) Pertaining to the division of the autonomic nervous system concerned with activities that, in general, inhibit or oppose the physiological effects of the sympathetic nervous system.

parathyroids (par''ă-thi'roidz) Small endocrine glands embedded on the posterior surface of the thyroid glands that are concerned with calcium metabolism.

parietal (pă-ri'ĕ-tal) Pertaining to a wall of an organ or cavity.

parietal pleura (ploor'ă) The thin serous membrane attached to the thoracic walls of the pleural cavity.

parotid gland (pă-rot'id) One of the paired salivary glands located on the side of the face over the masseter muscle just anterior to the ear and connected to the oral cavity by the parotid duct.

parturition (par''tyoo-rish'un) The process of giving birth.

pectoral (pek'to-ral) Pertaining to the chest region.

pectoral girdle The portion of the skeleton that supports the upper extremities.

pedicle (ped'i-k'l) The portion of a vertebra that connects and attaches the lamina to the body.

pelvic (pel'vik) Pertaining to the pelvis.

pelvic girdle The portion of the skeleton to which the lower extremities are attached.

pelvis (pel'vis) A basinlike bony structure formed by the sacrum and os coxae.

penis (pe'nis) The male organ of copulation, used to introduce spermatozoa into the female vagina and through which urine passes during urination.

pennate (pen'āt) Pertaining to a skeletal muscle fiber arrangement in which the fibers are attached to tendinous slips in a featherlike pattern.

perforating canal A minute duct through compact bone by which blood vessels and nerves penetrate to the central canal of an osteon; also called *Volkmann's canal.*

pericardium (per''ĭ-kar'de-um) A protective serous membrane that surrounds the heart.

perichondrium (per''ĭ-kon'dre-um) A toughened connective sheet that covers some kinds of cartilage.

perikaryon (per''ĭ-kar'e-on) The cell body of a neuron.

perilymph (per'ĭ-limf) A fluid of the inner ear that provides a liquid-conducting medium for the vibrations involved in hearing and the maintenance of equilibrium.

perimysium (per''ĭ-mis'e-um) Fascia (connective tissue) surrounding a bundle (fascicle) of muscle fibers.

perineum (per''ĭ-ne'um) The floor of the pelvis, which is the region between the anus and the scrotum in the male and between the anus and the vulva in the female.

perineurium (per''ĭ-noor'e-um) Connective tissue surrounding a bundle (fascicle) of nerve fibers.

periodontal membrane (per''e-o-don'tal) A fibrous connective tissue lining the dental alveoli.

periosteum (per''e-os'te-um) A fibrous connective tissue covering the outer surface of bone.

peripheral nervous system (pĕ-rif'er-al) The nerves and ganglia of the nervous system that lie outside of the brain and spinal cord.

peristalsis (per''ĭ-stal'sis) Rhythmic contractions of smooth muscle in the walls of various tubular organs by which the contents are forced onward.

peritoneum (per''ĭ-to-ne'um) The serous membrane that lines the abdominal cavity and covers the abdominal visceral organs.

Peyer's patches (pi'erz) (see mesenteric patches)

phalanx (fa'lanks), pl. *phalanges* A bone of a digit (finger or toe).

pharynx (far'ingks) The organ of the digestive system and respiratory system located at the back of the oral and nasal cavities that extends to the larynx anteriorly and the esophagus posteriorly; also called the *throat.*

photoreceptor (fo''to-re-sep'tor) A sensory nerve ending that responds to the stimulation of light.

physiology (fiz''e-ol'o-je) The science that deals with the study of body functions.

pia mater (pi'ă ma'ter) The innermost meninx that is in direct contact with the brain and spinal cord.

pineal gland (pin'e-al) A small cone-shaped gland located in the roof of the third ventricle.

pinna (pin'ă) The outer, fleshy portion of the external ear; also called the *auricle.*

pituitary gland (pi-too'ĭ-tar-e) A small, pea-shaped endocrine gland situated on the interior surface of the diencephalonic region of the brain, consisting of anterior and posterior lobes; also called the *hypophysis.*

pivot joint (piv'ut) A synovial joint in which the rounded head of one bone articulates with the depressed cup of another to permit a rotational type of movement.

placenta (plă-sen'tă) The organ of metabolic exchange between the mother and the fetus.

plantar (plan'tar) Pertaining to the sole of the foot.

plasma (plaz'mă) The fluid, extracellular portion of circulating blood.

platelets (plāt'letz) Small fragments of specific bone marrow cells that function in blood coagulation; also called *thrombocytes.*

pleural (ploor'al) Pertaining to the serous membranes associated with the lungs.

pleural cavity The potential space between the visceral pleura and parietal pleura.

pleural membranes Serous membranes that surround the lungs and provide protection and compartmentalization.

plexus (plek'sus) A network of interlaced nerves or vessels.

plexus of Auerbach (ow'er-bak) (see myenteric plexus)

plexus of Meissner (mīs'ner) (see submucosal plexus)

plicae circulares (pli'ce sur-kyŭ-lar'ēz) Deep folds within the wall of the small intestine that increase the absorptive surface area.

pneumotaxic area (noo''mo-tak'sik) The region of the respiratory control center located in the pons of the brain.

pons (ponz) The portion of the brain stem just above the medulla oblongata and anterior to the cerebellum.

popliteal (pop''li-te'al, pop-lit'e-al) Pertaining to the concave region on the posterior aspect of the knee.

posterior (pos-tēr'e-or) Toward the back; also called *dorsal.*

posterior root An aggregation of sensory neuron fibers lying between a spinal nerve and the dorsolateral aspect of the spinal cord; also called the *dorsal root* or *sensory root*.

posterior root ganglion (gang′gle-on) A cluster of cell bodies of sensory neurons, located along the posterior root of a spinal nerve; also called *sensory ganglion*.

postganglionic neuron (pōst″gang-gle-on′ik) The second neuron in an autonomic motor pathway. Its cell body is outside the central nervous system, and it terminates at an effector organ.

postnatal (post-na′tal) After birth.

preganglionic neuron (pre″gang-gle-on′ik) The first neuron in an autonomic motor pathway. Its cell body is inside the central nervous system, and it terminates on a postganglionic neuron.

pregnancy The condition in which a female is carrying a developing offspring within the body.

prenatal (pre-na′tal) Pertaining to the period of offspring development during pregnancy; before birth.

prepuce (pre′pyoos) A fold of loose, retractable skin covering the glans of the penis or clitoris; also called the *foreskin*.

prime mover The muscle most directly responsible for a particular movement.

pronation (pro-na′shun) A rotational movement of the forearm in which the palm of the hand is turned posteriorly.

proprioceptor (pro″pre-o-sep′tor) A sensory nerve ending that responds to changes in tension in a muscle or tendon.

prostate (pros′tāt) A walnut-shaped gland surrounding the male urethra just below the urinary bladder that secretes an additive to seminal fluid during ejaculation.

prosthesis (pros-the′sis) An artificial device to replace a diseased or worn body part.

protraction (pro-trak′shun) The movement of a body part, such as the mandible, forward on a plane parallel with the ground; the opposite of *retraction*.

proximal (prok′si-mal) Closer to the midline of the body or origin of an appendage; the opposite of *distal*.

puberty (pyoo′ber-te) The period of development in which the reproductive organs become functional.

pulmonary (pul′mo-ner″e) Pertaining to the lungs.

pulmonary circulation The system of blood vessels from the right ventricle of the heart to the lungs that transports deoxygenated blood and returns oxygenated blood from the lungs to the left atrium of the heart.

pulp cavity A cavity within the center of a tooth that contains blood vessels, nerves, and lymphatics.

pupil The opening through the iris that permits light to enter the vitreous chamber of the eyeball and be refracted by the lens.

Purkinje fibers (pur-kin′je) (see conduction myofibers)

pyloric sphincter (pi-lor′ik sfingk′ter) A modification of the muscularis tunic between the stomach and the duodenum that functions to regulate the food material leaving the stomach.

pyramid (pir′ă-mid) Any of several structures that have a pyramidal shape, including the renal pyramids in the kidney and the medullary pyramids on the ventral surface of the brain.

ramus (ra′mus) A branch of a bone, artery, or nerve.

raphe (ra′fe) A ridge or a seamlike structure.

receptor (re-sep′tor) A sense organ or a specialized distal end of a sensory neuron that receives stimuli from the environment.

rectouterine pouch (rek″to-yoo′ter-in) A pocket of parietal peritoneal membrane between the uterus and the rectum, forming the lowest point in the pelvic cavity; also called the *pouch of Douglas*.

rectum (rek′tum) The portion of the GI tract between the sigmoid colon and the anal canal.

red marrow (mar′o) A tissue that forms blood cells, located in the medullary cavity of certain bones.

red nucleus (noo′kle-us) An aggregation of gray matter of a reddish color, located in the upper portion of the midbrain. It sends fibers to certain brain tracts.

reflex (re′fleks) A rapid involuntary response to a stimulus.

reflex arc The basic conduction pathway through the nervous system, consisting of a sensory neuron, association neuron, and a motor neuron.

regional anatomy The division of anatomy concerned with structural arrangement in specific areas of the body, such as the head, neck, thorax, or abdomen.

renal (re′nal) Pertaining to the kidney.

renal corpuscle (kor′pus′l) The portion of the nephron consisting of the glomerulus and a glomerular capsule; also called the *malpighian corpuscle*.

renal cortex The outer portion of the kidney, primarily vascular.

renal medulla (mĕ-dul′ă) The inner portion of the kidney, including the renal pyramids and renal columns.

renal pelvis The inner cavity of the kidney that is formed by the expanded ureter and into which the major calyces open.

renal pyramid A triangular structure within the renal medulla, composed of the loops of the nephrons and the papillary ducts.

respiration (res″pi-ra′shun) The exchange of gases between the external environment and the cells of an organism.

respiratory center The structure or portion of the brain stem that regulates the depth and rate of breathing.

respiratory membrane A thin, moistened membrane within the lungs, composed of an alveolar portion and a capillary portion, through which gaseous exchange occurs.

rete testis (re′te tes′tis) A network of ducts in the center of the testis, site of spermatozoa production.

reticular formation (rĕ-tik′yoo-lar) A network of nervous tissue fibers in the brain stem that arouses the higher brain centers.

retina (ret′ĭ-nă) The inner layer of the eyeball that contains the photoreceptors.

retraction (re-trak′shun) The movement of a body part, such as the mandible, backward on a plane parallel with the ground; the opposite of *protraction*.

retroperitoneal (ret″ro-per″ĭ-tŏ-ne′al) Positioned behind the parietal peritoneum.

rhythmicity area (rith-mis′ĭ-te) A portion of the respiratory control center, located in the medulla oblongata. It controls inspiratory and expiratory phases.

ribosome (ri′bo-sōm) A cytoplasmic organelle composed of protein and RNA in which protein synthesis occurs.

right lymphatic duct (lim-fat′ik) A major vessel of the lymphatic system that drains lymph from the upper right portion of the body into the right subclavian vein.

rod A photoreceptor in the retina of the eye that is specialized for colorless, dim-light vision.

root canal The hollow, tubular extension of the pulp cavity into the root of the tooth that contains vessels and nerves.

rotation (ro-ta′shun) The movement of a bone around its own longitudinal axis.

round window (see cochlear window)

rugae (roo′je) The folds or ridges of the mucosa of an organ.

saccule (sak′yool) A saclike cavity in the membranous labyrinth inside the vestibule of the inner ear that contains a vestibular organ for equilibrium.

sacral (sa′kral) Pertaining to the sacrum.

sacral plexus (plek′sus) A network of nerve fibers that arise from spinal nerves L4 through S3. Nerves arising from the sacral plexus merge with those from the lumbar plexus to form the lumbosacral plexus and supply the lower extremity.

saddle joint A synovial joint in which the articular surfaces of both bones are concave in one plane and convex, or saddle shaped, in the other plane, such as in the distal carpometacarpal joint of the thumb.

sagittal (saj′ĭ-tal) A vertical plane through the body that divides it into right and left sides.

salivary gland (sal′ĭ-ver-e) An accessory digestive gland that secretes saliva into the oral cavity.

sarcolemma (sar″kŏ-lem-ă) The cell membrane of a muscle fiber.

sarcomere (sar′kŏ-mēr) The portion of a skeletal muscle fiber between the two adjacent Z lines that is considered the functional unit of a myofibril.

sarcoplasm The cytoplasm within a muscle fiber.

scala tympani (ska′lă tim′pă-ne) The lower channel of the cochlea that is filled with perilymph.

scala vestibuli (vĕ-stib′yŭ-le) The upper channel of the cochlea that is filled with perilymph.

Schwann cell (shwahn) (see neurolemmocyte)

sclera (skle′rah) The outer white layer of fibrous connective tissue that forms the protective covering of the eyeball.

scleral venous sinus A circular venous drainage for the aqueous humor from the anterior chamber; located at the junction of the sclera and the cornea; also called the *canal of Schlemm*.

scrotum (skro′tum) A pouch of skin that contains the testes and their accessory organs.

sebaceous gland (se-ba′shus) An exocrine gland of the skin that secretes sebum.

sebum (se′bum) An oily waterproofing secretion of sebaceous glands.

semen (se′men) The thick, whitish secretion of the reproductive organs of the male, consisting of spermatozoa and additives from the prostate and seminal vesicles.

semicircular ducts Tubular channels within the inner ear that contain the receptors for equilibrium; also called *semicircular canals*.

semilunar valve (sem″e-loo′nar) Crescent or half-moon-shaped heart valves, positioned at the entrances to the aorta and the pulmonary trunk.

seminal vesicles (sem′i-nal ves′i-k′lz) A pair of accessory male reproductive organs, lying posterior and inferior to the urinary bladder, that secrete additives to spermatozoa into the ejaculatory ducts.

seminiferous tubules (sem′i-nif′er-us) Numerous small ducts in the testes, where spermatozoa are produced.

senescence (sĕ-nes′ens) The process of aging.

sensory area A region of the cerebral cortex that receives and interprets sensory nerve impulses.

sensory neuron (noor′on) A nerve cell that conducts an impulse from a receptor organ to the central nervous system; also called *afferent neuron*.

septum (sep′tum) A membranous or fleshy wall dividing two cavities.

serous membrane (se′rus) An epithelial and connective tissue membrane that lines body cavities and covers visceral organs within these cavities; also called *serosa*.

Sertoli cells (ser-to′le) (see sustentacular cells)

serum (se′rum) Blood plasma with the clotting elements removed.

sesamoid bone (ses′ă-moid) A membranous bone formed in a tendon in response to joint stress (e.g., the patella).

shoulder The region of the body where the humerus articulates with the scapula; also called *omo*.

sigmoid colon (sig′moid ko′lon) The S-shaped portion of the large intestine between the descending colon and the rectum.

sinoatrial node (si″no-a′tre-al) A mass of specialized cardiac tissue in the wall of the right atrium that initiates the cardiac cycle; the SA node; also called the *pacemaker*.

sinus (si′nus) A cavity or hollow space within a body organ, such as a bone.

sinusoid (si′nŭ-soid″) A small, blood-filled space in certain organs, such as the spleen or liver.

skeletal muscle A specialized type of multinucleated muscle tissue that occurs in bundles, has crossbands of proteins, and contracts in either a voluntary or involuntary fashion.

small intestine The portion of the GI tract between the stomach and the cecum whose function is the absorption of food nutrients.

smooth muscle A specialized type of nonstriated muscle tissue, composed of fusiform, single-nucleated fibers. It contracts in an involuntary, rhythmic fashion within the walls of visceral organs.

soft palate (pal′at) The fleshy posterior portion of the roof of the mouth, from the palatine bones to the uvula.

somatic (so-mat′ik) Pertaining to the nonvisceral parts of the body.

spermatic cord (sper-mat′ik) The structure of the male reproductive system composed of the ductus deferens, spermatic vessels, nerves, cremasteric muscle, and connective tissue. The spermatic cord extends from a testis to the inguinal ring.

spermatogenesis (sper-mat″ŏ-jen′i-sis) The production of male sex gametes, or spermatozoa.

spermatozoon (sper-mat″ŏ-zo′on) A mature male sperm cell or gamete.

sphincter (sfingk′ter) A circular muscle that functions to constrict a body opening or the lumen of a tubular structure.

sphincter of ampulla The muscular constriction at the opening of the common bile and pancreatic ducts; also called the *sphincter of Oddi*.

sphincter of Oddi (o′de) (see sphincter of ampulla)

spinal cord (spi′nal) The portion of the central nervous system that extends downward from the brain stem through the vertebral canal.

spinal ganglion A cluster of nerve cell bodies on the posterior root of a spinal nerve.

spinal nerve One of the 31 pairs of nerves that arise from the spinal cord.

spinous process (spi′nus) A sharp projection of bone or a ridge of bone, such as on the scapula.

spiral organ The functional unit of hearing, consisting of a basilar membrane that supports receptor hair cells and a tectorial membrane located within the cochlea; also called the *organ of Corti*.

spleen (splēn) A large, blood-filled, glandular organ located in the upper left quadrant of the abdomen and attached by mesenteries to the stomach.

spongy bone A type of bone that contains many porous spaces; also called *cancellous bone*.

squamous (skwa′mus) Flat or scalelike.

stapes (sta′pēz) The innermost of the auditory ossicles of the ear that fits against the vestibular (oval) window of the inner ear; also called the *stirrup*.

statoconia Small hardened particles of calcium carbonate in the saccule and utricle of the inner ear, associated with the receptors of equilibrium; also called *otoliths*.

stomach A pouchlike digestive organ located between the esophagus and the duodenum.

stratified (strat′i-fīd) Arranged in layers, or strata.

stratum basale (ba′sal) The deepest epidermal layer, where mitotic activity occurs.

stratum corneum (kor′ne-um) The outer cornified layer of the epidermis of the skin.

stroma (stro′mă) A connective tissue framework in an organ, gland, or other tissue.

subarachnoid space (sub″ă-rak′noid) The space within the meninges, between the arachnoid and pia maters, where cerebrospinal fluid flows.

subdural space (sub-door′al) A space within the meninges, between the dura mater and arachnoid mater of the brain and spinal cord.

sublingual gland (sub-ling′gwal) One of the three pairs of salivary glands. It is located below the tongue and its duct opens to the side of the lingual frenulum.

submandibular gland (sub″man-dib′yŭ-lar) One of the three pairs of salivary glands. It is located below the mandible and its duct opens to the side of the lingual frenulum.

submucosa (sub″myoo-ko′să) A layer of supportive connective tissue that underlies a mucous membrane.

sulcus (sul′kus) A shallow impression or groove.

superficial (soo″per-fish′al) Toward or near the surface.

superficial fascia (fash′e-ă) A binding layer of connective tissue between the dermis of the skin and the underlying muscle.

superior Toward the upper part of a structure or toward the head; also called *cephalic*.

superior vena cava (ve′nă ka′vă) A large systemic vein that collects blood from regions of the body superior to the heart and returns it to the right atrium.

supination (soo″pi-na′shun) Rotation of the arm so that the palm is directed forward or anteriorly; the opposite of *pronation*.

surface anatomy The division of anatomy concerned with the structures that can be identified from the outside of the body.

surfactant (ser-fak′tant) A substance produced by the lungs that decreases the surface tension within the alveoli.

suspensory ligament (sŭ-spen′sŏ-re) **1.** A portion of the peritoneum that extends laterally from the surface of the ovary to the wall of the pelvic cavity. **2.** A ligament that supports an organ or a body part, such as that supporting the lens of the eye.

sustentacular cells (sus-ten-tak′yŭ-lar) Specialized cells within the testes that supply nutrients to developing spermatozoa; also called *Sertoli cells* or *nurse cells*.

sutural bone A small bone positioned within a suture of certain cranial bones; also called a *wormian bone*.

suture (soo′chur) A type of fibrous joint found between bones of the skull.

sweat gland A skin gland that secretes a fluid substance for evaporative cooling.

sympathetic (sim″pah-thet′ik) Pertaining to the division of the autonomic nervous system concerned with activities that, in general, inhibit or oppose the physiological effects of the parasympathetic nervous system; also called the *thoracolumbar division*.

symphysis (sim′fi-sis) A type of cartilaginous joint characterized by a fibrocartilaginous pad between the articulating bones, which provides slight movement.

symphysis pubis (pyoo′bis) A slightly movable joint anteriorly located between the two pubic bones of the pelvic girdle.

synapse (sin′aps) A minute space between the axon terminal of a presynaptic neuron and a dendrite of a postsynaptic neuron.

synarthrosis (sin″ar-thro′sis) A fibrous joint, such as a syndesmosis or a suture.

synchondrosis (sin″kon-dro′sis) A cartilaginous joint in which the articulating bones are separated by hyaline cartilage.

syndesmosis (sin″des-mo′sis) A type of fibrous joint in which two bones are united by an interosseous ligament.

synergist (sin′er-jist) A muscle that assists the action of the prime mover.

synovial cavity (si-no′ve-al) A space between the two bones of a synovial joint, filled with synovial fluid.

synovial joint A freely movable joint in which there is a synovial cavity between the articulating bones; also called a *diarthrotic joint*.

synovial membrane The inner membrane of a synovial capsule that secretes synovial fluid into the joint cavity.

system A group of body organs that function together.

systemic (sis-tem′ik) Relating to the entire organism rather than individual parts.

systemic anatomy The division of anatomy concerned with the structure and function of the various systems.

systemic circulation The portion of the circulatory system concerned with blood flow from the left ventricle of the heart to the entire body and back to the heart via the right atrium, in contrast to the pulmonary system, which involves the lungs.

systole (sis′tō-le) The muscular contraction of a heart chamber during the cardiac cycle.

systolic pressure (sis′tol′ik) Arterial blood pressure during the ventricular systolic phase of the cardiac cycle.

tactile (tak′til) Pertaining to the sense of touch.

taeniae coli (te′ne-e ko′li) The three longitudinal bands of muscle in the wall of the large intestine.

target organ The specific body organ that a particular hormone affects.

tarsal gland An oil-secreting gland that opens on the exposed edge of each eyelid; also called the *Meibomian gland.*

tarsus (tar′sus) Pertaining to the ankle; the proximal portion of the foot that contains the seven tarsal bones.

taste bud An organ containing the chemoreceptors associated with the sense of taste.

tectorial membrane (tek-to′re-al) A gelatinous membrane positioned over the hair cells of the spiral organ in the cochlea.

tendo calcaneus (ten′do kal-ka′ne-us) The tendon that attaches the calf muscles to the calcaneus bone; also called the *Achilles tendon.*

tendon (ten′dun) A band of dense regular connective tissue that attaches muscle to bone.

tendon sheath A covering of synovial membrane surrounding certain tendons.

tentorium cerebelli (ten-to′re-um ser″ĕ-bel′e) An extension of dura mater that forms a partition between the cerebral hemispheres and the cerebellum and covers the cerebellum.

teratogen (tĕ-rat′o-jen) Any agent or factor that causes a physical defect in a developing embryo or fetus.

testis (tes′tis) The primary reproductive organ of a male that produces spermatozoa and male sex hormones.

thalamus (thal′a-mus) An oval mass of gray matter within the diencephalon that serves as a sensory relay area.

thigh The proximal portion of the lower extremity between the hip and the knee in which the femur bone is located.

third ventricle (ven′tri-k′l) A narrow cavity between the right and left halves of the thalamus and between the lateral ventricles that contains cerebrospinal fluid.

thoracic (thō-ras′ik) Pertaining to the chest region.

thoracic duct The major lymphatic vessel of the body that drains lymph from the entire body, except for the upper right quadrant, and returns it to the left subclavian vein.

thorax (thor′aks) The chest.

thrombocyte (throm′bō-sīt) A blood platelet formed from a fragmented megakaryocyte.

thymus gland (thi′mus) A bilobed lymphoid organ positioned in the upper mediastinum, posterior to the sternum and between the lungs.

thyroid cartilage (thi′roid kar′tĭ-lij) The largest cartilage in the larynx that supports and protects the vocal cords; also called the *Adam's apple.*

tissue An aggregation of similar cells and their binding intercellular substance, joined to perform a specific function.

tongue A protrusible muscular organ on the floor of the oral cavity.

tonsil (ton′sil) A node of lymphoid tissue, located in the mucous membrane of the pharynx.

trabeculae (tra-bek′yu-le) A supporting framework of fibers crossing the substance of a structure, as in the lamellae of spongy bone.

trachea (tra′ke-ă) The airway leading from the larynx to the bronchi, composed of cartilaginous rings and a ciliated mucosal lining of the lumen; also called the *windpipe.*

tract A bundle of nerve fibers within the central nervous system.

transection (tran-sek′shun) A cross-sectional cut.

transverse colon (trans-vers′ ko′lon) A portion of the large intestine that extends from right to left across the abdomen between the hepatic and splenic flexures.

transverse fissure (fish′ur) The prominent cleft that horizontally separates the cerebrum from the cerebellum.

tricuspid valve (tri-kus′pid) The heart valve located between the right atrium and the right ventricle.

trigone (tri′gōn) A triangular area in the urinary bladder between the openings of the ureters and the urethra.

trochanter (tro-kan′ter) A broad, prominent process on the proximolateral portion of the femur.

trochlea (trok′le-ă) A pulley-shaped structure.

trunk The thorax and abdomen together.

tubercle (too′ber-k′l) A small, elevated process on a bone.

tuberosity (too′bĕ-ros′ĭ-te) An elevation or protuberance on a bone.

tunica albuginea (too′nĭ-kă al″byoo-jin′e-ă) A tough, fibrous tissue surrounding the testis.

tympanic membrane (tim-pan′ik) The membranous eardrum positioned between the outer and middle ear.

umbilical cord (um-bil′i-kal) A cordlike structure, containing the umbilical arteries and vein that connects the fetus with the placenta.

umbilicus (um-bil′i-kus) The site where the umbilical cord was attached to the fetus; commonly called the *navel.*

unipolar neuron (yoo′ni-po-lar noor′on) A nerve cell that has a single nerve fiber extending from its cell body.

upper extremity The appendage attached to the pectoral girdle, consisting of the shoulder brachium, elbow, antebrachium, and hand.

ureter (yoo-re′ter) A tube that transports urine from the kidney to the urinary bladder.

urethra (yoo-re′thra) A tube that transports urine from the urinary bladder to the outside of the body.

urinary bladder (yoo′ri-ner″e) A distensible sac situated in the pelvic cavity posterior to the symphysis pubis that stores urine.

urogenital triangle (yoo″ro-jen′i-tal) The region of the pelvic floor containing the external genitalia.

uterine tube (yoo′ter-in) The tube through which the ovum is transported to the uterus; the site of fertilization; also called the *oviduct* or *fallopian tube.*

uterus (yoo′ter-us) A hollow, muscular organ in which a fetus develops. It is located within the female pelvis between the urinary bladder and the rectum; commonly called the *womb.*

utricle (yoo′tri-kul) An enlarged portion of the membranous labyrinth, located within the vestibule of the inner ear.

uvula (yoo′vyu-lă) A fleshy, pendulous portion of the soft palate that blocks the nasopharynx during swallowing.

vacuole (vak-yoo′ol) Small spaces or cavities within the cytoplasm of a cell.

vagina (vă-ji-nă) A tubular organ that leads from the uterus to the vestibule of the female reproductive tract and receives the male penis during coitus.

vallate papillae (pă-pil′e) The largest of the papillae on the dorsal surface of the tongue. They are arranged in an inverted V-shaped pattern at the posterior portion of the tongue.

vasa vasorum (va-să va-so′rum) Tiny blood vessels that form a network through the tunica of larger blood vessels to supply them with nutrients.

vasomotor center (va-zo-mo′tor) A cluster of nerve cell bodies in the medulla oblongata that controls the diameter of blood vessels and therefore has an important role in regulating blood pressure.

vein A blood vessel that conveys blood toward the heart.

vena cava (ve′nă ka′vă) One of two large vessels that return deoxygenated blood to the right atrium of the heart.

ventral (ven′tral) Toward the front or belly surface; the opposite of *dorsal.*

ventricle (ven′tri-k′l) A cavity within an organ; especially those cavities in the brain that contain cerebrospinal fluid and those in the heart that contain blood to be pumped from the heart.

venule (ven′yool) A small vessel that carries venous blood from capillaries to a vein.

vermiform appendix (ver′mi-form ă-pen′diks) (see appendix)

vermis (ver′mis) The coiled, middle lobular structure that separates the two cerebellar hemispheres.

vertebral canal (ver′te-bral) The tubelike cavity extending through the vertebral column that contains the spinal cord; also called the *spinal canal.*

vestibular folds (vĕ-stib′yu-lar) The supporting folds of tissue for the vocal folds within the larynx; also called *false vocal cords.*

vestibular window An oval opening in the bony wall between the middle and inner ear, into which the footplate of the stapes fits; also called the *oval window.*

vestibule (ves′tĭ-byool) A space or cavity at the entrance to a canal, especially that of the nose, inner ear, and vagina.

villus (vil′us) A minute projection that extends outward into the lumen from the mucosal layer of the small intestine.

viscera (vis′er-ă) The organs within the abdominal or thoracic cavities.

visceral (vis′er-al) Pertaining to the membranous covering of the viscera.

visceral peritoneum (per″ĭ-tō-ne′um) A serous membrane that covers the surfaces of abdominal viscera.

visceral pleura (ploor′ă) A serous membrane that covers the surfaces of the lungs.

visceroceptor (vis″er-ō-sep′tor) A sensory receptor located within body organs that responds to information concerning the internal environment.

vitreous humor (vit′re-us hu′mor) The transparent gel that occupies the space between the lens and retina of the eyeball.

vocal folds Folds of the mucus membrane in the larynx that produce sound as they are pulled taut and vibrated; also called *vocal cords.*

Volkmann's canal (folk′manz) (see perforating canal)

vulva (vul′vă) The external genitalia of the female that surround the opening of the vagina; also called the *pudendum.*

white matter Bundles of myelinated axons located in the central nervous system.

wormian bone (wer′me-an) (see sutural bone)

yellow marrow (mar′o) Specialized lipid storage tissue within bone cavities.

zygote (zi′gōt) A fertilized egg cell formed by the union of a sperm cell and an ovulated secondary oocyte (ovum).

COMPUTERIZED STUDY GUIDE FOR
Human Anatomy
by Kent M. Van De Graaff
Written and compiled by Jeffrey and Karianne Prince

CUT STUDY TIME! • IMPROVE YOUR GRADES!

HOW DOES IT WORK?

- Take the computerized test for each chapter.
- Program presents you with feedback for each answer.
- Program creates a text page-referenced study plan for all incorrect answers.
- Look up the answers to the questions you missed.

✓ **FOCUS YOUR STUDY TIME ONLY ON THE AREAS WHERE YOU NEED HELP!**

- This easy to use Computerized Study Guide program is not a duplicate of the printed study guide that accompanies your text.

ORDER INFORMATION
Call Toll Free 1-800-748-7734
•••
Computerized Study Guide for
Human Anatomy
by Kent M. Van De Graaff

The Computerized Study Guide is available on
☐ **IBM PC/PC Compatible - 3½" Disk**
☐ **IBM PC/PC Compatible - 5¼" Disk**
☐ **Macintosh - 3½" Disk**
at $15.95 per Computerized Study Guide plus $2.95 for shipping and handling.
If not completely satisfied, simply return the program in 10 days for a full refund.

Payment Methods Available
☐ **Mastercard** ☐ **Visa** ☐ **Prepayment** ☐ **C.O.D.**

Have your credit card ready and
ORDER TOLL FREE TODAY 1-800-748-7734
Please be prepared to give the operator the following shipping information
• Name • Street Address • Telephone Number • City • State • Zip

Educational Software Concepts, Inc.   660 S. 4th Street   Edwardsville, KS 66113   Phone 1-800-748-7734

Illustrations

Chapter 1

Figure 1.2: From *A Short History of Anatomy and Physiology from the Greeks to Harvey* by C. Singer, 1957, Dover Publications, New York, NY. Reprinted by permission.
Figure 1.4: From J. Kevorkian, *The Story of Dissection.* Copyright © Philosophical Library, Inc., New York, NY. Reprinted by permission.

Chapter 2

Figure 2.8: From John W. Hole, Jr., *Human Anatomy and Physiology,* 6th ed. Copyright © 1993 Wm. C. Brown Communications, Inc., Dubuque, Iowa. All Rights Reserved. Reprinted by permission.

Chapter 3

Figure 3.3: From Stuart Ira Fox, *Human Physiology,* 4th ed. Copyright © 1993 Wm. C. Brown Communications, Inc., Dubuque, Iowa. All Rights Reserved. Reprinted by permission.
Figures 3.4, 3.12, 3.16b: From John W. Hole, Jr., *Human Anatomy and Physiology,* 6th ed. Copyright © 1993 Wm. C. Brown Communications, Inc., Dubuque, Iowa. All Rights Reserved. Reprinted by permission.

Chapter 4

Figure 4.7a: From John W. Hole, Jr., *Human Anatomy and Physiology,* 6th ed. Copyright © 1993 Wm. C. Brown Communications, Inc., Dubuque, Iowa. All Rights Reserved. Reprinted by permission.

Chapter 9

Figures 9.3, 9.12a: From John W. Hole, Jr., *Human Anatomy and Physiology,* 6th ed. Copyright © 1993 Wm. C. Brown Communications, Inc., Dubuque, Iowa. All Rights Reserved. Reprinted by permission.
Figures 9.7a, 9.8: From Stuart Ira Fox, *Human Physiology,* 4th ed. Copyright © 1993 Wm. C. Brown Communications, Inc., Dubuque, Iowa. All Rights Reserved. Reprinted by permission.

Figure 9.17a: Source: Richard T. Jolly (trans.), *The Color Atlas of Human Anatomy,* 1980.
Figure 9.51: Hollingsworth JW, Hashizume A, Jablon S: Correlations between Tests of Aging in Hiroshima Subjects: An Attempt to Define "Physiologic Age." *Yale Journal of Biology and Medicine,* 38:11–26, 1965. Reprinted by permission.

Chapter 11

Figures 11.2, 11.13, 11.14, 11.15, 11.22, 11.28, 11.38, 11.43, 11.46: From John W. Hole, Jr., *Human Anatomy and Physiology,* 6th ed. Copyright © 1993 Wm. C. Brown Communications, Inc., Dubuque, Iowa. All Rights Reserved. Reprinted by permission.
Figure 11.4: From John W. Hole, Jr., *Human Anatomy and Physiology,* 5th ed. Copyright © 1990 Wm. C. Brown Communications, Inc., Dubuque, Iowa. All Rights Reserved. Reprinted by permission.
Figures 11.6, 11.34, 11.40a: From Joan G. Creager, *Human Anatomy and Physiology,* 2d ed. Copyright © 1992 Wm. C. Brown Communications, Inc., Dubuque, Iowa. All Rights Reserved. Reprinted by permission.
Figures 11.7, 11.26, 11.27, 11.36: From Stuart Ira Fox, *Human Physiology,* 4th ed. Copyright © 1993 Wm. C. Brown Communications, Inc., Dubuque, Iowa. All Rights Reserved. Reprinted by permission.

Chapter 12

Figure 12.9: From John W. Hole, Jr., *Human Anatomy and Physiology,* 5th ed. Copyright © 1990 Wm. C. Brown Communications, Inc., Dubuque, Iowa. All Rights Reserved. Reprinted by permission.

Chapter 13

Figure 13.7: From Stuart Ira Fox, *Human Physiology,* 4th ed. Copyright © 1993 Wm. C. Brown Communications, Inc., Dubuque, Iowa. All Rights Reserved. Reprinted by permission.
Figure 13.8: From Joan G. Creager, *Human Anatomy and Physiology,* 2d ed. Copyright © 1992 Wm. C. Brown Communications, Inc., Dubuque, Iowa. All Rights Reserved. Reprinted by permission.

Chapter 14

Figures 14.4, 14.5, 14.13: From Stuart Ira Fox, *Human Physiology,* 3d ed. Copyright © 1990 Wm. C. Brown Communications, Inc., Dubuque, Iowa. All Rights Reserved. Reprinted by permission.
Figures 14.8a, 14.17, 14.22, 14.24: From John W. Hole, Jr., *Human Anatomy and Physiology,* 6th ed. Copyright © 1993 Wm. C. Brown Communications, Inc., Dubuque, Iowa. All Rights Reserved. Reprinted by permission.
Figure 14.14: From Eldon D. Enger, et al., *Concepts in Biology,* 6th ed. Copyright © 1991 Wm. C. Brown Communications, Inc., Dubuque, Iowa. All Rights Reserved. Reprinted by permission.
Figure 14.25: From John W. Hole, Jr., *Human Anatomy and Physiology,* 4th ed. Copyright © 1987 Wm. C. Brown Communications, Inc., Dubuque, Iowa. All Rights Reserved. Reprinted by permission.

Chapter 15

Figures 15.2, 15.20: From Joan G. Creager, *Human Anatomy and Physiology,* 2d ed. Copyright © 1992 Wm. C. Brown Communications, Inc., Dubuque, Iowa. All Rights Reserved. Reprinted by permission.
Figures 15.6a,b, 15.9, 15.12, 15.28, 15.32: From John W. Hole, Jr., *Human Anatomy and Physiology,* 6th ed. Copyright © 1993 Wm. C. Brown Communications, Inc., Dubuque, Iowa. All Rights Reserved. Reprinted by permission.
Figures 15.18, 15.21b, 15.24, 15.25, 15.33, 15.34, 15.36, 15.37 (left), 15.39b–d, 15.42: From Stuart Ira Fox, *Human Physiology,* 4th ed. Copyright © 1993 Wm. C. Brown Communications, Inc., Dubuque, Iowa. All Rights Reserved. Reprinted by permission.
Figure 15.37 (right): From J. R. McClintic, *Physiology of the Human Body,* 3d ed. Copyright © 1985 John Wiley & Sons, Inc., New York, NY. Reprinted by permission of John Wiley & Sons, Inc.
Figure 15.38: From Stuart Ira Fox, *Human Physiology,* 3d ed. Copyright © 1990 Wm. C. Brown Communications, Inc., Dubuque, Iowa. All Rights Reserved. Reprinted by permission.

Figures 15.40, 15.41a: From John W. Hole, Jr., *Human Anatomy and Physiology,* 4th ed. Copyright © 1987 Wm. C. Brown Communications, Inc., Dubuque, Iowa. All Rights Reserved. Reprinted by permission.
Figure 15.41b: From John W. Hole, Jr., *Human Anatomy and Physiology,* 2d ed. Copyright © 1981 Wm. C. Brown Communications, Inc., Dubuque, Iowa. All Rights Reserved. Reprinted by permission.

Chapter 16

Figure 16.2: From John W. Hole, Jr., *Human Anatomy and Physiology,* 5th ed. Copyright © 1990 Wm. C. Brown Communications, Inc., Dubuque, Iowa. All Rights Reserved. Reprinted by permission.
Figures 16.3, 16.6, TA 16.2, 16.20, 16.23a, 16.26, 16.33, 16.34, 16.40, 16.45: From John W. Hole, Jr., *Human Anatomy and Physiology,* 6th ed. Copyright © 1993 Wm. C. Brown Communications, Inc., Dubuque, Iowa. All Rights Reserved. Reprinted by permission.
Figures 16.4a, 16.47: From John W. Hole, Jr., *Human Anatomy and Physiology,* 4th ed. Copyright © 1987 Wm. C. Brown Communications, Inc., Dubuque, Iowa. All Rights Reserved. Reprinted by permission.
Figures 16.12, 16.49, 16.52, 16.53: From Stuart Ira Fox, *Human Physiology,* 4th ed. Copyright © 1993 Wm. C. Brown Communications, Inc., Dubuque, Iowa. All Rights Reserved. Reprinted by permission.
Figures 16.41, 16.46: From Joan G. Creager, *Human Anatomy and Physiology,* 2d ed. Copyright © 1992 Wm. C. Brown Communications, Inc., Dubuque, Iowa. All Rights Reserved. Reprinted by permission.

Chapter 17

Figures 17.9a,b, 17.10, 17.27: From John W. Hole, Jr., *Human Anatomy and Physiology,* 6th ed. Copyright © 1993 Wm. C. Brown Communications, Inc., Dubuque, Iowa. All Rights Reserved. Reprinted by permission.
Figure 17.17: From Stuart Ira Fox, *Human Physiology,* 4th ed. Copyright © 1993 Wm. C. Brown Communications, Inc., Dubuque, Iowa. All Rights Reserved. Reprinted by permission.

Exhibit II: From Joan G. Creager, *Human Anatomy and Physiology*, 2d ed. Copyright © 1992 Wm. C. Brown Communications, Inc., Dubuque, Iowa. All Rights Reserved. Reprinted by permission.

Chapter 18

Figures 18.6, 18.7, 18.11, 18.13, 18.19, 18.31: From John W. Hole, Jr., *Human Anatomy and Physiology*, 6th ed. Copyright © 1993 Wm. C. Brown Communications, Inc., Dubuque, Iowa. All Rights Reserved. Reprinted by permission.
Figure 18.23: From Stuart Ira Fox, *Human Physiology*, 4th ed. Copyright © 1993 Wm. C. Brown Communications, Inc., Dubuque, Iowa. All Rights Reserved. Reprinted by permission.

Chapter 19

Figures 19.3, 19.5, 19.8a, 10.10, 19.13b: From John W. Hole, Jr., *Human Anatomy and Physiology*, 6th ed. Copyright © 1993 Wm. C. Brown Communications, Inc., Dubuque, Iowa. All Rights Reserved. Reprinted by permission.
Figures 19.9b, 19.13a: From John W. Hole, Jr., and Karen A. Koos, *Human Anatomy*, 2d ed. Copyright © 1994 Wm. C. Brown Communications, Inc., Dubuque, Iowa. All Rights Reserved. Reprinted by permission.
Figure 19.21: Courtesy American Medical Systems, Inc., Minnetonka, MN.

Chapter 20

Figures 20.1, 20.5, 20.9a: From John W. Hole, Jr., *Human Anatomy and Physiology*, 6th ed. Copyright © 1993 Wm. C. Brown Communications, Inc., Dubuque, Iowa. All Rights Reserved. Reprinted by permission.
Figures 20.6, 20.7b, 20.8, 20.19: From Stuart Ira Fox, *Human Physiology*, 4th ed. Copyright © 1993 Wm. C. Brown Communications, Inc., Dubuque, Iowa. All Rights Reserved. Reprinted by permission.

Chapter 21

Figures 21.1, 21.19, 21.21: From John W. Hole, Jr., *Human Anatomy and Physiology*, 6th ed. Copyright © 1993 Wm. C. Brown Communications, Inc., Dubuque, Iowa. All Rights Reserved. Reprinted by permission.
Figure 21.2: From F. K. Shuttleworth, *Monograph of the Society for Research in Child Development*, Vol. 4, No. 3. Copyright © 1939 Society for Research in Child Development, Inc., Chicago, IL. Reprinted by permission of the University of Chicago Press.
Figure 21.8: From Stuart Ira Fox, *Human Physiology*, 3d ed. Copyright © 1990 Wm. C. Brown Communications, Inc., Dubuque, Iowa. All Rights Reserved. Reprinted by permission.

Chapter 22

Figures 22.8, 22.25: From John W. Hole, Jr., *Human Anatomy and Physiology*, 6th ed. Copyright © 1993 Wm. C. Brown Communications, Inc., Dubuque, Iowa. All Rights Reserved. Reprinted by permission.
Figure 22.27: R. E. Scammon and L. A. Calkins, *The Development and Growth of the External Dimensions of the Human Body in the Fetal Period*, 1929 University of Minnesota, xxiii & 367. Reprinted by permission of the University of Minnesota Press.

Photos

Chapter 1

Figure 1.1: Scala/Art Resource, NY; **Figure 1.3:** Published by permission of the Danish National Museum; **Figure 1.5:** The Trustees of the British Museum; **Figure 1.6:** Egyptian National Museum; **Figure 1.7, 1.8:** Bibliotheque Nationale; **Figures 1.9, 1.10:** Fratelli Alinari; **Figure 1.11:** Courtesy of the New York Academy of Medicine Library; **Figures 1.12, 1.13:** Reproduced by gracious permissions of Her Majesty Queen Elizabeth, Royal Art Collection, Great Britain; **Figure 1.14:** New York Academy of Medicine Library; **Figure 1.15:** Johan Maurita Van Nassau Mawitsuia: The Hague; **Figure 1.16:** Stock Montage, Inc.; **Figure 1.17a:** Photography by Debi Stansbaugh; **Figure 1.17b:** © George Mosil/Visuals Unlimited; **Figure 1.17c:** © CNRI/SPL/Photo Researchers, Inc.; **Figure 1.18a:** Courtesy of Kodak; **Figure 1.18b:** © Carroll H. Weiss, 1973/Camera M.D. Studios, Inc.; **Figure 1.18c:** Courtesy of Utah Valley Regional Medical Center, Department of Radiology; **Figure 1.19a:** © Lester V. Bergman and Associates; **Figure 1.19b:** From R. A. Robb, *Three-Dimensional Biomedical Imaging*, Vol. 1, 1985. Copyright CRC Press, Inc., Boca Raton, FL; **Figure 1.19c,d:** © Hank Morgan/Science Source/Photo Researchers, Inc.

Chapter 2

Figure 2.3b: From C. D. Haagensen, *Diseases of the Breast*, 2/E, 1974, W. B. Saunders Company; **Figure 2.4:** © Dr. Sheril Burton; **Figure 2.7a:** © Joan L. Cohen/Photo Researchers, Inc.; **Figure 2.7b:** © John D. Cunningham/Visuals Unlimited; **Figure 2.7c:** © James L. Shaffer; **Figure 2.7d:** © L. Lee Rue III/Photo Researchers, Inc.; **Figure 2.7e:** © Cyril Toker/Photo Researchers, Inc.; **Figure 2.7f:** © Tom Hollyman/Photo Researchers, Inc.; **Figure 2.3a,b:** Kent M. Van De Graaff; **Figure 2.11a:** © Dr. Sheril Burton; **Figure 2.12a:** Kent M. Van De Graaff; **Figure 2.12b:** © Arthur Glauberman/Photo Researchers, Inc.; **Figure 2.12c:** © Martin M. Rotker/Photo Researchers, Inc.; **Figures 2.13, 2.15a, 2.16a:** © Dr. Sheril D. Burton

Chapter 3

Figure 3.5a: Courtesy of Keith Porter; **Figure 3.5b:** Per H. Kjeldson, University of Michigan at Ann Arbor; **Figure 3.5c:** © John D. Cunningham/Visuals Unlimited; **Figure 3.6:** © BioPhoto Associates/Photo Researchers, Inc.; **Figure 3.7a:** Courtesy of Keith R. Porter; **Figure 3.8a:** © David M. Phillips/Visuals Unlimited; **Figure 3.9a:** © Keith Porter/Science Source/Photo Researchers, Inc.; **Figure 3.10a:** Courtesy of Keith Porter; **Figure 3.11a:** © David M. Phillips/Visuals Unlimited; **Figure 3.13a:** From Joseph Gall, *Journal of Cell Biology* 31(1966), Copyrighted permission of Rockefeller Univ. Press; **Figure 3.15:** © Dr. Don Fawcett/Penelope Goddum-Rosse/Photo Researchers, Inc.; **Figure 3.16a:** Courtesy Stephen L. Wolfe; **Figure 3.17:** © CNRI/Science Photo Library/Photo Researchers, Inc.; **Figure 3.21a-e:** © Edwin A. Reschke; **Figure 3.22a:** © SIU/Visuals Unlimited; **Figure 3.22b:** © John D. Cunningham/Visuals Unlimited

Chapter 4

Figure 4.1a: © Edwin A. Reschke; **Figure 4.1b:** Dr. Kerry Openshaw; **Figures 4.2b, 4.3b, 4.4b, 4.5b, 4.6b, 4.7b:** © Edwin A. Reschke; **Figure 4.8b:** Edwin A. Reschke/Peter Arnold, Inc.; **Figure 4.9b:** © Edwin A. Reschke; **Figure 4.11b:** © CNRI/Science Photo Library/Photo Researchers, Inc.; **Figure 4.14b:** © BioPhoto Associates/Science Source/Photo Researchers, Inc.; **Figure 4.15b, 4.16b:** © Edwin A. Reschke; **Figure 4.17b:** © Bruce Iverson, BSc; **Figures 4.18b, 4.19, 4.20b, 4.21b, 4.22b, 4.23b:** © Edwin A. Reschke; **Figure 4.24a,b:** Kent M. Van De Graaff; **Figures 4.25b, 4.27a-c:** © Edwin A. Reschke; **Figure 4.28b:** © John D. Cunningham/Visuals Unlimited; **Figure 4.28c:** © 1984 Martin M. Rotker/Photo Researchers, Inc.

Chapter 5

Figure 5.2: © Edwin Reschke/Peter Arnold, Inc.; **Figure 5.3:** © Dr. Jeremy Burgess/SPL/Photo Researchers, Inc.; **Figure 5.4:** Victor B. Eichler; **Figure 5.7:** © Dr. Sheril D. Burton; **Figure 5.8:** James Clayton; **Figure 5.9:** © Lester V. Bergman and Associates, Inc.; **Figure 5.10a:** World Health Organization; **Figure 5.10b:** © George Bogumil; **Figure 5.11a,b:** © Dr. Sheril D. Burton; **Figure 5.12a:** © Michael Abbey/Science Source/Photo Researchers, Inc.; **Figure 5.12b:** Dr. Kerry L. Openshaw; **Figure 5.13b:** © John D. Cunningham/Visuals Unlimited; **Figure 5.15:** Dr. Kerry L. Openshaw; **Figure 5.18:** © Martin M. Rotker; **Figure 5.19a:** © Dorte Groning/Tierbild Okapia/Photo Researchers, Inc.; **Figure 5.19b:** © James Stevenson/SPL/Photo Researchers, Inc.; **Figure 5.19c:** © John Radcliffe/SPL/Photo Researchers, Inc.; **Figure 5.21a-c:** Courtesy of Drs. David A. Kappel, E. Phillips Polack, and Marjorie L. Bush, Plastic Surgery Inc.; **Figure 5.22:** © Dan McCoy/Rainbow; **Figure 5.25:** © Jacques Jangoux/Peter Arnold, Inc.; **Figure 5.26:** (Child) from Science Year, *The World Book Science Annual*, 1973, Field Enterprise Educational Corp. By permission World Book, Inc.; (Woman) Black Star Publishing; **Figure 5.27:** © Norman R. Lightfoot/Photo Researchers, Inc.

Chapter 6

Figure 6.5a,b: © BioPhoto Associates/Photo Researchers, Inc.; **Figure 6.7a:** From R. G. Kessel and R. H. Kardon, *Tissues and Organs: A Text-Atlas of Scanning Electron Microscopy*, 1979, W.H. Freeman and Company; **Figure 6.7b:** © Edwin A. Reschke; **Figure 6.8:** © BioPhoto Associates/Photo Researchers, Inc.; **Figure 6.10:** © Edwin A. Reschke; **Figure 6.11:** Courtesy of Utah Valley Regional Medical Center, Department of Radiology; **Figure 6.18:** Kent M. Van De Graaff; **Figures 6.22a,b, 6.34a, 6.36a:** Courtesy of Utah Valley Regional Medical Center, Department of Radiology; **Exhibit 6.1b:** Ted Conde; **Figure 6.41:** Blayne L. Hirsche; **Figure 6.42:** © CNRI/SPL/Photo Researchers, Inc.; **Figure 6.43:** Carolina Biological Supply Company

Chapter 7

Figure 7.4: Kent M. Van De Graaff; **Figure 7.8b:** © Dr. Sheril D. Burton; **Figures 7.10, 7.12, 7.17, 7.19b:** Courtesy of Utah Valley Regional Medical Center, Department of Radiology; **Figure 7.19a:** © Dr. Sheril D. Burton; **Figure 7.21a:** Kent M. Van De Graaff; **Figure 7.21b:** Courtesy Blayne L. Hirsch; **Figure 7.22:** Kent M. Van De Graaff; **Figure 7.24e:** Kodak

Chapter 8

Figure 8.4: Courtesy of Utah Valley Regional Medical Center, Department of Radiology; **Figure 8.5:** © Paolo Koch/Photo Researchers, Inc.; **Figure 8.20a-h:** © Dr. Sheril D. Burton; **Figure 8.21a:** Kent M. Van De Graaff; **Figure 8.21b,c:** © Dr. Sheril D. Burton; **Figure 8.21d:** Kent M. Van De Graaff; **Figure 8.21e,f:** © Dr. Sheril D. Burton; **Figures 8.22a,b, 8.33a:** Kent M. Van De Graaff; **Figure 8.33b:** © Lester V. Bergman and Associates, Inc.; **Figure 8.34a:** SIU, School of Medicine, All Rights Reserved; **Figure 8.34b:** Kent M. Van De Graaff; **Figure 8.34c:** Courtesy of Dr. Julian Nemmers, D.D.S.; **Figure 8.34d:** SIU, School of Medicine, All Rights Reserved; **Figure 8.35:** Kent M. Van De Graaff

Chapter 9

Figure 9.7b: © John D. Cunningham/Visuals Unlimited; **9.10a-c:** From R. G. Kessel and R. H. Kardon, *Tissues and Organs: A Text-Atlas of Scanning Electron Microscopy*, 1979, Freeman and Company; **Figure 9.11a:** © Dr. H. E. Huxley; **Figure 9.12b:** © BioPhoto Associates/Photo Researchers, Inc.; **Figure 9.15a-h:** Dr. Sheril D. Burton; **Figure 9.52:** © John Bird

Chapter 10

Figure 10.1: Stock Montage, Inc.; **Figures 10.4a,b, 10.5, 10.6, 10.7, 10.8:** © Dr. Sheril D. Burton; **Figure 10.9a:** Kent M. Van De Graaff; **Figures 10.9b, 10.10a,b:** © Dr. Sheril D. Burton; **Figure 10.11, 10.12:** From S. D. Vidic and F. R. Suarez, *Photographic Atlas of the Human Body*, 1984 (pl 29, p. 38) St. Louis, The C. V. Mosby Co.; **Figure 10.13:** © Wm. C. Brown Communications, Inc./Karl Rubin, photographer; **Figure 10.14:** Kent M. Van De Graaff; **Figures 10.16, 10.17:** From S. D. Vidic and F. R. Suarez, *Photographic Atlas of the Human Body*, 1984 (pl 71, p. 62) St. Louis, The C. V. Mosby Co.; **Figures 10.18a,b, 10.19, 10.20, 10.21, 10.22:** Kent M. Van De Graaff; **Figures 10.23, 10.24:** From S. D. Vidic and F. R. Suarez, *Photographic Atlas of the Human Body*, 1984 (pl 115, p. 184) St. Louis, The C. V. Mosby Co.; **Figure 10.25:** Kent M. Van De Graaff; **Figure 10.26:** © Wm. C. Brown Communications, Inc./Karl Rubin, photographer; **Figures 10.27, 10.28:** From S. D. Vidic and F. R. Suarez, *Photographic Atlas of the Human Body*, 1984 (pl 40, p. 60) St. Louis, The C. V. Mosby Co.; **Figures 10.29, 10.30:** © Wm. C. Brown Communications, Inc./Karl Rubin, photographer; **Figure 10.31:** © Dr. Sheril D. Burton; **Figure 10.32:** © Wm. C. Brown Communications, Inc./Karl Rubin, photographer; **Figures 10.33, 10.34, 10.35, 10.36, 10.37, 10.38, 10.39, 10.40, 10.41a,b, 10.42:** From S. D. Vidic and F. R. Suarez, *Photographic Atlas of the Human Body*, 1984 (pl 129, p. 210) St. Louis, The C. V. Mosby Co.; **Figure 10.43:** From S. D. Vidic and F. R. Suarez, *Photographic Atlas of the Human Body*, 1984 (pl 130, p. 212) St. Louis, The C. V. Mosby Co.; **Figure 10.44:** Kent M. Van De Graaff; **Figures 10.45, 10.46:** © Dr. Sheril D. Burton; **Figures 10.47, 10.48, 10.49a–c, 10.50a–d:** Kent M. Van De Graaff; **Figure 10.51a,b:** Dr. Kenneth A. Johnson; **Figure 10.51c:** © Lester V. Bergman and Associates, Inc.; **Figure 10.52:** From S. D. Vidic and F. R. Suarez, *Photographic Atlas of the Human Body*, 1984 (pl 205, p. 336) St. Louis, The C. V. Mosby Co.; **Figure 10.53:** From S. D. Vidic and F. R. Suarez, *Photographic Atlas of the Human Body*, 1984 (pl 206, p. 338) St. Louis, The C.V. Mosby Co.; **Figures 10.54, 10.55, 10.56, 10.57, 10.58:** © Wm. C. Brown Communications, Inc./Karl Rubin, photographer; **Figure 10.59:** Dr. Blayne L. Hirsche.

Chapter 11

Figure 11.3: © John D. Cunningham/ Visuals Unlimited; **Figure 11.6b:** Courtesy Dr. John Hubbard; **Figure 11.8:** © Edwin A. Reschke; **Figure 11.17a–d:** Kent M. Van De Graaff; **Figure 11.18:** Monte S. Buchebaum, M.D.; **Figure 11.23a:** © Martin Rotker; **Figure 11.25b:** © William C. Brown Publishers/Karl Rubin; **Figure**

11.37a,b: Utah Valley Regional Medical Center, Department of Radiology; **Figure 11.40b:** Per H. Kjeldsen, University of Michigan at Ann Arbor

Chapter 12

Figure 12.2: From R. G. Kessel and R. H. Kardon, *Tissues & Organs: A Text-Atlas of Scanning Electron Microscopy*, 1979, W. H. Freeman and Company; **Figure 12.27:** Kent M. Van De Graaff; **Figure 12.32a–i:** Photographed by Dr. Sheril D. Burton, assisted by Dr. Douglas W. Hacking

Chapter 13

Figure 13.9a: © Wm. C. Brown Communications, Inc./Karl Rubin, photographer

Chapter 14

Figure 14.2b: © Edwin A. Reschke; **Figure 14.9:** © John D. Cunningham/ Visuals Unlimited; **Figure 14.10:** © Lester V. Bergman and Associates; **Figure 14.14b:** © SIU/Nawrocki Stock Photo, Inc.; **Figure 14.15:** © Fred Hossler/Visuals Unlimited; **Figure 14.18:** © BioPhoto Associates/Photo Researchers, Inc.; **Figure 14.21:** © Edwin A. Reschke; **Figure 14.23:** © R. Calentine/Visuals Unlimited; **Figures 14.26, 14.27:** © Lester V. Bergman and Associates

Chapter 15

Figure 15.6c: © Victor B. Eichler; **Figures 15.10, 15.13a–i:** © Dr. Sheril D. Burton; **Figure 15.14a,b:** © Carroll H. Weiss/Camera M.D. Studios, Inc.; **Figure 15.16c:** P. N. Farnsworth, Univ. of Medicine & Dentistry; New Jersey Medical School; **Figure 15.19b:** Per H. Kjeldsen, University of Michigan at Ann Arbor; **Figure 15.21a:** Copyright Thomas Sims; **Figure 15.29:** © Dr. Sheril D. Burton; **Figure 15.35:** From R. G. Kessel and C. Y. Shih, *Scanning Electron Microscopy in Biology*, Copyright 1976, Springer-Verlag; **Figure 15.39a:** Courtesy Dean E. Hillman; **Figure 15.43:** Courtesy Dr. Stephen Clark

Chapter 16

Figure 16.4b: © Bill Longcore/Science Source/Photo Researchers, Inc.; **Figures 16.7, 16.10:** Courtesy of Utah Valley Regional Medical Center, Department of Radiology; **Figure 16.11a:** © Carroll H. Weiss/Camera M.D. Studios, Inc.; **Figures 16.23b, 16.25a:** From *Practische Outleedkunde*, J. Dankmeyer, H. G. Lambers, and J. M. F. Landsmerr, Bohn, Scheltema & Holkkema; **Figure 16.42:** © Edwin A. Reschke; **Figure 16.44b:** © John D. Cunningham/Visuals Unlimited; **Figure 16.44c:** Kent M. Van De Graaff; **Figure 16.47b:** © John D. Cunningham/Visuals Unlimited; **Figure 16.48a–b:** Richard Menard; **Figure 16.51a:** Copyright Donald S.

Donald S. Baim, from Hurst et al. *The Heart*, 5/E, 1982, McGraw-Hill Book Company; **Figure 16.51b:** Zamboni from Weiss/Greep, *Histology*, 3/E, Elsevier Science Publishing Company, Inc.; **Figure 16.54a:** American Heart Association; **Figure 16.54b:** Lewis Lainey; **Figure 16.55b:** Courtesy of Utah Valley Regional Medical Center, Department of Radiology

Chapter 17

Figure 17.5: © CNRI/SPL/Photo Researchers, Inc.; **Figure 17.9c:** © CNRI/Phototake Inc.; **Figure 17.11a:** © John D. Cunningham/Visuals Unlimited; **Figure 17.11b:** © Edwin A. Reschke; **Figure 17.13:** © John Watney Photo Library; **Figure 17.15:** Courtesy of Kodak; **Figure 17.19:** © David M. Phillips/Visuals Unlimited; **Figure 17.24a,b:** From J. H. Comroe, Jr., *Physiology of Respiration*, 1974 Year Book Medical Publishers, Inc. Chicago; **Figure 17.26:** © Edward Lettau/Photo Researchers, Inc.; **Figure 17.30:** Edward C. Vasquez, RT., CRT/Dept. of Radiologic Technology, Los Angeles City College; **Figure 17.32a,b:** © Martin Rotker

Chapter 18

Figure 18.9: Kent M. Van De Graaff; **Figure 18.12a–c:** © BioPhoto Associates/Science Source/Photo Researchers, Inc.; **Figure 18.14:** © Edwin A. Reschke; **Figure 18.15:** Courtesy of Utah Valley Regional Medical Center, Department of Radiology; **Figure 18.16a,b:** Kent M. Van De Graaff; **Figure 18.18:** © Edwin A. Reschke; **Figure 18.22:** © Manfred Kagel/Peter Arnold, Inc.; **Figure 18.25:** Keith R. Porter; **Figure 18.27:** From W. A. Sodeman and T. M. Watson, *Pathologic Physiology*, 6/E, W. B. Saunders Company; **Figure 18.28:** © Lester U. Bergman and Associates; **Figure 18.30a:** © Edwin A. Reschke; **Figures 18.30b, 18.32a:** From R. G. Kessel and R. H. Kardon, *Tissues and Organs: A Text-Atlas of Scanning Electron Microscopy*, 1979, W. H. Freeman and Company; **Figure 18.33c:** © Victor B. Eichler; **Figure 18.36a:** © Carroll H. Weiss/Camera M.D. Studios, Inc.; **Figure 18.36b:** © Dr. Sheril D. Burton; **Figure 18.38:** Kent M. Van De Graaff; **Figure 18.39a–d:** John Van De Graaff, D.D.S. & Larry S. Pierce, D.D.S., both at Northwestern University

Chapter 19

Figure 19.2: © Science Photo Library/ Photo Researchers, Inc.; **Figure 19.4a,b:** Kent M. Van De Graaff; **Figure 19.6b:** © BioPhoto Associates/Photo Researchers, Inc.; **Figure 19.8b:** © David M. Phillips/Visuals Unlimited; **Figure 19.9a:** D. Friend/ Harvard Medical School; **Figure 19.11:** Per H. Kjeldsen, University of Michigan at Ann Arbor; **Figure 19.12:** Kent M. Van De Graaff; **Figure 19.15:** © Bruce Iverson/ Visuals Unlimited

Chapter 20

Figure 20.7a: © BioPhoto Associates/ Science Source/Photo Researchers, Inc.; **Figure 20.9b:** © Francis Leroy, Biocosmos/SPL/Photo Researchers, Inc.; **Figures 20.10, 20.11:** © Edwin A. Reschke; **Figures 20.12, 20.13:** © Manfred Kage/Peter Arnold, Inc.; **Figure 20.14:** © Edwin A. Reschke; **Figure 20.16a–c:** © Lennart Nilsson/A Child is Born/Dell Publishing Company; **Figure 20.2g,h:** Courtesy of Dr. Landrum Shettles; **Figure 20.21:** Center for Disease Control, Atlanta, GA

Chapter 21

Figure 21.4a,b: © Edwin A. Reschke; **Figure 21.5a:** From R. J. Blandau, *A Textbook of Histology*, 10/E, W. B. Saunders Company; **Figure 21.5b:** Courtesy of Dr. Landrum Shettles; **Figure 21.6:** From Bloom and Fawcett, *Histology*, 10/E, p. 859, W. B. Saunders Company; **Figure 21.7:** Courtesy of Dr. Landrum Shettles; **Figures 21.10, 21.14:** © Edwin A. Reschke; **Figure 21.17a,b:** © BioPhoto Associates/Photo Researchers, Inc.; **Figure 21.20:** Courtesy of Utah Valley Regional Medical Center, Department of Radiology; **Figure 21.25a,b:** SIU School of Medicine; **Figure 21.26a–f:** © 1986 SIU/Visuals Unlimited; **Figure 21.26g:** © Hank Morgan/Science Source/Photo Researchers, Inc.; **Figure 21.26h:** © Scott Camazine/Sue Trainor/ Photo Researchers, Inc.

Chapter 22

Figure 22.1c: © David Phillips 1987/ Visuals Unlimited; **Figure 22.2:** Luciano Zamboni from Roy Greep and Leon Weiss, *Histology*, 3/E, 1973, McGraw-Hill Book Company; **Figures 22.7a, 22.9a:** Copyright Roman O'Rahilly/Carnegie Laboratories of Embryology, University of California; **Figure 22.12a:** © Donald Yeager/ Carroll Weiss/Camera M.D. Studios, Inc.; **Figure 22.16e:** Courtesy of Dr. Landrum Shettles; **Figure 22.19a:** © Donald Yeager/Carroll Weiss/ Camera M.D. Studios, Inc.; **Figure 22.20a:** Courtesy of Dr. Landrum Shettles; **Figures 22.21, 22.23a,b:** © Donald Yeager/Camera M.D. Studios/Carroll Weiss; **Figure 22.24a:** © A. Tsiaras/Science Source/Photo Researchers, Inc.; **Figure 22.24b:** Kent M. Van De Graaff; **Figure 22.28a–c:** Courtesy of Utah Valley Regional Medical Center, Department of Radiology; **Figure 22.31:** From J. M. Tanner, *The Physique of the Olympic Athlete*, 1964; **Figure 22.32:** March of Dimes; **Figure 22.38:** © Lester V. Bergman and Associates; **Figure 22.39c:** Courtesy of Dr. Landrum Shettles; **Figure 22.41:** Gregory Dellore, M.D. and Steven L. Clark, M.D.

Index

Prefixes and Suffixes of Anatomical and Medical Terminology

Element	Definition and Example	Element	Definition and Example
a-	absent, deficient, without: *atrophy*	dipl-	double: *diploid*
ab-	off, away from: *abduct*	dis-	apart, away from: *disarticulate*
abdomin-	relating to the abdomen: *abdominal*	duct-	lead, conduct: *ductus deferens*
-able	capable of: *viable*	dur-	hard: *dura mater*
ac-	toward, to: *actin*	dys-	bad, difficult, painful: *dysentery*
acou-	hear: *acoustic*	e-	out, from: *eccrine*
ad-	toward, to: *adduct*	ecto-	outside, outer, external: *ectoderm*
af-	movement toward a central point: *afferent artery*	-ectomy	surgical removal: *tonsillectomy*
alba	pale or white: *linea alba*	ede-	swelling: *edema*
-alg	pain: *neuralgia*	-emia	pertaining to a condition of the blood: *lipemia*
ambi-	both: *ambidextrous*	end-	within: *endoderm*
angi-	pertaining to the vessels: *angiology*	entero-	intestine: *enteritis*
ante-	before, in front of: *antebrachium*	epi-	upon, after, in addition: *epigenesis*
anti-	against: *anticoagulant*	erythro-	red: *erythrocyte*
aqua-	water: *aqueous*	ex-	out of: *excise*
archi-	first: *archeteron*	exo-	outside: *exocrine*
arthr-	joint: *arthritis*	extra-	outside of, beyond, in addition: *extracellular*
-asis	condition or state of: *homeostasis*	fasci-	band: *fascia*
aud-	pertaining to the ear: *auditory*	febr-	fever: *febrile*
auto-	self: *autolysis*	-ferent	bear, carry: *efferent*
bi-	two: *bipedal*	fiss-	split: *fissure*
bio-	life: *biopsy*	for-	opening: *foramen*
blast-	generative or germ bud: *blastocyst*	-form	shape: *fusiform*
brachi-	arm: *brachialis*	gastro-	relating to the stomach: *gastrointestinal*
brachy-	short: *brachydont*	-gen	an agent that produces or originates: *pathogen*
brady-	slow: *bradycardia*	-genic	produced from, producing: *carcinogenic*
bucc-	cheek: *buccal cavity*	gloss-	tongue: *glossopharyngeal*
cac-	bad, ill: *cachexia*	glyco-	sugar: *glycosuria*
calc-	stone: *calculus*	-gram	a record, recording: *electroencephalogram*
capit-	head: *capitis*	gran-	grain, particle: *agranulocyte*
carcin-	cancer: *carcinogenic*	-graph	instrument for recording: *electrocardiograph*
cardi-	heart: *cardiac*	gravi-	heavy: *gravid*
cata-	lower, under, against: *catabolism*	gyn-	female sex: *gynecology*
caud-	tail: *cauda equina*	haplo-	simple or single: *haploid*
cephal-	head: *cephalic*	hema(o)-	blood: *hematology*
cerebro-	brain: *cerebrospinal fluid*	hemi-	half: *hemiplagia*
chol-	bile: *cholic*	hepat-	liver: *hepatic portal*
chondr-	cartilage: *chondrocyte*	hetero-	other, different: *heterosexual*
chrom-	color: *chromocyte*	histo-	web, tissue: *histology*
-cide	destroy: *germicide*	holo-	whole, entire: *holocrine*
circum-	around: *circumduct*	homo-	same, alike: *homologous*
co-	together: *copulation*	hydro-	water: *hydrocoel*
coel-	hollow cavity: *coelom*	hyper-	beyond, above, excessive: *hypertension*
-coele	swelling, an enlarged space or cavity: *blastocoele*	hypo-	under, below: *hypoglycemia*
con-	with, together: *congenital*	-ia	state or condition: *hypoglycemia*
contra-	against, opposite: *contraception*	-iatrics	medical specialties: *pediatrics*
corn-	denoting hardness: *cornified*	idio-	self, separate, distinct: *idiopathic*
corp-	body: *corpus*	ilio-	ilium: *iliosacral*
crypt-	hidden: *cryptorchidism*	infra-	beneath: *infraspinatus*
cyan-	blue color: *cyanosis*	inter-	among, between: *interosseous*
cysti-	sac or bladder: *cystoscope*	intra-	inside, within: *intracellular*
cyto-	cell: *cytology*	-ion	process: *acromion*
de-	down, from: *descent*	-ism	condition or state: *rheumatism*
derm-	skin: *dermatology*	iso-	equal, like: *isotonic*
di-	two: *diarthrotic*	-itis	inflammation: *meningitis*

Prefixes and Suffixes of Anatomical and Medical Terminology

Element	Definition and Example	Element	Definition and Example
labi-	lip: *levator labii superioris*	pod-	foot: *podiatry*
lacri-	tears: *nasolacrimal*	-poiesis	formation of: *hemopoiesis*
later-	side: *lateral*	poly-	many, much: *polyploid*
leuk-	white: *leukocyte*	post-	after, behind: *postnatal*
lip-	fat: *lipid*	pre-	before in time or place: *prenatal*
-logy	science of: *morphology*	prim-	first: *primitive*
-lysis	solution, dissolve: *hemolysis*	pro-	before in time or place: *prosect*
macro-	large, great: *macrophage*	proct-	anus: *proctology*
mal-	bad, abnormal, disorder: *malignant*	pseudo-	false: *pseudostratified*
medi-	middle: *medial*	psycho-	mental: *psychology*
mega-	great, large: *megakaryocyte*	pyo-	pus: *pyoculture*
meso-	middle or moderate: *mesoderm*	quad-	fourfold: *quadriceps femoris*
meta-	after, beyond: *metatarsal*	re-	back, again: *repolarization*
micro-	small: *microtome*	rect-	straight: *lateral rectus*
mito-	thread: *mitosis*	reno-	kidney: *renal*
mono-	alone, one, single: *monocyte*	rete-	network: *retina*
mons	mountain: *mons pubis*	retro-	backward, behind: *retroperitoneal*
morph-	form, shape: *morphology*	rhin-	nose: *rhinitis*
multi-	many, much: *multinuclear*	-rrhage	excessive flow: *hemorrhage*
myo-	muscle: *myology*	-rrhea	flow or discharge: *diarrhea*
narc-	numbness, stupor: *narcotic*	sanguin-	blood: *sanguiferous*
necro-	corpse, dead: *necrosis*	sarc-	flesh: *sarcoplasm*
neo-	new, young: *neonatal*	-scope	instrument for examining a part: *stethoscope*
nephro-	kidney: *nephritis*	-sect	cut: *dissect*
neuro-	nerve: *neurolemma*	semi-	half: *semilunar*
noto-	back: *notochord*	serrate-	saw-edged: *serratus anterior*
ob-	against, toward, in front of: *obturator*	-sis	process or action: *dialysis*
oc-	against: *occlusion*	steno-	narrow: *stenohaline*
-oid	resembling, likeness: *sigmoid*	-stomy	surgical opening: *tracheostomy*
oligo-	few, small: *oligodendrocyte*	sub-	under, beneath, below: *subcutaneous*
-oma	tumor: *lymphoma*	super-	above, beyond, upper: *superficial*
oo-	egg: *oocyte*	supra-	above, over: *suprarenal*
or-	mouth: *oral*	syn- (sym-)	together, joined, with: *synapse*
orchi-	testicles: *cryptorchidism*	tachy-	swift, rapid: *tachycardia*
ortho-	straight, normal: *orthopnea*	tele-	far: *telencephalon*
-ory	pertaining to: *sensory*	tens-	stretch: *tensor fascia lata*
-ose	full of: *adipose*	tetra-	four: *tetrad*
osteo-	bone: *osteoblast*	therm-	heat: *thermogram*
oto-	ear: *otolith*	thorac-	chest: *thoracic cavity*
ovo-	egg: *ovum*	thrombo-	lump, clot: *thrombocyte*
par-	give birth to, bear: *parturition*	-tomy	cut: *appendectomy*
para-	near, beyond, beside: *paranasal*	tox-	poison: *toxic*
path-	disease, that which undergoes sickness: *pathology*	tract-	draw, drag: *traction*
-pathy	abnormality, disease: *neuropathy*	trans-	across, over: *transfuse*
ped-	children: *pediatrician*	tri-	three: *trigone*
pen-	need, lack: *penicillin*	trich-	hair: *trichology*
-penia	deficiency: *thrombocytopenia*	-trophy	a state relating to nutrition: *hypertrophy*
per-	through: *percutaneous*	-tropic	turning toward, changing: *gonadotropic*
peri-	near, around: *pericardium*	ultra-	beyond, excess: *ultrasonic*
phag-	to eat: *phagocyte*	uni-	one: *unicellular*
-phil	have an affinity for: *neutrophil*	-uria	urine: *polyuria*
phlebo-	vein: *phlebitis*	uro-	urine, urinary organs or tract: *uroscope*
-phobe	abnormal fear, dread: *hydrophobia*	vas-	vessel: *vasoconstriction*
-plasty	reconstruction of: *rhinoplasty*	vermi-	worm: *vermiform*
platy-	flat, side: *platysma*	viscer-	organ: *visceral*
-plegia	stroke, paralysis: *paraplegia*	vit-	life: *vitamin*
-pnea	to breathe: *apnea*	zoo-	animal: *zoology*
pneumato-	breathing: *pneumonia*	zygo-	union, join: *zygote*